# Proceedings of

# THE THIRD IEEE CONFERENCE ON FUZZY SYSTEMS

## IEEE WORLD CONGRESS ON COMPUTATIONAL INTELLIGENCE

### June 26 - June 29, 1994

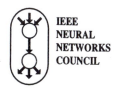

IEEE
NEURAL
NETWORKS
COUNCIL

### Walt Disney World Dolphin Hotel
### Orlando, Florida

IEEE

## Volume III
### Pages 1459 - 2118

# The Third IEEE Conference on Fuzzy Systems

IEEE Catalog Number: 94CH3430-6
ISBN Softbound: 0-7803-1896-X
Casebound: 0-7803-1897-8
Microfiche: 0-7803-1898-6
Library of Congress Number: 94-75609

Additional copies of this publication are available from

IEEE Service Center
445 Hoes Lane
P.O. Box 1331
Piscataway, NJ 08855-1331
1-800-678-IEEE

# FUZZ-IEEE TABLE OF CONTENTS

## VOLUME I

## IMAGE PROCESSING I

## HARDWARE I

## DATABASE/INFORMATION RETRIEVAL I

# VOLUME II

# VOLUME III

*FUZZY CONTROL*

# NEUROFUZZY SYSTEMS

# HARDWARE

# LEARNING

# GENETIC ALGORITHMS

# VISION

# STATISTICS

*EXPERT SYSTEMS*

# A FUZZY DECISION SYSTEM BASED ON STATISTICAL LEARNING FOR FAULT CLASSIFICATIONS

Dr. Yubao Chen
Dept. of Industrial and Manufacturing Systems Engineering
University of Michigan-Dearborn
Dearborn, MI 48128-1491, U.S.A.

## ABSTRACT

A Fuzzy Decision System(FDS) is proposed for condition monitoring of machining processes. The membership functions are established through a learning process based on test data, rather than being selected as *a priori*. The optimal partition and information gain weighting functions are also introduced in order to improve the robustness and reliability of this method. Experiment verification with an optimistic success rate of 97.5% was achieved.

## INTRODUCTION

Fuzzy systems have been used for process or system condition monitoring and diagnosis in numerous applications. For instance, a fuzzy pattern recognition technique was developed for the monitoring of drill wear [Li 1987]. The fuzzy set approach was also combined with the decision tree technique for the condition monitoring and fault classification of machining systems [Du, et. al. 1991, Zeng and Wang 1991, Kasool 1991]. However, in all these applications, the membership functions were determined subjectively as *a priori*. to the system. Due to high levels of uncertainty and complexity involved in a manufacturing system, the pre-sellected membership functions some time did not represent the true characteristics of the system or process under monitoring.

This paper presents a framework for an alternative diagnostic decision strategy with the membership functions learnt from statistical analysis. The fuzzy decision system(FDS) presented in this paper differs from previous research in the following aspects: 1) the partition of indices in the fuzzifier is optimized, unlike even selection as usually utilized in a fuzzy system; 2) the membership functions are learned from sample data based on information analysis of each index; and 3) a weighted multiple voting scheme is introduced with information gains as the weighting function. In such a way, groundwork is laid toward the construction of a robust diagnostics system for on-line implementation.

## FUZZY VARIABLE DEFINITION

A fuzzy variable is defined as a variable associated with a fuzzy set; it can assume only the values included in the fuzzy set. Three variables are used to construct the fuzzy decision system (FDS) for machine diagnosis: C, R, M. C is defined based on the class fuzzy set

$C = \{ (c_i, \mu_C(c_i) | c_i \in C \}$, where $C = \{ c_1, c_2, ..., c_{N_c} \}$, $c_i$ refers to the i-th class of machine conditions, and $N_c$ is the total number of classes under consideration. R is associated with the region fuzzy set: $R = \{ (r_j, \mu_R(r_j) | r_j \in R \}$, where $R = \{ r_1, r_2, ..., r_{NR} \}$, $r_j$ is the j-th region (interval) of index $y_k$, and $N_R$ the total number of regions in the span of index $y_k$. While $N_c$ is not necessarily equal to $N_R$, we assume $N_c = N_R$ here for simplicity and without losing the generality. M is corresponding to the two-dimensional fuzzy set representing the relationship between R and C: $M_{CR} = \{ (c_i, r_j), \mu_{CR}(c_i, r_j) | (c_i, r_j) \in C \times R \}$.

## DETERMINATION OF THE MEMBERSHIP FUNCTIONS

Membership functions play the most important role in the fuzzy logic inference mechanism. The membership function $\mu_R(r_j)$, i.e, the degree to which an index $y_k$ falls into region $r_j$, can be determined by observing index $y_k$. For machine diagnosis, we may define the membership function $\mu_R(r_j) = 1$ if $y_k \in r_j$ is observed, and 0 otherwise. Whereas the determination of the membership function $\mu_{CR}(c, r)$, which defines the relationship between the two fuzzy variables R and C, is more crucial for a successful FDS. $\mu_{CR}(c_i, r_j)$ actually represents the degree to which the machine condition belongs to $c_i$ when the observation of index $y_k$ is in the region $r_j$. This information can be extracted in the learning process by gathering a set of samples in the following format:

| Test No. | $y_1$ | $y_2$ | ... | $y_k$ | ... | $y_L$ | Class |
|----------|-------|-------|-----|-------|-----|-------|-------|
| 1 | $y_{11}$ | $y_{12}$ | | $y_{1k}$ | ... | $y_{1L}$ | $c_1$ |
| 2 | $y_{21}$ | $y_{22}$ | | $y_{2k}$ | ... | $y_{2L}$ | $c_3$ |

Based on these data, a partition over y can be formed by dividing its span into $N_R$ regions (intervals). The boundary of the regions, $b_i$, i=1, ..N, is determined according to its probability distributions. In a particular region, $r_j$, the conditional probability that the machine condition belong to $c_i$ can be estimated by

$$p(c_i | r_j) = \frac{n_{ij}}{N_j^R} \qquad \forall \; i,j \tag{1}$$

where $n_{ij}$ is the number of samples from class $c_i$ while the corresponding index value y falls into interval $r_j$, and $N_j^R$ is the total number of samples that fall into $r_j$, no matter which class it is from. The conditional probability can be used to define the membership function $\mu_{cR}(c,r)$. i.e.,

$$\mu_{CR}(c_i, r_j) = \frac{n_{ij}}{N_j^R} \qquad \forall \; i,j \tag{2}$$

Emphasis should be made here that the formation of partition R will affect a great deal of the membership function defined by Equation (4). The best relationship between R and C is that $\mu_{CR}(c_i, r_j) = 1$ for i=j and 0 otherwise, so that $M_{CR}(c,r) = \{\mu_{CR}(c_i, r_j)\}$ becomes a diagonal matrix. Although the diagonal matrix can hardly be obtained due to the high level of uncertainty which exists for a manufacturing process, an optimal partition of R can be achieved so as to diagonalize the matrix $M_{CR}(c, r)$ to a maximum degree. A practical way for searching such an optimal partition can be estimated by optimizing its total information gain [Chen 1990].

## DETERMINATION OF THE WEIGHTING FUNCTION

When multiple indices are used, the contribution of each index to the final decision varies to a large degree according to the real situation under which the diagnosis decision is to be made. Such contribution can be represented by a weighting function associated with each index. The determination of the weighting function is critical to a successful diagnosis decision making process. Instead of selecting the weighting function empirically based on experience or through a "blind" learning, a objective scheme for determining the weighting function is proposed based on information measures of each index to the decision making process.

In information theory, the entropy of a random variable indicates the uncertainty about the ensemble it is from. The entropy for the index space $Y \supset y_k$, H(Y), can be defined as the average self-information over all classes $\{c_i; i=1,..., N_c\}$ [Chen 1990, Devijver 1982]

$$H(\mathbf{Y}) = -\sum_{i=1}^{N_c} P(c_i) \log P(c_i) \tag{3}$$

The conditional entropy of all classes $\{c_i\}$ on region $r_j$ is defined as the average conditional entropy

$$H(\mathbf{Y}|r_j) = -\sum_{i=1}^{N_c} P(c_i | r_j) \log P(c_i | r_j) \tag{4}$$

where $P(c_i|r_j)$ is the conditional probability of $c_i$ given $r_j$. The regional information gain for index k, $G(r_j|y_k)$, is defined as the entropy for index space $\mathbf{Y}$ less the conditional entropy based on region $r_j$; that is,

$$G(r_j|y_k) = H(\mathbf{Y}) - H(\mathbf{Y}|r_j)$$

(5)

$G(r_j|y_k)$ represents the information gain carried by index $y_k$ over its region. In other words, $G(r_j|y_k)$ is the amount of uncertainty reduced after the fact that $y_k$ falls into a particular region $r_j$ is observed.

While eq. (5) explores the regional information gain, the information measure inherent in different classes has also to be included for a full evaluation of the contribution of index $y_k$. For a manufacturing process, different classes may have different probability distributions over the partitioned regions for an observed class. This distribution indicates variations in entropies with respect to classes $\{c_i\}$. Some indices may provide higher levels of information or lower entropies for certain classes. To incorporate this phenomenon into the weighting function, another information measure, $Q(c_i|y_k)$, is introduced:

$$Q(c_i|y_k) = -\sum_{j=1}^{N_R} P(r_j|y_k)\log P(r_j|y_k) + \sum_{j=1}^{N_R} P(r_j|c_i,y_k)\log P(r_j|c_i,y_k)$$

(6)

which is the entropy over all regions less the conditional entropy on class $c_i$ given index $y_k$. Finally, the weighting function for index $y_k$ can be established based on the two information measures, $G(r_j|y_k)$ and $Q(c_i|y_k)$. i.e.,

$$\mathbf{W}_k = \{Q(c_i|y_k) \cdot G(r_j|y_k)\} \qquad \forall\ i,j,k$$

(7)

## THE STRUCTURE OF THE FUZZY DECISION SYSTEM

The fuzzy decision system (FDS) is used to classify machine conditions into their appropriate classes based on the observation of all suitable indices $y_k$, k=1, .., L. In other words, the FDS is designed to process the mapping $R \rightarrow C$. More specifically, the decision-making process takes the following steps:

1) An input index vector $\mathbf{Y} = (y_1, y_2, ... y_L)$ is formed based on the samples of measurements, where $y_k$ is the value of k-th index, and L is the total number of indices involved.
2) The fuzzifier transfers the index vector $\mathbf{Y}$ into L fuzzy set vectors
3) The mapping: $R \rightarrow C$ is then processed in a parallel manner for each index $y_k$, according to the following fuzzy operation:

$$\mu^k_C(c_i) = \mu^k_R(r_j) o M^k_{CR} \qquad \forall\ i,j,k$$

(8)

where o is the fuzzy operator; $M^k_{CR}$ is the fuzzy relationship matrix estimated in the learning phase.

4) The membership function vector or the voting function, $\mu_C(c_i)$ is a weighted sum of $\mu^k_C(c_i)$ obtained separately for each index; i.e.,

$$\mu_c(c_i) = \sum_{k=1}^{L} \frac{w_k \mu^k_c(c_i)}{\sum_{k=1}^{c} w_k}$$

(9)

The weighting function $w_k$ represents the contribution of each index to the diagnosis process, which is obtained in the learning process based on eq. (9).

5) As the output of the defuzzifier, the final diagnosis decision is made by classifying the machine condition into class $c_{i*}$ such that

$$i^* = \text{Arg max } (\mu_c(c_i)|\ i=1, 2, ..., N_c)$$

## EXPERIMENTAL VERIFICATION

The Fuzzy Decision System (FDS) was designed and used for the diagnosis of a tapping process for validation of this approach. Tapping is a process to make threads inside a hole. The tap used in the operation is very fragile, and the process is a discontinuous type with very short time intervals. The tapping operation is performed on a machining center with a tap attached on the spindle head. The most common malfunctions of the tapping process include (1) tap wear; (2) hole undersize; (3) hole oversize; and (4) misalignment. The dynamic signals used for the diagnosis of the tapping process are the cutting forces, which include the tapping torque (T), the thrust force ($F_z$), and the lateral forces ($F_x$ and $F_y$). More details about this process can be found in [Chen 1990, Liu 1991].

From the measured four components of the cutting forces, eight indices were extracted and deemed suitable for the diagnosis purpose: A total of 80 tests were performed in the learning process, from which the fuzzy mapping matrix, $M_{CR}$, and the weighting function, $w_k$, were estimated for each index. The learned mapping matrix $M_{CR}$, and the weighting function, i.e., $G(r_j|y_k)$ and $Q(c_i|y_k)$, are also illustrated in Tables 1, 2, and 3.

Table 1 The fuzzy  mapping matrix $M_{CR}$ for index $y_1$

| | | | Region | | |
|---|---|---|---|---|---|
| Class | $r_1$ | $r_2$ | $r_3$ | $r_4$ | $r_5$ |
| $c_1$ | 0.0667 | 0.6111 | 0.2667 | 0.0000 | 0.0000 |
| $c_2$ | 0.0000 | 0.0000 | 0.0000 | 0.0000 | 1.0000 |
| $c_3$ | 0.0000 | 0.0000 | 0.0000 | 1.0000 | 0.0000 |
| $c_4$ | 0.9333 | 0.1111 | 0.0000 | 0.0000 | 0.0000 |
| $c_5$ | 0.0000 | 0.2778 | 0.7333 | 0.0000 | 0.0000 |

Based on the fuzzy membership functions established in the learning stage, the conditions of the tapping process were classified once a new set of observations of the selected indices was available. The classification decision made by the FDS was then compared to the true condition of the tapping process under which the data wereacquired. As the first test, each of the 80 sets of data used during learning was treated as a new condition. As expected,all the those 80 sets were correctly classified and, therefore, a 100% success rate was achieved for the learning data. In addition, 40 sets of new data (which were not used in the learning process) were acquired as test data for the FDS. Again, each set of data was fed through the FDS for condition classification. The FDS identified classes, together with the fuzzy voting function, $\mu_c(c_i)$, and original classes under which data were measured. A success rate of 97.5% was achieved.

## CONCLUSIONS

The membership functions in a fuzzy system can be established based on information gained directly from data in a learning process, instead of selecting a membership function as *a prior.* . In addition, weighting functions can be determined objectively based on information measures carried by each index. After all these being done, machine condition can be classified using a multiple voting scheme based on fuzzy composition principle with the information measures as the weighting function. These unique features embedded in the fuzzy system enhances its robustness and effectiveness in decision making, which is confirmed with experimental tests for tapping process. In this way, a fuzzy decision system (FDS) can be constructed for fault classification of a manufacturing process under high levels of uncertainty.

Table 2. Information measure  $G(r_j|y_k)$

| $y_k$ | $r_1$ | $r_2$ | $r_3$ | $r_4$ | $r_5$ |
|---|---|---|---|---|---|
| 1 | 1.1188 | 2.3219 | 2.3219 | 1.8768 | 1.4805 |
| 2 | 1.3264 | 2.3219 | 1.8662 | 1.5276 | 1.3255 |
| 3 | 1.3197 | 2.3219 | 2.3219 | 2.3219 | 1.3304 |
| 4 | 0.7832 | 0.8425 | 0.0390 | 0.7656 | 1.5807 |
| 5 | 1.4275 | 1.9500 | 0.6564 | 1.2984 | 1.0654 |
| 6 | 1.5107 | 2.3219 | 0.8654 | 1.4645 | 0.9817 |
| 7 | 0.7738 | 0.8382 | 0.1225 | 0.9231 | 1.1358 |
| 8 | 0.9802 | 0.6760 | 0.9339 | 0.4896 | 1.0012 |

Table 3　Information measure $Q(c_i|y_k)$

| $y_k$ | $c_1$ | $c_2$ | $c_3$ | $c_4$ | $c_5$ |
|---|---|---|---|---|---|
| 1 | 1.9686 | 1.0222 | 1.4853 | 2.3219 | 2.3219 |
| 2 | 1.9081 | 1.2169 | 0.5988 | 2.3219 | 2.3219 |
| 3 | 2.3219 | 1.3262 | 1.3239 | 2.3219 | 2.3219 |
| 4 | 0.5439 | 0.4086 | 0.7768 | 0.6817 | 1.6000 |
| 5 | 1.3244 | 0.8383 | 1.0686 | 1.1818 | 1.9846 |
| 6 | 1.4259 | 0.7934 | 1.2811 | 1.3219 | 2.3219 |
| 7 | 0.9022 | 0.9120 | 0.1399 | 0.8129 | 1.0265 |
| 8 | 2.3219 | 0.3919 | 0.0897 | 0.3265 | 0.9510 |

## REFERENCES

Bezdek, J.C., 1981, *Pattern Recognition with Fuzzy Objective Function Algorithm*, Plenum Press, New York.

Chen, Y.B., Sha, J.L., and Wu, S.M., 1990, "Diagnosis of the Tapping Process by Information Measure and Probability Voting Approach," *Trans. of ASME, J. of Engineering for Industry*,. Vol. 111, pp. 319-325.

Danai, K. and Chin, H., 1989, "Sensor-based Diagnostic Reasoning with Process Uncertainty," *ASME DSC*. Vol. 18, pp. 65-74.

Danai, K. and Ulsoy A.G., 1987, "An Adaptive Observer for On-line Tool Wear Estimation in Turning-Part I: Theory," *Mechanical Systems and Signal Processing*, Vol.1, No.2. pp.211-225.

Devijver, P.A., and Kittler, J., 1982, *Pattern Recognition: A Statistical Approach*, Prentice Hall.

Doyle, E.D., and Dean, S.K., 1974, "The Effect of Axial Forces on Dimensional Accuracy During Tapping," *Machine Tool Design and Research Journal*, Vol 14, pp. 525-333.

Du., R.X., 1989, Elbestawi, M.A., and L., Sam, 1991, "Tool Condition Monitoring in Turning Using Fuzzy Set Theory," *Int. J. of Machine Tool Manufacturing*. No.3.

Dyer, D., and Stewart, R.V., 1978, "Detection of rolling element bearing damage by statistical vibration analysis," *Journal of Mechanical Design*, Vol. 100, pp. 229-235.

Gmytrasiewicz, P., Hassberger, A., and Lee, J.C., 1990, "Fault Tree Based Diagnostics Using Fuzzy Logic," *IEEE Transaction on Pattern Analysis and Machine Intelligence*, Vol. 12. No.11. pp. 1115-1119.

Hayashi, Y., and Nakai, M., 1988, "Efficient Method for Multi-dimensional Fuzzy Reasoning and Its Application to Fault Diagnosis," *International Workshop on Artificial Intelligence for Industrial Applications*, pp. 27-32.

Kenarangui, Rasool, 1991, "Event-Tree Analysis by Fuzzy Probability," *IEEE Transaction on Reliability*, Vol. 40, No.1. pp. 120-124.

Klir, G.J., and Folger, T.A., 1988, *Fuzzy Sets, Uncertainty, and Information*, Prentice Hall.

Kosko, Bart, 1992, *Neural Networks and Fuzzy Systems*, Prentice-Hall, Inc.

Lee, C.C., 1990, "Fuzzy Logic in Control Systems: Fuzzy Logic Controller-Part I," *IEEE Transactions on Systems, Man, and Cybernetics*, Vol. 29, No.2, pp.404-418.

Li, P.G., and Wu, S.M., "Monitoring of Drill Wear States Using Fuzzy Pattern Recognition Technique," *Trans. of SAME, J. of Engineering for Industry*, Vol. 110, 1987, pp. 297-300.

Quinlan, J.R., 1983, "Induction of Decision Tree," *Machine Learning*, No.1, pp. 81-106.Surface Roughness and Pandit , S.M., and Wu, S.M., 1983. *Time Series and System Analysis with Applications*, John Wiley & Sons.

Ralston, P.A., and Ward, T.L., 1989, "Fuzzy Logic Control of Machining," *Manufacturing Review*, Vol. 3, No.3. pp. 147-154.

Zeng, L., and Wang, H.P. 1991, "Machine-Fault Classification: a Fuzzy-Set Approach," *International J. Advanced Manufacturing Technology*, Vol.6, pp. 83-94.

# Analysis of Fuzzy Controlled Systems Using Hyperstability Theory

Rolf Böhm

Institut für Regelungs- und Steuerungssysteme

Universität Karlsruhe

Kaiserstr. 12

D-76131 Karlsruhe

*In 1973 V. M. Popov introduced a criterion of hyperstability to analyse MIMO-systems as a generalization of the absolute stability problem [1]. In this paper a method to analyse a large class of fuzzy-controlled systems using hyperstability theory is presented.*

## 1. Introduction

In many cases fuzzy controllers are implemented because of a missing exact mathematical model. But even if exact models are available the use of a fuzzy controller seems reasonable, because often it is much more easy to understand and design fuzzy controllers than other non-linear controllers. However for every closed-loop system one has to guarantee stability. Methods from non-linear theory like the second method of *Ljapunov* are too complex to deal with systems of high order and non-linear fuzzy type controllers. Less complex methods like the criterion of absolute stability [5] or harmonic balance just deal with one-dimensional non-linearities. Therefore in this paper the use of hyperstability theory is presented to analyse closed-loop systems with fuzzy controllers.

## 2. Introductory Definitions

Consider the linear time-invariant multivariable system (Fig. 1)

$$\dot{x} = A \cdot x + B \cdot u$$
$$y = C \cdot x \tag{1}$$

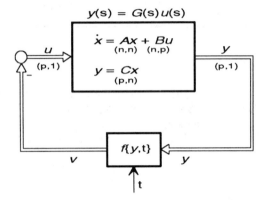

**Fig. 1** *Closed loop system under investigation*

where $x$ is a $n$-state vector, $u$ and $y$ are $p$-dimensional vectors. $A, B$ and $C$ are matrices of suitable dimension. $G(s) = C(sI - A)^{-1}B$ represents the transfer matrix. The pair $(A, B)$ is completely controllable, the pair $(A, C)$ completely observable.

The feedback block can be described by $v = f(y, t)$, $f(y = 0, t) = 0$, in this case the fuzzy controller.

From [2] one knows that the closed-loop system is asymptotically hyperstable, if

1. the inequation

$$\int_0^t v^T(\tau) \cdot y(\tau) \, d\tau \geq -\varepsilon_0^2 \quad \forall t \geq 0 \tag{2}$$

is satisfied and

2. the linear feedforward block $G(s)$ is strictly positive real.

If the feedforward block is not apriori strictly positive real one can add a fictive matrix $D$ using the Kalman-Yakubovich-Lemma

$$A^T \cdot P + P \cdot A = -LL^T = -Q \qquad (3\text{-}1)$$

$$L \cdot V = C^T - P \cdot B \qquad (3\text{-}2)$$

$$D + D^T = V^T \cdot V. \qquad (3\text{-}3)$$

Assuming a symmetric positive matrix $Q$, one can find a unique matrix $P$, to determine the $(m,p)$-matrix $V$ and the matrix $D$.

Then the modified system $\hat{G}(s) = C(sI - A)^{-1}B + D$ is strictly positive real.

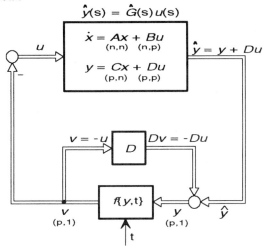

$$\hat{y}(s) = \hat{G}(s)\,u(s)$$

**Fig. 2** *Modified system with strictly positive real feedforward block*

To compensate this modification one has to add it to the non-linear feedback block as well [4]. Inequation (2) in this case reads

$$\int_0^t v^T(\tau)\hat{y}(\tau)\, d\tau \geq -\varepsilon_0^2 \quad \forall t \geq 0 \qquad (4)$$

with $\hat{y}(\tau) = y(\tau) + Du$.

### 3. Method to enlarge the admissible sectors

From Fig. 2 it is obvious that $v = f(y,t)$ and $u = -v$. With this one gets

$$\int_0^t \left[ f^T(y,\tau)y - f^T(y,\tau)Df(y,\tau) \right] d\tau \geq -\varepsilon_0^2. \quad (5)$$

Because in most cases we are not able to calculate the dynamic behaviour of $f(y,\tau)$ in

an analytic way, we demand as sufficient condition

$$\left[ f^T(y,\tau)y - f^T(y,\tau)Df(y,\tau) \right] \geq 0$$
$$\Rightarrow \sum_{j=1}^p \left[ f_j(y,\tau) \cdot \left( y_j - \sum_{i=1}^p d_{i,j} \cdot f_i(y,\tau) \right) \right] \geq 0. \quad (6)$$

(6) holds, if

$$f_j(y,\tau) \cdot \left( y_j - \sum_{i=1}^p d_{i,j} \cdot f_i(y,\tau) \right) \geq 0 \ \forall y_j. \quad (7)$$

In most cases the assumption

$$f_j(y_1,\ldots,y_p,\tau) \cdot y_j \geq 0 \ \ \forall y_j \qquad (8)$$

will be true. If so, one merely has to prove

$$y_j - \sum_{i=1}^p d_{i,j} \cdot f_i(y_1,\ldots,y_p,\tau) \geq 0 \ \ \forall \, y_j \geq 0$$
$$\text{and} \qquad\qquad\qquad\qquad\qquad\qquad (9)$$
$$y_j - \sum_{i=1}^p d_{i,j} \cdot f_i(y_1,\ldots,y_p,\tau) < 0 \ \ \forall \, y_j < 0$$

Now (9) has to hold for every $y$. Especially $d_{i,j} \neq 0$, $i \neq j$ cause couplings between $y_j$ and $f_i(y)$; thus (9) usually is not fulfilled. Therefore we assume $D$ as a diagonal matrix with $d_{i,j} = 0 \ \forall i \neq j$. From (9) then the following multidimensional sector condition

$$\frac{f_j(y,\tau)}{y_j} \leq \frac{1}{d_{j,j}}, \ \ \forall j \, , y \qquad (10)$$

is obtained, since $d_{j,j} \geq 0$ according (3-3). Condition (10) is quite easy to proof. It is obvious that the smaller the elements of $D$ are, the bigger are the admissible sectors for the controller $f$.

To fulfil (3) with $d_{i,j} = 0 \ \forall i \neq j$ one has to have a look at $L$ and $V$. From (3-3) we choose a matrix $V'$ as

$$V' = \begin{pmatrix} v'_{1,1} & \cdots & v'_{1,p} \\ \vdots & & \vdots \\ v'_{n,1} & \cdots & v'_{n,p} \\ v'_{n+1,1} & 0 & 0 \\ v'_{n+2,1} & & \vdots \\ \vdots & & 0 \\ v'_{m,1} & \cdots & v'_{m,p} \end{pmatrix} = \begin{pmatrix} v'_{1,1} & \cdots & v'_{1,p} \\ \vdots & & \vdots \\ v'_{n,1} & \cdots & v'_{n,p} \\ v'_{n+1,1} & 0 & 0 \\ v'_{n+2,1} & & \vdots \\ \vdots & & 0 \\ v'_{n+p,1} & \cdots & v'_{n+p,p} \end{pmatrix} \tag{11}$$

with $v_{i,j} = 0 \; \forall \, (i > n \wedge j > (i-n))$. With (11) all $v_{i,j} \neq 0$, $i > n$ can be chosen in such a way that $d_{i,j} = 0 \; \forall i \neq j$ holds.

To solve (3-1) we propose $L$ as

$$L = \begin{pmatrix} l_{1,1} & \cdots & l_{1,n} \\ \vdots & & \vdots \\ l_{n,1} & \cdots & l_{n,n} \end{pmatrix}, \tag{12}$$

but to fit dimensions without changing equation (3-2) one has to extend $L$ to

$$L' = \begin{pmatrix} l_{1,1} & \cdots & l_{1,n} & l_{1,n+1} & \cdots & l_{1,n+p} \\ \vdots & & \vdots & \vdots & & \vdots \\ l_{n,1} & \cdots & l_{n,n} & l_{n,n+1} & \cdots & l_{n,n+p} \end{pmatrix} \tag{13}$$

with $l_{i,j} = 0 \; \forall j > n$ to decouple the modification of $V$ from (3-2). As $L$ is regular, because $Q$ is a symmetric positive matrix [6], (3-2) can be solved by

$$V = L^{-1}(C^T - P \cdot B). \tag{14}$$

In (14) $V$ is a quadratic matrix. With $v_{i,j} = v'_{i,j} \; \forall i \leq n$ and $L'$ as extended $L$ (3-2) reads

$$L' \cdot V' = C^T - P \cdot B. \tag{15}$$

Now we are free to choose the elements $v_{i,j}$, $(i \leq n \wedge j \leq (i-n))$ in such a way $d_{i,j} = 0 \; \forall i \neq j$ is guaranteed.

To get the largest sectors we formulate an optimization criterion. As $D$ is a non-linear function of the elements of $L$: $D = D(l_{i,j})$. As from (10) $d_{j,j}$ has to become small the optimization task reads

$$\sum_{j=1}^{p} (g_j \cdot d_{j,j})^2 \to \min \tag{16}$$

as a weighted sum. The factors $g_j$ are used to give more importance to particular elements $d_{j,j}$.

The task is solved by a software routine using the optimization method of *Powell* [3].

For these calculations we have to use $L'$ and $V'$, because the elements of $V'$ which do not affect (3-2) take influence on (3-3) and $D$.

## 4. Example

Consider the linear time-invariant system

$$\dot{x} = \begin{pmatrix} -0.332 & 0.332 & 0 \\ 0.332 & -0.664 & 0.332 \\ 0 & 0.332 & -0.524 \end{pmatrix} \cdot x + \begin{pmatrix} 0.764 & 0 \\ 0 & 0 \\ 0 & 0.764 \end{pmatrix} \cdot u$$

$$y = \begin{pmatrix} 1 & 0 & 0 \\ 0 & 1 & 0 \\ 0 & 0 & 1 \end{pmatrix} \cdot x, \tag{17}$$

which describes a 3-tank-system, linearized at its operating point, with two inputs and three outputs (Fig. 3).

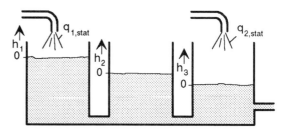

**Fig. 3** *3-tank-system*

The process-inputs $u_i$ are two bulk currents $q_i$, the output variables $y_j$ represent the filling level $h_j$ of each tank $j$. The first problem is to get same dimensions on $u$ and $y$. In this case its simple to extend the system (17) with respect to $u_i = -f_i(y)$. So one can write

$$\dot{x} = \begin{pmatrix} -0.332 & 0.332 & 0 \\ 0.332 & -0.664 & 0.332 \\ 0 & 0.332 & -0.524 \end{pmatrix} \cdot x$$
$$+ \begin{pmatrix} 0.764 & 0 & 0 \\ 0 & 0 & 0 \\ 0 & 0 & 0.764 \end{pmatrix} \cdot \begin{pmatrix} -f_1(y) \\ -f_2(y) = 0 \\ -f_3(y) \end{pmatrix} \quad (18\text{-}1)$$

$$y = \begin{pmatrix} 1 & 0 & 0 \\ 0 & 1 & 0 \\ 0 & 0 & 1 \end{pmatrix} \cdot x, \quad (18\text{-}2)$$

for the extended system.

Starting the optimization with a symmetric, regular $D_0$

$$D_0 = \begin{pmatrix} 0.251 & 0.25 & 0.25 \\ 0.25 & 0.251 & 0.25 \\ 0.25 & 0.25 & 0.251 \end{pmatrix}$$

and $g_1 = g_3 = 100$, $g_2 = 1$, because $d_{2,2}$ is less important than $d_{1,1}$ and $d_{3,3}$, finally it is obtained

$$D_{opt} = \begin{pmatrix} 0.002 & 0 & 0 \\ 0 & 1.248 & 0 \\ 0 & 0 & 0.002 \end{pmatrix}. \quad (19)$$

From (10) and (19) the non-linear feedback block has to satisfy three conditions

$$\frac{f_1(y)}{y_1} \le 500 \quad (20\text{-}1)$$

$$\frac{f_3(y)}{y_3} \le 500. \quad (20\text{-}2)$$

With (20-1) and (20-2) a fuzzy controller can be analysed or designed with respect to asymptotic hyperstability.

In this example the fuzzy controller is given by six rules R1-R6:

R1/R2: If $h_{1/3}$ negative then $q_{1/2}$ positive.

R3/R4: If $h_{1/3}$ zero then $q_{1/2}$ zero.

R5/R6: If $h_{1/3}$ positive then $q_{1/2}$ negative.

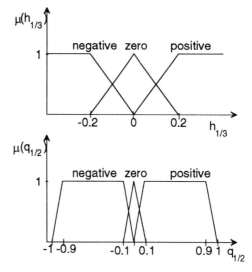

**Fig. 4** *Membership functions for* $h_{1/3}$ *and* $q_{1/2}$

The membership functions are shown in Fig. 4. Using the SUM-PROD-inference and the gravity method for defuzzyfication one can calculate the non-linear characteristics surface $u = f(y)$. Its easy to prove that this fuzzy-controller accomplishes (20-1) and (20-2). Simulations in Fig. 5 show the characteristics of $h_{1/2/3}$.

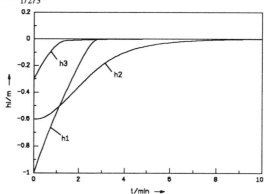

**Fig. 5** *Characteristics of* $h_{1/2/3}$

The results shown in Fig. 5 are better than results reached with *Riccati*-controllers, which use conrol inputs $q_{1/2}$ of comparable height.

## 5. Summary

In this paper hyperstability theory has been used to analyse fuzzy controlled systems. A new method to obtain simple criteria like the multi-dimensional sector criteria is presented which allows the optimization of the admissible sectors. Finally it should be mentioned that it is possible to use these criteria to design fuzzy controllers as well.

## Literature

[1] *Popov, V. M.:* Hyperstability of Control Systems. Springer-Verlag, Berlin, Heidelberg, New York, 1973.

[2] *Landau, Y. D.:* Adaptive Control - The Model Reference Approach. Marcel Dekker, New York, Basel, 1979.

[3] *Press, W. H. e. a.:* Numerical Recipes in C, 2nd ed. Cambridge University Press, Cambridge, 1992.

[4] *Föllinger, O.:* Nichtlineare Regelungen. Oldenbourg Verlag, Munich, 1993 (in German).

[5] *Popov, V. M.:* Absolute Stability of Nonlinear Systems of Automatic Control. Automat. Remote Control, 22 (8), 857-875, Plenum Publishing Corp, 1973.

[6] *Zurmühl, R., Falk, S.:* Matrizen und ihre Anwendungen Teil 1. 5. Aufl., Springer-Verlag, Berlin e.a., 1984.

# Fuzzy-Based Grasp-Force-Adaptation for Multifingered Robot Hands

Th. Dörsam, S. Fatikow and I. Streit

Institute for Real-Time Computer Systems and Robotics
Faculty for Informatics, University of Karlsruhe
Kaiserstr. 12, 76128 Karlsruhe, FRG
e-mail: doersam@ira.uka.de

## Abstract

A fuzzy logic approach for the on-line grasp-force-adaptation, which can be used for the control of fine manipulating with a multifingered robot hand, will be presented in this paper. The kernel of this approach consists of decision making logic which expresses the a-priory knowledge about the force behaviour inside the friction cones and the necessary reactions, regarding the criterion for grip stability. For the software realisation of the fuzzy control approach a Fuzzy-C Development System supporting the entire development process was used. Two corresponding fuzzy controllers have been designed: A finger controller and a grasp controller. During fine manipulations of an object the fuzzy controller interacts with an underlying conventional controller that receives, after defuzzyfication, the adapted force values which were applied. A computer based simulation system was developed to analyse the capabilities of the designed fuzzy controllers.

## 1. Introduction

Intelligent multifingered grippers are used for autonomously performing assembly tasks, like the peg-in-hole insertion. For the successful completion of such tasks a suitable controller is needed due to uncertainties. In spite of intensive research works on autonomous assembly and insertion tasks [1, 2, 3], there exists no general solution for these problems. One reason for this is that the tolerances of the objects, which have to be assembled in industry, are stricter than the accuracy of the used robot grippers.

The peg-in-hole insertion task is an example of such a tasks: One free standing cylindrical peg has to be inserted in a cylindrical, chamfered hole. This task can be successfully performed with a three-fingered robot hand. Compared to a conventional jaw-gripper multifingered hands have the advantage of making feasible various movements and fine-manipulations on the gripped object without moving the supporting arm of the robot. They also have a higher grip stability due to multi-contact points with the object and so more grasp positions on the object can be achieved. One of such robot grippers - the Karlsruhe Dextrous Hand - have been developed at the Institute of Real-time Computer Systems and Robotics (University of Karlsruhe). The hand mounted on a xyz-table can be used for performing the described insertion task because the peg and the hole can be located on different positions in different arrangements.

## 2. The Problematic

For the successful completion of the elementary peg-in-hole insertion task, a suitable control system is needed because uncertainties and noises can degrade the performance of the controller. Noise disturbs the signals of the sensors, thus the execution of the motion commands are inaccurate and the exact position of the grasped object between the fingers is uncertain. The common sources of uncertainty are the noisy signals of the position and force sensors and the inaccuracy of the relative position of gripper and environment on account of the uncertain knowledge of the environment. The uncertainty in the execution of motion commands is due to the dynamics and mechanics of the gripper itself: Back-lash, friction of gearboxes and toothed belts, etc. The third source of uncertainty is ignorance of the exact position of the grasped object between the fingers, since the object could have slipped because of gravity or collisions with the environment. There is also, at the beginning of the insertion task, a source of uncertainty in the grasping of the free standing peg. If the position is not well known the peg gets displaced and is not automatically centred between the fingertips, especially if the fingertips are soft and have a great adhesion coefficient.

The solution of these problems with an off-line planning is based on the assumption that all margins of uncertainty are available and so every movement and every reaction could be planned ahead. This kind of control has the disadvantage of inflexibility. The force control system must be able to handle all these uncertainties for a successful execution of the peg-in-hole task. It is clear that solving of this problem has to rely on the use of advanced methods of information processing, which allow decision making under faulty process states and optimal use of the a-priory information about the grasp force behaviour.

## 3. Solution Feasibilities

Before offering some solutions for the insertion task, the task has to be analysed. It can be decomposed into three phases: Firstly, the peg has to be grasped, compensating for gravity. Secondly, because of the position error due to the above mentioned forms of uncertainty, the reaction forces have to be compensated for by grasp-force-adaptation while inserting the peg into the hole. If these adaptations do not make a successful insertion, possibly a third phase has to initiate a movement of the gripper in consideration of the changed environment conditions.

In this paper we offer two solutions for the second phase of the insertion task: The grasp-force-adaptation. In this phase a control strategy is needed which handles the uncertain information about the state of a peg. All the forces in the contact points between the finger tips and the object are measurable with the help of the force sensors in the first joint of the fingers. But no exact statement could be mentioned about the force adaptation since the adhesive coefficient is not exactly known and the decomposition of the measured forces into a normal and a tangential component can be done roughly when there exists no a-priori-knowledge about the grasped object. The calculated outputs $\Delta N$ of the force adaptation algorithm are the desired values of the inner control loops which the Figure 1 shows.

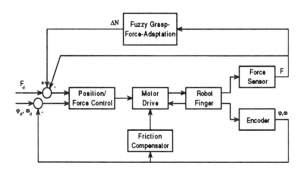

Fig. 1: Fuzzy-based control system

The armature current controller for the DC-motors with friction compensation is on the lowest control level. New research on friction compensation is presently in progress. The simple coulomb-friction-compensation should be improved by the consideration of all friction parts shown in the Stribeck-curve. Therefore fuzzy logic will also be an appropriate solution approach [4]. We have been used the ideas of the fuzzy logic referring to an understanding of the physical laws that describe the dynamic behaviour of grasp forces. The kernel of the approach consists of a decision making logic concerning the a-priory knowledge about the force change during manipulations

## 4. Development of the Fuzzy Controller

To model the grasp force behaviour during object manipulations a reaction-force simulator had to be used. This reaction-force simulator is based on the work of Whitney [5, 6] and considers one-point-contact as well as two-point-contact of the peg and the hole. It successively lowers the peg into the hole and determines the new position of the peg. As a model for the computation of the manipulation forces, a stiffness approach assuming a linear relationship between occurring forces and position changes has been used. The algorithms developed for force adaptation for a multi-fingered gripper are based upon the point contact with friction. The contact between a finger tip and the object is treated as a single point. Further it is assumed, that the finger is capable of only imposing forces on the object but no moments.

Two kinds of fuzzy controllers have been designed: A finger controller and a grasp controller. The finger controller works independently for every finger and calculates one output for every finger whereas the grasp controller considers all the forces on all the fingers and calculates only one output for all the fingers. The force adaptations are based on the following simple control strategy: The contact between the finger tips and the object must be saved. To grant this request the following friction condition has to be satisfied The tangential force must be less than the normal force multiplied by the friction coefficient $\mu$.

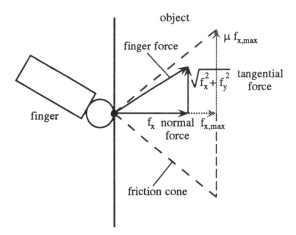

Fig. 2: Finger object contact

This means that the force in the contact point between the fingertip and the object has to be inside the friction cone (Figure 2). The friction cone is constructed by the maximal normal force $f_{x,max}$ that can be exerted by the finger itself. The finger forces acting on the contact point have to obey the conditions:

$$0 \le |f_x| \le f_{max} \, , \, \sqrt{f_y^2 + f_z^2} \le \mu \cdot |f_x|$$

Due to this condition, the set of the allowed finger forces which is called the friction cone is the following:

$$F := \left\{ f \;\middle|\; 0 \le |f_x| \le f_{max} \,,\; \sqrt{f_y^2 + f_z^2} \le \mu \cdot |f_x| \right\}$$

The force controller has to observe the condition under consideration that the minimal distance of the measured force from the lateral area and the bottomface of the friction cone is a maximum (Figure 3). If this condition is satisfied the grasp is called stable:

$$S(f) := \begin{cases} \min\!\big( |f - P(f)|, f_{max} - |f_x| \big) \,,\; \text{if } f \in F \\ 0 \,,\; \text{else} \end{cases}$$

With $P(f)$ being the projection of the finger force on the lateral area of the friction cone.

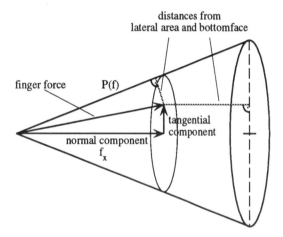

Fig. 3: Friction cone

The input values of the designed fuzzy controllers are the normal and tangential components of the forces in the contact-points, whereas the output value is the normal force adaptation $\Delta N$. This normal force adaptation is the desired value of the inner controller loops on the lower level of the controller system. In designing controllers for grasp force adaptation two routes were taken: A one finger control was designed, which can control every finger by itself and generates one $\Delta N$ for this finger. On the other hand a grasp controller was designed, which takes all contact forces of all the fingers into consideration and calculates one $\Delta N$ for all the fingers.

The finger controller has two input values (normal and tangential component of the contact force) and one output $\Delta N$. By designing this controller care has to be taken to ensure that $\Delta N$ is not to big for getting good adaptive behaviour of the finger. The resultant control strategy is that the resultant force in the con-

tact point must lie inside the friction cone and $\Delta N$ has to be as small as possible. The described a-priori knowledge has been transformed into the following fuzzy rule base:

IF normal_force IS medium THEN $\Delta N$=Null

IF normal_force IS large THEN $\Delta N$=Null

IF tangential_force IS very_small THEN $\Delta N$=Null

IF tangential_force IS small THEN $\Delta N$=Null

IF tangential_force IS medium THEN $\Delta N$=Null

IF tangential_force IS large THEN $\Delta N$=very_small

IF tangential_force IS very_large THEN $\Delta N$=small

IF tangential_force IS extrem_large THEN $\Delta N$=large

IF normal_force IS small AND tangential_force IS NOT very_large THEN $\Delta N$=Null

IF normal_force IS small AND tangential_force IS medium THEN $\Delta N$=small

IF normal_force IS medium AND tangential_force IS very_large THEN $\Delta N$=large

IF normal_force IS medium AND tangential_force IS very_large THEN $\Delta N$=small

The fuzzy controller using this fuzzy inference was implemented with the help of the Fuzzy-C Development System sold by the Togai InfraLogic, Inc. This multi-functional tool supports the entire software development process, from the fuzzy algorithms to the final board implementation on the fuzzy processors. The fuzzy knowledge base is described in the special Fuzzy Programming Language (FPL). For the process of creating a FPL fuzzy knowledge base we have selected a way using the TIL Shell editor, which allows one to enter and edit membership functions, fuzzy rules and variables graphically. As the result of the fuzzy knowledge base description, an output file is generated which can be used as input for either the software or the hardware realisation.

The grasp controller takes all the forces in all the contact points into consideration. Its design is specific to the number of fingers of the gripper - in our case for a three-fingered-gripper. For every other multifingered-hand you have to change the design. For the Karlruhe Dextrous Hand the controller has six input values (a normal and tangential components for each of the three fingers) and one output value $\Delta N$. The control strategy of the grasp controller is that the resultant force in the contact point must lie inside the friction cone and $\Delta N$ has not become to big.

The simulated results of the fuzzy controller show their performance in handling the uncertainties of the peg-in-hole task [7]. The graphical visualisation of this task displaying the finger forces inside the friction cones are shown in Figure 4:

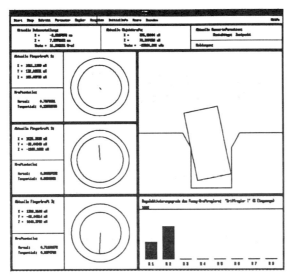

Fig. 4: Simulation of a peg-in-hole insertion

A comparison of the two designed controllers is shown in Table 1. A successful insertion was possible for the listed intervals of deviation of the exact position. The mentioned variables are shown in Figure 5. A look at the Table shows that the finger controller is better than the grasp controller. This result was expected for the finger controller, since it is more practical than the grasp controller. Also the finger controller has the advantage of being independent of the number of fingers. If the number of fingers of the gripper is changed you only have to add or remove finger controllers. Alternatively you have to design a new grasp controller because the number of input variables has changed. Another advantage of the finger controller is that it could be easily implemented on a single processor knot, supporting parallelism of the control system. This is an important aspect because of the design of an intelligent control system on parallel computers, will be done next.

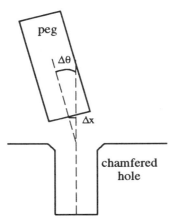

Fig. 5: Deviation of the peg

|  | $\Delta x$ ($\Delta\theta$=10°) | $\Delta\theta$ ($\Delta x$=2.5mm) |
|---|---|---|
| no control | [-1.2mm; 1.2mm] | - |
| finger contr. | [-4mm; 4mm] | [-12°; 12°] |
| grasp contr. | [-3mm; 3mm] | [-10.6°; 10.6°] |

Table 1: Comparison of the designed controllers

## 5. Conclusion

In this paper , we consider the problem of grasp force control for a three-fingered robot hand. Two fuzzy controllers were presented which are able to handle all the uncertainties of the peg-in-hole insertion. The controllers developed were tested by using a computer based simulation system and show good results. This allows force control and reliable operations of intelligent robots performing fine manipulations in unstructured environments.

## References

[1] V.Gullapalli, R.Grupen and A.Barto, "Learning reactive admittance control", Proc. of the IEEE Int. Conf. on Robotics and Automation, pp.1475-1480, 1992.

[2] M. Erdmann: "Multiple-Point Contact with Friction: Computing Forces and Motions in Configuration Space". Proc. of the IEEE/RSJ Int. Conf. on Intelligent Robots and Systems, IROS'93, Yokohama, pp.163-170, 1993.

[3] M. Umezu, K. Oniki: "Compliant Motion Control of Arm-Hand System". Proc. JSME Int. Conf. on Advanced Mechatronics, Tokyo, pp.429-432, 1993.

[4] T. Bertram, S. Svaricek: "Zur Kompensation der trockenen Reibung mit Hilfe der Fuzzy-Logik". Automatisierungstechnik 41 (1993), pp.180-184.

[5] D.E. Whitney: "Quasi Static Assembly of Compliantly Supported Rigid Parts". Journal of Dynamic Systems, Measurement, and Control, Vol. 104/65, 3/1992.

[6] D.E. Whitney: "Force Feedback Control of Manipulator Fine Motions". Journal of Dynamic Systems, Measurement, and Control, pp.91-97, 7/1977.

[7] I.Streit, "Development of a fuzzy-based approach to grasp force adaptation in a robot simulation system", Degree dissertation, University of Karlsruhe, Germany, 1993.

# A REFINED ON-LINE RULE/PARAMETER ADAPTIVE FUZZY CONTROLLER

**Zhuo Li[†], Shi-Zhong He[†] and Shaohua Tan[‡]**

[†]Department of Automation
Tsinghua University
Beijing 100084, China

[‡] Department of Electrical Engineering
National University of Singapore
10 Kent Ridge Crescent, Singapore 0511

## Abstract

A refined on-line rule/parameter adaptive fuzzy control scheme is presented in this paper. Modifying an early design in [1], the refined scheme has made improvements in three aspects. First, the integral action is rearranged to be in parallel with the fuzzy controller to improve steady-state response. Secondly, the rule adaptation is amended to be more reflective of human expert control experiences. Finally, an additional parameter adaptation mechanism is added to further improve on transient response. The refined scheme will be described in the paper along with the analysis justifying various refinements introduced. Simulation analysis is also included to assess the performance improvement and the robustness of the scheme.

## 1 Introduction

A rule-adaptive fuzzy control scheme has been introduced recently to improve the transient performance of a closed-loop fuzzy control system [1]. It has been shown that by using the adaptation mechanism described in the paper, the response can have both fast rise-time and small overshoot. However, it has been found that the steady-state performance may not always be desirable. In particular, due to the fuzzy quantization the integration action which is in series with the fuzzy controller does not seem to be able to compensate for the steady-state error adequately.

To overcome the difficulty, the present paper proposes a refinement of the previous scheme by rearranging the integration in parallel with the fuzzy controller. The analysis shows that such an arrangement is able to correct the steady-state error effectively. Besides, the paper also proposes to revise the $\alpha$ adaptation rule which is a key to realize the rule-adaptation in the original scheme, and to introduce an additional adaptation mechanism for adapting the controller output weighting factor $k_u$. Both ideas are intended to improve further on the transient response of the control system.

The paper is structured as follows. In the ensuring section, simple analysis is used to reveal the source of the steady-state error in the adaptive fuzzy control scheme used in [1], and shows how the proposed rearrangement of the parallel fuzzy integration action reduces the error. Section 3 briefly reviews the operation of the fuzzy-rule adaptation as in the original scheme, and discusses the amendment in adapting the parameter $\alpha$. Section 4 introduces the parameter adaptation mechanism for adapting $k_u$. Simulation analysis is conducted in Section 5 for comparison and performance evaluation, followed by conclusions in the final section.

## 2 Fuzzy integration action

For process control applications especially, the configuration for fuzzy controller appears to have gradually evolved into a relatively standarized form as shown in Fig. 1. Where in the figure, $k_e$ and $k_r$ are weighting factors for $e(t)$, respectively, $\dot{e}(t)$; $k_u$ the weighting factor for the control $u(t)$; and $E$, $R$ and $U$ are the linguistic variables of $e(t)$, $\dot{e}(t)$ and $u(t)$, respectively.

Observe in the configuration that there exists an integrator between $k_u$ and the process, which intends to eliminate the steady-state error. However, the fuzzy nature of $U$ dictates that $u(t)$ can not assume all the values within its operating range for fixed membership functions and defuzzification rule. Consequently, the integration can not guarantee $e(t) \to \infty$ when $t \to \infty$ in a rigorous sense.

In fact, upon assuming the sampling time to be $T$, the control $u(t)$ in the original scheme in [1] can be approximately written as $u(t) = \int_0^t k_u U dt = k_u T \sum_{i=0}^t U_i$, where $U_i$ is the value for $U$ at the $i$th sampling interval. Let the control to be some $u_0$ at the steady-state, then we need to have $k_u T \sum_{i=0}^t U_i = u_0$. This equality is not always true as the $U_i$'s are discrete in nature. Note that when $T$ is fixed, $k_u$ can be made small enough so that $k_u T \sum_{i=0}^t U_i$ falls into a small neighborhood of $u_0$. But small $k_u$ also slows down the system response and prolongs the transient response. Thus varying $k_u$ while keeping everything else unchanged does not entirely solve the problem.

The solution we propose is to rearrange the integral action in parallel with the fuzzy controller as shown in Fig. 2. When this is done, $u(t)$ becomes approximately

$$u(t) = k_u U + k_I T \sum_{i=0}^t E_i \tag{1}$$

in the modified configuration. This simple arrangement effectively decomposes the the control $u(t)$ into two parts: the first part $k_u U$ governs the steady-state response and the second part $k_I T \sum_{i=0}^t E_i$ the transient response. This decomposition allows the adjustment of $k_I$ to a small value to improve the steady-state response without significantly affecting the transient response. We can also take the advantage of the decomposition by modifying $k_u$ for improving the transient response without affecting the steady-state response. In the later part of the paper, one such on-line adaptation mechanism for adjusting $k_u$ will be discussed.

## 3  Fuzzy rule-adaptation

One of the key ideas for rule-adaptation used in [1] is to parameterize the fuzzy control table into the following form

$$U = < \alpha E + (1 - \alpha)R > \tag{2}$$

where $\alpha$ is a parameter. Obviously, the change of $\alpha$ changes the weighting proportion for both $E$ and $R$, thus the rules in the table. In [1], the on-line adaptation of $\alpha$ is devised according to the closed-loop system response profile. Referring to Fig. 3, the closed-loop response profile can be divided into 4 different stages defined by the signs of $e(t)$ and $\dot{e}(t)$. $\alpha$ is then adapted in such a way that it gives appropriate emphasis on $E$ or $R$ during various stages. It is found, however, that in the stages where there is divergent trend, an increase in $E$ feedback appears more appropriate. The design in [1] can also be improved upon whenever $e(t)$ is getting nearer to 0. Moreover, there is possibility to incorporate some sort of global trend into the $\alpha$ adaptation to complement its pure stage-dependent adaptation.

In what follows, we discuss the appropriate adaptation for $\alpha$ in the 4 stages, and the simple way for its implementation. In the OA stage ($e > 0, \dot{e} < 0$), as the system response tends to the set-point, $\alpha$ should be increased to fasten the system response. However, when $e$ is near 0, $\alpha$ should be decreased to avoid overshoot. Thus $\alpha$ in this stage should be greater initially, then gradually gets smaller.

In the AB stage ($e < 0, \dot{e} < 0$), the response diverges away from the set-point. To avoid big overshoot, there should be a large $\dot{e}$ feedback immediately after the system enters this stage, then a large $e$ feedback is required to prevent $e(t)$ from diverging further away. As a result, $\alpha$ changes gradually from the small to the large.

In the BC stage ($e < 0, \dot{e} > 0$), the error starts to reduce again, the system gradually approaches steady-state. $\alpha$ should gradually decrease to keep up with the trend.

Finally, in the CD stage ($e > 0, \dot{e} > 0$), the system gets into an undershoot. $\alpha$ can be maintained at a small value for the same reason given in the AB stage. If the undershoot is small, a small $\alpha$ can also facilitate the system to reverse its trend quickly.

The actual implementation of $\alpha$ is via the following updating equation

$$\alpha(t) = \alpha(t-1) + \gamma h(t) \tag{3}$$

where $\gamma$ is a positive constant governing the rate of change for $\alpha$. $\alpha(0)$ is normally chosen to be 0.9. $h(t)$ is the defuzzified version of a fuzzy variable $H(t)$, which itself is determined by a fuzzy table given in Table 1. In fact, the fuzzy table for $H(t)$ can be designed to generate an arbitrary $\alpha$ trend. This particular fuzzy map shown in the table is design to realize the desired $\alpha$ trend, and is not unique.

Actually, the $\alpha$ trend plotted in Fig. 3 is obtained using Table 1 and $\gamma = 1$. This trend obviously confirms to what we have discussed just now.

## 4  Fuzzy parameter-adaptation

In many situations, the adaptation of $\alpha$ can only effect marginal improvement in the quality of transient response. To improve further, we may consider to introduce parameter adaptation in conjunction with the adaptation of $\alpha$. As we have seen that the change of $k_u$ has a directly bearing on the transient performance. More precisely, the response is fast when $k_u$ is large, but excessive $k_u$ may lead to system oscillation; a small $k_u$ will slow down the response. Further, $k_u$ in general does not affect the steady-state response. Similar to the $\alpha$ adaptation, we can take advantage of these features and design an adaptation for $k_u$ for each of the 4 stages.

Referring to Fig. 3, we note that in the OA stage, $k_u$ should be increased. When the response approaches the point A, $k_u$ should be reduced to avoid overshoot. In the AB stage, to reverse the divergent trend, $k_u$ must be increased. In the BC stage, $k_u$ should be reduced to let the response not deviate away from the set-point, and so on. This analysis shows that the desired $k_u$ trend coincides with that of $\alpha$. Hence for the purpose of $k_u$ adaptation, we can use

$$k_u = k_{u0} \cdot \alpha, \quad \alpha \in (0, 1). \tag{4}$$

This allows the same fuzzy map to be used for the adaptation of both $\alpha$ and $k_u$. Here $k_{u0}$ is a constant factor. The entire system is shown in Fig. 4.

## 5  Simulation analysis

The 2nd-order delay system of the form

$$g(s) = \frac{K}{(T_1 s + 1)(T_2 s + 1)} e^{-\tau s} \tag{5}$$

is used in the simulation analysis. Such a process is representative of a wider range of industrial processes, and therefore qualifies to be a benchmark testing example for all kinds of control techniques. It should be noted that the interest is not so much in controlling this simple process, rather in depicting the ideas and drawing the comparison on a simple basis. The scheme with the proposed modifications is actually aimed at being developed into an effective methodology.

The purpose of the simulation is to highlight the performance gain when the adaptations are introduced. Another purpose is to assess the robustness in performance when certain parameters of the process (5) are varied. Both the set-point and the load disturbance responses are considered.

To show the performance gain due entirely to the $\alpha$ adaptation, we fix $k_u$ and let $\alpha$ adaptive. The closed-loop response of this system with that where $\alpha$ is fixed $\alpha$ (set to be 0.6) is then compared. In the two cases, the parameters are chosen to be $k_e = 10, k_r = 20, k_u = 0.03$ and $K_1 = \tau = 1, K_2 = 0.002$. The sampling time is set at $T = 0.1$sec. The responses are plotted in Fig. 5. We can see for the plot that when $\alpha$ is adaptive, the response has both shorter rise-time and smaller overshoot. Fig. 6 shows the responses when there is load disturbance in the system. Observe that the adaptation of $\alpha$ allows fast recovery from the disturbance.

In the second part of the simulation, we try to compare the response where both $k_u$ and $\alpha$ are adaptive with that where $k_u$ is adaptive and $\alpha$ is fixed. Fig. 7 shows the responses where both controllers are tuned to near optimal for the set of parameters $T_1 = 0.5, T_2 = 1, \tau = 1$ and the

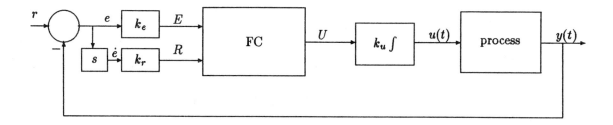

Figure 1: A typical fuzzy control configuration

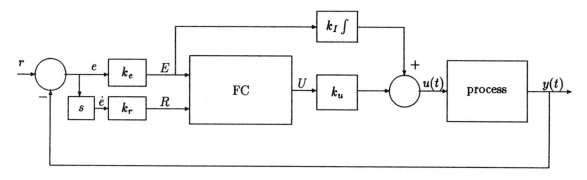

Figure 2: Modified fuzzy control configuration

sampling time $T = 0.1$sec. It can be seen clearly that the adaptation of $\alpha$ provides additional performance improvement even when $k_u$ is adaptive.

To see the robustness in performance for the preceding two cases, we fix all the parameters for both controllers and try to change the process parameters. Fig. 8 shows the case where $T_1, T_2$ are unchanged, but $\tau$ is changed to 4. Fig. 9 shows the case where $\tau$ and $T_2$ are unchanged, but $T_1$ is changed to 2. The plots show clearly that the proposed refinements can still keep up with its performance features even for changing processes.

## 6 Conclusions

This paper has introduced amendments into a previously proposed rule-adaptive fuzzy control scheme. The fuzzy integration is rearranged to improve the steady-state response, the adaptation on $\alpha$ is redesigned and an additional fuzzy adaptation mechanism is introduced to adapt the weighting factor $k_u$ for further response improvement. The analysis has been used to substantiate the changes introduced, and simulation analysis has verified the claims and clearly shows the advantages in adding these refinements.

## References

[1] S. Z. He, S. Tan, C. C. Hang and P. Z. Wang (1993), "Control of dynamical processes using an on-line rule-adaptive fuzzy control system," *Fuzzy Sets and Systems*, Vol. 54, pp. 11-22.

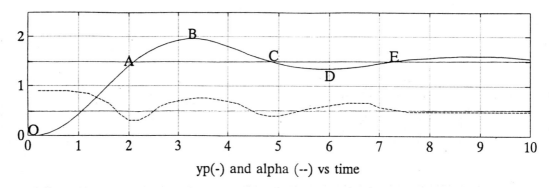

yp(-) and alpha (--) vs time

Fig.3. A typical set-point response and the trend of alpha
in relation to the process output

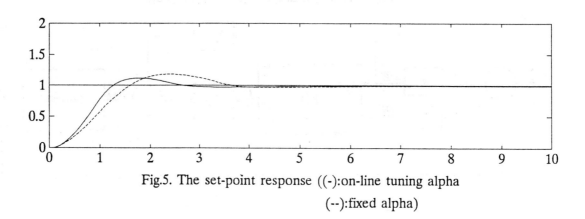

Fig.5. The set-point response ((-):on-line tuning alpha

(--):fixed alpha)

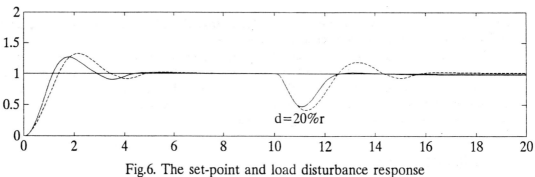

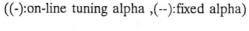

Fig.6. The set-point and load disturbance response

((-):on-line tuning alpha ,(--):fixed alpha)

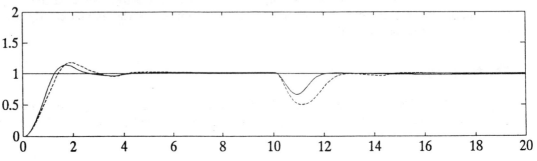

Fig.7. The set-point and load disturbance response

(-):on-line tuning both alpha and ku
(--):fixed alpha

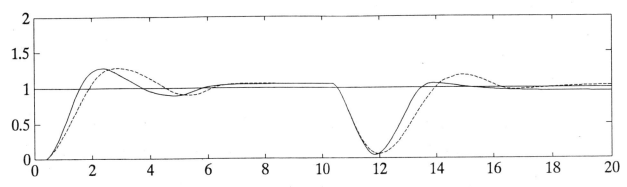

Fig.8. The set-point and load disturbance response

(-):on-line tuning both alpha and ku
(--):fixed alpha

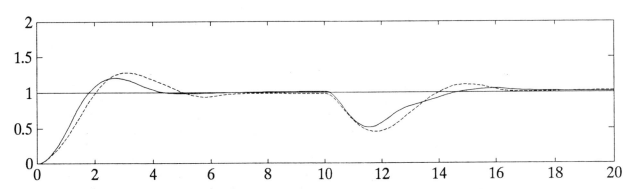

Fig.9. The set-point and load disturbance response

(-):on-line tuning both alpha and ku
(--):fixed alpha

table 1. An example of the fuzzy map to be used to update alpha

| H  EC=-6 | -5 | -4 | -3 | -2 | -1 | 0 | 1 | 2 | 3 | 4 | 5 | 6 |
|---|---|---|---|---|---|---|---|---|---|---|---|---|
| E=-6   2 | 2 | 2 | 2 | 2 | 2 | -.5 | 0 | 0 | 0 | 0 | -1 | -1 |
| -5     2 | 2 | 2 | 2 | 2 | 2 | -.5 | 0 | 0 | 0 | -1 | -1 | -1 |
| -4     2 | 2 | 2 | 2 | 2 | 1 | -.5 | 0 | -1 | -1 | -1 | -1 | -1 |
| -3     2 | 2 | 1 | 1 | 1 | 1 | -.5 | -.5 | -1 | -1 | -1 | -1 | -1 |
| -2     2 | 2 | 1 | 1 | 1 | 1 | -.5 | -.5 | -1 | -1 | -1 | -1 | -2 |
| -1     1 | 1 | 1 | 1 | 1 | 0 | 0 | -.5 | -1 | -1 | -1 | -2 | -2 |
| 0     -1 | -1 | -1 | -1 | -1 | 0 | 0 | 0 | -.5 | -1 | -1 | -1 | -1 |
| 1     -2 | -2 | -2 | -2 | -2 | -1 | 0 | 0 | 1 | 1 | 1 | 1 | 1 |
| 2     -2 | -2 | -2 | -2 | -1 | -1 | 0 | .5 | 1 | 1 | 1 | 2 | 2 |
| 3     -2 | -2 | -2 | -1 | -1 | -1 | 0 | 1 | 1 | 1 | 1 | 2 | 2 |
| 4     -2 | -2 | -1 | -1 | -1 | 0 | 0 | 1 | 1 | 2 | 2 | 2 | 2 |
| 5     -2 | -1 | -1 | 0 | 0 | 0 | 1 | 2 | 2 | 2 | 2 | 2 | 2 |
| 6     -1 | -1 | 0 | 0 | 0 | 0 | 1 | 2 | 2 | 2 | 2 | 2 | 2 |

# ON AN OBJECTIVE-CENTERED ADAPTIVE
# FUZZY CONTROL METHODOLOGY

**Shaohua Tan[†], Yu Lin[†], Yang-Ming Pok[‡], and Pei-Zhuang Wang[*]**

[†] Department of Electrical Engineering
National University of Singapore
10 Kent Ridge Crescent, Singapore 0511

[‡] Department of Electrical Engineering
Ngee Ann Polytechnic
535 Clementi Road, Singapore 2159

[*] Institute of System Sciences
National University of Singapore
Heng Mui Keng Terrace, Kent Ridge, Singapore 0511

## Abstract

This paper presents a new adaptive fuzzy control scheme that is formulated and constructed directly in the control objective space. The formulation on the basis of decomposition of closed-loop response profile is clarified first followed by a detailed description of the scheme. Unlike the existing adaptive fuzzy control methods, the fuzzy controller in the new scheme is fixed and the adaptation is done on the input and output weighting factors of the fuzzy controller. A state-space based approximate analysis technique is employed to analyze the stability of the closed-loop system. A simulation analysis is also conducted to evaluate the controller performance in regulating a structure varying process, and to illustrate the advantage of the scheme in controlling plants that can not be easily handled by other control approaches.

## 1   Introduction

*Plant, controller* and *control objective* are three basic elements in a control system. The traditional control theory appears to be centered at the plant. Within this framework, one designs a control system by first modelling the plant in terms of a suitable mathematical model, then constructing the controller based on the model to meet the control objective. It can be expected that the effectiveness of this paradigm (often termed model-based control) will, to a large extent, depend upon effective modelling.

Many real world plants are often too complex to be modelled effectively. A typical situation would be that a proposed mathematical model structure is unable to capture the relevant nature of a plant all the time. Consequently, the controller designed based on the inadequate model structure can not function as desired all the time. Effective modelling also implies simplicity. Sometimes, a faithful model of a plant is simply too complex to be of use for the controller design. These and other considerations have raised the need to seek for other control system design paradigms [1].

In many practical control situations, human experts often act as satisfactory controllers in certain control systems. In these cases, an alternative paradigm for the design of a controller would be to model directly the control actions by human experts, and code them in the form of discrete rules in a knowledge base. The controller should then provide appropriate mechanisms for invoking and interpolating the rules in the knowledge base in accordance with various closed-loop response conditions. This paradigm is obviously centered at controller rather than plant, and is commonly known as knowledge-based control. In fact, The conventional expert control and fuzzy linguistic control approaches are seen as special instances of this paradigm.

Observe the three ingredients that necessitate the use of the knowledge-based control paradigm: the availability of experts, the proper extraction of their knowledge and the appropriateness of the inference mechanism. Absence of any of them will render the paradigm inapplicable. Moreover, none of these ingredients can be easily quantified thus judged with precision. This explains the ad hoc nature of the knowledge-based control, and the amount of subjectivity associated with the designed control systems.

Effective within their own scopes, though, the aforementioned two paradigms still can not handle a large number of control problems that are both hard to model due to the plant complexity, and difficult to build a knowledge base for lack of experts. This motivates yet another paradigm that is centered on the control objective. One of the aims of the present paper is to clarify such an objective-centered paradigm that has appeared piecemeal in recent fuzzy control literature, and has yet to be identified as a coherent framework for control system design.

With such a framework in hand, the paper will propose a novel adaptive fuzzy control scheme as one of the concrete forms of this paradigm. Among other things, the detailed functional blocks in the scheme will be discussed in line with the objective-centered principle. One of the important results is that the control scheme is proven to function as desired using a state-spece technique. As an application of the scheme, the set-point change of a plant with sudden substantial structural variation is chosen as a benchmark test. Both the model-based control and knowledge based control will be shown to be completely ineffective, whereas the new scheme will be able to generate satisfactory control.

## 2  Objective-centered control methodology

Let us start by considering the following control problem: Given a dynamical plant with single input $u(t)$ and single output $y(t)$, where $u(t)$ and $y(t)$ are both real functions of time $t$. A controller needs to be constructed such that it generates an appropriate control $u(t)$ that drives $y(t)$ to a prescribed set-point $y_r$.

When such a control problem is handled within the framework of model-based control, the first step taken is to assume that there is a mathematical model that links $u(t)$ and $y(t)$. Then various assumptions will be made to facilitate the construction of the model. The controller design will then be entirely based on the model.

In the framework of knowledge-based control, the control design is done by appropriately extracting and representing the knowledge of an existing controller often in the form of human expert. The effectiveness of the design relies on such conditions as the existence of experts and the availability of effective interpolation techniques.

A less familiar framework for the control problem is targeted at the control objective. One possible formulation of such a framework is as follows. As the control objective is to drive $y(t)$ to the given set-point $y_r$, we can introduce an error term $e(t) = y_r - y(t)$ to measure how successful the control objective is accomplished. Obviously, $e(t)$ will have to assume one of the following three cases

$$e(t) > 0, \quad e(t) = 0, \quad e(t) < 0 \tag{1}$$

at any time instance $t$. Of the three, $e(t) = 0$ is what we desire. However, as the underlying plant is dynamical, the equivalence can not be achieved exactly in finite time, and the best we can hope for is $y(t)$ gets within some neighborhood of $y_r$. The size of such a neighborhood, however, needs to be specified explicitly for a specific control problem. With this consideration, (1) is modified to be

$$e(t) > \epsilon, \quad -\epsilon < e(t) < \epsilon, \quad e(t) < -\epsilon \tag{2}$$

where $\epsilon$ is a prescribed small positive real number. As $e(t)$ is a function of time, we can further distinguish whether $e(t)$ approaches to 0, and diverges from 0 in the first and the third cases by examining the sign of $\dot{e}(t)$. With all these considerations, we arrive at 5 different regions that indicate various possible positional and trend relationships between $y(t)$ and $y_r$: $e(t) > \epsilon, \dot{e}(t) < 0$ in Region I; $e(t) > \epsilon, \dot{e}(t) > 0$ in Region II; $e(t) < -\epsilon, \dot{e}(t) < 0$ in Region III; $e(t) < -\epsilon, \dot{e}(t) > 0$ in Region IV and $-\epsilon < e(t) < \epsilon$ in Region V.

Fig. 1 shows geometrically these regions on a typical response plot of $y(t)$ relative to $y_r$. Note that each of these regions has a clear interpretation in terms of the position and trends of $y(t)$ relative to $y_r$. For example, $y(t)$ in Region I is convergent from below to $y_r$, and $y(t)$ in Region II diverges away from $y_r$. As we shall make clear later, the transition between any two different regions will be gradual through the use of fuzzy transition.

It should be realized that the about partition into 5 different regions is only based on $e(t)$ and $\dot{e}(t)$. In fact, we can go to higher order derivatives of $e(t)$ and introduce more complex partitions. However, this partition seems to capture the human intuition best, yet simple enough to depict the essential ideas of the methodology.

Within each of the regions defined above, the original control objective reduces to a set of local control objectives that are relatively easy to realize. For instance, as $y(t)$ in Region I is convergent to $y_r$, the control should be so designed as to facilitate the convergence process. The regionalization is in effect a specific type of task-decomposition. Such a decomposition breaks up a complex task into manageable small tasks that can then be solved relatively easily. Decomposition is the essence of objective-centered control approaches. It should also be noted that the knowledge of the plant to be controlled enters into this framework not in the form of conventional dynamical equations, but in a qualitative form such as if $u(t)$ goes higher than $y(t)$ goes higher. Acquiring this latter qualitative type of knowledge about a plant is much less demanding than finding a dynamical model for it.

The idea of task decomposition has long been a familiar notion. It not only is one of the key ideas in the field of AI, but also has been used previously in many learning control schemes [2] [3]. However, their objective-centered nature is often obscured by their reliance on other approaches.

With the preceding discussion, we now proceed to elaborate on a specific adaptive fuzzy control scheme that will generate appropriate control for each of the regions. We shall first present the structure of the controller, then provide analysis and performance assessment for the control scheme.

# 3 Controller structure and operation

Referring to Fig. 1 for the decomposition of the objective space we have introduced, we need to design a controller that will generate appropriate control actions for each of the five regions. Specifically, the controlller will have to be such that as $y(t)$ in Regions I and III is moving towards $y_r$, the controller will have to facilitate such a trend; similarly, as $y(t)$ in Regions II and IV is moving away from $y_r$, the controller will have to reverse the trend; finally, as $y(t)$ in Region V is in the prescribed neighborhood of $y_r$, the controller will have to provide actions to let $y(t)$ home in quickly.

These intuitive thoughts have led us to develop an adaptive fuzzy controller that appear to be capable of generating these appropriate control actions. In what follows, we shall detail the structure of the controller.

Fig. 2 shows the general functional block diagram of the controller. It consists of two fuzzy controllers, called fuzzy PD-controller, and fuzzy I-controller, that are working in parallel. The former is reminiscent of a conventional PD controller, and the latter of an I controller. Of course, the similarities are only superficial as they are nonlinear controllers in nature. This arrangement is in recognition of the special requirement in Region V compared with other four regions. In this regard, the Fuzzy I-controller will only act in Region V, and the Fuzzy PD-controller will function in all five regions.

The structure of fuzzy I-controller is shown in Fig 3. The controller part is a familiar integrator whose action is further modified by an adaptive weight $k_i$. Because of the construction, the controller itself is not fuzzy. The fuzzy comes in when the parameter $k_i$ is adapted. This adaptation is done directly using a fuzzy table given in Table 1 which takes $e(t)$ and $\dot{e}(t)$ as inputs, and produces $k_i$ as its output. The precise nature of the adaptation is dependent upon the fuzzy table which has been constructed in such a way that it switches the controller on when in Region V. As this switching is done using a fuzzy mapping, the boundary of Region V becomes gradual rather than crisp. Another point to note is that the fuzzy table dictates both the size of Region V and the magnitude of $k_i$. It can be tuned to suit specific situations. Our experience shows that Table 1 is sufficiently general for many examples we have tested.

Fig 4 shows the detail of the fuzzy PD-controller. This controller consists of a standard fuzzy controller with inputs $e(t)$ and $\dot{e}(t)$ and the output $u(t)$ which are weighted by $k_e, k_r$ and $k_u$

respectively. The fuzzy control table is given in Table 2. The weights are adapted by fuzzy tables shown in Table 3. The underlying idea of the adaptation tables is to apply different weights in different regions. And the fuzzy tables naturally provide a mechanism for smooth transition among these regions. In all these fuzzy tables, the triangular membership functions and the compositional rules of inference are assumed.

Let us examine the operation of the controller in detail using the given fuzzy tables. When the closed-loop response falls into Region I and Region III, Table 3 produces the set of weights $k_e = 5, k_u = 5, k_r = 0.3$ which strengthen $e$ and $u$, and lower $k_r$. Consequently, the controller acts more like a bang-bang proportional error control with rapid but smooth transition between its control actions. This control action is designed to generate a fast response in the two Regions to meet the sub-objectives.

When the closed-loop response falls into Regions II and IV, Table 3 generates the set of weights $k_e = 0.3; k_u = 0.3; k_r = 5$ which strengthen $\dot{e}$ and lowers both $e$ and $u$. The effect of the control is like a derivative bang-bang control with rapid but smooth transition between its control actions. Together with a small $k_u$, this is designed to slow down and reverse the divergent trends by quickly getting out of these regions.

Finally, when the closed-loop response is in Region V, The weights generated by Table 3 are $k_e = 1, k_u = 1$ and $k_r = 1$ corresponding to a normal PD control action. The I-controller is also activated in this region producing the integral control action that then is combined with the PD control to form a complete PID control action.

## 4    Analysis and simulation

A state-space analysis is used to provide a theoretical basis for the control scheme described above. The details of the analysis will not be given for lack of space. Instead, we shall describe the essential elements in the analysis.

The key questions the analysis tries to answer are how the controller functions, and under what conditions such function is possible. To answer the questions, the controller action is examined in the state-space spanned by $e$ and $\dot{e}$. The adaptive nature of the controller and the fuzzy controller itself induces a partition in the state-space. In each of the partitioned regions, the adaptive fuzzy controller reduces approximately into a relatively simple linear controller. The fuzzy control formulation ensures that the transitions among these controllers are smooth, and the fuzzy adaptations attempt to modify the sizes of these regions to maintain $e(t)\dot{e}(t) < 0$ at almost all time instances $t$. Roughly speaking, the function of the adaptive fuzzy controller can be seen as minimizing the cost function $J = e^2(t)$. Such a minimization is done by maintaining the opposing signs for $e(t)$ and $\dot{e}(t)$. Note that the distinct feature here is that instead of going downhill to maintain $e\dot{e} < 0$ all the time, momentary uphill motion is allowed. If this happens, the adaptations will act to reverse such motion. The stability of the overall closed-loop system is thus maintained. The transient behaviour can also be modified by the way the adaptations are designed to update the sizes of the regions. As to the applicability of the scheme, it has been found that plants that have $\Delta y/\Delta u > 0$ and do not have considerable delay from $u$ to $y$ can be controlled relatively satisfactorily by the control scheme, whether they are linear or not, structure-varying or not.

To evaluate the performance of the control scheme, we apply it to regulate a structure-varying plant constructed as

$$g(s) = \frac{k_1 \alpha}{(T_1 s + 1)^2} + \frac{k_2(1 - \alpha)}{(T_2 s + 1)^3},\tag{3}$$

where, for the sake of definiteness, $k_1, k_2, T_1$ and $T_2$ are all set to 1. The sampling rate is chosen to be 0.1. $c_1 = 0.3, c_2 = 1, c_3 = 4.5$ are used for the adaptation tables for $k_e, k_u$; and $c_1 = 4.5, c_2 = 1, c_3 = 0.3$ are used for the adaptation of of $k_r$. Finally, $i_1 = 1.5$ and $i_2 = 2$ are used for the adaptation of $k_i$. Note that all these parameters are obtained after several rounds of fine-tuning. No knowledge of the plant is assumed in the tuning process.

The parameter $\alpha$ changes in such a way that it is 1 for the first 30 seconds, then drops to 0 for the next 30 seconds, and 1 again. The set-point change coincides with the change of $\alpha$. The closed-loop response of the adaptive fuzzy control scheme is shown in Fig. 5.

For the purpose of comparison, both a traditional PID controller and a fuzzy controller have been used to regulate the same plant with the responses shown in Fig. 6 and Fig. 7, respectively. Both controllers have been tuned with considerable effort, and their responses are hardly comparable to that of the adaptive fuzzy control scheme.

## 5  Discussions and conclusions

An adaptive fuzzy control scheme has been introduced and discussed in some detail in the light of objective-centered control framework. It is shown to be applicable to control problems that are considered to be hard in other controller design frameworks. Analysis also appears possible to understand the underlying mechanism of the scheme, though no elaboration has been done in the present paper. It should be understood, however, that the present scheme claims no universality, and does have its limitations. It is in its present form only suitable for SISO plants, and structurally not deep at all. It can not deal with plants that have a long time-delay or large non-minimum phase effect. Its advantage is that by relying on a non-model based formulation, it is able to regulate plants that do not fit into the descriptions of conventional model-based formulation. In this sense, it complements other control paradigms.

## References

[1] L. A. Zadeh (1973), "A rational for fuzzy control," *J. Dynamic Systems, Measurement and Control*, Vol.3, p.4.

[2] A. G. Barto, R. S. Sutton and C. W. Anderson (1983) "Neuronlike adaptive elements that can solve difficult learning control problems," *IEEE Trans. Syst., Man Cybern.*, Vol. SMC-13, pp. 834-846.

[3] Z. J. Nikolic and K. S. Fu (1966), "An algorithm for learning without external supervision and its application to learning control systems," *IEEE Trans. Autom. Control*, Vol. AC-11, pp. 414-422.

Table 1: Fuzzy I-adaptation table

|   | $K_I$ | \multicolumn{5}{c}{$Y$} |   |   |   |
|---|---|---|---|---|---|---|
|   | $K_I$ | -2 | -1 | 0 | 1 | 2 |
|   | 0 | 0 | 0 | 0 | 0 | 0 |
|   | 0 | 0 | $i_2$ | $i_2$ | 0 | 0 |
| $E$ | 0 | 0 | 0 | $i_1$ | 0 | 0 |
|   | 1 | 0 | 0 | $i_2$ | $i_2$ | 0 |
|   | 2 | 0 | 0 | 0 | 0 | 0 |

$i_1 = 1.5, i_2 = 2$

Table 2: Fuzzy PD control rules

|   | $U$ | \multicolumn{5}{c}{$E'$} |   |   |   |
|---|---|---|---|---|---|---|
|   | $U$ | -2 | -1 | 0 | 1 | 2 |
|   | -2 | -2 | -2 | -2 | -1 | 0 |
|   | -1 | -2 | -2 | -1 | 0 | 1 |
| $E'$ | 0 | -2 | -1 | 0 | 1 | 2 |
|   | 1 | -1 | 0 | 1 | 2 | 2 |
|   | 2 | 0 | 1 | 2 | 2 | 2 |

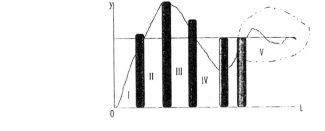

Figure 1: Division of five regions

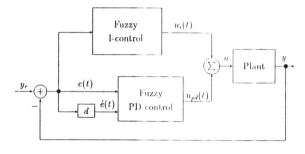

Figure 2: **The general block diagram of the adaptive fuzzy control scheme**

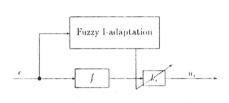

Figure 3: The functional blocks of the fuzzy I-control

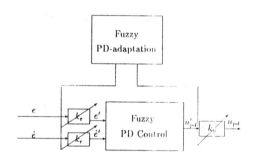

Figure 4: **The block diagram of the fuzzy PD control**

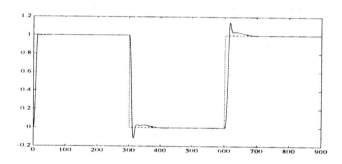

Figure 5: The set-point response of the adaptive fuzzy control scheme

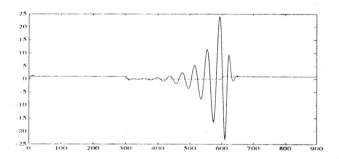

Figure 6: The set-point response of PID control

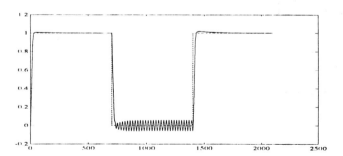

Figure 7: The set point response of the conventional fuzzy control

1483

# Fuzzy Control Used in Robotic Arm Position Control

*Zhang Nianzu, Zhu Ruhui, Fan Maoji*
*Dept. of Automatic control*
*Shanghai Jiao Tong University,*
*Shanghai, 200030, PR. China*

Abstract: Fuzzy control used in robotic arm position control is described in this paper. By the using of fuzzy control, the difficulty for control because of nonlinear factors, such as dead zone, resistant torques not equal in positive and neggative directions, is overcome. With adopting the self regulating factor, accuracy position control is obtained and overshoot is controlled effectively. At last the experimental results are presented in this paper.

## 1.Introduction and system structure

In robotic position control, PI and PID are used extensively. Although these control methods are more mature, there exist some limits. After PID's parameters are determined, the control is sensitive to the shift of system parameters. Also, it is difficult to overcome the affect of nonlinear factors in system.

What we work at is flexible robotic arm experimental equipment. For discussing the application of fuzzy control in position control, we consider the case in single link. The structure of the exprimental equipment can be seen in figure 1. The link is made of a aluminium square hollow beam, with 76cm in length.

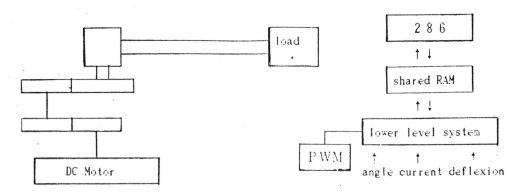

Fig.1 Robot arm structure        Fig.2 Control system structure

The load in endpoint weighs about 10 kilograms. The link is driven by DC servo motor through a gear box with the speed ratio 138/3. Control system is composed of two level computer systems[1].

The higher level is a PC 286 computer and the lower level is 8031 based system. The two levels communicate through a shared RAM. The realization of algorithm, digit analyses, and results outputing are finished in higher level. In the lower level, a 12 bit A/D convertor is used to sample the armature current of DC motor, deflextion and vibration of the flexible arm. From a photoelectric pulse generator, which connected with the axle of DC motor, acquires angle signal, with 10000 pulses per rotation of motor.

In the same time, the lower level is also resposible for the management of a DC PWM; its modulating frequence is 4KHZ. The pulse width coeffcient is from − 5000 to 5000 that correspond to armature voltage from –110V to 110V. The structure of the control system is shown in figure 2.

Because of the severe nonlinear factors in the system, it is difficult to describe the system with linear model and get accurate and smooth control with traditional control methods. In figure 3, there shows the response curve of angle to sine input. We can see the resistant torques in the different directions are not equal. Besides that, a severe dead zone exsits in system.

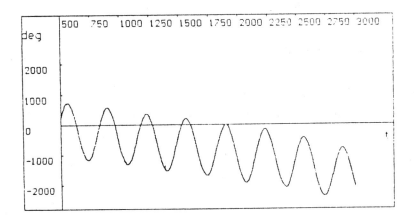

Figure 3. Response curve of angle to sine input

2. Control

As the robotic experimental equipment is for a flexible link, its control is divided into several periods: start, stable velocity, brake and position control[2]. When angle error is small to a some extent, the position control begins. This time, the velocity and acceleration are very small because of braking and the vibration has been not obvious.

In the position control, we once adopted PID control, but the results were not satisfactory. The first problem is that dead zone is too large. If the integral coeffcient is small, the control accuracy will be low. The control is not smooth because the motor may stop within the dead zone and start suddenly as the error integral and the control goes beyond the dead zone. If the integral coeffcient is large, a great overshout and oscillation may be resulted. The transition time is long and the control accuracy is hard to guarantee. The another problem is of less identity in experimental results; we may get a different control result in one position point from other position points, which brings great difficulty to PID paramaters deciding through experiment. All these problems made us have a try to use fuzzy control[3]. The control structure is shown in figure 4.

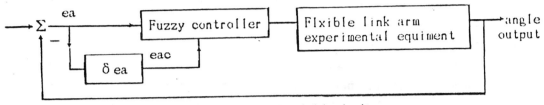

Fig.4    Fuzzy control block diagram

In figure 4, ea is the error of angle and eac is its increment. Suppose E is fuzzed varable of ea, EC is that of eac and U is that of control. The fuzzy set of these varables are

{ NB, NM-, NM, NM+, NS-, NS+, NO, O, PO, PS-, PS+, PM-, PM, PM+, PB},

that is { Negative Big, Negative Medium Minus, Negative Medium,

Negative Medium Plus, Negative Small Minus, Negative Small Plus,

Negative Zero, Zero, Positive Zero, Positive Small Minus,

Positive Small Plus, Positive Medium Minus, Positive Medium,

Positive Medium Plus, Positive Big }.

The whole fields of these varables are

{ -7, -6, -5, -4, -3, -2, -1, 0, 1, 2, 3, 4, 5, 6, 7 }.

In this experimental equipment, the sample value of angle is 10000 pulses per rotation. As the angle error equals to 1500 pulses,the position control begins. We use the rules below to fuzz the angle error.

$$\text{IF } ( \quad ea > 1000), \quad \text{THEN } E=PB \text{ AND } E=7;$$
$$\text{IF } (1000 > ea > 500 ), \quad \text{THEN } E=PM+ \text{ AND } E=6; \cdot$$
$$\text{IF } ( 500 > ea > 100 ), \quad \text{THEN } E=PM \text{ AND } E=5;$$
$$\text{IF } ( 100 > ea > 30 ), \quad \text{THEN } E=PM- \text{ AND } E=4;$$
$$\text{IF } ( 30 > ea > 15 ), \quad \text{THEN } E=PS \text{ AND } E=3;$$
$$\text{IF } ( 15 > ea > 5 ), \quad \text{THEN } E=PO+ \text{ AND } E=2;$$
$$\text{IF } ( 5 > ea > 1 ), \quad \text{THEN } E=PO- \text{ AND } E=1;$$
$$\text{IF } ( |ea| < 1 ), \quad \text{THEN } E=O \text{ AND } E=0;$$
$$\text{IF } ( -1 > ea > -5 ), \quad \text{THEN } E=NO+ \text{ AND } E=-1;$$
$$\text{IF } ( -5 > ea > -15 ), \quad \text{THEN } E=NO- \text{ AND } E=-2;$$
$$\text{IF } ( -15 > ea > -300), \quad \text{THEN } E=NS \text{ AND } E=-3;$$
$$\text{IF } ( -30 > ea > -100), \quad \text{THEN } E=NM+ \text{ AND } E=-4;$$
$$\text{IF } (-100 > ea > -500), \quad \text{THEN } E=NM \text{ AND } E=-5;$$
$$\text{IF } (-500 > ea > -1000), \quad \text{THEN } E=NM- \text{ AND } E=-6;$$
$$\text{IF } ( -1000 > ea), \quad \text{THEN } E=NB \text{ AND } E=-7.$$

The angle error increment is defined as dea=a(i)-a(i-1). Where a(i) is the present angle value and a(i-1) is angle value of the last time. The rules to fuzz the increment of the angle error are presented as below:

IF ( dea>100), THEN EC=PB AND EC=7;

IF (100>dea>65 ), THEN EC=PM+ AND EC=6;

IF ( 65>dea>40 ), THEN EC=PM AND EC=5;

IF ( 40>dea>20 ), THEN EC=PM– AND EC=4;

IF ( 20>dea>10 ), THEN EC=PS AND EC=3;

IF ( 10>dea>5 ), THEN EC=PO+ AND EC=2;

IF ( 5>dea>2 ), THEN EC=PO– AND EC=1;

IF ( |dea|<2 ), THEN EC=O AND EC=0;

IF ( –2>dea>-5), THEN EC=NO+ AND EC=-1;

IF ( -5>dea>-10), THEN EC=NO– AND EC=-2;

IF ( -10>dea>-20), THEN EC=NS AND EC=-3;

IF ( -20>dea>-40), THEN EC=NM+ AND EC=-4;

IF ( -40>dea>-65), THEN EC=NM AND EC=-5;

IF ( -65>dea>-100), THEN EC=NM– AND EC=-6;

IF ( -100>dea), THEN EC=NB AND EC=-7。

In the control, we adopt the control strategy with weight factor to regulate the weight of angle error and its increment in control.

U=Ke×Ue – Kec×Uec,

where $0 < ke < 1$, Kec = 1 – ke.

Because the resistant torque in positive is larger than the negative one, we adjust the control values accordingly.

Let U(*)={ U(-7), U(-6), U(-5), U(-4), U(-3), U(-2), U(-1), U(0), U(1), U(2),

U(3), U(4), U(5), U(6), U(7) }

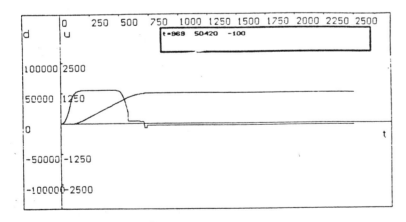

Figure 5. Response curve

={ -400, -300, -250, -200, -100, -80, -50, 0, 60, 90, 110, 210, 260, 310, 410},

where the values of U(*) are PWM pulse width coefficient. When Ke=0.7, the control result can be seen in figure 5. The position error exists due to the dead zone. In this case, the angle given is 50000 pulses.

1487

The values in figure 5 stand for time, angle and control. Here time is expressed with sample number; the sample frequence is 500HZ. Angle is expressed with pulses and control with pulse width coefficient. Because the the dead zone is too large and the control can not overcome it, the position error is 420 pulses. In order to make control accurate, we must increase the control value. Let

$$U(*)=\{ -520,-500, -450, -380, -350, -250, -150, 0,$$
$$170, 270, 370, 400, 470, 520, 540\}.$$

The control curve is shown in figure 6, from which we know that we have reach accurate position control; the error is one pulse. But we can also see that there is a great overshout appearing at 649, of 1075 pulses. To reduce the overshout, the weight to the increment of angle error should be increased properly. But on the other hand, if Ke is samller, the control accuracy will be lower.

For example, Ke=0.4, then Kec=0.6.

At this time, if Ue=2, Uec=0, then $U = Ke \times Ue - Kec \times Uec$

$$=2 \times 0.4 - 0 \times 0.6 = 0.8.$$

After integralizing, U=0. In this case, from the rules above, the position error will be 1 < ea < 5, that means the control accuracy has been lower. To solve the contridiction between accuracy and overshout, the weigh factor should be varible, which should increase with the decreasing of the increment of the angle error. The caculation of Ke shown as below:

$$Ke = 1 - Uec/10 .$$

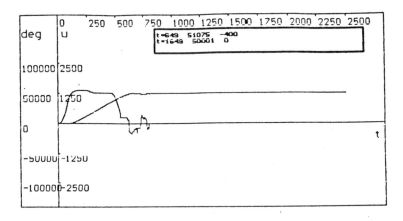

Figure 6. Response curve, the given angle is 50000

Figure 7a is the response curve when the given angle is 50000 pulses, from which we know the overshout is 12 pulse and the position error is one pulse. Figure 7b and 7c show the control results respectively when the given angles are 55000 and 60000 pulses. The control characteristics and identity are satisfactory.

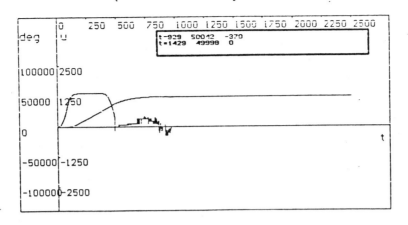

Figure 7.　a.Given angle 50000

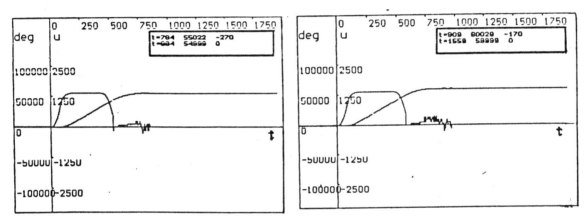

Figure 7. b.Given angle 55000　　　　Figure 7. c.Given angle 60000

Figure 7 Response curve of angle with self-regulating weight

3.Conclude

We adopt fuzzy control to the position control of a robotic arm, with self regulating weight factor, having overcome the severe nonlinear affect to system control. Experiment has verified the control characteristics are superior to PID and the controller parameters can be decided easily.

Reference

[1].Zhang Nianzu, Fan Maoji

A Distrubuted Computer System Used for Robot Ccontrol

Microcomputer Aplications, 1993, Vol.14, No. 3 ,pp.8-12

[2].Zhang Nianzu, Zhu Ruhui, Fan Maoji

Exprimental Research on One-link Flexible Robotic Arm Control

Journal of Fozhou University, 1993, Vol.21, No.5 pp.164-168.

[3].Li Shiyong

Fuzzy control and intelligent control theroy and applications

The Press House of Haerbin University of Technology

# Fuzzy–Variable Structure–Robust Model Reference Adaptive Combined Control

Ye Zhang      Zhong–Ren Liu      Shi–Zhong He

Department of Automation, Tsinghua University, Beijing 100084, China

Abstract: This paper presents a novel fuzzy–variable structure–robust model reference adaptive combined control scheme. The characteristic of this scheme lies in the improvement of the tracking performance of uncertain plants in smoothness, steady–state accuracy and robustness. The new controller not only maintains good output–tracking property, but also removes the oscillation brought out by variable structure control by replacing the sign function in old schemes with a fuzzy control output.

Keywords: Fuzzy control (FC), variable structure control (VSC),robust model reference adaptive control (RMRAC)

## 1. Introduction

To strengthen the robustness of control systems, several techniques have been proposed, including modification of conventional control laws[1,2], combination of different kinds of schemes[3,4,5] and introduction of intelligent concepts into system design[6,7], etc.

Robust adaptive control is an important approach for solving the control problem of uncertain plants. Many papers have been published in this field. Ioannou presented a general approach for designing robust adaptive control and the proof of its robustness[1]. But this scheme requires signal vector, as well as parameter vector which is adjusted by the adaptive law. When there exists a large initial error in parameter vector, it usually takes a certain amount of time to adjust the parameters to their matching values, so the dynamic performance of the control system is not satisfactory.

Fu proposed a new robust model reference control (RMRC) of multivariable unknown plants[5] based on variable structure control to overcome the above mentioned difficulty. The scheme which is a kind of null–estimation approach improved the dynamic performance while keeping the robustness and stability property. Because in VSC systems, oscillation usually occurs in steady state, therefore, the steady–state accuracy is relatively lower.

FC forms its control signal based on error($e$) and its differential($\dot{e}$), which is also the main characteristic of VSC. The dead zone of FC systems is a drawback to the improvement of its steady–state accuracy, but when integrated with VSC properly, FC can smooth the response. Furthermore, FC has more robustness compared with other non–intelligent control schemes.

In the fuzzy – variable structure – robust model reference adaptive control (F–VS–RMRAC) we proposed, robustness and stability is guaranteed by the RMRAC, and VSC improved rapidity of system. With the combination of VSC with FC, the former application is enlarged, the robustness is enhanced, meanwhile the control is easier to be realized.

In the section below, we first state the general scheme of RMRAC[1] and VS–RMRC[5]. In section 3, the analysis of F–VS–RMRAC is given. Some simulation results and remarks are given in section 4.

## 2. RMRAC

A general RMRAC scheme was proposed based on several modification ones[1]:
Plant:

$$y = G(s)u + \rho \qquad (2.1)$$
$$G(s) = G_0(s)(1 + \mu\Delta_m(s)) + \mu\Delta_a(s) \qquad (2.2)$$

$$G_0(s) = K_p \frac{Z_0(s)}{R_0(s)} \qquad (2.3)$$

where u,y are the input and output signals of plant, $\rho$ represents a bounded output disturbance. G(s) is a strictly proper transfer function; $G_0(s)$ represents the modelled part of the plant; $\mu\Delta_m(s)$, $\mu\Delta_a(s)$ are multiplicative and additive plant uncertainty; $\mu$ is a positive scalar representing the amplitude of unmodelled parts. And $K_p$,$Z_0(s)$,$R_0(s)$ denote the gain of plant, zeros polynomial and poles polynomial respectively.

Reference model:

$$y_m = W_m(s)r = K_m \frac{1}{D_m(s)} r \tag{2.4}$$

where r represents any bounded piecewise continuous reference input, $K_m$ is a positive constant, and $D_m(s)$ is a monic Hurwitz polynomial.

Assumptions:

1. $\mu\Delta_a(s)$ is a strictly proper transfer function.
2. $\mu\Delta_m(s)$ is a stable transfer function.
3. p is the stable domain of poles of $\mu\Delta_a(s-p)$ and $\mu\Delta_m(s-p)$, and p > 0; $p_0$ which is the lower bound of p, is known, $p_0 > 0$.
4. $R_0(s)$ is a monic polynomial of degree n.
5. $Z_0(s)$ is a monic Hurwitz polynomial of degree $m \leqslant (n-1)$.
6. The sign of $K_p$ is known.
7. The relative degree $n^* = (n-m)$ is known.

when the degree $[D_m(s)] = n^*$, the formula of control u can be expressed as:

$$u = g_1(s,\theta_1)\frac{1}{\Lambda(s)}u + g_2(s,\theta_2)\frac{1}{\Lambda(s)}y + \theta_3 y + c_0 r \tag{2.5}$$

where $\Lambda(s)$ is an arbitrary monic Hurwitz polynomial of degree n−1, $g_1(s,\theta_1)$, $g_2(s,\theta_2)$ are polynomials of degree n−2, which has coefficients of $\theta_1,\theta_2 \in R^{n-2}$, $\theta_3, c_0$ are scalars. It has been proved that $\theta_1^*, \theta_2^*, \theta_3^*$ and $c_0^*$ exist, satisfying the following matching condition[1]:

$$\frac{c_0^* K_p \Lambda(s) Z_0(s)}{[\Lambda(s) - g_1(s,\theta_1^*)]R_0(s) - K_p[g_2(s,\theta_2^*) + \theta_3^*\Lambda(s)]Z_0(s)} = W_m(s) \tag{2.6}$$

Define parameter vector as:

$$\theta = [\ \theta_1^T,\ \theta_2^T,\ \theta_3,\ c_0\ ]^T \tag{2.7}$$

Define signal vector as:

$$W = [\ W_1^T,\ W_2^T,\ y,\ r\ ]^T \tag{2.8}$$

where

$$W_1 = \frac{\begin{bmatrix} s^{n-2} \\ \vdots \\ 1 \end{bmatrix}}{\Lambda(s)}u, \qquad W_2 = \frac{\begin{bmatrix} s^{n-2} \\ \vdots \\ 1 \end{bmatrix}}{\Lambda(s)}y$$

thus (2.5) can be written as

$$u = \theta^T W \tag{2.9}$$

Different kinds of RMRAC have their own parameter adjusting rules.

VS−RMRC scheme was proposed based on the above RMRAC schemes, and the proofs of its robustness to $n^* = 1,2$ were presented[5]. As to single input single output (SISO) plants, the scheme can be simplified as:

$$u(t) = \theta^T(t)W(t) + v_p(t) \tag{2.10}$$

When $n^* = 1$,

$$v_p(t) = -sgn(e_0)(\beta_1\|W(t)\| + \beta_2) \tag{2.11}$$

$$e(t) = y(t) - y_m(t) \tag{2.12}$$

If $\theta$ is held constant, then there exist $\mu^*, \beta_1^*, \beta_2^* \geqslant 0$ such that for $\mu \in [\ 0,\mu^*)$, all signals inside the closed−loop system are uniformly bounded and the tracking error will converge to a residual set related to $\mu$ and $\rho$, as long as $\beta_i \in (\beta_i^*,\infty)$, $i = 1,2$[5].

The above principle is also suitable for plants with $n^* = 2$.

From the proof of the above scheme, we can find that $\theta(t) = 0$ is permitted. So (2.10) can also be written as:

$$u(t) = -sgn(e_0)(\beta_1\|W(t)\| + \beta_2) \tag{2.13}$$

From (2.13) we can find u(t) is a kind of variable structure control. To simplify the control scheme and improve its rapidity, the adjustment of $\theta$ is removed from control schemes. But there exists oscillation in steady–state behavior, which is an important drawback of VSC, thus usually leads to unsatisfactory steady–state performance.

## 3. F–VS–RMRAC

Comparing the VS systems with those FC ones of two order plants, it is easy to find that they both use the two inner state of plant to form their control. Further study shows their inner accordance.

### 3.1. Accordance of VSC and FC

The VSC of two order error model with bounded control signal can be generally described as[8]:

$$\dot{e} = Ae + bu, \qquad e \in R^2, A \in R^{2 \times 2}, b \in R^2 \tag{3.1}$$

$$s = c_1 e_1 + c_2 e_2 \tag{3.2}$$

$$u_{vs} = \begin{cases} u^+(e_1, e_2) = K_1, & if \ s < 0 \\ u^-(e_1, e_2) = K_2, & if \ s > 0 \end{cases} \tag{3.3}$$

where (3.1) is the error model; $e = [e_1, e_2]^T$ is an error state vector; s is the sliding mode and $u_{vs}$ represents the control signal.

In FC,

$$E = <e_1> \qquad EC = <e_2> \tag{3.4}$$

The FC rule with weighting factor can be described as[9]:

$$u_{FC} = - <\alpha E + (1 - \alpha) EC>, \qquad \alpha \in (0,1) \tag{3.5}$$

where E,EC are error and change of error respectively, $\alpha$ represents the weighting factor, $u_{FC}$ is the output of fuzzy controller.

When $c_1 = \alpha$, $c_2 = 1 - \alpha$, the accordance of VSC and FC can be easily found comparing (3.2),(3.3) with (3.4),(3.5). The main difference between them is that the sign of $u_{vs}$ is only determined by the sign of s, while that of $u_{FC}$ is related to E and EC, meanwhile the fuzzification and defuzzification exist in the process of calculating $u_{FC}$.

### 3.2. F–VS–RMRAC

From the above analysis, it is reasonable to replace VSC with FC. Setting $\theta(t) = 0$, we propose F–VS–RMRAC scheme of plants with $n^* = 1$ as the following:

$$u(t) = u_{FC}(\beta_1(\|W(t)\| + \beta_2) \tag{3.6}$$

where the definitions of $\beta_1, \beta_2, W(t)$ are the same as those in section 2; $u_{FC}$ represents the output of fuzzy controller in (3.5), $\|u_{FC}\|_{max} = 1$. The new scheme (3.6) maintains the control structure of (2.13) and replaces $-sgn(e)$ with $u_{FC}$.

Robustness:

When $\alpha = 1.0$, (3.6) is the same as (2.13), so its stability and robustness can be directly obtained from the proof of (2.13). But generally, $\alpha$ has its value less than 1.0 in order to fully use the information of E and EC, thus the control rules of FC can be simply stated as the right table:

The table shows the FC rules (3.5) corresponding to

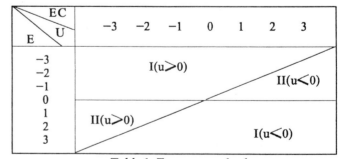

Table 1. Fuzzy control rules

different input pairs (E,EC), when $\alpha$ is chosen to be 0.5. From it, we can see that if the input pairs locate in area I, $u_{FC}$ has the same sign as $-sgn(e)$. While in area II, $u_{FC}$ has opposite sign of $-sgn(e)$, but the language variable of system is stable, so the system converges to its equilibrium.

Although $u_{vs}$ and $u_{FC}$ have different signs in this case, system is bounded and stable.

Smoothness:
Setting dead zone is an effective approach of reducing oscillation in VSC. Because dead zone exists naturally in FC, that is, $u_{FC} = 0$ when (E = 0, EC = 0), from (3.6) we can see that oscillation can be smoothed. Besides, the control signal of (2.13) indicates that the control is a kind of Bang–Bang control, which switches between maximum and minimum values with high frequency. If we set $\|u_{FC}\|_{max}$ = 1, then the control signal of (3.6) has different levels between the maximum and minimum, so the control effect of (3.6) is more smooth. The sign of VS control is determined by −sgn(e), while that of FC is related to E and EC, thus (3.6) use an additional information of EC other than (2.13). This would be helpful to improve system smoothness and reduce its overshot.

Rapidity:
To simplify the RMRAC scheme and acquire better dynamic performance, the parameter vector adjustment of adaptive control was removed in VS−RMRAC (2.13), which is an important improvement of the scheme. No complex algorithm is involved in F−VS−RMRAC, so the new control system also contains the characteristics of simplicity and rapidity.

## 4. Simulation results and remarks

Plant:

$$G_0(s) = \frac{0.36s^2 + 3.6s + 6}{s^3 + 11s^2 + 6s + 6}$$

$$\Delta_m(s) = \frac{1}{s + 2}$$

$$\Delta_a(s) = \frac{1}{s + 2}$$

Reference model:

$$G_m(s) = \frac{5}{s + 5}$$

$\Lambda(s) = s^2 + 10s + 100, \theta^T = 0$. Simulation results of the above system have been shown in figure 1−6, where ym represents the model output, yp1 represents the system output with controller (2.13), while yp2 is the system output with controller (3.6). Sample time T equals to 0.01 second.

When $\mu = 0$, $\rho = 0$, there is no uncertainty and disturbance in the plant, system output is indicated in figure 1, and figure 2,3 represent the control signals of (2.13) and (3.6) respectively. Figure 1 shows that the smoothness, steady−state accuracy of system were much improved while its rapidity was reduced a little. Comparing figure 2 with figure 3, we can find that VS control signal is difficult to be realized because of its large value and high switch frequency, while the realization of the control signal of FC is easy, since its amplitude and switch frequency are much lower.

When $\mu = 0.1$, $\rho = 0$, the corresponding system response is shown in figure 4, and the response when $\mu = 0$, $\rho = 0.1\sin(10t)$ is shown in figure 5. These simulation results demonstrate F−VS−RMRAC also guarantees the stability and robustness when there exist uncertainty and disturbance in the plant. Besides, it can offer a smooth response and high steady−state accuracy.

If a change of relative order occurs, for example, $G_0(s)$ changes to

$$G_0(s) = \frac{3.6s + 6}{s^3 + 11s^2 + 6s + 6}$$

and the simulation result without uncertainty and disturbance is shown in figure 6. We can see that when $n^*$ changes, the output performance of yp1 varies to a certain degree, while there is almost no change in yp2. So F−VS−RMRAC has better performance robustness than VS−RMRC.

## 5. Conclusion

After combined with FC, the novel control scheme offer an improvement in smoothness, steady−state accuracy and robustness without losing the simplicity and rapidity of the former. The new scheme basically solve the main control problem of uncertain plant with $n^* = 1$.

Reference:

[1] P.Ioannou and J. Sun, Theory and design of robust direct and indirect adaptive-control schemes, INT. J. Control, 1988, Vol.47, No.3, 775-813

[2] P.Ioannou, Adaptive systems with reduced models, Automatica, 20, 583, (1984)

[3] Hsu, L., Variable structure model-reference adaptive control (VS- MRAC) using only input output measurements: the general case, IEEE Trans. Automatic Control, AC-35, 1238-1243, (1990)

[4] Fu, L.C., A robust model reference control using variable structure adaptive for a class of plants, Int. J. Control, 53. 1359-1375, (1991)

[5] Chiang-ju Chien and Li-Chen Fu, A new robust model reference control with improved performance for a class of multivariable unknown plants, INT. J. of Adaptive Control and Signal Processing, Vol.6, 69-93, (1992)

[6] P.R.Chang and C.L. Tai, Model-reference neural color correction for HDTV systemn based on fuzzy information criteria, IEEE ICFS, 1383-1388, (1993)

[7] L.X. Wang, Training of fuzzy logic system using nearest neighborhood clustering, IEEE ICFS, 13-17, (1993)

[8] Wei-Bing Gao, Fundamentals of variable structure control, Scientific press, 1990

[9] Long Shengzhao and John G. Keifeldt, Human fuzzy control model and its application to fuzzy control system design, The 4th IFAC conference on Man-Machine Systems, (1989)

Simulation Results:

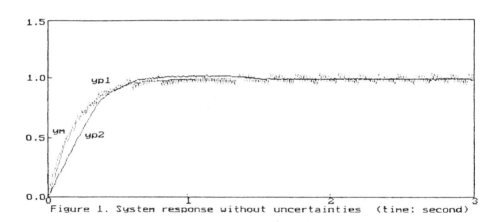

Figure 1. System response without uncertainties (time: second)

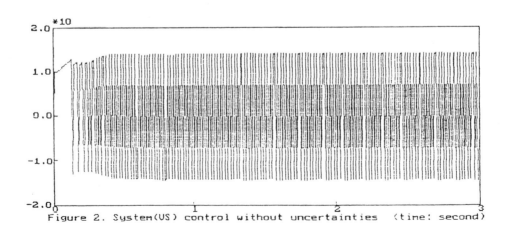

Figure 2. System(US) control without uncertainties (time: second)

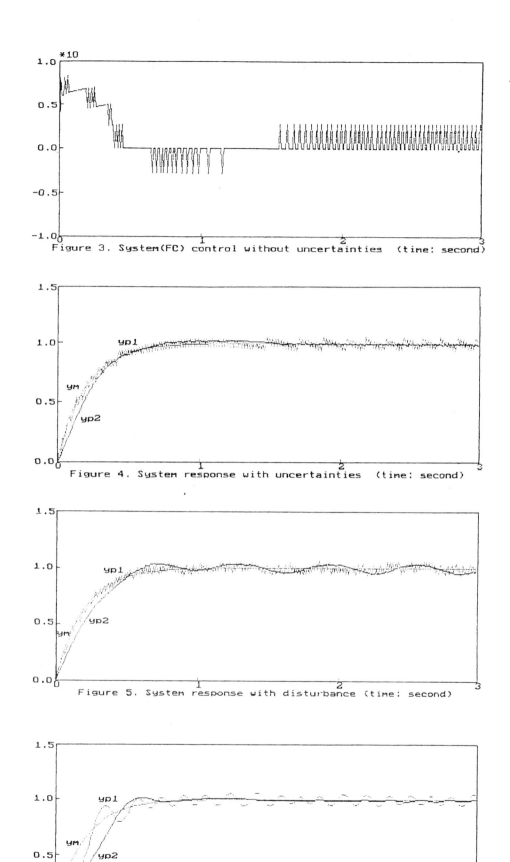

Figure 3. System(FC) control without uncertainties (time: second)

Figure 4. System response with uncertainties (time: second)

Figure 5. System response with disturbance (time: second)

Figure 6. System(n*=2) response without uncertainties (time:second)

# An Object-oriented Framework for Adaptive Fuzzy Circuit Control

Dorian Yeager
The University of Alabama
Department of Computer Science
Tuscaloosa, AL  35487

Don Stanley
Alex May
U.S. Bureau of Mines
P.O. Box L, Univ. of Alabama Campus
Tuscaloosa, AL  35486

**Abstract**:  A class library for process control is under development as a cooperative effort between the U. S. Bureau of Mines and the University of Alabama.  In this paper, some essential ingredients of such a library are described.  In particular, an adaptive fuzzy controller class and a minerals beneficiation circuit class are presented.  The fuzzy controller is a software object of type *fuzzyController* communicating with another software object of type *controllableObject* for the purpose of controlling the latter.  A circuit is an interconnected set of software objects of type *flow* and *process*.  The fuzzy controller model and the circuit model are unified by means of more specific derived classes which inherit properties as needed from the classes *controllableObject* and *process*.  An example is presented in which a model of an operating minerals beneficiation circuit was constructed, tuned with a genetic algorithm, and brought under control by fuzzy controllers in two different ways:  (a) by independently controlling the processes in the circuit, and (b) by directly controlling the entire circuit.  The study indicated that controlling the entire circuit was somewhat more effective than independently controlling the processes in the circuit.

## 1.  Introduction.

The U. S. Bureau of Mines has a long-term interest in the development of  software for process control.  The processing of ore to extract valuable  minerals is conducted on a large scale and is very costly.  Any technology which can be developed to yield a small improvement in a process has the potential to greatly increase the profits of the company employing that process.  There is, therefore, considerable potential benefit in a comprehensive approach to the control of minerals beneficiation circuits.  To realize this benefit, researchers at the U. S. Bureau of Mines and the University of Alabama are developing a comprehensive class library which accommodates fuzzy logic controllers, genetic algorithms, minerals beneficiation processes, and arbitrarily complex minerals beneficiation circuits.  This paper presents an important first step in constructing the desired class library, that of developing base classes and derived classes which constitute a usable framework for fuzzy control of a set of cooperating processes.

Adaptive control techniques integrating genetic algorithms and fuzzy logic are well developed when applied to individual controllable units.  Examples include a cart-pole system [3], pH control [4], control of an exothermic chemical reaction [6], and even control of a chaotic system [5].  However, all these examples involve a single process which can be controlled without affecting other related processes.  When multiple processes are involved, independent control of those processes may not be an acceptable strategy.  Applications abound for which sets of cooperating agents must be controlled in a coordinated fashion.  Such an application which holds significant interest for the U. S. Bureau of Mines is the simultaneous adaptive control of multiple processes in a minerals beneficiation circuit.

Valuable minerals often occur naturally as fine-grained particles of various types (species) locked together with large amounts of valueless minerals.  Extracting the desirable from the undesirable minerals (beneficiation)

1496

requires the use of a complex circuit of crushing, grinding, sizing, and separation processes. Control of a minerals beneficiation circuit requires the simultaneous coordination of these complex, mutually interdependent processes. Each process can be individually controlled, and control parameters may be different for the different processes. Control of the circuit as a whole has as its purpose reaching a stable performance level for the entire set of processes working as a unit. In order to make the approach to control as flexible and general as possible, a consistent control framework has been developed which allows the user of the class library to individually tailor a fuzzy controller for any one of the processes in the circuit, or for the circuit itself.

Let us use as an example the column flotation process. A flotation column is a large, usually cylindrical, vertical tank into which a mixture of water and finely ground ore is continuously fed. The feed is introduced at a level which is about two-thirds of the distance to the top of the column, and the mixture in the tank is continuously removed from the bottom. A sparger at the bottom of the tank introduces air into the mixture in the form of very fine bubbles. The bubbles float to the top and produce a foam, which is washed from above with a relatively small flow of wash water.

The column has two input flows and two output flows. The input flows consist of the feed and the wash water. The output flows consist of the tailings, which is the mixture of ore and water that is removed from the bottom of the column, and the concentrate, which is the foamy mixture of air, water, and ore that overflows the top of the column. Some of the minerals in the ore (called *hydrophobic* minerals) have a tendency to become attached to the bubbles and carried up to the foam; hence, the concentration of those minerals is higher in the concentrate than in the tailings. This separation of minerals is the purpose of the flotation column. In order to improve the separation characteristics of the column, chemicals which affect bubble size and particle-bubble attachment are mixed in with the feed.

A flotation column usually is combined with one or more additional separation processes in a circuit in order to recover more of the valuable mineral. For example, multiple flotation columns may be linked together in series, to form circuits (Figure 1). If the circuit is designed correctly, each successive column is able to improve the performance of the columns that precede it in the circuit. In Figure 1, three columns are connected "tailings to feed" and the concentrates are combined into a single output flow. In connection with the control of such a circuit it is relevant to ask, "To achieve stabilizing control of the circuit, is it adequate to simply control the individual columns, or is a circuit-wide control strategy necessary?"

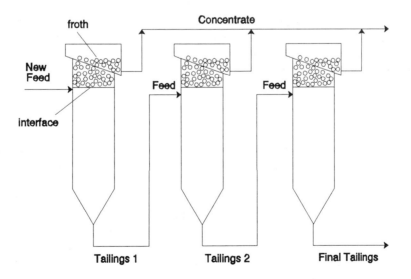

Figure 1. A multiple column flotation circuit.

The most fundamental aspect of the control of a flotation column is the level of the liquid-froth interface in the column. The circuit controller discussed in this paper is based on simultaneous *stabilizing control* of multiple flotation columns. Stabilizing control of a single column amounts to keeping the liquid-froth interface at a set level.

*Optimizing control* of a flotation column usually consists of maximizing the rate of recovery of the valuable mineral while maintaining an acceptable level of separation (grade). Since an optimizing controller should perform stabilizing control by default, development of a stabilizing controller is a necessary first step toward optimizing control. Both types of control require a well-tuned and accurate model. The flotation column model incorporated here is a well-developed, complex model attributable in large part to J. A. Finch and G. S. Dobby [2]. When properly tuned to reflect the operating characteristics of a specific column, the model is capable of satisfactorily reproducing the behavior of that column.

## 2. The Object-oriented Circuits Model.

There are four fundamental notions incorporated in our software model of minerals beneficiation circuits. Each of these has been implemented as a C++ class. The individual instances of these classes, called objects, are used as proxies for their real-world counterparts. The four classes are *species*, *flow*, *process*, and *circuit*. A *species* is an individual mineral present in a flow. Each such species has a density and a volumetric flow rate, as well as other attributes which describe the properties of the species which are relevant to the separation processes which must act on it. A *flow* is a collection of a variable number of species. Its attributes are largely derived from the corresponding attributes of the species which compose it, and include volumetric flow rate, volume fraction solids, weight fraction solids, and slurry density. For control purposes, a flow can be increased or decreased, in which case all the component species are increased or decreased proportionately. A *process* is a software object communicating with a collection of input and output flows. An object of type *process* can be *cycled*, which means that it examines its own internal state and the states of its flows, determines in some way what the contents and flow rates of its output flows should be, and sets those flows to the desired states. A *circuit* is a collection of flows and processes. Each flow is connected as an input or an output (or both) to one of the processes in the circuit. A flow is an input to the circuit as a whole if it is not an output of any of the processes in the circuit, and it is an output of the circuit as a whole if it is not an input of any of the processes in the circuit. Since a circuit is a special case of a process, it can be cycled. The cycling of a circuit consists of the ordered cycling of the individual processes in the circuit.

In addition to the circuit class, several other classes have been derived from the process class for the purpose of constructing software models of minerals beneficiation circuits. For example, the *mixer*, a simple class which merges two separate input flows into a single output flow, and the *column*, a class which has the essential capabilities of the flotation column discussed above.

## 3. The Fuzzy Controller Model.

A fuzzy controller needs two things: an object to control and a fuzzy rule base which gives the controller the knowledge needed for controlling the object. A controllable object allows the controller to request a series of decision variable values (sensor outputs, called *readings* in the discussion below) which define the current status of the object as understood by the controller. The controller supplies this information to the fuzzy rule base, which uses its inferencing mechanism to produce a set of control variable values (called the *adjustments* in what follows). The controllable object makes the recommended adjustments to its state when instructed to do so by the fuzzy controller. The fuzzy controller, controllable object, and fuzzy rule base have been implemented as the C++ classes *fuzzyController*, *controllableObject*, and *fuzzyRuleBase* (Figure 2). These classes provide the basis for control of circuits and circuit processes.

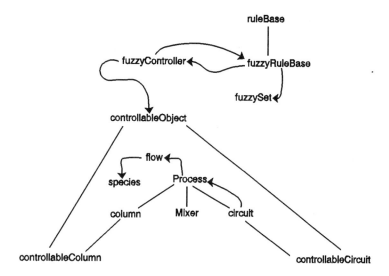

Figure 2. Class relationships in the controllable circuit system. Directed arcs indicate a client/server relationship between classes, whereas heavy undirected arcs indicate a base class/derived class relationship, with the base class positioned above the derived class.

Addition of control capabilities to the *column* and *circuit* classes results in the derived classes *controllableColumn* and *controllableCircuit*. When a circuit is first created, it is supplied with a single input flow. The circuit constructs itself by seeking out all processes which are fed by that flow, directly or indirectly, and all flows that are output by one of those processes or that feed into one of those processes. If the circuit is a controllable circuit, it will examine each of its component processes and determine which of them are controllable. For each controllable process, it will determine the number of readings provided and the number of adjustments required by that process. Once the circuit has performed this self-examination, it is ready to be controlled by a controller.

The C++ *fuzzyRuleBase* class is derived from a simpler *ruleBase* base class consisting of a collection of rules of the form

> IF readings are from sets $R_1$, $R_2$, ..., $R_m$
> THEN adjustments should be from sets $A_1$, $A_2$, ..., $A_n$

This rather abstract formulation takes a more intuitive form when rules are echoed by the *ruleBase* class itself. Following is an example of the output produced by the *echoRule()* member function:

> IF level is high
>         AND change in level is positive large
> THEN change in tailings is positive large.

The base class is capable of operating independently, in which case the sets are traditional sets and the inference method is simple logical inference. When the fuzzy rule base is used, however, the sets contained in the rule base are interpreted as fuzzy sets and de-fuzzification is achieved via the center-of-area method [7]. The fuzzy set concept is represented as the C++ class *fuzzySet*.

## 4. An example - the Mount Isa Mines column flotation circuit.

The class framework under discussion performed very well in controlling a model of the flotation columns which form a part of an operating circuit at Mount Isa Mines in Australia. The circuit separates zinc and lead. Data detailing its performance was published by R. A. Alford in [1]. The configuration of the circuit has three columns in series, as in Figure 1. The three columns have identical sizes, but flow rates and target interface levels differ. Using the published data from a specific test run, an object of type *controllableCircuit* was configured as nearly as possible to reproduce the test results. In the test run used, target interface levels differ by 20 cm from one column to the next, being set to 871 cm, 891 cm, and 911 cm respectively in the three columns. Configuration of the model to conform to the behavior of the actual circuit was achieved by means of a genetic algorithm set up to minimize the mean squared error between the published grades and flow rates and those produced by the model. The final tuned model reproduced most measured outputs (flow volume, percent zinc, percent lead, and percent iron) of the actual circuit to within two or three decimal places.

Two rule bases were constructed for stabilizing control. The first was for controlling a single column, and the second for controlling the entire circuit. The rule bases were obtained by a training process, in which the input flow to the column (circuit) was varied randomly and rules were chosen as needed to minimize the mean squared error between the resulting column (circuit) conditions and the desired conditions.

To test the performance of the single-column rule base, a separate controller was attached to each of the three column models, and a randomly varying flow was fed into the circuit. Each column began at a level 8 cm below the target level. To test the circuit-wide rule base, a single controller was attached to the entire circuit and an identical feed source was used. A direct comparison of the two control strategies is presented by the graph in Figure 3.

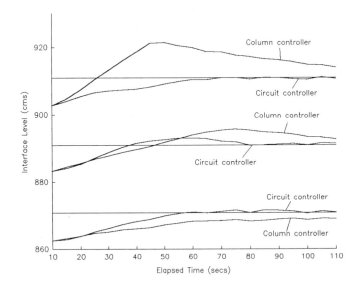

Figure 3. Stabilizing control results for the Mount Isa Mines circuit.

Examination of the graph reveals that for stabilizing control both the circuit-wide control strategy and the independent column control strategy worked satisfactorily. The circuit controller was clearly superior in achieving the set point more quickly and holding it more accurately. It is expected that when optimizing control strategies are tried the benefits of using a circuit-wide controller will be even more evident.

## 5. Conclusions.

The object-oriented approach to the software modeling of controllers and circuits presented in this paper is intuitive, flexible, and extendible. In the example presented, (1) multiple, individually configured copies of the flotation column model were combined into one software circuit model, (2) a genetic algorithm was used to fit the software model to published data, (3) the configured model was brought under stabilizing control using independent fuzzy logic controllers attached to the three columns, and (4) the same control was achieved, somewhat more effectively, using a single circuit-wide fuzzy logic controller. The example shows the benefits of both the object-oriented design approach and the circuit-wide approach to control. The use of base classes such as *process* and *controllableObject* along with inheritance means that future work which incorporates additional models of minerals beneficiation processes and their controllers will fit smoothly into the framework established here. The aim is ultimately to have an extensive class library which makes it possible not only to build an accurate software model of any existing circuit, but also to provide powerful tools for the design of new circuits.

## 6. References.

1. Alford, R. A., <u>Modeling and Design of Flotation Column Circuits</u>. Ph. D. thesis, Julius Kruttschnitt Mineral Research Centre, University of Queensland, March 1991.

2. Finch, J. A., and G. S. Dobby, <u>Column Flotation</u>. Oxford, Pergamon Press, 1990.

3. Karr, C. L., "Design of an Adaptive Fuzzy Logic Controller using a Genetic Algorithm", in <u>Proceedings of the Fourth International Conference on Genetic Algorithms</u>, ed. R. K. Belew and L. B. Booker, pp. 450-457. San Mateo, CA, Morgan Kaufmann, 1991.

4. Karr, C. L., and E. J. Gentry, "Real-time pH Control using Fuzzy Logic and Genetic Algorithms", paper presented at the Annual Meeting of the Society for Mining, Metallurgy, and Exploration (preprint number 92-49), Phoenix, AZ, February 1992.

5. Karr, C. L., and E. J. Gentry, "Control of a Chaotic System using Fuzzy Logic", submitted.

6. Karr, C. L., S. K. Sharma, W. Hatcher, and T. R. Harper, "Fuzzy Logic and Genetic Algorithms for the control of an Exothermic Chemical Reaction", submitted.

7. Larkin, L. I. "A Fuzzy Controller for Aircraft Flight Control", in <u>Industrial Applications of Fuzzy Control</u>, ed. M. Sugeno. New York, North-Holland, 1985.

# FUZZY CONTROLLER DESIGN TO DRIVE
# AN INDUCTION MOTOR

G. D'ANGELO ♦, M. LO PRESTI ♦, G.RIZZOTTO■

■ Corporate Advanced System Architectures
S.G.S. THOMSON Microelectronics
20041 Agrate Brianza (Mi) - Italy


♦ Fuzzy Logic Research Group
Co.Ri.M.Me. Consortium between
University of Catania and ST
Strada Statale 114, Torre Galiera
95121 Catania - Italy

## Abstract

This paper describes the simulation results of an Induction Motor Speed Control based on Fuzzy Logic Theory. Voltage Impress Control technique has been applied to medium power Induction Motor. An automatic approach to extract control rules allowS us to reduce the design time. Computer simulations have been carried out in order to test the performances of the whole control system. Then a hardware implementation of the Fuzzy Controller is proposed by means of a dedicated Fuzzy Processor: WARP (Weight Associative Rule Processor).

_Keywords:_ Fuzzy Logic, Induction Motor, Control System, FAM, Neural-Network, Coprocessor.

## 1. Introduction

In the few last years interest on Induction Motor Drive has increases, due to structure and performances that this type of motor guarantees with respect to other ones. Although electrical drives are well-known systems and are described by well established and accurate non-linear mathematical models, problems due to parametric variations do not guarantee robustness of traditional control [1]. Classical controls are based on knowledge of the model and when a parametric variation occurs, the behaviour of the system becomes unsatisfactory. Fuzzy Logic approach [2][3][4] can be usefully utilised to develop robust non-linear control, based on the designer experience, that works better than traditional regulators [5][6]. Speed and rotor position fuzzy controls of electrical drives, have been proposed mostly to improve the performances of conventional PI regulators [7]. Now a completely Fuzzy Control is proposed. The design of controller is carried out by using a Neuro-Fuzzy network that utilises a training set determinated with Cell-to-Cell procedure [8][9][10]. The design and simulation results of speed control are presented . These results show a very interesting and fruitful direction for future research work.

## 2. Fuzzy Regulator Preliminary

An Induction Motor Model is described by using a fifth order differential equation [11].

| Model | Parameters |
|---|---|
| $$\dot{\phi}_{sd} = -\frac{1}{\sigma T_s}\phi_{sd} + \omega_a\phi_{sq} + \frac{K_r}{\sigma T_s}\phi_{sd} + V_{sd}$$ $$\dot{\phi}_{sq} = -\frac{1}{\sigma T_s}\phi_{sq} - \omega_a\phi_{sd} + \frac{K_r}{\sigma T_s}\phi_{rq} + V_{sq}$$ $$\dot{\phi}_{rd} = \frac{K_s}{\sigma T_r}\phi_{sd} + (\omega_a - \omega)\phi_{rq} - \frac{1}{\sigma T_r}\phi_{rd}$$ $$\dot{\phi}_{rq} = \frac{K_s}{\sigma T_r}\phi_{sq} - (\omega_a - \omega)\phi_{rd} - \frac{1}{\sigma T_r}\phi_{rq}$$ $$\dot{\omega} = \frac{1}{J}(T_L - T_e)$$ $$\sigma = 1 - K_s K_r$$ | $$K_s = \frac{L_m}{L_s}; T_s = \frac{L_s}{R_s}$$ $$K_r = \frac{L_m}{L_r}; T_r = \frac{L_r}{R_r}$$ $$R_s = 1.5\Omega; R_r = 0.075\Omega;$$ $$L_s = 0.191mH; L_m = 0.0379mH;$$ $$J = 0.02Nms^2; P_n = 2.8KW$$ |

The control, generally, must guarantee the motor to run at a fixed speed with assigned set-point error, bounded overshoot and minum settling time. We impose the stator voltage (Vsd e Vsq in the d-q frame) at suitable frequency, generally proportional to voltage. This type of control is quite simple to implement with respect to other ones (Current Impress or Field Oriented) but introduces instability. Aim of this work is to implement a control that guarantees stability, robustness and appropriate performances. The scheme where the fuzzy controller device was applied is reported in Fig. 1. The controller, in corrispondence to inputs (error and error variation), gets the voltage value that the inverter (generally PWM) must impose to the

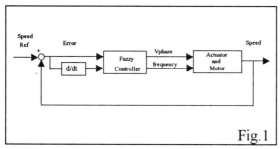

Fig.1

stator of the motor. The design procedure adopted in the implementation of the fuzzy controller makes use of neuro-fuzzy networks. This algorithm determines the suitable fuzzy rules for the control, fuzzy sets for each state variable and control variable.

The training phase was performed by using patterns obtained by means of the Cell-to-Cell Mapping Technique [12][13]. This procedure allows us to have optimal voltage, to apply at the motor, it assures performances and stability. In this way we obtain good performances with a simplified control. The input/output fuzzy sets of Fuzzy Controller are shown in the Fig.2,3,4, while examples of the rules, used to control system, are reported in Fig.5. In this way the regulator is completely designed.

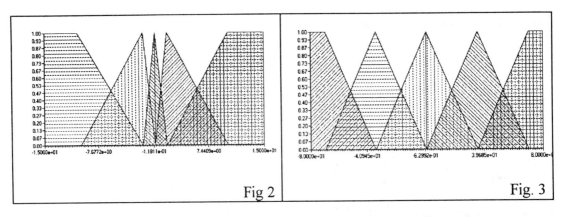

Fig 2    Fig. 3

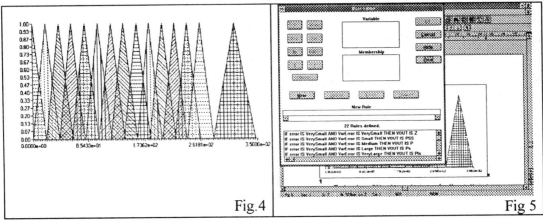

Fig.4    Fig 5

In next section  simulation tests are presented.


## 3. Fuzzy Regulator Results

The plot below shows some comparisons between a Fuzzy Controller and a traditional one (PID), the improvement obtained by using the fuzzy controller are clear. The model of WARP [14], able to process fuzzy rules hundred times faster than conventional microprocessor at a much lower cost (table I shows characteristic of this processor), has been utilised in the simulation algorithm. The reported simulation tests have shown that controllers implemented in this way are more robust than conventional regulators.

In particular,  in Fig.6 the outputs of the system (rotor speed)  in two different versions are reported. The fuzzy controller has better performances in terms of overshoot, settling time and steady-error, moreover the controller output (voltage) reduces the ratio  dV/dt  (see Fig.7) to impose to the switch devices of driver block, in this manner it is avoided additional burden. In Fig.8 the goodness of fuzzy control respect to traditional ones  when disturbs are present is underline; Fig.9 shows the output system when a load variation  (steady-state plus noise) is applied. If speed measure has additional noise, it becomes more difficult to control the system with PID control (Fig.10).

The Development Board to support WARP is shown in Fig.11. It has four inputs (two analogue inputs and two digital ones)  extendible to eight and two outputs (analogue and digital). Interface with PC  is available for system monitoring. The whole system, where the fuzzy regulator is applied, is reported in Fig.12.

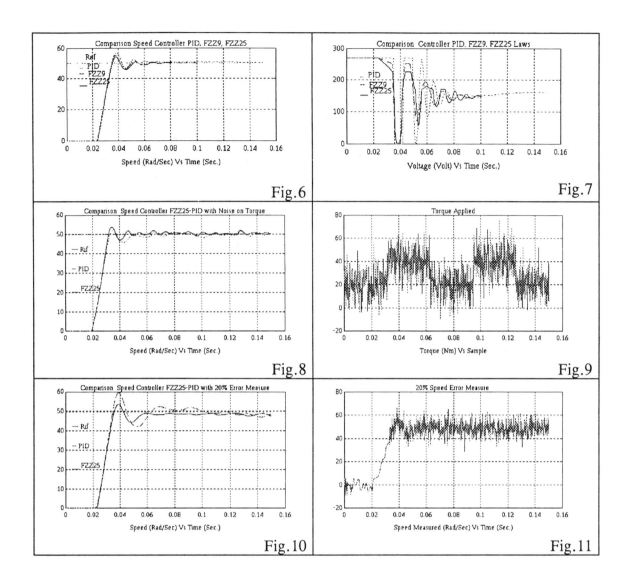

Fig.6

Fig.7

Fig.8

Fig.9

Fig.10

Fig.11

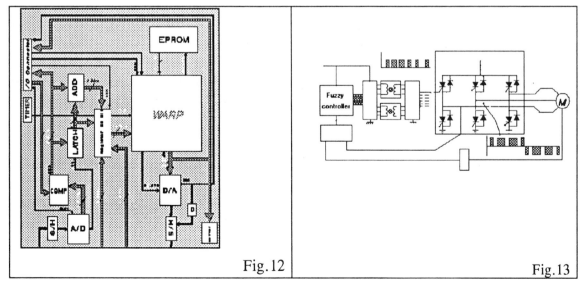

Fig.12

Fig.13

## 4. Conclusion

The speed control of an Induction Motor Drive by means of the Fuzzy Regulator has been investigated. Under reference speed trajectory the Fuzzy Controller has shown good performances. Moreover , when in the system some are disturbs present fuzzy control shows better performances with respect to traditional ones. In order to design the controller an automatic synthesis allows easy implementation. The availablity of WARP with its high speed computational power, gives us the possibility to realise real-time control.

| | |
|---|---|
| **8 bits Digital Inpu**<br>**High Speed Processing ($\approx$ 1.5 MFlips)**<br>**Up to 16 M.F.s for Input**<br>**Up to 16 Output Variables** | **In/Out easy Configurable**<br>**Defuzzification on chip**<br>**Max-Dot inference Method**<br>**Support S/W development tool** |

**Tab I**

## 5. REFERENCES

[1] R. Joetten and G. Maeder, " *Control Methods for Good Dynamic Performance Induction Motor Drives Based on Current and Voltage as Measured Quanties*". IEEE Transactions on Industry Applications, June 1983.

[2] D. Dubois and H. Prade," *Fuzzy Set and System: Theory and Applications*". New York: Academic Press, 1980.

[3] L.A. Zadeh , "*Fuzzy Set*", Information and Control, vol. 8, pp.338-352, 1965

[4] G. Klir, G. Folger, and T. Folger, "*Uncertainty and Information*". Englewood Cliffs, NJ: Prentice-Hall, 1988.

[5] P.J.King and E.H. Mamdami, "*The application of fuzzy controll systems to industrial process*",Automatica (1977), pp235-242

[6] Y.F. Li, C. Lau, "Application of fuzzy control for servo systems", IEEE International Conference on Robotics and Automation (1988), 1511-1519.

[7] I. Michi, Nagai, N. Nishiyama, S. Yamada, "*Vector Control of Induction Motor with PI controller*". IEEE Conference on Industry Application Society, Dearborn, Michigan, pp. 854-861.

[8] B.Kosko, "*Fuzzy Systems and Neural Networks*", Prentice Hall, 1992

[9] J. A. Freeman D. M. Skapura " *Neural Networks Algorithms, Applications and programming Thechinique*", Addison-Wesley

[10] C.S.Hsu, R. Guttalu, "*A Theory of Cell-to-Cell Mapping Dynamical Systems*", Asme Journal of Applied Mechanics, Vol.47, 1980, pp.931-939.

[11] P. Krause, *"Analysis of electric machinary"*, McGraw-Will Book [1988].

[12] L.Fortuna, G. Muscato, M. Lo Presti, K. Vinci, *"A parallel scheme for Cell-to-Cell based analisys of dynamic systems"*, annual Allerton Conference on communication, Control and Computing. Illinois (USA, (1992).

[13] M. Lo Presti, R. Poluzzi, GG Rizzotto, A. Zanaboni, *"FAM approach to design a Fuzzy Controller"*, IFSA 93, Seoul.

[14] K.D. Jee, R. Poluzzi, B. Russo, *"Memory Organitation for a Fuzzy Controller"*, IFSA 93, Seoul.

# A Self-Tuning Adaptive Resolution (STAR) Fuzzy Control Algorithm

Ho Chung Lui, Ming Kun Gu,
Tiong Hwee Goh and Pei Zhuang Wang
Institute of Systems Science, National U. of Singapore
Kent Ridge, Singapore 0511
Tel: (65)-772-2532, Fax: (65) - 774-4998
E-mail: luihc@iss.nus.sg

December 8, 1993

**Abstract**

A novel Self-Tuning Adaptive Resolution (STAR) fuzzy control algorithm is introduced in this paper. One of the unique features is that the fuzzy linguistic concepts change constantly in response to the states of input signals. This is achieved by modifying the corresponding membership functions. We use this adaptive resolution capability to realize a control strategy that attempts to minimize both the rise time and the overshoot. Simulation results on a simple inverted pendulum problem are presented. Its characteristics are compared with the classical PD controller. Finally, the algorithm is also realized to control a real inverted pendulum hardware. Experimental results show that the STAR controller is both robust and can minimize positional error with drastically reduced overshoot.

## 1    Introduction

Although there have been many successful applications of fuzzy control, control designers still need to face two major obstacles in implementing a fuzzy controller, namely:

1. The acquisition of fuzzy rules,

2. The search for optimal membership functions for the linguistic concepts.

Many researchers have attempted to alleviate these difficulties in the past. Wu et.al. [1] observed that the entries in the control rule table can be expressed as a function of the inputs. So they proposed a method to modify the functional form according to the input settings. He et.al.[2] extended their ideas and tested it on many experiments. Recently, Lin et.al.[3] discussed an alternative approach whereby both inputs and output of a fuzzy controller adapts to environmental changes so as to minimize overshoot and response time. Our approach is similar to theirs but address the two abovementioned issues from a slightly different angle.

In our approach, the control rule table will not change, but the definition of the linguistic concepts such as *POSITIVE LARGE* adapts constantly to environmental requirement. Our thesis is that the resolution (or granularity) of these variables need not be constant, but can change so that we can maintain fine resolution at the operating point but coarse granularity elsewhere. This paper discusses our approach for a simple two-input, one output fuzzy controller.

## 2 Background

Consider a conventional fuzzy controller with error ($e$) and rate-of-change-of-error ($\dot{e}$) as inputs, the control force ($u$) as output; a control rule

> if $e$ is $E$ and $\dot{e}$ is $DE$ then $u$ is $U$

where $E$, $DE$ and $U$ represent fuzzy linguistic variables, can be conveniently represented as an entry of the control table as shown in Table 1. The corresponding $(e, \dot{e})$ plane is shown in Fig. 1 (notice that the arrow of the $\dot{e}$ axis is going downward,

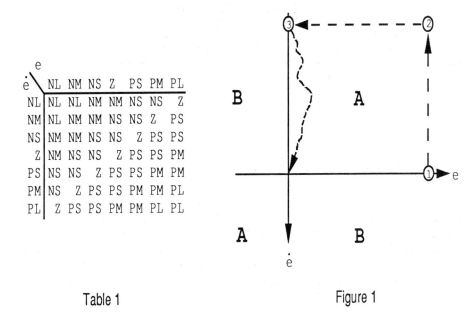

| $\dot{e}$ \ $e$ | NL | NM | NS | Z | PS | PM | PL |
|---|---|---|---|---|---|---|---|
| NL | NL | NL | NM | NM | NS | NS | Z |
| NM | NL | NM | NM | NS | NS | Z | PS |
| NS | NM | NM | NS | NS | Z | PS | PS |
| Z | NM | NS | NS | Z | PS | PS | PM |
| PS | NS | NS | Z | PS | PS | PM | PM |
| PM | NS | Z | PS | PS | PM | PM | PL |
| PL | Z | PS | PS | PM | PM | PL | PL |

Table 1

Figure 1

so that it corresponds to the control table orientation). By studying the input trajectory on the $(e, \dot{e})$ plane, we can partition the plane into two major regions - A and B as shown in Fig. 1. Region A corresponds to the case where $(e \cdot \dot{e} < 0)$ and B represents $(e \cdot \dot{e} > 0)$. It is easy to observe that region A represents condition that the error is reducing while the error in region B will actually increase ( the overshoot case). Hence we seek a method to confine the trajectory to region A as much as possible.

It is well-known that when triangular membership functions with 50% overlap are used (as shown in Fig. 2), the control action $u$ is a piece-wise linear function of the inputs $e$ and $\dot{e}$ when the conventional Center-of-Gravity defuzzification method is used. Moreover, there are only four control rules that will be activated at any instance.

By observing the membership function in Fig. 2, it is obvious that a linguistic concept such as $PM$ spans a range from $a_1$ to $a_3$. Its membership degree peaks at $a_2$. A closer observation reveals that all the membership functions are uniquely defined when $a_1$, $a_2$ and $a_3$ are given. So when these values are changed, the related membership functions will change. This means that what is regarded as $PM$ will also be changed. The STAR algorithm is our attempt to modify the linguistic definitions on-the-fly in order to force the $(e, \dot{e})$ trajectory to region A as much as possible.

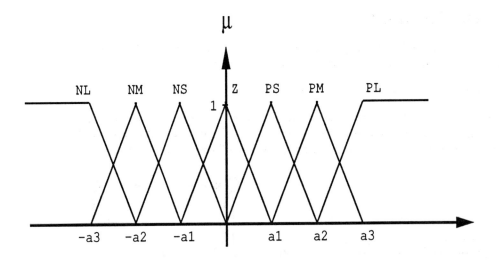

Figure 2

## 3 The STAR Control Strategy

Before we discuss how the membership functions are changed, let us first state the control strategy. With respect to table 1 and Fig. 1, the biggest control force $u$ that can output from region A is $PM$ in Table 1, which is indicated as State-1 in Fig. 1. To quickly minimize the error $e$, we want to force the trajectory to go to State-1, no matter what the intitial condition is. At State-1, the controller becomes a proportional controller (since $\dot{e}$ is ZERO), and it exerts the biggest possible $u$ within region A to minimize the error.

While this action can reduce the rise time, it increase the chance of overshoot. To avoid overshoot, the controller must at the appropriate time generate a $u$ in the opposite direction when the actual error is close to zero. This can be done by forcing the trajectory to State-3 in Fig. 1. Here the controller becomes a derivative controller since $e$ is ZERO. Before entering State-3, the controller can go to State-2, where the controller's output is ZERO. This represents the relaxation state.

If the controller is properly designed, then after State-3, both $e$ and $\dot{e}$ should be very small. Hence it can enter into State-4, which uses the conventional fuzzy control method to bring the close-loop system to the desire set-point. Hence the STAR control strategy is to force the fuzzy controller to traverse from State-1 to State-4.

## 4 Membership Function Adaptation

Since State-1 represents the control rule:

>   if $e$ is $PL$ and $\dot{e}$ is $Z$ then $u$ is $PM$

we need to change the definition of $PL$ with regard to $e$ and $Z$ with regard to $\dot{e}$ in order that the trajectory will stay at State-1. In other words, the controller must view that the error $e$ is positively large ($PL$), regardless what the actual value of $e$ is. This can be done by modifying $a_3$ on-the-fly. For example, we can choose

$$a_3(t) = e(t) \tag{1}$$

When $a_3$ is changed, $a_1$ and $a_2$ also change proportionally. In this way, the membership resolution (or granularity) expands or contracts as needed. Likewise, the definition of ZERO for $\dot{e}$ can be altered by modifying $a_1$ in the $\dot{e}$ axis in a similar manner.

By using the same method, we can force the trajectory to any of the first three states.

# 5 State Transition

To properly traverse the states, we devise the following criteria for state transitions:

1. State-1 will exert a large $u$ to reduce error. Hence the longer it stays in State-1, the smaller the rise time will be. However, if it stays here too long, then the chance of overshoot is high. We choose the criteria that

$$\text{if } (\mid e \mid > \alpha \mid \dot{e} \mid) \text{ then goto State-2}$$

2. State-2 is the relaxation state. The control output $u$ is zero. The error $e$ will continue to decrease because of the inertia ($\mid \dot{e} \mid > 0$). If the momentum is too strong, then overshoot will appear so it needs to enter State-3 (to apply the brake). In our experiments we choose

$$\text{if } (\mid e + \beta\dot{e} \mid) > thr_1 \text{ then goto State-3}$$

3. State-3 tries to prevent overshoot by sending a big control force in the opposite direction. At the appropriate time when both $\mid e \mid$ and $\mid \dot{e} \mid$ are small, it will enter into State-4. Our criteria is to check

$$\text{if } (\mid \ddot{e} \mid > \gamma \mid \dot{e} \mid) \text{ then goto State-4}$$

4. At State-4, the control action is the same as the conventional fuzzy controller. If the external disturbance is bigger than a certain threshold, then it can enter State-1 again and the cycle repeats itself.

The choices of constants and thresholds are determined empirically. We found that it is not difficult to find a good set of parameters.

# 6 Experimental Results

## 6.1 Software Simulation

We have simulated the STAR fuzzy controller to control the invertical pendulum problem in software. The dynamic equation given in [4] is adopted. The fuzzy controller takes in two inputs $\theta$ and $\dot{\theta}$ and outputs the control force $u$. Fig. 3a and 3b show the responses of the STAR fuzzy controller and the conventional PD controller. In both cases, the initial conditions are $\theta = 10°$ and $\dot{\theta} = 0$. Notice that in Fig. 3a, the control force $u$ clearly reflects how the controller traverses from State-1 to State-4 as discussed in Section 3.

## 6.2 Implementation on a Real Inverted Pendulum Apparatus

The algorithm is then used to control a real inverted pendulum. Here we need to control both the position offset ($x = 0$) while maintaining the pole upright ($\theta = 0$). Since there are four input variables ($x, \dot{x}, \theta, \dot{\theta}$), we use two two-input fuzzy controllers cascading together to handle this[5]. One controller is dedicated to balance the pole while the other uses $(x, \dot{x})$ as input, and outputs an angular setpoint for the former controller to track. The STAR adaptive algorithm is implemented in both controllers. Experimental results show that the cascaded controller is robust against external disturbances and uneven load. Moreover, when instructed to go from one $x$ position to another, the STAR algorithm, with its adaptive capability, can settle to the new position faster with less overshoot, as shown in Fig. 4a. Fig. 4b shows the corresponding angular deviation. Compared to the conventional fuzzy controller, the STAR approach reduces the positional overshoot and also the angular error in steady state.

## Conclusion

As mentioned in Section 1, conventional fuzzy controller design faces two obstacles. We have introduced the STAR fuzy control algorithm in an attempt to simplify the design and implementation process. In STAR, the fuzzy rule table is fixed (hence there is no need to determine the control rules), and the fuzzy concepts such as POSITIVE LARGE, MEDIUM, ..., etc. change according to the settings of the control inputs. In effect, the membership functions adapt to the environmental conditions on-the-fly. This will alleviate the designer's effort to search for good membership functions. We have also presented experimental results, both in software simulation and in controlling the real inverted pendulum, to demonstrate the characteristics of the STAR fuzzy control algorithm.

## Acknowledgement

We express our appreciation to Dr. Ming Bai and Dr. Meng Zhang of the Beijing Normal University for their insightful discussions with us.

## References

[1] Zhi Qiao Wu, Pei Zhuang Wang, and Hoon Heng Teh. A rule self-regulating fuzzy controller. *Fuzzy Sets and Systems*, 47, 1992.

[2] S.Z. He, S.H. Tan, C.C. Hang, and P.Z. Wang. Control of dynamical processes using an on-line rule-adaptive fuzzy control system. *Fuzzy Sets and Systems*, 54, 1993.

[3] Y. Lin, S.H. Tan, S.Z. He, and Z.R. Liu. Adaptive fuzzy control of structural and parameter varying processes. In *Proc. of First Asian Fuzzy Systems Symposium*, 1993.

[4] A.G. Barto, R. S. Sutton, and C. W. Anderson. Neuronlike adaptive elements that can solve difficult learning control problem. *IEEE Trans. on Systems, Man, and Cybernetics*, SMC-13(5):834 – 846, 1983.

[5] Ming Kun Gu, Ho Chung Lui, Tiong Hwee Goh, and Pei Zhuang Wang. A cascaded architecture of adaptive fuzzy controllers for inverted pendulums. In *submitted to the* Third IEEE Int'l Conf. on Fuzzy Systems., 1994.

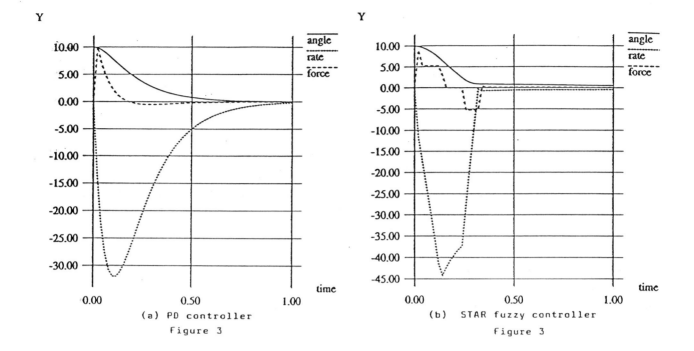

(a)  PD controller
Figure 3

(b)  STAR fuzzy controller
Figure 3

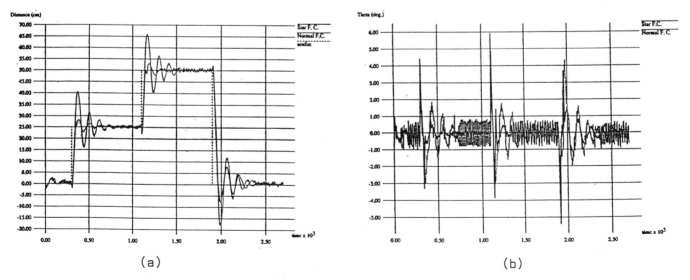

(a)

(b)

Figure 4: Comparison between the STAR and
Conventional Fuzzy Controllers in (a)
positional response and (b) angular
response.

# A Cascade Architecture Of Adaptive Fuzzy Controllers For Inverted Pendulums

Ming Kun Gu, Ho Chung Lui, Tiong Hwee Goh and Pei Zhuang Wang
Institute of Systems Science, National University of Singapore
Kent Ridge, Singapore 0511
E-mail: mkgu@iss.nus.sg, luihc@iss.nus.sg

## Abstract

This paper presents a cascade architecture design for the control of an inverted pendulum. A novel self-tuning adaptive resolution (STAR) fuzzy control approach is implemented in this control design. The controller consists of two sub-controllers: one of which is used to control the small cart's position and the other is used to balance the pole. Because of the unique architecture, the controller shows very good stability and robustness even under disturbances and uneven load conditions. The STAR fuzzy control provides excellent performances in controlling the cart's position and balancing the pole. In order to compare the performance of the STAR approach, a conventional fuzzy control method has also been implemented to control the same inverted pendulum.

## 1  Introduction

Fuzzy logic provides a means to deal with nonlinear functions. Suppose that $e$ and $\dot{e}$ are inputs and $u$ is control action. Then, in fuzzy logic, the relationship between inputs and output (i.e. control action) can be represented in linguistic variables as

if $e$ is E and $\dot{e}$ is DE then $u$ is U

where E, DE and U are fuzzy linguistic variables. The above description is called a control rule. All such rules can be expressed in a so-called rule table. The control rule table is a very important factor in fuzzy control. Another important factor is membership functions which define the fuzzy degree of an input, e.g. $e$, belonging to some fuzzy linguistic variable, e.g. E in the above rule.

In conventional fuzzy control methods, the control rule table and the membership functions are usually fixed and remain unchanged over all operation time once they are defined. However, such a fuzzy controller shows poor control performance in real-time controls because of unknown and time-varying models of plants to be controlled. To improve the performance and effectiveness of fuzzy controls, Wang et al. [1,2] proposed a method to adapt the fuzzy control rule while Tan et al [3] developed an adaptive fuzzy control by modifying the scale factors of inputs and the output. In this paper, a novel self-tuning resolution adaptive (STAR) fuzzy control [4], which adapts the membership functions for inputs and the output, is implemented to control the inverted pendulum.

For the control of an inverted pendulum mounted on a small cart, we need to control the cart's position and to balance the pole simultaneously. In this case, the control task can be viewed as two sub-tasks: position control and pole balance. Balancing the pole will cause the cart to deviate from the desired position since the two dynamics are coupled. In general, we always move the cart from one desired position to another position while assuring the pole upright. If we set a task priority respectively to these two sub-tasks, hence the priority of balancing the pole is higher than that of controlling the cart's position. Based on this consideration, we propose a cascade controller architecture to control the inverted pendulum. By using two simple fuzzy controllers with a well-defined control algorithm, we achieve an overall control system that is robust against disturbances.

The rest of this paper is organized as follows. A cascade controller architecture and system configuration used in our experiments are described first in Section 2. Section 3 briefs the STAR algorithm proposed by authors. Implementations of the proposed controller architecture and the STAR algorithm are presented in Section 4. Some conclusions are given in Section 5.

## 2 Cascade Controller Architecture

The proposed architecture is shown in Fig. 1. The fuzzy controller consists of two cascade fuzzy sub-controllers. The first sub-controllers is used to control the cart's position. According to the errors, $e_x$ and $\dot{e}_x$, of the cart's position and speed, this sub-controller produces an offset angle $\theta_x$ as the desired set point to the second sub-controller. The desired angular velocity $\dot{\theta}_x$ is approximately calculated through the first-order differentiation of $\theta_x$. This second sub-controller is used to balance the pole. It generates a control action, $u$, in terms of the errors, $e_\theta$ and $\dot{e}_\theta$, of angles and angular velocities. Then the control output is inputted to the motor driving system via the I/O interface of PC to drive the motor. The following shows how the controller works.

Firstly, let us consider the case without external disturbances exerted on the pole. In this case, if the cart is not located at the desired position, i.e. $x \neq x_d$, there is a position error $e_x$ and the position sub-controller will output a non-zero offset angle $\theta_x$ to the second sub-controller. When there is no external disturbances, the pole's angle $\theta$ and angular velocity $\dot{\theta}$ tend to be zero if the pole is balanced. Therefore, the offset angle $\theta_x$ and its corresponding velocity $\dot{\theta}_x$ produce non-zero errors $e_\theta$ and $\dot{e}_\theta$. Then the second sub-controller outputs a control action to move the cart to the desired position. When the cart is stable at the desired position and the pole is in balance, the errors $e_x$ and $\dot{e}_x$ of the cart's position and speed are zero. The angle $\theta$ and angular velocity $\dot{\theta}$ are also zero. The outputs, $\theta_x$ and $\dot{\theta}_x$, of the position sub-controller will be zero. Hence the zero inputs $e_\theta$ and $\dot{e}_\theta$ make the second sub-controller output a zero control action. The cart's position remains unchanged and the pole's angle is kept zero.

Now let us look at the cases where some external disturbances are exerted on the pole. Referring to Fig. 2, two types of external disturbances are considered here. One is an external force, $F$, exerted on the pole in Fig.2 (a). The other is an uneven load $G$ shown in Fig.2 (b). For the sake of simplicity, suppose that the desired cart's position is set to zero, i.e. $x_d = 0$ and the cart is at $x = 0$. There will be an offset angle $\theta$ when the disturbance is exerted on the pole. When the errors $e_x$ and $\dot{e}_x$ are zero and as such $\theta_x$ and $\dot{\theta}_x$ are zero too, the non-zero offset angle $\theta$ produces non-zero error $e_\theta$ and $\dot{e}_\theta$. Then the controller generates a control action $u$ to move the cart. However, the cart will no longer move further when it goes to somewhere, $x \neq 0$, such that the outputs, $\theta_x$ and $\dot{\theta}_x$, of the position sub-controller are the same as the pole's $\theta$ and $\dot{\theta}$. If the disturbance is exerted on the pole all the time, the cart-pole system will be stable with a steady position offset $x$ and angle offset $\theta$ as shown in Fig. 2. The values of these offsets are dependent on the disturbances.

Moreover, the above cascade architecture makes the controller more sensitive to the pole's angle than the cart's position. Hence the controller takes balancing the pole as its first task and controlling the cart's position as its second task. When the small cart is required to move from one desired position to another desired position, the controller will move the cart while assuring the pole balance.

# 3 Self-Tuning Adaptive Resolution (STAR) Fuzzy Control

In STAR strategy [4], the input trajectory on the $(e, \dot{e})$ plane is divided into two regions A and B. Region A represents $(e \cdot \dot{e} < 0)$ and the error is reducing in this region. The error in region B, where $(e \cdot \dot{e} > 0)$, will increase. Hence STAR is to confine the input trajectory to region A as much as possible. To this end, a 4-state control strategy has been proposed. The control actions in these four states and criteria for state transitions are briefly summarized as follows:

State-1 When the error is very large, e.g. $|e| > e_{max}$ where $e_{max}$ is a prescribed threshold, STAR goes to State-1. In this state, the controller produces a large control action $u$ to reduce error $e$. STAR will transit to State-2 if the input trajectory satisfies the criterion

$$|e| > \alpha|\dot{e}|.$$

where $\alpha$ is a constant.

State-2 This is a relaxation state. The control action $u$ is zero. The error $e$ will continue to reduce because of the inertia. When the following criterion is satisfied, STAR enters State-3

$$(|e| + \beta|\dot{e}|) > thr1$$

where $\beta$ and $thr1$ are some thresholds to be determined.

State-3 State-3 tries to prevent overshoot by exerting a large control action in the opposite direction. STAR goes into State-4 if $e$ is small, e.g. $|e| < e_{min}$ or the following criterion is satisfied:

$$(|\ddot{e}| > \gamma|\dot{e}|)$$

where $e_{min}$ is a threshold and $\gamma$ a constant.

State-4 The control action is the same as a conventional fuzzy controller. If the error is bigger than a certain threshold, e.g. $|e| > e_{max}$, STAR enter into State-1 again and the cycle repeats itself.

# 4 Implementation and Experimental Results

The experimental system is shown in Fig. 1. It consists of three parts: a fuzzy controller implemented on a personal computer (PC'486), a servo-motor driving system and an inverted pendulum.

The servo-motor driving system is based on pulse-width modulation (PWM). It receives an input voltage with ranges of -10 ~ +10 V from the D/A converter of I/O interface. After amplifying the input voltage, the driver outputs a large DC voltage to drive the motor. Fig. 3 illustrates the driving system.

There is only one motor mounted on the frame of the inverted pendulum to move the small cart along the track, called X-axis. Two encoders are mounted respectively on the motor's axis and the pole's axis. They are respectively used to measure the cart's position, $x$, from the center of the track and the angle, $\theta$, between the pole and the vertical axis. Position $x$ and angle $\theta$ are inputted to the fuzzy controller. They are also used to calculate the approximate speed, $\dot{x}$, of the cart and angular velocity, $\dot{\theta}$, of the pole through calculations of their first-order derivation respectively.

Two fuzzy control approaches, a conventional fuzzy control and STAR fuzzy control, have been implemented in the proposed cascade architecture to control the real inverted pendulum. The pole is a solid steel bar with a length of 48.5 cm and a diameter of 1.0 cm. The weight of the small cart is about 2.0 kg. For the sake of safety, the maximum allowable distance from the track's center is specified as -90 cm to +90 cm.

In both approaches, all inputs are divided into seven ranges: Negative Large (NL), Negative Medium (NM), Negative Small (NS), Zero (Z), Positive Small (PS), Positive Medium (PM) and Positive Large (PL). Triangular membership functions with 50% overlap are applied to inputs of the fuzzy controllers, as shown in Fig. 4. The same control rule table defined in Table 1 are used in two sub-controllers.

For the conventional fuzzy control, the parameters $(a_1, a_2, a_3)$ of membership functions are specified as follows:

Angle sub-controller:

inputs

$\theta$: $(a_1, a_2, a_3) = (3.0, 6.0, 10.0)$;

$\dot{\theta}$: $(a_1, a_2, a_3) = (1.0, 3.0, 5.0)$;

output

$u$: $(a_1, a_2, a_3) = (1.0, 3.0, 5.0)$.

Position sub-controller:

inputs

$x$: $(a_1, a_2, a_3) = (10.0, 20.0, 30.0)$;

$\dot{x}$: $(a_1, a_2, a_3) = (1.0, 5.0, 10.0)$;

output

$\theta_x$: $(a_1, a_2, a_3) = (1.0, 3.0, 5.0)$.

The center-of-gravity method is used to defuzzify the outputs $u$ and $\theta_x$ of two sub-controllers. All membership function parameters remain unchanged during overall operation time.

For STAR fuzzy control, the initial membership functions of inputs and outputs are the same as those of conventional fuzzy control. Its special parameters described in Section 3 are chosen as follows:

Angle sub-controller:

$e_{min} = 1.0$, $e_{max} = 5.0$, $\alpha = 0.25$, $thr1 = 0$, $\beta = -1.0$ and $\gamma = 1.0$;

Position sub-controller:

$e_{min} = 3.0$, $e_{max} = 10.0$, $\alpha = 0.02$, $thr1 = 0$, $\beta = -9.0$ and $\gamma = 1.0$.

Three experiments have been done to observe the stability and robustness of the controller and to compare the performances of the STAR algorithm with a conventional fuzzy control. In our first experiment, the initial desired cart's position is set to zero ($x_d = 0$). Then the cart is demanded to move in sequence to $x_d = 25cm, 50cm$ and return to $x_d = 0cm$. Figures 5 and 6 show the response curves of cart's positions and pole's angles. It can been seen from these figures that the STAR algorithm can move the cart to the new desired positions faster with less overshoots.

In the second and third experiments, a toy windmill and an uneven load are respectively mounted or put on the top of the pole and blown by a stand fan when the pole is upright. The experimental results show that the cascade controller is very stable and robust against external disturbances and uneven loads when both the conventional fuzzy control algorithm or the STAR algorithm are implemented.

## 5  Conclusions

The task of controlling an inverted pendulum has been divided into two sub-tasks: cart's position control and pole balance. The pole balance sub-task has a higher priority over the position control sub-task. Based on these priorities, a cascaded controller architecture has been proposed to control the inverted pendulum. Experiment results have shown that this controller is of very good stability and robustness against external disturbances and uneven loads. Moreover, a novel self-tuning adaptive fuzzy control algorithm has been implemented in this controller to further improve the control performance. As we have seen in Section 4, the new algorithm provided considerable improvements in reducing the position overshoots and balancing the pole.

## References

[1]  S. Z. Ron, P. Z. Wang, " Self-tuning issues of fuzzy control rules", *Fuzzy Mathematics*, No.3, Sept. 1982, pp.105-111. (in Chinese)

[2]  S. Z. He, S. Tan, C. C. Hang and P. Z. Wang, "Control of dynamical processes using an on-line rule-adaptive fuzzy control system", *Fuzzy Sets and Systems* 54 (1993), pp.11-22.

[3]  Y. Lin, S. H. Tan, S. Z. He and Z. R. Liu, " Adaptive fuzzy control of structural and parameter varying processes", *Proc. of First Asian Fuzzy Systems Symposium*, Singapore, Nov. 23-26, 1993.

[4]  H. C. Lui, M. K. Gu, T. H. Goh and P. Z. Wang, "A self-tuning adaptive resolution (STAR) fuzzy control algorithm", submitted to *Third IEEE Int. Conf. on Fuzzy Systems*, Orlando, Florida, June 26-29, 1994.

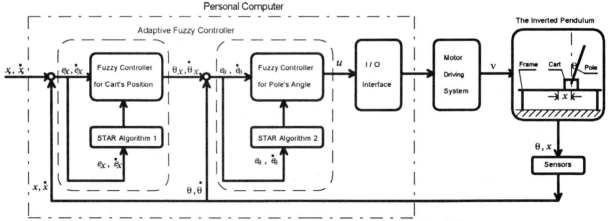

Figure 1.  The Control System for An Inverted Pendulum

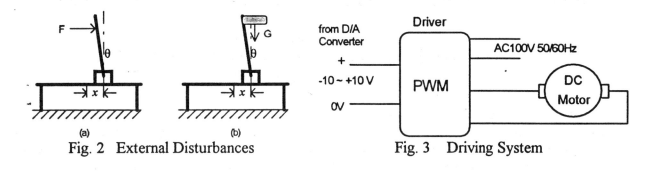

Fig. 2 External Disturbances

Fig. 3 Driving System

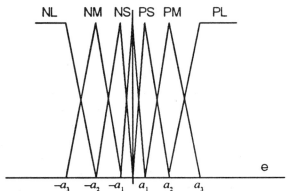

Fig. 4 Membership Functions

| $\dot{e}$ \ $e$ | NL | NM | NS | Z | PS | PM | PL |
|---|---|---|---|---|---|---|---|
| NL | NL | NL | NM | NM | NS | NS | Z |
| NM | NL | NM | NM | NS | NS | Z | PS |
| NS | NM | NM | NS | NS | Z | PS | PS |
| Z | NM | NS | NS | Z | PS | PS | PM |
| PS | NS | NS | Z | PS | PS | PM | PM |
| PM | NS | Z | PS | PS | PM | PM | PL |
| PL | Z | PS | PS | PM | PM | PL | PL |

Table 1. Control Rule Table

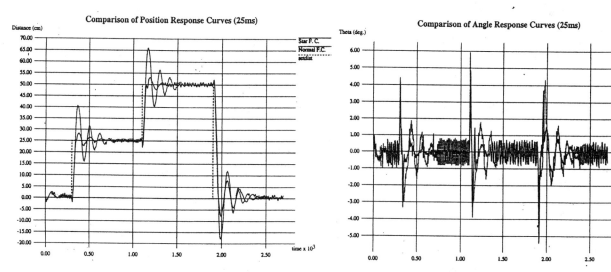

Fig. 5 Cart's Position Responses

Fig. 6 Pole's Angle Responses

# A Fuzzy Chip-Based Real-Time Fault Classifier in a Power Controller

Xing Wu and Chwan-Hwa Wu

Department of Electrical Engineering, 200 Broun Hall

Auburn University, AL 36849-5201

*Abstract* - A fuzzy chip-based electrical power faults classifier is presented in this paper. The system, which utilizes a fuzzy chip designed for the fuzzy rule base inference, detects the faults in the electrical power system in real time and activates the circuit control unit to take the appropriate actions. A set of features are extracted, and two sets of fuzzy inference rules are used to classify faults based on those features. The membership functions for all fuzzy variables are trained based on a supervised learning algorithm. Features extracted from structure properties of the patterns enable the classifier to rapidly detect the faults appearing in electrical power within 50µs. THe Fuzzy chip, in this fault classifier, provides speed and cost improvement over the existing general-purpose microprocesser technologies.

## 1. Introduction:

A higher quality electrical power would be beneficial to both industry and personal life; therefore, it is critical to design a real-time electrical power controller, especially for the machines based on properties of the electrical power, such as frequency, shape and other features. The faults that occur in the power may be disastrous for some equipment. Finding methods that can be used to quickly detect the faults and to compensate them in order to turn out high quality signals is an important research objective in electrical power system research, which is related to pattern recognition and control techniques. There are several approaches to this problem; such as expert system, statistics and neural network and fuzzy method. However, the neural network and fuzzy logic solutions are two prospective approaches among them because of the nonstationary nature of the problem. In this paper, a fuzzy chip-based power controller using a neural fuzzy classifier , which can rapidly detect faults in the power system in a very brief time (50µs) and activate the power generator control unit or system operator to take the appropriate actions, is implemented.

The subject of fuzzy sets, a mathematical representation of the human linguistic expression and inference, was introduced by Zadeh [1] in 1965. Now, fuzzy approaches to the control processes and pattern recognition have been successful, especially for solving the problems in the nonstationary situations. Since neural networks are designed to mimic the architecture and functions of human neurons, and humans are somewhat ambiguous in making judgments in complicated situations, the fuzzy set theory has been coupled with neural network models. Keller and Hunt [2] incorporated fuzzy membership functions into perceptron algorithm; Carpenter et al. [3] developed a fuzzy ART and a fuzzy ARTMAP architecture to embed fuzzy operations in their ART neural networks; S. G. Kong and B. Kosko [4] developed an adaptive fuzzy system for backing up a truck-and-trailer. In this neural fuzzy system, input-output associative data was used to generate the fuzzy associative memory (FAM) rules that solve the nonlinear behavior of practical problems. Simpson [5] developed a fuzzy min-max neural network to classify overlapped classes, Pal and Mitra [6] used fuzzy membership concepts for backpropagation algorithm for speech recognition, and W. Pedrycz [7] introduced a fuzzy neural network with reference neurons as pattern classifiers.

In this paper, a neural fuzzy-based classifier is introduced. A set of features based on 1/4 cycle of information of power signals' energy distribution and shapes makes up a feature space for the classification of the wave form of the power signals. Additional features, such as the angles around turn point from the transition state of the faulty pattern within a short period of time are extracted to quickly detect the faulty patterns appearing in the power system flow. The membership functions of all features are trained by a rapid learning algorithm. A fuzzy inference engine is set up for the classification and is implemented on a fuzzy chip, FC110 from TOGAI InfraLogic. All the fuzzy operations are realized on this special fuzzy logic chip, FDP ( Fuzzy Digital Processing Chip), which accomplishes the fuzzy operations at high speed. By fully utilizing the special FDP chip as the accelerator, we can classify the power signal waveforms and detect any faults in real time.

We introduce our neural fuzzy classifier in section 2 of this paper, . In section 3, a power controller using the neural fuzzy classifier is introduced, and extraction of transition structure properties features to quickly detect the faults is described as well. The implementation of a fault classifier system of the power signals on a fuzzy chip-based accelerator board is presented; and in the last section, we give a short conclusion about our neural fuzzy classification system.

## 2. Neural fuzzy classifier

Like the pattern classifier described by Duda and Hart [8], a five-layer structure is used to describe the NF system for a pattern recognition application. Based on the information provided by past engineering experiences and input-output data pairs, a feature vector $\mathbf{f}$ in a multidimensional feature space $(f_i, i = 1, 2, ..., n)$ is chosen as input to the system, and the domain of each input feature is divided into fuzzy subsets. The discriminant calculations are implemented with fuzzy membership functions $(m_j(f_i)$ where $i = 1, 2, ..., n$ and $j = 1, 2, ..., m)$. The fuzzy rules provide the fuzzy relations $(R_j, j=1, ..., m)$ and decide the connections between the input feature subsets and the fuzzy reasoning output classes. The final class output $(R_X, X = A, B, ..., Z)$ is the combination of the fuzzy rules (fuzzy relations) belonging to the same class $X$. The final decision is made by a maximum selector, which selects one class with the largest output active value.

A fuzzy IF-THEN rule provides a convenient way to express the input-output reasoning and recognition policy; the fuzzy classification rules are defined for the knowledge base of the fuzzy logic system as

$$If\{(f_1 \, is \, X_1),\; ;AND; (f_2 \, is \, X_2), ..., AND \, (f_n \, is \, X_n)\} \; then \; \{The \; output \; class \; is \; X \}$$

where $X_i$ is a fuzzy set in feature space $U$. The membership function of each fuzzy set is a membership located on the position $f_i$ of a universe of discourse $U_i$ to form a hyperbox in a multidimensional feature space. Four critical points, $X_i ll$, $X_l$, $X_r$, and $X_r rr$, are used to define a trapezoidal membership function for a fuzzy set $X_i$ in every feature space $U_i$.

A fuzzy classification rule, the antecedent with a linguistic "AND" is implemented by the Cartesian product in the product space. In a multidimensional feature space, every fuzzy rule $R$ is represented by a hyperbox with a membership function $\mathbf{R}$, given as:

$$\mathbf{R} = m_{X_1 \times ... \times X_n \to X}(\mathbf{f}) = m_{X_1 \times ... \times X_n}(\mathbf{f}) = m_{X_1} \bullet m_{X_2} \bullet \cdots \bullet m_{X_n}$$

Based on the easy and fast generalized delta rule, a developed on-line supervised learning algorithm uses pairs of input-output samples to generate and modify characterized membership functions. During the training scheme, every feature element $f_i$ of an input-output data pair $(\mathbf{f}, y_x)$ is used to compare the decision region of the desired output class. A decision region learning law is designed based on the fast-commit slow-recode [3] with learning factors $1 > Q1 > Q2 > 0$. $Q1$ and $Q2$ are the learning factors that dictate the amount of reinforcement required during the learning process. A large learning rate $Q1$ is used to quickly extend the right boundary of set $X$. A small learning rate $Q2$ is used to slowly contract the left fuzzy-set boundary. The initial membership function of a triangle shape will be trained to trapezoid shape. The algorithm is presened by following equations:

$$Case \; 1 : X_r = X_r^{old} + Q1(f - X_r^{old}) \quad X_r \angle f \qquad Case \; 2 : Xl = Xl^{old} + Q2(f - X_c^{old}) \quad f \angle Xl$$

$$Case \; 3 : Xl = Xl^{old} + Q1(f - Xl^{old}) \quad X_c \le f \le X_r \qquad Case \; 4 : X_r = X_r^{old} + Q2(f - X_c^{old}) \quad Xl \le f \le X_c$$

From the above discussion, a neural fuzzy classifier is set up and trained. Now, when the data is measured, the preprocessing module calculates all the features, then the input vector is sent to the fuzzy inference engine for fuzzy classification in order to obtain the classification results.

## 3. A real-time fault fuzzy chip-based classifier for power controller

We use the algorithm introduced in Section 2 to utilize a neural fuzzy classifier for the real-time classification of faults in three phase electricity power. A power controller system diagram is shown in Fig. 1. Its major parts are the power waveform generator, A/D converter, a neural fuzzy classifier for fault detection, output display and circuit control unit.

### a. Two sets of features extracted from power signals

There are seven kinds of basic patterns appearing in our experiments(as shown in Fig 1.b). They are sine wave, square wave, triangle wave, spike wave, chop wave, 2/3 duty wave and slow-rising and -falling square wave. Among these wave forms, the spike wave and chop wave are the faulty patterns of sine wave. The patterns appearing in electrical power are sine wave and its faulty patterns, spike and chop wavceforms; The other waveforms we add in the system are for demonsrating the versatility of our system.

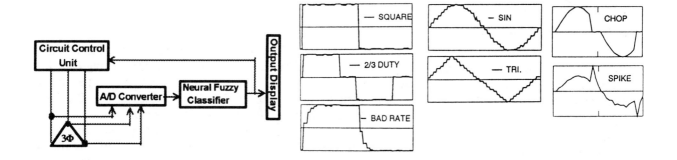

Fig. 1a   Three-phase electrical power controller          Fig. 1b   Seven experiment waveforms

A feature extractor, which extracts the specific features from the measuring data, can reduce the amount of data processing to an input feature vector in a multidimensional feature space.

Two sets of features are extracted for the fuzzy classification. The first set of four features is extracted from a 1/4 cycle of information of the power signal waveforms. The four features from the properties of the energy and shape information averaged and collected from a 1/4 cycle offer enough information for a classifier to classify the type of waveform the input signal is after the pattern appears for a 1/4 cycle. Whereas, for the 2/3 duty waveform, we need a 1/2 cycle to recognize it from a square wave. This set of features have to be calculated from normalized data of the input signals. They are the basic properties of periodic power signals.

The first set of features includes four features:

1. Feature $f_1$ is based on the average magnitude over 1/4 cycle.

2. Feature $f_2$ is the averaged magnitude difference between the input wave form and a predicted triangle waveform over 1/4 (positive) cycle.

3. Feature $f_3$ is the fraction of a 1/4 cycle waveform whose magnitude is around one.

4. Feature $f_4$ is the fracture of a 1/4 cycle waveform whose magnitude is around zero.

The fuzzy classifier can use the four features to distinguish the power signals. But, all four features are based on a 1/4 cycle of normalized information of the power signals. We have to know the cycle of the signal and starting point (the first positive point) in a 1/4 cycle. It needs at least a 1/4 cycle of data to extract the four features.

The second set of features is extracted on the short period transition state of the signal. The faulty patterns have large changes in the slope. There exists some structure angles around the critical turn points.

To get the new features, the turn points should be detected first, then we calculate the slopes of the signal around turn points in a small time window. Therefore, we can calculate the angles around the term points and critical slope values, and we can get all these following features without waiting for a 1/4 cycle. The procedure is expressed as the following:

We define:

$S(i,j) = M(i-1) - M(i+j)$,   $p(i,j) = M(i-j-1) - M(i)$ , where $M(i)$ is the data sequence acquired from A/D.

(1) Detect the turn points.

$$\frac{|s(i,0)|}{M} < \rho_1 \ and \ \frac{|s(i+1,0)|}{M} > \rho_2 \ then \ P_i \ may \ be \ a \ turn \ point$$

If $(S(i,0) = 0)$  $P_i$  is a turn point.

If $(s(i,0) \neq 0$ and $|s(i+1,0)/s(i,0)| > \rho_3)$  $P_i$  is a turn point

where M is the magnitude estimated from the former cycle, $P_i$ is the ith data point from A/D converter, $\rho_1$ $\rho_2$ and $\rho_3$ are vigilance parameters.

(2) Extract features of structure angels and critical slopes

$ns(i,j) = s(i,j)/max(M(j))$,   $np(i,j) = p(i,j)/max(M(j))$,  $j = i-3 \ldots i+3$,

$Angle1 = | \tan^{-1}(-ns(i,0)) + \tan^{-1}(ns(i,1)) |$,    $Angle2 = | \tan^{-1}(-ns(i+1,0)) + \tan^{-1}(ns(i,2)) |$,

$$\text{Angle3} = |\tan^{-1}(-np(i,2))+\tan^{-1}(ns(i,0))|, \quad \text{Angle4} = |\tan^{-1}(-ns(i,0))+\tan^{-1}(ns(i+1,4))|,$$
$$\text{slope1} = |np(i,2)+np(i,4)|/2, \quad\quad\quad \text{slope2} = |ns(i+1,2)+ns(i+1,4)|/2.$$

These features are extracted based only on several points around turn points. The other important advantage of the second set of features is that it is not required to detect the starting and ending point of the signal every 1/4 cycle in order to recognize the faulty signals. The delay between the presence of the fault and the recognition is about 50μs

### b. Fuzzy membership functions and inference rules

The domain of each input feature is divided into fuzzy sets which are related to linguistic variables. Each fuzzy set should be assigned the membership function to describe how the feature is linked with class decision region. All the membership functions are trapezoid. A fuzzy rule base is set up by experience to link input subsets from feature space to the fuzzy reasoning output classes. Based on the learning algorithm introduced in Section 2, the membership functions are set up by the first training pairs, then trained by more training pairs to "grow" and be "mature" (see Fig. 2). The fuzzy rules base on the second set of features can recognize the faulty pattern in a very short time. They can detect the single shot fault in the power system. The inference rules are represented as the mapping from hyperbox to the classification output. This hyperbox structure facilitates utilization of the special DFP fuzzy processor's fuzzy operation. Altogether we have about 45 fuzzy subsets and their membership functions from training, which are ready for the fuzzy engine's inference.

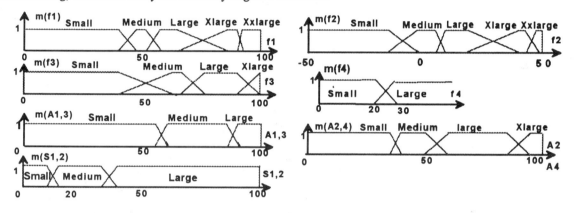

Fig. 2 The Trained membership functions for all fuzzy variables

Fuzzy rules are all expressed in the expressions of IF-THEN, which links the input fuzzy subsets in the features to the classification output space. A rule in rule base **R** is expressed as

$R_r$: *If($f_1$ is large or $f_2$ is large and $f_3$ is small) then it is a sine wave*

To recognize the faulty patterns quickly, another set of rule base is set up based on the second set of features. When a fault appears in the power system, such as a spike wave, it enters the short window of the turn point detecter. The detecter detects the turn points and extracts the second set of features, then the set of features are applied to the rule base for fault detection. This kind of fuzzy inference rule is drawn based on the second set of features, such as

*If(Angle1 is Large and Angle2 is Large and Slope1 is Medium and Slope2 is Small) then it is a chop wave*

### c. Power fuzzy classifier on a fuzzy-chip based module

In a fuzzy logic system, fuzzy operations include long vectors comparisons and products. The real time requirement needs a high-speed fuzzy processor to do the special fuzzy operations; hence, we use an ASIC fuzzy processor, FC110 from Togai InfraLogic, to calculate all the fuzzy operations in this fuzzy classifier.

The block diagram of the fuzzy chip-based power controller is shown in Fig. 3. It consists of a host 486 computer, a FC110AT fuzzy accelerator board, a display device and a control unit for a power generator. The accelerator board is embedded in the ISA bus. A/D channels on the board are used to receive the power signal. The host machine, a 486, is utilized to preprocess the data and calculate all features, then the host sends them to the fuzzy accelerator board through ISA bus. On the fuzzy accelerator board, KMB (knowledge Memory Base) restores all the fuzzy subsets' membership functions and fuzzy inference rules. When the host sends a vector of features to the FC110, the DFP chip runs evaluation and inference operations and offers outputs directly to the host. The host makes the final decision from the belief values for the premises of the rule base. The two different sets of features we extracted have different rule base, and they are stored and called at their relative areas on the KMB.

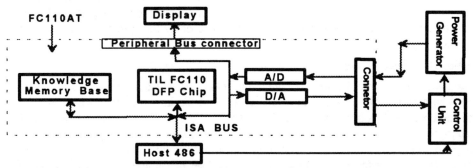

Fig. 3  Architecture of power controller based on fuzzy chip

The system works in the way described in Fig. 4.  There is a double recognition process:  (1) The host first checks the turn points in the signal flow.  If one turn point is found, then calculate the second set of features and call the second rule base to recognize the faulty patterns; (2) The system will calculate all first set features and call the basic rule base after every 1/4 cycle of the signal is obtained.  To recognize any faulty pattern appearing in the signal flow, the system offers some information from a 1/4 cycle for the next recognition process.  This double recognition process makes the system perform more reliable.

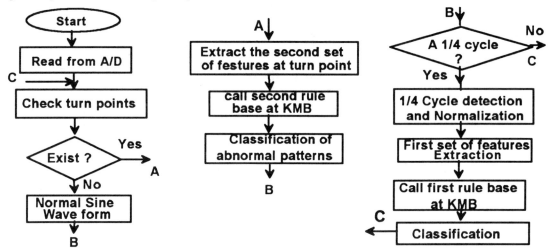

Fig. 4    Flow chart of the power control system

The fuzzy rule base is first written into a TIL file named power.til in FPL language designed by Togai InfraLogic, which can be further used by the Fuzzy-C system and the FC110 development system.  A portable C code is generated for the fuzzy rule base evaluation in Fuzzy-C.  We use software simulation to certify that our rule bases worke well.  Therefore, we use the FC110 compiler to compile the power.til in the FC110 developing system.  After they are linked, a machine code for the evaluation rule bases is obtained and is ready to run on the FC110 chip.  A driver program is made for preprocessing and driving the FC110 chip work.  The system works in the following way:  when a set of features is available, the driver program will send the input fuzzy variables to the chip by calling a function from the host processor, then the chip on the accelerator board operates all fuzzy operations in high speed using the special fuzzy instructions.  All the inference rules and evaluation information are used by the chip directly from KMB.  The classification results are sent to the host driver as returned results of a function.  One set of input features gets one classification in the real-time response.

The power.til for the fuzzy rule base in the power signal classifier written in Togai FPL language is similar to a linguistic expression.  In this file, we define all input and output fuzzy variables and their membership functions and set up the inference engine and all fuzzy rules.  It includes all the information needed for the fuzzy classification. In Fig. 5, we show how to implement fuzzy variables, their membership function and fuzzy rules for inferences in the power.til file ( $f_1$ is the feature described).

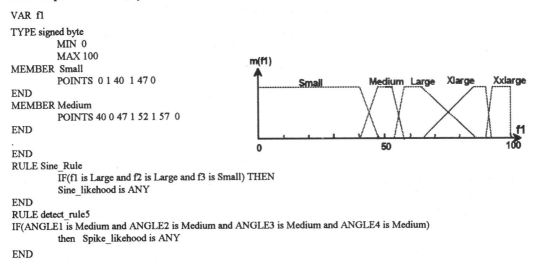

```
VAR f1

TYPE signed byte
        MIN 0
        MAX 100
MEMBER  Small
        POINTS 0 1 40  1 47 0
END
MEMBER Medium
        POINTS 40 0 47 1 52 1 57 0
END
.
END
RULE Sine_Rule
        IF(f1 is Large and f2 is Large and f3 is Small) THEN
        Sine_likehood is ANY
END
RULE detect_rule5
IF(ANGLE1 is Medium and ANGLE2 is Medium and ANGLE3 is Medium and ANGLE4 is Medium)
        then  Spike_likehood is ANY
END
```

Fig. 5    Part of the power.til file

The output variable is defined as ANY for classifier application, which gives the premise value in a fuzzy classifier application, which offers information needed to classify input signals.

### d. Results

The power controller based on the fuzzy chip using structure features extracted from transition state in power signals works stablely.  It can detect and recognize the faults in the power signal with a 100% accuracy. With the second rule base of the structure features, any faulty pattern appearing can be recognized within 50µs. Because of the implementation of the algorithm on a fuzzy chip, it provides real-time response with a low price cost.

## 4. Conclusions

In this paper, a neural fuzzy classifier is implemented on a Fuzzy chip-based module for power controller. A complete set of features are extracted from the power signal, which include basic-property features and short-time-transition-structure features. The learning algorithm of membership functions trains all the membership functions of the fuzzy subsets, and a fuzzy inference rule base is set up for the power signal classifier. The high speed, low price DFP chip implementation  recoginzes faults within 50µs. The system provides a 100% accuracy in classifying faults and the shapes of waveforms.

## Reference:

1.  L. A. Zadeh, "Fuzzy sets," *Information and Control*, Vol. 8, pp.338-353, 1965.
2.  J. Keller and D. Hunt, "Incorporating fuzzy membership functions into perceptron algorithm," *IEEE Tans. Patt. Anal.Amach.Intell.*, Vol.,7, pp.693-699, 1985
3.  G. A. Carpenter, S. Grossberg, and D.B. Rosen, "Fuzzy ART: fast stable learning and categorization of analog by an adaptive resonance system," Neural Networks, Vol. 4, pp. 759-771,1991
4.  S. G. Kong and B. Kosko, "Adaptive fuxxy system for backing up a truck-and-trailer," IEEE Trans. Neural Networks, Vol. 3, no.2, pp.211-223,1992
5.  P. K. Simpson, "Fuzzy Min-Max neural Networks-part1: Classification," *IEEE Trans. Neural Networks* vol. 3, no. 5, pp. 776-786, 1992
6.   S. K. Pal and S. Mitra, "Multilayer perceptron, fuzzy sets, and classification," *IEEE Trans. Neural Networks*, vol. 3, no. 5, pp. 683-697, 1992.
7.  W. Pedrycz, "Fuzzy neural networks with reference neurons as pattern classifier, " *IEEE Trans. Neural Networks,* Vol 3  pp.770-775, 1992
8.  R. O. Duda and P. E. Hart, *Pattern Classification and Scene Analysis*, Wiley, N. Y., 1973

# Fuzzy Control of Traffic Lights

Robert HOYER and Ulrich JUMAR

Institut für Automation und Kommunikation Magdeburg
Bahnhofstr. 27-28
D - 39179 Barleben
GERMANY
Tel.: +49 39203/ 61055
Fax: +49 39203/ 60938

## Abstract

The paper is dealing with a fuzzy control approach to urban road traffic lights. Starting with briefly reviewing some of the fairly rare publications to this topic special emphasis is laid on possible applications of fuzzy logic in traffic light control. Afterwards, the problem of controlling an intersection with 12 main direction traffic flows is tackled. Depending on traffic situations the fuzzy rule set decides on the activation of a two-state, three-state or four-state control. In this way the merits of each of these control schemes are combined. The state machine used, linguistic variables and essential ideas of fuzzy rule design are explained.

## 1. Introduction

Fuzzy control has been successfully applied to many tasks of automatic control over the last decades. A few theoretical papers on control of traffic systems using fuzzy statements have been published. PAPPIS and MAMDANI described in [PAP-77] a fuzzy logic controller for a single traffic junction of two one-way streets. The fuzzy controller uses three linguistic input variables and one linguistic output. Every ten seconds it decides on an extension of the current green period. The authors established five rules for every interval of ten seconds. The number of intervals is limited to five resulting in 25 fuzzy rules on the whole. Fuzzy variables of PAPPIS and MAMDANI were 1) passed time of the current interval, 2) number of vehicles crossed a detecting pad of the green direction, 3) length of queue in the red direction and 4) the value of the time extension calculated by fuzzy controller.

NAKATSUYAMA expanded the isolated one-way intersection to an one-way arterial road with two successive intersections [NAK-84]. First, there is applied a fuzzy logic controller (FLC) for the independent regulation of traffic flows of both intersections. However, co-ordination of the two intersections is not considered, thus enabling the system to be improved with respect to overall delays of crossing vehicles. Second, a newly developed fuzzy logic phase controller (FPC) determines the signal offset between successive intersections. A comparison of delays caused by FLC and FPC showed, that the superiority of FPC and FLC depends upon distribution and density of traffic flows. A further reduction of overall delays was achieved by a suitable co-operation of FPC and FLC. Switching of both controllers is handled by additional fuzzy statements. The interchanging strategy is a prospect method for a signal controller of arterial one-way roads.

Focusing the control of an isolated single intersection KELSEY [KEL-93] chooses the fuzzy variables 1) traffic density of the roads distinguishing closed and released directions, 2) time elapsed since the last signal change and 3) degree of change. The task is solved by 26 rules that are invoked every second. Investigations are not confined to one-way road systems, but the signal program has 2 states only, i.e. the consideration is restricted to 2 fixed main directions of traffic. The controller is unable to respond to changing and turning main traffic flows. A recently published paper dealing with the subject of a self-learning fuzzy traffic controller for an isolated single intersection is [SKO-93]. Details of this approach are given in [SHA-92].

## 2. Fuzzy logic and traffic light design

As far as it is known to the authors, no cases of fuzzy control practically applied to urban road traffic signal systems have been reported until now. Nevertheless, it seems to be beneficial to think about possibilities of practical use. Within the design of signal programs we distinguish states and state changes. During a state all traffic lights are fixed. The duration of a phase can be influenced by a vehicle actuated controller or a fuzzy logic controller. The state change is fixed by the design and it is not allowed to be modified by any controller because of traffic safety reasons. Furthermore, the designer has to keep in mind a minimal time of green light and a maximal time of red light. Therefore, it makes sense to investigate conventional approaches of control of traffic flows and to clarify their shortcomings.

Typically, three levels of control hierarchy, the operational, the tactical and the strategic level are assumed. The strategic or macroscopic level refers to the whole network or parts of it. It activates different signal plans for a longer time depending on a time schedule or on traffic density. The tactical or microscopic level is a subordinated level. It reacts more quickly to changing traffic situations. On the operational level the presence of every single car in front of the traffic light is considered.

Possible linguistic variables of fuzzy logic controllers for intersections or traffic networks are the traffic density, the traffic strength, the relative throughput, the gap of time between single cars, the congestion, the mean velocity, the rate of trucks, the requests of privileged public traffic, etc. Fuzzy logic will not provide an optimal solution with respect to a special criterion since no algorithm of optimisation is involved. Nevertheless, fuzzy logic control seems to be a promising approach to the

- description and analysis of traffic situations,

- calculation of control signals,

- evaluation and decision making

in all three levels of the control hierarchy. Mention should be made, that not all variables used in fuzzy control are strictly fuzzy. However, in between the framework of fuzzy rules a common treatment of fuzzy and not fuzzy-like crisp variables is convenient.

A typical problem offering a good chance to be solved by the help of fuzzy logic is time gap control. In order to decide, whether a green time has to be aborted or not, a measurement of the time gap of successive cars is commonly used. If the gap exceeds a fixed value, traffic light will change to red. A frequently used value in urban arterial roads are two seconds. Disregard of the acceleration behaviour of trucks and the disregard of various weather and roads conditions are disadvantages. The objective of time gap control in urban traffic networks is holding together the groups of cars in the "Green Wave". Fuzzy logic is able to support the identification of such groups by flexible time gap evaluation with only a few rules. For example, the time gap has to be longer than two seconds if the rate of trucks is higher than normal or the road is icy.

The acceleration of competing public traffic is another problem. In order to maintain optimised schedules preferential treatment in control of traffic lights is required. Especially, in case of many lines crossing large intersections it is difficult to determine the order of service of competing vehicles. Possible criteria to fix this order are the sequence of request, the kind of vehicle (bus or tram), proximity velocity, the rank of importance of the line, the schedule (being late or not), the number of transported people, etc. An evaluation of many criteria to determine the next vehicle requires a complex control algorithm which is impractical, or even impossible. Fuzzy logic could provide a solution to priority control.

A third problem worthwhile noting briefly is the selection of signal plans. This task occurs, when traffic light control has to be adapted to changing traffic flows in a traffic network. Optimised signal plans with fixed parameters (cycle times, phase lengths and delays) are designed for a given number of different situations. According to the present distribution of traffic flows a controller of the strategic level has to select the required signal plans of intersections. Again, the recognition of certain traffic situations could successfully be managed by fuzzy logic. Furthermore, notice that on certain conditions a few detectors are sufficient to map the traffic situation. Fuzzy control can support the completion of incompletely available traffic parameters. Additional information on weather, road conditions or degree of air pollution etc. can be taken into consideration.

## 3. The problem considered

The primary concern of our research is an extension of the approaches already mentioned in the brief literature review with respect to the following aspects:

- two-way arterial roads instead of one-way roads,

- multiple state control instead of two state control, and as a result of this

- consideration of turning main directions of traffic flows.

The structure of the studied fuzzy controller is depicted in figure 2. The controller operates with 10 fuzzy input variables and two output variables. As shown in figures 1 and 2 the traffic density of different lanes and elapsed time since last state change were chosen as inputs. Outputs are on the one hand the extension of time after the precalculated moment of state change (E), and on the other hand the selection of next state (NextPhase). The number of the controller's rule blocks was fixed to four in order to provide sufficient transparency. Rule block EX is fed by the traffic density of some red and green lanes (GA1, RA1...) as well as time (T). Variable T was included in this rule set for the sake of weighting. This is necessary to guarantee a minimal and maximal state length. Rule blocks GP and RP result in a preliminary decision on the next state. The task of the last rule block SUM is to balance signals GPhase and RPhase together with time T.

Looking at an intersection with four approaches including all turning traffic flows there exists a total of 12 possible main directions. However, in our state machine these 12 traffic flows are served by six states. According to usual design rules of traffic lights we distinguish two cases of left turning traffic control. In the first case (state 1 in figure 1) the left turning drivers have to observe the oncoming vehicles, in the second case oncoming traffic is stopped by traffic light (states 2 and 3). States 4, 5 and 6 are deduced from states 1 to 3 by rotation, thus reducing the number of necessary fuzzy statements. States 2 and 3 are introduced to be activated when heavy left turning traffic occurs. Again, the number of state changes is diminished by a common treatment of different situations within one fuzzy rule set. This is achieved by means of relative inputs mapping the actual detector signals of traffic density. The numbers nearby the arrows hint to the relative directions (angles) of released traffic flows during the next state. As a matter of course the state machine is not part of the fuzzy controller, and has to be implemented separately.

A total of 72 fuzzy rules was established to construct the fuzzy controller. Due to space limits of this paper we give only some of these rules in figure 3, which requires no detailed explanation. The column headed by the abbreviation DoS contains the degree of support of the rule. To provide some insight we outline the main ideas:

- If there is only weak left turning traffic the controller switches between states 1 and 4 only.

- In case of heavy left turning traffic decision has to be made whether state 2 or state 3 is next.

- An extension of green time is invoked when the ratio of released and stopped flows is large.

- Weighting of selected rules by time avoids unlimited extension of green time. In addition, this weighting - handled by the degree of support term (DoS) - ensures to keep a minimal green time.

A fuzzy controller of this kind is investigated at our institute at the moment. Its performance is assessed by extensive simulations based on microscopic and mesoscopic models [HOY-93]. Obtained results, which will be presented on the congress, are very promising. Of course, a comparison of the fuzzy approach and conventional traffic light controls will be of chief concern.

## 4. Concluding Remarks

It is expected, that fuzzy control will result in a more efficient and transparent design process of traffic light systems than design procedures in current use. We suppose a great potential in guaranteeing a high throughput of traffic by a flexible use of a two-state, three-state, and four-state control. Switching is accomplished by suitable fuzzy rules. The merit of two-state control lies in a minor number of state changes and loss of time, respectively. The merit of four-state control is the possibility of successive release of all lanes without conflicts. In this way, the advantages of both control schemes may be combined. Furthermore, with comparative ease terms like weather conditions, environmental aspects, etc. can be added to the fuzzy system. A lot of research work has to be done to verify the expected features by simulation and to study first practical solutions.

## Acknowledgements

This research is financially supported by the Ministry of Science and Research of Saxony-Anhalt in Germany under grant number FKZ 686 A 01312.

# References

[KEL-93] Kelsey, R.; Bisset, K.; Jamshidi, M.: *A Simulation Environment for Fuzzy Control of Traffic Systems.* 12th IFAC-World Congress, Sydney, Australia, 18. - 23. July 1993, Preprints, Vol. 5, pp. 553 - 556.

[NAK-84] Nakatsuyama, M.; Nagahashi, H.; Nishizura, N.: *Fuzzy Logic Controller for a Traffic Junction in the One-Way Arterial Road.* 9th IFAC-World Congress, Budapest, Hungary, 1984, Preprints pp. 13 - 18.

[PAP-77] Pappis, C.P.; Mamdani, E.H.: *A Fuzzy Logic Controller for a Traffic Junction.* IEEE Trans. Syst. Man. Cybern. 1977, SMC-7, pp. 707 - 717.

[SKO-93] Skowronski, W. M.; Shaw, L. S.: *Self-Learning Fuzzy Traffic Controller applied to an isolated Intersection.* First European Congress on Fuzzy and Intelligent Technologies, Aachen, Germany, 7. - 10. September 1993, Proceedings Vol. 2, pp. 751 - 761.

[SHA-92] Shaw, I. S.; Kruger, J. J.: *New fuzzy learning model with recursive estimation for dynamic systems.* Fuzzy Sets and Systems 48, pp. 217 - 229, North Holland 1992.

[HOY-93] Hoyer, R.: *Modellierung und Simulation fuzzy-gesteuerter Lichtsignalanlagen in einem Straßennetz.* In: Sydow, A. (Editor): Fortschritte in der Simulationstechnik. Bd. 6, S. 337 - 340, Braunschweig/Wiesbaden: Viehweg & Sohn Verlagsgesellschaft, 1993 (in German).

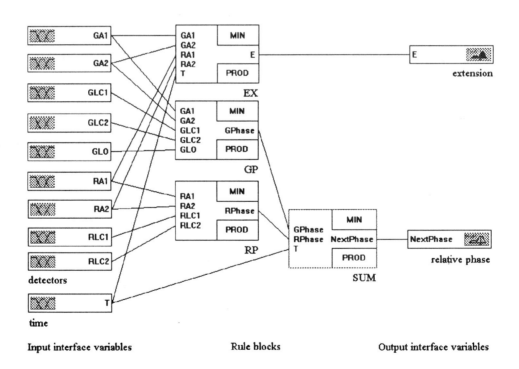

Figure 2. Structure of the fuzzy controller.

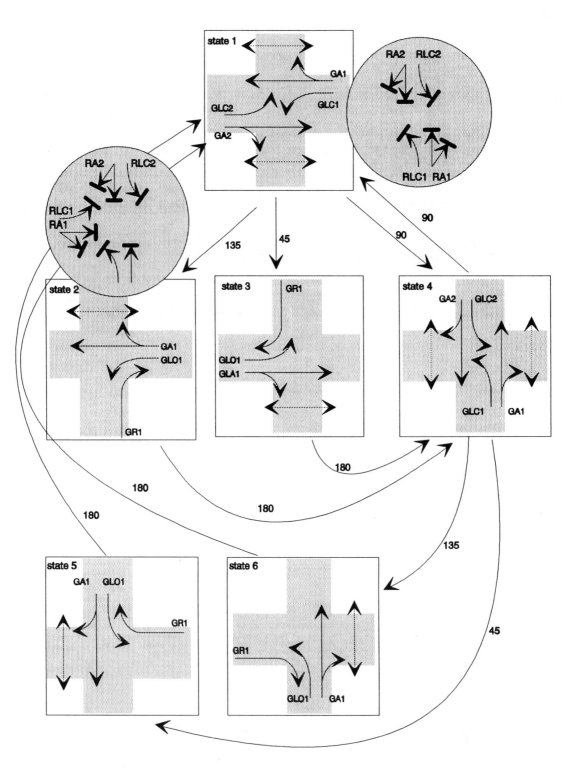

GLO: green direction left only  RA:  green direction across
GA:  green direction across     RLC: green direction left common
GLC: green direction left common
GR:  green direction right

Figure 1. State machine.

| Matrix / Utilities | GA1 | GA2 | RA1 | RA2 | T | DoS | E |
|---|---|---|---|---|---|---|---|
| 1 | high | | | | long | 1.00 | long |
| 2 | medium | | | | | 1.00 | medium |
| 3 | low | | | | | 1.00 | short |
| 4 | | high | | | long | 1.00 | long |
| 5 | | medium | | | | 1.00 | medium |
| 6 | | low | | | | 1.00 | short |
| 7 | | | high | | | 1.00 | short |
| 8 | | | medium | | | 1.00 | medium |
| 9 | | | low | | long | 1.00 | long |
| 10 | | | | high | | 1.00 | short |
| 11 | | | | medium | | 1.00 | medium |
| 12 | | | | low | long | 1.00 | long |
| 13 | | | | | long | 1.00 | short |
| 14 | | | | | medium | 1.00 | medium |
| 15 | | | | | short | 1.00 | long |
| 16 | | | | | | | |

| Matrix / Utilities | GA1 | GA2 | GLC1 | GLC2 | GLO | DoS | GPhase |
|---|---|---|---|---|---|---|---|
| 1 | high | | | | | 1.00 | Shift0 |
| 2 | medium | | | | | 0.50 | Shift0 |
| 3 | low | | | | | 1.00 | Shift90 |
| 4 | | | | | high | 1.00 | Shift0 |
| 5 | | | | | medium | 0.50 | Shift0 |
| 6 | | | | | low | 1.00 | Shift180 |
| 7 | | | high | | | 1.00 | Shift135 |
| 8 | | | medium | | | 0.50 | Shift135 |
| 9 | | | low | | | 1.00 | Shift180 |
| 10 | | high | | | | 1.00 | Shift0 |
| 11 | | medium | | | | 0.50 | Shift0 |
| 12 | | low | | | | 1.00 | Shift90 |
| 13 | | | | high | | 1.00 | Shift45 |
| 14 | | | | medium | | 0.50 | Shift45 |
| 15 | | | | low | | 1.00 | Shift180 |
| 16 | | | | | | | |

Figure 3. Part of the fuzzy rule set.

# Fuzzy Logic Attitude Control for Cassini Spacecraft *

*Richard Y. Chiang*

Jet Propulsion Laboratory
California Institute of Technology
4800 Oak Grove Dr
Pasadena, CA 91109-8099

*Jyh-Shing Roger Jang*

The MathWorks, Inc.
Cochituate Place
24 Prime Park Way
Natick, Mass. 01760

## Abstract

A fuzzy logic attitude controller has been developed for Cassini spacecraft. Feedback control issues such as tracking capability, thruster on/off time and cycle have been investigated and compared with conventional bang/bang control. A discrete nonlinear simulation was set up to assess the system performance with different controllers.

## 1 Introduction

Saturn, one of the most interesting planet in our solar system, will be visited by Cassini spacecraft in 2004. Cassini spacecraft will be launched in 1997 and arrive Saturn orbit in 2004 for a four-year mission of orbiting Saturn and flying by its largest moon Titan, which scientists believe containing materials just like earth at its primitive stage millions years ago.

Spacecraft attitude control system plays a crucial role in this mission. It has to stabilize the spacecraft attitude and track a set of complicated maneuver command profiles to within $\pm 2$ mrad pointing accuracy in the presence of external disturbance and internal plant uncertainty. Cassini spacecraft dynamics, disturbance and model have been introduced in [1], where a modern $\mathbf{H}^{\infty}$ controller was designed and compared to bang/bang control. In this paper, a fuzzy logic controller is developed for Cassini and compare to the same bang/bang controller.

---

*Presented at *1994 IEEE World Congress on Computational Intelligence, June 26 — July 2, 1994, Orlando, Florida*

Fuzzy logic has become one of the most active and fruitful areas of research and applications ([?],[2] and the references therein). MathWorks is currently developing a MATLAB$^{TM}$ toolbox that works with Simulink$^{TM}$ for designing fuzzy logic control system [3]. Design algorithm and simulation are presented here in detailed.

# 2 Conventional Bang/Bang Attitude Control

The conventional spacrcraft controller using on/off thrusters consists of a bang/bang relay, a deadband and a set of thruster mapping logics. It is called Reaction Control Subsystem (RCS). Basically, it takes both position and rate error signals and processes through these three components. In general, both position and rate signals are computed from the spacecraft attitude estimator. The additional rate feedback (in Cassini, rate gain = 3) can provide damping to the overall system. Figure 1 shows a block diagram of the RCS controller.

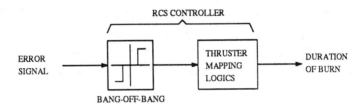

Figure 1: RCS Attitude Controller (Relay/Deadband + Thruster Logic).

A small deadband is also assigned to the controller so that the sensor noise can not trigger additional thruster activities. The minimum impulse bit defined in valve specification is about 7 mNs ± 35 %. Overall, the system will limit cycle with a peak-to-peak value approximately equal to the preset deadband. For Cassini spacecraft, the peak-to-peak deadband is currently set as 4 mrad to accommodate the pointing requirement.

### Thruster Mapping Logic

Thruster mapping logic is an essential part of the RCS attitude controller. As shown in Figure 2, there are two sets of thrusters: Y-facing and Z-facing, where Y-facing thrusters control Z-axis motion and Z-facing thrusters control X,Y-axes turns. Each thruster has a backup module to be used in case the primary one fails. When RCS is functioning, eight thrusters are available to stabilize spacecraft 3-axis dynamics and provide precise control of its commanded attitude.

Based on the above concept, a set of logics has been developed:

- If X Torque ¡ 0, then thruster 1 and 2 are on

- If X Torque ¿ 0, then thruster 3 and 4 are on

- If Y Torque ¡ 0, then thruster 2 and 3 are on

- If Y Torque ¿ 0, then thruster 1 and 4 are on

- If Z Torque ¡ 0, then thruster 5 and 7 are on

- If Z Torque ¿ 0, tehn thruster 6 and 8 are on

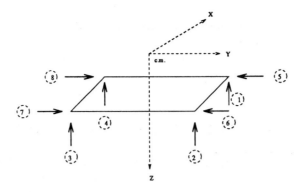

Figure 2: Thruster Locations and Firing Directions.

This set of simple logics seems very efficient to accomplish all of the Cassini spacecraft RCS maneuvers such as Titan flyby, sprint turn, probe release, reaction wheel momentum unload, and sun acquisitions, etc. The duration of burn for each thruster, $\Delta t$ sec, is defined by the control designer to set an upper bound on the overall RCS duty cycle. It is also directly related to the fuel consumption. Currently, the thruster on time $\Delta t$ is set to be fully open for the 125 msec rate group to reduce thruster cycles, which may end up using more fuel than it could have.

# 3   Fuzzy Attitude Control

One can substitute the regular thruster mapping logics with a set of fuzzy if-then rules. For instance, the fuzzy rules for thruster 4 would be

$$\begin{cases} \text{rule 1: } if\ tx \stackrel{\sim}{>} center,\ then\ out4 = 1, \\ \text{rule 2: } if\ ty \stackrel{\sim}{>} center,\ then\ out4 = 1, \\ \text{rule 3: } if\ tx \stackrel{\sim}{<} center\ and\ ty \stackrel{\sim}{<} center,\ then\ out4 = 0, \end{cases}$$

where $\stackrel{\sim}{>}$ and $\stackrel{\sim}{<}$ are the fuzzy versions for $>$ and $<$, respectively. However, the above rules do not take into consideration the constraint of no simultaneous diagonal fires. To accommodate this restriction, we obtain the following set of rules for thruster 4:

$$\begin{cases} \text{rule 1: } if\ tx \stackrel{\sim}{>} center\ and\ ty \stackrel{\sim}{>}\ - center,\ then\ out4 = 1, \\ \text{rule 2: } if\ tx \stackrel{\sim}{>}\ - center\ and\ ty \stackrel{\sim}{>} center,\ then\ out4 = 1, \\ \text{rule 3: } otherwise,\ out4 = 0. \end{cases}$$

This set of fuzzy rules are demonstrated in Figure 3, where (a) is rule 1, (b) is rule 2, and (c) is rule 3. The overall output of this fuzzy rule set is

$$output = \frac{w_1 f_1 + w_2 f_2 + w_3 f_3}{w_1 + w_2 + w_3},$$

where $w_i$ and $f_i$ are the firing strength and output, respectively, for rule $i$. If we choose the product as our T-norm operator, then

$$w_3 = (1 - w_1)(1 - w_2).$$

Also remember that $f_1 = f_2 = 1$ and $f_3 = 0$, therefore we have

$$output = \frac{w_1 + w_2}{1 + w_1 w_2}.$$

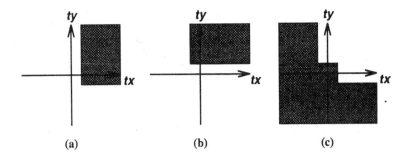

Figure 3: *Fuzzy rules for thruster 4: (a) rule 1; (b) rule 2; and (c) rule 3.*

The membership function (MF) we use to characterize $x \tilde{>} c$ is a generalized version of the common $s$ membership function with three parameters $c$, $s$ and $k$:

$$\mu_{x \tilde{>} c}(x) = \begin{cases} 0 & \text{if } x \leq c - s, \\ \frac{1}{2}[\frac{x-(c-s)}{s}]^{2k} & \text{if } c - s < x \leq c, \\ 1 - \frac{1}{2}[\frac{c+s-x}{s}]^{2k} & \text{if } c < x \leq c + s, \\ 1 & \text{if } c + s < x, \end{cases}$$

where $c$ determines the cross-over point ($\mu_{x \tilde{>} c}(c) = 0.5$); $s$ determines the spread of this MF ($0 < \mu_{x \tilde{>} c}(x) < 1$ whenever $c - s < x < c + s$); and $k$ (together with $s$) control the slope of the entire curve (for instance, the slope at the cross-over point is $\mu'_{x \tilde{>} c}(x)|_{x=c} = k/s$). Figure 4 shows the physical meanings of these parameters. Figure 5 (a) and (b) demonstrate the effects of changing parameter $s$ and $k$, respectively.

The MF for $x \tilde{<} - c$ is just an image of the MF for $x \tilde{>} c$ w.r.t Y-axis.

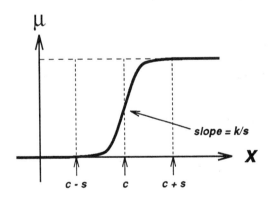

Figure 4: *Physical meanings of MF's parameters.*

# 4 Simulation and Result

A discrete nonlinear simulation was set up to evaluate both conventional and fuzzy attitude controllers (A & B). The sampling time is 125 msec. The spacecraft model consists only the rigid body Euler equations.

Figure 6 shows a comparison of a two degree step response in Y-axis for different RCS controller.

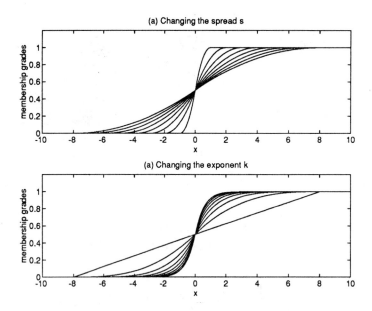

Figure 5: *The effects of changing MF's parameters: (a) change parameter s; (b) change parameter k.*

**Table 1: Total Thruster On-time and Cycles**

| Controller | Total On-time | Cycle |
|---|---|---|
| Bang/Bang | 77 Sec | 30 |
| Fuzzy A (Slope=1) | 70 Sec | 419 |
| Fuzzy B (Slope=500) | 75 Sec | 46 |

Some observations:

- Both fuzzy controllers can track better than the conventional bang/bang controller

- By approximating the relay/deadzone with a continuous membership function, fuzzy controller A can save about 10 % fuel but ends up higher thruster on/off cycles

- By increasing the membership function slope (from 1 to 500), fuzzy controller B approaches bang/bang structure with much less thruster cycle

# 5 Conclusion

Spacecraft attitude control problem has been well known for decades. In this paper, a modern fuzzy logic controller has been developed to compare with the conventional bang/bang control on the issues involving tracking capability, thruster on-time/cycle trade-offs etc.

It has been shown that the fuzzy logic controller can track the system command better but ends up higher thruster on/off cycles. The design process of such a fuzzy logic controller is straightforward and systematic as shown in Section 3. The theory of fuzzy logic control is interesting and easy-to-use.

Future study will focus on comparing other fundamental feedback issues such as robustness, disturbance rejection, and sensor noise, etc. with a more complicated spacecraft model with fuel slosh mode and flexible body.

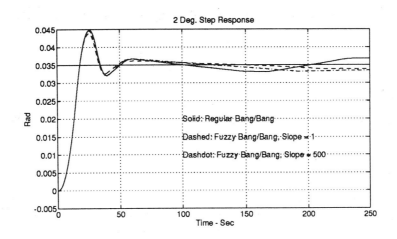

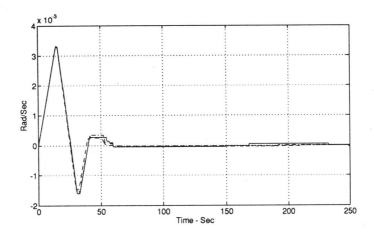

Figure 6: 2 Degree Step Response (Bang/Bang vs. Fuzzy).

# 6  Acknowledgement

The research described in this paper was partially carried out by the Jet Propulsion Laboratory, California Institute of Technology, under contract with the National Asronautics and Space Administration.

# References

[1] R. Y. Chiang, S. Lisman, E. Wong, P. Enright, W. Breckenridge, M. Jahan, "Robust Attitude Control for Cassini Spacecraft Flying By Titan," *Proc. of AIAA Guidance and Control Conf.,* Monterey, CA., Aug. 1993.

[2] L. A. Zadeh, "Fuzzy Sets," *Information and Control,* 8:338-353, 1965.

[3] J. S. Roger Jang, "Self-Learning Fuzzy Controller Based on Temporal Back-Propagation," *IEEE Trans. on Neural Networks,* vol. 3, no. 5, pp 714-723, Sept. 1992.

[4] J. S. Roger Jang and N. Gulley, *Fuzzy Logic Toolbox,* The MathWorks, Inc., to be released 1994.

# Fuzzy Logic Control of a Floating Level in a Refinery Tank

S. Galichet, L. Foulloy [*]
M. Chebre [**], J.P. Beauchêne [***]

LAMII [*]
41, Avenue de la plaine
BP 806
74016 Annecy Cedex
FRANCE

ELF ANTAR FRANCE [**]
Centre de Recherche de Solaize
BP 22
69360 Solaize
FRANCE

ELF ANTAR FRANCE [***]
Raffinerie de Donges
44480 Donges
FRANCE

**Abstract:** This paper presents the development of a fuzzy logic controller that maintains a floating level in a tank on top of the atmospheric distillation unit of a refinery (the Donges refinery of the ELF ANTAR FRANCE Company). This particular level regulation does not need any fixed setpoint as it is only required for the level to be kept between a minimal and a maximal bounds. The important constraint to take into account is to minimize the action range in order to stabilize the feed flow at the output of the tank. Besides, the system has to anticipate properly the different disturbances that can occur in the distillation unit (crude oil changes...).
At the beginning of the study, the tank level was controlled by an algorithmic controller that needed to be enhanced by introducing smooth recentering around the medium level in a nominal context without any disturbances. Furthermore, the system has also to be improved by introducing anticipation so as to avoid manual intervention during huge disturbances.
The fuzzy logic controller was installed six months ago and gives entire satisfaction to the refiners.
The development of the fuzzy logic controller is described step by step. The membership functions and the rulebase are first initialized in order to obtain a fuzzy logic system that is equivalent to the algorithmic controller. Some particular rules are then modified according to the expert knowledge and the particular specifications required. Finally, an adaptation of the rulebase is done to take into account an evolving context in order to reject smoothly important disturbances by anticipation.

## 1 Introduction

The aim of the presented fuzzy system is to maintain a floating level in a tank on top of the atmospheric distillation unit of a refinery. An important constraint to take into account is to minimize the action range to stabilize the feed flow at the output of the tank.

The purpose of this study is to replace an existing algorithmic controller that is not fully successful. In that framework, the following approach has been chosen. The existing algorithm is first translated into an equivalent Mamdani fuzzy controller ([7],[8]). This step is described in section 2. Then, some expert knowledge is introduced in the fuzzy controller in order to improve the controller behavior in particular cases. Section 3 examines this aspect of the problem. Finally, the section 4 deals with the adaptation of the rulebase when a disturbance of the feed flow can be predicted.

## 2 Modal equivalences principle

### 2.1 Description of the algorithmic controller

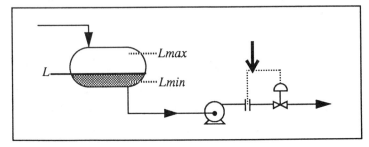

Fig. 1 : Floating level regulation

The level in the tank has to be maintained between $Lmin$ and $Lmax$. As shown in Fig. 1, the regulation action is realized on the output flow. The change in output flow is evaluated in order to stabilize the level before the time required for the level to reach the bound.

Let $L$ be the level, $\delta L$ the change in level and $\delta F$ the change in output flow. The following relationships are derived:

$$\delta F = K \cdot \frac{(\delta L)^2}{Lmax - L} \qquad \text{if} \qquad \delta L \geq 0 \tag{1}$$

$$\delta F = K \cdot \frac{(\delta L)^2}{Lmin - L} \qquad \text{if} \qquad \delta L < 0 \tag{2}$$

where $K$ is a constant that characterizes the tank and the sampling period.

A threshold $G$ is then applied to the change in output flow such as: $\qquad |\delta F| \leq G.$ \hfill (3)

## 2.2 Synthesis of an equivalent fuzzy controller

Our purpose is now to build a Mamdani type fuzzy controller equivalent to the presented algorithmic controller. The main idea is to express the fuzzy rules by using a general parametric function ([2],[9]) which has to be determined in order for the fuzzy controller to achieve the desired control law. The modal equivalences principle applied in ([5], [6]) for the synthesis of linear controllers is extended here to the nonlinear case. The developed approach consists in equating the output of the fuzzy controller with the output of the algorithmic controller at some particular input values, called modal values. The parametric function describing the rule base and the distribution of the membership functions can thus be deduced.

From equations (1) and (2), $L$ and $(\delta L)^2$ are chosen as inputs and $\delta F$ as output for the fuzzy controller. In a Mamdani type fuzzy controller the rules are expressed in a symbolic form. The choice of the input and output variables leads to a definition of the rules in the form:

**If** *level* is $A_i$ **and** *change in level* is $B_j$ **then** *change in output flow* is $C_{f(i,j)}$ \hfill (4)

For the sake of simplicity, the $(\delta L)^2$ variable is translated in terms of *change in level*. But, the correct expression should be *square to the change in level*. The terms $A_i$, $B_j$ and $C_{f(i,j)}$ represent the symbols respectively attached to $L$, $(\delta L)^2$ and $\delta F$. We assume that the membership functions of the meaning of these symbols [4] are triangular, regularly distributed with an overlap of 1 (i.e. providing a Bezdek partition of the universes of discourse [1]). Any such value where the membership function of the meaning of a symbol is equal to 1 is called a modal value. $\Delta a$ represents the distance between two consecutive modal values of the level. Similarly, $\Delta b$ and $\Delta c$ are respectively the distances between two consecutive modal values of the change in level and of the change in output flow. By assuming that the meaning of the symbol $A_0$ (respectively $B_0$ and $C_0$) has a membership function centered on $Lmin$ (respectively on the zero of the corresponding universe of discourse), we have:

$$L_i = Lmin + i\,\Delta a \tag{5}$$
$$(\delta L^2)_j = j\,\Delta b \tag{6}$$
$$\delta F_{f(i,j)} = f(i,j)\,\Delta c \tag{7}$$

where $L_i$ represents the $i^{th}$ modal value of the level, $(\delta L^2)_j$ the $j^{th}$ modal value of the "change in level" and $\delta F_{f(i,j)}$ the $f(i,j)^{th}$ modal value of the change in output flow.

Let us first examine the case when the change in level is negative. For input modal values $L_i$ and $(\delta L^2)_j$, the output of the algorithmic controller, $\delta F_{ALG}$, is derived from equation (2):

$$\delta F_{ALG} = K \cdot \frac{(\delta L^2)_j}{Lmin - L_i} \tag{8}$$

From expressions (4) and (5), the output of the algorithmic controller can be reformulated as:

$$\delta F_{ALG} = -K \cdot \frac{j\Delta b}{i\Delta a} \tag{9}$$

For the same modal values, the output of the Mamdani controller, $\delta F_{MAM}$, is: (see [5] or [6] for details)

$$\delta F_{MAM} = \delta F_{f(i,j)} = f(i,j)\,\Delta c \qquad (10)$$

By equating the values $\delta F_{ALG}$ and $\delta F_{MAM}$ derived in (8) and (9), the following relationship is obtained:

$$f(i,j)\cdot \Delta c = K\cdot \frac{\Delta b}{\Delta a}\left(-\frac{j}{i}\right) \qquad (11)$$

Since $f(i,j)$ represents the index of the conclusion symbol of the rule, the parametric function $f$ has to give an integer value. The easiest choice for $f(i,j)$ is given by:

$$f(i,j) = \alpha\left(-\frac{j}{i}\right) \qquad (12)$$

where $\alpha$ is the lowest integer such that $f(i,j)$ is an integer, that is the lowest common multiple of all possible values of $i$. In that case, the distribution of the membership functions must satisfy the relationship:

$$\Delta c = \frac{K}{\alpha}\cdot \frac{\Delta b}{\Delta a} \qquad (13)$$

The fuzzy controller generated according to the relationships (11) and (12) is not precisely equivalent to the algorithmic one. The equivalence is only guaranteed at the input modal values. When the distance between two consecutive modal values tends to zero, that is, when the number of rules tends to infinite, the fuzzy control law converges to the algorithmic one. The distribution of the input modal values is set up with the parameters $\Delta a$ and $\Delta b$.

In our case, precise equivalence has no sense. Indeed, the fuzzy controller will be modified according to the expert knowledge. This synthesis step is just an efficient way to translate the original controller into linguistic terms. For the sake of simplicity, the number of symbols attached to the level (respectively to the change in level) has been limited to 5 (respectively to 4), that is $i = 0, 1, 2, 3, 4$ and $j = 0, 1, 2, 3$. Thus, we finally obtain $\alpha = 12$. The automatically generated rule base may be reformulated as shown in Table 1.

Let us now examine the case when the change in level is positive. The same method is applied. As the behavior of the algorithmic controller is globally symmetrical, the rulebase described in Table 2 is obtained.

In order to regroup the two proposed fuzzy controllers in a single controller, we simply have to modify the «change in level» variable for it takes into account the sign of the change. The new input variable is finally chosen as $\mathrm{sign}(\delta L)*\delta L^2$. The corresponding rule base is described in Table 3. The two overlined squares in Table 3 correspond to discontinuities of the algorithmic control law.

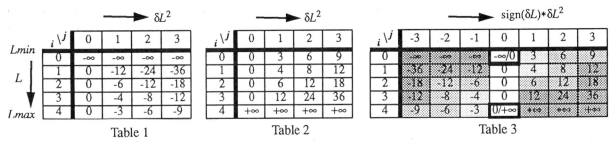

| | $i\backslash j$ | 0 | 1 | 2 | 3 |
|---|---|---|---|---|---|
| Lmin | 0 | $-\infty$ | $-\infty$ | $-\infty$ | $-\infty$ |
| L | 1 | 0 | -12 | -24 | -36 |
| | 2 | 0 | -6 | -12 | -18 |
| | 3 | 0 | -4 | -8 | -12 |
| Lmax | 4 | 0 | -3 | -6 | -9 |

Table 1

| $i\backslash j$ | 0 | 1 | 2 | 3 |
|---|---|---|---|---|
| 0 | 0 | 3 | 6 | 9 |
| 1 | 0 | 4 | 8 | 12 |
| 2 | 0 | 6 | 12 | 18 |
| 3 | 0 | 12 | 24 | 36 |
| 4 | $+\infty$ | $+\infty$ | $+\infty$ | $+\infty$ |

Table 2

| $i\backslash j$ | -3 | -2 | -1 | 0 | 1 | 2 | 3 |
|---|---|---|---|---|---|---|---|
| 0 | $-\infty$ | $-\infty$ | $-\infty$ | $-\infty/0$ | 3 | 6 | 9 |
| 1 | -36 | -24 | -12 | 0 | 4 | 8 | 12 |
| 2 | -18 | -12 | -6 | 0 | 6 | 12 | 18 |
| 3 | -12 | -8 | -4 | 0 | 12 | 24 | 36 |
| 4 | -9 | -6 | -3 | $0/+\infty$ | $+\infty$ | $+\infty$ | $+\infty$ |

Table 3

For a practical implementation of the fuzzy controller, the different parameters have now to be set up. Let [0,100] be the universe of discourse of $L$ and [-100,100] the universe of discourse of $\mathrm{sign}(\delta L)*\delta L^2$. With the number of symbols defined on the input variables, we obtain $\Delta a = 100/4$ and $\Delta b = 200/6$. From equation (12), the parameter $\Delta c$ can be deduced: $\Delta c = 1/12 . \Delta a/\Delta b = 1/9$. The implementation of the rules numbered "$\infty$" corresponds to the thresholding step introduced in the algorithmic controller (see equation (3)). For example, if $G = 10$, we have "$\infty$". $\Delta c = 10$, that is "$\infty$" = 90. The Fig. 2 shows the control surfaces obtained with the algorithmic controller and with the fuzzy controller. The fuzzy controller provides the same output as the algorithmic controller for the input modal values, distributed on the grid generated by the marked points. Between the input modal values, the output of the fuzzy controller is obtained by a quasi linear interpolation ([3], [9]). It is clear that the finer the partition of the universes of discourse, the better the equivalence between the two controllers.

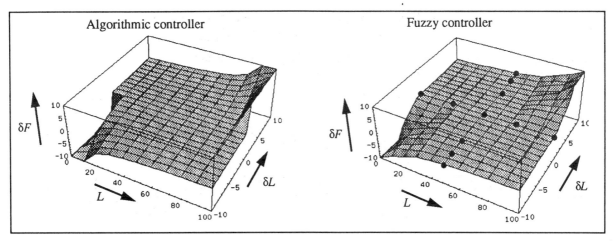

Fig. 2 : Control surfaces

# 3 Improving the rulebase

The first step consists in decreasing the number of symbols associated to the output variable. Seven classes of the control action, numbered from -3 to 3, can be distinguished by the expert. A regrouping of the output symbols is performed as shown by the different gray levels in Table 3. In spite of a significant degradation of the control law, the global behavior of the fuzzy controller is preserved. The condensed information is now easily understandable and local transformations of the rulebase can be imagined.

In order to correctly deal with abrupt changes in the input flow, it is important to be in a favorable situation when the disturbance occurs. Thus, a stabilization of the tank level around the middle level is a good compromise whatever the sense of a sudden variation. One way to avoid a stabilization of the tank level far away from the middle level is to modify the central column of the rulebase. Indeed, this column corresponds to a nearly zero change of level, that is a quasi steady state. In the basic rulebase, whatever the level, a zero action is applied when the level is stable. In order to force a recentering on the middle level, the central column has to be filled with the adequate actions, that is with negative actions when the current level is under the middle level, and with positive actions when the current level is above the middle level.

A special treatment of the limit cases may also be introduced in the fuzzy controller. When the level is in the neighborhood of the bounds *Lmin* or *Lmax*, an efficient control action must be applied in order to avoid an overshoot of the limit. Whatever the sign of the change in level, a high negative control action is generated when the current level is near *Lmin* and a high positive control action when the current level tends to *Lmax*. This improvement can be integrated in the fuzzy controller by simply modifying the first and the last lines of the rulebase. The final rulebase is given in Table 4.

# 4 Anticipation of the disturbances

The proposed rulebase provides a smooth recentering around the medium level in a nominal context without any disturbance. When a disturbance may be predicted, an adequate anticipation strategy can be applied in order to minimize the effects of the disturbance. For example, if a high positive disturbance is predicted, the tank level will rapidly increase. If the current level is already high, the tank has to be emptied before the occurrence of the disturbance. A high positive control action must thus be applied. On the other hand, that is, if the current level is relatively low, the normal control action leading to a filling of the tank may be significantly reduced. Indeed, the disturbance will naturally induce a filling of the tank.

One way to implement the anticipation strategies in the fuzzy controller consists in adding a new input variable describing the disturbance prediction. From expert knowledge, the disturbances can be distributed into seven fuzzy classes. The final rulebase is now composed of rules of the type:

**If** *level is $A_i$* **and** *change in level is $B_j$* **and** *predicted disturbance is $D_k$*
**then** *change in output flow is $C_{f(i,j,k)}$* (14)

For example, the Table 5 regroups the rules corresponding to the prediction of a very high positive disturbance. The original table (Table 4) is conserved for a zero disturbance prediction.

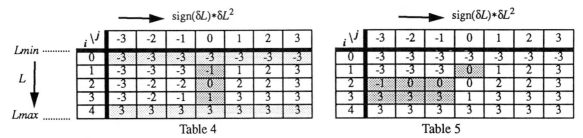

| Table 4 | | | | | | | | Table 5 | | | | | | |
|---|---|---|---|---|---|---|---|---|---|---|---|---|---|---|

$i \backslash j$ — Table 4 ($\text{sign}(\delta L)*\delta L^2$):

| $i \backslash j$ | -3 | -2 | -1 | 0 | 1 | 2 | 3 |
|---|---|---|---|---|---|---|---|
| 0 | -3 | -3 | -3 | -3 | -3 | -3 | -3 |
| 1 | -3 | -3 | -3 | -1 | 1 | 2 | 3 |
| 2 | -3 | -2 | -2 | 0 | 2 | 2 | 3 |
| 3 | -3 | -2 | -1 | 1 | 3 | 3 | 3 |
| 4 | 3 | 3 | 3 | 3 | 3 | 3 | 3 |

$i \backslash j$ — Table 5 ($\text{sign}(\delta L)*\delta L^2$):

| $i \backslash j$ | -3 | -2 | -1 | 0 | 1 | 2 | 3 |
|---|---|---|---|---|---|---|---|
| 0 | -3 | -3 | -3 | -3 | -3 | -3 | -3 |
| 1 | -3 | -3 | -3 | 0 | 1 | 2 | 3 |
| 2 | -1 | 0 | 0 | 0 | 2 | 2 | 3 |
| 3 | 3 | 3 | 3 | 1 | 3 | 3 | 3 |
| 4 | 3 | 3 | 3 | 3 | 3 | 3 | 3 |

# 5 Conclusion

The proposed fuzzy controller was installed six months ago in the Donges refinery of the ELF ANTAR FRANCE Company. Since its introduction, the operation staff has mentioned a reduction of the manual interventions and a stabilization of the output controlled flow. The crude oil changes are successfully taken into account by the fuzzy controller. The fuzzy logic has proved to be an efficient way to improve a classical control by integrating the human operator ability and some process expertise for anticipation. Indeed, by applying the modal equivalences principle, the analytic description of the algorithmic controller is translated in terms of understandable information. Heuristic knowledge can then be easily integrated in the same formalism, that is in the rulebase. Finally, a hybrid control is obtained based at one and the same time on the process control theory and on the heuristic reasoning.

The modal equivalences principle proposed for the initialization of the rulebase can be also applied when no mathematical equation of the control is available. In that case, a set of training points can be used as modal values.

# 6 References

[1]  J.C. Bezdek, *Pattern Recognition with Fuzzy Objective Function Algorithms*, New-York: Plenum, 1981.

[2]  F. Bouslama, A. Ichikawa, "Fuzzy control rules and their natural control laws", Fuzzy Sets and Systems, vol. 48, pp. 65-86, 1992.

[3]  J.J Buckley, "Fuzzy controller : Further limit theorems for linear control rules", Fuzzy Sets and Systems, vol. 36, pp. 225-233, 1990.

[4]  L. Foulloy, S. Galichet, "Fuzzy Controllers Representation", in *Proc. of the First European Congress on Fuzzy and Intelligent Technologies*, 1993, Aachen, pp 142-148.

[5]  S. Galichet, L.Foulloy, "Fuzzy Equivalences of Classical Controllers", in *Proc. 1st European Congress on Fuzzy and Intelligent Technologies*, 1993, Aachen, pp. 1567-1573.

[6]  S. Galichet, L.Foulloy, "Fuzzy Controllers : Synthesis and Equivalences", submitted to IEEE trans. on Fuzzy Systems (LAMII-CESALP Internal Report #9308, University of Savoie).

[7]  Lee C., "Fuzzy logic in control systems: Fuzzy logic controller - part I", IEEE trans. on Systems, Man and Cybernetics, vol. 20, No 2, 1990, pp. 404-418.

[8]  Mamdani E.H., "Application of fuzzy algorithms for control of simple dynamic plant", Proc. of the Institution of Electrical Engineers, Control and Science, vol. 121, No 12, 1974, pp. 1585-1588.

[9]  H. Ying, "General Analytical Structure of Typical Fuzzy Controllers and Their Limiting Structure Theorems", Automatica, vol. 29, No 4, 1993, pp. 1139-1143.

# An Estimator Based On Fuzzy If-Then Rules For The Multisensor Multidimensional Multitarget Tracking Problem

C.W. Tao[1], J.S. Taur[2], H.C. Kuo[1], J.C. Wu[1], and W.E. Thompson[3]

[1]Department of Electrical Engineering
National I-Lan Institute Of Agriculture And Technology
I-Lan, Taiwan, R.O.C.

[2]National Chung-Hsin University
Taiwan, R.O.C.

[3]Department of Electrical and Computer Engineering
New Mexico State University, Box 3-0
Las Cruces, New Mexico 88003

## Abstract

In this paper, an estimator based on fuzzy if-then rules are developed for multidimensional multitarget tracking with multisensor data taken in a cluttered environment. The clustering algorithm based upon a pseudo k-means algorithm and the Match-Agreement data technique designed in our previous paper are used here for clustering multisensor data from a clustered environment and data association problem in multitarget tracking. The estimator based on fuzzy if-then rules consists of Gaussian membership functions, min-"and" inference, and centroid defuzzification. Examples are presented to illustrate the comparisons between a Kalman estimator and the fuzzy estimator.

## I. Introduction

The problem of multisensor multidimensional multitarget tracking in clutter has been extensively studied [1,2]. Three important parts in this work are data clustering, data association and the design of estimators [3]. The clustering algorithm based upon a pseudo k-means algorithm and the Match-Agreement data association algorithm used here have been shown with the advantages of no required a priori information and computational efficiency, respectively, in our previous paper [4]. Therefore, this paper concentrates on the design of an estimator based on fuzzy if-then rules. As is well known, when the uncertainty may be represented statistically and the system dynamics are available, the Kalman filter has been very successful, according to its optimal property. However, when the system dynamics are unknown or the statistical model can not be properly applied to the uncertainty involved, the performance of the Kalman filter degrades. Moreover, the erroneous uncertainty model might cause divergence in a Kalman filter system [5]. In situations where the knowledge of the moving target dynamics and the statistical uncertainty model are not available, the estimator based on fuzzy if-then rules is a proper alternative.

The problem is formulated in section II. Section III presents the design of a Multidimensional multitarget fuzzy estimator. The results are shown in section IV, and the conclusion is in section V.

## II. Problem Formulation

The problem considered here is that of tracking multiple targets based upon multisensor measurements taken in a cluttered environment. It is assumed that the number of targets and the statistical uncertainty model are unknown, and that $n$ two-dimensional position measurements are available at time index k and are represented as

$$p_{iqf}(k)=\tan(\theta_{iq}(k)+\Theta_{iqa}(k)+\theta_{iq1}(k-1))T+p_{iqf}(k-1); \quad k>2$$

The following parts of this section give the details of the design of the $iq^{th}$ fuzzy estimator from a fuzzy partition of the universe of discourse of variables $(\theta_{iq}(k),\theta_{iqa}(k))$ to defuzzification.

## A. FUZZY PARTITION AND MEMBERSHIP FUNCTIONS

First, the universe of discourse of the variables $\theta_{iq}(k)$ and $\theta_{iqa}(k)$ needs to be defined. The closed range $[-\pi,\pi]$ is considered to be an universe of discourse $U_{iq1}$ for the input variable $\theta_{iq}(k)$. In order to create effective and stable (with respect to parameter $\sigma$) membership functions, a mapping function $f_{iq}$: $U_{iq1} \rightarrow U_{iq}$, specifically,

$$y_{iq}(k)=f_{iq}(\theta_{iq}(k))=(\theta_{iq}(k)/\gamma)a$$

$$\text{where} \quad a=2\sigma(\sqrt{2}\log 2), \quad \gamma=30$$

is applied on the universe of discourse $U_{iq1}$ which maps $U_{iq1}$ to an universe of discourse $U_{iq}=[-6a, 6a]$. Secondly, the $U_{iq}$ is fuzzily partitioned into a collection of fuzzy sets corresponding to those fuzzy sets in the fuzzy If-then rules. The membership function of each fuzzy set is defined by the bell-shaped function:

$$\mu_{iqyj}(y_{iq}(k))=\exp\{-(y_{iq}(k)-u_{iqyj})^2/2\sigma^2\} \; ; j=1, 2,...., 13$$

where $u_{iqyj}$ is the middle point of the support of each fuzzy set.

The universe of discourse $U_{iq2}$ for the variable $\theta_{iqa}(k)$ is defined to be the range $(-\infty,\infty)$. $U_2$ is fuzzily partitioned into a collection of 13 fuzzy sets ( pvvb, pvb, pb, pm, ps, pv, ze, nvs, ns, nm, nb, nvb, nvvb). Each of the fuzzy sets has the support $(-\infty,\infty)$ and the membership function with the form

$$\mu_{iq\theta j}(\theta_{iqa}(k))=\exp\{-(\theta_{iqa}(k)-u_{iq\theta j})^2 /2\sigma^2 \} \; ; j=1, 2,...., 13$$

where $u_{iq\theta j}$ is the number with membership value 1.

These letters p, n, ze, s, m, b, v in the names of the fuzzy sets correspond, respectively, to the meaning of positive, negative, zero, small, big, and very. For example, pvvb means "positive very very big". Actually, the name of each fuzzy set really means that the element with membership value 1 in this fuzzy set has the property which is described by the name of this fuzzy set.

## B. FUZZY RULE BASE

A fuzzy rule base consists of fuzzy if-then rules. There are several ways of deriving fuzzy if-then rules [6]. The fuzzy if-then rules in this paper are derived from the smoothness assumption. Each fuzzy if-then rule includes two variables $(y_{iq}(k), \theta_{iqa}(k))$, for example:

$$R1: \text{if} ( y_{iq}(k) \text{ is pvvb} ) \text{ then } (\theta_{iqa}(k) \text{ is nvvb})$$

## C. DEFUZZIFICATION

Let $\mu_{iqyj}(y_{iq}(k))$'s, j=1, 2,...., 13, be membership values of input fuzzy sets, nvvb, nvb, nb, nm, ns, nvs, ze, pv, ps, pm, pb, pvb, pvvb, respectively, with input $y_{iq}(k)$. Also, let $u_{iq\theta j}$, be the maximum of the membership functions $\mu_{iq\theta j}$, j=1, 2,...., 13, of output fuzzy sets, pvvb, pvb, pb, pm, ps, pv, ze, nvs, ns, nm, nb, nvb, nvvb, respectively. The centroid method, which is the standard technique in design of a fuzzy controller [7], is used for the defuzzification, that is,

$$\Theta_{iqa}(k) = \frac{\sum_{j=1}^{13} \mu_{iqyj}(y_{iq}(k))u_{iq\,\theta j}}{\sum_{j=1}^{13} \mu_{iqj}}$$

## IV. SIMULATION RESULTS

The Multisensor 2-dimensional Multitarget tracking situation involves three simulated crossing/intersecting trajectories. The multisensor measurements are corrupted by zero mean Gaussian noise with high variance of 0.2 and are shown in Figure 1. Application of the Pseudo K-means Clustering Algorithm yields the measurement cluster centers shown in Figure 2, and the Match Agreement Algorithm in combination with the fuzzy estimator and basic Kalman-filter algorithm yields the estimated trajectories shown by dashed lines in Figure 3 and Figure 4. The true trajectories are shown by solid lines in Figure 3 and Figure 4. The results indicate that when the variance of noise is high, only the fuzzy estimator obtains correct trajectories.

## V. CONCLUSIONS

This paper presents one estimator based on fuzzy if-then rules for multi-dimensional multisensor multitarget tracking. The simulation results indicate that this fuzzy estimator works even better for the multidimensional multitarget tracking when the variances of noises are high. Even though only simulation results for two dimensional target tracking is shown, similar simulations can be developed for m dimensional (m>2) target tracking problems.

## REFERENCE

[1]   Y. Bar-Shalom, "Tracking Methods in a Multitarget Environment," IEEE Transactions on Automatic Control, AC-23, 4 (Aug. 1978).

[2]   D. Sengupta, and R. Iltis, "Neural Solution to the Multitarget Tracking Data Association Problem," IEEE Transactions on Aerospace and Electronic Systems, AES-25, 1, (Jan. 1989).

[3]   E. Emre, and J. Seo, "A Unifying Approach to Multitarget Tracking," IEEE Transactions on Aerospace and Electronic Systems, AES-25, 4(July 1989).

[4]   W.E. Thompson, R. Parra-Loera, and C.W. Tao, "A Pseudo K-Means Approach to The Multisensor Multitarget Tracking Problem," Proceedings of the SPIE's International Symposium on Data Structures and Target Classification, Orlando, FL, April 1991, Vol. 1470, pp. 48-58.

[5]   F. H. Schlee and N. F. Toda, "Divergence in the Kalman Filter," AIAA J.,vol. 5,pp.1114-1120, June 1967.

[6]   C. C. Lee, "Fuzzy Logic in Control Systems: Fuzzy Logic Controller-Part I," IEEE Trans. Syst. Man Cybern., vol. SMC-20, no. 2, pp. 404-418, Mar./April 1990.

[7]   B. Kosko, Neural Networks and Fuzzy Systems. New Jersey: Prentice Hall, 1992.

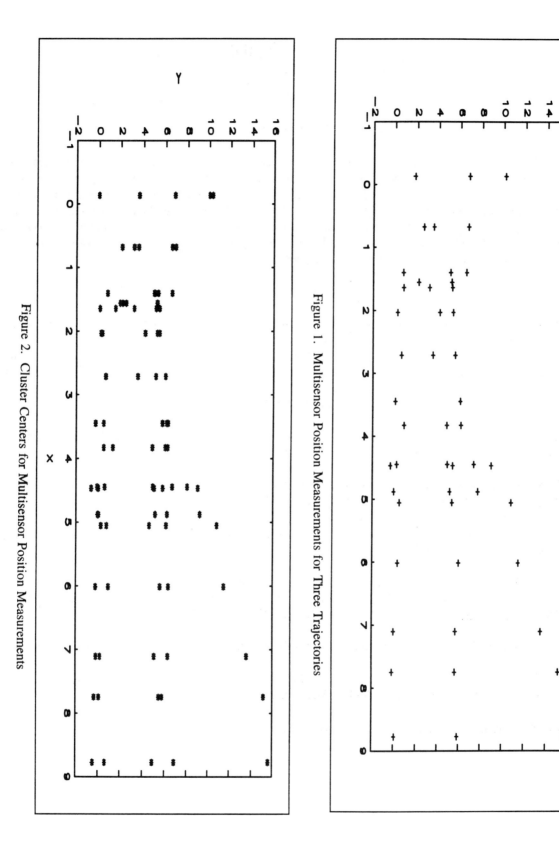

Figure 2. Cluster Centers for Multisensor Position Measurements

Figure 1. Multisensor Position Measurements for Three Trajectories

$$\{ (x_1(k) , y_1(k)) , (x_2(k) , y_2(k)) , \ldots , (x_n(k) , y_n(k)) \}$$

The basic problems involved here include measurement clustering, data association, and tracking. Measurement clustering involves grouping the measurements into a number of groups called clusters and determining a representative point for each cluster called cluster centers. Each cluster center is then viewed as a measurement for a target. The data association problem involves identifying each target measurement as belonging to a particular trajectory. The tracking problem involves utilizing target measurements to arrive at the "best" estimates for the trajectories. Because the pseudo k-means clustering algorithm and match-agreement data association algorithm in our previous paper [4] are used directly, only tracking problem is considered in this paper.

### III. Design Of A Multidimensional Multitarget Fuzzy Estimator

By utilizing the match-agreement algorithm for data association, each target measurement is associated to particular trajectories. These target measurements are used for the estimation of associated trajectories. The multidimensional multitarget fuzzy estimator developed here is constructed based on a simple fuzzy estimator. The simple fuzzy estimator (SFE) is an estimator for one target moving on one dimension. For the tracking problem of one target moving on an m dimensional space, the multi-dimensional fuzzy estimator (MFE) consists of m SFEs, and each SFE parallelly estimates one position coordinate of an m dimensional trajectory at each time index. Therefore, for the tracking problem of n targets moving on an m dimensional space, the multi-dimensional multitarget fuzzy estimator (MMFE) consists of n MFEs, and each MFE estimates the trajectory for one of n targets. The remainder of this section is the details of the design of an MMFE.

Based on common sense knowledge about moving targets with reasonable changes in velocity, it is assumed that moving targets (n targets) can not change their velocity dramatically in any one of the m dimensions, i.e., the difference $\theta_{iq}(k)$ of the angles $\theta_{iq1}(k)$ and $\theta_{iq1}(k-1)$, i=1,2, ...., m, q=1,2,....,n, shown in the following equations can not be very large:

$$\theta_{iq1}(k)=\tan^{-1}((p_{iq}(k)-p_{iqf}(k-1))/T)$$

$$\theta_{iq1}(k-1)=\tan^{-1}((p_{iqf}(k-1)-p_{iqf}(k-2))/T)$$

$$\theta_{iq}(k)=\theta_{iq1}(k)-\theta_{iq1}(k-1) ; T=\text{sampling period}$$

where k needs to be greater than 2. The input position variable $p_{iq}(k)$, k>2, is then transformed into an angle variable $\theta_{iq}(k)$ in order to utilize the smoothness assumption. The output position variable $p_{iqf}(k)$ of the fuzzy filter is assumed to be equal to the input position variable $p_{iq}(k)$ when k is less than or equal to 2.

The smoothness assumption described above formulates the fuzzy If-then rules for the fuzzy controller in the fuzzy filter, for example:

R1:    If the angle difference, $\theta_{iq}(k)$, is positive small, then the angle adjustment,
       $\theta_{iqa}(k)$, is negative small.

The "positive small" and "negative small" are not exact regions but are fuzzy sets which are defined by fuzzy membership functions (part A). For an input variable $\theta_{iq}(k)$, it is necessary to find the corresponding degree of "positive small" for the variables $\theta_{iq}(k)$ to be able to apply the fuzzy If-then rule R1. For each particular input variable $\theta_{iq}(k)$, every possible value of $\theta_{adj}(k)$ in its universe of discourse gives a degree of appropriateness for the control. In order to obtain the best $\Theta_{iqa}(k)$, which is a crisp value, to be the output of the fuzzy controller, a defuzzification technique is needed. Then the algorithm below is used to obtain the filtered target position.

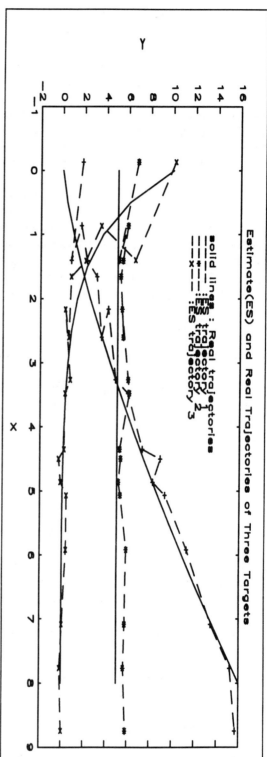

Estimate(ES) and Real Trajectories of Three Targets

solid lines : Real trajectories
— + — :ES trajectory 1
— * — :ES trajectory 2
— x — :ES trajectory 3

Figure 3. Estimated with fuzzy estimator and True Trajectories for Three Targets

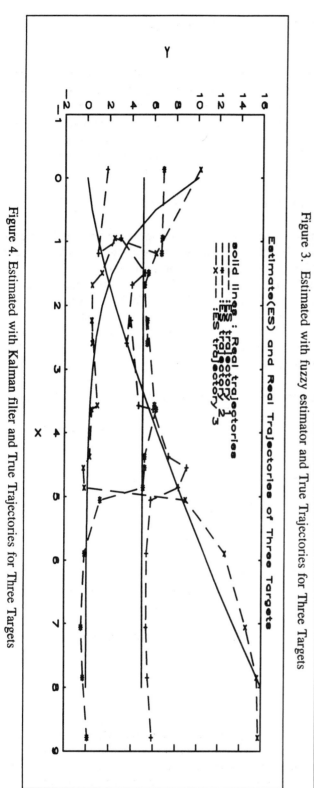

Estimate(ES) and Real Trajectories of Three Targets

solid lines : Real trajectories
— + — :ES trajectory 1
— * — :ES trajectory 2
— x — :ES trajectory 3

Figure 4. Estimated with Kalman filter and True Trajectories for Three Targets

# A Comparison of FAM and CMAC for Nonlinear Control

Arit Thammano and Cihan H. Dagli

Department of Engineering Management
University of Missouri-Rolla
Rolla, Mo. 65401

## Abstract

In the past, various neural network-based controllers are proposed to master the nonlinear control problems with different level of success. The recent trend is to incorporate fuzzy logic to this process. This article compares a neural network-based controller, both local and global networks, with Fuzzy associative memories (FAM) on a nonlinear problem. CMAC and FAM are chosen as representatives of local generalization networks. CMAC controller is trained off-line, therefore, it can response to the incoming input immediately. CMAC can intrapolate its memory and give a reasonable control signal even the input has not been trained on. Backpropagation is picked as a representative of global generalization networks. All three systems are studied on a simple simulated control problem. This preliminary research will be adapted later to control the laser cutting machine. A performance measure that depends on the transient response and the steady state response of the controlled system is used. The results indicate that CMAC and FAM are comparable.

## Introduction

During the last year neural network-based controllers are proposed in literature for modelling nonlinearities inherent on control problems. Various architectures are proposed. Learning algorithms adapted in these architectures depend local and global information captured from the control data. Recent trend is to incorporate fuzzy logic to the process. The quest for robust controller design is still continuing.

Most of the applications in the manufacturing process are nonlinear. However, many previous research show that the classical control concept cannot be used effectively to control nonlinear processes. Nonlinear mapping capabilities of neural networks are used extensively to solve control problems.

Previous research on control examined the use of fuzzy set theory and local networks. However, there are some niches to be penetrated. Some researchers [ L. Gordon Kraft and David P. Campagna, 1990 ], [Lichtenwalner, F. Peter (1993)] had shown the comparison between the neural network controller and the traditional adaptive control system. The former indicates that the neural network performs best when the plant is nonlinear, even it takes quite a long time to learn the process. The latter shows that for the fiber placement composite manufacturing process, the neurocontroller behave like a PI controller when the network receive an input which it has not experienced before. However, after learning from experience, the performance greatly improves and exceeds that of conventional methods. Miller, W. T. et al (1990) reviews the comparison between CMAC and Backpropagation and shows that CMAC can learn a large variety of nonlinear function in a fewer iteration with a little or no learning interference due to recent learning in remote parts of the inputs space. These advantages of CMAC are due to the local generalization at the expense of large memory. Another research [Lin, C. S. and Hyongsuk Kim (1991)] confirms that learned information is distributively stored in adaptive critic learning control and no memory capability is wasted on useless states. The adaptive critic method is a humanlike self-learning scheme that learns performance evaluation as well as control actions based on experience. In adaptive critic method, the user specifies a utility function to be controlled and an acceptable range of system response. An additional neural network, called a critic network, has been adapted to evaluate the progress that the system is making. The output signal of the critic network indicates whether the system status is getting better or worse. Therefore, the action network which outputs the actions to the process is adapted to maximize the utility function of the critic network. Christopher G. Atkeson et al (1990) shows the benefit of using table-based controller to control robots. Because more work is needed in the fuzzy-based controller area, this paper presents the direct comparison between CMAC neural network [Albus, J. S. (1979)] and FAM [Kosko, B. (1992)]. These two methods are of interest because of their powerful architectures. Both of them are local generalizer and behave as associative memory. Therefore, they can learn and response to the process pretty fast. Some differences between these two are (1) FAM stores rules in its memory and processes the incoming inputs in parallel on real time basis, but CMAC precalculates its look-up table off-line, so it can response to the incoming input immediately. (2) The system input-output characteristics of CMAC are continuous but those of FAM are discrete. This research tries to point out the effects of these differences. The objective of the controller is to minimize the error of the desired variable between the target and the actual and to reduce the rise time to the minimum value. In addition, The Backpropagation neural network, the well-known representative of the global network, is also included to compare the ability of the local and global network on the process which the input patterns are not in the same direction.

In the next section, the two different controllers are explained briefly. More details of these methods can be found in the references. Each of the methods are simulated on the same control process under the same conditions. The result of each are discussed in the sections that follow.

## Fuzzy Associative Memories

Proper fuzzy sets are the ones that violate the law of noncontradiction and excluded middle. Fuzzy set theory holds that all things are matters of degree. It reduces black and white logic to the mathematics of gray relationships. The fuzzy power set $F(2^X)$, which contains all fuzzy subsets of X, corresponds to the unit square when $X = \{x_1, x_2\}$. Figure 1 displays the fuzzy power set $F(2^X)$ in 2 dimensional unit hypercube. From Figure 1, the fuzzy subset A corresponds to the fit vector (1/4,3/4), therefore, A has membership degrees $m_A(x_1) = 1/3$ and $m_A(x_2) = 3/4$. The midpoint M of the unit cube has the maximum fuzziness.

In order to find the FAM bank as shown in Figure 2, since the dynamic equations of the process are known and the process is not complicated, FAM rules need to be identified.

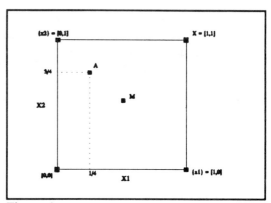

**Figure 1**

FAM bank can be randomly calculated from the process equations of motion. On the other hand, if the process equations of motion are not known, the differential competitive learning (DCL) has to be applied to estimate the FAM bank. The differential competitive learning law [Kosko, Bart (1992)] is shown below :

$$\dot{m}_{ij} = \dot{S}_j(y_j)[S_i(x_i) - m_{ij}] + n_{ij} = y_j[S_j - m_{ij}] - S_j[S_i - m_{ij}] + n_{ij}$$

where $y_j$ is the output signal of the $j^{th}$ neuron. When the $j^{th}$ neuron wins, $y_j = 1$ and it equals to 0 when the $j^{th}$ neuron loses. $S_j$ is the competitive signal, which is between zero and one. $m_{ij}$ is the synaptic weights of the connection matrix M. The $ij^{th}$ synapse is excitatory if $m_{ij} > 0$, inhibitory if $m_{ij} < 0$.

If the $j^{th}$ neuron continues to win, $S_j$ rapidly approaches unity, and learning ceases. The rapid burst of learning as $S_j$ approaches unity helps prevent the $j^{th}$ neuron wining too frequently. If this happens, it prematurely encodes a new synaptic pattern in $m_j$ at the expense of the current $m_j$ pattern. In differential competitive learning, the win signal $S_j$ rapidly stops changing once the $j^{th}$ neuron has secured its competitive victory. Differential competitive learning punishes losing with a sign change ( when $y_j(s) = 0$ ). Then $S_j$ rapidly falls to zero, and learning again ceases. Before $S_j$ reaches zero, the competitive learning law reduces to the anticompetitive law

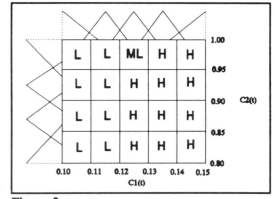

**Figure 2**

$$\dot{m}_{ij} = -S_j[S_i - m_{ij}] + n_{ij}$$

Note that the input and the output of the FAM system are the fuzzy sets and the output of the FAM, R, equals a weighted sum of the individual vectors $R_k'$:

$$R = \sum_{k=1}^{m} w_k R_k'$$

Because the output of the FAM is also a fuzzy set, therefore, the fuzzy centroid defuzzification scheme is introduced to produce a single numerical output.

$$Fuzzy\ centroid\ \bar{R} = \sum_{j=1}^{p} y_j m_R(y_j) \ / \ \sum_{j=1}^{p} m_R(y_j)$$

## Cerebellar Model Articulation Controller

The basic idea behind the CMAC approach is to create the look-up table from the input-output of the system. Then using the data in the table as feedforward information to calculate the appropriate control signal. In this case, the value of the system parameters were known, hence, all the value in the table were precalculated and stored in the memory. As the input are fed into the controller, CMAC would be able to look up in its memory and provide the appropriate controller output.

The CMAC algorithm maps any input it receives into a set of points in a large conceptual memory in such a way that two inputs that are close in input space will have their points overlap in the memory as shown in Figure 3, with more overlap for closer inputs. If two inputs are far apart in the input space, there will be no overlap in their sets in the memory, and also no generalization. With a built-in local generalization, input vectors that are close in the input space will give outputs that are close, even the input has not been trained on, as long as there has been training in that region of the state-space.

The method which was used in this article to improve the value in memory is through first-order learning law :

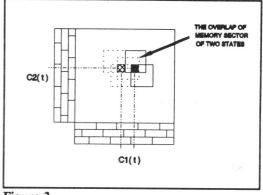

Figure 3

$$m_{ij}(k+1) = m_{ij}(k) + \mu[u(k) - m_{ij}(k)]$$

where m(k) is the present value of the memory location, m(k+1) is the updated value, u(k) is the desired output of the controller at time k , and $\mu$ is the learning rate which is between zero and one. If the memory contents m(k) is larger than u(k), then m(k) is corrected by subtracting a number proportional to the error.

The control signal generated by the network is found by summing the values in the system associated with the current inputs. This signal is then fed to the process to maintain the actual output at the target.

## Properties of CMAC and FAM

This section illustrates the similarities and differences between CMAC and FAM.
• Both CMAC and FAM are local generalizer. The input vectors that are close in the input space will provide the close outputs.
• Both of them use a look-up table method. Hence, they can be used appropriately to control the process because of their fast response.
• CMAC and FAM have the property that large network can be used and trained in reasonable time. This is because there is a small number of calculations per output.
• The input-output characteristics of CMAC are continuous but those of FAM are discrete. Therefore, CMAC uses more memory than FAM.
• CMAC has to be trained off-line before being used to control the process but FAM calculates the outputs on-line, hence, it uses more time to response. Since there is only a small number of calculations per output, the

difference between CMAC and FAM 's processing time is very small.

## Sample Process Description

Both control system algorithms were applied to the same process which is the bioreactor containing water, nutrients, and biological cells as shown in Figure 4. This problem has been suggested in *Neural Network for Control* by Anderson, Charles W. et al (1991). The state of this process is characterized by the number of cells and the amount of nutrients. The volume in the tank is maintained at a constant level by removing tank content at a rate equal to the incoming rate. This flow rate is the variable by which the bioreactor is controlled. The objective is to achieve and maintain a desired cell amount, $c_1^*(t)$, by altering the flow rate throughout a learning trial. In this article, $c_1^*(t)$ was setted at 0.1205. The initial conditions

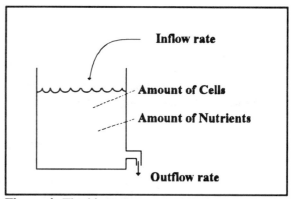

**Figure 4** The bioreactor

$c_1(0)$ is the random variable on the interval [0.10,0.14] and $c_2(0)$ is the random variable on the interval [0.8,1.0]. The system constraints are $0 \leq c_1, c_2 \leq 1$ and $0 \leq r \leq 2$. And the process equations of motion are :

$$c_1[t+1] = c_1[t] + 0.005(-c_1[t]r[t] + (1-c_2[t])e^{c_2[t]/0.48})$$

where
$c_1(t)$ : Amount of cells
$c_2(t)$ : Amount of Nutrients
$r$ : Flow rate

$$c_2[t+1] = c_2[t] + 0.005(-c_2[t]r[t] + (1-c_2[t])e^{c_2[t]/0.48}\frac{1.02}{1.02 - c_2[t]})$$

## Results and Conclusions

The result of the CMAC neural network system is shown in Figure 5. This CMAC controller was trained 15 iterations before being used to control the process. The system rise time was quite large but the system offset was very small.

The performance of the FAM method is plotted in Figure 6. The steady state performance was a little bit worse than CMAC but the rise time was very small. With these characteristics of the response, it indicates that FAM works as well as CMAC control system. The steady state response of FAM fluctuates because FAM's system input-output characteristics are discrete but those of the CMAC are continuous. There are the boundaries between each rules of FAM, hence, changing the rule from one to the other is not as smooth as CMAC. However, in the large process, which has many input, FAM is more favorable because FAM needs much smaller memory spaces than CMAC.

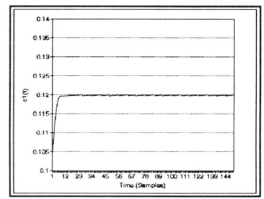

**Figure 5**

Finally, Backpropagation response, the global generalization neural network, is shown in Figure 7. The figure confirms that the global generalization neural network cannot be used to control the process, which the patterns of the inputs do not go in the same direction.

The CMAC and FAM based control system has been developed and implemented for this bioreator problem. These two methods have been chosen because of their fast learning characteristic and their local generalization background. Both of them give the favorable responses on this nonlinear process. They did track the target very well. CMAC gave slightly better steady state response than FAM. However, DCL will be applied to improve the steady state response of the FAM in the future research. Finally, the research shows that this type of process is not a good application for global generalization network.

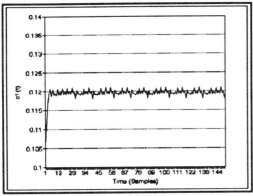

**Figure 6**

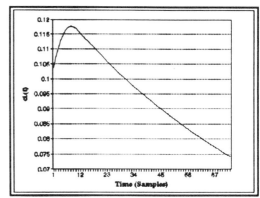

**Figure 7**

**References**

Albus, James S. (1979), "Mechanisms of planning and problem solving in the brain," *Mathematical Biosciences,* pp. 247-293.

Anderson, Charles W. and W. Thomas Miller, III (1991), "Challenging control problems," *Neural Networks for Control,* MIT press.

Atkeson, G. Christopher and David J. Reinkensmeyer (1990), "Using associative content-addressable memories to control robot," *Neural Networks for Control,* MIT Press.

Kosko, Bart (1992), *Neural Networks and Fuzzy Systems,* Prentice-Hall, New Jersey.

Kraft, L. Gordon and David P. Campagna (1990), "A comparison between CMAC neural network control and two traditional adaptive control systems," *IEEE Control Systems Magazine,* pp. 26-43, April 1990.

Lichtenwalner, F. Peter (1993), "An on-line learning neurocontroller for the fiber placement composite manufacturing process" in, *Intelligent Engineering Systems through Artificial Neural Networks,* Dagli, C. H., et. al. eds, pp. 529-534, ASME Press, New York.

Lin, Chun-Shin and Hyongsuk Kim (1991), "CMAC-Based adaptive critic self-learning control," *IEEE Transactions on Neural Networks,* pp. 530-532, September 1991.

Miller, W. Thomas, et. al.(1990), "CMAC : An associative neural network alternative to backpropagation," *Proceedings of the IEEE,* pp. 1561-1567, October 1990.

Sun, Kuo-Tung, et. al. (1993), "Sliding-mode-aided fuzzy control system design" in, *Intelligent Engineering Systems through Artificial Neural Networks,* Dagli, C. H., et. al. eds, pp. 517-522, ASME Press, New York.

# A Robust Fuzzy PI Controller for
# a Flexible-Joint Robot Arm with Uncertainties

## Weiming Tang  and  Guanrong Chen

Department of Electrical Engineering
University of Houston
Houston, TX 77204-4793

## Abstract

Fuzzy control strategy offers an alternative approach for many conventional control systems, which has certain advantages over the other techniques. For example, the *improved PI fuzzy controller* that we designed in this paper, which is based on an existing fuzzy PI controller, can control an uncertain flexible-joint robot arm to produce satisfactory tracking results. The improved fuzzy PI controller not only can control (stable and unstable) conventional linear systems, performing as well as the conventional PI controller, but also is capable of controlling many nonlinear systems such as the flexible-joint robot arm under investigation which contains uncertainties within ten percent tolerance of all the nominal system-parameter values. In this improved fuzzy PI controller, we used only three simple membership functions plus six simple fuzzy logic control rules. In this paper, we briefly describe the design principle of this improved fuzzy PI controller and its tracking performance in handling the nonlinearity, flexibilities and uncertainties of the flexible-joint robot arm system.

## 1 Introduction

Robot control is an important area of robotics engineering and a significant application of system control theory. A robotic system has very strong nonlinear characteristics and variable parameters in real environments. Recently, experimental evidence indicates that joint flexibility should be accounted for in both modeling and control of many engineering robotic manipulators [4]. Due to the nonlinearity and the joint flexibility, it is usually difficult to control the robot-link position (as well as the velocity and acceleration) in robot trajectory tracking.

Supported by the Institute of Space Systems Operations, University of Houston.

Fuzzy logic control has emerged as an alternative or complement to conventional control strategies in many engineering areas, from process control to robotics. Fuzzy control theory usually provides nonlinear controllers which are capable of performing various nonlinear control actions. If the parameters of the fuzzy controllers are chosen appropriately, it is also possible for them to work for uncertain nonlinear systems. In addition, fuzzy controllers are capable of handling many complex situations such as those control systems with large uncertainties in process parameters and/or systems structures [6].

In this research, we design an Improved Fuzzy PI (Proportional-Integral) Controller based on the successful Fuzzy PI Controller developed in [7] and [1] for a flexible-joint robot arm model, with both position and velocity involved in the feedback loop, to perform efficient control of the flexible-joint robot arm which can have relatively large uncertainties. The criterion is to force the position error tend to zero, or within a small range of tolerance. Due to the joint flexibility, the link may oscillate even under conventional control, and hence its position error may not be able to be eliminated completely. The crucial requirement is that the controller can handle the system nonlinearity, flexibility and uncertainty simultaneously.

Several researchers have used fuzzy control laws for rigid-joint robot model (e.g., [2] and [5]). Ying et al. [7] introduced a simple fuzzy PI controller (with simple fuzzy sets and membership functions) for conventional (linear and nonlinear) systems. In this paper, we improve this fuzzy PI controller, by introducing into it an additional sufficient condition, for the tracking control of the uncertain flexible-joint robot arm. In this design, we use only position and velocity feedbacks, avoiding sensing the acceleration (and jerk) of the link-angle output, while achieving a large robustness of the entire robot control system.

## 2 Design of the Improved Fuzzy PI Controller

In this section, we briefly describe the improved fuzzy PI controller: its mathematical principle and engineering design. We first establish a framework for the fuzzy PI controller of the robot system, and then describe the three main steps in its design: fuzzification, rule base establishment, and defuzzification. Finally, we introduce the additional sufficient condition for the control of the flexible-joint robot arm system.

### 2.1 The Fuzzy PI Controller

Recall from [7] and [1] that the fuzzy PI controller is described by

$$U(nT) = U(nT - T) + T\Delta U(nT), \qquad (1)$$

where $\tilde{k}_p = k_p - k_i \cdot T/2$ and $\tilde{k}_i = k_i$, with $k_p$ and $k_i$ being the conventional proportional and integral gains, respectively, and

$$\Delta U(nT) = \tilde{k}_p \cdot d(nT) + \tilde{k}_i \cdot e(nT), \qquad (2)$$

with

$$d(nT) = \frac{e(nT) - e(nT - T)}{T}.$$

Similarly, we can set up the fuzzy PI controller for a robot as shown in Fig. 1.

In Fig. 1, we use the following notation: $T$ is the sampling period, $\theta_d(nT)$ is the desired position, $\dot{\theta}_d(nT)$ is the desired velocity, $\theta(nT)$ is the actual position, $\dot{\theta}(nT)$ is the actual velocity, $\tilde{k}_p$ is the proportional gain, $\tilde{k}_i$ is the integral gain, $k_u$ is fuzzy control gain, and $\Delta U(nT)$ is the incremental control output of the fuzzy PI controller.

Next, denote the position error $e(nT)$ by $e_p(nT) = \theta_d(nT) - \theta(nT)$ and the velocity error $d(nT)$ by $e_v(nT) = \dot{\theta}_d(nT) - \dot{\theta}(nT)$, respectively. As can be seen from Fig. 1, we have

$$U(nT) = U(nT - T) + k_u \cdot \Delta U(nT). \qquad (3)$$

The constant gain $k_u$ can be chosen *a priori* or determined by the stability consideration etc. If we know the initial control U(0) (usually, zero), then we only need to find $\Delta U(nT)$ in the design. Since $k_u$ is constant, we can combine $k_u$ and $\Delta U(nT)$ together in order to simplify the design, namely, we can let $\Delta \tilde{U}(nT) = k_u \cdot \Delta U(nT)$ and consider $\Delta \tilde{U}(nT)$ as the incremental control output (to be determined).

### 2.2 Fuzzification, rule base, and defuzzification

Following the standard fuzzy controller design procedure, we have three parts in the design: (1) fuzzification, (2) fuzzy logic control rules, and (3) defuzzification.

STEP 1: Fuzzification

In our design, we use the simple fuzzy membership functions shown in Figs. 2-3.

The fuzzification algorithm, for scaled position error and velocity error, is shown in Fig. 2. The position error has two membership values: $e_p \cdot p$ (error positive) and $e_p \cdot n$ (error negative). Similarly, the velocity error has two membership values: $e_v \cdot p$ (error positive) and $e_v \cdot n$ (error negative). $L$ is some constant chosen by the designer according to the application and will be fixed after being selected.

The membership functions for the incremental control output $\Delta \tilde{U}(nT)$ is shown in Fig. 3. It has three membership values: $o \cdot p$ (output positive), $o \cdot z$ (output zero), and $o \cdot n$ (output negative). For simplicity, we choose the same constant value $L$ as that used for the inputs.

STEP 2: Set up fuzzy control rules

If $e_p > 0$ and $e_v > 0$, we see that $\theta_d > \theta$. In this case, the control law should lead the position output to increase. Since $\dot{\theta}_d > \dot{\theta}$, on the other hand, the control law should let this position-increase be fast. For this purpose, we should require the control output increase, which means that $\Delta \tilde{U}(nT)$ should be output positive ($o \cdot p$), because of the relation $U(nT) = U(nT - T) + \Delta \tilde{U}(nT)$. This yields the first control rule.

Similarly, we can find the control rules for the other three possible cases: $e_p > 0$ and $e_v < 0$; $e_p < 0$ and $e_v > 0$; and $e_p < 0$ and $e_v < 0$. The complete set of fuzzy control rules adopted in this design are summarized as follows:

$R1$ : *IF position error $= e_p \cdot p$*
  *AND velocity error $= e_v \cdot p$*
  *THEN output $= o \cdot p$;*
$R2$ : *IF position error $= e_p \cdot p$*
  *AND velocity error $= e_v \cdot n$*
  *THEN output $= o \cdot z$;*
$R3$ : *IF position error $= e_p \cdot n$*
  *AND velocity error $= e_v \cdot p$*
  *THEN output $= o \cdot z$;*

$$\times [\tilde{k}_i \cdot e_p(nT) + \tilde{k}_p \cdot e_v(nT)]. \quad (7)$$

Note that in regions $I\,3$ and $I\,4$, $e_v(nT)$ is positive, and so $e_v(nT) = |e_v(nT)|$. But in regions $I\,7$ and $I\,8$, the denominator will be $2(2L + \tilde{k}_p \cdot e_v(nT))$, where $e_v(nT)$ is negative, which can be rewritten as $2(2L - \tilde{k}_p \cdot |e_v(nT)|)$. This is the denominator in the above formula. For other regions, see Table 1.

## 2.3  Additional Sufficient Conditions

The design of the fuzzy PI controller described above assumes that the output of the process (plant) under control is proportional to the input, namely, when the input to the process is positive, then the process output will increase; when the input to the process is negative, then the process output will decrease. But for the flexible-joint robot arm model under investigation, due to the nonlinearity, this property is not satisfied. The situation is even worse if we take the system uncertainties into account. Hence, we need to derive some additional sufficient conditions to guarantee this proportional property.

Now, consider the well-known flexible-joint robot arm model shown in Fig. 5:

$$I_1 \cdot \ddot{\theta}_1 + Mgl \cdot sin(\theta_1) + k \cdot (\theta_1 - \theta_2) = 0, \quad (8)$$
$$I_2 \cdot \ddot{\theta}_2 + k \cdot (\theta_2 - \theta_1) = u, \quad (9)$$

where

| | |
|---|---|
| $I_1$ | is the link inertia, |
| $I_2$ | is the motor inertia, |
| $M$ | is the mass, |
| $g$ | is the gravity, |
| $l$ | is the length of the link, |
| $k$ | is stiffness, |
| $\theta_1$ | is the link position, |
| $\theta_2$ | is the motor position, |
| $u$ | is the torque input. |

Note that equation (8) can be rewritten as

$$0 = Y \cdot P + k(\theta_1 - \theta_2), \quad (10)$$

where

$$Y = [\ddot{\theta}_1 \quad sin(\theta_1)],$$
$$P = [I_1 \quad M \cdot g \cdot l].$$

If the desired link trajectory $\theta_{1d}$ and its velocity $\dot{\theta}_{1d}$ are both given, we can estimate the desired motor trajectories $\theta_{2d}$ and $\dot{\theta}_{2d}$ using (10) as follows:

$$\theta_{2d} = k^{-1} \cdot Y_d \cdot P + \theta_{1d}, \quad (11)$$
$$\dot{\theta}_{2d} = k^{-1} \cdot \dot{Y}_d \cdot P + \dot{\theta}_{1d}, \quad (12)$$

---

$R4$ : *IF position error* $= e_p \cdot n$
*AND velocity error* $= e_v \cdot n$
*THEN output* $= o \cdot n$.

Here, the "AND" is the Zadeh's logical "AND" [8] defined by

$$\mu_A \text{ AND } \mu_B = \min \{ \mu_A, \mu_B \} \quad (4)$$

for any two membership values $\mu_A$ and $\mu_B$ on the fuzzy subsets $A$ and $B$, respectively.

STEP 3: Defuzzification

The control rules shown above all employ the Zadeh "AND" for two signals: One is the scaled position error and the other is the scaled velocity error. Since the Zadeh "AND" is the minimum of two values, two different conditions arise for each rule: When the scaled position error is less than the scaled velocity error and when the scaled velocity error is less than the scaled position error. The twenty different combinations of the scaled position error and the scaled velocity error are shown in Fig. 4. From all these combinations, we can calculate the corresponding membership values according to the appropriate fuzzy control rules [7].

To this end, we can determine the fuzzy control $\Delta \tilde{U}(nT)$ using the following standard "center of mass" formula:

$$\Delta \tilde{U}(nT) = \frac{\sum (\text{member value})(\text{membership value})}{\sum \text{membership value}}. \quad (5)$$

Summarizing all the input combinations, we obtain the following incremental control output $\Delta \tilde{U}(nT)$ for the fuzzy PI controller:

• IF $\tilde{k}_p \cdot |e_v(nT)| \le \tilde{k}_i \cdot |e_p(nT)| \le L$, THEN

$$\Delta \tilde{U}(nT) = -\frac{L \cdot k_u}{2(2L - \tilde{k}_i \cdot |e_p(nT)|)}$$
$$\times [\tilde{k}_i \cdot e_p(nT) + \tilde{k}_p \cdot e_v(nT)]. \quad (6)$$

Note that in regions $I\,1$ and $I\,2$, $e_p(nT)$ is positive, and so $e_p(nT) = |e_p(nT)|$. But in regions $I\,5$ and $I\,6$, the denominator will be $2(2L + \tilde{k}_i \cdot e_p(nT))$, where $e_p(nT)$ is negative, which can be rewritten as $2(2L - \tilde{k}_i \cdot |e_p(nT)|)$. This is the denominator in the above formula.

IF $\tilde{k}_i \cdot |e_p(nT)| \le \tilde{k}_p \cdot |e_v(nT)| \le L$, THEN

$$\Delta \tilde{U}(nT) = -\frac{L \cdot k_u}{2(2L - \tilde{k}_p \cdot |e_v(nT)|)}$$

where

$$Y_d = [\ddot{\theta}_{1d} \quad sin(\theta_{1d})],$$
$$\dot{Y}_d = [\theta_{1d}^{(3)} \quad cos(\theta_{1d})].$$

Let

$$e_p(nT) = \theta_{2d}(nT) - \theta_2(nT),$$
$$e_v(nT) = \dot{\theta}_{2d}(nT) - \dot{\theta}_2(nT),$$

be the inputs to the fuzzy PI controller. If we know both $e_p(nT)$ and $e_v(nT)$, then we can find the incremental control $\Delta \tilde{U}(nT)$ according to the defuzzification algorithm described in the last section.

Now, integrating (9) gives

$$I_2 \int_0^t \ddot{\theta}_2 d\tau + k \int_0^t (\theta_2 - \theta_1) d\tau = \int_0^t u d\tau \quad (13)$$

or

$$I_2 \cdot (\dot{\theta}_2(t) - \dot{\theta}_2(0)) = \int_0^t u d\tau - k \int_0^t (\theta_2 - \theta_1) d\tau. \quad (14)$$

Assuming zero initial condition $\dot{\theta}_2(0) = 0$ (which is true for this robot arm model), we obtain

$$\dot{\theta}_2(t) = \frac{1}{I_2} \int_0^t u d\tau - \frac{k}{I_2} \int_0^t (\theta_2 - \theta_1) d\tau. \quad (15)$$

Thus, we see that a sufficient condition for the flexible-joint robot arm model to have the aforementioned proportional property should be:

- If $1/I_2 \int_0^t u d\tau > k/I_2 \int_0^t (\theta_2 - \theta_1) d\tau$, so that $\dot{\theta}_2(t) > 0$, then $\theta_2(t)$ should increase.

- If $1/I_2 \int_0^t u d\tau < k/I_2 \int_0^t (\theta_2 - \theta_1) d\tau$, so that $\dot{\theta}_2(t) < 0$, then $\theta_2(t)$ should decrease.

In summary, we not only need to determine the control law $U(nT)$ according to the fuzzy PI control rules, but also need to force the control $U(nT)$ to satisfy the sufficient condition described above. This modified version of fuzzy PI controller is called the *improved fuzzy PI controller.*

The overall improved fuzzy PI controller system is shown in Fig. 6.

## 3 Simulation Results

Usually, when a robot manipulator operates, the effects of the nonlinearities, time-varying parameters, and external uncertainties such as friction and damping, may cause the system very significant tracking errors. This is even worse when the robot joint has flexibility. The improved fuzzy PI controller is expected to be able to handle these problems to certain degree of success. The computer simulation results have convinced this belief, which will be discussed in this section. Here, we only show the robot arm system with ten percent uncertainty of tolerance in all system parameters.

Taking into account uncertainties, the flexible-joint robot arm model is described by

$$(I_1 + \delta I_1)\ddot{\theta}_1 + (M \cdot g \cdot l + \delta M)sin(\theta_1)$$
$$+ (k + \delta k)(\theta_1 - \theta_2) = 0, \quad (16)$$
$$(I_2 + \delta I_2)\ddot{\theta}_2 + (B + \delta B)\dot{\theta}_2$$
$$+ (k + \delta k)(\theta_2 - \theta_1) = u + \delta u, \quad (17)$$

where

$$|\delta I_1| \le C \cdot I_1,$$
$$|\delta I_2| \le C \cdot I_2,$$
$$|\delta M| \le C \cdot M \cdot g \cdot l,$$
$$|\delta k| \le C \cdot k,$$
$$|\delta u| \le C \cdot u,$$
$$B = 0.007 n \cdot m \cdot sec/rad,$$
$$|\delta B| \le C \cdot B,$$
$C$ is a varying constant (uncertainties).

Let $x_1 = \theta_1$, $x_2 = \dot{x}_1 = \dot{\theta}_1$, $x_3 = \theta_2$ and $x_4 = \dot{x}_3 = \dot{\theta}_2$. Let also $t_i$ be the $i$th time step, $t_{i-1}$ the $(i-1)$st time step, and $T$ the sampling period. Then, in this improved fuzzy PI controller, to obtain $\Delta \tilde{U}(t_i)$, the position error is

$$e_p(t_{i-1}) = \theta_{2d}(t_{i-1}) - \theta_2(t_{i-1})$$

and the velocity error is

$$e_v(t_{i-1}) = \dot{\theta}_{2d}(t_{i-1}) - \dot{\theta}_2(t_{i-1}).$$

Using this notation, the robot model can be rewritten as

$$x_4(t_i) = x_4(t_{i-1}) - \frac{T \cdot (k + \delta k)}{I_2 + \delta I_2} \cdot (x_3(t_{i-1}) - x_1(t_{i-1}))$$
$$+ \frac{T}{I_2 + \delta I_2} \cdot U(t_i) - \frac{B + \delta B}{I_2 + \delta I_2} \cdot x_4(t_{i-1}), \quad (18)$$
$$x_3(t_i) = x_3(t_{i-1}) + T \cdot x_4(t_i), \quad (19)$$
$$x_2(t_i) = x_2(t_{i-1}) - \frac{m \cdot g \cdot l + \delta m}{I_1 + \delta I_1} \cdot T \cdot sin(x_1(t_{i-1}))$$
$$- \frac{k + \delta k}{I_1 + \delta I_1} \cdot T \cdot (x_1(t_{i-1}) - x_3(t_i)), \quad (20)$$
$$x_1(t_i) = x_1(t_{i-1}) + T \cdot x_2(t_i). \quad (21)$$

We use the Improved Fuzzy PI Controller described in this paper to calculate the control action $U(t_i)$, and complete the simulations as shown below.

Fig. 7 is one of the many simulations, in which we compare the link positions for two cases: without system uncertainties and with system uncertainties (less than 10% tolerance of the nominal values).

## 4    Conclusions

Due to its strong nonlinear characteristics, joint flexibility and parameter variations in real environments, tracking control of a flexible-joint robot arm system is difficult. Many researchers in the field of robot control have been trying to use various control methods to resolve these problems, such as feedback linearization, adaptive control and the integral-manifold approach. Fuzzy control provides an alterative or complement for such robotic system tracking problems. The Improved Fuzzy PI Controller discussed in this paper provides such an example, which not only can control (stable and unstable) conventional linear systems, performing as well as (if not better than) the conventional PI controller, but also is capable of controlling many nonlinear systems such as the flexible-joint robot arm under investigation which contains uncertainties within up to ten percent tolerance of all the nominal system-parameter values. We have shown some computer simulation results on its tracking performance in handling the nonlinearity, flexibilities and uncertainties of the flexible-joint robot arm system.

## References

1. G. Chen and H. Ying, "Stability Analysis of Fuzzy PI Control Systems," *Proceedings of the 3rd International Workshop on Industrial Applications of Fuzzy Control and Intelligent Systems*, Houston, TX, Dec. 10-12, 1993, pp 128-133.

2. R. Palm, "Fuzzy Controller for a Sensor Guided Robot Manipulator," *Fuzzy Sets and Systems*, Vol. 31, pp 133-149, 1989.

3. M. Spong, "Modeling and Control of Elastic Joint Robots," *A.S.M.E. Journal of Dynamic Systems, Measurement and Control*, vol. 109, pp 310-319, 1987.

4. R. Steinvorth, *Model Reference Adaptive Control of Robots*, Technical Report, Rensselaer Polytechnic Institute, New York, March 1991.

5. M. Uragami, M. Mizumoto and K. Tanaka, "Fuzzy Robot Controls," *Journal of Cybernetics*, Vol. 6, pp 39-64, 1976.

6. L. Wang, "Stable Adaptive Fuzzy Control of Nonlinear Systems," *I.E.E.E. Transactions on Fuzzy Systems*, vol. 1, No. 2, pp 146-155, 1993.

7. H. Ying, W. Siler and J. J. Buckley, "Fuzzy Control Theory: A Nonlinear Case," *Automatica*, vol. 26, No. 3, pp 513-520, 1990.

8. L. Zadeh, "Fuzzy Sets," *Information and Control*, vol. 8, pp 338-353, 1965.

Table 1: Output of the Fuzzy Controller *

| Input Combinations | Output $\Delta U(nT)$ |
|---|---|
| $I$ 9 and $I$ 10 | $-[k_p \cdot e_v(nT) + L] \cdot k_u/2$ |
| $I$ 11 and $I$ 12 | $-[k_i \cdot e_p(nT) + L] \cdot k_u/2$ |
| $I$ 13 and $I$ 14 | $-[k_p \cdot e_v(nT) - L] \cdot k_u/2$ |
| $I$ 15 and $I$ 16 | $-[k_i \cdot e_p(nT) - L] \cdot k_u/2$ |
| $I$ 17 | $-L \cdot k_u$ |
| $I$ 18 | 0 |
| $I$ 19 | $L \cdot k_u$ |
| $I$ 20 | 0 |

(* when the scaled position and velocity errors are not within the interval $[-L, L]$)

Fig. 1.

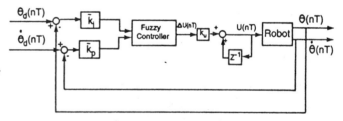

Fig. 2.

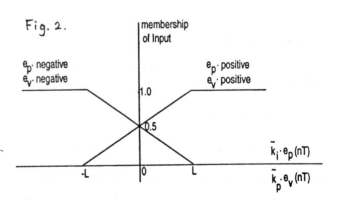

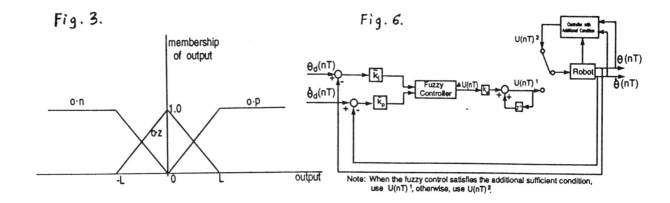

Fig. 3.

membership of output

o·n

1.0

o·z

o·p

-L          0          L          output

Fig. 6.

Note: When the fuzzy control satisfies the additional sufficient condition, use U(nT)¹, otherwise, use U(nT)².

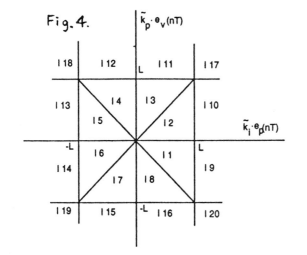

Fig. 4.

$\tilde{k}_p \cdot e_v(nT)$

| I 18 | I 12 | L I 11 | I 17 |
| I 13 | I 4 | I 3 | I 10 |
| | I 5 | I 2 | |
| -L | I 6 | I 1 | L |
| I 14 | I 7 | I 8 | I 9 |
| I 19 | I 15 | -L I 16 | I 20 |

$\tilde{k}_i \cdot e_p(nT)$

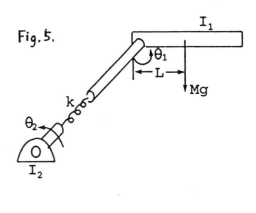

Fig. 5.

$I_1$

$\theta_1$

L

Mg

k

$\theta_2$

O

$I_2$

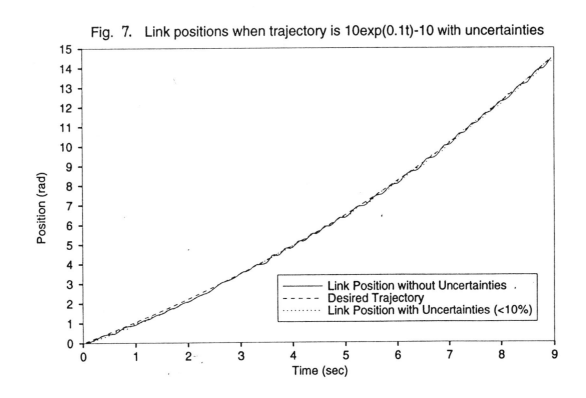

Fig. 7.  Link positions when trajectory is 10exp(0.1t)-10 with uncertainties

Position (rad)

—— Link Position without Uncertainties
--- Desired Trajectory
···· Link Position with Uncertainties (<10%)

Time (sec)

# A Self-Tuning Fuzzy Logic Controller for Temperature Control of Superheated Steam

Pekka Isomursu and Tapio Rauma
VTT (Technical Research Centre of Finland)
Computer Technology Laboratory
P.O. Box 201, FIN-90571 Oulu, Finland
Pekka.Isomursu@vtt.fi

*Abstract* - We first introduce a non-adaptive fuzzy logic controller (FLC) for the control of live steam temperature in a coal-fired power plant. For further enhancing performance, we introduce a self-tuning method for the FLC that modifies the scaling factor of one FLC output. To make the FLC more portable to other similar plants and more robust we add another self-tuning mechanism that runs on-line and modifies the membership functions of the fuzzy rule set. We have used the meta-rule approach in the tuning mechanisms. The performance of the FLC is compared to a cascade PI controller.

## I. INTRODUCTION

Fuzzy logic control appears very useful when linearity and time-invariance of the controlled process cannot be assumed, when the process lacks a well posed mathematical model, or when the human understanding of the process is very different from its model. However, constructing a fuzzy logic controller (FLC) poses some problems which are mainly related to the knowledge acquisition, tuning, and proving the optimality and stability of the control system.

Following the creation of a fuzzy rule base, the FLC typically needs further tuning before it reaches the desired level of performance. This tuning process is non-trivial, and can be time consuming. It is often done manually on a trial-and-error basis. The tunable parameters can be grouped as follows [1]:

❑ the scaling factors of the input and output variables
❑ the membership functions of the variables
❑ the set of control rules.

Although a non-adaptive FLC can be designed to cope with a certain amount of process nonlinearity, in a highly dynamic process a need may arise for on-line tuning of the properties of the FLC via an adaptation mechanism. A variety of methods have been proposed for tuning the FLC, e.g. [2,3] for tuning the scaling factors, [2,4] for the membership functions, and [3,5] for the set of control rules. The methods draw their ideas from neural network theory, artificial intelligence, mathematics, and other sources. However, many of these methods have been demonstrated only with simple control problems, such as pole balancing. It is not always clear how they scale up to controlling real-world industrial processes with complex nonlinearities and long time delays. [6]

In this paper we introduce an adaptive FLC for controlling the temperature of superheated steam. Using fuzzy meta-rules we have implemented two simple and practical mechanisms for self-tuning of the FLC. We compare the performance of the FLC to a cascade PI controller.

## II. PROBLEM STATEMENT AND PROCESS DESCRIPTION

The superheating process is depicted in Figure 1. To achieve the highest efficiency in the process the temperature of the superheated steam fed into turbines should be kept as high as possible. However, too high a temperature damages certain physical components of the system.

The superheating process is highly non-linear and has long dead times (greater than 30 seconds). Due to factors like wear and dirt, the process parameters change in time. The process consists of three stages -- primary, secondary, and tertiary superheating -- as shown in Figure 1. The temperature of steam at point C is controlled by two separate water sprayers (attemperators) V1 and V2. When the process is running normally the temperature can be controlled rather easily. However, quite frequently there are sudden irregular changes in process parameters, such as pressure changes in feed water and disturbances in fuel or air feed. A conventional controller -- in this case, a cascade PI -- works well under normal process conditions but cannot handle all the changes well. Due to disturbances the system may either oscillate a long time or the temperature may temporarily go much too high or low. Neither of these phenomena is desirable: both sudden temperature changes as well as high peak temperatures may damage the equipment. To prevent the damage, the process has to be run in a lower temperature and thus less cost-effectively.

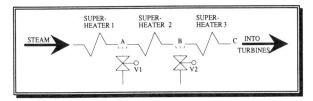

*Figure 1.* The superheating process.

## III. DESIGN OF THE NON-ADAPTIVE FLC

To control the process, we first developed a non-adaptive FLC that consists of five inputs and two outputs. Each rule has from one to three inputs and one output. The rules are of the form:

$R_m$: IF $X_1$ is $A_{1m}$
AND $X_2$ is $A_{2m}$
THEN $Y_1$ is $B_{1m}$

where $A_{im}$ and $B_{im}$ are the linguistic fuzzy labels for $X_i$ and $Y_i$ used in the rule $R_m$ respectively. The inference method used is Max-Product and the defuzzification method center of area (see [7] for more detailed discussion of the methods). The input variable membership functions are either triangular or trapezoidal whereas output variable membership functions are triangles of equal shape and size. We selected these design parameters because of their clarity. They are commonly used in industrial applications.

The rule base is divided into two groups:

1  Rules for controlling valve V1 (see Figure 1). The task of these rules is to keep the temperature in point B at such level that the position of valve V2 is not close to its extremes. The rules have two inputs: the change in temperature at B (caused by V2) and its first derivative. The number of these rules is 9 (for 3*3 input membership functions (mbfs)).

2  Rules for controlling valve V2. These rules keep the temperature in C as close to the set point as possible. The rules have three inputs: error in C, the change of error in C, and the change of temperature in B. The number of these rules is 27 (for 3*3*3 input mbfs).

## IV. CONSTRUCTION OF THE SELF-TUNING MECHANISMS

We first manually tuned the non-adaptive FLC so that it functions properly under normal operational conditions. The non-adaptive version already handled disturbances generally better than a tuned cascade PI controller (see Section V). However, to tune the FLC even further and to make it more portable and robust, we constructed two self-tuning mechanisms. These mechanisms use fuzzy meta rules for inferring possible adaptation actions in the FLC. The technique is based on the work of Maeda et al. [3] who used meta rules for tuning input and output scaling factors of an FLC. A

version of this approach is also introduced by Daugherity et al. [6] in their FLC of a simple water heating system where only the input scaling factors are tuned. In our approach, we modify the scaling factor of one FLC output and membership functions of the rule outputs.

The meta rules we have used are of the form:

$R_m$: IF PerformanceMeasure$_1$ is $A_{1m}$
AND PerformanceMeasure$_2$ is $A_{2m}$
THEN Adaptation is $B_m$

where PerformanceMeasure is an input from the process that describes its state, $A_{im}$ are linguistic labels describing these performance measures, and $B_m$ is a linguistic label describing the amount of adaptation needed.

The goal of the first self-tuning mechanism is to automate optimization of the scaling factors for the rule base (see Figure 2). Performance evaluation, in this mechanism, is based on the measurement of oscillation amplitude and frequency. We were able to achieve good tuning results with only four meta rules (one rule with one input and three rules with two inputs for 2*3 input mbfs). Although this mechanism could be used continuously on-line, we have chosen not to do so, because having it working at the same time as the second self-tuning mechanism may in some situations result in unpredictable operation of the FLC. Instead, the mechanism is used for optimizing the output scaling factor for the normal operational conditions by running it once, removing the mechanism, and setting the scaling permanently to the tuned value.

The second self-tuning mechanism alters the position of output fuzzy membership values of certain rules (Figure 2), leaving rules with extreme or zero output values intact. The main idea here is to damp oscillation when recovering from a deviation in temperature. Performance evaluation is based on the error in temperature at point C (see Figure 1) and its first derivative. In this rule group there are 11 rules (one rule with one input and three rules with two inputs for 3*5 mbfs).

One advantage of using the self-tuning mechanisms was that the rule base became modular, each module fulfilling a different function. In our case this approach led to a smaller total number of rules and their easier construction and tuning. Construction became easier because it could be divided into separate tasks: construction of rules for normal operation and separately tuning different functions such as normal operation and taking care of disturbances. We believe this idea of modularity can be applied and is often helpful also when constructing the rule base for a non-adaptive FLC.

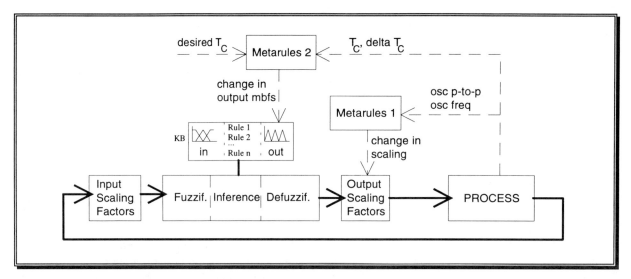

**Figure 2.** The FLC with its two self-tuning mechanisms.

## V. TEST RESULTS

We have compared the performance of the FLC with a cascade PI controller. This type of controller is generally used for controlling the temperature of superheated steam. Tuning of the PI controller was aided by control engineering experts from the Imatran Voima power company. While the cascade PI control works well under normal conditions and during certain disturbances, the tests we have run show that the self-tuning FLC better handles a wider range of disturbances.

For brevity, we only show the results of two typical test cases. The tests shown here have been run in the APROS [8] simulation environment. APROS (Advanced Process Simulator) was developed for very accurate simulation of various power plant processes, both nuclear and conventional. It has been widely used in industry. The test results are shown in Figures 3 to 8. The scaling in each figure is the same (temperature range 527.5 to 532°C, time range 0 to 700s). The set value for temperature is 530 °C (point C in Figure 1).

Figures 3 and 4 show test runs with cascade-type PI control and self-tuning FLC during a 5% change in steam pressure. The two controllers both handle the disturbance rather well, the FLC having a slightly higher temperature peak value but recovering faster.

Figures 5 to 8 show test runs during a 30% change in fuel feed. The change in fuel feed is shown with a dotted line. A solid line shows the temperature. From Figure 5 it can be seen that the cascade PI recovers from this change rather poorly and keeps the temperature about 1° C below set value for a long time. A manually tuned, non-adaptive FLC (Figure 6) already compensates the disturbance significantly better. Overshoot is 0.2°C and the oscillation is damped reasonably well. With the self-tuning method we have used for the output scaling factor

the temperature peak is clearly reduced (Figure 7). Notice that this result can be achieved manually via trial and error. In practice, however, finding the right parameter value can be very tedious and time-consuming. Automating the tuning process speeds it up.

Finally, adding the self-tuning mechanism for membership functions of rule outputs speeds up recovery back to the exact set temperature value and reduces temperature overshoot even further.

## VI. CONCLUSIONS

We have presented an adaptive FLC with two simple self-tuning methods. The methods are based on the use of fuzzy meta rules that modify parts of the basic FLC. Based on the frequency and amplitude of possible temperature oscillation, the scaling factor of one of the two FLC outputs is modified. The mechanism is run once before continuous operation of the FLC and then removed. During operation, membership functions of rule outputs are adjusted by another set of meta rules according to changes in temperature after the last superheating unit.

The self-tuning mechanisms have made the FLC more robust and portable. Modular structure also makes construction of the FLC easier. The test results we have shown demonstrate that the adaptive FLC handles disturbances better than the cascade PI generally used for controlling the temperature of superheated steam.

The time used for the construction of a controller is an important factor in real-world systems. For the superheating process, better controllers than the cascade PI do exist but their use is often limited by the large amount of work needed for their construction. The self-tuning features of the FLC speed up its tuning, making it more attractive for industrial use.

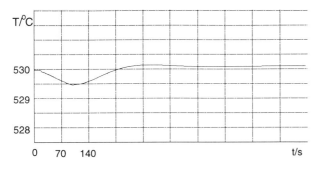

**Figure 3.** Cascade PI, 5% change in steam pressure.

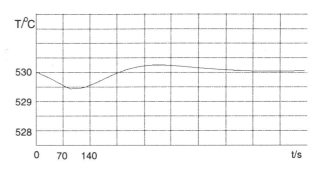

**Figure 4.** Self-tuning FLC, 5% change in steam pressure.

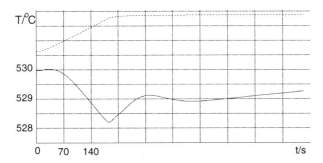

**Figure 5.** Cascade PI, 30%/3min change in fuel feed.

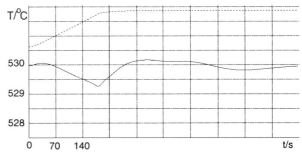

**Fig. 6.** Manually tuned FLC, 30%/3min change in fuel feed.

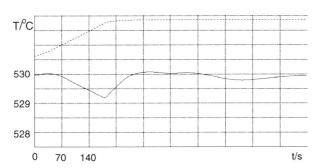

**Figure 7.** FLC with output scaling factor automatically tuned, 30%/3min change in fuel feed.

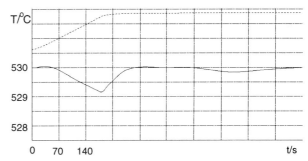

**Figure 8.** FLC with self-tuning of membership functions, 30%/3min change in fuel feed.

## REFERENCES

[1] Pedrycz W.: "Fuzzy Control and Fuzzy Systems", John Wiley & Sons, N.Y., 1989, 258 p.

[2] Burkhardt, D. and Bonissone, P.: Automated Fuzzy Knowledge Base Generation And Tuning. Proceedings, FUZZ-IEEE, San Diego, March, 1992, pp. 179-188.

[3] Maeda M., Murakami S.: A Self-Tuning Fuzzy Controller. Fuzzy Sets and Systems vol.51, 1992, pp. 29-40.

[4] Bartolini, G., Casalino, G., Davoli, F., Mastretta, M., Minciardi, R., and Morten, E.: Development of Performance Adaptive Fuzzy Controllers with Application to Continuous Casting Plants. In: R. Trappl (ed.), Cybernetics and Systems Research, Amsterdam, North-Holland, 1982, pp. 721-728.

[5] Procyk, T.J. and Mamdani, E.H.: A Linguistic Self-Organizing Process Controller, Automation, vol.15, no.1, 1979, pp. 15-30.

[6] Daugherity W., Rathakrishnan B., and Yen J.: "Performance Evaluation of a Self-Tuning Fuzzy Controller", Working Notes of the First International Workshop on Industrial Applications of Fuzzy Control and Intelligent Systems. College Station, TX, Nov 21-22, 1991.

[7] Lee, C. C.: Fuzzy Logic in Control Systems: Fuzzy Logic Controller, Parts I and II. IEEE Transactions on Systems, Man, and Cybernetics, vol.20, no.2, March/April 1990, pp. 404-435.

[8] Juslin K., Silvennoinen E., Kurki J., and Porkholm K.: APROS: An Advanced Process Simulator for Computer Aided Design and Analysis. 12th IMACS World Congress on Scientific Computation, Paris, 18-22 July, 1988, IMACS, Paris, 1988, 3 p.

# Modular Reconfigurable Controllers with Fuzzy Meta-Control

Alexander A. Lavrov

Institut für Informatik
Universität Hildesheim
31141 Hildesheim, Germany

*Abstract*—

This paper proposes an approach to implementing fuzzy controllers which is based on their modularity and a dynamic adaptation of a modular structure using internal fuzzy (meta-) control. The meta-level makes fuzzy decisions intended to adjust the total behavior of a controller by assigning an appropriate set of active components. The paper describes an underlying modular structure, presents a fuzzy-based description for "composite" control strategies, and reveals some special features of their implementation.

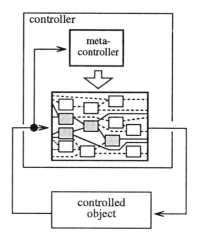

Figure 1: Two levels of control

## I. INTRODUCTION

The flexibility and effectiveness of fuzzy logic control results often in its preference over traditional PID-controllers technique. At the same time, with the further development and associated growing complexity of fuzzy controllers, a real return of their potential depends essentially on their design and internal operational logic. So, considering fuzzy controllers as complex systems, it would be natural to employ a fuzzy approach, with all its advantages, to direct the operation of the controllers themselves (which would in fact be a specific application of the extention principle).

In practice, keeping all the components (first of all, the rules) of a sophisticated controller in an active state can be superfluous, resulting in multiple futile reactions to insignificant data changes, in monitoring a multitude of presently inappropriate data, *etc*. A solution here can be found in a modular, distributed technique. The role of fuzzy meta-control in this case can consist of a dynamic construction of an appropriate current configuration from available functionally-specialized "building blocks" (fig.1). In fig.1, the meta-control flow is shown with a big arrow.

Section 2 describes an underlying modular structure of a controller; a fuzzy approach to its structural adaptation is presented in section 3.

## II. MODULAR STRUCTURE

The modular structure of a controller comprises a set $M = \{M_1, M_2, \ldots, M_n\}$ of interconnected modules (fig.2).

Two ways of implementing modules are possible: 1) each module is in fact a full fuzzy controller having its own fuzzifier and defuzzifier, and 2) each one is represented by a fuzzy rule inference component (a group of rules), while the fuzzifier and defuzzifier are implemented as separate components which process only input and ouput data of the entire modular structure. Without loss of generality, we assume in the following the second case, as it requires less intermediate data transformations. The internal design of a module is organized as a fuzzy rule-based inference network. A module can either be based on a separate chip or can virtually be represented by a particular part of a common one.

The controller receives a set $V^x = \{V(X_1), V(X_2), \ldots, V(X_k)\}$ of values of the input variables $X = \{X_1, X_2, \ldots, X_k\}$ from external sources (e.g. sensors). The fuzzifier transforms $V^x$ into a set of facts $F^x = \{F_1^x, F_2^x, \ldots, F_l^x\}$ (which are actually fuzzy sets defined on the values of input variables) with corresponding certainty factors (degrees of membership) $C^x = \{c_1^x, c_2^x, \ldots, c_l^x\}$ (e.g. "speed" is a variable, "80 m/h" is its value, "speed is high",0.8 and

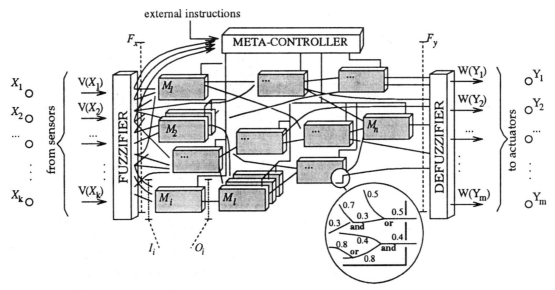

Figure 2: Modular structure of a controller

"speed is very low",0.2 are pairs "fact,certainty factor"). Defuzzifier transforms the sets of its input facts $F^y = \{F_1^y, F_2^y, \ldots, F_p^y\}$ and corresponding certainty factors $C^y = \{c_1^y, c_2^y, \ldots, c_p^y\}$ into a set $W^Y = \{W(Y_1), W(Y_2), \ldots, W(Y_m)\}$ of values of the controlled variables $Y = \{Y_1, Y_2, \ldots, Y_m\}$. Each module $M_i$ has a set of its input facts - $I_i = \{I_{i1}, I_{i2}, \ldots; I_{i\,l_i}\}$ - and a set of output facts - $O_i = \{O_{i1}, O_{i2}, \ldots; O_{i\,p_i}\}$. A module receives and produces the certainty values for corresponding facts.

The modular structure is built such that it respects the following conditions:

$$(F^x \cap (\cup_i I_i) \neq \emptyset) \wedge (F^y \cap (\cup_i O_i) \neq \emptyset),$$

$$(\forall M_i)(I_i \neq \emptyset) \wedge (O_i \neq \emptyset) \wedge$$
$$(I_i \subseteq F_x \cup (\cup_{j \neq i} O_j)) \wedge (O_i \subseteq F_y \cup (\cup_{j \neq i} I_j)).$$

Such a modular structure corresponds to a directed graph $\Gamma = (M, G)$, where $G = M \times M \to \{0, 1\}$ and $G(M_i, M_j) = 1 \leftrightarrow O_i \cap I_j \neq \emptyset$.

Each module (as, for example, $M_l$ in fig.2) can have its duplicates $(M_l', M_l'', \ldots; \quad I_l = I_l' = I_l'' = \ldots)$ which perform, on the same inputs, different kinds of processing (fig.3, cases (a) and (c)) or differ only "parametrically" within similar processing laws (fig.3 cases (a) and (b)).

(Examples in fig.3 illustrate the representation of input-output relations by fuzzy rules; separate intervals in each axis correspond to facts (fuzzy sets) representing the values of associated variables in a five-value logic: "very low", "low", "medium", "high", "very high".)

Duplicates can either possess their own chips, or have own parts in a general network on the common chip, or be represented virtually by a different

parametric tuning within the same part.

Each (inference) module has control connections to a meta-control module ("meta-controller") $M^{mc}$.

### III. FUZZY CONTROL OF A MODULAR STRUCTURE

We denote by $U^B$ and $U$ ($U^B \subseteq U$) the set of basic (fixed) strategies and the set of all executable strategies of a controller, respectively.

Each strategy $S^i \in U^B$ can be characterized by a set of active modules used by it, and by the degree of each module's contribution. A natural description can be found in terms of fuzzy sets: $S_i$ can be assumed to be a fuzzy subset of $M$ with a membership function:

$$\mu_{S_i} : M \to [0, 1] \qquad (1)$$

We denote $M^{S_i} = \{M_j \mid \mu_{S_i}(M_j) \neq 0\}$ - a set of "active" modules (under $S_i$); $\Gamma^{S_i} = (M^{S_i}, G^{S_i}) \subseteq \Gamma$ (where $G^{S_i}$ is a restriction of $G$ to the set $M^{S_i}$) - a corresponding subgraph.

Each strategy $S_i$ must be constructed whilst obeying the following requirements:

(C1)-"unambiguous causality":
$\quad \Gamma^{S_i}$ is acyclic,

(C2)-"connections to external world":
$$(F^x \cap (\cup I_j \mid_{M_j \in M^{s_i}}) \neq \emptyset) \wedge$$
$$(F^y \cap (\cup O_j \mid_{M_j \in M^{s_i}}) \neq \emptyset),$$

(C3)-"absence of hanging modules":
$$(\forall M_j \in M^{S_i})(I_j \cap (F^x \cup (\cup O_r \mid_{M_r \in M^{s_i}})) \neq \emptyset) \wedge$$
$$(O_j \cap (F^y \cup (\cup I_r \mid_{M_r \in M^{s_i}})) \neq \emptyset).$$

From (C1)-(C3) obviously follows the next property

(C4)-"existence of pure source modules":
$$(\forall S_i \in U^B)(\exists M_j \in M^{S_i}) :$$

1565

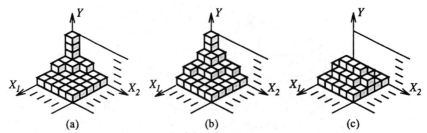

Figure 3: Simplified examples of different input-output characteristics of duplicate modules

$$(I_j \cap (\cup O_r \mid_{M_r \in M^{s_i}}) = \emptyset) \wedge (I_j \cap F^x \neq \emptyset).$$

(Note that (C4) allows for
$$I_j \cap (\cup O_z \mid_{M_z \in M \backslash M^{s_i}}) \neq \emptyset.)$$

Each strategy $S_i$ is activated, with a certainty factor $c(S_i)$, as the result of firing (in the meta-controller) a rule of type

**IF** $<antecedent>$ **THEN** $< S_i >$.

$c(S_i)$ is obtained as *min* of certainty factors of the set of preconditions of a rule. Simultaneous firings of several rules in $M^{mc}$ result in a composite strategy $S_e^\Sigma$ which actually represents a fuzzy subset of $U^B$ with a membership function

$$\mu_{S_e^\Sigma} : U^B \to [0, 1] \qquad (2)$$

where $\mu_{S_e^\Sigma}(S_i) = c(S_i)$.

The resulting membership function of modules $\tilde{\mu}_{S_e^\Sigma} : M \to [0, 1]$ can easily be obtained by combining (??) and (??):

$$\tilde{\mu}_{S_e^\Sigma}(M_j) = max_i\{\mu_{S_i}(M_j) \cdot \mu_{S_e^\Sigma}(S_i)\} \qquad (3)$$

A mixed (composite) strategy obtained with the help of (??), clearly, preserves properties (C2)-(C4), provided that the constituting basic strategies do so as well. Requirement (C1) can be met either be using an entirely acyclic general modular structure, or by introducing special priority assumptions.

In addition to monitoring current sensor data values, the meta-controller $M^{mc}$, while making decisions, takes into consideration long-term tendencies in data changes, external instructions (e.g. global aims), *etc.* Based on the decisions made, $M^{mc}$ "switches" corresponding modules (i.e. those belonging to $M^{S_e^\Sigma}$) and sends the values $\tilde{\mu}_{S_e^\Sigma}(M_j)$ via control connections. There are two ways of using the values of $\tilde{\mu}_{S_e^\Sigma}(M_j)$:

1) $\tilde{\mu}_{S_e^\Sigma}(M_j)$ impacts the resulting certainty value of each fact $F_k \in O_j$, so that the modules $\{M_l \mid F_k \in I_l \cap O_j\}$ receive a value $\varphi(\mu_{S_e^\Sigma}(M_j), c(F_k))$, where $\varphi$ is a composition function; in the simplest case, $\varphi(a, b) = a \cdot b$;

2) $\tilde{\mu}_{S_e^\Sigma}(M_j)$ impacts the resulting certainty factors of facts $F_k \in F^y \cap (\cup O_j \mid_{M_j \in M^{S_e^\Sigma}})$, while all internal informational exchange is fulfilled through actual values of modules' input/ouput.

In the latter case, it is reasonable to assign $\tilde{\mu}_{S_e^\Sigma}$ only to modules $M_l \in M^{S_e^\Sigma} \cap \hat{M}$ (where $\hat{M} = \{M_e \mid O_e \cap F^y \neq \emptyset\}$) while assuming
$$\forall M_x \in M_{S_e^\Sigma} \backslash \hat{M} : \quad \tilde{\mu}_{S_e^\Sigma}(M_x) = 1.$$

For a current strategy $S_e^\Sigma$, a particular attention must be paid to treating the facts $F_l \in ((\cup I_j \mid_{M_j \in M^{S_e^\Sigma}}) \cup F_y) \backslash ((\cup O_k \mid_{M_k \in M^{S_e^\Sigma}}) \cup F_x)$: the values $c(F_l)$ must be considered as unknown (rather than $c(F_l) = 0$), or standard optional values must be associated with them.

An advanced meta-controller $M^{mc}$ can perform its functions on a lower level: by manipulating separate modules, rather than (or better, in addition to) choosing among fixed groups of them. We extend the definition of a composite strategy to this case. Note that the manipulation by separate modules must be accompanied by special checking and correction procedures to prevent a violation of (C1)-(C4) (see, for example, comments to the algorithm *Alg.A* given below). A special case is represented by choosing a particular module from some duplicates. Of cource, it can be explicitly determined already by $S_i$. But it is in this case, that additional rules for module choice could fruitfully coexist with main strategy-level ones: in the scope of a general strategy from $U$ it would be possible to assign proper duplicate modules. The existence of duplicate modules provides a possibility to adjust a particular function which corresponds to a module, without going down to a rule level.

The change in a strategy must be synchronised with the operation of the modules. The average frequency of change should clearly be sufficiently lower than that of module occurences. The problems of allowable values here as well as problems related to dealing with the overlap resulted from different strategies could be the subject of a separate study.

A functioning of (an active part of) a modular

structure allows for parallel data processing. In the general case, an independent activation of modules, by changes of input data, is possible. But one of the most effective techniques can be a wave-like activization, when all nodes of a certain group process in parallel the outputs of their preceding modules, while the latter can be already occupied by the next portion of data and so on. The number $n_j$ of a group to which a module $M_j$ belongs is inductively defined by

$$(\forall M_j \in M_{source}^{S_i^{\Sigma}})(n_j = 1),$$

$$(\forall M_j \in M^{S_i^{\Sigma}})(\forall k > 1)((n_j = k) \Leftrightarrow$$
$$(\exists M_l \in M^{S_i^{\Sigma}})(n_l = k - 1) \wedge$$
$$(I_j \cap O_l \neq \emptyset)(\forall M_t \in M^{S_i^{\Sigma}}) \wedge$$
$$((n_t > k) \Rightarrow (I_j \cap O_t = \emptyset))),$$

where $M_{source}^{S_i^{\Sigma}} = \{M_j \in M^{S_i^{\Sigma}} \mid$
$(I_j \cap (\cup O_r \mid_{M_r \in M^{S_i^{\Sigma}}}) = \emptyset) \wedge (I_j \cap F^x \neq \emptyset)\}$-
the set of source modules of a strategy $S_i^{\Sigma} \in U$ (see (C4) ).

An algorithm for forming groups is given below. It is initiated every time a (set of) module is added or removed.

$Alg.A$:  $Group_1 := M^{S_j^{\Sigma}}$;
$\qquad j := 1$;
$\qquad$ **DO**
$\qquad\qquad Group_{j+1} := \{M_k \in Group_j \mid$
$\qquad\qquad\qquad \exists M_l \in G_j : O_l \cap I_k \neq \emptyset\}$;
$\qquad\qquad$ **IF**  $\mid Group_{j+1} \mid = \mid Group_j \mid$
$\qquad\qquad$ **THEN** { announce:
$\qquad\qquad\qquad$ "cycle(s)!"; **STOP** };
$\qquad\qquad Group_j := Group_j \backslash Group_{j+1}$;
$\qquad\qquad j := j + 1$;
$\qquad$ **UNTIL** $Group_j = \emptyset$.

After normal termination of the algorithm, the sets $Group_1 - ... - Group_{j-1}$ contain corresponding groups of modules. If $Alg.A$ terminates due to the detection of cycle(s), all modules which belong to the cycle(s) are contained in $G_j$ (as in $G_{j-1}$). In this case, for a module $M_x \in Group_j$ :
$\qquad M_x = argmin_{M_k \in Group_j}\{\mu_{S_i^{\Sigma}}(M_k)\}$
the standard optional values are assigned to all inputs from the set $I_x \cap (\cup O_t \mid_{M_i \in Group_j})$ and the corresponding connections are assumed to be removed. Then, $Alg.A$ is used again, beginning with the current $G_j$ and $j$, rather than with $M^{S_i^{\Sigma}}$ and 1, respectively. These steps are repeated until all modules from $M^{S_i^{\Sigma}}$ are separated into groups.

The approach outlined here can give more promising results in automatic creation of control strategies from small "general purpose" building blocks, each having specific input-output characteristics but not being strictly associated with the data content (that is with the application). The meta-module $M^{mc}$, having to its disposal some basic modules which differ in the type of input-output functions (i.e. in the kind of response surface) and allow parametrical adjustment, constructs a strategy $S_i^{\Sigma} \in U$. $M^{mc}$ properly connects input - output of "building blocks", based on knowledge concerning the type of processing needed.

## IV. Conclusions

A dynamic structural adaptation of modular fuzzy controllers suggests a proper field for fuzzy aproach application. After being equipped with an internal fuzzy meta-control component, modular fuzzy controllers can adjust their behavior (via changing configuration) to reach the most suitable overall functioning. The paper reveals structural features of modular controllers, discusses a fuzzy representation for forming a current control strategy, and touches on some implementation aspects.

## References

[1] B.Kosko, *Neural Networks and Fuzzy Systems*, Prentice Hall, Englewood Cliffs, N.J., 1992.

[2] M.-H Lee, "Fuzzy Logic in Control Systems", *IEEE Trans. on Systems, Man, and Cybernetics*, SMC, vol.20, No 2, 1990, pp.404-435.

[3] M.H.Lim and Y.Takefuji, "Implementing Fuzzy Rule-Based Systems on Silicon Chips', *IEEE Expert*, February 1990, pp.31-45.

Alternative Fuzzy Controller Logics

Brian Schott & Thomas Whalen / Georgia State University
Atlanta, GA / USA 30303 / qmdbms@gsusgi2.gsu.edu

## Abstract

We compare six fuzzy logics used to control a simulated inverted pendulum "plant." Among the six is the standard "Mamdani" fuzzy logic. All systems contain eleven rules and are optimized for fuel economy. Mamdani's logic fares very well in the comparisons, but serious challengers are identified.

## Fuzzy Controller

The purpose of this research is to study common alternative systems of fuzzy logic based on S- and R-implications, and especially to contrast them with the Mamdani logic standard. De Baets and Kerre performed an analytical study of generalized *modus ponens*.[1] In the introductory remarks of that study they say, "We cannot imagine why Mamdani's inference rule has been able to hold out for so long in certain application areas, while it is even not an extension of the [*modus ponens*] inference rule." The current study was initiated because of a similar belief. We study the robustness of S- and R-implication logics to varying physical attributes of the controlled system.

We consider traditional fuzzy controllers in which the knowledge is encoded as rules comprised of combinations of subrules. In operation the fuzzy controller is supplied the actual data values for the antecedent variables. As is usual in practice, these actual values are assumed to be crisp numerical singletons, in this research. Also the operational controller defuzzifies the rule's detached consequent value into a crisp numerical singleton which is employed to control the "plant," the system which is being controlled.

## Controller Design and Tuning

All fuzzy (linguistic) variables were represented as triangular fuzzy sets. The base values of the universes of discourse for all antecedent and consequent terms were discretized into 17 values, nominally labeled -8, -7, ... , 0 , ... ,8. The mapping from these nominal values to angles, angular velocity, and forces was optimized separately for each logic by an ordinal search technique. This process is referred to as "tuning." The scale of the all triangles on each universe of discourse were uniform relative to one another; but, the scales on different universes were independent.

In this study the "complex method" is employed for tuning the controllers[2]. We calibrated the scale of the axes of the three universes: $\theta$, $\dot{\theta}$, and $f$. The

criterion variable was the amount of fuel consumed during a fixed length of time (difference equations are used to model the cart-pole system for 10 seconds by 0.02 second increments). The less fuel consumed by the cart-pole system, the better the controller.

We assume the simulated inverted pendulum is carried atop a cart travelling on a 2 meter long track. (Actually, we trained the system on a 1 meter long track, but tested it on a 2 meter long track.) The cart-pole system is always initialized midway along the track. The standard values for the parameters used in the simulations are those given by Barto *et al*, except that their maximum pole angle is narrower, 12° *vs* our 16.6°, and their track is considerably wider, ±2.4*m*.[3] The controller does not steer the cart back to the starting position.

The fuzzy controller was constructed in a traditional manner[4]. The fuzzy subrules were aggregated into a single rule. To speed the simulated trials the possible push values were all precomputed, using the "center of area" method of defuzzification.

## Design Desiderata

The fuzzy controller was constructed with eleven sub-rules containing $\theta$ and $\dot{\theta}$ as antecedent variables and with $f$ as the consequent variable. The term set for the fuzzy values of these variables include NEGATIVE BIG, NEGATIVE, ZERO, POSITIVE, AND POSITIVE BIG. A typical subrule is, "If the error angle is NEGATIVE BIG and the angular velocity is ZERO, then the force of the push should be NEGATIVE BIG." The eleven sub-rules are summarized in Figure 1.

|  |  | Velocity | | | | |
|---|---|---|---|---|---|---|
|  |  | NB | N | Z | P | PB |
| A | NB |  |  | NB |  |  |
| n | N |  |  | N | N |  |
| g | Z | NB | N | Z | P | PB |
| l | P |  |  | P | P |  |
| e | PB |  |  | PB |  |  |

Figure 1. Eleven fuzzy subrules.

The design of the controllers in this study is

heavily influenced by the goal of comparing alternative fuzzy logics and by the authors' previous research regarding potential inference errors of fuzzy logics.[5] Our discussion of design issues are organized as follows.

a) Select key logics and a traditional "plant" to control.

b) Maintain each logic's internal consistency.

c) Construct a smooth control surface.

d) Avoid discretization error.

e) Optimize independently each logic's design and tuning, while maintaining comparability among logics.

**a)** The six logics listed in Table 1 were chosen to control an inverted pendulum. The Mamdani logic is used in most existing fuzzy controllers and is treated as the standard against which to compare others. The other five logics are the main exemplars of the t-norm families commonly referred to as S-implications and R-implications in the literature.

**b)** For each of the six logics every effort is made to construct the linguistic fuzzy terms and the logical connectives in an internally consistent manner. For example, depending on the logic's detachment operation, subrules are aggregated into a single rules based on coherent use of connectives such as AND, OR, and ELSE. Table 1 summarizes these factors for each logic in the columns labeled "Implication formula," "Subrule aggregation," and "Antecedent combination." The Mamdani logic system is unique in using an OR style combination "max" for combining subrules into a single rule. All others use an AND style combination "min". This is shown in the column "Subrule aggregation," in Table 1.

**c)** Every effort was made to construct and organize the subrules and linguistic terms so that a smooth control surface results. Smoothness of the control surface lead to carefully-shaped membership functions with adequate overlaps among the supports of subrules' fuzzy terms. The desire to achieve smoothness ultimately resulted in the fuzzy linguistic terms for each logic being constructed and tested in four different ways. The issues which lead to these four designs are discussed here. Because of space limitations only the results of

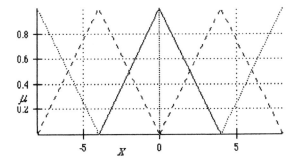

the "best" construction for each logic are reported later in this study. "Best" means most robust performance under the varying conditions in this study

The antecedent fuzzy terms in all cases are constructed from symmetrical, fixed width triangles. The exact width either spanned 8 or 16 base values, depending on the logic. The final choice of the width of the antecedent terms for each logic was based on robustness performance considerations.

Controlling the degree of overlap among the fuzzy set terms in each antecedent clause is especially critical

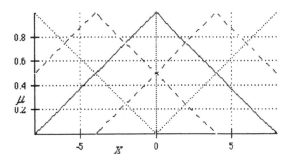

for the nonMamdani logics. For example, systems based on possibilistic logics (such as Mamdani's popular system) can function well with triangular shaped fuzzy terms with slight gaps between the cores of adjacent terms (in subrules).[6] But a system based on Lukasiewicz' logic requires fuzzy terms with broad, overlapping supports.[7,8] In this study this factor is magnified (relative to single variable antecedent subrules) because the contrast the t-norm operator $\max(0, a + a' - 1)$ yields a very narrow support set. The t-norm operator $a \cdot a'$ produces a support set which is intermediate to these extremes. To compensate for the nonliberal t-norm operator, the antecedent terms of the Lukasiewicz logic system were made especially broad; they span 16 discretized base values. Similarly, we anticipated that the Kleene-Diens logic system would benefit from broader antecedent terms to compensate for the especially restrictive OR operator which underlies its implication operation (A IMPLIES C is modeled NOT A OR C).

The Gödel (Brouwer) and the Goguen (quotient) logics encounter another consideration. For these two logics if the antecedent terms of two subrules overlap, then if the consequent terms of two corresponding subrules are disjoint, the resulting implication produces an aggregated rule which fires as "undecided" (all memberships equal zero) for certain singular inputs; the subrules are said to "eradicate" one another. This phenomenon is not common for systems with single antecedent subrules, because the overlaps are contained in a single dimension. But with multiple antecedent subrules such as those studied here, eradication is a delicate matter. To mitigate the possibility of this eventuality, the supports of the antecedent terms for these two logics were made narrow and spanned only 8 base values.

As the previous two paragraphs suggest, the widths of the **antecedent** fuzzy terms and the consequent fuzzy terms for 4 of the logic systems were pretty well predetermined by each logic's special nature. No such features clearly presented themselves to the authors for the remaining two logics, Reichenbach (probability) and Mamdani. The choice of the widths of the antecedent fuzzy terms for Reichenbach and Mamdani are performance based: in both cases the wider antecedent terms were more robust under the varying conditions, and so the wider terms were used. (The Mamdani logic tuned well–used less fuel–with the more narrowly defined antecedent terms, but then suffered somewhat in the tests of robustness.)

In fact, for the sake of equity, all six logics were tested under two antecedent width configurations.

i) The antecedent fuzzy terms are 8 base values wide and the consequent fuzzy terms are 16 base values wide.

ii) The antecedent fuzzy terms are 16 base values wide and the consequent fuzzy terms are 16 base values wide.

These tests confirmed most of our predispositions regarding the 4 logics described in the previous paragraphs. An exception was the performance of the Kleene-Diens logic system. As anticipated, wider antecedent terms outperformed narrower terms when the consequent terms were symmetric, but when the consequent terms were asymmetric (as described below for "cover triangles"), the expected improvement was not achieved.

The **consequent** fuzzy terms were also constructed in one of two ways. That is, for each of the two antecedent width configurations mentioned in the previous paragraph, the consequent fuzzy terms were constructed in two ways.

j) "symmetric triangles": symmetrical, fixed width triangles which span 16 base values.

jj) "cover triangles": triangles for which the support spans all the base values of the universe. These triangles are not generally symmetrical around their center.

We were unable to anticipate the best configuration with respect to the shape of consequent fuzzy terms. The choice of the shape of the consequent terms for each logic was based solely on performance considerations.

The final shapes and widths of the membership functions for each logic are given in Table 1 in the column "Antecedent widths and consequent shapes".

**d)** Discretization error has been shown to negatively influence the performance of fuzzy logic systems.[9] To grossly summarize, we attempted to minimize potential discretization error for all logics, and were able to completely eliminate this error for all but Goguen and Reichenbach logics.

**e)** The controllers were given an unlimited quantity of "fuel" with which to propel the cart. Then each simulation was run for 10 seconds or until the system failed by either the pole falling down or by the cart running off the track. The complex method was adapted to accommodate two level (lexicographic) tuning.[10] Any control system which failed, was inferior to any non-failed system. If the pole fell during the trial, or if the cart went off either end of the finite track, time stopped and the system failure was recorded.

The "complex" method for all controllers used the same six initial settings for each of the three controllable scale values. The system was initialized with the cart at the center of the track, with zero angular velocity, and with an angle of 0.05 radians. The complex method was halted after the search vector coefficient was less than $10^{-6}$.

Tuning

The results of the complex optimization are summarized in Table 2. The figures in Table 2 are limited to the one "best" of four previously mentioned separate fuzzy logic systems which were employed for each logic.

The optimal scale for push amounts (the force corresponding to +1 on the -8 to +8 base value scale) varies widely from 0.336 to 10.47$n$. The push scales are roughly inversely related to the corresponding "fuel"

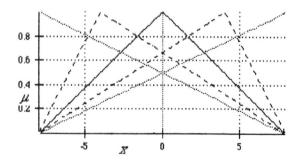

consumption discussed above. Perhaps systems with coarse push granularity produce relatively light "fuel" consumption, and systems with fine push granularity produce relatively heavy "fuel" consumption.

The optimal scale for pole angle varies from 0.00652 to 0.0189 *rad*, and the optimal scale for angular velocity varies from 0.0672 to 0.2815 *rad/s*. While these two scales do not show a clear relationship with one another, if the Lukasiewicz logic is excluded from consideration, there is a direct relationship between the angle scale values and the angular velocity scale values. When the angular velocity scale is finely scaled, the angle is finely scaled, and so on.

In Table 2 the "Maximum possible push ($n$)" is reported for each logic. These figures reflect the greatest force that each optimal controller could possibly apply in order to balance the pole. Except for the Mamdani logic, the ratio of each logic's "Maximum possible push ($n$)" to its "Push scale ($n$)" are very similar, between 2.6 and 3. The corresponding Mamdani logic ratio is 1.08. We wonder if this is an important distinction, and why.

The six control surfaces are displayed in Figure 2. For pole angles near zero angle and for angle speeds near zero, all six logics have control surfaces which are monotonically nondecreasing in push amount with respect to both pole angle and speed. In this central region, the gradient is greatest for the logics based on t-norm families $R_3$ and $S_3$, less for the logics based on t-norm families $R_2$ and $S_2$, and least for the logic based on t-norm families $R_1$ and $S_1$. The control surface for the Lukasiewicz logic displays much, much less gradation than the others. In a parallel manner, as the index of the t-norm implication family grows from small to large, the width of the antecedent terms decrease and the shape of the consequent terms change from "symmetric" to "cover."

## Robustness

Figure 3 displays several "disturbance" charts. Six "disturbance" charts for fuel used and six "disturbance" charts for angle overshoot (not shown) summarize the results of the robustness analysis; we "disturbed" each physical parameter relative to its original setting which had been employed during fine tuning.

For this system, changing the weight of the pole seemed to disturb the logics least of all. Only the Lukasiewicz logic had any difficulty with lighter or heavier poles. The Kleene-Diens, Mamdani, and Brouwer logics seemed to perform the best when pole weight was varied.

When the length of the pole is varied, the best logics are Brouwer, Mamdani, and Kleene-Diens. These three logics are very similar when the length is less different from the tuned length, but the Kleene-Diens logic degrades somewhat in the extremes.

Changing the weight of the cart seemed to disturb the logics most of all; the lines are shortest. In this situation the best logics are Kleene-Diens, Brouwer, and Mamdani.

The three strongest competitors seem to be Kleene-Diens, Mamdani, and Brouwer logics and it is difficult to choose between the first two. The Kleene-Diens logic seems to have slightly more ability to avoid failure and uses slightly less fuel when the disturbances are slight. The Mamdani logic seems stronger in the intermediate disturbance range. The three contenders are quite comparable with regard to angle overshoot, except that the Kleene-Diens logic degrades slightly more than the others when the disturbance gets extreme.

## Summary

Six different fuzzy logic systems were designed, objectively tuned, and evaluated in a simulated control environment, an cart and inverted pole. Each fuzzy logic controller was designed and tuned in four alternative configurations which varied the membership functions while maintaining the same eleven (sub)rules containing two antecedent variables and a single output variable. The controller was not guided.

In the design phase several issues with varying degrees of subtlety were confronted. These issues included: subrule aggregation operators; combining multiple antecedent variables into a single antecedent; consistent antecedent and consequent fuzzy memberships; and potential discretization error.

In the tuning phase we used the "complex" method of nonlinear optimization to scale the universes of discourse for pole error angle, pole angular velocity, and cart push force to minimize total fuel usage while disallowing system failure. Almost all logics were tunable in all four designs, and a single "best" design emerged for each logic. The selection of the "best" logic was based on both the optimal fuel usage and the behavior of the logic under robustness trials. We report and analyze various facts regarding the tuning process for the "best" design for each logic.

In the robustness phase we "disturbed" each of three parameters of the controlled system. We employed 20 values from one-tenth of the assumed parameter's value to eleven times the parameter's value. We compare and contrast the performance of the six logics.

## Table 1. Six Logic Systems

| Logic | T-norm Implication symbol | Implication formula | Subrule aggregation | Antecedent combination | Antecedent widths and consequent shapes |
|---|---|---|---|---|---|
| Lukasiewicz | $S_1, R_1$ | $\min(1, 1-a+c)$ | $\min(sr', sr)$ | $\max(0, a+a'-1)$ | 16, sym |
| Reichenbach (product) | $S_2$ | $1-a+ac$ | $\min(sr', sr)$ | $a \cdot a'$ | 16, sym |
| Goguen (quotient) | $R_2$ | 1, if $a \leq c$; $c/a$, otherwise | $\min(sr', sr)$ | $a \cdot a'$ | 8, sym |
| Kleene-Diens (Zadeh,Q) | $S_3$ | $\max(1-a, c)$ | $\min(sr', sr)$ | $\min(a, a')$ | 8, cover |
| Gödel (Brouwer) | $R_3$ | 1, if $a \leq c$; $c$, otherwise | $\min(sr', sr)$ | $\min(a, a')$ | 8, cover |
| Mamdani | na | $\min(a, c)$ | $\max(sr', sr)$ | $\min(a, a')$ | 16, cover |

## Table 2. Optimal Tuning Results

| Logic | Complex trials | "Fuel" consumed | Push scale (n) | Angle scale (rad) | Angular velocity (rad/s) | Max possible push (n) |
|---|---|---|---|---|---|---|
| Lukasiewicz | 129 | 249 | 0.336 | 0.00652 | 0.2815 | 1.008 |
| Reichenbach | 85 | 97 | 10.47 | 0.0189 | 0.1635 | 28.009 |
| Goguen | 69 | 233 | 2.6 | 0.01828 | 0.1872 | 7.8 |
| Kleene-Diens | 100 | 76 | 9.327 | 0.00793 | 0.0672 | 26.0 |
| Gödel | 54 | 74 | 9.347 | 0.01146 | 0.1087 | 29.3 |
| Mamdani | 72 | 109 | 9.578 | 0.00959 | 0.105 | 10.4 |

Figure 2. Control Surface Contours

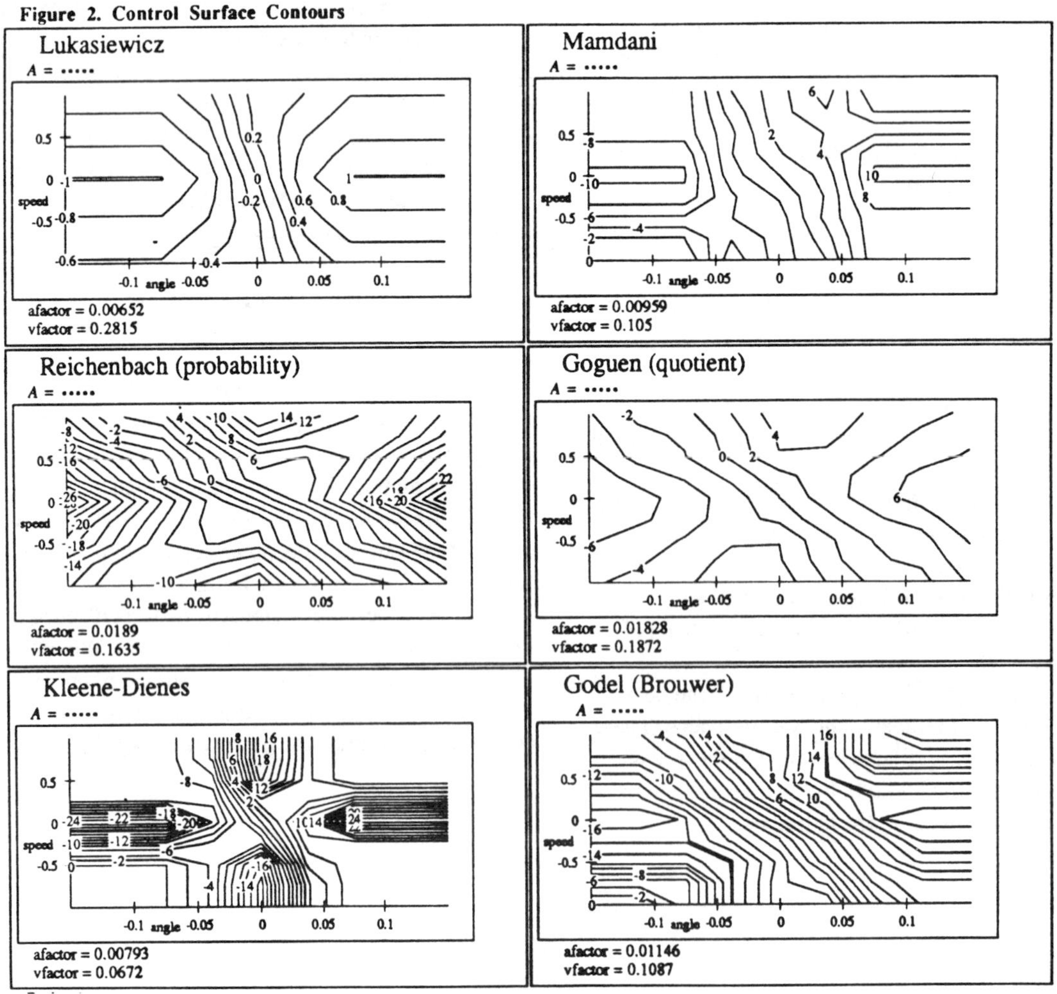

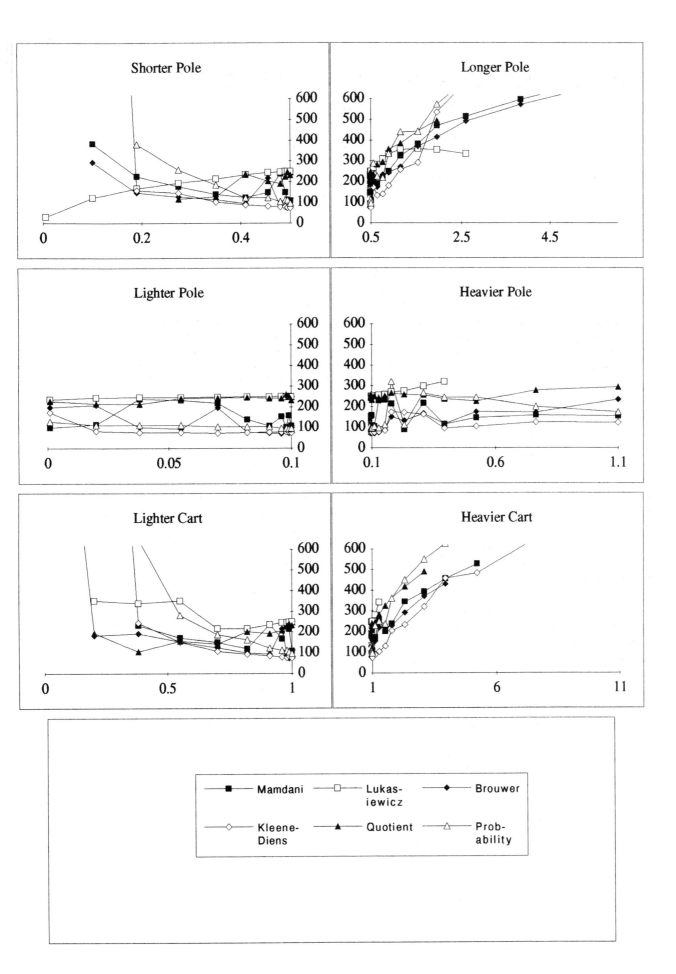

# MODEL BASED TUNING AND ADAPTION OF FUZZY LOGIC CONTROLLERS

Z. Papp, B.J.F. Driessen
TNO Institute of Applied Physics
P. O. Box 155, 2600AD Delft, The Netherlands
E-mail: papp@tpd.tno.nl

**Abstract**

This paper presents a fuzzy logic based control structure enhanced with supervised learning and/or adaption functionalities. Availability of at least a partial process model is assumed. Nonlinear process identification procedure is used to complete the partial model. Based on this process identification model using the techniques of systems sensitivity theory the necessary gradients are generated to guide the training process and thus to keep the training time (the number of observations) at minimum. The process identification and the controller tuning can run parallel, in this way the on-line adaption of the controller can be realized in a straightforward way. A supervisory robot control problem is shown to demonstrate the capabilities of the scheme proposed.

## 1 Introduction

Fuzzy logic based controllers (FCs) have been proved to be feasible tools to implement (nonlinear) controllers for complex, many times ill-defined processes. Beside the advantageous implementational features of FCs, the key of the success is that FCs provide a way to formalize operators' heuristics thus "expertise" can directly be involved in generating control commands. Unfortunately this feature is also the most serious limiting factor of the applications: to achieve reasonable (robust) control performance (i.e. to tune the FC parameters) in multidimensional systems is unmanageable many times.

According to this application driven need, a number of learning schemes have been developed. A considerable part of them follows a "model-free" approach based on the assumption that not even an approximate process model can be derived [1, 3, 9]. They use unsupervised and reinforcement learning schemes – which are observation demanding procedures – to improve control performance. The architecture in [1] can work with "weak" (e.g. binary) reinforcement signal and uses a neural network based evaluation network to "store" knowledge acquired about the process dynamics to speed up learning. The control scheme in [9] learns I/O relationships directly and it is capable to synthesize control rules using I/O space clustering and conflict resolution algorithms. Other approaches use process knowledge (formalized in some kind of model) to design and or tune FCs. In [2] a dynamic programming method based on the cell-space formalism [10] is described as a direct way to design FCs. There FC is considered as an approximator of the optimal control table. As a consequence of the "off-line" nature of the design procedure the control performance depends on the accuracy of the process model available and cannot be refined (adapted) during the operation. In [7] a direct control architecture is shown where an approximate process model is used for on-line training extended with a "supervisory" level to maintain global stability using Lyapunov synthesis. [8] provides a wide selection of possibilities for describing the process knowledge in a unified framework and the the adaptive network based FC can be trained using temporal back propagation algorithm to follow a predefined trajectory. The size of the trajectory adaptive network, the target of back propagation, linearly depends on the trajectory length and this can mean severe constraint in many control applications.

To overcome some of these difficulties the controller architecture proposed follows a model based approach in a different way: to optimize the nonlinear mapping of the FC – initially defined by the rule set supplied – the tuning procedure uses the approximate sensitivity model of the process to "guide" the optimization and thus minimize the number of training steps necessary to reach the control performance required.

The paper is organized as follows. First a "smooth" FC implementation is introduced, then the control structure and the learning/adaption algorithm are described. The next section shows a possible way to calculate the partial derivatives required by the learning algorithm. Then a robot supervisory control example follows to demonstrate some of the features of the control scheme.

## 2 A smooth fuzzy controller

As will be shown in the next session the FC serves as a parameterization of a nonlinear mapping in the control scheme proposed. The training process is basically a gradient descent optimization procedure to find optimal parameter values for the FC to satisfy the control goal. The gradient descent optimization procedure needs continuous partial derivatives, namely continuous $\partial C / \partial u$ and $\partial C / \partial \alpha$, where $C = C(u \mid \alpha)$ denotes the mapping realized by the FC ($u$ is the input vector, $\alpha$ is the parameter vector of the controller).

It is straightforward to proof that choosing singleton fuzzifier, product-operation fuzzy implication, algebraic product as t-norm, Gaussian-like membership functions and a sufficiently smooth defuzzifier (e.g. Center-Of-Area, Center-Average) results in continuous partial derivatives.

The implementation of the fuzzy mapping satisfying the previous conditions can be divided into two consecutive steps[1]:

- Calculating the rule firing strength: $\varphi_i = \exp \left( (\vec{x} - \vec{c_i})^T G (\vec{x} - \vec{c_i}) \right)$,

- Calculating the mapping output value $z$ using the firing strengths as input: e.g. using the Center-Average type of defuzzification $z = \vec{\varphi}^T S \vec{m} / \sum_{i=1}^R \varphi_i$,

where $\vec{\varphi} = (\varphi_1, \ldots, \varphi_N)$ is the firing strength of $i$th rule, $\vec{x} = (x_1, \ldots, x_N)$ is the normalized input vector, $\vec{c_i}$ is the vector containing the centers of the membership functions of the fuzzy sets involved in the rule left hand side, $G$ is a diagonal matrix defining the memberships' "spread", $\vec{m} = (m_1, \ldots, m_M)$ contains the centers of the fuzzy membership functions on the output side, and $S$ is an R by M matrix describing the *antecedent* → *consequent* structure of the rule set. $S_{ij} = 1$, if rule $i$ is a fuzzy implication having the fuzzy set with $m_j$ center on the consequent side; $S_{ij} = 0$ otherwise.

Though the mapping described above belongs to a wider class of fuzzy mappings, which – under certain conditions – are universal approximators of continuous nonlinearities [9], to keep the learning/adaption process feasible we fix the rule structure $S$ and use only a subset of $\alpha = \{G, \cup \vec{c_i}, \vec{m}\}$ for tuning. Practice shows that while the "input parameters" (i.e. $G$ and $\cup \vec{c_i}$) can be calculated from the experts' knowledge with confidence, choosing the parameters of the output membership functions ($\vec{m}$) is far not straightforward and the manual "optimization" is a trial-and-error process. On the other hand in most of the cases the parameters of the output membership functions provide sufficient freedom for tuning[2] ($\alpha = \{\vec{m}\}$).

## 3 The control structure

Our goal is to find an optimal parameter set for the fuzzy controller (C), which drives the process (P) to follow the desired trajectory ($y_d$) (Fig.1). The control performance is measured by the $J(t \mid \alpha) = 1/T \int_{t-T}^t e^T(\tau \mid \alpha) e^T(\tau \mid \alpha) d\tau$ performance index. According to the gradient descent optimization method the $\alpha$ update at discrete intervals can be calculated as

$$\alpha(t) = \alpha(t - T) - \eta \frac{2}{T} \int_{t-T}^t e(\tau) \frac{\partial e(\tau)}{\partial \alpha} d\tau.$$

---

[1]For simplifying the notation scalar output is assumed.

[2]It should be emphasized that involving the whole $\alpha$ in the tuning theoretically does not mean problem, but it would have application related consequences: in this case backpropagation type learning should be implemented, which would slow down the learning dramatically. In our case only one "layer" has to be trained (similar to the Radial Basis Function type neural networks [11], which exhibit very good learning capabilities).

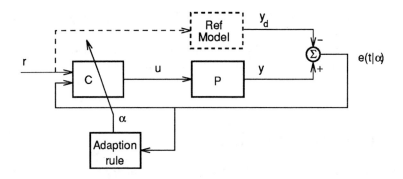

Figure 1: *The control structure*

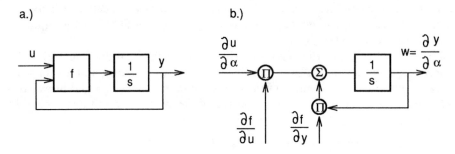

Figure 2: *A nonlinear process and its sensitivity model*

(It should be emphasized that for ensuring stability $\eta$ and $T$ have to be chosen carefully to keep the system in "quasi time-invariant" operating condition.) If the model of the globally stable process P is known $\partial e(t)/\partial \alpha = \partial y(t)/\partial \alpha$ as function of time can be calculated using the methodology of the systems sensitivity theory [5, 6]. For example assume that the process model is

$$\dot{x}(t) = f(x, u); \qquad y(t) = x(t)$$

and $f(.,.)$ is sufficiently "smooth" (Fig. 2/a). It is straightforward to proof that $w(t) = \partial y(t)/\partial \alpha$ is the solution of the following differential equation:

$$\dot{w} = \frac{\partial f}{\partial x} w + \frac{\partial f}{\partial u} \frac{\partial u}{\partial \alpha}$$

The corresponding sensitivity model is shown in Fig. 2/b. The same procedure should be followed to generate $\partial y/\partial \alpha$ for the closed loop system substituting $\partial u/\partial \alpha = \partial C/\partial \alpha + (\partial C/\partial y)(\partial y/\partial \alpha)$. As the figure shows to derive $\partial y/\partial \alpha$ the calculation involves $\partial C/\partial y$ , $\partial C/\partial \alpha$ , $\partial f/\partial u$ and $\partial f/\partial x$. The first two partial derivatives are easy to calculate from the FC structure introduced in the previous section. If the process model of P would completely be known deriving $\partial f/\partial u$ and $\partial f/\partial x$ would not pose problem. Because it is almost never the case it should be investigated how a process identification procedure for delivering the necessary partial derivatives can be integrated into the control structure.

## 4 Calculating the gradients

The architecture proposed can be considered as an indirect adaptive control scheme: to guide the adaption $\partial e(t)/\partial \alpha$ (as function of time) are calculated by the sensitivity model of the P process (Fig.3.); the sensitivity model is fed by the estimated gradient values provided by the system identification model of P (denoted by M).

The methodology in [4, 5] provides a relevant starting point for nonlinear on-line identification, where different architectures were investigated for identifying nonlinear systems using multilayer

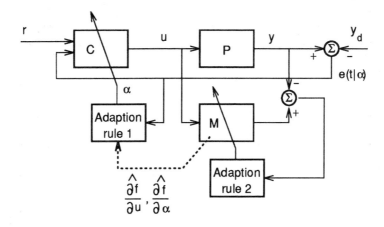

Figure 3: *The indirect adaptive control scheme*

neural networks as parameterization of static nonlinearities making possible to determine partial derivatives w. r. t. the inputs and network parameters. This framework can readily be extended to support process identification in wider sense: the neural network can be replaced any sufficiently smooth structurally known nonlinearity for which the partial derivatives can be calculated. Making use of this generalization there are several ways to involve a priori process knowledge in the controller tuning. The three most important issues are:

- Having knowledge about the the underlying operational phenomena of the process, more specific model with better identification properties can be built up (e.g. the input dimensionality of the nonlinearities can be kept low thus making the use of RBF neural networks feasible).

- Instead of neural networks other smooth nonlinear approximators with structural resemblance with the real nonlinearities can be applied, which, as a consequence, can significantly speed up the process identification.

- Based on the structural properties of the process a serial–parallel identification structure might be found, stability and convergency properties of which are superior to the common parallel structure.

These possibilities will be shortly demonstrated in the next section.

# 5  Robot supervisory control application

Due to the singular coordinate transformations in robot manipulator control systems it can happen that small change in position setpoints results in big setpoint change in motor coordinates. These wide range motions have to take place in a well conditioned way among different drive chains. Unfortunately the "secondary" parameters of the drive chains (e.g. frictions, belt losses) are different (and unknown) thus the traditional independent driver controller structures have difficulties with managing this problem. A possible solution for controlling robot arms around singular points using fuzzy supervisory controller is presented in [12]. Here the extension of that control structure with automatic tuning and adaption is presented.

A simplified model of the control problem is shown in Fig. 4. It is formalized as a model reference control structure where one of joints (J2) gives the desired trajectory. The two joints are identical except the "secondary" parameters. The shaded box in the figure contains a simplified dynamic model of the drive chain consisting of electric motor, transmission line and an arm segment. The speed dependent friction is modelled by the F nonlinearity. The joints are controlled independently by P/PI position/speed control loops (represented by $K_3$ and $C(s)$, respectively). The goal is to find a parameter set ($\alpha = \{\vec{m}\}$) for the S supervisory controller (which produces the $z$ "correction signal") to minimize the average position error ($1/T \int_{t-T}^{t} e_{\Omega}^2(\tau)d\tau$). Based on the sensitivity model

1577

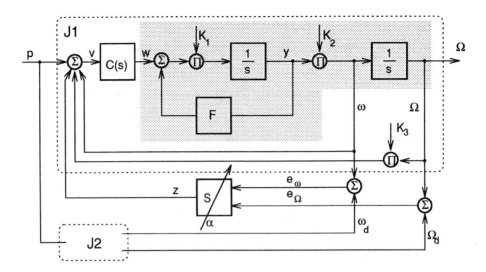

Figure 4: *The robot control structure*

formalism $\partial e_\Omega/\partial\alpha_i = \partial\Omega/\partial\alpha_i$ can be derived to guide learning. The following linear time variant system (the sensitivity model) generates the partial derivatives ($I(s) = 1/s$):

$$\frac{\partial\Omega}{\partial\alpha_i} = I(s)\frac{\partial\omega}{\partial\alpha_i}; \qquad \frac{\partial\omega}{\partial\alpha_i} = K_1 I(s)\left(\frac{\partial v}{\partial\alpha_i}K_2 C(s) + \frac{\partial F}{\partial y}\frac{\partial\omega}{\partial\alpha_i}\right)$$

$$\frac{\partial v}{\partial\alpha_i} = \left(1 + \frac{\partial S}{\partial e_\omega}\right)\frac{\partial\omega}{\partial\alpha_i} + \left(K_3 + \frac{\partial S}{\partial e_\Omega}\right)\frac{\partial\Omega}{\partial\alpha_i} + \frac{\partial S}{\partial\alpha_i}$$

Using the smooth FC showed in section 2, $\partial S/\partial e_\omega$, $\partial S/\partial e_\Omega$ and $\partial S/\partial\alpha_i$ can be calculated at a certain input value and controller parameter set. The $\partial F/\partial y$ is the partial derivative of the friction model block, which is unknown. The structure of the friction model is determined in advance based on the physics of the friction mechanism and a priori process knowledge (e.g. what friction types are relevant, what are negligible). Experiments showed that a friction model consisting of the weighted linear combination of three nonlinear functions provides reasonable accuracy. Using discrete time approximation a serial-parallel identification structure could be designed and applying the LMS method could deliver the optimal weights receiving a few hundred [$w,\omega$] observation pairs and could easily follow the the relatively slow friction changes due to "disturbances" (e.g. lubrication, temperature change, aging, etc.). This parameterized nonlinear function block (the friction identification model) can calculate $\partial F/\partial y$ for the sensitivity model. With this, the model based tuning scheme is completed.

# 6    Conclusions, further activities

In the paper an indirect adaptive control scheme has been presented to train/adapt fuzzy controllers. Model based approach has been followed to minimize learning time.

In this scheme the process identification model delivers the necessary partial derivatives for the sensitivity model to calculate the gradient on the error surface. It should be emphasized that – in contrast to the "traditional" indirect adaptive scheme, where the estimated process parameters are directly used in controller design – here the estimated derivatives just guide the optimization process, thus the inaccuracy of the gradients (i.e. the inaccuracy of the process model) does not deteriorate the control performance, but slows down the learning.

The method proposed can be used where partial (at least structural) process knowledge is available to set up an approximate process identification model (e.g. in mechatronics systems, where the structure, most of the parameters of the dynamics can be derived formally). The advantages of the method are more significant if conducting long experiments are expensive (or impossible) and/or collecting observations is a slow procedure. In case of complete knowledge about the process to be controlled, the control scheme can be considered as an efficient implementation tool for nonlinear controllers.

In the robot supervisory control example the scheme performed well, the position error could be reduced to the level reached by extremely time consuming manual tuning (below two degrees [12]) in a few training cycles provided the friction model parameters were identified in advance.

Further research is carrying out investigating the stability properties of the sensitivity model in simultaneous identification/control problems. The efficient real-time implementation of the combined identification/control scheme is under development.

# References

[1] H. B. Berenji, P. Khedkar, "Learning and Tuning Fuzzy Logic Controllers Through Reinforcements." *IEEE Trans. on Neural Networks*, vol. 3, pp. 724-740, 1992.

[2] S. M. Smith, D. J. Corner, "Automated Calibration of a Fuzzy Logic Controller Using Cell State Space Algorithm." *IEEE Control Systems Magazine*, vol. 2, pp. 18-28, August 1991.

[3] S.-G. Kong, B. Kosko, "Adaptive Fuzzy Systems for Backing up a Truck-and-Trailer." *IEEE Trans. on Neural Networks*, vol. 3, pp. 211-223, 1992.

[4] K. N. Narendra, K. Parthasarathy, "Identification and Control of Dynamical Systems Using Neural Networks." *IEEE Trans. on Neural Networks*, vol. 1, pp. 4-27, 1990.

[5] K. N. Narendra, K. Parthasarathy, "Gradient Methods for the Optimization of Dynamical Systems Containing Neural Networks." *IEEE Trans. on Neural Networks.* vol. 1, pp. 4-27, 1990.

[6] J. B. Cruz, Jr.(ed), *System Sensitivity Analysis.* (Benchmark Papers in Electrical Engineering and Computer Science). Dowden, Hutchinson and Ross, 1973.

[7] L. X. Wang, "Stable Adaptive Fuzzy Control of Nonlinear Systems." *IEEE Trans. on Fuzzy Systems*, vol. 1, pp. 146-155, 1993.

[8] J. R. Jang, "Self-Learning Fuzzy Controllers Based on Temporal Back Propagation." *IEEE Trans. on Neural Networks*, vol. 3, pp. 714-723, 1992.

[9] L. X. Wang, J. M. Mendel, "Generating Fuzzy Rules by Learning from Examples." *IEEE Trans. on Systems, Man, and Cybernetics*, vol. 22, pp. 1414-1427, 1992.

[10] C. S. Hsu, "A Discrete Method of Optimal Control Based upon the Cell State Space Concept." *J. of Optimization Theory and Applications*, vol. 46, pp. 547-569, 1985.

[11] T. Poggio, F. Girosi, "Regularization Algorithms for Learning That Are Equivalent to Multilayer Networks." *Science*, vol. 247, pp. 978-982, 1990.

[12] B. J. F. Driessen, Z. Papp, "Fuzzy Supervisory Controller for MANUS - A Wheelchair Mounted Manipulator." *Proceedings of the First European Congress on Fuzzy and Intelligent Technologies (EUFIT-93)*, vol. 2, pp. 894-899, 1993, Aachen, Germany.

# Robustness with a Digital Fuzzy Logic Controller

J.D. Winter, H.S. Tharp, M.K. Sundareshan

Electrical and Computer Engineering Department

University of Arizona, Tucson, AZ 85721

**Abstract:** Robustness characteristics associated with a digitally implemented fuzzy logic controller are illustrated using a second-order system. Two competing perfomance objectives (tracking and step response) are used to design the controller. A systematic design leads to a controller that is robust with respect to both the presence of a computational time-delay and the existence of parameter variations in the system. The controller is restricted to only measure the output error signal and must accommodate a saturating system input.

## I. Introduction

There has been a tremendous level of interest in the design and implementation of fuzzy logic controllers. To a large extent, this interest is a result of the ease with which fuzzy logic controllers (FLC) can be designed. For example, many successful fuzzy logic controller designs have been developed without an explicit model of the system being controlled. In addition, the understanding and the use of analytical control design techniques are usually unnecessary. This paper will highlight another important feature of a FLC which has not been appreciated in the literature, viz. a successful satisfaction of competing design objectives while being robust to changes in the system parameters.

Designing classical controllers to meet two competing performance objectives is challenging. Usually, the design ends up being an ad hoc combination of two independent designs that address each objective separately. This combination of two different designs can lead to a long trial-and-error design process. In this paper our interest is to investigate the possibility of designing a FLC for two competing objectives without the usual trial-and-error procedure that accompanies classical controller designs, and further to demonstrate the robustness properties of such a design.

The primary focus in our design is on a digital controller due to the simplicity it offers for a practical implementation. It must be emphasized that while there exist a number of attempts in the literature on the design of analog fuzzy controllers [1-3], the design of digital fuzzy controllers (which are amenable to microprocessor-based implementations) has not received significant attention. While the design details and the properties that we discuss hold in general, for the sake of motivating the discussion of the specific design objectives used, we shall consider the design of a tracking servo controller which is used to position the read/write head in a disk drive actuator system and to make it follow a specific track with negligible error.

## II. Problem Description

A standard second-order transfer function is used to represent the system being controlled as

$$G(s) = \frac{y(s)}{u(s)} = \frac{K}{s^2 + 2\zeta\omega_n s + \omega_n^2} . \quad (1)$$

There are some basic restrictions associated with this system. First, only an error signal can be measured directly, where

$$\bar{e}(t) = K_q e(t) = K_q(r(t) - y(t)) \quad (2)$$

with $r(t)$ a reference input, $y(t)$ the system output, and $K_q$ a scalar gain. Second, the dynamic range on the error signal, $\bar{e}$, is restricted to lie within a given set of upper and lower limits, e.g., $\pm L$. Any error signal that exceeds either the upper limit or the lower limit will be assigned a value equal to that limit. Third, the input into the system, $u(t)$, will saturate when the output position, $y(t)$, accelerates by more than $8g$'s, where one $g$ equals $9.8\ m/(sec)^2$.

Under the above constraints, the closed-loop system is supposed to satisfy two competing performance goals. First, achieve an acceptable steady-state error level when a 50 Hz sinusoid is used as the reference input. Second, the step response should have a quick rise time, a short settling time, and a reasonable percent overshoot.

The sinusoidal reference input that will be used is given by

$$r(t) = (75.0 \times 10^{-6}) \cdot \sin(100\pi t) . \quad (3)$$

An error, $e(t) = \bar{e}(t)/K_q$, of less that $9.0 \times 10^{-8}$ meters is considered acceptable when the reference input is the sinusoid in equation (3). The step input that will be used has an amplitude of $1.0 \times 10^{-6}$. For the step input, an overshoot less than 40%, a rise time less than 5 milliseconds, and a settling time less that 15 milliseconds are considered acceptable.

There is no constraint on how fast or how slow the controller can sample the error signal. However, if the time necessary to compute the control law is fixed based on the hardware used to realize the controller, then a slower sampling rate will result in a smaller percentage of the sample period being required for computation. In general, when a smaller percentage of the sample period is needed to compute the input into the system, the less impact the computation delay will have on the closed-loop system performance. For this reason, the sample rate should be selected as low as the performance of the system allows.

The performance specifications stated above describe the requirements for a digital tracking controller [4]. The desired tolerance on the steady-state error, viz. $9.0 \times 10^{-8}$ meters (which corresponds to 6% of the width of a track), reflects the precision tracking requirements usually demanded in this application. The control signal desired to be tailored is a current that drives a voice coil motor to provide the necessary torque to move the read/write head.

The following parameter values are used in the remainder of this paper: $K = 286.7$; $\zeta = 0.2$; $\omega_n = 125.6$ (rad/sec); $L = 0.45$ (volts); and $K_q = 1.5 \times 10^6$ (volts/meter).

### III. Fuzzy Logic Controller Design

As is well known, fuzzy logic controllers are typically required to perform three operations. First, the variables used as inputs to the FLC must be represented as fuzzy numbers. Second, these fuzzy numbers must be combined to generate a fuzzy output signal from the FLC. Finally, this fuzzy output signal must be defuzzified in

order for it to be used as an input into the system.

Initial controllers were attempted to be designed using only the error signal, $\bar{e}(t)$, as the FLC input signal. These controllers were unable to meet the required high precision performance objectives. More information was needed by the FLC, and to provide this additional information, it was decided to numerically approximate the derivative of the error signal using the measured error signal values. By using both $\bar{e}(t)$ and $\dot{\bar{e}}(t)$, more information was available to use in creating the fuzzy output signal of the FLC.

To develop an implementable controller, the three variables ($\bar{e}(t)$, $\dot{\bar{e}}(t)$, and $u(t)$) must be converted into fuzzy numbers. Two important issues are involved in converting these variables into fuzzy numbers. First, the amount of resolution for each variable must be determined. In other words, how many categories will be used to represent the variable? Should $\bar{e}(t)$, be resolved into five categories like large negative numbers, small negative numbers, zero, small positive numbers, and large positive numbers, or should $\bar{e}(t)$ simply be resolved into the two categories of negative and positive? Second, the numerical values that define the limits associated with each fuzzy category must be determined.

The error signal and its numerically approximated derivative were resolved into two possible categories. Each variable could be positive and/or negative. Note that either of these signals can belong to both the positive fuzzy number category and the negative fuzzy number category simultaneously.

The upper limit on the membership function associated with the output error signal ($\bar{e}/K_q$) was determined from the steady-state error specification. Based on a desired steady-state error of less than $9 \times 10^{-8}$ meters, the limits on the positive and the negative membership functions became 0.13, since

$$\frac{0.13}{K_q} = 8.667 \times 10^{-8} . \quad (4)$$

The membership function limits associated with the rate of change of the error signal were obtained through experimentation. Initially, the largest non-unity values in the membership functions were selected to be $\pm 3,000$.

The membership functions of the input to the system will, in general, depend on the two competing objectives. However, preliminary designs indicated that the steady-state error specification was the most constraining objective. Thus, we designed the membership functions of $u(t)$ using the steady-state error specification and investigated, after the design, the resulting step response performance.

Before completing the membership function design associated with $u(t)$, the number of rules associated with the error and its rate of change must be determined. Because both $\bar{e}(t)$ and $\dot{\bar{e}}(t)$ were fuzzified into two categories each, a total of four rules is all that is necessary to produce $u(t)$. See Table 1 for the four rule combinations.

With these four rules, the limits for each of the three categories associated with $u(t)$ were determined based on the sinusoidal reference input. The sinusoidal input peaks at 5 milliseconds. In order for the system to respond properly, a fuzzy mean (FM) of 0.08 was selected for the positive category, a FM of $-0.08$ was selected for the negative category, and the zero category had a FM of zero with its lower and upper limits equalling $-0.08$ and 0.08, respectively. The FM is obtained using standard fuzzy number calculation techniques [5].

A defuzzification procedure is needed to produce an input signal that can be applied to the system. The defuzzification procedure was accomplished using the formula

$$u(t) = \frac{N}{D} , \qquad (5)$$

where

$$N = DOF_1 \times FM_{neg} + DOF_2 \times FM_{zero}$$

$$+ DOF_3 \times FM_{zero} + DOF_4 \times FM_{pos}$$

and

$$D = DOF_1 + DOF_2 + DOF_3 + DOF_4 .$$

In the above equation, $FM_i$ stands for the fuzzy mean of the $i$-th category of the input's value and $DOF_i$ stands for the input's degree of fulfillment in that category. For example, when the input is positive it has a fuzzy mean of 0.08 and its degree of fulfillment of the positive category depends on the values of $\bar{e}(t)$ and $\dot{\bar{e}}(t)$. As

an example, if $\bar{e}(t) = -0.06$ then it is positive with a degree of membership equal to

$$\mu_{pos}(\bar{e}) = \frac{(-0.06 - (-0.13))}{0.26} = 0.269 \qquad (6a)$$

and if $\dot{\bar{e}}(t) = 700$ then it is positive with a degree of membership equal to

$$\mu_{pos}(\dot{\bar{e}}) = \frac{(700 - (-1000))}{2000} = 0.85 . \qquad (6b)$$

The degree of fulfillment of rule number four ($\bar{e}$ and $\dot{\bar{e}}$ both positive) is obtained from $\mu_{pos}(\bar{e})$ and $\mu_{pos}(\dot{\bar{e}})$ and for this example is

$$\begin{aligned} DOF_4 &= \mu_{pos}(\bar{e}) \times \mu_{pos}(\dot{\bar{e}}) \\ &= 0.269 \times 0.85 = 0.22865 . \end{aligned} \qquad (7)$$

As shown above, the $DOF_i$ associated with each rule is determined by multiplying the degree of membership of $\bar{e}$ and $\dot{\bar{e}}$. This combination technique has the advantage of giving weight to both the FLC inputs.

There are other possible ways to defuzzify the above data to yield alternative control strategies [6]. Using different defuzzification strategies may modify the overall results reported in this paper. Different defuzzification possibilities will not be considered further.

With the variables fuzzified, combined, and defuzzified, simulations can be generated to investigate the performance of the FLC, which is of particular significance to us in this paper.

### IV. Studies of Controller Robustness

Figure 1 shows that the FLC designed in the previous section satisfies the steady-state error specification. The sampling frequency for this controller was selected to be 50 kHz. The control effort, $u(t)$, necessary to produce Figure 1 never exceeded its saturation limit. It may be recalled that the input saturates when the position variable accelerates more than $8g$. An acceptable step response with this controller is shown in Figure 2.

With successful results using a 50 kHz sampling rate, a slower sampling rate of 30 kHz was then investigated. Without modifying the FLC in any way, the acceptable results were maintained with the lower sampling frequency. In the step response behavior the overshoot changed to 12% versus 10%, no perceivable difference was seen in the rise-time, and the settling time increased to 4 milliseconds from 3 milliseconds. Because the 30 kHz design was acceptable, the 30 kHz sampling frequency will be maintained throughout the remainder of the paper.

Time delay was introduced into the system to simulate the existence of the delay that is present in a physical realization of a digital FLC. Introducing $20\mu\text{sec}$ of delay (approximately 60% of a sample period for a 30 kHz sampling rate), caused unacceptable behavior using the previously designed FLC. In particular, the step response oscillated with an amplitude of $\pm10\%$ around its final value and the error due to the sinusoidal signal included much more high frequency content and also grew in amplitude to about $15.0 \times 10^{-8}$ meters. Figure 3 shows the unacceptable steady-state error behavior.

Due to the unacceptable performance in the presence of time delay, the FLC was modified by reducing the limiting values associated with the $\dot{e}(t)$ signal to $\pm1,000$ from its earlier value of $\pm3,000$. This setting for $\dot{e}(t)$ allowed for acceptable behavior of the tracking error signal with the time delay present (see Figure 4). Note that reducing the value for the $\dot{e}(t)$ setting can only improve the overshoot in the step response behavior. The overshoot, for this modified FLC, was lowered to about 5%. The settling time remained at 4 milliseconds, but the rise-time was lengthened to about 0.8 milliseconds.

With an acceptable FLC, the robustness of this system was investigated for system parameter variations. To modify the parameters, the system poles in the original system model were changed by $\pm20\%$. The resulting performance was almost unchanged and continued to meet all of the specifications.

As a final robustness check, a high frequency disturbance signal was superimposed on the 50 Hz sinusoidal reference signal. The disturbance signal was a positive pulse train with an amplitude of $9.4 \times 10^{-8}$ meters and a frequency of 1000 Hz. Figure 5 shows how the steady-state error signal behaves with this disturbance and $20\mu\text{sec}$ of delay. The input to the system that is required to produce the error signal in Figure 5 is shown in Figure 6.

## V. Discussion

The first parameter that was changed in the present study was the sampling frequency which was reduced from 50 kHz down to 30 kHz. The results were pretty insensitive to this change in the sampling frequency. This insensitivity is probably due to the high sampling rate relative to the closed-loop bandwidth. The closed-loop bandwidth of the system was approximately 500 Hz based on the observed rise time or time to peak [7]. Because the sampling rate was at or above 60 times the closed-loop bandwidth, the digital FLC was effectively performing like an analog controller. If we reduce the sampling rate, then the membership functions and the rules will need to be changed if there is time delay. Preliminary experiments show that when the sampling rate is 20 kHz with the present FLC, the performance is acceptable without computation delay. When the computation delay is $20\mu\text{sec}$, then the performance is unacceptable at a sampling rate of 20 kHz. An interesting study would be to determine the lower limit on the sampling frequency before the performance becomes unacceptable both with and without time delay.

Introducing the time-delay in the system caused unacceptable performance with the original FLC. Acceptable performance was recovered by reducing the limits on the $\dot{e}(t)$ signal. This reduction on the limits of $\dot{e}(t)$ has the effect of making the controller less aggressive. (The input into the system is less influenced by the rate of change of the error signal.) Because there is more lag present with computation delay, less emphasis should be placed on the prediction component of the controller, i.e., a smaller magnitude on the derivative of the error signal is necessary. Lowering these limits is like changing the gain on the derivative term in a PD-type controller.

Changing the system parameter values by $\pm20\%$ had little impact on the performance of the closed-loop system. The performance does change slightly in response to changes in the parameter values. This slight change indicates that by designing the FLC for a nominal system, this controller will be robust to parameter value changes. The FLC does not try to achieve the exact same performance independent of the parameter values. Instead, the FLC is a controller based on a nominal set of parameters that is robust to parameter variations around the nominal set.

The FLC does not explicitly handle disturbance rejection. Because the system gain decreases with increasing frequency, the disturbance will be attenuated slightly. If one closely examines Figure 5, one can see that the FLC is actually trying to track this disturbance signal. Some other type of compensation scheme may be required if the magnitude of this disturbance signal gets too large. The presence of time delay makes the response with the high frequency disturbance even worse.

Additional studies should be carried out to see how robust the FLC is to changes in the amplitude and the frequency of the sinusoidal reference input.

## VI. Conclusions

A digital FLC was designed from basic principles. This controller exhibited very robust behavior to changes in the sampling frequency, reduction in the time delay, and changes in the system parameters. The controller was designed to accommodate two competing objectives without a large amount of trial-and-error experimentation. Further work should investigate the robust nature of FLCs when the sampling frequency is reduced further. In addition, accomplishing disturbance rejection using FLC could be investigated.

## VII. References

[1] E.H. Mamdani, "Application of fuzzy algorithms for control of a simple dynamic plant," *Proc. IEE*, Vol. 121, pp. 1585-1588, 1974.

[2] P.J. King and E.H. Mamdani, "The application of fuzzy control systems to industrial processes," *Automatica*, Vol. 13, pp. 235-242, 1976.

[3] B. Kosko, *Neural Networks and Fuzzy Systems: A Dynamical Systems Approach to Machine Intelligenc*, Prentice-Hall, 1992.

[4] G.F. Franklin, J.D. Powell, and M.L. Workman, *Digital Control of Dynamic Systems*, Second Edition, Addison Wesley, 1990.

[5] A. Kaufmann and M.M Gupta, *Introduction to Fuzzy Arithmetic: Theory and Applications*, Van Nostrand Reinhold, 1991.

[6] C.C. Lee, "Fuzzy logic in control systems: fuzzy logic controller," *IEEE Trans. Syst., Man, Cyber.*, Vol. 20, No. 2, pp. 404-435, 1990.

[7] G.F. Franklin, J.D. Powell, and A. Emani-Naeini, *Feedback Control of Dynamic Systems*, Second Edition, Addison-Wesley, 1990.

**Table 1**

| | | error derivative | |
|---|---|---|---|
| | | Negative | Positive |
| **error** | Negative | Negative I | Zero II |
| | Positive | Zero III | Positive IV |

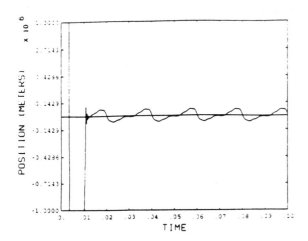

**Fig-1** Steady-state error performance of initial FLC with 50 kHz sampling rate.

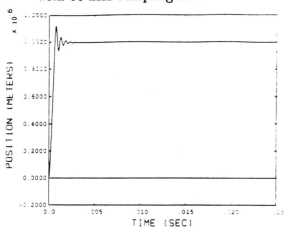

**Fig-2** Step response performance of initial FLC with 50 kHz sampling rate.

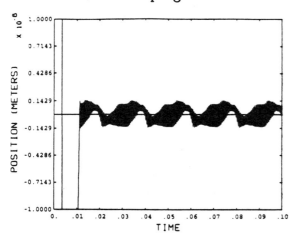

**Fig-3** Unacceptable steady-state error performance with 30 kHz sampling and a 20$\mu$sec delay.

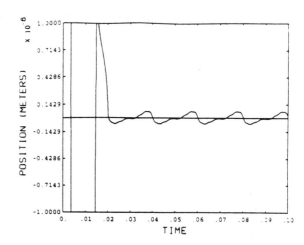

**Fig-4** Acceptable steady-state error performance with 30 kHz sampling and a 20$\mu$sec delay.

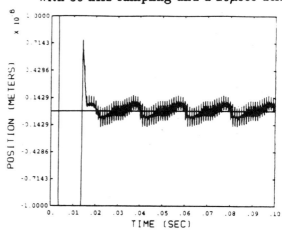

**Fig-5** Steady-state error performance with 20$\mu$sec delay and high frequency disturbance.

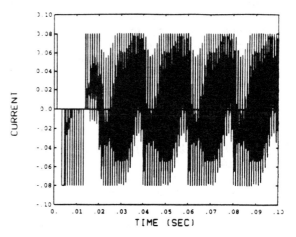

**Fig-6** System input response associated with Figure 5.

1585

# Membership Function-based Fuzzy Model
# And Its Applications To Multivariable Nonlinear Model-predictive Control

Renhong Zhao and Rakesh Govind
Department of Chemical Engineering, University of Cincinnati
Cincinnati, OH 45221

## 1. The Membership Function-based Fuzzy Model

Fuzzy control can incorporate the "rule of the thumb" experience in the design of the controller. In fuzzy control rules, usually two quantities, which are the measured control error and the change in control error, are used for each process input, and each quantity is assigned by following seven fuzzy subsets or fuzzy variables: positive large, positive medium, positive small, near nil, negative small, negative medium and negative large. So in the rule-based fuzzy control system, the complete control table for a single-input process should be consisted of $7^2=49$ rules, the complete control table for a two-input process should be consisted of $7^2 \times 7^2 = 2401$ rules. Although the number of rules can be reduced by using incomplete control tables or by using less fuzzy subsets, such kind of reduction can only result in the fall of control quality. Some study of reducing the number of rules through the multidimensional fuzzy reasoning and through the linear input-output relation has been reported.[1], [2].

Another difficulty of using rule-based fuzzy control is that there is no general method for the assignment of membership functions and control rules. The process-based experience usually cannot be used for other processes.

Membership functions proposed by Zadeh[3] were originally one-dimensional functions used to represent the partial belonging property. In this paper, the two-dimensional membership functions are proposed to represent the gradual deviation from the process known states. The proposed membership functions are not subjectively assigned by experts but can be obtained or approximated by using limited process response data. In this paper, the model-predictive control of a two-input, two-output nonlinear chemical reaction problem is used as an example of using the membership-based fuzzy models.

### 1.1 The membership function-based process step responses

Model-predictive control is a kind of digital model-based control in which the process models are incorporated directly in the control law as the basis for controller design. Process modeling of direct digital control with zero-order hold elements is a special kind of nonlinear modeling problems in which the process inputs are stair-step-like signals.

Since the process outputs generally do not change at the sampling instants, and the stair-step-like signals equal to the summation of successive step function signals, when the process additive property[4] is assumed, the process in a digital control system using zero-order hold element can be modeled by

$$Y(kT) = \sum_{n=0}^{\infty} \left\{ y\big[x(kT, nT)\big] - y\big[x(kT, nT+T)\big] \right\} \tag{1}$$

In equation (1), $T$ is the sampling time interval, $kT$ is the current sampling instant, $x(kT, nT)$ is the process step input beginning at $nT$ with the step size $x$, and $y\big[x(kT, nT)\big]$ is the open-loop process response with respect to the $x(kT, nT)$. In this paper, $x$, $y$ and $Y$ are deviation variables.

For linear systems, if $y\big[x(kT, nT)\big]$ is known, because of the homogeneous property[4], the process step responses to any other step size $x^*$ can be obtained from

$$\frac{y\big[x^*(kT, nT)\big]}{y\big[x(kT, nT)\big]} = \frac{x^*}{x} \tag{2}$$

In equation (2), we require that

$$x \neq 0 \quad \text{and} \quad y\big[x(kT, nT)\big] \neq 0 \tag{3}$$

By using the subscripts $p$ and $n$ to denote the positive and negative step changes respectively, by using the subscripts $max$ and $min$ to denote the allowed maximum size for the positive change and for the negative change respectively, from equation (2) we can have

$$\frac{y\big[x_p(kT, nT)\big]}{y\big[x_{max}(kT, nT)\big]} = \frac{x_p}{x_{max}} \tag{4}$$

and

$$\frac{y\big[x_n(kT,nT)\big]}{y\big[x_{min}(kT,nT)\big]} = \frac{x_n}{x_{min}} \qquad (5)$$

Because

$$0 \le x_p \le x_{max} \text{ and } x_{min} \le x_n \le 0 \qquad (6)$$

then from equations (4), (5) and (6) we have

$$0 \le \frac{y\big[x_p(kT,nT)\big]}{y\big[x_{max}(kT,nT)\big]} \le 1 \qquad (7)$$

and

$$0 \le \frac{y\big[x_n(kT,nT)\big]}{y\big[x_{min}(kT,nT)\big]} \le 1 \qquad (8)$$

For linear processes, both equalities (4), (5) and inequalities (7), (8) hold true. The models satisfying only the inequalities (7), (8) but not the equalities (4) and (5) can be used to represent nonlinear processes.

The range of inequalities (7) and (8) is [0, 1] which is exactly same as that of membership functions. We can define following two membership functions

$$\mu_p\big[x_p(kT,nT)\big] = \frac{y\big[x_p(kT,nT)\big]}{y\big[x_{max}(kT,nT)\big]} \qquad (9)$$

and

$$\mu_n\big[x_n(kT,nT)\big] = \frac{y\big[x_n(kT,nT)\big]}{y\big[x_{min}(kT,nT)\big]} \qquad (10)$$

The physical meaning of above two membership functions is a quantitative representation of the 'close degree' to the extreme step response $y\big[x_{max}(kT,nT)\big]$ or $y\big[x_{min}(kT,nT)\big]$. If the membership functions $\mu_p\big[x_p(kT,nT)\big]$ and $\mu_n\big[x_n(kT,nT)\big]$ are available, then from the known extreme step responses, the process step responses for any step size in $\big[x_{min}, x_{max}\big]$ can be derived directly from equations (9) and (10).

## 1.2 The approximation of two-dimensional membership functions from limited process data

For any specified value of $nT$, $\mu_p\big[x_p(kT,nT)\big]$ and $\mu_n\big[x_n(kT,nT)\big]$ are two-dimensional functions of the time $kT$ and the step size $x_p$ or $x_n$ when the process extreme step response are known.

For linear processes, when the process extreme step responses $y\big[x_{max}(kT,nT)\big]$ and $y\big[x_{min}(kT,nT)\big]$ are known, from equations (4), (5) and (9), (10), the membership functions $\mu_p\big[x_p(kT,nT)\big]$ and $\mu_n\big[x_n(kT,nT)\big]$ become one-dimensional functions of the step size only, and

$$\mu_p\big[x_p(kT,nT)\big] = \frac{x_p}{x_{max}} \qquad (11)$$

and

$$\mu_n\big[x_n(kT,nT)\big] = \frac{x_n}{x_{min}} \qquad (12)$$

For nonlinear processes, the membership functions defined by equations (9) and (10) are problem dependent and usually cannot been known exactly in practice. The proposed two-dimensional membership functions can be approximated by using some fixed structure functions which use limited process response data.

Since $x$ and $y$ are deviation variables, from equation (9),

$$\mu_p\big[x_p(kT,nT)\big]\Big|_{x_p=0} = \frac{y[0]}{y\big[x_{max}(kT,nT)\big]} = 0 \qquad (13)$$

and

$$\mu_p\big[x_p(kT,nT)\big]\Big|_{x_p=x_{max}} = \frac{y\big[x_{max}(kT,nT)\big]}{y\big[x_{max}(kT,nT)\big]} = 1 \qquad (14)$$

For open-loop stable processes, the model horizon[5] is defined as that minimal $t = mT$ such that for all $x_p$ in $\big[0, x_{max}\big]$ and for all $x_n$ in $\big[x_{min}, 0\big]$ the values of $y\big[x_p(kT,nT)\big]$ and $y\big[x_n(kT,nT)\big]$ become stable at all

1587

$kT \geq nT + mT$. Based on the definition of model horizon, from equation (9),

$$\mu_p\big[x_p(kT,nT)\big]\big|_{kT \geq nT+mT}$$

$$= \frac{y\big[x_p(nT+mT,nT)\big]}{y\big[x_{max}(nT+mT,nT)\big]} \qquad (15)$$

Equations (13) and (14) are the boundary conditions for step size $x_p$. Equation (15) is the boundary condition for time $kT$ and can be thought as the steady state part of $\mu_p\big[x_p(kT,nT)\big]$. By comparing equation (15) to equation (11), we can construct following approximation

$$\mu_p\big[x_p(kT,nT)\big]\big|_{\text{at steady state}}$$

$$= \frac{y\big[x_p(nT+mT,nT)\big]}{y\big[x_{max}(nT+mT,nT)\big]} = \left(\frac{x_p}{x_{max}}\right)^{\alpha_p} \qquad (16)$$

The advantage of using equation (16) is that: 1). For linear processes, $\alpha_p$ is not dependent on $x_p$ and $\alpha_p \equiv 1$; 2). Equation (16) satisfies the boundary conditions given by equations (13) and (14); 3). The deviation of $\alpha_p$ from unity is a quantitative representation of process steady state nonlinearity for the given step size $x_p$.

For nonlinear processes, $\alpha_p$ is dependent on the $x_p$. In practical applications, the value of $\alpha_p$ for the most often used step size can be used to approximate the values of $\alpha_p$ for other step sizes.

Extend from equation (16), which is an approximation expression for the steady state part of $\mu_p\big[x_p(kT,nT)\big]$, we can develop following approximation

$$\mu_p\big[x_p(kT,nT)\big] = \frac{y\big[x_p(kT,nT)\big]}{y\big[x_{max}(kT,nT)\big]}$$

$$= \left[\left(\frac{x_p}{x_{max}}\right)^{\alpha_p}\right]^{\left\{D(kT,nT,mT)^{\beta_p}\right\}} \qquad (17)$$

Equation (17) is an approximation expression for both the steady state and the dynamic state parts of $\mu_p\big[x_p(kT,nT)\big]$. In equation (17), $\alpha_p$ is a parameter for process steady state responses, $\beta_p$ is a parameter for process dynamic responses, and $D(kT,nT,mT)$ is defined as

$$D(kT,nT,mT)$$

$$= \begin{cases} \dfrac{kT-nT}{mT} & \text{for } nT < kT \leq nT+mT \\ 1.0 & \text{for } kT \geq nT+mT \end{cases} \qquad (18)$$

Similar to equation (16) we can approximate the steady state part of $\mu_n\big[x_n(kT,nT)\big]$ by

$$\mu_n\big[x_n(kT,nT)\big]\big|_{\text{at steady state}}$$

$$= \frac{y\big[x_n(nT+mT,nT)\big]}{y\big[x_{min}(nT+mT,nT)\big]} = \left(\frac{x_n}{x_{min}}\right)^{\alpha_n} \qquad (19)$$

and similar to equation (17), we can approximate both the steady state and the dynamic state parts of $\mu_n\big[x_n(kT,nT)\big]$ by

$$\mu_n\big[x_n(kT,nT)\big] = \frac{y\big[x_n(kT,nT)\big]}{y\big[x_{min}(kT,nT)\big]}$$

$$= \left[\left(\frac{x_n}{x_{min}}\right)^{\alpha_n}\right]^{\left\{D(kT,nT,mT)^{\beta_n}\right\}} \qquad (20)$$

The advantage of using equations (17) and (20) is that: 1). For linear systems, $\alpha_p$, $\alpha_n$, $\beta_p$ and $\beta_n$ are process independent, $\alpha_p$ and $\alpha_n$ are always equal to unity, $\beta_p$ and $\beta_n$ are always equal to zero; 2). The deviation of $\alpha_p$ and $\alpha_n$ from unity is a quantitative representation of steady state nonlinearity of the process step responses, the deviation of $\beta_p$ and $\beta_n$ from zero is a quantitative representation of dynamic nonlinearity of the process step responses; 3). Equations (17) and (20) also satisfy the boundary condition given by equations (13) and (14). 4). The parameters used by approximation expressions can be identified from limited process response data.

### 1.3 The identification of $\alpha_p$, $\alpha_n$, $\beta_p$ and $\beta_n$ in $\mu_n[x_n(kT,nT)]$ and $\mu_p[x_p(kT,nT)]$

Parameters of $\mu_n[x_n(kT,nT)]$ and $\mu_p[x_p(kT,nT)]$ are not subjectively determined but are identified by using the process reaction curve data. The required data can be obtained either from the limited process testing or from the analytical/numerical solutions of the valid differential/difference equation models.

Parameters $\alpha_p$ and $\beta_p$ can be identified by using: 1). the process known extreme step response $y[x_{max}(kT,nT)]$ where $n \leq k \leq k+m$; 2). the data from another step response $y[x_p*(kT,nT)]$ where $x_p*$ selected by user is a value between 0 and $x_{max}$.

First, the steady state parameter $\alpha_p$ can be identified from equation (16) by using the value of $y[x_p*(kT,nT)]$ at $k = nT+mT$,

$$\frac{y[x_p*(nT+mT,nT)]}{y[x_{max}(nT+mT,nT)]} = \left(\frac{x_p*}{x_{max}}\right)^{\alpha_p} \quad (21)$$

then the dynamic parameter $\beta_p$ can be identified from equation (17) by using the obtained $\alpha_p$ and the value from $y[x_p*(kT,nT)]$ at $k*T$ where $k*$ is a value between $n$ and $n+m$.

$$\frac{y[x_p*(k*T,nT)]}{y[x_{max}(k*T,nT)]} = \left[\left(\frac{x_p*}{x_{max}}\right)^{\alpha_p}\right]^{\{D(k*T,nT,mT)^{\beta_p}\}} \quad (22)$$

Similarly, $\alpha_n$ and $\beta_n$ can be identified by using: 1). the process known extreme step response $y[x_{min}(kT,nT)]$ where $n \leq k \leq k+m$; 2). the data from another step response $y[x_n*(kT,nT)]$ where $x_n*$ is a value between $x_{min}$ and 0..

The steady state parameter $\alpha_n$ can be identified by from equation (19) by using the value of $y[x_n*(kT,nT)]$ at $k = nT+mT$,

$$\frac{y[x_n*(nT+mT,nT)]}{y[x_{min}(nT+mT,nT)]} = \left(\frac{x_n*}{x_{min}}\right)^{\alpha_n} \quad (23)$$

then the dynamic parameter $\beta_n$ can be identified from equation (20) by using by using the obtained $\alpha_n$ and the value from $y[x_n*(kT,nT)]$ at $k**T$ where $k**$ is a value between $n$ and $n+m$.

$$\frac{y[x_n*(k**T,nT)]}{y[x_{min}(k**T,nT)]} = \left[\left(\frac{x_n*}{x_{min}}\right)^{\alpha_n}\right]^{\{D(k**T,nT,mT)^{\beta_n}\}} \quad (23)$$

## 2. The Using Of Membership-based Fuzzy Model For The Predictive Control Of A Chemical Reaction Process

Consider an isothermal continuous-stirred tank reactor(CSTR) with the following reactions:

$$A \underset{k2}{\overset{k1}{\rightleftarrows}} B \overset{k3}{\longrightarrow} C$$

where the reaction rates are given by

$$r_1 = k_1 C_A^{0.5}, \quad r_2 = k_2 C_B^{2.0}, \quad r_3 = k_3 C_B^{0.5} \quad (24)$$

In equation (24), $C_A$ and $C_B$ are outlet concentrations of A and B.

Assume that we want to control the outlet concentrations of A and B by adjusting the feed concentrations of A and B, by using the similar method like Ray[6], the given two-input and two-output nonlinear chemical reaction process can be represented by following differential equations:

$$\frac{dv_1}{dt} = (u_1 - v_1) - Da_1 v_1^{0.5} + Da_2 v_2^{2.0} \quad (25)$$

$$\frac{dv_2}{dt} =$$

$$(u_2 - v_2) + Da_1 v_1^{0.5} - Da_2 v_2^{2.0} - Da_3 v_2^{0.5} \quad (26)$$

In equations (25) and (26), $u_1$ and $u_2$ respectively represent the dimensionless quantities of the feed concentrations of A and B, $v_1$ and $v_2$ respectively represent the dimensionless quantities of the outlet concentrations of A and B. If we assume that $Da_1 = 2.0$, $Da_2 = 1.0$ and $Da_3 = 2.0$, if we also assume that at steady state $u_1 = 0.8$ and $u_2 = 1.0$, then from equations (25) and (26) we can obtain that the corresponding $v_1$ and $v_2$ at steady state are equal to 0.158 and 0.391 respectively.

We assume that the allowed extreme step sizes for $u_1$ and $u_2$ are $\pm 0.4$ and $\pm 0.5$ respectively. The step response for $u_1$ with $\pm 0.2$ step sizes and for $u_2$ with $\pm 0.25$ are used to identify the parameters of the membership functions. By using $x_1$ and $x_2$ to denote the deviation quantities of $u_1$ and $u_2$, by using $y_1$ and $y_2$ to denote the deviation quantities of the deviation of $v_1$ and $v_2$, then based on the step sizes specified above, we have

$$x_{1p}{}^* = 0.2, \quad x_{1n}{}^* = -0.2$$
$$x_{1max} = 0.4, \quad x_{1min} = -0.4$$
$$x_{2p}{}^* = 0.25, \quad x_{2n}{}^* = -0.25$$
$$x_{2max} = 0.5, \quad x_{2min} = -0.5$$

The sampling period $T$ is set as $T = 0.05$. The model horizon is set as $mT = 50 \times 0.05 = 2.5$. from the process responses for above eight step sizes.

By denoting the subscript $ij$ as the $i$th output variable corresponding to the $j$th input variable, and by using $k^* = k^{**} = n + 1$, from the method introduced in section 1.3 of this paper we can obtain following eight membership functions:

$$\mu_{p11} = \left[ \left( \frac{x_{1p}}{x_{1max}} \right)^{1.0886} \right]^{[D(kT,nT,mT)]^{0.0185}}$$

$$\mu_{n11} = \left[ \left( \frac{x_{1n}}{x_{1min}} \right)^{0.8378} \right]^{[D(kT,nT,mT)]^{-0.0484}}$$

$$\mu_{p21} = \left[ \left( \frac{x_{1p}}{x_{1max}} \right)^{0.9592} \right]^{[D(kT,nT,mT)]^{-0.0106}}$$

$$\mu_{n21} = \left[ \left( \frac{x_{1n}}{x_{1min}} \right)^{1.0404} \right]^{[D(kT,nT,mT)]^{0.0101}}$$

$$\mu_{p12} = \left[ \left( \frac{x_{2p}}{x_{2max}} \right)^{1.1964} \right]^{[D(kT,nT,mT)]^{-0.07189}}$$

$$\mu_{n12} = \left[ \left( \frac{x_{2n}}{x_{2min}} \right)^{0.7559} \right]^{[D(kT,nT,mT)]^{-0.1893}}$$

$$\mu_{p22} = \left[ \left( \frac{x_{2p}}{x_{2max}} \right)^{1.0131} \right]^{[D(kT,nT,mT)]^{0.00042}}$$

$$\mu_{n22} = \left[ \left( \frac{x_{2n}}{x_{2min}} \right)^{0.9551} \right]^{[D(kT,nT,mT)]^{-0.01175}}$$

For the example given by this paper, the step response for any step size in $[x_{min}, x_{max}]$ can be obtained from equations (9) or (10) by using the membership functions derived above, then the obtained process step responses can be used in equation (1) to predict the process output.

By using the membership function-based fuzzy model introduced in this paper, and by using the single-step control objective in the controller algorithm, for a first-order disturbance given by

$$Y_d(t)$$
$$= \begin{cases} 0 & \text{for } t < 0 \\ 0.157 \times 0.08 \times \left[ 1 - \exp^{(-t/0.4)} \right] & \text{for } t \geq 0 \end{cases}$$

assume that the $Y_d(t)$ is simultaneously added to the two outputs $y_1$ and $y_2$, the simulation result is given by Figure 1 and Figure 2 in which $Y1$ and $Y2$ are given by

$$Y1(t) = \frac{y_1(t) - \left[y_{1 \text{ at steady state}} = 0.158\right]}{0.157} \quad (27)$$

and

$$Y2(t) = \frac{y_2(t) - \left[y_{2 \text{ at steady state}} = 0.391\right]}{0.182} \quad (28)$$

In Figure 1 and Figure 2, the model predictive control using membership function-based fuzzy model is compared to the decoupled linear control. The models used by linear control are constructed by using the step change of $u_1$ from 0.8 to 1.0 and the step change of $u_2$ from 1.0 to 1.25. The PI controllers are turned by using the Ziegler-Nichols method[7].

### 3. Conclusions

The nonlinear processes in direct digital control systems can be modeled by the membership function-based fuzzy models proposed in this paper. The two-dimensional membership functions used by this paper are identified by using limited process response data. In stead of using membership functions to represent the belonging to a set, this paper uses the membership functions to represent the gradual deviation from the known states.

The membership function-based fuzzy models are effective nonlinear models which can be used for multivariable nonlinear predictive control in which the process interaction is used to enhance the control action rather than are decoupled like in linear control.

### 4. References

[1] M. Sugeno and T. Takagi, "Multi-dimensional fuzzy reasoning", Fuzzy Sets and Systems, Vol. 9, No.2, 1983.

[2] T. Takagi and M. Sugeno, "Fuzzy identification of Systems and its applications to modeling and control", IEEE Transactions on Systems, Man, and Cybernetics, Vol. SMC-15, No.1, pp.116-132, 1985.

[3] L. Zadeh, "Fuzzy sets", Information and Control, Vol.8, pp.338-353, 1965.

[4] H. Vanlandingham, Introduction to Digital Control Systems, Macmillan Publishing company, New York, 1985.

[5] D. Seborg, T. Edgar and Mellicamp, Process Dynamics and Control, John Wiley & Sons, 1989.

[6] W. Ray, Advanced Process Control, Mcgraw-Hill Book Company, New York, 1981.

[7] J. Ziegler and N. Nichols, Trans. ASME. Vol. 64, pp.759, 1942.

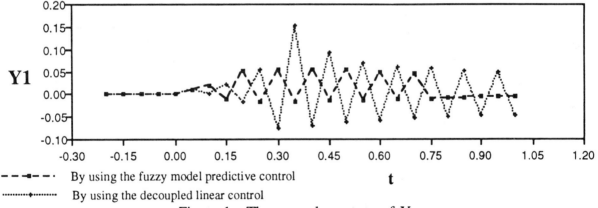

- - -■- - · By using the fuzzy model predictive control

·········◆········· By using the decoupled linear control

Figure 1   The control response of $Y_1$

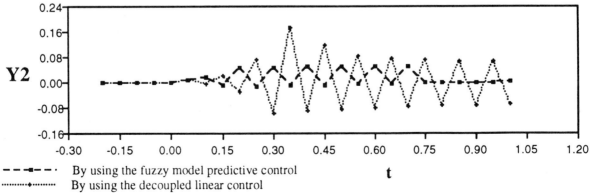

- - -■- - · By using the fuzzy model predictive control

·········◆········· By using the decoupled linear control

Figure 2   The control response of $Y_2$

# Nonlinear System Control with Fuzzy Logic Design

Chu Kwong CHAK and Gang FENG

Department of Systems and Control, School of Electrical Engineering

University of New South Wales, Kensington, NSW 2033, Australia

*Abstract: In this paper, we present a new control design method for nonlinear systems. In this method a set of control laws will be designed based on the linear models about some points of state space of nonlinear systems. The controller output will be the result of applying fuzzy logic theory to manipulate the given set of control laws. By this approach, we ensure the good performance as expected when the state space trajectory is around or about the points where linear models are based to find the control laws and the reasonable performance in other regions. This method can be applied to the nonlinear systems which is not feedback linearisation applicable. Some simulation results of a $4^{th}$ order nonlinear system, ball and beam system, are presented to demonstrate the feasibility of this method.*

## 1. Introduction

To design a smooth nonlinear control system, a number of approaches have been addressed. For some slightly nonlinear systems, designers will normally determine the control law based on a linearised model about some point where may normally be the centre of state space trajectory. The performance will be acceptable if the state space trajectory is nearly around that point. For better performance designers normally employ feedback linearisation technique to obtain a resulting equivalent system which is linear. Then linear control system design theory can be applied to design a controller for the nonlinear system. However, feedback linearisation requires a very accurate plant model so that the resulting system is fully linearised. Unfortunately, in practical situation, to find a accurate model of a plant is not always possible due to external disturbances, measurement noise, etc. Then the resulting model based on which the control law is designed will not give satisfactory performance. Besides it is not a global design technique applicable to all nonlinear systems because the conditions for feedback linearisation of nonlinear systems are quite restrictive. For example the ball and beam system [9] violates the integrability condition. Thus it is not possible to fully linearise the ball and beam system. One may consider that the nonlinear system can be piecewise linearised about some points and the corresponding control laws are designed. The control laws are fired sequentially according to the state space trajectory. But this approach will produce the discontinuity of controller output and hence the discontinuity in the description of the system behaviour. Nevertheless, this drawback can be overcome if some smoothing techniques are applied to approximate the controller output based on the output of all linear control laws. If the trajectory is around the points where linear models are formed the closed loop dynamics approach to what we expect since powerful linear control theory is used to design the control law about that point. If the trajectory is away from the points, the closed loop dynamics are only approximated. It is expected that more linear models are formed, more accurate closed loop dynamics are obtained.

Of the approximation techniques, multidimensional interpolation and fuzzy logic are the candidates for the approximations. Fuzzy logic control systems provide more robustness in noise. It has shown to give better results in approximation reasoning and in many respects than those obtained with conventional control techniques [6]. Some papers [3][7][8] has addressed fuzzy logic control systems with fuzzy rules the premise condition of which concerns with the state variables. It is based on the concept that for a fuzzy system, due to the conditions on the premise parts, each fuzzy if-then rule can be viewed as local description of system under consideration. In paper [3], the local description of system is linear and the global description of system is of "fuzzy combination" of a set of local linear systems. Paper [7][8] extends this concept in a more systematic way that the state space is partitioned into a number of cells called fuzzy cells. Each cell is associated with a local dynamical system. The global dynamics can then be described by the local dynamical systems which in each fuzzy cell activates locally. This framework can readily suit our approximation requirements since each linearised system about a point can be regarded as a local dynamical system stated in paper [7][8]. In this paper, we will address this approach and design the controller the output of which is the result of fuzzy logic operation on state variables with fuzzy rules. The fuzzy rules are determined by the state space feedback design of the linearised systems.

In section 2, we will discuss how to find linearised systems with the given plant model. Then the local control law design will be presented in section 3. With this set of control laws, we will propose the new method based on fuzzy logic, to determine the controller output in section 4. Some simulation results are presented to show the feasibility of this approach in section 5 and finally a conclusion will be drawn in section 6.

## 2. The Linearised Systems of a Nonlinear System

Consider a smooth nonlinear system

$$\frac{dx}{dt} = f(x,u) \tag{1}$$

where $f(x,u) = \begin{bmatrix} f_1(x,u) & f_2(x,u) & \dots & f_n(x,u) \end{bmatrix}^T$

Let $k = (k_1, k_2, \dots, k_n)$ and $l = (l_1, l_2, \dots, l_m)$ be an n-tuple of non-negative integers (a "multi-index"). We have $x^k = \begin{bmatrix} x_1^{k_1} & x_2^{k_2} & \dots & x_n^{k_n} \end{bmatrix}^T$ and $u^l = \begin{bmatrix} u_1^{l_1} & u_2^{l_2} & \dots & u_m^{l_m} \end{bmatrix}^T$. There may exist a set of linearised dynamic models about the point $(x^k, u^l)$, $k, l \in N_+$. Using Taylor's method, we can find the corresponding Jacobian matrices $A^{kl}$ and $B^{kl}$ such that

$$\frac{d\delta x}{dt} = A^{kl}\delta x + B^{kl}\delta u$$

where $\delta x = x - x^k$ and $\delta u = u - u^l$. It can be further written into

$$\frac{dx}{dt} = A^{kl}x + B^{kl}u + d^{kl}$$

where $d^{kl} = -(A^{kl}x^k + B^{kl}u^l)$ is constant disturbance

## 3. The Design of Linear Control Laws

For each linearised dynamic models about the point $(x^k, u^l)$, we design the corresponding feedback control law

$$v^{kl} = -K^{kl}x$$

where $K^{kl}$ is feedback gain matrix. With these feedback laws, the dynamics of the closed loop systems about the point $(x^k, u^l)$ is

$$\frac{dx}{dt} = A_c^{kl}x + d^{kl}$$

where $A_c^{kl} = A^{kl} - B^{kl}K^{kl}$ is a constant matrix whose eigenvalues depend on the design specifications.

## 4. Nonlinear Control Law Formulation (Fuzzy Logic Controller Design)

In practical consideration, the set of $(x^k, u^l)$ is finite and thus the number of corresponding control laws is finite. Since state space trajectory of the system is continuous, these discrete control laws cannot achieve an accurate control performance, some approximation techniques can be utilised to help us to determine a nonlinear control law with the given control laws or determine controller outputs with the outputs of the set of control laws. In other words, we approximate a function h(.)

$$v = h(\{v^{kl}, k=1,2,\dots,K, l=1,2,\dots,L\})$$

with the given $v^{kl} = -K^{kl}x$. Fuzzy logic is used to approximate the function h(.) in the following sections. We design the control laws for each given linearised model of the nonlinear plant. With these control laws, we construct a number of linguistic rules to describe a global control law in fuzzy framework. Then the global closed loop system can be regarded as the combination of the linear closed loop systems which are formed by linearised models and their control laws. These linear closed loop systems describe the dynamics of global system locally. The combination effect of the local linear system dynamics can represent the dynamics of the global system. Fuzzy logic can readily be applied to the design of controller output.

For a general fuzzy logic controller design, please refer to papers [4][5] for details. In this paper, we describe the structure and mechanism of a fuzzy logic controller for a general MIMO nonlinear system (1). The fuzzy controller aims to approximate the control output based on the given linear control laws and functions as a nonlinear control law. From another point of view, this problem can be regarded as a special case of fuzzy modelling which uses linguistic rules to model a nonlinear controller according to a set of linear control laws. Figure 1 shows the fuzzy logic controller design. In this framework, fuzzy sets are introduced to provides a basis for a systematic way for the manipulation of vague and imprecise concepts and hence they are employed to represent state variables $x_i$ and input variable $u_j$ in linguistic value. The linguistic values of state variables $x_i$ and input variable $u_j$ are defined in linguistic terms. Each term of linguistic variables is characterised as fuzzy sets associated with a membership function.

We describe the fuzzy system in more precise mathematical terms. Let $x_i$ be state variable in the universal of discourse $X_i$ and input variable $u_j$ in the universal of discourse $U_j$. For $x \in R^n$ and $u \in R^m$, let $x = \begin{bmatrix} x_1 & x_2 & \dots & x_n \end{bmatrix}^T$ be the state vector, $u = \begin{bmatrix} u_1 & u_2 & \dots & u_m \end{bmatrix}^T$ be the input vector and $\overline{X}$ be the vector

1593

space spanned by $x$ and $u$. Then $\bar{x} = \begin{bmatrix} x & u \end{bmatrix}^T$ is a point of the state space $\bar{X}$. Let $k = (k_1, k_2, ..., k_n)$ and $l = (l_1, l_2, ..., l_n)$ be an n-tuple of non-negative integers (a "multi-index"). We have $x^k = \begin{bmatrix} x_1^{k_1} & x_2^{k_2} & ... & x_n^{k_n} \end{bmatrix}^T$ and $u^l = \begin{bmatrix} u_1^{l_1} & u_2^{l_2} & ... & u_m^{l_m} \end{bmatrix}^T$. Let $X_i^{k_i}$ be a fuzzy set in a universe of discourse $X_i$ characterised by membership function $\mu_{X_i^{k_i}} : X_i \rightarrow [0,1]$ which indicates its grade of membership function: $X_i^{k_i} = \{(x_i, \mu_{X_i^{k_i}}(x_i) | x_i \in X_i\}$. Similarly let $U_j^{l_j}$ be a fuzzy set in a universe of discourse $U_j$ characterised by membership function $\mu_{U_j^{l_j}} : U_j \rightarrow [0,1]$ which indicates its grade of membership function: $U_j^{l_j} = \{(u_j, \mu_{U_j^{l_j}}(u_j) | u_j \in U_j\}$. Then $P_{X_i} = \{X_i^1, X_i^2, ..., X_i^K\}$ is a collection of fuzzy partitions of $X_i$ and $P_{U_j} = \{U_j^1, U_j^2, ..., U_j^L\}$ is a collection of fuzzy partitions of $U_j$. Both $X_i^{k_i}$ and $U_j^{l_j}$ represent linguistic terms and the related membership functions $\mu_{X_i^{k_i}}(x_i)$ and $\mu_{U_j^{l_j}}(u_j)$ which satisfy normal, complete and convex conditions.

We now consider a space partition $\bar{X}^p = X_1^{k_1} \times X_2^{k_2} \times ... \times X_n^{k_n} \times U_1^{l_1} \times U_2^{l_2} \times ... \times U_m^{l_m}$ in the product space $X_1 \times X_2 \times ... \times X_n \times U_1 \times U_2 \times ... \times U_m$ where $p = p(k_1, k_2, ..., k_n, l_1, l_2, ..., l_m)$. Then $\bar{X}^p$ is a fuzzy set in the universal of discourse $\bar{X}$ associated with a membership function $\mu_{\bar{X}^p}(\bar{x})$

$$\mu_{\bar{X}^p}(\bar{x}) = \mu_{X_1^{k_1}}(x_1) \wedge \mu_{X_2^{k_2}}(x_2) \wedge ... \wedge \mu_{X_n^{k_n}}(x_n) \wedge \mu_{U_1^{l_1}}(u_1) \wedge \mu_{U_1^{l_1}}(u_1) \wedge ... \wedge \mu_{U_m^{l_m}}(u_m).$$

Then the space partition $\bar{X}^p$ and $\bar{X}^q$ in the universal of discourse $\bar{X}$ satisfies normal, complete and convex conditions. From the conditions, in each space fuzzy partition $\bar{X}^p$ in $\bar{X}$, there exists a point $\bar{x} = \begin{bmatrix} x & u \end{bmatrix}^T$ where $x = \begin{bmatrix} x_1 & x_2 & ... & x_n \end{bmatrix}^T$ and $u = \begin{bmatrix} u_1 & u_2 & ... & u_m \end{bmatrix}^T$ such that $\mu_{\bar{X}^p}(\bar{x}) = 1$, that is $\mu_{X_i^{k_i}}(x_i) = 1$, $i = 1, 2, ..., n$ and $\mu_{U_j^{l_j}}(u_j) = 1$, $j = 1, 2, ..., m$. We then find a linearised model and the corresponding control law about that point.

The fuzzy system is then characterised by the set of linguistic rules derived by the control laws based on the linearised models. Using the linear laws, we construct the following fuzzy control rules

$R_{kl}$:   If $\bar{x}$ is $\bar{X}^p$ then $v = v_p$   or   If $x_1$ is $X_1^{k_1}$ and $x_2$ is $X_2^{k_2}$ and ... and $x_n$ is $X_n^{k_n}$ and $u_1$ is $U_1^{l_1}$ and $u_2$ is $U_2^{l_2}$ and ... and $u_n$ is $U_m^{l_m}$ then $v = v_{kl}$

where $v_{kl} = -K_{kl} x$ is the linear control law output about the point $(x^k, u^l)$, $x^k = \begin{bmatrix} x^{k_1} & x^{k_2} & ... & x^{k_n} \end{bmatrix}^T$ and $u^l = \begin{bmatrix} u^{l_1} & u^{l_2} & ... & u^{l_m} \end{bmatrix}^T$

to form a rule base.

Obviously, it is a Sugeno type fuzzy reasoning [1] because the consequence of the rules is a function of input linguistic variable ($x_i$ and $u_j$). Since the inputs are measured by sensors and are crisp. These crisp values are treated as fuzzy singletons. The rules can be easily manipulated as a fuzzy implication (relation) $R_{kl}$. The firing strength $\alpha_{kl}$ of $kl^{th}$ control rules may be expressed as

$$\alpha_{kl} = \mu_{X_1^{k_1}}(x_1) \wedge \mu_{X_2^{k_2}}(x_2) \wedge ... \wedge \mu_{X_n^{k_n}}(x_n) \wedge \mu_{U_1^{l_1}}(u_1) \wedge \mu_{U_2^{l_2}}(u_2) \wedge ... \wedge \mu_{U_m^{l_m}}(u_m)$$

where $\wedge$ is composition operator and $\mu_{X_i^{k_i}}(x_i)$ and $\mu_{U_j^{l_j}}(u_j)$ play the role of the degrees of partial match between the input data and the data in the rule base. The inferred value of the control action from the $kl^{th}$ rule is $\alpha_{kl} v_{kl}$. Correspondingly, a crisp control action is given by

$$v = \frac{\sum_{kl=0} \alpha_{kl} v_{kl}}{\sum_{kl=0} \alpha_{kl}}$$

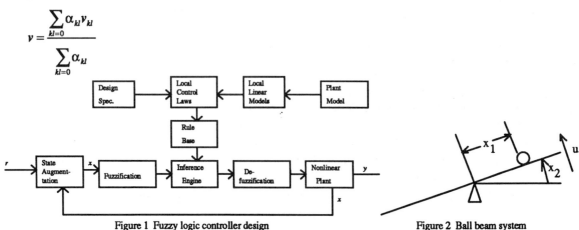

Figure 1 Fuzzy logic controller design                    Figure 2 Ball beam system

## 5. Simulation Examples

In this section, we will consider a $4^{\text{th}}$ order nonlinear system, ball and beam system [9], for the simulation. The ball and beam system is shown in figure 2. The beam is made to rotate in a vertical plane by applying a torque at the centre of rotation and the ball is free to roll along the beam. The ball remains in contact with the beam. Let $x = \begin{bmatrix} x_1 & x_2 & x_3 & x_4 \end{bmatrix} := \begin{bmatrix} r & dr/dt & \theta & d\theta/dt \end{bmatrix}$ be the state of the system, $y = x_1$ be the output of the system and $u$ be the input to the system. Then, from [9], the system can be represented by the state-space model

$$\begin{bmatrix} dx_1/dt \\ dx_2/dt \\ dx_3/dt \\ dx_4/dt \end{bmatrix} = \begin{bmatrix} x_2 \\ B(x_1 x_4^2 - G\sin x_3) \\ x_4 \\ 0 \end{bmatrix} + \begin{bmatrix} 0 \\ 0 \\ 0 \\ 1 \end{bmatrix} u$$

where $u$ is the input to the system, and $B$ and $G$ are system parameters. We find Jacobian matrix about $x^k = \begin{bmatrix} x_1^{k_1} & x_2^{k_2} & x_3^{k_3} & x_4^{k_4} \end{bmatrix}^T$

$$\begin{bmatrix} d\delta x_1/dt \\ d\delta x_2/dt \\ d\delta x_3/dt \\ d\delta x_4/dt \end{bmatrix} = \begin{bmatrix} 0 & 1 & 0 & 0 \\ B(x_4^{k_4})^2 & 0 & -BG\cos x_3^{k_3} & 2Bx_1^{k_1}x_4^{k_4} \\ 0 & 0 & 0 & 1 \\ 0 & 0 & 0 & 0 \end{bmatrix} \begin{bmatrix} \delta x_1 \\ \delta x_2 \\ \delta x_3 \\ \delta x_4 \end{bmatrix} + \begin{bmatrix} 0 \\ 0 \\ 0 \\ 1 \end{bmatrix} u$$

For each linear system, we design the corresponding control law. We consider the linear control laws in discrete time mode

$$v^k(t) = -K^k x(t)$$

where $t \in \{p\tau | p \in N_+\}$ and $\tau$ is sampling period. We choose sampling time $\tau = 0.1s$ and the closed loop poles $= \{0.93, 0.92, 0.91, 0.90, 0.89\}$. Using MATLAB functions, we obtain the equivalent discrete time models. An integrator is augmented into the model. The feedback matrix $K$ is calculated using MATLAB functions. The purpose of control is to determine $v(x)$ such that the closed-loop system output $y$ will converge to zero from certain initial conditions.

We simulated each of two methods, fuzzy logic and multidimensional interpolation for two initial conditions: $x(0) = \begin{bmatrix} 1.6 & 0.05 & -0.6 & -0.05 \end{bmatrix}^T$ and $\begin{bmatrix} 2.4 & -0.1 & 0.6 & 0.1 \end{bmatrix}^T$. In the simulations, we solved the differential equations using the second/third order Runge-Kutta method. It is obvious that the linguistic variables of the system is $x_i \in \{$Ball Position, Ball Velocity, Beam Angle and Beam Angular Velocity$\}$. We select the universe of discourse $U$ to be $U = \begin{bmatrix} -10 & 10 \end{bmatrix} \times \begin{bmatrix} -5 & 5 \end{bmatrix} \times \begin{bmatrix} -\pi/4 & \pi/4 \end{bmatrix} \times \begin{bmatrix} -1.6 & 1.6 \end{bmatrix}$. Let $x_1^{k_1} \in \{-10, -\frac{10}{3}, \frac{10}{3}, 10\}, x_2^{k_2} \in \{-5, -\frac{5}{3}, \frac{5}{3}, 5\},$ $x_3^{k_3} \in \{-\frac{\pi}{4}, -\frac{\pi}{12}, \frac{\pi}{12}, \frac{\pi}{4}\}$ and $x_4^{k_4} \in \{-1.6, -\frac{1.6}{3}, \frac{1.6}{3}, 1.6\}$, $k_i \in [1, 4]$, $i \in [1, 4]$. We divide universe of course of each linguistic variables into four terms (Fuzzy partitions) $X_i^k \in \{$large negative, small negative, small positive and large positive$\}$. Each term of linguistic variables is characterised as fuzzy set whose membership function is chosen to be

$$\mu_{x_i^k}(x_i) = \frac{(b_i^k - |x_i - x_i^{k_i}|)}{b_i^k} \qquad \text{and} \qquad \mu_{x_i^k}(x_i) = \exp(-\frac{|x_i - x_i^{k_i}|^2}{(b_i^k)^2})$$

(triangular membership function)        (Gaussian membership function)

The composition operator for connective '*and*' is product and '*or*' is sum. Hence the control action is

$$v = \frac{\sum\limits_{k=1}^{256} \alpha_k v_k}{\sum\limits_{k=1}^{256} \alpha_k}$$

where $v_k = -K_k x$ and $\alpha_k = \prod\limits_{i=1}^{4} \mu_{x_i^k}(x_i)$ is the firing strength.

Figure 3, 4, 5, 6, 7 and 8 show the performance of the closed loop system with different methods for same initial condition $x(0) = \begin{bmatrix} 1.6 & 0.05 & -0.6 & -0.05 \end{bmatrix}^T$. Figure 9, 10, 11, 12, 13 and 14 show the performance of the closed loop system with different methods for another initial condition $x(0) = \begin{bmatrix} 2.4 & -0.1 & 0.6 & 0.1 \end{bmatrix}^T$. Simulation results show that fuzzy logic controllers generate more smooth control output than that of multidimensional interpolation controller. The implementation of multidimensional interpolation controllers results in control chattering which is always undesirable. Chattering can excite high frequency, unmodeled dynamics in the system and lead to degradation in performance. Furthermore, mechanical actuator will be worn due to chattering control. In this simulation, two different membership functions are used. We find that the performance of them are different. This shows that the shape of membership function is one of factors for the controller design. It is expected that fine tuning the membership function will result a better performance. The position error will go zero as shown in the graphs since an integrator is added for the position control.

## 6. Conclusion

A new control design method for smooth nonlinear systems is presented. This design method is shown to be effective through some simulation results of controlling a 4th order nonlinear system which is not feedback linearisation applicable. The controller output is derived from the outputs of linear control laws which are designed based on a set of linearised models of the nonlinear system. The approximation method, fuzzy logic, is proposed to determine the controller output. Although we discuss pole placement state space design theory for linear control laws, any linear control theory can be applied to the local linear law design. It is expected that the overall controller output will be beneficial from more sophisticated linear control theory applied to the local linear systems. eg. robust design.

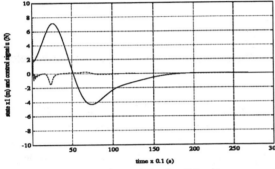

Figure 3 The state $x_1$ and control signal $u$ of closed loop system using fuzzy logic (Gaussian membership function) with initial condition
$$x(0) = \begin{bmatrix} 1.6 & 0.05 & -0.6 & -0.05 \end{bmatrix}^T$$

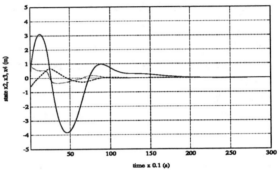

Figure 4 The state $x_2$, $x_3$ and $x_4$ of closed loop system using fuzzy logic (Gaussian membership function) with initial condition
$$x(0) = \begin{bmatrix} 1.6 & 0.05 & -0.6 & -0.05 \end{bmatrix}^T$$

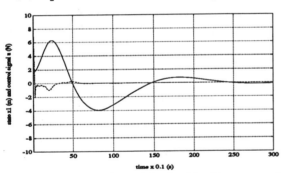

Figure 5 The state $x_1$ and control signal $u$ of closed loop system using fuzzy logic (triangular membership function) with initial condition
$$x(0) = \begin{bmatrix} 1.6 & 0.05 & -0.6 & -0.05 \end{bmatrix}^T$$

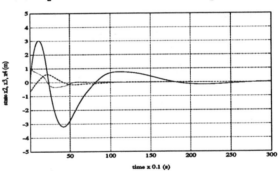

Figure 6 The state $x_2$, $x_3$ and $x_4$ of closed loop system using fuzzy logic (triangular membership function) with initial condition
$$x(0) = \begin{bmatrix} 1.6 & 0.05 & -0.6 & -0.05 \end{bmatrix}^T$$

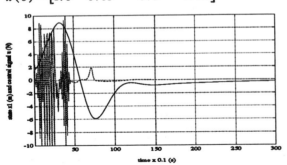

Figure 7 The state $x_1$ and control signal $u$ of closed loop system using multidimensional interpolation with initial condition
$$x(0) = \begin{bmatrix} 1.6 & 0.05 & -0.6 & -0.05 \end{bmatrix}^T$$

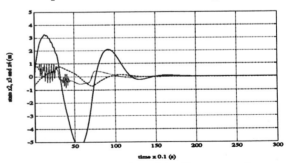

Figure 8 The state $x_2$, $x_3$ and $x_4$ of closed loop system using multidimensional interpolation with initial condition
$$x(0) = \begin{bmatrix} 1.6 & 0.05 & -0.6 & -0.05 \end{bmatrix}^T$$

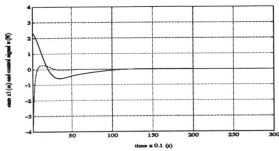

**Figure 9** The state $x_1$ and control signal $u$ of closed loop system using fuzzy logic (Gaussian membership function) with initial condition
$$x(0) = \begin{bmatrix} 2.4 & -0.1 & 0.6 & 0.1 \end{bmatrix}^T$$

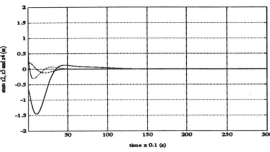

**Figure 10** The state $x_2$, $x_3$ and $x_4$ of closed loop system using fuzzy logic (Gaussian membership function) with initial condition
$$x(0) = \begin{bmatrix} 2.4 & -0.1 & 0.6 & 0.1 \end{bmatrix}^T$$

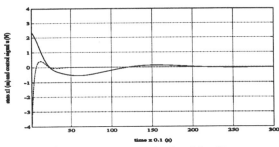

**Figure 11** The state $x_1$ and control signal $u$ of closed loop system using fuzzy logic (triangular membership function) with initial condition
$$x(0) = \begin{bmatrix} 2.4 & -0.1 & 0.6 & 0.1 \end{bmatrix}^T$$

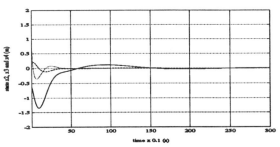

**Figure 12** The state $x_2$, $x_3$ and $x_4$ of closed loop system using fuzzy logic (triangular membership function) with initial condition
$$x(0) = \begin{bmatrix} 2.4 & -0.1 & 0.6 & 0.1 \end{bmatrix}^T$$

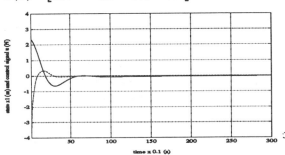

**Figure 13** The state $x_1$ and control signal $u$ of closed loop system using multidimensional interpolation with initial condition
$$x(0) = \begin{bmatrix} 2.4 & -0.1 & 0.6 & 0.1 \end{bmatrix}^T$$

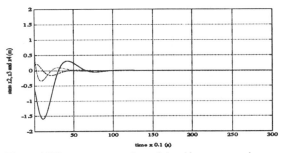

**Figure 14** The state $x_2$, $x_3$ and $x_4$ of closed loop system using multidimensional interpolation with initial condition
$$x(0) = \begin{bmatrix} 2.4 & -0.1 & 0.6 & 0.1 \end{bmatrix}^T$$

**Reference**

[1] T. Takagi and M. Sugeno,"Fuzzy identification of systems and its applications to modeling control", *IEEE Tran. Sys., Man, and Cyb.*, **SMC-15**, Jan/Feb 1985.

[2] J.J. Buckley, "'Sugeno type controllers are universal controller",Fuzzy Sets and Systems,1992,**53**,299-303.

[3] K. Tanakaand K. Sugeno, "Stability analysis and design of fuzzy control system", *Fuzzy Sets and Systems*, **45**, 135-156, 1992.

[4] C.C. Lee, "Fuzzy logic in control systems: fuzzy logic controller, part I", *IEEE trans. Syst. Man Cybern.*, **20**, (2), 404-415, March/April 1990.

[5] C.C. Lee, "Fuzzy logic in control systems: fuzzy logic controller, part II", *IEEE trans. Syst. Man Cybern.*, **20**, (2), 419-435, March/April 1990.

[6] E.H. Mandani, J.J. Ostergaard and E. Lembesis, "Use of fuzzy logic for implementating rule-based control of industrial processes", in Advances in Fuzzy Sets, Possibility Theory, and Applications, edit by Wang, 1983, 307-323, Plenum Press, New York.

[7] L.M. Jia and X.D. Zhang, "Identification of mltivariable fuzzy systems through fuzzy cell mapping", *12th World Congress IFAC*, **9**, 389-393, 1993.

[8] L.M. Jia and X.D. Zhang, "Fuzzy cell mapping approach of complex system modelling and control", *Singapore Internation Conference on Intelligence Control and Instrumentation*, 1380-1385, 1992.

[9] J. Hauser, S. Sastry and P. Kokotovic, "Nonlinear control via approximatie input-output linearization: the ball and beam example", *IEEE Trans. Auto. Contr*, **37**, (3), 392-398, 1992.

# CALLIGRAPHIC ROBOT BY FUZZY LOGIC

Shigehiro Masui, Hosei University, Kajino-cho, Koganei, Tokyo 184, Japan
Toshiro Terano, LIFE, Siber Hegner Bld., 89-1 Yamashita, Naka,Yokohama

## 1.   Introduction

Most of the research of robot or AI aim to imitate the logical thought of human. However, human thought is not only logical but also sensuous. The research, whose goal is to make computer have the sensuous ability, is being increased. The authors have developed a flower arrangement robot[1] which could create a beauty.   In this paper, they also suggest a new robot which writes a beautiful "Kanji" (Chinese letter) automatically.

Japanese calligraphy is a manner how to write Kanji beautifully by using wrighting brush. Of course, this has real purpose of writing a letter, but it has been evaluated as an abstract art for long time. Until now the demonstration of writing Kanji by robot has been done very often. But these are performed by master-slave type robot. That is, robot follows the action of a human calligrapher automatically.

The robot suggested here is constructed so as to have a basic knowledge of generating beautiful Kanji.   The most essential features of beautiful Kanji are extracted from the typical copies and represented as fuzzy rules. This is a difficult work. Another problem is to realize writing motion by arm robot. Calligraph requires very delicate operation of writing brush. The results are effected sensitively by the touch, speed and control of writing brush. But more important factor of evaluation is the balance or harmony of whole Kanji. The rules of precise writing motion or the sensuous evaluation are represented by natural language. Therefore, it is achieved only by fuzzy logic. Fig.1 shows the structure of this system.

## 2. Structure of Kanji

There are two types of Kanji. One of them is a basic character which is transformed from original hieroglyph and has its own meaning. The other has more complex structure, that is, a combination of some basic character. We call these component characters as "Bushu" (element). The element is also composed of some parts such as line(horizontal, vertical), dot, sweep up and wield.

Now we consider the arrangement of elements. Some types of the combination of elements are shown in Fig.2. In this paper we treat only the type A ~ G. These knowledge are stored in "Bushu file". If we input some Kanji to the robot, neccessary elements are automatically retreived from the file.

## 3. Fuzzy Rules of Calligraph

Through the practice of calligraph we see that there are basic rules for brush control when we write parts. For example, when we draw a horizontal line, we first put down the brush obliquely and press it for a time, then move it smoothly to horizontal direction. We stop the brush at the end of line and press it again. By this way, we can draw a strong and sprightly line as shown in Fig.3. The same kinds of basic operation, such as dotting, sweeping up or wielding, are initially programmed and stored in a file. Any parts can be written by combining these basic operations.

It must be pointed out that the robot can't write a beautiful Kanji if the above rules are applied faithfuly. Calligraph is different from the print. We must modify line by warping or inclining in accordance with its size and position in order to draw a beautiful line. Generally speaking, the whole configuration of Kanji should be composed so as to slightly right-side up, and the width of line or the combination of dot and line should be well harmonized. Though the evaluation of harmony is emphirical, the results of evaluation by master calligraphers agree well each other. This means that there exist some common rules.

## 3.1 Fuzzy Rules for Parts

The modification of line is inclination and warp whose degree depends on the position and length of line. Some examples of modification rule are as follows.

As to the horizontal line:

$R_1$ : if the position is upper and length is short, then inclination is big and warp is negative medium.

$R_9$ : if the position is lower and length is long, then inclination is small and warp is positive big.

As to the vertical line:

$R_{10}$: if the position is left and length is short, then inclination is left.

$R_{18}$: if the position is right and length is long, then inclination is zero and warp is negative small.

Fig.4 shows the membership functions of $R_1 \sim R_9$ as an example. As for the vertical line, the warp is very small and always convex to left. Therefore, the value of NS is fixed on 0.05. As to the sweep, standard angle is 45 degree to left and right, but it is changed according to the aspect ratio of the element. In our experiment, the size of Kanji is 140 × 140 mm.

## 3.2 Fuzzy Rules of Combining Element

It is very important to harmonize the size and distance of each element. If the number of parts are many, the element becomes large generally, and vice versa. Beside, there exists an optimum distance among the elements which depends on the size of elements. This size is conveniently represented by the aspect ratio in a standard frame. That is, if the aspect ratio is large, relative size of the element is small. Therefore, we obtain the following rules.

$R_a$ : if the number of parts is few, then the aspect ratio is large.

$R_c$ : if the number of parts is many, then the aspect ratio is small.

Relating to the optimum distance, the rules will be expressed as follows.

$R_{19}$: if the aspect ratio of left element is small and that of right is small , then the distance is large.

$R_{23}$: if the aspect ratio of left element is large and that of right is large , then the distance is small.

These rules are applicable to both lateral and vertical types of combination of elements.

From the result of experiment, we can see that the rules described above are useful and beautiful Kanji can be written when the composition is simple . However, the balance is bad for complex Kanji. Because the distance is too narrow or too wide. Therefore, more precise modification is needed. This modification is realized by adjusting the longitudinal and cross size of the element slightly. The degree of adjustment is determined experimentally. The following is example of typical modification rule.

For the combination of type B ~ D

$R_{24}$: if the aspect ratio of left element is medium and that of right is small, then the longitudinal size of left element is decreased and that of right element is slightly decreased and the cross size of left element and of right element are both slightly decreased.

$R_{29}$: if the aspect ratio of left and right element are both large, then the longitudinal size of left and right is not changed and the cross size of left element is not changed but that of right is slightly decreased.

For the combination of type E ~ G

$R_{30}$: if the aspect ratio of upper element is small and that of lower is small, then the longitudinal size of upper element is slightly decreased and that of lower is not changed and the cross size of upper element is slightly decreased and that of lower is decreased.

$R_{35}$: if the aspect ratio of upper element is medium and that of lower is large, then the longitudinal size of upper element is slightly decreased and that of lower is not changed and the cross size of upper element is slightly increased and that of lower is slightly decreased.

The robot can write good Kanji by these rules. However, final evaluation by human depends on the personel preference. Therfore, the results can be corrected through the human interaction as follows.

$R_{36}$: slightly decrease (or increase) the distance between the elements.

$R_{39}$: increase (or decrease) the size of an element.

These interaction are performed by pushing a key on the board. If an user operates a key once, the distance or size are changed by a given step.

## 4. Experiment

The flow chart of this system is shown in Fig.5. When we input a Kanji, its type is recognized and the elements are retrieved from data file. Next, the size of each element is determined generally by $R_a$ ~ $R_c$. The distances between the elements are calculated from $R_{19}$ ~ $R_{22}$ by approximate reasoning. Then, basic arrangement of elements is obtained. This result is refined by approximate reasoning from $R_{23}$ ~ $R_{34}$. After that, the size and the position of each element are fixed. Next, robot makes a plan of writing parts which are retrieved from data file. The detailed motion of robot is determined by approximate reasoning from $R_1$ ~ $R_{18}$. Fig.6 shows the whole structure of rules which are developed to generate beautiful Kanji.

Fig. 7 and 8 are the examples of Kanji which the robot writes by itself. These Kanji are not so bad. However, another example shown in Fig.9 has a rather complicated structure and its balance is bad. In such a case, human interacts and refines it as shown in Fig.10.

## 5. Conclusion

Our robot is different from the master-slave type which follows the human motion faithfully, but has an intelligence similar to human. Therefore , it can write many Kanji by small number of fuzzy rules. The result is inferior to that written by a master calligrapher, but almost same or little higher level compared with that of average Japanese. One of the problems remained here is automatic evaluation of Kanji. This is very difficult, because it depends on the personel feeling and the traditional culture. Perhaps, special learning process will be neccessary for this purpose. Anyway future robot must have not only logical intelligence but also sensuous feeling. Fuzzy engineering will be a powerful tool for this goal.

References
[1]T.Terano, S.Masui: Flower Arrangement by Fuzzy Robot, Proc. 3rd IFSA Cong . 1/4, Seattle, Aug. 1989
[2]Murakami, Mishima: Calligraphy (Japanese), Geijutu Shinbunsha, 1993
[3]Calligraphy Text Book for middle school (Japanese), Gakko Tosho Ltd.
[4]T.Terano, K.Asai, M.Sugeno: Fuzzy Systems and Applications, Academic Press, 1993

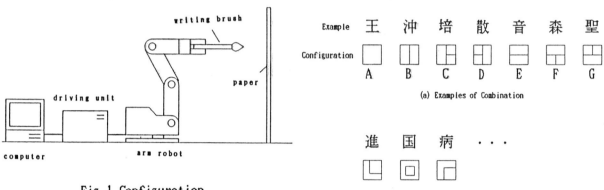

Fig.1 Configuration

Fig.2 Configuration of Elements

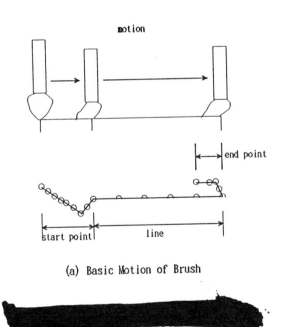

(a) Basic Motion of Brush

(b) Experimental Result

Fig.3 Basic Motion of Writing

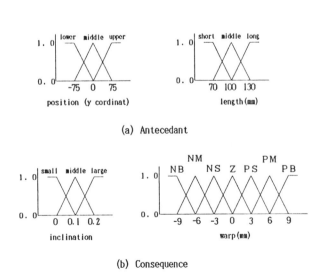

(a) Antecedant

(b) Consequence

Fig.4 M.F. of Modification Rule for Horizontal Line

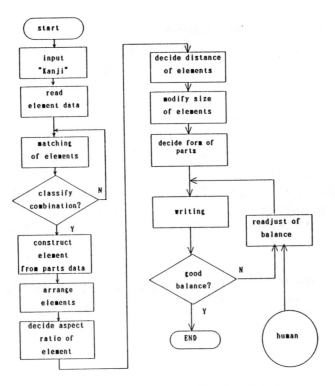

Fig.5 Flow Chart of Writing Robot

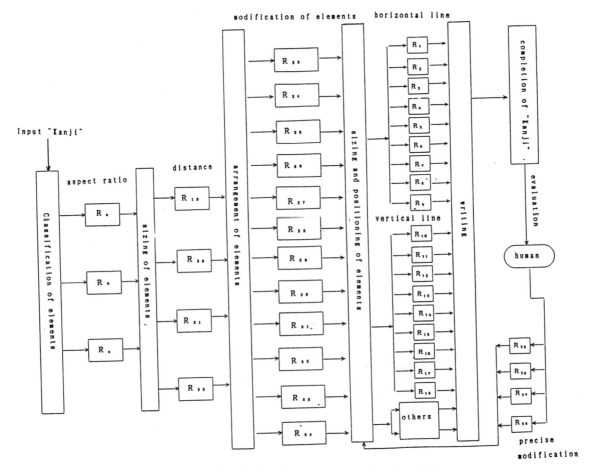

Fig.6 Configuration of Writing Rules

Fig.7 Example of Result(1)

Fig.8 Example of Result(2)

Fig.9 Unbalanced Example

Fig.10 Modification by human

# Mapping alpha-cut borders: classification and PID realization

L. Reznik and A. Stoica
Dept. of Electrical and Electronic Engineering
Victoria University of Technology
PO Box 14428 MMC, Melbourne 3000, Australia
Email: leonid or adrian@cabsav.vut.edu.au

*Abstract*—A new method for fuzzy processing Mapping Alpha-Cut Borders was previously introduced [3] and demonstrated on a control problem. We investigate here some relations with other processing methods. Different variations of the method are being discussed in terms of general properties of fuzzy controllers. The possibility of implementing classic PID controllers with this method is demonstrated.

Keywords: fuzzy reasoning, PID controllers, alpha-cut

## I. INTRODUCTION

Nowadays fuzzy rule-based systems employ different methods of fuzzy processing. A recent comprehensive description of common methods and a comparison between them can be found in [1]. Most of the classic methods operate in three steps: fuzzification (mapping of crisp (fuzzy) inputs into fuzzified inputs), fuzzy inference (mapping fuzzified inputs into fuzzy output sets, according to the rules in the rule-base (or directly into crisp values according to the formulas, i.e. Sugeno's method) followed by a combination of contributions from different rules, and defuzzification (mapping output fuzzy sets to crisp outputs). In [3] a new method for fuzzy processing oriented to fuzzy control applications was proposed. The method, based on mapping of alpha-cut borders, is faster than the methods defuzzifing an output fuzzy set, making it competitive for real-time control applications.

## II. BRIEF DESCRIPTION OF THE MACB METHOD

Determine a fuzzy set **A** as the set of ordered pairs [5], $A = \{(x, \mu_A(x))/x \in X\}$, the alpha-cut $A_\alpha = \{x \in X | \mu_A(x) \geq \alpha\}$, and the prototype for a normalized convex set **A** as the value $x_p$, for which the degree of membership equals unity, $\mu_A(x_p) = 1$.

A first difference between Mapping of Alpha-Cut Borders (MACB) and other methods of fuzzy processing lies in the fuzzification step. Fuzzification in MACB associates an input value not only with a degree of membership to a fuzzy set, but also with the position of the input relative to the prototype (in the set of ordered pairs). An input value x corresponds to one of the two borders of an alpha-cut of an input set $InSet_i$. The fuzzification process specifies the level $\mu$ of the alpha-cut and which of the two borders of the alpha-cut is determined by x, e.g. by indicating the neighbour fuzzy set $InSet_{i\pm1}$ the side on which x stands as relative to the prototype of $SET_i$. For example, if $x > x_p(InSet_i)$ then the set chosen is $InSet_{i+1}$. Thus the notation $\{\mu_1, InSet1_i, InSet1_{i\pm1}\}$ indicates the border of $\{\mu_1$-cut of set $InSet_i$ on the side of the prototype of $InSet_{i\pm1}$. Let InSet1, InSet2 be input sets, OutSet the output set.

Classic fuzzification
$x \rightarrow \{\mu, InSet_i\}$
MACB fuzzification
$x \rightarrow \{\mu, InSet_i, InSet_{i\pm1}\}$

The MACB inference is based on the interpretation that alpha-cuts map according to the rules mapping the fuzzy sets.

Classic inference:

$$\{\mu_1, InSet1_i\} \wedge \{\mu_2, InSet2_j\} \rightarrow \{\mu_0, OutSet_k\}$$

MACB inference:

$$\begin{aligned}
&\{\mu_1, InSet1_i, InSet1_{i\pm1}\} \wedge \\
&\{\mu_2, InSet2_j, InSet2_{j\pm1}\} \rightarrow \\
&\{\mu_0, OutSet_k, OutSet'_k\}
\end{aligned} \qquad (1)$$

$\mu_0 = \mu_1 t \mu_2$, where $t$ is a t-norm and $OutSet_k = rule[InSet1_i, InSet2_j]$ - the set on which $InSet1_i$ and $InSet2_j$ map according to the rules. $OutSet_k = rule[InSet1_i, InSet2_j]$ the set on which $InSet1_i$ and $InSet2_j$ map according to the rules.

Each rule gives one crisp value $z^i$, obtained from the result produced by (1) either
- as a $\mu_0$ cut border of $OutSet_k$ on the side of $OutSet'_k$, if $OutSet_k$ and $OutSet'_k$ do not coincide,
- as the average of the two $\mu_0$ if the two, $OutSet_k$ and $OutSet'_k$ coincide.

"Defuzzification" is determined here by the inverse of the membership function, determining the alpha-cut of the output set and selecting of the borders. The method respects a unique convention for fuzzification and defuzzification, expressed by
$x \rightarrow \{\mu, SET_i, SET_{i\pm1}\} \rightarrow x$
The final result is given by

$$u = \frac{\sum_{k=1}^n z^i \mu_i}{\sum_{k=1}^n \mu_i} \qquad (2)$$

where n is the number of fired rules, $z^i$ is the crisp value contributed by rule i. Weighting can be determined by the importance degree assigned to each rule by the expert, or simply by the $\mu_i$-cut of each rule. If rules have equal weights, the average of the values contributed by each rule can also be considered in determining the final result (as in fact the degree of membership to the output sets was effectively used once in determining the correspondent $\mu_i$-cut), $u = \frac{\sum_{k=1}^n z^i}{n}$

## III. COMPUTATIONAL EFFICACY OF MACB

The proposed method does not require defuzzification of a final output set, which is usually the most computation consuming step in fuzzy processing. The computation time can be evaluated by the number of rules fired by any particular input. If a threshold level is introduced, which determines the minimal value of membership degree necessary for firing a rule, the number of fired rules can be reduced (sometimes significantly, such as in the case of gaussian membership functions). In this way rules with small activation without a significant contribution to the output, are removed from computational procedure. A level of 0.1 (10% of the maximum) can be considered reasonable. For an ordinary controller, it leaves at maximum two rules to be fired by any input. The output value is determined by "strong" rules only.

## IV. ON SOME PROPERTIES OF FUZZY CONTROLLERS BASED ON MACB

Generally, like other fuzzy controllers, those using MACB method are nonlinear controllers, which can produce discontinuity in the output. Discontinuity in the output can be regulated by changing the weights of different values in (2) and using continuous membership functions, or changing threshold level imposed for firing the rules. Continuous control surface is produced in limited regions. The ability of this method to realize different types of control such as PD and PID is demonstrated further.

## V. FORMAL SIMILARITIES WITH OTHER METHODS

### A. Tsukamoto's method

A simplified method of fuzzy reasoning, operating on fuzzy sets for which the membership functions are monotonic was proposed by Tsukamoto, [4]. Our approach is related, we use with partly monotonic function, i.e. monotonic on both sides of the prototype. The way that each of the monotonic functions is mapped is determined by the double access of the rule table.

### B. Sugeno's method

In Sugeno's method the IF part of the rule has fuzzy sets associated to the linguistic terms, the THEN part is given by a formula. The output is a combination of individual outputs

1605

given by fired rules. If the particular formula used in THEN part can be expressed on the base of the inverse of the membership function of an output set selected by the rule than MACB can be considered a particular case of Sugeno's method. If the expert provides the local control function by formula, then Sugeno's method is the appropriate choice. If the expert provides a membership function for an output set, then MACB can be considered. A combination of the two can be imagined, if the knowledge is in hybrid format: for some regions formulas are known, while for other regions fuzzy sets can be provided. As the expert can provide a membership function for the fuzzy output, the choice of its inverse can be considered reasonable to be used as a formula in the THEN part of the rule.

## C. Height Method

Let us compare the last step of MACB (2) with the height method. The Height Method is similarly based on a weighted combination of peaks. In MACB the peaks (corresponding to the prototypes) are replaced by a border of the h-cut, or by the center of the h-cut, (when uncertainty in selecting one of the borders occurs). In some sense the MACB can be regarded as an extension of a method having for defuzzification the Height Method.

## VI. PID REALIZABLE

Mizumoto [2] analyzed some fuzzy processing methods in respect of the possibility of implementing classic PI, PD, PID controllers with fuzzy controllers based on these methods. For a simple case in which an error, a change in error, and an integral of error have associated fuzzy sets with linear membership functions as presented in the figure he concluded that "min-max-gravity" method could not be used for constructing a PD or PID controller while "product-sum-gravity" and "simplified fuzzy reasoning" could. We will demonstrate that MACB also supports the implementation of a classic PID controller under the same assumptions and following [2] in our discussion.

## A. PD controller

Consider the fuzzy sets associated with an error and a change in error as in the figure, where $e_1$ is the minimal and $e_2$ is the maximum value of the error. Let e be the input value for error, de the input value for the change in error. Consider also the output given by fuzzy sets having as prototypes $u_1 = \alpha e_1 + \beta de_1$, $u_2 = \alpha e_1 + \beta de_2$, $u_3 = \alpha e_2 + \beta de_1$, $u_4 = \alpha e_2 + \beta de_2$.

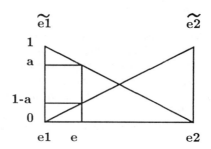

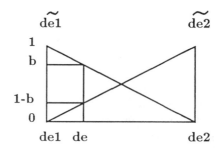

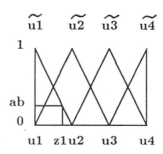

The rules are given in the table.

| $de/e$ | $\tilde{e}_1$ | $\tilde{e}_2$ |
|--------|---------------|---------------|
| $\tilde{de}_1$ | $\tilde{u}_1$ | $\tilde{u}_3$ |
| $\tilde{de}_2$ | $\tilde{u}_2$ | $\tilde{u}_4$ |

*Table 1:* Rule table

## A.1 Fuzzification

$$e \to \{a, \widetilde{e_1}, \widetilde{e_2}\} \quad e \to \{1 - a, \widetilde{e_2}, \widetilde{e_1}\}$$
$$de \to \{b, \widetilde{de_1}, \widetilde{de_2}\} \quad de \to \{1 - b, \widetilde{de_2}, \widetilde{de_1}\}$$

$$a = \frac{e_2 - e}{e_2 - e_1}$$

$$b = \frac{de_2 - de}{de_2 - de_1}$$

## A.2 Inference

**As a particular t-norm in the following we choose the product.**

$$R1 : \{ab, \widetilde{u_1}, \widetilde{u_4}\} \to z_1$$
$$R2 : \{a(1 - b), \widetilde{u_2}, \widetilde{u_3}\} \to z_2$$
$$R3 : \{(1 - a)b, \widetilde{u_3}, \widetilde{u_2}\} \to z_3$$
$$R4 : \{(1 - a)(1 - b), \widetilde{u_4}, \widetilde{u_1}\} \to z_4$$

**The following relations are valid (see below an example for $z_1$)**

$$z_1 = u_2 - abu_2 + abu_1$$
$$z_2 = u_3 - a(1 - b)u_3 + a(1 - b)u_2$$
$$z_3 = u_2 - (1 - a)bu_2 + (1 - a)bu_3$$
$$z_4 = u_3 - (1 - a)(1 - b)u_3 + (1 - a)(1 - b)u_4$$

**Example for obtaining z1.** In the triangle associated with the membership function of $u_1$, the following relation is valid,

$$\frac{u_2 - z_1}{u_2 - u_1} = \frac{ab}{1}$$

$$z_1 = u_2 - abu_2 + abu_1.$$

## A.3 Contribution of all fired rules

$$u = \frac{z_1 + z_2 + z_3 + z_4}{4}$$

**Substituting and grouping the terms the following relation describing a PD controller, is obtained:**

$$u = Ae + Bde + C$$

**where**

$$A = \frac{\alpha}{4}$$

$$B = \frac{2\beta(de_2 - de_1) - \alpha(e_2 - e_1)}{4(de_2 - de_1)}$$

$$C = \frac{\beta(de_2^2 - de_1^2) + 2\alpha(de_2 e_2 - de_1 e_1) + \alpha(de_2 e_1 - de_1 e_2)}{4(de_2 - de_1)}$$

## B. PID controller

Replacing the change of error de with the integral of error ie, we can similarly prove the possibility of the PI controller realization by MACB method. A corresponding set of coefficient $A', B', C'$ can be determined for PI

$$u = A'e + B'ie + C'$$

As the MACB method realizes both PD and PI controllers, our conclusion can be expanded to PID realization. Thus a PID controller realized through a summation of the outputs of a PD and PI controllers would be determined according to

$$u = (A + A')e + Bde + B'ie + C + C'$$

## VII. CONCLUSION

Relations between a recently proposed method, Mapping of Alpha-Cut Borders and some classic methods for fuzzy information processing were investigated. The method was also demonstrated to allow implementation of classic PID controllers.

## REFERENCES

[1] D. Driankov, H. Hellendoorn, and M. Reinfrank. *An Introduction to Fuzzy Control.* Springer Verlag, 1993.

[2] M. Mizumoto. Realization of pid controls by fuzzy control methods. In *Proc. of the 1st IEEE Conference on Fuzzy Systems*, pages 709–715. IEEE, 1992.

[3] A. Stoica. Fuzzy processing based on alpha-cut mapping. In *Proc. of the 5th IFSA Congress*, pages 1262–1265, 1993.

[4] Y. Tsukamoto. *An approach to fuzzy reasoning method.* In Advances in Fuzzy Set Theory and Applications, Eds: M.M. Gupta , R. K. Ragade and R. R. Yager. 1979.

[5] H. J. Zimmermann. *Fuzzy Set Theory and Its Applications.* Kluwer-Nijhoff Publishing, 1991.

# A HYBRID NEURO-FUZZY POWER SYSTEM STABILIZER

A. M. Sharaf
Department of Electrical Engineering
The University of New Brunswick
Fredericton, NB
Canada

T. T. Lie
School of Elect. and Elec. Engineering
Nanyang Technological University
Nanyang Avenue
Singapore 2263

**Abstract** — The paper presents a novel artificial intelligent (AI) neuro-fuzzy hybrid power system stabilizer (PSS) design for damping electromechanical modes of oscillations and enhancing power system synchronous stability. The hybrid PSS comprises a front end conventional analog PSS design, an artificial neural network (ANN) based stabilizer, and a fuzzy logic post-processor gain scheduler.

The stabilizing action is controlled by the post-processor gain scheduler based on an optimized fuzzy logic excursion based criteria ($J_0$). The two PSS stabilizers, conventional and neural network have their damping action scaled on-line by the magnitude of $J_0$ and its rate of change ($dJ_0$). The ANN feed forward two layer based PSS design is the curve fitted nonlinear mapping between the damping vector signals and the desired optimized PSS output and is trained using the bench-mark analog PSS conventional design. The fuzzy logic gain scheduling post-processor ensures adequate damping for large excursion, fault condition, and load rejections. The parallel operation of a conventional PSS and a neural network one provides the optimal sharing of the damping action under small as well as large scale generation-load mismatch or variations in external network topology due to fault or switching conditions.

**Keywords:** Hybrid Power System Stabilizer, Neural Networks, Fuzzy Logic Gain Scheduling.

## 1 Introduction

Power system stabilizers (PSS) are usually used to enhance the power system synchronous stability [1 - 3] and damp the electromechanical oscillatory modes such as machine local, intra-area, and inter-area modes of oscillations. These electromechanical modes of oscillations are usually characterized by natural frequencies in the range (0.1 - 3 Hz.).

Most PSS used in the electric utility industry are analog type with lead-lag compensators, washout, and amplifier gains. The additional damping is introduced through the excitation system by the extra damping electric torque modulation.

Most PSS utilize speed deviation ($\Delta\omega$), accelerating power ($\Delta P_a$), and actual generator active and reactive powers ($\Delta P_G, \Delta Q_G$) or current ($I_g$) as effective modulating signals. The design and selection of the best PSS structure, transfer function, and gains is a complex, iterative process which is usually optimized for a given power system topology structure and for given loading conditions. Such PSS design method usually result in a fixed structure, fixed parameter type PSS, which is limited and may lose its effective damping robustness for any network topology variations, loading conditions, large excursions such as short circuit faults and load rejections.

New techniques [4 -8] such as self tuning, on-line adaptive control, Lyapunov methods result in effective PSS designs which can be made adaptive to any changes in the network or loading conditions. These new techniques require extensive knowledge of the power system dynamics and require estimators which are rather slow to implement on-line and less effectiveness due to incorrect state identification under noisy measurements and system nonlinearities. New techniques [9 - 12] start emerging for the use of expert system, rule based, fuzzy logic, and neural network PSS designs.

This paper presents a novel supplementary ANN based PSS design with a fuzzy switching criteria to allow independent, sequential, or parallel operation in conjunction with the fixed 'optimized' PSS structure. The paper is structured in the following sequence:

1. Introduction

2. Power System Model

3. ANN Based PSS Design

4. Fuzzy Logic Gain Scheduler

5. Sample Simulation Results

6. Conclusions

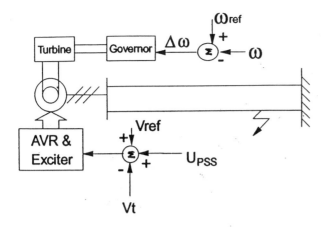

Figure 1: Sample One Machine Infinite Bus System

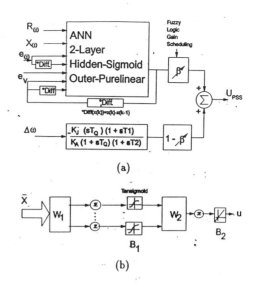

Figure 2: (a) ANN Based PSS and (b) ANN Structure

# 2 Power System Dynamic Model

Figure 1 depicts the sample one machine infinite bus system with the excitation and speed control blocks and the configurations of the conventional and ANN PSS structures with the fuzzy logic gain scheduling post-processor. Figure 2 depicts the detailed ANN based PSS with output $(U_{NPSS})$ using the input modulating damping vector $\overline{X}$

$$\overline{X} = [e_\omega, de_\omega, e_v, R_\omega, X_\omega, dU_{NPSS}.d(dU_{NPSS})]^T$$

and ANN-PSS output

$$y = [U_{NPSS}]$$

where $d$ is the difference on rate operator $dx = x(k) - x(k-1)$ and

$$
\begin{aligned}
e_\omega &= \Delta\omega = \omega - \omega_0 \\
de_\omega(new) &= \omega(new) - \omega(old) \\
R_\omega &= \sqrt{e_\omega{}^2 + de_\omega{}^2 + e_v{}^2} \\
X_\omega &= e_\omega \cdot de_\omega
\end{aligned}
$$

$e_\omega, de_\omega$ define the speed error deviation $\Delta\omega$ and its incremental rate. $e_v$ is the incremental 'error' change in machine terminal voltage. $U_{NPSS}$ is the output of the ANN based PSS block. The system was simulated using the MATLAB software package with sampling period of 10 ms. $R_\omega, X_\omega$ are the synthesized additional damping signals representing excursion vector length $(R_\omega)$ and momentum measure $(X_\omega)$. These additional signals are utilized to improve the curve fitting and training of the ANN based PSS design.

## 2.1 The Dynamic Model

The generator is adapted as a third first order differential equations given below:

$$\dot{\delta} = \omega_0\omega \qquad (1)$$

$$\dot{\omega} = \frac{1}{M}(P_M + G + K_d\omega - P_e) \qquad (2)$$

$$\dot{E}_q' = \frac{1}{T_{d0}'}(E_{fd} - (x_d - x_d')i_d - E_q') \qquad (3)$$

where

$$
\begin{aligned}
i_d &= \frac{E_q - E\cos\delta}{x_e + x_q} \\
E_q &= E_q' + (x_q - x_d')i_d \\
P_e &= \frac{E_q' E \sin\delta}{x_e + x_d'}
\end{aligned}
$$

For the AVR and exciter, the following dynamic model is adapted.

$$\dot{E}_{fd} = \frac{K_A}{T_A}(V_{ref} - V_t + U_{PSS}) - \frac{E_{fd}}{T_A} \qquad (4)$$

where

$$V_t = \sqrt{V_d{}^2 + V_q{}^2}$$

and

$$
\begin{aligned}
V_d &= \frac{x_q E \sin\delta}{x_e + x_q} \\
V_q &= E_q' - x_d' i_d
\end{aligned}
$$

For the governor, the following transfer function is considered.

$$G = \left[a + \frac{b}{1 + sT_g}\right]\dot{\delta} \qquad (5)$$

For the conventional PSS, the following transfer function is considered.

$$U_{PSS} = -\frac{K_J}{K_A}\left[\frac{sT_Q}{1 + sT_Q}\right]\left[\frac{1 + sT_1}{1 + sT_2}\right]\dot{\delta} \qquad (6)$$

# 3 ANN Based PSS Design

The best conventional analog PSS bench-mark stabilizer data was utilized to train and validate the ANN PSS. The feed forward neural network [13] structure was utilized with one hidden layer of seventeen (17) neurons and tansigmoid activation functions. The output layer has one (1) neuron with a purelinear activation function. The ANN network was trained using off-line simulation data correlating the input vector $(\overline{X})$ and output $(U_{NPSS})$ for different disturbances such as external network switching, generation-load mismatch, and fault conditions. The ANN block behaves as a nonlinear function approximator correlating the selected input vector $\overline{X}$ to the output $U_{NPSS}$. The error goal was (0.001) while the learning rate was selected at (1e-5). To ensure better function approximation, the speed error difference and PSS output rates (first and second change) were utilized as extra input variables. Additional signals $R_\omega$ and $X_\omega$ were added to enhance the function approximation.

The ANN feed forward structure is shown in Figure 2b.

# 4 Fuzzy Logic Gain Scheduler

Figure 3 depicts the fuzzy logic gain scheduling structure comprising the fuzzification, decision table, and defuzzification stages. The two fuzzy variables $\|J_0\|$ and its rate of change $d\|J_0\|$ defined as fuzzy input variables defined as follows:

$$\|J_0(k)\| = \sqrt{e_\omega{}^2(k) + de_\omega{}^2(k) + e_v{}^2(k)} \quad (7)$$

$$d\|J_0(k)\| = \|J_0(k)\| - \|J_0(k-1)\| \quad (8)$$

Each fuzzy variable is assigned four membership classes, namely {small, medium, large, very large}. Also, the fuzzy output scaling $\beta$ is assigned the same four membership classes {small, medium, large, very large}. Membership functions as shown in Figure 3. The decision table (rule base) is given in Table 1 as follows:

The on-line gain scaling $\beta$ is assigned the following numerical weights corresponding to the four classes {small(0.1), medium(0.6), large(0.8), very large(1.0)}. So, gain $\beta_j$ vector has the following form discrete weights, [0.1, 0.6, 0.8, 1.0] depicting the possible relative sharing levels between the conventional and ANN based power system stabilizers.

The equivalent weighting $\overline{\beta}$ is obtained using the modified center of area (COA) criteria as follows:

$$\overline{\beta} = \frac{\sum_j \omega_j{}^k \cdot \beta_j}{\sum_j \omega_j{}^2} , \quad k = 0.8 - 1.8 \quad (9)$$

and

$$\omega_j = \min\{\mu_{J0}, \mu_{dJ0}\} \quad (10)$$

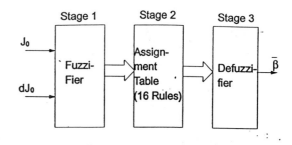

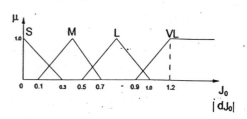

Figure 3: Post-Processor Gain Scheduler ($\overline{\beta}$ varies from 0 - 1.0 pu)

Table 1: Assignment Table (Decision Rule)

| $J_0$ \ $|dJ_0|$ | S | M | L | VL |
|---|---|---|---|---|
| S | S $^{1}$ | S $^{5}$ | M $^{9}$ | L $^{13}$ |
| M | S $^{2}$ | M $^{6}$ | M $^{10}$ | L $^{14}$ |
| L | M $^{3}$ | M $^{7}$ | L $^{11}$ | L $^{15}$ |
| VL | M $^{4}$ | L $^{8}$ | L $^{12}$ | VL $^{16}$ |

The modified COA criteria with $k$ around 1.5 results in better damping, less overshoot/undershoot in the power system underfault and excursion conditions.

## 5 Sample Simulation Results

To validate the hybrid neuro-fuzzy PSS damping effectiveness, two large excursions $F_1, F_2$ were introduced. A 0.5 step change increase in $P_m$ was applied at time $t = 1.5 sec.$ and a 0.6 step change increase in external system equivalent impedance $x_e$ was also applied at time 1.5 sec. Figures 4 and 5 depict the system response without and with the conventional PSS. Figures 6 and 7 depicts the performance with the ANN-PSS design and the hybrid neuro-fuzzy PSS design.

## 6 Conclusions

The paper presents a novel hybrid PSS design for damping electromechanical modes of oscillations and enhancing power system stability. The design utilizes a hybrid conventional PSS, an ANN based PSS, and a gain scheduling fuzzy logic postprocessor.

The proposed hybrid PSS is robust and effective PSS with an on-line scaled damping under small scale and large scale disturbances.

## 7 Acknowledgement

The Authors wish to acknowledge the support of Nanyang Technological University, Singapore and The University of New Brunswick, Canada.

## References

[1] F. P. deMello and C. Concordia, "Concepts of synchronous machine stability as affected by excitation control," *IEEE Trans. on PAS-88* (1969) 316 - 329.

[2] H. A. M. Moussa and Y. N. Yu, "Dynamic interaction of multimachine system and excitation control," *IEEE Trans. on PAS-94* (1974) 1150 - 1158.

[3] R. Fleming, M. A. Mohan, and K. Parvatisam, "Selection of parameters of stabilizers in multimachine systems," *IEEE Trans. on PAS-100* (1981) 2329 - 2333.

[4] S. Lefebvre, "Tuning of stabilizers in multimachine power systems," *IEEE Trans. on PAS-102* (1983) 290 - 299.

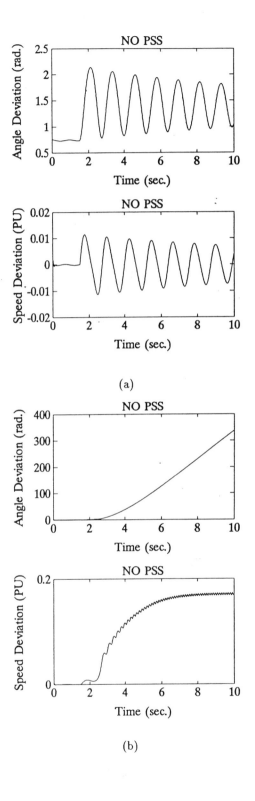

Figure 4: Response to (a) a 0.5 pu Step Change Increase in $P_M$ and (b) a 0.6 pu Step Change Increase in $x_e$ without Conventional PSS

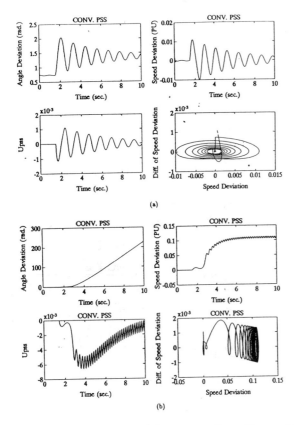

Figure 5: Response to (a) a 0.5 pu Step Change Increase in $P_M$ and (b) a 0.6 pu Step Change Increase in $x_e$ with Conventional PSS

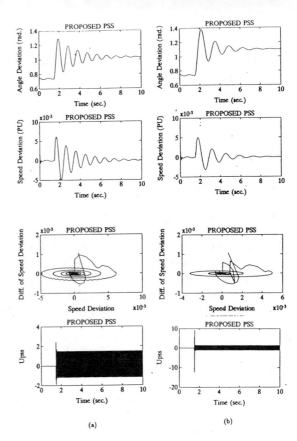

Figure 7: Response to (a) a 0.5 pu Step Change Increase in $P_M$ and (b) a 0.6 pu Step Change Increase in $x_e$ with Hybrid Neuro-Fuzzy PSS

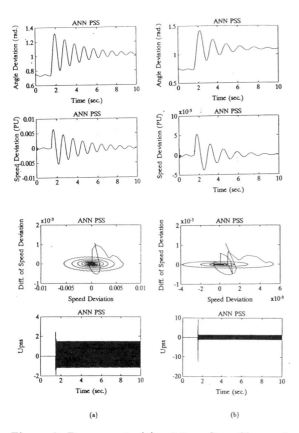

Figure 6: Response to (a) a 0.5 pu Step Change Increase in $P_M$ and (b) a 0.6 pu Step Change Increase in $x_e$ with ANN-PSS

[5] Y. N. Yu, C. M. Lim, and S. Elangovan, "Direct power system stabilizer design for optimal performance of multimachine power systems," *Proc. IFAC Symp. Power Systems and Power Plant Control*, Beijing, China, 1986, pp. 107 - 112.

[6] C. M. Lim and S. Elangovan, "A new stabilizer design technique for multimachine power systems," *IEEE Trans. on PAS-104* (1985) 2393 - 2400.

[7] S. J. Cheng, Y. S. Chow, O. P. Malik, and G. S. Hope, "An adaptive synchronous machine stabilizer," *IEEE PES Joint Power Generation Conf.* Milwaukee, WI, USA, 1985, Paper No. 85 JPGC 601-0.

[8] S. J. Cheng, O. P. Malik, and G. S. Hope, "A self-tuning stabilizer for a multi-machine power system," *Proc. Inst. Electr. Eng.*, Part C, 133 (1986).

[9] A. M. Sharaf, J. Heydeman, and G. Hondred, "Application of Regression analysis in novel power system stabilizer design," *Electric Power Systems Research*, Vol. 22, 1991, pp. 181 - 188.

[10] T. Hiyama, T. Sameshima, and C. M. Lim, "Fuzzy logic stabilizer with digital compensator for stability enhancement of multi-machine power system," *Proc. of $3^r d$ Symp. on Expert System Application of Power System*, 1991, pp. 455 - 461.

[11] Y. Y. Hsu and C. R. Chen, "Tuning of power system stabilizer using artificial neural network," *IEEE Trans. on EC*, Vol. 6, No. 4, 1991, pp. 612 - 619.

[12] T. Hiyama, "Application of neural network to real time tuning of digital type PSS," *Proc. of IPEC'93*, Singapore, March 1993, pp.

[13] D. E. Rumelhart, G. E. Hinton, and R. J. Williams, Learning Internal Representation By Error Propagation, Vol. 2, Chap. 8, *M. I. T. Press*, Cambridge, MA, 1986, pp. 318 - 362.

[14] A. M. Sharaf, T. T. Lie, and H. B. Gooi, "Neural Network Based Power System Stabilizers", *Accepted for presentation and publication in ANNES '93*, New Zealand, 1993.

[15] A. M. Sharaf, T. T. Lie, and H. B. Gooi, "A Neuro-Fuzzy Switchable Power System Stabilizer," *Proc. of IASTED*, Pittsburgh, USA, May 1993.

[16] A. M. Sharaf, T. T. Lie, and H. B. Gooi, "A Neuro-Fuzzy Adaptive Power System Stabilizer," *Accepted for presentation and publication in AUPEC '93*, Australia, 1993.

## Disturbances

$F_1$ – A 0.5 pu step change increase in $P_M$ was applied at time t = 1.5 sec.
$F_2$ – A 0.6 pu step change increase in $x_e$ was applied at time t = 1.5 sec.

## List of Symbols

$\delta$ - rotor angle corresponding to the infinite bus system

$\omega$ - deviation of rotor speed from synchronous speed $\omega_0$

$M$ - effective inertia $(2 * H)$

$K_d$ - equivalent damping factor

$E$ - amplitude of the infinite bus voltage

$E'_q$ - transient q-axis voltage

$E'_{fd}$ - transient excitation voltage

$T'_{d0}$ - equivalent transient rotor time constant

$x'_d$ - d-axis transient reactance

$x_d$ - d-axis reactance

$x_q$ - q-axis reactance

$P_M$ - mechanical power

$P_e$ - electrical power

$V_t$ - terminal bus voltage

$U_{PSS}$ - PSS control signal

$K_A$ - amplifier gain constant

$T_A$ - amplifier time constant

# Appendix

## Synchronous Machine Parameters (in PU)

$\omega_0 = 314.16$   M = 6.92   $K_d$ = -5.0
$x_d = 1.24$   $x_q = 0.743$   $x'_d = 0.32$
$x_e = 0.6$   $P_M = 0.75$   $E = 1.0$
$T'_{d0} = 5.0$

## AVR and Excitation System Parameters (in PU)

$K_A = 400$   $T_A = 0.02$   $V_{ref} = 1.0$

## Governor System Parameters (in PU)

a = -0.001238   b = -0.17   $T_g = 0.25$

## Conventional PSS Parameters (in PU)

$K_J = 40$   $T_Q = 2.5$   $T_1 = 0.1$   $T_2 = 0.03$

# Optimization of a water-treatment system with fuzzy logic control

Th. Froese, Foxboro Deutschland GmbH, Düsseldorf, Federal Republic of Germany,
C. v. Altrock, INFORM GmbH, Aachen, Federal Republic of Germany / Chicago, Illinois,
St. Franke, INFORM GmbH, Aachen, Federal Republic of Germany / Chicago, Illinois.

*Abstract:* Fermentation sludge obtained in the course of producing penicillin as a by-product contains microorganisms and remnants of nutrient salts. It is the basic material for a high quality fertilizer that can be sold. To gain the fertilizer the sludge is concentrated in a decanter and then cleared of the remaining water in a vaporizer. In order to reduce energy costs of the vaporizing process the separation of water and dry substance in the decanter has to be optimized. Due to complexity, caused by long deadtimes, the proportioning process of precipitants of a decanter is often controlled manually. From an economical point of view installations are normally operated suboptimal to avoid endangering the process. Sludge draining of a fermentation applied by a company in Austria producing penicillin demonstrates how the process can be optimized by use of fuzzy technologies.

## 1. Introduction

Restrictive ecological regulations and rising operating expenses are forcing operators of sewage works to apply economical methods to optimize their processes. In this context proven techniques of control engineering are often taken into account first. In chemical industry conventional control systems have been successfully applied especially to continous processes [3, 4, 6, 8].

The major drawback of these methods is the discrepancy between a high amount of time that is necessary for deriving an exact mathematical model of the process and the economical advantages offered by a high degree of automation [5]. The expense required for optimizing complex processes is mostly within the range of some man-years, hence break-even may take a long time.

Due to the problems mentioned above many plants are operated manually. In this case is Fuzzy logic a good method to optimize this proces with very low costs.

In the case presented in this paper, the proportioning process of precipitants of a decanter, it turned out that the use of fuzzy logic made it possible to solve a problem normally requiring complex algorithms within a very short development time. Using Inform's *fuzzy*TECH design software and Foxboro's I/A-Series-System an efficient controller was implemented within only two man-weeks.

## 2. Description of the process

The application presented in this paper was realized in cooperation with one of the largest producers of oral penicillin. The penicillin is obtained by a fermentation process and then extracted from fermentation sludge by solvents. The remaining slurry contains dead microorganisms and remnants of nutrient salt, and therefore has to be regarded as biological waste which has to be disposed of in an economical and ecological justifiable way. For further utilization the slurry is dried. The water precipitated in the course of this process is clarified in a purification plant and then introduced into surface water. The remaining dry substance is a high quality fertilizer that is being sold.

The dry substance and the liquid are separated in two steps. First the slurry is concentrated as high as possible by a decanter. Then, in a second step, the thickened slurry is cleared of remaining water in a vaporizer.

The sewage out of the fermenters contains about 2 % dry substance. In a first step the sewage is neutralized by milk of lime. Second it is biologically degraded in a fermenter and conducted to a large reactor where bentonite is added. This results in a pre-coagulation of the sludge. Now the slurry is enriched with cationic polymer before it is separated from water in the decanter. The cationic polymer discharges the surface charge of the sludge particles and hence leads to coagulation.

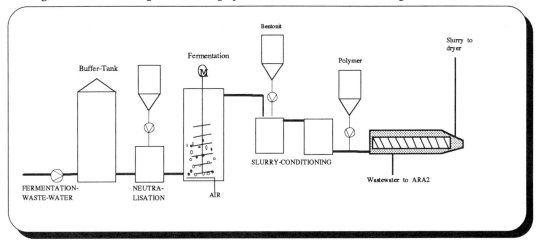

Fig. 1: Process diagram of sludge draining

Bentonite and polymer addition exert a fundamental influence on the drainage quality obtained in the decanter. In order to achieve a high drainage quality, the gradation of chemicals has to be optimized.

A feed-forward control of the input of chemicals is not possible because no data is available on which judgement about the requirement of chemicals of the fluid slurry can be based. Drainage quality and the requirement of chemicals is determined later by two values:

· The percentage purity of the water is measured by a turbidimeter. The higher the turbidity, the higher is the remaining solids content of the water.
· The draining degree of the slurry is measured by conductimetry. The higher the conductability of the slurry, the higher is its water content.


### 3. Objectives of the Project

Objective of the optimization is a highly economical realization of the drying process, in particular:

· The amount of precipitants should be minimized. Particularly the reduction of polymer is a point of interest due to its high cost.
· Sludge draining ought to be carried out to the greatest possible extent in order to reduce the cost of vaporizing the remaining water.
·The biomass content of the sewage water coming out of the decanter should range about 0.7 g/l. The value of 1.5 g/l must not be exceeded because otherwise the operatability of the clarification plant might be reduced. Exceeding the limit can result in a breakdown of the next sewage stage because of its limited capacity to degrade biomass.

Measurement of biomass content by degrees of turbidity (TE/F) enforces to define a limit in TE/F that must not be exceeded. Due to inaccuracy in turbidimetry a large security range has to be obeyed. Accuracy of the tubidimeter is only guaranteed up to 3,500 TE/F, so that this value has been defined as a limit. In case of exceeding this limit the controller has to reduce turbidity without respect to economical aspects.

Below a critical biomass content of the sewage water only the first two objectives are of relevance. A rough cost estimation shows that the most effective way to reduce the expenses of the process is to minimize the energy used in the evaporaters. If the dry substance content of the thick slurry is increased by 1% (this corresponds to a change of conductability of about 0.5 mS/cm), the costs of energy (used in the evaporator) could be reduced by 120$ per day. A reduction of bentonite addition by 10% saves about 25$ per day, and a decrease of polymer addition by 10% reduces costs by 38$ per day. A small decrease of chemical gradation can result in a significant reduction of dry substance contents in thick slurry. Therefore the main objective of optimization is to obtain the best possible result of draining with least use of chemicals.

## 4. Correlations and Rules for the Design of a Fuzzy Controller

Increase of polymer addition in principle results in an increase of drainage grade; the degree of turbidity of the sewage water (degree of suspended matter) and conductability of the thick slurry (water content) is hence reduced. As soon as an optimal polymer gradation is exceeded, further increase of polymer addition leads to a decline of the separative power in the centrifuge. The objective of a controller is to find the optimum of the unknown function which desrcibes this dependency.

As no mathematical model for this plant exists and a shift of the optimum is also influenced by exogenous causes in its position as well as in its size, its actual position can only be infered from an evaluation of the systems reaction to an arbitrary change in polymer gradation. If the draining of thick slurry improves, the conclusion will be that this decision was correct, and it can be repeated, if necessary. If the decision results in a decline of the process, the fuzzy controller must pursue the opposite strategy. The stronger the system's reaction to a change in polymer addition, the greater is the system's distance to the optimum.

Likewise an increase in bentonite addition leads to an increase in separative power of the decanter, until an optimum is reached. If more bentonite is added beyond this point, neither improvement nor decline will occur. Thus an increase in bentonite addition beyond the optimal point is less critical than an increase in polymer addition. Yet bentonite addition should be kept as low as possible to lower the operating expenses.

According to the works manager's observations it is probable that there exists a correlation between the required quantities of polymer and bentonite. Although the degree of this correlation is only of subordinate importance it should be considered in the development of the controller.

With respect to this correlation it was agreed on an independent regulation of bentonite and polymer addition, except until a relation of polymer to bentonite (measured in $kg/m^3$) falls below 0.06 or exceeds 0.09. In operation there has been positive experience within this variation. Usually the composition was changed manually so that the relation of polymer to bentonite was about 0.075.

Regarding the biomass content of the sewage water, a gradation controller must under any circumstances prevent exceeding the critical value of about 1.5 g/l. This presumably corresponds to about 5,000 turbidity units (TE/F). The turbidimeter can only measure a maximum of 4,000 TE/F and a set-value of 0.7 g/l (i.e. about 2,500 TE/F) is recommended. For these reasons neither polymer nor bentonite addition should be reduced if about 3,000 TE/F are exceeded. From a value of 3,500 TE/F upwards the quantity of precipitants even should be increased because the measured value should be within a measurable range to exclude exceeding the value of 5,000 TE/F at all events.

## 5. Design of the Fuzzy Logic Controller

A change in bentonite addition has much more a long-term effect than a change in polymer addition, due to the higher dead time. As a correlation between bentonite and polymer addition exists, it is sensible to change the polymer addition in dependence of the separative quality in the short term and to let the bentonite controller run with delay.

Sewage quality is only restricted by an upper limit, whereas slurry quality is of great economical importance (see above). For these obvious reasons polymer addition is controlled in a first set of rules in correlation to the slurry quality, and thereby aims at the optimum. An absolute point of reference cannot be chosen because of the problem with the changing optimum of polymer addition, and for these reasons absolute input or output values cannot be used. The controller can determine its position in the objective function from a prior change in polymer addition and the resulting reaction of conductability. Due to these considerations an algorithm for optimization can be developed.

The first rule block of the fuzzy controller works on the targeting. Input values for this rule block are a recommendation of the fuzzy controller, taken out of the previous cycle pr_decision (increment on the set-point previous cycle) and the first differentiation of conductability d_Leitf. The output of the rule block is an increment (d_poly) on the set-point of the polymer controller. Control strategy of this rule block is represented by the matrix shown below. The markings in the matrix represent the operational signs of the output increment d_poly.

| d_conduct. / pr_decision | negativ | null | positiv |
|---|---|---|---|
| negativ | negativ | 0 | positiv |
| null | 0 | 0 | positiv (1) |
| positiv | positiv | negativ | negativ |

Fig. 2: Matrix representing the control strategy of polymer proportion

A positive change in conductability implies a decline of the plant's operating state. If the controller in the previous cycle (pr_decision) recommended "negative" (reduce polymer addition) and conductability increased as a reaction to this, the controller must now operate in the opposite direction and increase polymer addition, etc..

The turbidity of the sewage water forms a restriction that in some cases forbids the reduction of the precipitant addition. Therefore the result of the first rule block must be verified by a second rule block, which takes into account the turbidity of the sewage water. This second rule block takes the output of the first rule block for its input, and the degree of turbidity (measured in TE/F) for a second input. As long as turbidity units are below 2,500 TE/F, the second rule block transfers the first block's recommendation to the output. At about 3,000 TE/F all recommendations of the first rule block are transferred, except recommendations to reduce polymer addition. If about 3,500 TE/F are exceeded the second rule block always recommends an increase of flocculent addition. Only the rate of increase of flocculent addition is changed, dependent on the recommendation of the first rule block.Fig. 3: The structure of the Fuzzy Controller.

The controller was implemented on a Foxboro I/A-Series-System. The Fuzzy-Controller ist connected to the conventional control-blocks. To reduce disturbances by noise, all input-signals are conditioned by filter-modules. The complete structure of the generated controller is shown in Fig.4.

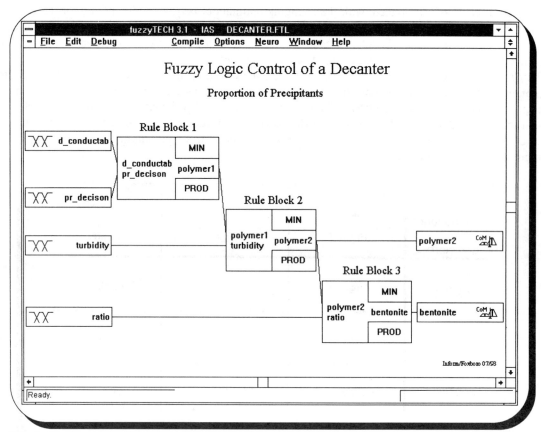

Fig. 3: The structure of the Fuzzy- Controller

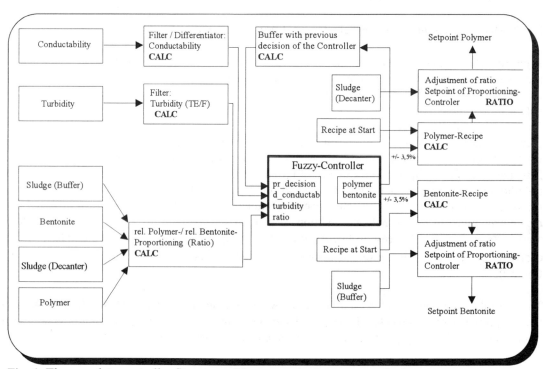

Fig. 4: The complete controller-Structure of the proportioning process of a decanter

## 6. Results

The project was completed to the development of the above described prototype in about two man-weeks. Most of the time was used to design the filters through respective CALC-blocks, to process them, and to document the development work. The prototype hardware used to integrate the fuzzy logic controller into the I/A-Series system also contributed to the time consumed (e.g. programming, compilation). The result of this project can be illustrated by the development of expenses.

After first problems in software adaption were solved expenses of the plant operated by the fuzzy controller reduced considerably, compared to manual operating, resulting in a ROI of about three months.

## References:

[1]     von Altrock, C., Krause, B., Limper, K. und Schäfers, W., "Regelung einer Müllverbrennungsanlage mit Fuzzy Logic", in Zimmermann/v. Altrock (Hsg.), "Fuzzy Logic - Anwendungen", Oldenbourg Verlag, München, 1993.

[2]     Assilian, S. and Mamdani, E. H., "An experiment in Linguistic Synthesis with a Fuzzy Logic Controller", Intern. J. Man-Machine Stud. 7, pp. 1-13, 1975.

[3]     Froese, Th., "Die Optimierung einer C2-Hydrierung", in Zimmermann/v. Altrock (Hsg.), "Fuzzy Logic - Anwendungen", Oldenbourg Verlag, München 1993.

[4]     Hanakuma, Y. et al., "Ethylen plant destillation column bottom temperature control", Keisi Vol. 32, No. 8 1989, pp. 28-39.

[5]     Müller-Nehler, U., 1992. Werkzeug und Erfahrungen in der Hoechst AG beim Einsatz von Fuzzy Control. IIR-Fuzzy-Conference Munich, 05/1992.

[6]     Roffel, B. and Chin, P.A., "Fuzzy control of a polymerisation reactor", Hydrocarbon Processing, June 1991, S. 47 - 50.

[7]     Tobi, T. and Hanafusa, T., "A practical application of fuzzy control for an air-conditioning system", International Journal of Approximate Reasoning 5 (1991), S. 331 - 348.

[8]     Yagishita, O., Ioth, O., Sugeno, M., 1985. Application of fuzzy reasoning to the water purification process. In Sugeno, M., ed. Industrial Applications of Fuzzy Control. Amsterdam, New York.

[9]     Zadeh, L. A., "Outline of a New Approach to the Analysis of Complex Systems and Decision Processes", IEEE Transactions on Systems, Man and Cybernetics, Vol. SMC-3, No. 1, pp. 28-44, 1973.

[10]    Zimmermann, H.-J., Fuzzy Set Theory - and its Applications, 2nd rev. Ed., Kluver, Boston, 1991.

[11]    Zimmermann, H.-J. and Zysno, P., "Latent Connectives in Human Decision Making", FSS 4, pp. 37-51, 1980.

# Fuzzy Sliding Mode Control of Nonlinear System*

Chen-Sheng Ting, Tzuu-Hseng S. Li†, and Fan-Chu Kung
Control System Laboratory, Department of Electrical Engineering
National Chen-Kung University, Tainan, 70101, Taiwan, R.O.C.

## Abstract

The unmodeled dynamics of nonlinear control system is studied by using fuzzy control approach. Based on variable structure system(VSS) theory, this paper presents two fuzzy control schemes. As the upper bound of system uncertainty is known, it can be used as a clue to form fuzzy control logic. The fuzzy inference provides an estimation for the unknown upper bound. Examples are given to illustrate the application of the proposed methods.

## 1 Introduction

The fuzzy control system (FCS) introduced by Mamdani [1] has been applied in the industry for years. Generally it is called a heuristic control because the control rules are formed mostly dependent on the expert knowledge. For an ill-defined and complex system fuzzy control perhaps provides another way to the design. However, the major drawback of FCS is lack of adequate analysis and design technique. Most of its control rules are human dependent, i.e., different expert will give different rules in spite of the same performance of the system. Hence the research on the systematic analysis and design of FCS becomes an important subject of fuzzy control theory.

In this paper, we present two analytical models of FCS which are based on sliding mode control (SMC) of variable structure system theory[2]. The sliding mode control is a nonlinear control approach which garantees an effective control strategy in the face of modeling imprecisions. The stability is attained by driving system states to the sliding surface. The modeling imprecisions, comparing with fuzziness, seem to have some common natures. As the upper bound of system uncertainty is known ( or crisp) , the control output will be determined according to the distance from sliding hyperplane and magnitude of this bound. Usually it is very large and discontinuous, by which the chattering may be induced. We adopt the intrinsic quality of sliding mode control as a basic structure of FCS. The upper bound can directly be used as control rules or estimated by fuzzy inference. The stability analysis are also examined.

## 2 Fuzzy Control in Sliding Mode

Consider a single-input single-output (SISO) nth-order nonlinear system described by

$$x^n = f(X) + u, \qquad y = x \tag{1}$$

*This work was supported by National Science Council of the Republic of China under grant No. NSC83-0404-E006-002.
†Author for correspondence.

where $X = [x \; \dot{x} \cdots x^{n-1}]^T$ denotes the state veccter and $f(X)$ is an unknown continuous function. The control objective is to force the state to track the desired trajectory $X_d = [x_d \; \dot{x}_d \cdots x_d^{n-1}]^T$ in the presence of imprecision on $f(X)$.

Let the tracking error be

$$\tilde{X} = X - X_d \tag{2}$$

i.e., $\tilde{x} = x - x_d, \dot{\tilde{x}} = \dot{x} - \dot{x}_d, \cdots, \tilde{x}^{n-1} = x^{n-1} - x_d^{n-1}$. Define a time-varying surface $S(t)$ in the state-space $R^n$ by the scalar equation $s(X,t) = 0$ [3], where

$$
\begin{aligned}
s(X,t) &= (\frac{d}{dt} + \lambda)^{n-1}\tilde{x} \\
&= \tilde{x}^{n-1} + a_1\tilde{x}^{n-2} + \cdots + a_{n-1}\tilde{x}
\end{aligned}
\tag{3}
$$

and $\lambda$ is a strictly positive constant. As the states are maintained on $S(t)$, the invariancy of the sliding mode to the uncertainty or disturbance is achieved. Since $s(X,t)$ is a Hurwitz polynomial, $\tilde{x}$ will approach zero with given initial condition $\tilde{x}(0)$, and control object is completed. Furthermore, $s(X,t)$ also represents the distance from the surface $S(t)$ while its value is not zero.

The sufficient condition for $S(t)$ being reached is $s(X,t)\dot{s}(X,t) < 0$ [2]. Differentiating $s(X,t)$ with respect to time yields

$$
\begin{aligned}
\dot{s}(X,t) &= \tilde{x}^n + a_1\tilde{x}^{n-1} + \cdots + a_{n-1}\dot{\tilde{x}} \\
&= x^n - x_d^n + a_1\tilde{x}^{n-1} + \cdots + a_{n-1}
\end{aligned}
\tag{4}
$$

By substituting the system dynamics (1) into (4) the control u is then appeared

$$\dot{s}(X,t) = f(X) + u - x_d^n + a_1\tilde{x}^{n-1} + \cdots + a_{n-1} \tag{5}$$

If the upper bound of $f(X)$ is known, control u is chosen as

$$u = -f^u(X,t)sgn(s) + x_d^n - a_1\tilde{x}^{n-1} - \cdots - a_{n-1}\dot{\tilde{x}} \tag{6}$$

This gives

$$\dot{s}(X,t) = f(X,t) - f^u(X,t)sgn(s) \tag{7}$$

and the absolute distance from $S(t)$ is diminished. In general, control u has to be discontinuous across $S(t)$, which will cause an undesirable chattering. Beside, in order to account for the uncertainty, u is very large and implementation cost is increased.

To improve those inadequacy, we propose the fuzzy control scheme as follows. Let system control u be composed of $u_s$ and $u_f$, i.e., $u(t) = u_s(t) + u_f(t)$, where

$$
\begin{aligned}
u_s(t) &= x_d^n - a_1\tilde{x}^{n-1} - \cdots - a_{n-1}\dot{\tilde{x}} \\
u_f(t) &= determined \; by \; fuzzy \; control \; rules.
\end{aligned}
$$

It is assumed that the system state $X$ and $s(X,t)$ can be measured. Then, $s(X,t)$ and $u_f(t)$ form the input space $\Sigma$ and output space $U$ of fuzzy controller respectively. The fuzzy control rules are of the form:

$$R_j : if \; s(t) \; is \; \tilde{\Phi}_j, \; then \; u(t) \; is \; \tilde{U}_{-j}, \; j = -n,\cdots,0,\cdots,+n.$$

The popular triangular shape is used to represent fuzzy set $\tilde{\Phi}_j$ as shown in Fig. 1. To complete the proof of system stability, we introduce the following assumptions:

1. $\tilde{\Phi}_j$ is uniformly spaced over the universe of discourse $\Sigma$, and so is $\tilde{U}_j$.

2. For any $s(X,t) \in \Sigma$, there exists j such that $\mu_{\tilde{\phi}_j}(s) > 0$.

3. Given $U_j$, the center of $\tilde{U}_j$, there exists a positive constant $\kappa$ such that $|U_j| \leq \kappa |S_j|$, where $S_j$ is the center of $\tilde{\Phi}_j$ [4].

4. $\tilde{U}_n = f^u(X) > U_{n-1} > \cdots > U_0 = 0$, and $\tilde{U}_{-n} = -f^u(X) < U_{-n+1} < \cdots < U_{-1}$.

Moreover, based on center of area (COA) strategy, the defuzzification value of $u_f$ is defined as

$$
\begin{aligned}
u_f(t) &= \frac{\mu_{\tilde{\phi}_j}(s)U_{-j} + \mu_{\tilde{\phi}_{j+1}}(s)U_{-(j+1)}}{\mu_{\tilde{\phi}_j}(s) + \mu_{\tilde{\phi}_{j+1}}(s)} \\
&= \frac{\mu_j U_{-j} + \mu_{j+1}U_{-(j+1)}}{\mu_j + \mu_{j+1}} \qquad -n \leq j \leq n-1
\end{aligned}
\tag{8}
$$

From Fig.1 and the described fuzzy control rules, it is clear that $u_f(t) = -f^U(X)sgn(s)$ as $|s(X,t)| > \Phi$ will assure the stability constraint $s(X,t)\dot{s}(X,t) < 0$. When s(X,t) lies within the boundary layer $|\Phi|$, the surface dynamics becomes

$$
\dot{s}(X,t) = f(X,t) + \frac{\mu_j U_{-j} + \mu_{j+1}U_{-(j+1)}}{\mu_j + \mu_{j+1}}
\tag{9}
$$

Since the fuzzy set $\tilde{\Phi}_j$ is uniformly spaced (Assumption 1), let $\Delta = \frac{\Phi}{n}$, $S_j = \Delta \cdot j$, $S_0 = 0$, $S_n = \Phi$, and $S_{-n} = -\Phi$. For $S_j \leq s(X,t) \leq S_{j+1}$, and $0 \leq j \leq n-2$,

$$
\mu_j = 1 - \frac{s(X,t) - S_j}{\Delta}
\tag{10}
$$

$$
\mu_{j+1} = \frac{s(X,t) - S_j}{\Delta} = 1 + \frac{s(X,t) - S_{j+1}}{\Delta}
\tag{11}
$$

consequently

$$
\mu_j + \mu_{j+1} = 1
\tag{12}
$$

Substituting (12)-(12) into (8) yields

$$
\begin{aligned}
u_f(t) &= (1 - \frac{s - S_j}{\Delta})U_{-j} + \frac{s - S_j}{\Delta}U_{-(j+1)} \\
&= U_{-j} - \frac{\delta}{\Delta}s + \frac{\delta}{\Delta}S_j
\end{aligned}
\tag{13}
$$

where $s(X,t)$ is succinctly replaced by $s$, and $\delta$ represents the absolute distance between centres of fuzzy set $\tilde{U}_j$ and $\tilde{U}_{j+1}$. The derivative of $s(X,t)$ can be put in the form

$$
\begin{aligned}
\dot{s} &= -\frac{\delta}{\Delta}s + U_{-j} + \frac{\delta}{\Delta}S_j + f(X) \\
&= -\eta s + f(X) + U_{-j} + \eta S_j \\
&= -\eta s + (\ perturbation\ \Delta S)
\end{aligned}
\tag{14}
$$

which implies that the variable s can be viewed as the output of the first order filter and the perturbation $\Delta S$ caused by system uncertainty is as the input.

An important characteristics has been verified in [3]

$$|\tilde{x}^i(t)| \leq (2\lambda)^i \varepsilon \qquad i = 0, \cdots, n-1 \tag{15}$$

where $\varepsilon = \frac{\Phi}{\lambda}$. With this specific property, we may conclude that $f(X)$ is bounded. Assumption 3 shows $U_{-j}$ is limited. Therefore, $\Delta S$ is bounded and $s(X,t)$ will approach sliding surface finally.

Another aspect we should consider is $s(X,t)$ lying in the interval of $S_{n-1}$ and $S_n$ ( or $S_{-n+1}$ and $S_{-n}$. The $u_f$ is defuzzified as

$$u_f(t) = (1 - \frac{s - S_{n-1}}{\Delta})U_{-n+1} + \frac{s - S_{n-1}}{\Delta}U_{-n} \tag{16}$$

and surface dynamics is

$$\begin{aligned} \dot{s} &= f(X) + (1-b)U_{-n+1} - bf^u(X) \\ &= f(X) - [(1-b)U_{n-1} + bf^u(X)] \end{aligned} \tag{17}$$

where $b = \frac{s - S_{n-1}}{\Delta}$, $0 \leq b \leq 1$, and $U_{n-1}$ is positive. With proper selection of $U_{n-1}$, i.e., expanding boundary layer thickness, negation of $\dot{s}$ might be achieved and system is asymptotically stable. The property in (15) may provide an estimation of $\Phi$.

## 3 Adaptive Fuzzy Control for SMC

For the design of SMC, the upper bound of uncertainty is usually assumed to be obtainable. However, this condition sometimes may not be satisfied due to extraneous disturbance or system complexity. The estimation on perturbation of SMC has been discussed in [5]. Currently, we give a diffenrent estimation model according to the theory of fuzzy reasoning [6], then incorporate estimated result in SMC. Comparing with the conventional estimation, fuzzy estimation can use prior expert knowledge to achieve better performance[7].

Before proceeding some derivations, we give the following statement. For the system described in (1), there exists a positive constant vector $\theta^*$ such that

$$|f(X)| \leq \theta^* W \tag{18}$$

where $W$ is called fuzzy basis function [7] and will be defined latter. Then, use fuzzy implications and reasoning to approximate $\theta^*$. Based on the Lyapunov function, the adaptation law is proposed for regression of a parameter vector in fuzzy inference.

The previous fuzzy input $u_f(t)$ in (4) is redefined as

$$u_f(t) = -K_f sgn(s) \tag{19}$$

and $K_f$ is determined by fuzzy implications. Again, assume that $s(t)$ is measurable but $\dot{s}(t)$ can not be obtained directly. The approximation is made for this value

$$\dot{\tilde{s}}(t) = \frac{s(t) - s(t-l)}{l} \tag{20}$$

where $l$ is a delay constant. With fuzzy implications,

$$R^i : \ if \ s(t) \ is \ \tilde{A}^i \ and \ \dot{\tilde{s}} \ is \ \tilde{B}^i, \ then \ K_f \ is \ \tilde{C}^i, \ i = 1, \cdots, m$$

$K_f$ is calculated as[6]

$$
\begin{aligned}
K_f &= \frac{\sum_{i=1}^{m} \mu_{\tilde{A}^i} \mu_{\tilde{B}^i} \cdot C^i}{\sum_{i=1}^{m} \mu_{\tilde{A}^i} \mu_{\tilde{B}^i}} \\
&= \theta^T W
\end{aligned} \tag{21}
$$

where $\theta = (C^1, \cdots, C^m)^T$ is an adjustable parameter vector, and $W = (w_1, \cdots, w_m)$ is a regressor vector called fuzzy basis function. Note that $C^i$ denotes the center value of fuzzy set $\tilde{C}^i$, and $w_i$ is defined as

$$
w_i = \frac{\mu_{\tilde{A}^i} \cdot \mu_{\tilde{B}^i}}{\sum_{i=1}^{m} \mu_{\tilde{A}^i} \mu_{\tilde{B}^i}} \tag{22}
$$

Let $\tilde{\theta}$ denote the parameter error, $\theta - \theta^*$ and choose the following Lyapunov function,

$$
V = \frac{1}{2}(s^2 + \frac{1}{\gamma}\tilde{\theta}^T \tilde{\theta}) \tag{23}
$$

then differentiating V with respect to time yields

$$
\dot{V} = s\dot{s} + \frac{1}{\gamma}\tilde{\theta}\dot{\theta} \tag{24}
$$

Substituting (19) and (21) into (24), we have

$$
\begin{aligned}
\dot{V} &= s(f(X) + u_f) + \frac{1}{\gamma}\tilde{\theta}^T \dot{\theta} \\
&= sf(X) - \theta^{*T}W|s| + \frac{1}{\gamma}\tilde{\theta}^T(\dot{\theta} - \gamma W|s|)
\end{aligned} \tag{25}
$$

If the adaptive law is chosen as

$$
\dot{\theta} = \gamma W|s| \tag{26}
$$

then (25) becomes

$$
\dot{V} = sf(X) - \theta^{*T}W|s| \tag{27}
$$

which is obviously non-positive for the given condition (18). Hence, $V \to 0$ as $t \to \infty$, that is, $s(X, t)$ will approach zero and asymptotic stability is satisfied. In conclusion, the fuzzy inference can be used as an estimation mechanism for any real continuous function on the compact set [7].

## 4  Examples

Consider a nonliner system $\ddot{x} + a(t)\dot{x}^2 cos(3x) = u$, which was adopted as an example in [3], is used to examine the proposed schemes. The control objective is to follow the desired trajectory $x_d(t) = sin(\frac{\pi}{2}t)$ with the given initial condition $x(0) = -0.4$. The value of $a(t)$ in simulation is $|sin(t)| + 1$ and $f^U(X) = 5x_2^2$. The $\lambda$ in (3) is 20. There are seven fuzzy subsets spaced over $\Phi = 10$ and $\delta = \frac{20}{3}$ in the first methodology. Next, in adaptive fuzzy control, we define five fuzzy subsets, PB, PS, Z, NS, and NB for $s$. The regression rate $\gamma$ and delay constant $l$ are set to 1.0 and 0.01 respectively. The fuzzy sets of $\dot{s}$ are chosen as P(ositive), Z(ero), and N(egative). Hence, we have 15 implications for estimation of $k_f$. The simulation results are shown in Fig.2.

# 5   Conclusion

For unmodeled dynamics systems, this paper introduces analytical design methodologies of fuzzy control. The chattering and large control input of VSS can be avoided by the proposed fuzzy control schemes simultaneously. The adaptive fuzzy control will modify inappropriate control rules and achieve better performance. Simulation shows both strategies can be implemented for the design.

# References

[1] P.J. King and E.H. Mamdani, "The application of fuzzy control systems to industrial processes," *Automatica*, vol. 13, pp. 235–242, 1977.

[2] V.I. Utkin, "Variable structure systems with sliding modes," *IEEE Trans. Automat. Contr.*, vol. AC-22, no. 2, pp. 212–222, 1977.

[3] J.-J. Slotine and W. Li, *Applied Nonlinear Control.* Englewood Cliffs, NJ: Prentice Hall, 1991.

[4] G. Langari and M. Tomizuka, "Stability of fuzzy linguistic control systems," *Proc. IEEE Conf. Decision and Control*, 1990, pp. 2185–2190.

[5] D.Y. Yoo and M.J. Chung, "A variable structure control with simple adaptation law for upper bounds on the norm of the uncertainties," *IEEE Trans. Automat. Contr.*, vol. 37, no. 6, pp.860–865, 1992.

[6] C.C. Lee, "Fuzzy logic in control systems: Fuzzy logic controller, parts I and II," *IEEE Trans. Syst., Man, Cybern.*, vol. 20, no. 2, pp. 404–435, 1990.

[7] L.X. Wang, "Stable adaptive fuzzy control of nonlinear systems," *IEEE Trans. Fuzzy Syst.*, vol. 1, no. 2, pp. 146–155, 1993.

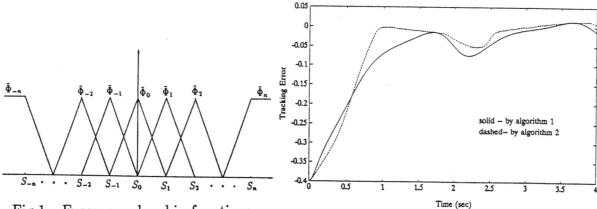

Fig.1.  Fuzzy membership functions

Fig.2.  Control performance

# Fuzzy Logic-Based 'Perception-Action' Behavior Control

# of a Mobile Robot in Uncertain Environments

Wei Li

National Laboratory of Intelligent Technology and Systems,
Department of Computer Science, Tsinghau University, Beijing (100084), China

## Abstract

This paper presents a method for fuzzy logic-based 'perception-action' behavior control of a mobile robot in uncertain environments. A key problem in 'perception-action' behavior control is to coordinate and integrate more reactive behaviors when the mobile robot executes tasks in complex environments. The main idea of the paper is to formulate 'perception-action' behaviors and to coordinate their conflicts and competitions by fuzzy sets and fuzzy rules. An advantage of this method is that the coordination of more reactive behaviors is very robustness (nearly independent of dynamic environments). The simu-lation results show that the proposed method, only using dynamic information acquired by ultrasonic sensors, can perform robot navigation in complex and uncertain environments by efficiently weighting reactive behaviors, such as avoiding obstacle, follow-ing edge, and moving to the target and so on.

## 1 Introduction

A key issue in autonomous robot is robot navigation in uncertain and complex environments. If a mobile robot moves among unknown obstacles to reach a specified target without collisions, sensors must be used to acquire information about real world. Obviously, using such information it is very difficult to build a precise and entire world model in real-time for preplanning a collision-free path. On the basis of reactive behaviors, 'perception-action' behavior control [1][2][3] has been proposed to realize robot navigation. For example, robot wandering behavior can be described by the following two simple perception-actions: if obstacles are located to the left side, the robot turns to the right; if obstacles are located to the right side, the robot turns to the left. Since this method does not need building an entire world model and complex reasoning process, it is suitable for robot navigation in dynamic environments. In practice, more perception-action behaviors must be supplemented, so a key problem in the method is how to coordinate these perception-action behaviors efficiently. In [1], the coordination of more perception-action behaviors is done by inhibiting those reactive behaviors with lower levels according their priority. However, this strategy is not very efficient when a mobile robot executes tasks in complex environments. The example in Fig.1 shows that the robot must efficiently weight more reactive behaviors according to dynamic information, such as avoiding obstacle, following edge, and moving to the target ect., when it reaches a target inside a U-shaped object. The usual approach for implementing perception-action behaviors is artificial potential field [4][5][6]. A drawback to the approach is that during preprogramming much effort must be made to test and to adjust some thresholds regarding potential field for avoiding obstacle, wandering, and moving to target ect. Besides, these thresholds frequently depend on environments.

This paper presents a new method for 'perception-action' behavior control based on fuzzy logic [10]. Unlike 'perception-action' behavior control based on artificial potential field [1][2][3], this method is to compute weights of more reactive behaviors in dynamic environments by the fuzzy logic algorithm rather than simply to inhibit those reactive behaviors with lower levels according their priority. This method also differs from the fuzzy control approaches for obstacle avoidance in [7][8][9] since perception and decision units in this method are integrated in one module by the use of the idea of 'perception-action' behaviors and are directly oriented to a dynamic environment to improve real-time response and reliability. To demonstrate the effectiveness and robustness of the proposed method, we report a lot of simulation results on robot navigation in uncertain environments, such as moving obstacle avoidance in real-time, decelerating at curved and narrow road, escaping from a U-shaped object and moving to target and so on.

## 2 Fuzzy Logic Control Scheme for a Mobile Robot

In order to acquire information about dynamic environments, 15 ultrasonic sensors are mounted on the THMR-II mobile robot with 1.0m length and 0.8m width. These ultrasonic sensors are divided into three groups to detect obstacles to the left, front, right locations, as shown in Fig.2, respectively. The THMR-II mobile robot is equipped with two driving wheels and one driven wheel. The velocities of the driving wheels are controlled by a motor drive unit. The input signals to fuzzy logic navigation algorithm are distances between the robot and obstacles to the left, front, and right locations as well as the heading angle between the robot and a specified target, denoted by *left_obs*, *front_obs*, *right_obs* and *head_ang*, respectively, as shown in Fig.3(a). When the target is located to the left side of the mobile robot, a heading angle *head_ang* is defined as negative; when the target is located to the right side of the mobile robot, a heading angle *head_ang* is defined as positive, as shown in Fig.3(b). According to acquired dynamic information, reactive behaviors are weighted by the fuzzy logic algorithm to control the velocities of the two driving wheels of the robot, denoted by *left_v* and *right_v*, respectively. The linguistic variables *far*, *med* (*medium*) and *near* are chosen to fuzzify *left_obs*, *front_obs* and *right_obs*. The linguistic variables *P* (*positive*), *Z* (*zero*) and *N* (*negative*) are used to fuzzifiy *head_ang*; the linguistic variables *fast*, *med*, and *slow* are used to fuzzify the velocities of the driving wheels *left_v* and *right_v*. In analogy to artificial potential field, distances between the robot and obstacles serve as a repulsive force for avoiding obstacle, while the heading angle serves as an attractive force for moving to target.

# 3 Description of 'Perception-Action' Behaviors Using Fuzzy Logic

In order to reach a specified target in a complex environment, the mobile robot must at least have the following reactive behaviors: 1. obstacle avoidance; 2. following edges; 3. target steer; 4. decelerating at curved and narrow roads. Because a real world is very complex, using ultrasonic sensors it is very difficult to acquire precise information about dynamic environments. In this case, a set of fuzzy logic rules is used to describe the perception-action behaviors mentioned above. Now, we list parts of fuzzy rules from the rule base to explain, in principle, how these the perception-action behaviors are realized (in fact, much more fuzzy logic rules have been used in our navigation algorithms).

## 3.1 Obstacle Avoidance and Decelerating at Curved and Narrow Roads

When the acquired information from the ultrasonic sensors shows that there exist obstacles nearby robot or the robot moves at curved and narrow roads, it must reduce its speed to avoid obstacles. In this case, its main reactive behavior is decelerating for obstacle avoidance. To realize this behavior, we use such fuzzy logic rules as follows:

*If* (*left_obs is near and front_obs is near and right_obs is near and head_ang is any*) *Then* (*left_v is fast and right_v is slow*)
*If* (*left_obs is med and front_obs is near and right_obs is near and head_ang is any*) *Then* (*left_v is slow and right_v is fast*)
*If* (*left_obs is near and front_obs is near and right_obs is med and head_ang is any*) *Then* (*left_v is fast and right_v is slow*)
*If* (*left_obs is near and front_obs is med and right_obs is near and head_ang is any*) *Then* (*left_v is med and right_v is med*)

Such fuzzy rule represents that the robot only pays attention for obstacle avoidance and moves slowly when it is very close to obstacles or at curved and narrow roads.

## 3.2 Following Edge

When the robot is moving to a specified target inside a room (Fig.1) or escaping from a U-shaped obstacle, it must reflect following edge behavior. In order to describe this behavior, we use the following fuzzy rules:

*If* (*left_obs is far and front_obs is far and right_obs is near and head_ang is P*) *Then* (*left_v is med and right_v is med*)
*If* (*left_obs is near and front_obs is far and right_obs is far and head_ang is N*) *Then* (*left_v is med and right_v is med*)
*If* (*left_obs is far and front_obs is med and right_obs is near and head_ang is P*) *Then* (*left_v is med and right_v is med*)
*If* (*left_obs is near and front_obs is med and right_obs is far and head_ang is N*) *Then* (*left_v is med and right_v is med*)

These fuzzy rules show that the robot shall follow an edge of an obstacle when the obstacle is very close to the left (or right) of the robot and the target also is located to the left (or right).

## 3.3 Target Steer

When the acquired information from the ultrasonic sensors shows that there are no obstacles around robot, its main reactive behavior is target steer. Here, we use the following fuzzy rules to realize this behavior:

If (*left_obs is far and front_obs is far and right_obs is far and head_ang is Z*) Then (*left_v is fast and right_v is fast*)

If (*left_obs is far and front_obs is far and right_obs is far and head_ang is N*) Then (*left_v is slow and right_v is fast*)

If (*left_obs is far and front_obs is far and right_obs is far and head_ang is P*) Then (*left_v is fast and right_v is slow*).

These fuzzy logic rules show that the robot mainly adjusts its motion direction and quickly moves to the target if there are no obstacles around the robot.

## 3.4 Coordination of Reactive Behaviors

In 'perception-action' behavior control based on artificial potential field [1][2][3], the velocities of the driving wheels *left_v* and *right_v* are controlled by a reactive behavior that is determined by inhibiting reactive behaviors with lower levels according their priority. In doing this, much effort must be made to test and to adjust some thresholds during preprogramming. Besides, these thresholds usually depend on environments. In our 'perception-action' behavior control, reactive behaviors are formulated by fuzzy rules and the velocities of the driving wheels *left_v* and *right_v* are determined by weighting all reactive behaviors. This is done by the Min-Max inference algorithm and the centroid defuzzification method. Obviously, this strategy for the coordination of reactive behaviors is not to inhibit reactive behaviors with lower levels according their priority, so this is more reasonable for robot navigation.

## 4. Simulations

To demonstrate the effectiveness and robustness of the proposed method, here we report several simulation results on robot navigation in dynamic environments using ultrasonic sensors, such as avoiding obstacle in real-time, decelerating at curved and narrow roads, escaping from a U-shaped object and moving to target and so on.

## 4.1 Moving to a Target inside a U-Shaped Object

Fig.1 illustrates robot motion to a target inside a U-shaped object. At start stage, the robot moves to the target with a high speed since the "moving to target"

behavior is strong due to the large free space around the robot. When the robot approaches to the U-shaped object, it is decelerating by automatically reducing the weight of the "moving to target" behavior and increasing the weight of the "avoiding obstacle" and "following edge" behaviors. When the robot finds out the entry of the U-shaped object, it slowly reaches the target by integrating the "avoiding obstacle" and "moving to target" behaviors.

## 4.2 Escaping from a U-Shaped Object

Fig.4 shows a robot start position is located to the entry side of the U-shaped object and a target position is located to the back side of the U-shaped object. In this case, using artificial potential field the robot is usually trapped inside the U-shaped obstacle due to local minimum. Using our navigation algorithm, the robot moves to the target with a high speed at start stage since there is a large free space around the robot . When it is trapped inside the U-shaped object, the robot is moving along the edge of the U-shape object by increasing the weight of the "following edge" behavior as so to escape the U-shaped object. When the robot goes around the U-shaped object, it drives to the target with a high speed again.

## 4.3 Moving in a Cluttered Environment

Fig.5 shows robot motion in a cluttered environment. We choose at random several targets that are located among different obstacle distribution. Path 1 in Fig.5 represents robot motion from the start position to target 1 located in a narrow road;  Path 2 in Fig.5 represents robot motion from target 1 to target 2 that is behind more obstacles; and path 3 represents robot motion from target 2 to target 3 that is placed in the region where start position is located. It can be observed that, only using ultrasonic sensors to acquire dynamic information, the robot can successfully reach all targets by efficiently weighting more 'perception-action' behaviors using the proposed fuzzy logic navigation algorithm.

## 4.4 Following Wall Edges

In some applications, a mobile robot should be able to move from a room to another room. Fig.6 shows that a start position and a target position are located in different rooms. Using artificial potential field, it is difficult for the robot to reach the target in absence of complete information about the environment. Using our navigation algorithm, however, the robot can automatically act "following edge" behavior (in our algorithm the right-oriented principle is implemented) as so to reach the target when it "hits" the wall.

## 4.5 Decelerating at Curved and Narrow Roads

When the mobile robot operates in outdoor environments, it should be able to tack roads to reach a target. The example in Fig.7 shows robot navigation at curved and narrow roads. The robot begins from its start position and is automatically decelerating at the first curved road with 90°. Then it moves into a very narrow road with a slow speed. At the following curved roads with 90°, the robot automatically makes turns to keep on the roads. Finally, the robot gets the road where the target is located and move to the target with obstacle avoidance by the use of distance information between the robot and obstacles and their heading angle.

## 4.6 Moving Obstacle Avoidance

Fig.8(a)-(d) shows the simulation of robot motion in an uncertain environment with avoiding a moving obstacle. In the example, we set a moving obstacle nearby the target whose speed is lower than that of the mobile robot. Its motion direction is along the wall and just blocks the robot motion to the target in Fig.8(a). In this case, the robot only pays attention for avoiding this obstacle by making right turn, as show in Fig.8(a)-(b). After the robot goes round the moving obstacle, it moves directly to the target in Fig.8(c)-(d)

## 5. Conclusions

In this paper, we use fuzzy logic to realize the perception-action behavior control for robot navigation. Since this method is to weight more reactive behaviors by the fuzzy logic algorithm rather than simply to inhibit those reactive behaviors with lower levels according their priority, it may be more efficient than traditional perception-action behavior control. The navigation algorithm has better reliability and real-time response since perception and decision units are integrated in one module and are directly oriented to a dynamic environment. The simulation results show that the proposed method, only using information acquired by ultrasonic sensors, can perform robot navigation in complex and uncertain environments by weighting more reactive behaviors, such as avoiding obstacles, decelerating at curved and narrow roads, escaping

from a U-shaped object, and moving to target and so on. However, this method does not guarantee to reach a target in some case since complete information is not available. This can be improved by adding a vision system.

## References

[1] R.A. Brooks, "A robust layered control system for a mobile robot", IEEE J. of Robotics and Automation, RA-2, pp. 14-23, April 1986

[2] Ronald C. Arkin, and Robin R. Murphy, "Autonomous navigation in a manufacturing environment", IEEE Tran. on Robotics and Automation, vol.6, no.4, pp.445-454, Aug. 1990

[3] M.D. Adams, Housheng Hu and P.J. Probort, "Towards a real-time architecture for obstacle avoidance and path planning in mobile robot", Proc. IEEE Int. Conf. on Robotics and automation, pp.584-589, March 1990

[4] B.H. Krogh, "A generalized potential field approach to obstacle avoidance control", SME-RI Technical Paper MS84-484, 1984

[5] O. Khatib, "Real-time obstacle avoidance for manipulators and automobile robots", Int. J. of Robotics Research, vol.5, no.1, 1986

[6] T. Wang and B. Zhang, "Time-varying potential field based 'perception-action' behaviors of mobile robot", Proc. IEEE Int. Conf. on Robotics and Automation, pp.2549-2554, May 1992

[7] M. Sugeno and M. Nishida, "Fuzzy control of model car", Fuzzy Sets and Systems vol.16, pp.103-113, 1985

[8] T. Takeuchi; Y. Nagai and N. Enomoto, "Fuzzy control of a mobil robot for obstacle avoidance", Information Science, vol.45, pp.231--248, 1988

[9] M. Maeda; Y.Maeda and S. Murakami, "Fuzzy drive control of an autonomous mobile robot", Fuzzy Sets and Systems, vol.39, pp.195--204, 1991

[10] W. Li. "Robot obstacle avoidance and navigation using fuzzy logic in uncertain environments", technical report, Department of Computer Science, Tsinghua University, 1993

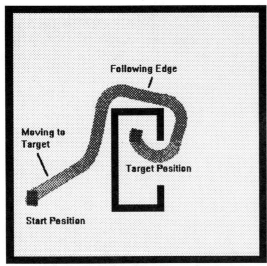

Fig.1: Robot motion to reach a target inside a U-shaped object

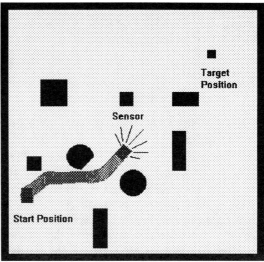

Fig.2: Ultrasonic sensor-based robot motion

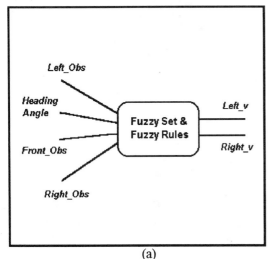

(a)

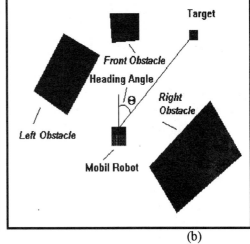

(b)

Fig.3(a)-(b): Fuzzy logic scheme for perception-action behavior control

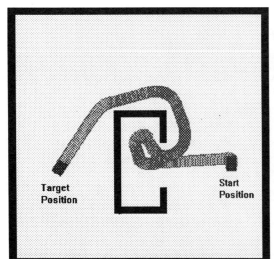

Fig.4: Robot motion to a target with escaping from the U-shaped object

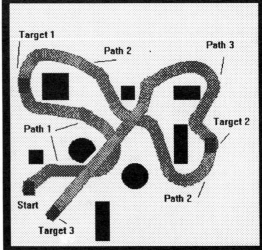

Fig.5: Robot motion to reach more targets in a cluttered environment

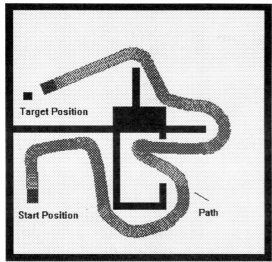

Fig.6: Robot motion to reach a target by following edge behavior

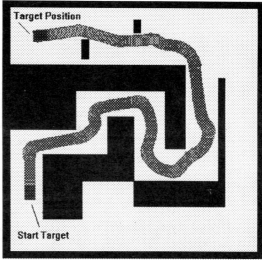

Fig.7: Robot motion with lower speed at curved and narrow roads

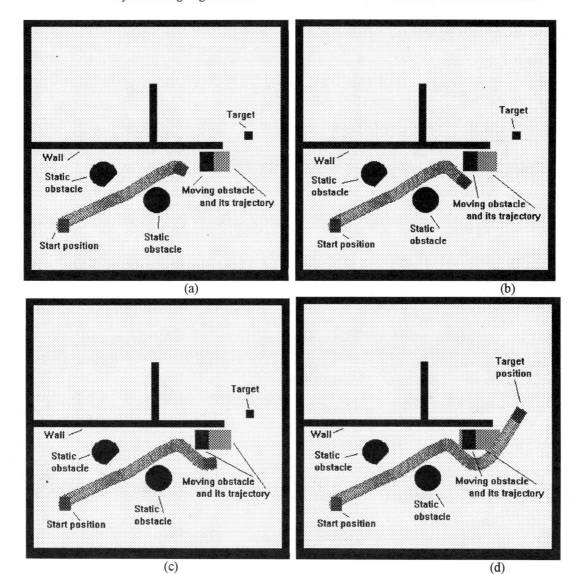

(a)

(b)

(c)

(d)

Fig.8(a)-(d): Robot motion with avoiding a moving obstacle

# Fuzzy Control for Platoons of Smart Cars

Julie Dickerson, Hyun Mun Kim, and Bart Kosko
Department of Electrical Engineering—Systems
Signal and Image Processing Institute
University of Southern California
Los Angeles, California 90089-2564

**Abstract**—An additive fuzzy system can control the throttle of cars in single lane platoons. The system uses fuzzy controllers for velocity control and gap control. Fuzzy controllers create, maintain, and divide platoons on the highway. Each car's controller uses data from its car and the car in front of it. Cars drop back during platoon maneuvers to avoid the "slinky effect" of tightly coupled platoons. Some car and engine types need their own fuzzy rules and sets. A hybrid neural fuzzy system can learn the fuzzy rules and sets from input-output data. We compute new fuzzy rules and sets for a truck velocity controller. The learned system controls the velocity of the truck with no overshoot or slow response. We tested the fuzzy gap controller first with a car model and then with a real car on highway I-15 in California. Each car's controller used data only from sensors on the car. We used this controller to drive the "smart" car on the highway in a two-car platoon.

## I. PLATOONS OF SMART CARS

Traffic clogs highways around the world. Platoons of cars can increase the flow and mean speed on freeways. Platoons are high-speed groups of smart cars in single lanes on freeways of the future. Electronic links tie the cars together. Computer control speeds response times to road hazards so that cars can travel more safely on their own or in groups.

A platoon is a group of cars with a lead car and one or more follower cars that travel in the same lane. The lead car plans the course for the platoon. It picks the velocity and car spacing and picks which maneuvers to perform. Platoons use four maneuvers: merge, split, velocity change, and lane change [5]. A merge combines two platoons into one. A split splits one platoon into two. A lane change moves a single car into an adjacent lane. Combinations help cars move through traffic.

Standard control systems use an input-output math model of the car and its environment. Fuzzy systems do not use an input-output math model or exact car parameters. A fuzzy system is a set of fuzzy rules that maps inputs to outputs [4]. The fuzzy platoon controller uses rules that act like the skills of a human driver. The rules have the form "If input conditions hold to some degree, then output conditions hold to some degree" or "If $x$ is **A**, then $y$ is **B**" for fuzzy sets **A** and **B**. Each fuzzy rule defines a fuzzy patch or a Cartesian product **AxB** in the input-output state space **XxY**. To approximate the function the fuzzy system covers its graph with fuzzy patches and averages patches that overlap [3].

The fuzzy platoon controller (FPC) is a distributed control system for future freeways that drives a car in or out of a platoon. The FPC includes an integrated maneuver controller (IMC) for course selection and an individual vehicle controller (IVC) for throttle, brake, and steering control as in Figure 1.

We designed a fuzzy throttle controller (FTC) for velocity and gap control in smart platoons [1]. The controller uses the throttle only and has thre subsystems for control of velocity and gap distance as in Figure 2. The FTC gets information from its own sensors, the car ahead, and from the platoon goals. We tested the fuzzy gap controller on the highway. The controller got information from its own sensor and from the platoon goals. The controlled car followed the lead car as it changed speed and went over hills. The system performed smoothly in all cases.

Platoons travel at high speeds and need precise longitudinal control for safety. We used a hybrid neural-fuzzy system that finds the ellipsoidal rule-

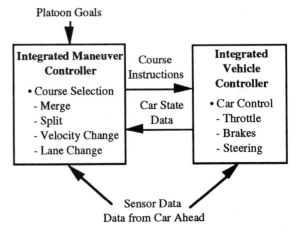

Figure 1. Block diagram of the fuzzy platoon leader system.

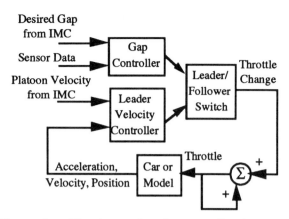

Figure 2. The fuzzy throttle controller has two subsystems that maintain the platoon. The gap controller keeps the cars at a constant distance from one another. The velocity controller keeps the platoon leader at a constant speed.

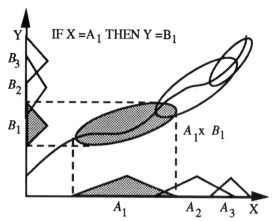

Figure 3. The fuzzy rule patch "If X is fuzzy set $A_1$, then Y is fuzzy set $B_1$" is the fuzzy Cartesian product $A_1xB_1$ in the input-output product space XxY.

patches with unsupervised and supervised learning. The hybrid system uses unsupervised learning to quickly pick the first set of ellipsoidal fuzzy rules. Then supervised learning tunes the rules using gradient descent. Each rule defines a fuzzy subset or connected region of state space [2] and thus relates throttle response, acceleration, and velocity. [2] describes the hybrid ellipsoidal learning system. We used ellipsoidal learning to find the fuzzy rules and sets for a truck velocity controller.

## II. ADDITIVE FUZZY SYSTEMS WITH ELLIPSOIDAL RULES

Additive fuzzy systems can uniformly approximate continuous [4] or measurable [5] functions with fuzzy rule patches. It gives a model-free estimate of a continuous function since it does not use an input-output math model of the function. A fuzzy system contains a set of rules of the form "If input conditions hold, then output conditions hold" or "If X is **A**, then Y is **B**" for fuzzy sets **A** and **B**. Each fuzzy rule defines a fuzzy patch or a Cartesian product AxB as in Figure 3. To approximate the function the fuzzy system covers its graph with fuzzy patches and averages patches that overlap. Centroidal defuzzification with correlation product inference [4] gives the output value $y_k$ at time $k$:

$$y_k = \frac{\int_{D_y} y\, m_B(y)\,dy}{\int_{D_y} m_B(y)\,dy} = \frac{\sum_{j=1}^{r} c_{y_j}\, I_j\, m_{A_j}(x_k)}{\sum_{j=1}^{r} I_j\, m_{A_j}(x_k)} \quad (1)$$

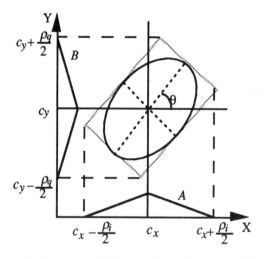

Figure 4. A positive definite matrix **A** defines an ellipsoid about the center **m** of the ellipsoid. The eigenvalues of **A** define the length of the axes. The projections of the ellipsoid onto the axes define the input and output fuzzy sets.

$$= \frac{\sum_{j=1}^{r} \text{Area}\left(B_j'\right) \text{Centroid}\left(B_j'\right)}{\sum_{j=1}^{r} \text{Area}\left(B_j'\right)}$$

$I_j$ is the area of the $j$th output set. $c_{y_j}$ is the centroid of the $j$th output set. $m_{B_j}(x_k)$ scales the output set $B_j$. $r$ is the number of output fuzzy sets $B_j$.

A fuzzy patch can take the form of an ellipsoid [2]. A positive-definite matrix **A** defines an ellipsoid in the $q$-dimensional input-output state space, where $q=n+p$ (Figure 4), $n$ is the number of inputs to the fuzzy system, and $p$ is the number of outputs. The ellipsoid is the locus of all **z** that satisfy [7]

$$\alpha^2 = (\mathbf{z} - \mathbf{m})^T \mathbf{A} (\mathbf{z} - \mathbf{m}) \qquad (2)$$

where $\alpha$ is a positive real number and $\mathbf{m}$ is the center of the ellipsoid in $R^q$. The eigenvalues of $\mathbf{A}$ are $\lambda_1, \ldots, \lambda_q$. The eigenvectors define the ellipsoid axes. The Euclidean half-lengths of the axes equal $\alpha / \sqrt{\lambda_1}, \ldots, \alpha / \sqrt{\lambda_q}$. The projections of the rules onto the input axes form the fuzzy sets.

## III. FUZZY THROTTLE CONTROLLER

In a platoon each car tries to travel at the desired platoon velocity and maintain a fixed gap between cars. The IMC system in the leader car chooses the desired platoon velocity and gaps between cars. When the platoon travels at a constant velocity each car uses its own velocity controller to maintain the desired platoon velocity. These systems use the velocity and acceleration data of the controlled car. Each car measures the velocity and acceleration data of the car in front of it. The system output is the change in throttle angle.

The fuzzy throttle controller (FTC) performs longitudinal maneuvers for the platoon. Separate fuzzy subsystems for velocity and gap control change the throttle angle for cars in the platoon as in Figure 2. The velocity controller controls the speed of the platoon leader. The gap controller controls splits, merges, and changes in spacing for follower cars. The subsystems work together to control the platoon.

The gap controller corrects the distance error when it is too large. The gap controller for platoon followers uses the differences in acceleration and velocity between cars and the distance error to achieve or maintain a constant spacing. The distance error $\Delta d_i(t)$ is the difference between the desired gap between the cars and the actual gap. A range-finding system on each car in the platoon measures the distance between the cars. Equations 3, 4, and 5 give the inputs for the $i$th car [1]:

$$\Delta d_i(t) = d_{gap} - d_i(t) \qquad (3)$$

$$\Delta v_i(t) = v_{i-1}(t) - v_i(t) \qquad (4)$$

$$\Delta a_i(t) = a_{i-1}(t) - a_i(t) \qquad (5)$$

In this analysis the follower cars only get data from their own sensors. The sensor measures the distance and the velocity difference between a car and the car in front of it. We estimate the acceleration input in (5) with the difference of the velocity measurements:

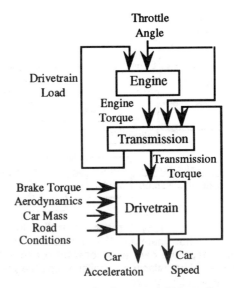

Figure 5. Longitudinal car model block diagram.[6]

$$\Delta a_i(t_k) = \text{sign}(\Delta v_i(t_k) - \Delta v_i(t_{k-1})) \cdot c_a \qquad (6)$$

The acceleration measurements are noisy so we use only the scaled sign of the acceleration. $c_a$ equals the inverse of the system update rate $\tau_s^{-1}$. We used $\tau_s = 0.05$ seconds.

The output of the fuzzy controller in the $i$th car is $\Delta \theta_i$. We use a low pass filter to smooth the throttle input to the car [9]. The input to the low pass filter $\theta_i(t_k)$ is

$$\theta_i(t_k) = \theta_i^{LP}(t_{k-1}) + \Delta \theta_i(t_k) \qquad (7)$$

$\theta_i^{LP}(t_{k-1})$ is the previous output of the filter. The smoothed throttle input is

$$\theta_i^{LP}(t_k) = \frac{2}{3} \theta_i^{LP}(t_{k-1}) + \frac{1}{3} \theta_i(t_k)$$
$$= \theta_i^{LP}(t_{k-1}) + \frac{1}{3} \Delta \theta_i(t_k) \qquad 8)$$

We stored the fuzzy controller as a look-up table based on the fuzzy sets and rules. We scale the output of the look-up table by 1/3. Scaling decreases round-off errors since the microprocessor in the car only uses integer operations.

## IV    GAP CONTROL TEST

We tested the gap controller on highway I-15 in Escondido, California. First we tested small platoons with a realistic car model [6]. Then we put the controller in a real car from VORAD Incorporated. In

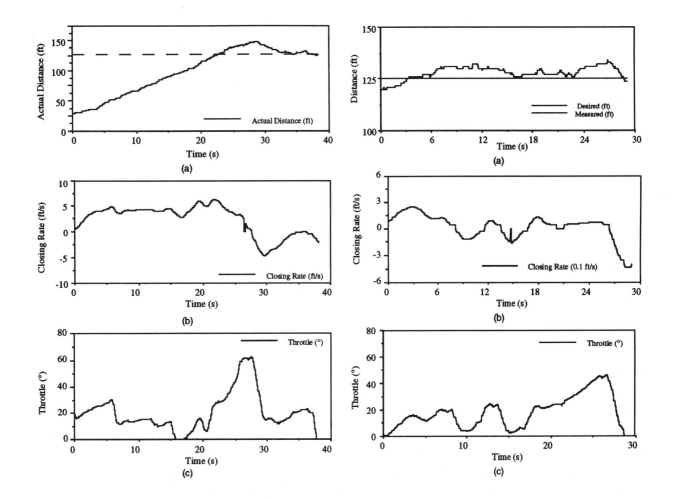

Figure 6. Follower car data as a two-car platoon accelerates on the highway. (a) The distance between the lead and follower cars. (b) Closing rate of the velocity differences between the cars. (c) Throttle values for the follower cars.

Figure 7. Follower car data as a two-car platoon climbs a hill. (a) The distance between the lead and follower cars. (b) Closing rate of the velocity differences between the cars. (c) Throttle values for the follower cars.

this test the follower cars got data from only their sensors. Figure 5 shows the basic subsystems of the validated car model that we used to test our controller. The inputs to this model are the throttle angle and the brake torque. The outputs are the speed and acceleration of the car. This study looks only at throttle control. We set the brake torque to zero. The system has a mechanical delay of 0.25 seconds.

We used a pulse-doppler radar system. The radar measures the distance and the velocity difference between the computer controlled car and the car ahead. The radar has a measurement delay of 0.05 seconds. The radar locks on a target on a car ahead.

We tested the gap controller in a two-car platoon on highway I-15. The desired gap between the cars was 125 feet. Figure 6 shows the platoon as it accelerates onto the highway. The cars started with a separation of 10 feet. The fuzzy controller started when the follower car reached 25 miles per hour. The cruise control does not work below this speed. The platoon

accelerated to 55 miles per hour in 20 seconds. Figure 6 a shows the follower car gap as the platoon accelerates. Figure 6 b shows the closing rate between the cars. Figure 6 c shows the throttle value as the car accelerates. The follower car slowly falls back to the desired gap as the platoon reaches the desired velocity. The follower car overshot the desired gap and then converged to the proper spot.

In the second test the platoon went up and down hills. The desired gap is 125 feet. The follower car drops back as the platoon starts up the hill. Figure 7 a shows the gap distance as the platoon goes up a hill. The follower drops back and then moves to the proper gap. Figure 7 b shows the closing rate between the lead and follower cars. The "spike" at 15 seconds shows the radar sensor losing the lead car in the platoon. When this happens the follower car maintains a constant throttle until the sensor detects a new target. Figure 7 c shows the throttle values as the car ascends the hill.

## V. HYBRID ELLIPSOIDAL LEARNING

We tested a hybrid system that combined unsupervised and supervised learning. The hybrid system used unsupervised learning to quickly pick the first set of ellipsoidal fuzzy rules—it picks them in shape, orientation, and number. Then supervised learning tuned the rules. The first choice of ellipsoid parameters dictates how well the supervised system performs and how fast it converges.

### A. Unsupervised Rule Estimation

First and second order statistics of the data can estimate the fuzzy patches as ellipsoids. Vector quantizers can use competitive learning to estimate the local conditional covariance matrix for each pattern class [4]. The covariance matrix defines an ellipsoid in the $q$-dimensional input-output space centered at the quantization vector or centroid where $q$ is the combined number of inputs $n$ and outputs $p$ of the fuzzy system. Regions of sparse data lead to large ellipsoids and thus less certain rules. The quantizing vector "hops" more as it matches or quantizes more diverse data.

Adaptive vector quantization (AVQ) systems adaptively cluster pattern data in a state space. An autoassociative AVQ system combines the input $\mathbf{x}$ and the output $\mathbf{y}$ of the data to form $\mathbf{z}^T = [\mathbf{x}^T \mid \mathbf{y}^T]$. The $j$th neuron "wins" if $\mathbf{z}$ is closest in Euclidean distance.

Competitive learning estimates the first and second order statistics of the data with the stochastic difference equations for the winning neuron [4]

$$\mathbf{m}_j(k+1) = \mathbf{m}_j(k) + c_k \left[ \mathbf{z}_k - \mathbf{m}_j(k) \right] \tag{9}$$

$$\mathbf{K}_j(k+1) = \mathbf{K}_j(k) + d_k \left[ \left( \mathbf{z}_k - \mathbf{m}_j(k) \right) \left( \mathbf{z}_k - \mathbf{m}_j(k) \right)^T - \mathbf{K}_j(k) \right] \tag{10}$$

The coefficients $c_k$ and $d_k$ must satisfy the convergence conditions in [4]. In practice they decrease linearly in time. The initial choices of $\mathbf{m}_j(0)$ should equal sample data $\mathbf{z}$.

### B. Supervised Ellipsoidal Learning

Supervised learning can also find and tune the ellipsoidal fuzzy patches. The "backpropagation" algorithm[4] learns the ellipsoidal patches by locally minimizing the *mean-squared error* function. The error is the desired output minus the output of the additive fuzzy system. It performs stochastic gradient descent on the instantaneous mean-squared error $SE_k$ [4]:

$$SE_k = \frac{1}{2} \cdot \left( d_k - y_k(\mathbf{x}_k) \right)^2 \tag{11}$$

$d_k$ is the desired output of the system. $y_k(\mathbf{x}_k)$ is the output of the additive fuzzy system at time $k$ for the input $\mathbf{x}_k$ in (1). Gradient descent estimates the eigenvalues, rotation angles, and centroids of the ellipsoidal patches [2]. The partial derivatives and update equations are found in [2].

## VI. LEARNING OF FUZZY RULES

We learned the fuzzy rules for the leader velocity controller using the car model in [8]. The leader velocity controller in [1] generated 7500 training samples in 200 trajectories for a sport utility vehicle. The training vectors ($a$, $\Delta v$, $\Delta \theta$) defined points in the three dimensional input-output space. Unsupervised ellipsoid covariance learning clustered the data and computed its local statistics. The AVQ system had 450 synaptic vectors or local pattern classes. The sum of the ellipsoid projections onto each axis of the state space gave a histogram of the density of the pattern classes. We chose seven sets in each of the input dimensions. The center of each fuzzy rule set matched a peak in the histogram.

We then partitioned the state space into a grid of possible rule patches. We found the rules by counting the number of synaptic vectors in each cell. Clusters of synaptic vectors in fuzzy rule cell defined the rules [4]. The sets for the truck are larger and coarser than those for the smaller and lighter cars.

Next we used supervised learning to tune the rules. We optimized the rules to minimize mean-squared error for the training data. There were 49 rules. The supervised system took 30,000 iterations to refine the data.

Figure 8 shows the results of changing the platoon velocity for the truck using the unsupervised

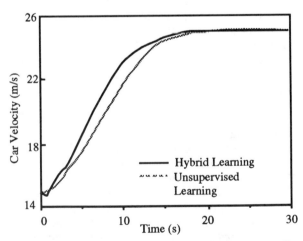

Figure 8. Comparison of the velocity controller performance after unsupervised and hybrid learning. The hybrid controller gives a quick response with no overshoot.

and hybrid controllers. Both controllers had no overshoot when the car sped up. The hybrid controller performed better when the car slowed down. The hybrid controller had no overshoot at the desired speed.

## VII. CONCLUSION

We designed and tested a fuzzy throttle controller for a car platoon. The next phase of the FPL will add the brake and steering controllers so the platoon can maneuver on the highway.

This controller worked well for coupled systems where a series of objects must track and predict the object in front of it. Networks of these controllers could control the rate of message or car traffic flow through electronic and physical intersections. The coupled systems can differ. The distributed structure of the fuzzy controller could apply to factory assembly lines or to robotic limb control.

Unsupervised ellipsoidal learning tuned the fuzzy rules and sets for cars of different sizes and engine types. This gives a new way to find a fuzzy system using only data from a human driver or other controller. Ellipsoidal learning works for any control system when input-output data is available as in the control of biological or economic processes. Future research will compare how well supervised and unsupervised learning find the optimal ellipsoids and rules for fuzzy systems. On-line adaptive fuzzy control systems with ellipsoidal learning can adapt the system over time as engine parameters change.

## ACKNOWLEDGMENT

The Caltrans PATH Program supported this research (Agreement #20695 MB). VORAD Incorporated provided the car and radar sensor for the gap controller test. The authors thank Jim Rowland and John Olds of VORAD for their help in testing this controller.

## REFERENCES

[1] Dickerson, J.A., Kosko, B., "Ellipsoidal Learning and Fuzzy Throttle Control for Platoons of Smart Cars," in *Fuzzy Sets, Neural Networks, and Soft Computing*, edited by R.R. Yager, L. Zadeh, Van Nostrand Reinhold, 1994.

[2] Dickerson, J.A., Kosko, B., "Fuzzy Function Approximation with Supervised Ellipsoidal Learning," *World Congress on Neural Networks (WCNN '93)*, Volume II, 9-17, 1993.

[3] Kosko, B., "Fuzzy Systems as Universal Approximators," *IEEE Transactions on Computers*, 1993; an early version appears in *Proceedings of the 1st IEEE International Conference on Fuzzy Systems(FUZZ-IEEE FUZZ92)*, 1153-1162, March 1992.

[4] Kosko, B., *Neural Networks and Fuzzy Systems*, Prentice Hall, Englewood Cliffs, 1992.

[5] Hsu, A., Eskafi, F., Sachs, S., Varaiya, P., "The Design of Platoon Maneuver Protocols for IVHS," *PATH Research Report*, UCB-ITS-PRR-91-6, April 20, 1991.

[6] Ioannou, P., Xu, T., "Throttle and Brake Control for Vehicle Following," *32nd Control Design Conference (32nd CDC)*, 1993.

[7] Strang, G., *Linear Algebra and Its Applications*, Second Edition, Academic Press, 1980.

[8] Sheikholesam, S., Desoer, C.A., "Longitudinal Control of a Platoon of Vehicles with No Communication of Lead Vehicle Information: A System Level Study," *PATH Technical Memorandum*, 91-2,1991.

[9] Kuo, B.C., *Automatic Control Systems*, Fourth Edition, Prentice Hall, Englewood Cliffs, 1982.

# Automatic Generation of Application Specific Fuzzy Controllers for Rapid-Prototyping

Saman K. Halgamuge, Thomas Hollstein, Andreas Kirschbaum, Manfred Glesner

*Abstract*—Automatic generation of application specific fuzzy controllers from a description in an user friendly high level language is sucessfully implemented and tested with a real world application. The compiler FUZ2LCA generates net listings to be down loaded for FPGAs for rapid prototyping. Several new aspects in designing hardware for fuzzy systems are discussed giving more emphasis to the storing membership functions and the defuzzification.

## I. INTRODUCTION

Realtime fuzzy solutions can be realised using a number of commercially available processors. Most of them are ideally suited for rapid prototyping of online solutions with certain limitations. However, application specific designs are more suitable for time critical solutions.

Either programming a standard processor to function as a fuzzy system or to use a processor specially constructed for efficient fuzzy processing may be the commercially available solution [T91], [Gmb92].

Standard processors such as microcontrollers and digital signal processors can be programmed to function as fuzzy systems. The reverse driving support of truck and trailer described in [HRG94] is driven by a NEC$\mu$PD78310 microcontroller programmed as a fuzzy controller. Although an application specific design generated from the FUZ2LCA tool described in this paper is much faster, the microcontroller based solution is sufficient due to the moderate time limitations of the application.

## II. APPLICATION SPECIFIC DESIGNS

Application specific hardware implementations of fuzzy systems perform much faster though the development time required is longer compared to other alternatives. The main objective of creating the compiler FUZ2LCA is to allow rapid prototyping of high

Mailing Address: Darmstadt University of Technology, Institute of Microelectronic Systems, Karlstr. 15, D-64283 Darmstadt, Germany, Tel.: ++49 6151 16-4337, Fax.: ++49 6151 16-4936
Email: saman@microelectronic.e-technik.th-darmstadt.de

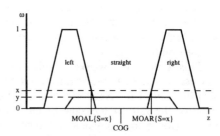

Figure 1: Variations of MOA Defuzzification

performance application specific designs. It automatically creates module based designs for down-loading in Xilinx FPGAs.

In the preliminary version of the compiler certain limitations were defined. Membership functions are stored as piece wise linear functions in external RAMs. An effective method of approximating the nonlinear membership functions with pies wise linear functions is described in section IV–$B$. Among numerous inference/composition schemes, the most popular Min/Max method is selected.

Midpoint of Area (MOA), also known as Center of Area (COA), Mean of Maxima (MOM), are the standard defuzzification methods ([DHR93]) that can be implemented efficiently in hardware. The well known Center of Gravity (COG) method is not appropriate in some conflicting situations, c.g. the situation where $y = 0$ in Fig. 1. In this case two rules are active to the left and the right with equal strengths. One can imagine this special case as moving a vehicle towards two equally possible paths in the left and the right with an obstacle in the middle. COG defuzzification delivers the most inappropriate solution in the middle, where as MOA delivers either the right corner of the "left" membership function or the left corner of the "right" membership function. Even MOA delivers the same inappropriate crisp output as COG, if the obstacle is replaced by a swampy path, where $0 \leq y \leq x$ in Fig. 1 and x is defined as the "swamp constant", that can be set by the user. Even though

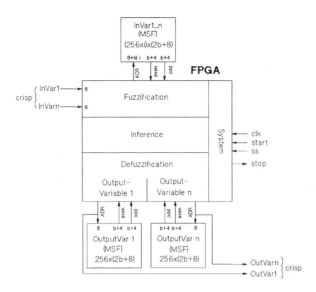

Figure 2: Generated Fuzzy Hardware

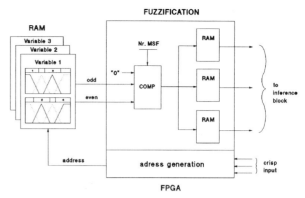

Figure 3: Fuzzification Block

the vehicle can move to th straight in the swampy path the desirable way would be to the left or to the right. To overcome these undesirable properties of standard defuzzification methods authors implemented modified strategies MOAL and MOAR.

In both strategies the membership value of the defuzzified crisp output must be at least equal to the swamp constant. The left and right limitation is the only difference in these modified methods as shown in Fig. 1.

## III. FUZ2LCA COMPILER

Fuzzy systems written in Togai's FPL or C language can be synthesized and converted to Xilinx Netlist Format (XNF). This enables user to define the fuzzy system in a problem specific manner. Problems arising in converting large fuzzy systems can be solved by effectively partitioning the design into several FPGAs.

Each fuzzy design consists of three modules or functional units: fuzzification, rule inference, and composition and defuzzification. All three modules have their own local controllers allowing them to function independently. User can set parameters depending on the availability of hardware resources and the required speed so that a highly parallel design, a completely sequential design or a design in between them is created. Due to high time consumption of many commonly used methods the defuzzification unit should normally operate in parallel to fuzzification and inference units. The system controller supports depending on user selectable parameters both sequential and pipe-line modi. Due to the modularity

FUZ2LCA can be easily extended by adding alternative modules. In addition to the FPGAs generated automatically memories are needed for storing antecedent and consequent membership functions (Fig. 2).

## IV. FUZZIFICATION UNIT

Piece-wise linear membership functions can be easily stored using two different external RAM blocks. All the odd numbered membership functions are stored in odd-RAM block while the even numbered membership functions are stored in even-RAM block (Fig. 3). The restriction in this method is that maximum of two membership functions can overlap.

### A. Automatically Generated Membership Functions

In many neuro-fuzzy approaches, where fuzzy systems are automatically generated [HPG93], [LL92], the resulting membership functions are bell shaped functions. The typical approach utilizing look-up tables for fuzzification is inefficient in those cases, because of the requirement of huge fuzzification memory. One solution to this problem is the approximation of membership functions using straight lines. Each membership function form is represented with maximum of 8 straight lines reducing the memory capacity. In case of implementing a sigmoidal membership function 256 bytes were needed for simple look-up-table approach where as only 24 bytes were needed for the approach with membership function approximation, without reducing the speed compared to the first effort (see also Fig. 4). Each line $Y = a_i(X - X_i) + Y_i$ is characterized by the three parameters: the tangential coefficient $a_i$, and the coordinates of the left most position of the line $X_i$ and $Y_i$. Three short memories are used for storing those parameters. The fuzzifier works in two phases. The process of searching the proper approximation line by

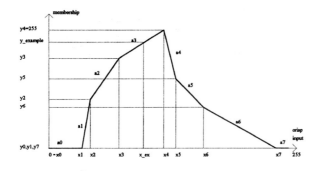

Figure 4: Membership Function Approximation

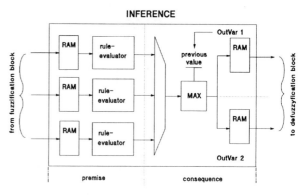

Figure 5: Rule Inference

bisection searching ([Knu73]) is the first phase. The search always last three clock cycles and results in the index used for addressing parameter memories. Then they are processed in the multiplier/adders delivering the crisp output. Searching and computation are performed independently and concurrently.

### B. Rule Inference

Inference is the process where the evaluation of the premise and the consequent of a single rule occurred (Fig. 5) where as in the composition the inference outcome of many rules are combined (left part of the (Fig. 6)).

Three different types of rule evaluators can be generated:

- simple evaluators that can either read or negate the membership value of an input
- Min/Max rule evaluators for rules with less complexity
- complex rule evaluators with maximum of 16 Min/Max operations

The premises are evaluated by using a single rule evaluator or many of them depending on the complexity. The initially implemented but easily extendable inference/composition method is Min/Max. The outcome of the composition is stored in internal RAMs so that the output can be obtained after parallel driven defuzzification.

### C. Defuzzification

Defuzzification unit is normally the most time consuming module, specially if a method such as COG is implemented. Two steps were taken to overcome this problem:

- defuzzication module is always generated as a parallel module
- less time consuming methods are generated with efficient hardware structures

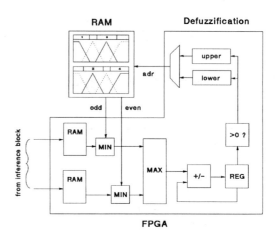

Figure 6: Composition and MOA Defuzzification

MOA method and its modifications MOAL, MOAR can be generated as defuzzification units. The hardware implementation of MOA method is depicted in Fig. 6. The pointers upper and lower covers the output range starting from the lower and the upper limit respectively and moving stepwise towards the middle. After composition of the inference outputs, the area underneath the fuzzy output is added to a register as the pointer lower moves and subtracted from the register as the pointer upper moves. Comparison of the content of register with zero allows the effective moving of pointers so that the midpoint of the area could be reached. After number of iterations the crisp output can read from the position of the pointers. Since operations such as multiplication or division are not involved, this method is much faster than COG.

### D. Results and Discussion

After several tests, the compiler is sucessfully applied in generating a fuzzy controller for the fuzzy truck and trailer mentioned before [HWG94]. In this appli-

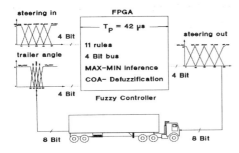

Figure 7: Fuzzy Truck Control

cation the fuzzy controller supports continuous fast reverse driving of a truck with a long trailer without taking the risk of bending the trailer so that no reverse driving is possible. The fuzzy controller generated contains 11 fuzzy rules, 2 inputs and 1 output, each with 5 membership functions, employs Max-Min inference/composition and Mid Point of Area defuzzification. A 4-bit version of the generated fuzzy controller, of which the accuracy was good enough for the application, could be implemented in a XC4006-FPGA and only $42\mu S$ are needed for calculating a new output. This result can be compared with standard solutions such as DSP-TMS320 (150 $\mu S$), and special fuzzy solutions of Togai FC110 (32 $\mu S$).

### REFERENCES

[DHR93]  D. Driankov, H. Hellendoorn, and M. Reinfrank. *An Introduction to Fuzzy Control.* Springer-Verlag, USA, 1993.

[HPG93]  S. K. Halgamuge, W. Poechmueller, and M. Glesner. A Rule based Prototype System for Automatic Classification in Industrial Quality Control. In *IEEE International Conference on Neural Networks' 93,* San Francisco, USA, March 1993.

[HRG94]  S. K. Halgamuge, T. A. Runkler, and M. Glesner. A Hierarchical Hybrid Fuzzy Controller for Realtime Reverse Driving Support of Vehicles with Long Trailors. In *IEEE International Conference on Fuzzy Systems' 94, (submitted),* Orlando, USA, June 1994.

[HWG94]  S. K. Halgamuge, T. Wagner, and M. Glesner. Validation and Application of an adaptive transparent Defuzzification Strategy for Fuzzy Control. In *IEEE International Conference on Fuzzy Systems' 94, (submitted),* Orlando, USA, June 1994.

[Inf92]  INFORM GmbH. *FUZZY-166 Hybrid Fuzzy Processor.* Aachen, Germany, 1992.

[Knu73]  D. E. Knuth. *The Art of Computer Programming, Vol. 3: Sorting and Searching.* Addison-Wesley Publishing Company, USA, 1973.

[LL92]  C. T. Lin and C. S. G. Lee. Real-Time Supervised Structure/Parameter Learning for Fuzzy Neural Network. In *IEEE International Conference on Fuzzy Systems,* San Diego, USA, 1992.

[Tog91]  Togai Infralogic Inc. *FC110 Togai Fuzzy Processor.* Irvine, U.S.A., 1991.

# Validation and Application of an Adaptive Transparent Defuzzification Strategy for Fuzzy Control

Saman K. Halgamuge, Tilman Wagner, Manfred Glesner

*Abstract*—**An extended parametrizable defuzzification method is implemented as a special transparent neural network, which can be considered as a global defuzzification approximater. The customization of the method to different applications and the analysing capability of the trained solution are the key features here. The network is validated by showing its convergence to various existing defuzzification methods. The results of an application example and a number of less known defuzzication methods are also discussed.**

## I. INTRODUCTION

The Center of Gravity (COG) and the Mean of Maxima (MOM) are well known defuzzification methods described in literature [DHR93]. There are also methods such as Midpoint of Area (MOA) that can be implemented in software and hardware efficiently. Let us denote U as the finite set of possible normalized output values of a controller: $U = \{U_1, U_2.....U_n\}$, where $\forall i, 0 \leq U_i \leq 1$. The output of the rule block is a fuzzy output that must be defuzzified to get the crisp control output $U_{out} \epsilon U$. The method MOA can be defined as:

$$\int_0^{U_{out}^{MOA}} \mu_i \cdot di = \int_{U_{out}^{MOA}}^n \mu_i \cdot di$$

A new method described here is Center of Maxima (COM):

$$\int_0^{U_{out}^{COM}} \mu_i^{Max} \cdot di = \int_{U_{out}^{COM}}^n \mu_i^{Max} \cdot di$$

where $\mu_i^{Max}$ is the maximum of $\mu_i$ $\forall i$ and $n$ denotes the *netwidth*.

The method MOM can be considered as a special case of COG, where as COM is a special case of MOA.

Mailing Address: Darmstadt University of Technology, Institute of Microelectronic Systems, Karlstr. 15, D-64283 Darmstadt, Germany, Tel.: ++49 6151 16-4337, Fax.: ++49 6151 16-4936 Email: saman@microelectronic.e-technik.th-darmstadt.de

Many research has been reported on learning fuzzy systems, i.e., learning rules and fine tuning of membership functions. Few papers could be found in describing the extraction of complete fuzzy systems from sample data. And serious efforts in adapting defuzzification to the specific nature of an application as reported in [YF93] is not very common. The application of different defuzzification methods may lead to completely different crisp outputs depending on the shape of the fuzzy outputs. Therefore, tuning to an application or customization of the defuzzication strategy is an interesting alternative to the conventional trial and error methods used. In a previous publication ([HG93]), the concept of **C**ustomizable **BA**sic **D**efuzzification **D**istributions was presented. The CBADD is an extension of well known BADD method proposed by Yager et al [FY91].

This paper describes the CBADD network in section 2 presenting simulation results in section 3 and showing the applicability in a real world problem in section 4.

## II. CBADD TRANSPARENT NEURAL NETWORK

CBADD uses a special transparent neural network with *netwidth* (n) number of inputs, each of them feeding a $\mu_i$, the $i$th discrete value of the fuzzy output, where $0 \leq i \leq n$ to approximate the crisp output for $U_{out}$. It consists of several consecutive layers of

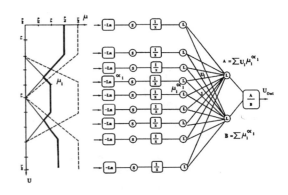

Figure 1: CBADD Transparent Neural Network

neurons, each layer consisting of different types of neurons concerning the calculation of *net input* and *activation function* (Fig. 1). The first layer consists of neurons having logarithmic activation functions, and the connections to the next layer are weighted by the variable confidence, $\alpha_i$. The confidence on the fuzzy output values can be divided into different regions as shown in Fig. 2. We define the set *variable confidence measure* $\alpha_I$: $\alpha_I = \{\alpha_1, \alpha_2 ... \alpha_n\}$. They are the only trainable weights in the CBADD network. The following neurons have an exponential activation function which partly eliminates the logarithm of the first layer and delivers $\mu^{\alpha_i}$. Within the first two layers there are no cross-connections but only connections among the neurons of the same net index. In the second last neuron layer, all outputs are fed into two neurons that build a weighted sum. The incoming connections to the neuron $(A)$ is weighted by $U$, the discrete values of the possible range of outputs, and it has a linear activation function The neuron $(B)$ has unity incoming weights and an inverting activation function. In the last layer the resulting sum of neuron $A$ is multiplied with the inverted sum calculated in neuron $B$ by the using a Π-Neuron. The network has a special structure which can be resolved mathematically to match the CBADD equation as described in [HG93].

$$U_{out}^{CBADD} = \frac{\sum_{i=1}^{n} \mu_i^{\alpha_i} \cdot U_i}{\sum_{i=1}^{n} \mu_i^{\alpha_i}}$$

The use of the network is as follows: The fuzzy variable to be defuzzified is applied to the inputs of the network. Each input receives a value in the range from `ALMOST_ZERO` to 1.0. `ALMOST_ZERO` is a constant value defined as the lower border of the input range and set to $10^{-6}$ in the simulation examples. The input must be greater than zero as the first layer of neurons calculate the logarithm. The output $U_{out}^{CBADD}$ of the trained network represents the defuzzified value of the pattern.

The network, containing Σ-Neurons with summed net inputs and a single Π-Neuron with multiplied net inputs, can be trained by gradient descent learning. The use of Π-Neurons requires different $\delta$ values [RM86] for backpropagation learning whenever the neurons following the actual neuron layer are Π-Neurons. In the case of Σ-Neurons $A$ and $B$ are in the actual layer and a single Π-Neuron is in the next layer as in the CBADD network, the equations reduce to

$$\delta_A^{\Pi} = (\delta_{ps} \cdot w_{sb} \cdot o_{pb} \cdot w_{sa}) \cdot f_a'(net_{pa})$$

$$\delta_B^{\Pi} = (\delta_{ps} \cdot w_{sa} \cdot o_{pa} \cdot w_{sb}) \cdot f_b'(net_{pb})$$

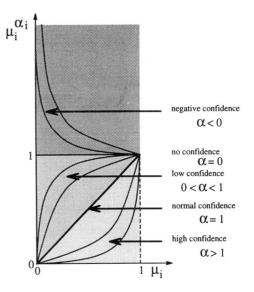

Figure 2: Different Regions in the Confidence Chart

where $\delta_A^{\Pi}$, $\delta_B^{\Pi}$ are error measures to be back propagated from the neurons A and B, $\delta_{ps}$ is the error measured at the output, $w_{sa}$ and $w_{sb}$ are weights from neuron A and B to the Π-Neuron, $o_{pa}$ and $o_{pb}$ are outputs of the neurons A and B, and $f_a'(net_{pa})$ and $f_b'(net_{pb})$ are the partial derivatives with respect to the net inputs of activation functions in the neurons A and B respectively.

The CBADD network was trained to perform different defuzzification strategies, using data sets created by some of the standard methods, and tested with the different test data sets. Since this is a transparent network, the learned weights of the set $\alpha_I$ can be analysed for the validation of the network.

### III. VALIDATION OF THE CBADD NETWORK

The validation of the CBADD Network is done by training the network with pattern sets generated with the defuzzification strategies COG and MOM. It can be mathematically shown, that $U_{out}^{CBADD}$ can be reduced to $U_{out}^{COG}$ by setting all the elements in set $\alpha_I$ to 1, and it can be reduced to $U_{out}^{MOM}$ by setting those elements virtually to $\infty$. We expect to approximate this mathematically exact solution after the training phase and take it for a validation for the convergence of the system.

The pattern sets used have $n = 32$ inputs in the range from `ALMOST_ZERO` $= 10^{-6}$ to 1.0 and one output range from 0.0 to `maxu` = 31.0. We use 2000 different patterns that are applied over 1000 sweeps, that is 1000 times. Furthermore we use single mode backpropagation and update the weights after each single pattern. Each pattern represents one epoch.

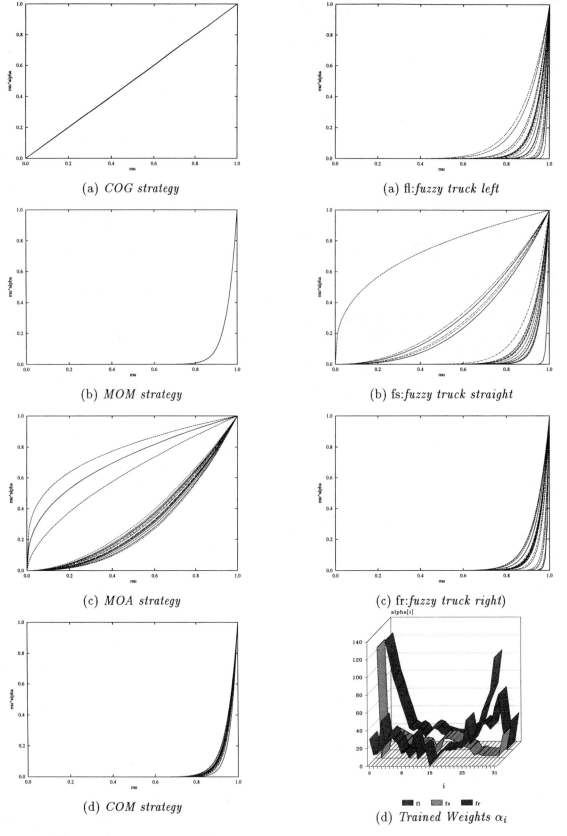

(a) COG strategy

(a) fl:fuzzy truck left

(b) MOM strategy

(b) fs:fuzzy truck straight

(c) MOA strategy

(c) fr:fuzzy truck right)

(d) COM strategy

(d) Trained Weights $\alpha_i$

Figure 4: Confidence Chart $\mu^{\alpha}$ versus $\mu$ for Pattern Type 2

Figure 5: Confidence Chart $\mu^{\alpha}$ versus $\mu$ for Fuzzy Truck Patterns

The system has been tested with six different pattern sets, each defuzzified with the strategy of interest. This paper reports the simulation results for the pattern sets of type 1 and 2, which are similar to real fuzzy output variables. The randomly generated patterns in the set 1 are evenly distributed and, therefore have their $U_{out}^{COG}$ and $U_{out}^{MOA}$ values often close to the middle of the output range. Pattern set 2 has patterns composed of 5 random heights that are bounded horizontally by the limits of the output range and vertically by two randomly selected edge values. The resulting pattern sets can be of a rectangular shape as well as of a triangular or multi-trapeze shape. Therefore $U_{out}^{COG}$ and $U_{out}^{MOA}$ values are almost randomly distributed.

## A. Validation with COG

The training of patterns defuzzified with *Center Of Gravity* (COG) strategy converges successfully, leading to weights $\alpha_I$ in the range of $1.0 \pm 0.01$ for pattern type 1(Fig. 3(a)) and type 2 (Fig. 4(a)). Pattern type 1 shows a slower convergence than pattern type 2, the first shows an RMS-error of 0.004% whereas the second has an RMS-error of 0.0002% after 1000 sweeps. The mathematically correct solution for $\alpha_I$ is 1.0. Therefore we see that CBADD approximates the set *variable measure* $\alpha_I$ for the COG in the training phase. When trained over a longer period of time (10000 sweeps) the tendency for the weights is towards an asymptote at 1 and an RMS-error of 0.

## B. Validation with MOM

When trained with pattern sets defuzzified by MOM strategy the CBADD network shows $\alpha$-weights in the range of $33.0 \pm 2.0$ for pattern type 1 (Fig. 3(b)) and weights in the range of $21.5 \pm 0.2$ for pattern type 2 (Fig. 4(b)). The RMS-error for training the first set is 1.46%, the RMS-error for the second is 0.49%. In this case the mathematically correct solution for $\alpha_I$ is $+\infty$, a value that can hardly be reached in a computer simulation. We therefore conclude that CBADD can correctly represent MOM, since confidence values are well above 20.

As shown by the simulation results CBADD can successfully represent the defuzzification strategies *Center Of Gravity* (COG) and *Mean Of Maxima* (MOM) and the implementation has herewith shown to be correct. The trained CBADD approximations for MOM and COG are sucessfully tested with different data sets.

## IV. Application

### A. Approximation of MOA

The resulting weights for the MOA defuzzification strategy can not be mathematically obtained as for COG and MOM. The *variable confidence measure* $\alpha_I$ reaches values varying from 0.0 at both ends to 2.5 in the middle range. Pattern type 1 (Fig. 3(c)) shows an additional peak in the middle of the input range, the RMS-error for pattern type 1 is 2.24%, the RMS-error for pattern type 2 (Fig. 4(c)) is 2.20%.

### B. Approximation of COM

The training of *Center Of Maxima* (COM) defuzzified patterns leads to weights for $\alpha_I$ between 23.0 and 41.0 for pattern type 1 (Fig. 3(d)) and to weights between 19.0 to 29.0 for pattern type 2 (Fig. 4(d)). The RMS-error for the first type is 9.96%, the RMS-error for the second is 3.53%. The weights for $\alpha_I$ are somewhat similar to those of MOM, which is not a surprise as a single block of maximum height will have the same crisp output with both strategies. But the tests conducted show that COM test test have the least error with the COM trained network and not with the MOM network. Therefore the resulting weights represent the new defuzzification strategy COM.

### C. Application Oriented Data

The training of application data obtained from a fuzzy controlled model truck [HRG94] for three drives: reverse turn to left (fl), straight line reversing (fs) and reverse turn to right (fr) show a new distribution of weights similar to COM (Fig. 5). The truck is on-line fuzzy controlled in order to enhance the reverse dynamics, so that even an unexperienced driver would encounter hardly any problems in fast reverse driving the truck with a trailer. Inputs to the fuzzy controller are the driver command to the wheels of truck at the front and the current rare wheel position of the trailer. The output is the fuzzy controlled command to the front wheels of the truck.

## V. Discussion

After the validation of the CBADD network for MOM and COG, it is shown that the system needs different weights $\alpha_i$ for different areas of the output range to represent the defuzzification strategies MOA and COM, which is possible due to the extension of single parameter $\alpha$ of BADD to the vector $\alpha_I$ in CBADD. The strategies MOA and COM can successfully be represented by CBADD although they can not be represented by BADD. Therefore CBADD is a powerful extension of the BADD defuzzification strategy.

The CBADD strategy serves as a global defuzzification approximater and it is an alternative way of finding the most appropriate defuzzification method. The confidence charts described above play a significant role in analysing the results. CBADD network is integrated as a part of the FuNe II fuzzy neural system. This system generates a fuzzy controller including fuzzy rules, from training data, tunes parameters in membership functions and approximates the defuzzification strategy. Future publications will reveal some of the applications of FuNe II, with the CBADD network and the analog hardware implementation of the CBADD resulting network.

### REFERENCES

[DHR93] D. Driankov, H. Hellendoorn, and M. Reinfrank. *An Introduction to Fuzzy Control.* Springer-Verlag, USA, 1993.

[FY91] D. P. Filev and R. R. Yager. A Generalized Defuzzification Method via Bad Distributions. *International Journal of Intelligent Systems*, 6, 1991.

[HG93] S. K. Halgamuge and M. Glesner. The Fuzzy Neural Controller FuNe II with a New Adaptive Defuzzification Strategy Based on CBAD Distributions. In *European Congress on Fuzzy and Intelligent Technologies'93*, Aachen, Germany, September 1993.

[HRG94] S. K. Halgamuge, T. A. Runkler, and M. Glesner. A Hierarchical Hybrid Fuzzy Controller for Realtime Reverse Driving Support of Vehicles with Long Trailors. In *IEEE International Conference on Fuzzy Systems' 94, (submitted)*, Orlando, USA, June 1994.

[RM86] D. E. Rumelhart and J. L. McClelland. *Parallel Distributed Processing: Explorations in the Microstructure of Cognition.* MIT Press, USA, 1986.

[YF93] R. R. Yager and D. P. Filev. SLIDE: A Simple Adaptive Defuzzification Method. *IEEE Transactions on Fuzzy Systems*, 1(1), February 1993.

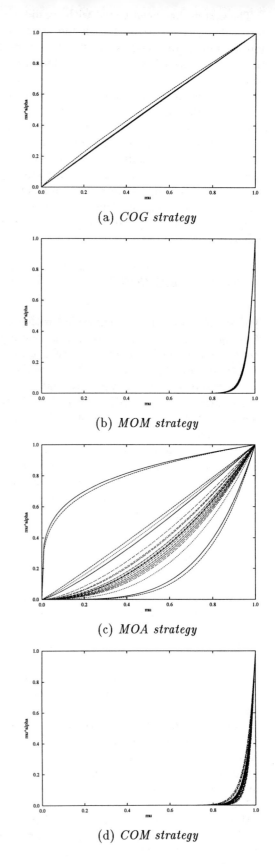

(a) *COG strategy*

(b) *MOM strategy*

(c) *MOA strategy*

(d) *COM strategy*

Figure 3: Confidence Chart $\mu^\alpha$ versus $\mu$ for Pattern Type 1

# Defuzzification Based on Fuzzy Clustering

Harald Genther, Thomas A. Runkler and Manfred Glesner

*Abstract*— We develop a modified fuzzy clustering algorithm for parametric defuzzification in fuzzy rule base systems. Using examples and basic defuzzification properties we compare defuzzification by clustering with the standard defuzzification methods COG and MOM. Concerning fuzzy sets with forbidden zones the new method proves to be superior. We present how heuristic preprocessing and quality measures are used for appropriate parameter selection.

## I. INTRODUCTION

Since they were first mentioned [13], fuzzy sets were applied to a wide range of applications, which we divide into *set oriented* approaches like fuzzy data analysis methods [1] and *rule base oriented* approaches like fuzzy control [3] and expert systems [14].

In fuzzy data analysis we use fuzzy sets to model the fuzziness of the statistical examinations with only a finite number of (usually noisy) examples. We obtain *fuzzy statistical information* like positions and shapes of fuzzy cluster from non optimal data sets, where similar data sets produce similar results.

Fuzzy rule base systems realize algorithms specified using fuzzy IF ... THEN ... rules, which usually perform forward chaining. Internally, these systems calculate truth values and fuzzy sets using operations of the well–defined families of t–norms and t–conorms [11]. The final result of the rule base evaluation being fuzzy, a *defuzzification* [12] step is necessary realizing the retransformation to the (usually crisp) real world, which turns out to be the most crucial operation in a fuzzy rule base system.

Because of some undesirable properties of the standard defuzzification algorithms [9, 8, 7] we are looking for new methods appropriate for the defuzzi-fication task, i.e. methods, which can be used to localize "characteristic" values with respect to a given fuzzy set.

We will show that fuzzy data analysis, especially fuzzy clustering algorithms, provide a method for the determination of crisp characteristic values. We use fuzzy clustering as a parametric defuzzification method.

We introduce the defuzzification and its standard methods in section II. The Fuzzy c–Means algorithm used for fuzzy clustering is described in section III. Proceeding from an algorithm for grayscale and color image clustering we develop a modified fuzzy clustering algorithm, which we use for the purpose of defuzzification (Section IV.). We show examples for defuzzification with fuzzy clustering, examine its properties, and describe the parameter selection process using heuristic preprocessing and quality measures. Conclusions are given in section V.

## II. DEFUZZIFICATION

Fuzzy controllers realize nonlinear control algorithms specified using fuzzy linguistic expert knowledge represented in IF ... THEN ... rules. They consist of a *fuzzification* unit transforming crisp input to membership degrees using the premise membership functions specified, an *inference* and *composition* unit calculating the fuzzy output following the fuzzy rule base, and a *defuzzification* unit transforming the fuzzy result into a crisp steering action.

We express defuzzification with the operator $\mathcal{F}^{-1}$ mapping fuzzy sets to "characteristic" or "significant" crisp elements of the universe of discourse:

$$\mathcal{F}^{-1} : \mathcal{F}(x) \to \mathcal{X}, \qquad (1)$$

or, equivalently, mapping membership functions (representing fuzzy sets) to crisp universe elements:

$$\mathcal{F}^{-1} : \{\mu(x)\} \to \mathcal{X}. \qquad (2)$$

Widely used standard defuzzification methods are

- **Mean of Maxima (MOM)** calculating the

This work is a part of the doctoral research programme of H. Genther and T. A. Runkler within the Graduiertenkolleg "Intelligent Systems for Information and Automation Technology" at Darmstadt University of Technology.

Mailing Address: Darmstadt University of Technology, Institute of Microelectronic Systems, Karlstrasse 15, 64283 Darmstadt, Germany, Tel. (+49)6151/16–4337, Fax (+49)6151/16–4936, Email haraldg@microelectronic.e-technik.th-darmstadt.de.

mean of the values with maximum membership:

$$\mathcal{F}_{\mathrm{MOM}}^{-1}\left(\mu\left(x\right)\right) := \frac{\int_{\mu(x)=\sup_{x\in\mathcal{X}}(\mu(x))} x \; dx}{\int_{\mu(x)=\sup_{x\in\mathcal{X}}(\mu(x))} dx}, \quad (3)$$

- **Center of Gravity (COG)** calculating the centroid of the area under the membership function:

$$\mathcal{F}_{\mathrm{COG}}^{-1}\left(\mu\left(x\right)\right) := \frac{\int_{x_{\inf}}^{x_{\sup}} \mu(x)\cdot x \; dx}{\int_{x_{\inf}}^{x_{\sup}} \mu(x) \; dx}, \quad (4)$$

and

- **Center of Area (COA)** calculating the position of the divisive axis of the area under the membership function:

$$\int_{x_{\inf}}^{\mathcal{F}_{\mathrm{COA}}^{-1}(\mu(x))} \mu(x) \; dx = \int_{\mathcal{F}_{\mathrm{COA}}^{-1}(\mu(x))}^{x_{\sup}} \mu(x) \; dx. \quad (5)$$

Due to various undesirable properties of the standard defuzzification algorithms extended methods like parametric methods (e.g. *BADD* [4], *CBADD* [6] and *XCOA* [8]), decision based methods (e.g. *CDD* [7]), and approximate methods (e.g. *DECADE* [10]) were developed.

### III. Fuzzy Clustering

Fuzzy Clustering is a method of partitioning a set of input data into subgroups. Fuzzy membership values to the subgroups called *clusters* are assigned to each input datum. A well known method for calculating the cluster centers is the Fuzzy c–Means (FCM) algorithm [2]. Using $c$ clusters and $n$ input vectors, an objective functional given by

$$\mathcal{J}(U,\vec{v}) = \sum_{i=1}^{n}\sum_{j=1}^{c} u_{ij}^{m} d(\vec{x}_i,\vec{v}_j)^2 \quad (6)$$

is minimized by a Picard iteration using

$$\mu_{ij} = \frac{1}{\sum_{k=1}^{c}\left(\frac{d(\vec{x}_i,\vec{v}_j)^2}{d(\vec{x}_i,\vec{v}_k)^2}\right)^{\frac{1}{m-1}}}; \quad \begin{array}{l} 1 \le j \le c; \\ 1 \le i \le n, \end{array} \quad (7)$$

$$\vec{v}_j = \frac{\sum_{i=1}^{n} u_{ij}^{m}\vec{x}_i}{\sum_{i=1}^{n} u_{ij}^{m}}; \quad \forall \; j. \quad (8)$$

$m$ is a weighting factor for the membership grades $u_{ij}$, that influences the fuzziness of the resulting partition. We use $m = 2$ in our simulations. FCM uses the Euclidean distance to calculate $d(i,j)$. This works well for data that can be described in vector form, but leads to problems when data described by multiple features or multiple vectors are given. FCM can be extended for these problems, an application to grayscale and colour image processing is described in [5].

### IV. Defuzzification by Fuzzy Clustering

For our defuzzification application we consider a distance function of the location $x$ and the associated membership value $\mu(x)$. Using these two features, we write the distance function

$$D_{ij} = \sqrt{\alpha\frac{(x_i-v_j)}{(x_i-v_j)_{max}} + (1-\alpha)\frac{(\mu(x_i)-w_j)}{(\mu(x_i)-w_j)_{max}}}. \quad (9)$$

We use $\alpha$, the weighting factor, as a parameter for defuzzification representing the sensitivity for the values of $x$, and calculate the cluster centers using (7), (8) and

$$\vec{w}_j = \frac{\sum_{i=1}^{n} u_{ij}^{m}\mu(x_i)}{\sum_{i=1}^{n} u_{ij}^{m}}; \quad \forall \; j. \quad (10)$$

We obtain $c$ cluster centers with associated membership values $(v_j, w_j)$, choose the cluster center $k$ with $w_k = \max_j(w_j)$ and take $v_k$ as the result of the defuzzification.

#### A. Examples

In Fig. 1 and 2 we show a typical piecewise linear fuzzy output of a fuzzy controller. Using three cluster centers and $\alpha = 0.6$, we receive the result shown in Fig. 1.

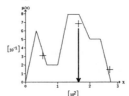 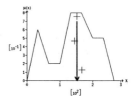

Fig. 1: Clustering Results ($\alpha = 0.6$)    Fig. 2: Clustering Results ($\alpha = 0.1$)

The output is defuzzified to 170, which is a reasonable result.

Fig. 2 shows the result obtained with $\alpha$ set to 0.1. Although the clustering result is mainly influenced by the membership values, more or less neglecting the location, the defuzzified value is still reasonable. In this example the method is robust against changes of $\alpha$.

Our next example is the defuzzification of a bimodal function shown in Figs. 3 to 5.

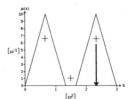 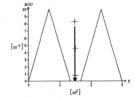

Fig. 3: Clustering Results ($c = 3; \alpha = 0.5$)    Fig. 4: Clustering Results ($c = 3; \alpha = 0.1$)

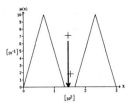

Fig. 5: Clustering Results ($c = 2; \alpha = 0.5$)

Defuzzification with three clusters and $\alpha = 0.5$ gives a good result, one of the cluster centers near the vertices of the triangles is chosen for the defuzzified value (Fig. 3). When $\alpha$ is set to a very low value (e.g., 0.1, Fig. 4) or only two cluster centers are chosen (Fig. 5), poor results are obtained. To achieve acceptable defuzzification results, appropriate parameter selection is necessary.

### B. Properties of the Clustering Defuzzification

We examine the Fuzzy Clustering Defuzzification concerning the rational static properties proposed in [9].

### B.1. Problem Parity: The Zero Element

Because of symmetry considerations unimodal rectangular membership functions (*zero elements*, Fig. 6) have to be defuzzified to their symmetry axis. We defuzzify random rectangles and determine the average distances between the defuzzified and the symmetry point (Fig. 7). Low distances (errors) are achieved with odd numbers of clusters, even cluster numbers result in high errors.

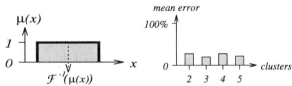

Fig. 6: Zero Element     Fig. 7: Error (Zero Element)

Obviously, unimodal rectangular membership functions specify an *odd problem* requiring a defuzzification with an odd number of clusters. Generally membership functions can be separated in *odd* and *even problems*, depending on the number of appearing modes.

### B.2. α Parameter: The One Element

Rational considerations lead us to require singleton membership functions (*one elements*, Fig. 8) to be defuzzified to their non zero value. We defuzzify singletons with random position and height and calculate the mean deviation from the non zero value. For increasing numbers of clusters the deviation decreases, because the resolution of the $x$ coordinate becomes more precise. The error also depends on the weighting factor $\alpha$ (Fig. 9): Low $\alpha$ (e.g. $\alpha \leq \frac{1}{4}$ for $c = 3$ clusters) result in high concentration on

the membership values $\mu$ instead of the position $x$, which increases the average error. Obviously, $\alpha$ has to be chosen appropriately with respect to the fuzziness of the membership function to be defuzzified. Rather crisp membership functions (singletons) require high $\alpha$ values, rather fuzzy membership functions require low $\alpha$.

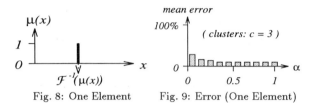

Fig. 8: One Element     Fig. 9: Error (One Element)

### B.3. Maxima Behavior: Fuzzy Number

Fuzzy numbers (Fig. 10) are widely used in fuzzified algebraic computations. They are considered as fuzzifications of numbers represented by the mean values $m$ with membership $\mu(m) = 1$. Rationally, defuzzification of a fuzzy number is required to restore the crisp mean value $m$.

Defuzzified random fuzzy numbers are the nearer to $m$, the mean value, the larger $c$, the number of clusters. By using a large number of clusters a high sensitivity for high membership values can be achieved. Defuzzification with many clusters is similar to the *mean of maxima* defuzzification.

### B.4. Problem Quantity: Forbidden Zones

Many applications (e.g. robotics) use membership values with legislative semantics, i.e. the higher $\mu(x)$, the membership of an output $x$, the more the output is *allowed* to become equal to $x$. Zero or low (i.e. *sub−α*) membership values specify forbidden zones (Fig. 11).

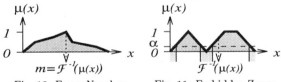

Fig. 10: Fuzzy Number     Fig. 11: Forbidden Zones

We applied clustering defuzzification to random piecewise linearities with forbidden zones and calculated $\overline{m} = \overline{\mu(\mathcal{F}^{-1}(\mu(x)))}$, the average membership of the defuzzified value as a measure of the defuzzification quality concerning the forbidden zones (*statistical examination* [7]).

The best results achieved were $\overline{m_{cluster}} \approx 0.7$, between $COG$ and $MOM$ ($\overline{m_{COG}} \approx 0.5$ and $\overline{m_{MOM}} \approx 0.9$). For each membership function a specific cluster number $c_{opt}$ leading to maximum $\overline{m_{max}}$ can be found indicating the *specific quantity of the problem*. Typically this quantity is equal to the number of modes contained in the membership function.

## C. Heuristic Preprocessing and Quality Measures

The considerations of the previous sections lead to a mechanism for the selection of the parameters $c$ and $\alpha$ (*heuristic preprocessing*, Fig. 12, left). We analyze the number of modes contained in the membership function to estimate the problem parity and quantity, which we use for the appropriate selection of $c$, the cluster number. To determine an appropriate weighting factor $\alpha$ we use fuzziness measures estimating the needed sensitivity concerning the membership values.

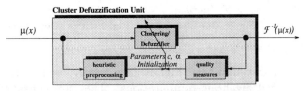

Fig. 12: Heuristics and Quality Measures

Further iterations can be performed with other initializations and improved parameter selections considering quality measures of the clustering process (Fig. 12, right).

## V. RESULTS

Proceeding from a modification of the Fuzzy c-Means algorithm for clustering grayscale and colour images we have developed a defuzzification method based on fuzzy clustering.

Using characteristic examples and considering various basic defuzzification properties we have shown that defuzzification by clustering successfully determines characteristic values of fuzzy sets.

Concerning fuzzy sets with forbidden zones (as frequently used in robotics applications, e.g.), defuzzification using fuzzy clustering leads to much better results than *COG*.

We have shown, how heuristic preprocessing determining characteristics of the fuzzy set can be used for appropriate parameter selection, as well as quality measures quantifying the success of the fuzzy clustering process itself.

Thus, defuzzification by fuzzy clustering is an interesting alternative, where standard defuzzification methods fail.

We are currently working on real world applications of fuzzy controllers using defuzzification by clustering.

## REFERENCES

[1] H. Bandemer and W. Näther. *Fuzzy Data Analysis*. Kluwer Academic Publishers, Dordrecht, 1992.

[2] James C. Bezdek. *Pattern Recognition with Fuzzy Objective Algorithms*. Plenum Press, New York, 1981.

[3] D. Driankow, H. Hellendoorn, and M. Reinfrank. *An Introduction to Fuzzy Control*. Springer-Verlag, Berlin, 1993.

[4] D. P. Filev and R. R. Yager. A generalized defuzzification method via bad distributions. *International Journal of Intelligent Systems*, 6:687–697, 1991.

[5] Harald Genther and Manfred Glesner. Fuzzy Clustering Verfahren für Grauwert- und Farbbilder. In *3. Dortmunder Fuzzy-Tage*, Dortmund, June 1993.

[6] S. K. Halgamuge and M. Glesner. The Fuzzy Neural Controller FuNe II with a New Adaptive Defuzzification Strategy Based on CBAD Distributions. In *European Congress on Fuzzy and Intelligent Technologies*, Aachen, September 1993.

[7] T. A. Runkler and M. Glesner. Defuzzification as crisp decision under fuzzy constraints – new aspects of theory and improved defuzzification algorithms. In *Workshop "Fuzzy-Systeme"*, pages 156–164, Braunschweig, October 1993.

[8] T. A. Runkler and M. Glesner. Defuzzification with improved static and dynamic behavior: Extended center of area. In *European Congress on Fuzzy and Intelligent Technologies*, pages 845–851, Aachen, September 1993.

[9] T. A. Runkler and M. Glesner. A set of axioms for defuzzification strategies — towards a theory of rational defuzzification operators. In *Second IEEE International Conference on Fuzzy Systems*, pages 1161–1166, San Francisco, March/April 1993.

[10] T. A. Runkler and M. Glesner. DECADE — Fast centroid approximation defuzzification for real time fuzzy control applications. In *ACM Symposium on Applied Computing (SAC'94)*, Phoenix AZ, March 1994.

[11] B. Schweizer and A. Sklar. Associative functions and statistical triangle inequalities. *Publicationes Mathematicae Debrecen*, 8:169–186, 1961.

[12] R. R. Yager and D. P. Filev. On the issue of defuzzification and selection from a fuzzy set. Technical Report MII–1201, Machine Intelligence Institute, Iona College, 1991.

[13] L. A. Zadeh. Fuzzy sets. In *Information and Control 8*, pages 338–353, 1965.

[14] H. J. Zimmermann. *Fuzzy Sets, Decision Making, and Expert Systems*. Kluwer Academic Publishers, Boston, 1987.

# REALIZATION OF
# A MODIFIED FUZZY LOGIC CONTROLLER

S. Tandio, N.N. Bengiamin, J. Hootman, and N. Swain

Electrical Engineering Department
University of North Dakota
Grand Forks, ND USA

*Abstract-* **A novel method for realizing Fuzzy Logic controllers is presented in this paper. The basic concept of fuzzy logic is modified to accommodate desirable feedback control implementation features. Moreover, TUTSIM is used to introduce a flexible, economical, and effective tool for studying Fuzzy Logic based control algorithms. Fuzzification, fuzzy inference, and defuzzification are addressed. The developed methodology is applied to a typical control system and its effectiveness is demonstrated when compared to PID control. The developed fuzzy control module has value as a readily available tool for analysis and design.**

## I. INTRODUCTION

Controls is a prime area for fuzzy logic applications [1,2]. Fuzzy control theory has been proven successful in numerous applications varying from fast processes in motion control; e.g. robotics, inverted pendulum, disk drives, and machine tools; to relatively slower systems for heat exchange, traffic control, machine automation, and camera focus. The unique feature of Fuzzy Logic methodologies is in emulating human experience and intuition in dealing with indefiniteness. This feature improves the adaptiveness capability of closed loop systems such that modeling uncertainties are tolerated without much loss in performance. Consequently, Fuzzy Controllers are very flexible and they provide a strong competitor to the long accepted classical PID controller. Economy is another incentive for adopting fuzzy controllers as they can be realized cheaply. Special chips are commercially available to implement a wide variety of industrial fuzzy controllers [3]. However, while fuzzy controllers appear to be simple for single-output systems, their complexity increases rapidly for multiple input/output applications.

Fuzzy controllers are essentially rule-based where the continuous feedback input signals are assigned certain validity grades corresponding to their analog level. The grades are determined using logic membership functions (fuzzification process). The inference process that follows is to apply predetermined rules of action. These rules are usually coded on the basis of intuition and general understanding of the controlled process. The rules and logic membership functions are often determined in a heuristic way, initially, then tuned on a simulator. Self learning and dynamic rule-set schemes facilitate more systematic design approaches [6]. The output of the inference process gets defuzzified to produce a continuous control signal. The defuzzification process is also selected in a nonorthodox way and sometimes is determined by common sense and for convenience. Computer simulation is effective in facilitating fuzzy logic variables evaluation.

Computer aided analysis and design tools for fuzzy control are still in their early stages of development. A few rigorous software packages are commercially available at a significant cost [3]. Attempts have been made to demonstrate the usefulness of some general purpose simulation programs as a fuzzy control simulation tool [4]. TUTSIM is an economical fully developed dynamic systems computer simulation program with a comprehensive functions library [5]. It is user friendly and its controls oriented features favored its adoption in this work. This paper demonstrates the usefulness of TUTSIM in analyzing and synthesizing fuzzy control systems. Practical realization aspects are considered and a computer simulation module for the controller is provided. Interested users may use this module directly in their particular application for expedient realization of controllers. This module is suitable for both motion control and slow dynamics applications.

## II. THE FUZZY CONTROLLER

The objective is to realize a two-input-single-

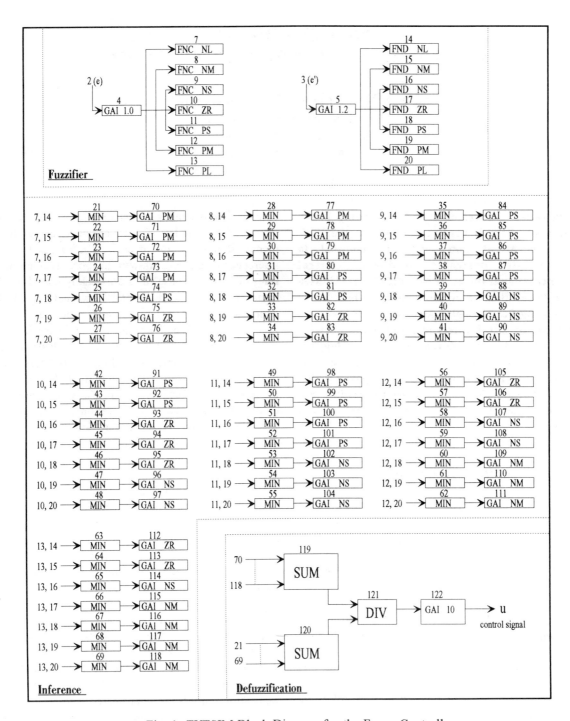

Fig. 1 TUTSIM Block Diagram for the Fuzzy Controller

output fuzzy controller with sufficient flexibility to change rules, membership functions, and control gains. General familiarity with TUTSIM and fuzzy logic basics are assumed in the following development. The fundamental approach is to use the error signal (e) and its derivative (e') as analog inputs to the fuzzy controller, as depicted in Figs. 1 and 3. The Fuzzifier is constructed of seven inverted V-shaped overlapping membership functions (blocks 7-13 for the error and 14-20 for the derivative error), Fig. 1. The input to the fuzzifier is a scaled signal with adjustable gain to be tuned for an acceptable performance (blocks 4 and 5). This process assigns the input to a certain class of membership, namely negative large (NL), negative medium (NM), negative small (NS), for negative inputs, and zero (ZR) for a zero input. A similar pattern is used for positive values. This process, in a sense, quantifies the analog input. Increasing the number of membership functions at the input would increase the quantification of the input signal into the membership rules. With the proper processing of these quantified signals the Fuzzy logic controller should approximate an analog controller. For this study seven membership functions are considered an acceptable number. The shape and centroid of the membership functions can be altered easily by changing the parameters of the function blocks.

One of the fundamental difficulties in implementing fuzzy logic controllers is in the required scaling of the base width of membership functions to accommodate the full range of input signals. In motion control systems, the inputs (e, e') vary over a wide range based on the mode of operation. Fixing the scale to a wide base is a mixed blessing. On the one hand it covers a wider range of inputs, and on the other hand it reduces the resolution of quantization for small input modes of operation. One of the commonly used methods to reach a compromise is to narrow the base of some of the membership functions in a discriminatory fashion. This approach is semiheuristic and it provides a compromised solution. The other option is to increase the number of membership functions and their corresponding inference rules, resulting in a complicated controller.

In this work, the **MAJOR MODIFICATION** to classical fuzzy logic is in assigning the membership functions. Traditionally, membership functions are selected such that they assign positive validity grades only, for a limited base at zero membership level. Based on the degree of overlap between these functions and the input value of the signal, a finite number of functions will contribute to the inference process at one time, points f and g of input 3 in Fig. 2. The input to these functions has to be within the range of their base. Unlike classical fuzzy logic, **the membership functions in this realization are extended into the negative range of assigned grading**, points a-e of Fig. 2. This modification gives the following advantages:

1. All membership functions are contributing to the output at all times. The grade points are marked a-g.

2. Inputs outside the range of regular fuzzy functions are handled without rescaling, input 4.

3. Less effort would be devoted to discriminating between width of membership functions for special applications.

4. All fuzzy logic rules will be active at all times producing a less noisy defuzzification output.

5. Loss of information due to quantization is reduced.

Blocks 21-118 of Fig.1 realize the 49 fuzzy rules given in Table 1. These rules cover all possible combinations of processing the outputs of the fuzzifier and, therefore, represent the most flexible realization. The logical product (minimum) is adopted in this work. The gains of blocks 70-118 determine the centroid of the output functions. These gains can be used at the designer's discretion to change the rules by shifting the centroids, or to eliminate some of the unneeded rules when set to zero. Many applications require fewer rules than the maximum of 49 without any noticeable loss in performance. One important observation relative to Table 1 is that the rules seem to defy

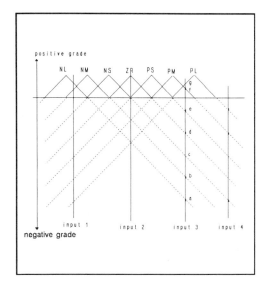

Fig. 2  Membership Functions

| e \ e' | 14 NL | 15 NM | 16 NS | 17 ZR | 18 PS | 19 PM | 20 PL |
|---|---|---|---|---|---|---|---|
| 7 NL | 70 PM | 71 PM | 72 PM | 73 PM | 74 PS | 75 ZR | 76 ZR |
| 8 NM | 77 PM | 78 PM | 79 PM | 80 PS | 81 PS | 82 ZR | 83 ZR |
| 9 NS | 84 PS | 85 PS | 86 PS | 87 PS | 88 NS | 89 NS | 90 NS |
| 10 ZR | 91 PS | 92 PS | 93 ZR | 94 ZR | 95 ZR | 96 NS | 97 NS |
| 11 PS | 98 PS | 99 PS | 100 PS | 101 PS | 102 NS | 103 NS | 104 NS |
| 12 PM | 105 ZR | 106 ZR | 107 NS | 108 NS | 109 NM | 110 NM | 111 NM |
| 13 PL | 112 ZR | 113 ZR | 114 NS | 115 NM | 116 NM | 117 NM | 118 NM |

Table 1  Inference Rules

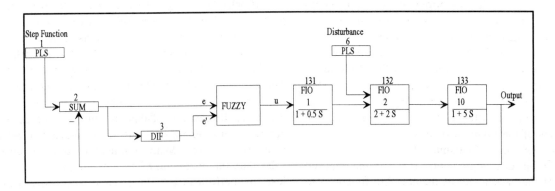

Fig. 3 Control System

intuitive logic, i.e. logically reversed. Assuming a desired positive output in a feedback control loop, (Fig. 3), a PL error (e) means that the current output is too small or negative. Therefore, the control signal should be positive. Table 1, however, shows negative for the control signal because the modified fuzzy logic controller assigns higher grade to the opposite membership function, e.g. point "a" for input 3 of Fig.2.

The weighted-average method is used to defuzzify the control signal, blocks 119-122 of Fig. 1. The implemented formula is such that the summation of the logical product (minimum) times the output function centroid, for each rule, is divided by the summation of the minimums.

Blocks 122, 4, and 5 are used for tuning the controller. Their effect is to scale each of the output (block 122) and the input signals according to their relative significance (blocks 4 and 5).

## III. SIMULATION RESULTS

The block diagram of Fig. 3 depicts a typical feedback control loop for a third order plant withexternal disturbances. The objective of this study is twofold: first to demonstrate that the modified fuzzy controller is viable and it provides an option to PID; second, to examine the sensitivity of the controller to the tuning gains.

Fig. 4 illustrates the superiority of fuzzy control over PID when they both respond to sudden changes in input, disturbances, and plant parameter variations. A step disturbance of magnitude of 0.5 is injected at 8 sec and a change of 100% in the time constant of block 133 occurred at time 13 sec. The PID controller exhibited greater deviation from the desired reference input than did the fuzzy controller. This demonstrates

the proper functionality and effectiveness of the developed TUTSIM model.

Ying, et. al, proved that fuzzy controllers are similar in their performance to a nonlinear PI controller with gains changing with the error and its derivative [7]. They derived direct relationships between PI gains and the tuning gains for classical fuzzy logic controllers. The modified fuzzy controller presented here differs from classical controllers in two respects. First, change in gain in classical controllers beyond a certain threshold causes a switch from one membership function to another resulting in use of different inference rules. Consequently, the useful range of membership functions would change and an incremental change in performance results. The modified controller, however, maintains contributions from all membership functions. Therefore, the effect of changing the tuning gains follows a linear pattern that reduces the effect of quantization and provides better understanding of the tuning process. Figures 5, 6, and 7 show the effect of change in tuning parameters on the performance of the closed loop system.

## IV. CONCLUSIONS

The modified fuzzy logic controller is an extension of the classical logic controller. It requires no scaling to accommodate a wide range of input signals. Its gains adjusting follows a linear pattern that makes the tuning process more systematic. It seems to provide a solution to several limitations encountered in classical fuzzy logic control. The modified controller is presently under further studies by the authors.

The simulation module for TUTSIM is an effective tool for studying fuzzy controllers. This module can be used with any plant while the first 122 blocks are reserved for the fuzzy controller.

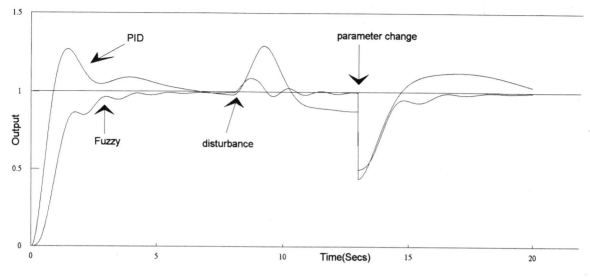

Fig. 4   Time Response

## V. REFERENCES

[1]   David I. Brubaker, " Fuzzy-Logic System Solves Control Problem," EDN, June 1992, pp. 121-127.

[2]   Omron Electronic, Inc., "An Introduction to Fuzzy Logic and its Applications in Control Systems," 1991.

[3]   Doug Conner, "Fuzzy-Logic Control System," EDN, March 1993, pp. 77-88.

[4]   Charles Hymowitz, "SPICE Simulates Fuzzy Logic Motion Control Systems," PCIM, September 1993, pp. 60-63.

[5]   Twente University of Technology SIMulation (TUTSIM) program, Applied i, Palo Alto, California.

[6]   Jeffery R. Layne, et. al., "Fuzzy Learning Control for Anti-Skid Breaking Systems," IEEE proceedings of the 31st Conference on Decision and Control, December 1992, pp. 2523-2528

[7]   Hao Ying, et. al., "Fuzzy Control Theory: A Nonlinear Case," Automatica , vol. 26, no. 3, 1990, pp. 513-520.

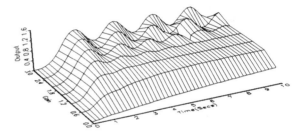

Fig. 5   Sensitivity to Error Gain

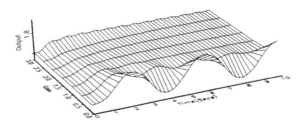

Fig. 6   Sensitivity to Derivative Gain

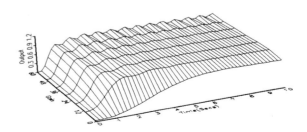

Fig. 7   Sensitivity to Output Gain

# A Comparison of Robustness:
# Fuzzy Logic, PID, & Sliding Mode Control

Charles P. Coleman and Datta Godbole

Department of Electrical Engineering and Computer Sciences

University of California at Berkeley

Berkeley, CA 94720, USA

*Abstract*— This work is performed in order to foster and promote unbiased and accurate comparison of fuzzy logic control and classical control design methodologies. We are motivated to execute this study by the demonstrated robustness to plant perturbations of the fuzzy logic controller given in [1].

Robust fuzzy logic, PID, and sliding mode controllers are designed to control the speed of a nominal third order linear time-invariant model of a motor. The step response performance of each controller, applied to the nominal and two perturbed motor plants, is presented.

We conclude fuzzy logic control can be a useful tool for the control engineer. We encourage more benchmark comparisons of fuzzy logic control with classical control techniques for the benefit of the practicing control engineer.

## I. INTRODUCTION

### A. New Control Tools, New Enemies?

The successful use of fuzzy logic controllers has greatly expanded in the last twenty years ([2]–[5]). This expansion has prompted much comparison to classical control techniques. The ensuing discussions have not always been amicable. Nor have they necessarily lead to manifest results and conclusions for the practicing control engineer.

### B. Debate and Dilemma

At times the debate over the use of fuzzy logic control techniques versus classical control techniques has become quite heated with ardent detractors on both sides ([6]–[11]).

Some control engineers have been placed in a dilemma by this debate, and have been left without an appeasing study or comparison upon which to formulate a useful opinion.

### C. Well-Being of the Control Engineer

In the process of this raging and seemingly growing battle, it appears that the best interest of the control engineer has been occluded. The practicing control engineer usually has little interest in ideological debates. It is her interest to produce a working controller, with efficacy, making use of any and all techniques at her disposal.

### D. A Call for Peace or at Least a Cease Fire! and Negotiations

Fortunately, some have called for control engineers to consider inclusion of fuzzy logic control into their control enginering toolboxes, and for the utilization of both fuzzy logic control and classical control when and where appropriate ([12] [13]). This appears to be a reasonable request. At the very least, the fuzzy logic tool can be compared to classical control tools. Should the fuzzy logic control tool prove useful and acceptable to the control engineer, it should become welcome addition to her control toolbox.

### E. Robust Controllers

The control engineer is concerned about such properties as the robustness of a controller to plant perturbations and uncertainty. We take the perspective of the control engineer who has several control tools available for the synthesis of robust controllers, and who is interested in determining the capabilities of each control tool to synthesize a robust controller.

To assess the usefulness of each tool in designing a robust controller, we choose a benchmark problem, and then engage in a comparison of the control designs which result from the application of each control design method.

## II. CHOICE OF CONTROL TOOLS

Fuzzy logic control, PID control, and sliding mode control tools are chosen for our investigation. Fuzzy logic control is chosen because of its empirically demonstrated robustness properties shown in [1]. PID control is chosen because it is one of the most commonly used controllers in industry, and because it has plant perturbation robustness properties which can be mathematically analyzed. We choose a sliding mode controller because it is a robust non-linear control method whose robustness properties can also be mathematically analyzed.

We do not claim to investigate and present every robust control technique. Indeed, the popular and successful $H_\infty$ and adaptive control techniques are noticeably absent from this study. We felt it was necessary to restrict ourselves to fuzzy logic control and two other control methods, in order to keep our study brief, intelligible, and meaningful.

## III. Control Problem

### A. Motor Speed Control

Motivated by the work in [1] and [14], and encouraged by the robustness results presented for fuzzy logic control, we consider the robust speed control of a rotating motor, modeled by a third order linear time-invariant transfer function. The general configuration for the motor speed control problem is given in Figure 1.

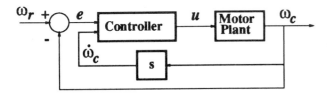

Figure 1: Motor Speed Control Problem

We use the nominal plant, perturbed motor plants, and the performance criteria given in [1] to design and compare robust fuzzy, PID, and sliding mode controllers.

### B. Restrictions on Controller Inputs and Outputs

As shown in Figure 1, the reference step input speed $\omega_r$ and the output motor speed $\omega_c$ are available for comparison. Each controller only has access to the motor speed error $e = \omega_r - \omega_c$ and the motor angular acceleration $\dot{\omega}_c$ as inputs. Each controller produces only one output $u$.

### C. Nominal Plant

The nominal motor plant is modeled by the following transfer function:

$$G_1(s) \;=\; \frac{5}{s(s+1)(s+2)} \tag{1}$$

### D. Perturbed Plants

The two perturbed motor plants are given by the following transfer functions:

$$G_2(s) \;=\; \frac{15}{s(s+1)(s+2)} \tag{2}$$

$$G_3(s) \;=\; \frac{5}{s^2(s+2)} \tag{3}$$

### E. Performance Criteria

The designed controllers must be robust to variations of system parameters. Specifically, for the fuzzy logic, PID, and sliding control designs, one single controller must render acceptable the closed loop unit step responses of the nominal plant and the two perturbed plants. To be acceptable, the closed loop step responses must simultaneously have short rise time and no overshoot.

### F. Design Approach

We design the fuzzy, PID, and sliding mode controllers based on the nominal plant given in equation (1). Each designed controller is then applied to all three transfer functions, and numerical simulations are performed to analyze the controller's robustness.

The design of the fuzzy logic controller is given in Section IV. The design of the PID controller is given in Section V. The design of the sliding mode controller is given in Section VI. Plots of each controller's closed loop step response and control effort are shown in Appendix A.

## IV. Analysis and Reconstruction of a Fuzzy Robust Controller

We implement the robust fuzzy controller given in [1] and [14]. The inputs to the fuzzy controller are $e$ and $\dot{\omega}_c$. The output of the fuzzy controller is $u$. The universes of discourse of $e, \dot{\omega}_c$, and $u$ are partitioned into seven fuzzy sets:

- NB - negative big
- NM - negative medium
- NS - negative small
- ZE - zero
- PS - positive small
- PM - positive medium
- PB - positive big

Each fuzzy set is represented by a Gaussian membership function. The rule base of the fuzzy logic controller contains forty-nine rules which are tabulated in Figure 2. The output of each rule is determined by min-inference. The crisp output $u$ of the fuzzy logic controller is generated by centroid defuzzification.

|  |  | $e$ | | | | | | |
|---|---|---|---|---|---|---|---|---|
|  |  | NB | NM | NS | ZE | PS | PM | PB |
|  | PB | NB | NB | NB | NB | NM | NS | ZE |
|  | PM | NB | NB | NB | NM | NS | ZE | PS |
|  | PS | NB | NB | NM | NS | ZE | PS | PM |
| $\dot{\omega}_c$ | ZE | NB | NM | NS | ZE | PS | PM | PB |
|  | NS | NM | NS | ZE | PS | PM | PB | PB |
|  | NM | NS | ZE | PS | PM | PB | PB | PB |
|  | NB | ZE | PS | PM | PB | PB | PB | PB |

Figure 2: Fuzzy Logic Controller Rule Base

The step response of this fuzzy logic controller applied to the nominal and perturbed plants is shown in Appendix A. The step responses for all three plants have short rise times and no overshoot. Thus, they meet the specified robust performance criteria. The control effort generated by the fuzzy logic controller is also shown in Appendix A.

## V. ANALYSIS AND SYNTHESIS OF A ROBUST PID CONTROLLER

The fuzzy logic controller designed in [1, 14] is similar to a nonlinear PD controller [15]. We design a linear PID controller to obtain satisfactory step response for all three transfer functions.

A linear PID controller can be characterized by the following transfer function:

$$\bar{C}(s) = K_p + K_d\, s + \frac{K_i}{s} \qquad (4)$$

Because, it is very difficult to realize a pure differentiator, we use the following transfer function for implementation.

$$C(s) = K_p + \frac{K_d\, s}{\tau\, s + 1} + \frac{K_i}{s} \qquad (5)$$

To have a fair comparison, we use a small value for $K_i = 0.001$, just enough to keep steady state error zero. We use a small value for $\tau = 0.01$, so that its effect on the dynamics is minimal, but the derivative controller is still realizable.

The selection of $K_p$ and $K_d$ is based on the root locus of $C(s)G_1(s)$. The root locus gives us the locations of closed loop poles. The proportional and derivative gain values are chosen such that the closed loop pole locations are in the left half complex plain for all three plants. Another criterion for selection of these gains is the step response of the closed loop system.

We select the following gains:
$K_p = 2, K_d = 5$
With these gains, the closed loop system is stable for the open loop gain up to 51. Thus in particular, the above PID controller stabilizes the systems given by the transfer functions (1,2). For the model of (3), we have closed loop stability for the open loop gain upto 30. The step responses have no overshoot and are critically damped.

## VI. ANALYSIS AND SYNTHESIS OF ROBUST SLIDING MODE CONTROLLER

To design a sliding mode controller, we convert the transfer function model into state space format. The controllable canonical form realization of the model of (1) is given by:

$$\dot{x} = \begin{bmatrix} 0 & 1 & 0 \\ 0 & 0 & 1 \\ 0 & -2 & -3 \end{bmatrix} x + \begin{bmatrix} 0 \\ 0 \\ 5 \end{bmatrix} u$$

$$y = \begin{bmatrix} 1 & 0 & 0 \end{bmatrix} x \qquad (6)$$

We consider the following sliding surface for this design:[1]

$$S = (\ddot{y} - \ddot{y}_d) + \lambda_1(\dot{y} - \dot{y}_d) + \lambda_2(y - y_d) \qquad (7)$$

The choice of sliding surface is based on the following considerations:

- The relative degree of the system (6) with $S$ as output should be 1. This ensures that the input $u$ appears explicitly on the right hand side of $\dot{S}$ equation.
- System dynamics on the sliding surface should be stable. (This requires $\lambda_1, \lambda_2 > 0$)

$y_d(t)$ specifies the reference trajectory to be tracked by output. If one knows the entire trajectory of $y_d$, then the derivative information $\dot{y}_d(t), \ddot{y}_d(t)$ can be extracted off line and used in the design to improve the performance. If the information is not available (or as in this case, $y_d(t)$ being unit step, is nondifferentiable) we can assume $\dot{y}_d$ and $\ddot{y}_d$ to be zero and let the robustness property of the sliding mode controller take care of the mismatch.

With the definition of sliding surface as above, we have reduced the design requirement from tracking $y_d(t)$ to being on the surface $S = 0$. Once on the sliding surface, the dynamics (equation 7) is exponentially stable and asymptotic trajectory tracking is achieved. The control $u$ is designed to make the surface $S = 0$ attractive and to reach the surface in finite time.

Consider the following lyapunov function:

$$V = \frac{1}{2}\, S^2 \qquad (8)$$

Its time derivative is given by

$$\dot{V} = S\, \dot{S} \qquad (9)$$

As the system has relative degree 1 with $S$ as output, we can solve for $u$ from the equation

$$\dot{S} = -K\ \text{sgn}(S) \qquad (10)$$

This results in a negative definite $\dot{V}$ and also guarentees finite time convergence to the sliding surface. But, the controller will result in high frequency chattering near sliding surface. To avoid this, we use the following expression to solve for $u$.[2]

$$\dot{S} = -K\, S \qquad (11)$$
$$\Rightarrow \dot{V} = -K\, S^2 \qquad (12)$$

[1]Refer to [16] for details of sliding mode controller design.
[2]With this controller, the trajectory is not guaranteed to reach the sliding surface in finite time.

Expanding both sides of the equation, we get

$$\dot{S} = -2x_2 - 3x_3 + 5u + \lambda_1 x_3 + \lambda_2 x_2$$
$$= -K(x_3 + \lambda_1 x_2 + \lambda_2 x_1 - \lambda_2 y_d) \text{ Desired}$$

The $u$ obtained from this equation will be used for the control.

This controller needs access to all three states, whereas the fuzzy logic controller of [1, 14] make use of only $x_1$ and $x_2$. To have a fair comparison, we will construct an observer to get an estimate of the state variable $x_3$.

Standard Luenburger observer is designed to get an estimate of $x_3$. We use the model of (1) to design the observer as follows:

$$\dot{\hat{x}} = A\hat{x} + bu + L(y - \hat{y}) \qquad (13)$$
$$\hat{y} = c\hat{x}$$

where, the matrices $A$, $b$ and $c$ correspond to the model in equation (1). The matrix $L$ is chosen such that the eigenvalues of observer error dynamics (A-LC) are in the left half plane and the error dynamics is faster than the dynamics on the sliding surface.

The closed loop step responses with this controller are shown in Appendix A. The controller is robust to input gain changes as well as to the changes in system dynamics while keeping the input magnitude small.

Note: With the use of equation (11), the controller is essentially a linear state feedback controller. It results in closed loop eigenvalues of -1.3392 & -4.33±j 1.91 for model (1). The same controller, when applied to other plants will give the following closed loop eigenvalue locations:
Model of (2) $\Rightarrow$ $-1.32, -3.55, -19.12$
Thus, by design, the increase in gain makes the closed loop system more stable...
Model of (3) $\Rightarrow$ $-1.4, -3.79 \pm j2.62$

## VII. CONCLUSIONS

### A. Comments on Controller Designs

The control engineer proficient in PID and sliding mode control techniques can readily synthesize robust controllers to perform the task dictated by this example. As demonstrated in [1] and verified in this study, the control engineer utilizing fuzzy control techniques can readily achieve the same goal. Thus, all three control tools are available to the control engineer for use in synthesizing an acceptable robust controller for this control problem.

With the appropriate software tools available, the fuzzy logic controller for this problem was relatively simple to implement, and it provided satisfactory results. Software tools were also available for the design and simulation of the PID and sliding mode controllers for this problems, and so these two controllers were easy to implement.

### B. Analytical Tools for the Analysis of Fuzzy Logic Control

A control engineer adding the fuzzy logic control tool to her toolbox already full of classical control tools, might ask "Why does this method demonstrate robustness?", and "When will it fail?" These questions often have mathematically analytical answers for classical control tools. To satisfy this inquiry, we call for the continued development of mathematically analytical tools to answer such questions to the satisfaction of the control engineering community.

### C. Fuzzy Logic Control: A Useful Tool

From our brief study, we conclude that fuzzy logic control should have a place in the control engineer's toolbox. Only time, experience, and further analysis will determine whether fuzzy logic control becomes a prominent tool in the control toolbox used by devotees of classical control techniques.

## VIII. FUTURE WORK

We encourage more extensive fair and unbiased benchmark comparisons of fuzzy logic control with classical control techniques for the benefit of the practicing control engineer, and we hope to become a part of this effort.

In the future, we hope to contribute to the creation of new analytical tools for the analysis of fuzzy logic controllers. We also hope to be able to contribute to the continuing effort to identify classes of problems where rule based fuzzy logic control techniques have advantages over and are more appropriate than classical control techniques.

## IX. ACKNOWLEDGMENTS

The authors would like to thank Professor S. Shankar Sastry for his gracious support and encouragement. The authors would also like to thank Dr. Shahram M. Shahruz for helpful discussions.

### REFERENCES

[1] H. T. Nguyen, C.-W. Tao, W. E. Thompson, "An Empirical Study of Robustness of Fuzzy Systems", *Proc. of 2nd IEEE Intl. Conf. on Fuzzy Systems*, pp. 1340–1345.

[2] R. M Tong, "An Annotated Bibliography of Fuzzy Control", in *Industrial Applications of Fuzzy Control*, M. Sugeno, Ed., North-Holland, Amsterdam, Holland, 1985, pp249–269.

[3] T. Terano, K. Asai, M. Sugeno. *Fuzzy Systems Theory and Its Applications*. Academic Press, Boston, MA, USA.

[4] W. Pedrycz. *Fuzzy Control and Fuzzy Systems*, 2nd extended ed. John Wiley & Sons, New York, NY, USA.

[5] H. Berenji, "Fuzzy Logic Controllers", in *An Introduction to Fuzzy Logic Applications in Intelligent Systems*, R. R. Yager and L. A. Zadeh, Eds., Kluwer Academic Publishers, Boston, MA, USA, 1992, pp 69–96.

[6] S. Chiu, S. Chand, D. Moore, A. Chaudhary, "Fuzzy Logic for Control of Roll and Moment for a Flexible Wing Aircraft", *IEEE Control Systems Magazine*, pp 42–48, Vol 11, No 4, June 1991.

[7] A. L. Schwartz, "Comments on Fuzzy Logic Control of Roll and Moment for a Flexible Wing Aircraft", *IEEE Control Systems Magazine*, pp 61-62, Vol 12, No 1, February 1992.

[8] S. Chiu, "Author's Reply", *IEEE Control Systems Magazine*, pp 62–63, Vol 12, No 1, February 1992.

[9] E. Cox, "Adaptive Fuzzy Systems", *IEEE Spectrum*, pp 27–31, Vol 30, No 2, February 1993.

[10] C. J. Herget, Ed., "Reader's Forum", *IEEE Control Systems Magazine*, pp 5–7, Vol 13, No 3, June 1993.

[11] M. Athans, "Control - The Adventure Continues", Bode Lecture, *32nd IEEE Conference on Decision and Control*, San Antonio, TX, USA, December 15–17, 1993.

[12] E. H. Mamdani, "Twenty Years of Fuzzy Control: Experiences Gained and Lessons Learnt", *Proc. of 2nd IEEE Intl. Conf. on Fuzzy Systems*, pp. 339–344.

[13] M. Tomizuka, "Fuzzy Control in Control Engineer's Tool Box", Lecture, *ARO/NASA Workshop on Formal Models for Intelligent Control*, MIT, Cambridge, MA 02139, USA, September 30 – October 2, 1993.

[14] C.-W. Tao, R. Mamlook, W. E. Thompson, "Reduction of Complexity for a Robust Fuzzy Controller", *Proc. of 2nd IEEE Intl. Conf. on Fuzzy Systems*, pp. 1346–1349.

[15] H. Ying, W. Siler, J. J. Buckley, "Fuzzy Control Theory: A Nonlinear Case", *Automatica*, Vol. 26, No. 3, pp 513–420, 1990.

[16] J.-J. E. Slotine and W. Li, *Applied Nonlinear Control*. Prentice Hall, 2nd ed., 1991.

## A  CONTROLLER PERFORMANCE RESULTS

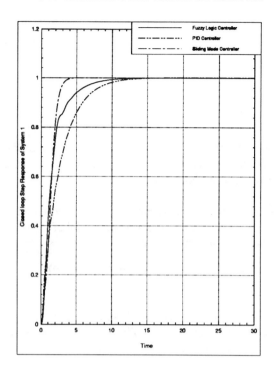

Figure 3: Closed Loop Step Response of nominal plant

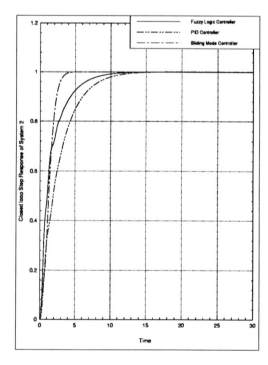

Figure 4: Closed Loop Step Response of the perturbed model $G_2(s)$

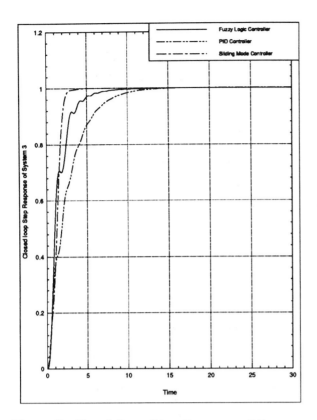

Figure 5: Closed Loop Step Response of the perturbed model $G_3(s)$

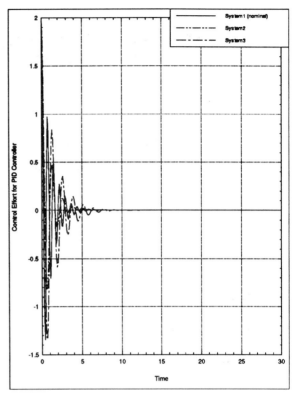

Figure 7: PID Controller Control Effort

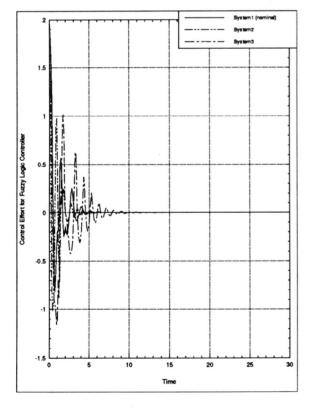

Figure 6: Fuzzy Logic Controller Control Effort

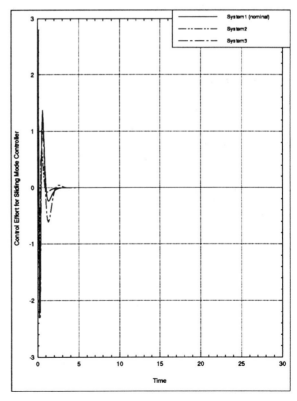

Figure 8: Sliding Mode Controller Control Effort

# APPLICATION OF SELF-ORGANIZING FUZZY LOGIC CONTROLER IN 2-D MOTION TRACKING CONTROL

Jenn-Terng Gau and W. F. Lu

Department of Mechanical and Aerospace Engineering and Engineering Mechanics

University of Missouri-Rolla

Rolla, MO 65401-0249

## ABSTRACT

In this study, Self-organizing Fuzzy Logic Controller (SOFLC) is used in 2-D motion tracking control. SOFLC avoids the need of the prior knowledge of the system, and is able to create and modify the control rules by learning based on a given performance criterion. Simulation results compare the performance of SOFLC with simple Fuzzy Logic Controller. It is found that the SOFLC is able to lower the tracking error below a desired set value in a short time for 2-D tracking control.

## 1. INTRODUCTION

Many successful applications of Fuzzy Logic Controllers have been reported[1,2,3,4] since the pioneer work of Mamdani and his colleagues on fuzzy logic control[5,6,7] motivated by Zadeh's paper on the theory of fuzzy sets[8]. Fuzzy logic controllers (FLC) are rule-based systems that use fuzzy linguistic variables to model human rule-of-thumb approached to problem solving. Due to their inherent characteristics of capturing the approximate, inexact nature of a real world, the FLC are capable to merge the experience, heuristic and intuition of expert operations. Thus FLC can perform well in spite of noise or sensor failures in the system, and has the ability to handle nonlinearities without degradation of the system.

While FLC is gaining its popularity, it is not without any drawbacks in the implementation. The control rules are usually derived from experience of experts or operators, but these are not always available or easily formulated.

Furthermore, the derived control rules might not cope with the new environment if the change occurs in the process. Based on these problems, an attractive solution is provided by the Self-Organizing Fuzzy Logic Controller(SOFLC) proposed by Procyk and Mamdain[9]. The SOFLC is capable to develop and improve the control rules based upon the evaluation of their performance with varying process dynamics without human expertise. SOFLC has been used successfully in several applications [10,11,12,13].

In this paper, we attempt to study the application of SOFLC for 2-D motion tracking control since 2-D motion is widely used in industry automation. With the aim of high quality product and high production rate, the tracking ability and traveling time of the system must be improved. Therefore, it is important to limit the tracking error in 2-D motion within a tolerable value corresponding to the product tolerance and less traveling time. Self-paced Fuzzy Tracking Controller(SPFTC) proposed by Huang and Tomizuka[14] is used here to adjust the tracking speed according to the condition of the desired contour. SOFLC is then utilized to generate and modify fuzzy control rules to limit the tracking error.

## 2. BASIC IDEA OF SELF-ORGANIZING FUZZY LOGIC CONTROLLER

The structure of the SOFLC is shown in Figure 1. It is obvious that the controller is an extension of the 'simple' FLC that incorporates self-organizing mechanism or performance feedback. Since the overall idea of simple FLC has been presented by many researchers [1,2,3,4], only the structure of SOFLC is briefly described in this paper.

## 2.1 Self-organizing Mechanism

Self-organizing mechnism consists of three components which are performance index , correction of the control action, and fuzzy rules modification.

### 2.1.1. Performance Index

The performance index measures the deviation of actual output of plant from the desired output and emits the appropriate changing commands which are required at the output. The performance index rules are formulated in terms of linguistic statement as shown in Table 1 which is adopted form [10][11]. They are written to create control rules starting from a controller containing no rules. In this paper, the two measurments are deviation of desired tracking error, EE, and change of deviation of desiried tracking error, CEE.

The performance index rules can be transformed into a look-up table of output commands using the standard techniques of fuzzy calculus. This look-up table used in this study is derived from [10] as shown in Table 2.

### 2.1.2. Correction of the Control Action

In simple input single output (SISO) systems, the correction of control action is

$$Pi(nT) = G*r(nT)$$

where =Pi is the output of performance index, and G is the gain which is one in this study, and r is the changing commands required at the controller output.

For multi-input multi-output processes [9,10]

$$\underset{\sim}{p}(nT)=[M]\underset{\sim}{r}(nT)$$

where $\underset{\sim}{p}(nT)$ is a output vector of performance index, and $\underset{\sim}{r}(nT)$ is the vector of changing commands required at controller output. [M] can be regarded an incremented model of the process. The elements of M martix are between +1 and -1.

### 2.1.3. Fuzzy Rules Modification

They are three forms of modifying the components of the fuzzy controller. This paper uses fuzzy rules modification. Assuming that a process has time delay of d samples, the control action at (n-d)T has most contributed to the systems performance at nT.

The original implication

$$EE[(n-d)T] \rightarrow CEE[(n-d)T] \rightarrow U[(n-d)T]$$

has to be modified into

$$EE[(n-d)T] \rightarrow CEE[(n-d)T] \rightarrow U[(n-d)T]+r(nT)$$

where r(nT) is the changing commands.

Shihuasng Shao [12] has developed a new algorithm to modify the fuzzy rules with less computational time and memory.

## 3. SPFTC PLUS SELF-ORGANIZING FUZZY LOGIC CONTROLLER FOR 2-D MOTION

In 2-D motion tracking control, two steps are used in this study. First, SPTFC proposed in [14] is employed here to adjust the tracking speed according to the shape of the path. In step 2, self-organizing mechanism is used to modify the control rules in fuzzy controller.

The tracking speed from SPTFC is the input to the SOFLC in Figure 1. Figure 2 shows the structure of the plant in this study. It consists of DC motor, and leadscrew and the parameters of the plant are shown in Table 3.

### 3.1 Using SPTFC to Adjust Tracking Speed

Here, we defined the tracking error as $[(X_a(n)-X_d(n))^2 + ((Y_a(n)-Y_d(n))^2]^{1/2}$ where $(X_d(n),Y_d(n))$ is the desired coordinate at the sampling time n and $(X_a(n),Y_a(n))$ is the actual coordinate.

In SPFTC, current curvature C, change of curvature CC and tracking speed, V, are selected as the linguistic variables in the control rules. C and CC are defined as

C = current curvature*G1
CC =(previewed curvature-current curvature) *G2

where G1 and G2 are scale gains. The consequent of each rule is defined in terms of the change of tracking speed CV(n). Once CV(n) is determined, the tracking speed at sampling time (n+1) can be updated as V(n+1)=V(n)+CV(n).

In this paper, a set of fuzzy control rules for SPFTC is proposed as shown in Table 4. There are three promise linguistic variables, V, C, CC, and one consquence linguistic output, CV in the talbe. P, N, S, M, and B means positive, negative, small, medium, and big respectively. Table 4 tabulates the linguistic fuzzy rules in the form of

If F is A and C is B and CC is D then CF is E

Their membership functions are also shown in Figures 3 and 4.

## 3.2 Design of SOFLC

In most of the automated production, the tracking error of the motion is normally set to be less than a desired value in order to assure the quality of the product. In this study, a desired tracking error corresponding to the path tolerance is defined and use SOFLC to generate and modify fuzzy control rules to assure that the tracking error is within the set threshold value.

First of all, we defined two linguistic variable, EE and CEE as follows.

a). The performance index
$$=[(X_a(n)-X_a(n-1))^2+(Y_a(n)-Y_a(n-1))^2]^{1/2} -$$
$$[(X_d(n) - X_d(n-1))^2+(Y_d(n)-Y_d(n-1))^2]^{1/2}.$$

If the index is greater then zero, then

$$EE(n)=GEE*(E_{real}(n)-E_{set}); \text{ otherwise}$$
$$EE(n)=GEE*(E_{set}-E_{real}(n)).$$
where
$$E_{real}(n)=[(X_a(n)-X_a(n-1))^2+(Y_a(n)-Y_a(n1))^2]^{1/2}$$
and $E_{set}$ = desired tracgking error .

also, $(\bullet)_a$ indicates actual position ;

and $(\bullet)_d$ represents desire position.

b). CEE=GCEE*[EE(n)-EE(n-1)].

Next , the operation procedures of SOFLC are summarized.

Step1. Simple fuzzy logic controller issues U[(n-d)T] command to plant at nT [1][9].

Step 2. Calculate the error of tracking error EE(nT) and the change error of tracking error CEE(nT) at nT.

Step 3. Determine the changing command r(nT), r(nT)=Pi(nT) for SISO, from EE(nT) and CEE(nT) from the performance look-up table.

Step 4. Modify the fuzzy control rules.[1][9].

Step 5. Determine the control action from the new fuzzy control rules.

## 4. SIMULATION RESULTS

In this section, the performance of both simple fuzzy logic controller and self organizing fuzzy logic controller are studied by simulations of 2-D motion tracking control. The objective of the contour following tasks is to limit the tracking error to within a set threshold value. A desired motion path as shown in Figure 5 was selected to ensure that a diverse rules would be generated. The motion started from point 'a' in an anti-clockwise direction. In order to avoid the unbound of tracking speed, the speed is restricted to a range between Vmin =0.2 cm/sec and Vmax =2.0 cm/sec.

In these simulation, SPFTC is first used to adjust the tracking speed based on the curvature of the path. Gains and parameters chosen are G1=10, G2=2, Gcf=0.0055, sampling time T=0.004 sec, and preview steps=max(-100*current curvature+50, 20). Using the fuzzy control rules in Table 4, the tracking speed are found varied with the shape of the path as shown in Figure 6. Once the tracking speeds are determined, both simple FLC and SOFLC are employed to simulate the tracking error of the motion.

## 4.1 Simple Fuzzy Logic Controller (FLC)

The structure of a simple FLC can be found in the portion of Figure 1. Since both tracking error, E, and change of tracking error, CE, are important in determining the control input U, they are selected as linguistic variables to be used in the premise of control rules. The control input can be seen as the mapping from E and CE through control rules and the fuzzy reasoni.ng. The membership fuctions of linguistic values E, CE, and U are shown in Figures 3 and 4. These linguistic variables are defined as

E = GE* (desired position - actual position)
CE =GCE* (current tracking error- previous tracking error)

1662

Table 5 is the control rules of this simple fuzzy logic controller. In this simulation, the defuzzification method [1] is used to determine U.

$$U= \{ \sum_{i=1}^{n} u_{C_i}(w_i) \bullet w_i \}/\{ \sum_{i=1}^{n} u_{C_i}(w_i) \}*GU.$$

where GE=8, GCE= 6, GU=0.005 are chosen. The simulated result is showen in figure 7. In this case, the tracking error depends only on the control rules.

## 4.2 Self-organizing Fuzzy Logic Controller

In this simulation, assuming the tracking error is limited to 0.0008cm. Using simple FLC, it is not gurantee that the tracking errors will be below the desired value of 0.0008 cm as shown in Fig. 7. SOCFL is adopted here. We utilize Table 2 as the performance look-up table and let the control rules empty at the begining.

In the design of SOFLC, delay d=1, GEE=1600, GCEE=600, and GU=23 are chosen. The simulated results are shown in Figure 8 with the combination of three runs. At the beginning of the first run, the tracking error went beyond the desired value since no control rules are available at the beginning. Once more control rules were created and modified based on the performance index, the tracking errors are kept below the threshold value. By self-organizing and self-learning, enough rules are already generated and modified in runs 2 and 3, the tracking error are controlled below the set value. Since our performance index is the tracking error= 0.0008 cm, there is a tendency for the controller to move the tracking error close to this value as can be seen in the later portion of runs 2 and 3. Comparing Fig. 7 and Fig. 8, it is clear that simple FLC can not gurantee that the tracking error is kept below a desired value since the tracking error depends only on curvature of the path. The simple FLC uses only fixed rules provided and can not learn from the changing environmnet.

## 5. CONCLUSIONS

In this paper, an application of Self-organizing Fuzzy Logic Control to a 2-D motion tracking is considered. Combining with SPFTC, SOFLC has shown its superiority and flexibility over simple FLC from the simulated results. The controller is capable of self learning from its feedback of performance index and keeping the tracking error below a desired value. It is believed that this study can be extended to several machining problems. Currently, a 2-D end milling process is under studied.

## REFERENCES

1. Chuen Chien Lee " Fuzzy logic in control systems : Fuzzy logic control Part I and Part II ", IEEE Trans. Sys. Man Cyber. , vol. 20 , no. 2 , pp. 404 - 435 , Mar./ Apr. 1990.

2. Michio Sugeno " An Introductory survey of fuzzy control " Information sciences , 36, pp. 59-83,1985.

3. R. M. Tong " A control engineering review of fuzzy systems " Automatica , vol. 13 , no.6 , pp. 559-569 , Nov. 1977.

4. P. J. King and E. H. Mamdani " The application of Fuzzy system to Industry processes ," Automatica , vol. 13, pp. 235-242, 1977.

5. E. H. Mamdani, "Application of fuzzy algorithms for simple dynamic plant," Proc. IEE, vol. 121, no 12, pp. 1585-1588, 1974.

6. E. H. Mamdani and S. Assilian, " An experiment in linguistic sythesis with a fuzzy logic controoler," Int. J. Man Mach. Studies, vol 7, no 1, pp. 1-13, 1975.

7. E. H. Mamdani, "Advances in the linguistic sythesis of fuzzy controllers," Int. J. Man Mach. Studies, vol 8, no 6, pp. 669-678, 1976.

8. L. A. Zadeh " Fuzzy Sets, " Information and Control, vol. 8, no.3, pp.338-353, June 1965.

9. T. J. Procyk & E. H. Mamdani , " A Linguistic Self-Organizing Process Controller, " Automatica, vol.. 15 , pp. 15-30, 1979.

10. S Daley & K. F. Gill " A design study of a self-organizing fuzzy logic controller, " procee. Inst. Mech. Engers., vol. 200, pp. 59-69, 1985.

11. S DaLey & K. F. Gill " Attitude control of spacecraft using an extended self-organizing fuzzy logic controller, " procee. Inst. Mech. Engers. ,vol. 201, ND. C2, pp. 97-105, 1987.

12. Shihuang Shao, " Fuzzy-organizing controller and Its application for dynamic process, " Fuzzy Sets and System, 26, pp.151-164, 1988.

13. B. A. M. Wakileh and K. F. Gill " Robot control using self-organizing Fuzzy logic, " Computer in Industry , 15, pp. 175-186, 1988.

14. L.-J. Huang and M. Tomizuka " A self-paced fuzzy tracking controller for a two-dimensional motion control, " IEEE Trans. Sys. Man and Cyber., vol. 20, no. 5, pp. 115-1124, 9/10, 1990.

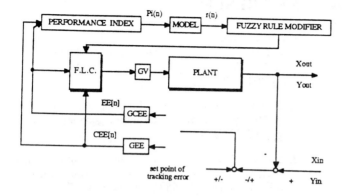

Figure 1 Structure of SOFLC

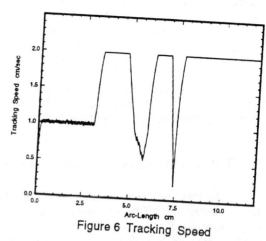

Figure 6 Tracking Speed

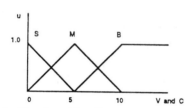

Figure 2 Plant ( DC Motor & Leadscrew)

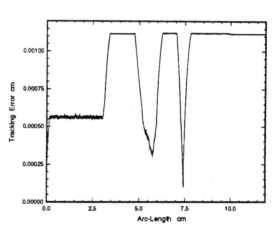

Figure 7. Tracking Error of SPFTC plus FLC

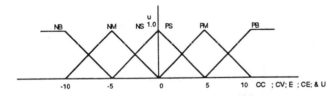

Figure 3 Membership Functions of V and C

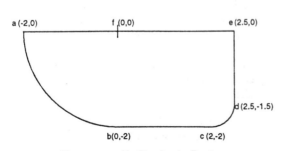

Figure 4 Membership Functions of CC, CV, E, CE, & U

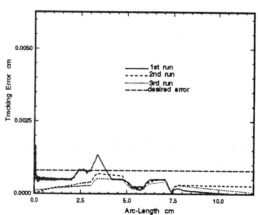

Figure 8 Tracking Error of SPFTC plus SOFLC

Figure 5 Desired Part

## Table 1  Performance Index Rule

| EE \ CEE | PB | PM | PS | ZO | NS | NM | NB |
|---|---|---|---|---|---|---|---|
| PB | NB | NB | NB | NB | NM | NS | ZO |
| PM | NB | NB | NB | NM | NS | NS | PS |
| PS | NB | NB | NM | NS | NS | ZO | PM |
| PO | NB | NM | NM | NS | ZO | ZO | PB |
| NO | NM | NM | NS | ZO | ZO | PS | PB |
| NS | NS | NS | ZO | ZO | PS | PM | PB |
| NM | NS | ZO | PS | PM | PM | PB | PB |
| NB | ZO | PS | PM | PM | PB | PB | PB |

Table 1  Performance Index Rule

## Table 2  Performance Look-up Table

| C EE \ EE | -6 | -5 | -4 | -3 | -2 | -1 | 0 | 1 | 2 | 3 | 4 | 5 | 6 |
|---|---|---|---|---|---|---|---|---|---|---|---|---|---|
| -6 | 7.0 | 6.5 | 7.0 | 6.5 | 7.0 | 7.0 | 4.0 | 4.0 | 4.0 | 3.0 | 1.0 | 0.0 | 0.0 |
| -5 | 6.5 | 6.5 | 6.5 | 5.0 | 6.5 | 6.5 | 4.0 | 4.0 | 4.0 | 2.5 | 1.5 | 0.0 | 0.0 |
| -4 | 7.0 | 7.0 | 7.0 | 5.0 | 4.0 | 4.0 | 4.0 | 1.0 | 1.0 | 1.0 | 0.0 | -1.5 | -1.0 |
| -3 | 6.5 | 6.5 | 6.5 | 5.0 | 4.0 | 4.0 | 4.0 | 0.0 | 1.5 | 1.0 | -1.0 | -1.0 | -1.5 |
| -2 | 7.0 | 6.5 | 7.0 | 4.0 | 1.0 | 1.0 | 1.0 | 0.0 | 0.0 | -1.0 | -1.0 | -4.0 | -4.0 |
| -1 | 6.5 | 6.5 | 6.5 | 4.0 | 1.5 | 1.5 | 0.0 | 0.0 | 0.0 | -1.0 | -1.5 | -4.0 | -4.0 |
| -0 | 7.0 | 6.5 | 4.0 | 3.0 | 1.0 | 1.0 | 0.0 | 0.0 | -3.0 | -3.0 | -4.0 | -6.5 | -7.0 |
| +0 | 7.0 | 6.5 | 4.0 | 3.0 | 0.0 | 0.0 | 0.0 | -1.0 | -3.0 | -4.0 | -4.0 | -6.5 | -7.0 |
| 1 | 4.0 | 4.0 | 1.5 | 1.0 | 0.0 | -1.0 | -1.5 | -1.5 | -1.5 | -4.0 | -6.5 | -6.5 | -6.5 |
| 2 | 4.0 | 4.0 | 1.0 | 1.0 | 0.0 | 0.0 | -1.0 | -1.0 | -1.0 | -4.0 | -7.0 | -6.5 | -7.0 |
| 3 | 1.5 | 1.0 | 0.0 | -1.0 | -1.5 | -1.0 | -4.0 | -4.0 | -4.0 | -5.0 | -6.5 | -6.5 | -6.5 |
| 4 | 1.0 | 1.5 | 0.0 | -1.0 | -1.0 | -1.0 | -4.0 | -4.0 | -4.0 | -5.0 | -7.0 | -6.5 | -7.0 |
| 5 | 0.0 | 0.0 | -1.5 | -2.5 | -4.0 | -6.5 | -4.0 | -6.5 | -6.5 | -5.0 | -6.5 | -6.5 | -6.5 |
| 6 | 0.0 | 0.0 | -1.0 | -3.0 | -4.0 | -7.0 | -4.0 | -7.0 | -7.0 | -6.5 | -7.0 | -6.5 | -7.0 |

Table 2  Performance Look-up Table

## TABLE 3  Parameter Table

| parameter | value |
|---|---|
| J | 0.19 in-lb-sec$^2$ |
| Ra | 0.75 ohms |
| L | 0.01 H |
| Kt | 10.27 in-lb-sec$^2$-ohm/v |
| Kb | 1.16 v/rad/sec |
| Lp | 20mm |
| B | 0.0 in-lb/rad/sec |
| Kg | 5/6 |

TABLE 3  Parameter Table

## Table 4  Fuzzy Rules For SPFTC

| V | S | | | M | | | B | | |
|---|---|---|---|---|---|---|---|---|---|
| C | S | M | B | S | M | B | S | M | B |
| CC |  |  |  |  |  |  |  |  |  |
| NS | PB | PM | PS | PM | PS | NS | PS | NS | NS |
| NM | PB | PM | PM | PM | PM | PS | PM | PS | PS |
| NB | PM | PB | PB | PB | PB | PS | PS | PS | PS |
| PS | PS | PS | NS | PS | PM | NM | NM | NM | NM |
| PM | NS | NS | NS | NS | PM | NM | NM | NM | NB |
| PB | NS | NS | NS | NS | NM | NB | NB | NB | NB |

Table 4  Fuzzy Rules For SPFTC

## Table 5  Fuzzy Control Rules

| CE \ E | NB | NM | NS | PS | PM | PB |
|---|---|---|---|---|---|---|
| PB | PB | PS | NB | NB | NB | NB |
| PM | PB | PS | NM | NM | NB | NB |
| PS | PB | PS | NS | NS | NM | NB |
| NS | PB | PM | NS | NS | NS | NB |
| NM | PB | PB | PM | NS | NS | NB |
| NB | PB | PB | PB | PS | NS | NB |

Table 5  Fuzzy Control Rules

# Input Scaling Factors in Fuzzy Control Systems

Yung-Yaw Chen, *Member IEEE* and Chiy-Ferng Perng, *Student Member IEEE*

*Abstract—* The input scaling factors in a fuzzy control system are commonly used to conduct proper transformations between the real input data and the pre-specified universe of discourses of the fuzzy input variables in the system. Theoretically, they are constant parameters. However, it is noticed that, sometimes, they are also used to fine-tune the performance of the system in a similar way to the tuning of a PID controller. In this paper, the roles of the input scaling factors in a fuzzy control system are discussed. The firing of the rules in a fuzzy controller is shown with different values of the scaling factors to demonstrate the fact that the adjustment of the factors is equivalent to the re-construction of the membership functions in the rule-base. The results also suggest that such an adjustment should be done carefully if the linguistic meaning of the rule-base is to be preserved. An inverted pendulum system is used to verify the points raised.

## I. Introduction and Motivation

Since the introduction of fuzzy set by Zadeh[7], the concept has been extended to many fields. One of the successful examples, fuzzy control, has found many applications[6] and caught the eyes of the world in recent years. A number of commercial implementations by Japan, e.g. the detection of load and control of the washing cycle of a washing machine, the focusing of the video camera, etc, have promoted fuzzy theory from academic research to production lines. In Japan, the word "fuzzy" has even become a point of sales with great popularity in the market.

The basic idea of fuzzy control is to make use of the knowledge and experience of the experts to form a rule-base with linguistic *if-then* rules. Proper control actions are then derived from the rule-base which can be considered as an emulation of the behavior of the human operators. Different from other control theories, fuzzy control does not involve complex mathematical operations and models of the plants. From one point of view, fuzzy control is actually a form of control rather than a control algorithm.

The design of a fuzzy controller usually involves the determination of the control strategies in the rule-based form and the construction of the inference engine. To begin with, a number of fuzzy terms, such as *small* and *large*, are defined within the universe of discourse for each variable. Sometimes the fuzzy terms can also be normalized and be defined within the range of [ 0 , 1 ]. Then, scaling factors are used to conduct proper transformations between the actual range of parameters variations and the interval [ 0 , 1 ]. The input scaling factors usually seen in a fuzzy control system are $G_E$ ($E$ stands for the error signal) and $G_{CE}$ ($CE$ stands for the change of error), while the output scaling factor is $G_U$ ($U$ stands for the output).

Procyk and Mamdani[5] discussed the effects of the scaling factors in their linguistic self-organizing controller by exhaustively changing the values of the scaling factors. Maeda and Murakami[4] proposed a self-tuning algorithm on improving the system performance by adjusting the scaling factors. Braae and Rutherford[3] tracked the system trajectory in the linguistic space to determine if the scaling factor $K_1$ and $K_2$ should be adjusted. Bare, Mulholland and Sofer[2] studied the tuning of the scaling factors for the fuzzy controller of a gasoline refinery catalytic reformer.

Basically, the tuning of the scaling factors in the literature can be divided into two catagories. One is in the learning controller design, such as in [5] and [4]. In those cases, there are many tunable parameters in the system. The input scaling factors are simply part of them and, perhaps, the most likely ones to start with. Since there are no known or meaningful values of the input scaling factors, their tunings are reasonable and may result in an improvement of the rule-base with a better system performance. On the other hand, when a rule-base is constructed by experts, the scaling factors are defined to be the transformation between the values of the real inputs

The authors are with the Electronic Research Laboratory, University of California, Berkeley, CA 94720, USA and Department of Electrical Engineering, National Taiwan University, Taipei, Taiwan, R.O.C.

and the usually specified [ 0 , 1 ]. In those cases, the tuning of scaling factors still will affect the sysytem performances, as in [3] and [2]. However, the linguistic meaning of the rule-base will not be preserved if the scaling factors are changed.

In this paper, we show that the adjustments of the input scaling factors can be interpreted as the change of the universe of discourse for input variables. The result is that the tuned fuzzy controller can no longer preserve its linguistic meaning of the rule-base since the definitions of the membership functions of the input variables are altered. Such an outcome may not be the intention of the experts who supplied the rules in the first place and will also prohibit the incorporation of new knowledge into the rule-base in the future.

The organization of the paper is as follows. The basic structure of a fuzzy control system and the input scaling factors are introduced in section 2. The effects of the tuning of the scaling factors are shown in section 3 with an example of the inverted pendulum. Detailed discussions are given in section 4. Finally, the conclusion is summarized in the last section.

## II. Fuzzy Control System and Its Scaling Factors

The basic structure of a fuzzy control system is shown in Figure 1. The dashed block represents a

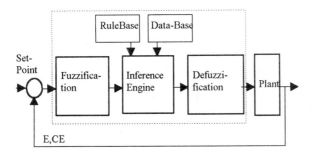

Figure 1: Basic Structure of Fuzzy Control Systems

fuzzy logic controller which can be described by five different functional blocks, i.e. *fuzzification, rule-base, data-base, inference engine,* and *defuzzification.* (Detailed description of each functional blocks can be found in [1] and will not be described in this paper.)

Since the inputs and the outputs of a fuzzy controller must be real numbers in order to match the sensors and the actuator requirements, fuzzification of input variables and defuzzification of output variables are necessary. The purpose of fuzzification is

to transform the real sensor data into fuzzy linguistic terms so that further fuzzy inferences can be performed according to the rule-base. The premise of the rules in the rule-base usually are fuzzy conditional statement, such as:

If $x_1$ is *Small*, and $x_2$ is *Medium*

The linguistic terms above are defined based on the interpretations of the experts, which require the determination of the universe of discourse for different variables and the membership functions for different fuzzy terms. A commonly used set of fuzzy terms are shown in Figure 2. To simplify the notation,

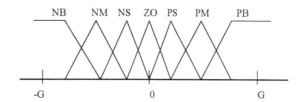

Figure 2: Membership Functions of Fuzzy Terms

the fuzzy linguistic terms in the premise of the rules in the rule-base are sometimes defined within the range of [ 0 , 1 ]. As a result, it is necessary to normalize the actual variations of the sensor inputs into the interval of [ 0 , 1 ]. The input scaling factors, $G_E$ and $G_{CE}$, are determined by the experts or designers so that the universe of discourse of the input variables are mapped into the unity interval. As shown in Figure 3, it can be easily seen that an input scaling factor of $G_1$ and a normalized set of linguistic terms are equivalent to a set of linguistic terms with the universe of discourse between $[ -\frac{1}{G_1}, \frac{1}{G_1} ]$.

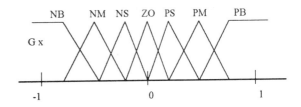

Figure 3: Normalized Liguistic Terms

Now supposed that the input scaling factors $G_E$ and $G_{CE}$ are altered during a tuning process and have become $G'_E$ and $G'_{CE}$. There much exist real numbers $k_E$ and $k_{CE}$ such that

$$G'_E = k_E \times G_E$$

$$G'_{CE} = k_{CE} \times G_{CE}$$

and the fuzzy controller can be represented as shown in Figure 4.

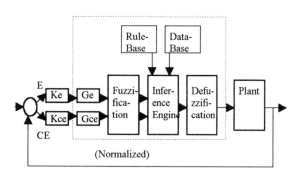

Figure 4: Tuning Paramters of Input Scaling Factors

## III. EFFECTS OF TUNINGS OF INPUT SCALING FACTORS

As shown in the previous section, the input scaling factors are the coefficients between the universe of discourse of the input variables and the unity interval, which are supposedly constant if the range of input variations are approximately known. For an auto-tuning or learning controller design, most of the parameters are not known and the tuning of a set of parameters according to a learning scheme, such as the reinforcement learning and the genetic algorithm, may be able to improve the system performance and to derive a better controller. The set of tuning parameters involved may include the input scaling factors.

However, if the relation between the ranges of inputs and the unity interval are known, tuning of the scaling factors becomes meaningless because those parameters are supposed to be constant coefficients. We shall show in this section that tunings of the input scaling factors is, in fact, equivalent to the reconstruction of the rule-base. An example of the well-known inverted pendulum is used to demonstrate the effects of the tunings of the input scaling factors in the following.

Consider the inverted pendulum system in Figure 5. The parameters of the inverted pendulum system are as follows:
length of the pole $l = 0.5m$,
mass of the pole $m_p = 0.3kg$,
mass of the cart $m_c = 0.5kg$,
friction coefficient of pole $\mu_p = 0.000002$,
friction coefficient of cart $\mu_c = 0.0005$.
The dynamical equation of the pendulum and the cart can be found in [1]. Without loss of generality,

only the pole angle control is considered. A fuzzy controller is constructed to control the pole angle position so that a vertical upright position is maintained. Suppose that the domains of interest of the

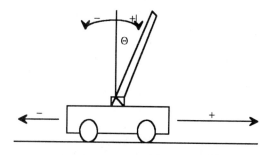

Figure 5: Inverted Pendulum System

pole angle $\theta$ is between $-12$ *degrees* and $12$ *degrees* (i.e. $\pm 0.21$ *rad*) and the pole angular velocity is between $\pm 5$ *rad/sec* (i.e. $\pm 286.5$ *deg/sec*). Then the system diagram of the inverted pendulum system can be described by Figure 4 with $G_E = \frac{1}{0.21} = 4.76$, $G_{CE} = \frac{1}{0.873} = 1.145$, $k_E = 1$, and $k_{CE} = 1$. A rule-base can then be constructed as shown in Table 1 to balance the pole.

Table 1: Rule-Base of the Fuzzy Controller

| deg/sec | deg | | | | | | |
|---|---|---|---|---|---|---|---|
| | NL | NL | NM | NS | NS | ZE | ZE |
| | NL | NM | NM | NS | ZE | ZE | PS |
| | NL | NM | NS | NS | ZE | ZE | PS |
| | NM | NS | NS | ZE | PS | PS | PM |
| | NS | ZE | ZE | PS | PS | PM | PL |
| | NS | ZE | ZE | PS | PM | PM | PL |
| | ZE | ZE | PS | PS | PM | PL | PL |

Now, assuming that the result of a tuning process renders a new value $G'_E$ which is equivalent to $k_E = 4$. Let's compare the behaviors of the two fuzzy controllers by considering the firing of the rules between the two rule-bases. There are two possible interpretations:

(a) Assume that the rule-base does not change, then the input values to the rule-base of the tuned fuzzy controller will be twice as large as the original one. As a result, the rules fired will be farther away from the origin than the old controller. Such an effect is not linear in any sense since fuzzy controller is a nonlinear operator.

(b) Assume that $G'_E$ is included into the definitions of the membership functions of the linguistic terms in the rule-base, then the new rule-base shall have the universe of discourse in the interval of $[-\frac{G_E}{4}, \frac{G_E}{4}]$. However, the input values shall remain in the same range of $[-G_E, G_E]$. Similarly, we can see that the firings of the new rules with the same input value will also be farther away from the origin compared with the old rule-base. Besides, it is more likely for the modified rule-base to have the rules in the boundary fired and the combined effect is not a linear function of $k_E$.

Figure 6 and 7 demonstarted the output surfaces formulated by the two different rule-bases in the inverted pendulum system. It is obvious that the original and the new output surfaces are different and the liguistic meanings of the rule-bases are also different. The saturation region around the bound-

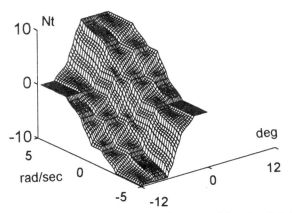

Figure 6: Output Surface Generated by the Original Fuzzy Controller

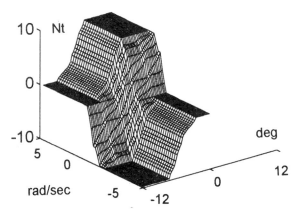

Figure 7: Output Surface Generated by the Modified Fuzzy Controller

ary of the universe of discourse plays an important role in our comparison between the original and the tuned rule-bases. Mostly because of this saturation effect, the tuned rule-base loses its relationship with the original one.

## IV. Discussions

The fine-tuning of a fuzzy controller has always been a difficult problem for controller designers. There are so many parameters involved that it is virtually impossible to choose a certain one manually to optimize the system performance. It is therefore intuitive to seek for a certain set of parameters for further adjustments when the expert-provided rule-base does not meet the specifications of design. An easy choice of parameters are the input and the output scaling factors, which bear a resemblance to the $k_p$, $k_i$, and $k_d$ in conventional PID control. If the sole purpose of such a tuning process is to improve the system performance without the intention to maintain the liguistic essence of a fuzzy controller, then the tuning of any set of parameters will serve the purpose in a certain degree. However, it is our intention to show in this paper that such an approach may not be appropriate if there is such a concern. The resulting rule-base after tuning may well lose its original linguistic meaning and prohibit itself from incorporating new knowledge in the future. As a matter of fact, the new rule-base can completely lose its linguistic form, as seen in many learning control design, that the derived fuzzy controller can only be regarded as a nonlinear function of the input variables.

## V. Conclusions

In this paper, we have shown that the effects of tuning the input scaling factors in a fuzzy control system is equivalent to the scalings of the domain of interest of the input variables in a nonlinear way. The resulting fuzzy controller after the tuning, i.e. for $k_E \neq 1$ or $k_{CE} \neq 1$, is practically different from the original fuzzy controller, not only numerically but also linguistically. Consequently, it is inappropriate to tune the input scaling factors when the rule-base is constructed by experts. The fine-tuning of such kind of fuzzy controllers seems to be better achieved by membership tunings or individual rule modifications so that the linguistic meanings of the rule-base can be preserved.

## References

[1] Y.-Y. Chen, "Global anaysis of fuzzy dynamical systems", Ph.D. thesis, University of California, Berkeley, 1989.

[2] W. Bare, R. Mulholland, and S. Sofer, "Design of a self-tuning rule based controller for a gasoline refinery catalytic reformer", *IEEE Trans. of Automatic Control*, Vol. 35, pp. 156-164, 1990.

[3] M. Braae and D.A. Rutherford, "Selection of parameters for a fuzzy logic controller", *Fuzzy Sets and Systems*, Vol. 2, pp. 185-199, 1979.

[4] M. Maeda and S. Murakami, "A self-tuning fuzzy controller", *Fuzzy Sets and Systems*, Vol. 51, pp. 29-40, 1992.

[5] T. Procyk and E. Mamdani, "A Linguistic Self-Organizing Process Controller", *Automatica*, Vol. 15, pp. 15-30, 1979.

[6] M. Sugeno ed., *Industrial applications of fuzzy control*, 1986.

[7] Zadeh, L.A., "Fuzzy Sets", *Information and Control*, pp. 338-353, 1965.

# A Study on Fuzzy Tracking of Maneuvering Aircraft Using Real Data

Peter Ott          Keith C.C. Chan
Dept. of Electrical and Computer Engineering
Ryerson Polytechnic University
350 Victoria St.
Toronto, ON, Canada
M5B 2K3

Henry Leung
Surface Radar Section, Radar Division
Defense Research Establishment Ottawa
3701 Carling Ave, Ottawa, ON, Canada
K1A 0K2

*Abstract*— **This article presents a method for introducing fuzzy logic into a conventional $\alpha$-$\beta$ tracker in an attempt to improve its tracking ability for maneuvering targets. To evaluate the performance of the fuzzy tracker, we compare its tracking ability with that of the original $\alpha$-$\beta$ tracker. Both simulated and real radar data is used for the comparison. Results show that the fuzzy tracker performs better.**

## I. Introduction

Target tracking is the ability to predict the future trajectory of an object given its past history. This subject has plagued engineers and scientists for years and many methods have been developed which attempt to cope with the associated problems. The classical solution is known widely as $\alpha$-$\beta$ tracking and employs a fixed-coefficient filter to perform both the smoothing and prediction. Though it is old, this method is still accepted and recognized as an important tool in the target tracking community. In this paper we examine the $\alpha$-$\beta$ tracker to determine if the inclusion of fuzzy logic into the algorithm will improve its tracking ability on maneuvering targets.

The main problem with tracking schemes is computational overhead. If, for example, we take the case of multiple target tracking in the context of air-traffic control, it can be shown that the difficulty level of finding a particular track is proportional to the square of the number of objects being considered [1]. Since any single target tracking algorithm will no doubt be used in a multi-target environment, it makes sense to be concerned about computational efficiency. The age-old $\alpha$-$\beta$ tracker works very quickly and is very efficient, making it a desirable tracking base on which to build.

The standard $\alpha$-$\beta$ tracker is not adaptive and is, consequently, not going to perform in any optimal way. Expert choices of the tracking coefficients, $\alpha$ and $\beta$, will only ensure accurate tracking only when the target is moving in a straight, constant velocity trajectory. The fuzzy logic approach has the immediate advantage of being adaptive and should provide good tracking, especially during target maneuvers.

## II. The $\alpha$-$\beta$ Tracker

The alpha-beta tracker is a straight forward, easy to implement tracking algorithm using only positional $x$-$y$ information. It is computationally efficient and works very quickly, making it one of the most widely used fixed parameter trackers.

Equations 1 through 4 define the $\alpha$-$\beta$ tracker.

$$X_s(k) = X_p(k|k-1) + \alpha(X_m(k) - X_p(k|k-1)) \quad (1)$$

$$Vx_s(k) = Vx_p(k|k-1) + \frac{\beta}{qT}(X_m(k) - X_p(k|k-1)) \quad (2)$$

$$X_p(k+1|k) = X_s(k) + T \cdot Vx_s(k) \quad (3)$$

$$Vx_p(k+1|k) = Vx_s(k) \quad (4)$$

where:

- $X_s(k)$ is the smoothed (averaged) $X$ position for iteration $k$
- $X_m(k)$ is the actual measured $X$ position for iteration $k$
- $X_p(k|k-1)$ is the predicted $X$ position value for iteration $k$ given past values up to iteration $k-1$.
- $Vx_s(k)$ is the smoothed (averaged) $X$ velocity for iteration $k$
- $Vx_p(k|k-1)$ is the predicted $X$ velocity for iteration $k$ given past values up to iteration $k-1$.
- $T$ is the time for the radar to complete 1 revolution
- $q$ is a value associated with the number of consecutive missed scans
- $\alpha$ and $\beta$ decide how much of the positional error is used in the smoothed position and velocity calculations.

These quantities are calculated for both the $X$ and $Y$ coordinates. Only the $X$ equations are given here since they are the same for both. It has been shown in [3] that the tracker is stable for ranges of $(0 < \alpha < 2)$ and $(\beta > 0)$.

Due to its simplicity, the fixed coefficient alpha-beta tracker is quite fast, but it is not adaptive and will not work well with maneuvering targets. A low value of $\alpha$ causes the smoothed positional value to become more related to the predicted value. This will lead to larger maximum errors because any abrupt change in target direction will cause an underdamped response in the predicted values. The convergence rate will be slower as the value of $\alpha$ gets closer to zero. Setting $\alpha = 1.0$ will give acceptable tracking behaviour, but is not a realistic value since the tracking algorithm will eventually be used in the context of a multi-target tracking environment; setting $\alpha = 1.0$ essentially states that we are *sure* that the measured value $X_m(k)$ is the point associated with the target.

The value of $\beta$ is critical to the number of lost tracks (or detections out-of-gate) the algorithm causes over the course of a complete track. A good choice of $\beta$ will enable the smoothed positional values $X_s$ and $Y_s$ to "fill in" the gaps caused by single missing points, avoiding a possible lost track.

This leads us to the primary design problem for this tracker: we must choose $\alpha$ and $\beta$ such that $X_p(k|k-1)$ is calculated to cause an underdamped response to abrupt target maneuvers. It is also desirable that target is not lost due to one missed detection. Because the $\alpha$-$\beta$ tracker is not adaptive and target maneuvers are highly non-linear, the parameter choices that are based on the above criteria will still give only average results.

Given the above discussions, we are forced to ask the question: can we design an adaptive mechanism which will choose these two coefficients as they are required? The fuzzy tracker attempts to address this.

### III. The Fuzzy Tracker

The fuzzy tracker is essentially an $\alpha$-$\beta$ tracker with $\alpha$ and $\beta$ constantly being recalculated based on prompting from a fuzzy decision maker.

The fuzzy algorithm operates in four stages:

1. Calculating $E_{dist}$ and $E_{vel}$
2. Deciding how much of a member of ZE, SP, MSP, MP, and LP each of the errors is, by indexing them into the membership functions (this results in a 5x1 vector).
3. Cycling through the rules to find which are true and what contributions each should make to the final decision.
4. De-fuzzifying the final decision to find actual values for both $\alpha$ and $\beta$.

The decision is ultimately based on the mean squared distance and velocity prediction errors over one scanning interval. Mathematically this amounts to:

$$E_{dist} = \sqrt{(X_m(k) - X_p(k-1))^2 + (Y_m(k) - Y_p(k-1))^2}$$

$$B_{vel} = \sqrt{(\frac{X_m(k) - X_m(k-1)}{T} - V_{xp})^2 + (\frac{Y_m(k) - Y_m(k-1)}{T} - V_{yp})^2}$$

As with the original $\alpha$-$\beta$ tracker, the decision is made based only on the values taken from a single scan interval.

The 25 rules that were formulated and used in the fuzzy tracker are given below.

Table 1: Fuzzy decision rules for $\alpha$

$E_{dist}$

| $E_{vel}$ | ZE | SP | MSP | MP | LP |
|-----------|----|----|-----|----|----|
| ZE | LP | LP | MP | MP | MP |
| SP | LP | LP | MP | MP | MP |
| MSP | LP | LP | MP | MP | MP |
| MP | LP | LP | MP | MP | MP |
| LP | LP | LP | MP | MP | MP |

Table 2: Fuzzy decision rules for $\beta$

$E_{dist}$

| $E_{vel}$ | ZE | SP | MSP | MP | LP |
|-----------|-----|-----|-----|-----|-----|
| ZE | LP | LP | LP | LP | LP |
| SP | MP | MP | MP | MP | MP |
| MSP | MSP | MSP | MSP | MSP | MSP |
| MP | SP | SP | SP | SP | SP |
| LP | SP | SP | SP | SP | SP |

where:

- $E_{dist}$ and $E_{vel}$ are the distance and velocity errors, respectively.
- ZE, SP, MSP, MP, and LP represent zero, small, medium small, medium, and large positive, respectively.

The rules in Table 1 determine what $\alpha$ should be given a distance and velocity error. Table 2 is the equivalent for the value of $\beta$. It was decided during the research that, if the distance error is small, the $\alpha$ and $\beta$ values should be changed to force $X_p$ and $Y_p$ to be equal to $X_m$ and $Y_m$, thus nullifying the error before it gets a chance to increase. The value of $\alpha$ should never need to be less than MP since this would cause the smoothed values to approach the predicted values. This has the effect of increasing the time required for the predicted position to converge to the actual.

The memberships of $E_{dist}$ and $E_{vel}$ were chosen to be triangular in shape and are shown in Figure 1. The spans of the memberships define what area ZE, SP, MSP, MP, and LP cover and depend on the maximum velocity of the targets that are to be tracked.

The de-fuzzification is performed by finding the weighted average of the contributions of each rule and the actual values of ZE, SP, MSP, MP, and LP obtained in number 2 above. Mathematically,

$$\alpha = \frac{(rule\ contributions) * (actual\ values\ from\ indexing\ E_{dist})}{rule\ contributions}$$

$$\beta = \frac{(rule\ contributions) * (actual\ values\ from\ indexing\ E_{vel})}{rule\ contributions}$$

### IV. Experiments

To compare the performance of the fuzzy and $\alpha$-$\beta$ trackers, both simulated and actual radar scan data are employed. The simulated data gives us the ability to test the algorithms under extremes that the real data

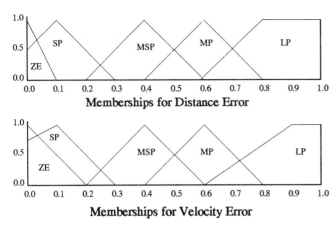

Figure 1: The Fuzzy Membership Functions

may not currently possess and lend insight into where the algorithm may fail. The actual data draws an important link between theory and practicality and should be included in any practical testing.

The simulated data is shown in Figure 2. In designing it, we require that the sequence involved as many different position range transitions as possible. The sequence was chosen to be:

SSS-LLL-SSS-MMM-LLL-SSS-MMM-SSS-LLL-MMM.

Here SSS-LLL means: take three small distance steps and then take three large distance steps. A curve was also placed in the data to add another dimension to the data non-linearity. This data spans the operating range of the trackers and will give an accurate assessment of their ability to function under extremes.

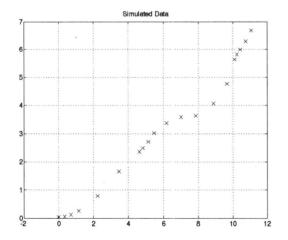

Figure 2: The Simulated Data
Used for Tracker Testing

The real data, shown in Figure 3, is the hexagonal shaped flight path of an F-18 aircraft taken from data gathered during the Radar Acquisition and Tracking Trials (RATT) in 1990. The original data contained information for six F-18 fighters. One target was selected

and its track was extracted from the data.

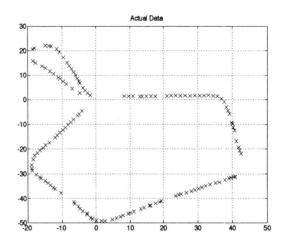

Figure 3: The Real Data Used
for Tracker Testing

The tracking performance is measured by evaluating the number of track terminations, the maximum mean squared positional error, and the total error over the course of an entire track. A target is considered terminated if there is no radar detection within a predefined elliptical region surrounding the current predicted position. This region is commonly referred to as the gate of the target and its size depends on the maximum speed of the aircraft. The ability for a tracker to be able to avoid track terminations is very important. In noise and clutter there are many possible targets and it becomes of prime concern not to lose track of a possibly hostile target due to a single missing point or non-standard maneuver; these important issues make the number of tracker terminations a good criteria to use in evaluating tracking performance. Care must be taken when basing the performance on the positional and cumulative errors. Gating will result in a lower total error because several detections are not used while the tracking algorithm resets itself to begin another track. Thus, all criteria must be used together to evaluate performance.

### V. Results and Discussion

Several different combinations of $\alpha$ and $\beta$ where used during the testing of the $\alpha$-$\beta$ tracker. There is no optimal set of coefficients since one set will work well for one batch of data, while some other set will not, leaving the final choice to be almost arbitrary. By using several sets of coefficients, we cover the basis for many of the possible choices.

Tables 3a and 3b contain the results obtained by using the $\alpha$-$\beta$ tracker with the simulated and actual data, respectively.

1673

Table 3a: $\alpha$-$\beta$ Tracker $\rightarrow$ Simulated Data

| $\alpha$ | $\beta$ | # Term. | Max. Error | Total Error |
|---|---|---|---|---|
| 0.25 | 0.25 | 5 | 1.3495 | 6.1954 |
| 0.5 | 0.5 | 5 | 1.2713 | 5.6041 |
| 0.75 | 0.75 | 3 | 1.1951 | 5.8731 |
| 1.0 | 0 | 5 | 1.2713 | 5.6176 |
| 0 | 1.0 | 5 | 1.2713 | 5.5947 |

Table 3b: $\alpha$-$\beta$ Tracker $\rightarrow$ Actual Data

| $\alpha$ | $\beta$ | # Term. | Max. Error | Total Error |
|---|---|---|---|---|
| 0.25 | 0.25 | 3 | 13.1318 | 160.8611 |
| 0.5 | 0.5 | 2 | 11.8804 | 131.5874 |
| 0.75 | 0.75 | 2 | 11.2001 | 119.3587 |
| 1.0 | 0 | 3 | 12.3591 | 172.2582 |
| 0 | 1.0 | 3 | 13.7572 | 309.3514 |

Table 4 contains the results obtained by using the fuzzy tracker with the simulated and actual data.

Table 4: Fuzzy Tracker $\rightarrow$ Simulated and Actual Data

| Data Type | # Term. | Max. Error | Total Error |
|---|---|---|---|
| Simulated | 3 | 1.1361 | 5.7518 |
| Actual | 2 | 10.8518 | 127.8132 |

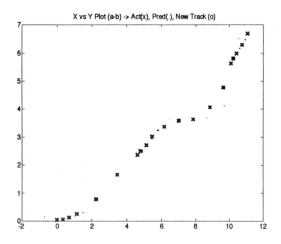

Figure 3: $\alpha$-$\beta$ Tracker Response
to the Simulated Data

From the tables we see that the fuzzy tracker performs better than the $\alpha$-$\beta$ for many of the possible coefficient choices. At its worst, the tracking response of the fuzzy algorithm is approximately equal to that of the $\alpha$-$\beta$ tracker utilizing the best choices for $\alpha$ and $\beta$ for a given track. Since this *best choice* of coefficients can not be known before hand, we can conclude that the introduction of fuzzy logic into the $\alpha$-$\beta$ tracker has made it

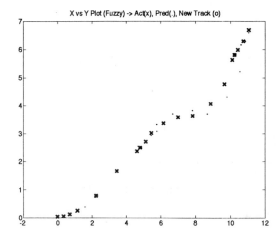

Figure 4: Fuzzy Tracker Response
to the Simulated Data

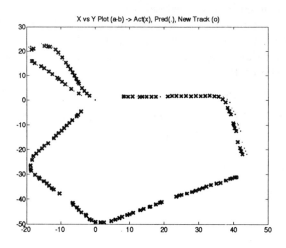

Figure 5: $\alpha$-$\beta$ Tracker Response
to the Actual Data

better. Acceptable tracking is obtained without concern for the type of target being tracked.

The adaptive nature of the fuzzy tracker has decreased the maximum error in every case. This, in turn, helps to decrease the number of lost tracks and enhance overall tracking ability.

## VI. Conclusions

We have presented a method for introducing fuzzy logic into the conventional $\alpha$-$\beta$ tracker. Through our experimental results, we have shown that the fuzzy tracker is superior to the original in many cases. By testing the algorithm on both simulated and real radar scan data, we have proven that this technique has not only theoretical implications, but practical as well.

The fuzzy tracker presented in this article is based entirely on the $\alpha$-$\beta$ algorithm. Consequently, it cannot be further enhanced until this algorithm is modified. The

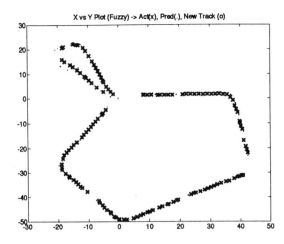

Figure 6: Fuzzy Tracker Response
to the Actual Data

next stage of the research should focus on modifying the basic algorithm to better deal with missing scans and noisy points (essentially equivalent problems). Several ways to achieve this goal are currently under consideration.

### REFERENCES

[1] J.K. Uhlmann. "Algorithms for Multiple-Target Tracking". *American Scientist*, Vol 80, March-April 1992, pp 128-141.

[2] S.S. Blackman. "Multiple-Target Tracking with Radar Applications". Artech House, 1986, pp 21-47.

[3] K. Kim and J. Hansen. "Development of Fuzzy Algorithm For Tracking of Maneuvering Targets". *Proceedings of the IEEE Conf. on Decision and Control*, 1992, pp 803-808.

[4] Porter, Priebe, and Jones. "Multiple Target Tracking in Clutter with Fuzzy Logic". *Acquisition, Tracking, and Pointing VI*, Vol 1697, 1992.

[5] Ott, Chan, and Leung. "Enhanced Maneuvering Target Tracking Using Fuzzy Logic". (Future Paper by given authors).

# A Novel Implementation of Fuzzy Logic Controller Using New Meet Operation

Mamoru Sasaki and Fumio Ueno
Dept. of Electrical Engineering and Computer Science,
Kumamoto University, Kumamoto, 860 JAPAN

*Abstract*— The authors have proposed a simple fuzzy logic hardware controller, in which the weighted average operation can be simplified to the weighted sum operation by introducing a new meet operation [1]. However, the characteristics of the edges of the fuzzy membership functions in the controller vary with temperature and processing variations. In this paper, to overcome this limitation, we propose two bias circuits and a variable current mirror. The circuits can improve the characteristics of the controller not to be influence on temperature and processing variations without sacrificing the circuit simplicity. The high performance of the improved hardware controller, for example 200ns total delay time, 1.0% maximum operation error, 3.0V low supply voltage and 150$\mu$W low power dissipation per one rule, was confirmed by SPICE simulation.

## I. INTRODUCTION

One of the most successful applications of the fuzzy logic and the fuzzy algorithm is in the field of process control. The fuzzy-logic rule-based system allow natural translations of domain knowledge into fuzzy rules. The inference procedure derives effective control actions using the fuzzy rules. However, the computations required in the fuzzy inference procedure are usually too sophisticated to meet the real-time requirements of process controllers. So, fuzzy logic implementation which can realize high-speed operation and small size, is expected as a new application area well suited to analog circuits [2], [3].

The authors have proposed a simple fuzzy logic hardware controller based on the singleton fuzzy controller [1]. In the controller, the weighted average operation which is executed to obtain a final inference result, can be simplified to the weighted sum operation by introducing a new meet operation. Therefore, the hardware controller can be simply implemented with current mode CMOS circuits, because the weighted sum operation requires no divider but only adders which can be easily imple-

mented by wired connections in current mode. The validity of the fuzzy logic controller with the new meet operation has been confirmed by applying to regulation of ĆUK converters [4] and inverted pendulum control.

However, the characteristics of the edges of the fuzzy membership functions in the hardware controller vary with temperature and processing variations. In this paper, to overcome this limitation, we propose two bias circuits and a variable current mirror. The variable current mirror can change the current gain according to the control voltages. The circuits can improve the characteristics of the hardware controller not to be influence on temperature and processing variations without sacrificing the circuit simplicity. One bias circuit generates the bias voltage for the membership function circuits by referring a unit input-voltage, which will be given from outside of the VLSI chip. The other bias circuit generates one of the control voltages for the variable current mirror by referring a unit output-current, which will be also given from outside of the VLSI chip. An overview of this paper is as follows. The inference algorithm with the new meet operation is explained in Section 2, and many of the details on the implementation of the controller are discussed in Section 3. Results of SPICE simulation are given to confirm the circuit performance of the proposed hardware controller in Section 4. Finally, some conclusion remarks are given in Section 5.

## II. AN ALGORITHM SUITABLE FOR HARDWARE IMPLEMENTATION

### A. A singleton fuzzy controller

The basis of the proposed hardware controller is the singleton fuzzy controller. The singleton fuzzy controller can be designed using the following if-then fuzzy rules.

if $x$ is $A_1$ and $y$ is $B_1$ then $z$ is $C_1$
if $x$ is $A_2$ and $y$ is $B_2$ then $z$ is $C_2$
$$\vdots \qquad \vdots \qquad \vdots$$
if $x$ is $A_n$ and $y$ is $B_n$ then $z$ is $C_n$

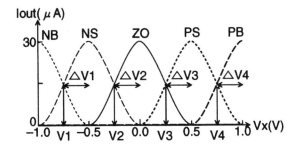

Figure 1: Membership functions

where $x$ and $y$ are inputs, $z$ is output, $A_i$, $B_i$ and $C_i$ are linguistic values $(i = 1, 2, \cdots, n)$, n is the number of rules.

The linguistic values $A_i$ and $B_i$ are represented with typical membership functions as shown in Fig.1. While, the linguistic value $C_i$ is a singleton membership function defined as follows:

$$C_i(z) = \begin{cases} 1 & \text{if } z = z_i \\ 0 & \text{otherwise} \end{cases} \quad (1)$$

Since the conclusions in the rules are represented with the singleton membership functions, the inference algorithm is simpler than the typical fuzzy controllers.

The maximum/minimum operations are used as the inference operations in the many applications. In the case, when the inputs $x'$ and $y'$ are given, the output $z'$ of the singleton fuzzy controller can be calculated as follows.

$$z' = \frac{\sum_i g_i \cdot z_i}{\sum_i g_i} \quad (2)$$

$$g_i = A_i(x') \wedge B_i(y') \quad (3)$$

where $g_i$ is an adapting degree of the $i$th fuzzy rule, $A_i(\cdot)$ is a membership function representing the linguistic value $A_i$, $\wedge$ is minimum operation.

If the singleton fuzzy controller is implemented with the analog circuits, however, the circuitry is too large and the operation speed is low because a divider should be implemented for the weighted average operation.

### B. An algorithm of the proposed hardware controller

In this section, a new meet operation is introduced to overcome the above limitation. The meet operation $\cap$ is,

$$x \cap y \overset{\text{def}}{=} \frac{(x \wedge y) + (x \odot y)}{2} \quad (4)$$

where $\odot$ is the bounded product which is one of the fuzzy logic operations. The bounded product is defined as follows,

$$x \odot y \overset{\text{def}}{=} \begin{cases} (x + y) - 1 & \text{if } (x + y) \geq 1 \\ 0 & \text{otherwise} \end{cases} \quad (5)$$

Here, let us assume that the membership functions in the antecedent parts satisfy the following two conditions,

1. $\forall x \; \sum_j A_j(x) = 1$ (orthogonal condition) where the subscript $j$ indicates kinds of the membership functions.
2. The number of the membership functions overlapping each other is less than equal two.

In this condition, "$\sum_i g_i = 1$" can always hold by using the meet operation instead of the minimum operation in (3). Where "$\sum_i g_i$" is the denominator in the weighted average operation (2). Therefore, the output $z'$ can be obtained by the weighted sum operation as follows,

$$z' = \sum_i g_i \cdot z_i \quad (6)$$

$$g_i = A_i(x') \cap B_i(y') \quad (7)$$

The membership functions shown in Fig.1 satisfy the above conditions. The authors have given the proof of "$\sum_i g_i = 1$" in the previous paper [1].

### III. A CIRCUIT CONFIGURATION

From the definition (4) of the introduced meet operation, the output of $i$th rule can be represented by the following function.

$$r_i(x, y) = R_i \{ A_i(x) \wedge B_i(y) + A_i(x) \odot B_i(y) \} \quad (8)$$

where $R_i$ is a coefficient $(R_i = \frac{z_i}{2})$. In what follows, we will implement the function $r_i(x, y)$ with the current mode CMOS circuits.

### A. A membership function circuit

To solve a problem of the current mode circuits which is restriction of the fan-out number, a voltage-input and current-output membership function circuit has been introduced [5]. The circuit is shown in Fig.2.

The circuit can realize one of the membership functions with cross point parameters $V_j$ and $V_{j-1}$ and width parameters $\Delta V_j$ and $\Delta V_{j-1}$ shown in Fig.1. A source coupled type OTA (operational transconductance amplifier) plays an important role in the circuit. When the MOS FETs operate in saturation region, the drain currents $Ia_j$ and $Ib_j$ in

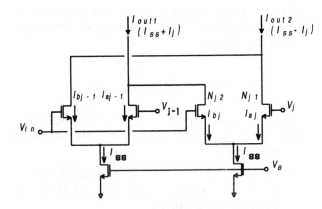

Figure 2: A membership function circuit

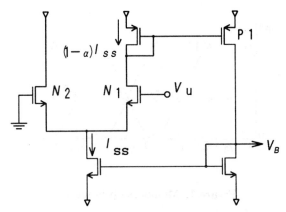

Figure 3: A bias circuit (1)

Fig.2 can be expressed as follows:

1) when $V_{in} < V_j - \Delta V_j$

$$Ia_j = 0 \qquad (9)$$

$$Ib_j = Iss \qquad (10)$$

2) when $V_j - \Delta V_j \leq V_{in} \leq V_j + \Delta V_j$

$$Ia_j = \frac{1}{2}Iss + I_v \qquad (11)$$

$$Ib_j = \frac{1}{2}Iss - I_v \qquad (12)$$

$$I_v = \frac{1}{2}K(\frac{W}{L})_j(V_{in} - V_j)\sqrt{\frac{2Iss}{K(\frac{W}{L})_j} - (V_{in} - V_j)^2} \qquad (13)$$

3) when $V_j + \Delta V_j < V_{in}$

$$Ia_j = Iss \qquad (14)$$

$$Ib_j = 0 \qquad (15)$$

where $K$ is the gain parameter of the MOSFET, $(\frac{W}{L})_j$ is the W/L ration of both MOSFETs $Nj_1$ and $Nj_2$.

By substituting "$Ia_j = 0$" and Eq.(13) into Eq.(11), the width parameter $\Delta V_j$ can be expressed as follows :

$$\Delta V_j = \sqrt{\frac{Iss}{K(\frac{W}{L})_j}} \qquad (16)$$

The $V_B$ in Fig.2 is the bias voltage and it can generated by a bias circuit shown in Fig.3.

We assume that the unit input-voltage $V_u$ is given. The bias circuit can generate the bias voltage $V_B$ using the unit input-voltage $V_u$.

In Fig.3, due to the channel length modulation effect of the MOSFET $P1$, the drain current $I_D$ of the MOSFET $N1$ is :

$$I_D = (1 - \alpha)Iss \qquad (17)$$

where $1 \gg \alpha > 0$. And, from Eqs. (11) and (13), $I_D$ can be expressed :

$$I_D = \frac{1}{2}Iss + I_v \qquad (18)$$

$$I_v = \frac{1}{2}K(\frac{W}{L})_u(V_u-0)\sqrt{\frac{2Iss}{K(\frac{W}{L})_u} - (V_u - 0)^2} \qquad (19)$$

Hence, from Eqs. (17), (18) and (19),

$$\begin{aligned} \Delta V_u &= V_u - 0 \\ &= \sqrt{\frac{(1-\alpha)Iss}{K(\frac{W}{L})_u}} \\ &\simeq \sqrt{\frac{Iss}{K(\frac{W}{L})_u}} \end{aligned} \qquad (20)$$

where $(\frac{W}{L})_u$ is the W/L ratio of both MOSFETs N1 and N2.

From (16) and (20),

$$\frac{\Delta V_j}{\Delta V_u} = \sqrt{\frac{(\frac{W}{L})_u}{(\frac{W}{L})_j}} \qquad (21)$$

As shown in (21), the $\Delta V_j$ can be set by only the ratio of the W/L ratios of the MOS FETs. Thus, the characteristic of the membership function circuit can not be influence upon temperature and processing variations through the gain parameter $K$. As shown Fig.2, the simple circuit generates two output currents $Iout_1$ and $Iout_2$ as follows, for convenience of the following building blocks which are min and bounded product circuits.

$$Iout_1 = Ib_{j-1} + Ia_j \ (= Iss + I_j) \qquad (22)$$

$$Iout_2 = Ia_{j-1} + Ib_j \ (= Iss - I_j) \qquad (23)$$

Where $I_j$ represents the membership function with cross-point parameters $V_{j-1}$ and $V_j$, and $Iss$ corresponds to the logical value 1.

When $V_j - \Delta V_j \leq V_{in} \leq V_j + \Delta V_j$,

$$\begin{aligned}
\sum_j I_j &= I_j + I_{j+1} \\
&= (Iss - Ia_j) + (Iss - Ib_j) \\
&= 2Iss - (Ia_j + Ib_j) \\
&= Iss \quad (24)
\end{aligned}$$

Therefore, the orthogonal condition of the membership function is satisfied.

*B. A min operation circuit*

De Morgan's law can hold between the maximum operation and the minimum operation as follows,

$$\begin{aligned}
\min(I_x, I_y) &= \overline{\max(\overline{I_x}, \overline{I_y})} \\
&= Iss - \quad (25) \\
&\quad \max(Iss - I_x, Iss - I_y)
\end{aligned}$$

Hence, a min operation can be realized by one max circuit and one complement circuit using the outputs "$Iss - Ix$" and "$Iss - Iy$" from the membership function circuits. In the min operation circuit, we use the max circuit implemented with current mirrors and an OR circuit [6].

*C. A bounded product circuit*

From the definition (5), the bounded product can be expressed as follows.

$$BP(I_x, I_y) = \begin{cases} I_x + I_y - Iss & \text{if } I_x + I_y \geq Iss \\ 0 & \text{otherwise} \end{cases}$$

$$= \begin{cases} (Iss + I_x) + (Iss + I_y) - 3 \cdot Iss \\ \quad \text{if } I_x + I_y \geq Iss \quad (26) \\ 0 \quad \text{otherwise} \end{cases}$$

The bounded product can be realized with the outputs "$Iss + I_x$" and "$Iss + I_y$" from the membership function circuits, using a diode-connected MOS FET and a current source.

*D. A coefficient circuit*

A coefficient circuit can be implemented with a variable current mirror shown in Fig.4.

In the circuit, the MOSFETs N1 and N2 operate in linear region and the other transistors operate in saturation region. The cross-coupled source follower consists of the MOSFETs N3 - N6. From Fig.4,

$$V_1 = V_i - (\Delta V_3 + \Delta V_6) \quad (27)$$

$$V_2 = V_i - (\Delta V_4 + \Delta V_5) \quad (28)$$

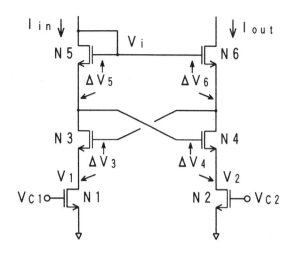

Figure 4: A variable-gain current mirror

Assuming that the transistor N3 and N4 are matched to N5 and N6 respectively,

$$\Delta V_3 + \Delta V_6 = \Delta V_4 + \Delta V_5 \quad (29)$$

Thus,

$$V_1 = V_2 \quad (30)$$

Assuming $Vc_1 \gg V_1$ and $Vc_2 \gg V_2$,

$$I_{in} = K(\frac{W}{L})_1 (Vc_1 - Vth) V_1 \quad (31)$$

$$I_{out} = K(\frac{W}{L})_2 (Vc_2 - Vth) V_2 \quad (32)$$

Where $K$ and $Vth$ are gain parameter and threshold voltage of the MOSFETs respectively. $(\frac{W}{L})_1$ and $(\frac{W}{L})_2$ are the W/L ratios of the MOS FETs N1 and N2, respectively.
From (30), (31) and (32),

$$I_{out} = \frac{(\frac{W}{L})_1}{(\frac{W}{L})_2} \frac{Vc_1 - Vth}{Vc_2 - Vth} \cdot I_{in} \quad (33)$$

A bias circuit generating the control voltage $Vc_1$ and $Vc_2$ is shown in Fig.5. We assume that the unit output-current $I_u$ is given. The circuit can generate the control voltage $Vc_2$ using the unit output-current $I_u$. The circuit adjusts the control voltage $Vc_2$ so that the following equation is satisfied :

$$R \cdot I_u = G \cdot Iss \quad (34)$$

where R is the coefficient and G is the current gain. We can set the coefficient R by only the ratio of the W/L ratios of the transistors constituting the PMOS current mirror in Fig.5. In virtue of this mechanism, even if the sink current $Iss$ of the OTAs changes by temperature and processing variations,

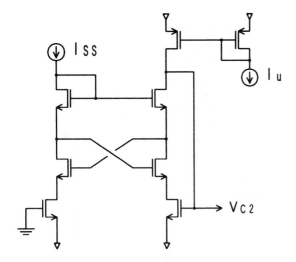

Figure 5: A bias circuit (2)

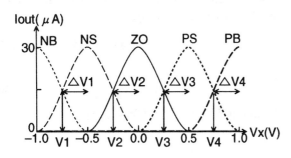

Figure 6: DC analysis 1

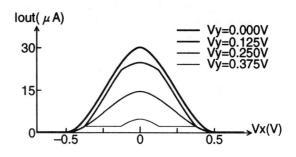

Figure 7: DC analysis 2

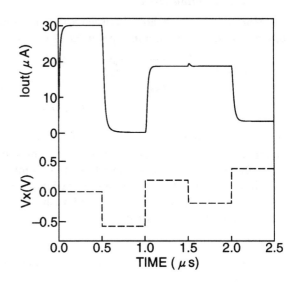

Figure 8: Transient analysis

the output current of the controller can be stabilized.

The function $r_i(x, y)$ can be composed of the four building blocks explained in Section $A$, $B$, $C$ and $D$. The fuzzy logic hardware controller can be easily constructed by connecting the output terminals of the rule blocks because of the current outputs.

## IV. SPICE SIMULATIONS

We designed the rule block using $2\mu$m CMOS design rule. In the circuit, the cascode current mirrors and the folded circuits are used on purpose to reduce the channel length modulation effect and to connect the building blocks, respectively. We have simulated the circuit using SPICE2. The electrical parameters are $VDD = 1.0V$, $VSS = -2.0V$ and the unit input-voltage $V_u$ and the unit output-current $I_u$ are set to 0.25V and 10.0 $\mu$ A, respectively. The coefficient $R$ is 3.0. The result of the DC analysis is shown in Figs.6 and 7.

In Fig.6, we set $Vy_1 = -0.25V$, $Vy_2 = +0.25V$, $Vy = 0.0V$, $Vx_2 = Vx_1 + 0.5V$ and we use $Vx$ as the input and $Vx_1$ as the parameter. From Fig.6, the membership functions were confirmed. In Fig.7, we set $Vx_1 = Vy_1 = -0.25V$, $Vx_2 = Vy_2 = +0.25V$ and we use $Vx$ as the input and $Vy$ as the parameter. From Fig.7, the proposed meet operation was confirmed. The DC characteristics agree well with the theoretical prediction. The maximum error of the DC characteristic was within 1.0% to the full scale current.

The result of the transient analysis is shown in Fig.8.

In the transient analysis, we set $Vx_1 = Vy_1 = -0.25V$, $Vx_2 = Vy_2 = +0.25V$, $Vy = 0.0V$ and $Vx$ is the pulse voltage as shown by dotted line in Fig.8. From the transient analysis, the rise time and the fall time to within 1.0% of final value were less than 200ns. The power dissipation was $150\mu$W in the above condition.

1680

## V. Conclusions

A fuzzy logic hardware controller has been proposed and it was implemented with the current mode CMOS circuits. The performance of the circuit was confirmed by SPICE simulation. The advantages of the proposed circuit are as follows,

1. The simple circuit configuration.
2. The high-speed operation
3. The low supply voltage and the wide input range.
4. The low power dissipation.
5. The characteristic of the circuit is independent of temperature and processing variations.
6. The circuit can be compatible with standard MOS fabrication process.

## References

[1] M.Sasaki and F.Ueno, "A fuzzy logic function generator (FLUG) implemented with current mode CMOS circuits," *Proc. of the 21st IS-MVL*, pp.356-362, May 1991.

[2] Y.P.Tsividis, "Analog MOS integrated circuits - certain new ideas, trends, and obstacles," *IEEE J.Solid-State Circuits*, vol.SC-22, pp.317-321, Jun. 1987.

[3] T.Yamakawa, "A fuzzy inference engine in non-linear analog mode and its application to a fuzzy logic control," *IEEE Trans. on Neural Networks*, vol. 4, No. 3, pp.496-522, May 1993.

[4] F.Ueno, T.Inoue, I.Oota and M.Sasaki, "Regulation of ĆUK converters using fuzzy controllers," *Proc. of the INTELEC'91*, pp.261-267, Nov. 1991.

[5] M.Sasaki, N.Ishikawa, F.Ueno and T.Inoue, "Current-mode analog fuzzy hardware with voltage input interface and normalization locked loop," *IEICE Trans. Fundamentals*, vol.E75-A, No. 6, pp.650-654, 1992.

[6] F.Ueno, T.Inoue, T.Morimoto and M.Sasaki "A fuzzy inference engine using current-mode max-min composition circuits," *Proc. of the 3rd IFSA Congress*, pp.639-642, Aug. 1989.

# HYBRID FUZZY LOGIC PID CONTROLLER

Thomas Brehm and Kuldip S. Rattan

Department of Computer Science & Engineering
Wright State University
Dayton, OH 45435

Department of Electrical Engineering
Wright State University
Dayton, OH 45435

Abstract --This paper investigates the design of a fuzzy logic PID controller that uses a simplified design scheme. Fuzzy logic PD and PI controllers are effective for many control problems but lack the advantages of the fuzzy PID controller. Fuzzy controllers use a rule base to describe relationships between the input variables. Implementation of a detailed rule base, such as a PID controller, increases in complexity as the number of input variables grow and the ranges of operation for the variables become more defined. We propose a hybrid fuzzy PID controller which takes advantage of the properties of the fuzzy PI and PD controllers. The effectiveness of the PID fuzzy controller implementation is illustrated with an example.

## I Introduction

Fuzzy logic controllers (FLCs) demonstrate excellent performance in numerous applications such as industrial processes [8] and flexible arm control [5]. Mamdami's [3] work introduced this control technology that Zadeh pioneered with his work in fuzzy sets [9]. Unlike "two valued" logic, fuzzy set theory allows the degree of truth for a variable to exist somewhere in the range [0,1]. For example, if pressure is a linguistic variable that describes an input, then the terms low, medium, high and dangerous describe the fuzzy set for the pressure variable. If the universe of discourse for pressure is [0, 100], then low could be defined as "close to 10", "medium" could be "around 40", and so on. For control applications, linguistic variables describe the control inputs for dynamic plants and rules define the relationships between the inputs. Thus, precise knowledge of a plant's transfer function is not necessary for design and implementation of the controller. The thrust of earlier efforts involved replacing humans in the control loop by describing the operators' actions in terms of linguistic rules.

There are three processes involved in the implementation of an FLC; fuzzification of inputs, a rule base or an inference engine, and defuzzification to obtain a "crisp output." Fuzzification involves dividing each input variables' universe of discourse into ranges called fuzzy sets. A function applied across each range determines the membership of the variable's current value to the fuzzy sets. The value at which the membership is maximum is called the peak value. Width of a fuzzy set is the distance from the peak value to the point where the membership is zero. Linguistic rules express the relationship between input variables. Table I is an example of a matrix of rules that covers all possible combinations of fuzzy sets for two input variables. The rules describe a proportional-plus-derivative FLC (PDFLC). The rule matrix is just a convenience and still represents all the rules in "English" of the form:

*$R_N$ : If error is $E_i$ and change in error is $\Delta E_j$ then output is $U_{ij}$*

where $1 \le i \le$ number of sets for error, $1 \le j \le$ number sets for change in error and $1 \le N \le$ (number of sets for error multiplied by the number of sets for change in error). $E_i$ and $\Delta E_j$ are fuzzy sets for error and change in error, respectively and $U_{ij}$ are the output fuzzy sets. In this case, each variable has seven fuzzy sets that gives a total of 49 rules. The notation PB means positive big; PM means positive medium; PS means positive small; ZO means zero; NS means negative small; NM means negative medium; and NB means negative big. The defuzzification process determines the "crisp output" by resolving the applicable rules into a single output value.

Table I  Rule Matrix for PDFLC

|  |  | **Error** | | | | | | |
|---|---|---|---|---|---|---|---|---|
|  |  | NB | NM | NS | ZO | PS | PM | PB |
| **Change** | NB | NB | NB | NB | NB | NM | NS | ZO |
| **in** | NM | NB | NB | NB | NM | NS | ZO | PS |
| **Error** | NS | NB | NB | NM | NS | ZO | PS | PM |
|  | ZO | NB | NM | NS | ZO | PS | PM | PB |
|  | PS | NM | NS | ZO | PS | PM | PB | PB |
|  | PM | NS | ZO | PS | PM | PB | PB | PB |
|  | PB | ZO | PS | PM | PB | PB | PB | PB |

Recent research into fuzzy control has applied classical techniques to stability analysis [1] and design [6,10]. The operation of a fuzzy controller behaves similar to a classical proportional-plus-derivative (PD) or proportional-plus-integral (PI) controller [1,6]. For a classical PD controller, the position and derivative gains remain constant for all values of input. However, for an FLC, the gains depend on the range where the control variables exist at any instant. The piecewise linearity of the FLC provides better system response than a classical controller [2,6,7].

Design concepts for FLCs involve manipulating the fuzzy sets and rules to obtain the desired effective controller gain. Section II discusses steady-state error reduction using a proportional FLC (PFLC). Drawbacks of the PD and PI controllers are discussed in Section III. As the number of control variables increase, implementation of the design of an FLC becomes more complicated. A hybrid proportional-plus-integral-plus-derivative FLC (PIDFLC) that uses a simplified design and implementation is proposed in Section IV. Simulations to demonstrate the effectiveness of the PID fuzzy controllers are given. Some conclusions are drawn in Section V.

## II  Steady-State Error Reduction

For the classical controller, the steady-state error is dependent on the system type for a specific input and the proportional gain. For example, the steady-state error for a type-0 system with a unit step input is $\frac{1}{1+K_p}$. The larger the proportional gain, the smaller the steady-state error. For a classical controller, the proportional gain is fixed but the effective proportional gain of the PFLC can be increased around zero.

The PFLC is able to minimize the steady-state error by the decreasing the width of the "around zero" fuzzy sets. In terms of linguistic interpretation, this is equivalent to saying "the range of error in which the control action is zero should be smaller". Mathematically, the steady-state error can be approximated by using the rate of change of the control action to calculate the effective gain, $K_{p\text{-eff}}$, and the steady-state error will be $\frac{1}{1+K_p}$. $K_p$ is the effective gain multiplied by the PFLC output gain. The peak values for the error and output fuzzy sets must give at least an effective gain for the desired steady-state.

To obtain the optimum $K_p$ output gain and $K_{p\text{-eff}}$, effective gain of the controller that minimizes steady-state error, a three rule PFLC test needs to be conducted for various values of output gains. This test involves using a PFLC with three input fuzzy sets and three corresponding rules as shown in Table II. PB and NB for the output fuzzy set peak values are the maximum normalized output (i.e. ±1). ZO for the error and the output fuzzy set peak value is zero. PB and NB for the error fuzzy set peak values are ±a which is varied from ±1 to near zero. Figure 1 shows the controller output for PB and NB equal to ±a respectively. Figure 2 shows the corresponding system output for a type-0 system. As "a" becomes closer to 0, the steady-state error is reduced.

Table II  Control rule matrix for determining PFLC output.

|        | NB | ZO | PB |
|--------|-----|----|----|
| Error  | -a  | 0  | a  |
| Output | -1  | 0  | 1  |

Design of a PDFLC is an extension of PFLC design. The PFLC design provides a response that meets the steady-state error specifications. With further manipulation of the input and output fuzzy sets, the overshoot can be minimized. The change in error input provides additional overshoot reduction. By manipulation of the change in error fuzzy sets, the overshoot can be minimized without significantly affecting the rise time.

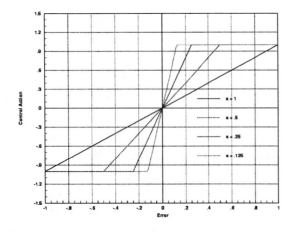

Figure 1  Controller output for PB and NB equal to ±a.

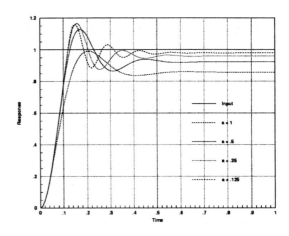

Figure 2  System response for PB and NB equal to ±a.

## III  Controller Limitations

Like the classical PD controller, its fuzzy counterpart can not eliminate steady-state error. The fuzzy control action drives the plant output to the ranges for the zero set for both the error and change in error (the center rule of the rule matrix). However, since the zero set is defined as a range, it is possible to have a small error when the controller action is near zero. As described in the previous section, to reduce the steady-state error, ZO for the error term can be made smaller so that there is more contribution from the control action of adjacent rules of the matrix. Narrowing the ZO range corresponds to increasing the gain of the controller around the zero error point. If the change in error is zero, and ZO for the error term decreases to a width of zero, the control action is discontinuous.

Figure 3 shows the controller output of a PDFLC for a step input to a closed loop system with a plant that has the transfer function $G_p(s) = \dfrac{45}{s^2 + 18s + 45}$. All fuzzy sets are equally spaced except for PS and NS which are near zero. The large effective gain of the controller causes the system response to continually oscillate as shown in Figure 4. Therefore, while the steady-state error can be reduced by manipulating the effective gains, it can not be eliminated.

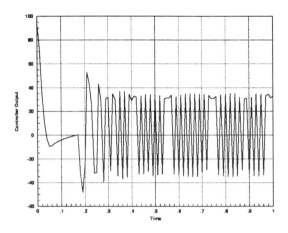

Figure 3  Controller output of example system.

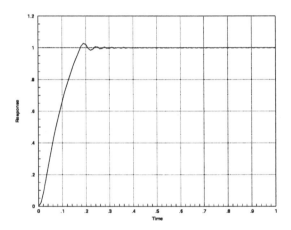

Figure 4  Response of example system

The PI fuzzy controller has a slower response due to the integral error control variable. The controller response maps to a path through the matrix of control rules. Movement from the currently applied rule to the next rule is dependent on the current and next set of control variable. Movement from set to set based on the integral error control variable is slow because the integral error variable changes slowly, thus slowing down the fuzzy controller response.

In terms of classical control, the PI controller has the effect of adding a pole at the origin. The PI controller acts as a low pass filter and reduces the bandwidth of the frequency response. A lower bandwidth

corresponds to a slower rise time. However, the main advantage of the PI controller is that it increases the type of the system which is used to reduce or eliminate steady-state error.

## IV Hybrid PIDFLC Design

Since neither the PD nor PI controllers are capable of individually meeting design criteria of steady-state error, overshoot and rise time, the PID controller is used. For PIDFLC design, there are three inputs which makes the design and implementation more complex.

The number of rules to cover all possible input variations for a PDFLC is the number of sets for the error multiplied by the number of sets for the change in error. The rule matrix for a controller that has seven sets for error and seven sets for change in error has 49 rules. Similarly, the PI control rule matrix has 49 rules. The addition of another control variable significantly increases the number of rules. For example, if the integral error control variable with seven sets is added to the PD controller, there would be 343 control rules. Design of such a large rule base would be a tedious task.

The following section describes a method of PIDFLC design and implementation. The goal is to obtain a PID like response with a FLC without significantly increasing design complexity.

### Reduced Rule PID Controller

Logically, it should be possible to divide the action of the PID controller into two separate control actions: PD controller for fastest response and PI controller for the elimination of the steady-state error. The plant must be capable of being compensated by a PI controller. Figure 5 is the implementation of the proposed hybrid fuzzy PID control scheme. Similar to separate control rule tables for "coarse" and "fine" control [2], a PD controller provides the "coarse" control and the PI controller gives the "fine" control. Logic is required to determine the switching between the two control actions.

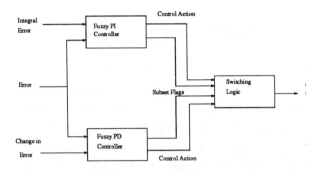

Figure 5  Proposed hybrid PIDFLC design.

The PDFLC is active when the error is large and should be reduced fast. On the other hand, the PI portion activates only when the PD portion reduces the error and change in error to where both are in the ZO fuzzy set range. Therefore, at any instant, calculation of the control action involves only four control rules whereas an FLC with three input variables (i.e. PID) requires eight.

For a PIDFLC with three control variables of seven sets each, 21 sets would have to be checked to determine applicable rules. For the hybrid controller, only a maximum of sixteen sets would be checked. The rule search first checks the two ZO sets for the PD portion and then checks at most all fourteen of the PI portion sets. If the check on the two ZO sets is negative then the remaining 12 sets would be checked.

For the hybrid fuzzy PID controller, the PD and PI portions are designed separately and logic controls when to switch between the two controllers. The logic switches to the PI portion when both change in error and error are in the ZO range. The PD portion must not be re-enabled until the error variable moves out of the ZO range, regardless of the change in error variable. The PI portion in the process of reaching steady-state obviously creates a change in error that might be out of the PD's

ZO range and should not reactivate the PD portion.

## Numerical Example

The hybrid fuzzy PID controller was used to control the plant that has the transfer function $G_p(s) = \dfrac{45}{s^2 + 18s + 45}$.  The controller gains were $K_p = 19$ and $K_d = .5$.  The peak values of the fuzzy sets is given in Table III.  The PI portion was designed with $K_i/K_p$ set to .1, $K_p$ is 100, and seven fuzzy set with all peak values equally spaced.  Figure 6 shows that after a slight overshoot, the ramp response settles to the desired output.

Table III

|  | NB | NM | NS | ZO | PS | PM | PB |
|---|---|---|---|---|---|---|---|
| Error | -1 | -.35 | .03 | 0 | .03 | .35 | 1 |
| ΔError | -1 | -.35 | -.10 | 0 | .10 | .35 | 1 |
| Output | -1 | -.75 | -.3 | 0 | .3 | .75 | 1 |

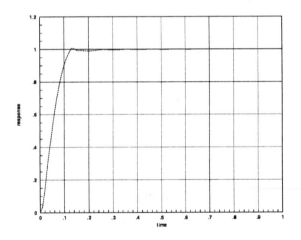

Figure 6  Response for example system controlled by hybrid PIDFLC

## IV Conclusions

PD and PI fuzzy controllers have the same design disadvantages as their classical counterparts. Therefore, in some cases a fuzzy

PID controller maybe required. The fuzzy PID controller entails a large rule base which presents design and implementation problems. A reduced rule fuzzy PID scheme was implemented to take advantage of both PD and PI control actions. Some further research is required for the process of switching between the control actions. Results from the simulation demonstrates the effectiveness of the PID controllers.

1. Gholamresa Langari, *A Framework for Analysis and Synthesis of Fuzzy Linguistic Control Systems*, Ph.D. Thesis, University of California at Berkeley, December 1990.

2. Y. Li and C Lau, "Development of Fuzzy Algorithms for Servo Systems," *IEEE Control Systems Magizine*, pp 65-71, April, 1989.

3. E.H. Mamdani, "Application of Fuzzy Algorithms for Control of Simple Dynamic Plant," *Proc. IEE* 121 Vol. 12 pp. 1585-1588, 1974.

4. M. Mizumoto, "Realization of PID Controls By Fuzzy Control Methods," IEEE 1992.

5. Kuldip S. Rattan, B.Chiu, V Feliu and H.B. Brown Jr., "Rule Based Fuzzy Control of a Single-Link Flexible Manipulator in the Presence of Joint Friction and Load Changes," *American Control Conference*, Pittsburgh, PA, June 1989.

6. D. Sabharwal and K. Rattan, Design of a Rule Based Fuzzy Controller for the Pitch Axis of an Unmanned Research Vehicle," *NAECON Proceedings*, Dayton, 1992.

7. D. Sabharwal and K. Rattan, Design of a Rule Based Fuzzy Controller for the Pitch Axis of an Unmanned Research Vehicle," *NAECON Proceedings*, Dayton, 1991.

8. M. Sugeno, *Industrial Applications of Fuzzy Control*, North-Holland, 1985.

9. L. A. Zadeh, "Outline of a New Approach to the Analysis of Complex Systems and Decision Processes," *IEEE Trans. Systems, Man Cybernetics*, Vol 3, pp. 28-44, 1973.

10. L. Zheng, "A Practical Guide to Tune of Proportional and Integral (PI) Like Fuzzy Controllers," IEEE 1992.

# Fuzzy linearization for nonlinear systems: A preliminary study[1]

Jin Yaochu,  Zhu Jing  and   Jiang Jingping

Electrical Engineering Department, Zhejiang University

Mailing address: Electrical Engineering Department, Zhejiang University
Hangzhou, 310027, P. R. of China

Abstract: This paper proposes a novel control diagram for nonlinear systems, namely fuzzy linearization. On the basis of fuzzy reasoning, we build a set of fuzzy linear subsystems to linearize the original nonlinear system. Consequently, we design an optimal controller for every linear subsystem using the mature linear control theory. The control effect of each subsystem is composed via fuzzy reasoning to control the nonlinear system. Therefore, the design of any nonlinear systems can be simplified to the control problem of linear time-invariant systems. Compared to the existing methods such as the Taylor expansion and piecewise linearization, the proposed approach exhibits higher precision, better control performances and stronger robustness to system uncertainties.
Key words: Linearization, fuzzy reasoning, neural network, optimal control

## 1. Introduction

The control of nonlinear systems remains to be one of the most difficult problem in control engineering. As it is well known, the basic technique for controlling nonlinear systems is to linearize the given system and then use the linear control strategies. Generally speaking, there are two main approaches for system linearization, namely, approximate linearization and exact linearization. The most widely used approximate linearization method is to linearize the original system around its operating point using Taylor expansion. After that, either a conventional linear controller ( for example, the PID controller[1] ) is applied or a self-tuning controller is constructed by virtue of on-line identification[2]. Another approximate linearization method is to replace the original system with  a group of piecewise linear systems[3]-[4]. However, for a class of substantially nonlinear systems, all the approximate linearization methods exist large approximation error and therefore satisfying control performances can not be  obtained.

The differential geometry theory that was developed since 1970s  provided an alternative way  for nonlinear system control. By taking advantage of diffeomorphism transformation and nonlinear state feedback, a class of nonlinear systems is successfully and exactly linearized[5]. Despite that, there are many problems remained to be solved to implement the algorithm in control engineering. For example, it  is strongly dependent on the accuracy of the mathematical model built for the system. Therefore, the robustness of the controller is very weak.

Fuzzy reasoning  is an effective methodology  for modeling and control of complex systems[6]. It combines human heuristic reasoning and logic thinking with control systems and forms a particular  tool for   system modeling and control. The authors[7] have proposed a new  way of adaptive fuzzy reasoning  and successfully realized the modeling and control for a class of complex systems.

This paper extends the fuzzy modeling  method of [6] to a class general nonlinear systems. Using the adaptive fuzzy reasoning developed in [7], we build a set of fuzzy linear subsystems  to exactly[2]  model the given nonlinear system. A linear optimal controller is designed for each subsystems and  at last is synthesized to control the nonlinear system via fuzzy composition. In th[3]is way, the problem of nonlinear control is transferred to linear system control. Simulation is carried out on a single link manipulator, which demonstrates that the proposed methodology is feasible.

## 2. Fuzzy linearization and fuzzy model identification
### 2.1  Fuzzy linearization

Without loss of generality, we discuss the following  nonlinear system:

$$x_k = f(x_{k-1}, u_{k-1}) \qquad (1)$$

[1]This work is financially supported by National Science Foundation of China.
[2]Theoretically, we can exactly model the nonlinear system with the linear subsystems if the system satisfies some conditions. In practice, we approximate the nonlinear system to a certain precision.

where, $x_k$ and $x_{k-1} \in R^{n \times 1}$, are the state vector at instant k and k+1 respectively, $u_{k-1} \in R^{m \times 1}$ is the control of the system. If system (1) is expanded at $x_{k-1} = x_{k-1}^*$ and $u_{k-1} = u_{k-1}^*$ with Taylor series, then we have:

$$\Delta x_k = A_{k-1} \Delta x_{k-1} + B_{k-1} \Delta u_{k-1} \tag{2}$$

where, $A_{k-1} = \partial f / \partial x_{k-1}|_*$, $B_{k-1} = \partial f / \partial u_{k-1}|_*$. Here, even if system (1) has no uncertainties, the linear system $(A_{k-1}, B_{k-1})$ is time-varying. If system (1) is enough smooth, $(A_{k-1}, B_{k-1})$ can be thought as constant within a period of time and therefore the following piecewise LTI models are obtained(refer to Fig. 1):

$$
\begin{aligned}
x_k &= A_1 x_{k-1} + B_1 u_{k-1} & &, \text{ if } a_1 \leq x_{k-1} \leq a_2 \\
x_k &= A_2 x_{k-1} + B_2 u_{k-1} & &, \text{ if } a_2 \leq x_{k-1} \leq a_3 \\
&\vdots & &\vdots \\
x_k &= A_p x_{k-1} + B_p u_{k-1} & &, \text{ if } a_p \leq x_{k-1} \leq a_{p+1}
\end{aligned} \tag{3}
$$

In this situation, $(A_j, B_j)$ $(j=1,2, \ldots, P)$ are LTI systems. According to the piecewise linear models in equation (3), linear controller can be designed for each subsystem to control the nonlinear model. However, we can see from Fig. 1 that the bounds for each linear model $[a_1, a_2], [a_2, a_3], \ldots, [a_p, a_{p+1}]$ are all definite, which unavoidably causes large errors near the bounds. As a matter of fact, we argue that the original system does not have these bounds inherently. If it has, the bounds are fuzzy but not concrete. To this end, we take the fuzzy bounds as in Fig.2, which has proved to greatly reduce the approximating error. In such case, the nonlinear system (1) is fuzzily but much more exactly linearized compared to the piecewise method. We notice that the definite ranges $[a_1, a_2], [a_2, a_3], \ldots, [a_p, a_{p+1}]$ have become fuzzy partitions $[F_1, F_2], [F_2, F_3], \ldots, [F_p, F_{p+1}]$ with certain membership functions. On the basis of the above discussion, the nonlinear system (1) can the described using the following P fuzzy linear models:

$R^j$:  If $x_{k-1}$ is $F^j$ and $u_{k-1}$ is $G^j$, then

$$x_k^j = A_{k-1}^j x_{k-1} + B_{k-1}^j u_{k-1}^j \tag{4}$$

where, $R^j$ means the jth fuzzy rule, $(A_{k-1}^j, B_{k-1}^j)$ represents the jth fuzzy linear subsystem. If system (1) has no uncertainties, then the subsystems $(A_{k-1}^j, B_{k-1}^j)$ are time-invariant. So we have transferred the control of nonlinear system (1) to the control of fuzzy system (4). According to reference [5], given the system state $X_{k-1} = (x_1^0(k-1), x_2^0(k-1), \ldots, x_n^0(k-1))^T$ the whole fuzzy system can be expressed by:

$$x_k = \sum_{j=1}^{L}(w^j x_{k-1}^j)/\sum_{j=1}^{L} w^j = [\sum_{j=1}^{L}(w^j A_{k-1}^j)/\sum_{j=1}^{L} w^j] x_{k-1} + [\sum_{j=1}^{L}(w^j B_{k-1}^j)/\sum_{j=1}^{L} w^j] u_{k-1} \tag{5}$$

where, $w^j$ is the true value of the jth rule, which satisfies $w^j \geq 0$, $\sum w^j > 0$ and

$$w^j = t \, F_j^j(x_{k-1}^0) \times G^j(u_{k-1}) \tag{6}$$

where, 't' called t norm[8], is a fuzzy operator, $F_j^j$ and $G^j$ are membership functions of Gaussian or triangular type. Generally, they can be described by a feedforward neural network.

2.2 Identification of the fuzzy linear models

In paper [6], structure and parameter identification are introduced. In our paper, we suppose the structure of the system is known, what we need to identify is the premise and consequence parameter. By premise parameter identification, we mean that the partition of fuzzy subspaces and its membership functions of each premise variable need to be established or modified. That is to say, at the beginning of the identification process, we only give a few fuzzy rules and we try to modify the membership function of each fuzzy subset to acquire a good performance. If it fails, the fuzzy system automatically produces new fuzzy rule until desired precision is obtained. Explicitly, consequence parameter identification means to update the parameters of the linear models.

Sugeno and Tanaka[6] proposed a general identification method for fuzzy models. Jin et al [7] proposed an adaptive fuzzy identification scheme based on a hybrid Pi-sigma neural network. The neural network architecture we proposed is shown in Fig.3. In the figure, '$\Lambda$' represents the fuzzy node that carries out the fuzzy operation defined in equation (6), '$\Sigma$' and '$\prod$' mean the summing node and the multiplication node respectively. Suppose

$$A^j = \begin{bmatrix} a_{11}^{\ j} & a_{12}^{\ j} & ...... & a_{1n}^{\ j} \\ a_{21}^{\ j} & a_{22}^{\ j} & ...... & a_{2n}^{\ j} \\ : & : & ...... & : \\ a_{n1}^{\ j} & a_{n2}^{\ j} & ...... & a_{nn}^{\ j} \end{bmatrix} \qquad B^j = \begin{bmatrix} b_{11}^{\ j} & b_{12}^{\ j} & ...... & b_{1m}^{\ j} \\ b_{21}^{\ j} & b_{22}^{\ j} & ...... & b_{2m}^{\ j} \\ : & : & ...... & : \\ b_{m1}^{\ j} & b_{m2}^{\ j} & ...... & b_{mm}^{\ j} \end{bmatrix}$$

For convenience of derivation, we note:

$$c_{1,1}^{\ j} = a_{11}^{\ j}, \ ... , \ c_{1,n}^{\ j} = a_{1n}^{\ j}, \ c_{1,n+1}^{\ j} = b_{11}^{\ j}, \ ... , \ c_{1,n+m}^{\ j} = b_{1m}^{\ j}$$
$$c_{2,1}^{\ j} = a_{21}^{\ j}, \ ... , \ c_{2,n}^{\ j} = a_{2n}^{\ j}, \ c_{2,n+1}^{\ j} = b_{21}^{\ j}, \ ... , \ c_{2,n+m}^{\ j} = b_{2m}^{\ j}$$
$$\vdots \qquad\qquad \vdots \qquad\qquad \vdots \qquad\qquad \vdots$$
$$c_{n,1}^{\ j} = a_{n1}^{\ j}, \ ... , \ c_{n,n}^{\ j} = a_{nn}^{\ j}, \ c_{n,n+1}^{\ j} = b_{n1}^{\ j}, \ ... , \ c_{n,n+m}^{\ j} = b_{nm}^{\ j}$$

$$z_1 = x_1(k-1), \ ... , \ z_n = x_n(k-1), \quad z_{n+1} = u_1(k-1), \ ... , \ z_{n+m} = u_m(k-1)$$

Thus, the fuzzy system (4) can be rewritten by:

$R^j$: If $z_1$ is $F_1^j$, ... , $z_n$ is $F_n^j$, $z_{n+1}$ is $G_1^j$, ... , $z_{n+m}$ is $G_m^j$, Then

$$x_1^j(k) = c_{1,1}^{\ j} x_1(k-1) + c_{1,2}^{\ j} x_2(k-1) + \cdots + c_{1,n}^{\ j} x_n(k-1)$$
$$+ c_{1,n+1}^{\ j} u_1(k-1) + c_{1,n+2}^{\ j} u_2(k-1) + \cdots + c_{1,n+m}^{\ j} u_m(k-1)$$

$$\vdots \qquad\qquad \vdots \qquad\qquad\qquad\qquad\qquad (7)$$

$$x_n^j(k) = c_{n,1}^{\ j} x_1(k-1) + c_{n,2}^{\ j} x_2(k-1) + \cdots + c_{n,n}^{\ j} x_n(k-1)$$
$$+ c_{n,n+1}^{\ j} u_1(k-1) + c_{n,n+2}^{\ j} u_2(k-1) + \cdots + c_{n,n+m}^{\ j} u_m(k-1)$$

and

$$x_i(k) = \sum_{j=1}^{L} (w_i^j \sum (c_{i,r}^{\ j} z_r)) / \sum_{j=1}^{L} w_i^j \qquad\qquad (8)$$

We are now in the position of discussing the selection of the membership function. As mentioned before, if we use the Gaussian function as the membership function, its shape is greatly limited and therefore lack of flexibility. In our paper, we replace it with a sub network having the following form:

$$\mu_i^j(v) = g(\sum_{k=1}^{H} l_{j,k} h(d_k v - s_k)) \qquad\qquad (9)$$

where, $g(.)$ and $h(.)$ are nonlinear activation functions in the output layer and hidden layer respectively. We found in practice that if both $g(.)$ and $h(.)$ take the usual sigmoid function, the sub network is awkward to realize the non smooth membership functions such as the triangular type. Hence, we select the wavelet mother function as the activation function $h(.)$[9]. To keep the conservation of the membership function represented by the sub network, the activation function of the output layer takes the following sigmoid function:

$$g(v) = (1 + \exp(-(5.0(v-0.5))))^{-1} \qquad\qquad (10)$$

It can be easily checked that $g(v)$ is a smooth saturation function, which both ensures the membership between [0,1] and satisfies the updating of the neural network.

Define the error function of the hybrid Pi-sigma neural network as follows:

$$E_i = 0.5[x_i(k) - x_i^d(k)]^2 \qquad (i = 1, 2, ..., n) \qquad\qquad (11)$$

Using the gradient method, we have[7]

$$\Delta c_{i,r}^{\ j} = \lambda_1 [x_i(k) - x_i^d(k)] w_i^j z_r / \sum_{j=1}^{L} w_i^j \qquad\qquad (12)$$

As to the updating of the membership sub network, the learning varies slightly according to the definition of the fuzzy operation in equation (6). If we use the minimum as the fuzzy operation, the learning algorithm for its output layer as follows:

$$\Delta l_{j,k} = \lambda_2 [x_i(k) - x_i^d(k)][x_i^j / \sum_{j=1}^{L} w_i^j - \sum_{j=1}^{L} (w_i^j x_i^j) / (\sum_{j=1}^{L} w_i^j)^2] \cdot g'(.) \cdot h(d_k z_k - s_k) \qquad\qquad (13)$$

The learning algorithm for the weight $d_k$ and $s_k$ can be obtained through the principle error backpropagation.

It should be pointed out that when the fuzzy linear systems take the standard controllable form, the number of the parameters to be identified will be significantly reduced.

## 3. Optimal controller design

For the sake of simplicity, consider the optimal regulator design problem for the linear single input system. Suppose each linear subsystem has the same control vector, therefore, equation (4) and (5) can be written by:

$$x_k^j = A_{k-1}^j x_{k-1} + B_{k-1} u_{k-1}^j \qquad (14)$$

$$x_k = [\sum_{j=1}^{L} (w^j A_{k-1}^j)/\sum_{j=1}^{L} w^j] x_{k-1} + B_{k-1} u_{k-1} \qquad (15)$$

where, $(A_{k-1}^j, B_{k-1})$ is a controllable pair. Choose the performance index as:

$$J^j = \sum_{k=1}^{N} [(x_{k-1})^T V^j x_{k-1} + (u_{k-1})^T W^j u_{k-1}] \qquad (16)$$

where $V^j$ and $W^j$ are positive definite matrices. If the state vector $\{x_k^j\}$ possesses the Markov feature, then we can use the dynamical programming technique to solve the optimal control law that minimizes the above performance index. The optimal feedback gain is recursively calculated as follows:

$$u_{k-1}^j = \Lambda_{N-k+1}^j A_{N-k+1}^j x_{N-k} \quad , k=1, 2, ..., N \qquad (17)$$

$$\Lambda_{N-k+1}^j = [(B_{N-k})^T V_{N-k+1}^j B_{N-k} + W_{N-k}]^{-1} (B_{N-k})^T V_{N-k+1}^{j0} \qquad (18)$$

$$V_{N-k+1}^{j0} = V_{N-k+1}^j + (A_{N-k+1}^j)^T \underline{V}_{N-k+2}^j A_{N-k+1}^j \qquad (19)$$

$$\underline{V}_{N-k+2}^j = V_{N-k+1}^j + B_{N-k} \Lambda_{N-k+1}^j \qquad (20)$$

When k=1, $\underline{V}_{N+1}^j = 0$, $V_N^{j0} = V_N^j$. In equation (17), $L_{N-k+1}^j = \Lambda_{N-k+1}^j A_{N-k+1}^j$ is the optimal feedback gain. The control effect acted on the nonlinear system is composed from each subsystem in the following form:

$$u_{k-1} = \{\sum_{j=1}^{L} [(w^j)^2 L_k^j] + 2\sum_{i=1}^{L}\sum_{j=i+1}^{L} [w^i w^j \Lambda_k^i A_{k-1}^j]\}/\sum (w^j)^2 x_{k-1} \qquad (21)$$

**Conclusion 1:** If $B_{k-1}^j (j=1,2,...,m)$ are all equal and the control action is composed as in equation (21), then the control problem of the whole fuzzy system (15) is equivalent to the control of each subsystem(14).

**Proof:**

For system (15), we have

$$u_{k-1} = -\Lambda_k [\sum(w^j A_{k-1}^j)/\sum w^j] x_{k-1} = -[\sum(w^j \Lambda_k^j)/\sum w^j][\sum(w^j A_{k-1}^j)/\sum w^j] x_{k-1}$$

$$= \{\sum[(w^j)^2 L_k^j] + 2\sum_{i=1}^{L}\sum_{j=i+1}^{L} [w^i w^j \Lambda_k^i A_{k-1}^j]\}/\sum (w^j)^2 x_{k-1}$$

Thus the conclusion is arrived.

Substitute $u_{k-1}^j = -L_k^j x_{k-1}$ into (14), we have

$$x_k^j = A_{k-1}^j x_{k-1} + B_{k-1} u_{k-1}^j = (A_{k-1}^j - B_{k-1} L_k^j) x_{k-1} = G_{k-1}^j x_{k-1} \qquad (22)$$

The whole fuzzy system is :

$$x_k = [\sum(w^j G_{k-1}^j)/\sum w^j] x_{k-1} \qquad (23)$$

**Conclusion 2:** For each subsystem in equation (22), if

(i) There exist two constants a , b and a symmetric matrix $P_k$, which satisfies:

$$0 < aI \leq P_k \leq bI < \infty, \text{ for all } k \qquad (24)$$

(ii) There exist some matrix $H_k$, which satisfies

$$(G_{k-1}^j)^T P_k G_{k-1}^j - P_k = -H_k H_k^T \qquad (25)$$

(iii) There exist a constant C and a function Sn(k), for a some n and all k, the following relations hold:

$$S_n(k) = \sum_{i=1}^{N-1} \Phi^T(k+i,k) H(k+i) H^T(k+i) \Phi(k+i,k) \geq CI > 0 \qquad (26)$$

where $\Phi(k,j)$ is the transfer matrix of system (22). Then system (23) is exponentially stable at $x_k=0$. The proof is omitted here for space limit.

4. Simulation study

Simulation is carried out on a single link robot whose dynamics is expressed by:

$$u=( ml_c^2+I)\ddot{q}+\gamma\dot{q}+ml_c^2g \cos(q) \tag{27}$$

where $l_c=0.5\ l$, $I=(1/12)\ ml^2+m/4R^2$. Let $x_1=q$, $x_2=\dot{q}$, then we have

$$\dot{x}_1=x_2$$
$$\dot{x}_2=[u-\gamma\dot{q}-ml_c^2g \cos(q)]/( ml_c^2+I)$$

where, $m=15.91$ kg, $l=0.432$m, $\gamma=0.1$, $R=0.05$m.

**Simulation 1:** Training example. We take the following as the training sample:

$$q=2.0t +\sin(2\pi t/20), \ t=0{\sim}20.$$

In this case, q and q are divided into three fuzzy subspaces, namely, positive small, positive medium and positive large. The control variable is divided into six fuzzy subspaces, namely NL, NM, NS, PS, PM,PL, where P stands for positive, N stands for negative and therefore, there are fifty-four sub models in the fuzzy system. After 5000 times of training, the fuzzy system approximates the nonlinear satisfactorily, as shown in Fig. 4.

**Simulation 2:** Control example. In this example, we let the system to track the following trajectory:

$$q=0.5\pi(1.0-\exp(-t)), \ t=0{\sim}20.$$

Since all the input variables of the fuzzy system are positive, they are all partitioned into three subspaces and the fuzzy system has twenty-seven sub models. After 2000 times of training, the approximation result is given in Fig. 5. Then we design a optimal controller for each linear system and the whole control action is obtained following the algorithm given in equation (21). We find that the system behaves quite well, as demonstrated in Fig. 6.

5. Conclusion

In this paper, a new control diagram for nonlinear system is developed. The simulation results show that the proposed is feasible. The further work remained to be done is to applied the method to more complex system, such as the MIMO systems. Another problem we need to solve is to reduce the number of the fuzzy sub models. One possible solution to this problem is to build the fuzzy reasoning machine hierarchically.

References:
1. S. Q. Zhu, Approximating analysis for nonlinear system. Education Press, Beijing, 1980(in Chinese)
2. C. S. G. Lee et al . An adaptive control strategy for mechanical manipulators. IEEE Trans. on AC, Vol.29,1984
3. J.S. Shamna and M. Athans, Analysis of gain scheduled control for nonlinear plants. IEEE Trans. on AC, Vol.35,1990
4. R.A. Jacos and M.I. Jordan, Learning piecewise control strategy in a modular neural network architecture. IEEE Trans. on SMC, Vol.23, 1993
5. A. Isidori, Nonlinear control system: an introduction. Springer Verlag, Berlin,1985
6.M. Sugeno and K Tanaka, Successive identification of a fuzzy model and its application to prediction to prediction of complex system. Fuzzy Set and Systems, Vol.42, 1991
7. Y. C. Jin, J. Zhu and J. P. Jiang, Adaptive fuzzy modeling and identification with applications. International Journal of Systems Science, to be published, 1993
8.Q. Zhang, and A. Benveniste, A wavelet network. IEEE Trans. on NN, Vol. 3, 1992
9. S Tanaka and M. Sugeno, Stability analysis and design of fuzzy control systems. Fuzzy Set and Systems, Vol.45, 1992

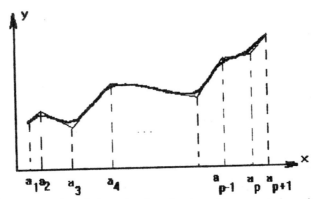

Fig. 1 Illustration of piecewise linearization with concrete bounds

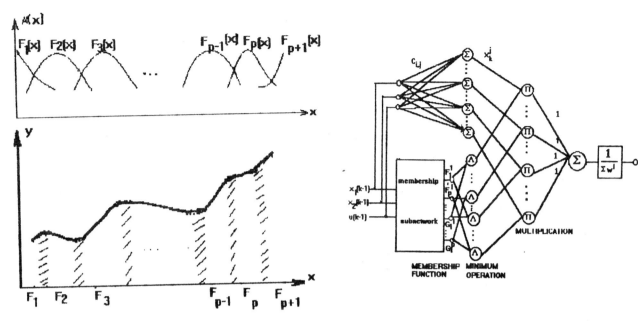

Fig. 2 Fuzzy linearization with fuzzy bounds

Fig.3 Hybrid Pi-sigma neural networks

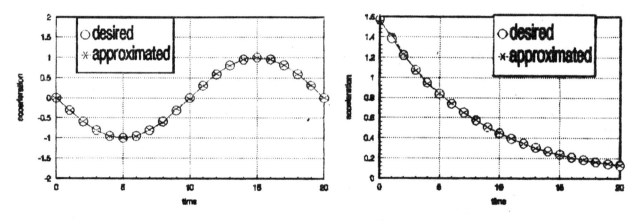

Fig. 4 Approximation result : example 1

Fig. 5 Approximation result: example 2

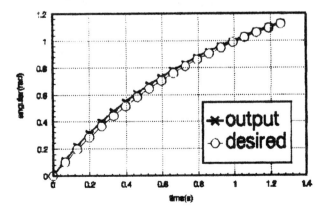

Fig.6 Desired and real system response

# Fuzzy Inverse Incremental Model as Tracking Controller for SISO Systems

Mahesh V. Joshi, P.G. Poonacha, B.Seth

Indian Institute of Technology, Bombay, INDIA.

### Abstract

The problems encountered in using fuzzy logic based single neuron controller (SNC) for tracking control of nonlinear SISO systems is shown to be overcome by the use of Fuzzy Inverse Incremental Model (FIIM) of the same process as the tracking controller. The proposed method of tracking control uses on-line tuning of the universes of discourse and on-line identification of FIIM. Three different algorithms for the linguistic/fuzzy modeling of SISO systems are proposed. The comparative results of using these algorithms for tracking control of some nonlinear systems are shown.

keywords : fuzzy variable, linguistic value, universe of discourse, membership function.

## 1 Introduction :

The last two decades have witnessed the successful use of fuzzy logic and artificial neural networks (ANNs) in controlling complex industrial processes [1], [2], [3]. The fuzzy logic and ANNs represent two aspects of artificial human intelligence. The fuzzy logic tries to formalize the imprecise qualitative nature of human thought processes, whereas the ANNs try to ape the learning and fast processing ability of human brain. Integration of these two to achieve the automatic control systems which will imitate the human operator's capability to control the complex processes, was done to evolve a fuzzy logic based Single Neuron Controller (SNC) [4].

The SNC is a neuron trained for the decision table of a fuzzy logic controller. Some default rules and membership functions are used to develop the SNC, which can be modified implicitly by changing some gain paramemters and the slopes of the sigmoidal nonlinearity, for tuning purposes. It can also explain its actions in terms of rules. This SNC works well for the regulatory control of almost all classes of SISO systems and tracking control of linear systems. But it faced problems while doing the tracking control of nonlinear processes. This was observed to be due to the complete absence of the knowledge regarding the process.

The default rules of SNC are formed on the basis of what is the desired change in output for the given values of the error and change in error, based on common sense of an average human operator. It is assumed that the change in process input is directly proportional to the change in process output, which is true only in case of linear systems. For the nonlinear systems SNC may supply overestimated or underestimated value of change in process input, for required change in process output. This is because for different values of output, the proportion between the change in input and corresponding change in output is different. Here arises the need for inverse incremental identification of the process, which can be done either quantitatively, using some existing identification technique or qualitatively as the human operator will form the rules regarding the process. The later way leads to the proposed idea of fuzzy modeling.

## 2 Identification of Fuzzy Model of the System :

The Process can be represented by a set of fuzzy rules :

$$\text{IF } X = A_i \text{ Then If } Y = B_i \text{ Then } Z = C_i , i = 1 \text{ to } n. \qquad \ldots (1)$$

where, X and Y are the fuzzy variables for the inputs and Z is the fuzzy variable for the output. $A_i$, $B_i$, and $C_i$ are the linguistic values of these variables respectively. $A_i \epsilon T_X$, where $T_X$ is the set of all linguistic variables of X, called the term set. Similarlly $B_i \epsilon T_Y$ and $C_i \epsilon T_Z$. One example of the term set can be,

$T_X$ = { negative large, negative medium, negative small, negative zero, positive zero, positive small, positive medium, positive large} or in an alternate way of representation, $T_X = \{-3, -2, -1, -0, +0, +1, +2, +3\}$.

The range of the numerical values of the fuzzy variables is called the universe of discourse (UD), over which linguistic values are defined. It can be assumed without loss of generality that $T_X$, $T_Y$, $T_Z$ are fixed and the UD for each of the X, Y and Z is kept variable, to implicitly change the definitions of the terms in the term sets. Then the set of rules can be represented as a two dimensional rule table being indexed by all the entries of the term sets $T_X$ and $T_Y$. The entries in the rule table will be the terms in $T_Z$. Thus each location will represent one rule. The identification of the process model will now get transformed into filling the locations in the rule table with appropriate terms from $T_Z$.

The training data for identification will be the numerical values of the fuzzy variables X, Y, and Z, say $x_t$, $y_t$, $z_t$ respectively. The identification should be done such that when $x_t$ and $y_t$ are given as inputs to the fuzzy system represented by the rule table, the deterministic output will lie closer to $z_t$. Here, the need arises to know how the fuzzy system makes its decision.

## 2.1 Conceptual Framework of Fuzzy Modeling :

Following assumptions are made before proceeding further :

    1. The linguistic terms of all the variables are triangular in shape. The overlap between the two alternate terms in a given term set is only at one point. This is shown in Fig.1. This implies that for any value on the universe of discourse, there exist only two linguistic terms in which its membership is nonzero.

    2. The input values are fuzzified to singletons. (Unity membership functions at the input value and Zero membership functions everywhere else)

    3. The Mamdani's Minimum Implication is used for rule representation and Zadeh's Max-Min Compositional Rule of Inference is used for evaluating the rules.

    4. The Centre of Gravity method is used for defuzzification.

Assume the the set of rules as given in (1). For inputs $x_t$ and $y_t$ to the fuzzy system, the result of the $i^{th}$ rule will be a fuzzy subset characterized by the membership function of the form

$$\mu_{Ci'}(z) = \max(\mu_{A'}(x), \mu_{B'}(y), \mu_{Ri}(x,y,z)), \text{ over all } x \in X, y \in Y \text{ and for all } z \in Z.$$

Here $R_i$ is the relational equivalent of the $i^{th}$ rule using Mamdani's Implication. Using the singleton nature of A' and B', this can be simplified to

$$\mu_{Ci'}(z) = \mu_{Ri}(x_t, y_t, z), \text{ for all } z \in Z.$$

Using Mamdani's Implication,

$$\mu_{Ci'}(z) = \min(\mu_{Ai}(x_t), \mu_{Bi}(y_t), \mu_{Ci}(z)), \text{ for all } z \in Z.$$

This says that Ci' is the intersected set Ci at the value,

$$\min(\mu_{Ai}(x_t), \mu_{Bi}(y_t)).$$

The Contribution of this rule will be nonzero if the values $\mu_{Ai}(x_t)$ and $\mu_{Bi}(y_t)$ are nonzero. The net infered output fuzzy set will be the union of all such Ci' s. The centre of gravity of this fuzzy set will be the output of the fuzzy system.

With the assumption 1 above $\mu_{Ai}(x_t)$ will be nonzero for maximum two values of Ai. The same is true for $\mu_{Bi}(y_t)$. Hence maximum four rules will get fired, for any input values.

While Identifying, for given training pattern $(x_t, y_t, z_t)$, with the knowledge of the UDs, maximum of four locations in the rule table will be filled with the appropriate terms from $T_Z$. All the terms in $T_Z$ are eligible candidates for filling these locations, as far as the centre of gravity of the net infered set lies closer to $z_t$. *The most immediate and simple candidates that one can think of are those terms in which $z_t$ has nonzero membership. According to the assumption there will be only two such terms. So, the infered set will be the union of maximum of two adjacent terms from $T_Z$ each one intersected at different value. The centre of gravity of this union will always lie in the close viscininty of $z_t$. See Fig.2.*

In summary, fuzzy identification process reduces down to the filling of four locations in the rule table with two possible values at each location.

## 2.2 Algorithms for Identification :

The nomenclature that will be used in describing the algorithms is :

$A_i$ , $A_j$ : The Linguistic terms in which $x_t$ has nonzero membership.

$B_k$ , $B_l$ : The Linguistic terms in which $y_t$ has nonzero membership.

$C_m$ , $C_n$ : The Linguistic terms in which $z_t$ has nonzero membership.

$C_{ik}$ : The entry in the rule table at the location corresponding to $A_i$ and $B_k$.

$\tau_{ik}$ : Closeness Measure of the entry $C_{ik}$ , which is a positive quantity.

Before the modeling algorithms are invoked, the rule table is filled with the default rules and all the locations in the closeness measure table are flagged ( with say the value -1).

The Step 0 in all the three algorithms is identical, in which the four locations are located by finding $A_i$ , $A_j$ , $B_k$ and $B_l$ , knowing $x_t$ and $y_t$ ; and the two terms $C_m$ and $C_n$ , knowing $z_t$.

### Algorithm 1 : Four Location Maximum Closeness Modification (FLMCM) :

1. Since four locations are to be filled with two terms, there are in all 16 possible combinations of doing so. Let the combinations be numbered 0 to 15. Select combination 0.
2. Find Out the defuzzified output, $z_o$ for the current combination.
3. Calculate the closeness of $z_o$ to $z_t$ as the membership of $z_o$ in the fuzzy set *close*, which is centred around $z_t$, and its shape is similar to the terms in $T_Z$.
4. If the closeness is greater than that for the previous combination then select current combination as valid combination.
5. Select next combination, and repeat steps 3 to 5. At the end, the combination with maximum closeness will be known.
6. For the location addressed by $A_i$ and $B_k$, let the proposed entry be $C_p$ , which will either be $C_m$ or $C_n$. Let the closeness of $C_p$ be $\tau_p$.

    If ( $\tau_{ik}$ is flagged) or ( $\tau_p$ is greater than $\tau_{ik}$ ) then
        Replace $C_{ik}$ with $C_p$ and $\tau_{ik}$ with $\tau_p$.

    Do this for all other locations $C_{il}$ , $C_{jk}$ and $C_{jl}$ .
7. Go to the next training pattern $(x_t, y_t, z_t)$.

### Algorithm 2 : Single Location Maximum Membership Modification (SLMMM) :

Let $C_{max}$ be the term (either $C_m$ or $C_n$) in which $z_t$ has the maximum membership value.

It was found in the FLMCM algorithm that the combination with $C_{max}$ at all locations is always among the combinations those give the maximum closeness. (Note that there will be more than one combinations doing so, because of the nature of evaluation of the rules, which can be verified).

Here, the way in which a human operator will form the rules is taken into consideration. He will not modify all the four rules simultaneously, rather he will find only one rule out of any combination. This rule will correspond to that location where the $x_t$ and $y_t$ will be closer in meaning to the linguistic terms. To state it clearly, let $A_{max}$ be the term from among $A_i$ and $A_j$ with maximum membership of $x_t$, and $B_{max}$ be that term from $B_k$ and $B_l$ ahich is the better linguistic description of $y_t$, that is with maximum membership of $y_t$. Then the human operator will tend to modify only that location addressed by $A_{max}$ and $B_{max}$.

Based on this concept, the SLMMM algorithm will be as follows :

1. Find $A_{max}$ , $B_{max}$ and $C_{max}$. Let $\tau_{max}$ be the membership of $z_t$ in $C_{max}$. $\tau_{max}$ will be the closeness measure of this term.
2. At the location addressed by $A_{max}$ and $B_{max}$ which is say $C_{mm}$;

If ( $\tau_{mm}$ is flagged) or ( $\tau_{max}$ is greater than $\tau_{mm}$ ) then

    Replace $C_{mm}$ by $C_{max}$ and $\tau_{mm}$ by $\tau_{max}$ .

3. Go to the next training pattern ($x_t$ , $y_t$ , $z_t$ ).


**Agorithm 3 : Single Location Weighted Average Modification (SLWAM) :**

    In FLMCM and SLMMM, the previous entry in the rule table was replaced with the proposed entry if the closeness measure of the proposed term was greater.

    Here again the way a human operator may tend to work out the new entry at the proposed location can be looked at. He will tend to form the new rule by taking the weighted average of the existing and the observed rule. So this algorithm is proposed.


1. Same as Step 1 in SLMMM.
2. At the location addressed by $A_{max}$ and $B_{max}$ , which is say $C_{mm}$;

    If ( $\tau_{mm}$ is flagged) then

        Replace $C_{mm}$ by $C_{max}$ and $\tau_{mm}$ by $\tau_{max}$.

    Else

        Compute C = ( $C_{max} \tau_{max}$ + $C_{mm} \tau_{mm}$ ) / ( $\tau_{max}$ + $\tau_{mm}$ ).

        Compute $\tau$ = ( $\tau_{max}$ + $\tau_{mm}$ ) / 2.

        Replace $C_{mm}$ by C and $\tau_{mm}$ by $\tau$.

3. Go to the next training pattern ($x_t$ , $y_t$ , $z_t$ ).


Now, the use of these algorithms for the identification of the Fuzzy Inverse Incremental Model (FIIM) of the nonlinear processes will be seen in the tracking control applications.


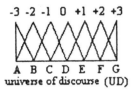

Fig. 1 Linguistic Terms of the Fuzzy Variable

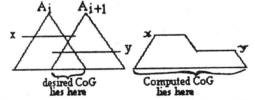

Fig.2 Centre of Gravity (CoG) for two Adjacent Terms

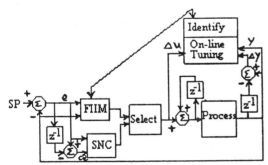

Fig.3 Tracking Control System with FIIM.

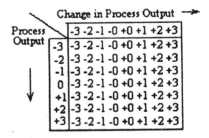

Fig.4 Default Model of a Process


# 3 Fuzzy Inverse Incremental Model (FIIM) for Tracking Control :

The Inverse Incremental Model of the process in terms of the rules is given by,

    IF *output is* $A_i$ THEN IF *desired change in output is* $B_i$ THEN *input should be changed by* $C_i$ , (i = 1 to n)

This model can be identified in an on-line fashion and can be simultaneously used for controlling a SISO process. The block diagram of the resulting control system is shown in Fig.3. The following definitions are used ( the linguistic terms are indicated using numerical values) :

$T_X$ = {-3,-2,-1,0,+1,+2,+3}, where X is Process Output .

$T_Y$ = {-3,-2,-1,-0,+0,+1,+2,+3}, where Y is Change in Process Output.

$T_Z$ = {-6,-5,-4,-3,-2,-1,-0,+0,+1,+2,+3,+4,+5,+6}, where Z is Change in Process Input.

The implementation algorithm for sinusoidal setpoint tracking is as follows :

    **a.** Initialize the rule table to that shown in Fig.4.

    **b.** Read the Process Output (PO).

    **c.** For the current timepoint evaluate setpoint (SP); Error(E) = SP-PO; and Change in error(CE) = current_error-previous_error.

    **d.** If (time < N) then activete SNC with inputs E and CE Else activate the Inverse Model as Controller with inputs E and PO, because E is the desired change in output.

    **e.** Increment the Process Input by the output of the controller and Simulate/Run the process.

    **f.** Record actual values of change in input, change in output and previous process output.

    **g.** Modify the UDs of X,Y and Z. This is done by calculating the maximum values of the change in output, output, and change in input observed so far.

    **h.** Identify the inverse incremental model knowing $x_t$=*output*, $y_t$=*change in output*, and $z_t$=*change in input* using the selected algorithm for identification, and then go back to step b.

This algorithm uses on-line identification of FIIM and on-line tuning of the UDs. The time N needs to be at least one cycle long (of the sinusoid of setpoint variation). During this time N, SNC takes care of controlling the process while producing the training data for identification.

# 4 Results and Discussions :

The above control algorithm was simulated for sinusoidal tracking of three types of nonlinear systems taken from Narendra's work [5]. For a setpoint variation of amplitude 7 and frequency of 100 time points, the simulation was done for seven cycles of setpoint variation.

The first process chosen was a mixed nonlinear process (having nonlinearity in both input and output). The discrete time model of the process was :

$$y(k+1) = y(k) / ( 1 + y(k)^2) + u(k)^3.$$

Fig.5a shows two parts. Upto 2 cycles SNC was controlling. After that the default model of the process took over with *no* on-line identification but with on-line tuning of the UDs. Fig.5b to Fig.5d show the tracking response with FIIM controlling after taking over at the end of two cycles from SNC. The on-line identification as well as on-line tuning were used. The models formed at the end of 7 cycles are given in Tables 1 to 3. For these models, following were the universes of discourse, at the end of 7 cycles, given as [ lower_limit , upper_limit ], in the order of Output, Change in Output and Change in Input : [-7.0 , 7.0] , [-2.49 , 2.49] , [0.679 , 0.679].

The similar sinusoidal tracking responses for following nonlinear processes were observed.

Output Nonlinear Process : $y(k+1) = u(k) + [ y(k) y(k-1) (y(k) + 2.5) ] / [ 1 + y(k)^2 + y(k-1)^2 ]$,

Input Nonlinear Process : $y(k+1) - 0.3 y(k) - 0.6 y(k-1) = u(k)^3 + 0.3 u(k)^2 - 0.4 u(k)$.

The results are shown in Fig.6 and Fig.7 respectively.

Following conclusions could be drawn from the results observed.

    - The on-line modification of the UDs to span all the possible values of that particular variable helps in tuning the controller performance.

    - The SLMMM and SLWAM algorithms work well for on-line inverse incremental identification of general nonlinear process, at least in the sinusoidal tracking control applications. The SLMMM algorithm seems to be better in some cases.

    - The tracking performance for input nonlinear process was not observed to be very good, which might be due to the fact that the FIIM takes into account only the output nonlinearity ( $\Delta u = f ( \Delta y, y)$ ).

# 5 References :

[1] Zadeh L.A., "Outline of new approach to the analysis of complex processes", IEEE Trans. on Systems, Manand Cybernetics, vol.SMC-3, no.1, pp.25-44, Jan 1973

[2] Lee C.C., "Fuzzy Logic in Control Systems : Fuzzy Logic Controller - Parts I andII", IEEE Trans. on Systems, Man and Cybernetics, vol.SMC-20, no.2, pp.404-435, Mar/Apr 1990.

[3] Lippmann R.P., "Introduction to Computing with Neural Nets", IEEE ASSP Magazine, pp.4-22, Apr 1987.

[4] Mahesh V. Joshi, P.G.Poonacha, B.Seth, "Fuzzy Logic Based Single Neuron Tracking Controller for SISO Systems", Proc. of International Joint Conference on Neural Networks '93 - Nagoya, pp. 2885-2888, Oct 1993.

[5] Narendra K.S., et.al., "Identification and Control of Dynamic Systems using Neural Networks", IEEE Trans. on Neural Networks, vol.1, no.1, pp.4-27, Mar 1990.

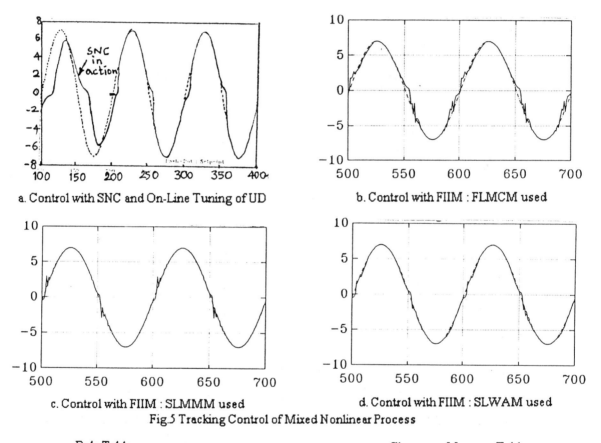

a. Control with SNC and On-Line Tuning of UD

b. Control with FIIM : FLMCM used

c. Control with FIIM : SLMMM used

d. Control with FIIM : SLWAM used

Fig.5 Tracking Control of Mixed Nonlinear Process

Rule Table

| op \\ cop | -3 | -2 | -1 | -0 | +0 | +1 | +2 | +3 |
|---|---|---|---|---|---|---|---|---|
| -3 | -3 | -0 | -0 | -0 | +0 | +0 | +1 | +2 |
| -2 | -4 | -4 | -1 | -0 | -0 | +2 | +3 | +3 |
| -1 | -3 | -4 | -2 | +3 | +5 | +4 | +2 | +2 |
| 0 | -5 | -3 | -3 | +3 | +4 | +4 | +5 | +4 |
| +1 | -3 | -2 | -2 | +0 | +0 | +0 | +4 | +3 |
| +2 | -2 | -2 | -2 | -0 | -0 | +1 | +3 | +3 |
| +3 | -2 | -1 | -2 | -0 | +0 | +0 | +2 | +2 |

op : output
cop : change in output

Closeness Measure Table

| op \\ cop | -3 | -2 | -1 | -0 | +0 | +1 | +2 | +3 |
|---|---|---|---|---|---|---|---|---|
| -3 | 0.959 | 0.986 | 0.997 | 1.000 | 1.000 | 0.996 | 0.986 | 0.970 |
| -2 | 0.997 | 0.997 | 0.997 | 1.000 | 1.000 | 0.999 | 0.999 | 0.998 |
| -1 | 0.997 | 0.997 | 0.997 | 0.999 | 1.000 | 1.000 | 0.999 | 0.998 |
| 0 | 0.987 | 0.997 | 0.997 | 0.999 | 1.000 | 1.000 | 0.995 | 0.978 |
| +1 | -1.00 | 0.998 | 0.998 | 0.996 | 0.996 | 0.995 | 0.997 | 0.997 |
| +2 | 1.000 | 1.000 | 0.998 | 1.000 | 1.000 | 0.997 | 0.997 | 0.997 |
| +3 | 1.000 | 1.000 | 0.998 | 1.000 | 1.000 | 0.997 | 0.995 | 0.995 |

Table 1 : Model due to FLMCM

## Table 2: Model due to SLMMM

**Rule Table**

| op \ cop | -3 | -2 | -1 | -0 | +0 | +1 | +2 | +3 |
|---|---|---|---|---|---|---|---|---|
| -3 | -3 | -2 | -1 | -0 | +0 | +1 | +2 | +3 |
| -2 | -3 | -2 | -1 | -0 | +0 | +1 | +2 | +3 |
| -1 | -6 | -6 | -3 | -0 | +0 | +2 | +4 | +3 |
| 0 | -6 | -6 | -6 | -6 | +6 | +6 | +6 | +5 |
| +1 | -3 | -2 | -1 | -0 | +0 | +2 | +4 | +3 |
| +2 | -3 | -2 | -1 | -0 | +0 | +1 | +2 | +3 |
| +3 | -3 | -2 | -1 | -0 | +0 | +1 | +2 | +6 |

op : output
cop : change in output

**Closeness Measure Table**

| op \ cop | -3 | -2 | -1 | -0 | +0 | +1 | +2 | +3 |
|---|---|---|---|---|---|---|---|---|
| -3 | -1.00 | -1.00 | 0.981 | 0.918 | 0.932 | 0.849 | -1.00 | -1.00 |
| -2 | 0.763 | 0.938 | 0.815 | 0.975 | 0.973 | 0.934 | 0.925 | -1.00 |
| -1 | 1.000 | 1.000 | 0.998 | 0.817 | 0.884 | 0.999 | 0.973 | -1.00 |
| 0 | 1.000 | 1.000 | 1.000 | 1.000 | 1.000 | 1.000 | 0.974 | 0.951 |
| +1 | -1.00 | 0.988 | 0.999 | 0.884 | 0.817 | 0.998 | 0.944 | -1.00 |
| +2 | -1.00 | 0.997 | 0.934 | 0.973 | 0.975 | 0.815 | 0.938 | -1.00 |
| +3 | -1.00 | -1.00 | 0.961 | 0.937 | 0.918 | 0.829 | 0.993 | 1.000 |

## Table 3 : Model due to SLWAM

**Rule Table**

| op \ cop | -3 | -2 | -1 | -0 | +0 | +1 | +2 | +3 |
|---|---|---|---|---|---|---|---|---|
| -3 | -3 | -2 | -1 | -0 | +0 | +0 | +2 | +3 |
| -2 | -4 | -3 | -1 | -0 | -0 | +1 | +2 | +3 |
| -1 | -6 | -4 | -2 | -1 | +0 | +2 | +3 | +3 |
| 0 | -6 | -4 | -4 | -4 | +2 | +5 | +4 | +5 |
| +1 | -3 | -4 | -2 | -1 | -0 | +1 | +3 | +3 |
| +2 | -3 | -3 | -2 | -0 | +0 | +1 | +2 | +3 |
| +3 | -3 | -2 | -1 | -0 | +0 | +0 | +2 | +3 |

op : output
cop : change in output

**Closeness Measure Table**

| op \ cop | -3 | -2 | -1 | -0 | +0 | +1 | +2 | +3 |
|---|---|---|---|---|---|---|---|---|
| -3 | -1.00 | -1.00 | 0.646 | 0.537 | 0.705 | 0.681 | -1.00 | -1.00 |
| -2 | 0.703 | 0.694 | 0.813 | 0.775 | 0.814 | 0.622 | 0.667 | -1.00 |
| -1 | 0.938 | 0.812 | 0.856 | 0.703 | 0.747 | 0.846 | 0.840 | -1.00 |
| 0 | 0.890 | 0.854 | 0.850 | 0.771 | 0.817 | 0.890 | 0.582 | 0.797 |
| +1 | -1.00 | 0.849 | 0.689 | 0.784 | 0.798 | 0.746 | 0.799 | -1.00 |
| +2 | -1.00 | 0.774 | 0.621 | 0.867 | 0.761 | 0.813 | 0.675 | -1.00 |
| +3 | -1.00 | -1.00 | 0.718 | 0.720 | 0.537 | 0.597 | 0.664 | 0.825 |

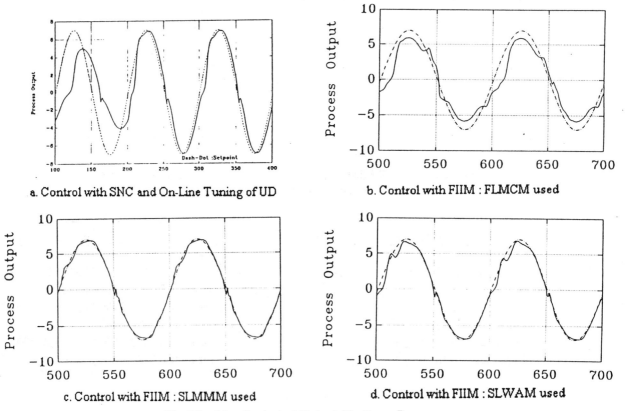

a. Control with SNC and On-Line Tuning of UD

b. Control with FIIM : FLMCM used

c. Control with FIIM : SLMMM used

d. Control with FIIM : SLWAM used

Fig.6 Tracking Control of Output Nonlinear Process

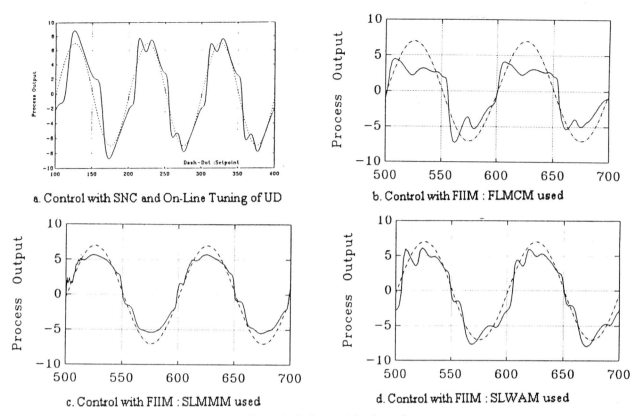

a. Control with SNC and On-Line Tuning of UD

b. Control with FIIM : FLMCM used

c. Control with FIIM : SLMMM used

d. Control with FIIM : SLWAM used

Fig.7 Tracking Control of Input Nonlinear Process

# Modeling Fuzzy Logic Controllers Having Noisy Inputs

Ronald R. Yager and Dimitar P. Filev
Machine Intelligence Institute, Iona College, New Rochelle, NY 10801

**ABSTRACT.** We suggest a new approach for reasoning with fuzzy models that have noise in the inputs. It is inspired by a Dempster-Shafer belief structure interpretation of the probabilistic signal. We obtain result that are computationally effective and can be useful in the applications of fuzzy controllers in a noisy environment.

## 1. Introduction

Fuzzy systems models requires a current reading of the input to be used in the inference process. In fuzzy systems models of the type used in fuzzy logic controllers it is usually assumed that the input variables take deterministic values. In a more realistic setting the sensor measurements are contaminated by noise.[1]. One approach to modeling this noise is to sample the input signal multiple times for each required reading. Thus the measured value of a required input is a a collection of values and it can be modelled by a histogram which defines the density function of the input to the model to be used . This type of model of the signal has a probabilistic nature and its accommodation in the environment of the reasoning mechanism of the fuzzy systems is not an evident task. In this paper we discuss the problem of reasoning in fuzzy systems that use this approach to handle noisy inputs. We provide a solution to this problem inspired by using the framework of Dempster-Shafer theory of evidence[$]. We also provide a method for the implementation of the results in engineering applications.

## 2. Representation of Noisy Inputs

Assume we have a single-input, single-output system described by a fuzzy model consisting of m rules of the form

**IF U is $B_i$ THEN V is $D_i$,**

where $B_i$ and $D_i$ are fuzzy subsets of the universes X and Y. We consider the case where input variable U includes noise and has been sampled. Thus each input value is not a single number, but a histogram as shown in Fig.1a.

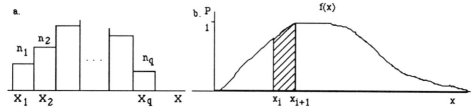

**Fig.1. Histogram (a.) and probability density function (b.) of the input signal.**

The histogram is essentially a manifestation of an underlying probability distribution. The information about the input signal resides in the probability density function, $f(x) = \{n_1/n, n_2/n, ..., n_q/n\}$, $n = \Sigma_i n_i$ , Fig.1b. We recall that the area under the curve is equal to one. Furthermore, since $P(x_i \leq x \leq x_{i+1}) = \int_{x_1}^{x_2} f(x)\, dx$ then the area under the curve between $x_i$ and $x_{i+1}$ is the probability that the input value x lies between $x_i$ and $x_{i+1}$. (See Fig. 1b.). One can obtain an approximation to this probability density function by its partitioning (Fig.2a). Using a disjoint partitioning we obtain a set of intervals $Q_1, Q_2, ..., Q_q$ and an set associated probabilities $p_1, p_2, ..., p_q$ where $\Sigma_i p_i = 1$, $p_i \in [0, 1]$, where $p_i$ is the probability that the input lies in the interval $Q_i$. An alternative representation can be obtained by a fuzzy partitioning of the support set of the probability density function. This fuzzy approximation consists of a collection of fuzzy subsets of real line, $Q_1, Q_2, ..., Q_q$ and associated probabilities $p_1, p_2, ..., p_q$. Thus $p_i$ is the probability fuzzy subset $Q_i$ will occur. In particular $p_i$ is the area under the curve indicating the probability that the input value x lies in the fuzzy interval $Q_i$. Essentially we are formally saying $P_i = P(x \in Q_i) = \int_{Q_i} f(x)\, dx.$

Using this formulation for the input noise it appears very natural to use the framework of the Dempster-Shafer [$] theory to represent the probabilistic input. The probabilistic input can be seen as a belief structure N with focal elements $Q_1, Q_2, ..., Q_q$ and weights $p_1, p_2, ..., p_q$. In the case of disjoint partitioning of the probability density function these focal elements are intervals, while in the case of fuzzy partitioning they are fuzzy sets. In both cases the focal elements, the $Q_j$'s, are characterized by their membership functions $Q_j(x)$ (in the case of disjoint partitioning these membership functions are rectangular with unit heights), and by their weights $p_i$'s. Figures 2b and 2c show the approximation of the probability density function by a belief structure with crisp or fuzzy focal elements. However, the belief structure representation of the probabilistic input is only an approximation of the probability density function of the input signal. This approximation is obtained by a disjoint or fuzzy quantization of the probability density function.

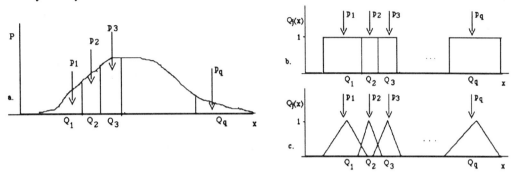

Fig.2. Probability density function of the input signal (a.) approximated as a belief structure; crisp (b.) and fuzzy (c.) partitioning with interval, respectively fuzzy focal elements Q1, ..., Qq and weights p1, ..., pq.

We now can use the concept of a partitioning of the input signal to express the output inferred by a fuzzy model assuming that the input is probabilistic i.e. described by a probability distribution, rather than by a single deterministic value. We shall demonstrate in the next section that the problem of reasoning in the fuzzy models that have inputs of this type can be described in the framework of the Dempster - Shafer theory of evidence.

## 3. Fuzzy Models with Noisy Inputs

In the following we shall look at this how the probabilistic input signal, approximated by the belief structure N, affects the output of the system described by a fuzzy model. Each rule of the rule-base of the fuzzy model,

**IF U is $B_i$ THEN V is $D_i$,**

when fired by by the belief structure N generates as its output a pseudo belief structure $M_i$. The focal elements of $M_i$ are denoted as $F_{ij}$, j = 1 to q, and the weights are the weights of the focal elements of N, $m_i(F_{ij}) = p_j$. The focal elements of $M_i$ are defined as

$$F_{ij} = \tau_{ij} \wedge D_i,$$

where $\tau_{ij}$ denotes the level of matching the ith antecedent $B_i$ with the $Q_j$th component of the input signal. We can calculate $\tau_{ij}$ by using the conditional possibility; $\tau_{ij} = Poss(B_i|Q_j) = Max_x (B_i(x) \wedge Q_j(x))$. In the case when the $Q_j$'s are intervals then $\tau_{ij}$ simply becomes the maximum membership grade of $B_i$ in the interval $Q_j$. Thus each focal element is an output of the rule for a particular $Q_j$.

To obtain the overall output F inferred by the fuzzy model we have to take a union of outputs of each of the individual rules, belief structures $M_i$,

$$F = M = \cup_i M_i$$

We recall that $M_i(F_{ij}) \in [0, 1]$ and $\sum_j M_i(F_{ij}) = 1$. In [4, 5] we applied a similar approach to obtain a belief structure interpretation of the fuzzy rules.

In [6-8] we discussed the process of combining of belief structures under various operations we draw upon this work for the following discussion. Because each of the $M_i$ are generated from the same belief structure, N, rather then being independent, as is usually assumed in work with belief structures, these belief structures are highly interrelated. This interrelationship leads to simpler combination rule for the formulation of the focal elements of M. Denoting the focal elements of M as $H_j$, j = 1 to q, we obtain that

$$H_j = \cup_i F_{ij} = \cup_i (\tau_{ij} \wedge D_i)$$

with weight $M(H_j) = p_j$.

It is easy to see that the jth focal element $H_j$ is actually the fuzzy output inferred by the model if we consider as input simply the jth input focal element $Q_j$.

As has become prevalent in most of the works on fuzzy logic modeling we shall use summation operator rather then the *max* to implement the union operator and the multiplicative operator rather then the *min* operator $\wedge$ to obtain an analytical expression for the system output $y^*$. Obviously the summation brings the output of the fuzzy model F out of the unit interval. However, as we shall see it doesn't have effect on the defuzzified value due to the normalization when calculating the centroid. In [9] we describe this as a kind of soft union. Using this operator we get

$$H_j(y) = \Sigma_i \, F_{ij}(y) = \Sigma_i \, \tau_{ij} \, D_i(y)$$

with weights $M(H_j) = p_j$.

Thus as a result of this third step we obtain a Dempster-Shafer structure M as our output of the fuzzy system model where the focal elements of M are the subsets $H_j$ with weights $M(H_j) = p_j$.

The next step in the procedure is to apply the defuzzification process to M to obtain the singleton output $y^*$. The procedure used first obtains the defuzzified value for each focal element of M. Let us denote the defuzzified value of the fuzzy set $H_j$ by $d_j$. Then the final step is to obtain the overall system output $y^*$ as the expected value of these defuzzified values,

$$y^* = \sum_{j=1}^{q} d_j \, M(H_j) = \sum_{j=1}^{q} d_j \, p_j \qquad (I)$$

Using the simplified (RDFR) method of reasoning for each focal element $H_j$ we calculate its defuzzified value $d_j$ as follows

$$d_j = \frac{\sum_{i=1}^{m} \tau_{ij} \, y_i^*}{\sum_{i=1}^{m} \tau_{ij}}$$

where $y_i^*$ is the centroid of the consequent of the ith rule, the $D_i$. We then use these defuzzified values $d_j$'s to obtain the defuzzified output of the system $y^*$ as

$$y^* = \sum_{j=1}^{q} d_j \, M(H_j) = \sum_{j=1}^{q} \frac{\sum_{i=1}^{m} \tau_{ij} \, y_i^*}{\sum_{i=1}^{m} \tau_{ij}} \, p_j. \qquad (II)$$

Thus $y^*$ is essentially the expected value of the defuzzified values of the focal elements of M over probabilities $p_j$'s.

We shall illustrate the above reasoning process in the following example.

**Example 1.** Assume a single-input single-output fuzzy model consisting of two rules

**IF** U is $B_1$ **THEN** V is $D_1$

**IF** U is $B_2$ **THEN** V is $D_2$

The input probability density function is approximated by partitioning into three intervals $Q_1$, $Q_2$ and $Q_3$ with probabilities $p_1$, $p_2$ and $p_3$ (Fig.3); this yields a belief structure N with focal elements $Q_1$, $Q_2$ and $Q_3$ and weights $p_1$, $p_2$ and $p_3$.

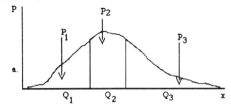 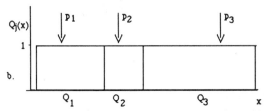

**Fig.3. Probability density function (a.) of the input and its belief structure approximation (b.). Focal elements $Q_1$, $Q_2$, $Q_3$ are intervals.**

Applying this belief structure to the system model induces two output belief structure $M_1$ and $M_2$ with focal elements $F_{ij}$

$M_1$: $F_{11} = \tau_{11} \, D_1$, $M_1(F_{11}) = p_1$; $F_{12} = \tau_{12} \, D_1$, $M_1(F_{12}) = p_2$, $F_{13} = \tau_{13} \, D_1$, $M_1(F_{13}) = p_3$.

$M_2$: $F_{21} = \tau_{21} \, D_2$, $M_2(F_{21}) = p_1$; $F_{22} = \tau_{22} \, D_2$, $M_2(F_{22}) = p_2$; $F_{23} = \tau_{23} \, D_2$, $M_2(F_{23}) = p_3$.

where $\tau_{1j} = \text{Max}_x \, (B_1(x) \wedge Q_j(x))$ and $\tau_{2j} = \text{Max}_x \, (B_2(x) \wedge Q_j(x))$, j=1, 2, 3.

The aggregation of these outputs leads to a belief structure M with three focal elements $H_1 = \text{Agg}(F_{11},$

$F_{21}$), $H_2 = \text{Agg}(F_{12}, F_{22})$, $H_3 = \text{Agg}(F_{13}, F_{23})$. By application of the summation operator for calculating the union of the focal elements followed by defuzzification of the $H_j$'s via the center of area method we obtain:

$$H_1(y) = F_{11}(y) + F_{21}(y); \qquad d_1 = \frac{\tau_{11} y_1^* + \tau_{21} y_2^*}{\tau_{11} + \tau_{21}} \qquad \text{with } M(H_1) = p_1$$

$$H_2(y) = F_{12}(y) + F_{22}(y); \qquad d_2 = \frac{\tau_{12} y_1^* + \tau_{22} y_2^*}{\tau_{12} + \tau_{22}} \qquad \text{with } M(H_2) = p_2$$

$$H_3(y) = F_{13}(y) + F_{23}(y); \qquad d_3 = \frac{\tau_{13} y_1^* + \tau_{23} y_2^*}{\tau_{13} + \tau_{23}} \qquad \text{with } M(H_3) = p_3$$

where $y_1^*$ and $y_2^*$ are the centroids of the sets $D_1$ and $D_2$.

The final step is to obtain the overall system output, $y^*$, as the expected value of these individual rule outputs

$$y^* = p_1 d_1 + p_2 d_2 + p_3 d_3 = \frac{\tau_{11} y_1^* + \tau_{21} y_2^*}{\tau_{11} + \tau_{21}} p_1 + \frac{\tau_{12} y_1^* + \tau_{22} y_2^*}{\tau_{12} + \tau_{22}} p_2 + \frac{\tau_{13} y_1^* + \tau_{23} y_2^*}{\tau_{13} + \tau_{23}} p_3$$

## 4. Multiple Probabilistic Inputs

In the following we deal with the case when the system has multiple inputs. For notational simplification we assume a system with two input $U_1$ and $U_2$ and a fuzzy model consisting of the family of rules:

**IF** $U_1$ is $B_{i1}$ **AND** $U_2$ is $B_{i2}$ **THEN** V is $D_i$, $\qquad i = (1, m)$

Let $\{Q_1, Q_2, ..., Q_q\}$ and $\{s_1, s_2, ..., s_q\}$, and $\{R_1, R_2, ..., R_r\}$ and $\{t_1, t_2, ..., t_r\}$, be the focal elements and the weights of the belief structures N and L approximating the probability density functions of the inputs $U_2$ and $U_2$. Applying these two independent belief structures as inputs to the ith rule induces as output the belief belief structure $M_i$ with focal elements $F_{ij}$ and weights $p_j$ expressed as follows.

$$\tau_{i1} = \text{Poss}(B_{i1}|Q_1) \wedge \text{Poss}(B_{i2}|R_1) \qquad F_{i1} = \tau_{i1} \wedge D_i \qquad \text{with } p_1 = s_1 t_1$$
$$\tau_{i2} = \text{Poss}(B_{i1}|Q_1) \wedge \text{Poss}(B_{i2}|R_2) \qquad F_{i2} = \tau_{i2} \wedge D_i \qquad \text{with } p_2 = s_1 t_2$$
$$. . .$$
$$\tau_{ir} = \text{Poss}(B_{i1}|Q_1) \wedge \text{Poss}(B_{i2}|R_r) \qquad F_{ir} = \tau_{ir} \wedge D_i \qquad \text{with } p_r = s_1 t_r \qquad \text{(III)}$$
$$\tau_{ir+1} = \text{Poss}(B_{i1}|Q_2) \wedge \text{Poss}(B_{i2}|R_1) \qquad F_{ir+1} = \tau_{ir+1} \wedge D_i \qquad \text{with } p_{r+1} = s_2 t_1$$
$$. . .$$
$$\tau_{ig} = \text{Poss}(B_{i1}|Q_q) \wedge \text{Poss}(B_{i2}|R_r) \qquad F_{ig} = \tau_{ig} \wedge D_i \qquad \text{with } p_g = s_q t_r$$

where $g = q\,r$ is the product of the cardinalities q and r of the partitioning the probability densities of $U_1$ and $U_2$. The output of the fuzzy model is a belief structure M which is a union of the belief structures $M_i$'s:

$$M = \cup_i M_i$$

The belief structure M will have $g = q\,r$ focal elements $H_k$'s defined as follows:

$$H_k = \cup_i F_{ik} = \cup_i (\tau_{ik} \wedge D_i) \quad \text{with weight } M(H_k) = p_k.$$

Let $d_k$ be the centroid of the focal element $H_k$. Then the overall system output $y^*$ as the expected value of these centroids $d_k$'s:

$$y^* = \sum_{k=1}^{g} d_k M(H_k) = \sum_{k=1}^{g} d_k p_k \qquad \text{(IV)}$$

We can see that the result obtained for the case of multiple inputs is formally the same as that derived for the case of a single input (I). Analogously, by replacing of the *max* interpretation of the union by the summation and the *min* operator by the mutiplicative operator $\wedge$, we obtain the same expressions for the focal elements $H_k$'s of the belief structure M

$$H_k(y) = \Sigma_i F_{ik}(y) = \Sigma_i (\tau_{ik} D_i(y)) \quad \text{with weight } M(H_k) = p_k.$$

Similarly by using the RDFR method of reasoning we get for the output inferred by the fuzzy system:

$$y^* = \sum_{k=1}^{g} d_k M(H_k) = \sum_{k=1}^{g} \frac{\sum_{i=1}^{m} \tau_{ik} y_i^*}{\sum_{i=1}^{m} \tau_{ik}} p_k \qquad \text{(V)}$$

The following example illustrates the technique just described.

**Example 2.** Consider a fuzzy model consisting of two rules:

$$\textbf{IF } U \text{ is } B_{11} \textbf{ AND } U_2 \text{ is } B_{12} \textbf{ THEN } V \text{ is } D_1$$
$$\textbf{IF } U \text{ is } B_{21} \textbf{ AND } U_2 \text{ is } B_{22} \textbf{ THEN } V \text{ is } D_2$$

The input probability density function of the first input $U_1$ is approximated by partitioning into two intervals $Q_1$, $Q_2$ with probabilities $s_1$, $s_2$; analogously the density function of the second input $U_2$ is partitioned into two intervals $R_1$, $R_2$ with probabilities $t_1$, $t_2$. These two belief structures fire the rules and induce belief structures $M_1$ and $M_2$ with focal elements $F_{1j}$, $F_{2j}$ and weights $p_j$ as follows:

$$\tau_{11} = Poss(B_{11}|Q_1) \wedge Poss(B_{12}|R_1) \qquad F_{11} = \tau_{11} \wedge D_1 \qquad \text{with } p_1 = s_1\, t_1$$
$$\tau_{12} = Poss(B_{11}|Q_1) \wedge Poss(B_{12}|R_2) \qquad F_{12} = \tau_{12} \wedge D_1 \qquad \text{with } p_2 = s_1\, t_2$$
$$\tau_{13} = Poss(B_{11}|Q_2) \wedge Poss(B_{12}|R_1) \qquad F_{13} = \tau_{13} \wedge D_1 \qquad \text{with } p_3 = s_2\, t_1$$
$$\tau_{14} = Poss(B_{11}|Q_2) \wedge Poss(B_{12}|R_2) \qquad F_{14} = \tau_{14} \wedge D_1 \qquad \text{with } p_4 = s_2\, t_2$$

$$\tau_{21} = Poss(B_{21}|Q_1) \wedge Poss(B_{12}|R_1) \qquad F_{21} = \tau_{21} \wedge D_2 \qquad \text{with } p_1 = s_1\, t_1$$
$$\tau_{22} = Poss(B_{21}|Q_1) \wedge Poss(B_{12}|R_2) \qquad F_{22} = \tau_{22} \wedge D_2 \qquad \text{with } p_2 = s_1\, t_2$$
$$\tau_{23} = Poss(B_{21}|Q_2) \wedge Poss(B_{12}|R_1) \qquad F_{23} = \tau_{23} \wedge D_2 \qquad \text{with } p_3 = s_2\, t_1$$
$$\tau_{24} = Poss(B_{21}|Q_2) \wedge Poss(B_{12}|R_2) \qquad F_{24} = \tau_{24} \wedge D_2 \qquad \text{with } p_4 = s_2\, t_2$$

The belief structure $M$ associated with the overall output will have 4 focal elements $H_k$'s:

$$H_k = F_{1k} \cup F_{2k} = (\tau_{1k} \wedge D_1) \cup (\tau_{2k} \wedge D_2) \quad \text{with weight } M(H_k) = p_k.$$

Let $d_k$ be the centroid of the focal element $H_k$. Then the overall system output $y^*$ is the expectation of these centroids $d_k$'s over the probabilities (weights) $p_1, ..., p_4$:

$$y^* = d_1\, p_1 + d_2\, p_2 + d_3\, p_3 + d_4\, p_4$$

## 5. Simplified Reasoning with Probabilistic Inputs.

We observe that according to expressions (IV) or (V) to calculate the output of a fuzzy model in the presence of two probabilistic inputs whose probability density functions are partitioned into q and r fuzzy or crisp intervals we need to repeat the calculation qr times for all possible combinations of the input components and then to take the expected value of these qr output values. The number of calculations drastically increases with increasing the number of input variables. For this reason we shall look for a simplification of the results (I), (II) and -(IV), (V).

In both cases of single and multiple inputs we found that the focal elements $H_j$'s of the belief structure $M$ are $H_j(y) = \sum_i F_{ij}(y) = \sum_i (\tau_{ij}\, D_i(y))$ where $M(H_j) = p_j$. At first we shall discuss the single input case. Instead of calculating the crisp output of the fuzzy model as the expected value of the defuzzified values $d_j$'s associated with each of these focal elements $H_j$'s, we shall calculate first the weighted average of all the $H_j$'s over the weights $M(H_j) = p_j$ and then we shall perform the defuzzification step on this weighted average. The weighted average $H$ of the focal elements is:

$$H = \sum_j H_j\, p_j = \sum_j \sum_i F_{ij}\, p_j = \sum_j \sum_i \tau_{ij}\, D_i\, p_j = \sum_i \sum_j \tau_{ij}\, p_j\, D_i = \sum_i \bar{\tau}_i\, D_i$$

where $\bar{\tau}_i$

$$\bar{\tau}_i = \sum_j \tau_{ij}$$

is the expected value of all firing levels $\tau_{ij}$'s of the ith rule considering each focal element $Q_j$ of the input belief structure. Then the centroid defuzzified value of $H$ is obtained by the application of the RDFR method of reasoning:

$$y^* = \frac{\displaystyle\sum_{i=1}^{m} \bar{\tau}_i\, y_i^*}{\displaystyle\sum_{i=1}^{m} \bar{\tau}_i} \tag{VI}$$

The above result will be extended to the multiple input case. In addition to the assumptions made before, we shall assume the the firing strength of the rules is determined by the product of the levels of matching the antecedent fuzzy sets by the focal elements, the $Q_j$'s, rather than by their minimum, $\tau_{ik} = Poss(B_{i1}|Q_j) \cdot Poss(B_{i2}|R_p)$, where index k takes values from 1 to g, index j - from 1 to q and p - from 1 to r (see the family of expressions (III) describing the belied structures $M_i$'s for details). In a similar manner we obtain the weights $p_i$'s (see also (III)), $p_k = s_j\, t_p$. Therefore for each focal element $F_{ik}$ of the belief structure $M_i$ we have in the case of two probabilistic inputs:

$$F_{ik} = Poss(B_{i1}|Q_j) \cdot Poss(B_{i2}|R_p)\, D_i \quad \text{with weight } p_k = s_j\, t_p$$

where indexes k, j and p take the same values as in the above two expressions. The focal elements $H_k$'s of the belief structure M are

$$H_k = \Sigma_i\ F_{ik} = \Sigma_i\ \tau_{ik}\ D_i = \Sigma_i\ Poss(B_{i1}|Q_j) \cdot Poss(B_{i2}|R_p) \cdot D_i \quad \text{with weight } p_k = s_j\ t_p$$

Their weighted average is H:

$$H = \Sigma_k\ H_k\ p_k = \Sigma_k\ \Sigma_i\ F_{ik}\ p_k = \Sigma_k\ \Sigma_i\ \tau_{ik}\ D_i\ p_k = \Sigma_i\ \Sigma_k\ \tau_{ik}\ p_k\ D_i$$

$$= \Sigma_i\ \Sigma_j\ \Sigma_p Poss(B_{i2}|Q_j) \cdot Poss(B_{i2}|R_p)\ s_j\ t_p\ D_i = \Sigma_i\ \Sigma_j\ \Sigma_p Poss(B_{i1}|Q_j)\ s_j \cdot Poss(B_{i2}|R_p)\ t_p\ D_i$$

$$= \Sigma_i\ (\Sigma_j\ Poss(B_{i1}|Q_j)\ s_j) \cdot (\Sigma_p\ Poss(B_{i2}|R_p)\ t_p)\ D_i = = \Sigma_i\ \overline{\tau}_i\ D_i$$

where $\overline{\tau}_i$ is the firing strength of the ith rule:

$$\overline{\tau}_i = (\Sigma_j\ Poss(B_{i1}|Q_j)\ s_j) \cdot (\Sigma_p\ Poss(B_{i2}|R_p)\ t_p)$$

We see that the firing strength of a rule $\overline{\tau}_i$ in the case of two probabilistic inputs is calculated as a product of the expectations of the levels of matching the antecedents fuzzy sets $B_{i1}$ and $B_{i2}$ by the focal elements $Q_1, Q_2, ..., Q_q$ and $R_1, R_2, ..., R_r$ over the respective probabilities $s_1, s_2, ..., s_q$ and $t_1, t_2, ..., t_r$.

The crisp output is calculated as a centroid of the H which yields expression (VI).

Using the above result we obtain for the output y* considered in the Example 2:

$$y^* = (\overline{\tau}_1\ y_1^* + \overline{\tau}_2\ y_2^*) / (\overline{\tau}_1 + \overline{\tau}_2)$$

where

$$\overline{\tau}_1 = [Poss(B_{11}|Q_1)\ s_1 + Poss(B_{11}|Q_2)\ s_2] \cdot [Poss(B_{12}|R_1)\ t_1 + Poss(B_{12}|R_2)\ t_2]$$

$$\overline{\tau}_2 = [Poss(B_{21}|Q_1)\ s_1 + Poss(B_{21}|Q_2)\ s_2] \cdot [Poss(B_{22}|R_1)\ t_1 + Poss(B_{22}|R_2)\ t_2]$$

Generally, when we have r probabilistic inputs $U_1, U_2, ..., U_r$ whose probability density functions are partitioned into q intervals $Q_{11}, ..., Q_{1q}; Q_{21}, ..., Q_{2q}; ...; Q_{r1}, ..., Q_{rq}$ with probabilities $p_{11}, ..., p_{1q}; p_{21}, ...,$ $p_{2q}; ...; p_{r1}, ..., p_{rq}$ we calculate the output via expression (VI):

$$y^* = \frac{\sum_{i=1}^{m} \overline{\tau}_i\ y_i^*}{\sum_{i=1}^{m} \overline{\tau}_i}$$

where the firing levels $\overline{\tau}_i$'s are obtained from the expectations of the levels of matching the antecedents by the respective intervals to which the inputs are partitioned:

$$\overline{\tau}_i = (\Sigma_j\ Poss(B_{i1}|Q_{1j})\ p_{1j}) \cdot (\Sigma_j\ Poss(B_{i2}|Q_{2j})\ p_{2j}) \cdot ... \cdot (\Sigma_j\ Poss(B_{ir}|Q_{rj})\ p_{rj})$$

## 6. Conclusion.

We derived a new approach for reasoning in fuzzy models that have probabilistic inputs. It utilizes a belief structure interpretation of the probabilistic signal. We obtained result that is computationally effective and can be performed in on-line calculations as well.

## 7. References

[1]. Berenji, H.R., "Fuzzy logic controllers", In: An Introduction to Fuzzy Logic Applications in Intelligent Systems, edited by Yager, R.R. and Zadeh, L.A., John Wiley & Sons, New York, 69-96, 1991

[2]. Dempster, A. P., "Upper and lower probabilities induced by a multi-valued mapping," Ann. of Mathematical Statistics 38, 325-339, 1967.

[3]. Shafer, G., A Mathematical Theory of Evidence, Princeton University Press: Princeton, N.J., 1976.

[4]. Yager, R. R. and Filev, D. P., "Including Probabilistic Uncertainty in Fuzzy Logic Controller Modeling Using Dempster-Shafer Theory," Technical Report # MII-1309, Machine Intelligence Institute, Iona College, New Rochelle, NY, 1993.

[5]. Yager, R. R. and Filev, D. P., "Template based fuzzy systems modeling" Technical Report # MII-1310, Machine Intelligence Institute, Iona College, New Rochelle, NY, 1993.

[6]. Yager, R. R., "Arithmetic and other operations on Dempster-Shafer structures," Int. J. of Man-Machine Studies 25, 357-366, 1986.

[7]. Yager, R. R., "Quasi-associative operations in the combination of evidence," Kybernetes 16, 37-41, 1987.

[8]. Yager, R. R., "On the Dempster-Shafer framework and new combination rules," Information Sciences 41, 93-137, 1987.

[9]. Yager, R. R. and Filev, D. P., "Fuzzy logic controllers with flexible structures," Proceedings Second International Conference on Fuzzy Sets and Neural Networks, Iizuka, Japan, 317-320, 1992.

# Design of Sophisticated Fuzzy Logic Controllers Using Genetic Algorithms

Kim Chwee Ng  and Yun Li
Department of Electronics and Electrical Engineering
University of Glasgow, Rankine Building
Glasgow  G12 8LT, Scotland, U.K.

*Abstract -* Design of fuzzy logic controllers encounters difficulties in the selection of optimized membership functions and fuzzy rule base, which is traditionally achieved by a tedious trial-and error process. This paper develops genetic algorithms for automatic design of high performance fuzzy logic controllers using sophisticated membership functions that intrinsically reflect the nonlinearities encounter in many engineering control applications. The controller design space is coded in base-7 strings (chromosomes), where each bit (gene) matches the 7 discrete fuzzy value. The developed approach is subsequently applied to design of a proportional plus integral type fuzzy controller for a nonlinear water level control system. The performance of this control system is demonstrated higher than that of a conventional PID controller. For further comparison, a fuzzy proportional plus derivative controller is also developed using this approach, the response of which is shown to present no steady-state error.

## I. INTRODUCTION

Modern control theory has been successful for well defined, either deterministically or stochastically, systems. This approach, however, encounters problems in many engineering applications where systems to be controlled are difficult to model, have a strong nonlinearity or are embedded in a changing environment with uncertainty. With the development of modern information processing technology and computational intelligence, an alternative solution to these problems has been to incorporate human intelligence directly into automatic control systems. These intelligent control schemes tend to imitate the way of human decision making and knowledge representation and have received increasing attention widely across the control community in the world. It has been shown, in applications such as robot control, automotive systems, aircraft, spacecraft and process control, that they offer potential advantages over conventional control schemes in (a) less dependency on quantitative models; (b) natural decision making; (c) learning capability; (d) a greater degree of autonomy; (e)

ease of implementation and (f) friendly user interface [1~3].

Incorporating the uncertainty and abstract nature inherent in human decision making into intelligent control systems, a fuzzy logic controller (FLC) offers a more accurate and efficient approach, which tends to capture the approximate and qualitative *boundary conditions* of system variables (in contrast to the probability theory that deals with random *behaviour*) by fuzzy sets with a membership function. Such a system flexibly implements functions in near human terms, i.e. IF-THEN linguistic rules, with reasoning by fuzzy logic, which is a rigorous mathematical discipline. It has been demonstrated that fuzzy logic control systems are reliable and robust and are straightforward to implement [1~3].

The crux of designing an FLC lies, however, in the selection of high-performance membership functions that represents the human expert's interpretation of the linguistic variables, because different membership functions determine different extent to which the rules affect the action and hence the performance. The existing iterative approaches for choosing the membership functions are basically a manual trial-and-error process and lack learning capability and autonomy. Therefore, the more efficient and systematic genetic algorithm (GA) [4], which acts on the survival-of-the-fittest Darwinian principle for reproduction and mutation, has recently been applied to FLC design for searching the poorly understood, irregular and complex membership function space with improved performance. Successful application of this approach has been demonstrated in spacecraft rendezvous [5], cart-pole balancing [6], linear motion of cart [7], three-term control [8] and pH value control [9], where the FLC membership functions are represented by internally linear triangular/trapezoidal shapes which are mainly encoded in binary numbers for optimal searching by GAs.

This paper develops genetic algorithms for designing fuzzy logic controllers using sophisticated membership functions that intrinsically reflect the nonlinearity encountered in many engineering control problems. Since the coding parameters are increased in these FLCs and decision making of most FLCs are based on seven discrete values in one dimension, the base-7 coding is used for the coding process. The developed approach is subsequently applied to design two FLCs for a nonlinear coupled tank water level control system and their performances are compared with conventional PID control schemes.

## II. FUZZY LOGIC CONTROLLER

A schematic of a fuzzy control system for a regulation control task is shown in fig 1. The fuzzy logic controller relates the control variables (error and change_in_error) being translated by a component known as a condition interface into fuzzy linguistic terms which are specified by the membership functions of the fuzzy sets. In order to reflect the fuzzy nature of the seven linguistic classifications and to allow for the best options for high performance design, this paper uses the symmetrical exponential membership functions given by :

$$\mu_{\pm i}(x) = \exp\left(-\frac{|x \mp \alpha_i|^{\beta_i}}{\sigma_i}\right) \quad (1a)$$

$$\forall \, x \in [\text{- large}, \text{+ large}]$$

where

$i = \{\text{zero}, \pm \text{small}, \pm \text{medium}, \pm \text{large}\}$,

and

$$\mu_{+\text{large}} = 1 \quad x > \alpha_{+\text{large}} \quad (1b)$$

$$\mu_{-\text{large}} = 1 \quad x < \alpha_{-\text{large}} \quad (1c)$$

Fig. 1  A fuzzy controller

An example of the shapes of the membership functions of the error and change_in_error-error variable is shown in fig 2. Here, $\alpha_i$ is the position parameter which describes the centre point of the membership function along the universe of discourse, $\beta_i \in [1.5, 5.0]$ is the "shape parameter" which resembles the shapes from a triangular to a trapezoidal, and $\sigma_i \in [0.1, 3.0]$ is the "scale parameter" which modifies the base-length of the membership functions and determines the amount of overlapping. Note that a small overlapping is necessary for a distinctive fuzzy decision making. Note also that in (1a), $\beta_i$ is not included in the power of $\sigma_i$ and this will allow for a gradual and consistent change in the base-length.

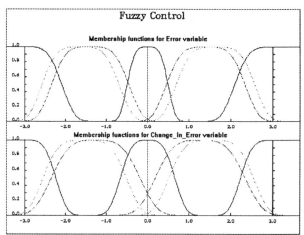

Fig. 2  Symmetrical exponential membership functions

In fuzzy control, the linguistic terms which are defined on the appropriate universe of discourse evaluate the control rules using the compositional rules of inference. The result of application of the rules is a fuzzy set defined on the universe of possible control actions and in turn, an appropriately computed control action is determined and then reconverted to the crisp value to regulate the process. Therefore, the essential design steps in designing fuzzy controllers include 1) defining input and output variables, 2) specifying all the fuzzy sets and their membership functions defined for each input and output variable, 3) converting the input variables to fuzzy sets, 4) compilation of an appropriate and complete set of heuristic control rules that operate on these fuzzy sets, i.e. formulating the fuzzy rule-base, 5) designing the computational unit that accesses the fuzzy rules and computes for the fuzzy control action, 6) devising a transformation method for converting fuzzy control action to crisp value.

The major task in the design of a fuzzy controller lies in the optimal choice of the membership functions, or the $\alpha$, $\beta$ and $\sigma$ parameters in the case of membership functions given in (1). In manual design, these functions and the rule-sets are usually obtained by a tedious trial-and-error process which after does not go through the entire possible solution space and therefore does not result in an optimal design. This is also the reason that manually design FLCs compromise accuracy with simplicity by using pure triangular and trapezoidal membership functions.

## III. DESIGN OF SOPHISTICATED FLCs USING GAs

The genetic algorithms developed by Holland [4] is to simulate the natural evolution process that operates on chromosomes. The simple genetic algorithm that yields satisfactory results in many practical problems is done by using three operators: (1) reproduction; (2) crossover; and (3) mutation in the following process.

> Make initial population
> **REPEAT**
> > Choose parents from the population;
> > Selected parents produce children with the number weighted by their individual fitness;
> > Extend the population with the children;
> > Select fittest elements of the extended population to survive for the next cycle
> **UNTIL** satisfactory generation found
> Decode the optimum population (and form the final FLC)

By coding the coefficients ($\alpha$, $\beta$ & $\sigma$) of the membership functions, the fuzzy logic rule-set and the gains of error and change_in_error into a decimal-number string, FLC design can be developed and optimised by using GAs. The coded FLC design population can be found by the entire space of the strings termed "chromosomes", each of which was randomly generated "bits", termed "genes". Then the GA process is used to reproduce and select the "fittest" individual, i.e., the "optimal" solution to designing FLCs. A FLC design process is usually complicated, nonlinear and poorly understood and GAs has been proven to be an extremely efficient searching tool for such a process. It is also shown that such a searching technique is robust and convergies faster than conventional searching algprithms.

Small alterations had been made to Goldberg's general-purpose GA [10] to include an adaptive mutation method and the selection of the fittest elements from the new and previous generations to survive. In the adaptive mutation method, an identical string of chromosome is prevented in the new generation by increasing the mutation rate based on the similarity compared to their parents after the process of crossover. With this conception, the maximum mutation rate is limited to 0.2 and the lowest is at 0.03.

With a concatenated , mapped, unsigned coding method, the coefficients ($\alpha$, $\beta$ and $\sigma$) and the gains of the error and change_in_error ($K_1$ and $K_2$) are coded with values mapped form a minimum value $C_{min}$ to a maximum value $C_{max}$ using an $n$-bit, unsigned base-7 integer starting from 0. The decoding mapping is given by

$$C = C_{\min} + \left(C_{\max} - C_{\min}\right) \times \frac{string\_val}{7^n - 1} \qquad (2)$$

where the *string_val* is the base-7 value represented by an *n*-bit string, and $C$ is the decimal (real) value being coded. The choice of the specific number of bits $n$ used to represent each sub-string variable is dependent on the resolution required in its variation. An example of a complete chromosome string is shown in fig 3, where sub-string groups A, B, C, D and E represent the rule-set for the fuzzy rule-base, scaling parameters ($\sigma$), position parameters ($\alpha$), gains ($K$) and shape parameters ($\beta$).

Fig. 3 A coded chromosome string and it's partitioned sub-strings

The fuzzy rule-set focuses 7x7 possible control actions corresponding to values in input error and change_in_error and therefore 49 bits are used in string group A to form the look-up table, where a single bit represents each control action. This illustrated in Fig 4.

In sub-string B, each 2 bit group represents the value of $\sigma_{\pm B}$, $\sigma_{\pm M}$, $\sigma_{\pm S}$ and $\sigma_{ZO}$ to be used for modifying the scaling factors of the error and change_in_error membership functions, respectively. The next sub-string of 8 integers are coded for defining the positions of the fuzzy sets "small" and "medium" along the universe of discourse, whilst the positions of "Big" and "Zero" are fixed, i.e. $\alpha_B = 3.0$ and $\alpha_{ZO} = 0.0$. Again, each $\alpha$ parameter will require two bits. Sub-string D represents $K_1$ and $K_2$ used as gains of the error and change_in_error, with three bits assigned to each. The final group of 8 integer characters is coded for the shape coefficients $\beta$ of both the error and change_in_error variables, requiring one bit for each parameter.

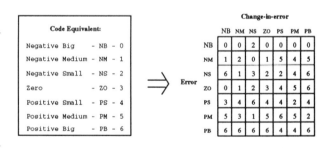

Fig. 4 Fuzzy rule-set look-up table for control actions

For each individual chromosome (a complete string) in the population, it is necessary to establish a measure of its fitness, $f(x)$, that is often used to accurately evaluate the performance of the controller and will be used to generate a probability according to which the individual in question will be selected for reproduction. However, the task of defining a fitness function is always application specific. In this paper, the objective of the controller is to drive the output of the process to the desired set-point in the shortest time possible and to maintain the output at the desired set-point, which is evaluated by:

$$x = \sum_{n=0}^{finish\_time} \left\{ n\,e_n^2 + n(\Delta e_n)^2 \right\}$$

$$f(x) = \exp\left( -\sqrt{\frac{x}{finish\_time}} \right) \qquad (3)$$

where $n$ is the time index, $e$ the error, $\Delta e$ the change_in_error and the *finish_time* in the following implementation is 300.

For a given set of membership functions, error energies were calculated with the intent of using the GA to minimise it. This fitness function provides a mean for evaluating the performance of each FLC using different fuzzy membership functions and rules-base being selected, so that an optimised FLC would be developed.

### IV. IMPLEMENTATION ON A NONLINEAR SYSTEM

The approach developed in the previous section is programmed in Pascal and is then used to design an FLC (of proportional plus integral type) for a nonlinear twin-tank coupled water level control system. The membership functions designed by the GA are shown in Fig. 2, whose parameters are given below:

|          | For error | For change in error |
|----------|-----------|---------------------|
| Position | $\alpha_S = 1.40$ | $\alpha_S = 1.27$ |
|          | $\alpha_M = 1.63$ | $\alpha_M = 1.53$ |
| Shape    | $\beta_{ZO} = 4.0$ | $\beta_{ZO} = 3.85$ |
|          | $\beta_S = 4.0$ | $\beta_S = 3.58$ |
|          | $\beta_M = 4.0$ | $\beta_M = 4.0$ |
|          | $\beta_B = 4.0$ | $\beta_B = 4.0$ |
| Scale    | $\sigma_{ZO} = 0.09$ | $\sigma_{ZO} = 0.33$ |
|          | $\sigma_S = 1.11$ | $\sigma_S = 1.88$ |
|          | $\sigma_M = 1.11$ | $\sigma_M = 1.51$ |
|          | $\sigma_B = 0.98$ | $\sigma_B = 0.50$ |

It can be inferred that manually designed membership functions could not be as sophisticated as those shown in Fig. 2 and would not ultimately lead to optimised results by trial-and-error. The response of the water level of one tank to a step with amplitude 75mm is given by curve (1) in Fig. 5, using this PI type FLC designed automatically by a genetic algorithm. For comparison, the performance of a conventional PID controller is shown in curve (2), whose gains were

initially determined by the Ziegler-Nichols rule and further manually tuned to their best performance. As can be seen, the performance achieved by the genetic FLC is apparently superior to that obtained from this PID controller, with both the overshoot and decay rate being improved.

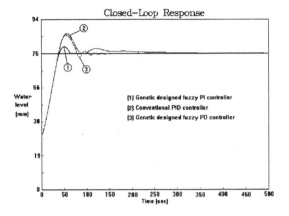

Figure 5 Performance comparison between GA designed PI and PD FLCs and manually designed PID controller

The GA approach developed in the previous section can also be extended to other designs. For comparison purpose, it has been used to design an FLC in a traditional and simplest way, where the control action generated obeys the PD instructions, rather than the PI instructions. The performance is shown in curve (3) of Fig. 5, which is slightly better than that of the manually tuned PID controller. It is interesting to note that there is no steady-state error resulting from this fuzzy PD controller, whilst this is not the case for a conventional PD controller (results not shown in Fig. 5).

### V. CONCLUSION AND FUTURE WORK

GAs have been proven to be an extremely efficient and robust searching tool for complicated and poorly understood processes. It is also shown in the literature that such a searching technique converges intelligently and much faster than conventional learning means.

Genetic algorithms for automatic design of fuzzy logic controllers have been developed, using sophisticated membership functions that intrinsically reflect the nonlinearity encountered in many engineering applications. Utilising these sophisticated membership functions, the control laws can be implemented in a simple format, such as PI or PD scheme. The sophistication obtained by the machine based automatic design could not be reached by manual design which is exclusively based on a painstaking trial-and-error process. The genetic design approach discussed in this paper offers a convenient and complete way to design a fuzzy controller in the shortest time.

Further work underway includes on-line design of adaptive FLCs. For such a genetic-fuzzy controllers, parallel architectures are currently studied in order to provide a high throughput rate for the control signals with short system latency, whilst performing the adaptation tasks.

## ACKNOWLEDGEMENT

Mr. Ng is grateful for the University of Glasgow and CVCP for their support in the form of a Postgraduate Scholarship and Overseas Research Scheme award. The authors would like to thank their colleague, Dr. Ken Sharman, for useful discussions on evolutionary algorithms.

## REFERENCES

1. Special Issue on Intelligent Control, *IEEE Control Systems*, vol.13, no.3, June 1993.
2. K.C. Ng and Y. Li, *Application of Genetic Algorithms to Design of Fuzzy Logic Controllers*, Internal Report, Department of Electronics and Electrical Engineering, University of Glasgow, Aug. 1993.
3. E. Rogers and Y. Li., Eds., *Parallel Processing in a Control Systems Environment*, London: Prentice Hall International, May 1993.
4. J.H. Holland, "Genetic algorithms," *Scientific American*, pp.44-50, July 1992.
5. C.L. Karr, L.M. Freeman and D.L. Meredith, "Genetic algorithm based fuzzy control of spacecraft autonomous rendezvous," *Proc. 5th Conf. on Artificial Intelligence for Space Applications*, 1990, ch62 3073, pp.43-51.
6. C.L. Karr, "Design of an adaptive fuzzy logic controller using a genetic algorithm," *Proc. 4th Int. Conf. on Genetic Algorithms*, 1991, pp.450-457.
7. A. Homaifar and E. McCormick, "Full design of fuzzy controllers using genetic algorithms," *Proc. SPIE Conf. on Neural and Stochastic Methods in Image and Signal Processing*, 1992, vol.1766, pp.393-404.
8. P. Wang and D.P. Kwok, "Optimal fuzzy PID control based on genetic algorithm," *Proc. 1992 Int. Conf. on Industrial Electronics, Control, Instrumentation and Automation*, 1992, ch286, vol.3, pp.977-981.
9. C.L. Karr and E.J. Gentry, "Fuzzy control of pH using genetic algorithms," *IEEE Trans. Fuzzy Systems*, vol.1, no.1, pp.46-53, Jan. 1993.
10. D.E. Goldberg, *Genetic algorithms in search, optimization and machine-learning*, Reading MA: Addison-Welsey, 1989.

# CUSTOMER-ADAPTIVE FUZZY LOGIC CONTROL OF HOME HEATING SYSTEM

C. v. Altrock, Inform Software Corp., 1840 Oak Avenue, Evanston, IL 60201
H.-O. Arend, Viessmann Werke GmbH & Co., D-3559 Allendorf(Eder)
B. Krause and C. Steffens, INFORM GmbH, Pascalstrasse 23, D-52076 Aachen

*Abstract*: To maximize both heating economy and comfort of a private home heating system, fuzzy-logic control has been used by a German company in a new generation of furnace controllers. The fuzzy-logic controller ensures optimal adaptation to changing customer heating demands while using one sensor less than the former generation. Both the fuzzy-logic controller and the conventional control system were implemented on a standard 8-bit microcontroller. The design, optimization and implementation of the fuzzy controller was supported by the software development system *fuzzy*TECH.

## I.  INTRODUCTION

Most European houses have a centralized heating system which uses furnace for diesel-type fuel to heat the water supply (boiler). From the boiler, the hot water is distributed by a pipe system to individual radiators in the rooms of the house. To meet the different needs of customer heating habits, the temperature of the furnace-heated water must constantly be adjusted in relation to the outdoor temperature (heat characteristic). To measure the outdoor temperature, a sensor is installed at the outside of the house. Fig. 1 depicts the basic structure of such a system.

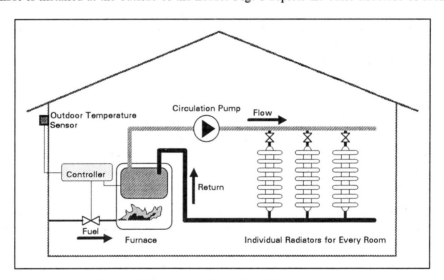

*Fig. 1: Scheme of a Centralized Heating System*

The basic structure of a controller for this system is shown in Fig. 2. The controller itself realizes a on-off characteristic. If the water temperature in the furnace drops to 2 Kelvin below the set temperature, the fuel valve opens and the ignition system starts the burning process. When the water temperature in the boiler itself rises to 2 Kelvin above the set temperature, the fuel valve closes. This on-off control strategy involving hysteresis minimizes the number of starts while assuring that the boiler temperature remains within the desired tolerance in contrast to a on-off control strategy without hysteresis.

Although the structure of this control loop is quite simple, the task of determining the appropriate set boiler temperature is not. Since the maximum heat dissipation of the room radiators depends on the temperature of the incoming water (approximately the boiler temperature), the set point for the water temperature in the boiler must never be set so low that it cannot warm the house when necessary. On the other hand, an excessively high setting of the boiler temperature would result in energy loss in both the furnace and the piping system. Thus the set boiler temperature needs to be carefully set to ensure both user comfort and energy efficiency.

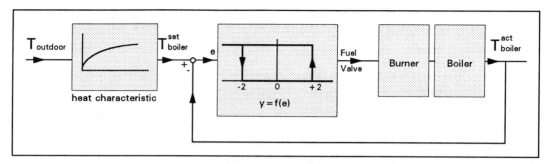

*Fig. 2: Block scheme of the Conventional Furnace Controller*

In the 1950´s, the German Electrical Engineering Society (VDE) defined the following procedure for this: based on the assumption that the maximum amount of heat, required by the house, depends on the outdoor temperature ($T_{outdoor}$), a parameterized function $T^{set}_{boiler}=f(T_{outdoor})$ was defined to adjust the set boiler temperature in relation to the outside temperature (heat characteristic) [3]. Parameters are the insulation coefficient of the house and a so-called "comfort parameter". The physical model of this is one in which the maximum amount of available heat equals the amount of heat disposed by the house plus some excess energy to compensate occasional door and window opening.

Back in those days when most houses had only poor thermal insulation, the assumption, that the energy to be delivered by a heating system was largely outdoor-temperature dependent, was appropriate. Today, this is obsolete. Due to rising energy costs and environmental concerns, modern houses are built with improved insulation. Therefore, to achieve high efficiency, the outdoor temperature is not the only parameter which reflects the required energy amount. Other factors, such as ventilation, door/window openings and personal lifestyle, have to be considered as well.

## II. THE FUZZY CONTROLLER

Basically, there are two approaches for determining the appropriate set boiler temperature for a well-insulated house:
- extensive use of sensors (i.e. temperature sensors in every room) and use of a mathematical model.
- definition of engineering heuristics to determine the set boiler temperature; based on a knowledge-based evaluation of existing sensor data.

Since the use of extensive sensors is expensive and the construction of a comprising mathematical model is of overwhelming complexity, the second approach has been chosen for realizing the new generation of heating system controllers.

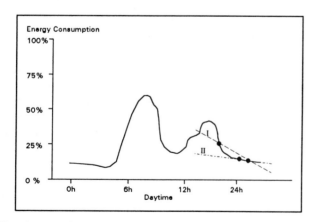

*Fig. 3: Actual Energy Consumption of the House (draft)*

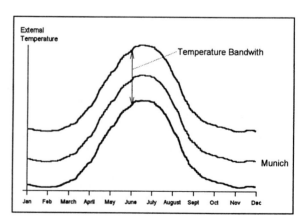

*Fig. 4: Average Outside Temperatures in Munich*

The most important criteria about individual customer heat demand patterns comes from the actual energy consumption curve of the house, which is measured by the on/off-ratio of the burner. An example of such a curve is given in Fig. 3. From this curve, four describing parameters are derived:

1. Current energy consumption, indicating current load.
2. Medium term tendency (I), indicating heating-up and heating-down phases.
3. Short term tendency (II), indicating disturbances like door/window openings.
4. Yesterday average energy consumption, indicating the general situation and house heating level.

These parameters were used to heuristically form rules for the determination of the appropriate set boiler temperature. To allow for the formulation of plausibility rules (such as "temperatures below thirty degrees Fahrenheit are rare in August") the appropriate average outdoor temperature for that season is also a system input parameter. These curves are plotted in Fig. 4. Since the average temperature curves are given, no outdoor temperature needs to be measured. Hence, the outdoor temperature sensor can be eliminated.

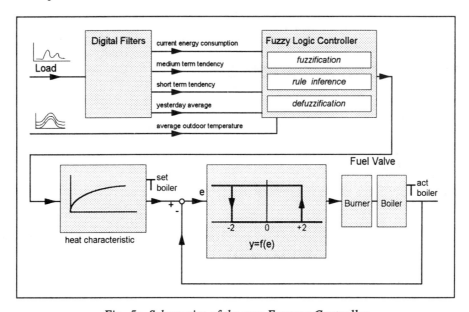

*Fig. 5: Schematics of the new Furnace Controller*

The structure of the new furnace controller is shown in Fig. 5. The fuzzy controller uses a total of five inputs: four of which are derived from the energy consumption curve using conventional digital filtering techniques; the fifth is the average outdoor temperature. This input comes from a look-up table within the system clock. The output of the fuzzy system represents the estimated heat requirement of the house and corresponds to the $T_{outdoor}$ value in the conventional controller (Fig. 2).

## III. DEVELOPMENT OF THE SYSTEM

The objective of the fuzzy controller is to estimate the actual heat requirement of the house. For this, IF-THEN rules were defined to express the engineering heuristics of this parameter estimation:

|       |                                           |       |                                        |
|-------|-------------------------------------------|-------|----------------------------------------|
| IF    | current_energy_consumption IS low         | AND   | medium_term_tendency IS increasing     |
| AND   | short_term_tendency IS decreasing         | AND   | yesterday_average IS medium            |
| AND   | average_outside_temperature IS very_low   | THEN  | estimated_heat_requirement IS medium_high |

In total, 405 rules were defined for the parameter estimation. To develop and optimize such a large system efficiently, *fuzzy*TECH's matrix representation was used. This technique enables rule bases to be viewed and defined graphically rather than in text form. Fig. 6 shows a screen shot of such a rule matrix. In this representation, all linguistic labels of two selected linguistic variables (established heating requirement and yesterday's average energy consumption) are displayed. All other variables (medium term tendency) are kept at a selected label. The matrix may be browsed to show the entire rule base by selecting other terms for these variables.

Within the matrix, a white square indicates rule plausibility whereas a black square indicates rule implausibility (not existent in the rule base). For instance, the highlighted rule in Fig. 6 is valid. Its textual representation (in the lower part of the window) can be read as:

IF        medium_term_tendency IS stable
AND       yesterday_avg IS medium
THEN      est._heat_req. IS medium.

For the formulation of these IF-THEN rules, an initial systems prototype was built. During system optimization, however, it became apparent that some rules were more important than others and that mere rule addition/deletion was too inexact of a system-tuning method. Thus the inference strategy had to be extended to allow rules to be associated with a "degree of support", a number between 0 and 1 which expresses the individual importance of each rule with respect to all other rules. The degree of support for each rule is indicated in the matrix by a gray-shaded square. This allows for the expression of rules like:

IF        medium_term_tendency IS stable
AND       yesterday_avg IS very_high
THEN      est_heat_req IS between high and very_high, rather more high.

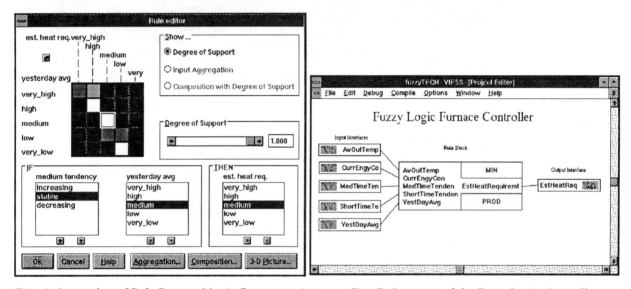

*Fig. 6: Screenshot of Rule Base as Matrix Representation*          *Fig. 7: Structure of the Fuzzy Logic Controller*

The inference method used to represent individual degrees of support is based on approximate reasoning and Fuzzy Associative Map (FAM) techniques: after fuzzification, all rule premises are calculated using the minimum operator for the representation of the linguistic AND and the maximum operator for the representation of the linguistic OR [11]. Next, the premise's degree of validity is weighted with the individual degree of support of the rule, resulting in the degree of truth for the conclusion [2,9,10]. In the third step, all conclusions are combined using the maximum operator. The result of this is a fuzzy set. The Center-of-Maximum defuzzification method is used to arrive at a real value from a fuzzy output [5].

The entire structure of the fuzzy controller is shown in Fig. 7. In this screen shot, the large block in the middle represents the previously-described rule base while the small blocks represent input and output interfaces. The icons denote the fuzzification/defuzzification methods used in the respective interfaces.

## IV. IMPLEMENTATION AND OPTIMIZATION

After completion of the design of the fuzzy controller and the definition of linguistic variables, membership functions and rules, the system was compiled to the target hardware, i.e. to 8051 assembly language. With this technology, the fuzzy controller only uses 2.1 Kbyte of the internal ROM area. Once the fuzzy controller had been linked to the entire furnace controller code, the system was optimized.

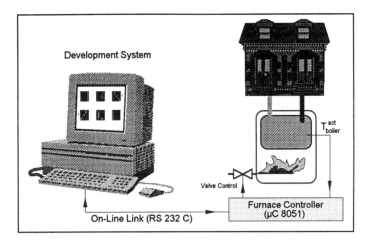

*Fig. 8: Optimization using the "Online" Technique allows for Cross-Debugging and "on-the-fly" Modifications*

To achieve the most efficient system optimization, *fuzzy*TECH's online module was used and the target hardware (8051-based) was connected to the developer's workstation (Windows-PC). The online technique allows for the graphical visualization of the information flow while the system is running and all fuzzification, defuzzification and rule inference steps can be graphically cross-debugged in real-time. In addition, the fuzzy controller can be modified and optimized "on-the-fly" during run-time using the graphical editors.

During optimization, the fuzzy logic controller was connected to a real heating system. This enabled the optimization of the system robustness against process disturbances such as:
- preparation of hot water (e.g. for a bath tub)
- opening of windows
- extended departure, like for vacation

## V. PERFORMANCE

To evaluate system performance, both the conventional controller and the fuzzy controller were connected to a test house. One such example is shown in Fig. 9. Over a period of 48 hours, three graphs were plotted:
- Optimal boiler temperature (calculated from the external/internal house condition).
- Set boiler temperature, as derived from the conventional controller (considering outdoor temperature).
- Set boiler temperature, as derived from the fuzzy controller.

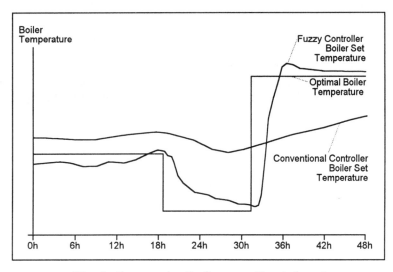

*Fig. 9: Comparative Performance Test (scheme)*

The result of the comparative performance tests showed that the fuzzy controller was highly responsive to the actual heat requirement of the house. It was very reactive to sudden heat demand changes like the return of house inhabitants from vacation. In addition to this, the elimination of the outdoor temperature sensor saved about $30 in production costs and even more in installation costs. By setting the set boiler temperature beneath the level typically used by a conventional controller in low-load periods, the fuzzy controller actually saved energy. Long-term studies collecting statistical data for quantifying exactly how much energy per house could be saved annually are currently investigated. In Addition to this, the two knobs parameterizing the heat characteristic for the individual house (cf. Fig. 2) used by conventional heating systems, are not necessary with the fuzzy logic controller any more. This eases the use of the heating system, since the parameterisation of the heating curves requires a expertise most home owners do not have.

With these new generation of fuzzy logic heating controller, we achieved:
- Improved engergy efficiency, since the fuzzy controller reduces heat production at low heat demand periods.
- Improved comfort, due to the detection of sudden heat demand peaks.
- Easy setup, since the heat characteristic does not need to be parameterized manually.
- Savings both in production and installation costs.

Taking into account the benefits of introducing engineering heuristics, formulated using fuzzy logic technologies, the price was rather low. In the product, the fuzzy logic controller only requires 2 KB of ROM. Using matrix rule representation and online development technology, the optimization of a complex fuzzy logic system containing 405 rules was done efficiently.

## VI. LITERATURE

[1]    H.-J. Zimmermann, Fuzzy Set Theory - and its Applications, 2nd rev. Ed. (Kluver, Boston, 1991)
[2]    C. v. Altrock, B. Krause and H.-J. Zimmermann, Advanced Fuzzy Logic Control in Automotive Applications, IEEE Conf. on Fuzzy Systems (1992) 835-842
[3]    DIN 32729, Teil1, Meß-, Steuer- und Regeleinrichtungen für Heizungsanlagen. Witterungsgeführte Regelung der Kesselwasser- und Vorlauftemperatur, 1992
[4]    M. M. Gupta and J. Qi, Design of Fuzzy Logic Controllers based on generalized T-Operators. FSS 40 (1991), 473-489.
[5]    fuzzyTECH MCU-51 Edition Manual, Inform Software Corporation, Evanston, IL, 1992
[6]    E. H. Mamdani and S. Assilian, An experiment in linguistic synthesis with a fuzzy logic controller, Internat. J. Man-Machine Stud. 7 (1975) 1-13
[7]    Mitsumoto and H.-J. Zimmermann, Comparison of Fuzzy Reasoning Methods. FSS 8 (1992), 253-285
[8]    A. Nafarieh and J. M. Keller, A new Approach to Inference in Approximate Reasoning. FSS 41 (1991), 17-37
[9]    C. v. Altrock, B. Krause and H.-J. Zimmermann, Advanced Fuzzy Logic Control of a Model Car in Extreme Situations. FSS 48 (1992) 41-52
[10]   C. v. Altrock and B. Krause, Online Development Tools for Fuzzy Knowledge-Based Systems of Higher Order. Proc. of the 2nd Int'l Conf. on Fuzzy Logic and Neural Networks Iizuka, Japan 1992, 269-272
[11]   B. Kosko, Neural Networks and Fuzzy Systems (1992) Prentice-Hall, Englewood Cliffs, New Jersey
[12]   L. A. Zadeh, Outline of a New Approach to the Analysis of Complex Systems and Decision Processes, IEEE Transactions on Systems, Man, and Cybernetics, Vol. SMC-3, No. 1 (1973) 28-44

# GENERATING FUZZY RULES FOR A NEURAL FUZZY CLASSIFIER

Chihwen "Chris" Li
Department of Electrical Engineering
National I-Lan Institute of Agriculture and
Technology
I-Lan 26015, Taiwan, R.O.C.

Chwan-Hwa "John" Wu
Department of Electrical Engineering
Auburn University
AL 36849, U.S.A.

Abstract— It is difficult to design a classifier when overlap problems occur between the decision regions of different classes. A top-down learning procedure trains the NF system from global to local views of the overlap decision region and generates nested IF-THEN rules. With the nested IF-THEN rules, the NF system can correctly separate similar classes within the overlap decision region. Two operation examples of the NF system are given.

## 1. GENERATING FAM RULES

### 1.1 NF Classifier

Like the pattern classifier described by Duda and Hart, a five-layer structure is used to describe the NF system for a pattern recognition application, as shown in Fig. 1. Based on the information provided by past engineering experience and input-output data pairs, a feature vector $f$ in a multi-dimensional feature space ($f_i$, $i = 1, 2, ..., n$) is chosen to input the system, and the domain of each input feature is divided into fuzzy sets. The discriminant calculations are implemented with fuzzy membership functions ($m_X(f_i)$ where $i = 1, 2, ..., n$ and $X = A, B, ..., Z$). The fuzzy rules provide the fuzzy relations ($R_j$, $j=1, ..., m$) and decide the connections between the input feature subsets and the fuzzy reasoning output classes. The final class output ($R_X$) is the combination of the fuzzy rules (fuzzy relations) belonging to the same class $X$. The final decision is made by a maximum selector, which selects one class with the largest output active value.

For solving overlap problems, a hierarchical NF system with nested IF-THEN rules is developed. With a basic module as shown in Fig. 1, a multi-layered network consisting of a cascaded connection of many layers of module units is feasible for massive pattern classification and is used to separate overlapping decision regions, as shown in Fig. 2.

### 1.2 General Learning Procedure for Generating FAM Rules

A general procedure for generating fuzzy associative memory (FAM) rules for an adaptive NF system consists of five steps:(1) setting up the input features; (2) generating fuzzy rules from the first set of input-output data pairs; (3) quickly expanding and finely contracting membership functions by fast-commit slow-recode learning rules; (4) detecting overlap; and (5) collecting the same class components and generating fuzzy rules.

### 1.3 Hierarchical NF System

One level of fuzzy rules generated from one iteration of the general learning procedure can be used for a classification application without an overlap problem. For solving overlap problems, a hierarchical NF system is developed as shown in Fig. 2. Once we know where the decision region overlaps, the training for the decision region can focus on the overlapping region and generate a more precise linguistic description for defined the next level of fuzzy IF-THEN rules. This procedure will be repeated if overlap is detected inside the new level of fuzzy rules. With multi-level fuzzy rules, a hierarchical NF system is formed by this recursive procedure going through the overlap regions.

### 1.3.1 Overlapping Region for Recursive Training

In an $n$-dimensional feature space, $n \geq 2$, $I_{XY} = 0$, there is an overlapping $n$-dimensional hyper-box $O$ between classes $X$ and $Y$, represented by

$$O = (XY) = \left\{ (\mathbf{f}, m_{XY}(\mathbf{f})) \mid \mathbf{f} \in U \right\}. \tag{1}$$

As shown in Fig. 3, for five different types of overlap, the boundaries of the decision region of a new generated fuzzy set $(XY)$ within the overlapping region in every feature space are

*Type 1: $Xl < Yl < Xr < Yr$*

$$(XY)r = Xr \qquad (XY)l = Yl \tag{2}$$

*Type 2: $Yl < Xl < Yr < Xr$*

$$(XY)r = Yr \qquad (XY)l = Xl \tag{3}$$

*Type 3: $Yl < Xl < Xr < Yr$*

$$(XY)r = Xr \qquad (XY)l = Xl \tag{4}$$

*Type 4: $Xl < Yl < Yr < Xr$*

$$(XY)r = Yr \qquad (XY)l = Yl \tag{5}$$

*Type 5: $Xl \approx Yl < Yr \approx Xr$*

$$(XY)r = \mathrm{MAX}(Xr, Yr) \qquad (XY)l = \mathrm{MIN}(Xl, Yl). \tag{6}$$

### 1.3.2 Adaptive Learning Unit for Recursive Learning Procedure

Figure 4 illustrates the recursive learning procedure with an adaptive learning unit, which controls the training of the hierarchical NF system. The training procedure is form a global view of a universe of discourse (level $\lambda=1$) to a local view of small overlap regions ($\lambda=\lambda+1$). Therefore, in a higher level of training, only the training data located inside the overlapping regions $O^\lambda$ is used for the training.

The steps 1-5 of the general learning procedure are the basic procedure for all of the classification application. When there is an overlapping region ($I_{xy}=0$), the second level of training is active when the ratio of the overlapping region is larger than a user-defined tolerance T, given as

$$\frac{(XY)r_i - (XY)l_i}{\text{the collecting data range of the universe of discourse } U_i} > T. \tag{7}$$

In the next level of training (level $\lambda$, $\lambda>1$), the general training procedure is repeated from Step 2 to Step 5, and the training region is focused on the overlapped fuzzy set $(XY)$ of the previous level.

A limited size of hyper-box cluster method similar to Simpson's [1] is adapted to solve the overlap problem. An adaptive learning threshold is necessary to limit the expansion size of the decision region for the next level of training. The learning threshold is bounded by overlapping regions of the feature space $i$, given as

$$0 < \theta_{(XY)i} \leq [(XY)r_i - (XY)l_i].\qquad\qquad \textbf{(8)}$$

In simulations, an initial learning threshold is set with the maximum bounded value $\theta_i = [(XY)l_i, (XY)r_i]$. When classes $X$ and $Y$ are overlapped in type 5 (Eq. 6) in all $n$-dimensional feature spaces, the maximum bounded value of a learning threshold cannot separate this kind of overlap problem. To solve this problem, the learning threshold is changed to a smaller value, $\theta_i = q\,[(XY)l_i, (XY)r_i]$, $q < 1$, which makes the NF system generate more precise fuzzy rules in the next level learning, $\lambda = \lambda + 1$, and more numbers of small hyperboxes (clusters) will be generated to do the more precise separation.

### 1.3.3 Rules Recall

With the top-down training, level $\lambda = 1, 2, ..., m$ and cluster $i = 1, 2, ..., n$, every fuzzy rule $R^\lambda_i$ and overlap region $O^\lambda$ is a hyper-box in a multi-dimensional space $U$. Let $\mathbf{R}^\lambda_i$ denote the expanded membership functions, $\mathbf{O}^\lambda_{ij}$ ($j = 1, 2, ..., n; j \neq i$) denote the membership functions of the overlap hyper-box ($\mathbf{O}^\lambda_{ij} \subset \mathbf{R}^\lambda_i$), and $\mathbf{Y}^\lambda_{ij}$ denote the output membership function of the NF system. In the recall scheme, the excitatory membership functions $\mathbf{R}^\lambda_i$ are inhibited by the membership functions $\mathbf{O}^\lambda$. With an input feature vector $\mathbf{f}$, the output is

$$\mathbf{Y(f)}^\lambda_{ij} = \max\!\left[\mathbf{R(f)}^\lambda_i - \mathbf{O(f)}^\lambda_{ij},\, 0\right].\qquad\qquad \textbf{(9)}$$

Since the overlapping decision regions are inhibited by the overlap hyper-box, there is no overlap between membership functions and only one hyper-box will be activated by an input $\mathbf{f}$. With a maximum selector, the classification $\mathbf{C}$ is decided by the hyper-box with maximum output value, given as

$$\mathbf{C} = \max_{\lambda = 1,2,...,m;\; i = 1,2,...,n} \mathbf{Y(f)}^\lambda_i.\qquad\qquad \textbf{(10)}$$

## 2. RESULTS OF THE NF CLASSIFIER

Two operation examples are presented. Example one is a NF classifier in a parallel power controller. Without overlapping problem, the FAM rules of the NF classifiers are generated by one iteration of the learning algorithm (Steps 1-5). Example two is the classification of three subspecies of irises[2]. The measured data are overlapped in four-dimensional feature spaces. For solving this overlap problem, we used the recursive training procedure to generate a hierarchical neural fuzzy classifier with nested fuzzy rules.

### 2.1 Classifier in a Parallel Power Controller Application

The NF power controller system is a magnitude-, frequency-, noise-, and time-invariant system [3]. A waveform generator is the input module of the NF system and generates seven basic waveforms, as shown in Fig. 5. Using the normalized waveform magnitude, the NF system can classify the input waveform independent of its magnitude. With the feature extraction algorithm [3] and developed training method, the NF system can classify input waveforms within a wide range of frequencies (30-85 Hz) and has the generalization to detect the temporally variant chop and spike occurring within a sine waveform. In Fig. 6, some normalized spike and chop samples, which are deformed by different magnitudes, shifted by different time positions, and coupled with random noises, are classified correctly by the NF system.

The membership functions in five feature spaces ($f_1$-$f_5$ [3]) for the seven-waveform classifier are trained by the developed learning algorithm, as shown in Fig. 7. In Fig. 7, the two-third duty, sine, triangle, square, bad-rate square, spike, and chop waveforms are represented with the classes $A$, $B$, $C$, $D$, $E$, $F$, and $G$, respectively.

## 2.2 Hierarchical NF Classifier for Overlap Iris Data

The iris data [2] is a three-class, four-feature classification problem that is commonly used to evaluate classification algorithms. There are 150 data vectors in a four-dimensional feature space $U$, which are measured in centimeters, including sepal length $U_1$, sepal width $U_2$, petal length $U_3$, and petal width $U_4$. Three subspecies of iris are setosa, versicolor, and virginica irises, which are represented with the classes $A$, $B$, and $C$, respectively. Each subspecies of iris has 50 data vectors [2]. The iris data of the classes $B$ and $C$ are overlapped in all of the four feature spaces, which makes classes B and C nonseparable.

The results presented in this section are produced by a hierarchical NF classifier, which is trained by the recursive learning procedure and classifies classes with nested fuzzy IF-THEN rules. With four levels and nine fuzzy IF-THEN rules, a four-layered NF classifier, where each layer has one module unit (Fig. 1), successfully separates the iris data. The decision regions of the classes $A$, $B$, and $C$ in the four feature spaces of level 1-4 are illustrated by Fig. 8(a)-(d), respectively. For testing the training and classification algorithms of the NF classifier, 75 input-output data pairs, 25 data vectors per class randomly selected from each iris subspecies, are used for the recursive training procedure. The other 75 data vectors are used to test the classification performance. Table 1 compares the performance of the hierarchical NF classifier with the fuzzy min-max classifier [1], the perceptron, the fuzzy perceptron, the $k$-nearest neighbor (KNN)[4,5], the fuzzy KNN[4], the Bayes classifier[5], the Fisher ratios[5], and the Ho-kashyap[5]. The fuzzy min-max classifier used 48 clusters in [1]. The classic perceptron algorithm requires iterative training and is limited by linearly separable problems. The fuzzy perceptron reduces the influence of the nonseparable data vectors in determining a decision boundary and makes the training converge in fewer iterations. The NF system has 1 misclassifications and the fuzzy min-max network has 2 misclassifications [1]. Moreover, the NF system used 9 clusters, which is much smaller than 48 clusters used in [1].

## 3. CONCLUSIONS

To solve the overlap problem in multi-dimension feature spaces of large-scale classification applications, a hierarchical NF system can be generated by a newly developed general learning procedure. With the recursive learning procedure, nested fuzzy IF-THEN rules are generated for the hierarchical NF system. Moreover, the nested fuzzy rules can be easily implemented by parallel machines with parallel or two-way searching methods. In test example two, the hierarchical NF system correctly separates the iris data [2] and performs better than other classic, fuzzy, and neural network classifiers in the iris data. Although there is no standard way to perform fuzzy logic in classification applications, we have developed a general procedure for generating a hierarchical NF system.

## REFERENCES

[1]    P. K. Simpson (1992), *IEEE Trans. Neural Networks*, **3**, 776.

[2]    R. Fisher (1936), *Annals of Eugenics*, **7**, 179.

[3]    C. Li and C. Wu (1993), *Proceedings of World Congress on Neural Networks (WCNN)*, Portland, Oregon, **2**, 39.

[4]    J. Keller, M. Gray, and J. Givens (1985), *IEEE Trans. Syst. Man, Cybern.*, **SMC-15**, 580.

[5]    S. Fahlman and C. Lebiere (1990), Carnegie Mellon University, School of Computer Science, Tech. Rep. **CMU-CS-90-100**.

Table 1  A performance comparison between the hierarchical NF classifier
and other classic, fuzzy, neural network classifiers.

| Classification Algorithm | No. Misclassification |
|---|---|
| Hierarchical NF system | 1 |
| Fuzzy min-max network [1] | 2 |
| Perceptron | 3 |
| Fuzzy perceptron | 2 |
| KNN [4] | 5.9 |
| KNN [5] | 4 |
| Fuzzy KNN [4] | 4.7 |
| Bayes classifier [5] | 2 |
| Fisher ratios [5] | 3 |
| Ho-Kashyap [5] | 2 |

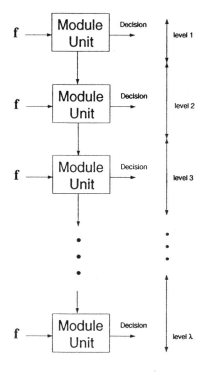

Fig. 2  With a basic module as shown in Fig. 1, a multi-layered network
consists of a cascaded connection of many layers of module units.

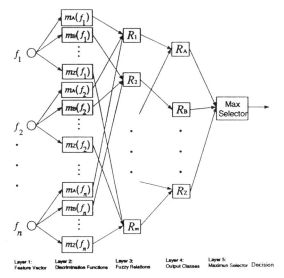

Layer 1:        Layer 2:                    Layer 3:           Layer 4:          Layer 5:
Feature Vector  Discrimination Functions    Fuzzy Relations    Output Classes    Maximum Selector   Decision

Fig. 1  A five-layer structure is used to describe the NF system for a pattern
recognition application.

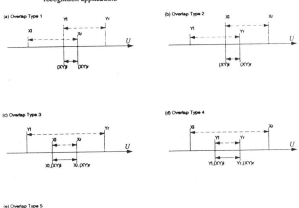

Fig. 3   Five different types of overlap.

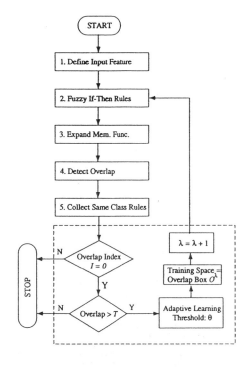

Fig. 4  The recursive learning procedure with an adaptive learning unit is used
to control the training of the hierarchical NF system.

1723

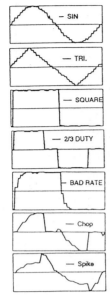

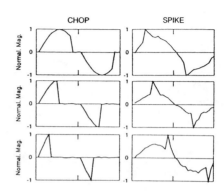

Fig. 5 The seven basic waveforms, which are provided by Westinghouse Electric Corporation, are sine, triangle, square, two-third duty, bad-rate square, chop, and spike.

Fig. 6 some normalized spike and chop samples, which are deformed by different magnitudes, shifted by different time positions, and coupled with random noises, are classified correctly by the NF system.

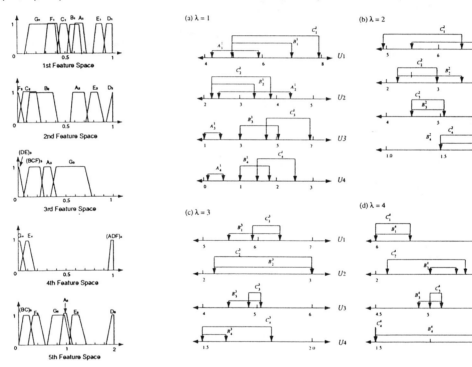

Fig. 7 The final membership functions in five feature spaces ($U_1$-$U_5$) for the seven-waveform classifier. The two-third duty, sine, triangle, square, bad-rate square, spike, and chop waveforms are represented with the classes $A$, $B$, $C$, $D$, $E$, $F$, and $G$, respectively. The membership functions for the four-waveform power classifier are also shown in this diagram with only the classes $A$, $B$, $F$, and $G$.

Fig. 8 The decision regions of the classes $A$, $B$, and $C$, in the four feature spaces of level 1-4.

# AN APPLICATION OF FUZZY ARTMAP NEURAL NETWORK TO REAL-TIME LEARNING AND PREDICTION OF TIME-VARIANT MACHINE TOOL ERROR MAPS

Narayan Srinivasa  and  John C. Ziegert

Machine Tool Research Center
Department of Mechanical Engineering
University of Florida
Gainesville, FL-32611.

## ABSTRACT

The problem of real-time learning of thermal error maps in machine tools is investigated. This problem is treated as an incremental approximation of a functional mapping between thermal sensor readings and the associated positional errors at each location of the cutting tool. The Fuzzy ARTMAP is used as a tool to achieve this approximation in real-time. Experimental measurements of the positional errors for a turning center were performed using a laser ball-bar over two separate thermal duty cycles. The Fuzzy ARTMAP was trained on-line using the data collected during the first duty cycle. Data from a new duty cycle is used to test the performance of the trained network. Results show that the Fuzzy ARTMAP is not only able to learn thermal errors in real-time but also make accurate predictions of the test data.

## 1. INTRODUCTION

### 1.1 Background

Machine tools are serial-link, open loop, kinematic chains. At one end of the chain is the part to be machined or measured, and at the other is the cutting tool. One of the major functional requirements of this class of mechanisms is the ability to position the cutting tool with absolute positioning accuracy on the order of 0.001% of the total working volume. Improvement of machine tool accuracy is an essential part of quality control in manufacturing processes. In today's industries there is a continuous demand for precision machined parts. Hence, methods for error compensation before and during machining process are continually being investigated.

The positioning errors of a machine tool can be divided into three components, *geometric*, *elastic* and *thermal* errors. The geometric errors arise from imperfect construction of the slides and joints of the machine. Elastic errors result from non-rigid body effects in the kinematic elements of the positioning chain. They may be quasi-static, due primarily to gravitational and/or clamping effects in the work piece and moving elements

of the machine; or dynamic(vibrational), resulting from cutting forces and inertial effects. Thermal errors arise due to thermal deformations of the machine elements caused by an extremely complex temperature field. These heat sources exist within the structure including leadscrew bearings and nuts, axis drive motors, spindle, friction on the way surfaces and heat generated by the cutting process. These errors are time-variant in nature as they cause the positional errors to grow with time. They are generally thought to be the largest contributor to overall machine inaccuracy, contributing as much as 70% of the total positioning error. In this paper, we are interested in learning and predicting of these time-variant thermal errors for the entire workspace of the machine (ie. error map of the machine tool).

Thermal errors can be mitigated in two different ways. The first method minimizes the problem by thermally stabilizing the machine[1,2]. This is accomplished by using good design practice, which isolates the major heat sources from the positioning elements, and then operating the machine in a thermally controlled environment. While this method can be adopted for new machines, it is neither cost-effective nor practical to re-design all the machine tools that already exist in the industry. The second method attempts to predict the thermal errors of the machine and to compensate for them. Numerous researchers have addressed this problem[3-8]. The fundamental approach in all these methods is to model the errors in machine elements using rigid body kinematics with each machine element having six error motions(three translational and three rotational). These errors referred to as *parametric errors* are considered as functions of the direction of motion of the machine element. Laser interferometry and capacitance probe gages are standard devices to measure these functions. The characterization of time-variant thermal errors of a machine tool requires each parametric error function to be measured at various thermal states of the machine. Typically, this is done by operating the machine in between brief periods of heating, for several hours, until the machine reaches a thermal equilibrium. Then the machine is stopped and the parametric errors are measured during the cooling period. Since machine tools are typically massive with high heat storage capacity, the time required for a thermal duty cycle may be as much as 12 hours. For a three axis machine, this corresponds to a machine downtime of 216 hours(ie.,6(errors/axis) x 3(axes) x 12(hours)). Hence, this method while accurate is highly time consuming and not economical for machine tool industries.

## 1.2 Motivation

Our main motivation behind this work is to provide a fast and accurate method for calibration of machine tools. As a first step, a new measuring device called the *Laser Ball-Bar( LBB)* was designed and tested at the Machine Tool Research Center[9, 10]. This device is capable of directly measuring the positioning error at the tool instead of deriving it from a kinematic model containing the individually measured parametric errors of the machine as described above. The operating principles of the LBB will be described later on. In order to hasten the process further, it was decided to utilize the learning abilities of neural networks to predict the total positioning error at any given thermal state and location of the cutting tool from knowledge of that error at some specified points in the workspace as measured by the LBB. This combination of the LBB and neural networks can effectively achieve our goal of fast and accurate calibration of machine tools.

In our previously published work, we have shown that a neural net using the back-propagation algorithm [11] is capable of approximating error maps of machine tools[12]. However, training of these networks can proceed only after the entire experimental data is collected. The training time is prohibitively long and the network architecture has to be decided on a trial and error basis. These drawbacks cause a sufficient delay before actual error compensation can proceed. This nullifies the advantage gained in real-time measurement of positioning errors using the LBB.

In this paper, we propose a Fuzzy ARTMAP neural network[13] based learning and prediction of thermal error maps. The network learns each input as it is received on-line from the LBB. The architecture is very simple to implement and the learning is extremely fast. The paper is organized as follows. Section II outlines the Fuzzy ARTMAP algorithm. Section III describes the working principle of the LBB and the technique for positioning error measurement of a two axis turning center using it. Section IV discusses the experiments conducted using the Fuzzy ARTMAP and the LBB combination and the results obtained. Concluding remarks are given in section V.

## 2. THE FUZZY ARTMAP ALGORITHM

The Fuzzy ARTMAP is a neural net architecture used for incremental learning of an associative map in response to arbitrary sequences of binary/analog input vectors. This net contains a pair fuzzy adaptive resonance theory modules ($ART_A$ and $ART_B$) as shown in Figure 1. During learning, $ART_A$ receives a vector A, which is the input to the mapping, and $ART_B$ receives a vector B, which is the correct output of the mapping corresponding to A. These vectors are coded into recognition classes in their respective fuzzy ART modules as follows. For convenience, this process is outlined for fuzzy $ART_A$ alone and is identical for fuzzy $ART_B$.

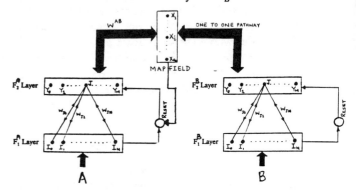

Figure 1. The Fuzzy ARTMAP Architecture.

The $ART_A$ module contains two fields: a field $F_1^A$ that receives both the bottom-up input and the top-down input from field $F_2^A$ where $F_2^A$ represents the category field. The bottom-up input is denoted by A = $(A_1,...,A_{2M})$ where $A_i$ is normalized within [0, 1] and M is the number of features in A. The other M features of the A vector represent the complement of the first M features. This is called *complement coding* and is used to prevent the system from a category proliferation problem[13]. The $F_1^A$ and $F_2^A$ activity vectors are denoted by $x^A = (x_1^A,....,x_{2M}^A)$ and $y^A = (y_1^A,.....,y_N^A)$ respectively where N denotes the number of category nodes in $F_2^A$. All the category nodes in $F_2^A$ have an adaptive weight $w_j^A = (w_{j1}^A,.....,w_{j2M}^A)$ that connects them to activity pattern $x^A$. Initially, $w_{j1}^A = .... = w_{j2M}^A = 1.0$ for j = 1 to N. For each input A and $F_2^A$ node j, the function $T_j(A)$ is defined as,

$$T_j(A) = \frac{\sum_{i=0}^{2M} \min(A_i, w_{ji})}{\alpha + \sum_{i=0}^{2M} w_{ji}} \qquad (1)$$

where $\alpha$ is a choice parameter that helps to minimize the re-coding process during learning. It is normally set as $\alpha = 0.001$. The system makes the category choice node J where

$$T_J = \max(T_j : j = 1, , N) \quad (2)$$

In case of a tie, the node with the lowest index is chosen as the active node. When the Jth category is chosen, $y_J = 1$ and $y_j = 0$ for $j \neq J$. If this chosen category meets the *vigilance criterion* defined as

$$\frac{\sum_{i=0}^{2M} \min(A_i, w_{Ji})}{\sum_{i=0}^{2M} A_i} \geq \rho^A \quad (3)$$

where $\rho^A$ is the vigilance parameter, then category J is the node that codes the input vector A. It should be noted that the higher the vigilance, the more sensitive the system is to small changes in the input features. If the condition in equation (3) is not met, then *reset* occurs and $T_J$ is set to 0 for the duration of the present input. A new index J is then chosen using equation (2) and the search continues until the chosen node J satisfies equation (3). Once the search ends, the weight vector $w_J$ is modified using the following learning rule:

$$w_{Ji}^{new} = \min(A_i, w_{Ji}^{old}) \quad i = 1, 2M \quad (4)$$

Hence, the inputs A and B are coded into category fields $F_2^A$ and $F_2^B$ of fuzzy $ART_A$ and $ART_B$ respectively. These fields are linked together via an inter-art module $F^{AB}$, called a *map field*. The map field is used to form predictive associations between the category fields $F_2^A$ and $F_2^B$ as follows. Let $x^{AB} = (x_1^{AB},....,x_L^{AB})$ denote the $F^{AB}$ output vector and let $w_j^{AB} = (w_{j1}^{AB},....,w_{jL}^{AB})$ denote the weight vector from the jth $F_2^A$ node to $F^{AB}$ where L is the number of nodes in $F_2^B$. If node K in $F_2^B$ is active, then the node K in $F^{AB}$ is activated by 1-to-1 pathways between $F_2^B$ and $F^{AB}$. The $F^{AB}$ output vector $x^{AB}$ obeys $x_i^{AB} = \min(y_i^B, w_i^{AB})$ for $i = 1$ to L. If $x^{AB} \neq 0$, then the prediction by $ART_A$ is confirmed by $ART_B$. If not, a *mismatch* event occurs which triggers an $ART_A$ search for a better category as follows. Let the $ART_A$ vigilance at the start of each input presentation be $\rho_0^A$ called the *baseline vigilance*. Let the map field vigilance parameter be denoted by $\rho^{AB}$. Due to the *mismatch* event, the $ART_A$ is made more vigilant by slightly increasing $\rho^A$ as

$$\rho_a > \frac{\sum_{i=0}^{2M} \min(A_i, w_{Ji}^A)}{\sum_{i=0}^{2M} A_i} \quad (5)$$

When this occurs, $ART_A$ search leads to activation of either an $ART_A$ category that correctly predicts B (ie satisfies $X^{AB} \neq 0$) or to a previously uncommitted category node. The weights $w_{jk}^{AB}$ in $F_2^A ->$ $F^{AB}$ paths initially satisfy $w_{jk}^{AB} = 1$ for all j and k. Once node J of $ART_A$ learns to predict node K of $ART_B$, then $w_{JK}^{AB} = 1$ and $w_{jk}^{AB} = 0$ for $j \neq J$ and $k \neq K$. This constitutes the working principle of the Fuzzy ARTMAP. For further details on this neural net, the reader is referred to [13].

## 3. METHODOLOGY

### 3.1 Working Principle of the LBB

As mentioned before, direct measurement of the positioning errors is achieved using LBB. The LBB consists of a laser interferometer aligned between two precision spheres by a telescoping tube as shown in Figure 2.

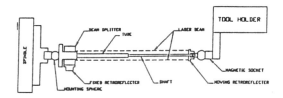

Figure 2. Schematic of the Laser Ball-Bar.

The light from the laser is collected in a polarization preserving fiber optic cable and carried to the polarization beam splitter(PBS) where the light is separated into two components: one component passes through the PBS(reference beam) and is reflected in the retroreflector attached to the PBS and returned to the fiber-optic pick-up; the other component travels along one side of the telescopic tube to the retroreflector attached to the opposite end. The retroreflector returns the beam along the other side of the telescoping tube where it re-enters the PBS and it is reflected to fiber optic pick-up and combined with the reference beam. This recombined beam is carried to a receiver where the displacement information can be extracted from the phase relationship of their combined beat frequency relative to the reference frequency of the laser. An anti-rotation tube is provided to restrict the rotation of the measurement retroreflector relative to the PBS and also limit the travel of the telescoping element. By itself the LBB can only measure changes in length. To obtain positional measurements, its absolute length must be known (as will be explained in the next section). This is accomplished by a self-initialization procedure as described in [9].

## 3.2 LBB Based Measurement Technique

In all our experiments, the positioning errors of the cutting tool of a MAZAK Quick Turn 28N two axis turning center is to be measured using the LBB. The turning center consists of a carriage(Z axis motion), a cross-slide(X axis motion) and a turret attached to the cross-slide on one end of the kinematic chain, and, the spindle on the other end of it as shown in Figure 3(a).

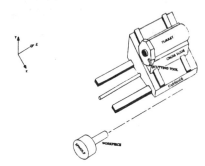

Figure 3(a). Mazak Quick Turn 28N Turning Center

To determine the position of the cutting tool relative to a reference coordinate frame, the method of triangulation is adopted as shown in Figure 3(b). A grid of points were selected which would fit within the LBB's range.

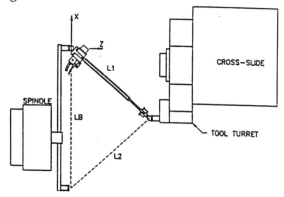

Figure 3(b). Set-up of LBB on turning center[9].

The LBB is first placed between the base sockets and the length, $L_B$, is recorded. Next the LBB is placed for measurement at each grid point i and the machine records the length $L_{1i}$ for all these points. The procedure is repeated for $L_{2i}$ and the coordinates $(X_i, Z_i)$ as measured by the LBB for each grid point is calculated as:

$$x_i = \frac{L_{2i}^2 - L_{1i}^2 - L_B^2}{2 L_B} \qquad (6)$$

$$z_i = \sqrt{L_{1i}^2 - x_i^2} \qquad (7)$$

These values of $x_i$ and $z_i$ are transformed into the coordinate system of the machine tool to obtain the total positional errors for all the grid points.

## 4. EXPERIMENTAL RESULTS

In order to train the fuzzy ARTMAP on-line, a thermal duty (TDC1) was formulated. It consists of a series of measurement periods with a warming period in between each of them until the machine warms up to thermal equilibrium (ie., for 6 hours). During each measurement period, the x and z positioning errors of the cutting tool was measured for a grid of ninety eight (7 X 14) points using the LBB. At each grid point, these errors are fed as inputs to the $ART_B$ module in the complement coded form. Temperatures from nine different locations in the machine and the machine coordinates of the cutting tool are fed as inputs to the $ART_A$ module in the complement coded form. These temperatures are measured using thermocouples at the following locations of the carriage and cross-slide: its two slide ways, the leadscrew nut and the leadscrew bearing housing. In addition the temperature of the ambient air was also used as an input as the room is not temperature controlled. During the warming period, the carriage and cross-slide are exercised to move along a diagonal to their respective ranges of travel(480mm for the carriage and 192mm for the cross-slide) at a constant speed of 1300mm/min for 25 minutes. Typical examples of training data are shown in Figures 4.

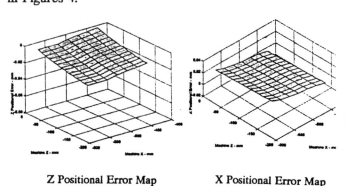

Z Positional Error Map     X Positional Error Map

Figure 4. Error maps of x and z at cold-state.

The inputs to ART$_A$ and ART$_B$ must be normalized in the range [0, 1]. However, it is not possible to know *a priori*, the range of the thermocouple readings or the range of the positional errors. There are two solutions to this problem. The first solution evaluates the minimum and the maximum value for each input after each measurement period and if these values are different from that calculated during the previous measurement period, it is normalized with respect to the new minimum and maximum values. The Fuzzy ARTMAP is then made to forget all previously learned data (by setting $w^A = w^B = w^{AB} = 1.0$) and re-learn the entire data after re-normalization of all inputs have occurred. The re-learning takes place during the warming period as it requires sufficient time for processing. While this solution is portable to any machine tool, the learning is not truly on-line. The second solution sets the ranges of all inputs based on its previous history. This solution results in on-line learning and is feasible in most industries as there is a sufficient history of each machine readily available. This solution is adopted in this paper.

The training set consists of 1960 input vectors for TDC1. The baseline vigilance of ART$_A$ module was set at 0.95; the vigilance parameter of the map field can be set to any value between 0 and 1 for fast-learn conditions; The vigilance of ART$_B$ module was set to 1.0 to achieve a excellent approximation of the mapping. By setting the vigilance of ART$_B$ to 1.0, the system categorizes the inputs to ART$_B$ module into categories separated by the resolution of the machine.(ie., 1 micron). This is very desirable as the machine cannot compensate for errors below its resolution and at the same time we also require a very fine associative mapping. The training resulted in 526 ART$_A$ categories and 85 ART$_B$ categories.

In order to test the performance of the system, a new thermal duty cycle (TDC2) was constructed as shown by Table 1. Five materials were selected ranging from mild to alloy steel and the KENNAMETAL tables were used to find recommended feed rates for each material for a constant spindle speed of 1300 rpm. During each warming period, a material and corresponding feed rate was randomly selected to lie between 10% to 100% of the entire ranges of X and Z travel. The warming period was randomly selected to be between 15 and 30 minutes. At each grid point location, the ART$_A$ module of the Fuzzy ARTMAP is presented with the nine thermocouple readings and cutting tool coordinates as inputs and the ART$_B$ module predicts a corresponding x and z positional error. This prediction is compared with the actual measurement made by the LBB in order to characterize the performance of the proposed technique.

A general measure of the performance of the

Table 1: Data for the test thermal duty cycle.

| Trial | Mat. AISI# | Feed mpm | X (mm) | Z (mm) | Time secs |
|-------|-----------|----------|--------|--------|-----------|
| 1 | 1026 | 950 | 165 | 395 | 1305 |
| 2 | 1045 | 310 | 43 | 243 | 1053 |
| 3 | 1095 | 1145 | 178 | 378 | 1789 |
| 4 | 4340 | 770 | 105 | 52 | 961 |
| 5 | 4023 | 580 | 72 | 372 | 1444 |
| 6 | 1095 | 1145 | 135 | 135 | 1567 |
| 7 | 4340 | 770 | 111 | 391 | 1605 |
| 8 | 1026 | 950 | 127 | 327 | 1472 |
| 9 | 1045 | 310 | 63 | 323 | 1565 |
| 10 | 1026 | 950 | 40 | 300 | 1066 |
| 11 | 4340 | 1145 | 145 | 230 | 1606 |

Fuzzy ARTMAP, R$_{eP}$, is computed. It is defined as the mean value over the number of measured points, N, of the absolute difference between the positional error P of the each axis (P = X or Z) measured by the LBB and the Fuzzy ARTMAP predicted value of the same error:

$$R_{eP} = \frac{\sum_{i=0}^{N} |P_{i,laser} - P_{i,net}|}{N} \qquad (8)$$

For TDC2, R$_{eX}$ and R$_{eZ}$ were found to be 1 and 3 microns respectively. These results are extremely good considering that the test set is completely different from the training set and the resolution of the machine is 1 micron. The results for some of the test trials are shown below in Figure 5(a) and 5(b). The performance could be further improved if the network is trained on a larger training set. However, the limit chosen here gave satisfactory results and is sufficient to illustrate the technique. Also, an off-line voting scheme[12] can further improve its predicting capabilities.

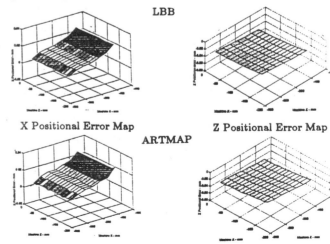

Figure 5(a). Error map comparison between the LBB and the Fuzzy ARTMAP for trial#3 of the test set.

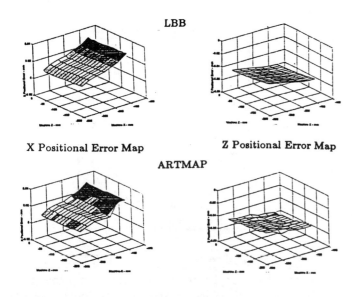

LBB

X Positional Error Map     Z Positional Error Map

ARTMAP

Figure 5(b). Error map comparison between the LBB and Fuzzy ARTMAP for trial#9 of test set.

## 5. CONCLUSIONS

The problem of real-time learning of thermal error maps in machine tools using a Fuzzy ARTMAP neural net and LBB combination is investigated. The positional errors of a two axis turning center were collected at a grid of points during two different thermal duty cycles. The first thermal duty cycle consisted of six hours of warming. The Fuzzy ARTMAP was trained on data collected during the first duty cycle. Inputs to the $ART_A$ module were the cutting tool coordinates and a set of nine thermocouples which provide information of the thermal condition of the machine. The x and z positional errors were inputs to the Fuzzy $ART_B$ module. The vigilance parameters of the $ART_A$ and $ART_B$ modules were set to very high values in order to obtain a very accurate mapping of the error map. A second thermal duty cycle was generated randomly and consisted of a range of realistic machining feed rates over randomly generated paths for random time periods. The Fuzzy ARTMAP was used to predict the positional errors collected during the second thermal duty cycle. The results show that these predictions are accurate and comparable to actual measurements taken by the LBB over the same duty cycle.

## ACKNOWLEDGEMENT

This material is based on work supported by the National Science Foundation under Award No. DDM-9017293.

## REFERENCES

1) Weck, M., and Zangs, L., "Computing the Thermal Behavior of Machine Tools Using the Finite Element Method- Possibilities and Limitations, *Proceedings of the 16th MTDR Conference*, 1975.

2) Okushima, K., et al., "An Optimal Design of Machine Tools for Thermal Deformations," *Bulletin of Japan Society of Precision Engineering*, vol. 7, no. 2, 1973.

3) Tlusty, J., and Mutch, G. F., "Testing and Evaluating Thermal Deformations in Machine Tools," *Proceedings of the 14th MTDR Conference*, 1973.

4) Hocken, R. J., "3-Dimensional Metrology," *CIRP Annals*, vol. 26, no. 1, 1977.

5) Donmez, M. A., "A General Methodology for Machine Tool Accuracy Enhancement: Theory, Application and Implementation," *Ph. D. Dissertation*, Purdue University, 1985.

6) Venugopal, R., and Barash, M., "Thermal Effects on the Accuracy of Numerically Controlled Machine Tools," *CIRP Annals*, vol. 35, no. 1. pp. 255-258, 1986.

7) Duffie, N. A., and Malmberg, S. J., "Error Diagnosis and Compensation Using Kinematic Models and Position Error Data," *CIRP Annals*, vol. 36, no. 1, 1987.

8) Chen, J. S., "Real-time Compensation of Time-varying Volumetric Error on a Machining Center," *Ph. D. Dissertation*, University of Michigan, 1991.

9) Mize, C. D., "Design and Implementation of a Laser Ball-Bar Based Measurement Technique for Machine Tool Calibration," *Master's Thesis*, University of Florida, 1993.

10) Ziegert, J. C. and Mize, C. D., "Laser Ball-Bar: A New Instrument for Machine Tool Metrology," *Precision Engineering* (accepted for publication).

11) Srinivasa, N., Ziegert, J. C., and Smith, S., "Prediction of Positional Errors of a Three Axis Machine Tool Using a Neural Network," *Proceedings of the Japan/USA Symposium on Flexible Automation*, San Fransisco, pp. 203-209, 1992.

12) Rumelhart, D. E., Hinton, G. E., and Williams, R. J., "Learning Internal Representations by Error Propagation," *Parallel Distributed Processing*, MIT Press, Cambridge, MA, vol. 1, pp. 318-362, 1986.

13) Carpenter, G. A., et al., "Fuzzy ARTMAP: A Neural Network Architecture For Incremental Supervised Learning of Analog Multidimensional Maps," *IEEE Transanctions on Neural Networks*, vol. 3, no. 5, pp. 698-712, 1992.

# A Hybrid Fuzzy/Neural System Used to Extract Heuristic Knowledge from a Fault Detection Problem

Paul V. Goode
pvgoode@eos.ncsu.edu
IEEE Student Member

Mo-yuen Chow
chow@eos.ncsu.edu
IEEE Senior Member

Dept. of Electrical and Computer Engineering
North Carolina State University
Raleigh, NC 27695
USA

*Abstract*

Artificial neural networks have proven to be capable of solving the motor monitoring and fault detection problem using an inexpensive, reliable, and non-invasive procedure. The neural network, unfortunately, cannot provide heuristic knowledge about the motor or the fault detection process. This paper introduces a novel hybrid fuzzy/neural fault detector that will use the learning capabilities of the neural network to detect if a motor has an incipient fault. Once the fuzzy/neural fault detector is trained, heuristic knowledge about the motor and the fault detection process can also be extracted. With better understanding of the heuristics through the use of fuzzy rules and fuzzy membership functions, we can have a better understanding of the fault detection process of the system, thus we can design better motor protection systems. The electric motors in industry are exposed to a wide variety of environments and conditions. These factors, coupled with the natural aging process of any machine, make the motor subject to incipient faults. These incipient faults, if left undetected, contribute to the degradation and eventual failure of the motors. With proper monitoring and fault detection schemes, the incipient faults can be detected; thus, maintenance and down time expenses can be reduced while also improving safety. In this paper, motor bearing faults in single phase induction motors will be used to illustrate this novel system. This illustration will demonstrate successful training of a hybrid fuzzy/neural system that can provide accurate fault detection and give the heuristic reasoning for the fault detection procedure.

## I. Introduction

The use of electric motors in industry is extensive. These motors are exposed to a wide variety of environments and conditions. These factors, coupled with the natural aging process of any machine, make the motor subject to incipient faults [1-6]. These incipient faults, if left undetected, contribute to the degradation and eventual failure of the motors. With proper monitoring and fault detection schemes, the incipient faults can be detected; thus, maintenance and down time expenses can be reduced while also improving safety [1].

Several methods have been proposed to monitor the motor condition [1,7-8]. Many of these, however, are expensive to implement or require expert system knowledge in the form of precise mathematical models [1-6]. Others even require an invasive procedure [7] or an off-line motor fault detection procedure [1]. Artificial neural networks (ANN), however, have proven to be very capable of performing motor fault detection [3,5,8-9] while avoiding many of these unpleasant requirements [10-12]. The ANN can inexpensively perform this function in a non-invasive manner without the need for any mathematical models of the motor. Furthermore, it can adapt itself to learn arbitrarily complicated continuous nonlinear functions which will enable it to give the exact solution to a particular problem. Different advantages of using ANNs instead of other fault detection techniques are discussed in [10-12].

While the ANN can correctly monitor the condition of the motor, it cannot provide general heuristic, qualitative information about what contributes to a fault; i.e., under what conditions do a fault occur? This inability is due to the inherent "black box" feature of the ANN. Even though the neural network can perform the correct input-output relationship for the given problem; it cannot perform this function in a manner which makes heuristic sense. Furthermore, it is unable to provide qualitative information about the circumstances under which faults may occur; i.e., it cannot provide expert knowledge about the motor condition in heuristic terms which humans prefer.

The inability to extract this type of knowledge from the ANN is a result of its complex architecture. The ANN consists of neurons (summing junctions) and numerical weight values assigned to the interconnection strengths between these neurons. In standard feedforward neural network configuration, heuristics cannot be easily extracted from the neurons and weights. However, if these heuristics could be extracted, it would not only give us more insight into the fault detection process, but also provide more knowledge about the actual system.

This problem can be solved by incorporating the use of fuzzy logic with the ANN structure. It is well known that fuzzy logic has the capability of transforming heuristic and linguistic terms into numerical values for use in complex machine computations via fuzzy rules and membership functions [13-15]. It can also provide a heuristic output as a result of those same complex computations by quantifying the actual numerical data into heuristic and linguistic terms [13-15]. For these reasons, fuzzy logic can be used to provide a general heuristic solution to a particular problem with the use of general heuristic knowledge about the problem. Unfortunately, *a priori* knowledge about the system is necessary to develop the fuzzy rules and membership functions.

This paper will merge these two technologies to be used as a novel motor fault detection technique. From this synergy of technology, a fuzzy/neural motor fault detector will be obtained that does not need a mathematical model, or *a priori* knowledge, of the motor, but will still be able to provide qualitative expert knowledge of the motor through valid heuristics and give exact quantitative solutions to detect the motor faults. This paper will address bearing wear as one of the common causes of failure as an illustration of the fuzzy/neural system. As a result, expert knowledge of the motor can be extracted from the fuzzy/neural fault detector in the form of fuzzy logic membership functions and fuzzy logic rules. The knowledge obtained will aid in the understanding of motor bearing as well as the fault detection process. In addition, this fault detector will be able to learn the bearing faults and the conditions under which they occur through an inexpensive and non-invasive procedure.

Section II of this paper will discuss the fuzzy/neural fault detector architecture. Section III will provide the illustration of bearing wear as an application to the fuzzy/neural motor fault detector. This example illustration will adhere to the system architecture described in Section II and will also be explained in detail.

## II. Fuzzy/neural System for Fault Detection

The fuzzy/neural architecture takes into account both fuzzy logic and neural network technologies. The system is a neural network structured upon fuzzy logic principles, which enables the fuzzy/neural system to provide such fuzzy parameters as fuzzy membership functions and fuzzy rules. This is done by constructing the fault detector using two modules: the fuzzy membership function module (module 1) and the fuzzy rule module (module 2). The use of these modules for the motor fault detection problem is discussed below. The fuzzy/neural motor fault detector is shown in Figure 2.

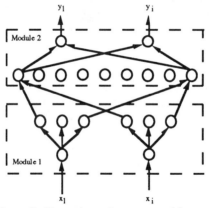

Figure 2. Fuzzy/neural system architecture

### Module 1

The purpose of the fuzzy membership function module is to provide the fuzzy membership functions of the inputs. For the illustration of motor fault detection, these membership functions will provide qualitative heuristic knowledge of the motor current and rotor speed. This knowledge will be in the form of grades of membership that indicate, for example, what range of current is considered *low current* and what range of speed is considered *high speed*, etc. From these linguistic terms, fuzzy rules can be expressed that give qualitative descriptions of the motor; i.e., 'when the current is *low* and the speed is *high*, then the bearing condition is *good*.'

The fuzzy membership function module is composed of independent 'sub-networks.' The inputs to these 'sub-networks' are the system inputs (which are first normalized). The function of the 'sub-network' is to partition the normalized values into fuzzy membership function space and provide these as outputs of the module. The information for the fuzzy membership functions is contained in the weights of the 'sub-networks', which determine the shape of the membership functions of interest. Sub-networks are used because they allow for representation of very complex membership functions [16] which are more flexible to adaptation for decision classification.

The form of the fuzzy membership functions of do not need to be known because the fuzzy/neural system will adaptively determine these membership functions. However, a good initialization of these 'sub-networks' will aid in training of the fuzzy/neural system by giving it a better starting point. This starting point is important because most of the changes made during training will occur in module 2. This is evident by the generalized delta rule backpropagation training algorithm. For more information on this algorithm, the reader should refer to [17].

Therefore, a good partitioning of the fuzzy sets will aid in the learning of the fuzzy rules done by module 2 (this will be explained later). An example is shown in Figure 3.

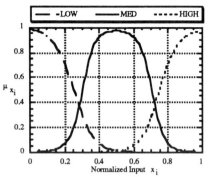

Figure 3. Example of initial membership functions for one input

These vague heuristics serve to build the universes of discourse, $X$ and $Y$, for each of the input spaces of $I$ and $\omega$, respectively. These are represented in the standard notation used in [18]:

$$X = [\,\mu_{\text{low}}(I)\,,\,\mu_{\text{medium}}(I)\,,\,\mu_{\text{high}}(I)\,]\,,\,I \in X, \tag{1}$$

$$Y = [\,\mu_{\text{low}}(\omega)\,,\,\mu_{\text{medium}}(\omega)\,,\,\mu_{\text{high}}(\omega)\,]\,,\,\omega \in Y, \tag{2}$$

where $\mu_i(\lambda)$ = the grade of membership of $\lambda$ in $\mathcal{I} = \{\,i \in \mathcal{I}\,|\,\text{low, medium, high}\,\}$, $\lambda \in X, Y$. The universes of discourse represent the operating range of the inputs. For the motor fault detection problem, the motor current operating range was from 3 to 14 amps, and the motor speed operating range was from 140 to 188.5 rad/sec. Both sets of these values were normalized to [0,1].

As mentioned above, the actual fuzzy membership function information is contained in the weights of the 'sub-network' in module 1 after training. These final fuzzy membership function values are extracted by looking at the outputs of the 'sub-networks.' Each output node represents a fuzzy membership set. For example, the 'sub-networks' of Figure 3 would each have three output nodes, one each for: *low*, *medium*, and *high* . After training, the 'sub-networks' are evaluated by inputting a set of incremented values between [0,1] and recording the outputs. The outputs will represent the final form of the fuzzy membership functions.

*Module 2*

The fuzzy rule module provides the antecedent-consequence statements of fuzzy logic. These statements provide the condition of the fault being monitored given the linguistic operating range of the inputs. For example, 'when the current is *low* and the speed is *high*, then the bearing condition is *good*.' The structure of the module is a two layer feedforward ANN. The inputs to this module are the outputs of the fuzzy membership function module (membership nodes), while the outputs of this module are the classification conditions (classification nodes) of the motor condition.

As with the fuzzy membership function module, the information for the antecedent-consequences are in the weights connecting the membership nodes to the classification nodes. These rules are determined through training of the fuzzy/neural system. No rule initialization is needed; i.e., prior knowledge of what the rules should be is not necessary. The rule extraction is based upon the following mathematics.

In a standard feedforward ANN, each node performs a function on the sum of its inputs. An input is represented as the output of a node from the previous layer, where $m$ represents the membership function node number, multiplied by the weight value which connects it to the $n$th node. The total input to a node can be expressed as:

$$s_n = \sum_{m=1}^{m=k} x_m w(y_n, x_m) + b_n, \tag{3}$$

where $b_n$ represents the bias weight value which connects the $n$th node to the bias node and $k$ is the number of inputs from the previous layer. Because the bias node generally has an output of one [17], $b_n$ is sufficient to represent the bias input. For further details on standard feedforward ANN, the interested reader should refer to [17]. The output of the classification node, $y_n$, is expressed as $\qquad y_n = f(s_n),$ (4)

where $f$ is the sigmoid activation function. For the fuzzy/neural system, the bias weights for the classification nodes are set to zero. Using closest neighborhood classification with the properties of the sigmoid activation function, we obtain

$$y_n = \begin{cases} \geq 0.5 \text{ if } s_n \geq 0 \ \Rightarrow \ \text{TRUE} \\ < 0.5 \text{ if } s_n < 0 \ \Rightarrow \ \text{FALSE} \end{cases}. \tag{5}$$

Therefore, it becomes clear that the classification output is a function of the sign $s_n$. Thus,

$$y_n = f(\text{sgn}(s_n)). \tag{6}$$

1733

If the sign of $s_n$ is positive, then $y_n$ is the proper consequence for the set antecedents. If the sign of $s_n$ is negative, then $y_n$ is not the proper consequence for the set antecedents. Therefore, referring to (3) and (5), the fuzzy rules are extracted by assuming that $x_m = 1$ for the antecedents (the membership nodes) of the rule while $k$ is the number of antecedents for the rule, and taking the sign of $s_n$, the sum of the weights connecting the antecedent nodes to a particular consequence (classification) node.

For example, referring to Figures 2 and 3, to determine whether the fuzzy rule indicates *good* for the case of $I=low$ and $\omega=high$, the following classification node output is evaluated

$$s_{good} = w(good, lowI) + w(good, high\omega) \tag{7}$$

If the result of $s_{good}$ in (7) is positive, then, by (5), $y_{good}$ is TRUE and considered to be the correct consequence. Therefore, under these conditions, the fuzzy rule would be: if $I=low$ and $\omega=high$ then *condition=good*.

Because $x_m \in [0,1]$, it is clear that the sign of $s_n$ is dictated by the sign of $w(y_n, x_m)$. Using the generalized delta rule to backpropagate the error, the weights are changed to minimize a measure of the network's error. The change in weights between an antecedent node, $b$ and a consequence node, $a$ is $\Delta w_{ba}$ and is calculated each iteration by:

$$\Delta w_{ba} = \eta(t_b - o_b)(o_b - o_b^2)o_a, \tag{8}$$

where $\eta$ is the learning rate and is greater than zero, $t_b$ is the target output of a consequence node for a given pattern, $o_b$ is the actual output of a consequence node for a given pattern, and $o_a$ is the output of the antecedent node for a given training pattern. Because the sigmoidal activation function is used, the outputs of the antecedent and consequence nodes are bounded between zero and one. Therefore:

$$o_a, o_b, t_b \in [0,1]. \tag{9}$$

Because $t_b$ is either zero or one for classification problems and $(o_b - o_b^2)o_a > 0$, then

$$(t_b - o_b) \ is \begin{cases} \leq 0 \ if \ t_b = 0 \\ \geq 0 \ if \ t_b = 1 \end{cases} \Rightarrow \Delta w_{ba} \ is \begin{cases} \leq 0 \ if \ t_b = 0 \\ \geq 0 \ if \ t_b = 1 \end{cases}. \tag{10}$$

Therefore, the weights which connect the correct consequence for a set of antecedents will continue to increase positively in value whereas the weights which connect the incorrect consequence for an antecedent will continue to decrease negatively in value. Using this procedure, there should exist only one consequence per set of antecedents. If there are no training patterns belonging to that particular antecedent set, then that set of antecedents will provide more than one consequence and is therefore not a valid rule.

### III. Bearing Wear Fault Detection

As previously mentioned, the two most popular faults, bearing wear and insulation failure, are used here to illustrate the proposed fuzzy/neural system. The fuzzy/neural motor fault detector is first trained to learn the bearing wear faults. The bearing condition was classified into three categories: *good, fair,* and *bad*. These classifications were made based upon motor efficiency under increasing bearing wear conditions [6]. For example, rotational losses, $P_{rotational\ loss}$, due to bearing wear generally contributes five to ten percent of all power losses, $P_{loss}$, experienced by an average healthy motor. Considering this, data classification for bearing wear was constructed based upon the following heuristics:

$$P_{loss} = (P_{stator\ loss} + P_{rotor\ loss} + P_{rotational\ loss})$$

$$IF \quad P_{rotational\ loss} \leq 0.05\ (P_{loss}) \quad THEN\ GOOD$$

$$IF \quad 0.00\ (P_{loss}) < P_{rotational\ loss} \leq 0.10\ (P_{loss}) \quad THEN\ FAIR \tag{11}$$

$$IF \quad 0.10\ (P_{loss}) < P_{rotational\ loss} \quad THEN\ BAD$$

Training data for bearing wear was obtained using these same heuristics. A plot of the training data obtained (121 data points) is shown in Figure 4.

Observe that the bearing condition is more dependent on motor speed than motor current, but is still a function of both inputs. Also note the absence of data from the low current, low speed and high current, high speed regions. As bearing wear increases, the rotor speed is reduced. This reduction in rotor speed causes an increase in slip, thus increasing motor current. Therefore, it is physically impossible for the motor to maintain a low current while decreasing rotor speed. This condition nullifies the possibility of data belonging to a low current, low speed region. Furthermore, the motor cannot maintain a high speed with increased bearing wear. As mentioned, the rotor speed will reduce and the motor current will increase. This condition nullifies the possibility of data belonging to a high current, high speed region.

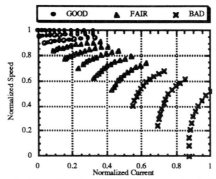

Figure 4. Distribution of bearing wear fault data

As mentioned previously, the vague heuristics of *low*, *medium*, and *high* are used to initialize the membership function module as shown in Figure 3. This gives the module a better starting point for classification. The fuzzy rules are assumed to be unknown and will therefore be randomly initialized. The fuzzy/neural motor fault detector was trained on the bearing wear data and achieved 100% classification accuracy for both the training data and the testing data.

The membership functions were extracted using the procedure outlined in the previous section; i.e., evaluate the output nodes of module 1. The resulting membership functions are shown in Figure 5. These membership functions indicate the actual regions of *low*, *medium*, and *high* for each of the respective regions. For example, to classify $I=ligh$ and $\omega=low$ , the range of *high* current covers roughly the top one-third of the input range of current; whereas, the range of *low* speed covers roughly the bottom two-thirds of the input range of motor speed.

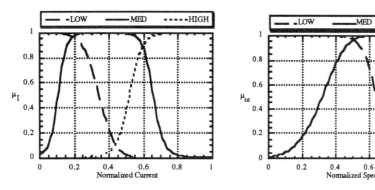

Figure 5. Bearing wear membership functions after training

The fuzzy rules extracted from module 2 are done following the procedure outlined in the previous section; i.e., evaluate the sum of weights entering each classification node for each input condition. The resulting fuzzy rules were found to be analogous to the fault data distribution plot of Figure 5. For example, the case for $I=low$ and $\omega=high$ will be analyzed. From Figure 5, when $I=low$ , the range a normalized motor current it strongly activates is {0,0.35} while $\omega=high$ activates {0.9,1.0}. Looking at these ranges on Figure 1(a) ($I$ is the x axis, $\omega$ is the y axis) shows that the data points that are in that area are *good*. The rest of the rules are shown in Table I.

| If $I$ = | and $\omega$ = | Then $B_c$ is: |
|----------|----------|----------|
| low | medium | fair |
| low | high | good |
| medium | low | bad |
| medium | medium | fair |
| medium | high | good |
| high | low | bad |
| high | medium | bad |

Table I. Fuzzy rules for bearing wear

Notice that there are no rules for the cases $I = low$, $\omega = low$ and $I = high$, $\omega = high$. This was determined by multiple output classifications (*good* and *bad*) for these conditions as mentioned in the previous section. The non-use of these rules is verified when comparing the training data of Figure 4 with the resulting membership functions of Figure 5. There were no data inputs for the case of $I = high$, $\omega = high$. while there did exist two data inputs for the case $I = low$, $\omega = low$. However, it should be noted that these two data inputs only partially activated (approximately 50%) this rule while they fully activated the rule $I = medium$, $\omega = medium$. Therefore, the latter rule is the dominant rule and the former rule can be ignored without any impact on the system.

## IV. Conclusion

A fuzzy/neural system has been proposed to perform bearing wear fault detection in single phase induction motors. The fuzzy/neural system can provide quantitative descriptions of the motor faults under different operating conditions as well as qualitative heuristic explanations of these operating conditions and the fault detection procedures. This knowledge is gained through the use of a non-invasive fault monitoring scheme that requires *essentially* no expert knowledge of the motor to be monitored; however, a minimal amount of knowledge about the motor is required. The resulting fuzzy rules and membership functions did make heuristic sense while the fuzzy/neural motor fault detector achieved 100% classification accuracy in the fault detection application. These facts validate the procedure as a means to extract knowledge of motor faults and fault detection.

## References

[1] Peter J. Tavner, James Penman, <u>Condition Monitoring of Electrical Machines</u>, Research Studies Press Ltd. John Wiley & Sons Inc.

[2] S. Cambrias and S.A. Rittenhouse, *Generic Guidelines for Life Extension of Plant Electrical Equipment*, Electric Power Research Institute Report EL-5885, July 1988.

[3] Mo-yuen Chow, Sui Oi Yee, "Application of Neural Networks to Incipient Fault Detection in Induction Motors," *Journal of Neural Network Computing*, Auerback, Vol. 2, No. 3, pp. 26-32, 1991.

[4] D.R. Boothman, E.C. Elgar, R.H. Rehder, and R.J. Woodall, "Thermal tracking-A rational approach to motor protection," *IEEE Transactions Power Apparatus and Systems*, Sept./Oct. 1974, pp. 1335-1344.

[5] Mo-yuen Chow, "Artificial Neural Network Methodology in Real-Time Incipient Fault Detection of Rotating Machines," in *Proc. National Science Foundation Workshop on Artificial Neural Network Methodology in Power Systems Engineering*, Clemson University (Clemson, SC), April 8-10, 1990, pp. 80-85.

[6] R.W. Smeaton, *Motor Application and Maintenance Handbook*, 2nd ed. New York: McGraw-Hill, 1987.

[7] James E. Timperly, "Incipient Fault Detection Through Neutral RF Monitoring of Large Rotating Machines," *IEEE Transaction on Power Apparatus and Systems*, Vol. PAS-102, No. 3, March 1983.

[8] Mo-yuen Chow, Sui Oi Yee, "Methodology For On-Line Incipient Fault Detection in Single-Phase Squirrel-Cage Induction Motors Using Artificial Neural Networks," *IEEE Transactions on Energy Conversion*, Vol. 6, No. 3, pp. 536-545, Sept., 1991.

[9] Mo-yuen Chow, Peter M. Mangum, Sui Oi Yee, "A Neural Network Approach to Real-Time Condition Monitoring of Induction Motors," *IEEE Transactions on Industrial Electronics*, Vol. 38, No. 6, pp. 448-453, Dec. 1991.

[10] Arun K. Sood, Ali Amin Fahs, Naeim A. Henein, "Engine Fault Analysis Part I: Statistical Methods," *IEEE Transaction on Industrial Electronics*, Vol. IE-32, No. 4, Nov. 1985.

[11] A. Keyhani, S.M. Miri, "Observers for Tracking of Synchronous Machine Parameters and Detection of Incipient Faults," *IEEE Transaction on Energy Conversion*, Vol. EC-1, No. 2, June 1986.

[12] James C. Hung, B.J. Doran, "High Reliability Strapdown Platforms Using Two-Degree-of-Freedom Gyros," *IEEE Transaction on Aerospace and Electronic Systems*, Vol. AES-9, no. 2, pp. 253-259, March 1973.

[13] R.M. Tong and P.P. Bonissone, "A Linguistic Approach to Decision Making with Fuzzy Sets," *IEEE Transactions on Systems, Man, and Cybernetics*, Vol. SMC-10, pp. 716-723, 1980.

[14] L.A. Zadeh, "Fuzzy Sets," *Information and Control*, Vol. 8, pp. 338-353, 1965.

[15] L.A. Zadeh, The Concept of a Linguistic Variable and its Application to Approximate Reasoning," Parts 1 and 2, *Information Sciences*, Vol. 8, pp. 199-249, 301-357, 1975.

[16] C. Lin and C.S.G. Lee, "Neural-Network-Based Fuzzy Logic Control and Decision System," in *IEEE Transactions on Computers*, Vol. 40, No. 12, pp. 1320-1336, December, 1991.

[17] D.E. Rumelhart and J.L. McClelland, *Parallel Distributed Processing* Cambridge, MA: The MIT Press, 1986.

[18] G.J. Klir and T.A. Folger, *Fuzzy Sets, Uncertainty, and Information*. New Jersey: Prentice Hall, 1988.

# Solving fuzzy relational equations by max-min neural networks[1]

A. Blanco, M. Delgado, I. Requena
Departamento de Ciencias de la Computación e Inteligencia Artificial de la Universidad de Granada.
Facultad de Ciencias 18071 Granada España.

**Abstract** *The problem of identifying a fuzzy system has been faced from several points of wiew which include statiscal methods, neural networks and relational equations solving approaches.*

*In this paper we present the use of a neural networks without activation function in order to identify a fuzzy system through the solution of a fuzzy relational equation from a set of examples. The main contribution is to define a "smooth derivative" to be use in the minimization of the energy function which drives the learning procedure.*

*Some examples show the effectiveness of this new approach.*

*Keywords:* Neural network, smooth derivative, max-min composition.

## 1  Introduction

Any fuzzy system can be represented by a fuzzy relational equations system, and thus to identify it forces us to solve equations like $X \oplus R = Y$, where $X$ and $Y$ are inputs and outputs respectively, and where the composition operation $\oplus$ is generally a combination t-conorm / t-norm.

This equations and systems of equations of this type has been studied by several researches, to obtain different methods of resolution [1], [9], [10], [2], [6], [5], [3].

One of the most recent methods is that of fuzzy neural networks which has been developed and extensively used by authors such Pedrycz [7], Mukaidono [8].

In this paper we will present a new formulation of the learning procedure for max-min neural networks. This new methods attemp to overcome some drawbacks that the currently learning procedures present.

The paper will be organized into four sections. In the next one we present the problem, the drawbacks of used methods and our alternative solution. Section 3 devoted to some examples which show the effectiveness of our approach. In the section 4 we present the conclusions.

## 2  Identification of fuzzy relational equations by fuzzy neural networks without activation function

### 2.1  The problem

Our objective is to identify a fuzzy system through solving a fuzzy relational equation by a max-min fuzzy neural network. We will assume the fuzzy relational equations is $X \oplus R = Y$, $X \in [0,1]^r$ , $Y \in [0,1]^s$, $R \in [0,1]^{r \times s}$. We will limit ourselves to the case $\oplus = max - min$ and we suppose that we have a set of examples $[X^i, Y^i : i = 1...p]$ to solve $R$, and we will use a neural network max-min for the identification, using these examples, to train the neural network.

Obviously, the problem is to establish, the design the neural network and the learning method.

---

[1]This work has been developed under project PB 92-0945 of DGICIT. MADRID

## 2.2  Net topology

We are going to consider a fuzzy neural network with the following topology:

The input-output pairs are $(x_1, \cdots, x_i, \cdots, x_r)$ and $(Out_1, \cdots, Out_j, \cdots, Out_s)$, where $Out_j$ is determined by $Out_j = max[min(x_i, w_{ij})]$, the $w_{ij}$ being the elements of the weight matrix $W$ that assess the synaptic conexions.

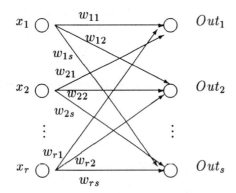

Let observe no activation function is considered here. The activation function has a double objective in an artificial neural: on the one hand, to adapt the output to prefix range, on the other to fix a threshold. In our case, we use the $max - min$ operation and obviously, the output range is fixed by the operation, it is $[0, 1]$, so the first objective is reached.

On the other hand, the operation $min$ is a threshold function, for each input $x_j$ and each weight $w_{ij}$, which represent the saturation level

$$min(x_j, w_{ij}) = \begin{cases} x_j & si \ x_i \leq w_{ij} \\ w_{ij} & si \ x_i > w_{ij} \end{cases}$$

Thus using any activation function, is not needed after to apply $max - min$ to the input, since it is underlying in the own process.

So, we are considering a neural network without hidden layer, where the inputs are values $X \in [0, 1]^r$ and the outputs $Y \in [0, 1]^s$ are obtained by $Y = max(min(W, X))$ being $W$ the weight matrix. If $X = (x_1, x_2, \ldots x_r)$, $Y = (Out_1, y_2, \ldots Out_s)$ and the elements of $W$ matrix, we note by $w_{ij}$, the outputs are obtained such that:

$$Out_1 = max[min(x_1, w_{11}), min(x_2, w_{21}) \ldots min(x_r, w_{r1})]$$
$$\vdots$$
$$Out_s = max[min(x_1, w_{1s}), min(x_2, w_{2s}) \ldots min(x_r, w_{rs})]$$

## 2.3  Learning whit smooth derivative.

The objective of training the network is to adjust the weights so that application of a set of inputs produces the desided set of outputs.

This is driven by minimizing the square of the difference between the desired output $T_j$ and the actual $O_j$, for all the patterns to be learnt so, $E = 1/2 \sum (T_j - O_j)^2$, where $O_j = max_i(min(x_i, w_{ij}))$

$$\text{It is obvious that} \qquad \frac{\partial E}{\partial w_{ij}} = \frac{\partial E}{\partial O_j} * \frac{\partial O_j}{\partial w_{ij}} \qquad\qquad (1)$$

We are going to expand the second factor of ( 1)

$$\frac{\partial O_j}{\partial w_{ij}} = \underbrace{\frac{\partial Max(Min(x_i, w_{ij}))}{\partial Min(x_s, w_{sj})}}_{P1} * \underbrace{\frac{\partial Min(x_s, w_{sj})}{\partial w_{sj}}}_{P2} \qquad\qquad (2)$$

$$P1 = \frac{\partial Max(Min(x_s, w_{sj}), \underset{i \neq s}{Max}(Min(x_i, w_{ij})))}{\partial Min(x_s, w_{sj})}$$

After that
$$P1 = \begin{cases} 1 & Min(x_s, w_{sj}) \geq \underset{i \neq s}{Max}(Min(x_i, w_{ij})) \\ 0 & Min(x_s, w_{sj}) < \underset{i \neq s}{Max}(Min(x_i, w_{ij})) \end{cases} \tag{3}$$

$$P2 = \frac{\partial Min(x_s, w_{sj})}{\partial w_{sj}} = \begin{cases} 1 & si \ x_s \geq w_{sj} \\ 0 & si \ x_s < w_{sj} \end{cases} \tag{4}$$

By combining P1 and P2 we will obtain the value of ( 2)

$$\frac{\partial O_j}{\partial w_{ij}} = \begin{cases} x_s < w_{sj} \begin{cases} x_s \geq \underset{i \neq s}{Max}(Min(x_i, w_{ij})) & \longrightarrow & C1 \\ x_s < \underset{i \neq s}{Max}(Min(x_i, w_{ij})) & \longrightarrow & C2 \end{cases} \\ x_s \geq w_{sj} \begin{cases} w_{sj} \geq \underset{i \neq s}{Max}(Min(x_i, w_{ij})) & \longrightarrow & C3 \\ w_{sj} < \underset{i \neq s}{Max}(Min(x_i, w_{ij})) & \longrightarrow & C4 \end{cases} \end{cases}$$

So, the value of ( 2), say C, will 0 in the cases C1, C2 ó C4, and 1 in C3.

Now we are going to expand the first factor of ( 1), for that we will denote $-\frac{\partial E}{\partial O_j}$ by $\delta_j$ $\frac{\partial E}{\partial O_j} = -(T_j - O_j)$;

Therefore, $-\frac{\partial E}{\partial w_{ij}} = \delta_j C$. Finally, the changes for the weights will be obtained by the expression: $\triangle w_{ij} = \mu \delta_j C$, from which the learning rule is stated as:

$\triangle w_{ij} = \mu \delta_j C$, where $\delta_j = (T_j - O_j)$

By applying this learning process, it is not guaranteed the network to learn, obviously because the value of C is null in three of the four cases C1-C4.

If we observe the pictorial representation of the function $min(x, w) = g(x)$

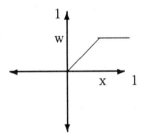

we see this function has a "crisp" behavior, to change its form and value in the threshold w, that produces the values of C in cases C1-C4. Thus to improve the perfomance of the learning process, we are interesting to change this "crisp" behavior by one "fuzzy" able to capture the real meaning of $min(x, w)$ in a vague context. Takin onto account that $min(x, w)$ measures the relative position of $x$ with respect to $w$, we propose to measure for each $x$ the inclusion degree $w$ in $x$, which we will denote $\|w \subset x\|$

On its turn, to assess we can apply any implication function to get the inclusion degree but our experiences have shown that the Gödel implication is the most suitable one.

$$\|w \subset x\| = w \overset{G}{\longrightarrow} x = \begin{cases} 1 & si \ w \leq x \\ x & si \ w > x \end{cases}$$

On the other hand, similar to before when dealing with $max\{a, b \ldots c\}$, if we denote "$max1 = max\{a, b \ldots c\}$" and "$max2 = max\{\{a, b \ldots c\} - \{max1\}\}$" we are interesting to know the inclusion degree of $max1$ in $max2$, $\|max1 \subset max2\|$.

By the Gödel implication

$$\|max1 \subset max2\| = max1 \xrightarrow{G} max2 = \begin{cases} 1 & si\ max1 \le max2 \\ max2 & si\ max1 > max2 \end{cases}$$

So, when we are using the inclusion degree, P1 and P2 have not only 0 or 1 values but values in $[0, 1]$.

In ( 4), we observe that P2 is zero, when $x_s < w_{sj}$, but using the inclusion degree of $w_{sj}$ in $x_s$ $\|w \subset x\|$, a new value of P2 is obtained to be.

$$P2 = \begin{cases} 1 & si\ w \le x \\ x & si\ w > x \end{cases} \tag{5}$$

By the same process with P1 and to denoting $max2 = \underset{i \ne s}{\text{Max}}\ (Min(x_i, w_{ij}))$

P1 being zero, obviously, when $min(x_s, w_{sj}) < max2$, so, we are going to take the inclusion degree of $max2$ in $min(x_s, w_{sj})$:

$$\|max2 \subset min(x_s, w_{sj})\| = \begin{cases} 1 & si\ max2 \le (min(x_s, w_{sj})) \\ min(x_s, w_{ij})) & si\ max2 > (min(x_s, w_{sj})) \end{cases}$$

Fynaly, we obtain the P1 value:

$$P1 = \begin{cases} 1 & si\ max2 \le (min(x_s, w_{sj})) \\ min(x_s, w_{ij})) & si\ max2 > (min(x_s, w_{sj})) \end{cases} \tag{6}$$

By combining ( 5) and ( 6), we will obtain:

$$\frac{\partial O_j}{\partial w_{ij}} = \begin{cases} x_s < w_{sj} \begin{cases} x_s \ge \underset{i \ne s}{\text{Max}}\ (Min(x_i, w_{ij})) \longrightarrow C1 = x_s \\ x_s < \underset{i \ne s}{\text{Max}}\ (Min(x_i, w_{ij})) \longrightarrow C2 = x_s * x_s \end{cases} \\ x_s \ge w_{sj} \begin{cases} w_{sj} \ge \underset{i \ne s}{\text{Max}}\ (Min(x_i, w_{ij})) \longrightarrow C3 = 1 \\ w_{sj} < \underset{i \ne s}{\text{Max}}\ (Min(x_i, w_{ij})) \longrightarrow C4 = w_{sj} \end{cases} \end{cases}$$

Obviously, the values obtaining from $\frac{\partial O_j}{\partial w_{ij}}$ are depend of the implication to choosing. We have making several trials whit all implications, except with those, that it give us an inclusion degree null.

Remark After the training, the network is able to reproduce the behabiour of the equation. But according to the topology and perfomance of the network, it is obvious that $W$ is just a solution for $R$. Then this method does not only identifie the system but solves the relational equation.

## 3  Examples

In this section we will illustrate our former developments with an example. It will constructed according to the following steps.

S1.- A fuzzy relational matrix $R$ is fixed arbitrarily to be

$$R = \begin{pmatrix} 0.6 & 0.5 & 0.8 & 0.3 & 0.2 \\ 0.4 & 0.1 & 0.9 & 0.6 & 0.4 \\ 0.1 & 0.1 & 0.9 & 0.8 & 0.5 \\ 0.9 & 0.2 & 0.9 & 0.1 & 0.5 \\ 0.4 & 0.5 & 0.3 & 0.8 & 0.9 \end{pmatrix}$$

**S2-.** For a suitable set of $X_t \in [0, 1]$ the associated $Y_t$ are obtained by max-min composition $Y_t = X_t \oplus R$
$t = 1 \cdots 5$

$X_1 = (1, 0, 0, 0, 0)$  $Y_1 = (0.6, 0.5, 0.8, 0.3, 0.2)$
$X_2 = (0, 1, 0, 0, 0)$  $Y_1 = (0.4, 0.1, 0.9, 0.6, 0.4)$
$X_3 = (0, 0, 1, 0, 0)$  $Y_1 = (0.1, 0.1, 0.9, 0.8, 0.5)$
$X_4 = (0, 0, 0, 1, 0)$  $Y_1 = (0.9, 0.2, 0.9, 0.1, 0.5)$
$X_5 = (0, 0, 0, 0, 1)$  $Y_1 = (0.4, 0.5, 0.3, 0.8, 0.9)$

**S3.-** The network with the topology we have presented in the section 2 is trained with the pairs $\{(X_t, Y_t)$
$t = 1 \cdots 5\}$

The following table and graphic show the behavior of the training process against different learning rates.

| $\mu$ | Iteration | Eror achieved |
|-------|-----------|---------------|
| 0.2 | 57 | 7.22990E-11 |
| 0.5 | 19 | 2.93584E-11 |
| 0.8 | 8 | 5.28875E-11 |
| 1 | 1 | 1.11669E-23 |

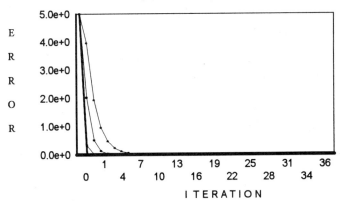

Let observe the obtained weight matrix $W$ coincides wiht $R$

$$\mu = 0.5 \text{ Iteration } 5$$
$$\begin{pmatrix}
0.603125 & 0.503906 & 0.801562 & 0.305468 & 0.206250 \\
0.409375 & 0.114062 & 0.901565 & 0.606240 & 0.409375 \\
0.114062 & 0.114062 & 0.901565 & 0.803125 & 0.507812 \\
0.901565 & 0.212500 & 0.901565 & 0.114062 & 0.507812 \\
0.409375 & 0.507812 & 0.319375 & 0.803125 & 0.901565
\end{pmatrix}$$

$$\mu = 0.5 \text{ Iteration } 10$$
$$\begin{pmatrix}
0.600097 & 0.500122 & 0.800048 & 0.301708 & 0.200195 \\
0.400292 & 0.100439 & 0.900048 & 0.600195 & 0.400292 \\
0.100439 & 0.100439 & 0.900048 & 0.800097 & 0.500244 \\
0.900048 & 0.200390 & 0.900048 & 0.100439 & 0.500244 \\
0.400292 & 0.500244 & 0.300341 & 0.800097 & 0.900048
\end{pmatrix}$$

$$\mu = 0.5 \text{ Iteration } 19$$
$$\begin{pmatrix}
0.600000 & 0.500000 & 0.800000 & 0.300001 & 0.200001 \\
0.400001 & 0.100001 & 0.900000 & 0.600000 & 0.400001 \\
0.100001 & 0.100001 & 0.900000 & 0.800000 & 0.500000 \\
0.900000 & 0.200001 & 0.900000 & 0.100001 & 0.500000 \\
0.400001 & 0.500000 & 0.300001 & 0.800000 & 0.900000
\end{pmatrix}$$

$$\mu = 1 \text{ Iteration } 1$$
$$\begin{pmatrix}
0.600000 & 0.500000 & 0.800000 & 0.300001 & 0.200000 \\
0.400000 & 0.099999 & 0.900000 & 0.600000 & 0.400000 \\
0.099999 & 0.099999 & 0.900000 & 0.800000 & 0.500000 \\
0.900000 & 0.200000 & 0.900000 & 0.099999 & 0.500000 \\
0.400000 & 0.500000 & 0.300000 & 0.800000 & 0.900000
\end{pmatrix}$$

# 4  Conclusions

**1-.** Any fuzzy system described by a fuzzy relational equation may be identified by using a max-min fuzzy neural network.

**2.-** The introduction of the so called "smooth derivative" improves the effectiveness of the training process.

**3-.** According to the topology and of the network and its perfomance rules it happens that the weight matrix after the training is a possible solution for $R$. So our method not only identifies the system but slves the associated fuzzy relational equation.

# References

[1] Czogala, E., Drewniak, J. Pedrycz, W. (1982). Fuzzy relation equations on a finite set. *Fuzzy Sets and Systems* 7, 275-285.

[2] Bour, L., Lamotte, M. (1986). Determination d'un operateur de maximalisation pour la resolution d'equations de relation floues. *BUSEFAL* 25, 95-106.

[3] Bour, L., Lamotte, M. (1988). Equations de relations floues avec la composition Conorme-Norme triangulaires. *BUSEFAL*, 34 86-94.

[4] Draper, N., Smith, H. (1966). *Applied Regression Analysis*. Wiley, New York.

[5] Dubois, G., Lebrat, E., Lamotte, M., Bremont, J. (1992). Solving a system of fuzzy relation equations by using a hierarchical process. *Procedings of the IEEE International Conference of Fuzzy Sets and Systems*, 679-686.

[6] Di Nola, A., Pedrycz, W. Sessa, S., Wang, P. (1984). Fuzzy relation equations under a class of triangular norms: a survey and new results. *Stochastica*, 2 99-145.

[7] Pedrycz, W. (1990). Relational Structures in fuzzy sets and neurocomputation. *Proceedings of International Conference on Fuzzy Logic and Neural Networks*,, Vol. (pp. 235-238) IIZUKA'90.

[8] Saito, T., Mukaidono, M. (1991). A learning algorithm for max-min network and its application to solve fuzzy relation equations. *Proceedings of the 2nd International Conference on Fuzzy Logic and Neural Networks* , Iizuka'92 184-187.

[9] Sánchez, E. (1984). Solution of fuzzy equation with extended operations. *Fuzzy Sets and Systems*, 12, 237-248.

[10] Sessa, S. (1984). Some results in the setting of fuzzy relation equation theory. *Fuzzy Sets and Systems*, 14, 281-297.

[11] Zadeh, L. (1973). *The Concept of a Linguistc Variable and its Application to Approximate Reasoning*, American Elsevier, New York.

# An ART Network with Fuzzy Control for Image Data Compression

C. J. Wu, A. H. Sung, and H. S. Soliman
Department of Computer Science
New Mexico Tech.
Socorro, NM 87801-4682

**Abstract - In this paper, the application of a simplified ART1 (adaptive resonance theory) network with a fuzzy controller to image data compression is presented. The unique feature of a vigilance parameter of ART allows the direct control of trade-off between compression ratio and image quality; the fuzzy controller can be used to adjust vigilance to seek a better compromise automatically. Furthermore, the decision table of this fuzzy controller is designed to guarantee convergence. Therefore, the network is insensitive to the given initial vigilance values. Simulations are performed and the results indicate that this fuzzy-control-equipped simplified ART1 network provides a promising technique for image data compression.**

## I. INTRODUCTION

The large amount of time and storage required to transmit pictorial data brings about the need of image data compression. In [1], the application of Adaptive Resonance Theory 1 (ART1) [2], [3], to image data compression was studied, and its performance was compared with CPN's and we showed that ART1 networks can be a promising alternative. In this work, another important issue for image data compression in real-time environment is studied. In general, the combination of compression and distortion ratios is a function of pictures and techniques used in the compression process; experimentally, some of the images will have profound effect on the compression process. It is rather difficult to obtain the balance between those two ratios for various pictures in real-time environment. For this reason, we usually use trial-and-error method to search for the better compromise of distortion and compression ratios.

Fuzzy logic has been successfully used in many applications to control the process automatically. In our work, the incorporation of fuzzy controllers in the design of simplified ART1 networks is investigated. The simplified ART1 networks has several advantages over the original ART1 networks including less implementation cost and processing time. The rest of the paper is organized as follows. Section II describes the architecture and the learning algorithms of simplified ART1 networks. The mechanism of fuzzy controller is introduced in section III. Simulation results are given in Section IV. At last, conclusions and discussions are given in section V.

## II. THE ARCHITECTURE AND LEARNING ALGORITHMS OF SIMPLIFIED ART1

The advantage of using adaptive resonance theory (ART1) in image data compression is that it has an external control mechanism--the vigilance parameter--to control directly the trade-off between compression and distortion ratios. However, the ART1 has some disadvantages for this specific application as well [1]. One disadvantage is the possible large number of iterations which slow down the process of encoding each input presented to the network. In addition, we found out that the function of bottom-up weight matrix is just to decide the retrieval order of categories; therefore, we can compare the input with existing prototypes directly and eliminate the bottom-up weight matrix. The simplified ART1 (SART1) has two immediate advantages. Since the most similar prototype will be picked first, there is no need to check on other prototypes; hence, it runs much faster than the ordinary ART1 networks. The other advantage is less implementation cost since the SART1 only uses one weight matrix instead of two in the original ART1.

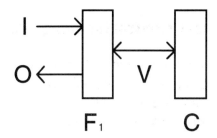

Figure 1. The architecture of simplified ART1.

### *The Learning Algorithm Of SART1*

Step 1: Initialize the weight matrix and vigilance parameter as follows:

$$V = [1] \tag{1}$$

$$0 \leq \rho \leq 1 \tag{2}$$

where $V$ and $\rho$ are the weight matrix and vigilance parameter, respectively.

Step 2: Present the binary (unipolar) input vector $I$ to $F_1$. The winning node, say $j$, is the node with the weight vector ($V_j$) most similar to input $I$ in terms of the Hamming distance ($\sum_{i=1}^{M} |I_i - V_{ij}|$) in layer $C$ where $M$ is the dimension of $I$. In case of tie, one of them is to be selected arbitrarily.

Step 3: Test the similarity between the prototype of winner $j$ and the input pattern $I$ by computing the following ratio:

$$\gamma = \frac{(M - \sum_{i=1}^{M} |I_i - V_{ij}|)}{M} \tag{3}$$

Step 4: If $\gamma \geq \rho$, go to Step 5; otherwise, a new category (node) will be added to layer $C$ unless there is no new node available. In that case, the operation terminates.

Step 5: Update only the weight vectors associated with the winner $j$ as follows.

$$V_{ij}(t+1) = I_i V_{ij}(t) \quad \text{for } 1 \leq i \leq M \tag{4}$$

Step 6: If no new input vector, terminate the process; otherwise, get the next input vector and go back to Step 2.

## III. THE CONTROL MECHANISM OF FUZZY CONTROLLER

Fuzzy logic has been successfully applied to automatic control [4]. Therefore, we propose to incorporate a fuzzy controller into the SART to achieve the automatic control during the compression process. In the algorithm above, the compromise of those two ratios is closely related to $\rho$. A high vigilance leads to low distortion and compression ratios. The relation of SART1 and fuzzy controller is depicted in Figure 2. The outputs of SART1 is the following two ratios:

(1) Generated Compression Ratio (GCR):

$$GCR = \frac{NF}{C \cdot N + F \cdot \log_2 C} \tag{5}$$

where $N$ is the dimension of the input vectors or subimages, $F$ is the total number of subimages, and $C$ is the total number of categories formed during training.

(2) Generated Distortion Ratio (GDR):

$$GDR = \left( \sum_{j=1}^{F} \sum_{i=1}^{N} (I_{ij} - I'_{ij})^2 \Big/ NF \right) \times 100 \tag{6}$$

where $I_{ij}$ is the value in the original image and $I'_{ij}$ is the corresponding value in the reconstructed image. Since an accepted distortion ratio is our primary concern, we use the difference between the target and the actual distortion ratios ($\Delta E$) to be one of the inputs from SART1 network to the fuzzy controller. The other input ($C\rho$) used by the fuzzy controller is the change of vigilance to detect the possible oscillation occurred in the process.

$$\Delta E = TDR - GDR \tag{7}$$

$$C\rho = \rho_t - \rho_{t-1} \tag{8}$$

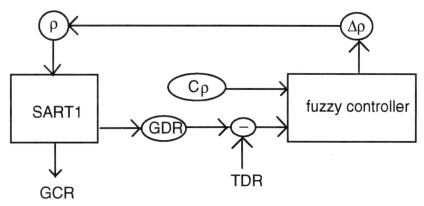

GCR: generated compression ratio
GDR: generated distortion ratio
TDR: target distortion ratio

Figure 2. The schematic diagram of SART1 with fuzzy controller.

In general, the membership function of a fuzzy set $A$, $\mu_A$, is defined as follows: [5]

$$\mu_A : X \to [0,1] \tag{9}$$

where $X$ is the universal set. In this paper, seven fuzzy sets are used--PB (positive big), PM (positive medium), PS (positive small), ZE (zero), NS (negative small), NM (negative medium), and NB (negative big). Since we use the triangular type of membership functions (see Eq. 10), except fuzzy sets NB (see Eq. 11) and PB (see Eq. 12), the membership function of fuzzy sets can be characterized by three values--(left boundary, central value, right boundary). Hence, for any crisp value $x$, the membership values are decided as follow:

$$\mu_A(x) = \begin{cases} (x-L)/(C-L) & L \le x \le C \\ (x-R)/(C-R) & C \le x \le R \\ 0 & otherwise \end{cases} \tag{10}$$

$$\mu_{NB}(x) = \begin{cases} (x-R)/(C-R) & C \le x \le R \\ 1 & x \le C \\ 0 & otherwise \end{cases} \tag{11}$$

$$\mu_{PB}(x) = \begin{cases} (x-L)/(C-L) & L \le x \le C \\ 1 & x \ge C \\ 0 & otherwise \end{cases} \qquad (12)$$

where C, L, and R are the central value, left boundary, and right boundary respectively and A={NM, NS, ZE, PS, PM}. For the input $\Delta E$, let us choose the membership functions of the fuzzy sets to be as follows:

NB     (-100, -6, -3)      NM     (-5, -3, -1)      NS     (-2, -1, 0)

ZE     (-0.5, 0, 0.5)

PS     (0, 1, 2)      PM     (1, 3, 5)      PB     (3, 6, 100)

For the input $C\rho$ and output $\Delta\rho$, the membership functions of fuzzy sets are defined as follows:

NB     (-1, -0.25, -0.1)      NM     (-0.15, -0.1, -0.05)      NS     (-0.1, -0.05, 0)

ZE     (-0.025, 0, 0.025)

PS     (0, 0.05, 0.1)      PM     (0.05, 0.1, 0.15)      PB     (0.1, 0.25, 1)

The decision table used by fuzzy controller is designed as follow to guarantee the convergence.

Table 1. The decision table of fuzzy controller.

| | | NB | NM | NS | ZE | PS | PM | PB |
|---|---|---|---|---|---|---|---|---|
| | NB | PM | PS | ZE | ZE | PS | PM | PB |
| | NM | PS | PS | ZE | ZE | PS | PM | PM |
| | NS | PS | PS | ZE | ZE | PS | PS | PS |
| $\Delta E$ | ZE | ZE | ZE | ZE | ZE | ZE | ZE | ZE |
| | PS | NS | NS | NS | ZE | ZE | NS | NS |
| | PM | NM | NM | NS | ZE | ZE | NS | NS |
| | PB | NB | NM | NS | ZE | ZE | NS | NM |

*(Column header spanning top: C $\rho$)*

The convergence of this fuzzy controller is guaranteed by the following three special design criteria and one constraint to avoid possible infinite oscillations:

    (1) $\Delta\rho$ should not be bigger than $C\rho$ in terms of fuzzy sets.

    (2) Whenever the oscillation is observed (through the decision table), $\Delta\rho$ should be smaller than $C\rho$ in terms of fuzzy sets. For example, if $\Delta E$ is PB and $C\rho$ is PB, then $\Delta\rho$ is at most jump back NM.

    (3) If $\rho$ is equal to zero and GDR is smaller than TDR, the process is terminated since it makes no sense have negative $\rho$ in ART networks.

    (4) $0 \le TDR \le 1$

However, the convergence does not mean that the system will always find the proper $\rho$, which generates the TDR, if it exists.

    The membership value of entry $ij$ in the decision table is decided as follow:

$$M_{ij} = \min(M_i, M_j) \qquad (13)$$

where $M_i$ and $M_j$ are the membership values of $\Delta E$ and $C\rho$, respectively. The crisp output $\Delta\rho$ of fuzzy controller is decided as follow: [6]

$$\Delta\rho = \frac{\sum_{i=1}^{K}\sum_{j=1}^{N}f(M_{ij})C_{ij}}{\sum_{i=1}^{K}\sum_{j=1}^{N}f(M_{ij})} \tag{14}$$

where N is the number of fuzzy sets, $f(M_{ij})$ and $C_{ij}$ are the membership and central values of entry $ij$ in the decision table respectively. The process will be terminated when $\Delta\rho$ reaches 0.

## IV. SIMULATION RESULTS

Experiments are performed on two pictures of $256^2$ black-and-white pixels using the simplified ART1 network with fuzzy controller. Simulation results of ten randomly generated initial vigilance values using 8x8 subimages are listed in Table 2. As shown, the system is insensitive to the given initial $\rho$'s and the final $\rho$'s are staying in the target range. The reconstructed images of "goat" and "monkey" along with their original pictures are shown in Fig. 3.

(a)

(b)

(c)

(d)

**Figure 3: The comparison between original and reconstructed images (a) the original image of the goat (b) the reconstructed image of the goat (c) the original image of the monkey (d) the reconstructed image of the monkey.**

Table 2. Simulation results of simplified ART1 with fuzzy controller (DDQ=18.5%)

| | | Initial $\rho$ | | | | | | | | | |
|---|---|---|---|---|---|---|---|---|---|---|---|
| | | 0.38 | 0.58 | 0.13 | 0.15 | 0.51 | 0.27 | 0.1 | 0.19 | 0.12 | 0.86 |
| Goat | final $\rho$ | 0.667 | 0.670 | 0.667 | 0.667 | 0.686 | 0.661 | 0.689 | 0.640 | 0.663 | 0.710 |
| | GCQ | 10.240 | 10.240 | 10.240 | 10.240 | 10.039 | 10.240 | 9.942 | 10.240 | 10.240 | 9.846 |
| | GCE(%) | 21.013 | 21.013 | 21.013 | 21.013 | 18.292 | 21.013 | 17.165 | 20.764 | 21.013 | 17.038 |
| Monkey | final $\rho$ | 0.665 | 0.657 | 0.665 | 0.667 | 0.660 | 0.670 | 0.703 | 0.704 | 0.705 | 0.610 |
| | GCQ | 9.309 | 9.309 | 9.309 | 9.309 | 9.309 | 9.309 | 9.225 | 9.225 | 9.225 | 9.309 |
| | GCE(%) | 19.136 | 19.136 | 19.136 | 19.136 | 19.136 | 19.136 | 17.899 | 17.899 | 17.899 | 19.136 |

where DDR, TDR, and GCR are the desired distortion ratio, target distortion ratio, and generated compression ratio, respectively.

## V. CONCLUSIONS AND DISCUSSIONS

Since we can not predict the proper vigilance value in advance, with the help of fuzzy controller, the ART1 network can be insensitive to a given initial vigilance values. Nevertheless, our computer simulations show that the final vigilance will converge to the target range. In addition, the simplified ART1 (SART1) used in this work has two advantages over the original one--less processing time and implementation cost. Hence, the image data compression using SART1 with fuzzy controller in the real-time environment is promising. The on-going research is to extend this architecture to accommodate the gray scale and color pictures as well as tune-up the fuzzy controller to keep the final vigilance as close as possible to the target value.

## REFERENCES

[1] C. J. Wu, A. H. Sung, and H. S. Soliman, "Image Data Compression Using ART Networks," in *Proc. Artificial Neural Networks In Engineering*, 1993, pp. 417-422.

[2] S. Grossberg, "Adaptive Pattern Classification and Universal Recording: I. Parallel Development and Coding of Neural Feature Detectors," *Biological Cybernetics*, vol. 23, pp. 121-134, 1976.

[3] G. A. Carpenter and S. Grossberg, "A Massively Parallel Architecture for a Self-Organizing Neural Pattern Recognition Machine," *Computer Vision, Graphics, and Image Processing*, vol. 37, pp. 54-115, 1987.

[4] M. Sugeno, *Industrial Applications of Fuzzy Control*. North Holland, Amsterdam, 1985.

[5] G. J. Klir and T. A. Folger, *Fuzzy Sets, Uncertainty, and Information*, Englewood Cliffs, NJ: Prentice- Hall, 1988.

[6] B. Kosko, *Neural Networks and Fuzzy Systems: A Dynamical Systems Approach to machine Intelligence*, Englewood Cliffs, NJ: Prentice-Hall, 1992.

# A Fuzzy Finite State Machine Implementation Based on a Neural Fuzzy System

**Fatih A. Unal**          **Emdad Khan**

**National Semiconductor, Embedded Systems Division**
**2900 Semiconductor Dr., Santa Clara, CA, 95051**

**Abstract** — The outputs of a feedforward neural network depend on the present inputs only. Difficulties arise when a solution requires memory in such applications as speech processing, seismic signal processing, language processing, and spatiotemporal signal processing. For such applications, the outputs are not only the functions of the present inputs but the present states (or the past inputs and the outputs) as well. The fuzzy finite state machines can be effectively used in these applications. The aim of this study is to show that a fuzzy finite state machine can be realized using our neural fuzzy system. In a fuzzy finite state machine, the output and the next state depend on the input and the present state which in turn is a function of the previous inputs. To accommodate the memory requirement, the feedforward structure of the neural fuzzy system is changed to a recurrent architecture by adding a feedback loop from the output layer to the input layer during the recall mode. The validity of the approach is verified with a temporal pattern matching experiment.

## 1. Introduction

In our previous works, it has been shown that the artificial neural networks (ANN) can be combined with fuzzy logic systems to achieve high performance and low cost solutions [1]-[3]. In this work, the feedforward structure of our neural fuzzy system (NeuFuz) is changed to a recurrent architecture by adding a feedback loop from the output layer to the input layer during the recall mode. With this modification, we demonstrate that NeuFuz can be used to implement a Fuzzy Finite State Machine (FFSM).

A FFSM is an extension of a conventional FSM. The difference is that the inputs, the present states and the next states are represented using fuzzy sets in contrast with the crisp inputs and the states of a traditional crisp FSM. The output function and the state transition function of the FSM are replaced by the recurrent fuzzy rules which determine the outputs and govern the transitions of the FFSM between the fuzzy states. The fuzzification of FSM results in data reduction (less memory to implement the FFSM) as well as more robust operation (less susceptible to system parameter changes or to noise) as stated in [5].

Also, with this approach, writing a new program is not required since the ANN of NeuFuz is trained with the state transition table and implementation of a new FFSM is achieved directly with the training data. This increases the reliability. Furthermore, the suggested approach facilitates the design and implementation of a FFSM using a microcontroller.

The FFSM concept is introduced in the next section. In Section 3, a brief description of NeuFuz is given. An illustrative temporal pattern matching experiment is carried out in Section 4 to verify the validity of our approach. Section 5 contains the conclusions.

## 2. A Fuzzy Finite State Machine

A FFSM is a synchronous sequential machine and it can be defined as a quintuple following the crisp FSM definition in [4]:

$$FFSM = (I, O, S, f, g) \qquad (1)$$

where I, O, and S are finite, nonempty sets of fuzzy inputs, fuzzy outputs, and fuzzy states respectively, and f: I X S → S is the state transition function, g is the output function such that g: I X S → O. The Cartesian product I X S contains all pairs of fuzzy elements (u(t), x(t)). The state transition function f maps each fuzzy pair (u(t), x(t)) onto a fuzzy next state x(t+1) in S, and the output function g maps each fuzzy pair (u(t), x(t)) onto a fuzzy output y(t) in O in accordance with

$$x(t+1) = f(x(t), u(t)) \qquad (2)$$

and

$$y(t) = g(x(t), u(t)) \qquad (3)$$

where x(t) is the fuzzy present state, u(t) is the fuzzy input, and x(t+1) is the fuzzy next state. The FFSM is defined only at discrete times t = 0, 1, 2, ... in this work.

A Moore type of FFSM is employed in this study since it is more convenient to use it with our neural fuzzy system (NeuFuz has one output at this time). The output of a Moore machine is a function of present state only

$$y(t) = g(x(t)) \qquad (4)$$

Note that, in a Mealy machine, the output is a function of both present state and the input. For both machine types, the next state depends on both the input and the present state as seen in equation (2).

In general, the state transition function f, and the output function g (given in equations (2) and (3) respectively) can be represented using fuzzy rules in the following format

If *Input = u(t)* and *PresentState = x(t)*
   then *NextState = x(t + 1)* $\qquad (5)$

If *Input = u(t)* and *PresentState = x(t)*
   then *Output = y(t)* $\qquad (6)$

and the membership functions $\mu_{u(t)}$, $\mu_{x(t)}$, $\mu_{x(t+1)}$ are defined for the fuzzy input u(t), fuzzy present and the next states x(t) and x(t+1) respectively as usual.

Note that the derivation of the fuzzy rules (given in equation (5) and (6)) and the associated membership functions is not trivial. The task is to transform the crisp FSM into the corresponding FFSM. This requires the partitioning of the crisp states and the crisp inputs into the clusters (fuzzy sets) and generate the associated membership functions using the state transition table that corresponds to the crisp FSM. This is achieved by means of the ANN in NeuFuz.

### 3. NeuFuz Neural Fuzzy System

A detailed description of our neural fuzzy system can be found in [1] - [3]. Here, a brief description will be given. The neural fuzzy system consists of four components: A feedforward ANN, a fuzzy rules and membership function generator (FRMFG), a fuzzy rule verifier and optimizer (FRVO) and an automatic code converter (ACC). A supervised learning is employed to train the ANN. A modified backpropagation algorithm is used to learn the input output relationship for a given system. The ANN has four inputs, one output and five hidden layers. The architecture of the ANN explicitly maps to the fuzzy logic structure, and the connection structure along with the weights directly define the fuzzy rules and the membership functions. The FRMFG extracts the fuzzy rules and the membership functions when the ANN converges to a minimum energy state. The validity of the solution can be verified by the FRVO and the solution can be optimized by eliminating the fuzzy rules that do not contribute to the result significantly. This way the output error can be controlled and kept at tolerable levels which is not a trivial task when an ANN is not utilized to obtain the fuzzy system. Finally, the ACC generates a program which corresponds to the fuzzy system.

During the training phase, one input of the ANN is used for the external input U(t), and another input is used for the present state X(t). The output of the ANN is used for the next state X(t+1) as shown in Figure 1. There is no need to use separate outputs for the next state X(t+1) and the external output Y(t) of the FSM, since a Moore type of FFSM is implemented.

The output X(t+1) is fed back to the present state input X(t) through a unit delay D, once the training is completed. Thus, the architecture of the neural fuzzy system is changed from a feedforward structure to a recurrent architecture during the recall mode as depicted in Figure 2. There is a comparator at the output to determine whether the system is at the final

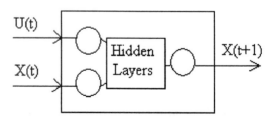

Legend:
U(t): Input, X(t): Present State, X(t+1): Next State

Figure 1: NeuFuz in training mode

1750

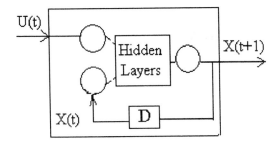

Legend:
D: Unit Delay

Figure 2: NeuFuz in recall mode

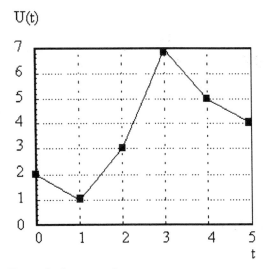

Figure 3: A temporal pattern

state. Note that the output y(t) is zero in all states except the final state.

The defuzzification in NeuFuz can be summarized as

$$X(t+1) = \sum_{i=1}^{M} \mu_{i,u}(U(t)) \cdot \mu_{i,x(t)}(X(t)) \cdot \mu_{i,x(t+1)}(X(t+1)) \qquad (7)$$

where M is the total number of fuzzy rules, $\mu_u$, $\mu_{x(t)}$, $\mu_{x(t+1)}$ are the membership functions for the fuzzy input u(t), fuzzy present state x(t) and the fuzzy next state x(t+1) respectively. The capital U, X(t) and X(t+1) represent the crisp values. In this work, singletons are used for the next state. The addition and multiplication are used for the disjunction and conjunction operations as shown in equation (7).

## 4. Experimental Results

To show the validity of the proposed approach, a Moore type of FFSM is implemented to recognize a temporal pattern shown in Figure 3. First, a state transition diagram (depicted in Figure 4), and a corresponding state transition table (tabulated in Table 1) are derived to recognize the given temporal pattern. The FFSM changes from a fuzzy state to another fuzzy state depending on the temporal pattern sample values arriving at discrete instances. There is an initial state S0 at which the system waits for the specified temporal pattern. The initial state is resumed each time the correct temporal sequence is broken. The final state S6 is reached when the entire temporal pattern is received. The FFSM stays in S6 indefinetely until it is reset.

Table 1: State Transition Table

| INPUTS | PRESENT STATE | NEXT STATE |
|---|---|---|
| U(t) | X(t) | X(t+1) |
| U(0) | S0 | S1 |
| Else | S0 | S0 |
| U(0) | S1 | S1 |
| U(1) | S1 | S2 |
| Else | S1 | S0 |
| U(0) | S2 | S1 |
| U(2) | S2 | S3 |
| Else | S2 | S0 |
| U(0) | S3 | S1 |
| U(3) | S3 | S4 |
| Else | S3 | S0 |
| U(0) | S4 | S1 |
| U(4) | S4 | S5 |
| Else | S4 | S0 |
| U(0) | S5 | S1 |
| U(5) | S5 | S6 |
| Else | S5 | S0 |
| All | S6 | S6 |

Table 2: A Procedure to Generate the Training Data

1. Devise the state diagram
2. Generate the state transition table from the state diagram
3. Assign numerical values to the states
4. Generate the "Else" inputs within the input range along with the corresponding states

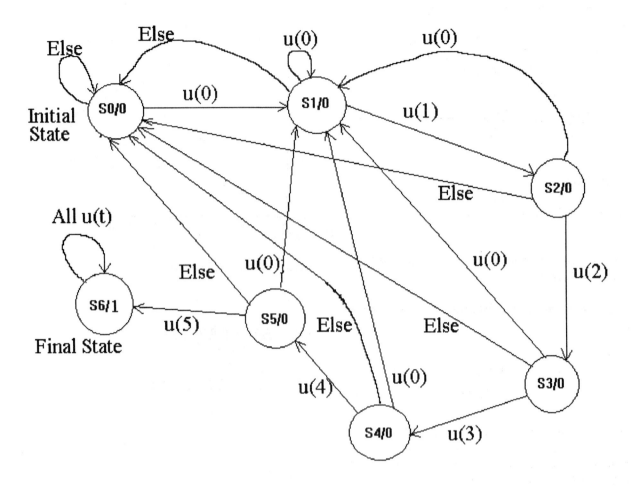

Figure 4: State transition diagram to recognize the temporal pattern in Figure 3.

The next step is the training of the ANN in NeuFuz. Table 2 contains the procedure to generate the training data set. Following the procedure, the numerical values are assigned to the states as listed in Table 3. The numerical assignments can be arbitrary. However, to facilitate the convergence of the ANN in NeuFuz, the assignment in Table 3 is chosen to minimize the variations between the next state values

Table 3: Numerical state assignments

| STATE | NUMERIC VALUE |
|-------|---------------|
| S0 | 3 |
| S1 | 4 |
| S2 | 5 |
| S3 | 6 |
| S4 | 2 |
| S5 | 7 |
| S6 | 1 |

X(t+1) corresponding to the consecutive next states and inputs as can be seen in Table 4.

The parameters shown in Table 5 are used for the training. It takes 109,286 iterations for the ANN to converge with these parameters. The generated membership functions for the input u(t) and the present state x(t), and the fuzzy rules are given in matrix format in Figure 5a, 5b and Table 6 respectively. The NeuFuz generates an assembler program code automatically. The required amount of memory to implement this FFSM is 894 bytes (in National's COP8 Microcontroller ).

The recognition rate of FFSM is 100 % when the input sample values are within the specified error range ($\varepsilon = 0.01$). A detailed analysis of recognition performance is not carried out as the major focus is to show the feasibility of implementing the FFSM using NeuFuz.

Table 4: Training Data.

| U | X(t) | X(t+1) |
|---|------|--------|
| 0 | 3 | 3 |
| 1 | 3 | 3 |
| 2 | 3 | 4 |
| 3 | 3 | 3 |
| 4 | 3 | 3 |
| 5 | 3 | 3 |
| 6 | 3 | 3 |
| 7 | 3 | 3 |
| 0 | 4 | 3 |
| 1 | 4 | 5 |
| 2 | 4 | 4 |
| 3 | 4 | 3 |
| 4 | 4 | 3 |
| 5 | 4 | 3 |
| 6 | 4 | 3 |
| 7 | 4 | 3 |
| 0 | 5 | 3 |
| 1 | 5 | 3 |
| 2 | 5 | 4 |
| 3 | 5 | 6 |
| 4 | 5 | 3 |
| 5 | 5 | 3 |
| 6 | 5 | 3 |
| 7 | 5 | 3 |
| 0 | 6 | 3 |
| 1 | 6 | 5 |
| 2 | 6 | 4 |
| 3 | 6 | 3 |
| 4 | 6 | 3 |
| 5 | 6 | 3 |
| 6 | 6 | 3 |
| 7 | 6 | 3 |
| 0 | 2 | 3 |
| 1 | 2 | 3 |
| 2 | 2 | 4 |
| 3 | 2 | 3 |
| 4 | 2 | 3 |
| 5 | 2 | 7 |
| 6 | 2 | 3 |
| 7 | 2 | 3 |
| 0 | 7 | 3 |
| 1 | 7 | 3 |
| 2 | 7 | 4 |
| 3 | 7 | 3 |
| 4 | 7 | 1 |
| 5 | 7 | 3 |
| 6 | 7 | 3 |
| 7 | 7 | 3 |

Table 4 cont'd.

| 0 | 1 | 1 |
|---|---|---|
| 1 | 1 | 1 |
| 2 | 1 | 1 |
| 3 | 1 | 1 |
| 4 | 1 | 1 |
| 5 | 1 | 1 |
| 6 | 1 | 1 |
| 7 | 1 | 1 |

Table 5: Training Mode Parameters
Legend:
$\varepsilon$: Tolerated error, LR: Learning rate, LF: Learning factor, MF: Membership functions, u: Fuzzy input, x(t): Fuzzy present state

| $\varepsilon$ | 0.01 |
|---|---|
| LR | 0.4 |
| LF | 0.08 |
| # MF u | 7 |
| # MF x(t) | 7 |
| # Rules | 49 |

Table 6: Fuzzy Rules
Legend:
u: Fuzzy input, x(t): Fuzzy present state, singleton values are for next state x(t+1), VLM: Very Low Match, LM: Low Match, MLM: Medium Low Match, MM: Medium Match, MHM: Medium High Match, HM: High Match, VHM: Very High Match. The numerical values are truncated to two digits after the decimal point.
Example: If u(t) is VH and x(t) is VHM then x(t+1) is 2.89

| u \ x | VLM | LM | MLM | MM | MHM | HM | VHM |
|-------|------|------|------|------|------|------|------|
| VL | 0.74 | 2.66 | 2.20 | 3.18 | 2.66 | 2.20 | 2.91 |
| L | 0.68 | 2.33 | 2.49 | 6.77 | 1.17 | 6.05 | 1.62 |
| ML | 0.16 | 3.48 | 4.66 | -1.27 | 4.85 | -2.69 | 4.93 |
| M | 1.00 | -0.49 | -0.96 | 6.32 | 3.71 | 5.59 | -1.60 |
| MH | 0.51 | 3.78 | 6.47 | -0.11 | 0.97 | -0.37 | 1.85 |
| H | -1.14 | 8.14 | -4.02 | 2.52 | 1.30 | 1.92 | 0.86 |
| VH | 0.74 | 2.61 | 3.27 | 3.17 | 2.64 | 2.19 | 2.89 |

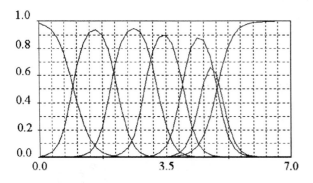

Figure 5a: Membership Functions $\mu_u$ (U(t)) for the Fuzzy Input u(t)

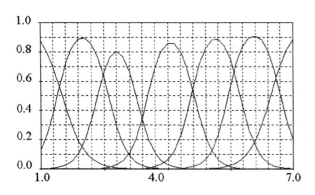

Figure 5b: Membership Functions $\mu_{x(t)}$ (X(t)) for the fuzzy present state x(t)

## 5. Conclusions

The main objective of this work is to show that a FFSM can be implemented using a neural fuzzy system, namely, NeuFuz. The experimental results show that the proposed approach can be effectively used especially when the cost is a primary concern. The main advantage of this approach is that the fuzzification of the states and the input result in the reduction of memory size that is required to implement the FFSM in a microcontroller. Besides, there is no need to write a program (or design hardware) to implement the FFSM. The training data corresponding to the state transition table can be used to implement the FFSM directly. This ensures the reliability of the operation. Also, additional flexibility (to control the transitions between the states) can be achieved by modifying the training set (adding, deleting and changing training data). For instance, changing noise characteristics can be considered to update the training data set when a FFSM is employed as an adaptive

signal detection subsystem. Furthermore, the suggested approach facilitates the design and implementation of a FFSM using a microcontroller and provides more robust response (less susceptible to system parameter changes or to noise) as stated in [5]. The results obtained encourage us to apply our method to the automotive problems such as airbag deployment and antilock braking system as well as other spatiotemporal signal processing applications. Further studies are underway to determine the applicability of this approach to speech processing.

## References

[1] E. Khan and P. Venkatapuram, "NeuFuz: Neural Network Based Fuzzy Logic Design Algorithms," FUZZ-IEEE'93 Proceedings, vol. 1, pp. 647-654, San Francisco, California, March 28 - April 1, 1993.

[2] E. Khan, "Neural Network Based Algorithms For Rule Evaluation & Defuzzification in Fuzzy Logic Design," WCNN'93 Proceedings, vol. 2, pp. 31-38, July 11-15, 1993.

[3] E. Khan, "NeuFuz: An Intelligent Combination of Fuzzy Logic with Neural Nets," IJCNN'93 Proceedings, vol. 3, pp. 2945-2950, Nagoya, Japan, October 25-29, 1993.

[4] Z. Kohavi, Switching and Finite Automata Theory. New York, NY: McGraw-Hill Publishing Company, 1978.

[5] E. H. Mamdani, "Twenty years of Fuzzy Control: Experiences Gained and Lessons Learnt," FUZZ-IEEE'93 Proceedings, vol. 1, pp. 339-344, San Francisco, California, March 28 - April 1, 1993.

# Design of a Fuzzy Controller Mixing Analog and Digital Techniques

J.Tombs, A.Torralba and L.G.Franquelo
Dpto. de Ingeniería Electrónica, de Sistemas y Automática
Escuela Superior de Ingenieros
Avda. Reina Mercedes, s/n, SEVILLA–41012 (Spain)
Tlf: 34–5–4556849, Fax: 34–5–4629205
e–mail: torralba@gtex02.us.es

## I. Introduction

In spite of its early introduction by L.A.Zadeh [1], only in the recent years the interest of the researchers has been turned into fuzzy–logic. Fuzzy inference systems have usually been implemented in software on digital computers or microprocessors. Although this approach can cope with most of the actual problems, real–time systems often require very short time responses, and a hardware implementation becomes the only solution.

The first fuzzy chips were due to Togai and Watanabe and [2]–[4], marked the beginning of the hardware approach using digital circuits. Later, Yamakawa [5]–[7], developed a fuzzy–logic controller using analog circuits. Since then, several fuzzy–logic chips have been proposed using both, analog and digital techniques [9]–[26].

Analog circuits are attractive for fuzzy VLSI, as they allow an easy implementation of the arithmetic functions which are needed in a fuzzy controller. Analog circuits are efficient in area allowing parallelism to be employed. However, they suffer from noise disturbances and interferences. Besides, they are susceptible to process variations and mismatching. On the other hand, digital circuits are robust and easy to design and process. However, arithmetic functions (such as multiplication and division) require large digital circuits.

To overcome the above mentioned drawbacks, pulse stream techniques have been recently proposed [27]–[29]. Pulse stream techniques use digital signals to carry information and control analog circuitry. Among the existing pulse stream techniques (PWM, PFM, PDM), pulse width modulation (PWM) has been selected here due to its simplicity. This paper presents the architecture of a fuzzy logic controller using PWM techniques. A prototype circuit is being designed using a standard CMOS process.

## II. Structure of the Controller

Figure 1 depicts the block diagram of the controller. It uses Larsen's product operation rule [30]. The connective *and* is implemented by means of the MINIMUM operator

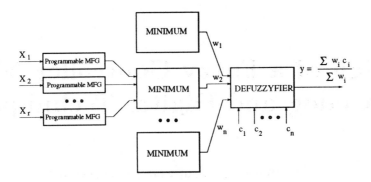

*Figure 1:* Block Diagram of the Controller

rule $i$: IF $x_1$ is $A_{1i}$ and ... and $x_r$ is $A_{ri}$ THEN $y$ is $C_i$

$$\mu_{C_i'} = MINIMUM(\ \mu_{A_{1i}}(x_{1k}), ..., \mu_{A_{ri}}(x_{rk})\ ) \quad \times \quad \mu_{C_i} = w_i \times c_i \tag{1}$$

where the output membership functions $\mu_{C_i}$ are singletons. The defuzzifier unit uses the *center-of-gravity* method,

$$y = \frac{\sum_1^n w_i \times c_i}{\sum_1^n w_i} \tag{2}$$

Figure 2 depicts the internal structure of the Membership Function Generator (MFG), that uses the current mode approach [31]. In figure 2, $I_{in}$ is substracted from the midpoint current $I_{mid}$. The difference is rectified using the circuit proposed in [32] and the result is amplified and substracted from the reference current $I_{max}$. As $I_{mid}$ is a digitally programmable current, the circuit of figure 2 can generate a set of triangular membership functions. A similar approach was previously used in [33]. The output current $I_{out}$ drives a capacitor whose voltage is compared to a reference voltage $V_{ref}$ to obtain its PWM codification.

As the inputs to the MINIMUM block are PWM signals, a simple digital AND gate performs the minimum operation. Note that for a large number of rules, the inputs of the MINIMUM blocks can be multiplexed using a digital or analog multiplexer.

Figure 3 shows the multiplication circuit that uses a transconductance amplifier adapted for PWM. The magnitude of the output current $I_1$ in figure 3 is proportional to $V_{gi}$, for the duration of the driving pulse $V_{wi}$. Transistor $M4$ in figure 3 reduces disturbances in the power lines due to simultaneous current switching.

Figure 3 also shows the adder circuit, that uses a dynamically biased opamp and an integrator capacitor. At the end of a clock cycle, the sampled output voltage is proportional to $\sum g_i c_i$. In figure 3, the first cycle of a conversion is distinguished from the rest. In the first cycle, $V_{gi} = V_{ci}$ and the sampled output $V_{wc} = K \sum w_i c_i$. In the following cycles, the same input $V_y$ is applied to the gates of all the M1 transistors, i.e. $V_{gi} = V_y \quad i = 1...n$, and the sampled output $V_z = K y \sum w_i$. Capacitor $C$ has to be a good linear capacitor. The product and addition circuit, first proposed in [28], has been previously used in a neural network design. A test circuit implemented in 1.5 $\mu$m CMOS technology [34]–[35], showed that a high precision can be achieved wilst using a reduced area.

Figure 4 shows the divider circuit that uses the product–and–addition circuitry of figure 3. In this circuit, $m+1$ clock cycles are necessary to obtain the controller output, where $m$ is the desired output

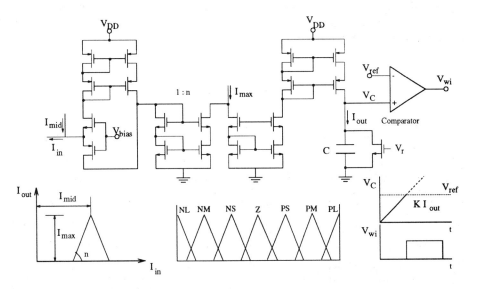

Figure 2: Membership Function Generator

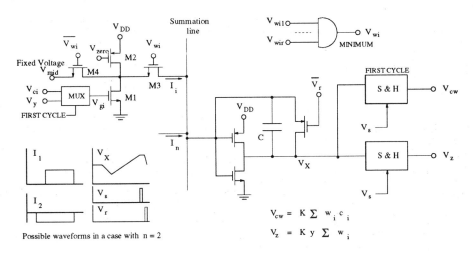

$$V_{cw} = K \sum w_i c_i$$
$$V_z = K y \sum w_i$$

Figure 3: Product and addition circuit.

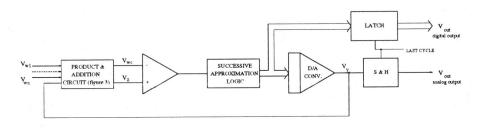

Figure 4: Divider circuit

precision. Successive aproximation logic and a $D/A$ converter (a capacitor array) generate a sequence of $V_y$ analog signals which converges in $m$ clock cycles to the output controller $y$ (expression (2)). Note that the output does not depend on the exact value of capacitor $C$ in figure 3, provided that $V_X$ is in the active comparator input range.

## REFERENCES

[1] L.A.Zadeh. "Fuzzy Sets". *Inform.Contr.*, vol. 8, pp. 338–358, 1965.

[2] M.Togai and H.Watanabe. "Expert system on a chip: an engine for real–time approximate reasoning". *IEEE Expert Syst. Mag.*, vol. 1, pp. 55–62, 1986.

[3] M.Togai and S.Chiu. "A fuzzy accelerator and a programming environment for real–time fuzzy control" in *Proc. 2nd IFSA Congress*, Tokyo, Japan. July 1987, pp. 147–151.

[4] H.Watanabe, W.Dettloff and E.Yount. "A VLSI Fuzzy Logic Inference Engine for Real–Time Process Control". *IEEE Journal of Solid–State Circuits*, vol. 25, no. 2, pp. 376–382, April 1990.

[5] T.Yamakawa and T.Miki. "The current mode fuzzy logic integrated circuits fabricated by the standard CMOS process", *IEEE Trans. Computers*, vol. C–35, no. 2, pp. 161–167, 1986.

[6] T.Yamakawa and K.Sasaki. "A simple fuzzy computer hardware system employing min and max operations. A challenge to 6th generation computer", in *Proc. 2nd IFSA Congress*, Tokyo, Japan, July 1987.

[7] T.Yamakawa. "Fuzzy microprocessor rule chip and defuzzifier chip". *Proc. of Int. Workshop on Fuzzy Syst. Appl.*, Aug. 1988, pp. 51–52.

[8] M.Sasaki, T.Inone, Y.Shirai and F.Ueno. "Fuzzy multiple–input maximum and minimum circuits in current mode and their analysis using bounded–difference equations". *IEEE Trans. on Comput.*, vol. C–39, no. 6, June 1990, pp. 768–774.

[9] H.Watanabe, J.R.Symon, W.Dettloff, and E.Yount. "VLSI Fuzzy Chip and Inference Accelerator Board Systems". *Proc. of IEEE Int. Symp. on Multi–Valued Logic*, pp. 120–127, May 1991.

[10] H.Watanabe. "RISC Approach to Design of Fuzzy Proc. Architecture". *Proc. Int. Conf. Fuzzy Systems*, San Diego, March 1992, pp. 431–443.

[11] M.Morisue, Y.Kogure. "A Superconducting Fuzzy Processor". *Proc. Int. Conf. Fuzzy Systems*, San Diego, March 1992, pp. 443–450.

[12] T.Yamakawa. "A Fuzzy Programmable Logic Array (Fuzzy PLA)". *Proc. Int. Conf. Fuzzy Systems*, San Diego, March 1992, pp. 459–468.

[13] M.Sasaki, N.Ishikawa, F.Ueno and T.Inone. "Current Mode Analog Fuzzy Hardware with Voltage Input Interface and Normalization Loop". *Proc. Int. Conf. Fuzzy Systems*, San Diego, March 1992, pp. 451–457.

[14] T.Yamakawa. "A Fuzzy Programmable Logic Array (Fuzzy PLA)". *Proc. Int. Conf. Fuzzy Systems*, San Diego, March 1992, pp. 459–468.

[15] H.Ikeda, N.Kisu, Y.Harimoto and S.Nakamura. "A Fuzzy Inference coprocessor using a flexible Active–Rule–Driven Architecture". *Proc. Int. Conf. Fuzzy Systems*, San Diego, March 1992, pp. 537–544.

[16] D.Ishizuke, K.Tanno, Z.Tang and H.Matsumoto. "Design of a Fuzzy Controller with Normalization Circuits". *Proc. Int. Conf. Fuzzy Systems*, San Diego, March 1992, pp. 1303–1308.

[17] H.Eichfeld, M.Lohner and M.Muller. "Architecture of a Fuzzy Logic Controller with Optimized memory organization and operator design". *Proc. Int. Conf. Fuzzy Systems*, San Diego, March 1992, pp. 1317–1323.

[18] T.Tsukano, T.Inone, F.Ueno. "A Design of Current–Mode Analog Circuits for Fuzzy Inference Hardware Systems", *Proc. Int. Conf. Fuzzy Systems*, San Diego, March 1992, pp. 1385–1388.

[19] T.Miki, H.Matsumoto, K.Ohto, T.Yamakawa. "Silicon Implementation for a Novel High–Speed Fuzzy Inference Engine: Mega–FLIPS Analog Fuzzy Processor". *Journal of Intelligent & Fuzzy Systems*, vol.1, no. 1, 1993, pp. 27–42.

[20] F.Colodro, A.Torralba and L.G.Franquelo. "A Fuzzy–logic chip using stochastic–logic: the defuzzifier". *Proc. of the Int. Conf. on Fuzzy Control Applications, IFCA'93*, Tarrasa, February 1993.

[21] A.P.Ungering, K.Thuerner and K.Goser. "Architecture of a Fuzzy Logic Controller with Pipelining and Optimized chip area". *Proc. 2nd. Int. Conf. Fuzzy Systems*, San Francisco, March 1993, pp. 447–452.

[22] J.L.Huertas, S.Sánchez–Solano, A.Barriga, I.Baturone. "A Fuzzy Controller using Switched–Capacitor techniques". *Proc. Int. Conf. Fuzzy Systems*, San Francisco, March 1993, pp. 516–520.

[23] H.Watanabe, D.Chen. "Evaluation of Fuzzy Instruction in a RISC Processor", *Proc. of the 2nd. IEEE Int. Conf. on Fuzzy Syst.*, San Francisco, March 1993 pp. 521–526.

[24] M.Sasaki, F.Ueno and T.Inoue. "7.5 MFLIPs Fuzzy Microprocessor Using SIMD and Logic–in–Memory Structure". *Proc. of the 2nd IEEE Int. Conf. on Fuzzy Syst.*, San Francisco, March. 1993, pp. 527–533.

[25] T.Kelter, K.Schumacher, K.Goser. "Realization of a Monolitic Analog Fuzzy Controller". *Proc. of the 20th European Solid–State Circuits Conf.*, Sevilla, Sept. 1993, pp. 47–49.

[26] F.Colodro, A.Torralba and L.G.Franquelo. "A Digital Fuzzy–logic controller with a simple architecture". Submitted for presentation at the *Int. Symp. Circ. and Syst., ISCAS'94*.

[27] A.F.Murray. "Pulse Arithmetic in VLSI Neural Networks". *IEEE Micro Mag.*, Dec. 1989, pp. 64–74.

[28] A.F.Murray, D.del Corso and L.Tarassenko, "Pulse–Stream VLSI Neural Networks mixing Analog and Digital techniques". *IEEE Trans. on Neural Networks*, vol. 2, no. 2, March 1991.

[29] A.Hamilton, A.F.Murray, D.J.Baxter, H.Martin Reeki and L.Tarassenko. "Integrated pulse stream Neural Networks: Results, Issues and Pointers". *IEEE Trans. on Neural Networks*, vol. 3, no. 3, May 1992, pp. 385–393.

[30] C.C.Lee. "Fuzzy Logic in Control Systems: Fuzzy Logic Controller–Parts I and II". *IEEE Trans. Syst. Man Cybern.*, vol. 20, no.2, pp. 404–435, Mar./Apr. 1990.

[31] C.Tomazou. F.J.Lidgey and R.Haigh. *Analog IC Design: the Current–Mode Approach*. Peter Peregrinus Ltd., 1990.

[32] Z.Wang. *Current–Mode Analog Integrated Circuits and Linearization Techniques in CMOS Technology*. Hartung–Gorre Verlag, Konstanz. Series in microelectronics; vol. 7, 1990.

[33] I.Baturone, A.Barriga, S.Sánchez–Solano, J.L.Huertas. "Una implementación Hardware de Circuitos Básicos de Lógica Difusa". *Actas del VII Congreso de Microelectrónica*, pp. 407–412, Toledo, 1992 (in spanish).

[34] J.Tombs, *Multilayer Neural Networks and their Implementation in Analogue VLSI*, D.Phil. Thesis Trinity Term 1992, Oxford University Engineering Department.

[35] J.Tombs, L.Tarassenko and A.Murray. "A Novel Analogue VLSI Design for Multi–Layer Networks". *IEE Special Issue on Artificial Neural Networks*, 1992.

# A Fuzzy–logic Controller with On–chip Learning, employing Stochastic Logic

A.Torralba, F.Colodro and L.G.Franquelo
Dpto. de Ingeniería Electrónica, de Sistemas y Automática
Escuela Superior de Ingenieros,
Avda. Reina Mercedes, s/n, SEVILLA–41012 (Spain)
Tlf: 34–5–4556849, Fax: 34–5–4629205
e–mail: torralba@gtex02.us.es

## I. INTRODUCTION

In the recent years there has been an increasing interest in the development of efficient fuzzy controller hardware, capable to cope with the requirements of real–time systems. Togai and Watanabe developed the first fuzzy logic chip in 1985 [2]–[4]. Later, Yamakawa [5]–[7], developed a fuzzy–logic hardware using analog techniques. Since then, several fuzzy–logic chips have been proposed using both, analog and digital techniques [9]–[25].

This paper presents a new hardware implementation of a digital fuzzy controller that uses stochastic logic to implement the arithmetic functions involved in the defuzzification and learning processes. Stochastic logic systems use binary random signals whose average can be viewed as an analog value in the range [0,1]. Using stochastic logic has a number of advantages over other analog and digital implementations, such as multiplication using a simple AND gate. Stochastic logic has been successfully applied to different fields including neural processing [26]–[30].

A proper selection of the different parts of the controller, and the use of stochastic logic leads to a simple digital architecture with a short response time (less than 21 $\mu s$ for a 7 bit stochastic precision and a 12 $MHz$ system clock). A main feature of the proposed controller is its learning capability, obtained by adjusting the position of the output singletons. Presently, a prototype of the controller is being designed using a 1.5 $\mu m$ CMOS technology.

## II. ARCHITECTURE OF THE CONTROLLER

Figure 1 depicts the block diagram of the controller. Triangular–shaped membership function are used to define the linguistic terms of the input variables. The linguistic terms of the output variable are singletons, defined by their position $c_i$ in the $x$–axis. Fuzzy implication follows the Larsen's product operation rule [31]. The $MINIMUM$ operator implements the connective *and*,

rule $i$: IF $x_1$ is $A_{1i}$ and ... and $x_r$ is $A_{ri}$ THEN $y$ is $C_i$

$$\mu_{C_i'} = MINIMUM(\ \mu_{A_1 i}(x_1 k),...,\mu_{A_n i}(x_n k)\ )\ \times\ \mu_{C_i} = w_i \times c_i \tag{1}$$

The algebraic addition implements the connective *also*. The defuzzification process is carried-out by means of the *Center-of-Gravity* method

$$y = \frac{\sum_1^n w_i \times c_i}{\sum_1^n w_i} \tag{2}$$

where $n$ is the number of control rules.

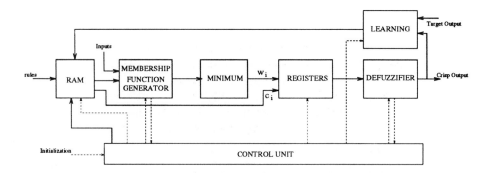

Figure 1: Block Diagram of the controller

In the prototype here presented, the controller has three input variables and one output variable and it is able to process 16 fuzzy rules. To reduce the response time, the controller uses a pipeline structure: while the rule implication module computes the values $w_i$ corresponding to time step $k$ (by means of the expression (1)), the defuzzifier module obtains the crisp output value corresponding to time step $k-1$ (following the expression (2)). The values $w_i$ $(i = 1, 2, .., 16)$, are sequentially obtained in order to reduce the complexity of the circuit. This does not increase the response time as the defuzzification process is the slowest part of the controller.

**Rule implication module**

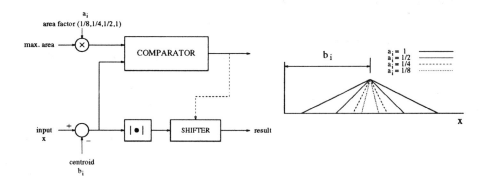

Figure 2: Membership function evaluator

The module that computes the membership functions of the input variables is depicted in figure 2. The input value is substracted from the membership centroid $b_i$, and the result is shifted to account for an area factor $a_i$. The values $w_i$ are computed by means of the expression (1). Only those $w_i$ whose values are greater than zero are stored in the intermediate registers for defuzzification. In the case of three input variables, there are no more than 8 non-null $w_i$, provided that the overlap degree

1760

of the input variable membership functions are 2; that is, only the membership functions in direct neighborhood are allowed to overlap.

**The defuzzifier**

Computing the numerator of the expression (2) requires the addition of $n$ product terms. The product $w_i \times c_i$ can be obtained by means of a logical AND of two stochastic signals, which pulse with a probability which is proportional to $w_i$ and $c_i$, respectively. The addition $\sum_1^n w_i \times c_i$ can be achieved by accumulating the stochastic product pulses in a counter. In each cycle, the signal $s_i$ that pulses with the probability of the $i$-th product term $w_i \times c_i$ increments or not a counter depending on its binary value (figure 3). In this way, $n$ cycles are required to update each counter.

The circuit in figure 3 also computes the division of equation (2). In this circuit, the upper counter accumulates the denominator value. When this counter reaches a reference value (e.g., $2^q$), the most significant bits of the numerator counter contain the desired result.

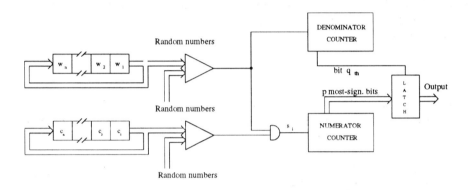

*Figure 3:* Defuzzifier circuit

This circuit deserves further attention. To obtain a precise result in a short period of time, previous denominator scaling is convenient. To this purpose, the $w_i$ terms are right-shifted until the maximum of them is in the range $[2^{q-1}, 2^q]$, where $q$ is the number of bits of the random number stream.

In spite of its simplicity, the proposed hardware is capable of a very high speed. For instance, with a 7 bits resolution in the denominator counter, 256 cycles are required in the worst case. If we use the architecture of figure 3, $8 \times 256 = 2048$ clock cycles are required in all. That is, 62 $\mu s$ using a 33 MHz sytem clock.

This delay can be further reduced if the pulse signals of the 8 input terms are accumulated in parallel, as proposed by the autors in [30]. In this case, the defuzzifier speed becomes independent on the number of rules. With the same figures of the example above, $7.8\mu s$ are expended in the computation of the controller output, allowing more than 125000 FLIPs. Note that a more accurate result can be obtained at the expense of a higher response time.

### III. LEARNING PROCESS

On-chip learning circuits allow a fast and low cost controller design. However, as far as the authors know, there are not on-chip learning implementations for fuzzy-logic controllers. The fuzzy-logic controller here proposed implements hardware learning using the Least Mean Square (LMS) method. This method attemps to minimize the squared error for training sample input–output pairs; for every training sample, it modifies the position of each output singleton $c_i$ using the partial derivative of the squared error with respect to $c_i$.

Figure 4 depicts the full block diagram of the defuzzifier circuit including the learning part. In normal mode, the target output is fixed to zero and, when the denominator counter attains the reference value, the numerator counter contains the crisp output value of the controller, as it was explained in the previous section.

In learning mode, the numerator counter accumulates the error

$$error = \frac{\sum_1^n w_i \times target\_output}{\sum_1^n w_i} - \frac{\sum_1^n w_i \times c_i}{\sum_1^n w_i} = target\_output - controller\_output \tag{3}$$

Now, when the denominator counter attains the reference value, the numerator counter contains the error between the target output and the controller output. Computing the error in this way reduces the negative effects of the random noise, due to the compensation that exists between the two terms in expression (3). Finally, a second learning phase is started where the singletons are updated in the bottom part of the figure 4, following the expression

$$\Delta c_j = \eta \; \frac{w_j}{\sum_1^n w_i} \; error \tag{4}$$

Once again, the denominator of this expression is $\sum_1^n w_i$. Hence, when the denominator counter attains the reference value, counters $C_1$ through $C_n$ contain the updated singleton values.

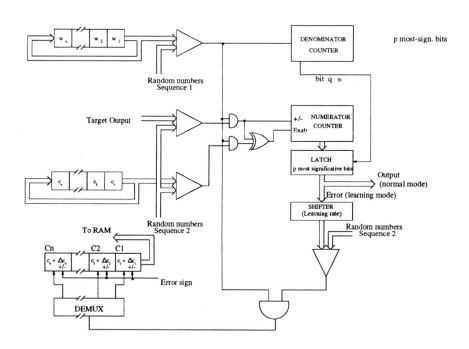

Figure 4: Defuzzifier circuit including learning

To study the behavior of the proposed controller and its learning algorithm, the whole circuit has been simulated using a hardware description language, and the results of these simulations are depicted in figures 5 and 6. In figure 5 the controller has learnt the $1 + \cos x$ function, and the controller response after learning is superimposed over the original function. In figure 6a, the original function is $(1 + \cos x)(1 + \cos y)$. Figure 6b and 6c show the controller response after learning with 16 and 36 rules, respectively, showing the effects of a small number of rules in the controller response.

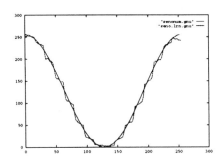

*Figure 5:* Results obtained after learning (1 input variable).

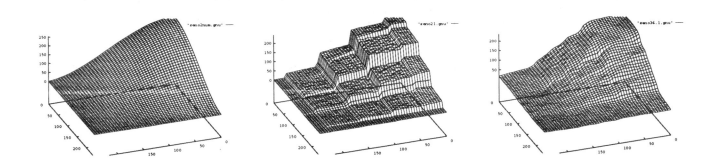

*Figure 6:* Results obtained after learning (2 input variables).

## REFERENCES

[1] L.A.Zadeh. "Fuzzy Sets". *Inform.Contr.*, vol. 8, pp. 338–358, 1965.

[2] M.Togai and H.Watanabe. "Expert system on a chip: an engine for real–time approximate reasoning". *IEEE Expert Syst. Mag.*, vol. 1, pp. 55–62, 1986.

[3] M.Togai and S.Chiu. "A fuzzy accelerator and a programming environment for real–time fuzzy control" in *Proc. 2nd IFSA Congress*, Tokyo, Japan. July 1987, pp. 147–151.

[4] H.Watanabe, W.Dettloff and E.Yount. "A VLSI Fuzzy Logic Inference Engine for Real–Time Process Control". *IEEE Journal of Solid–State Circuits*, vol. 25, no. 2, pp. 376–382, April 1990.

[5] T.Yamakawa and T.Miki. "The current mode fuzzy logic integrated circuits fabricated by the standard CMOS process", *IEEE Trans. Computers*, vol. C–35, no. 2, pp. 161–167, 1986.

[6] T.Yamakawa and K.Sasaki. "A simple fuzzy computer hardware system employing min and max operations. A challenge to 6th generation computer", in *Proc. 2nd IFSA Congress*, Tokyo, Japan, July 1987.

[7] T.Yamakawa. "Fuzzy microprocessor rule chip and defuzzifier chip". *Proc. of Int. Workshop on Fuzzy Syst. Appl.*, Aug. 1988, pp. 51–52.

[8] M.Sasaki, T.Inone, Y.Shirai and F.Ueno. "Fuzzy multiple–input maximum and minimum circuits in current mode and their analysis using bounded–difference equations". *IEEE Trans. on Comput.*, vol. C–39, no. 6, June 1990, pp. 768–774.

[9] H.Watanabe, J.R.Symon, W.Dettloff, and E.Yount. "VLSI Fuzzy Chip and Inference Accelerator Board Systems". *Proc. of IEEE Int. Symp. on Multi–Valued Logic*, pp. 120–127, May 1991.

[10] H.Watanabe. "RISC Approach to Design of Fuzzy Proc. Architecture". *Proc. Int. Conf. Fuzzy Systems*, San Diego, March 1992, pp. 431–443.

[11] M.Morisue, Y.Kogure. "A Superconducting Fuzzy Processor". *Proc. Int. Conf. Fuzzy Systems*, San Diego, March 1992, pp. 443–450.

[12] T.Yamakawa. "A Fuzzy Programmable Logic Array (Fuzzy PLA)". *Proc. Int. Conf. Fuzzy Systems*, San Diego, March 1992, pp. 459–468.

[13] M.Sasaki, N.Ishikawa, F.Ueno and T.Inone. "Current Mode Analog Fuzzy Hardware with Voltage Input Interface and Normalization Loop". *Proc. Int. Conf. Fuzzy Systems*, San Diego, March 1992, pp. 451–457.

[14] T.Yamakawa. "A Fuzzy Programmable Logic Array (Fuzzy PLA)". *Proc. Int. Conf. Fuzzy Systems*, San Diego, March 1992, pp. 459–468.

[15] H.Ikeda, N.Kisu, Y.Harimoto and S.Nakamura. "A Fuzzy Inference coprocessor using a flexible Active–Rule–Driven Architecture". *Proc. Int. Conf. Fuzzy Systems*, San Diego, March 1992, pp. 537–544.

[16] D.Ishizuke, K.Tanno, Z.Tang and H.Matsumoto. "Design of a Fuzzy Controller with Normalization Circuits". *Proc. Int. Conf. Fuzzy Systems*, San Diego, March 1992, pp. 1303–1308.

[17] H.Eichfeld, M.Lohner and M.Muller. "Architecture of a Fuzzy Logic Controller with Optimized memory organization and operator design". *Proc. Int. Conf. Fuzzy Systems*, San Diego, March 1992, pp. 1317–1323.

[18] T.Tsukano, T.Inone, F.Ueno. "A Design of Current–Mode Analog Circuits for Fuzzy Inference Hardware Systems", *Proc. Int. Conf. Fuzzy Systems*, San Diego, March 1992, pp. 1385–1388.

[19] T.Miki, H.Matsumoto, K.Ohto, T.Yamakawa. "Silicon Implementation for a Novel High–Speed Fuzzy Inference Engine: Mega–FLIPS Analog Fuzzy Processor". *Journal of Intelligent & Fuzzy Systems*, vol.1, no. 1, 1993, pp. 27–42.

[20] F.Colodro, A.Torralba and L.G.Franquelo. "A Fuzzy–logic chip using stochastic–logic: the defuzzifier". *Proc. of the Int. Conf. on Fuzzy Control Applications, IFCA'93*, Tarrasa, February 1993.

[21] A.P.Ungering, K.Thuerner and K.Goser. "Architecture of a Fuzzy Logic Controller with Pipelining and Optimized chip area". *Proc. 2nd. Int. Conf. Fuzzy Systems*, San Francisco, March 1993, pp. 447–452.

[22] J.L.Huertas, S.Sánchez–Solano, A.Barriga, I.Baturone. "A Fuzzy Controller using Switched–Capacitor techniques". *Proc. Int. Conf. Fuzzy Systems*, San Francisco, March 1993, pp. 516–520.

[23] H.Watanabe, D.Chen. "Evaluation of Fuzzy Instruction in a RISC Processor", *Proc. of the 2nd. IEEE Int. Conf. on Fuzzy Syst.*, San Francisco, March 1993 pp. 521–526.

[24] M.Sasaki, F.Ueno and T.Inoue. "7.5 MFLIPs Fuzzy Microprocessor Using SIMD and Logic–in–Memory Structure". *Proc. of the 2nd IEEE Int. Conf. on Fuzzy Syst.*, San Francisco, March. 1993, pp. 527–533.

[25] T.Kelter, K.Schumacher, K.Goser. "Realization of a Monolitic Analog Fuzzy Controller". *Proc. of the 20th European Solid–State Circuits Conf.*, Sevilla, Sept. 1993, pp. 47–49.

[26] D.E. van der Bout, T.K.Miller III, "A Digital Architecture Employing Stochasticism for the Simulation of Hopfield Neural Nets". *IEEE Trans. Circuits Syst.*, vol. 36, pp. 732–738, May 1989.

[27] M.S. Tomlimson, D.J.Walker, "DNNA: A Digital Neural Network Architecture". *Proc. Int. Neural Networks Conf.* (INNC–90), vol. 2, pp. 589–592, 1990.

[28] M.S.Melton, T.Phan, D.S.Reeves, D.E. van der Bout, "The TInMANN VLSI Chip". *IEEE Trans. Neural Networks*, vol.3, no. 3, May 1992.

[29] Y.Kondo Y.Sawada, "Functional Abilities of a Stochastic Logic Neural Network". *IEEE Trans. Neural Networks*, vol. 3, no. 3, May 1992.

[30] A.Torralba, F.Colodro. "Towards a fully parallel stochastic Hopfield neural netwotk". *Proc. of the Int. Symp. on CAS*, Chicago, May 1993, pp. 2741–2743.

[31] C.C.Lee. "Fuzzy Logic in Control Systems: Fuzzy Logic Controller–Parts I and II". *IEEE Trans. Syst. Man Cybern.*, vol. 20, no.2, pp. 404–435, Mar./Apr. 1990.

# A HIGH SPEED PARALLEL ARQUITECTURE FOR FUZZY INFERENCE AND FUZZY CONTROL OF MULTIPLE PROCESSES

**Andrés Jaramillo-Botero**

Faculty of Engineering, Pontificia Universidad Javeriana, Calle 18 No. 118-250 Via a Pance, Cali, Colombia.
e-mail: ajaramil@ujccol.bitnet

**Yoichi Miyake**

Faculty of Engineering, Chiba University, 1-33 Yayoi-cho, Inage-ku, Chiba 263, Japan.
e-mail: miyake@ics.tj.chiba-u.ac.jp

## Abstract

A hardware processing arquitecture and an associated graphical compiler for Fuzzy inference and control of multiple antecedents and output variables is presented. The hardware arquitecture, running as a Fuzzy co-processor for IBM PC (or compatible) platforms, uses as core elements the FP9000 rules chips and FP9001 defuzzifier chips proposed by Yamakawa et al[1]. The claimed response for these processors is in the order of 0.714 μs per fuzzy inference without defuzzification, and 1.6 μs per fuzzy inference including center of gravity defuzzification. The arquitecture proposed permits fuzzy inference simulation under PC master control, as well as direct fuzzy control of output processes. The system relies on the partial flexibility of the fuzzy chips (fuzzy rule parameters are programmable through a digital interface), their embedded parallel processing, and the analog processing speeds, in order to achieve high speed simulation and real-time control response. An object oriented graphical compiler designed to permit intuitive definitions and tuning of linguistic labels within membership functions, as well as plant control systems modeling, is currently under development. The compiler interacts directly with the PCs i80X86 and the Fuzzy parallel engine (in a master-slave configuration), and generates, at users request, intermediate flat ANSI C type code for portability.

## Introduction

Human knowledge and the mechanisms involved in reasoning with accumulated knowledge is typically represented in the relational form causes-to-effect, or IF-THEN clauses containing fuzzy values. These fuzzy terms (usually linguistic in nature) describe the degree or level at which an event occurs, not event occurrence [2]. This degree can be defined by what is known as a membership function, a function (or functions, including piece-wise linear functions) that depicts a relation between fuzzy terms and how well, or not, are given deterministic input values represented by the set of fuzzy functions. Typically several fuzzy rules can account for an experts (human quantitative and qualitative) solution to a given problem, where antecedents are separate pieces of information he knows, or perceives, as influencing factors in the decision, and consequent/s correspond to the summarization of all these "reasons" into one subjective (seldom deterministic) assessment. In practical applications, fuzziness of the antecedents eliminates the need for precise matching with input facts[3]. It is also true that most common actuators, and transducers require/provide deterministic inputs/outputs to work properly within artificially controlled environments, so any fuzzy, or non-fuzzy, controller must take this into consideration.

### The FP9000 rule chip and the FP9001 defuzzifier chip

Although the reader is advised to refer directly to the original papers on the subject [1,4,5], a brief introduction will follow in order to point out the main arquitectural design decisions taken into account, due to the FP9000 series fuzzy processors internal hardware structure.

The fuzzy engine is implemented in a parallel arquitecture, where the results of all the fuzzy rules in response to the deterministic antecedent/s, defined and programmed internally, are aggregated by an analog OR construct and combined to produce a deterministic output value by the defuzzifier. The internal processing, in both FP9000 & FP9001, takes place in analog voltage mode as opposed to other implementations [6,7,8,9], although a digital interface has been included in the latest version (predecessors were completely analog) in order to define-modify (write), and read the parameters of each fuzzy rule, quickly and easily. In other words, center and bandwidth values can be adjusted in digital mode, as opposed to the cumbersome external trimmer adjustments required in the all-analog versions (which by the way, also make it quite difficult to implement adaptive fuzzy controllers).

## The Rule chip

Each unit consists of an antecedent block, a consequent block, and a rule memory to store fuzzy rule sets. It supports up to four fuzzy if-then rules simultaneously, each with 3 antecedent variables and one consequent variable. All three antecedents are logically ANDed by a MIN circuit defining the consequents' degree of membership. Only S, or Z functions, or combinations of both, are permitted as membership functions. An 8 bit opcode determines, the center of the fuzzy label within the universe of discourse (5 MSBs), and the slope of the membership function (3 LSBs) (figure 1). This means that, each membership function circuit can produce up to six different alternative type of functions, at a total of thirty one (31) different center positions within the universe of discourse, and at least at eight different bandwidth alternatives.

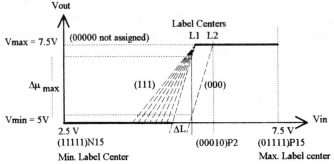

**_Figure 1._** Slope and center coding for an example S function.

In order to determine the matching grade from the deterministic facts and the membership functions, a MIN operation takes place between the S function and Z function circuits.

This MIN operation is followed by a MAX operation in the consequent block. The consequent block has 4 demultiplexer circuits (one for each rule in memory) to decode the single consequent label defined for each fuzzy rule. A three bit opcode, provides 7 possibilities (code 000 not assigned) labeled NL, NM, NS, ZR, PS, PM, and PL, for the definition of consequent center values within the universe of discourse.

The rule memory provides the appropriate digital interface for fuzzy rule definition and application. It is a two stage (buffered to permit writing while execution of a fuzzy inference) memory consisting of 24 8 bit registers (3 duplicated antecedents, and 4 inference engines) and four 3 bit registers for the consequent of each engine.

## The Defuzzifier Chip

The defuzzifier chip accepts a set of singletons, one from the consequent of each rule defined, to produce a single deterministic value as final conclusion. It uses a deterministic "singleton" consequent in order to reduce hardware implementation complexity, and increase inference speed. Singletons were also preferred for the consequent, instead of fuzzy terms, since they gives rise to a linear interpolation between antecedents and consequent [1]. Since a crisp output is required (most practical applications), the Center of Gravity method is used to achieve *defuzzification*:

$$C.G = \frac{\int i\mu(i)di}{\int \mu(i)di} \tag{1}$$

for continuous expressions of a membership function, where ($i$) is the grade of $i$th element. It represents a division of the weighted sum of consequents over the normal sum of consequents, and it is achieved via current weighted summation and normal current summation circuitry cascaded with an analog divider. The assignment of weights for each singleton output, is defined as follows:

$$\mathbf{NL_W} = 0.5, \ \mathbf{NM_W} = 0.67, \ \mathbf{NS_W} = 0.83, \ \mathbf{ZR_W} = 1, \ \mathbf{PS_W} = 1.17, \ \mathbf{PM_W} = 1.33, \ \mathbf{PL_W} = 1.5.$$

## The Proposed Arquitecture

The design implementation introduced by the FP series fuzzy processors provides a good basis for construction of simple, but not so flexible, single-output fuzzy control systems. On the other hand, additional support hardware must be added in order to contemplate a more programmable, and flexible environment, for simulation, monitoring, and control of robust systems with multiple inputs-outputs. The proposed arquitecture contemplates the possibility of having more than 3 antecedents, and more than 1 controlled variable, as well as several levels of decision. More "clearly defined" fuzzy linguistic terms can be achieved by a multi level decision tree (similar to the grammatical constructs permitted by PROLOG). It takes into account that additional information can be obtained from the combination of deterministic facts (analog or binary valued), context or time variant fuzzy operators, or other external and internal signals. It has been designed in order to allow rules like:

**IF (FOR** (*in* **from** 1 **to** 5 **by** 1) *in*=OFF **end_for**), **XOR(NOT**(*a*),*b*)=HIGH, average(*b*)=LOW **THEN** *c*=HIGH

where,

> *a*, *b*, and *in* are deterministic input variables (external or internal), and *c* a singleton consequent for internal (multi-level inferences) or external use (actual control of some actuator).
> OFF is a linguistic term representing a binary variable (ex: state of some external contact switches),
> HIGH, and LOW are fuzzy linguistic terms,
> and bold tokens correspond to control language constructs.

In other words inputs can be analog, binary, or heuristically fuzzy terms defined by the user as influencing variables to alter or compliment antecedent facts. Figure 3 represents a block diagram of the fuzzy processing units composed of 3 rule chips, 1 defuzzifier, and additional support hardware. Analog switches at the output conclusion and at the antecedent facts Xin, Yin, and Zin, plus A/D and D/A converter circuits have been added to the main controller in order to permit software processed facts, multi-level decision trees before actual output control activation, and more than 3 antecedent facts. Figure 4 shows a more global block diagram of the complete system, including several fuzzy units running as slaves to a PC platform.

Flash converters are used to increase conversion speeds. We are currently testing discrete wired component converters, and PWM techniques to replace the D/A circuitry. From the hardware implementation of the FP series processors, where the slope of the membership functions can have 7 different values and the label centers a total of 31 possibilities, the following analysis suggests an appropriate reduction in D/A converter circuits (for Xin, Yin, & Zin) requirements without detrimental loss of precision (refer also to figure 1):

$$\Delta\mu_{max}(f) = \mu_{L1}(f) - \mu_{L2}(f) \qquad \text{at slope = (000)} \qquad (2)$$

$$\Delta\mu_{min}(f) = \mu_{L1}(f) - \mu_{L2}(f) \qquad \text{at slope = (111)} \qquad (3)$$

$$\Delta L_{min} = L2 - L1 \Rightarrow {}^{5V}\!/_{31} = 0.1613V \quad \text{for label centers, L1 and L2 represent neighboring labels} \qquad (4)$$

$$\frac{\Delta\mu(f)}{\Delta L_{min}} = \text{measured slope} \qquad (5)$$

where,

> $\Delta L$ is the defined minimum increment in label centers (a total of 31).
> $\Delta\mu_{max}(f)$ is the maximum change in degree of membership for a fixed fact $f$ within $\Delta L$,
> and, $\Delta\mu_{min}(f)$ is the minimum change in degree of membership for a fixed fact $f$ within $\Delta L$,

therefore,

> $$\Delta\mu_{min}(f) = \Delta L_{min} \times \text{min. measured slope (around 1.2) = aprox. 0.191 Volts} \qquad (6)$$
> $$\Delta\mu_{max}(f) = \Delta L_{min} \times \text{max. measured slope (around 5.6) = aprox. 0.9 Volts (i.e worst case)} \qquad (7)$$

When the slope of the membership function is high, any small change in input fact values (Xin, Yin, & Zin) will produce a big change in the grade of membership for a given antecedent (equation 7). For a precision of around 55mV on the grade (for biggest change of $\mu$ in $\Delta L$ at biggest slope) we need at least an 8 bit D/A for the antecedent facts, but since labels are intuitively assigned this precision is not so strict (6 bits) since even the $\Delta L_{min}$ can be enough to better tune the fuzzy rules.

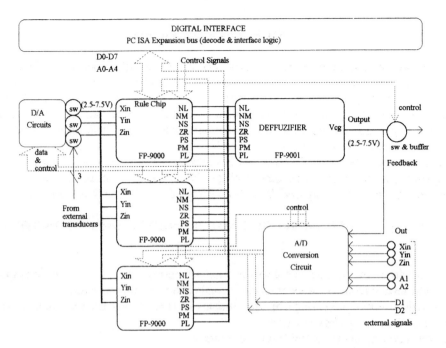

**Figure 3.** Block diagram of one process output unit (fuzzy unit)

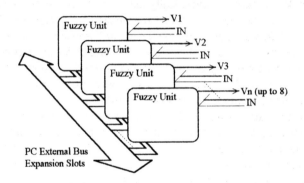

**Figure 4.** Block diagram of Fuzzy slave units under PC control

## The Compiler

The following code represents the grammar constructs (BNF notation) of the proposed language, defined for fuzzy processing and control, at the core of the graphical compiler. It works on PC platforms under Windows 3.0™ (Microsoft Corp.) or higher versions. An initial version has been implemented in Borlands Turbo C++ compiler, and development still continues. The main data structures are lists (includes support of polymorphic lists). The graphical interface will allow, at users request, the generation of C (Draft Proposed American National Standard for Information Systems - Programming Language C.[10]) code corresponding to the control system schemes defined.

```
<Program>              ::=<Block_Declarations>        <Block_Includes>      ::=#includes
                          <Block_Processes>                                    <List_Includes>
<Block_Declarations>   ::= <Block_Includes>                                  | ()
                          <Block_Types>              <List_Includes>       ::=<List_Includes>
                          <Block_Constants>                                    fileName
                          <Block_Fuzzy_Defs>                                 | fileName
                          <Block_Prototypes>         <Block_Types>         ::= #type_Def <List_Types>
                          <Block_GlobalVars>                                 | ()
```

| | |
|---|---|
| &lt;List_Types&gt; | ::=&lt;List_Types&gt; &lt;Record_Def&gt;<br>\| &lt;Record_Def&gt; |
| &lt;Record_Def&gt; | ::=**record**ID '=' &lt;List_Vars&gt; **end** ID |
| &lt;List_Vars&gt; | ::=&lt;List_Vars&gt;';'<br>&lt;Group_Vars&gt; ∈<br>&lt;Type&gt;&lt;Unit_Measure&gt;<br>\|&lt;Group_Vars&gt; ∈&lt;Type&gt;<br>&lt;Unit&gt; |
| &lt;Group_Vars&gt; | ::= &lt;Group_Vars&gt; ',' ID<br>\| ID |
| &lt;Block_Constants&gt; | ::=**#constants**<br>&lt;List_Constants&gt; \| () |
| &lt;List_Constants&gt; | ::=&lt;List_Constants&gt;';' ID '='<br>&lt;Value_Unit&gt;<br>\| ID '=' &lt;Value_Unit&gt; |
| &lt;Value_Unit&gt; | ::=Number &lt;Unit_Measure&gt;<br>\|Constant_Character<br>\| Constant_String |
| &lt;Value&gt; | ::= Number<br>\| Constant_Character<br>\| Constant_String |
| &lt;Block_Fuzzy_Defs&gt; | ::=**#fuzzy_Def**<br>&lt;Fuzzy_Defs&gt;<br>\| () |
| &lt;Fuzzy_Defs&gt; | ::=&lt;List_Labels&gt;<br>&lt;List_Rules&gt; |
| &lt;List_Labels&gt; | ::=&lt;List_Labels&gt;<br>&lt;Label_Def&gt;<br>\| &lt;Label_Def&gt; |
| &lt;Label_Def&gt; | ::= ID'='**shape** ':' &lt;Shape&gt; ','<br>**bandwidth** ':' &lt;Value&gt; ','<br>**center** ':' &lt;Value&gt; ';' |
| &lt;Shape&gt; | ::= **Z-Shape** \| **S-Shape**<br>\| **U-Shape** \| **Π-Shape**<br>\| **V-Shape** \| **Δ-Shape** |
| &lt;List_Rules&gt; | ::=&lt;List_Rules&gt;<br>&lt;Rules_Def&gt;<br>\| &lt;Rules_Def&gt; |
| &lt;Rules_Def&gt; | ::= ID **is** ID **if** &lt;Antecedent&gt;<br>\| ID (ID) **if** &lt;Antecedent&gt; |
| &lt;Antecedent&gt; | ::= '(' &lt;Antecedent&gt; ')'<br>\|&lt;Antecedent&gt;**and**<br>&lt;Antecedent&gt;<br>&lt;Antecedent&gt;**or**<br>&lt;Antecedent&gt;<br>\|**not** &lt;Antecedent&gt;<br>\| ID **is** ID \| ID **is_not** ID<br>\| '(' **if** &lt;Antecedent&gt; ')' |
| &lt;Block_Prototypes&gt; | ::=**#prototypes**<br>&lt;List_Prototypes&gt;<br>\| () |
| &lt;List_Prototypes&gt; | ::=&lt;List_Prototypes&gt;<br>&lt;Prototypes&gt;<br>\| &lt;Prototypes&gt; |
| &lt;Prototypes&gt; | ::=**Def** ID '=' &lt;Inputs&gt;<br>&lt;Outputs&gt; |
| &lt;Inputs&gt; | ::= **Inputs** ':' &lt;Vars&gt; ';'<br>\| () |
| &lt;Outputs&gt; | ::= **Outputs** ':' &lt;Vars&gt; ';'<br>\| () |
| &lt;Vars&gt; | ::=&lt;Vars&gt;';'&lt;Group_Vars&gt;<br>∈&lt;Type_Def&gt;<br>\|&lt;Group_Vars&gt;<br>∈&lt;Type_Def&gt; |
| &lt;Type_Def&gt; | ::= &lt;Type&gt; &lt;Unit&gt;<br>\| TID |
| &lt;Type&gt; | ::= **Z** \| **Z+** \| **Zs** \| **Zs+** \| **Zl**<br>\| **Zl+** \| **R** \| **R+** \| **D** \| **D+**<br>\| **Char** \| **Str** \| **Fuzzy** \| **Bool**<br>\| **list of** &lt;Type&gt; |
| &lt;Unit_Measure&gt; | ::= **en** ID &lt;Unit&gt;<br>\| () |
| &lt;Unit&gt; | ::=Units (Kg, Lbs, m/s, etc.) |
| &lt;Block_Processes&gt; | ::=&lt;Processes&gt;**main**<br>&lt;Statements&gt; **end_main** |
| &lt;Processes&gt; | ::= &lt;Processes&gt; &lt;Function&gt;<br>\| &lt;Function&gt; |
| &lt;Function&gt; | ::= ID ':' &lt;LocalVars&gt;<br>&lt;Statements&gt; **end** ID |
| &lt;LocalVars&gt; | ::= **Var** &lt;Vars&gt;<br>\| () |
| &lt;Statements&gt; | ::= &lt;Statements&gt; ';'<br>&lt;Statement&gt;<br>\| &lt;Statement&gt; |
| &lt;Statement&gt; | ::= **for** ID **from** &lt;Value&gt;<br>**to** &lt;Value&gt; **by** &lt;Value&gt;<br>&lt;Statements&gt; **end_for**<br>\| **while** '('&lt;Expression&gt;')'<br>&lt;Statements&gt; **end_while**<br>\| **repeat** &lt;Statements&gt;<br>**until** '(' &lt;Expression&gt; ')'<br>\| &lt;S_If&gt; **end_if**<br>\| &lt;S_If&gt; **else** &lt;Statements&gt;<br>**end_if**<br>\| &lt;S_Case&gt;<br>\| ID '=' &lt;Expression&gt; |
| &lt;S_If&gt; | ::= **if** '(' &lt;Expression&gt; ')'<br>**then** &lt;Statements&gt; |
| &lt;S_Case&gt; | ::= &lt;Case_Header&gt;<br>**end_case**<br>\| &lt;Case_Header&gt;<br>**default** ':' &lt;Statements&gt;<br>**end_case** |

```
<Case_Header>      ::= case '(' <Expression> ')'        | ID '(' <Groups_Vars> ')'
                   of <List_Options>                   | ID <Qualifier_Op>
<List_Options>     ::=<List_Options><Value>            | <Value>
                   : <Statements> break   <Qualifier-Op>  ::= '.' ID <Qualifier_Op>
                   |<Value>:<Statements>                | ()
                   break                  <OpBin>       ::= AND | OR | XOR | <
                                                        | <= | > | >= | == | <= | + | -
<Expression>       ::='(' <Expression> ')'             | * | / | mod
                   | <Expression> <OpBin>  <OpUnary>   ::= NOT | + | -
                   <Expression>
                   | <OpUnary> <Expression>
```

## Concluding remarks

The presented arquitecture and control language form an ideal system for simulation and actual control using fuzzy reasoning. The fuzzy unit shown in figure 3 has been tested as a stand-alone controller, using a PCs parallel printer port as data interface, with optimal results for real-time control applications. A simple PID fuzzy algorithm has been tested for control of small DC motors (< 1 Amp), with faster settling times than classical PID algorithm implementations under critically damped conditions.

One of the intended practical uses of the system is in "target-tracking" trajectory planning of mechanical manipulators, by approximating targets position relative to the robots end-effector using data obtained from 2 dimensional images of the workspaces as partial parameters for fuzzy control rules. Work is being done to test the language, and arquitecture, for real-time control of a 6 D.O.F. mechanical arm by calculating the inverse kinematics and motor control parameters, based only on image understanding and using the combined fuzzy neural model proposed by Jaramillo et al [11].

## Acknowledgments

The authors wish to thank sincerely Dr. Takeshi Yamakawa for supplying the FP9000 and FP9001 units, and for his most enlightening comments. Thanks are also due to Mr. Gabriel Tamura, and Mr. Andrés Dorado, for their help on the compilers design and implementation stages.

## References

[1]   T. Yamakawa, "Silicon Implementation for a novel high-speed fuzzy inference engine: mega-flips analog fuzzy processor". Journ. of Intell. and Fuzzy systems, Vol 1 (1), 1993.

[2]   B. Kosko. "Neural Networks and Fuzzy Systems: A dynamical systems approach to machine intelligence." Prentice-Hall, Inc., 1992.

[3]   Zadeh, L.A., "Fuzzy Sets", Information and Control 8:338-353, 1965.

[4]   T. Yamakawa, "A Fuzzy Logic Controller", Journal of Biotechnology, 24, 1992.

[5]   T. Yamakawa, "FP9000 and FP9001 technical notes", 93/3/11.

[6].  M. Togai and H. Watanabe, "An inference enginefor real-time approximate reasoning: Toward an expert on a chip." IEEE E 1 (3) 55-62 (1986).

[7]   H. Watanabe and Wayne Dettloff, "Reconfigurable fuzzy logic processor: A full custom digital VLSI." Proceedings of the InternationalWorkshop on Fuzzy System Applications, Iizuka, Japan, 1988, pp. 49-50.

[8]   H. Arikawa and K. Hirota, "Fuzzy inference engine by address-look-up and paging method." Proceedings of the InternationalWorkshop on Fuzzy System Applications, Iizuka, Japan, 1988, pp. 45-46.

[9]   K. Shimizu, M. Osumi, F. Imae, "Digital fuzzy processor FP-5000." Proceedings of the International Conference on Fuzzy Logic and Neural Networks, Iizuka, Japan, 1988, pp. 49-50.

[10]  A. Holub, Compiler Design in C, Prentice-Hall, Inc.,1990.

[11]  A. Jaramillo, K. Yamaba, "An Artificial Neural Network for Classification of Color Images." Proceeding of the First IS&T/SID Color Imaging Conference: Transforms and Transportability of Color, Scottsdale, Arizona, U.S., 1993, pp. 167-173.

# A Tool for Automatic Synthesis of Fuzzy Controllers

A. Costa‡, A. De Gloria‡, P. Faraboschi‡, A. Pagni†

*Abstract*— We present a two step synthesis approach to the design of VLSI Fuzzy Controllers. First, we derive a VHDL description of the ASIC from the problem specifications, the hardware constraints and the performance requirements. Then, we map the VHDL description to gate level description with a standard logic synthesis tool. The process is repeated until the resulting design is tuned to the global requirements of the control device in terms of performance and cost.

## I. INTRODUCTION

Fuzzy Logic has recently gained a growing interest for its capability of simply modeling human experience [1, 6]. This feature make Fuzzy Logic particularly suited to control applications where the system to be controlled can only be described by a nonlinear mathematical model.

Nowadays Fuzzy Logic covers a very wide set of industrial applications such as servo controllers for optical disk drives, control of blood pressure, camera autofocus, etc. [2, 3].

As a consequence of this application spread, it is necessary to study efficient implementations of the Fuzzy inference mechanism in embedded systems.

Software implementations represent the easiest and most flexible way to perform a fuzzy control. Most software applications adopt a general-purpose micro-controller and an high-level programming language such as C. However, software implementations may be expensive in terms of component requirements with respect to a fully integrated solution, for instance if we are looking at mass volume consumer electronics.

The research for performance/cost balance has recently lead to the development of architectures with specific support, and several accelerator coprocessors dedicated to fuzzy logic have been proposed [4, 5]. Dedicated hardware obviously represents the best solution in terms of performance but at the same can only cover a limited range of applications.

Despite its lack of flexibility, the choice of a fully dedicated hardware solution may represent an effective way to support Fuzzy Logic, in particular for those applications whose large number of rules has a very high computational burden. Fuzzy logic specific hardware allows to reach a better cost/performance ratio by exploiting the particular features of a single application. Dedicated hardware responds to the demand of an improvement in computing power through the introduction of special purpose units and the exploitation of parallelism. Moreover, this solution can be cost effective as it only implements the hardware resources needed by the application and the availability of logic synthesizers allows also non-VLSI specialists to design an ASIC, correct-by-construction, in a short time.

Unfortunately, currently available synthesis tools do not support high-level descriptions and the generation of an appropriate architecture from an algorithm specification still remains a major research challenge.

The need to reach an optimal price/performance ratio requires to explore tradeoffs both in the algorithm and in the physical design space. This pushes to the development of tools that can perform this task. Even if a complete solution to the general problem is not near to be reached, we can develop specific tools dedicated to specific applications. From these considerations, the realization of a tool that automates the design phase of a Fuzzy Controller starting from high level specifications supplied by the user appears suitable.

In this paper we present an automatic synthesis approach that derives a complete circuit definition from a high level description of the fuzzy rules and the system requirements.

## II. THE SYNTHESIS SYSTEM

The proposed synthesis system implements a particular type of Fuzzy Controller, with fixed methods for inference and defuzzification. The reference fuzzy controller (assumed according to [1]) operates on the following hypotheses:

†Dept. of Biophys. and Elect. Eng. (DIBE), University of Genova V. Opera Pia 11a, 16145 Genova, Italy

‡SGS-THOMSON Microelectronics, Centro Direzionale Colleoni Palazzo Andromeda 3, 20041 Agrate Brianza (MI), Italy

on the following hypotheses:

- 'IF $x_1$ IS $a_1$ AND ...$x_n$ IS $a_n$ THEN $y$ IS $b$' rule format. This choice requires a software pre-processing of the expert rules, in order to reduce them in the required format.
- Correlation-product inference to produce fuzzy output.
- Defuzzification performed using fuzzy centroid computation.
- Memory Oriented approach in membership function evaluation.

Our synthesis procedure is structured in three main steps:

- Rule Definition
- VHDL Description Generation
- Netlist Synthesis

There are different reasons for the choice of VHDL as the description language: VHDL is a standard *de facto* and allows an easy design exchange in industrial and research environments. VHDL also enables a unified description for simulation, synthesis and documentation. Finally, VHDL covers the description from architectural to gate level and it saves the user from low level implementation details.

VHDL Generation and Synthesis are recursive and correlated steps: (Figure 1) from the VHDL Generator we derive a RTL description of the architecture and from the Synthesizer we obtain the design netlist and an indication about area, that drives the successive iterations of the procedure. The goal is to obtain an optimal trade-off between required performance and hardware constraints.

## A. The Rule Formalizer

The Rule Formalizer represents the user interface for the description of the fuzzy system.

The first task to be accomplished is the definition of the controller input parameters. In this initial phase it is important to guarantee an user friendly interface to allow a simple requirements definition. For this purpose the user can specify the inference rules in a linguistic way according to the format IF - AND ... AND - THEN. Our system allows the definition of the parameters in interactive mode (the user provides the design parameters from standard input) and batch mode (the user specifies the parameters in a *.param* file with a predefined template). The user interface will be integrated within the SGS-Thomson environment [8] in the near future.

The user-provided rules are then processed and converted in a mathematical format suitable for the VHDL Automatic Generator.

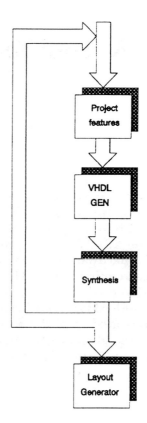

Figure 1: Design Flow Chart

## B. The VHDL Generator

The key tool of the system is the VHDL generator. The generator accepts two kinds of constraints:

- Application Constraints, related to the characteristics of the controlled system, such as the number of rules $NR$, the number of inputs $NI$, the number of outputs $NO$ and the number of fuzzy subsets $NS$ for each input and output.
- Hardware Constraints, related to the hardware resources available for the synthesis step, such as the bit resolution or the number and the characteristics of adders, multipliers, registers, memory banks.

The first task performed by the VHDL Generator is the evaluation of the number of pins of the controller to be synthesized. The parameters involved belong both to Application and to Hardware Constraints. For instance, if we assume single memory bank, the number of I/O ports is given by $N_{in} = NI \cdot (NS + 1) + 2 \cdot NO \cdot NS + 2$ and $N_{out} = 2 \cdot NO + 1$.

The controller input and output data are:

- the crisp inputs $(in_i)$;
- the base memory address for each subsets $(sval_{ij})$;

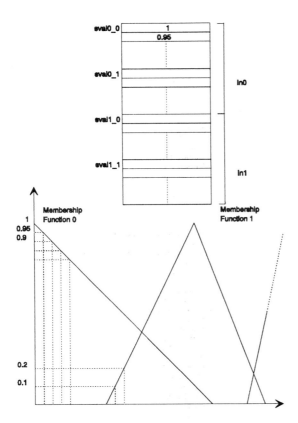

Figure 2: Organization of membership memory

- the parameters for centroid ($k_{ij}$) and area ($h_{ij}$) of the output fuzzy subset;
- the address bus to external memory;
- the data bus from external memory;
- the values for denominator ($den_i$) and numerator ($num_i$) of the crisp output

The data flow of the Architecture defined by the VHDL generation tool reflects the three main tasks to be accomplished by a Fuzzy Controller:

- **Fuzzification**: the inputs received by the controlled system are used to address the external membership memory. The controller computes the addresses of the proper values adding the crisp input to the base addresses of each membership function (figure 2).

  The number of cycles required to read the fuzzy values depends on the characteristics of the external memory and on the available adders. In case of multiple memory banks (one per input), the parallel computation of the proper addresses may allow the concurrent acquisition of multiple fuzzy values.

- **Inference**: the second step of a Fuzzy Control is rule application. From an hardware point of view, this corresponds to the execution of a set of minimum operations per rule.

- **Defuzzification**: the final steps performs the computation of the crisp outputs from their fuzzy values. The controller performs this task according to the fuzzy centroid calculation. This operation is the most area expensive and time consuming as it requires several multiply-accumulate steps.

We can define a basic resource set as the minimum set of units necessary to the controller design. In our case, it is composed of a Multiplier, three Adders (one for the evaluation of the membership functions and two for the computation of output numerators and denominators), a *minimum* unit and an external memory for the membership look-up tables. According to the affordable hardware resources, we can increase the number of units of each type, to decrease the number of control steps (cycles) necessary for the computation.

The realization of the units is strictly related to the application features. The system can use different types of adder (carry select, carry lookahead, etc.) and multipliers (parallel, sequential) according to performance and area constraints. For instance, the use of a parallel multiplier leads to a considerable performance improvement, but there are classes of applications without strict time constraint where it would be probably more efficient to use a sequential (*add-and-shift*) structure.

An example of generated VHDL description for $NI = 2$, $NO = 1$ and $NS = 3$, and a Basic Resource Set is shown in Figure 3.

*C. Synthesis*

Once obtained the VHDL description, we synthesize it to gate level by means of a standard logic synthesizer (Mentor Graphics AutoLogic [7]). AutoLogic associates a cost parameter to each basic element of the target cell library to determine the best way to optimize the design. The cost of a high level component (such as an adder) is computed by considering the cost of its low level components and an estimate of the interconnection load.

### III. EXPERIMENTS AND RESULTS

In order to verify the validity of our approach, we have applied the described system to the synthesis of controllers dedicated to different problems.

We have considered several parameters regarding system specifications (number of inputs $NI$, outputs $NO$, subsets $NS$, rules $NR$ and bit resolution) and hardware resources (number of adders $NA$, multipliers $NM$, comparators $NC$). The obtained VHDL descriptions have been mapped to a $1.0\mu$ CMOS standard cell library and, for each synthesized circuit, we have evaluated hardware complexity (num-

```vhdl
use IEEE.std_logic_1164.all;

ENTITY Fuzzy_Controller IS PORT(
    in0,in1:   IN STD_ULOGIC_vector(7 downto 0);
    sval0_0,sval0_1,sval0_2,sval1_0,sval1_1,sval1_2:
               IN STD_ULOGIC_vector(7 downto 0);
    k0_0,k0_1,k0_2: IN STD_ULOGIC_vector(7 downto 0);
    h0_0,h0_1,h0_2: IN STD_ULOGIC_vector(7 downto 0);
    data:  IN STD_ULOGIC_vector(7 downto 0);

    ck: IN STD_ULOGIC;

    address:   OUT STD_ULOGIC_vector(7 downto 0);
    num0:   OUT STD_ULOGIC_vector(7 downto 0);
    den0:   OUT STD_ULOGIC_vector(7 downto 0)
    );
END Fuzzy_Controller;

ARCHITECTURE behave OF Fuzzy_Controller IS
    TYPE states IS ( sstart,st1,st2,st3,st4,st5,st6,st7,st8,
    st9,st10,st11,st12,st13,st14,st15,st16,st17,st18,st19,
    st20,st21,st22,st23,st24,st25,st26,st27,st28,st29,st30,
    st31,st32,st33,st34,st35,st36,st37,st38,st39,st40,
    st41,st42,st43,st44,st45);
    SIGNAL state: states;
    SIGNAL m0: STD_ULOGIC_vector(7 downto 0);
    SIGNAL r: STD_ULOGIC_vector(7 downto 0);
    SIGNAL Inum0,Iden0: STD_ULOGIC_vector(7 downto 0);
    SIGNAL o1: STD_ULOGIC_vector(7 downto 0);
    SIGNAL o2: STD_ULOGIC_vector(7 downto 0);

BEGIN
min:o1<=m0 WHEN m0<=r ELSE r;

mul:PROCESS(m0,r)
    BEGIN
        o2<=to_std_ulogic(to_integer(m0)*to_integer(r),8);
    END PROCESS mul;

fsm: PROCESS(ck)
    BEGIN
    IF(ck='1' AND ck'LAST_VALUE='0') THEN
        CASE state IS
            WHEN sstart => -- RULE 1
                address <= sval0_0 + in0; -- load membership 00
                state <= st1;
            WHEN st1 =>
                m0 <= data;
                address <= sval1_1 + in1; -- load membership 11
                state <= st2;

            WHEN st2 =>
                r<=m0;
                m0<= data;
                state <= st3;
            WHEN st3 =>
                r<=o1;  -- r=min(r,m0)
                m0<=k0_1;
                state <= st4;
            WHEN st4 =>
                Inum0<=Inum0+o2; -- Inum += r*k01
                m0<=h0_1;
                state<=st5;
            WHEN st5 =>
                Iden0<=Iden0+o2; -- Iden += r*h01
                address <= sval0_1+in0;
                state <= st6;
                        .
                        .

            WHEN st41 => -- RULE 9
                m0<=data;
                address<=sval1_2+in1; -- load membership 12
                state<=st42;
            WHEN st42 =>
                r<=m0;
                m0<=data;
                state<=st43;
            WHEN st43 =>
                r<=o1; -- r = min(r,m0)
                m0<=k0_0;
                state<=st44;
            WHEN st44 =>
                Inum0<=Inum0+o2; -- Inum += r*k00
                m0<=h0_0;
                state<=st45;
            WHEN st45 =>
                Iden0<=Iden0+o2; -- Iden += r*h00
                den0<=Iden0;
                num0<=Inum0;
                state<=sstart; -- loop back

        END CASE;
    END IF;
    END PROCESS fsm;
END behave;
```

Figure 3: Example of VHDL description of a fuzzy controller

| Bits | Gates |
|------|-------|
| 8    | 3085  |
| 10   | 5346  |
| 12   | 6560  |

Table 1: Hardware complexity and bit resolution

ber of gates) and performance (number of cycles).

A first experiment aims at estimating the relationship between hardware complexity and bit resolution. In particular, we have considered a fixed problem described by $NI = 2$, $NO = 1$, $NS = 3$ and $NR = 9$ and a basic set of resources ($NA = 3$, $NM = 1$, $NC = 1$) and have compiled the description for different datapath widths. The results (Table 1) show a linear dependence.

A second experiment aims at estimating the relationship between hardware complexity and problem parameters, such as number of rules and inputs. In

this case $NI$ ranges from 2 to 5, $NR$ ranges from 2 to 50, $NO = 1$, $NS = 3$, and the resolution is 8 bits. In this case we considered a basic resource set with $NA2$, $NM = 1$, $NC = 1$ and the increase in complexity is given by the growing decode part (to multiplex data) and control part to specify the algorithm steps. The obtained results are shown in Figure 4 and indicate that hardware complexity increases with order $O(NR \cdot NI \cdot log_2(NR \cdot NI))$.

Finally, we have computed the time complexity of the algorithm, that is linear with respect to the number of rules, inputs and outputs (with fixed resources), as it is shown in figure 5. If we consider a cycle time of 40ns (for a $1.0\mu$ CMOS technology and 16 bit accuracy), we can see that a typical problem (3 inputs, 1 output and 20 rules) can be executed in 161 cycles, that corresponds to 6 $\mu s$ of input sampling rate (166KHz) and is sufficient to support most control applications.

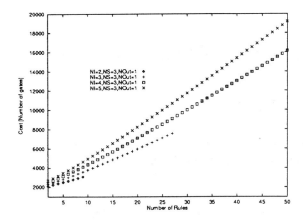

Figure 4: Complexity of a Fuzzy Controller as a function of number of inputs and rules

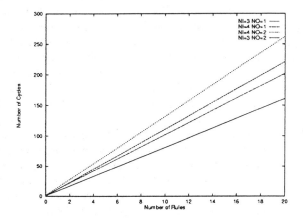

Figure 5: Time performance of a Fuzzy Controller as a function of number of inputs,outputs and rules

## IV. CONCLUSIONS

As complexity of custom integrated circuits increases, the corresponding design and verification effort required to map high level project specifications to silicon implementations grows dramatically.

The proposed approach overcomes the limitations of currently available synthesis tools and allows to reach a gate-level description of a dedicated fuzzy logic control unit starting from a high-level specification and according to a predefined set of resources. The choice of a VHDL interface enables the system to interface with standard simulation and synthesis tools, and shows important advantages:

- the design is portable without modifications over a wide range of technologies, such as PLD for fast prototyping, or ASIC for volume production
- the fuzzy logic controller can be seen as a macro-block to be included in a more complex circuit, yet within a unified design environment
- it is possible to perform an early system sim-

ulation including a precise model of the fuzzy controller, to evaluate the actual system behavior

Future work involves the development of a more flexible and optimized resource scheduling algorithm and the integration of the toolset within the SGS-Thomson user interface.

### REFERENCES

[1] L.A. Zadeh, 'Outline of a New Approach to the Analysis of Complex Systems and Decision-Making Approach', IEEE Transactions on Systems, Man and Cybernetics, Vol. SME-3, No. 1, pp. 28–45, January 1973

[2] C. Von Altrock, B. Krause, H.J. Zimmermann, 'Advanced Fuzzy Logic Control Technologies in Automotive Applications', IEEE International Conference on Fuzzy Systems, pp. 835–842, March 1992

[3] L.P. Holmbald, J.J. Ostergaard, 'Control of a Cement Kiln by Fuzzy Logic', Fuzzy Information and Decision Process, North-Holland Publishing Co., pp. 389–399, 1982

[4] O. Ishizuka, K. Tanno, Z. Tang, H. Matsumoto, 'Design of a Fuzzy Controller with Normalization Circuits', IEEE International Conference on Fuzzy Systems, pp. 1303–1308, March 1992

[5] H. Watanabe, W.D. Dettloff, K.E. Yount, 'A VLSI Fuzzy Logic Controller with Reconfigurable, Cascadable Architecture', IEEE Journal of Solid State Circuits, vol. 25. No. 2, pp. 376–382, Apr. 1990

[6] B. Kosko, 'Neural networks and fuzzy systems: a dynamical approach to machine intelligence' (Prentice Hall, Englewood Cliffs, N.J., 1992)

[7] Mentor Graphics Corporation: 'AutoLogic User Guide', 1992.

[8] A. Pagni, R. Poluzzi, G. Rizzotto, 'Integrate development environment for fuzzy logic applications', SGS-Thomson technical report. 1993.

# Architecture of a 64-bit Fuzzy Inference Processor

Ansgar P. Ungering, Karl Goser

University of Dortmund, LS-BE, Emil-Figge-Str. 68, 44227 Dortmund, Germany
Fax: + 49 231 7554450    E-mail : ungering@luzi.e-technik.uni-dortmund.de

*Abstract* — **The architecture of a 64-bit Fuzzy Inference Processor (FIP) will be presented. A fuzzy system consisting of the FIP and a 64-bit microprocessor speeds-up the inference by up to 10 times. In addition we present an optimized inference algorithm which achieves a 50 fold acceleration for the calculation of the rule base. The FIP will be used only for the inference while the fuzzification and defuzzification will be done by the microprocessor ($\mu$P), which will also do the controlling of the FIP. This results in a simple architecture and low hardware requirement. We use the MIN/MAX algorithm and an internal resolution of 8-bit. Up to 8 membership functions can be used for every input and output. A prototype (with 32-bit) was simulated on FPGA's and needed 180ns for the calculation of one rule with 8 inputs and 2 outputs.**

## I. INTRODUCTION

Today most fuzzy logic systems are implemented as software on microprocessors ($\mu$P). If there is a need for higher operation speed, we can add a fuzzy processor (FP) like the FC110 from Togai. For 8- and 16-bit systems the FP speeds-up the system. But when using a 32- or 64-bit microprocessor, an optimized fuzzy controller algorithm implemented as software is faster than the FC110. The advantage of the FC110 is the powerful arithmetic logical unit (ALU). For the defuzzification a fast multiplication and division are needed. 8- and 16-bit $\mu$P's are not optimized in these instructions. If using a 32- or 64-bit $\mu$P with a powerful ALU, the advantages of the FC110 are only the fast minimum- and maximum-operations. Some investigations of the fuzzy controller algorithm show that there is usually only a small part of the whole rule base which has to be calculated [1, 2]. In this case, a fast 32- or 64-bit $\mu$P with an optimized fuzzy algorithm is faster than the FC110. We presented a fuzzy processor for 64-bit systems which can be used in combination with an optimized fuzzy algorithm[1]. For high operation speed and low hardware requirement we used the MIN/MAX algorithm and singletons for the conclusion part.

[1]This architecture can be used for 32-bit systems with some modifications, too.

## II. ANALYSIS OF THE FUZZY CONTROLLER ALGORITHM

Fig. 1 shows the average percentage of operation time for fuzzification, inference and defuzzification. The fuzzy controller algorithm is implemented as "software" on a 64-$\mu$P. We used 8 inputs with 8 membership functions (MF) each, 4 outputs with 8 singletons each, an 8-bit resolution and 50 rules. Although the rule base is not very complex, the inference takes the longest time (83%). The high data width of the 64-bit $\mu$P provides no advantages for the 8-bit fuzzy operations in comparison to e.g. 8-bit $\mu$P's. This is caused by the serial calculation and the emulation of the minimum and maximum operations with standard $\mu$P instructions. The fuzzification is very fast and takes only 1% of the whole operation time, because it is only a memory access. For the defuzzification the $\mu$P only needs about 16% of the whole operation time because of the powerful ALU.

## III. POSSIBILITIES FOR AN OPTIMIZATION

In principle there are two different possibilities to optimize a fuzzy system: An optimization of the software or an optimization of the hardware. The aim of the optimized software is to calculate only a small part of the rule base, because of serial calculation of the rule base. If the degree of overlap of the MF's is limited to 2 (Fig. 2), most of the rules have a truth value equal to zero. The aim is to find those rules which have a truth value greater than zero. A fuzzy controller with e.g. 4 inputs has up to 16 rules with a truth value greater than zero [1]. It is simple to select all these rules, but for a fuzzy controller with e.g. 8 inputs the

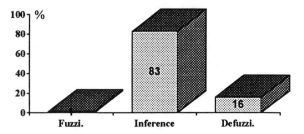

Fig. 1. Percentage of processing time

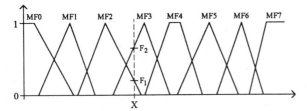

Fig. 2. Input with 8 MF's, up to 2 MF's at the same time have a truth value greater than zero

operation time for selecting these rules (up to 256 rules) and the required memory for storing the rule base (up to 16 MByte) will be very high. In this case it is more suitable to use a not so optimized algorithm which selects more rules (as well as rules with a truth value equal to zero), but with a moderate memory requirement and a low operation time for selecting these rules. In the following chapter such an algorithm will be presented.

Basically the aim of the optimized hardware is to calculate the whole or big parts of the rule base in parallel and implement the minimum- and maximum operations. This can be succeeded in an expansion of a standard µP or a special fuzzy ASIC. The expansion of a standard µP is suitable for 8-bit µP's, but 64-bit µP's are too complex and an expansion will be very difficult and costly. A special fuzzy ASIC for supporting the fuzzy logic operations will be the simplest way.

In our work we used a 64-bit µP with an optimized software algorithm in combination with a fuzzy ASIC. The software algorithm is so optimized that there is no need for a parallel calculation of the rule base. Only one rule in parallel has to be calculated.

## IV. OPTIMIZED INFERENCE ALGORITHM

Usually only a small part of the whole rule base has to be calculated if the degree of overlap of the MF's is limited [1]. In most fuzzy systems the degree of overlap can be limited to 2 without any restrictions. In the following the degree of overlap[1] is limited to 2.

The optimized algorithm consists of the two parts `sorting the rule base' and `calculation of the block address'. If the rule base is sorted according to the first input (Fig. 3a), it is clear that only the rules of one of the three blocks (I, II and III) have to be calculated because of the limited degree of overlap. Only up to 2 directly neighboured MF's of input A have a truth value greater zero. In this simple example only 50% of the whole rule base has to be calculated. The `starting addresses' and the length of the 3 blocks are stored in a look-up table. The lowest number of the `hidden' MF's

[1]The optimized algorithm could also be used for a higher degree of overlap but with lower effectiveness.

points directly at the position of the look-up table where the starting address of the block is stored. Using 8 MF's for the first input, only 25% (in the best case) of the whole rule base has to be calculated.

It is also possible to sort the rule base according to N inputs. In this case the rule base must be devided into additional blocks. These blocks are build in the following manner:

*1)* Sort the rule base according to the first input. Build blocks by collecting those rules which use the same or the direct neighboured MF's of the first input. We get M-1 blocks, when M is the number of MF's for the input. The blocks which are directly neighboured overlap each other. Fig. 3a gives an example with the following blocks: Block I with A1 and A2, block II with A2 and A3 and block III with A3 and A4 . Block I and block II as well as block II and block III overlap.

*2)* Sort the received main blocks (I, II and III) of (*1*) according to the second input. Build sub blocks of these main blocks by collecting those rules which use the same or the direct neighboured MF's of the second input (Fig. 3b). Some of the rules have to be used two times because the main blocks now don't overlap. Only the sub blocks of the main blocks overlap. So the rule base needs more memory to be stored. But the maximum expansion of the rule base will be less than factor 2.

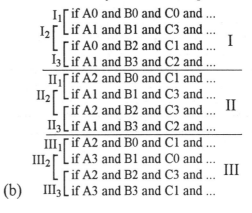

Fig. 3. Sorted rule base by one input (a) and two inputs (b)

1777

*3)* Do the same procedure like (2) for the other inputs. For every additional input we get a new level of sub blocks. The size of the rule base grows (less than factor 2 for every additional input) because some rules are multiply used.

As the number of sub blocks increases, the memory for the look-up table where the starting address and length of the sub blocks are stored grows. The number S of sub blocks can be calculated by the following formula:

$$S = (A'-1)*(B'-1)*(C'-1)*...$$  (1)

Where A', B' and C' are the number of MF's of the inputs A, B and C. The pointer P of the look-up table could be calculated by using the numbers $A_X$, $B_X$ and $C_X$, of the lowest 'hidden' MF's of the inputs:

$$P = A_X*1 + B_X*(B'-1) + C_X*(B'-1)*(C'-1) + ...$$  (2)

If A1 and B2 in the example given by Fig. 3 are the lowest numbers of the 'hidden' MF's, the value of the pointer will be 1*1 + 2*3 = 7. The starting address of the sub block is stored at this location of the look-up table. Notice that the address pointer to the look-up table (7) must not be equal to the number of the sub block (6).

The advantage of the described optimization is the splitting of the rule base into small sub blocks. Only the rules of one of these sub blocks and the starting address of this sub block have to be calculated. Depending on the number of sorted inputs the size of the sub blocks will be reduced. In the same way the needed memory for storing the expanded rule base and the look-up table increases. Although more rules have to be calculated than selecting only those rules with a truth value greater than zero, the described method is very fast because the selection of rules takes a high operation time.

The best results will be reached using those inputs for the sorting of the rule base which use a lot of MF's. Especially for large rule bases this optimization will be very helpful. For a rule base with e.g. 2500 rules, 8 MF's per input and sorted by 4 inputs less then 10 rules in average have to be calculated (best match). The needed memory for storing the expanded rule base is up to 16 times higher than without expansion. Also about 2400 memory locations are needed for the look-up table. In real applications some of the MF's of one input are used in a lot of rules - others not. So a calculation of 50 rules in average will be a realistic result. This is equal to a 50 fold speed-up in comparison to a non optimized inference.

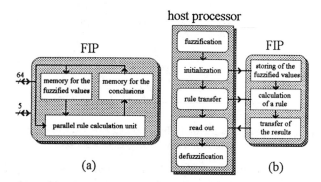

Fig. 4. Architecture of the FIP (a) and controlling of the FIP by the host processor (b)

## V. ARCHITECTURE OF THE FUZZY INFERENCE PROCESSOR

The Problems of most fuzzy processors (FP) are the high costs and the needed additional components like memory, a clock generator, latches, and others. In additional most of the components of the host processor like ALU, MMU, control unit, and others are implemented in the FP, too. Such systems are not very effective but very costly. Our idea of a 64-bit fuzzy system was to minimize the needed hardware requirement for the fuzzy processor by using already existing components of the 64-bit µP and implement only those components which speed-up the fuzzy algorithm. As shown in chapter II the fuzzification and defuzzification are very fast - this means that there is no possibility for an optimization. But the inference takes the longest time because of the bottleneck of serial processing and not implemented minimum and maximum instructions. An optimization of the inference yields the highest speed-up and can easily be realized. In this case the FP consists only of an inference unit and will be called fuzzy inference processor (FIP). The developed FIP calculates one rule in parallel with up to 8 inputs and outputs. If using the optimized inference algorithm described in chapter III, it is not necessary to calculate more rules in parallel. To minimize the number of system components and the

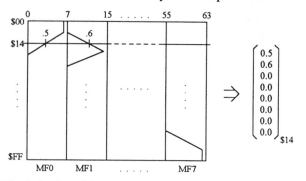

Fig. 5. Storing of the MF's MF0, ..., MF7 in 256 64-bit words

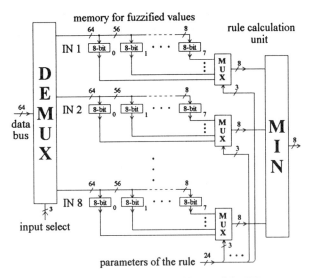

Fig. 6. Architecture of the if-part of the FIP

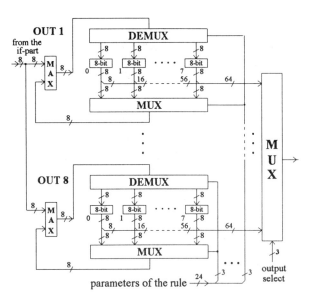

Fig. 7. Architecture of the than-part of the FIP

chip area, the FIP has a behaviour like an intelligent memory. The FIP will be fully controlled by the host μP and all I/O operations will be done by the μP. Therefore the FIP needs no MMU, control unit and ALU. Furthermore there is no system clock needed because the FIP works asynchronously. The FIP consists only of two memory blocks and a parallel rule calculation unit (Fig. 4a). The advantages of this architecture are the low hardware requirement for the FIP, no additionally system components and the high operation speed. Furthermore there are no problems with the sharing of memory with the host μP because the host μP controls all I/O operations of the FIP. A calculation of the rule base will be split into 5 parts as shown in Fig. 4b. First the host μP fuzzifies the input values. With only one memory access all MF's of one input are fuzzified (Fig. 5). In the second part, the host μP stores the fuzzified values in the FIP and calculates the starting address of the sub block of the rule base. The next step is the transfer of the rules of the sub block to the FIP, which will be done rule by rule. Using 8 inputs and outputs with 8 MF's for each input and output, one rule is coded with 48 bit (8*3 + 8*3 = 48). Therefore one rule is stored in one 64-bit word and is read by the 64-bit μP in one memory read cycle. In the meantime while, the host μP is fetching a new rule, the FIP calculates the rule in parallel. This is the reason why the FIP calculates only one rule in parallel because the rules are transmitted serially into the FIP. A parallel calculation of a sub block has no advantages. After all rules are transferred into the FIP, the host μP reads out the results of the inference. In the last part, the host μP defuzzifies the results of the inference.

The internal architecture of the FIP is described in the following paragraphs.

*Architecture of the if-part:*

Fig. 6 depicts the architecture of the if-part. There are 64 8-bit latches (for every input 8), where the fuzzified values will be stored. Depending on the input address which will be given by the host μP, the fuzzified values will be stored in the latches. Depending on the parameters of the rule one of the 8 latches of every input will be selected and switched to the MIN-gate. The output of the MIN-gate is the truth value of the rule. Only one step is needed to calculate the whole if-part of a rule. For the minimum gate we used the architecture described in [6].

*Architecture of the then-part:*

The architecture of the then-part is shown in Fig. 7. There are 64 8-bit latches (for every output 8), where the height of the MF's are stored [6]. After the truth value of a rule has been calculated this value is compared with the already stored values in the corresponding latches, which are selected by the parameters of the rule. The highest values will be stored back in the latches. When all rules were calculated, the host μP reads out the results of the inference. Depending on the address given by the host μP all 8 MF's of an output are read out at the same time.

The rule calculation unit doesn't exist as an extra unit. It is integrated in the if- and then-part.

The FIP doesn't limit the degree of overlap of the MF's. With a limited degree of overlap, the architecture can be simplified and some latches will be saved.

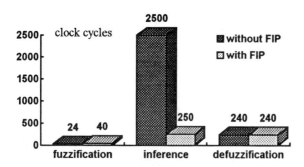

Fig. 8. Execution time in clock cycles

## VI. RESULTS

Because of the limitations of the FPGA's only a 32-bit version of the FIP was realized. The number of inputs is limited to 8 and the number of outputs is limited to 2. So it is possible to code one rule in 32 bit using 8 MF's for the inputs and outputs. Also the degree of overlap of the MF's is limited to 2. In this case only 3 latches are needed to store the 2 fuzzified values greater than zero and the number of the lowest `hidden' MF's of one input. The if-part is not changed, only the number of outputs is limited. While the data bus has a width of only 32-bit, 2 memory accesses are needed to read out the results of one output. The FIP is realized on two FPGA's (XILINX 3090) and uses about 6500 gate equivalents. Simulations of the FPGA implementation show a delay of about 180ns for the calculation of one rule. This is fast enough for using the FIP in a 20 MHz system because the host μP needs some clock cycles for fetching a new rule. The usage of the FIP speeds the inference up to a factor of about 10 (Fig. 8). For the whole fuzzy controller algorithm including the fuzzification, inference and defuzzification we achieve a 5 fold speed-up in comparison to Fig. 1.

## VII. SUMMARY

The presented architecture of the fuzzy inference processor (FIP) is very simple and needs only a small chip area, because only the inference of the fuzzy controller algorithm is supported by the FIP. The fuzzification and defuzzification, as well as the controlling of the FIP are done by the host microprocessor. Another advantage is the universal use in every 64-bit microprocessor system because the FIP behaves like a memory. The inference was accelerated to a factor of 10. This results in a complete speed-up including the fuzzification, inference and defuzzification to a factor of 5. In future works this architecture will be implemented into an ASIC.

In addition we presented an optimized organization of the rule base for a fast inference. Only one block with a small number of rules has to be calculated. Using the FIP and the new optimized organization of the rule base, a 64-bit fuzzy system is nearly as fast as special fuzzy ASIC's.

### REFERENCES

[1] H. Eichfeld, M. Löhner and M. Müller, "Architecture of a Fuzzy Logic Controller with optimized memory organisation and operator design," FUZZ-IEEE '92, San Diego, March 1992, pp. 1317-1323

[2] H.N. Teodorescu and T. Yamakawa, "Architectures for Rule-Chips Number Minimizing in Fuzzy Inference Systems," Proceedings of the 2nd International Conference on Fuzzy Logic & Neural Networks, Iizuka, Japan, July 1992, pp. 547-550

[3] H. Watanabe, "A RISC Approach to Design of Fuzzy Processor Architecture," FUZZ-IEEE '92, San Diego, March 1992, pp. 431-440

[4] Ansgar P. Ungering, Karsten Thuener, Karl Goser, "Architecture of a PDM VLSI Fuzzy Logic Controller with Pipelining and Optimized Chip-Area," FUZZ IEEE'93, San Francisco, March 1993, pp. 447-452

[5] H. Watanabe, W.D. Dettloff and K.E. Yount, "VLSI Fuzzy Logic Controller with Reconfigurable, Cascadable Architecture," IEEE Journal of Solid-State Circuits, vol. 25, no. 2, April 1990, pp. 376-382

[6] Ansgar P. Ungering, K. Goser, "Architecture of a PDM VLSI Fuzzy Logic Controller with an Explicit Rule Base," IFSA '93, Seoul, Korea, July 1993, pp. 1386-1389

[7] Ansgar P. Ungering, Karl Goser, "Chip-Area Optimised Storing of Membership Functions," EUFIT '93, Aachen, Germany, September 1993, pp. 98-103

[8] F. Deffontaines, A. Ungering, V. Tryba and K. Goser, "The concept of a RISC architecture for combining fuzzy logic and a Kohonen map on an integrated circuit," Neuro Nimes 92, November 1992, pp. 555-564

# CUSTOM DESIGN OF A HARDWARE FUZZY LOGIC CONTROLLER

Donald L. Hung

Department of Electrical Engineering
Gannon University, Erie, PA 16541
Tel: (814) 871-7527; E-mail: hung@knight.gannon.edu

*Abstract* - **The paper describes a custom designed hardware fuzzy logic controller (FLC) for high speed real-time control applications. With a pipelined parallel architecture, the FLC can operate at very high speeds. The FLC can also gain the ability of on-line adaptation by connecting itself to a supervisory microprocessor so that its performance can be constantly monitored and its knowledge base can be updated at run time. A two inputs, one output prototype of the FLC has been implemented with a Xilinx XC4008-6 FPGA and a separate EPROM. With the FLC's control unit operating at a clock speed of 20MHz, the FLC can produce 9 million control actions per second.**

## I. INTRODUCTION

In the recent years, there has been increased use of fuzzy logic in control applications. For most of the reported applications, algorithms of the FLCs are implemented in software and executed on a standard microprocessor or microcontroller, or even on a general purpose computer. Although software-based FLCs are in general more economical and flexible, they often have difficulty in dealing control systems that require very high processing and I/O handling speed, such as real-time controlled robots. Due to the development of ASIC technology and the fact that standard miroprocessors/microcontrollers with the conventional von Neumann architecture are inherently inappropriate for implementing fuzzy inference algorithms, custom designed hardware FLCs become a natural solution for speed-demanding control applications [1-9].

The main features of the digital hardware FLC discussed in this paper are: 1) high speed; 2) on-line adaptability. The high speed of this FLC is due to its pipelined parallel architecture and the elimination of the multiplication and division hardware based on iterative algorithms. The on-line adaptability is obtained by providing a communication link between the FLC and a supervisory microprocessor. The supervisory processor may monitor the FLC's performance and update the FLC's knowledge base at run time when dealing with changing environment and plant characteristics.

Discussion in the rest of the paper will focus on a two-input, one-output FLC as illustrated by Fig. 1. This simplification will not limit the generality since generalized modus ponens (GMP) based multiple-input multiple-output (MIMO) FLCs can be partitioned into a set of multiple-input single-output (MISO) sub FLCs with two-input, one-output as a special case. The algorithm, architecture, building blocks, as well as implementation and testing results of the FLC will be introduced in the following sections.

## II. ALGORITHM AND DESIGN CONSIDERATIONS

Like most fuzzy logic controllers, the FLC discussed here adopts the GMP for fuzzy inference, the "mini" operator for fuzzy implication, the "max-min" operators for fuzzy composition, and the center of gravity (COG) method for defuzzification. The inputs and output (linguistic variables) of the FLC's inference engine (please refer to Fig. 1) are characterized by their linguistic values as follows:

$$E = \{E1, E2, E3\}, D = \{D1, D2, D3\},$$
$$C = \{C1, C2, ..., C9\},$$

where E1, ..., E3, D1, ..., D3, C1, ..., C9 are fuzzy sets. There are possibly nine distinct control rules as listed below:

R1: if $x$ is E1 and $y$ is D1 then $z$ is C1
R2: if $x$ is E1 and $y$ is D2 then $z$ is C2
R3: if $x$ is E1 and $y$ is D3 then $z$ is C3

  ...   ...   ...
  ...   ...   ...

R9: if $x$ is E3 and $y$ is D3 then $z$ is C9

Since the "mini" and "max-min" operators are used for fuzzy implication and composition respectively, the inference procedure for a specific control rule

If $x$ is E$i$ and $y$ is D$j$ then $z$ is C$k$

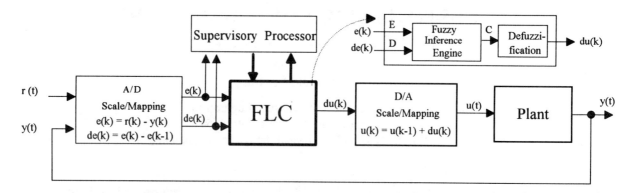

Figure 1. The fuzzy logic controller in a control system

can be expressed as [10]:

$$C'k = (E'i, D'j) \circ (Ei \times Dj \rightarrow Ck)$$
$$= \min [E'i \circ (Ei \rightarrow Ck), D'j \circ (Dj \rightarrow Ck)]$$

where $E'i$ and $D'j$ are the observed input fuzzy sets and $C'k$ is the inferred output fuzzy set. The membership function values of $C'k$ can be computed from

$$w_k = \mu_{c'k}(z) \qquad (1)$$
$$= \min \{\min [(\max \min (\mu_{E'i}(x), (\mu_{Ei}(x)),$$
$$(\max \min (\mu_{D'j}(y), (\mu_{Dj}(y)))], \mu_{Ck}(z)\}$$

With a pair of observed inputs, multiple control rules may be activated. To produce a crisp value as the FLC's output, the defuzzification algorithm based on the COG method requires the following computation:

$$\Delta u = z_0 = \frac{\sum w_i z_i}{\sum w_i} \qquad (2)$$

For simplification, $z_i$'s in (2) are often chosen as the center values of each activated membership function, in the universe of discourse of the output space. The hardware algorithm for implementing fuzzy inference computations described by (1) and (2) is listed below:

1. Perform the "max-min" operation between each observed input value and each of the membership functions in the correspondent input space.
2. Perform the "mini" operation among values obtained from Step 1 and relate to the same control rule.
3. For all weights obtained from Step 2, perform
   a) the "summation" operation $\sum w_i$,
   b) the "multiplication" operation $w_i z_i$.
4. Perform the "summation" operation $\sum w_i z_i$ for all $w_i z_i$'s obtained from Step 3b.
5. Perform the "division" operation $\sum w_i z_i / \sum w_i$ using the results obtained from Step 3a and Step 4.

Note that he five steps listed above must be executed sequentially, but operations inside each step can be executed concurrently. Step 3 through Step 5 are for the COG-based defuzzification and involves multiplication and division. Since their correspondent arithmetic algorithms are complicated and require significant amount of iterative operations, hardware fuzzy controllers employ multiplier and divider in their COG circuitry have the disadvantage of higher complexity and lower overall processing speed.

From practical considerations and experience in applying fuzzy logic to control systems, the following constraints can be applied to simplify the hardware FLC design and improve the FLC's performance:

1. The observed inputs of the FLC are crisp and digitized.
2. For each input variable, the overlapping degree of its membership functions is limited to two (i.e., no more than two membership functions in the same input space should be overlapped) and the membership function values at the crossover points will not exceed 50% of their full range.
3. Membership functions in the output space have symmetric shapes.

With the first constraint, the "max-min" operation in Step 1 of the hardware algorithm can be simplified as a table look-up operation. Values of the input membership functions can be prestored in *input look-up tables* and the observed crisp inputs serve as the addresses for accessing to the table contents. The second constraint means that for an $n$-input FLC, a sample of observed inputs can activate at most $2^n$ control rules. This consideration will reduce the required hardware, also limit the maximal length of the summations $\sum w_i$ and $\sum w_i z_i$ in Step 3a, 4 and 5 of the hardware algorithm. With the third constraint, each membership function in the output space can be treated as a fuzzy singleton situated at the center point $z_i$ of an output membership function, therefore computation for the COG-based defuzzification can be significantly reduced. Notice that membership functions for practical FLCs usually require limited resolution, which suggests the following approaches for eliminating the multiplication and division required by the COG algorithm:

1. For a fuzzy output singleton situated at $z_i$, all possible $w_i z_i$ values can be precalculated and stored in a *product look-up table* with values of $w_i$ serving as the addresses for accessing to the table contents.
2. All possible values of the division $\Sigma w_i z_i / \Sigma w_i$ can be precalculated and stored in a *defuzzification look-up table* with values of the concatenated $\Sigma w_i z_i$ and $\Sigma w_i$ serving as the addresses for accessing to the table contents.

Since the FLC supports all possible control rules, its control surface can be manipulated by changing the contents of the input and product look-up tables, while the defuzzification look-up table should be fixed since it is independent to the rest of the FLC.

With discussions in this section, the hardware architecture and building blocks of the designed FLC are introduced in the next section.

## III. HARDWARE

The generalized data processor of the designed FLC is shown in Figure 2. Building blocks of the data processor and the control unit of the FLC will be introduced individually in this section.

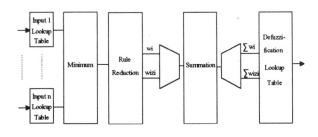

Figure 2. Data processor of a MISO FLC

### A. The Input Look-up Tables

The input look-up tables are used for the implementation of Step 1 of the hardware algorithm discussed in the previous section. The designed FLC has two inputs with each characterized by three membership functions. Both of the membership functions and the digitized inputs have 4-bit resolution. To facilitate on-line adaptation, twelve $16 \times 4$ bit SRAMs are used to store the input membership function values. Six of them for normal operation, another six are secondary copies for on-line adaptation. Each of the $16 \times 4$ SRAM can be accessed by both the related crisp input and the supervisory processor. The bus arbitration is controlled by the FLC's control unit. At any time, a sample of observed inputs can read from six SRAMs concurrently; at the same time, the supervisory processor can write to the other six SRAMs. Figure 3 illustrates the configuration of a pair of $16 \times 4$ bit SRAMs serving as the look-up tables for input membership function E1.

### B. The Minimum Module

This module is for the implementation of Step 2 of the hardware algorithm discussed in the previous section. The module contains nine identical "mini" blocks that execute in parallel. Each of the "mini" blocks selects the minimum from its two 4-bit inputs, as shown in Figure 4.

### C. The Rule Reduction Module

With Constraint 2 introduced in the previous section, the FLC under discussion cannot have more than four activated control rules with a given input sample. The Rule Reduction module produces weights ($w_i$'s) as well as the products ($w_i z_i$'s) that correspondent to the activated control rules. The rule reduction criterion is shown in Figure 5a. A part of the module which selects the weight and product from four possible control rules is illustrated by Figure 5b. The $64 \times 8$ ROM in Fig. 5b is for the four product look-up tables corresponding to the four possible control rules. It can be replaced by two pieces of SRAMs to allow on-line adaptation, and the implementation is similar to the example shown in Fig. 3. Note that the Rule Reduction module also operates in a parallel fashion.

### D. The Summation Module

This module is used for producing the two sums $\Sigma w_i$ and $\Sigma w_i z_i$. The module consists two levels of adders in a tree structure. Note that there is a partial pipeline: when the Summation module is summing up the weights ($w_i$'s), the Rule Reduction module is producing the products ($w_i z_i$'s) based on the same weights. The sums $\Sigma w_i$ and $\Sigma w_i z_i$ are latched at the output of the Summation module and proceed to the Defuzzification module.

### E. The Defuzzification Module

The Defuzzification module is a lookup table which stores the precalculated values of $\Sigma w_i z_i / \Sigma w_i$. The concatenated $\Sigma w_i z_i$ (9 bits) and $\Sigma w_i$ (5 bits) serve as the address lines to access the table contents. As discussed in the previous

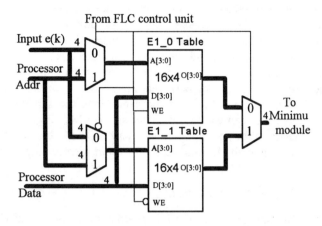

Figure 3. Example of the input lookup table

section, this module replaces the iterative division hardware and is independent to the rest of the FLC.

### F. The Control Unit

The control unit of the FLC is a finite state machine which generates the following basic control signals with appropriate timing so that the data processor operates in a fully pipelined fashion:

- latch the FLC inputs
- latch the Minimum module inputs
- latch the Rule Reduction module inputs
- latch the $w_i$ inputs to the Summation module
- latch the $\Sigma w_i$ outputs from the Summation module
- latch the $w_i z_i$ inputs to the *Summation* module

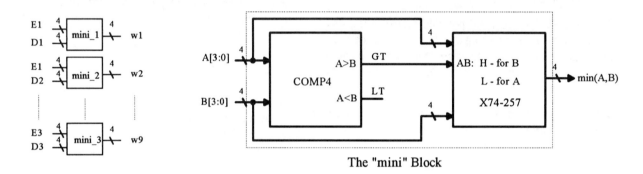

Figure 4. The Minimum module

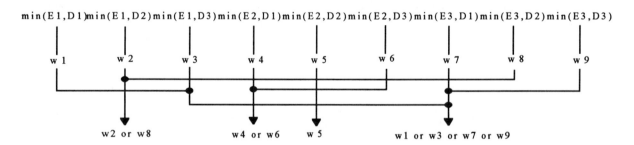

Figure 5a. The rule reduction criterion

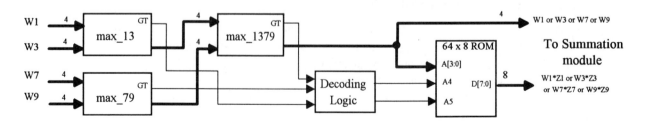

Figure 5b. Part of the Rule Reduction module

- latch the $\Sigma w_i z_i$ outputs from the *Summation* module
- read the *Defuzzification* lookup table

The control unit also controls the FLC's communication with the supervisory processor. A semaphore is used for this purpose. During normal operation, the semaphore is cleared and the supervisory processor can write to the FLC's secondary input or product look-up tables freely in a burst mode. When an adaptation is considered necessary, the supervisory processor sets the semaphore. The FLC's control unit constantly watches the semaphore and, when it detects that the semaphore is set, it immediately switches the FLC to use the newly updated look-up tables and then clears the semaphore.

## IV. IMPLEMENTATION

The FLC discussed in the previous sections is implemented with a Xilinx XC4008PG191-6 FPGA and a Cypress CY7C251 16K x 8bit EPROM. The EPROM (50 ns access time) is used for implementing the Defuzzification module. The rest parts of the FLC are all on the FPGA. The on-line adaptation feature for the product ($w_i z_i$) look-up table is not implemented due to the resource limitation of the FPGA package used. Two versions of the FLC are tested. The first version runs with a system clock up to 40 MHz without pipelining, and produces over 1.7 million control actions per second. The second version is fully pipelined. It runs with a system clock up to 20 MHz and produces over 9 million control actions per second.

## IV. CONCLUSIONS

A custom designed hardware fuzzy logic controller is discussed and the testing results are reported. Its high processing speed and the capability of on-line adaptation make it suitable for handling high-speed control tasks in distributed complex control systems, or serving as an autonomous controller when backed up by a microprocessor.

## REFERENCES

[1]   Masaki Togai and Hiroyuki Watanabe. "Expert system on a chip: An engine for real-time approximate reasoning," *IEEE EXPERT* (Fall 1986):56-58.

[2]   T. Yamakawa, "High speed fuzzy controller hardware systems," Proceedings of the 2nd Fuzzy System Symposium, Japan, 1986, pp. 122-130.

[3]   Masaki Togai and S. Chiu, "A fuzzy logic chip and a fuzzy inference accelerator for real-time approximate reasoning," *Proceedings of the 17th International Symposium on Multiple-Valued Logic*, pp.25-29, 1987.

[4]   T. Yamakawa, "Fuzzy microprocessors - Rule chip and defuzzifier chip," *International Workshop on Fuzzy System Applications*, Iizuka, Japan, Aug. 1988, pp. 51-52.

[5]   W.D. Detloff, K.E. Yount and H. Watanabe, "A fuzzy logic controller with reconfigurable, cascadable architecture," *Proceedings of 1989 IEEE International Conference on Computer Design: VLSI in Computers and Processors*, pp. 474-478, 1989.

[6]   Hiroyuki Watanabe, Wayne Dettloff and Kathy Yount, "A VLSI fuzzy logic controller with reconfigurable, cascadable architecture," *IEEE Journal of Solid-State Ciruits*, Vol. 25, No. 2, pp. 376-382, April 1990.

[7]   H. Watanabe, W.D. Detloff, K.E. Yount, "VLSI fuzzy chip and inference accelerator board systems," *Proceedings of the 21th International Symposium on multiple-valued logic*, pp. 120-127, 1991.

[8]   Sujal Shah and Ralph Horvath, "A Hardware Digital Fuzzy Inference Engine Using Standard Integrated Circuits," *Proceedings of the 1st International Conference on Fuzzy Theory and Technology*, Durham, NC, 1992, pp. 109-114.

[9]   A. P. Ungering, K. Thuener and K. Goser, "Architecture of a PDM VLSI fuzzy logic controller with pipelining and optimized chip area," *Proceeding of the 2nd IEEE International Conference on Fuzzy Systems*, 1993, Vol.1, pp. 447-452.

[10]  Chuen Chien Lee, "Fuzzy Logic in Control Systems: Fuzzy Logic Controller, Part II," *IEEE Trans. on Systems, Man and Cybernetics,* Vol.20, No.2, pp. 419-435, March/April 1990.

# Fuzzy control design system based on DSP

Nobukazu Iijima*, Kouichi Koizumi**, Hideo Mitsui*
and Mototaka Sone*

*1-28-1 Tamazutsumi, Setagaya-ku, Tokyo 158 Japan
Musashi Insutitute of Technology
** 2000 Showacyo Tsukijiarai, Nakakoma-gun
Yamanashi-ken 409-38 Japan
KITO Corp.

## ABSTRACT

Now, one of the most significant problems and requirements about Fuzzy Control is the establishment of tuning method of parameters concerned about Fuzzy control. And it seems that the suitable tuning method which clear this demands hasn't been developed. Therefore the Fuzzy control system design tool that will be able to perform a systematic experiment for tuning the parameters in brief time is required.

This paper proposes a new Fuzzy control design system that is realized by using Digital Signal Processor(DSP). This system can perform the systematic experiment for tuning, the repetitive of changing the parameters and carrying on the Fuzzy control, in extremely short time because of the merits of DSP.

The use of DSP in Design system can calculate the fuzzy calculations in high speed and can change the parameters of fuzzy system easily by software itself. So it seems that this design system is available to decide the optimum parameter set at the time of development of a new fuzzy control system, especially requires the extremely short control period.

## 1.INTRODUCTION

Fuzzy control has the following characters that is basically different from it of other usual control which is popular and basic control method, a manner of handling differential equations.
1)treat the subject qualitative
2)control by rules
3)rules are expressed by language
4)robust control

Because of these characters, Fuzzy control is effective to the humanistic system and systems of the same complexity that the usual control is comparatively not suited for. Recently it is expected that Fuzzy control is useful on adaptive control and robust control too and various fuzzy systems(expert systems) have been realized.

Now, in Fuzzy control the tuning method has not been established yet and it becomes one of the most important problem. So at the practical application of Fuzzy control, Fuzzy control design system that is able to perform a systematic experiment for tuning parameters is required seriously. Specifically speaking, a design system that carry on the repetitive of changing parameters and perform the fuzzy control under this condition systematically in high speed.

And if the design system aims to certificate the result of the fuzzy control system as the practical application, the design system will require the still faster operation for accomplishing the real time control of a high speed subject.

Fuzzy control is a digital control and has generally high cost of fuzzy calculation on fuzzy inference. The calculation of fuzzy inference must be fast for developing a high speed Fuzzy design system. The most effective means for realizing high speed Fuzzy design system is a hardware, using fuzzy chips for fuzzy inference. This hardware has a remarkable result concerning the calculation speed but it is short of flexibility, because it needs a down load processing of parameters that are obtained by simulation carried out in other machines. So it is good for the practical fuzzy control system, but not good for Fuzzy design system.

This paper shows a new Fuzzy design system based on Digital Signal Processor(DSP). Fuzzy design system using by DSP has a merit in high speed fuzzy calculation and flexibility obtained by software. In this paper we try to consider the availability of this design system on an example, a position control by an AC servo motor.

## 2.FUZZY CONTROL DESIGN SYSTEM
## 2.1 EXPERIMENT BY AC SERVO MOTOR

The Block diagram of experimental system is shown in Fig.1 and the free running characteristic of used AC servo motor is Fig.2. AC servo motor as a subject is very simple to control but it is good for discussing the high speed ability of the design tool. The used servo motor is RA-13M produced by TOSHIBA Corp..

The separated block by left side dashed line is the Fuzzy design system that consist of personal computer (PC386G: produced by EPSON) and DSP(MSP86220:produced by MITEC) , main chip is MB86220 by FUJITSU. This DSP's performance is shown in Table.1.

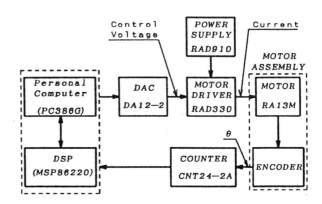

Fig.1 block diagram

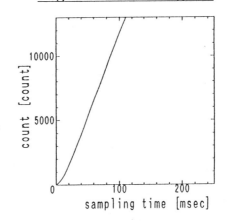

Fig.2 free running

Table.1 performance of DSP

| processor | 24bits floating point DSP MB86220 [FUJITSU] |
|---|---|
| process ability | Max 12MFLOPS |
| machine cycle | 12 MHz |
| operation time | 83.3 [nsec] |
| internal RAM | 512 [word] |
| external RAM | 64 [kword] 4 pages [at 1 wait] |

Fuzzy controller designed adaptively by the result of the experiment can control the motor instead of missing the Fig.2 character.

## 2.2 FUZZY INFERENCE

The main parts of fuzzy inference are two, composition and defuzzification. These conditions which are carried on are shown in Table.2. And a typical membership functions and rules are shown in Fig.3.

In this experiment, we change the number of rules and shape of membership function, practically change the base of triangle etc.

### Table.2 fuzzy inference

| method | composition | defuzzfication |
|--------|-------------|----------------|
| T | | height |
| M | | area |
| D | product—addition | center of gravity |
| K | MIN—addition | center of gravity |
| J | MIN—MAX | center of gravity |
| DM | product—MAX | center of gravity |

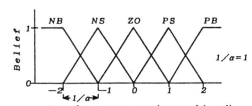

(a) input membership function

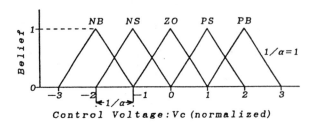

(b) output membership function

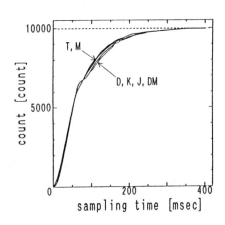

(c) rule matrix(9 rules)
Fig.3 fuzzy rules

## 3.RESULT AND DISCUSSION
## 3.1 EFFECTIVITY OF DSP AT SPEED

The result of control that are performed according to the 6 composition-defuzzfication methods are shown in Fig.4 as compared with Fig.5 that shows the result without DSP (personal computer only).

About the result of this Fuzzy control design system, all 6 methods have completely finished. And each control period is shown in Table.3. At the result only two simple methods(T,M method) that don't use the center of gravity at difuzzfication have finished in Fig.5. The others are in failure, because these control period are too long to control this AC servo motor.

Fig.4 responses by DSP

From the result the control period obtained by Fuzzy design system with DSP has about 10 times short of it without DSP.

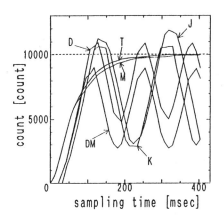

Fig.5 responses by C-language

Table.3 control period

| method | control period [msec] | | period rate (C-langage/DSP) |
| --- | --- | --- | --- |
| | C-langage | DSP | |
| T | 1.2 | 0.42 | 2.9 |
| M | 1.4 | 0.45 | 3.1 |
| D | 21.4 | 3.2 | 6.7 |
| K | 29.5 | 2.8 | 10.5 |
| J | 31.1 | 2.9 | 10.7 |
| DM | 23.3 | 3.3 | 7.1 |

It seems that if an adequate control period is got in spite of inference method, these control result shall be the same under the same rules. And even if we can obtain more short control period, we can't expect a control result to be better.

Thus, it seems that the parameters (such as rules etc.) have direct effects upon the control result under this circumstances.

In the simple inference method(T,M) the control period are about 0.4[msec]. So if the adequate period is about 5[msec], we can increase the number of fuzzy space(the number of input variables of Fuzzy system). The maximum number of variables will be about 200 by this design system with DSP.

## 3.2 EFFECT OF THE NUMBER OF RULES

In this section the number of rules is changed. This meaning is the change of sensitivity and the number of membership functions that takes part in the inference doesn't change.

The result of control is shown in Fig.6. All of them are controlled by T-method.

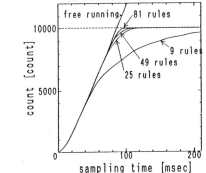

Fig.6 responses for each rules

It shows that the control result is better according to the increase of rules at the speed especially. It is clear from the comparison with this motor free running in the same figure. At the number of rules(25, 49, 81) the control period is the same at the 9 rules.

Thus the parameter(the number of rules) has direct effect on the control result.

### 3.3 CHANGE OF THE BASE

Fig.7 shows the control result when the base of membership function. The change of the base is the change of the number of membership functions which takes part in the inference at a sampling time.

These result are performed by the 9 rules and T-method. The effects of the change of the base is small to the control result. But in the case that a change of the base inhomogeneously over the whole fuzzy space that is shown in Fig.8. And the result is shown in Fig.9.

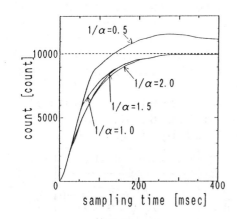

Fig.7 responses for the base

These result are carried on by the same condition of Fig.7 obtained.

From these result the control result is affected considerably by the fuzzy parameters, especially the sensitivity. The sensitivity is changed significantly by the change of the number of rules and the base.

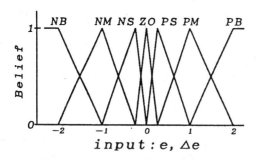

(not uniform)
Fig.8 change the base

So when one fuzzy controller becomes practical use, many tries of tuning parameters will be performed systematically and repetitively.

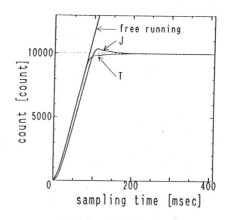

Fig.9 responses(by Fig.8)

## 4.CONCLUSION

A new Fuzzy control design system is proposed. This design system is developed by Digital Signal Processor technique.

In an experiment of an AC servo motor control, the availability of this design system is clear. The characteristics of this new system are the speed and the flexibility.

From the result one significant point becomes clear. If an adequate control period for the subject, the inference method doesn't influence the control result greatly under the same rule. And the quality of rule has direct effects to the control result. So when one fuzzy controller becomes practical use, many tries of tuning parameters will be performed systematically and repetitively.

By this design system about 400 times tuning processing is accomplished within 30 minutes.

# Importance of Membership Functions: A Comparative Study on Different Learning Methods for Fuzzy Inference Systems

A. Lotfi & A. C. Tsoi

Department of Electrical and Computer Engineering
University of Queensland, Brisbane, Qld. 4072 Australia
Email: lotfia@s1.elec.uq.oz.au   act@s1.elec.uq.oz.au

*Abstract*— **This paper investigates different adaptive structures for fuzzy inference systems. We examine the effect of membership functions on reasoning process when the number of rules is fixed. Three commonly used membership function shapes have been employed in this study. It has been shown that membership functions have the dominant effect on reasoning process rather than number of rules or inference mechanism. We compare our adaptive membership function scheme with two already proposed by others.**

## I. INTRODUCTION

It is obvious that altering the parameters of knowledge-base i.e.:

1. the membership function of linguistic values,
2. fuzzy reasoning methods,
3. the number of rules,

will affect the overall input-output mapping. Assignment of a membership function (MF) to each fuzzy value for Fuzzy Inference System (FIS)is problematic. As we will show, altering the MF has the dominant effect on the two others. It can be said that for a fixed number of rules in production rules, changing the membership function can achieve the same input-output mapping, regardless of the fuzzy reasoning method. Alternatively, for a fixed fuzzy reasoning method we can attain the same input-output mapping with different number of rules and different membership function.

The first attempt to provide a general theory for the realization of a fuzzy inference system was proposed by Jang [2] who introduced a General Neural Networks based fuzzy inference systems. The network was able to adjust its membership functions parameters such that the error between the desired and the actual decision surface was reduced. The fuzzy inference method used was based on a specific type of fuzzy inference system introduced earlier by Takagi and Sugeno [9]. The consequent of each rule is assumed to be a crisp, rather than fuzzy value. We will use the term *semi-fuzzy model* in comparison with fuzzy model where both the antecedent and the consequent are fuzzy values.

Some researches has been conducted using triangular membership functions and a simple reasoning structure [7] [11]. As it will be explained, this model is a simplified configuration of a semi-fuzzy model. The technique proposed in [11] can be used as an advisor that makes suggestions about how to tune membership functions so as to make the fuzzy system approach the desired behavior.

We have already proposed an adaptive membership function scheme (AMFS) for FIS's [5]. The scheme that we have proposed is to employ the same type of network introduced earlier by Jang [2] except that we will use a fuzzy model instead of a semi-fuzzy model for FIS.

We have employed three different and commonly used MF shapes with three above mentioned adaptive structures for FIS. This comparison has been performed for a particular nonlinear control problem, viz. truck backer-upper control.

The organization of the rest of this paper is as follows; after introducing fuzzy inference systems in section II, membership function adaptation schemes and membership function tuning will be introduced. Application to the truck backer-upper control is employed to illustrate this comparative study. Pertinent conclusions will be drawn.

## II. FUZZY INFERENCE SYSTEMS

A general inference mechanism, which is often used in representation of human reasoning, can be presented as a set of $n$ implication rules, combined together by a connective operator *else*.

$P^i$:If $\mathcal{X}$ is $\mathcal{A}^i$ then $\mathcal{Y}$ is $\mathcal{B}^i$ , else
Q:    $\mathcal{X}$ is $\mathcal{A}'$

---

∴                    $\mathcal{Y}$ is $\mathcal{B}'$

The antecedent vector $\mathcal{X} = [X_1, ..., X_m]$is an $m$ vector with elements which are linguistic variables in the universe of $\mathcal{U} = [U_1, ..., U_m]$. The consequent vector $\mathcal{Y} = [Y_1, ..., Y_k]$ is a $k$ vector with elements which are linguistic variables in the universe of $\mathcal{V} = [V_1, ..V_k]$.Vectors $\mathcal{A}^i = [A_1^i, ..A_j^i, ..A_m^i]$ and $\mathcal{B}^i = [B_1^i, ..., B_j^i, ..., B_k^i]$ are vectors of linguistic values referring to vector of fuzzy variables $\mathcal{X}$ and $\mathcal{Y}$ respectively. Vector $\mathcal{A}'$ is fuzzy/crisp observation vector and vector $\mathcal{B}'$ is fuzzy/crisp conclusion vector.

Vector $\mathcal{X}$ contains $m$ linguistic variables that are connected together by liaison operator *and*. The consequent vector $\mathcal{Y}$ comprises $k$ linguistic variables. It is reasonable to assume that there is no relationship between a linguistic variable $y_j$ and the other linguistic variables $y_l$, $l \neq j$. Therefore such an inference might be decomposed into $k$ inference with antecedent vector $\mathcal{X}$ and consequent linguistic variable $Y_j$  $(j = 1, 2, ..., k)$ individually. Without loss of generality, we can assume that the consequent premise is just one variable, i.e. $k = 1$.

We further assume that the universe of antecedent and consequent i.e. $\mathcal{U}$ and $\mathcal{V}$ are limited to a specific domain interval defined as indicated below:

$$U_j = [U_j^- \ U_j^+] \qquad V_j = [V_j^- \ V_j^+] \qquad (1)$$

Three commonly used symmetric membership functions have been considered in our study. They are defined for antecedent fuzzy values as follows

$$A_j^i = exp(-\left(\left(\frac{u_j - \sigma_{ij}}{\rho_{ij}}\right)^2\right)^{\beta_{ij}}) \qquad (2)$$

$$A_j^i = \frac{1}{1 + \left(\left(\frac{u_j - \sigma_{ij}}{\rho_{ij}}\right)^2\right)^{\beta_{ij}}} \qquad (3)$$

$$A_j^i = 1 - \frac{2|u_j - \sigma_{ij}|}{\rho_{ij}}, \quad \sigma_{ij} - \frac{\beta_{ij}}{2} < u_j < \sigma_{ij} + \frac{\beta_{ij}}{2} \qquad (4)$$

and consequent fuzzy values are defined by

$$B^i = exp(-\left(\left(\frac{v - \hat{\sigma}_i}{\hat{\rho}_i}\right)^2\right)^{\hat{\beta}_i}) \qquad (5)$$

$$B^i = \frac{1}{1 + \left(\left(\frac{v - \hat{\sigma}_i}{\hat{\rho}_i}\right)^2\right)^{\hat{\beta}_i}} \qquad (6)$$

$$B^i = 1 - \frac{2|v - \hat{\sigma}_i|}{\hat{\rho}_i} \quad \hat{\sigma}_i - \frac{\hat{\beta}_i}{2} < v < \hat{\sigma}_i + \frac{\hat{\beta}_i}{2} \qquad (7)$$

where $\sigma_{ij}, \rho_{ij}, \beta_{ij}, \hat{\sigma}_i, \hat{\rho}_i,$ and $\hat{\beta}_i$ are unknown constant parameters. As will be shown subsequently, these parameters can be adjusted on-line using a gradient descent algorithm.

To make an inference "Y is $\mathcal{B}'$" from a set of rules $P$ and observation $Q$, different methods of reasoning have been examined. Moreover, numerous methods of defuzzification have been proposed [4]. Among these methods, the *centroid* method has been shown to be most effective.

In our comparative study, we consider three methods of inference and defuzzification that have been used intensively in recent research [10] [8] [1]. The first method uses a fuzzy model, and other two methods use semi-fuzzy models for system under consideration.

**A)** Pacini and Kosko [8] have proven that if correlation product inference determines the output fuzzy values, the global centroid $F_1$ can be computed from local consequent premise centroids. i.e.

$$F_1 = \frac{\sum_{i=1}^n w_i \bar{C}_i I_i}{\sum_{i=1}^n w_i I_i} \qquad (8)$$

where $w_i, \bar{C}_i,$ and $I_i$ are rule firing weights, local centroid and area of consequent premise respectively; they are defined as follows;

$$w_i = \prod_{j=1}^m A_j^i(X_j) \qquad (9)$$

$$\bar{C}_i = \frac{1}{I_i(v)} \int_{V-}^{V+} B^i(v)dv \qquad (10)$$

$$I_i(v) = \int_{V-}^{V+} B^i(v)dv \qquad (11)$$

**B)** Another type of fuzzy if-then rule was proposed by Takagi and Sugeno [9]. In this method, the fuzzy values are involved only in the antecedent part of premises and the consequent is a linear function of its inputs.

$$F_2 = \sum_{i=1}^n \bar{w}_i f_i \qquad (12)$$

where

$$\bar{w}_i = \frac{w_i}{\sum_{i=1}^n w_i} \qquad f_i = \hat{\sigma}_i + \sum_{j=1}^m q_{ij} X_j \qquad (13)$$

where $q_{ij}$ and $\hat{\sigma}_i$ are constant parameters that must be determined; $w_i$ is defined in Equation (9).
**C)** The third method of inference which has been used extensively [10] is a simplified model of (13) when $q_{ij} = 0$ for i=1...n, and j=1...m.

$$F_3 = \frac{\sum_{i=1}^n w_i \hat{\sigma}_i}{\sum_{i=1}^n w_i} \qquad (14)$$

Wang and Mendel [10] have proven that fuzzy representation in Equation 14 is a universal approximator.

It is to be noted that in the last two methods, $\hat{\sigma}_i$ is assumed to be the center of consequent membership function.

### III. Membership Function Adaptation Schemes

It is still unknown how to to assign a MF to each fuzzy value in FIS. However, there has been some research in this direction. Jang [2] introduced a Generalized Neural Network structure based fuzzy inference system. The developed model for the inference mechanism was based on the second method ($F_2$) presented in the previous section. The network contains a multilayer feedforward network in which each node performs a particular function (node function). There are two types of nodes:

- nodes with fixed parameters (they are called *circle node functions* by Jang [1])
- nodes that depend on a set of parameters specific to the node (Jang [1] called this type of node a *square node function*)

The performance of the node (and consequently the performance of the system) depends on these parameters. The structure of network for a rule base that containing 2 rules and 2 inputs is depicted in Figure (1).

Nomura et.al [7] and Zheng [11] proposed learning methods for FIS by a gradient descent method. They used triangular membership functions in the antecedent premises and the reasoning approach was based on the third method ($F_3$) presented in the previous section. The membership

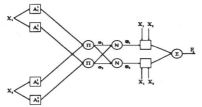

Figure 1: Adaptive network based fuzzy inference systems

function parameters in the antecedent, and the real number $\hat{\sigma}_i$ in consequent part are tuned by means of the descent method. The arrangement of adaptive inference system for a rule base containing 2 rules and 2 inputs is depicted in Figure (2).

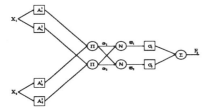

Figure 2: Simplified adaptive network

To have a general inference mechanism that can tackle systems with fuzzy values in the antecedent as well as the consequent part, we have already proposed an adaptive membership function scheme[5]. This method uses the first method ($F_1$) of inference presented in Section 2. Our proposed algorithm uses a neural network which contains four layers, with 2 circle and 2 square layers. The first and third layers (containing only square nodes) represent the membership functions given in fuzzy values $A_j^i$ and $B^i$ respectively. The other two layers contain only circle nodes. The node function of the second layer is a simple multiplication and the node function of the fourth layer is the actual output of the system governed by Equation (8).

The structure of the adaptive network based fuzzy inference system, for a system with 2 inputs ($m = 2$), and 2 rules ($n = 2$) is shown in Figure (3).

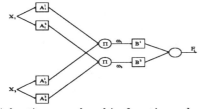

Figure 3: Adaptive membership function scheme network

### IV. MEMBERSHIP FUNCTION TUNING

The membership functions parameters are tuned to minimize the objective function $E$ which is defined as

$$E = \sum_{p=1}^{P} E^p = \sum_{p=1}^{P} (T^p - F^p)^2 \qquad (15)$$

where $E^p$ is the square of the difference between the actual output $F^p$ and the desired output $T^p$ of the system for the $p^{th}$ training data. We assume the number of exemplars in the training data set is $P$.

The parameters of membership functions in the antecedent and the parameters of consequent premises are defined as $\Theta_{ij} = [\sigma_{ij}, \rho_{ij}, \beta{ij}]$ and $\hat{\Theta}_i$, where

$$\hat{\Theta}_i = \begin{cases} [\hat{\sigma}_i, \hat{\rho}_i, \hat{\beta}i] & Model \ F_1 \\ [\hat{\sigma}_i, q_{i1}, q_{i2}, ..., q_{in}] & Model \ F_2 \\ [\hat{\sigma}_i] & Model \ F_3 \end{cases}$$

To update the parameters, we use a steepest descent gradient method to minimize the cost function $E$. The values $\triangle\Theta_{ij}$ and $\triangle\hat{\Theta}_i$ at $(t+1)^{th}$ instant, where $\triangle\Theta_{ij}(t) = \Theta_{ij}(t) - \Theta_{ij}(t-1)$, and $\triangle\hat{\Theta}_i(t)$, are defined in a similar fashion. It is given as a function of the values at $t^{th}$ instant as follows:

$$\triangle\Theta_{ij}(t+1) = -\eta\nabla E_{ij}^p + \alpha\triangle\Theta_{ij}(t) \qquad (16)$$

$$\triangle\hat{\Theta}_i(t+1) = -\eta\nabla\hat{E}_i^p + \alpha\triangle\hat{\Theta}_i(t) \qquad (17)$$

where $\bigtriangledown\hat{E}^p$, $\bigtriangledown E^p$, and $\eta$, are gradients of the parameters, and learning rate, which can be expressed as follows:

$$\nabla E_{ij}^p = \left[\frac{\partial E^p}{\partial\sigma_{ij}}, \frac{\partial E^p}{\partial\rho_{ij}}, \frac{\partial E^p}{\partial\beta_{ij}}\right] \qquad (18)$$

$$\nabla\hat{E}_i^p = \begin{cases} \left[\frac{\partial E^p}{\partial\hat{\sigma}_i}, \frac{\partial E^p}{\partial\hat{\rho}_i}, \frac{\partial E^p}{\partial\hat{\beta}_i}\right] & Model \ F_1 \\ \left[\frac{\partial E^p}{\partial\hat{\sigma}_i}, \frac{\partial E^p}{\partial q_{1i}}, \frac{\partial E^p}{\partial q_{2i}}, \cdots \frac{\partial E^p}{\partial q_{ni}}\right] & Model \ F_2 \\ \frac{\partial E^p}{\partial\hat{\sigma}_i} & Model \ F_3 \end{cases} \qquad (19)$$

$$\eta = \frac{k}{\sqrt{\sum_{i,j}\left(\nabla E_{ij}^p\right)^2 + \sum_i\left(\nabla\hat{E}_i^p\right)^2}} \qquad (20)$$

The constant parameter $\alpha$ is the *momentum* of the gradient descent and the constant $k$ is the step size of the gradient descent. The gradients defined in equations (19) and (20) are analytically available (see appendix) and this makes the presented networks realizable.

### V. ILLUSTRATIVE EXAMPLE

The truck backer-upper control system is used as an example to illustrate this comparative study. Backing a truck to a loading dock is a nonlinear control problem which can involve extensive computation time to steer the truck to a prescribed loading zone.

For truck backer-upper control, Nguyen and Widrow [6] use expert exemplars to train an artificial neural network based controller. Kong and Kosko [3] propose a fuzzy logic controller with 35 expert rules, and they compare their results obtained from FLC with results achieved by using a neural network controller. FLC has been shown to give more appropriate tracking results. A neural network controller only uses numerical data, and FLC employs linguistic rules inquired from expert drivers explicitly. To combine the two above methods, Wang and Mendel [10] utilize numerical-fuzzy approach with almost the same rules as [3] but with different membership functions for fuzzy values.

The truck used in our simulation is the cab part of [6] and the same truck for [3] and [10] except the size of the yard and the definition of steering and azimuth angle. Our study is performed in simulation, therefore the dynamic of truck backer-upper is required. We used the following approximate kinematics [10].

$$
\begin{aligned}
x(t+1) &= x(t) + v\left(cos[\phi(t)+\theta(t)] + sin[\theta(t)]sin[\phi(t)]\right) \\
y(t+1) &= y(t) + v\left(sin[\phi(t)+\theta(t)] - sin[\theta(t)]cos[\phi(t)]\right) \\
\phi(t+1) &= \phi(t) - v\left(sin^{-1}\left[\frac{2sin[\theta(t)]}{\ell}\right]\right) \quad (21)
\end{aligned}
$$

where $(x, y)$, and $\phi$ are rear center of truck coordinate and azimuth angle of truck in yard, respectively. They can be considered as state variables of the system which indicate position and direction of the truck in yard at any instant of time. $\theta$ is the steering angle to direct the truck to the loading zone $x_f$ and $y_f$. Constant parameters $v$ and $\ell$ are truck speed and length of the truck, respectively. The control goal is to steer the truck from any initial position to prespecified loading dock with a right azimuth angle ($\phi_f = 90$) and correct rear position. The steering angle $\theta$ is the control action which is provided by the designed fuzzy controller. Since we presuppose adequate clearance between the truck and the loading dock, state variable $y$ can be abandoned for the reason that it becomes a dependent variable. Therefore the inputs to the controller are $x$ and $\phi$. The range of variables for simulated truck and controller are as follows;

$$
\begin{aligned}
x &\in [x^-, x^+] = [0, 100] \\
\phi &\in [\phi^-, \phi^+] = [-90, 270] \\
\theta &\in [\theta^-, \theta^+] = [-30, 30]
\end{aligned}
$$

We used a rule base that contains 4 rules for backing-up the truck to loading zone. In our comparative study, we considered three different membership functions and three adaptive inference mechanisms explained earlier. The initial membership functions were almost with the same appearance. The reader can refer to [5] for more details about MF parameters. The initial consequent parameter $\hat{\sigma}_i$ was unique for the three different methods of inference (initial $q_{ij} = 0$ for inference $F_2$). The desired control surface which we used was the same control surface achieved from 35 rules given in [8]. We tuned the rule base with 4 rules such that the error achieved from control surface with 35 rules and 4 rules was minimized.

In order to evaluate the performance of different adaptive networks, we define *average percentage error* as follows [1];

$$
\bar{\varepsilon} = \frac{\sum_{p=1}^{P}(T^p - F^p)^2}{\sum_{p=1}^{P}(T^p)^2} \times 100\% \quad (22)
$$

|              | F1    | F2    | F3    |
|--------------|-------|-------|-------|
| Gaussian MF  | .690  | 3.983 | .8051 |
| Cauchy MF    | 2.177 | 3.364 | .744  |
| Triangular MF| 5.053 | 3.610 | 6.486 |

Table 1: Average percentage error $\bar{\varepsilon}$, for 9 possible case stud

where P=209.

Table (1) lists the average percentage error $\bar{\varepsilon}$ after 200 epochs of training for 3 membership functions and 3 inference mechanisms. Initial trajectory tracking from initial position $[x, y, \phi] = [20, 20, 90]$ for Gaussian, Cauchy and triangular MFs along with three adaptive inference mechanisms, are depicted in Figure (4-12)(a). Trajectory tracking after 200 epochs of training also are illustrated in Figure (4-12)(b) for all possible cases.

The simulation has been repeated for 9 rules in the rule base. The achieved results of tracking are confirming obtained results with 4 rules.

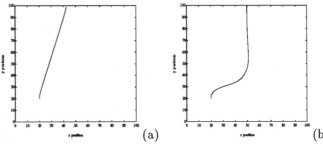

Figure 4: Initial and final trajectory tracking for Gaussian membership function and inference mechanism $F_1$

## VI. CONCLUSIONS

In this paper, through an illustrative example, it has been shown that;

- Achieving more or less the same control task is possible, regardless number of rules or inference mechanism, by tuning membership functions. As it has been shown, the control task obtained with 4, 9 or 36 rules can be identical.

- Using triangular MFs in adaptive networks in comparison with Gaussian or Cauchy bell shape membership functions, is not a good choice of MF. They can be used only for fine MF justification.

- The simplified model of Takagi and Sugeno ($F_3$) is an effective and satisfactory model for semi-fuzzy model instead of using model $F_2$.

### ACKNOWLEDGMENT

The first author would like to thank David Lovell for his valuable comments when preparing this manuscript.

1794

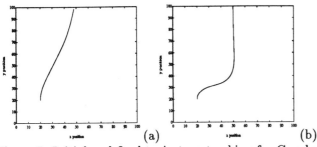

Figure 5: Initial and final trajectory tracking for Cauchy membership function and inference mechanism $F_1$

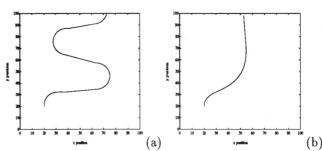

Figure 9: Initial and final trajectory tracking for triangular membership function and inference mechanism $F_2$

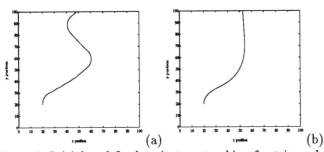

Figure 6: Initial and final trajectory tracking for triangular membership function and inference mechanism $F_1$

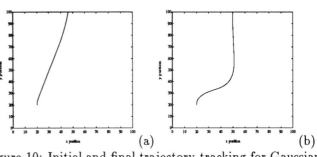

Figure 10: Initial and final trajectory tracking for Gaussian membership function and inference mechanism $F_3$

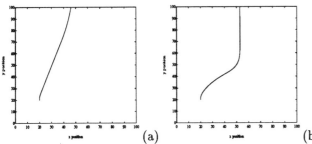

Figure 7: Initial and final trajectory tracking for Gaussian membership function and inference mechanism $F_2$

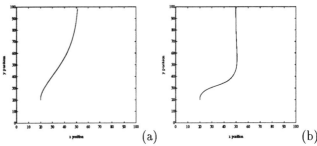

Figure 11: Initial and final trajectory tracking for Cauchy membership function and inference mechanism $F_3$

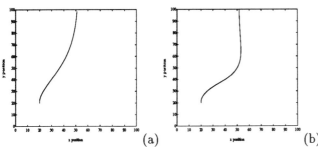

Figure 8: Initial and final trajectory tracking for Cauchy membership function and inference mechanism $F_2$

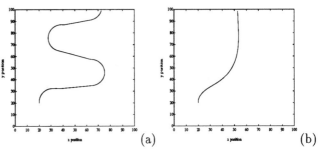

Figure 12: Initial and final trajectory tracking for triangular membership function and inference mechanism $F_3$

$$VVp = -2\hat{\sigma}_i V^+ + \hat{\sigma}_i^2 + V^{+2}$$
$$VVn = -2\hat{\sigma}_i V^- + \hat{\sigma}_i^2 + V^{-2}$$

## APPENDIX

In this appendix the gradient of learning parameters for Gaussian membership functions and the first method of inference ($F_1$) is derived.

$$\frac{\partial E^p}{\partial \sigma_{ij}} = \frac{(T^p - F^p)(F^p - \bar{C}_i)I_i \, w_i \, \beta_{ij}}{(u_j - \sigma_{ij})A_j^i(u_j)\sum_{i=1}^{n} w_i \, I_i}\left(\left(\frac{u_j - \sigma_{ij}}{\rho_{ij}}\right)^2\right)^{\beta_{ij}}$$
$$4e^{-\left(\left(\frac{u_j - \sigma_{ij}}{\rho_{ij}}\right)^2\right)^{\beta_{ij}}}$$

$$\frac{\partial E^p}{\partial \rho_{ij}} = \frac{(T^p - F^p)(F^p - \bar{C}_i)I_i \, w_i \, \beta_{ij}}{\rho_{ij} \, A_j^i(u_j)\sum_{i=1}^{n} w_i \, I_i}\left(\left(\frac{u_j - \sigma_{ij}}{\rho_{ij}}\right)^2\right)^{\beta_{ij}}$$
$$4e^{-\left(\left(\frac{u_j - \sigma_{ij}}{\rho_{ij}}\right)^2\right)^{\beta_{ij}}}$$

$$\frac{\partial E^p}{\partial \beta_{ij}} = \frac{-(T^p - F^p)(F^p - \bar{C}_i)I_i \, w_i}{A_j^i(u_j)\sum_{i=1}^{n} w_i \, I_i}\left(\left(\frac{u_j - \sigma_{ij}}{\rho_{ij}}\right)^2\right)^{\beta_{ij}}$$
$$2ln\left(\frac{u_j - \sigma_{ij}}{\rho_{ij}}\right)^2 e^{-\left(\left(\frac{u_j - \sigma_{ij}}{\rho_{ij}}\right)^2\right)^{\beta_{ij}}}$$

$$\frac{\partial E^p}{\partial \hat{\sigma}_i} = \frac{2(T^p - F^p)w_i}{\sum_{i=1}^{n} w_i \, I_i}\left[(F^p - \bar{C}_i)\frac{\partial I_i}{\partial \hat{\sigma}_i} - I_i\frac{\partial \bar{C}_i}{\partial \hat{\sigma}_i}\right]$$

$$\frac{\partial E^p}{\partial \hat{\rho}_i} = \frac{2(T^p - F^p)w_i}{\sum_{i=1}^{n} w_i \, I_i}\left[(F^p - \bar{C}_i)\frac{\partial I_i}{\partial \hat{\rho}_i} - I_i\frac{\partial \bar{C}_i}{\partial \hat{\rho}_i}\right]$$

$$\frac{\partial E^p}{\partial \hat{\beta}_i} = \frac{2(T^p - F^p)w_i}{\sum_{i=1}^{n} w_i \, I_i}\left[(F^p - \bar{C}_i)\frac{\partial I_i}{\partial \hat{\beta}_i} - I_i\frac{\partial \bar{C}_i}{\partial \hat{\beta}_i}\right]$$

For the sake of simplicity we will consider the case when $\hat{\beta}_i = 1$ and this variable is not be changed during the training process. Therefore we have a relatively simple gradient for the other variables.

$$\frac{\partial I_i}{\partial \hat{\sigma}_i} = -e^{-\left(\frac{\hat{\sigma}_i - V^+}{\hat{\rho}_i^2}\right)^2} + e^{-\left(\frac{\hat{\sigma}_i - V^-}{\hat{\rho}_i^2}\right)^2}$$

$$\frac{\partial I_i}{\partial \hat{\rho}_i} = \frac{-\sqrt{\pi}}{2}erf(\frac{\hat{\sigma}_i - V^+}{\hat{\rho}_i}) + \frac{(\hat{\sigma}_i - V^+)}{\hat{\rho}_i}e^{-\left(\frac{\hat{\sigma}_i - V^+}{\hat{\rho}_i^2}\right)^2}$$
$$+\frac{\sqrt{\pi}}{2}erf(\frac{\hat{\sigma}_i - V^-}{\hat{\rho}_i}) - \frac{(\hat{\sigma}_i - V^-)}{\hat{\rho}_i}e^{-\left(\frac{\hat{\sigma}_i - V^-}{\hat{\rho}_i^2}\right)^2}$$

$$\frac{\partial \bar{C}_i}{\partial \hat{\sigma}_i} = \frac{1}{I_i}\left((\hat{\sigma}_i + V^+)e^{-\frac{VVp}{\hat{\rho}_i^2}} - (\hat{\sigma}_i + V^-)e^{-\frac{VVn}{\hat{\rho}_i^2}}\right)$$
$$+\frac{\hat{\sigma}_i \, \bar{C}_i}{I_i} + 1 - \frac{\bar{C}_i}{I_i^2}\frac{\partial I_i}{\partial \hat{\sigma}_i}$$

$$\frac{\partial \bar{C}_i}{\partial \hat{\rho}_i} = \frac{1}{I_i^2}\left(\frac{\hat{\rho}_i^2}{2}e^{-\frac{VVp}{\hat{\rho}_i^2}} - \frac{\hat{\rho}_i^2}{2}e^{-\frac{VVn}{\hat{\rho}_i^2}} - I_i\right)\frac{\partial I_i}{\partial \hat{\rho}_i} + \frac{\hat{\sigma}_i}{\hat{\rho}_i}$$
$$+\frac{1}{I_i}\left(-\hat{\rho}_i e^{-\frac{VVp}{\hat{\rho}_i^2}} + \frac{\hat{\sigma}_i(\hat{\sigma}_i - V^+)}{\hat{\rho}_i}e^{-\left(\frac{\hat{\sigma}_i - V^+}{\hat{\rho}_i}\right)^2}\right)$$
$$+\frac{1}{I_i}\left(\hat{\rho}_i e^{-\frac{VVn}{\hat{\rho}_i^2}} - \frac{\hat{\sigma}_i(\hat{\sigma}_i - V^-)}{\hat{\rho}_i}e^{-\left(\frac{\hat{\sigma}_i - V^-}{\hat{\rho}_i}\right)^2}\right)$$
$$+\frac{1}{I_i}\left(-\frac{VVp}{\hat{\rho}_i}e^{-\frac{VVp}{\hat{\rho}_i^2}} + \frac{VVn}{\hat{\rho}_i}e^{-\frac{VVn}{\hat{\rho}_i^2}}\right)$$

## REFERENCES

[1] J. S. R. Jang, "Fuzzy Modeling Using Generalized Neural Networks and Kalman Filter Algoorithm," Proc. of the 9th National Conference on Artificial Intelligent, pp. 762–767, July, 1991.

[2] J. S. R. Jang, "Self-Learning Fuzzy Controllers based on Temporal Back Propagation," IEEE Trans. on Systems, Man, and Cybernetics , 3, no. 5, pp. 714–723, Sep., 1992.

[3] S. G. Kong and B. Kosko, "Comparision of Fuzzy and Neural Track Backer-Upper Control System," Proc. of Int. Joint Conf. on Neural Networks (IJCNN-90), 3, pp. 349–358, June, 1990.

[4] C. C. Lee, "Fuzzy logic in control systems; Fuzzy logic controller Part I / Part II," IEEE Trans. Systems, Man, and Cybernetics, 20, no. 2, May, 1990.

[5] A. Lotfi and A. C. Tsoi, "Redundant Rule Reduction Using Adaptive Membership Function Scheme Through Expert Knowledge and Exemplar," Proc. of IEEE Australia and New Zealand Inteligent Information Systems, Perth, Dec., 1993.

[6] D. H. Nguyen and B. Widrow, "Neural networks for self-learning control systems," IEEE Control Systems Magazine, pp. 18–23, April, 1990.

[7] H. Nomura, I. Hayashi and N. Wakami, "A learning Method of Fuzzy Inference Rules by Descent Method," IEEE Int. Conf. on Fuzzy Systems, pp. 203–210, March, 1992.

[8] P. J. Pacini and B. Kosko, "Adaptive fuzzy systems for target tracking," Intelligent Systems Engineering, pp. 3–21, July, 1992.

[9] T. Takagi and M. Sugeno, "Fuzzy identification of systems and its application to modeling and control," IEEE Trans. on Systems, Man, and Cybernetics, 15, no. 1, pp. 116–132, Jan., 1985.

[10] L. X. Wang and J. M.Mendel, "Generating Fuzzy Rules by Learning From Examples," IEEE Trans. on Systems, Man, and Cybernetics , 22, no. 6, pp. 1414–1427, Dec., 1992.

[11] L. Zheng, " A Practical Computer-Aided Tuning Technique for Fuzzy Control," Proc. of IEEE Conference on Fuzzy Systems, 2, pp. 702–707, April, 1993.

# A Study on Apportionment of Credits of Fuzzy Classifier System for Knowledge Acquisition of Large Scale Systems

Ken Nakaoka, Takeshi Furuhashi, Yoshiki Uchikawa
Department of Information Electronics, Nagoya University
Furo-cho, Chikusa-ku, Nagoya, 464-01, Japan
Tel.+81-52-781-5111 ext.2792 Fax.+81-52-781-9623
E-mail: nakaoka@bioele.nuee.nagoya-u.ac.jp

*Abstract*—This paper studies a knowledge acquisition of large scale systems using a Fuzzy Classifier System (FCS). A new method for apportionment of credits is proposed in this paper. The new method makes it possible to fully utilize the feature of the genetic algorithm, i.e. the effects of the crossover operator.

Simulations to obtain avoiding rules of a ship are done to show that the FCS can acquire fuzzy rules for complex tasks.

## 1. INTRODUCTION

Fuzzy controls described in linguistic IF-THEN rules have been widely used in the industry for its high performance in human-computer interactions. A demand for fuzzy inference system which can describe complex multi-input/output systems is growing. The methods used in [1][2] are effective in the case where there are input/output data of the systems in advance. However, if these data are not available, the capability of these methods is limited. An application of the genetic algorithm (GA) to automatic acquisition of fuzzy control rules has been reported[3]. The methods used in [3] has a problem that the sizes of the chromosomes become very large to describe the complex multi-input/output systems.

Valenzuela-Rendon proposed Fuzzy Classifier System (FCS)[4] by introducing fuzzy rule base and fuzzy inference into the Classifier System (CS)[5]. The FCS can handle continuous variables. The number of fuzzy rules used in the FCS increases linearly with the number of input variables. Moreover, the designer of the inference system has only to judge the final performance of the system. However, it was shown in [4] that the FCS was effective to handle continuous variables by applying the FCS to an approximation of a single input-single output function. Studies on apportionment of credits to fuzzy rules for describing the large scale systems have not been done.

The authors have studied application of the FCS to knowledge acquisition of large scale systems[6]. This paper presents a new method for apportionment of credits. The new method has the following two steps: 1)Each fuzzy rule receives credits in proportion to the number of membership functions having grades larger than zero; 2)The rule receives credits for the degrees of contribution to success/failure of simulations. The new method enables to fully utilize the feature of the GA that is the effects of the crossover operator. With this method, fuzzy rules which describe knowledge in complex multi-input/output systems can be obtained. Simulations are done to show that the FCS acquires the fuzzy rules for collision avoidance in steering a ship.

## 2. FUZZY CLASSIFIER SYSTEM

### 2.1 Problem formulation

This paper studies application of the FCS to a knowledge acquisition of a multi-input system. For acquiring avoiding rules in steering a ship, a new method for apportionment of credits is proposed.

The problem of this ship is just to steer the operator's ship to go into the goal avoiding two other ships. The velocities of the operator's ship and the other ships are $V$, $V_0$, respectively, and both are set to be constant. Fig.1 shows the denotations.

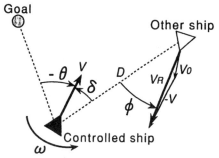

Fig.1 Denotations of parameters

The distance between the operator's ship and the

other ship is denoted by $D$. The angle between the direction of operator's ship and the direction of the other ship viewed from operator's ship is denoted by $\delta$. In the same way, the angle between the direction of operator's ship and the goal is $\theta$. $V_R$ is the relative velocity between the two ships. The angle between the relative velocity $V_R$ and the direction of operator's ship from the other ship is $\phi$. When $\phi$ is nearly zero, it means that the other ship comes toward operator's ship. Each angle is set to be counterclockwise positive viewed from the reference line in Fig.1.

2.2 Application of fuzzy classifier system

The FCS is a learning system having the following four blocks: rule generation mechanism, fuzzy rule base, fuzzy inference system and apportionment of credit. The configuration of the FCS is shown in Fig.2. In this paper, the functions of the blocks are as follows:

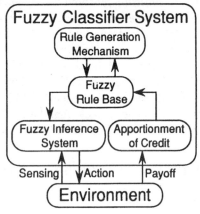

Fig.2 Configuration of Fuzzy Classifier System

(1)fuzzy rule base

The inputs of the FCS are the relative distance $D_1$, $D_2$, the angle $\phi_1$, $\phi_2$, the relative angle $\delta_1$, $\delta_2$, the relative angle $\theta$, and the angular velocity $\omega$. The output of the FCS is the steering angle $u$. The subscripts 1, 2 means the variables between the operator's ship and the other No.1, 2 ships, respectively. Two membership functions (Small (S), Big (B)) are used for the distance $D_1$, $D_2$, and five (Negative Big (NB), Negative Small (NS), Zero (ZO), Positive Small (PS), Positive Big (PB)) are used for other variables. The rules of the FCS are represented with nine bits as in Fig.3.

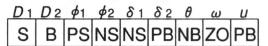

Fig.3 An example of cording of
input/output variables

Each bit corresponds to one of the membership

functions of each variable. The rule in Fig.3 can be read as follows:

> IF $D_1$ is S, $D_2$ is B, $\phi_1$ is PS, $\phi_2$ is NS,
> $\delta_1$ is NS, $\delta_2$ is PB, $\theta$ is NB, $\omega$ is ZO
> THEN $u$ is PB.

Fig.4 shows the membership functions for each variable.

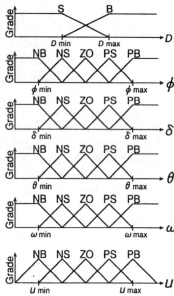

Fig.4 Membership functions

(2)fuzzy inference system

Fuzzy inference is done here using fuzzy rules in the fuzzy rule base. The product-sum-center of gravity method is used. The steering of the ship is continued until the controlled ship collides with either of the two other ships, or the controlled ship goes out of the screen on the CRT, or the ship goes into the goal avoiding the two other ships.

(3)apportionment of credit

The number of possible rules $n_{pos} = 2^2 \times 5^6 = 62,500$ in this system, whereas most of these $n_{pos}$ rules are anticipated not to be used for controlling the ship. It is easily expected that a small number of rules relative to $n_{pos}$ is sufficient to fulfill the steering task. On the contrary, with the small number of randomly generated rules, the number of rules which have the truth values larger than zero is expected to be very small and the steering of the ship becomes impossible. Some special considerations should be given to the apportionment of credit system for generating proper fuzzy rules. The authors propose to use the following two kinds of payoffs:

(a)Ones are directly given to the fuzzy rules of which some of the membership functions in the antecedent have a certain amount of grades at each sampling time of the simulation. For every generation, e.g. the $k$-

th($k = 1, 2, \cdots$) generation, the credit $\beta_{jk}$ ($j = 1, 2, \cdots$, $n_{chr}$) of each fuzzy rule is accumulated in proportion to the number of membership functions having grades larger than zero. $n_{chr}$ means the number of initial rules. At the early generations, there are few rules which have the truth values. This credit $\beta$, together with the crossover operator, is effective for searching valid combinations of membership functions in the antecedent.

(b)The other payoffs are given to the apportionment of credit system on success/failure of the steering. Success means that the controlled ship goes into the goal, and failure means that the ship goes out of the screen or the ship collides with one of the two other ships. The FCS receives positive payoff in case of success, and receives negative payoff in case of failure. The feature of this system is that the FCS acquires the rules using such a simple standard of evaluation without being taught detailed rules by operators. The way of delivering the other credits $\alpha$ to each fuzzy rule in the rule base is as follows:

Each rule has the initial credit of $\alpha_0$. Suppose the rules at the $k$-th($k = 1, 2, \cdots$) generation are used. The steering of the ship is done $n_{tri}$ times. The controlled ship starts from the positions at the wall opposite to the goal. The initial angular velocity is 0 rad/s. The truth value and the output steering angle $u$ of each fuzzy rule at each sampling time from the start to the goal/failure are memorized. The FCS receives the payoff $\mu_S$ (>0) in case of success or the payoff $\mu_F$ (<0) in case of failure. The apportionment of credit system assigns these payoffs to each fuzzy rule as the credit $\Delta\alpha_{jk}$ ($j = 1, 2, \cdots, n_{chr}$). The mean values of the positive and negative steering angles $u$ for each steering trial are obtained, respectively. Then the rules with the output $u$ bigger than the | mean value of the steering angle | $\times\gamma_{thr}$ receives the following credit:

$$\Delta\alpha_{jk} = \sum_l \mu_i \times \text{(the truth value of the fuzzy rule)}_l$$
$$(i = S, F, \quad j = 1, 2, \cdots n_{chr})$$

where $l$ means the $l$-th sampling time.

The rules with the output $u$ smaller than the | mean value of the steering angle | $\times\gamma_{thr}$ receives no credit $\Delta\alpha_{jk} = 0$. Thus the credits of the rules at the ($k$+1)-th generation $\alpha_{jk+1}$ is renewed by $\alpha_{jk+1} = \alpha_{jk} + \Delta\alpha_{jk}$. The rules which has never had the truth value bigger than 0 pay the tax $\Delta\alpha_{tax}$ (<0) as follows:

$$\alpha_{jk+1} = \alpha_{jk} + \Delta\alpha_{tax}.$$

The credit $\alpha$, together with the crossover and mutation operators, is effective for finding valid rules for the steering.

(4)rule generation mechanism

In this block, fuzzy rules are selected and reproduced using the genetic algorithm (GA). In this paper, the crossover and mutation operators are used.

(a)crossover

The crossover is used especially for searching fuzzy rules which have the truth values at each phase of the steering operations. The $n_{sel}$ rules with less credits $\beta_{jk+1}$ are selected. Among the selected $n_{sel}$ rules, there may be some rules having large credits of $\alpha_{k+1}$. Let $\zeta$ denote the credit $\alpha_{k+1}$ which is the $n_{less}$-th value from the least. In the $n_{sel}$ rules, the rules whose credits $\alpha_{k+1}$ are larger than $\zeta$ are replaced with the rules having less credits of $\alpha_{k+1}$. The $n_{sel}$ rules selected in this way are screened out. New rules are randomly reproduced from the remaining rules and one-point crossover is applied to the new rules.

(b)mutation

The mutation operation which changes the label of membership functions is used for the following two purposes in this paper: One is applied to the antecedent part of the newly generated rules in case that the antecedent part of the new rules coincide with that of the existing rules for avoiding the uniformity of the rules. The other is applied to the consequent part of the existing rules with the probability of 0.5 for improving the actions.

After these genetic operations are done, the credits of the new rules and the credits of the rules whose credits are less than $\alpha_0$ are renewed to be $\alpha_0$, and the other credits of all rules $\beta$ is reset to be zero. This is to give the new rules a equal chance to survive.

The rules at the ($k$+1)-th generation are produced. The simulation restarts at (2)fuzzy inference system.

## 3. SIMULATIONS

The rules in the FCS were first randomly generated. The number of rules $n_{chr} = 100$, the number of trials $n_{tri} = 5$, initial credit $\alpha_0 = 100$, the payoff in case of success $\mu_S = 3.0$, the payoff in case of failure $\mu_F = -1.0$, the threshold of the steering angle $\gamma_{thr} = 1.5$, the tax $\Delta\alpha_{tax} = -0.1$, the number of rules which are screened out $n_{sel} = 20$. The number $n_{less}$ which determines the threshold value $\zeta$ is set as $n_{less} = 30$. When the controlled ship goes into the goal in every trial or the 100th generation is over, the simulation is stopped. Two other ships appear from the left hand side wall and right hand side wall, and go across in front of the controlled ship. The ship which goes across from right to left is denoted as No.1 and the other from left to right is No2. Fig.5 shows an example of tracks of ships at the initial fuzzy rules.

The controlled ship went straight without steering because there existed few rules which have the truth values. Fig.6 shows the tracks of ships at the 10th generation. The controlled ship was steered a little. However, the steering was not sufficient to go into the goal avoiding the other ships yet. Fig.7 shows the tracks of ships at the 69th generation. With the improved fuzzy rules, the ship was steered well for every trial.

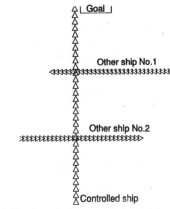

Fig.5 Tracks of ships with initial fuzzy rules

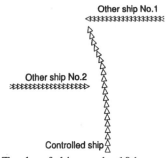

Fig.6 Tracks of ships at the 10th generation

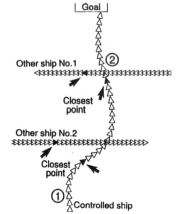

Fig.7 Tracks of ships at the 69th generation

At the point ① in Fig.7, the fuzzy rule

IF $D_1$ is B, $D_2$ is B, $\phi_1$ is ZO, $\phi_2$ is NS,
$\delta_1$ is PS, $\delta_2$ is NS, $\theta$ is PS, $\omega$ is PS
THEN $u$ is NB

is used and the rule means to avoid the collision with the No.2 ship by steering the controlled ship to the right. At the point ② in Fig.7, the fuzzy rule

IF $D_1$ is B, $D_2$ is B, $\phi_1$ is PS, $\phi_2$ is NS,
$\delta_1$ is NB, $\delta_2$ is NB, $\theta$ is PS, $\omega$ is NS
THEN $u$ is NB

is used and it means to go into the goal by steering the ship to the right.

## 4. CONCLUSIONS

This paper studied application of the FCS to knowledge acquisition of large scale systems. For acquiring avoiding rules in steering a ship, a new method for apportionment of credits was proposed. It was shown that with the two kinds of payoffs the FCS acquired complex fuzzy rules.

## REFERENCES

[1]Iokibe, T. "Fuzzy Clustering Method by Discretionary Fuzzy Performance Function -Automatic Rule and Membership Function Generation-", J. of Japan Society for Fuzzy Theory and Systems Vol.4, No.2 pp.334-343 (1992)

[2]Tsutsui, H., Kurosaki, A. "Fuzzy & Topological Case Based Modeling and its Application", 9th Fuzzy System Symposium (Sapporo, Japan) pp.249-252 (1993)

[3]Takahama, T., Miyamoto, S., Ogura, H., Nakamura, M. "Acquisition of Fuzzy Control Rules by Genetic Algorithms", 8th Fuzzy System Symposium (Hiroshima, Japan) pp.241-244 (1992)

[4]Valenzuela-Rendon, M. : The Fuzzy Classifier System: A Classifier System for Continuously Varying Variables, Proceeding of 4th International Conference on Genetic Algorithm, pp.346-353 (1991)

[5]Goldberg, D. E. : Genetic Algorithm in Search, Optimization and Machine Learning, Addison Wesley, (1989)

[6]Furuhashi, T., Nakaoka, K., Morikawa, K., Uchikawa, Y., Maeda, H. "A Study on Knowledge Acquisition of Large Scale Systems Using Fuzzy Classifier System" submitted to the J. of Japan Society for Fuzzy Theory and Systems.

# Fuzzy Approach in Model-based Object Recognition

Popović, D and Liang, N

*University of Bremen, FRG*

**Abstract.** *A fuzzy logic approach to pattern recognition is proposed along with the corresponding model-based problem solving algorithm suitable for recognition in intelligent robotics, where a good visual orientation is required for space orientation of a working robot. For simplified pattern recognition the angle-of-sight signature is used to represent the features of the object image. The features, defined in this way, are then used for building a reference model base. In addition, the membership function of the reference modes was defined in order to structure the demarcation rule base. Finally, using the model base built and the rule-based algorithm proposed, the stored image of the "seen" object is classified as pertaining to the reference one or not. Some simulation results are included.*

**Keywords.** *Pattern Recognition, Fuzzy Set, Artificial Intelligence, Robotics Control, Pattern classification.*

**Contact Author:** Prof. Dr. Dobrivoje Popovic, Dipl.-Ing. Nan Liang, Institut für Automatisierungstechnik, Fachbereich 1, Universität Bremen, Postfach 330440, W-2800 Bremen 33, Germany.
Telephone: 49-421-218-3580, Fax: 49-421-218-3601

# I. Introduction

Image data processing in computer vision-based robotics is primarily concerned with the object recognition problems. Here, the main impact is due to the temptation to simulate the capability of a human being to recognize relatively complex patterns, that still cannot be efficiently handled by the computers(Watanabe,1985), like the recognition of a human face, of the patient's status, etc. taken in the presence of a high-level distortion.

A way of enhancing the pattern classification power of intelligent machines would be to teach them to demarcate the given patterns, i.e. to simplify or approximate them, like a human being does when recognizing some well-defined objects. This practically means to search for some simplifed but relatively robust classification schemes. Such schemes are likely to be bound by implementing the relevant fuzzy logic approaches, as it will be pointed out here.

The fact, that we want to teach the computer to apply some rather vague but robust object identification criteria guide us to the use of fuzzy logic approach for mathematical formulationof the problem to be solved. Such approach has generally proved to be highly useful for pattern recognition(Zimmermann,1987), this mainly becuase the fuzzy sets have become the most adequate theoretical background for modeling cognitive processes of the human being, especially those related to the object recognition and the successive object classification problems.

This was our motivation for solving the image recognition problem in vision-based robotics using a fuzzy approach, based on on storage of all pattern classes of the objects to be expected in the "world" or work space of a robot. The stored patterns are then analyzed by the computer, i.e. the object features extracted and stored in the model-base built in so called AOS(angle-of-sight) signature form. At this point we suppose that the range of patterns to be identified is relatively limited. Usually, a robot to which our results should be applied will face only a relatively limited number of work objects, or tools, when carrying out its tasks. This, in turn, implies a spacely relatively limited model base. Therefore, we define the grade membership function or degree membership functions to represent the membership within the fuzzy sets mapped as referent image clusters. The larger the membership function value the closer the element (object image) to the corresponding fuzzy set (referent image). Finally, in accordance with the expert's experence(or the heuristic knowledge) we determine some decision threshold values for individual image features in the membership function value gap, seperating the object patterns into two distinct groups with the mean values above or below the threshold value. Based on this demarcation concept, we then set up a rule base of individual decisions. In the process of the object recognition we then, in order to achieve the demarcation to be stored in the model base, apply the decision rules, stored in the rule base. Thus, the rule base and the model base are the main supports of our classification algorithms.

## II. Elementary Individual Pattern Fuzzy Recognition

Let in the uniuerse $U$ the fuzzy subsets

$$A_{\sim 1}, A_{\sim 2}, ..., A_{\sim n},$$

represent a set of different pattern classes. For an element $u_j$ in $U$ we can define corresponding membership functions $\mu_{A_{\sim i}}(u_j)$, designating the grade to which the $u_j$ belongs to $A_{\sim i}$ ( $A_{\sim i} = \{(u, \mu_{A_{\sim i}}(u))|u \in U, i = 1, 2, ..., n)$. As we known, when $\mu_{A_{\sim i}}(u_j) = 1$, $u_j$ belongs to fuzzy set $A_{\sim i}$ definitely. And $\mu_{A_{\sim i}}(u_j) = 0$ means $u_j$ does not belong to fuzzy set $A_{\sim i}$. Generally, $\mu_{A_{\sim i}}(u_j)$ is between 1 and 0.

Assuming, $u_0$ is an element in $U$. There must be a set $A_{\sim k}$ $(k = 1, 2, ..., n)$, which makes $\mu_{A_{\sim k}}(u_0) = max\{\mu_{A_{\sim 1}}(u_0), \mu_{A_{\sim 2}}(u_0), ..., \mu_{A_{\sim n}}(u_0)\}$. i.e.$\mu_{A_{\sim k}}(u_0)$ is the largest one among the membership functions $\mu_{A_{\sim 1}}(u_0) \sim \mu_{A_{\sim n}}(u_0)$.

Our principle in the process is: because $\mu_{A_{\sim k}}(u_0)$ is largest, $u_0$ should be classified as a member of fuzzy set $A_{\sim k}$.

## III. Object Pattern Recognition for Robotic Vision

Let the features of a two-dimensional image $A_i$ are discribable by an angle-of-sight signature, and assume that the point $O$ is the origin of the reference polar coordinates system and the centre of gravity of the image itself(Fig.1(a)). Using the angle $\alpha$ we can now define

$$P(\alpha) =| OP |_\alpha \qquad 0° \leq \alpha \leq 360°$$

as the distance between $O$ and the corresponding cross point $P$. Here,

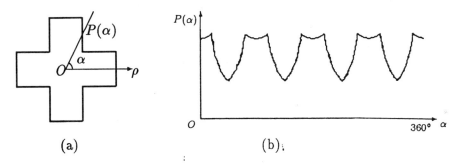

(a)                              (b).

*Fig. 1. (a) Example image $A_1$ and a polar coordination set for signating the structure properties of the image.*
*(b) Angle-of-sight signature of this example image $A_1$*

in fact, $P(\alpha)$ represents the AOS signature of $A_i$(Fig.1(b)). Some further examples of the images and the corresponding AOS signatures are given in Fig.2.

The problem now is to carry out the pattern matching. i.e. to establish weather an object, say $B_j$, should be classified as one of references , say $A_i$. In this way the original object recognition problem is reduced to the equivalent shape recognition problem.
The closeness of object's shape is reduced to the closeness of their signature lines. Denoting the AOS signature of the both images by $P_{jA}(\alpha)$ and $P_{iB}(\alpha)$, we can use the minimum least squares error

$$S = \tfrac{1}{n}\sqrt{\overline{P_{iA}(\alpha_1)P_{jB}(\alpha_1)}^2 + \cdots + \overline{P_{iA}(\alpha_k)P_{jB}(\alpha_k)}^2 + \cdots + \overline{P_{iA}(\alpha_n)P_{jB}(\alpha_n)}^2}$$

to quantify the distance between them, where $\overline{P_{jB}(\alpha_k)P_{iA}(\alpha_k)}$ is the distance between the points of $P_{jB}(\alpha_k)$ and $P_{iA}(\alpha_k)$ on the signature lines.

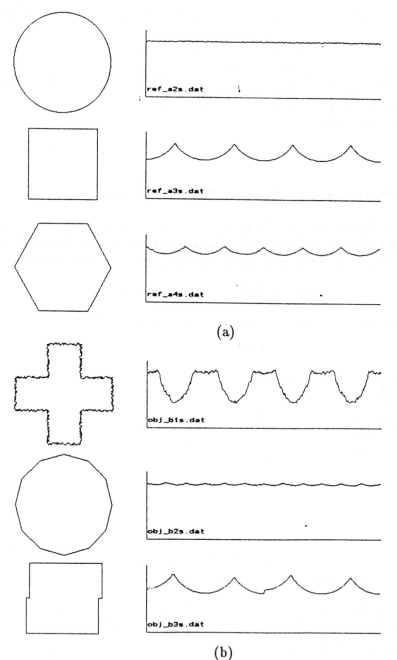

(a)

(b)

Fig.2. Images and their AOS signatures

Alternatively the discrete form of Mahalanofis disface

$$S_m = \sum_{\gamma=1}^{n} \mid P_{iA}(\alpha_\gamma) - P_{jb}(\alpha_\gamma) \mid$$

can be used.

It is obvious that $S = 0$ means that the AOS signatures of $A_i$ and $B_j$ are coincident. Based on the squares error concept and, supposing that the values of individual functions are within the range [0,1], we can now define the corresponding membership functions. Defined in this way, the function expresses the grade of similarity of $A_i$ and $B_j$, expressed by

$$\mu_{A_i(B_j)} = \left\{ \begin{array}{cc} 1 - S/S_t & 0 \leq S \leq S_t \\ 0 & otherwise \end{array} \right.$$

where $S$ is the minimum least squares error between AOS signatures $A_i$ and $B_j$, and $S_t$ is the bound value of the function concerned.

In order to apply the same approach to the pattern classification, we assume a threshold value of the membership function $\theta$, based on our heuristic knowledge. When the value of the membership function $\mu_{A_i}$ is larger than its threshold value, the objcet image $B_j$ should be classified as belonging to the class $A_i$. Building the membership functions of all reference images and selecting their largest value $max\{\mu_{A_1}, \mu_{A_2}, \cdots\}$, we can now check wheather the condition $max\{\mu_{A_1}, \mu_{A_2}, \cdots\} \geq \theta$ or $max\{\mu_{A_1}, \mu_{A_2}, \cdots\} < \theta$ holds. In the first case there are two possibilities:

1. $max\{\mu_{A_1}, \mu_{A_2}, \cdots\}$ is unique, then we identify the class corresponding to the maximum value and classify the image accordingly

2. $max\{\mu_{A_1}, \mu_{A_2}, \cdots\}$ is not unique, then we difine a priority among the $\mu_{A_1}$, $\mu_{A_2}$, ... and classify the object image accordingly

Otherwise, the object image is rejected as not belonging to any of the classes.
The following example should close explain the classification algorithm proposed.

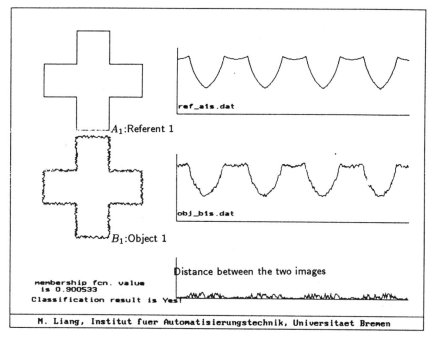

Fig. 3. To class the object image as the referent one, since the corresponding membership function value is larger than the threshold

Let, as shown in Fig. 3, $A_1$ be one of the referent images and $B_1$ the image of an object produced by a vision system. Let, in addition, $P_{A_1}(\alpha)$ and $P_{B_1}(\alpha)$ are their AOS. The minimum least squares error between the signatures is

$$S_1 = \frac{1}{n}\sqrt{\overline{P_{A_1}(\alpha_1)P_{B_1}(\alpha_1)}^2 + \cdots + \overline{P_{A_1}(\alpha_n)P_{B_1}(\alpha_n)}^2}$$

Membership function $\mu_{A_1(B_1)}$, giving the grade of membership of $B_1$ to $A_1$, is expressed as

$$\mu_{A_1(B_1)} = \left\{ \begin{array}{ll} 1 - S_1/S_t & 0 \leq S_1 \leq S_t \\ 0 & otherwise \end{array} \right.$$

with $S_t = 1.3$ for $\theta = 0.8$. this result in $\mu_{A_1(B_1)} = 0.900533 > \theta$, that corresponds to the $max\{\mu_{A_1}, \mu_{A_2}, \cdots\} > \theta$. Therefore, $B_1$ should be classified as $A_1$.

Table 1 shows some further results of the shape-recogniton process, apllied to the examples from Fig.2.


## IV.Conclusion

A pattern classificaiton approach is proposed for simplified solution of shape recognition problem for vision-based robotic systems. The simulation results have shown that the approach is particularly efficient when the shapes of the objects to be recognized by the robot are of a relatively limited range.

| | Referent Images | | | |
|---|---|---|---|---|
| Mem. function value $\mu$ | ✚ | ◯ | ▢ | ⬡ |
| ✚ | 0.900533 | 0.000000 | 0.000000 | 0.155623 |
| ◯ | 0.000000 | 0.907658 | 0.317461 | 0.734628 |
| ▢ | 0.000000 | 0.293665 | 0.864411 | 0.524671 |

*Table 1. Image recognition results*

In the mean time the approach was further improved by

- Using in the process of signature production an efficient algorithm for determination of the centres of gravity and for correct selection of the start point of the signature on the boundary of the image to be recognized.

- Making the results of calculation of the grade of similarify $\mu_{A_i}$ robust against the absolute measurements of the vision system.

# References

[1] W. Pedrycz, "Fuzzy Sets in Pattern Recognition: Methodology and Methods", *Pattern Recognition*, Vol. 23, No. 1/2, pp.121-149,1990.

[2] Silvana Dellepiane, Giovanni Venturi and Gianni Vernazza,"Model Generation and Model matching of Real Images by a Fuzzy Approach", *Pattern Recognition*, Vol.25,No.2,pp.115-137,1992.

[3] I. Gath and A.B.Geva,"Unsupervised Optimal Fuzzy Clustering", *IEEE Transaction on Pattern Analysis and Machine Intelligence*, Vol.11,NO.7,pp.773-781,July 1989.

[4] Michael J. Sabin, "Convergence and Consistency of Fuzzy c-means/ISODATA Algorithms", *IEEE Transaction on Pattern Analysis and Machine Intelligence*, Vol.PAMI-9, No.5, SEPTEMBER 1987.

[5] T.L.Huntsberger, C.L.Jacobs and R.L.Cannon,"Iterative Fuzzy Image Segmentation", *Pattern Recognition*, Vol.18, No.2, pp.131-138, 1985.

[6] Azriel Rosenfeld, "The Fuzzy Geometry of Image Subsets", *Pattern Recognition Letter* 2(1984), pp.311-317.

[7] Azriel Rosenfeld, "The Periveter of A Fuzzy Set", *Pattern Recognition*, Vol.18, No.2, pp.125-130,1985.

[8] Fang-Hsuan Cheng, Wen-hsing Hsu and Chien-An Chen, "Fuzzy Approach to Solve The Recognition Problem of Handwritten Chinese Characters", *Pattern Recognition*, Vol.22,No.2,pp.131-141,1989.

[9] Hans-Jürgen Zimmermann, "Fuzzy Sets in Pattern Recognition", *Pattern Recognition, Theory and Applications*, pp.383-391, 1987.

[10] Yuhua Yao, Weixin Xie, Shanrong Dai and Wu Gang, "X-ray Image Processing Unsing Fuzzy Technology", CH2614-6/88/0000/0730, 1988 IEEE.

[11] K.P.Chan and Y.S.Cheung, "Fuzzy-Attribute Graph with Application to Chinese Character Recognition", *IEEE Transaction on System, Man, and Cybernetics*, Vol.23, No.1, January/February 1992.

[12] King Ngi Ngan and Sing Bing Kang, "Fuzzy Quaternion Approach to Object Recognition Incorporating Zernike Moment Invariants", CH2898-5/0000/0288,1990 IEEE.

[13] H.J.Zimmermann, "Fuzzy Set Theory and Its Applications",1991.

[14] I.Tchoukanov, R.Safaee-Rad, K.C.Smith and B.Benhabib, "The Angle-of-Sight Signature for Two-Dimensional Shape Analyses of Manufactured Objects", *Pattern Recognition*, Vol.25,No.11, pp.1289-1305,1992.

[15] E.K.Wong, "Model Matching in Robot Vision by Subgraph Isomorphism",*Pattern Recognition*, Vol.25,No.3,pp.287-303,1992.

[16] Kandel, Abraham, "Fuzzy Techniques in Pattern Recognition", 1982.

[17] Kashipati Rao, Gerard Medioni, Huan Liu, and George A.Bekey, "Shape Description and Grasping for Robot Hand-Eye Coordination", *IEEE Control System Magazine 1989*.

[18] Wen-Hsiang Tsai and Cheng-lin Chou, "Detection of Generalized Pricipal axes in Rotationally Symmetric Shapes", *Pattern Recognition*, Vol.24,No.2,pp.95-104,1991.

[19] W.Drotning, B.Christensen, and S.Thunborg, "Graphical Model Based Control of Intelligent Robot Systems", 0272-1708/92/. 1992 IEEE.

# Fuzzy Feature Extraction on Handwritten Chinese Characters

Hahn-Ming Lee and Chiung-Wei Huang

Department of Electronic Engineering
National Taiwan Institute of Technology
Taipei, Taiwan
E-mail: hmlee@et.ntit.edu.tw

## Abstract

In this paper, a new feature extraction method based on the fuzzy set theory is proposed. Our feature extraction focuses on decomposing the structure of handwritten characters. Due to the Chinese characters are composed of strokes, features adopted here are all strokes. During extraction, extracted strokes are assigned various degrees for each stroke type. Also, some flexible fuzzy rules are given to deal with the strokes combination. Since the strokes of handwritten Chinese character are fuzzy in nature, our approach can obtain a better result on depressing the writing variants and distortions which are caused by thinning. Moreover, current simulation results encourage us to do further investigation.

## 1. Introduction

Due to the characteristics of handwritten Chinese characters, i.e., large character sets of characters (at least 5401s for daily use), very complex in structure, and writings vary from writer to writer, stroke extraction and stroke combination are the key issues in the recognition of handwritten Chinese characters. A handwritten basic stroke may have various degrees on different stroke types. For example, a stroke may obtain degree 0.4 as horizontal line and degree 0.7 as right-slanting line. That is, we may treat this stroke as a horizontal line or a right-slanting line. Therefore, there are flexible choices for the latter processing, stroke combination. In this way, the result of the feature extraction will be more reasonable and reliable. Besides, fuzzy rules are used to govern the stroke combination. Therefore, the capability of the system can be enhanced by using fuzzy rules appropriately. Due to hierarchical rule sets are provided, the computation effort is not heavy. That is, we use the spirit of fuzzy set theory to reduce the possibility of mis-recognition caused by writing variants as well as distortions caused by thinning.

## 2. System Architecture

The architecture of the Fuzzy Feature Extractor is shown in Fig. 1. It is composed of the Primitive Feature Extractor, Compound Feature Extractor, and a Stroke Feature Pool. Also there are some fuzzy rules given to combine strokes. Inputs to the system must be thinned [1]. The Primitive Feature Extractor aims at extracting basic strokes of handwritten Chinese characters from the input. Also it assigns degrees to the extracted strokes according to the membership definition of each stroke type, and stores these information into the Stroke Feature Pool. In the meantime, the Compound Feature Extractor combines the basic strokes appropriately into some complex strokes. Finally, the stroke features stored in the Stroke Feature Pool can be fed into a classifier to be recognized.

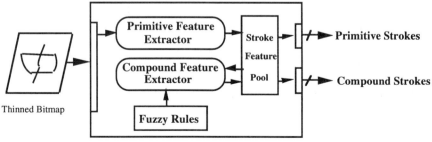

Fig. 1 Fuzzy Feature Extractor.

During extraction, a 3x3 window is used to trace the skeleton of strokes. The tracing direction on input bitmap is from top to bottom and from left to right. Using a tracing window to trace input bitmap will take the advantage of enduring distortions caused by thinning. For example, a thinned handwritten stroke may have some discontinuous points inside its skeleton. If we apply a 3x3 window and adopt some flexible tracing rules [3] on tracing strokes, a more reasonable result will be obtained. Fig. 2(a) shows a tracing window. Also an example of thinned stroke and its extracted result are shown in Fig. 2(b) and Fig. 2(c), respectively. Although in Fig. 2(b) there are 4 break points inside the stroke, they are not feature points, i.e., not corner or crossing point. Thus, the tracing strategy would not regard it as five short strokes but a single slanting line stroke. After this step, the extracted stroke will be stored into the Stroke Feature Pool waiting for the chance to be combined with others. Besides, following the order of extraction may keep the combining information and may save much computation effort for combination. This is because neighboring or crossing strokes have greater opportunity to be combined together. Also, in order to keep the tracing strategy as simple as possible, we delete those strokes which have been traced.

Fig. 2 (a) tracing window (b) thinned pattern (c) extracted stroke.
Based on a 3x3 tracing window, the next movement of point P may have
flexible choices. Though the thinned pattern, (b), is not perfect, the
tracing result, (c), will be much better.

Since the complex structure and large character sets of Chinese characters, it is difficult to cope with them at once. Based on the structure of the Chinese characters, characters can be classified into indivisible characters, left-right characters, top-bottom characters, and special characters [2]. For Example, "仁" is a left-right character, and "兄" is a top-bottom one. In this study, divisible characters, most characters belong to, are used to investigate our approach. Furthermore, some superior methods developed by previous researchers are adopted to retain high performance of our Fuzzy Feature Extractor. They are thinning algorithm [1], stroke extraction [3], and radical segmentation [4].

Tab. 1. Primitive Strokes.
Six primitive strokes of the Chinese characters used in our system.

| Stroke Shape | Stroke Type | Abbreviation |
|---|---|---|
| — | Horizontal | HL |
| \| | Vertical | VL |
| \ | Left-Slanting | LS |
| / | Right-Slanting | RS |
| ﹅ | Dot | DT |
| ㄱ , ㄴ | Corner-Stroke | CS |

## 2.1 Primitive Feature Extractor

It is well known that the Chinese character has very complex structure in its shape. Many researchers have spent much effort on how to decompose its structure. So there are many key-in methods based on the structure decomposition for Chinese characters, e.g., Changjei [5], Dayi, Wuhsiami, ..., etc. In Taiwan, Changjei key-in method is widely used. It almost provides a unique way for every Chinese character. Based on its decomposition on Chinese character structure, we choose Horizontal, Vertical, Left-Slanting, Right-Slanting, Dot, and Corner-Stroke, as our primitive features. Tab. 1 shows all the primitive strokes defined in the Primitive Feature Extractor, and the membership definition for each stroke type is given as follows.

(1) Horizontal: $\mu_{\widetilde{HL}(x)} = \begin{cases} 1- |m(x)|, & |m(x)|<1 \ ; \\ 0, & \text{others} \end{cases}$

, where m(x) denotes the slope of line x.

(2) Vertical: $\mu_{\widetilde{VL}(x)} = \begin{cases} 1- \left|\dfrac{1}{m(x)}\right|, & |m(x)| >1 \ ; \\ 0, & \text{others} \end{cases}$

, where m(x) denotes the slope of line x.

(3) Left-Slanting: $\mu_{\widetilde{LS}(x)} = \begin{cases} 1- \left|\dfrac{\theta(x) - 135^\circ}{45^\circ}\right|, & 122.5^\circ < \theta(x) \le 157^\circ \ ; \\ 0, & \text{others} \end{cases}$

, where $\theta(x)$ denotes the theta between line x and the horizontal line.

(4) Right-Slanting: $\mu_{\widetilde{RS}(x)} = \begin{cases} 1- \left|\dfrac{\theta(x) - 45^\circ}{45^\circ}\right|, & 22.5^\circ < \theta(x) \le 67.5^\circ \ ; \\ 0, & \text{others} \end{cases}$

, where $\theta(x)$ denotes the theta between line x and the horizontal line.

(5) Dot: $\mu_{\widetilde{DT}(x)} = \begin{cases} 1-\min(1, \dfrac{4*len(x)}{3*max\_len}), & \dfrac{1}{8}*max\_len < len(x) < \dfrac{3}{4}*max\_len \ ; \\ 0, & \text{others} \end{cases}$

, where len(x) and max_len denotes the length of line x and the maximum length of stroke of that character, respectively.

(6) Corner-Stroke:

The corner stroke defined here is a consecutive stroke which has a corner inside. That is, there is a direction change in the whole stroke, for example ∟, ⊃ . For convenience, the substroke before the corner is called the former substroke, and the other one is called the latter substroke. The important features of corner stroke are its corner point and the length of latter substroke. One stroke without corner point is undoubtedly not a corner stroke. Thus, we define the membership function as follows:

$\mu_{\widetilde{CS}(x)} = \begin{cases} \dfrac{len(\text{latter substroke})}{len(\text{former substroke})}, & \text{line x has a corner point} \ ; \\ 0, & \text{others} \end{cases}$

, where len() denotes the length of the specified stroke.

## 2.2 Compound Feature Extractor

The Compound Feature Extractor combines primitive features into complex ones. Fuzzy rules are used to carry out strokes combination. It completely retains the fuzzy characteristics of the handwritten Chinese characters. During extraction, the Primitive Feature Extractor extracts out primitive strokes. At the same time, the Compound Feature Extractor takes a look at the connectivity and the order of extracted strokes to check if they can be combined together or not. The CONNECTIVITY denotes the degree of connectivity of two specified strokes. Because the writing habits vary from writer to writer, a connected stroke may be written as a crossing stroke or two independent strokes, such as ∫ and ↖. For retaining fuzzy characteristics, the CONNECTIVITY is very important in the Fuzzy Feature Extractor. Its definition is shown as follows:

$$\text{CONNECTIVITY}(S1, S2) = 1 - \frac{min\_distance(S1, S2)}{max(len(S1), len(S2))}$$

, where len() denotes the length of the specified stroke and min_distance() represents the minimum distance between two specified strokes.

Once strokes get the opportunity to be combined, its degree must be reevaluated for the new stroke type. The used expression is:

$$\text{CombinationDegree}(S1,S2) = \min(\deg(S1),\deg(S2)) * \text{CONNECTIVITY}(S1,S2)$$
, where deg() means the degree of the specified stroke for a special stroke type.

Furthermore, another important parameter, named TYPE, is used to define the connection type of connected stokes. TYPE = (MEET, CROSS), where MEET represents strokes meet together, such as ↑, and CROSS means strokes are crossed, such as ↑. Also, a special α level set degree, named REGULARITY, is used to decide the regularity of handwritten characters. If we set the degree high, it only accepts characters which are written in a very regular manner. That is, strokes will not be combined if their stroke type degrees are low. Otherwise, if the degree of α level set is low, more cursive writings will be accepted. However if the degree set too low, it would possibly return a wrong combination of a character.

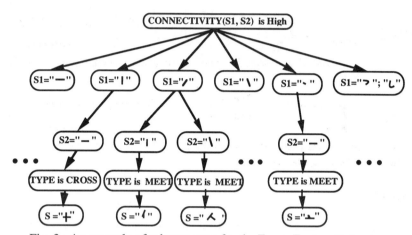

Fig. 3  An example of rule structure for the Fuzzy Feature Extractor.

Due to the complex structure of the Chinese characters, hierarchical rule sets are used to deal with the combination. Otherwise, there must be a lot of rules and takes much effort on inference. Each rule set represents a unique condition for strokes combination. If we treat the rule structure as a hierarchical tree, the root of each subtree stands for the first stroke of a special compound stroke. An example of the hierarchical structure of the rule system is shown in Fig. 3. According to the radical segmentation and tracing strategy, the primitive strokes will be extracted out in an appropriate order. Assume that S1 is the first extracted stroke and S2 is the second one. If the stroke type degrees of S1 and S2 do not reach the α level, they will not be considered to be combined. Otherwise, the Compound Feature Extractor checks the connectivity of them. If CONNECTIVITY(S1,S2) is also greater than the REGULARITY, they get the chance to be combined using fuzzy rules. Through the hierarchical rule set, it is very easy to find that if they can be combined together or not. If so, it will find out the most promising compound stroke quickly, then records the CombinationDegree(S1,S2) as the degree of the combined stroke. Next, check other stroke type candidates of S1 and S2 to see if they also get the chance to be combined. If so, it treats them as the candidates for the combination of S1 and S2. Of course, all of the candidates of S1 and S2 must meet the requirement of REGULARITY, otherwise the further combination would not be proceeded. Finally, a set of possible strokes of the input character are obtained. In particular, for convenience to the classifier, stroke sets are grouped under their possible combinations. Since the rules can be treated as partitioned rule sets, the computation effort is reduced a lot.

As mentioned in the beginning of this section, since Chinese characters have very complex structure, the divisible characters [2] are focussed in this study. The compound features adopted here also follow the Changjei key-in method [5]. They are, totally 24 categories, "日 "，"月 "，"金 "，"木 "，"水 "，"火 "，"土 "，"竹 "，"戈 "，"十 "，"大 "，"中 "，"一 "，"弓 "，"人 "，"心 "，"手 "，"口 "，"尸 "，"廿 "，"山 "，"女 "，"田 "，"卜 ".

In what follows, we give a rule example used in the Fuzzy Feature Extractor.

If CONNECTIVITY(S1, S2) is High then

    If S1 is VL, S2 is HL, and TYPE is CROSS then
        combine_S1_and_S2 as " **十** "           /* Changjei code = " 十 " */

    If S1 is RS, S2 is VL, and TYPE is MEET then
        combine_S1_and_S2 as " **丨** "           /* Changjei code = " 人 " */

    If S1 is RS, S2 is LS, and TYPE is MEET then
        combine_S1_and_S2 as " **入** "           /* Changjei code = " 人 " */

    If S1 is DT, S2 is HL, and TYPE is MEET then
        combine_S1_and_S2 as " **⊥** "           /* Changjei code = " 卜 " */

In short, we can summarize extraction procedure as follows:

    Step 1. Extract one primitive stroke from the input pattern, and calculate the degrees for
          each type of stroke.
        The extracted strokes are deleted.
        (Tracing direction: from left to right, from top to bottom.)

    Step 2. Check if there are strokes in the feature pool able to be combined.
          If not, goto step 1.

    Step 3. Under the constraints of REGULARITY and fuzzy rules,
        combine the most possible type, i.e., the type of highest degree, of the specified strokes and
          calculate their combination result.
        Then check if the specified strokes have other stroke type candidates whose degrees are greater than
          REGULARITY. If so, do the possible combination.

    Step 4. If there are any strokes left in the input, goto Step 1.

    Step 5. Output results in the Stroke Feature Pool.

## 3. Examples

We use the following two examples to illustrate how the extractor extracts and combines strokes:
Let REGULARITY = 0.7,

*(1) An example of left-right divisible character:* 仁

1. " ╱ " is extracted out. The degree of RS is 0.93, and is 0.11 for VL.
2. " 丨 " is extracted out. The degree of VL is 0.91.
3. Combines previous extracted strokes as " 亻 ", the combination degree is 0.91.
    Another combination for VL(0.11) with second VL(0.91) is impossible, because the degree for first VL is
      below the REGULARITY and there is no such combination rule.
4. " ── " is extracted out. The degree of HL is 0.88.
5. " ── " is extracted out. The degree of HL is 0.79.
6. Since the strokes cannot be combined again,
    the results are " 人 ", " ─ ", " ─ ".

*According to this concept, the following characters can be decomposed correctly:*

仁 , 仨 .

*(2) An example of top-bottom divisible character:*

1. " / " is extracted out. The degree of VL is 0.83.
2. " ⁊ " is extracted out. The degree of CS is 0.9.
3. Combines previous extracted strokes as " ⊓ ", the combination degree is 0.8.
4. " — " is extracted out. The degree of HL is 0.75.
5. Combines previous extracted strokes as " ⊔⊃ ", the combination degree is 0.73.
6. " ╱ " is extracted out. The degree of VL is 0.57, and is 0.51 for RS.
7. " ∪ " is extracted out. The degree of CS is 0.73.
8. Since the strokes can not be combined again, the results are " 口 "," 竹 ", " 山 " .

   Note that the " 竹 " denotes " ╱ " and " 山 " represents " ∪ " in Changjei key-in method.

*According to this concept, the following character can be decomposed correctly:*

只 .

Also, it has another candidate: " 口 "," 竹 ", " 人 " .  That is the " 只 " .

## 4. Conclusion

A new method for extracting strokes of handwritten Chinese characters has been proposed. It takes the advantage of fuzzy set theory, and meets the needs of dealing with handwritten Chinese characters whose strokes are of fuzzy in nature. Some examples have been given to demonstrate the workings of our system. Besides, the capability of the whole system can be enhanced by using fuzzy rules appropriately. Also, it not only depresses the writing variants and distortions caused by thinning, but also provides candidates selection for the classifier. Therefore, the mis-recognition rate will be reduced a lot. In order to make the extractor for practical application, a handwritten Chinese character database, named HCCRBASE provided by CCL of ITRI in Taiwan, which contains 5401x250 daily-used characters is currently used to confirm our workings.

## References

[1] Y. S. Chen and Wen-Hsing Hsu, " A Modified Fast Parallel Algorithm for Thinning Digital Patterns," *Pattern Recognition Letter*, vol. 7, pp. 99-106, 1988.
[2] Fang-Hsuan Cheng and Wen-Hsing Hsu, " Research on Chinese OCR in Taiwan," *International Journal of Pattern Recognition and Artificial Intelligence*, vol. 5, no. 1&2, pp. 139-164, 1991.
[3] H. Ogawa and K. Taniguchi, " Thinning and Stroke Segmentation for Handwritten Chinese Character Recognition," *Pattern Recognition*, vol. 15, no. 4, pp. 299-308, 1982.
[4] Fang-Hsuan Cheng and Wen-Hsing Hsu, " Radical Extraction of Handwritten Chinese Characters by Background Thinning Method," *Trans. IEICE E71*, pp. 88-98, 1988.
[5] Hong-Lien Shen, " Handbook of The Fifth Generation Changjei Key-In Method," *Sung-Kang Computer Book Co., Ltd., Taipei, Taiwan, 1993.* (in Chinese)

# Fuzzy Control for Forecasting and Pattern Recognition in a Time Series

Andrew Zardecki

Los Alamos National Laboratory, MS E541, Los Alamos, NM 87545

## Introduction

Starting with pioneering work of Lapedes and Farber,[1] the neural networks have been advantageously applied to prediction and modeling of time series, in domains as distinct as chaotic dynamics and corporate bond rating prediction.

For most real-world control and signal processing problems, the information concerning design and evaluation can be classified into two kinds: numerical information obtained from sensor measurements, and linguistic information obtained from human experts. Generally, neural control is suited for using numerical data pairs (input-output pairs), whereas fuzzy control is an effective approach to utilizing linguistic rules. When fuzzy rules are generated from numerical data pairs, the two kinds of information are combined into a common framework.[2]

As compared to neural networks, the fuzzy controllers can operate in real time; their learning process does not require many iterations to converge. For this reason fuzzy controllers deserve their legitimacy in time series forecasting, where the real time detection and identification of trends is sought. From the standpoint of mathematics, both neural networks and fuzzy controllers stand on a solid footing: they can be viewed as universal approximators.[3]

The paper describes an object-oriented implementation of the algorithm advanced by Wang and Mendel.[2] Numerical results are presented both for time series with seasonal changes and time series corresponding to chaotic situations, such as encountered in the context of strange attractors. In the latter case, the effect of noise on predictive power of the fuzzy controller are explored. In addition, by introducing a distance between an observed and predicted data, one can apply the results of this study to a pattern recognition of temporal signatures.

## Generating Fuzzy Rules

Following Ref. 2, we summarize the algorithm that allows us to generate fuzzy rules from numeric data. The algorithm consists of five steps, the first of which simply divides the input and output spaces into fuzzy regions and assigns to each region a fuzzy membership function. The shape of each membership function is triangular; however, a different shape, e.g. trapezoidal, would not change the essence of the method.

The second step, in which fuzzy rules are generated from given data pairs, is crucial. Suppose we are given a set of desired input-output data pairs:

$$\left( x_1^{(1)}, x_2^{(1)}, x_3^{(1)} ; y^{(1)} \right), \quad \left( x_1^{(2)}, x_2^{(2)}, x_3^{(2)} ; y^{(2)} \right), \ldots \tag{1}$$

where $x_1$, $x_2$, and $x_3$ are inputs, whereas $y$ is the output. We generate a set of fuzzy rules from the input-output pairs of Eq. (1), and use these fuzzy rules to determine a mapping $f: (x_1, x_2, x_3) \rightarrow y$.

To resolve possible conflicts in rules definition, in the third step, we assign a degree to each rule. This is a numerical value equal to the product of degrees of individual rule's members. When some a priori information about the data is available, we can modify the rule degree by an additional multiplicative factor reflecting this information.

In the fourth step, we create a combined fuzzy rule base. If $M_1$, $M_2$, $M_3$, and $N$ are the numbers of quantized levels each rule member can take, then the number $N_R$ of possible rules is given by $N_R = M_1 \times M_2 \times M_3 \times N$. The arising four-dimensional parameter space can be viewed as consisting of $N_R$ cells. If a linguistic rule is an "and" rule, it fills only one cell; an "or" rule fills all the cells in the rows or columns corresponding to the regions of the IF part.

The final fifth step defines the centroid defuzzification scheme. For inputs $x_1$, $x_2$, and $x_3$, let $m^i_I(x_1)$, $m^i_I(x_2)$, and $m^i_I(x_3)$ denote the input membership values of the $i$-th fuzzy rule. Using product operation, we determine the degree of the output, $m^i_O$, corresponding to $x_1$, $x_2$, and $x_3$ as

$$m^i_O = m^i_I(x_1)\, m^i_i(x_2)\, m^i_I(x_3) \tag{2}$$

If the total number of rules is K, then the following centroid defuzzification formula defines the output,

$$y = \frac{\sum_{i=1}^{K} m^i_O \langle y^i \rangle}{\sum_{i=1}^{K} m^i_O}, \tag{3}$$

where $\langle y^i \rangle$ denotes the center value of the output region corresponding to $i$-th rule.

## Object Oriented Implementation

The implementation of the Wang-Mendel algorithm is based on three separate classes dealing with the input-output data arrangement, rule evaluation, and the defuzzification of outputs. The two classes that are relevant to describe both the input and output data are *mfType* (membership function type) and *ioType* (input-output type). In contrast to the approach of Viot,[4] we use arrays, not lists, to handle these data structures, since the number of elements in each remains fixed during a given control process. There is an instance (object) of ioType class corresponding to each member if the if-then rule. If we consider three-input one-output rules, we thus have four instances of the ioType class. Each instance, in turn, has an array of fuzzy sets (instances of the mfType) class attached to it. The rule data base class *rule_base* has been implemented as a list, in which each list element points to the *if* and *then* parts of a rule. Fig. 1 shows schematically this data structure modeled on Ref. 4.

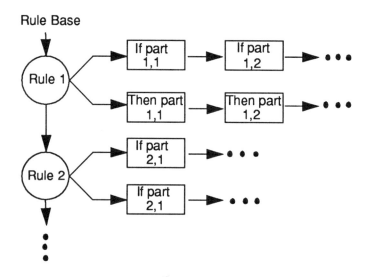

**FIGURE 1. Rule-base structure. Example: if part 1.1 is the first antecedent of rule 1.**

Rules are thus implemented by the sets of two pointers. The first set indicates which antecedent values are used to determine the rule's strength, and the second set points to output locations where the strength is to be applied.

The centroid defuzzification scheme described earlier has been somewhat simplified in the actual numerical implementation, by employing the center of gravity method instead.

## Numerical results

To illustrate our considerations, we apply our algorithm to two sample problems. In both cases we deal with a discrete time series $z(k)$, $k = 1, 2, \ldots$, which is simply a mapping from a set of whole numbers to reals. The prediction (forecasting) problem consists in finding $z(k)$, given $z(k-3)$, $z(k-2)$, and $z(k-1)$. The length of three of the input pattern is only taken as an example; any other length is admissible.

Steps 1-4 of the algorithm are used to generate the fuzzy rule base. Typically, we use between 100 and 200 input-output pairs to generate the rules. In both cases we use seven levels to quantize the input and output parameter space.

A pseudo-random time series with an upward trend has been used to emulate the special nuclear material inventory during 200 time periods. In contrast to the popular ARIMA model, we do not form a differenced time series to account for nonstationarity. Figure 2 shows the results of our simulation.

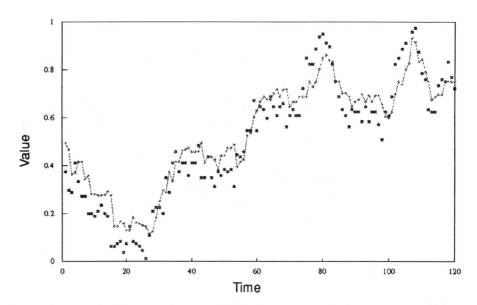

**FIGURE 2. Pseudorandom data points and their fit based on fuzzy controller.**

The training process covering 80 time period leads to a set of about 60 rules. As can be seen, the fuzzy controller correctly predicts the up and down trends of the data. The same data set can be used to train a backpropagation neural network. With little increase in predicting power, the cost is an increase in the computer time by two orders of magnitude.

The second example deals with a chaotic dynamic system whose bounded subset has a noninteger fractal dimension. Such subsets are termed strange attractors.[5] The two-dimensional mapping studied by Henon has the form

$$x_{n+1} = 1 - cx_n^2 + y_n, \tag{4}$$

$$y_{n+1} = \beta x_n. \tag{5}$$

Starting from an initial point $x_0 = 0.631$, $y_0 = 0.189$, a strange attractor structure is developed if we set $c = 1.4$ and $\beta = 0.3$. In the $x$-$y$ plane, we train the fuzzy controller by considering the Euclidean distance of the attractor point from the origin. The remaining points, shown in Fig. 3, are the results of the recall phase.

Considering the chaotic nature of the Henon attractor, it is remarkable we can forecast the 100 points after using only the first part of data as our training set.

In the presence of noise, at fine resolution below the noise magnitude, the self-similar features of the attractors are truncated. The fractal dimension can then be shown to measure the properties of noise, and not of the dynamical system.[6] As the additive noise is introduced, the fuzzy controller looses its predictive power.

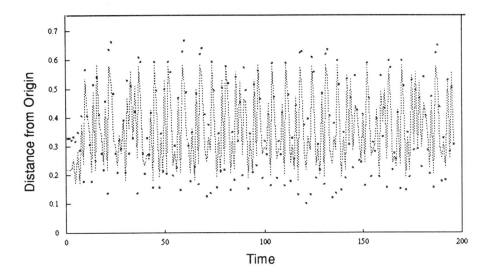

**FIGURE 3. The data point and their forecast for Henon attractor.**

To apply the developed theory to pattern recognition, we introduce a distance of two time series. For example, the Euclidean distance would be based on the sum of differences squared at the same value of time moment. A given temporal signature is accepted, if its distance from a fiducial shape does not exceed a chosen threshold value. This simple principle has been applied to pattern recognition of time series having some simple shapes. For example, a sinusoidal pattern is easily distinguished from a pattern of rectangular pulses.

## Conclusions

We have applied the algorithm of Wang and Mendel to predict the time series corresponding both to seasonal and chaotic systems. As the noise level increases, the predictive properties of the fuzzy controller are donwgraded. A distance function allows us to apply the theory to temporal pattern recognition.

## References

1. A. Lapedes and R. Farber, "Nonlinear signal processing using neural networks: prediction and system modeling," Los Alamos National Laboratory report LA-UR-87-2662, July 1987.

2. L. X. Wang and J. M. Mendel, "Generating fuzzy rules by learning from examples," IEEE Trans. Systems, Man and Cybernetics, **22**, 1414–1427 (1992).

3. K. Hornik, M. Stinchcombe, and H. White, "Multilayer feedforward networks are universal approximators," Neural Networks, **2**, 359–366 (1989).

4. G. Viot "Fuzzy logic: Concepts to constructs," AI Expert, **8**, 26–33 (1993).

5. E. Ott, "Strange attractors and chaotic motions of a dynamical system," Rev. Mod. Phys. **43**, 655–671 (1981).

6. A. Zardecki, "Noisy Ikeda Attractor," Phys. Lett. **90A**, 274–277 (1982).

# Handwritten Character Recognition by an Adaptive Fuzzy Clustering Algorithm

Amit Sharan and Sunanda Mitra

Computer Vision and Image Analysis Laboratory
Department of Electrical Engineering
Texas Tech University, Lubbock, Texas 79409

**Abstract** — Unconstrained handwritten characters pose a serious challenge to the development of a recognition algorithm. Many approaches have been studied over the years for such a recognition algorithm. We use an adaptive neuro-fuzzy clustering algorithm for classification and recognition of handwritten characters of a variety of styles and investigate the effectiveness of Fourier coefficients as representative features of handwritten characters in the presence of noise. Our results indicate that the adaptive clustering algorithm outperforms k-means clustering in handwritten character recognition for the same data representation. However some misclassifications cannot be avoided due to inherent problems associated with large variability in handwriting styles and the presence of excessive noise in practice.

## I. INTRODUCTION

Character recognition has always been an integral part of pattern recognition, since characters as patterns were relatively easy to obtain. An enormous amount of research has been conducted in this area [1, 2, 3, 4, 5]. While most work has concentrated on multi-font and fixed font reading, the most challenging area of research lies in recognition of handwritten characters. Machines that read fixed font, such as those on checks and credit cards are very reliable due to the constraints of using magnetic ink and the fixed style of letters used. Operating on a much less ideal case, a multi-font reader has to deal with noise from paper, printing quality, and type alignment. The ability to recognize handwritten characters automatically is a difficult task. The issue is made more complex by the existing noise problem and the large variability in handwriting styles. Noise arises from the writing instrument used, the quality of the paper, or the pressure applied when writing. Variability is something that has to be taken as granted, since even the same person's handwriting changes depending on the conditions and the nature of the writing. The problem facing researchers can be appreciated from the fact that even humans, which the computers are being made to emulate, have an error rate of 4% while reading handwriting without context. The automatic handwriting recognition success rate is not even remotely close to human ability in most cases.

This paper explores one approach to improve the current character recognition success rate. We used Fourier coefficients as features to classify handwritten characters using a modified Adaptive Fuzzy Leader Clustering (AFLC) algorithm [6, 7, 8]. The research was specifically aimed at finding out how effective Fourier coefficients were as a set of features.

## II. CLUSTERING FOR CHARACTER RECOGNITION

The approach of using clustering for character recognition has been used for quite some time. Apart from the neocognitron [9] network which has been specifically designed to recognize handwritten characters, two other unsupervised methods seen in recent literature for character recognition are the ART classifier [10, 11] and the nearest neighbor clustering algorithm [12]. Both of these have problems in recognition in the presence of noise. In this paper, we use a relatively new algorithm which overcomes a number of the earlier problems of other clustering approaches, namely noise and a priori knowledge of the number of classes.

## III. METHODOLOGY

The adaptive fuzzy leader clustering (AFLC) algorithm developed by Newton et al. [7] incorporates an incremental updating rule into the ART-1 architecture. This incremental updating rule is based on the FCM equations. It does not update all cluster centroids as in the FCM model but updates only the centroid of the winner cluster. This algorithm does not assume the number of clusters a priori but updates the number of clusters during processing, each category being represented by a cluster centroid. This algorithm uses a ratio of the Euclidean distance of the current input sample from the winner cluster centroid to the average Euclidean distance of all input patterns belonging to that category as the similarity measure for the vigilance criterion.

In the modified AFLC, updating process is followed by a verification procedure whose function is to check if the previous classification is still valid. The location of the samples which join the cluster at a later stage can often cause the prototype (centroid) of the cluster to shift in a particular direction. The extent and direction of the shift depend on the order in which the data is fed to the algorithm. As a result, the distance of some of the data samples from the new centroid of the cluster to which they originally belonged will increase drastically, and hence the vigilance condition might not be satisfied anymore. This could result in misclassifications if these samples are found to be closer to another neighboring cluster and satisfy the vigilance condition with respect to it. The modified AFLC avoids this problem by means of the procedure : a sample that does not satisfy the original classification condition is reclassified by selecting the cluster prototype with maximum value to be the winner.

However, classification of noisy data sets often poses a difficult problem as the noise sample points can severely bias the clustering procedure. Isolation of random samples has been achieved by the modified AFLC by introducing a noise cluster that contains all the outliers which cannot be matched to form a separate data cluster. To avoid the danger of treating a couple of neighboring noise samples as a separate valid cluster, the user can define the minimum number of samples required to form a proper cluster. Besides, if the noise points are in small groups, these groups can be merged and separated based on an additional feature like pixel intensity, in a typical digital image processing application. The details of the AFLC and its modified version can be found in [6, 7, 8].

We have chosen the lower-case letters of the English alphabet as input data set to this modified algorithm. Digitized images of the characters were generated by different methods, and were of varying characteristics to make the database more representative of real world situations. The database is not exhaustive, but is an effort to test the idea of using Fourier coefficients for character recognition, and the network's efficiency for this problem. Whole words were also used and noise was also introduced to test the validity of this approach for more difficult conditions. The image files were processed and Fourier coefficients

were computed. Next the desired coefficient set from each character was taken as the chosen features and run through the AFLC program to cluster the data. Figure 1 shows the flowchart of the algorithm used in this research.

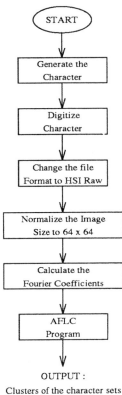

Figure 3.1 Flowchart of the algorithm

We generated 9 sets of lower case letters. These were of differing sizes, shapes, and with varying orientation, thickness and distortion. No constraints were placed on the writer, so as to give a more general flavor to the database. We used these various methods to try to ensure that we could deal with all the formats available and to find out if we had problems dealing with data generated from various sources and in differing formats and sizes.

To make a set of data which was homogeneous with respect to size of files and the gray-scales, we cropped the images and scaled them down to a pixel size of 64 x 64 to make the initial runs through the clustering algorithm less time consuming. The image files were all converted to a 2 gray-level binary format with values $0$ or $1$. In later stages we worked with characters of various gray-levels and sizes.

Once we had the individual character image data, the next step was to generate the features to be used by the AFLC algorithm. We selected the low-frequency Fourier coefficients as the features used for this classification. We use the symmetry of the Fourier transform to reduce the number of coefficients needed. Also, since we need only the lower frequency terms, we collected data for the lower harmonics for each character. For example, for the second harmonic we have 5 terms (-2 to dc to +2) in the x-direction and 5 terms (-2 to dc to +2) in the y-direction, giving us 25 real and 25 imaginary coefficients. According to the symmetry of the Fourier transform, there are 13 unique real terms and 12 unique imaginary terms (since the imaginary dc term is always zero). We took the first half of the real coefficients as the first half of the feature matrix, disregarding the second. In the second half of the feature matrix, we included the lower half of the imaginary coefficient matrix, thus getting all the unique coefficients from the Fourier transform, without any redundancy or loss of information. We computed the unique Fourier coefficients for the first, second and fifth harmonics. This gave us 9 coefficients for the first harmonic, 25 coefficients for the second, and 121 coefficients for the fifth harmonic.

Once the coefficients are collected, these are used as the feature vector to classify the characters. This means that each character is represented by the Fourier coefficient matrix, which in turn represents a point in an $n$-dimensional feature-space. For the case of 25 coefficients, $n$ is 25, and each character is a point in a 25-dimensional hyperspace. Thus, we have a set of samples, each with a feature vector of its own, which is then classified using the modified AFLC algorithm. One of the advantages of using this network is that we can easily modify the system to allow it to accept new data as it is generated, which means that this system has the capability to be used in a real-time situation. Various experiments were conducted to test the validity of this algorithm under various conditions.

## IV. EXPERIMENTS AND RESULTS

We conducted experiments on a wide range of characters to see how well the system reacted to variations. We created 9 different sets of the 26 alphabets in differing formats and 4 sets of characters by scanning them by a camera and digitizing the image captured. These were stored as binary, as well as TIFF file format. Then we created 4 sets of characters on the Paintbrush utility on Windows. These gave us

characters in BMP and PCX file formats. The last set of data was computer generated. We used the last set of data just in the initial runs of the programs to get an idea of the various parameters that we had to deal with. The different character sets were dissimilar not only with respect to their file formats, but also in size, shape, placement on the page, orientation, thickness and in the style of writing.

The characters were normalized to the same format and to the same size specification. The images were first cropped, i.e., the extra white space around the letter was cut off, leaving only the character within the image boundary. The files were scaled to the same size, 64 x 64, to make all the letters comparable in size and picture intensity. If the images were not binary, they were binarized to a 0/1 format.

### A. Number of Harmonics Needed

For the first experiment, we collected the Fourier coefficients for the normalized characters and stored the data for the first, second, fifth and seventh harmonic. The objective was to see how many coefficients were needed to classify the data to an acceptable degree of accuracy. The experiment was conducted initially with 6 sets of the characters $a$ and $h$ and then with 6 sets of the characters $a$, $b$, $c$, $d$ and $e$. The results in Table 1 show that the first harmonic itself has almost enough information to classify the letters. The coefficients from the second harmonic made the results totally error free. But as we increase the number of harmonics, thereby increasing the number of coefficients, the accuracy of the classification does not increase; on the contrary we see cases of misclassification. This shows that the lower frequency components of the Fourier transform contain most of the information regarding the shape of the character, and if we take higher frequency coefficients, distortions in the image and the variations in shape have to be taken into account. Just the second harmonic is enough to classify the data to the best degree of accuracy expected. So, for the rest of the experiments, we calculated only upto the second harmonic giving us 25 terms per character.

### B. Effect of Varying the Characters

The second experiment was to see how well the algorithm worked under variations in thickness, size, shape and orientation of the characters on the page. Here we took the original size of the images, and tested them with the images of the same character sized to 64 x 64 pixel size. The results show that scaling of the images does not affect the classification at all, and the characters are all recognized correctly even when the sizes are different, or when they are not cropped. The uncropped images also had the letters in different positions on the page, showing that translation did not affect the coefficients.

The experiment also showed that the algorithm was not very sensitive to shape variations. The characters were quite distorted when generated on paintbrush, since the resolution was not very good. These images were recognized to the same degree of accuracy as the characters generated by scanning the characters by the camera. The other area where the algorithm failed to group the characters properly was when the characters were written in two or more different styles for the same letter. For some of the characters, the character was classified into the noise cluster, which is what is expected.

The next experiment was to test how far this approach was successful in classifying characters which were not the same, but looked very much alike. For this experiment, we took all 9 sets of character coefficients and ran them through the AFLC program. After going through the results, we see that here is a shortcoming of this approach. An $f$ written like a $g$ are sometimes classified together, $e$ and $l$, and $e$ and $c$ if written in a way to create ambiguity did cause misclassifications. The only feasible solution is to have instructions for the writer so that we avoid any ambiguities. But this limits the application of character recognition to very specific cases.

*C. Rotational Invariance*

Rotational invariance in character recognition can be achieved by using the rotational property of the Fourier transform and correcting the angular shift [13].

*D. Comparison with K-Means Clustering*

In this experiment, we compared the performance of the AFLC algorithm with that of the K-means algorithm. The confusion matrices are shown in Tables 2 and 3. These show that the AFLC classifies most characters that are written legibly. There were some characters which were not classified properly. The AFLC classification put these characters difficult to recognize in the zeroeth cluster. There were a few

misclassifications, but that was expected when letters are very similar to each other.

## V. CONCLUSIONS

This research has shown that the modified AFLC can be used as an effective tool for character recognition. This algorithm worked extremely well under very difficult situations. The approach works with file formats of different kinds, making it free from the restriction of generating character data. The algorithm works very well for characters written in different ways, and though there are the natural problems of understanding illegible and unique characters, the overall performance is remarkable. The premise that Fourier transforms could be used to classify characters has been tested and found to be valid for handwritten characters. However, recognition of whole words, in general, was not as successful using Fourier coefficients as representative features due to large variability in the word arrangement with respect to spacing, size and orientation of the component characters. Future work will employ the use of a preprocessor such as morphological filters prior to feature extraction to improve recognition.

## REFERENCES

[1] C. S. Suen, M. Berthod and S. Mori, "Automatic recognition of handprinted characters - The state of the art," *Proceedings of the IEEE*, Vol. 68, No. 4, pp. 469-487, April 1980.

[2] V. K. Govindan and A. P. Shivaprasad, "Character recognition - a review," *Pattern Recognition*, Vol. 23, No. 7, pp. 671-683, 1990.

[3] S. Impedovo, L. Ottaviano and S. Occhinegro, "Optical character recognition - a survey," *International Journal of Pattern Recognition and Artificial Intelligence*, Vol. 5, No. 1 & 2, pp. 1-24, 1991.

[4] R. H. Davis and J. Lyall, "Recognition of handwritten characters - a review," *Image and Vision Computing*, Vol. 4, No. 4, pp. 208-218, Nov., 1986.

[5] S. Mori, C. Y. Suen and K. Yamamoto, "Historical review of OCR research and development," *Proceedings of the IEEE*, Vol. 80, No. 7, pp. 1029-1058, July 1992.

[6] S. Pemmaraju, *Adaptive Fuzzy Clustering of Noisy Data,* M.S.E.E. Thesis, Texas Tech University, December 1992.

[7] S. C. Newton, S. Pemmaraju and S. Mitra, "Adaptive Fuzzy Leader Clustering of complex data sets in pattern recognition," *IEEE Transactions on Neural Networks,* Vol. 5, No. 5, pp. 794-800, 1992.

[8] S. Mitra and Y. S. Kim and S. Pemmaraju, "Adaptive pattern recognition by self-organizing neural networks," in *Neural and Fuzzy Systems,* SPIE Institute Series, Eds. S. Mitra and M. M. Gupta (in press).

[9] K. Fukushima, "Neocognitron : A hierarchical neural network model capable of visual pattern recognition," *Neural Networks,* Vol. 1, No. 2, pp. 119-130, 1988.

[10] A. Farci, *The Art of Character Recognition,* M. S. Thesis, University of Texas at Arlington, 1989.

[11] K. W. Gan and K. T. Lua, "Chinese character classification using an adaptive resonance network," *Pattern Recognition,* Vol. 25, No. 8, pp. 877-882, 1992.

[12] M. A. O'Hair, *A Whole Word and Number Reading Machine Based on Two-Dimensional Low-Frequency Fourier Transforms,* Ph.D. Dissertation, Air Force Institute of Technology, December 1990.

[13] D. J. Lee, S. Mitra and T.F. Krile, "Analysis of sequential complex images using feature extraction and 2-D cepstrum techniques," *JOSA A,* Vol. 6, pp. 863-870, 1989.

Table 2 Confusion Matrix for the K-Means classification using 7 sets of characters

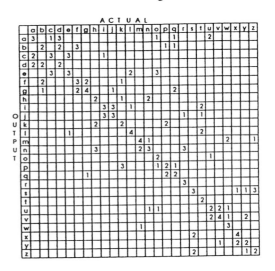

Table 3 Confusion Matrix for the AFLC classification using 7 sets of characters

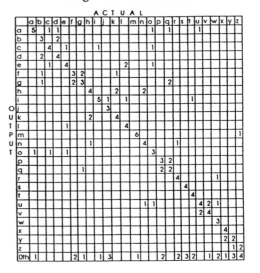

Table 1 Number of characters classified correctly out of 6 with increasing number of harmonics

For *a* and *h* :

| Character | 1st Harmonic | 2nd Harmonic | 5th Harmonic | 7th Harmonic |
|---|---|---|---|---|
| a | 6/6 | 6/6 | 6/6 | 6/6 |
| h | 6/6 | 6/6 | 5/6 | 6/6 |

For *a, b, c, d,* and *e* :

| Character | 1st Harmonic | 2nd Harmonic | 5th Harmonic |
|---|---|---|---|
| a | 6/6 | 6/6 | 6/6 |
| b | 6/6 | 6/6 | 5/6 |
| c | 5/6 | 6/6 | 5/6 |
| d | 6/6 | 6/6 | 6/6 |
| e | 5/6 | 6/6 | 6/6 |

# Pattern Recognition of the Polygraph Using Fuzzy Classification

Shahab Layeghi, Mitra Dastmalchi, Eric Jacobs, and R. Benjamin Knapp
Department of Electrical Engineering
San Jose State University, San Jose, California 95192-0084

Abstract- Polygraph tests are a widely used method to distinguish between truth and deception. Polygraph charts are usually analyzed by human interpreters. However, computer algorithms are now being developed to score the tests or verify the results. These methods are based on statistical classification techniques. In this study a number of time, frequency and correlation domain features were selected and used. The fuzzy K-nearest neighbor algorithm was used to classify the polygraph charts, a correct classification of ninety-one percent was obtained for a set of one hundred case files supplied by the NSA.

## I. Introduction

Polygraph examinations are the most widely used method to distinguish between truth and deception. In a polygraph examination a person is connected to a special instrument called a Polygraph which records several physiological signals such as electrocardiogram, galvanic skin response, and respiration. During the polygraph examination, the subject is asked a set of questions by an examiner. The examiner analyzes the graphs to determine the reactions of the subject to the questions for evidence of truth or deception.

Different formats are used for polygraph examinations. A given polygraph test format is an ordered combination of relevant, irrelevant and control questions. Relevant questions are questions about a specific issue. Control questions are not directly related to the specific issue under question but are designed to make the subject uncomfortable and provide a physical response for comparison. Irrelevant questions are very general questions that are not related to the issue but provide a response to comparison [1][4]. The rational for scoring the tests is that a deceptive subject will be more threatened by the relevant questions than by the control questions while a non deceptive subject will be more threatened by the control questions than the relevant questions.

Three general types of test formats are in use today. These are Control Question Tests, Relevant-Irrelevant Tests, and Concealed Knowledge Tests. Each of the general test formats may have a number of more specific variations. Each test consists of two to five charts containing a prescribed series of questions. The test format that is used in an examination is determined by the test objective [3][4].

A control question test is often used in criminal investigations. The control questions are compared to the relevant questions and if the responses to the relevant questions are greater, the subject is usually classified as deceptive. Irrelevant questions are used as buffers.

The problem with human classification of polygraph tests is that the outcome depends on the examiner's experience and judgment. As a result, automatic scoring systems to classify polygraph tests are being developed to overcome this problem. Several methods for polygraph classification have been studied which are mostly based on statistical classification techniques [1] [2]. This project, however, is focused on using fuzzy classification rather than statistical methods.

## II. Methods

Digitized polygraph data used in this project were collected from various police stations by the National Security Agency (NSA). The data files were organized according to the test format used and were decoded to ASCII format so they can be read by the mathematical computation package, MATLAB. All preprocessing and feature extraction routines were implemented in MATLAB.

Classification of polygraph charts like any other pattern recognition problem can be divided into two major sections, feature extraction and classification. The methods used for each one of these sections are explained in the following section.

1825

## A. Feature Extraction

Polygraph data consists of signals from four different channels: Galvanic Skin Response (GSR), electrocardiogram, higher respiration, and lower respiration. Before actual feature extraction was done, the data was preprocessed. The electrocardiogram signal was decomposed into a high frequency component showing heart pulse, and a low frequency component showing blood volume. The derivative of the blood volume was also used as a preprocessed channel. In order to eliminate any noise and trend, these six derived signals were detrended and filtered.

A broad range of features that are the best indicators of truth or deception were chosen based on previous work and on interviewing polygraph examiners[5][6]. In general, features are divided into three main groups, time-domain features, frequency-domain features and correlation features. Time-domain features involved standard statistical characteristics such as the mean, the standard deviation, and the median for each of the six channels. Other channel-specific time-domain features such as the ratio of inhalation over exhalation and the auto-regressive parameters of a tenth order AR filter model for the heart pulse were also considered as features. Frequency-domain features for each of the six channels included the fundamental frequency, the magnitude of the power spectral density at the fundamental frequency, and the coherency at the fundamental frequency. To extract each feature for each question a time fragment of each signal was selected starting several seconds after a question was asked and continuing for a number of seconds. The exact time frame used was dependent on the channel being measured.

A total of ninety-nine different features were extracted for each question in each chart. Each feature was extracted for each relevant, irrelevant, and control question in the test. In order to classify subjects using the difference between the control and relevant responses similar features for these question were combined. Seven methods were used to then combine each control and relevent features into one common feature. (The irrelevant features were not used.) The first method was subtracting the average response normalized to the control question from average response normalized to the relevant question. The second method was to use maximum response in place of average response. The other methods of combining control and relevant features were max - min, min -

max, min - min, dividing the averages, and using normalized averages.

## B. Classification

It was decided to use the K-nearest neighbor (KNN) classifier in this project because the distribution of the samples of deceptive and non-deceptive classes were not known beforehand, and the KNN classifier does not explicitly use the distribution of the samples.

One of the characteristics of the conventional KNN classification method is that it assigns each input to one of the possible classes (crisp classification). The way that humans think and classify objects is fundamentally different. Each object can be considered to belong to more than one class at the same time, and there are degrees of membership for each class. This is the basic idea that is followed in fuzzy logic. It was decided to use a modified version of KNN algorithm which uses fuzzy logic concepts [7] [8]. In this way the output will be the *possibility* of deception and thus give a continuous measure of truth versus deception rather than a discrete choice.

The first step in the fuzzy KNN algorithm is the same as first step in crisp classifier. In both cases K nearest neighbors of the input are found. In the crisp classifier, the majority class of the neighbors is used to assign the input to a class. In the fuzzy classifier, the membership of the input to each class is found. In order to do so, the membership vector of each neighboring sample is combined to obtain the membership vector of the input. If the samples are crisply classified, membership vectors should be assigned to them. One method to do so is to assign the membership of 1 to the class that it belongs to, and membership of 0 to other classes. Other methods assign different memberships to the samples according to their distance from the mean of the class, or the distances from the nearby samples of its own class and the other classes.

When the membership vectors of the labeled samples are specified, they are combined to find the membership vector of the unknown class. This procedure is done in a way that samples that are closer to the input have more effect on the resultant membership function. The following formula uses the inverse distance to weigh the membership functions. $x$ is the input to be classified, $x_j$ is the j*th* nearest neighbor and $u_{ij}$ is the membership of the j*th* nearest neighbor of the input in class i. $D(x,y)$ is a distance measure between the vectors x and y. Euclidean distance has been used as the distance measure in this project.

$$u_i(x) = \frac{\sum_{j=1}^{K} u_{ij}(1/D(x,x_j)^{\frac{1}{m-1}})}{\sum_{j=1}^{K}(1/D(x,x_j)^{\frac{1}{m-1}})}$$

where m is a parameter that changes the weighing effect of the distance.

When $m \gg 1$, all the samples will have the same weight. When $m \rightarrow 1$, nearest samples have much more effect on the membership value of the input [7].

The feature extraction mentioned in section II.A created 669 features for each chart. This number of features was larger than could be practically used by fuzzy KNN classifier. It was decided to reduce the number of features to 30 at this step. Two different methods were chosen to test the features one at time to find the best 30 features. The first method was using the fuzzy KNN classifier to classify the data files using one feature at a time. The classifier parameters such as K and threshold were changed to find the best classification results. The value 5 was selected for K because it gave better classification results. Also, a threshold of 0.5 was used to defuzzify the output of the classifier. The second method was using the scatter criterion presented below:

$$J = \frac{(m_1 - m_2)^2}{s_1^2 + s_2^2} \tag{1}$$

where $m_i$ is the mean of the class $i$, and $s_i$ is the standard deviation of class $i$.

This criterion measures the distance between the means of the two classes, normalized over the sum of the variances. Therefore, the more compactly the samples in each class are separated, the higher will the value of J.

The results of KNN and scatter criterion were averaged for three sets of data. Thirty features that showed the best performance in both methods or had a special significance to the polygraph examiner were selected.

Better classification was achieved by combining several features. The most basic way of finding the best combination is the exhaustive search method. That is trying all the combinations for these features. This is not practical when the number of features is large.

All combinations of two features out of the best 30 features were tried. Then, the 20 combinations with the best accuracy rate were selected and combined with other features of the best 30 set to build combinations of three. The same procedure was followed for combinations of three and four with the best selected set of features. This procedure was continued until adding features did not improve the classification results significantly.

## III. Results

The classification results were improved by increasing the number of combined features from 1 to 4. Using single features the best result was 70 percent by using the mean of GSR signal. When combinations of two features were used the best result was obtained using the difference between maximum and minimum of the GSR and the maximum of derivative of the low cardio signal. The average result was 73 percent. By combining three features the best result was 78 percent by using the maximum of the GSR, the maximum of the upper respiratory, and the frequency of the maximum integrated spectral difference of the control-relevant pair in the GSR. The combinations of four features that showed the best classification results are shown in Table 1. These results were obtained by using K=5 and defuzzification threshold of 0.5 in the fuzzy K-nearest neighbor classifier. The feature set that showed the best result on the average was used for further experiments. These features were the maxima of the GSR, the high cardio and the upper respiratory signals and the difference between the maximum and the minimum of the high cardio signals.

Different values for K and the defuzzification constant were tried to optimize the classifier. The best result was obtained using K = 6 and the defuzzification constant of 0.6. The average result for three sets was 81.6 percent.

Another experiment that was performed was combining the results of several charts that are used in a polygraph test. Usually a polygraph test is composed of two to four charts that contain the same questions. Previously, the charts were classified independently. The outcome of classifying every chart in a test was added and the whole test was classified accordingly. Correct classification results for sets one to three were 85.7, 80.0, and 91.4 percent.

# IV. Conclusion and Discussion

| Set | Features | | | | Accuracy |
|---|---|---|---|---|---|
| Set 1 | GSR(max) | HC(max-min) | LR(max) | UR(max) | 81.0 |
| | GSR(max) | HC(min) | LR(max) | UR(max) | 80.2 |
| | GSR(max) | LR(max) | UR(max) | GSR(isd) | 74.4 |
| Set 2 | GSR(max) | DLC(mean) | UR(max) | GSR(isd) | 81.0 |
| | GSR(max) | HC(min) | LR(max) | UR(max) | 79.4 |
| | GSR(max) | LR(max) | UR(max) | GSR(isd) | 79.0 |
| Set 3 | HC(max-min) | DLC(mean) | UR(max) | GSR(isd) | 87.4 |
| | GSR(max) | LR(max) | UR(max) | GSR(isd) | 86.6 |
| | GSR(max) | HC(max-min) | LR(max) | UR(max) | 82.5 |
| Average | GSR(max) | HC(max-min) | LR(max) | UR(max) | 81.0 |
| | GSR(max) | LR(max) | UR(max) | GSR(isd) | 80.0 |
| | GSR(max) | HC(min) | LR(max) | UR(max) | 79.8 |

GSR=Galvainc Skin Response. HC=High Cardio, LC=Low Cardio, DLC=Derivative of Low Cardio. UR=Upper Respiratory, LR=Lower respiratory isd=integrated spectral density.

**Table 1. classification results with combining 4 features**

The classification results improved consistently by increasing the number of features in the combination from one to four. The feature set that showed the best classification result only included simple time-domain features which were the maximums of GSR, lower respiratory, upper respiratory, and the difference between maximum and minimum of the high cardio signal. It is notable that these features come from different channels. It is possible to conclude that adding the information in different physiological channels is a good way of finding the evidence of deception in polygraph tests. Although these features showed the best results, some other features appeared in other combinations with approximately the same results. In future work combinations of more than four features could be studied in order to find an optimum feature set that uses all possible features.

The primary advantage of using the fuzzy KNN classifier is that it gives the possibility of deception rather than just classifying the person as deceptive or non-deceptive. This gives the examiner the ability to have a continuous measure of deception. Also using a fuzzy classifier whose membership functions could be trained during a polygraph exam may direct the examiner toward a specific line of questioning.

[This work was supported by a grant from the National Security Agency.]

# References

[1] Dale E. Olsen, et. al., "Recent developments in polygraph testing: A research review and evaluation - A technical memorandum, " Washington DC: US Government Printing Office 1983.

[2] John C. Kircher and David C. Raskin, "Human versus computerized evaluations of polygraph data in a laboratory setting, " Journal of Applied Psychology, Vol.73, 1988 No 2, pp. 291-308.

[3] John E. Reid and Fred E. Inbau, "Truth and Deception: The Polygraph ( Lie Detector ) Technique", The Williams & Wilkins Company, Baltimore, Md., 1966

[4] Michael H. Capps and Norman Ansley, "Numerical Scoring of Polygraph Charts: What Examiners Really Do", Polygraph, 1992, 21, 264-320

[5] Brian M. Duston, " Statistical Techniques for Classifying Polygraph Data ", Naval Command Control and Ocean Surveillance Center, RDT&E Division

[6] Howard W. Timm, " Analyzing Deception From Respiration Patterns " , Journal of Police Science and Administration, 1982, 1, 47 - 51.

[7] J.M. Keller, M.R. Gray and J.A. Givens, "A Fuzzy K Nearest Neighbor Algorithm", IEEE Trans. on Syst. Man. Cybernetics, vol SMC-15, no. 4 (1989)

[8] J.C. Bezdek and Siew K. Chuah, "Generalized K-Nearest Neighbor Rules, Fuzzy Sets and Systems vol. 18 (1986)

# Fuzzy Clustering as Blurring

Yizong Cheng

University of Cincinnati, Cincinnati, OH 45221-0008

yizong.cheng@uc.edu

*Abstract*— A family of fuzzy clustering algorithms that are Picard iterations based on alternate membership evaluations and cluster center shifts are compared with the blurring process, a deterministic dynamic system that moves data points to weighted means in their neighborhoods. It is shown in this paper that when the initial cluster centers are assigned as data points themselves, some fuzzy clustering algorithms, particularly the maximum-entropy clustering, become blurring processes. Some basic results obtained in the blurring process thus can be applied to these special runs of fuzzy clustering, and may serve as counterexamples for fuzzy clustering in general.

## I. Introduction

Fuzzy clustering is often credited to Ruspini, Dunn, and Bezdek [1]. A recent development named *maximum-entropy clustering* [6] adds a new algorithm to this paradigm, which is characterized by the optimization iterate consisting of two Picard phases, one optimizing the objective function with respect to the membership values while having the cluster centers fixed, another optimizing the objective function with respect to the cluster centers while having the membership values fixed. A degenerate case of this paradigm is the $k$-means clustering algorithm. These algorithms are often probabilistic and reach a local minimum of the objective function. The Picard iteration is similar to gradient descent in that the initial positions of the cluster centers determines the final result, because both can only reach local minima. It differs from gradient descent in that no choice on step size is needed. The Picard iteration also differs from stochastic approximation clustering algorithms like the competitive learning and the self-organizing map, in that it takes the data set as a batch.

The purpose of this paper is to show that some specific runs of the probabilistic fuzzy clustering algorithms can be treated as instances of the *blurring process*, a deterministic iterative process, and results obtained about the blurring process can be used on some of the fuzzy clustering algorithms.

In Section II, relevant fuzzy clustering algorithms will be presented. Section III reviews the blurring process, and an attempt to identify specific fuzzy clustering runs with the blurring process is in Section IV. Section V contains an example and concluding remarks.

## II. Fuzzy Clustering

Clustering is often studied as an optimization problem, in which a set of data or a data space is to be partitioned into *clusters*. A common feature in the fuzzy clustering algorithms discussed in this paper is that a Picard iteration is performed to update the membership value of each data point to each cluster and the cluster centers alternately. Each iterate contains two steps, the update of the membership value $v_i(s)$ of data point $s$ to the $i$th cluster, and the update of the cluster centers $z_i$'s. During the first step, the current cluster centers are fixed, and the membership values that optimize the objective function is found. During the second step, the membership values are fixed, and the best cluster centers are found. Hence, the clustering algorithm is completely determined by the objective function and the constraints on the parameters.

The most straightforward objective function is the so-called *within-group sum of squared errors*, or,

$$f(\mathbf{v}, \mathbf{z}) = \sum_{i=1}^{k} \sum_{s \in S} v_i(s)|s - z_i|^2, \qquad (1)$$

where $S$ is the data set, $z_i$ is the center of the $i$th cluster, $v_i(s)$ is the membership value of $s$ to the $i$th cluster, and $k$ is the number of clusters. A common constraint on the membership values is the they are values in $[0, 1]$ and $\sum_{i=1}^{k} v_i(s) = 1$ for every $s$. Because $f$ is a concave function in terms of $\mathbf{v}$, the Picard step on updating $\mathbf{v}$ causes $v_i(s)$ to assume its extreme values (0 or 1), and thus the algorithm is equivalent to the ordinary non-fuzzy *k-means clustering*, described as follows.

1. Start with initial cluster centers $z_1, \ldots, z_k$ with $k$, the number of clusters, determined by the user. Normally, $z_i$'s are randomly generated points in $R^n$.

2. Evaluate the membership values for each $s \in S$:

$$v_i(s) = \begin{cases} 1 & i = \mathrm{argmin}(|s - z_i|) \\ 0 & \text{otherwise} \end{cases} . \quad (2)$$

3. Shift each cluster center to the centroid of the cluster:

$$z_i \leftarrow \frac{\sum_{s \in S} v_i(s)s}{\sum_{s \in S} v_i(s)}. \quad (3)$$

If any of $z_i$'s is different from the previous one, goto step 2.

The algorithm converges to a local minimum of the objective function, which depends on the initial assignment of the cluster centers. Cluster centers do not merge or disappear. Finally, $R^n$ is partitioned into $k$ clusters according to (2).

To generate truly fuzzy clusters, either the objective function or the constraints on membership values have to be altered. Suppose the constraints on the membership values hold, but the objective function becomes

$$f(\mathbf{v}, \mathbf{z}) = \sum_{i=1}^{k} \sum_{s \in S} v_i^2(s)|s - z_i|^2. \quad (4)$$

The following is the Picard iteration that minimizes this objective function [1].

1. Start with initial cluster centers $z_1, \ldots, z_k$ with $k$, the number of clusters, determined by the user. Normally, $z_i$'s are randomly generated points in $R^n$.

2. Evaluate the membership values for each $s \in S$:

$$v_i(s) = \frac{|s - z_i|^{-2}}{\sum_{j=1}^{k} |s - z_j|^{-2}}, \quad (5)$$

and when $s = z_i$, $v_i(s) = 1$ and other $v_j(s)$'s are zero.

3. Shift each cluster center to the centroid of the cluster:

$$z_i \leftarrow \frac{\sum_{s \in S} v_i^2(s)s}{\sum_{s \in S} v_i^2(s)}. \quad (6)$$

If any of $z_i$'s is different from the previous one, goto step 2.

Rose, Gurewitz, and Fox [6] use the within-group sum of squared errors as the constraint and the entropy as the objective function. The following is the resulting Picard iteration [6].

1. Start with initial cluster centers $z_1, \ldots, z_k$ with $k$, the number of clusters, determined by the user. Normally, $z_i$'s are randomly generated points in $R^n$.

2. Evaluate the membership values for each $s \in S$:

$$v_i(s) = \frac{e^{-\beta|s - z_i|^2}}{\sum_{j=1}^{k} e^{-\beta|s - z_j|^2}}. \quad (7)$$

3. Shift each cluster center to the centroid of the cluster:

$$z_i \leftarrow \frac{\sum_{s \in S} v_i(s)s}{\sum_{s \in S} v_i(s)}. \quad (8)$$

If any of $z_i$'s is different from the previous one, goto step 2.

The final cluster center values $z_i$'s will be used to compute the membership values according to (7). Some of the centers will merge and the number of cluster centers so generated may be less than $k$.

In general, the paradigm of fuzzy clustering contains Picard iterative optimization algorithms. These are probabilistic algorithms, in the sense that the initial cluster centers must be arbitrarily chosen, and the choice determines which local optimum the objective function will eventually reach. Another system parameter that must be chosen is $k$, the initial number of clusters. These algorithms are better than gradient descent and stochastic approximation algorithms in the sense that there is no need to select another system parameter, the step size.

## III. The Blurring Process

The *blurring process* is a deterministic iterative process formalized in [3]. It is a generalization of the deterministic clustering algorithms proposed in [5] as the "mean-shift procedure" and in [2] as a conceptual clustering method.

The blurring process consists of *blurring steps* which maps each data point $s \in S$ to its "blurred image":

$$b(s) = \frac{\sum_{t \in S} K(s - t)P(t)t}{\sum_{t \in S} K(s - t)P(t)}, \quad (9)$$

where $K$ is a *kernel* in the form of $K(x) = k(|x|^2)$ with $k : [0, \infty] \rightarrow [0, 1]$ a nonincreasing function, and $P(s)$ is the *weight* of data point $s$. When the kernel is a density function ($\int K(x) = 1$), it has been shown in [3] that $b(s) - s$ is in the direction of the gradient at $s$ of the density estimation $q(s) = \sum_{t \in S} K_1(s-t)P(t)$ with $K_1(x) = k_1(|x|^2)$ and

$$k_1(r) = \alpha - \beta \int_0^r k(t)dt \quad r \geq 0. \qquad (10)$$

$\alpha$ and $\beta$ are constants to meet the density requirements. For example, when $K$ is the flat kernel specified by

$$k(r) = \begin{cases} c & \text{if } 0 \leq r \leq 1 \\ 0 & \text{if } r > 1 \end{cases}$$
$$K(x) = \begin{cases} c & \text{if } |x| \leq 1 \\ 0 & \text{if } |x| > 1 \end{cases} \qquad (11)$$

the corresponding $K_1$ is specified by

$$k_1(r) = \begin{cases} c(1-r) & \text{if } 0 \leq r \leq 1 \\ 0 & \text{if } r > 1 \end{cases}$$
$$K_1(x) = \begin{cases} c(1-|x|^2) & \text{if } |x| \leq 1 \\ 0 & \text{if } |x| > 1 \end{cases}. \qquad (12)$$

When $K$ is the *Gaussian kernel*

$$K(x) = \alpha e^{-\beta|x|^2} \qquad (13)$$

or the *truncated Gaussian kernel*

$$K(x) = \begin{cases} \alpha e^{-\beta|x|^2} & \text{if } \beta|x| \leq \gamma \\ 0 & \text{if } \beta|x| > \gamma \end{cases}, \qquad (14)$$

where $\alpha$, $\beta$, and $\gamma$ are positive constants, the two kernels are identical, $K_1 = K$, and

$$b(s) - s = \frac{1}{2\beta} \frac{\nabla_s q(s)}{q(s)} = \frac{1}{2\beta} \nabla_s \log q(s). \qquad (15)$$

This shows that a blurring step with the Gaussian kernel is hill-climbing on the density surface estimated with the same Gaussian kernel, truncated or not. Moreover, this is not ordinary hill-climbing. Logarithm applied on a density surface will make plateaus disappear, and there is no need to worry about a step size that may be too small or too large.

The blurring process is built on the blurring step in one of two versions, the "blurring with shifting data" version and the "blurring with fixed data" version. For the purpose of this paper, only the latter version is presented.

1. The initial finite set of data $S \in R^n$ along with their weights $P$ are provided. A kernel $K$ is chosen. The initial "set of images" $Z = S$.

2. The blurring step. For each $z \in Z$ do

$$b(z) = \frac{\sum_{s \in S} K(z-s)P(s)s}{\sum_{s \in S} K(z-s)P(s)} \qquad (16)$$

3. Update the set of images.

$$b(Z) = \{b(z); z \in Z\} \qquad (17)$$

If $b(Z) = Z$, terminate. Otherwise, goto Step 2.

A typical kernel will be the Gaussian one (13) with a chosen variance. In the case when data points are equally weighted, the blurring step is

$$b(z) = \frac{\sum_{s \in S} e^{-\beta|z-s|^2} s}{\sum_{s \in S} e^{-\beta|z-s|^2}} \qquad (18)$$

When $\beta = 0$, $Z$ becomes a singleton. When $\beta$ is very large, $Z$ contains as many elements as the number of distinct data points in $S$. Intermediate sizes for $Z$ can be obtained using intermediate $\beta$ values. Each element in $Z$ can be considered as the center of a cluster, and each $s \in S$ can be assigned to a cluster by tracing the images through the blurring steps.

The blurring clustering algorithm based on the Gaussian kernel is free from any arbitrariness of the selection of system parameters, including number of clusters, step size, even the objective function, except $\beta$, the span of the kernel. It is a deterministic algorithm and the result does not depend on any random step in the algorithm. It provides a unique clustering result for a each data base. It is an excellent mode-seeking clustering method. In the following section, I will show that, by initializing the cluster centers to be the data themselves, and by abandoning various objective functions, one can cast most fuzzy clustering algorithms in the mold of the blurring process.

## IV. Fuzzy Clustering as Blurring

When the initial cluster centers are assigned with the original data set, or, $Z = S$, the maximum-entropy clustering algorithm (8) can be put in the form of (16):

$$b(z) = \frac{\sum_{s \in S} K(z-s)A(s)s}{\sum_{s \in S} K(z-s)A(s)}, \qquad (19)$$

with

$$K(x) = \alpha e^{-\beta|x|^2} \qquad (20)$$

and

$$A(s) = \frac{1}{\sum_{t \in S} K(t-s)} \qquad (21)$$

Now, this special run of the maximum-entropy clustering algorithm becomes "blurring with fixed data".

Mathematically, the term $A(s)$ can be treated as $P(s)$, the weight, in the blurring process. Therefore, all conclusions about the blurring process are still valid with this extra multiplier. Conceptually, it can also be treated as part of the kernel, and this makes the kernel *context-dependent* or *adaptive*. In the case of the maximum-entropy clustering algorithm, $A(s)$ is inversely proportional to the "density" of the data at the point $s$. The effect of this context-dependent multiplier is that the density surface on which the cluster centers are climbing is made smoother.

Bezdek's prescription of the membership value, (5), also has a multiplicative context-dependent term. However, this term is more integrated with the core of the kernel, because the infinity both would reach when they got separated. Thus, it is more difficult to treat it in the mold of the blurring process.

When the initial cluster centers are the data points in $k$-means clustering, no merge and nontrivial clustering take place.

However, as long as the update step for cluster center is the mean weighted by the membership values and the membership value is a function of $|s - z_i|^2$ with a possible context-dependent multiplier, the blurring process can shed some light on the behavior of the fuzzy clustering algorithm. First, because this a special run of the probabilistic algorithm, it provides counter examples to general claims on the algorithm. Secondly, although there is no proof about whether assigning initial cluster centers to data points will make fuzzy clustering algorithms reach global optima of the objective functions, it seems that this very special assignment outweighs the controversial goal of optimizing an objective function, and it might serve the purpose of clustering better when the algorithm becomes deterministic.

## V. Fuzzy Hierarchies

The results from runs of the blurring process/fuzzy clustering are intriguing. Figure 2 is a "fuzzy" hierarchy generated from a sequence of eleven runs with $\beta$ values $0, 0.1, \ldots, 0.9, 1$ on the "animal" data provided by R. Forsyth [4]. The original data contains 101 points in the 17-dimensional space with many points shared by more than one animal. Therefore, there are only 59 distinct data points. Weights representing repetitions of points were used as $P(s)$. Only the first 16 attributes were used. The modified data is shown in Figure 1, where the 17th column is the repetition or $P(s)$ of each data point. A truncated Gaussian kernel (14) with $\gamma = 1$ was used along with a context-dependent term $A(s)$ (21) de-

Figure 1: Modified animal data from R. Forsyth.

```
1 0 0 1 0 0 0 1 1 1 0 0 4 1 0 1 6  antelope
0 0 1 0 0 1 1 1 1 0 0 1 0 1 0 0 5  bass
1 0 0 1 0 0 1 1 1 1 0 0 4 0 0 1 2  bear
1 0 0 1 0 0 0 1 1 1 0 0 4 1 1 1 4  calf
0 0 1 0 0 1 0 1 1 0 0 1 0 1 1 0 1  carp
1 0 0 1 0 0 0 1 1 1 0 0 4 0 1 0 1  cavy
0 0 1 0 0 0 1 0 0 0 0 0 0 0 0 0 1  clam
0 0 1 0 0 1 1 0 0 0 0 4 0 0 0 0 1  crab
0 1 1 0 1 0 1 0 1 1 0 0 2 1 0 0 2  crow
0 0 0 1 0 1 1 1 1 1 0 1 0 1 0 1 2  dolphin
0 1 1 0 1 0 0 0 1 1 0 0 2 1 1 0 3  dove
0 1 1 0 1 1 0 0 1 1 0 0 2 1 0 0 1  duck
0 1 1 0 1 0 0 0 1 1 0 0 2 1 0 1 1  flamingo
0 0 1 0 0 0 0 0 1 0 0 6 0 0 0 2  flea
0 0 1 0 0 1 1 1 1 1 0 0 4 0 0 0 1  frog1
0 0 1 0 0 1 1 1 1 1 1 0 4 0 0 0 1  frog2
1 0 0 1 1 0 0 1 1 1 0 0 2 1 0 0 2  fruitbat
1 0 0 1 0 0 1 1 1 1 0 0 2 0 1 1 1  girl
0 0 1 0 1 0 0 0 1 0 0 6 0 0 0 1  gnat
1 0 0 1 0 0 0 1 1 1 0 0 2 0 0 1 1  gorilla
0 1 1 0 1 1 1 0 1 1 0 0 2 1 0 0 3  gull
0 0 1 0 0 1 0 1 1 0 0 1 0 1 0 0 3  haddock
1 0 0 1 0 0 0 1 1 1 0 0 4 1 1 0 1  hamster
1 0 0 1 0 0 0 1 1 1 0 0 4 1 0 0 2  hare
1 0 1 0 1 0 0 0 1 1 0 6 0 1 0 1  honeybee
0 1 1 0 0 0 1 0 1 1 0 0 2 1 0 0 1  kiwi
0 0 1 0 1 0 1 0 0 0 0 6 0 0 0 1  ladybird
0 1 1 0 1 0 0 0 1 1 0 0 2 1 0 0 4  lark
0 0 1 0 0 1 1 0 0 0 0 6 0 0 0 2  lobster
1 0 0 1 0 1 1 1 1 1 0 0 4 1 0 1 1  mink
1 0 1 0 1 0 0 0 1 0 0 6 0 0 0 2  moth
0 0 1 0 0 1 1 1 1 1 0 0 4 1 0 0 1  newt
0 0 1 0 1 1 0 0 0 0 0 8 0 0 1 1  octopus
1 0 0 1 0 0 1 1 1 1 0 0 4 1 0 0 2  opossum
0 1 1 0 0 0 0 0 1 1 0 0 2 1 0 1 1  ostrich
0 1 1 0 0 1 1 0 1 1 0 0 2 1 0 1 1  penguin
0 0 1 0 0 0 1 1 1 1 0 0 1 0 0 0 1  pitviper
1 0 1 1 0 1 1 0 1 1 0 0 4 1 0 1 1  platypus
1 0 0 1 0 0 1 1 1 1 0 0 4 1 1 1 1  pussycat
0 1 1 0 0 0 1 0 1 1 0 0 2 1 0 1 1  rhea
0 0 0 0 0 1 0 0 1 1 0 8 1 0 0 1  scorpion
1 0 0 1 0 1 1 1 1 1 0 1 0 0 0 1 1  seal
1 0 0 1 0 1 1 1 1 1 0 1 2 1 0 1 1  sealion
0 0 0 0 0 1 1 1 1 0 1 0 0 1 0 0 1  seasnake
0 0 1 0 0 1 1 0 0 1 0 0 0 0 0 1  seawasp
0 0 1 0 0 0 1 1 1 0 0 0 1 0 0 1  slowworm
0 0 1 0 0 0 0 0 1 0 0 0 0 0 0 2  slug
1 0 0 1 0 0 0 1 1 1 0 0 2 1 0 0 1  squirrel
0 0 1 0 0 1 1 0 0 0 0 5 0 0 0 1  starfish
0 0 1 0 0 1 1 1 1 0 1 1 0 1 0 1 1  stingray
0 1 1 0 1 1 0 0 1 1 0 0 2 1 0 1 1  swan
0 0 1 0 0 1 0 1 1 1 0 0 4 0 0 0 1  toad
0 0 1 0 0 0 0 0 1 1 0 0 4 1 0 1 1  tortoise
0 0 1 0 0 0 1 1 1 1 0 0 4 1 0 0 1  tuatara
0 0 1 0 0 1 1 1 1 0 0 1 0 1 0 1 3  tuna
0 1 1 0 1 0 1 0 1 1 0 0 2 1 0 1 1  vulture
1 0 0 1 0 0 0 1 1 1 0 0 2 1 0 1 1  wallaby
1 0 1 0 1 0 0 0 0 1 1 0 6 0 0 0 1  wasp
1 0 0 1 0 0 1 1 1 1 0 0 4 1 0 1 10 wolf
```

fined with the same kernel. Integer data values were treated as real numbers.

The simulation was done on a SUN SPARCstation. For each $\beta$ value, the process entered a kind of oscillating steady state, with change of cluster center values well-below $10^{-10}$, after 50 to 130 iterations. I do not know the exact nature of the oscillation. But when blurring with shifting data or blurring with fixed data, without the context-dependent term, was used on the same data set, oscillation occurred much less frequently. For example, when blurring with shifting data was used, the process usually entered in about 20 iterations to a steady state where no single bit changed on all double-precision values. Each data point was traced to the final destination of its "image", or the shifting cluster centers. These centers merged and data points sharing the same final image were considered belonging to the same cluster.

It was generally true that the larger the $\beta$ value is, the more distinct final images will be. However, there were exceptions, particularly when the kernel was not quite truncated (for instance with $\gamma = 5$). A cluster generated with a $\beta$ value may be included in a cluster generated with a smaller $\beta$ value, but exception abounds. Figure 2 shows the complicated inclusion relation between clusters generated with different $\beta$ values. Some hierarchical inclusions of clusters are obvious, while others are fuzzy and uncertain. Merged clusters may split at a much smaller *beta* value.

I hope the addition of this fuzzyness in hierarchical clustering may be useful to show the validity of certain hierarchies, and thus to preserve more information about the data set in the clustering result, which often functions as an exploratory and summarizing representation of the data.

**Acknowledgement.** This work was supported by NSF grant IRI-9010555.

# References

[1] J.C. Bezdek, *Pattern Recognition with Fuzzy Objective Function Algorithms*. New York: Plenum, 1981.

[2] Y. Cheng and K.-S. Fu, "Conceptual clustering in knowledge organization", *IEEE Trans. Pattern Anal. Machine Intell.*, vol. PAMI-7, pp.592-598, 1985.

[3] Y. Cheng and Z. Wan, "Analysis of the blurring process", to appear in *Computational Learning Theory and Natural Learning Systems*.

[4] R. Forsyth. "zoo.data", data file available at the machine learning database at the University of California at Irvine, via the Internet.

[5] K. Fukunaga and L.D. Hostetler, "The estimation of the gradient of a density function, with applications in pattern recognition", *IEEE Trans. Information Theory*, vol. IT-21, pp.32-40, 1975.

[6] K. Rose, E. Gurewitz, and G.C. Fox, "Statistical mechanics and phase transitions in clustering", *Physical Review Letters*, vol. 65, pp.945-948, 1990.

Figure 2: A fuzzy hierarchy of clusters with the animal data in Figure 1. See Section V text for details.

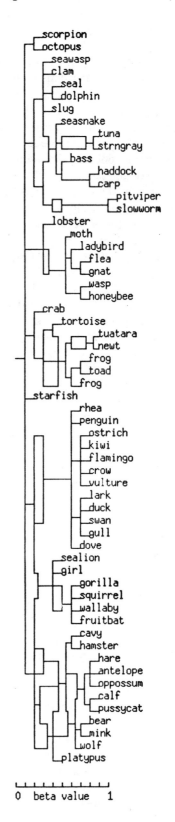

# An Error Convergence Simulation Study of Hard vs. Fuzzy C-Means Clustering

Michael E. Brandt, *Member, IEEE*, and Yezdi F. Kharas, *Student Member, IEEE*

*Abstract*— We have previously demonstrated that the fuzzy c-means (FCM) algorithm is effective for separating cerebrospinal fluid (CSF), white and gray matter tissue clusters in brain magnetic resonance images of children with and without hydrocephalus. In this paper we report results of some simulation studies comparing the hard c-means (HCM) algorithm, FCM and a variant of FCM referred to as SFCM. We show that under certain conditions of cluster shape, size and overlap, the two fuzzy algorithms are more stable than HCM in the sense that the error decreases more monotonically as a function of iteration number. We also demonstrate that the second error difference should be used as a stopping criterion for FCM. Finally we show that maximizing the sum of squared memberships is a better indicator of the number of clusters present in the data than a criterion based on both minimizing the intracluster distance and maximizing the intercluster distance.

## I. INTRODUCTION

We have studied the effectiveness of the fuzzy c-means (FCM) algorithm [1,2] for estimating the volumes of various brain tissues in pediatric populations from magnetic resonance images (MRI). In MRI, one can image a brain slice having a certain fixed thickness from one to several millimeters. Since a finite volume of brain is imaged in a given slice, more than a single tissue type is likely to be included in each imaged voxel. This leads to partial volume averaging of tissue. Hard clustering techniques will assign voxels exclusively to a particular tissue type. Fuzzy clustering methods can be used to estimate the partial tissue memberships in each voxel. For example, a given voxel may be composed of 40 percent gray matter and 60 percent white matter. We found that a modified version of the FCM algorithm is particularly effective in distinguishing children with hydrocephalus from normals [3]. In the hydrocephalic brain there is often an increase in brain cerebrospinal fluid (CSF) on the order of 10 to 15 percent with a comparable decrease in myelin (white matter). Insertion of a brain shunt in these children is often necessary to reduce intracerebral pressure caused by increased amounts of CSF. We have previously shown that decreased brain white matter content in hydro-

This work is supported in part by a grant from the National Institute of Neurologic Disorders and Stroke (NINDS), number NS25368.

M. E. Brandt is with the Department of Psychiatry and Behavioral Sciences, University of Texas Medical School, Houston, TX 77030-1501.

Y. F. Kharas is with the Department of Electrical Engineering, University of Houston, Houston, TX 77204-4793.

cephalic children correlates with neurobehavioral and cognitive deficits, thus highlighting the importance of early detection of such brain injury [4]. In other degenerating brain disorders such as early onset Alzheimer's, the tissue volume differences as compared to normals may be much smaller (on the order of a few percent).

We performed several simulations in order to determine if fuzzy clustering will ultimately be useful in distinguishing small brain tissue volume differences for use in clinical MRI studies. In a recent conference proceeding paper [5] we compared FCM, and hard c-means clustering (HCM) with several variants of Krishnapuram and Keller's [6] possibilistic clustering method (PCM). We showed that PCM is ineffective at separating even moderately overlapped clusters and that for certain geometrical configurations HCM outperforms FCM despite a high degree of cluster overlap.

In this paper we report results of further simulation studies using HCM, FCM, and a variant of FCM, which we will refer to as SFCM in what follows. We will show that under certain conditions of cluster geometry and overlap, the two fuzzy algorithms are more stable than HCM in the sense that the error decreases more monotonically as a function of iteration number. We also demonstrate that it is better to use the second error difference as a stopping criterion. Finally, we show that maximizing the sum of squared memberships is a better indicator of the number of clusters present in the data than a criterion based on both minimizing the intracluster distance and maximizing the intercluster distance.

## II. METHODS

The simulated data was designed to be similar in form to bispectral MRI data. MRI intensity data was simulated by generating pseudorandom Gaussian distributed numbers between 0 and 255 (8-bit grayscale resolution) [7] to create two-dimensional data scattergrams (two random numbers were produced for each image voxel for the $x$ and $y$ axes). Each dimension represented a separate MRI spectral channel. Clustering is performed in this two-dimensional feature space [8]. In the case of real data, the dimensions might represent a $T_2$-weighted image and a proton density (PD)-weighted image of a given anatomic slice. While normal distributions are convenient for study, it is unlikely that real brain MRI intensity values are Gaussian distributed. The FCM algorithm estimates the probability distribution from the data itself and therefore does not assume any *a priori* data distribution [9].

We studied how the three algorithms separate three equal sized and equidistant circular clusters that are each expanded by increasing their standard deviations (SDV) equally in the $x$ and $y$ directions. In the simulation set, cluster centers were positioned a fixed number of units from each other (132), with equal number of data points (6,787). Each cluster therefore shared one third of the total data space. We simulated cluster overlap by increasing the variance of each cluster equally in both the $x$ and $y$ directions. We refer to this as Method 1 in what follows. Fig. 1 shows the three clusters in this configuration each having an SDV along the $x$ and $y$ axes of 30 units. We also studied an arrangement with three elongated clusters along the $x = y$ diagonal. We kept their centroid locations fixed and expanded the clusters in the diagonal direction. We refer to this as Method 2 in the remainder of this paper. Fig. 2 shows this configuration with an SDV of 20 units.

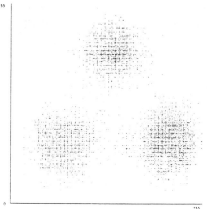

Fig. 1. Simulation of 3 equispaced, equal sized clusters with SDV in $x$ and $y$ directions of 30 units (Method 1).

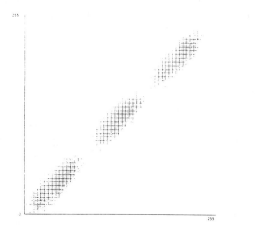

Fig. 2. Simulation of 3 equal sized clusters with SDV of 20 units along the $x = y$ direction (Method 2).

FCM makes use of a "fuzzification" parameter, $m$, which determines the amount of overlap between clusters. The choice of value for this parameter is determined empirically from the actual problem domain. It is used to weight both the distance and membership values. We have found that a value of $m = 4/3$ works very well for actual brain MRIs and we have used that value in what follows for the FCM and SFCM procedures. The use of $m = 1$ reduces the algorithm to the HCM clustering procedure in which objects are assigned 100 percent membership to the nearest cluster centroid.

FCM is straightforward to implement on the computer and basically consists of four calculations: 1) distances between pixel vectors and cluster centroids, 2) cluster membership values between 0 and 1 for each pixel and each cluster, 3) cluster centroid location, and 4) iterative convergence (sum squared difference between membership matrix in previous iteration and the current one). In FCM the memberships are constrained by three conditions: 1) each has a value between 0 and 1 (0 or 1 in HCM), 2) their sum across clusters for each object is 1.0, and 3) their sum for a given cluster across objects is less than the total number of such objects ($N$). The procedure convergences to a solution in the least mean squares sense [8] although incorrect clusters are infrequently arrived at.

In addition to HCM and FCM, we studied a variant of FCM. In the FCM procedure the error convergence (EC) criterion is:

$$\text{If } \|U^{(l)} - U^{(l-1)}\| < \varepsilon \text{ then stop,} \quad (1)$$

where $U^{(l)}$ is the membership matrix computed in current iteration l. That is, the sum of squared differences between all matrix elements ($N$ pixel values by $C$ clusters) in the current and previous iterations must be computed and compared to $\varepsilon$. In some cases, the error measured by (1) converges to a finite value larger than $\varepsilon$. It is therefore better to compute the second successive absolute difference of $U$ and compare with $\varepsilon$. In the HCM and SFCM procedures we use: if $\sum_{i=1}^{C} \|v_i^{(l)} - v_i^{(l-1)}\|^2 < \varepsilon$ then stop. That is, the squared differences in centroid locations are added and compared to $\varepsilon$ in each iteration. We used $\varepsilon = 0.1$ in the following results. Note that the usual FCM criterion specifies convergence of the membership values, yet this also implies convergence of the centroid locations. SFCM may represent a rather large savings in computation time. For example, in a typical MR brain image there are about 20,000 pixels with 5 main clusters (CSF, white and gray matter, air, and bone). FCM would require 100,000 subtractions, multiplies (squaring), and adds per iteration. SFCM would only require 5 subtractions, multiplies, and adds per iteration. If one were to analyze the whole brain at once (about 50 slices) the time savings over the FCM criterion would be quite substantial. We have found that SFCM produces convergent solutions on real and simulated data.

We also studied the behavior of two computations for determining the number of clusters using SFCM. The first one seeks to maximize the sum of the total memberships squared [10]:

$$F(c) = \frac{1}{N} \sum_{i=1}^{C} \sum_{j=1}^{N} (\mu_{ij})^2 \qquad (2)$$

where $c$ is the iteration number, and the other minimizes the difference between intra and intercluster distance [11]:

$$S(c) = \sum_{i=1}^{C} \sum_{j=1}^{N} (\mu_{i,j})^m |x_k - \nu_i| - |\nu_i - \overline{x}| \qquad (3)$$

where $\overline{x}$ is the grand mean of all pixel vector values.

### III. SECOND ERROR DIFFERENCE

Using simulation Method 1 with the conventional FCM algorithm, we found that the error of convergence often reaches a finite limit greater than $\varepsilon$ as a function of iteration number (#IT). Table I (next page) shows EC as a function of #IT for FCM, SFCM, and HCM for Method 1 with SDV equal to 30 units in the $x$ and $y$ directions. After the fourth iteration of FCM, the EC is stable at 0.180. Using the second EC difference detects when the first difference stops changing. In the remaining results we used second EC differences for all three algorithms.

### IV. ERROR CONVERGENCE AND MONOTONICITY

In general, HCM converges faster (with less iterations) than SFCM and the latter converges faster than FCM. This can be seen clearly in Table II (next page) for Method 1 with $x$ and $y$ SDV of 40 units, and in Fig. 3 for Method 2 with SDV along the diagonal of 40 units. Notice how the error for HCM shown in Fig. 3 does not decrease monotonically as opposed to the fuzzy methods. This can also be seen in Fig. 4 when the SDV is increased to 70 units, and in Fig. 5 for Method 1 with $x$ and $y$ SDV of 70 units. Also, there was a slight error increase at iteration number 16 of the FCM solution in Fig. 5. The error for SFCM decreased monotonically for all iterations shown in Fig. 5.

### V. DETECTING THE CORRECT NUMBER OF CLUSTERS

Table III displays results for the calculation of $F(c)$ (3) for Method 1 and SFCM as a function of increasing number of clusters and increasing $x$ and $y$ SDV. The maximum value of $F(c)$ is supposed to be a good predictor of the number of clusters present in the data using FCM [10]. This appears to be true (asterisk marks maximum value of $F(c)$) up to an SDV of 90 units which represents a very high degree of cluster overlap.

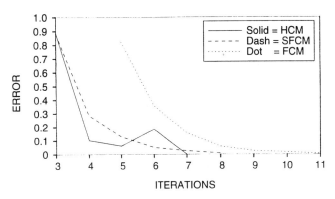

Fig. 3. Error vs. iteration number for Method 2 with SDV of 40 units.

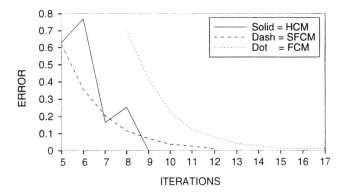

Fig. 4. Error vs. iteration number for Method 2 with SDV of 70 units.

Table IV displays results for the calculation of $S(c)$ (4) for Method 1 and SFCM as a function of increasing number of clusters and increasing $x$ and $y$ SDV. The minimum value of $S(c)$ is supposed to be a good predictor of the number of clusters present in the data using FCM [11]. This does not appear to be true for any SDV for this cluster configuration.

### VI. CONCLUSIONS AND DISCUSSION

In this report we have studied the effects of HCM, FCM, and a modified FCM procedure using a computationally simpler error convergence criterion. We performed clustering procedures on simulated data in the context of MRI intensity pixel values. Certain assumptions have been made which may not hold in the case of real MRIs. We made use of Gaussian pseudorandom cluster distributions of equal size. In simulation Method 1 we made use of three equispaced, overlapped circular clusters. In Method 2 we simulated three overlapping, elongated clusters with major diagonals along the $x = y$ line. Our goal was to study the computational performance of these algorithms with respect to error convergence and, in the case of SFCM, to determine which of two relationships (3,4) could most effectively be used to properly estimate the correct number

## TABLE I

First and Second Error Convergence (EC) Differences for FCM, SFCM, and HCM. The Standard Deviation (SDV) in the $x$ and $y$ Directions is 30 Units. Equal Number of Data Points in Each Equispaced Cluster (Method 1). #IT is the Number of Iterations to Convergence.

| #IT | FCM ($m = 4/3$) | | SFCM ($m = 4/3$) | | HCM ($m = 1$) | |
|---|---|---|---|---|---|---|
| | 1ST DIFF | 2ND DIFF | 1ST DIFF | 2ND DIFF | 1ST DIFF | 2ND DIFF |
| 1 | 19190.000 | — | 5.621 | — | 7.611 | — |
| 2 | 1.639 | 19189.310 | 0.107 | 5.514 | 0.119 | 7.429 |
| 3 | 0.234 | 1.404 | 0.004 | 0.103 | 0.001 | 0.118 |
| 4 | 0.182 | 0.052 | — | 0.004 | — | 0.001 |
| 5 | 0.180 | 0.002 | | | | |
| 6 | 0.180 | 0.000 | | | | |

## TABLE II

Error of Convergence (EC) for FCM, SFCM, and HCM as a Function of #IT. The Standard Deviation (SDV) in the $x$ and $y$ Directions is 40 Units. Equal Number of Data Points in Each Equispaced Cluster (Method 1).

| #IT | FCM | SFCM | HCM |
|---|---|---|---|
| 2 | 18426.730 | 20.764 | 31.032 |
| 3 | 6.629 | 0.907 | 0.698 |
| 4 | 0.883 | 0.129 | 0.194 |
| 5 | 0.124 | 0.018 | 0.005 |

## TABLE III

Values of $F(c)$ for Method 1 Using SFCM ($m = 4/3$) with Increasing cluster number (c) and Increasing SDV.

| SDV | $F(c = 2)$ | $F(c = 3)$ | $F(c = 4)$ | $F(c = 5)$ | $F(c = 6)$ |
|---|---|---|---|---|---|
| 10 | 0.972 | 1.000* | 0.955 | 0.910 | 0.865 |
| 20 | 0.948 | 0.994* | 0.948 | 0.945 | 0.902 |
| 30 | 0.912 | 0.963* | 0.918 | 0.880 | 0.853 |
| 40 | 0.880 | 0.926* | 0.888 | 0.861 | 0.842 |
| 50 | 0.873 | 0.902* | 0.875 | 0.852 | 0.835 |
| 60 | 0.869 | 0.889* | 0.869 | 0.850 | 0.844 |
| 70 | 0.866 | 0.880* | 0.868 | 0.849 | 0.848 |
| 80 | 0.864 | 0.877* | 0.857 | 0.854 | 0.849 |
| 90 | 0.881* | 0.877 | 0.872 | 0.857 | 0.855 |

## TABLE IV

Values of $S(c)$ for Method 1 Using SFCM ($m = 4/3$) with Increasing cluster number (c) and Increasing SDV.

| SDV | $S(c = 2)$ | $S(c = 3)$ | $S(c = 4)$ | $S(c = 5)$ | $S(c = 6)$ |
|---|---|---|---|---|---|
| 10 | -79536.0 | -1621605.0 | -1605030.0 | -1590115.0 | -1626813.0 |
| 20 | 22667.0 | -1296531.0 | -1304601.0 | -1318568.0 | -1354195.0 |
| 30 | 107519.0 | -979403.0 | -1004437.0 | -1044091.0 | -1219224.0 |
| 40 | 137347.0 | -728999.0 | -737808.0 | -1020766.0 | -993884.0 |

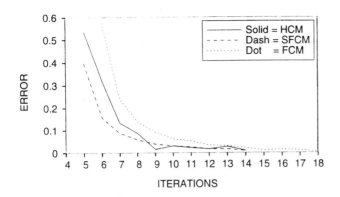

Fig. 5. Error vs. iteration number for Method 1 with SDV of 70 units.

of clusters in the simulated data. The following conclusions can be made:

1. The second difference with respect to iteration number should be used for the fuzzy procedures to avoid the problem of converging to an error value which is larger than $\varepsilon$ (Table I).

2. In general, HCM converges faster than SFCM which in turn converges faster than FCM (Table II, Fig. 3). Furthermore, SFCM is computationally faster than FCM depending on $N$ and $C$.

3. Error convergence of HCM is less monotonic (and therefore less stable) than the fuzzy algorithms (Figs. 3–5).

4. For SFCM, use of $F(c)$ is a good predictor of the number of clusters in the data with high degree of cluster overlap (Table III), whereas $S(c)$ was not found to be useful (Table IV).

We found previously [5] that SFCM performs just as well as FCM using simulated data with respect to the number of iterations to convergence, final centroid locations, and cluster size. Given the findings reported in the present paper, SFCM is a good computational choice for fuzzy c-means clustering. We also reported in [5] that HCM can be more effective than fuzzy clustering depending on the geometrical configuration of the data (cluster size, shape, relative location and degree of overlap).

### REFERENCES

[1] J. Bezdek, *Pattern Recognition with Fuzzy Objective Function Algorithms,* New York: Plenum Press, 1981, pp. 65-79.

[2] Y. Pao, *Adaptive Pattern Recognition and Neural Networks,* Reading, Mass.: Addison-Wesley, 1989, pp. 73-76.

[3] M. Brandt, T. Bohan, L. Kramer, and J. Fletcher, "Estimation of CSF, white and gray matter volumes in hydrocephalic children using fuzzy clustering of MR images," *Computerized Medical Imaging and Graphics,* in press.

[4] J. Fletcher, T. Bohan, M. Brandt, B. Brookshire, S. Beaver, D. Francis, K. Davidson, N. Thompson, and M. Miner, "Cerebral white matter and cognition in hydrocephalic children," *Archives of Neurology,* vol. 49, pp. 818-824, 1992.

[5] M. Brandt and Y. Kharas, "Simulation studies of fuzzy clustering in the context of brain magnetic resonance imaging," *Proceedings of the Third International Conference on Industrial Fuzzy Control and Intelligent Systems,* pp. 197-203, 1993.

[6] R. Krishnapuram and J. Keller, "A possibilistic approach to clustering," *IEEE Trans. Fuzzy Systems,* vol. 1, pp. 98-109, 1993.

[7] L. Baker, *More C Tools for Scientists and Engineers,* New York: McGraw Hill, 1991, pp. 167-185.

[8] B. Kosko, *Neural Networks and Fuzzy Systems,* Englewood Cliffs, NJ: Prentice Hall, 1992, pp. 13-14.

[9] M. Vannier, R. Butterfield, D. Rickman, D. Jordan, W. Murphy, and P. Biondetti, "Multispectral magnetic resonance image analysis," *CRC Critical Reviews in Biomedical Engineering,* vol. 15, pp. 117-144, 1987.

[10] J. Bezdek, *Pattern Recognition with Fuzzy Objective Function Algorithms,* New York: Plenum Press, 1981, p. 100.

[11] J. Bezdek, "Pattern Recognition with Fuzzy Objective Function Algorithms," in M. Adler (ed.) *Advanced Applications in Pattern Recognition,* New York: Plenum Press, 1981, pp. 65-85.

# A STUDY OF REPETITIVE TRAINING WITH FUZZY CLUSTERING

Hiroyuki Asaka    Hirokazu Takahashi    Mototaka Sone    Nobukazu Iijima

Musashi Institute of Technology

1-28-1 Tamatsutsumi Setagaya-ku Tokyo 158, Japan

## ABSTRACT

The back-propagation algorithm [1] [2] needs some couples of training data and supervised signals. Generally, the more training data there are, the more certain recognition result is obtained.

However, a plenty number of training data is not always gotten on the training stage. Thus, this paper propose a new repetitive training algorithm based on fuzzy logic. [3] This algorithm modify the network on the recognition stage so that the network would output more certain recognition result.

As the result, it comes to clear that this algorithm is effective for getting more certain recognition result.

Keywords: artificial neural network, fuzzy clustering, repetitive training, hidden node value.

## INTRODUCTION

It has been recently reported that an A.N.N. is effective for pattern recognition such as image, character and voice recognition. The structure of the networks have been proposed so far, and the most general structure is layer structure because of the B.P. training algorithm. This algorithm is for the layered network and so powerful that a good recognition result could be obtained.

The B.P. algorithm needs couples of input data and output data, and the network is modified so that the network would respectively output a supervised signal when the training datum coupled with the supervised signal is input. It is generally considered that the more training data there are, the more certain recognition result would be obtained. So, it is very important to use more training data on training stage. However, a plenty number of training data is not always gotten. Therefore, If it is possible to assign suitable supervised signals to the recognition data, the recognition rate would be higher because recognition data can be used as training data.

However, it is generally known that the huge number of training data make the training convergency worse. Thus, the number of training data must be as less as possible. So, This paper proposes a new repetitive training algorithm with a feature of the network and a classification method based on fuzzy logic. It creates new training data which has almost the same effect to the assignment supervised signals to the recognition data, but the number of new created training data is as less as possible.

## RECOGNITION USING NEURAL NETWORK

Fig. 1 shows an output state of a three-layered network trained with B.P. rule, which has two input nodes, six hidden nodes (including one bias node) and one output node. Training parameters are shown in Table 1.

■ indicate training data coupled with supervised signals which values are 0.0, and ● are those coupled with supervised signals which values are 1.0.

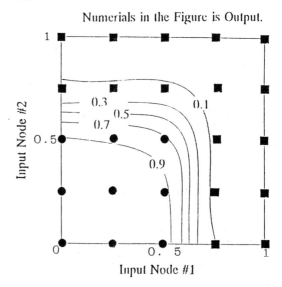

Fig. 1 Output State of Network

Table 1 Training Parameters

| Learning Rate | 0.2 |
|---|---|
| Momentum | 0.9 |
| Initial Weight | Random |
| Initial Weight Range | $-1.0 \sim +1.0$ |

In this figure, a training datum or a recognition datum is indicated as a point in the input space. In near region around training data in the input space, the output is equal to the supervised signals coupled with the training data.

The outputs on being input a similar datum to one of training data are almost correspond to the supervised signal. The outputs of the network are calculated by the hidden node values and the weights between hidden nodes and output nodes. Therefore, the network can correctly recognize fluctuated data because of the existence of the hidden nodes. Thus, the hidden node values are paid attention to.

Fig. 2 shows the the hidden node values of the same network in Fig. 2 on being input recognition data. Fig. 2–(a) indicates those on being input the recognition data marked in Fig. 2–(b).

These figures show that all the outputs on being input a datum similar to a training datum are almost the same. It is estimated that the recognition data which can be correctly recognized by the network causes almost the same hidden node value to the one caused by one of training data.

## ASSIGNMENT SUPERVISED SIGNALS TO RECOGNITION DATA

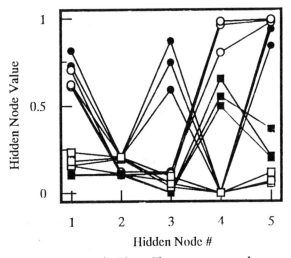

Marks in These Figures correspond.

(a) Hidden Node Value

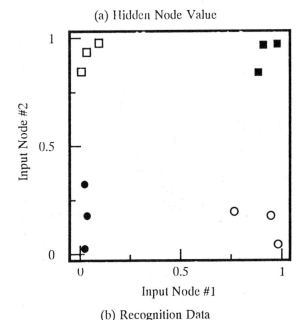

(b) Recognition Data

Fig. 2 Recognition Data and Hidden Node Value

Network can Only Determine Region uncertainly.

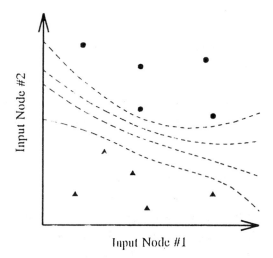

(a) Small Training Data Number

Network can Determine Region minutely.

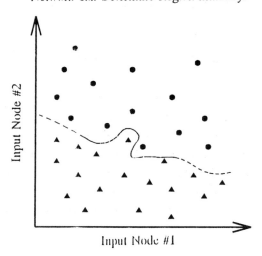

(b) Large Training Data Number

Fig. 3 Regions Made by A.N.N.

As described above, the recognition data which have almost the same hidden node values to the one of training data are certain to be correctly recognized by the network. And the network can correctly recognize slight different data from training data.

Thinking of training and recognition data as vectors, the network make some numbers of region in the input space by training data and supervised signals as Fig. 3-(a) shows. So, the network can roughly determine the region with a small number of training data and supervised signals. Therefore, Adding suitable supervised signals to recognition data which are certain to be correctly recognized by the network and using them as training data, the network can determine the regions more minutely as Fig. 3-(b) shows. Thus, this paper proposes a new repetitive training method as following.

1) Training is carried out with an initial training data group.

2) The hidden node values of the A.N.N. caused by training data and those caused by recognition data are compared.

3) If the hidden node values of a recognition datum are similar to the hidden node value caused by one of training data, the supervised signal coupled with the training datum is added to the recognition datum, and the recognition datum is added to the training data group.

4) Training is carried out with a new training data group.

5) Repeat the procedure 2) ~ 4).

With this repetitive training algorithm, recognition data can be used as training data because suitable supervised signals are assigned to recognition data. Therefore, the network can recognize more certainly because the more number of training data used on training stage grows in accordance with the number of iteration of the procedures.

## FUZZY CLUSTERING

The proposed repetitive training algorithm needs to know How similar the hidden node value caused by a recognition datum to the hidden node value caused by a training datum. Fuzzy clustering is a method to classify a group of data into some number of exclusive groups which consist of similar data. [4][5]

For instance, there are ten two-dimensional data as shown in Fig. 4-(a). These data will be classified by fuzzy clustering into several groups as Fig. 4 shows. As these figures show, several number of data which are similar to each other can be picked up with fuzzy clustering.

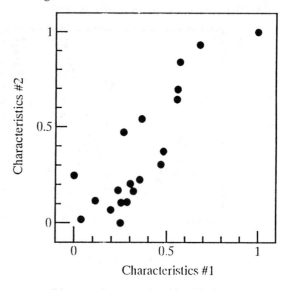

(a) Data Group to be Classified

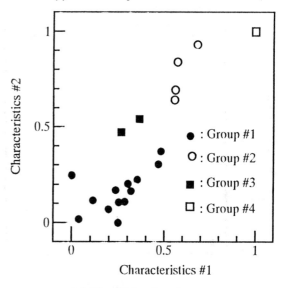

(b) Classifying Result

Fig. 4 Fuzzy Clustering

## REDUCING TRAINING DATA NUMBER

With the procedures described above, it is estimated that the more certainly the network can recognize, the more number of training iteration. However, the number of training data would increase in accordance with the number of iteration of procedure 2) ~ 4), and the increasing of the number of training data worsen the training convergency. In the worst case, training can not be completed. So, the increasing of the number of training data has to be avoided.

The neural network can correctly recognize slight different data from training data. In another words, recognition data which exist in a region around a training data can be correctly recognized by the network with regarding the recognition data and training data as vectors like Fig. 5-(a) shows. So, the region made by some numbers of extremely similar training data corresponds to the one made by a average data of the training data as Fig. 5-(b) shows. Through this averaging process, the increasing of training data number is as less as possible and the training convergency is kept good.

Therefore, the procedures described above are added to the repetitive algorithm. As the result, the repetitive training algorithm is as shown in Fig. 6.

Region Where Network Recognize Correctly.

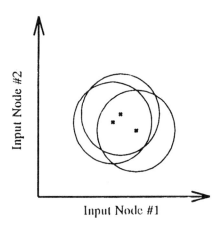

(a) Similar Training Data

Region Where Network Recognize Correctly.

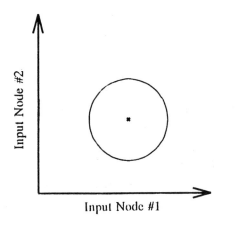

(b) Average Training Data
Fig. 5 Region Where Network can Recognize Correctly

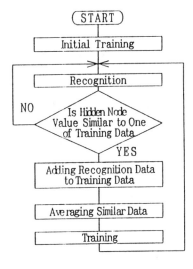

Fig. 6 Repetitive Training Algorithm

## SIMULATION OF REPETITIVE TRAINING

A simulation of repetitive training was carried out with a three-layered network which has two input nodes, six hidden nodes (including one bias node), and four output nodes. Training parameters are shown in Table 1. This simulation is in order to recognize in which region of the four regions shown in Fig. 7 the recognition data exists. Training data and recognition data are shown in Fig. 8 (the total number is 30x4=120), which are fluctuated to present the fluctuation included in real data such as noise and individual gaps.

The training with all of the data shown in Fig. 8 with supervised signals assigned by human was not able to be completed out. So, five data of each supervised signal (the total number 5x4=20) are picked up and used for training. And the training was completed.

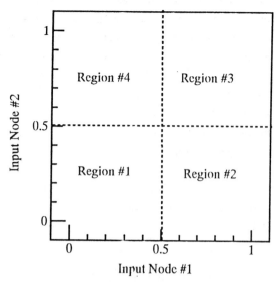

Fig. 7 Four Regions for Recognition

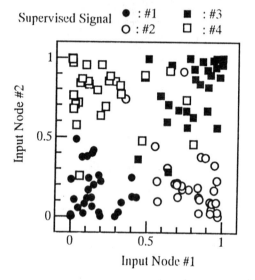

Fig. 8 Sample Data

Fig. 9(a)–(d) show the output state of the network of each output node. The output states of each output node ideally correspond to Fig. 7, but the real output state is so different especially output node #1 and #4.

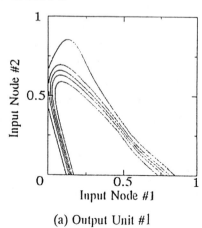

(a) Output Unit #1

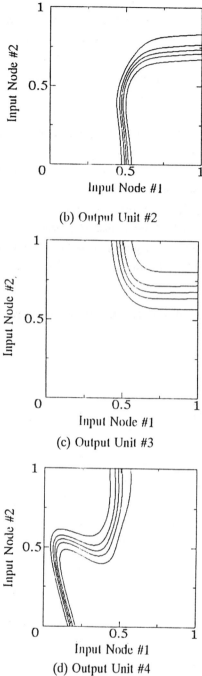

(b) Output Unit #2

(c) Output Unit #3

(d) Output Unit #4

Fig. 9    Output State (Initial)

Fig. 10(a)–(d) show the output after the repetitive training proposed by this paper. The output state of the network is improved well, and the output state is more similar to the ideal one (Fig. 7) than Fig. 9. Therefore, the proposed repetitive training algorithm is able to modify the network to obtain more certain recognition result. Indeed, the number of the training data is one hundred. Therefore, this repetitive training algorithm can modify the network so that the network would obtain more certain recognition

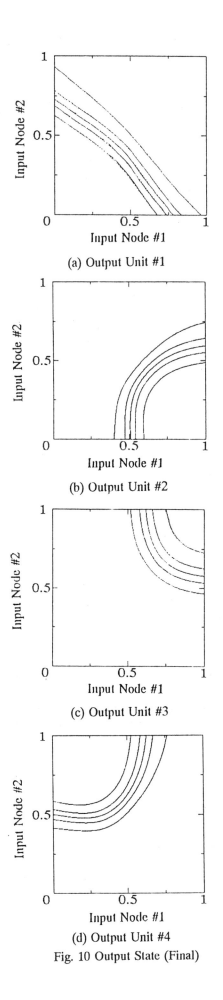

(a) Output Unit #1

(b) Output Unit #2

(c) Output Unit #3

(d) Output Unit #4

Fig. 10 Output State (Final)

result, creating new training data as less as possible from recognition data.

## CONCLUSION

This paper proposed a repetitive training algorithm based on fuzzy logic. The algorithm is very effective to obtain more certain recognition result.

## REFERENCES

[1] Rumelhart D.E. and McClelland J.L. : "Parallel Distributed Processing", Explorations in the Micro Structure of Cognition", MIT Press (1986).

[2] Lippmann R.P. : "An Introduction to computing with Neural Nets", IEEE ASSP Mag. pp.4–22 (April 1987).

[3] Zadeh L.A. : "Fuzzy Sets", Inform. Contr., vol.8 pp.338–353 (1965)

[4] Yamashita H. : "Fuzzy Clustering and Ordering in Instructional Analysis", Proc. of NAFIPS Congress VII pp.291–295 (1988)

[5] Yamashita H. : "Approximation Algorithm of a Fuzzy Graph and its Application", Proc. of IFSA Congress III pp.492–495 (1989)

# Alternative Membership Function for Sequential Fuzzy Clustering

John Sum and Lai-Wan Chan

Dept. of Computer Science, Chinese University of Hong Kong

Shatin, N.T., Hong Kong

December 31, 1993

**Abstract**

This paper presents an alternative membership function for fuzzy c mean. According to this membership function and Bezdek's definition, we derive two sequential algorithms for fuzzy c mean. Both of them are stochastic gradient descent algorithms which minimize Bezdek's objective functional. Analytical result indicates that both algorithms are actually compatible with each other. The convergence properties of both algorithms are studied. As the update equations are so simple, these sequential algorithms are embedded into neural network to form a class of fuzzy neural network analogue to unsupervised type neural network such that competitive learning is a special case.

## 1 Introduction

Recently, many attempts have been made to merge fuzzy logic and neural network to form a class of fuzzy neural network. Some of them are related to fuzzy competitive learning network (FCL) [2][3][4] which suffered from two serious designing problems: *(i) the algorithms are complicated and (ii) the algorithms showed no relation with fuzzy c-mean or hard c-mean*[1].

This paper presents two sequential fuzzy clustering (SFC) algorithms. They are different in the definition of the membership function. We embed them into a simple competitive learning network (CL), namely the fuzzy competitive learning network. It is easy to show that CL is just a special case of FCL. Intuitively, both algorithms share the same properties. They are simple in form and their convergence must be guaranteed. Theoretically, the new membership function covers the deficiency of Bezdek's definition.

In the next section, the two membership functions leading to sequential fuzzy clustering are derived. The mechanism of the corresponding fuzzy competitive learning will be described in section 3. Simulation and evaluation on these two membership functions are compared. The result is described in section 4. Finally, a conclusion follows in section 5.

## 2 Two membership functions leading to SFC

As our discussion is akin to fuzzy c mean, let us have a brief review. Given a set of unlabeled data containing finite number of elements, each element is denoted by $x$, and a set of representative vectors denoted by $v_i$, fuzzy c mean can be stated as a constraint optimization problem as follows.

**Definition 1: Fuzzy c mean is an optimization problem which minimizes the following function.**

$$J_m = \sum_{k=1}^{c} \sum_{x} \mu_k^m(x)\|x - v_k\|^2 f(x) \tag{1}$$

**subject to (i) $\sum_{k=1}^{c} \mu_k(x) = 1$ and (ii) $0 \le \mu_k(x) \le 1$. $f(x)$ is the pdf of $x$.**

---

[1] Jou has attempted to link fuzzy c-mean with competitive learning network CL, but he did not provide a vigorous treatment on his algorithm.

Consider the above problem, Bezdek derived the following membership function [1].

$$\mu_k(x) = \begin{cases} 1 & \text{if } x = v_k \\ 0 & \text{if } x = v_j \text{ ,and } j \neq k \\ \left[ \sum_{j=1}^{c} \left( \frac{\|x - v_k\|^2}{\|x - v_j\|^2} \right)^{1/(m-1)} \right]^{-1} & \text{otherwise} \end{cases} \tag{2}$$

However, *his algorithm is limited to two situations: (i)the input set is finite and the update must be in batch-mode, (ii) m cannot be set to between* $(0, 1]$. To solve the first problem, an intuitive method is to derive a gradient descent algorithm based on the continuous version of (1) [6][7].

$$J_m = \sum_{k=1}^{c} \int \mu_k^m(x) \|x - v_k\|^2 f(x) dx \tag{3}$$

In case of the second situation, $m$ is theoretically bounded.

## 2.1 Criteria of membership function

To define appropriate membership function leading to an effective sequential fuzzy clustering algorithm, we suggest the following criteria.

**a.** *For all $x$, $\sum_{k=1}^{c} \mu_k(x) = 1$.*

**b.** *$\mu_k(x; p)$ should be a function of $x$ and $p$, where $p \in R$ controlling the degree of fuzziness[2]. Besides, there exists $p_0$ such that $\mu_k(x; p)$ is a crisp membership as $p \to p_0$.*

**c.** *The corresponding SFC, which membership function is $\mu_k(x; p)$, must be simple in the sense that the form is comparable to simple competitive learning [5].*

**d.** *Convergence of SFC must be guaranteed.*

So, we restrict our concern on a general membership function defined as below.

**Definition 2: The general membership function is defined as**

$$\mu_k(x; p) = \begin{cases} 1 & \text{if } x = v_k \\ 0 & \text{if } x = v_j \text{ ,and } j \neq k \\ \left[ \sum_{j=1}^{c} \left( \frac{\|x - v_k\|^2}{\|x - v_j\|^2} \right)^{1/p} \right]^{-1} & \text{otherwise} \end{cases} \tag{4}$$

It is easy to show that (4) has the following properties.

**P1:** $\mu_k(x; p)$ **exists as** $p \to 0$, $x \to v_k$ **or** $x \to v_j$.
**P2:** $\sum_{k=1}^{c} \mu_k(x; p) = 1$ **for all** $p \in (0, \infty)$.
**P3: When** $p \to 0$, **(4) is reduced to the following equation.**

$$\mu_k(x; p \to 0) = \begin{cases} 1 & \text{if } \|x - v_k\| < \|x - v_j\| \text{ for all } j \neq k \\ 0 & \text{otherwise} \end{cases} \tag{5}$$

**P4: For all** $p > 0$, $\mu_k(x; p) > \mu_j(x; p)$ **iff** $\|x - v_k\| < \|x - v_j\|$.

---

[2]More precise, the parameter $m$ in (3) will be defined as the degree of fuzziness of objective.

## 2.2 General Stochastic Gradient Descent Algorithm

(5) is just the crisp membership function appeared in competitive learning. Observed that **P2** and **P3** fulfills (a) and (b). Our next step is to find $p$ in terms of $m$ such that (c) and (d) can also be satisfied. For simplicity, we write $\mu_k(x; p)$ as $\mu_k$ in the following text. Taking derivative of (4) with respect to $v_i$, we obtain the following equations.

$$\frac{\partial \mu_k}{\partial v_i} = -\frac{2}{p} \mu_k^2 \left( \frac{\|x - v_k\|^2}{\|x - v_i\|^2} \right)^{1/p} \frac{(x - v_i)}{\|x - v_i\|^2} \tag{6}$$

$$\frac{\partial \mu_i}{\partial v_i} = \frac{2}{p} \mu_i^2 \left( \sum_{j \neq i} \frac{\|x - v_i\|^2}{\|x - v_j\|^2} \right)^{1/p} \frac{(x - v_i)}{\|x - v_i\|^2} \tag{7}$$

Taking the derivative of (3), we obtain

$$\frac{\partial J_m}{\partial v_i} = \int \left( \sum_{k=1}^{c} m \mu_k^{m-1} \|x - v_k\|^2 \frac{\partial \mu_k}{\partial v_i} - 2\mu_i^m (x - v_i) \right) f(x) dx \tag{8}$$

$$= \int \left\{ -\frac{2m}{p} \sum_{k=1}^{c} \mu_k^{m+1} \left( \frac{\|x - v_k\|^2}{\|x - v_i\|^2} \right)^{1+1/p} + 2\frac{m-p}{p} \mu_i^m \right\} (x - v_i) f(x) dx \tag{9}$$

*With the help of (9), we can simply define a family of stochastic gradient descent algorithm for minimizing (3) as following. For $p > 0$,*

$$v_i(t+1) = v_i(t) + \alpha(t) \left\{ -\frac{2m}{p} \sum_{k=1}^{c} \mu_k^{m+1} \left( \frac{\|x - v_k\|^2}{\|x - v_i\|^2} \right)^{1+1/p} + 2\frac{m-p}{p} \mu_i^m \right\} (x - v_i(t)). \tag{10}$$

## 2.3 Two Special Cases

Though, we can define a family of algorithm based on different value of $p$. We wish to check whether there is any relationship between $p$ and $m$ such that (10) can be as simple as possible[3].

### 2.3.1 Algorithm 1: $p = m$

By inspection, if we put $p = m$, equation (10) is reduced to a simple form as. The second term will be vanished. Let us consider the term of

$$\sum_{k=1}^{c} \mu_k^{m+1} \left( \frac{\|x - v_k\|^2}{\|x - v_i\|^2} \right)^{1+1/m} = \sum_{k=1}^{c} \left[ \sum_{j=1}^{c} \left( \frac{\|x - v_k\|^2}{\|x - v_j\|^2} \right)^{1/m} \right]^{-m-1} \left( \frac{\|x - v_k\|^2}{\|x - v_i\|^2} \right)^{1+1/m}$$

$$= \sum_{k=1}^{c} \left[ \sum_{j=1}^{c} \left( \frac{\|x - v_k\|^2}{\|x - v_j\|^2} \right)^{1/m} \right]^{-(m+1)} \left[ \left( \frac{\|x - v_k\|^2}{\|x - v_i\|^2} \right)^{1/m} \right]^{(m+1)}$$

$$= c \left[ \sum_{j=1}^{c} \left( \frac{\|x - v_i\|^2}{\|x - v_j\|^2} \right)^{1/m} \right]^{-(m+1)} \tag{11}$$

Hence, it can be derived that (9) is reduced to the following equation.

$$\frac{\partial J_m}{\partial v_i} = -\int 2c \mu_i^{m+1}(x; m)(x - v_i) f(x) dx \tag{12}$$

---

[3]Note that $\mu_k$ specified in case 1 is different from case 2.

where $c$ is the total number of representative vectors and $m \in (0, \infty)$. The corresponding stochastic gradient descent algorithm can be defined as the following equation provided that $m > 0$.

$$v_i(t+1) = v_i(t) + c\alpha(t) \left[ \sum_{j=1}^{c} \left( \frac{\|x - v_i\|^2}{\|x - v_j\|^2} \right)^{1/m} \right]^{-(m+1)} (x - v_i(t)) \tag{13}$$

It minimizes the following objective function.

$$J_m = \sum_{k=1}^{c} \int \left[ \sum_{j=1}^{c} \left( \frac{\|x - v_k\|^2}{\|x - v_j\|^2} \right)^{1/m} \right]^{-m} \|x - v_k\|^2 f(x) dx \tag{14}$$

### 2.3.2   Algorithm 2: $p = m - 1$

While $p$ in (9) is substituted by $m - 1$, we obtained the following equation.

$$\frac{\partial J_m}{\partial v_i} = \int \left\{ -\frac{2m}{p} \sum_{k=1}^{c} \mu_k^{m+1} \left( \frac{\|x - v_k\|^2}{\|x - v_i\|^2} \right)^{m/(m-1)} + \frac{2}{m-1} \mu_i^m \right\} (x - v_i) f(x) dx$$

Consider the first term inside the bracket.

$$\sum_{k=1}^{c} \mu_k^{m+1} \left( \frac{\|x - v_k\|^2}{\|x - v_i\|^2} \right)^{m/(m-1)} = \sum_{k=1}^{c} \left[ \sum_{j=1}^{c} \left( \frac{\|x - v_k\|^2}{\|x - v_j\|^2} \right)^{1/(m-1)} \right]^{-(m+1)} \left[ \left( \frac{\|x - v_k\|^2}{\|x - v_i\|^2} \right)^{1/(m-1)} \right]^m$$

$$= \left[ \sum_{j=1}^{c} \left( \frac{\|x - v_i\|^2}{\|x - v_j\|^2} \right)^{1/(m-1)} \right]^{-(m+1)} \sum_{k=1}^{c} \left( \frac{\|x - v_k\|^2}{\|x - v_i\|^2} \right)^{1/(m-1)}$$

$$= \left[ \sum_{j=1}^{c} \left( \frac{\|x - v_i\|^2}{\|x - v_j\|^2} \right)^{1/(m-1)} \right]^{-m} \tag{15}$$

Substitute the above result into (9), we obtain,

$$\frac{\partial J_m}{\partial v_i} = \int \left\{ -2 \frac{m}{m-1} \mu_i^m + \frac{2}{m-1} \mu_i^m \right\} (x - v_i) f(x) dx \tag{16}$$

It brings out the following stochastic gradient descent algorithm provided that $m > 1$.

$$v_i(t+1) = v_i(t) + \alpha(t) \left[ \sum_{j=1}^{c} \left( \frac{\|x - v_i\|^2}{\|x - v_j\|^2} \right)^{1/(m-1)} \right]^{-m} (x - v_i(t)) \tag{17}$$

It minimizes the following objective function.

$$J_m = \sum_{k=1}^{c} \int \left[ \sum_{j=1}^{c} \left( \frac{\|x - v_k\|^2}{\|x - v_j\|^2} \right)^{1/(m-1)} \right]^{-m} \|x - v_k\|^2 f(x) dx \tag{18}$$

Observed that (16) and (12) are similar to that of competitive learning algorithm.

### Remark
**(I).** We constraint the membership function in the form of equation (4). One may argue whether case 1 ($p = m$) can lead to a global minima of (3). The answer is 'no'. In order to ease the discussion, we denote $J(m, p)$ as the general objective function[4] (3) which membership function is $\mu_k(x; p)$, (4). Bezdek has already proven that $J(m; p)$ is globally minimized only if $p = m - 1$ (see chapter 3 of [1]), i.e.

$$inf\, J(m, m - 1) \leq inf_p \{inf\, J(m; p)\} \;\; \forall m \geq 1.$$

---

[4] That is to say, $m$ is corresponding to the degree of fuzziness of the objective and $p$ is the degree of fuzziness of the membership.

Therefore, we still suggest the form (14) as well as algorithm 1. Moreover, algorithm 1 allows us to set $0 < m \leq 1$ but it is not possible for (18) and algorithm 2.

**(II).** By inspecting equation (13) and (17), it can be found that algorithms 1 and 2 are related in the following ways.

$$c\frac{\partial J(p+1,p)}{\partial v_i} = \frac{\partial J(p,p)}{\partial v_i} \tag{19}$$

for all $p > 0$. That is to say, the solution set of minimizing[5] $J(p+1;p)$ is identical to that of minimizing $J(p;p)$ if both algorithms start with the same initial state. In another ward, the two algorithms are compatible to each other even though they are solving different problems.

**(III).** Algorithm 1 and 2 can easily be embedded into competitive learning network. The constitution is as follow. We assign the weight vectors as $v_i$ and replace the output of each unit by $\mu_k^{p+1}$. For each of the input $x$, the network will output the corresponding membership value $\mu_k(x)$ of $x$ with respect to each of the class, from 1 to $c$. Based on the current output and the current weight vectors, the network is trained by (17) (or (13) respectively). Parameter $m$ (or $p$ corresponding) is treated as a control of fuzziness. So, we can further define certain decreasing schedule[6] for $m$ in order to decrease the effect due to fuzziness.

# 3 Simulation: Mechanism, Result and Convergence Rate

We consider the case $m = 2$. Be fixing $m$, we apply both algorithms to evaluate the cluster centers. The SFC mechanism is summarized as following.

**S1.** Select randomly one sample, $x$, from the butterfly data set.

**S2.** For all $i = 1, 2, \ldots, c$. Update $v_i$ based on (13) or (17) which is dependent on the algorithm chosen.

**S3.** Goto step **S1**.

Figure(1) shows the trajectories of the two algorithms. The cluster centers obtained by either cases are similar. The cluster centers obtained by $p = m$ is closer than that of $p = m - 1$. While each of the values of the cluster center is plotted against number of iterations in Figure(2), it can be observed that their convergence rate are more or less the same.

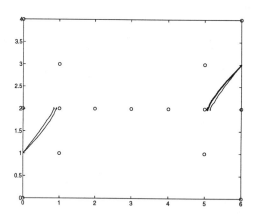

Figure 1: The trajectories of both algorithms. The lines which converge closer to $(1, 2)$ and $(5, 2)$ are corresponding to algorithm 1. The further ones are obtained by algorithm 2. The small circles are corresponding to the Butterfly Data.

---

[5] Based on the definition of (3).

[6] For instance, Bezdek et.al. [2] suggested $m = m_0 - \delta mt$, where $m \to 1$ as $t \to \infty$.

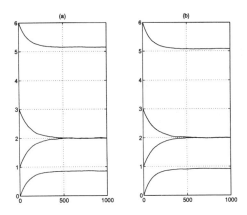

Figure 2: Convergence curve of algorithm (a) $p = m - 1$, (b) $p = m$.

## 4  Conclusion

This paper accounts for an alternative membership function for sequential clustering. The derivation is simply started by considering a family of membership function in the form of (4). Then we suggest two relationships between $p$ and $m$. Both relationships elicit elegant update equations (17) and (13) comparable with competitive learning. The second alogorithm is just the sequential version of Fuzzy c Mean. Whatever, both sequential update algorithms can be defined as mentioned in section 3. Furthermore, both algorithms are compatible with each other (see remark II in section 2.3). Both algorithms can be reduced to hard c mean (or competitive learning) if the parameter $p$ is set to zero.

In practice, we can put these algorithms into forms of neural network to form Fuzzy Competitive Learning Algorithm. In theory, the new relationship provides an essential extension on Bezdek's definition, i.e. $m$ can be set to be less than 1. We successfully derive a kind of fuzzy neural network and an alternative membership function which can cover the limitation of Bezdek's fuzzy c mean. However, further work has to be done on the investigation of their convergence rate and their application.

## References

[1] J.Bezdek, *Pattern Recognition with Fuzzy Objective Function Algorithms.* NY:Plenum,1981.

[2] J.Bezdek, E.Tsao and N.Pal, "Fuzzy Kohonen clustering networks,", *Proceeding of First IEEE Conf. on Fuzzy Systems*, 1035-1043, 1992.

[3] J.Bezdek, "A Review of Probabilistic, Fuzzy, and Neural Models for Pattern Recognition," *Journal of Intelligent and Fuzzy Systems*, VOL.1(1), 1-25, 1993.

[4] D.Zhang, M.Kamel and M.I.Elmasy, "Fuzzy Clustering Neural Network (FCNN) Using Fuzzy Competitive Learning," *Proceeding of World Congress in Neural Network*, VOL.II, 22-25, 1993.

[5] T. Kohonen, *Self-Organization and Associative Memory.* Springer-Verlag, 3rd Ed., 1989.

[6] C.C.Jou, "Fuzzy Clustering Using Fuzzy Competitive Learning Network," *Proceeding of IJCNN'92 Baltimore*, VOL.II, 714-719.

[7] C.C.Jou, "Fuzzy CounterPropagation Networks," Presented in the *Workshop on the Future Directions of Fuzzy Theory and Systems*, organized by faculty of engineering, CUHK, 27 Oct, 1993.

[8] E.Ruspini, "Numerical methods for fuzzy clustering," *Information Science*, VOL.2, 319-350, 1970.

[9] J.Sum, "An Alternative Membership Function for Sequential Fuzzy Clustering and Fuzzy Competitive Learning," unpublished manuscript.

# Qualitative Modeling Based on Fuzzy Relation and Its Application

Yingwu CHEN  Manshu ZHOU  Hao WANG
Department of Systems Engineering and Mathematics
National University of Defense Technology
Changsha, Hunan 410073
P. R. CHINA

## ABSTRACT

This paper describes the method of modeling of multivariable dynamical systems based on fuzzy relation equation solving. Its application to economic cycle research and macro economic early warning is disscused.

Keywords: Qualitative Modeling, Fuzzy Relation, Economic Cycles, Economic Early Warning.

## INTRODUCTION

Modeling, simulation and control of dynamical systems are important aspects of system research. There are bsically two kinds of conventional methods of modeling, i.e., mathematical modeling and experimental modeling[3]. Both have been developed very well and successed greatly in applications. In general, mathematical modeling starts from the understanding of problems or systems, and then the mathematical formulations are deduced, (such as differencial equations or difference equations). Rather, experimental modeling estimates the model representation of the system from experimental data or statistical data. Morever, from the modern point of view, modeling should consist of quantitative modeling and qualitative modeling. The methods of quantitative modeling, we think, overlap the conventional methods. They use quantitative data, information, varibles, states and relations between them in the process of modeling. Qualitative modeling is a modern modeling and analysis method developed as well as the advances of AI, fuzzy sets and system science. There are three representative branches:
- Qualitative Reasoning from AI [1]
- Fuzzy Modeling and Simulation [2,3]
- Qualitative Theory of Dynamical Systems

The method provided in this paper is most related to the first two. This paper discusses mainly the qualitative experimental modeling method. The detailed modeling procedures of system with multivariables and multioutputs and estimation algorithm are provided. Furthermore, we disscuss the qualitative modeling of economic cycles. and Macro Economic Early Warning (MEEW).

## FUZZY QUALITATIVE MODELING OF DYNAMICAL SYSTEM

Assume system state variable $x(t)$ is a fuzzy variable defined on $(p_0, p_1 \cdots p_m)$ or$(p_{-m}, \cdots p_0 \cdots p_m)$, So, we have:

$$x(t) = \mu_0(t)/p_0 + \mu_1(t)/p_1 + \cdots + \mu_m(t)/p_m$$

or

$$x(t) = \mu_{-m}(t)/p_{-m} + \mu_{-m+1}(t)/p_{-m+1} + \cdots + \mu_0(t)/p_0 + \cdots + \mu_{m-1}(t)/p_{m-1} + \mu_m(t)/p_m$$

here, $t \in R^+$. The above two fuzzy sets are correspond to two kinds of human intuitions:
- pure positive: (Very small, Small, Medium, Large, Very Large)
- positive and negative: (Negative Large(NL), Negative Medium(NM), Negative Small (NS), Zero(ZO), Positive Small(PS), Positive Medium(PM), Positive Large (PL))

They are two kinds of universe of discourse. For simplicity, the second will be considered here only, and it takes alsothe form as: $(P_1, P_2, \cdots P_m)$

Now, we discuss a qualitative multivariable system. Let the system state vector be $X(t) = (x_1(t), \cdots, x_n(t))$, and:

$$x_i(t) = (\mu_{i1}(t), \cdots, \mu_{im}(t)) \qquad (i = 1, \cdots n) \tag{1}$$

And the system output vector be $Y(t) = (y_1(t), \cdots, y_p(t))$, and:

$$y_i(t) = (v_{j1}(t), \cdots, v_{ji}(t)) \qquad (j = 1, \cdots, p) \tag{2}$$

Here, Card(X)=m, Card(Y)=$l$, and the system has n state variables and p outputs. The dimensions of X and Y are $1 \times nm$ and $1 \times pl$ respectively, Thus, we get the autonomous system model:

$$\begin{cases} X(t+1)=X(t) \circ R \\ Y(t) \quad =Y(t) \circ S \end{cases} \tag{3}$$

In equation (3), " o " is a Max-Min or Max-Product operator.For practical use, the Max-Product operator will be considered. Assume, there are K statistical samples, then the state equation and output equation have K-1, and K ordered pairs, which satisfy fuzzy relation equations (3) respectively. We get:
the state equation

$$(x_1(t),\cdots,x_n(t))=(x_1(t-1),\cdots,x_n(t-1)) \circ R \qquad (t=1,\cdots,K-1) \tag{4}$$

and the output equation

$$(y_1(t),\cdots,y_p(t))=(x_1(t),\cdots,x_n(t)) \circ S \qquad (t=0,\cdots,K-1) \tag{5}$$

Here, $R=[r_{ij}]$ $(i,j=1,\cdots,nm)$, $S=[s_{ij}]$ $(i=1,\cdots,nm,j=1,\cdots,pl)$. From (4), (5), we see, system modeling is to solve fuzzy relation equations. Although many efforts have been made for the solution of fuzzy relation equations, but analytical solving is still very difficult in the following aspects:
- whether the equations have solutions or not, because the conditions in which the solutions of a FRE exist are very severe;
- How to dertermine the solutions of a FRE, because if a FRE has one solution, then it may have more than one;
- How to calculate, because solving a FRE has been proved a NP-complete problem [9].

Therefore, we attempt to use some results of [5,6], and extend them to multivariable systems, and to utilize numerical estimation method to find the approximate solutions of (4) and (5). Output equation (5) is more general than (4), so we discuss it first as an example. From (4),(5), we know X and Y are defined on different universe of discourse. Our method is to make the estimated system outputs fit the samples best, thus it is necessary to define a performance index to measure the fitness. Now, we use Euclidean distance as the performance index, i.e.,

$$Q=\frac{1}{2}\sum_{t=1}^{K-1} \sum_{h=1}^{p} |\hat{y}_h(t)-y_h(t)|^2 \tag{6}$$

Here $\hat{y}_h(t)'s$ are estimated values. $y_h(t)$'s are samples. Hence,

$$Q=\frac{1}{2}\sum_{t=0}^{k-1} \sum_{h=1}^{p} \sum_{g=1}^{l} ( \underset{1 \le k' \le nm}{Max} \ s_{k'q} \bullet \mu_{uv}(t)-v_{hg}(t))^2 \tag{7}$$

where:

$$q=(h-1)\bullet l+g$$

$$u=u(k')=((k'-1) \ div \ m)+1$$

$$v=v(k')=((k'-1) \ mod \ m)+1$$

(div is the operator for exactly division, and mod for mode). Thus, we have the algorithm for $s_{ij}$:

$$s_{ij}^{(k-1)}=s_{ij}^k-\epsilon^{(k)} \partial Q/\partial s_{ij}|s_{ij}=s_{ij}^{(k)} \tag{8}$$

k is the number of iterations. From (7), we can get:

$$\partial Q/\partial s_{ij}=\sum_{t=0}^{K-1} (\hat{v}_{w(j)z(j)}(t)-v_{w(j)z(j)}(t)) \bullet P_{ij}(t) \tag{9}$$

where
$$w(j)=((j-1) \ div \ l)+1$$

$$Z(j)=((j-1) \ mod \ l)+1$$

Also, we have:

$$\partial Max f(s_{ij})/\partial s_{ij} = \partial Max f(s_{ij})/\partial f(s_{ij})/\partial s_{ij} \qquad (10)$$

here, $f(s_{ij})=S_{ij} \bullet \mu_{uv}(t)$ , and:
And

$$\partial Max f(s_{ij})/\partial f(s_{ij}) = \begin{cases} 1 & \text{if } s_{ij}=Max f(s_{ij}) \\ 0 & \text{otherwise} \end{cases} \qquad (11)$$

$$\partial f(s_{ij})/\partial s_{ij} = \mu_{uv}(t) \qquad (12)$$

From (11) and (12), we get:

$$P_{ij}(t) = \begin{cases} \mu_{u(j)v(j)}(t) & \text{if } s_{ij} \bullet \mu_{u(j)v(j)} \geq \underset{\substack{1 \leq k' \leq nm \\ k' \neq j}}{Max} \ s_{k'q}\mu_{u(k')v(k')}(t) \\ 0 & \text{otherwise} \end{cases} \qquad (13)$$

From (9) and (13), we get

$$\partial Q/\partial s_{ij} = \sum_{t=0}^{k-1} (\underset{1 \leq k' \leq nm}{Max} \ s_{k'j} \bullet \mu_{u(k')v(k')}(t) - v_{w(j)Z(j)})P_{ij}(t) \qquad (14)$$

According to [5], we set:

$$e^{(k)} = 1/(K+k^{\beta})$$

$\beta$ is an experimental parameter, and it should ensure good convergency of the algorithm and avoid oscillations.

The above algorithm is derived from output equation (8), but for the state equation (7), it can be done by replacing l with n, and $v$ with $\mu$, and let $t=1,\cdots,k-1$. The details are omitted here.

## ARCHITECTURE FOR MODELING OF ECONOMIC CYCLES

Because of the ecnomic reform and opendoor policy of China, our economy is changing from the central planned economy to market economy and is being in high speed growth period, economic fluctuation and instability can easily occur. Thus, macro economic forecasting and early warning become the most important task of acroeconomic management.

Economic cycle research has been continued for almost 100 years, but this problem has not been solved completely till now, especially forecasting the cyclical turning points is an important problem [8]. But the current approaches are based on precise forecasting method, such as econometric models, ARIMA model etc.. The common characteristics of these models is that the human factors are excluded from them. This results in its weakness in economic cycle forecasting.

Let's review the conventional method for economic cycle research briefly. According to economic cycle theory, any economic cycle indictors (ECI) is composed of four parts: Trend(T), Seasonality(S), Cyclical factor(C) and Irregular factor(I). Therefore, utilizing different models, the indictor series Y can be represented as:

addition model: $\qquad Y=T+S+C+I$

or multiplication model: $\qquad Y=T \times S \times C \times I$

Then through seasonal adjustment (such X-11 method), S and I can be removed, and we can get TC series. Depending on the basic cycle selected and with some constraints, the peaks and troughs of each ECI can be

1854

determind. Based on the above results, all ECI's can be classified into three groups of leading, roughly coincident and lagging indictors. Forecasting cyclical turning points often uses the relation between leading and coincident indictors. According to the average periods of past cycles, and the current situation, the future turning point can be determind. Some times, an ARIMA model should be used for extrapolation.

In China, the above method is the basic one at present. But because the practical effects are very limit, it makes economic early warning research faces severe difficulties. We think, the reasons for that are:
(1) Understanding of economic system is rather limit
(2) Statistical data is inexact and incomplete.
(3) Qualitative information is excluded from the forecasting method, especially the economic survey data cann't be applied together with statistical data.
(4) The method for forecasting cyclical turning points is too rough, and its functions are too weak.

Because of the above drawbacks, we provide an architecture for MEEW using qualitative modeling of dynamical system (Fig.1). We don't plan to give the details here, because it can be seen easily from Fig.1. However, how to obtain and express qualitative and quantitative information for modeling, and how to interpret simulation results to decision maker, will be discussed below.

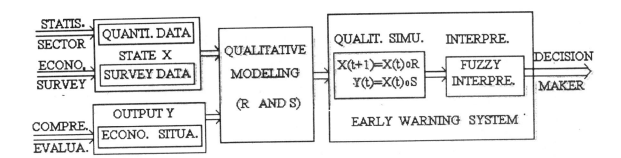

Fig.1 Macro Economic Early Warning

Assume system state variables are a set of ECI's, composed of two parts. One is quantitative data (generally statistical data) obtained from government statistical sector, such as: Industrial Production Growth Rate(IPGR), Total Investments (TI) and Saving (S) etc.. The other is survey data. Economic cycle survey (ECS) is another efficient approach for economic cycle research [7]. ECS's ask questions similar to economic statistics but mainly in qualitative term. In addition, these surveys focus on the coverage of the following kinds of information:
- Judgements (e.g., assessment of an enterprise's business situation or a household's financial situation).
- Anticipation (Plans and expectations of businessmen and consumers).

This type of information can be expressed by fuzzy sets almost as the same as the output given below.

Quantitative data must be qualitativized and expressed by fuzzy sets. This fuzzification is derived from the comcept 'Order of Magnitude' [1]. Given $x_i^*$ for any $x_i(t) \in X(t) \in R^n$, $x/x^*$ is used to calculate the membership of $x_i(t)$ on (NL, NM, NS, ZO, PS, PM, PL). (Fig.2).

Here, $x_i^* = \underset{0 \leq t \leq K-1}{Min} |x_i(t)|/2$ . or it is selected directly by experts.

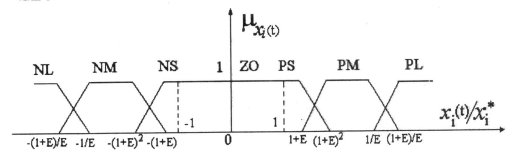

Fig.2 Membership of Quantitative Data

The output Y, a single variable, represents the situation of the entire economy. Of course, its historical values should be a set of values of comprehensive evaluations by some experts or by certain approach. More concretely, y is a fuzzy set defined on {Greatly Decrease (GD), Decrease (MD), Slightly Decrease (SD), Invariant (IN), Slightly Increase (SI), Moderately Increase (MI), Greatly Increase (GI)}. Moreover, we use the more generalized universe defined in section 2, i.e., (NL, NM, NS, ZO, PS, PM, PL). Therefore, if we get the value of y at time $t^*$:

$$Y(t^*)=(0, 0.1, 0.1, 0.2, 0.2, 0.2, 0.6, 0.1)$$

we can say the entire economy increases moderately rapidly.

After building the qualitative model of economic cycles based on fuzzy relations using the method described in section 2, Macro Economic Early Warning (MEEW) is a qualitative simulation (Fig.2). But the results y's are also fuzzy sets, which shouldn't be ouput to users directly, and should be interpreted before. It can be realized by a carefully designed interpretation block and need further research.

## SUMMARY

Qualitative modeling is a challenge direction for morden system research, and MEEW is a challenge field for qualitative modeling. There are two aspects of importances for applying qualitative modeling to MEEW, i.e.,
- Improve the method of MEEW
- Promote the human understanding of economic cycles.

Of course, there are many things to do.

## References

[1] D. S. Weld and de Kleer, eds, Readings in Qualitative Reasoning about Physical Systems, Los Altos, CA, 1990, Morgan Kaufmann

[2] W. Pedrycz, Fuzzy Modelling: Fondamentals, Construction and Evaluation, Fuzzy Sets And Systems, Vol.41, No.1, May.6, 1991

[3] P. A. Fishwick, Fuzzy Simulation: Specifying and Identifying Qualitative Models, Int. J. General Systems, Vol.19, No.3, 1991

[4] T. Söderström and P. Stoica, System Identification, Prentice Hall, 1989

[5] W. Pedryz, Numerical And Applicational Aspects of Fuzzy Relational Equations, Fuzzy Sets and Systems, Vol.11, No 28. 1983

[6] N. Ikoma and K. Hirota, Nonlinear Autoregressive Model Based on Fuzzy Relation, Information Sciences, Vol.71, 1993

[7] W. H. Strigel, Business Cycle Surveys: A New Quality in Economic Statistics, in P. A. Klein eds., Analyzing Morden Business Cycles, 1991

[8] A. H. Westlund. ed., Special Is sue on Business Cycle Forecasting, Journal of Forecasting, Vol.12, No.3-4, April, 1993

[9] Fen Jichen etc. eds., Fuzzy Pattern Recognition, Hebei Science and Technology Publishing Co., 1992(In CHINESE)

# Fuzzy Interpolation Of Hydro Power Sales Data In Simulink

Jan Jantzen
Technical University of Denmark
Building 325, DK-2800 Lyngby, DENMARK
Bitnet/earn: stardust@vm.uni-c.dk


Bo Eliasson
Sydkraft AB
S-205 09, Malmö, SWEDEN

*Abstract*—The problem in this case study can be described as a multi-dimensional surface fit to a given set of data. The data are sales figures in MWH/H for a hydro-thermal power generation system. The data are incomplete and not totally reliable. A model with ten fuzzy rules fits the data with a total error of 19%; with twenty-six rules the error would be 0%. The model is used in the long term planning concerning operating costs. It is further intended as a short term decision support for operators when negotiating power interchange contracts.

## I. INTRODUCTION

Before the high load season the water reservoirs in the Nordic countries for power generation are usually full. From winter to spring the water level gradually sinks to a minimum of, say, 15% to 30%. When the snow in the mountains melt, the spring flood sets in, and the water rises to its peak level sometime in the autumn. There are two main goals of water reservoir management: 1) to prevent the reservoirs from drying out prior to the spring flood, and 2) to fill the reservoirs in the autumn without spilling water. Excess water, especially in a "wet year" with plenty of rain or snow storage, may be converted to power at virtually no cost except taxes and sold to other power companies, neighbouring countries, or, in the near future, on a European "power exchange".

The energy stored in reservoirs is measured frequently — statistics for Norway, Finland and Sweden, for instance, are published quarterly — and the purpose of a model is to predict the amount of power for sale depending on variables like storage and precipitation. Such a prediction is especially useful in the long term planning of how to operate generating units (nuclear, fossil, gas, hydro, purchase) in order to meet the predicted consumer load ("unit commitment").

A local expert has supplied typical sales figures based on historical data and subjective judgment. The estimated sales figures depend on reservoir level (measured in percentages of full scale), month of the year, rain- and snowfall (in percentages above or below normal precipitation), and home power consumption (in MWH/H above or below normal). The estimates are precise numbers, but incomplete and not totally reliable.

To build a model the most straight forward way would be to use linear interpolation. The estimates could be placed in an array and missing elements filled in. The array would be multi-dimensional, and the interpolation procedure would be accordingly complex. A second possibility would be statistical or model identification methods, hoping to find some sort of correlation between the input and output variables. A third possibility would be a mathematical model based on first principles and problem insight. A fourth possibility might be to feed historical data into a neural network, and have it identify the relationship between the five variables.

The multivariable problem and the incompleteness of the data motivated a fuzzy rule-based model. The other possibilities mentioned above have not been investigated since we were looking for a demonstration problem for fuzzy logic. For maximal end-user control of the design phase, a prototyping approach was adopted using the simulation environment Simulink (Mathworks, 1993).

## II. INTERPOLATION MODEL

Fuzzy rules are appropriate in a number of cases within the problem area. One case is in the evaluation of the home power consumption compared to normal. An example of a set of rules is

If consumption is *above normal* then sell 100 MWH/H less
If consumption is *below normal* then sell 100 MWH/H more
$$(1)$$

The shape of the fuzzy membership functions, related to the fuzzy terms in italics, directly determine each rule's contribution to the sales measured in mega-watt-hours per hour (MWH/H). In a conventional program, the relationship between consumption on the one hand and sales on the other would probably be implemented as a piece-wise linear table function with interpolation. The major advantage of using fuzzy rules here is the rule format which is usually considered more user-friendly.

Another case is in the evaluation of the hydro for sale depending on water level and the time to spring flood:

If level is *above reference* and time to spring flood is *large* then sales are 200 MWH/H
If level is *above reference* and time to spring flood is *small* then sales are 800 MWH/H
... (2)

In this case there are two input variables to the rule block: *level* and *time to spring flood*. There is one output variable: *sales*. A fuzzy program will handle all possible combinations of input data and *interpolate* the sales between 0 and 800 MWH/H. This multivariable relationship is much harder to implement in a non-fuzzy program.

A third case is in the evaluation of the rain- and snowfall (precipitation). Several decisions depend on whether it is a *wet* year with up to 15 % more precipitation than usual, or not. The accumulated energy in the snow is a good indicator for wet and dry years. The rainfall is generally estimated.

The model is a four-input-one-output system. It contains eight rather similar rules and a rule block which takes care of the home power consumption (1). The latter relationship is simple because it is independent of the three other inputs, and the outcome is just added to the sales (Fig. 1). The rule block is named "Rules2", and it can be opened for further study by double clicking on the block (Fig. 2).

The rest of the model concerns the rather complex relationship between month, level, precipitation and sales. The overall design principle is that the relationship is mainly determined by eight characteristic cases (TABLE I). Each case holds original sales data for a combination of month and precipitation. The model is able to interpolate between the eight cases in the 'solution space', because the rules are fuzzy.

To be specific, the model assumes that the sales in January, May, June, and December are characteristic; the rest of the year is modelled as a blend of these, two-by-two. January is characteristic because sales are low here. This is probably due to fear of drying out the reservoir before the spring flood sets in. May is characteristic, because sales are rather high. This is because the spring flood is very close, and the risk of drying out is small. June is also characteristic, because sales are very reluctant, partly due to the risk of not filling the reservoir, partly due to government regulations concerning the environment. December is characteristic

because sales are very high; this is probably due to fear of spilling water.

The eight rules of the model are specified in a compact manner in TABLE 1. Each column represents a variable and each row a rule. In the familiar if-then format the first rule, for instance, would read:

If the *month* is January and it is not a *wet* year then sales would be [some number depending on **level**] (3)

### TABLE I
#### SALES STRATEGY

| MONTH | WET | LEVEL [%] SALES [MWH/H] | | |
|-------|-----|------|------|------|
| Jan | No | 70 0 | 75 210 | 85 200 |
| Jan | Yes | 75 400 | 85 600 | |
| May | No | 35 0 | 40 300 | 55 900 |
| May | Yes | 35 225 | 40 500 | |
| Jun | No | 40 0 | 60 100 | 70 900 |
| Jun | Yes | 40 100 | 60 500 | |
| Dec | No | 80 0 | 85 250 | 90 800 |
| Dec | Yes | 80 200 | 85 300 | 90 900 |

The variables MONTH and WET (year) are regarded as independent, and SALES as the dependent variable. The sales are assumed linearly dependent on level.

The resultant sales value is computed in a manner similar to the inference in a fuzzy controller (see, e.g., [1]), only slightly modified. The contribution of each rule is computed as

$$\text{JAN}(t) \text{ AND NOT WET}(p) = w_1$$
$$\text{JAN}(t) \text{ AND WET}(p) = w_2$$
$$\text{MAY}(t) \text{ AND NOT WET}(p) = w_3$$
$$... \qquad (4)$$

Here $t$ is the time of year, $p$ is the precipitation, and the $w_i$'s are the weights, or firing strengths, of the rules. The sales number $s_i$ from the $i$th rule is

$$s_i = \mathbf{f}_i(level) \qquad (5)$$

where $\mathbf{f}_i$ is a linear interpolation function using the $i$th interpolation table in the right column of TABLE I. The resulting sales number $s$ is the weighted average of all contributions, i.e.,

$$s = (\Sigma w_i \, s_i) \, / \, \Sigma w_i \qquad (6)$$

For example, if the input is $t = 1$ for *january* and $p = 0$ %, then the first rule will contribute, but none of the others will. The values of the MONTH variable are defined as

**jan** = 1 0.8 0.5 0.3 0 0 0 0 0 0 0 0
**may** = 0 0.2 0.5 0.7 1 0 0 0 0 0 0 0
**jun** = 0 0 0 0 0 1 0.9 0.5 0.4 0.2 0 0
**dec** = 0 0 0 0 0 0 0.1 0.5 0.6 0.8 1 1

If $t = 3$ for *March* and $p = 10$ %, more rules will contribute: Since the month is neither January nor May the model decides that it is January to the degree 0.5 and May to the degree 0.5. Similarly, it is a wet year in the degree 0.75 and *not* a wet year in the degree 0.25. In fact, all of rules 1 - 4 will contribute. The fuzzy membership curves for the MONTH variable have been handtuned to get the best fit to the original data, more or less. The membership curve for WET is a standard s-curve.

In the "Rules1" block, the input *level* is immediately compared to a reference curve on which the sales are zero (Fig. 3). Instead of using the absolute level the model computes the surplus relative to the reference curve, a number between 0 and 30 %, and uses this in further calculations. This is a normalisation that makes it easier to compare and interpolate sales in different months.

Some of the fuzzy membership values are being combined using fuzzy operations. These have been implemented as blocks (for example **AND, OR, NOT**) using the Simulink facilities. All fuzzy blocks are collected in a library (Fig. 4). The first row of blocks are various ways to implement fuzzy sets. The blocks have generic names like 's-curve', 'pi-curve', etc. to make them general. The names are easily changed into, say, 'low' and 'ok' by clicking on the names (not the block). The middle row of blocks concerns operations on sets. The common set operations 'and', 'or', 'not' are implemented quite easily in Simulink. For completeness the modifiers 'very' and 'rather' have been included. The last row contains some auxiliary functions that the model uses. They are not fuzzy operations, but necessary functions that did not exist in the standard Simulink library.

## III. Test Results And Discussion

Since the problem at hand is essentially a data fitting problem, it is possible to evaluate the accuracy by comparing the model's output against the original data. These are conveniently arranged in two matrices: one for a normal year and one for a wet year. Since they are incomplete there are a lot of empty cells.

The non-empty cells have been compared with the model's output. Summing all differences numerically, the model is off by a total error of 19%. The average error for each of the 55 non-empty cells is 67 MWH/H.

This result could perhaps be improved by optimizing the fuzzy membership curves, but probably only slightly. It will definitely help to increase the number of rules instead. The extreme case of 2 x 12 rules, one for each month in the wet / normal cases, plus the two rules for home consumption (1) will result in 0% error.

To get a picture of how the interpolation performs, Fig. 5 is a plot of an interpolated curve against some tabulated points. The plot concerns a fixed surplus of water (10 % over the reference), a fixed precipitation (0% precipitation over normal), and a fixed home consumption (0 % of normal), so the sales vary with the time of the year only. The plot shows that the interpolation is anchored in the values for January, May, June and December. The values in between are determined by the blending of two neighbouring anchor point values with a blending ratio governed by the fuzzy membership functions.

The model is fairly easy to use because of the graphical layout. Nevertheless, unskilled users have to be trained somewhat in Simulink and perhaps also Matlab. Basic knowledge of fuzzy sets is necessary also.

Some useful extensions are possible, including

- a dynamic model of the reservoirs including alternative regulation policies;
- integration with optimization programs (unit commitment, power flow).

Knowledge based decision systems for operators, e.g., voltage collapse constraints, monitoring of certain cut sets, are multi-input systems where the fuzzy techniques are of interest. The model is also useful for teaching purposes.

## IV. Conclusion

The case study shows that it is rather easy to model the complex relationship between the sales numbers and the four independent variable. The accuracy of the fit depends on the number of rules. Since a conventional linear interpolation program would provide a fit with zero error, the benefit of the fuzzy solution is not in the performance, but in the rule-based structure: it is user-friendly, and it is easy to add new rules — even new variables are quite easy to add.

ACKNOWLEDGEMENT

Mårten Eriksson, Sydkraft, has supplied the data (although they have been changed for publication) and expert knowledge. His insight and participation is gratefully appreciated.

REFERENCES

[1]     D. Driankov, H. Hellendoorn, and M. Reinfrank, *An Introduction To Fuzzy Control*. Berlin: Springer-Verlag, 1993.

[2]     Mathworks, *Simulink User's Guide*. The Mathworks, Inc., Cochituate Place, 24 Prime Park Way, Natick, MA 01760, USA, 1993.

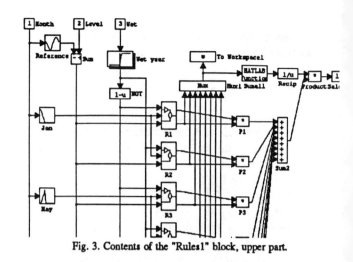

Fig. 3. Contents of the "Rules1" block, upper part.

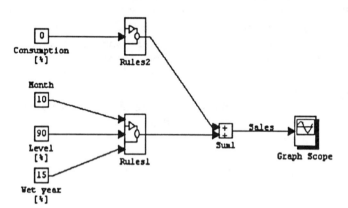

Fig. 1. Overall structure of the interpolation model.

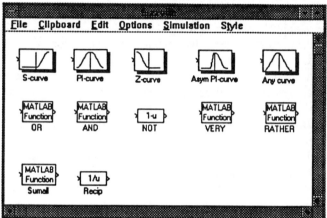

Fig. 4. Library of fuzzy blocks.

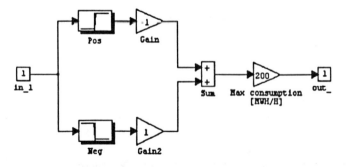

Fig. 2. Contents of the "Rules2" block concerning home consumption.

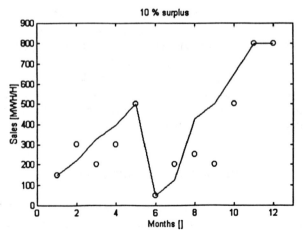

Fig. 5. Interpolated sales curve compared to tabulated points (circles).

# MODELING OF A FUZZY SYSTEM IN GEOTECHNICAL ENGINEERING

TUAN D. PHAM

School of Civil Engineering, University of New South Wales
P.O. Box 1, Kensington N.S.W. 2033, Australia

ABSTRACT: The modeling of fuzziness in geotechnical problems is addressed herein using the finite element method and the theory of fuzzy sets. The operations on fuzzy numbers are introduced in the formulation of finite element calculations. Fuzzy finite element solutions will be presented with lower and upper bounds or possibility distributions. These possibility distributions are investigated using fuzzy entropies. Fuzzy measure is also applied to introduce an estimate of uncertainty propagation in the fuzzy geotechnical system.

## 1. INTRODUCTION

Soil properties are not completely known in practice; hence the sense of exactness is doubtful in the analysis of geotechnical structures. Since the statistical data of soils are hardly obtained and regardless of the magnitude of testing programs, engineering judgment certainly plays an important role in the design process. Subjective estimates of geotechnical parameters such as elastic modulus and Poisson's ratio are assessed and fuzzified in the finite element analysis of an elastic soil medium subjected to a load at the surface. The results will be presented in terms of possibility distributions which show the lower and upper bounds of displacements and stresses at certain membership grades.

The fuzzy behavior of the soil medium is further investigated by considering the fuzzy entropies of the possibility distributions. Finally, Sugeno's fuzzy measure is used to estimate the error propagation in the finite element solution, which may be caused by impreciseness in choosing the geotechnical parameters and in the mesh discretization.

## 2. FUZZY-SET MODELING OF GEOTECHNICAL PARAMETERS

Following the formulations of the membership functions proposed by Zadeh (1976), the elastic modulus and Poisson's ratio are modelled as fuzzy sets whose membership functions may take the form either triangular function or $\pi$-function (Valliappan & Pham 1993a). In this paper, the elastic modulus and the Poisson's ratio of soil are characterized by convex fuzzy sets with triangular membership functions which can be defined by

$$\mu(x) = \begin{cases} 0 & for \ x \leq L' \\ \dfrac{x-L'}{P-L'} & for \ L' \leq x \leq P \\ \dfrac{H'-x}{H'-P} & for \ P \leq x \leq H' \\ 0 & for \ x \geq H' \end{cases} \tag{1}$$

where

$$L' = \begin{cases} P-2(P-L) \ for \ P \geq 2(P-L) \\ 0 \qquad\qquad for \ P \leq 2(P-L) \end{cases}$$

$$H' = P + 2(H-P)$$

in which L, P and H are expert estimates of low, most likely and high values respectively.

## 3. FUZZIFICATION OF THE ELASTICITY MATRIX

In finite element analysis, the load-displacement relation is written as

$$[K]\,\{\delta\} = \{P\}$$

(2)

where $[K]$ is the stiffness matrix, $\{\delta\}$ is the nodal displacement vector, and $\{P\}$ is the nodal force vector.

In two-dimensional finite element analysis, the stiffness matrix of an element is expressed as

$$[k] = \iint [B]^T [D][B]\ t\ dx\ dy$$

(3)

where $[B]$ is the strain-displacement matrix, $[D]$ is the elasticity matrix and t is the element thickness.

For plane-strain idealization in an isotropic material, $[D]$ will be fuzzified to give the lower and upper bounds of displacements using the operations of fuzzy numbers (Kaufmann & Gupta 1985) at an $\alpha$-level as follows.

$$[D]_L^\alpha = \frac{E}{(1+v_L^\alpha)(1-2v_R^\alpha)}\begin{bmatrix} 1-v_L^\alpha & v_R^\alpha & 0 \\ v_R^\alpha & 1-v_L^\alpha & 0 \\ 0 & 0 & \dfrac{1-2v_R^\alpha}{2} \end{bmatrix}$$

(4a)

$$[D]_R^\alpha = \frac{E}{(1+v_R^\alpha)(1-2v_L^\alpha)}\begin{bmatrix} 1-v_R^\alpha & v_L^\alpha & 0 \\ v_L^\alpha & 1-v_R^\alpha & 0 \\ 0 & 0 & \dfrac{1-2v_L^\alpha}{2} \end{bmatrix}$$

(4b)

where L and R stand for the left and the right values of a fuzzy number respectively. E and $v$ are elastic modulus and Poisson's ratio respectively. And with the understanding that the elements $D_{33}$ in the above two matrices in Eqs.(4a&b) are derived to yield fuzzy shear moduli.

## 4. FUZZY ENTROPIES OF FUZZY FINITE ELEMENT SOLUTIONS

Some of the procedures for measures of fuzziness are applied herein to show insight into the behavior of fuzzy finite element solutions. Different procedures are presented as follows.

Fuzzy entropy or measure of fuzziness of a fuzzy set, which was introduced by De Luca and Termini (1972) is defined by

$$d(A) = K \sum_{h=1}^{N} S(\mu(x_h))$$

(5)

where $d(A)$ is the fuzzy entropy of a fuzzy set A, K is a constant and $S(\mu(x_h))$ is expressed using Shannon's function:

$$S(\mu(x_h)) = -\mu(x_h)\ \ln \mu(x_h) - (1-\mu(x_h))\ \ln\ (1-\mu(x_h))$$

(6)

with the assumption that $0 \ln 0 = 0$.

The normalized version of $d(A)$ is obtained by the functional

$$d_n(A) = \frac{1}{N} \sum_{h=1}^{N} S(\mu(x_h))$$

(7)

where $S(\mu(x_h))$ is now expressed in logarithm in base 2.

Another measure of fuzziness is also suggested in terms of the Minkowski class of metric distances (Klir &

Folger 1988):

$$d_p(A) = |X|^{1/p} - D_p(A, \bar{A}) \tag{8}$$

where $|X|$ denotes the cardinality of the universal set X (x $\varepsilon$ X), p $\varepsilon$ [1,$\infty$] and

$$D_p(A, \bar{A}) = \left[ \sum_{x \in X} \delta_A^p(x) \right]^{1/p} \tag{9}$$

where

$$\delta_A(x) = |\mu_A(x) - \mu_{\bar{A}}(x)|$$

It is noted that when p= 1 or 2, Eq.(8) involves the Hamming distance or the Euclidean distance respectively. The measure of fuzziness $d_p(A)$ can be normalized as

$$d_{n,p}(A) = 1 - \frac{D_p(A, \bar{A})}{|X|^{1/p}} \tag{10}$$

## 5. FUZZY MEASURE OF ERROR PROPAGATION

### 5.1. *Fuzzy measures*

Some relevant definitions of Sugeno's fuzzy measures (Sugeno 1977) are presented as follows.

Let X be an arbitrary set and $\mathfrak{B}$ a Borel field of X. For all A, B $\varepsilon$ $\mathfrak{B}$ and A $\cap$ B = $\varnothing$, then a $\lambda$-rule which measures the union of A and B is defined by

$$g(A \cup B) = g(A) + g(B) + \lambda\ g(A)g(B)\ ,\quad \lambda > -1 \tag{11}$$

Let X = {$x_1$, ..., $x_n$} be a finite set, and $g^i = g(\{x^i\})$ for i=1,n be the values of g over the set of singletons of X. $g^i$ will also be called a fuzzy density function of $g_\lambda$. Assume that A = {$x_{i1}$, ..., $x_{im}$} $\subseteq$ X, then g(A) can be expressed as

$$g(A) = \sum_{j=1}^m g^{ij} + \lambda \sum_{j=1}^{m-1} \sum_{k=j+1}^m g^{ij} g^{ik} + ... + \lambda^{m-1} g^{il}\ ...\ g^{im} = \frac{1}{\lambda} \left[ \prod_{x_i \in A} (1+\lambda g^i) - 1 \right],\ \lambda \neq 0 \tag{12}$$

With g(A)=1 and if the fuzzy densities $g^i$ are known, then the $g_\lambda$-fuzzy measure can be constructed by solving the equation:

$$\lambda + 1 = \prod_{i=1}^n (1+\lambda g^i) \tag{13}$$

### 5.2. *Estimate of error propagation*

The three main factors which govern the reliability of the fuzzy finite element method of an elastic soil medium may be listed as:

1. Reliability in expert estimate of E and its shape of membership function.
2. Reliability in expert estimate of $\nu$ and its shape of membership function.
3. Reliability in mesh discretization.

If other factors can be negligible and the weights of contribution to the accuracy of finite element solutions for the above three factors are assigned in such a way:

$$w(E) + w(\nu) + w(M) = 1 \tag{14}$$

where

w(E): weight of contribution relating to E,  w(v): weight of contribution relating to v.

w(M): weight of contribution relating to mesh discretization.

Let R={E, v, M} and $\wp$(R)={$\varnothing$, {E}, {v}, {M}, {E,v}, {E,M}, {v,M}, {E,v,M}} be the power set of R.

If the value of each weight can be determined either subjectively or objectively, then our interest is to establish the fuzzy measures, which imply the "grades of efficiency" to the solution accuracy, on $\wp$(R), ie. $g(\varnothing)$, $g_\lambda$({E}), $g_\lambda$({v}), $g_\lambda$({M}), $g_\lambda$({E,v}), $g_\lambda$({E,M}), $g_\lambda$({v,M}) and $g_\lambda$({E,v,M}). And these are called the grades of ideal efficiency. However, in real applications, the grades of actual efficiency are usually less than those of the ideal owing to the errors in the subjective estimates of material properties and in the mesh discretization, ie. $w_E + w_v + w_M < 1$. From this standpoint, the error rate of subjective estimates denoted by $\xi$ is defined by the following equation.

$$\xi = \frac{\sum_{i=1}^{r} w(F_i)\left( g(F_i) - g(F_i)^\bullet \right)}{\sum_{i=1}^{r} w(F_i)\, g(F_i)} \qquad (15)$$

where

$F_i \subseteq \wp$(R) excluding $\varnothing$ and {E,v,M},

$r = |\wp(R)| - 2 = 2^q - 2$, $q = |R|$ (q=3 in this case),

$w(F_i)$ is the weight of a subset F in $\wp$(R).

$g(F_i)$ is the grade of ideal efficiency of $F_i$, $g(F_i)^\bullet$ is the grade of actual efficiency of $F_i$, and $g(F_i) \geq g(F_i)^\bullet$.

The value of $\xi$ is set in the limit: $0 \leq \xi \leq 1$. $\xi = 0$ means that the estimates of E, v and M are totally reliable, while the increase of $\xi$ implies the magnitude of the error affecting the result accuracy with reference to the estimates of E, v and M.

## 6. NUMERICAL EXAMPLE

Consider the half symmetrical finite element mesh of a soil medium which consists of two layers with a soft soil being on top of a stiffer layer as shown in Fig.1. The estimates of soil properties in the form of low (L), most likely (P) and high values (H) are as follows.

1. First layer (soft soil):

$E_1$ (kPa)= 1500 (L), 2000 (P), 2600 (H). $v_1$= 0.2 (L), 0.3 (P), 0.4 (H).

2. Second layer (stiff soil):

$E_2$ (kPa)= 8000 (L), 10000 (P), 13500 (H). $v_2$= 0.2 (L), 0.3 (P), 0.4 (H).

The strip load is assumed to be 240 kN/m.

The membership functions of $E_1$, $v_1$, $E_2$ and $v_2$ are constructed using Eq.(1) (see Figs. 2-4).

Fig.5 shows the possible vertical displacements at the center along the depth for the membership grades of 0.8, 0.9 and 1. The displacement variations are greater in the top (softer soil) than the second layer. Fig.6 shows the variations of the vertical stresses about the center along the soil depth. The vertical stresses in the top soil layer are not much affected by the variations of the fuzzy soil parameters. The vertical stress variations in the stiffer soil are also not considerable.

Fig.7 and Fig.8 show the normalized fuzzy entropies of the vertical displacements and vertical stresses respectively in the soil mass where the measures of fuzziness using the Euclidean distance (EN), the Hamming distance (HG) (Eq.(10)) and De Luca & Termini's entropy (D&T) (Eq.(7)) may be taken as the lower bound, median and upper bound respectively.

To illustrate how $g_\lambda$-fuzzy measure is applied to calculate the rate of error propagation as discussed in Section 5, we consider the following case.

For the problem of foundation settlement, the determination of the displacements is of most importance. Thus, the geotechnical engineer would rank the contributions of E, v and M to the accuracy of the displacements obtained from the finite element analysis with different weights as w(E) = 0.5, w(v) = 0.2 and w(M) = 0.3. If all the weights are assumed to be equal to the fuzzy densities, then $g^1 + g^2 + g^3 = 1$; therefore, $\lambda$ is zero (calculated from Eq.(13)). Now according to the degree of experience in estimating the soil parameters, the intrinsic uncertainty of the membership-function shape, and the degree of the mesh accuracy (with reference to displacement error only); the engineer decides to reduce the fuzzy densities to the new values: $g^1 = 0.3$, $g^2 = 0.1$ and $g^3 = 0.2$. Again using Eq.(13) a new $\lambda$ can be obtained: $\lambda = 3.109$. Then new fuzzy measures $g_{-3.109}$ of all subsets $A_j$ can be calculated using Eq.(11). Finally the estimate of error propagation $\xi = 0.258$ which is calculated from Eq.(15).

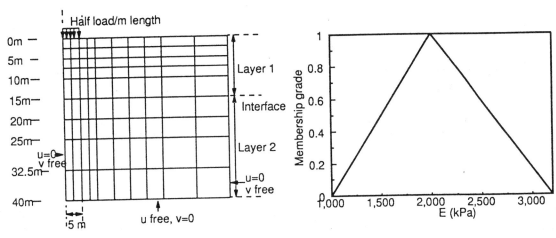

Fig.1 - A half symmetrical finite element mesh of a two-layered soil mass.

Fig.2 - Membership function of E for first soil layer

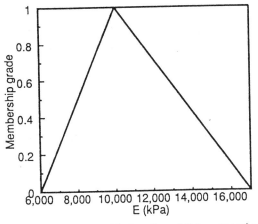

Fig.3 - Membership function of E for second soil layer

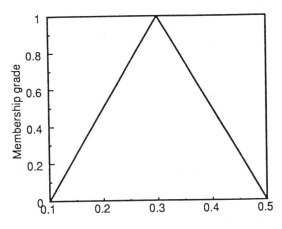

Fig.4 - Membership function of $\nu$ for both soil layers

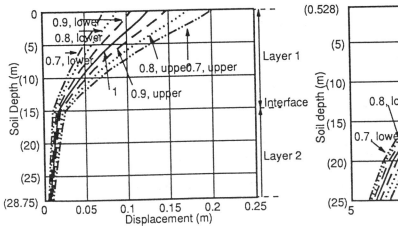

Fig. 5 - Vertical displacements below the center line of a strip load in a two-layered soil mass at different membership grades.

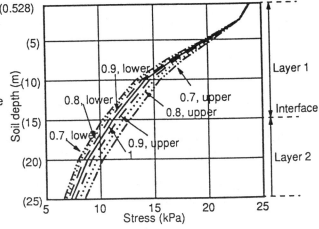

Fig.6 - Vertical stresses at Gaussian points below loading cente in a 2-layered soil mass at different membership grades.

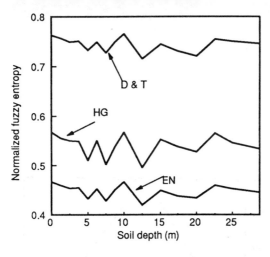

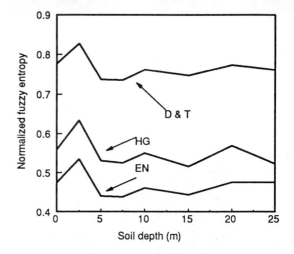

Fig.7 - Measures of fuzziness
of vertical displacements

Fig.8 - Measures of fuzziness
of vertical stresses

## 7. CONCLUSION

In the preceding sections of this paper, the mathematical aspects of fuzzy-set theory have been applied to the finite element analysis of an elastic soil medium in which the geotechnical parameters are subjected to imprecision. The results of displacements and stresses are presented with the possibility distributions which help as a guide-line to engineering design. In addition to the basic formulation of the fuzzy elastic matrix in the finite element method, we have attempted to elaborate the framework of the fuzzy finite element method by exploring the concepts of fuzzy entropies and Sugeno's fuzzy measures to study the possibility distributions and estimate the error propagation in the fuzzy geotechnical system respectively.

Fuzzy mathematical modeling has also been applied in elasto-plastic finite element analysis by Valliappan & Pham (1993b). In the light of this pilot study, the fuzzy finite element method would be pervasive in the analysis of complex civil engineering systems.

## ACKNOWLEDGEMENT

This work, under the direction of Professor S. Valliappan at the University of New South Wales, was financially supported by the Australian Research Council.

## REFERENCES

De Luca, A. & S. Termini 1972. A definition of a nonprobabilistic entropy in the setting of fuzzy sets theory. *Information and Control* 20: 301-312.

Kaufmann, A. & M.M. Gupta 1985. *Introduction to fuzzy arithmetic*. New York: Van Nostrand Reinhold.

Klir, G.J. & T.A. Folger 1988. *Fuzzy sets, uncertainty, and information*. Englewood Cliffs, NJ: Prentice-Hall.

Sugeno, M. 1977. Fuzzy measures and fuzzy integrals: A survey. In M.M. Gupta, G.N. Saridis & B.R. Gaines (eds.), *Fuzzy automata and decision processes*. Amsterdam, North-Holland.

Valliappan, S. & T.D. Pham (1993a). Fuzzy finite element analysis of a foundation on an elastic soil medium. To be published in *Int. J. Numerical and Analytical Methods in Geomechanics*.

Valliappan, S. & T.D. Pham (1993b). Elasto-plastic finite element analysis using fuzzy logic. Submitted to *Int. J. Numerical Methods in Engineering*.

Zadeh, L.A. 1976. A fuzzy-algorithmic approach to the definition of complex or imprecise concepts. *Int. J. Man-Machine Studies* 8: 249-291.

# The aggregation of information by examples via fuzzy sensors

Gilles Mauris, Eric Benoit, Laurent Foulloy

Laboratoire d'Automatique et de MicroInformatique Industrielle
LAMII / CESALP, Université de Savoie,
41 Avenue de la Plaine, BP 806,
74016 ANNECY CEDEX,
Tel : 50.67.56.11; Fax 50.57.48.85

*Abstract* **The problem of the aggregation of complementary information is a crucial point in the monitoring of large intelligent systems. This paper deals with the cases in which we do not have an analytical mathematic model to derive new information, nor of a rule-based model at our disposal, but only a few examples expressed in a linguistic manner by basic fuzzy sensors. In the first section, we explain within the frame of the fuzzy subset theory how the numeric-linguistic conversion is carried out inside a sensor, called then a fuzzy sensor. Next, the second section presents a method to extract rules from the examples studied, and how to obtain in this way a linguistic description of comfort from linguistic measurements of temperature and humidity. Finally, the advantages of our approach are pointed out as well as the developments to come to improve the process of linguistic modelling from examples.**

## I. INTRODUCTION

During the last ten years, the interest in multisensor intelligent systems has increased rapidly [1][2]. The reason for this is that the traditionally used single sensors cannot be used in a particular environment due to their physical limitations (e.g., acoustic sensors in space), while others are limited due to either technical or economic factors. The addition of extra sensors, providing redundant and complementary information, offers the capability of resolving most complex situations, and will lead to a richer description of the world. The use of a multisensor-system is particularly necessary in applications where the requirement is for the system to interact with and operate in an unstructured environment without the complete control of a human operator. Furthermore, existing multisensors are applied in the following areas of applications [3]: industrial tasks like equipment handling, part manufacturing, assembly inspection, military command and control, mobile robot navigation and control, ... .

If the benefit of multisensor systems seems promising (improved accuracy, increased detection range, enhanced reliability, tasks impossible to perform with individual sensors, shorter acquisition time, lower costs, ...), the ways to monitor such systems lead to many problems [2][4] : fusion of information, sensor monitoring, system issues.

The problem of the aggregation considered here is how to obtain a high level information from basic single measurements. To determine the operations leading to the acquisition of new information, we must know:

- first, what type of information we obtain from the elementary sensors,

- second, what type of information we want for the information resulting from the combination of the basic data.

- third, how the link between the basic information and the resulting information is expressed.

In fact, the process of aggregation is largely directed by the last aspect, which determines the method to use in order to combine the elementary information, according to the level of knowledge at our disposal. If we know an analytical relationship between the input and output information (i.e. a high level knowledge), we could use conventional statis-

---

. Manuscript received December 10, 1993.

tical methods [5][6]. When we have only sampled input-output numeric data pairs, neural methods are efficient to approximate the relation between them [7][8]. Our field of research is concerned with the situations where linguistic knowledge is involved in the process of aggregation. For these cases, artificial intelligence approaches based on the fuzzy subset theory are available [9] [10]. Previously, we have developed a rule based formalism to combine linguistic entities linked by a set of rules, generally obtained by human experience [11]. As regards the situations where the inputs are of numeric type and the output of a linguistic type, we have presented an interpolation method based on a fuzzy partition of the numerical multi-dimensional space of the basic features [12]. All of these approaches are summarized in the following table.

**Table 1: various approaches for the aggregation problem**

| inputs | outputs | relation | method |
|--------|---------|----------|--------|
| numeric | numeric | numeric | statistical |
| numeric | linguistic | linguistic | rule-based (symbolic view) |
| linguistic | linguistic | linguistic | rule-based (symbolic view) |
| numeric | numeric | linguistic | rule-based (numeric or symbolic view) |
| numeric | numeric | examples | interpolation (neural methods) |
| numeric | linguistic | examples | interpolation (fuzzy partition) |
| **linguistic** | **linguistic** | **examples** | **linguistic interpolation (fuzzy method)** |

This paper is concerned with the cases where the sampled input-output are both of a linguistic form (last line of the table). This situation often arises when the output is a not-measurable abstract feature (e.g. comfort, danger, shape, ...) and when the inputs, even though corresponding to a measurable physical entity, are also expressed in a linguistic manner. This occurs when they are provided by a human cue or by an intelligent component (e.g. a fuzzy sensor) that has already made a processing on the numeric information in order to give concise and relevant information to the central unit. After having recalled in the first section how a fuzzy sensor performs a linguistic description of a physical phenomenon from a numeric measurement inside itself, we present in section two a case-based reasoning method [13][14] to build a fuzzy knowledge base from a few examples. This fuzzy linguistic model of the relationship between the inputs and the output, thus obtained, is then used to derive the linguistic description of the output for any new measurements. As an example, the acquisition of comfort information from temperature and humidity observations is described.

## II. FUZZY SENSORS

The object of sensors is to select a particular physical quantity from the lot of received information. Generally, the description of a property or a feature of an object of the real world is made by the assignment of a numerical value. The numerical representation is precise, compact and based on an objective operational procedure. It provides a way of establishing scientific laws to explain complex phenomena. Nevertheless, in several cases, using a numerical description is not relevant, either because of problems of acquisition or storage of information, or because of the multidimensional nature of the analysed property, or because it concerns attributes describing human behaviour. In such cases, one is led to make a qualitative description of the observed phenomena with words of the natural language [15][16]. At first sight, linguistic symbolization seems less interesting than numerical symbolization. It is less precise and is subjective in the sense that it can not be experimentally checked in an objective operational procedure (i.e. independent from the observer). But the linguistic qualitative description is compact, easy to understand by human beings, and has a richness that the numeric measurement has not, i.e. background knowledge upon the global goal that started the measurement process. In fact, through the linguistic description, the data provided by the sensor move through ever more abstract interpretations, and therefore could easily be integrated at the reasoning level with the symbolic representation of tasks and actions.

Our approach is to implement these intelligent capabilities directly at the sensor level, where each numeric datum is related to a primitive logical concept, in order to reduce the computation load at the central unit. Moreover, in our opinion, the use of linguistic information at the lower level will improve a cooperative resolution of the overall problem using the solution of the subproblems available at autonomous sensor nodes. One way to provide a numeric-symbolic conversion is to associate a fuzzy set to each sensor in order to obtain a fuzzy measurement of some physical quantities. The interest of the fuzzy subset approach is twofold : it allows to take into account the continuous nature of the handled variables, inside the symbolic representation that is itself discrete, and it provides an effective way to process and to propagate measurement imprecisions and so to validate the information given back. Our view has led to the concept of *fuzzy symbolic sensors* or simply *fuzzy sensors* [17][18][19].

Hereafter, we briefly recall the mechanism of the numeric-linguistic conversion made in fuzzy sensors. Let us consider a set of linguistic terms $\mathcal{L}$. Let $\mathcal{E}$ be the set of all possible measurements. A fuzzy mapping, called a *meaning* [20], associates any term to a fuzzy subset of measurements. The fuzzy meaning of a symbol $L$ is characterized by its membership function noted as $\mu_{\tau(L)}(x)$. Examples of the membership function of the fuzzy meaning for many terms of temperature and humidity are given in figure 1.Conversely, a fuzzy description can be defined as a fuzzy mapping from the measurement set $\mathcal{E}$ in the set of the fuzzy subsets of symbols, so the fuzzy description is characterized by its membership function noted as $\mu_{\iota(x)}(L)$. There is a fundamental relation between the fuzzy description and the fuzzy meaning, that allows to deduce the linguistic description of a measurement from the fuzzy meanings located in the knowledge base of a fuzzy sensor (fig. 1) : $\mu_{\iota(x)}(L)=\mu_{\tau(L)}(x)$.

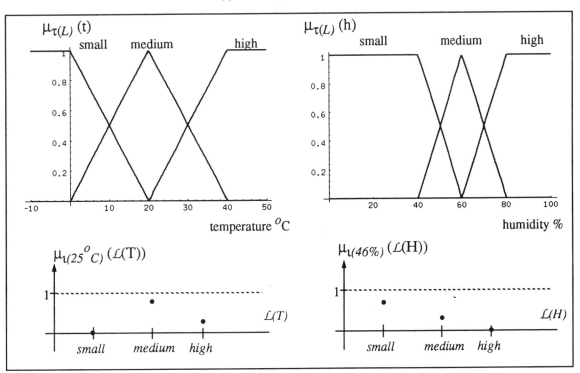

*Figure 1 : Examples of fuzzy meanings and fuzzy descriptions.*

## III. AGGREGATION FROM LINGUISTIC EXAMPLES

Let us recall that we consider the situations, where there is no mathematical relation between the inputs and the output, and also no linguistic rule base, but only the knowledge of a few linguistic examples provided by fuzzy sensors. Our approach consists in generating fuzzy rules from the examples in a manner similar as the one described in [20], and then in building a linguistic graph linking the input terms to the output terms. Next, the description of a new situation is obtained by computing the image of the new input with the graph using the combination/projection principle.

For the sake of simplicity, let us illustrate our approach on the case of comfort description from temperature and humidity measurements, with the following items for the considered linguistic sets : $\mathcal{L}$(T)={low, medium, high}; $\mathcal{L}$(H)={low, medium, high}; $\mathcal{L}$(C)={frosty, pleasant, muggy}. Let us consider the examples described in table 2.

### Tableau 2 : Examples used to create the fuzzy rule base

| Temperature | Humidity | Comfort |
|---|---|---|
| 0,75/medium + 0,25/high (25$^o$C) | 0,7/medium + 0,3/high (66%) | 0,8/pleasant + 0,2/muggy |
| 0,5/medium + 0,5/high (30$^o$C) | 0,25/medium + 0,75/high (75%) | 0,3/pleasant + 0,7/muggy |
| 0,8/low +0,2/medium (20$^o$C) | 1/low (40%) | 0,9/frosty +0,1/pleasant |

To generate a rule from an example, we propose for instance to consider for each input and output variable only the term with the highest degree. With this method, examples corresponding to close measurements lead to a unique rule. For the cases where we have the same IF part but a different THEN part, we keep only the rule with the highest validity, which is defined as the product of the membership degrees to any of the terms with the highest degree in the premise and in the conclusion. In this way, we lose a little information, but the number of rules is greatly reduced. So, for the preceding examples, we obtain the four following rules, because the second example leads to two rules.

•If the temperature is medium and the humidity is medium then the atmosphere is pleasant.

•If the temperature is medium and the humidity is high then the atmosphere is muggy.

•If the temperature is high and the humidity is high then the atmosphere is muggy.

•If the temperature is low and the humidity is low then the atmosphere is frosty.

This linguistic rule base could be represented by a graph on the cartesian product $\mathcal{L}$(T) $\times$ $\mathcal{L}$(H) $\times$ $\mathcal{L}$(C).

### Tableau 3 : Graph linking the linguistic inputs and output

| Hum.\Temp. | Low | Medium | High |
|---|---|---|---|
| **Low** | Frosty | | |
| **Medium** | | Pleasant | |
| **High** | | Muggy | Muggy |

Now, we compute the image F of a fuzzy subset (E, E') of $\mathcal{L}$(T)$\times$ $\mathcal{L}$(H) by using the combination/projection principle (also referred to as generalized modus ponens) with the formula (1) where $S$ is a co-norm and $T_1$ a norm.

$$\mu_F(W)=S_{(V,V')\in \mathcal{L}(T)\times\mathcal{L}(H)} \ (\mu_{(E,E')}(V,V') \ T_1 \ \mu_\Gamma(V,V',W)). \tag{1}$$

It remains now to define $\mu_{(E,E')}(V,V')$ . If we make the assumption that the two variables are independent, $\mu_{(E,E')}(V,V')$ can be decomposed and we could write with $T_2$ an operator of intersection (i.e. a t-norm):

$$\mu_{(E,E')}(V,V')=\mu_E(V) \ T_2 \ \mu_{E'}(V'). \tag{2}$$

If we have at our disposal two numeric measuremnts $t$ and $h$ :

$$\mu_{(E,E')}(V,V')= \mu_{(t,h)} \ (V,V')=\mu_{(t)} \ (V) \ T_2 \ \mu_{(h)} \ (V')=\mu_{\tau(V)} \ (t) \ T_2 \ \mu_{\tau(V')} \ (h). \tag{3}$$

For example, let us consider in details the expression of $\mu_F$(muggy). In This case $\mu_\Gamma(V,V',W)$ is not equal to zero only for (V, V')=(medium, high) and (V, V')=(high, high) for whose $\mu_\Gamma$(V,V', muggy)=1. Thus :

$$\mu_F(\text{muggy})=S \ (\mu_{(E,E')}(\text{medium, high}) \ T_1 \ 1; \ \mu_{(E,E')}(\text{high, high}) \ T_1 \ 1 \ ). \tag{4}$$

So, $\mu_{(t,h)}$(muggy)=$S \ ((\mu_{\tau(\text{medium})} \ (t) \ T_2 \ \mu_{\tau(\text{high})} \ (h)) \ T_1 \ 1; \ (\mu_{\tau(\text{high})} \ (t) \ T_2 \ \mu_{\tau(\text{high})} \ (h)) \ T_1 \ 1 \ ). \tag{5}$

Hereafter the meanings of the comfort terms, built with the proposed method, are plotted with $S(a,b)=\min(a+b, 1)$ and $T_1(a,b)=T_2(a,b)=a.b$, and with the meanings described fig. 1 for the terms of temperature and humidity.

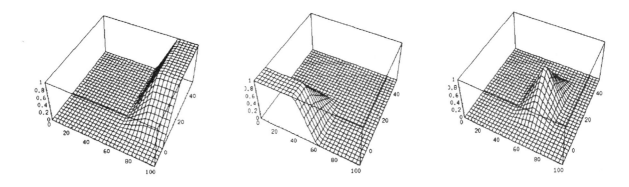

Fig. 2 Meanings of the obtained terms of comfort {muggy,frosty, pleasant}.

The results obtained provide a linguistic description for every point in the basic feature multi-dimensional space. But, in some areas, the degree of membership to any of the terms is zero. This corresponds to areas, where the considered examples are not active. The description could be improved by considering examples covering these areas, or if they are not available, by using directly rules provided by experts. This last opprotunity is simple to implement, because it consists only in filling in a box in the graph on $\mathcal{L}(T) \times \mathcal{L}(H) \times \mathcal{L}(C)$ with the corresponding rule.

## IV. CONCLUSION

In this paper, we have addressed the problem of the acquisition of high level information, i.e. properties or attributes related to more than one physical quantity in a non-analytical manner, only by using a few linguistic examples provided by fuzzy sensors. The proposed mechanism consists in generating a set of rules from the examples, and in building a graph linking the considered linguistic terms. The computation of the description of a new situation is then deduced from the combination/projection principle. The main features and advantages of the method presented are :

•it enables us to aggregate basic information when we only know a few linguistic examples,

•it is a simple procedure, that does not require a high computation load at the combination node, because it works on a small number of variables, since the numeric treatments are made at the sensor level,

•there is a possibility to simply take new examples or rules generated from expertsinto account.

Nevertheless, the way to create a rule from an example may seem abrupt, because we only keep the term with the maximum degree. An other way consists in keeping all the information included in each example, this time leading to the creation of a fuzzy graph. This promising method is being developed at present in our laboratory.

## V. REFERENCES

[1] Ren C. Luo, M.H. Lin, R.S. Scherp, *"Dynamic multisensor data fusion system for intelligent robots, "* IEEE Journal of robotics and automation, Vol. 4, No 4, Aug. 1988, pp. 386-396.

[2] J.K. Aggarwal, Y.F. Wang, *"Sensor fusion in robotics systems"*, Control and dynamic systems, Vol. 39,1991, pp. 435-462.

[3] Ren C. Luo, Michael G. Kay, *"Multisensor integration and fusion in intelligent systems, "* IEEE trans. on Systems, Man, and Cybernetics, Vol. 19, No 5, Oct. 1989, pp. 901-931.

[4] T.C. Henderson, E. Shilcrat, *"Logical sensor systems, "* Journal of robotic systems, 1984, pp. 169-193.

[5] S.S. Iyengar, R.L. Kayshyap, R.N. Madan, *"Distributed sensor networks - introduction to the special section, "* IEEE trans. on Systems, Man, and Cybernetics, Vol. 21, No 5, Sept-Oct. 1991, pp. 1027-1031.

[6] A. Lauber, *"Intelligent multi-sensor fusion"*, Int. Conf. on Fault Diagnosis, Toulouse (France), April 1993, pp. 140-144.

[7] X. Cui, K.G. Shin, *"Direct control and coordination using neural networks"*, IEEE trans. on Systems, Man, and Cybernetics, Vol. 23, No 3, Oct. 1989, pp. 686-697.

[8] T. Fukuda, K. Shimojima, F. Arai, H. Matsuura, *"Multi-sensor integration system based on fuzzy inference and neural network for industrial application"*, IEEE Int. Conf. on fuzzy systems, San Diego, USA, March 1992, pp. 907-914.

[9] T.D. Garvey, *"A survey of AI approaches to the the integration of information"*, Proc. of SPIE, Vol. 782, 1987, pp. 68-82.

[10] Foulloy L., Benoit E., Mauris G., *"Applications of fuzzy sensors"*, Proc. of the European Workshop on Industrial Fuzzy Control and Applications (IFCA 93), Terrassa, Spain, April 93.

[11] Foulloy L., Galichet S., *"Fuzzy Controllers Representation"*, Proc. of the 1st European Congress on Fuzzy Intelligent Technologies (EUFIT 93), Aachen, Germany, Sept. 93, pp. 142-148.

[12] E. Benoit, L. Foulloy, S. Galichet, G. Mauris, *"Fuzzy sensor for the perception of colour"*, submitted to the Third IEEE Int. Conf. on fuzzy systems, Orlando, USA, June 1994.

[13] P. Bonissone, S. Ayub, *"Similarity measures for case-based reasoning systems"*, Proc. of the Int. Conf. IPMU'92, Palma, Spain, July 1992, pp. 483-487.

[14] J.F. Baldwin, R. Ribeiro, *"Fuzzy reasoning by case for decision support systems"*, Proc. of the Int. Conf. IPMU'92, Palma, Spain, July 1992, pp. 479-481.

[15] Finkelstein L. *"Representation by symbol systems as an extension of the concept of measurement"*, Kybernetes, Vol. 4, pp. 215-223, 1975.

[16] Zingales G., Narduzzi C., *"The role of artificial intelligence in measurement"*, 8th Int. Symp. on Artificial Intelligence based Measurement and Control, Sept. 1991, Ritsumeikan University, Kyoto, Japan. pp. 3-12.

[17] Benoit E., Foulloy L., *"Symbolic sensors : one solution to the numerical-symbolic interface"*, Proc. of the IMACS International Workshop on Qualitative Reasoning and Decision Support Systems, Toulouse, France, March 1991.

[18] Foulloy L., *"Fuzzy Sensor for Fuzzy Control"*, Proc. of the Int. Conf. IPMU'92, Palma, Spain, July 1992, pp. 759-768.

[19] Mauris G., Benoit E., Foulloy L., *"Fuzzy symbolic sensors : From concept to applications"*, to appear in the Int. Journal of Measurement in 1994.

[20] L.X. Wang, J.M. Mendel, *"Generating fuzzy rules by learning from examples, "* IEEE trans. on Systems, Man, and Cybernetics, Vol. 22, No 6, Nov-Dec. 1992, pp. 1414-1427.

# MODELING DYNAMIC SOCIAL AND PSYCHOLOGICAL PROCESSES WITH FUZZY COGNITIVE MAPS

Philip Craiger
Organizational Systems Department
Navy Personnel Research and Development Center
San Diego, CA

Michael D. Coovert
Department of Psychology
University of South Florida
Tampa, FL

## Abstract

A fuzzy cognitive map (FCM) is a graphical means of representing arbitrarily complex models of interrelated concepts, the implications of which can be calculated simply via matrix algebra. This paper discusses the use of FCMs for providing qualitative information about complex models in social and behavioral science research; specifically we show how FCMs can be used to reveal implications of models comprised of dynamic social and psychological processes.

## I. Models of Social and Behavioral Processes

One of the most widely used statistical techniques in the social and behavioral sciences is structural equation modeling. Structural equation modeling allows researchers to define and test statistical models represented as networks of relationships among sets of latent (unobservable, hypothetical) variables (see Appendix 1). It is quite common, however, to encounter serious problems during the modeling process. For instance, non-convergence of solutions and impossible parameter estimates, such as negative error variances and standardized correlations greater than unity, are not uncommon, and their occurrence can often be attributed to problems with model identification. *Identification* concerns the question of whether the model and data constraints are sufficient to determine a unique set of parameter estimates[2]. *Empirical* underidentification occurs when the data are such that a unique set of parameter estimates is unobtainable, whereas *theoretical* underidentification occurs when the model specified does not allow for a determination of a unique set of parameter estimates[1]. One cause often associated with underidentification are reciprocal directional influences[4].

Below we describe fuzzy cognitive maps (FCM) and illustrate their application in the modeling of social and psychological processes. FCMs are particularly attractive as an alternative to classical statistical methods because they do not suffer from the limitations described above[3], and they facilitate the modeling of arbitrarily complex, dynamic processes. We argue that FCMs may be used in conjunction with classical statistical modeling techniques, as a precursor in the initial modeling building phase, or in place of these methods, as when it is desirable to explore and reveal model implications, but conditions are such that traditional modeling techniques are impractical.

## II. Fuzzy Cognitive Maps

Developed by Kosko[7-8], a FCM is a graphical means of representing directional influences among concepts (variables). FCMs can be used to evaluate the time-evolving effects of a model's directional influences on the global state of a model. The "fuzzy" indicates that FCMs are often comprised of concepts that can be represented as *fuzzy sets*. Fuzzy set theory was originally developed by Zadeh[10] as a means of representing and reasoning with vague and ambiguous information in mathematical terms. The reader is referred to Zadeh[10] and Kosko[8] for more information on fuzzy set theory.

A FCM is comprised of nodes (variables, concepts) and edges (directional arrows). Appendix 2 illustrates a FCM representing a model relating individual, organizational and extra-organizational variables to job turnover.[9]

1873

FCMs lack "traditional" statistical parameter estimates; real numbers estimated from sample data that indicate the strength of relationships between concepts. In simple FCMs directional influences take on trivalent values {-1, 0, +1}, where a -1 indicates a negative relation, 0 no relation, and +1 a positive relation. Because the directional influences are represented as all-or-none relationships, FCMs provide qualitative as opposed to quantitative information about relationships.

### III. Model Complexity and Implications

Model implications are determined by the *patterns* of the relationships represented in a model. To illustrate a simple case, the necessary implication of the model $A \rightarrow B \rightarrow C$ is that a change in $A$ results in a change in $B$ and a change in $C$. Now consider the FCM in Appendix 2. What are the implications of an increase in person/organization fit? What are the implications of simultaneous increases in stress and commitment, and decreases in satisfaction and involvement? It is very difficult to determine the effects because of the reciprocal (bi-directional) relationships and the number of relationships represented, the two ingredients that determine a model's complexity. It may be difficult, if not impossible, to determine the consequences of the spreading activation of "causality" in complex and dynamic models using ubiquitous "soft" science methodologies such as linear statistical models. In general, there is a negative relationship between the complexity of a model and our ability to determine the implications of the model; however, FCMs provide the ability to determine implications of a model regardless of the model's complexity[8]. In FCM nomenclature, model implications are revealed by *clamping* variables and using an iterative vector-matrix multiplication procedure to assess the effects of these "perturbations" on the state of a model. As such, an analogy can be drawn between clamping nodes and manipulating independent variables in the traditional psychological experiment.

If we designate independent variables as the rows and the dependent variables as the columns, we can transform a FCM into a *connection matrix*. A connection matrix is a specification of the relationships among concepts represented in matrix form; a square matrix with the number of cells equal to the number of latent variables squared. Appendix 3 represents the connection matrix for the turnover FCM.

The inference equation is:

$$I_{[i+1]} = O_{[i]} = \sum_{n=1}^{k} \sum_{m=1}^{k} I_{[m]} C_{[m.n]} \quad (1)$$

where $I$ is an input vector of $k$ elements, $O$ is an output vector of $k$ elements, and $C$ is a connection matrix of order $k \times k$. This equation calculates the inner (dot) product for each column of the connection matrix. The dot product defines the cosine of the angle between the vectors, which in statistical terms, is the correlation between the (normalized) vectors.

It is possible for the computed elements of the output vector to lie outside the range of valid bit values {-1, 0, 1}. To transform an out-of-range output vector element into an acceptable value for the input (bit) vector we apply the following threshold function:

$$I_{[m]} = \begin{cases} -1 & if \ x < 0 \\ 0 & if \ x = 0 \\ 1 & if \ x > 0 \end{cases} \quad (2)$$

Accordingly, to transform the output vector from iteration $i$ into an acceptable input (bit) vector for iteration $i+1$ requires a two-step process: (1) clamp any nodes turned on in the initial input vector, and (2) transform elements outside {-1, 0, 1} using the threshold function.

### IV. Hidden Patterns = Model Implications

In FCM nomenclature, a model implication or inference is a global stability, an equilibrium in the state of the system.[8] Repeating patterns can be *hidden patterns* or *limit cycles*. A hidden pattern is a single recurring pattern, whereas a limit cycle is a sequence of (multiple) repeating patterns. Hidden patterns and limit cycles reveal model inferences. Note that this equilibrium can be likened to the global minima of the maximum likelihood estimation procedure commonly employed in classical statistical methods. Hidden patterns, limit cycles, and global minima reveal the minimum energy state of the global system.

To illustrate the procedure, say we wish to evaluate the effect of family problems on our turnover model. We represent the effect of family problems by the following bit vector:

$$I_{[1]} = \{1 \ 0 \ 0 \ 0 \ 0 \ 0 \ 0 \ 0 \ 0 \ 0 \ 0 \ 0 \ 0 \ 0 \ 0\}$$

Each element of the vector represents the *state* of one of the concepts in our model.

1874

Thus, $I_{[1]} = 1$ indicates that the family problem variable is clamped, $I_{[2]} = 0$ and $I_{[3]} = 0$ indicate that job alternatives and economic conditions are off, and so on

Applying Eq. 1 results in the output vector:

$$O_{[1]} = \{0\ 0\ 0\ 0\ 0\ 0\ 1\ 1\ 1\ 0\ 0\ 0\ 0\ 0\ 0\ 0\}$$

An output vector can be interpreted as a "snapshot" of the global state of the model; this vector suggests that the initial effect of family problems is to increase absence, lateness, and stress.

Clamping node 1 and reapplying Eq. 1 results in:

$$O_{[2]} = \{0\ 0\ 0\ 0\ 0\ 0\ 1\ 1\ 1\ \text{-}1\ 0\ \text{-}1\ \text{-}1\ 0\ 0\ 0\}$$

The output vector now indicates that, in addition to increases in absence, lateness, and stress, we find decreases in job satisfaction, job performance, and involvement. The next step of the process reveals the spreading activation of family problems in the model:

$$O_{[3]} = \{0\ 0\ 0\ 0\ 0\ 0\ 1\ 1\ 1\ \text{-}1\ 0\ \text{-}1\ \text{-}1\ \text{-}1\ 1\ 0\}$$

The effect of family problems has now spread deeper into the model, now causing a decrease in job commitment, and an increase in leave intentions. Employing this process for two more iterations reveals a fixed point in phase space; a global stability in our model indicated by the following hidden pattern:

$$O_{[4]} = \{0\ 0\ 0\ 0\ 0\ 0\ 1\ 1\ 1\ \text{-}1\ \text{-}1\ \text{-}1\ \text{-}1\ \text{-}1\ 1\ 1\}$$

$$O_{[5]} = \{0\ 0\ 0\ 0\ 0\ 0\ 1\ 1\ 1\ \text{-}1\ \text{-}1\ \text{-}1\ \text{-}1\ \text{-}1\ 1\ 1\}$$

The procedure reveals the model's equilibrium state on the fourth iteration by the hidden pattern $O_{[4]}$: family problems results in decreases in job satisfaction, expectations, job performance, job involvement, and job commitment, and concomitant increases in lateness, absence, stress, leave intentions, and willingness to turnover. The hidden pattern is repeated *ad infinitum*. Note that it required four iterations to find an equilibrium. Research indicates that FCMs quickly come to an equilibrium, regardless of the complexity of the model[8].

V. Competing Concepts Model
The simple example above involving the clamping of a single variable illustrates the procedure by which FCMs reveal model

implications. A more interesting and useful example is one in which model variables "compete." For example, the effect of clamping family problems alone results in negative effects, such as increased stress, decreased job satisfaction, and decreased job performance. What would be the effect of clamping simultaneously a second variable that has positive implications, say, person/organization fit? Because the results are not readily apparent, clamping the two antithetical concepts allows the researcher to test the implications of one or more competing concepts.

To illustrate, we begin by clamping nodes one and four and apply Eq. 1, resulting in the following output vector:

$$O_{[1]} = \{0\ 0\ 0\ 0\ 0\ 0\ 1\ 1\ 0\ 1\ 0\ 1\ 0\ 1\ 1\ 0\ 0\ 0$$

The first iteration reveals the initial effects of family problems and person/organization fit are increases in absence and lateness, and increases in job satisfaction, job performance, and job involvement.
The equilibrium state is revealed by the hidden pattern resulting from the second iterative application of Eq 1:

$$O_{[2]} = \{0\ 0\ 0\ 0\ 0\ 0\ 0\ 0\ 0\ 1\ 1\ 1\ 1\ 1\ \text{-}1\ \text{-}1\}$$

$$O_{[3]} = \{0\ 0\ 0\ 0\ 0\ 0\ 0\ 0\ 0\ 1\ 1\ 1\ 1\ 1\ \text{-}1\ \text{-}1\}$$

$O_{[2]}$ suggests that the concomitant existence of family problems and person/organization fit causes increases in job satisfaction, job performance, involvement, and commitment, and decreases in leave intention and willingness to turnover.

Interestingly, our competing concept model reveals that the addition of a second competing variable (person/organization fit) essentially "negates" the effect of the first (family problems). As revealed in the first example, we expect family problems to elicit negative effects. The implications for the competing model, however, suggests positive results from clamping family problems and person/organization fit. Why might this be? Evidently, person/organization fit has a *greater effect on the state of the model by virtue of the way we have represented the relationships among the variables*. The differential impact of person/organization fit is not obvious because of the complexity of the model. The implication is that the effect of family problems is *moderated* by (depends

1875

on) person/organization fit. Thus, FCMs have the ability of modeling *both* mediator and moderator effects.

## VI. Discussion

FCMs exhibit a number of desirable properties that make it attractive as a supplemental process in model building for social and behavioral scientists:

- FCMs provide qualitative information about the (hidden, nonapparent) inferences in complex social and psychological models,
- providing a means of quickly mapping out and testing the implications of a model,
- can represent an unlimited number of reciprocal (bi-directional) relationships,
- do not require data,
- facilitate "thought experiments" by allowing the interactive manipulation of independent variables and evaluation of concomitant effects on the global state of a model,
- relationships need not be linear,
- can model both mediator and moderator relationships,
- there are no problems with empirical or theoretical identification,
- can be easily calculated by hand, and perhaps of most importance,
- FCMs facilitate the modeling of dynamic, time-evolving phenomena and processes.

A limitation is that FCMs provide only qualitative information regarding the predictions of emerging patterns; no real-value parameter estimates or statistical tests of significance are available. Thus, for the traditional methodologist, FCMs may not be an attractive alternative. Social and behavioral scientists should find, however, that FCMs are ideal for prototyping models, and for assessing the global implications of complex and dynamic social and behavioral processes.

[1] Brannick, M.T., & Spector, P. (1990). Estimation problems in the block-diagonal model of the multitrait-multimethod matrix. Applied Psychological Measurement, 14, 325-339.

[2] Coovert, M.D., Penner, L.A., & MacCallum, R. (1989). Covariance structure modeling in personality and social psychological research: An introduction. In C. Hendrick and M. Clark (Eds.), Review of Personality and Social Psychology: Research Methods in Personality and Social Psychology (Vol. II, pp. 291-330). Newbury Park: CA: Sage.

[3] Craiger, J. P. (April, 1994). Fuzzy cognitive maps and causal modeling. Paper to be presented at the Ninth Annual Meeting of the Society for Industrial and Organizational Psychology, Nashville, TN.

[4] Hayduk, L. (1987). Structural Equation Modeling with LISREL: Essential and Advances. Baltimore, MD: Johns Hopkins Press. 57, 239-251.

[5] Jöreskog, K. (1969). A general approach to maximum likelihood factor analysis. Psychometrika, 34, 183-202.

[6] Jöreskog, K. (1970). A general method for analysis of covariance structures. Biometrika, 57, 239-251.

[7] Kosko, B. (1986). Fuzzy cognitive maps. International Journal of Man-Machine Studies, 34, 65-75.

[8] Kosko, B. (1992). Neural Networks and Fuzzy Systems: A Dynamical Systems Approach to Machine Intelligence. Englewood Cliffs, NJ: Prentice Hall.

[9] Koslowsky, M. (1987). Antecedents and consequences of turnover: An integrated systems approach. Genetic, Social, and General Psychology Monographs, 113, 271-292.

[10] Zadeh, L. (1965). Fuzzy sets. Information and Control, 8, 338-353.

The underlying statistical model for structural equation modeling is[5-6]:

$$\eta = \beta\eta + \Gamma\xi + \zeta$$
$$x = \Lambda_x\eta + \Theta_\varepsilon$$
$$y = \Lambda_y\xi + \Theta_\delta$$

The first equation specifies how latent endogenous variables ($\eta$), latent exogenous variables ($\xi$), and errors in equations ($\zeta$) combine to form a causal network. The second and third equations are confirmatory factor analysis models that partition each measured variable into common or true variance ($\eta$ and $\xi$, modified by factor loadings $\Lambda_x$ and $\Lambda_y$, respectively) and associated error of measurement ($\theta_\varepsilon$ and $\theta_\delta$).

## Appendix 2

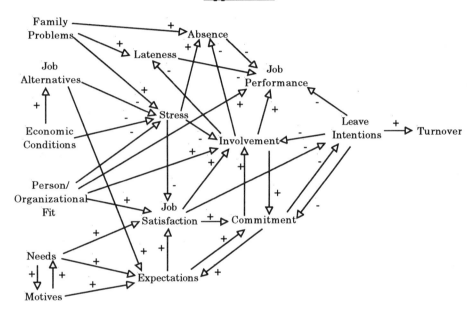

## Appendix 3

|  |  | 1 | 2 | 3 | 4 | 5 | 6 | 7 | 8 | 9 | 10 | 11 | 12 | 13 | 14 | 15 | 16 |
|---|---|---|---|---|---|---|---|---|---|---|---|---|---|---|---|---|---|
|  | 1 Family Problems | 0 | 0 | 0 | 0 | 0 | 0 | + | + | + | 0 | 0 | 0 | 0 | 0 | 0 | 0 |
|  | 2 Job Alternatives | 0 | 0 | 0 | 0 | 0 | 0 | 0 | 0 | − | 0 | + | 0 | 0 | 0 | 0 | 0 |
|  | 3 Economic Conditions | 0 | + | 0 | 0 | 0 | 0 | 0 | 0 | − | 0 | 0 | 0 | 0 | 0 | 0 | 0 |
|  | 4 Person/ Organ. Fit | 0 | 0 | 0 | 0 | 0 | 0 | 0 | 0 | − | + | 0 | + | + | 0 | 0 | 0 |
|  | 5 Needs | 0 | 0 | 0 | 0 | 0 | + | 0 | 0 | 0 | + | + | 0 | 0 | 0 | 0 | 0 |
|  | 6 Motives | 0 | 0 | 0 | 0 | + | 0 | 0 | 0 | 0 | 0 | + | 0 | 0 | 0 | 0 | 0 |
|  | 7 Absence | 0 | 0 | 0 | 0 | 0 | 0 | 0 | 0 | 0 | 0 | 0 | − | 0 | 0 | 0 | 0 |
|  | 8 Lateness | 0 | 0 | 0 | 0 | 0 | 0 | 0 | 0 | 0 | 0 | 0 | − | 0 | 0 | 0 | 0 |
| c = | 9 Stress | 0 | 0 | 0 | 0 | 0 | 0 | + | 0 | 0 | − | 0 | 0 | − | 0 | 0 | 0 |
|  | 10 Job Satisfaction | 0 | 0 | 0 | 0 | 0 | 0 | 0 | 0 | 0 | − | 0 | 0 | + | + | + | 0 |
|  | 11 Expectations | 0 | 0 | 0 | 0 | 0 | 0 | 0 | 0 | 0 | + | 0 | 0 | 0 | + | 0 | 0 |
|  | 12 Job Performance | 0 | 0 | 0 | 0 | 0 | 0 | 0 | 0 | 0 | 0 | 0 | 0 | 0 | 0 | 0 | 0 |
|  | 13 Involvement | 0 | 0 | 0 | 0 | 0 | 0 | − | − | 0 | 0 | 0 | + | 0 | + | 0 | 0 |
|  | 14 Commitment | 0 | 0 | 0 | 0 | 0 | 0 | 0 | 0 | 0 | 0 | + | 0 | + | 0 | − | 0 |
|  | 15 Leave Intentions | 0 | 0 | 0 | 0 | 0 | 0 | 0 | 0 | 0 | 0 | 0 | − | − | − | 0 | + |
|  | 16 Job Turnover | 0 | 0 | 0 | 0 | 0 | 0 | 0 | 0 | 0 | 0 | 0 | 0 | 0 | 0 | 0 | 0 |

# A Fuzzy-based Approach to Numeric Constraint Networks

**Kyeongtaek Kim and Pyung Dong Cho**
Electronics and Telecommunications Research Institute
P.O.Box 8, Daeduk Science Town, Taejon, Korea
ktkim@dooly.etri.re.kr, pdcho@dooly.etri.re.kr

## Abstract

Current numeric constraint propagation systems accepting exact numeric values or intervals have a limitation on representing the preference among values. To represent and handle the preference on the values, the notion of fuzzy numeric constraint networks is introduced in this paper. After defining the notion of fuzzy-consistency, we show that fuzzy-consistency can be represented by an extension of interval-consistency. Using this relation, we propose a propagation algorithm for solving fuzzy numeric constraint networks.

## 1. Introduction

Constraints have been used to represent the relations among several variables. A lot of research has been focused on Constraint Satisfaction Problem(CSP) [Waltz 1975, Montanari 1974, Mackworth 1977, Davis 1987]. A CSP is defined by a set of variables, the possible values that each variable can take, and a set of constraints on the variables.

Current numeric constraint systems can be classified into 3 groups on the basis of their input: exact value-based systems, label-based systems, and interval-based systems.

Exact value-based systems [Sussman and Steele 1980, Lelher 1988, Konopasek and Jayaraman 1984] accept only an exact value for input and correspondingly produce only exact values. However, exact value-based systems reveal several limitations in many applications such as dealing with inexact data, solving over-constrained problems, solving under-constrained problems, and fixing variable types [Hyvonen 1992].

Label-based systems accept a set of possible discrete values as input. During propagation, the constraints are used to restrict these sets. Several consistency techniques such as arc consistency [Mackworth 1977, Mohr and Henderson 1986, Deville and van Hentenryck 1991] and path consistency [Montanari 1974, Mackworth 1977, Mohr and Henderson 1986] have been successfully applied to the CSP. However, in some applications it is required to represent continuous values.

Interval-based systems use intervals to restrict domains. Both interval analysis [Moore 1966] and consistency techniques are used for interval propagation algorithms. They can be applied to either finite domains such as CHIP[Dincbas et al. 1988], or continuous domains such as LIC [Ward et al. 1990], TP[Hyvonen 1992], and ICN [Kim 1993]. Interval-based systems have been successfully applied either to under-constrained CSP or to CSP to restrict discrete domains rapidly. Usually the result from interval propagation is not a final solution for given problem. The selection of one or some solution sets among the solution sets identified as a result of applying interval propagation, is followed for further investigation or propagation. However, the result of applying interval propagation does not include any information on values within a interval to help the selection process. If no further objective information is provided, the selection process cannot help depending on the user's subjective judgment. Although the user is uncertain about the final value of a variable, he or she usually has a preference or desire for choosing certain values over others, even during constraints are formulated.

In this paper, fuzzy sets are used to represent and manipulate the preference or desire. The preference or desire is represented by assigning membership values to particular variable values based on user's preference or desire [Wood 1989]. Fuzzy extension principle is applied to determine the resulting membership values of fuzzy operations. In addition to representational merit, this approach has another advantage that the selection of a specific value at a particle membership level restricts values of the other variables to the values with same membership level.

In Section 2, we review numeric constraint networks and interval-based numeric constraint networks. Section 3 begins with describing several definitions on fuzzy-based numeric constraint networks and fuzzy-consistency. Then, we show that fuzzy-consistency can be accomplished by an extension of interval-consistency. Using this relation, we propose a propagation algorithm that produces the fuzzy-consistent solution for a fuzzy-based numeric constraint network.

## 2. Background

A constraint network is a declarative structure which expresses relations among variables. A constraint network consists of variables, which have some particular values, known or unknown, connected by constraints [Davis 1987]. A constraint represents a relation among the values of the variables it connects. A numeric constraint network (NCN) is a constraint network in which variables can take only numeric values. Formal definitions on NCN and its solution are:

### Definition 1: Numeric Constraint Network (NCN)
A numeric constraint network $N = <D, V, C>$ is defined by
- a set of value domains $D = \{D_1, ..., D_n\}$ where a value domain, $D_i = \{d_{i1}, ..., d_{ik}\}$ represents a set of exact values that a variable $V_i$ can take,
- a set of numeric variables $V = \{V_1, ..., V_n\}$,
- a set of numeric constraints $C = \{C_1, ..., C_n\}$.

### Definition 2: A Solution to an NCN
A solution to an NCN $N = <D, V, C>$ is an instantiation of the variables, i.e., $<d_{s1}, ..., d_{sn}>$ such that $d_{si} \subseteq D_i$, for $j = 1$ to $n$, and $<d_{s1}, ..., d_{sn}>$ satisfies all constraints.

In the case of over-constrained NCNs with continuous domains, it is impossible for exact value-based systems to identify the fact that it is over-constrained. In the case of under-constrained NCNs with continuous domains, infinite number of solutions exists. Since it is impossible to enumerate all solutions, intervals are adopted to express all the values within ranges. Specifying an interval for a variable indicates that it can only take on values within that interval. The values within the interval are completely "possible". Values outside the interval are "impossible". An interval-based numeric constraint network (INCN) is an NCN where each domain is an interval. Formal definitions on INCN and its solution are:

### Definition 3: Interval-based Numeric Constraint Network (INCN)
An interval-based numeric constraint network $N_I = <I, V, C>$ is defined by
- a set of domains $I = \{I_1, ..., I_n\}$ where a domain, i.e., an interval, $I_i = [d_{i1}, d_{i2}]$ represents the minimum value $d_{i1}$ and the maximum value $d_{i2}$ that $V_i$ can take,
- a set of numeric variables $V = \{V_1, ..., V_n\}$,
- a set of numeric constraints $C = \{C_1, ..., C_n\}$.

### Definition 4: A Solution to an INCN
A solution to an INCN $N_I = <I, V, C>$ is an instantiation of the variables, i.e., $<[d_{s1}, d_{s1}], ..., [d_{sn}, d_{sn}]>$ such that $[d_{si}, d_{si}] \subseteq I_i$, for $j = 1$ to $n$, and $<[d_{s1}, d_{s1}], ..., [d_{sn}, d_{sn}]>$ satisfies all constraints.

Each solution $<[d_{s1}, d_{s1}], ..., [d_{sn}, d_{sn}]>$ to an INCN corresponds to a solution $<d_{s1}, ..., d_{sn}>$ to an NCN, and vise versa. However, the main purpose of applying intervals to NCN is to reduce domains without excluding any solution rather than to find exact solutions. For this purpose interval-consistency is defined as follows:

**Definition 5: consistent INCN**

An INCN $N_{I'} = <I', V, C>$ is consistent if and only if $\forall I'_i \in I'$, $\forall d_{ij} \in I'_i$,

$\exists C(d_1, ..., d_{I-1}, d_{ij}, d_{i+1}, ..., d_n)$ is satisfied,

$d_k \in I'_k$,

$k = 1, ..., i-1, i+1, ..., n$.

**Definition 6: interval-consistent $I'$**

$I'$ is interval-consistent to $N_{I'}$ if and only if an INCN $N_{I'} = <I', V, C>$ is consistent.

**Definition 7: interval-consistent solution**

$I'$ is the interval-consistent solution to given $N_I = <I, V, C>$ if and only if

an INCN $N_{I'} = <I', V, C>$ is consistent,

$I' \subseteq I$, and

$I'$ does not exclude any solution to $N_I = <I, V, C>$.

The interval propagation algorithm such as TP[Hyvonen 1992] or ICN[Kim 1993] produces the interval-consistent solution to given interval-based constraint network.

## 3. Fuzzy-based Numeric Constraint Network

A fuzzy-based numeric constraint network (FNCN) is a NCN in which variable values have their membership values. If the membership value of each possible value is set to 1 and others are set to 0, any NCN or INCN can be represented by an FNCN. Thus, FNCNs are superset of NCN and INCN. The formal definitions on FNCN and its solution are:

**Definition 8: Fuzzy Numeric Constraint Network (FNCN)**

A fuzzy numeric constraint network $N_F = <F, V, C>$ is defined by

– a set of fuzzy sets $F = \{F_1, ..., F_n\}$ where a fuzzy set $F_i = \{(d_{I1}, \mu_{Fi}(d_{I1})), ..., (d_{Ik}, \mu_{Fi}(d_{Ik}))\}$ is a set of 2-tuples, a numeric value $d_{ij}$ that $V_i$ can take, and its membership value in which $\mu_{Fi}$ represents the membership function,

– a set of numeric variables $V = \{V_1, ..., V_n\}$,

– a set of numeric constraints $C = \{C_1, ..., C_n\}$.

**Definition 9: A Solution to an FNCN**

A solution to an FNCN $N_F = <F, V, C>$ is $\{(d_{s1}, \mu_s), ..., (d_{sn}, \mu_s)\}$ where

1) an instantiation of the variables, i.e., $<d_{s1}, ..., d_{sn}>$, such that

$d_{sj} \in D_i$, for $j = 1$ to $n$,

$<d_{s1}, ..., d_{sn}>$ satisfies all constraints,

2) an associated membership value $\mu_s$ such that $\mu_s = \min_i \{\max \mu_{Fi}(d_{si})\}$.

Similar to the case of INCN, the main purpose of applying fuzzy concept to NCN is not to find solutions. Rather, it is both to reduce domains without excluding any solutions and to rate solution candidates within resulting domains. For this purpose fuzzy-consistency is defined as follows:

**Definition 10: consistent FNCN**

An FNCN, $N_{F'} = <F', V, C>$, is consistent if and only if $\forall F'_i \in F'$, $\forall (d_{ij}, \mu_{F'i}(d_{ij})) \in F'_i$,

$\exists C(d_1, ..., d_{I-1}, d_{ij}, d_{i+1}, ..., d_n)$ is satisfied,

$\mu_{F'k}(d_k) \geq \mu_{F'i}(d_{ij})$,

$(d_k, \mu_{F'k}(d_k)) \in F'_k$,

$k = 1, ..., i-1, i+1, ..., n$.

## Definition 11: fuzzy-consistent

A set of fuzzy set $F'$ is fuzzy-consistent to $N_{F'}$ if and only if
an FNCN, $N_{F'} = <F', V, C>$, is consistent.

## Definition 12: fuzzy-consistent solution

$F'$ is the fuzzy-consistent solution to given $N_F = <F, V, C>$ if and only if
an FNCN $N_{F'} = <F', V, C>$ is consistent,
$F' \subseteq F$, and
$F'$ does not exclude any solution to $N_F = <F, V, C>$ .

Now we will show that fuzzy-consistency can be defined by interval-consistency. Let $(F')^\alpha$ be the $\alpha$-cut set of $F'$, i.e., $(F')^\alpha = \{(F'_1)^\alpha, ..., (F'_n)^\alpha\}$ where $(F'_i)^\alpha = \{d_{ij} \mid (d_{ij}, \mu_{Fi}(d_{ij})) \in F'_i, \mu_{F'i}(d_{ij}) \geq \alpha\}$ and $0 < \alpha \leq 1$. Then, the following proposition corresponds to definition 10.

## Proposition 1.

An FNCN $N_{F'} = <F', V, C>$ is consistent if and only if $\forall (F'_i)^\alpha \in F'$, $\forall d_{ij} \in (F'_i)^\alpha$,

$\exists C(d_1, ..., d_{I-1}, d_{ij}, d_{i+1}, ..., d_n)$ is satisfied,
$d_k \in (F'_k)^\beta$,
$0 < \alpha \leq \beta \leq 1$,
$k = 1, ..., i-1, i+1, ..., n$,

The proposition 1 corresponds to the following proposition by $(F'_i)^\alpha \supseteq (F'_i)^\beta$.

## Proposition 2:

An FNCN, $<F', V, C>$, is consistent if and only if every $(F')^\alpha$ is interval-consistent.

Using the above proposition and the properties of interval propagation, the fuzzy consistent solution can be obtained by the following proposition.

## Proposition 3:

$F'$ is the fuzzy-consistent solution to an FNCN $N_F = <F, V, C>$ if and only if every $(F')^\alpha$ is the set of intervals resulting from applying interval propagation to an INCN, $< (F)^\alpha, V, C>$.

---

Let $N_F = <F, V, C>$ be a given fuzzy-based constraint network and $\alpha i$, $i=1$ to $m$, be a given $\alpha$-cut level such that $\alpha 1 = 0$ and $\alpha i \geq \alpha j$ for $i > j$ .

Step 1. Set $(F')^{\alpha j}$ to $\phi$, for $j = 1$ to $m$.
Step 3. Set $i$ to 1.
Step 4. Set $I$ to $F^{\alpha i}$.
Step 5. Solve an INCN $N_I = <I, V, C>$ using interval propagation.
       Let $I'$ be a set of resulting intervals.
Step 6. If $I' = \phi$, go to step 9.
Step 7. Set $(F')^{\alpha i}$ to $I'$.
Step 8. If $i < n$, increment i by 1 and go to step 4.
Step 9. If $(F')^{\alpha l} = \phi$, then there is no fuzzy-consistent solution for the given problem,
       else, $F'$ is the fuzzy-consistent solution for $N_F = <F, V, C>$.

---

**Figure 1. A Fuzzy Propagation Algorithm**

By the above proposition, a fuzzy propagation algorithm that produces the fuzzy-consistent solution can be defined using interval propagation algorithm. The proposed fuzzy propagation algorithm is shown in Figure 1. The capability and performance of the proposed algorithm mainly depend on the used interval algorithm.

## 4. Conclusion

In this paper we have defined fuzzy-based numeric constraint networks, and shown that a consistent FNCN can be represented by some consistent INCNs. Using this relationship, we have proposed a fuzzy propagation algorithm that produces the fuzzy-consistent solution for FNCN.

There are several advantages to this approach. First, the preference or desire of the users can be represented more accurately since users can specify his preference or desire on particular values using membership values. Second, any NCN or INCN can be represented by an FNCN. Third, the result from fuzzy propagation can assist the user to select values within intervals for further investigation.

## Acknowledgements

We are very grateful to Dr. Sehyeong Cho for his valuable comments. The work in this paper was supported by Korea Telecom.

## References

Davis, E. (1987), "Constraint Propagation with Interval Labels," *Artificial Intelligence*, Vol. 32, pp. 281-330.

Deville, Y., and van Hentenryck, P. (1991), "An Efficient Arc Consistency Algorithm for a Class of CSP Problem," *Proceedings of the 12th International Joint Conference on Artificial Intelligence*, Sydney, Australia, pp. 325-330.

Dincbas, M., Van Hentenryck, P., Simonis, H., Aggoun, A., Graf, T., and Berthier, F. (1988), "The Constraint Logic Programming Language CHIP," *Proceedings of the International Conference on Fifth Generation Computer Systems*, ICOT.

Hyvonen, E (1992), "Constraint Reasoning Based on Interval Arithmetic: the Tolerance Propagation Approach," *Artificial Intelligence*, Vol. 58, pp. 71-112.

Kim, K. (1993), *An Interval-based Approach for Concurrent Engineering Under Imprecision*, Ph.D. Dissertation, North Carolina State University, Raleigh, NC.

Konopasek, M., and Jayaraman, S. (1984), The TK!Solver Book, McGraw-Hill, Berkeley, CA.

Lelher, W. (1988), *Constraint Programming Languages: Their Specification and Generation*, Addison-Wesley, Reading, MA.

Mackworth, A.K. (1977), "Consistency in Networks of Relations," *Artificial Intelligence*, Vol. 8, pp. 99-118.

Mohr, R., and Henderson, T.C. (1986), "Arc and Path Consistency Revisited," *Artificial Intelligence*, Vol. 28, pp. 225-233.

Montanari, U. (1974), "Networks of Constraints: Fundamental Properties and Applications to Picture Processing," *Information Sciences*, Vol. 7, pp. 95-132.

Moore, R.E. (1966), *Interval Arithmetic*, Prentice-Hall, Englewood Cliffs, New Jersey.

Sussman, G.J., and Steele, G.L. (1980), "CONSTRAINTS-A Language for Expressing Almost-Hierarchical Descriptions," *Artificial Intelligence*, Vol. 14, pp. 1-39.

Ward, A.C., Lozano-Perez, T., and Seering, W.P. (1990), "Extending the constraint propagation in intervals," *Artificial Intelligence in Engineering Design, Analysis, and Manufacturing*, 4(1), pp. 47-54.

Wood, K.L. (1989), *A Method for Representing and Manipulating Uncertainties in Preliminary Engineering Design*, Ph.D. Dissertation, California Institute of Technology, Pasadena, CA.

Waltz, A. (1975), "Understanding Line Drawings of Scene with Shadows," in *The Psychology of Computer Vision* (ed. P.H. Winston), McGraw-Hill, New York.

# CONSTRAINTS WITH INTERVAL COEFFICIENTS AND STRICTNESS PROBABILITY INDEX

Yozo Nakahara
Shizuoka Seika College
549, Honnakane, Yaizu, 425, Japan

Mitsuo Gen
Ashikaga Institute of Technology
268-1, Omae, Ashikaga, 326, Japan

**Abstract**

Recently, Nakahara *et al* have proposed the probability index of an inequality between intervals and the treatment of the constraint based on the ranking fuzzy intervals which is set by the lower bound of the index. In the treatment, a decision maker (DM) can choose the strictness of constraint by setting the lower bound of the index, because DM can choose the strictness of the ranking interval fuzzy numbers by setting the lower bound. But the constraint becomes equivalent to the non-linear inequality.

In this paper, we compare the rankings of interval fuzzy numbers, to demonstrate the usefulness of the rankings by Tanaka *et al* and that by Nakahara *et al*. We also compare the constraints of a linear programming problem with interval coefficients (an ILP problem) or a goal programming problem with interval coefficients (an IGP problem) set by using the above two rankings, and constraints created by regarding the interval fuzzy numbers as random variables with uniform distribution on the intervals and by using the chance-constrained programming. It showes of which the advantage and the disadvantage we have by using the ranking defined by Nakahara *et al* instead of the ranking defined by Tanaka *et al*. The statement of the theorem is also given which lightens the disadvantage(non-linearlity).

**Key words:** Decision making; mathematical programming; constraints; interval coefficients; probability index; linear approximation

## 1 Introduction

This paper considers the constraints of a kind of linear programming (LP) problems with interval coefficients (ILP problems) or goal programming (GP) problems with interval coefficients (IGP problems) as

$$\sum_{j=1}^{n} A_{ij} x_j \leq B_i, \qquad i = 1, 2, \cdots, m \quad (1)$$

$$x_j \geq 0, \qquad j = 1, 2, \cdots, n \quad (2)$$

where $A_{ij}$ and $B_i$ are interval fuzzy numbers represented by $A_{ij} = [a_{ij}{}^L, a_{ij}{}^U]$ and $B_i = [b_i{}^L, b_i{}^U]$ for $i = 1, 2, \cdots, m; j = 1, 2, \cdots, n$. Let us consider the constraints of a kind of LP porblems with imprecise coefficitns restricted by possibilistic distributions (PLP problems) as

$$\sum_{j=1}^{n} \widetilde{a_{ij}} x_j \leq \widetilde{b_i}, \qquad i = 1, 2, \cdots, m \quad (3)$$

$$x_j \geq 0, \qquad j = 1, 2, \cdots, n \quad (4)$$

where $\widetilde{a_{ij}}$ and $\widetilde{b_i}$ are fuzzy numbers for any $i$, $j$. The constraints of another kind of PLP problems is as follows:

$$\sum_{j=1}^{n} \widetilde{a_{ij}} x_j \lesssim \widetilde{b_i}, \qquad i = 1, 2, \cdots, m \quad (5)$$

$$x_j \geq 0, \qquad j = 1, 2, \cdots, n \quad (6)$$

where $\widetilde{a_{ij}}$, $\widetilde{b_i}$, $\forall i, j$, are fuzzy numbers and $\lesssim$ is the soft inequality [2], [3], [7], [15].

Various methods for ranking fuzzy numbers have been developed or used in order to solve the constraints (3), (4) or (5), (6) [1], [2], [3], [6]~[10], [15], [19]~[21]. As shown in the following section, some of these methods are reduced to a few rankings of interval fuzzy numbers. Although there are many other methods for ranking fuzzy numbers [6], almost no method can be reduced to other rankings of interval fuzzy numbers than the above or can be applied toward solving constraints(1). These rankings of interval fuzzy numbers are very useful, but there are cases where DM wants to use another ranking. Because of this, Tanaka *et al* [11] proposed the ranking of interval fuzzy numbers using 'the degree of inequality between interval fuzzy numbers' which defined by them. Using this ranking, the constraints (1) become equivalent to the linear inequalities. In [17], we proposed a new ranking interval fuzzy numbers using a probability degree (or index) of the inequality between interval fuzzy numbers, defined by using $P(\bullet)$ which represents the probability of $\bullet$. Using this ranking, (1) is equivalent to the following auxiliary inequalities:

$$P(\langle \sum_{j=1}^{n} A_{ij} x_j \rangle \leq \langle B_i \rangle) \geq q_i, \qquad i = 1, 2, \cdots, m$$

$$(7)$$

where $q_i$ are the lower bounds of the probability set by the DM and for any fuzzy interval $I$, $\langle I \rangle$ represents the random variable uniformly distributed on the support of $I$. In (7), the DM can reflect his own delicate or subtle request to the strictness of constraints (1) by setting the lower bounds $q_i, \forall i$, from the probability index. (7) is different from the constraints created by regarding the interval fuzzy numbers of (1) as random variables with uniform distribution on the intervals and by using the chance-constrained programming developed by Charnes, Cooper and Symonds [4], and Charnes and Cooper [5]. In fact, if we regard the $A_{ij}$ and $B_i$ as $\langle A_{ij} \rangle$ and $\langle B_i \rangle$ respectively for any $i,j$, (1) is regarded as the following condition by the chance-constrained programming with the lower bounds $q_i, \forall i$:

$$P(\sum_{j=1}^{n} \langle A_{ij} \rangle x_j \leq \langle B_i \rangle) \geq q_i, \qquad i = 1, 2, \cdots, m$$

$$(8)$$

The disadvantage of this ranking is that the constraints (7) has become equivalent to non-linear inequalities as (8) did, although the non-linearity is not so much as (8). We have already shown in [17] that the set of solutions for (the region expressed by) (7) and (2) can be approximated from the inside by their region expressed by corresponding linear inequalities. We have suggested that we could use this region as a substitution for the region expressed by (7) and (2), and that it was especially useful when the DM hoped to seek solutions necessarily belonging to the region expressed by (7) and (2) (the region (7) and (2)). In [18], we have shown that, under the reasonable condition of $q \in (1/2, 1]$, there was a region expressed by linear inequalities which gave a better approximation from the inside of the region (7) and (2).

In this paper, we will compare the rankings of interval fuzzy numbers which is defined by many researchers, to demonstrate the usefulness of the rankings by Tanaka et al and that by Nakahara et al. We will also compare the constraints of an ILP problem or an IGP problem set by using the above two rankings, and constraints (8), to show of which the advantage and the disadvantage we

have by using the ranking defined by Nakahara et al instead of the ranking defined by Tanaka et al. The statement of the theorem will be also given which lightens the disadvantage(nonlinearlity). The theorem will give other useful estimates about the approximation of the region expressed by (7) and (2) to show that the proposed method for constraints (1) can be used practically,. More precisely, it will show that there is a region expressed by linear inequalities which is an approximation from the outside of the region (7) and (2). This region is useful when the DM wants to know the limit of the objective function values in region (7) and (2). It will also show that when we have $q \in (1/2, 1]$, there are two regions expressed by linear inequalities, one of which gives a better approximation from the inside and another of which gives a better approximation from the outside. Moreover it will show that the region which approximates from the inside converges on the region which approximates from the outside, $q$ tending to $1/2$. The proof will be given in another paper, for there is no space for it in this paper. By these estimates, the DM can gain much information on any solution of a mathematical programming problem where the constraints are treated by the probability index.

## 2   Methods for ranking fuzzy numbers

Many methods for ranking fuzzy numbers have been developed or used in order to solve the constraints (3) and (4) or to solve constraints (5) and (6) [1], [2], [3], [6]~[10], [15], [19]~[21]. In this section, some expressions resulting from their application to ranking intervals will be given. We will also show two ranking intervals, proposed by Tanaka et al and has suggested by Nakahara et al. Let us define $I_f$ (the family of intervals) as

$$I_f = \{S \mid S; \; interval, \; supp S = [s_L, s_U],$$
$$s_L, s_U \in R, \; s_L \leq s_U\}, \qquad (9)$$

where $supp S$ means the support of $S$ and $[s_L, s_U]$ is defined as $[s_L, s_U] = \{s \in R \mid s_L \leq s \leq s_U\}$ for any $s_L, s_U \in R$. We can obtain the following ranking intervals from the corresponding ranking fuzzy numbers:

(a) By use of Ramik and Rimanek's ranking [19], we have

$$\forall [a_L, a_U], [b_L, b_U] \in I_f,$$
$$[a_L, a_U] \leq [b_L, b_U]$$
$$\iff a_L \leq b_L, \quad a_U \leq b_U. \quad (10)$$

(b) Tanaka, Ichihashi and Asai's ranking [20], [2], [7] was defined for fuzzy numbers with symmetric triangular possibilistic distributions, but we can easily extend it for arbitrary fuzzy numbers. By use of the ranking, we have

$$\forall [a_L, a_U], [b_L, b_U] \in I_f,$$
$$[a_L, a_U] \leq [b_L, b_U]$$
$$\Longleftrightarrow a_L \leq b_L, \quad a_U \leq b_U. \quad (11)$$

(c) By use of Adamo's ranking [1], [2], [3] using $k$-preference index, we have

$$\forall [a_L, a_U], [b_L, b_U] \in I_f,$$
$$[a_L, a_U] \leq [b_L, b_U]$$
$$\Longleftrightarrow a_U \leq b_U. \quad (12)$$

(d) By use of Yager's first index [21], [2], [3], we have

$$\forall [a_L, a_U], [b_L, b_U] \in I_f,$$
$$[a_L, a_U] \leq [b_L, b_U]$$
$$\Longleftrightarrow \frac{1}{2}(a_L + a_U) \leq \frac{1}{2}(b_L + b_U). \quad (13)$$

(e) By use of Yager's second index [21], [2], [3], we have

$$\forall [a_L, a_U], [b_L, b_U] \in I_f,$$
$$[a_L, a_U] \leq [b_L, b_U]$$
$$\Longleftrightarrow \frac{1}{2}(a_L + a_U) \leq \frac{1}{2}(b_L + b_U). \quad (14)$$

(f) By application of Fuller's approach [10], we have

$$\forall [a_L, a_U], [b_L, b_U] \in I_f,$$
$$[a_L, a_U] \leq [b_L, b_U]$$
$$\Longleftrightarrow \frac{1}{2}(a_L + a_U) \leq \frac{1}{2}(b_L + b_U). \quad (15)$$

(g) By application of Dubois's k-very weak feasibility concept [8], we have

$$\forall [a_L, a_U], [b_L, b_U] \in I_f,$$
$$[a_L, a_U] \leq [b_L, b_U]$$
$$\Longleftrightarrow a_L \leq b_U. \quad (16)$$

(h) By application of Dubois's k-middle weak feasibility concept [8], we have

$$\forall [a_L, a_U], [b_L, b_U] \in I_f,$$
$$[a_L, a_U] \leq [b_L, b_U]$$
$$\Longleftrightarrow a_U \leq b_U. \quad (17)$$

(i) By application of Dubois's k-middle strong feasibility concept [8], we have

$$\forall [a_L, a_U], [b_L, b_U] \in I_f,$$
$$[a_L, a_U] \leq [b_L, b_U]$$
$$\Longleftrightarrow a_L \leq b_L. \quad (18)$$

(j) By application of Dubois's k-very strong feasibility concept [8], we have

$$\forall [a_L, a_U], [b_L, b_U] \in I_f,$$
$$[a_L, a_U] \leq [b_L, b_U]$$
$$\Longleftrightarrow a_U \leq b_L. \quad (19)$$

As shown in the above, these rankings are reduced to 6 rankings of intervals. Although there are other methods for ranking fuzzy numbers [6], almost no method can be reduced to other rankings of intervals than the above or can be applied toward solving constraints(1). These rankings of intervals are very useful, but there are cases where DM wants to use another ranking. Because of this, Tanaka *et al* [11] defined, for any $q \in [0, 1]$ set by DM, the ranking as

$$\forall [a_L, a_U], [b_L, b_U] \in I_f,$$
$$[a_L, a_U] \leq [b_L, b_U]$$
$$\Longleftrightarrow P_2([a_L, a_U] \leq [b_L, b_U]) \geq q, \quad (20)$$

where $P_2(A \leq B)$ for any fuzzy interval $A$ and $B$ is defined as the following:

**Definition 1 ([17])** *For any fuzzy interval* $A = [a_L, a_U]$ *and* $B = [b_L, b_U]$, *we define* $P_2(A \leq B)$ *as*

$$P_2(A \leq B) = \frac{(b_U - a_U) + (a_U - a_L)}{a_U - a_L + b_U - b_L}. \quad (21)$$

*We call* $P_2(A \leq B)$ *degree of* $A \leq B$ [11].

For any $q \in [0, 1]$, we have

$$P_2([a_L, a_U] \leq [b_L, b_U]) \geq q$$
$$\Longleftrightarrow q a_U + (1 - q) a_L \leq q b_L + (1 - q) b_U. \quad (22)$$

Hence, from this and (20) we have for any $q \in [0, 1]$

$$\forall [a_L, a_U], [b_L, b_U] \in I_f,$$
$$[a_L, a_U] \leq [b_L, b_U]$$
$$\Longleftrightarrow q a_U + (1 - q) a_L \leq q b_L + (1 - q) b_U. \quad (23)$$

These include the ranking (13) $\sim$ (16), (19) and many other rankings. Table 1 shows any $q \in [0, 1]$ with which (20) is equivalent to some of (10) $\sim$

1885

| $q$ | the rankings equivalent to (20) with $q \in [0,1]$ |
|-----|---------------------------------------------------|
| 0   | (16)                                              |
| 1/2 | (13), (14), (15)                                  |
| 1   | (19)                                              |

Table 1: The relation between the rankings (10) $\sim$ (19) and the rankings (20)

| $q$ | the rankings equivalent to (26) with $q \in [0,1]$ |
|-----|---------------------------------------------------|
| 0   | (16)                                              |
| 1/2 | (13), (14), (15)                                  |
| 1   | (19)                                              |

Table 2: The relation between the rankings (10) $\sim$ (19) and the rankings (26)

(19), and the rankings equivalent to (20) with such $q$. Using one of these rankings with $q \in [0,1]$ fixed by DM, the constraints (1) become equivalent to linear inequalities. $P_2(A \leq B)$ is the degree of $A - B \leq 0$ rather than the degree of $A \leq B$, because $P_2(A \leq B)$ represents the rate

$$\frac{0 - (A-B)^L}{(A-B)^U - (A-B)^L}, \qquad (24)$$

where $(A-B)^U$ and $(A-B)^L$ means the upper bound of $A - B$ and the lower bound of $A - B$, respectively.

In [17], Nakahara *et al* defined for any $q \in [0,1]$ set by DM the new ranking as

$$\forall [a_L, a_U], [b_L, b_U] \in I_f,$$
$$[a_L, a_U] \leq [b_L, b_U]$$
$$\iff PI([a_L, a_U] \leq [b_L, b_U]) \geq q, \quad (25)$$

where $PI(A \leq B)$ for any fuzzy interval $A$ and $B$ is defined as the following:

**Definition 2 ([17])** *For any fuzzy interval $A$ and $B$, we define $PI(A \leq B)$ as*

$$PI(A \leq B) = P(\langle A \rangle \leq \langle B \rangle) \qquad (26)$$

*where, for any fuzzy interval $S$, $\langle S \rangle$ represents the random variable uniformly distributed on the support of $S$, and $P(\bullet)$ represents the conventional probability of $\bullet$ . We call $PI(A \leq B)$ a probability of $A \leq B$. Moreover, we call $PI$ the strictness probability index of ranking between fuzzy intervals.*

These proposed rankings also include the rankings (13) $\sim$ (16), (19) and many other rankings. Table 2 shows any $q \in [0,1]$ with which (20) is equivalent to some of (10) $\sim$ (19), and the rankings equivalent to (20) with such $q$. We clearly have the following relation from Definition 1 and 2:

$$\forall A, B \in I_f, \quad P_2(A \leq B) = PI(A - B \leq 0). \qquad (27)$$

Using one of the Nakahara *et al*'s rankings with $q \in [0,1]$ fixed by DM, the constraints (1) become equivalent to non-linear inequalities.

**Remark 1** *Using one of the rankings (20) with $q_i \in [0,1]$, $i = 1, \cdots, m$, fixed by DM, the constraints (1) becomes as following:*

$$P(\langle \sum_{j=1}^{n} A_{ij} x_j - B_i \rangle \leq 0) \geq q_i, \qquad i = 1, 2, \cdots, m$$
$$(28)$$

*We can see that the above constraints (28) and constraints (7) approximate the constraints (8) as following, respectively:*

$$\sum_{j=1}^{n} \langle A_{ij} \rangle x_j \leq \langle B_i \rangle$$
$$\underset{\sim}{} \langle \sum_{j=1}^{n} A_{ij} x_j - B_i \rangle \leq 0, \qquad (29)$$

$$\sum_{j=1}^{n} \langle A_{ij} \rangle x_j \leq \langle B_i \rangle$$
$$\underset{\sim}{} \langle \sum_{j=1}^{n} A_{ij} x_j \rangle \leq \langle B_i \rangle. \qquad (30)$$

*Note that the multiplication of a random variable by a real number and the summation(addition) of random variables in the left-hand side are defined usually, and that the multiplication of an interval fuzzy number by a real number and the summation(addition) or subtraction of interval fuzzy numbers in the right-hand side are defined by using the extension principle.*

*We can say that, in general, the constraints (7) approximate the constraints (8) more than the constraints (28) do. This is one of the advantages of using the ranking defined by Nakahara et al instead of using the ranking defined by Tanaka et al.*

*On the other hand, the constraints (28) are equivalent to the linear inequalities of $x_j, j = 1, \cdots, n$, the constraints (8) are equivalent to the non-linear inequalities of $x_j, j = 1, \cdots, n$, and the constraints (7) are also equivalent to the linear inequalities of $x_j, j = 1, \cdots, n$, although they have not so much non-linearity as the non-linear inequalities equivalent to the constraints (8). This is the disadvantage of using the ranking defined by Nakahara et al instead of using the ranking defined by Tanaka et al.*

# 3    Main results

At first, we define some regions:

**Definition 3** *For arbitrary intervals* $[a_j{}^L, a_j{}^U]$; $j = 1, \ldots, n$, $[b_L, b_U]$, *and any number* $q \in [0, 1]$, *we define* $r_x(q)$, $r_x(q)_1 \sim r_x(q)_5$ *as follows:*

$$r_x(q) = \{\mathbf{x} \in R^{+^n} \mid$$
$$PI(\sum_{j=1}^{n} [a_j{}^L, a_j{}^U] x_j \leq [b_L, b_U]) \geq q\}, \tag{31}$$

$$r_x(q)_1 = \{\mathbf{x} \in R^{+^n} \mid$$
$$q \sum_{j=1}^{n} a_j{}^U x_j + (1-q) \sum_{j=1}^{n} a_j{}^L x_j \leq b_L\}, \tag{32}$$

$$r_x(q)_2 = \{\mathbf{x} \in R^{+^n} \mid$$
$$q \sum_{j=1}^{n} a_j{}^U x_j + (1-q) \sum_{j=1}^{n} a_j{}^L x_j$$
$$\leq \frac{1}{2}(b_U + b_L) - \frac{1}{2}(2q-1)^2(b_U - b_L),$$
$$\sum_{j=1}^{n} a_j{}^L x_j \leq b_L\}, \tag{33}$$

$$r_x(q)_3 = \{\mathbf{x} \in R^{+^n} \mid$$
$$q \sum_{j=1}^{n} a_j{}^U x_j + (1-q) \sum_{j=1}^{n} a_j{}^L x_j$$
$$\leq \frac{1}{2}(b_U + b_L) - \frac{1}{2}(2q-1)(b_U - b_L)\}, \tag{34}$$

$$r_x(q)_4 = \{\mathbf{x} \in R^{+^n} \mid$$
$$q \sum_{j=1}^{n} a_j{}^U x_j + (1-q) \sum_{j=1}^{n} a_j{}^L x_j \leq b_U\}, \tag{35}$$

$$r_x(q)_5 = \{\mathbf{x} \in R^{+^n} \mid$$
$$q \sum_{j=1}^{n} a_j{}^U x_j + (1-q) \sum_{j=1}^{n} a_j{}^L x_j$$
$$\leq \frac{1}{2}(b_U + b_L)\}. \tag{36}$$

$r_x(q)_3$ is same as the region(7) proposed by Ishibuchi and Tanaka as the feasible region of (4) when $q \in (1/2, 1]$.

In this paper, we show the following Main Theorem:

**Main Theorem 1** *Let* $[a_j{}^L, a_j{}^U]$; $j = 1, \ldots, n$, $[b_L, b_U]$ *be arbitrary intervals such as* $b_L < b_U$ *and* $a_j{}^L < a_j{}^U$ *for some* $j \in \{1, \cdots, n\}$. *Then, if* $q \in (0, 1]$, *we have*

$$r_x(q)_1 \subset r_x(q) \subset r_x(q)_4, \tag{37}$$

*and if* $q \in (1/2, 1]$, *we have*

$$r_x(q)_2 \subset r_x(q), \tag{38}$$
$$r_x(q)_1 \subset r_x(q)_3 \subset r_x(q) \subset r_x(q)_5 \subset r_x(q)_4 \tag{39}$$

*and*

$$r_x(q)_3 \longrightarrow r_x(q)_5, \quad when \ q \longrightarrow 1/2 \tag{40}$$

*which means*

$$\forall q \in (1/2, 1], \forall \Omega \subset R^{+^n}$$
$$s.t. \ r_x(q)_3 \subset \Omega \subset r_x(q)_5$$
$$and \ \Omega \neq r_x(q)_5, \ \exists q_0 > 1/2;$$
$$q' \leq q_0 \Longrightarrow r_x(q')_3 \supset \Omega. \tag{41}$$

**Remark 2** *Although this Main theorem is stated for the constraint as (4), similar theorem holds for any constraint as the following:*

$$\sum_{j=1}^{n} f_j(\mathbf{x}) \leq B, \tag{42}$$
$$f_j(\mathbf{x}) = [f_j{}^L(\mathbf{x}), \ f_j{}^U(\mathbf{x})],$$
$$j = 1, \ldots, n, \tag{43}$$
$$f_j{}^L, f_j{}^U : R^{+^n} \longrightarrow R, \ f_j{}^L \leq f_j{}^U,$$
$$j = 1, \ldots, n, \tag{44}$$
$$B = [b_L, b_U]. \tag{45}$$

*The theorem for such constraint can be gained by exchanging* $A_j x_j$ *for* $f_j(\mathbf{x})$, $a_j{}^L x_j$ *for* $f_j{}^L(\mathbf{x})$ *and* $a_j{}^U x_j$ *for* $f_j{}^U(\mathbf{x})$ *in the stated Main Theorem.*

We showed in [17] the right-hand side of (37) and we showed (38) in [18]. Moreover, $r_x(q)_1 \subset r_x(q)_3$, $r_x(q)_5 \subset r_x(q)_4$ and (40) with $q \in (1/2, 1]$ hold clearly. Hence, remaining estimates to be proved are $r_x(q)_3 \subset r_x(q)_1 \subset r_x(q)_5$ with $q \in (1/2, 1]$ and $r_x(q) \subset r_x(q)_4$ with $q \in (0, 1]$. For there is no space for the proof in this paper, we omit that. We will show the proof in another paper.

# 4    Conclusion

In this paper, we have compared the rankings of interval fuzzy numbers which is defined by many researchers, to demonstrate the usefulness of the rankings by Tanaka *et al* and that by Nakahara *et al*. We have also compared the constraints of an ILP problem or an IGP problem set by using the above two rankings, and constraints (8) created by regarding the interval fuzzy numbers as random variables with uniform distribution on the intervals and by using the chance-constrained programming. It has shown of which the advantage and the disadvantage we have by using the ranking defined by Nakahara *et al* instead of the ranking defined by Tanaka *et al*. The statement of the theorem has been also given which lightens the disadvantage(non-linearlity). The theorem has shown that, under some reasonable condition, there is the region which is the approximation from the outside of the feasible region. This region is useful when DM wants to know the limit of the objective function values in the feasible region. It has also shown that when $q$ belongs $(1/2, 1]$ there is the region which is always included by the feasible region and which, $q$ tending to $1/2$, converges, in a sense, to the feasible region which gives the approximation from the outside. By these estimate, DM can gain much information of any solution of a mathematical programming problem with the constraints treated by the probability index.

# References

[1] J. M. Adamo, Fuzzy decision trees, *Fuzzy Sets and Systems* **4** (1980) 207-219.

[2] L. Campos and J. L. Verdegay, Linear programming problems and ranking of fuzzy numbers, *Fuzzy Sets and Systems* **32** (1989) 1-11.

[3] L. Campos, Fuzzy linear programming models to solve fuzzy matrix games, *Fuzzy Sets and Systems* **32** (1989) 275-289.

[4] A. Charnes, W. W. Cooper and G. H. Symonds, Cost horizons and certainty equivalents: an approach to stochastic programming of heating oil, *Management Science* **4** (1958) 235-263.

[5] A. Charnes and W. W. Cooper, Chance-constrainted programming, *Management Science* **5** (1959) 73-79.

[6] S. J. Chen and C. L. Hwang, *Fuzzy Multiple Attribute Decision Making* (Springer-Verlag Berlin Heidelberg, 1992).

[7] M. Delgado, J. L. Verdegay and M. A. Vila, A general model for fuzzy linear programming, *Fuzzy Sets and Systems* **29** (1989) 21-29.

[8] D. Dubois and H. Prade, Ranking Fuzzy Numbers in Setting of Possibility Theory, *Information Science* **30** (1983) 183-224.

[9] D. Dubois, Linear Programming with Fuzzy Data, in: *Analysis of Fuzzy Information, Volume 3: Applications in Engineering and Science* (CRC Press, Boca Raton, FL, 1987) 241-263.

[10] R. Fuller, On stability in fuzzy linear programming problems - a short communication, *Fuzzy Sets and Systems* **30** (1986) 511-519.

[11] H. Ishibuchi and H. Tanaka, Formulation and Analysis of Linear Programming Problem with Interval Coefficients, *J. of Japan Indust. Manage Assoc.* **40** (1988) 320-329, in Japanese.

[12] H. Ishibuchi and H. Tanaka, Interval 0-1 Programming Problem and Produced-Mix Analysis, *J. Oper. Res. Soc. Jap.* **32** (1989) 352-369, in Japanese.

[13] H. Ishibuchi and H. Tanaka, Multiobjective Programming in Optimization of the Interval Objective Function, *Euro. J. Oper. Res.* **48** (1990) 219-225.

[14] M. Inuiguchi, H. Ichihashi and H. Tanaka, Fuzzy Programming: A Survey of Recent Developments, in: R. Slowinski and J. Teghemi, Ed., *Stochastic versus Fuzzy Approaches to Multiobjective Mathematical Programming under Uncertainty* (Netherlands, 1990) 45-68.

[15] Y. Y. Lai and C. L. Hwang, *Fuzzy Mathematical Programming* (Springer-Verlag Berlin Heidelberg, 1992).

[16] Y. Nakahara, M. Sasaki, K. Ida and M. Gen, A Method for Solving 0-1 Linear Programming Problem with Interval Coefficients, *J. of Japan Indust. Manage Assoc.* **42** (1991) 345-351, in Japanese.

[17] Y. Nakahara, M. Sasaki and M. Gen, On the Linear Programming Problems with Interval Coefficients, *Int. J. of Comp. and Indust. Engg.* **23** Nos 1-4 (1992) 301-304.

[18] Y. Nakahara and M. Gen, On the Feasible Region of Linear Programming Problems with Interval Coefficients, Proceedings of the Korea-Japan Joint Conference on Fuzzy Systems and Engineering (1992) 245-248.

[19] J. Ramik and J. Rimanek, Inequality relation between fuzzy numbers and its use in fuzzy optimization, *Fuzzy Sets and Systems* **16** (1985) 123-138.

[20] H. Tanaka, H. Ichihashi and K. Asai, A formulation of fuzzy linear programming problem based on comparison of fuzzy numbers, *Control and Cybernetics* **13** (1984) 185-194.

[21] R. R. Yager, A procedure for ordering fuzzy subsets of the unit interval, *Information Sciences* **24** (1981) 143-161.

# Fuzzy Petri nets to Control Vision System and Robot Behaviour under Uncertain Situations within an FMS cell

## M. Hanna, A. Buck and R. Smith

School of Electrical, Electronic and Systems Engineering
University of Wales, College of Cardiff,
School of ELSYM, Newport Rd,
P.O Box 917, Cardiff CF2 1XH,
Cardiff, Wales, UK
e-mail: Hannam@uk.ac.cardiff

**Abstract:** The Petri net approach to modelling, monitoring and control of the behaviour of an FMS cell is shown. The FMS cell described comprises a pick and place robot, vision system, computer numerical control (CNC) drilling machine, two conveyors and a bin. The work demonstrates how a Petri net approach can be used to describe events at different levels of abstraction. Next, by utilising Fuzzy Logic Control and Petri Nets, a modelling approach using fuzzy rules based on a knowledge-based system illustrates how vision system decisions and robot behaviour within an FMS cell can be determined. This methodology can be used as a graphical modelling tool to monitor imprecise and uncertain situations, and determine the quality of the output products of an FMS cell.

## 1. INTRODUCTION

Flexible manufacturing systems (FMS) can share many resources such as computer numerical control (CNC) machinery, robots and vision systems. Such systems fall into a category of the type which exhibits asynchronicity and concurrency. Petri nets have been applied to the modelling, analysis, control and simulation of FMS's[1], [2], [3], [9] and can model and monitor events at different levels of activity, thereby leading to an increase in system efficiency and flexibility. In section 2, the FMS cell control Petri net illustrates that all activities are modelled and simulated at each level of a hierarchy[9].

Recently, Fuzzy logic with Petri nets (Fuzzy Petri nets) has been used to model imprecise and uncertain situations which can arise within FMS's[5], [6]. In general Fuzzy Petri nets can model vagueness and uncertainty by using fuzzy rules based on a knowledge-based system[7]. In Fuzzy Petri nets (similar to ordinary Petri nets); places are denoted by circles, transitions are denoted by bars and the routes from places to transitions and from transitions to places are denoted by arcs. Firing a transition in a Fuzzy Petri net depends upon the fuzzy conditions associated with the transition. The object of this work is to produce a Fuzzy Petri net which defines how a vision system might make a crisp decision suitable for activating a robot (section 3). Also, the paper shows that with Fuzzy Petri nets it is essential to define fuzzy marking, fuzzy variables and fuzzy firing sequences.

## 2. THE FMS CASE STUDY

### 2.1 Petri net approach

A Petri net can be defined as a four-tuple (P, T, I, O), where $P=\{P_1,....,P_n\}$ is the set of P places, $T=\{t_1,....,t_m\}$ is the set of T transitions, I is an input function, and O is an output function. A marking **v** of a Petri net indicates the number of tokens in the places of that net. The marked Petri net consists of circles representing places, bars for transitions and solid dotes for tokens inside the circles (places) of the net. A transition T represents an event that may occur and the places P represents possible states which may become active within the system. A Petri net executes by firing its transitions. A transition fires by removing one token from each of its input places and depositing one token in each of the output places. Firing a transition will enable the passage from a terminated state to a new state and usually change the marking of the Petri net **v** to new marking **v'**.

## 2.2 A Petri net Control for the FMS Cell

The FMS cell example used in this paper can be decomposed into the following elements: 1) robot, 2) CNC-drilling machine, 3) computer vision system, 4) conveyor systems C1 and C2, 5) bin for discarded workpiece. The operation of the FMS cell ( Fig. 1) is as follows:

An incoming conveyor C1 transports the workpieces to the manufacturing cell; the robot picks the incoming material from the conveyor and feeds the CNC-drilling machine in the facility. The robot then returns to its normal position waiting for the CNC-drilling machine to complete its cycle. When the CNC-drilling machine has finished its task, the robot arm is extended to the design position where it picks up the workpiece from the CNC-drilling machine and puts it on the inspection table within the vision system for inspection. The robot then places the workpiece on the conveyor C2 or in the bin according to the decision taken by the fuzzy logic control architecture (section 3).

The initial marking after the PLC (Programmable logic controller) power's up of the FMS cell control Petri net (Fig. 2) is (1, 1, 0, 0, 0, 0, 0, 0, 0, 0, 0, 0, 1)$^T$. The behaviour of the FMS cell control Petri net is as follows:

1- Ti fires automatically after the PLC powers up based on the assumption that all flags in the PLC are reset on power up, depositing a token in Pi and Ps.

2- Ts fires when all the s/w elements within the cell have been initialised and informs the cell control net that the s/w initialisation is complete.

3- Th fires when all the h/w elements within the cell have been initialised and informs the cell control net that the h/w initialisation is complete.

4- Tr fires when all the cell elements (robot, vision system, CNC-drilling machine and conveyor C1, C2) are in a safe condition and the start button is pressed.

5- T1 fires when the robot picks the workpiece from conveyor C1 and has placed it in the design position on the bed of the CNC-drilling machine.

6- T2 fires when the CNC-drilling machine has completed its cycle.

7- T3 fires when the workpiece has been placed on the vision table for inspection.

8- T4 fires when the vision system has inspected the workpiece and the decision made to place the rejected workpiece in the bin. The methodology used to accomplish this stage is based on the Fuzzy Petri net (section 3).

9- T5 fires when the vision system has inspected the workpiece and the decision made to place the accepted workpiece on conveyor C2. The methodology used to accomplish this stage is based on the Fuzzy Petri net (section 3).

10- T6 fires when the robot has placed the rejected workpiece in the bin

11- T7 fires when the robot has placed the accepted workpiece on the conveyor C2.

12- T8 fires if a demand comes from the cell controller to repeat the cycle (robot picks another workpiece and places it on CNC-drilling machine.

13- T9 fires if a demand comes from the cell controller to return to $P_r$ (cell ready).

## 3. FUZZY PETRI NET

### 3.1 Why fuzzy logic control architecture?

This section illustrates how the decision taken by the vision system can lead to an uncertain situation which may affect the FMS cell performance. The vision system within the cell proceeds in the following manner:

Once the robot has placed the workpiece on the inspection table a signal is sent to the vision camera to it. The vision algorithm decides if the depth and diameter of the hole made by the drilling tool in the workpiece is to be accepted or rejected and the vision system outputs a signal directing the robot to put the workpiece in the bin or on the conveyor C2. An uncertain situation can arise when the robot receives ill-known knowledge from the vision system. Under these circumstances the robot may fail to identify the correct path decision i.e. put the workpiece on conveyor C2 or in the bin. To deal with this kind of uncertainty, a Fuzzy Petri net is used to monitor and control the vision system decision and determine subsequent robot action. In order to achieve this objective, fuzzy logic control architecture is constructed within the vision control Petri net (Fig. 4) in order to control the robot action. In this net, P2 represents the fuzzy logic control algorithm which:

1- Converts the input data to a fuzzy form (Fuzzification).

2- Evaluates the fuzzy control rules (Inference).

3- Computes an output value from the fuzzy value determined from 1 and 2 (Defuzzification).

With fuzzy logic[4] it is possible to interpret the fuzzy modelling in terms of a qualitative model which describes the system behaviour using a natural language such as "small", "medium" or "large", similar to the way human observation can be expressed. Unlike binary logic, which can evaluate two states of events, fuzzy logic allows intermediate states to be represented and thus an object can be more than just small or large. The fuzzy logic control architecture consists of:

1- Fuzzy set: Contains the elements $x_i$ with the degree of membership $\mu_i$ belonging to them.

2- Linguistic variable: A variable word used to describe fuzzy sets for example, "small", "medium".

3- Membership function: A function $\mu_A(x)$, $\mu_B(x)$, $\mu_C(x)$ describing the members of an element x in fuzzy sets A, B and C.

4- Degree of membership: Describes the association between an element x and a fuzzy set C. For example, $\mu_C(x_1) = 0.70$.

5- Fuzzy operators: Combines degree of membership in logical expressions, for example, OR/AND.

6- Knowledge-based rule: A method similar to that used by an expert system [8] (IF - THEN).

7- Fuzzification: Conversion of a physical input variable to a fuzzy form (Fig. 3).

8- Fuzzy inference: Derivation of fuzzy information based on IF-THEN rules (Fig. 3).

9- Defuzzification: The procedure to obtain a deterministic value of the output y from a fuzzy value (Fig. 3).

The defuzzification procedure is a centre of gravity method, which can be calculated using the following equation[4], [6]:

$$z = \frac{\sum_{i=1}^{n} z_i \cdot \mu(z_i)}{\sum_{i=1}^{n} \mu(z_i)}$$

The input to the fuzzy logic control is defined by x and y (where x represents the depth of the hole and y represents the diameter of the hole as a result of the drilling operation on the workpiece) and the output from the fuzzy logic control is defined by z (where z represents the quality of the machining operation, i.e., high or low ).

For example, if z is $\geq$ 70% this implies the quality of the workpiece is high and consequently, the robot places it on conveyor C2.

The fuzzy model on which the vision system decision is based uses two rules in this example:

- **IF** x is small and y is small **THEN** z is low (Fig. 3 rule 1).
- **IF** x is medium and y is small **THEN** z is high (Fig. 3 rule 2).

### 3.2 Fuzzy Petri net

The aim of this section is to model the vision system decision making mechanism and the robot action using a Fuzzy Petri net (Fig. 5). For this net, the initial marking $(1, 0, 0, 0, 0)^T$ implies that the vision system including a workpiece on the inspection table are available. Only one transition T1 is initially enabled since its input place P1 has a token. Firing T1 removes the token from P1 and deposits it into P2. Place P2, being marked, implies that the vision system is inspecting the quality of the workpiece. Firing T2 removes the token from P2 and deposits it into P3. Place P3, being marked (fuzzy marking), implies that the quality of the workpiece is evaluated based on fuzzy logic rules (IF....THEN). Firing T3 or T4 removes the token from P3 (fuzzy marking) and deposits a token into P4 or P5 if the fuzzy conditions attached to T3 or T4 are satisfied. When the robot has responded to the fuzzy firing reasoning sequences according to the possible token locations in places P4 or P5, a token is deposited in P7 (Fig. 2) instructing the cell control net to continue the cycle or return to Pr.

With the fuzzy Petri net described above it is essential to identify clearly the meaning of three elements:

### 3.2.1 Fuzzy markings

In Petri net theory, a Petri net marking **v** *is* defined as a number of tokens attached to the places in the net where the vector v is defined as follows[6]:

$$\mathbf{v} = (v1, v2,......., v_i)$$

The execution of a Petri net is controlled by the movement of tokens through the net. The rule to the firing of a transition is established by removing a token from each of its input places and placing a token in each of the output places.

Fuzzy marking can be defined as a possibility of token location in the output places after firing of fuzzy transitions. In Fig. 5, after firing T2 a token deposited in P3 represents all the fuzzy rules which can be interpreted by the membership functions, for example $\mu(x)$ and $\mu(y)$. T3 fires when the fuzzy condition associated with it is satisfied. This is interpreted by the fuzzy rules such that if x is small and y is small then z is low. T4 fires when the fuzzy condition associated with it is satisfied. This is interpreted by the fuzzy rules such that if x is medium and y is small then z is high.

### 3.2.2 Fuzzy variables
The input variables <x, y> are attached to the arc going from T2 to place P3 (Fig. 5) and represent the depth and the diameter of the hole on the workpiece evaluated by the vision system. The Place P3 converts the input variables <x, y> to a fuzzy form by using linguistic model such as "small" and "medium" (Fig. 3). The output variable <z> is crisp following defuzzification and represents the decision taken by the fuzzy logic control algorithm and implies that the workpiece quality is high or low.

### 3.2.3 Fuzzy firing sequences
The fuzzy firing sequences illustrated by transitions T3 and T4(Fig. 5) contain fuzzy reasoning information to describe the robot action in the output places P4 and P5. The crisp output value following the defuzzification stage provides information which determines the robot action to be taken (P4 or P5).

### Conclusion
This paper presents a Petri net approach for the modelling and control of an FMS cell. The methodology illustrates how different levels of activity such as sequential operations and concurrency can be monitored in a graphical manner by means of Petri nets. The paper emphasises how a Fuzzy Petri net can be used to determine a vision system decision in order to identify robot action within an FMS cell. The technique described offers an opportunity to model decision making in imprecise and vague situations such as those where quality is concerned . This methodology can be used to monitor uncertain situations and make effective decisions when choices and conflicts arise within such a system. The paper has also identified the meaning of fuzzy marking,

fuzzy variables and fuzzy firing sequences; these are essential elements in describing a Fuzzy Petri net for such systems. More work is required to investigate the application of Petri nets with Fuzzy logic to manufacturing systems in order to demonstrate how the technique can improve flexibility in design, system efficiency, cost effectiveness and the quality of products.

### References
1- MengChu Zhou, Frank DiCesare and Alan A. Desrochers "A Hybrid Methodology for Synthesis of Petri Net Models for Manufacturing Systems" IEEE Trans. on Robotics and Automation, Vol. 8, NO. 3, June 1992.

2- J. A. Cecil, K. Srihari and C. R. Emerson, "A review of Petri-net applications in manufacturing", The International Journal of Advanced Manufacturing Technology, vol. 7, pp 168-177, 1992.

3- Carolyn L. Beck and Bruce H. Krogh " Models for Simulation and Discrete Control of Manufacturing Systems" IEEE Trans on System, Man and Cybernetics, pp 305- 310, 1986.

4- Chen Chien Lee "Fuzzy logic in control systems: Fuzzy Logic Controller-Part II", IEEE Trans on System, Man and Cybernetics, Vol. 20, No. 2, pp 419-433, March 1990.

5- JC. Pascal, R. Valette " A Petri net based fuzzy PLC for linear interpolation between control steps", IEEE International workshop on Emerging Technology and Factory automation, Australia, August 1992.

6- R. Valette "Advanced Practical Tutorial" International Conference on Petri Nets, Chicago, 1993.

7- Carl G. Looney "Fuzzy Petri Nets for Rule-Based Decision making" IEEE Trans. on System, Man, and Cybernetics, Vol. 18, No. 1, January/February 1988.

8- D. T. Pham and P. T. N. Pham "Expert systems in mechanical and manufacturing engineering", The International Journal of Advanced Manufacturing Technology, 3(3), 3-21, 1988.

9- M. Hanna, A. Buck and R. Smith "Modelling Safety Requirements of An FMS using Petri-Nets", SPIE's Int. Symp. on: Optical Tools for Manufacturing and Advanced Automation. Vol. 2064(Boston, USA), pp 348-357, September 1993.

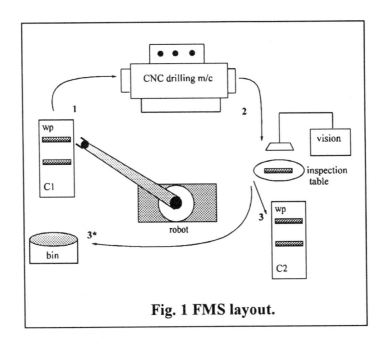

**Fig. 1 FMS layout.**

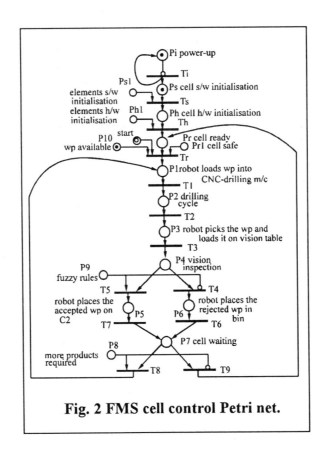

**Fig. 2 FMS cell control Petri net.**

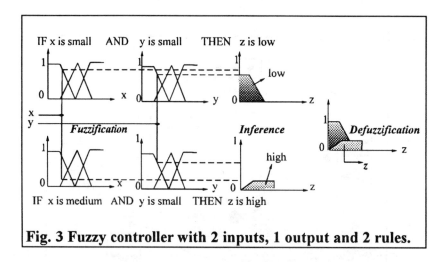

**Fig. 3 Fuzzy controller with 2 inputs, 1 output and 2 rules.**

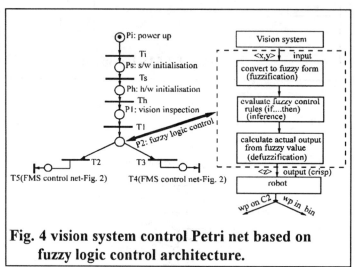

**Fig. 4 vision system control Petri net based on fuzzy logic control architecture.**

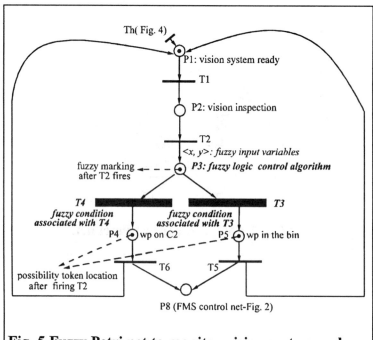

**Fig. 5 Fuzzy Petri net to monitor vision system and robot behaviour.**

# A Fuzzy Modeling of Very Large Scale System Using Genetic Algorithm and a Multiple-Representing Method

Tomohiro Yoshikawa, Takeshi Furuhashi, Yoshiki Uchikawa
Department of Information Electronics, Nagoya University
Furo-cho, Chikusa-ku, Nagoya, 464-01, Japan
Tel.+81-52-781-5111 ext.2792 Fax. +81-52-781-9263
E-mail: yoshi@bioele.nuee.nagoya-u.ac.jp

*Abstract-This paper presents a new approach to fuzzy modeling of very large scale systems using genetic algorithm. A new method for obtaining multiple phenotypes from a chromosome, i.e. a multiple-representing method, for expanding exploratory space is also proposed in this paper. The new method maintains the diversity of the phenotypes and is effective to explore vast space of solutions.*

*A simulation of fuzzy modeling to identify multi-inputs nonlinear system is done to show the effectiveness of the proposed method.*

*Keywords: Fuzzy modeling, Genetic algorithm, Large scale system*

## I.INTRODUCTION

Fuzzy inference has a high performance in human-computer interactions. Fuzzy modeling[1] is a method to describe systems with fuzzy rules and is effective for describing complex nonlinear systems. However, if the system is very large having many inputs and outputs, the fuzzy modeling becomes very difficult. There are many large scale systems such as chemical plant, human thought, etc. and the needs to describe these systems are growing.

Genetic algorithm (GA)[2][3] is one of the promising tools for solving combinatorial problems. The problems with many variables have a vast search space. The GA does a multi-point search and is suitable for solving this kind of problems. One of the difficulties of the GA is the premature convergence[2]. Diploidy and imposition of geography are said to be effective for this difficulty.

This paper presents a new approach to the fuzzy modeling of very large scale systems using the GA. This paper also presents a new method for obtaining multiple phenotypes from a chromosome for slowing done the premature convergence and for expanding exploratory space. The authors call the method a multiple-representing method. The chromosomes are selected with multiple fitness functions, and the diversity of chromosomes are maintained. The method is effective in searching vast space and is efficient to find out many pseudo-optimal solutions.

Simulations using the genetic algorithm with the proposed method to obtain a fuzzy model of a complex (ten inputs and one output) nonlinear equation are done. It is shown that the method is effective for searching a huge space in which the number of possible combinations of the fuzzy rules is several millions.

## II.MULTIPLE-REPRESENTING METHOD

Chromosomes consisting of 0, 1 bits or symbols are changed into solutions through some sort of rules. The authors call these rules representing methods. If there are multiple representing methods, multiple phenotypes are produced from a chromosome. Figure 1 shows a conventional representing method for a knapsack problem. There are eight packages numbered 1-8 and the loci of the chromosome correspond to these packages. If the locus has "1", the corresponding package is put into the knapsack. Figure 2 shows the multiple-representing method proposed in this paper. The figure shows the case where there are two

Fig.1 Representing method for knapsack problem

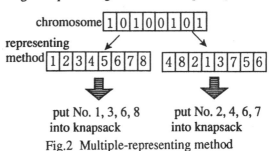

Fig.2 Multiple-representing method

representing methods for one chromosome. From the left hand side representing method, the packages 1, 3, 6, 8 are selected and from the right hand side, 2, 4, 6,7 are selected. One chromosome has the two combinations of packages in this case.

The features of this method are as follows:
(1)Multiple phenotypes are produced from a chromosome.

From $n$ chromosome and $m$ representing methods, $n \times m$ phenotypes are obtained.
(2)Diversity of chromosomes are maintained.

The chromosomes are grouped in subpopulations with the multiple-representing method through the selections, reproductions.
(3)Local and global search are done.

When the genetic operations do a local search under one representing method, at the same time, they do a global search from the point of view of the other representing method. Crossovers between the subpopulations also do global searches.

## III. FUZZY MODELING

The proposed representing method is applied to a search problem in a vast space. The problem is to obtain a fuzzy model of a complex nonlinear equation given by

$$y = (1 + x_1^{0.5} + x_2^{1} + x_3^{1.5} + x_4^{2.5} + x_5^{1.5})^2$$
$$+ (1 + x_6^{0.5} + x_7^{1} + x_8^{1.5})^2 \qquad (1)$$
$$+ 20 \sin (x_9 + x_{10}).$$

The fuzzy rule for this fuzzy modeling can be the form as

$$R^i: \text{If } x_1 \text{ is } A_{i1} \text{ and } x_2 \text{ is } A_{i2} \text{ and} \cdots x_{10} \text{ is } A_{i10}$$
$$\text{then } y = B_i \qquad (2)$$
$$(i = 1, 2, \cdots, n)$$

where $R^i$ is the $i$-th fuzzy rule and $A_{ij}$, $B_i$ are fuzzy variables. $n$ is the number of fuzzy rules. The min-max-center of gravity method is used to obtain the crisp output.

Figure 3 depicts an example of the inference method with three inputs $x_1$ - $x_3$, and three fuzzy rules $R^1$ - $R^3$. The triangular shapes in the figure are examples of membership functions. The truth value for each rule is derived as the minimum of the grades of the membership functions in the antecedent and the membership in the consequent is cut with the truth value. The inferred value $y^*$ is obtained by taking the envelope (maximum) of the trapezoidal membership functions in the consequent and by calculation the center of gravity of the envelope shape.

Since the minimum operation is used for deriving the truth value of the fuzzy rule, the whole grades of the membership functions $A_{ij}$ ($j$ = 1, 2, $\cdots$, 10) in the antecedent should have a certain amount of value for the fuzzy rule being activated, i.e. the truth value being greater than zero.

There are $(5^{10} =)$ 9,765,625 possible combinations of rules in the search space. This is a very large scale problem.

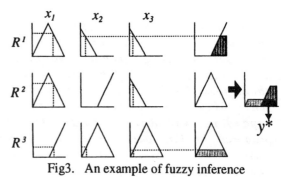

Fig3. An example of fuzzy inference

## IV. APPLICATION OF MULTIPLE-REPRESENTING METHOD

### 4.1 Coding

The chromosome for the fuzzy modeling in the previous chapter consists of 0, 1 bits. Five bits are allocated for each input/output variable. Each of the five bits correspond to either of the five membership functions Small(S), Medium Small(MS), Medium(M), Medium Big(MB), and Big(B). For simplicity, only one bit among the five bits for a variable is allowed to be "1". The length of the chromosome is then (10-inputs, 1-output) $* 5 = 55$ bits. Figure 4 shows an example of the chromosome and the corresponding membership functions. There are two representing methods in this case. One has a sequence of the membership functions as S, MS, M, MB, B as shown on the left hand side. The other has an opposite sequence as B, MB, M, MS, S. One chromosome has two different fuzzy rules.

### 4.2 Fitness value

Two types fitness values for each chromosome are used. One of the fitness values is given in proportion to the accumulated number of membership functions which have grades larger than zero. For a very large space search, it is not probable with the randomly generated chromosomes that the fuzzy rules can be activated, i.e. having the truth values larger than zero. This fitness value together with the crossover operator is effective to search activated fuzzy rules.

The other fitness value inversely proportional to the error of the rule is given to the rule. The error of the fuzzy rules $\Delta y$ are defined as

1896

$$\Delta y = |y^* - y| / (\max(y) - \min(y)) \quad (3)$$

where $y^*$ is the inferred value with the rule. $y$ is the output in eq. (1). When no rule is activated, $\Delta y = 1.0$. Once the rules having a certain amount of truth values are found, these fitness values are given to these rules to search valid rules with less errors.

Each chromosome has two rules in the case in Fig.4. Each rule has initial amount of fitness value. This fitness value is increased/decreased with the above two types of fitness values. For every one input/output pair of data, the fitness values of every chromosome is updated. The modifying amount of fitness value is made smaller in proportion to the number of activated rules for avoiding convergence of rules to particular data.

### 4.3 Selection, reproduction

Two groups of chromosomes are formed in the case in Fig.4. The chromosomes are grouped in view of the representing methods. If the roulette wheel selection[2] is used, the chromosomes are converged very prematurely. For maintaining the diversity of the chromosomes, eight tenth of the chromosomes are selected in the order of higher fitness values. If there are multiple chromosomes having the same strings, only one is selected. The rest of two tenth are selected with the roulette wheel selection. If there are 100 chromosomes and 2 representing methods, 40 chromosomes are selected in the order of the fitness values with one representing method and another 40 are selected in the order with another representing method. 20 chromosomes for each representing method are selected by the roulette wheel selection.

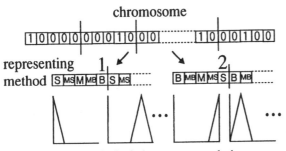

Fig.4 Coding method of chromosome and phenotypes

### 4.4 Crossover, mutation

Figure 5 shows an example of the crossover operation. Multi-point crossover is used. The crossover points are limited to the borders between the variables as shown in Fig.5. The probability of the crossover is $p_c$. The crossovers are done between the chromosomes in the same subpopulation. Between the two groups, one from each is selected and the crossover is done to them. This crossover constitutes a global search.

Mutation is applied to the chromosomes with the probability of $p_m$. The bit "1" for an input/output variable is changed to another locus in the same variable with the mutation operation.

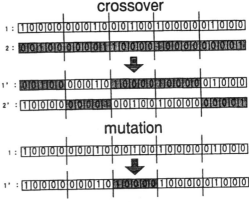

Fig.5 Crossover, Mutation

## V. SIMULATION

Simulations were done. Table 1 shows the input/output pairs of data used for the simulations. The data were randomly generated. The table shows the case where the number of data pairs N = 100. In a generation, all the pairs of data were used once. The ranges of the input variables are 1.0-5.0 for $x_1$-$x_4$ and $x_6$-$x_8$, 0.1-0.9 for $x_5$, and 0.0-1.5 for $x_9$ and $x_{10}$. The following two conditions were compared: (a) 180 chromosomes, 2 representing methods, (b) 360 chromosomes, 1 representing method. Other conditions of initial chromosomes, selection, reproduction, crossover, mutation, were made as equal as possible. The probabilities of crossover and mutation were set at $p_c = 0.5$, $p_m = 0.05$, respectively. For (a), the representing method 1 used the sequence S, MS, M, MB, B and the representing method 2 used that of M, MB, M, MS, S. Simulations up to the 30-th generation were done 50 times under each condition. Figure 6(a) shows the average number of activated rules at each generation. Figure 6(b) shows the average error. Since the number of possible combinations of the fuzzy rules was $5^{10} = 9,765,625$, there were few activated rules at the early generations. Approximately three tenth of the data had activated rules at the 30-th generation. The numbers were saturated because the search proceeded to decrease the errors than to find out activated rules in the midst of the generations. The performance of (a) was better than that of (b). Table 2 shows the ratios of effective crossovers for finding activated fuzzy rules under the condition (a). The crossovers (i) in a group of either of the representing method, (ii) in a group viewed from the different representing method, (iii) between the chromosomes each from the different group, were effective. (ii) and

(iii) share approximately a half of the effective crossovers and the proposed multiple representing method is found feasible.

Figure 7 shows the case where N = 1000, (a) 500 chromosomes, 2 representing method, (b) 1000 chromosomes, 1 representing method. Simulations up to 10-th generation were done 50 times under each condition. The average number of activated fuzzy rules are shown in Fig.7. The multiple-representing method (a) shows another high performance.

## VI.CONCLUSIONS

This paper presented a new fuzzy modeling method for very large scale systems using genetic algorithm. The multiple representing method was also proposed in this paper. The proposed method was effective for searching vast space of solutions.

## REFERENCES

[1] T.Takagi and M.Sugeno, "Fuzzy Identification of Systems and its Applications to Modeling and Control," IEEE Trans. Syst., Man, Cybern., vol.SMC-15, no.1, pp.116-132, (1985).

[2]D. E. Goldberg, "Genetic Algorithm in search, Optimization and Machine Learning", Addison Wesley (1989)

[3]L. Davis(Editor), "Handbook of Genetic Algorithm", Van Nostrand Reynold (1989)

Table 1  Input/output pairs of data

| No. | $x_1$ | $x_2$ | $x_3$ | $\cdots$ | $x_{10}$ |
|-----|-------|-------|-------|----------|----------|
| 1   | 3.40  | 2.18  | 1.15  |          | 1.33     |
| 2   | 3.81  | 1.54  | 3.27  |          | 0.35     |
| 3   | 1.90  | 1.54  | 2.63  |          | 1.25     |
| 4   | 1.59  | 4.93  | 3.06  |          | 0.56     |
| 5   | 3.03  | 2.67  | 1.09  |          | 0.45     |
| ⋮   |       |       |       |          |          |
| 99  | 1.76  | 2.82  | 1.33  |          | 0.84     |
| 100 | 1.72  | 2.68  | 4.45  |          | 0.15     |

Table 2  Effective crossovers

| i | ii | iii |
|-----|-----|-----|
| 54.3% | 40.5% | 5.1% |

number

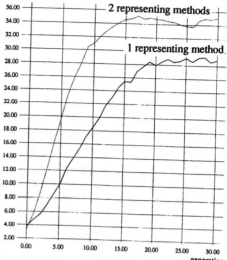

(a)  Average numbers of activated rules

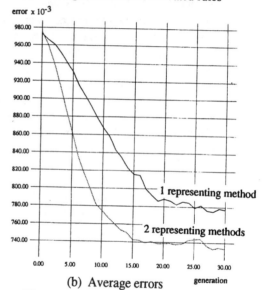

(b)  Average errors

Fig.6  Simulation results (N = 100)

number

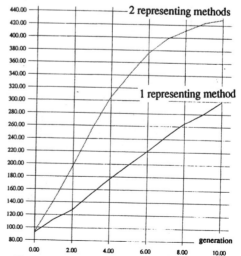

Fig.7  Simulation results (N = 1000)

# Fuzzy Fault Tree Analysis

**David P. Weber**
**GE Aircraft Engines**
**One Neumann Way MD T25**
**Cincinnati, OH 45215-6301 USA**

## SUMMARY & CONCLUSIONS

Reliability of products is frequently a prime safety consideration. Interpretation of reliability is both quantitative and qualitative. Extensive quantitative analysis employing probablistic risk assessment has been widely performed to provide predicted hazard or accident minimization. Weibull probability data and information is a vital tool of these quantitative risk assessments, but so are qualitative methods such as Fault Tree Analysis. Qualitative aspects of product risk contribution are subjective and contain many uncertainties. Risk analysis with all the inherent uncertainties is a prime candidate for Fuzzy Logic application. Most risk analyses do not have a method of expressing, propagating, or interpreting uncertainty. A Fuzzy Logic method employing Weibulls to represent membership functions for a set of fuzzy values (and fuzzy intervals) has been developed. (Ref. 1) Mixtures of "crisp" values with less-certain values do not "spoil" the analysis. The method can address subjective, qualitative, and quantitative uncertainties involving risk analysis.

## 1. INTRODUCTION

This paper will describe a new fuzzy logic software for risk analysis. A complete description of this new application with parallels to previous methods was presented by Weber (Ref. 1). The model discussed in this paper is called Fuzzy Fault Tree (FFT). The FFT employs "Fuzzy Weibull" membership (FWM) functions. A prototype system is programmed in the C programming language for Apple Macintosh® platforms.

Fault Trees study the avoidance of hazards. A hazard is an attribute of a product or a system of products used jointly, having capacity for harmful interaction with users. A risk on the other hand is the likelihood (0 to 1 scale) of some conceivable, unwanted, or undesired outcome (i.e., it is a probability of occurrence of that outcome). In product risk analysis, the objective is to determine the risk or probability of occurrence of outcomes affecting consumers. In the past, Fault Tree methods have been restricted to Boolean logic.

## 2. FUZZY FAULT TREE

Weber (Ref. 1) introduces a new graphical representation and computing paradigm, Fuzzy Fault Tree (FFT). FFT is a computer fuzzy logic program that automates the risk analysis of components used in any product. It has a Weibull graphical membership function for automating probability selection that transcends a wide probability scale (from $10^{-11}$ to $10^{-2}$). FFT contains a generic fault tree analysis system containing the means for addressing subjective qualitative and quantitative uncertainties involving risk analysis. FFT is a "C" program with graphs. It is written in the Symantec THINK C$^{TM}$ programming language for the Apple Macintosh® computer.

FFT provides graphical representation of risk analysis parametrics. FFT can provide rapid risk analysis of reliability data. It can identify areas of uncertainty and reflect uncertainty measurements in the results of the analysis. It can be used to check a basic component design by predicting its contribution to product risk. FFT will improve qualitative risk analysis and product reliability through superior analysis. FFT is based on Fuzzy Logic, it can handle fuzzy arithmetic with the application of fuzzy set theory (Refs. 1-6). FFT is not designed to produce crisp or finely tuned results. The resulting hazard potential is revealed with some measure of the uncertainty propagated to its level or state.

## 3. UNCERTAINTY IN RISK ANALYSIS

Even the most sophisticated, precise, and well-constructed quantitative model may give misleading results if uncertainties are not treated at some level. Uncertainty in risk analysis can range from modeling uncertainties, to incomplete and unreliable information. Data uncertainties are a major source. Some means of ranking incertitude relative to risk contribution is desired. Any system under study has dominant risk contributors in addition to the dependent failures usually studied. (Ref. 7) These can be categorized as either external events or human errors. Rankings can also range from "dominant" to "insignificant." As analysts of our own system we tend to treat only its inherent reliability and not its susceptibility to external causes and events. We often ignore human error in operation and maintenance as a major contributor to uncertainty. FFT was designed to handle these problems in incertitude.

## 4. CONVENTIONAL FAULT TREE METHODS

Fault tree analysis (FTA) is a systematic, deductive procedure. At its apex is a pre-defined undesired system event (the hazard to be analyzed). The analysis consists of determining the subevents, or combinations of subevents, within a system that will result in the pre-defined system event. The possible combinations and sequences of events that could contribute to this undesired event are graphically depicted by use of logic symbols. The logic sequence flows upward, toward the top of the tree. (See Figure 1) FTA defines the sequence of events required for the failure to progress to a potentially hazardous condition. The fault tree uses a set of standard logic symbols to represent the functional relationships that exist in the hardware or system. The symbols represent gates that are usually "AND," "OR," "IF," etc. By proper connection of these symbols and identification of the unique gate functions that each symbol represents, a logic diagram is produced. This diagram is a pictorial representation of the failure modes of the equipment and the sequence of events of each indenture level leading to the top undesired effect at the system level. Boolean logic is used for the probability evaluation of undesired events. The resulting Boolean logic diagram is then evaluated to estimate the probability of occurrence of the top level events.

## 5. SYMBOLIC PROCESSING FAULT TREE

For complex fault trees, application of this rigorous procedure requires an extensive quantity of "reasonably founded," detailed data for the probability solution. For this reason the author developed an interval method of fault tree analysis that uses symbolic processing or symbolic math. (Ref. 1) This much-simplified analytical approach processes symbols at gates in place of probability values. The Fault Tree (with probability scale) for this method is shown in Figure 1. The symbols in the middle match the semantic descriptions (probability terms) at the bottom of the figure. The uncertainty intervals used are those shown in Figure 1. The symbolic math is accomplished in this method using lookup tables for gate operations such as the "AND" gates, the "IF" (INHIBIT) gates and "OR" gates.

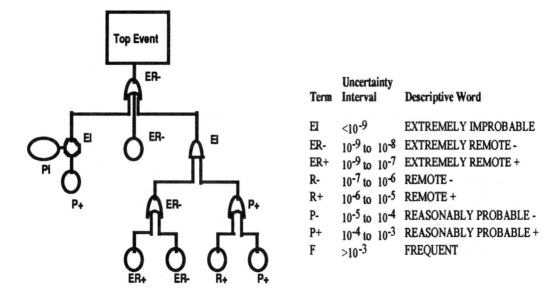

| Term | Uncertainty Interval | Descriptive Word |
|------|---------------------|------------------|
| EI | $<10^{-9}$ | EXTREMELY IMPROBABLE |
| ER- | $10^{-9}$ to $10^{-8}$ | EXTREMELY REMOTE - |
| ER+ | $10^{-9}$ to $10^{-7}$ | EXTREMELY REMOTE + |
| R- | $10^{-7}$ to $10^{-6}$ | REMOTE - |
| R+ | $10^{-6}$ to $10^{-5}$ | REMOTE + |
| P- | $10^{-5}$ to $10^{-4}$ | REASONABLY PROBABLE - |
| P+ | $10^{-4}$ to $10^{-3}$ | REASONABLY PROBABLE + |
| F | $>10^{-3}$ | FREQUENT |

**Figure 1 Fuzzy Fault Tree with Probability Scale**

This lookup table method works very well for symbolic calculation. The results of this type of symbolic math processing can be seen in the successive values achieved in the fault tree in Figure 1. Building upon this earlier method, the Fuzzy Fault Tree (FFT) method has been developed using Fuzzy Logic calculation in place of Boolean logic in lookup tables.

## 6. FUZZY SET THEORY APPLIED TO FAULT TREES

Fuzzy Logic describes and handles "vagueness" on equal terms with "crispness." (Refs. 5-6) If a properly constructed expert system has a built-in calculi of uncertainty and the means to handle these calculi in the reasoning process then qualitative risk assessment issues can be addressed somewhat superior to crisp Boolean calculations. (Ref. 8) Research was performed utilizing some of the uncertainties described in reference 1. Figure 2 is the representation of a fuzzy number. This graph is called a "T-norm" which means triangular "normal" type fuzzy number. The normal doesn't refer to a distribution but does imply that the apex of the triangle reaches 1.0.

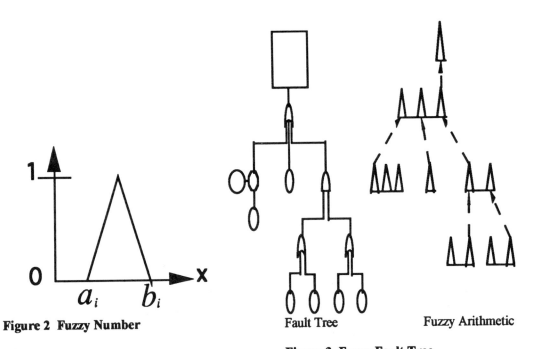

**Figure 2  Fuzzy Number**                Fault Tree                Fuzzy Arithmetic

**Figure 3  Fuzzy Fault Tree**

Fuzzy Logic provides a remarkably simple way to draw definite conclusions from vague, ambiguous or imprecise information. It is unlike Boolean logic which requires a deep understanding of a system, exact equations, and precise numeric values. Fuzzy Logic deals with vague statements and allows for partial membership in a set. It allows modeling complex systems using a higher level of abstraction originating from the analyst's knowledge and experience. Fuzzy Logic incorporates an alternative way of reasoning. FFT allows expressing this knowledge with subjective concepts.

A further extension of Fuzzy Logic itself has been made formulating a fuzzy membership function. The ordinate in Figure 2 is the membership function. This may be thought of as the "level of presumption" or "presumption level". (Ref. 10) The abscissa of Figure 2 is the interval of confidence. (Ref. 10)

In conventional set theory we use crisp intervals and something proposed is either a member of one interval or the other. In the fuzzy set, membership functions can overlap. Computationally, the semantic terms are input to the model and the operations are accomplished with equations that represent the T-norms involved.

Figure 3 illustrates how the fuzzy fault tree handles the fuzzy arithmetic of fuzzy gate calculation. Each gate operation is a fuzzy set operation of "OR," "AND" and "IF" combinations. Figure 4 shows four ways in which gates can be represented in the model. They can be represented as crisp numbers, crisp intervals, fuzzy numbers and as fuzzy intervals. The values read on the horizontal scale are used in the model as 4-tuple values. Computationally, the FFT model performs fuzzy arithmetic on these 4-tuples based on a method described by Bonissone (Ref. 8) as shown in Figure 4. Lopez de Mantaras, et al. have designed a fuzzy expert system shell called "MILORD" (Ref. 11) which also uses this system of fuzzy calculation.

Worst case studies with FTAs are sometimes performed where the failure rates and probabilities for components on the bottom of the tree are multiplied by 10 (or some multiple) to examine the net effect on the hazard potential at the top. With FFT fewer of these types of studies should be required in the future.

The matter of independence of events is handled by the type of fuzzy arithmetic operator (Ref. 8) used in the operation. For instance for "AND" gate operation the following rules of operator selection apply:

## "AND" Gate Computation

THE LOWER EXTREME BOUNDS

- if A and B are completely independent, then:
  A and B are two events
  $$P(AB) = P(A) \, P(B)$$

THE UPPER EXTREME BOUNDS

- if on the otherhand they may not be independent, then:
- the joint probability of the two events is:
  $$P(AB) = P(A) \, P(B|A) = P(B) \, P(A|B)$$
  since: $P(B|A) \leq 1$ and $P(A|B) \leq 1$
  then: $P(AB) \leq P(A)$
  $\qquad \leq P(B)$
  $$P(AB) \leq \text{Min} \, [\, P(A), \, P(B) \,]$$

A similar set of rules applies for "OR" gate operators where probabilities with dependencies between events are used as opposed to no dependency between events.

### Membership Distributions

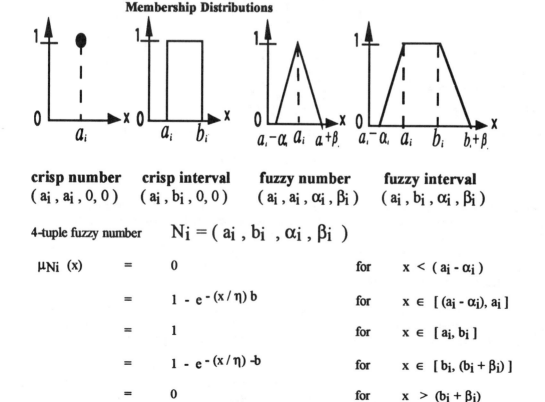

| crisp number | crisp interval | fuzzy number | fuzzy interval |
|---|---|---|---|
| $( a_i, a_i, 0, 0 )$ | $( a_i, b_i, 0, 0 )$ | $( a_i, a_i, \alpha_i, \beta_i )$ | $( a_i, b_i, \alpha_i, \beta_i )$ |

4-tuple fuzzy number $\qquad N_i = ( a_i, b_i, \alpha_i, \beta_i )$

$$
\begin{aligned}
\mu_{N_i}(x) &= 0 & \text{for} & \quad x < ( a_i - \alpha_i ) \\
&= 1 - e^{-(x/\eta)\, b} & \text{for} & \quad x \in [ (a_i - \alpha_i),\, a_i ] \\
&= 1 & \text{for} & \quad x \in [ a_i, b_i ] \\
&= 1 - e^{-(x/\eta)\, -b} & \text{for} & \quad x \in [ b_i, (b_i + \beta_i) ] \\
&= 0 & \text{for} & \quad x > (b_i + \beta_i)
\end{aligned}
$$

**Figure 4  Fuzzy Arithmetic With 4-tuple Numbers**

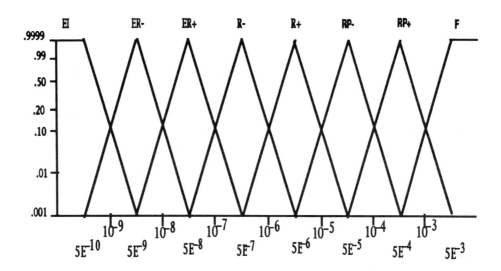

**Figure 5 Fuzzy Weibull**

## 7. FUZZY WEIBULLS

A unique Fuzzy Logic method has been developed employing the Weibull probability function to represent membership functions for sets of fuzzy values (and fuzzy intervals) along with crisp values. Since the membership function is similar to a cumulative density function (cdf), why not use the Weibull cdf? Weibulls have been used in risk assessment for years. Abernethy teaches courses in Weibull risk analysis and in measurement uncertainty handling with Weibulls. (Refs. 13-14) The author completed another research project that used Weibulls in an expert system for reliability design review. (Refs. 15, 16)

Because most risk analyses have to span a very wide probability scale (from $10^{-11}$ to $10^{-2}$), the concept that this could be represented on a Weibull scale was born. Semantic scale word inputs can be represented as fuzzy Weibull membership representations (FWMs) for fuzzy arithmetic computation. The Weibull function is asymptotic and therefore, never reaches either 0 or 1. FFT avoids discontinuity approaching 0 or 1 by calculating 0 at < .001 and 1 at > .9999. A 3-parameter Weibull could be used to represent the FWMs, but doesn't fit with the fuzzy arithmetic used.

An important fuzzy Weibull logic concept is the Weibull degree-of-membership function (FWM), which takes a value for an individual event and returns a value between 0.001 and 0.9999. (The Weibull cdf is asymptotic so 0 or 1 are only approached) The FWM indicates the extent to which that event belongs to a set. FFT uses terms that have shades of meaning between membership and non-membership in a set. In FFT the uncertainty calculi (linguistic statements of likelihood) are shown in Figure 1. The semantics and scales used for these intervals may vary. (Ref. 9) In FFT instead of intervals, Weibull T-norms (triangular normals as in Figure 2) are used. These are not linear, however, as can be seen in Figure 5 they are a Weibull graph. The x-scale is a natural log scale and the vertical scale is Weibull cdf (lnln). This is an extension of Fuzzy Set Theory (Refs. 1, 4, 8) which is, in turn, an extension of Set Theory. (Ref. 17) The uncertainty values can be chosen linguistically in FFT. The fuzzification in FFT is made using the semantic scale word inputs represented as their fuzzy Weibull representations for fuzzy arithmetic computation as shown in Figure 5.

## 8. ACKNOWLEDGEMENT

I wish to thank my daughter, Amy Weber, for her creative interpretation and programming of the Fuzzy Fault Tree prototype. All findings, opinions, and conclusions are those of the author alone. They do not represent the official position of the author's affiliation or of any organization or group.

# REFERENCES

1. Weber, David P., 1994, "Fuzzy Weibull for Risk Analysis," 94RM-150, *PROCEEDINGS Annual Reliability and Maintainability Symposium*, Anaheim, California.

2. Zadeh, Lotfi, 1965, "Fuzzy Sets," *Information and Control*, 8, (3), June, pp 338-353.

3. Bonissone, Piero P., 1991, "Approximate Reasoning Systems: A Personal Perspective," *Keynote paper at AAAI-91, The Ninth National Conference on Artificial Intelligence*, American Association for Artificial Intelligence, Anaheim, California.

4. Kosko, Bart, 1992, *Neural Networks and Fuzzy Systems: A Dynamical Systems Approach to Machine Intelligence*, Prentice-Hall, Inc., Englewood Cliffs, New Jersey.

5. Kosko, Bart, 1993, *Fuzzy Thinking: The New Science of Fuzzy Logic*, Hyperion, New York.

6. McNeill, Daniel and Friberger, Paul, 1993, *Fuzzy Logic*, Simon & Schuster, New York.

7. Vesely, William E., 1983, "The Facade of Probilistic Risk Analysis: Sophisticated Computation Does Not Necessarily Imply Credibility," *PROCEEDINGS Annual Reliability and Maintainability Symposium*.

8. Bonissone, Piero P. and Decker, K.S., 1986, "Selecting Uncertainty Calculi and Granularity: An Experiment in Trading-Off Precision and Complexity," *Uncertainty in Artificial Intelligence*, L. N. Kanak, J.F. Lemmer (Eds.), North Holland Publishing Company.

9. Fragola, Joseph R., 1993, "Designing for Success: Reliability Technology in the Concurrent Engineering Era," *PROCEEDINGS Annual Reliability and Maintainability Symposium*.

10. Kaufmann , Arnold and Gupta, Madan M., 1991, "Introduction to Fuzzy Arithmetic: Theory and Applications," Van Nostrand Reinhold, New York.

11. Lopez de Mantaras, R., Agusti, J., Plaza, E., and Sierra, C., 1991, "MILORD: A Fuzzy Expert Systems SHELL," Chapter 15, in *Fuzzy Expert Systems*, Kandel, Abraham, Editor, 1991, CRC Press.

12. Zadeh, Lotfi, 1992, "The Calculus of Fuzzy If/Then Rules," *IAI Expert Magazine*, March.

13. Abernethy, Robert B., Breneman, J.E., Medlin, C.H., Reinman, G.L., 1983, "AFWAL-TR-83-2079, Weibull Analysis Handbook," Aero Propulsion Laboratories, Airforce Systems Command, Wright-Patterson AFB, OH.

14. Abernethy, Robert B., 1989, "Workbook for Weibull Analysis-RMS Engineering Workshop," Presented at GE Aircraft Engines, Evendale, OH.+

15. Weber, David P., 1991, "Reliability Design Review ExPert (RDRx), *Workshop Notes from the AAAI-91 Workshop on "Knowledge-Based Construction of Probabilistic and Decision Models*. American Association for Artificial Intelligence, Anaheim, California.

16. Weber, David P., 1992, "Reliability Design Review Expert (RDRx) 92-WA/DE-13, *ASME 113th Winter Annual Meeting*, ASME, Anaheim, California.

17. Stoll, Robert R., 1979, "Set Theory and Logic," Dover Publications, Inc., New York.

+ This reference available from Dr. Robert B. Abernethy.

# BIOGRAPHY

David Weber is a Staff Engineer in Commercial Reliability Engineering. Mr. Weber has over 33 years experience in the engineering management of and performance of the design, evaluation test, and reliability analysis of electronic/mechanical components, computer models, computer software, missiles, space launch vehicles, and aircraft engines. Progressively responsible experience as manager of large and small groups of engineering professionals. Individual contributions have led to completion of many successful projects and programs. Mr. Weber has 31 years of experience in Reliability Engineering, with many original contributions of mathematical modeling for reliability, severity of operation, safety, maintainability, life cycle cost and fault analysis. He recently was responsible for development of Reliability and Maintainability methods and computer modeling software, including IR&D Expert Systems and AI Applications. Mr. Weber is a recognized expert in Weibull Probability Analysis modeling, simulation modeling and risk assessment. Mr. Weber served on both the USAF System Safety Groups for the C5-A and B1-B aircrafts as the engine representative. He performed the safety certification analysis (for FAA) on several GE Aircraft Engine's commercial engine programs. Senior Member of AIAA. Also member of ASME (27 years), and AAAI. Former member of ASQC, IES and SAE. Served on SAE Committee G-11, Reliability, Maintainability and Supportability (RMS) and on the Institute of Environmental Sciences' (IES) Reliability Growth Committee. Mr. Weber has patents in "Cold Welding" encapsulation of transistors and was named in the first edition of "Who's Who in Aviation and Aerospace" - US Edition, 1983. His BS in Mechanical Engineering is from the University of Evansville in 1959.

# FUZZY CALCULUS APPLIED TO REAL TIME SCHEDULING[*]

## François TERRIER[•], Ziqiang CHEN[+]

[•]E-mail : terrier @ albatros.Saclay.cea.fr
Phone : +33 (1) 69 08 62 59; Fax: (1)69 08 83 95
**Commissariat à l'Énergie Atomique**
**(French Atomic Energy Agency)**
Centre d'Études de Saclay
LETI (CEA - Technologies Avancées) DEIN
CE/S F91191 Gif sur Yvette Cedex France

[+]Phone: +33 (1) 45 35 46 06

**Department of Computer Science**
Roskilde University
DK-4000 Roskilde
Denmark

[*] This work has been supported by the European EUREKA project 711, IRTC (Intelligent Real Time Control).

**Abstract:**

Real time systems are more and more complex and, increasingly, involve large applications where real time constraints can be only roughly defined by use of vague linguistic terms. Generally, real time scheduling algorithms provide optimal solution when the application context is well known and its behavior completely mastered. But in practice, both the real time constraints and the tasks characteristics are imprecisely known. This results in sub-optimal behaviors generating failures which should be avoided. This paper proposes to apply the Fuzzy Calculus to real time tasks scheduling in order to allow both more realistic knowledge representations and definitions of more flexible real time scheduling algorithms.

**Keywords :**    Real time, fuzzy calculus, Earliest Deadline First (EDF), scheduling algorithms.

## 1    INTRODUCTION

Fuzzy set theory and logic, as a means of both capturing human expertise and dealing with uncertainty, have been applied to various domains such as industrial control, medical diagnosis, management and decision making systems. Some attempts have been made to process and manage temporal uncertainty ([DUB 92], [DUT 88], [CHE 91], [CHE 93]) and some of them have been applied in tasks scheduling, mainly for manufacturing system monitoring, for example:

- Fuzzy temporal windows have been used to solve a scheduling problem in [DUB 89].
- A fuzzy timed Pétri nets has been modeled and applied to monitoring in [VAL 89].

An other application domain concerned by fuzzy temporal information management is real time task scheduling. Real time domain covers a large set of applications going from very small and critic embedded systems (such as in-board missile command systems) to very large and complex in-line applications with a relatively low level of criticality (such as general purpose Man-Machine Interfaces). This kind of applications is different from those previously mentioned: a large part of the tasks are periodic and period concept has to be fully integrated to the scheduling.

In this paper, first we will define the application context: what is real time and how the scheduling algorithms work. Secondly, the role and the use of a fuzzy model in real time scheduling will be discussed. Finally, we will point out in the conclusion the remaining difficulties and work direction to follow in order to integrate fully a fuzzy scheduling model in industrial operating systems.

## 2    REAL TIME SCHEDULING

Real time applications can be defined as in [SPU 93]:

*An application is a real time one when it has to respect some time constraints .*

With this meaning the main point is the **respect of time constraints** and not *short execution times*. This distinction is very important because it is not a fast execution of the tasks but the guarantee of the time constraints respect which will ensure that a given real time application fits the users' needs. Real time task scheduling can be tackled by distinguishing two kinds of real time [ISO 93]:

- **Hard real time**: in this case, a failure of time constraints respect entails the whole application failure and application failure are not allowed for all the possible situations. *It is never allowed to fail.*

- **Soft real time**: in this case, a failure of the time constraints respect entails only the failure of some application tasks and causes no high damage. *Even if it is requested not to fail, it is allowed to fail.*

In fact, soft real time is more general than hard real time because it allows to deal with a bigger set of applications. It is on this context we would like to extend the classical scheduling algorithms by use of fuzzy constraints and values. Usually real time constraints correspond to two correlated aspects:

(A)• time dates or events associated to tasks execution;

(B)• task synchronization allowing task communication.

Real time scheduling algorithms are foccussed mainly on the management of the first kind of constraints:

- deadline, (**d**>0): the task must be completed before its deadline;
- ready time, (**r**≥0): the task can not start before its ready-time;
- period, (**p**>0): the task has to be periodically started, its ready time and deadline are inside each period.

No assumption is made on task communications, but practically, to achieve the scheduling, ready times and deadlines can be exploited only if the task execution time is also known. The knowledge of the task execution times often assumes that the tasks are quite independent and that no unpredictable communication delays are involved while the task execution. So, the role of the operating system is then to manage the tasks competing to the same resources, typically for the execution time aspects: the CPU resource. It is at this level that the connection is done between time constraints and execution time: the time used by a task to perform its operations depends of the CPU performance and of its availability at the request time. Generally, operating systems manage tasks competition from priority values defined by the developer or by scheduler, computing the tasks priorities depending on their real time constraints.

There is a very large set of scheduling algorithms adapted to different possible cases [LIU 73]. In the following, a simple, usual and general algorithm is described in order to have a precise idea of the existing solution before to analyze the difficulties of extending them to fuzzy information management.

**Earliest Deadline First : a typical scheduling algorithm**

Detailed definition and analysis of the EDF' algorithm can be found in [CHE 89] and [CHE 90]. Here, we consider execution plan (**schedule**) built only from periodic tasks with ready times and deadlines as real time constraints. Each execution of a periodic task can be seen by the scheduler as one particular task instance to schedule. This algorithm proposes both a schedulability criteria and a procedure for building a satisfactory solution. The **schedulability criteria** is defined by:

$$\sum_{i=1}^{n} e_i / p_i \leq 1 \quad \text{where } \mathbf{n} \text{ is the tasks number, } e_i \text{ the execution time of the task } T_i \text{ and } p_i \text{ its period.}$$

## 3  THE FUZZINESS IN REAL TIME CONSTRAINTS

The discussion presented hereafter tries to precise the bases of the problem of "Fuzzy scheduling". For that, the first point is to study and exploit the schedulability criteria which is in practice one of the main characterization feature of a task set. After this, the use of EDF algorithm in fuzzy context is briefly tackled.

## 3.1  APPLICATION BEHAVIOR DESCRIPTION

The fuzziness can be introduced at several levels:

- **Task execution time** evaluation.
- **Real time constraint** definition.

**Task execution time** represents a *measure* of a particular task characteristic and it is one of the most concrete and obvious place where imprecision has to be considered. It is often assumed that the task execution time is a priori and precisely known in the scheduling algorithms. But unfortunately, this usually needed hypothesis is often unrealistic and sometimes, completely false. In this case, the real time constraints guaranty provided by the algorithms becomes a "non-sense". Interesting approaches have been proposed based on the model called *variable depth resolution* [SHI 91]. But actually, these approaches do not correspond to the general case because they are applicable only when a task corresponds to an activity providing progressively more and more precise results by performing successive or iterative execution steps.

The nature of the **Real time constraints** is completely different from the previous one because the aim is not to measure a task characteristic, but is to define the context and the way to execute the task. We can often assume that these constraints are precise, but in order to be more general we can also consider that the real time constraints can be express by use of linguistic terms such as "*2 second periods*", "*less than 0.1 second response time*", etc. For large and complex application, the hypothesis of constraints precision is often arbitrary. For example, when the period is "*2 seconds*" it is generally accepted that little fluctuations append arround this value, for example "*2 seconds ± 0.1 %*" is set as correct. In other task the imprecision of these values can be very more important, for example a display task with a period of 10 seconds can support wide perturbations if globally the average of the display period value remains at 10 seconds. Here typically the size of the fluctuations is subjective and a good solution could be to capture this size by introducing the notion of fuzzy periods.

Another aspect of the real time constraints imprecision is due to the fact that we can accept that, sometimes (not often), some task execution are not performed (or lost). This aspect tends to associate to the time constraint another dimension: the relative task importance constraints. This concept of importance (criticality, etc.) is the base of a wide set of works ([JEN 93], [MAR 91], [BUR 91]) and will introduce too much complexity to be treated in this preliminary reflection.

## 3.2   EXTENSION OF EDF SCHEDULABILITY CRITERIA

Two elements have to be redefined for the use of fuzzy information: the schedulability criteria and the algorithm itself. For the schedulability criteria, the formulae used with precise information is, with the previous notations:

$$\sum_{i=1}^{n} e_i / p_i \leq 1 \qquad \text{what can be rewritten:} \qquad G \leq 1, \text{ by setting } G = \sum_{i=1}^{n} e_i / p_i$$

The way to manage fuzzy arithmetical expression has been deeply studied and in our case there is no particular theoretical difficulty. The usual results can be applied ([ZAD 68]). Two measures will be used to describe this criteria: *Possibility,* $\Pi ( G \leq 1 )$ and *Necessity* $N ( G \leq 1 )$. These two measures depend on the possibility distribution function of the variable G, $\pi_G(g)$:

$$\Pi ( G \leq 1 ) = \underset{g \leq 1}{\text{Sup}} \ ( \pi_G(g) ) \quad \text{and}$$

$$N ( G \leq 1 ) = 1 - \Pi(G > 1) = 1 - \underset{g > 1}{\text{Sup}} \ ( \pi_G(g) )$$

### 3.2.1   Fuzzy execution times and precise constraints

We assume here that only the task execution times are imprecise. Then we associate to each task execution time a possibility distribution specifying its fuzziness. For a task $T_i$ with an execution time $e_i$ its possibility distribution function depends on the time variable and is noted: $\pi_{e_i} ( t )$. We have now to define the possibility distribution of the criteria G from these of the task execution times. G appears here as a fuzzy multi-criteria relation depending on the fuzzy variables $e_1, ..., e_n$. So, by taking into account the minimum of specificity criteria for the definition of the G's possibility distribution function, we can set:

$$\pi_G ( g ) = \underset{t_1, ..., t_n, / \sum t_i / p_i = g}{\text{Sup}} \ ( \min ( \pi_{e_1}(t_1), \pi_{e_2}(t_2), ... , \pi_{e_n}(t_n) ) )$$

**Computation examples**

Let us consider now the previous example except that the task execution times here are imprecise and associated to triangular possibility distributions, for example:

$$T_1 (e_1 = 1 \pm .2 ; p_1 = 2) \quad T_2 (e_2 = 1 \pm .5 ; p_2 = 3) \quad T_3 (e_3 = 2 \pm 1 ; p_3 = 12)$$

The possibility distribution of the schedulability criteria variable G is:

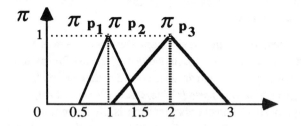 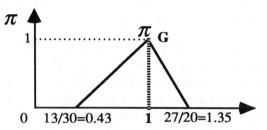

The possibility and necessity measures of this criteria are:

$$\Pi ( G \le 1 ) = \underset{g \le 1}{\text{Sup}} ( \pi_G(g) ) = 1 \quad \text{and} \quad N ( G \le 1 ) = 1 - \underset{g > 1}{\text{Sup}} ( \pi_G(g) ) = 0$$

It means practically that the criteria can be respected, but also that it is not guarantied. It is due to the fact that the central value of the imprecise execution time corresponds to the limit value of the criteria: **1**. More interesting cases for the criteria interpretation are the followings.

Central value greater than the criteria limit value:

$$T'_1 (e_1 = 1 \pm .2 ; p_1 = 2) \quad T'_2 (e_2 = 1 \pm 0.5 ; p_2 = 2) \quad T'_3 (e_3 = 2 \pm 1 ; p_3 = 8)$$

which gives the possibility and necessity values: $\Pi ( G \le 1 ) = 0.5$, $N ( G \le 1 ) = 0$. Here, the possibility to satisfy the scheduling criteria is low and indicates a bad real time constraints guaranty level.

Central value smaller than the criteria limit value:

$$T''_1 (e_1 = 1 \pm .2 ; p_1 = 3) \quad T''_2 (e_2 = 1 \pm .8 ; p_2 = 4) \quad T''_3 (e_3 = 2 \pm 1 ; p_3 = 12)$$

which gives the possibility and necessity values: $\Pi ( G \le 1 ) = 1$, $N ( G \le 1 ) \approx 1 - 0.4 = 0.6$. Now the good level of the necessity value gives a good level of the real time constraints guaranty.

### 3.2.2 Fuzzy constraints and general case

We use now the two possibility distributions: one for the task execution time, $\pi_{e_i} ( t )$, and one for the task periods, $\pi_{p_i} ( t )$. From them, we can construct for each task a characteristic variable by calculating the fuzzy ratio between the task execution time, $e_i$ and its period, $p_i$: $e_i/p_i$, called hereafter $x_i$. The possibility distribution of this ratio is then given by:

$$\pi_{X_i} ( x ) = \pi_{e_i / P_i} ( x ) = \underset{t_2 > 0,\ t_1 / t_2 = x}{\text{Sup}} ( \min ( \pi_{e_i}(t_1), \pi_{p_i}(t_2) ) )$$

The possibility distribution of the G variable can be obtained in the same way as previously:

$$\pi_G ( g ) = \underset{z_1, ..., z_n \sum z_i = g}{\text{Sup}} ( \min ( \pi_{x_1}(z_1), \pi_{x_2}(z_2), ... , \pi_{x_n}(z_n) ) )$$

## 3.3  EDF ALGORITHM EXTENSION

The extension of the scheduling algorithm has to start by the analysis of the operations involving fuzzy elements. For this, we have to list all used parameters : **t** (current time), **ST** (scale time), $r_{ij}$ (ready time of the $j^{th}$ occurrence of $T_i$), $d_{ij}$ (deadline of the $j^{th}$ occurrence of $T_i$), $e_i$ ($T_i$'s execution time) and finally, the **Schedule** itself because it is built based on other parameters.

We can see that the task periods do not appear explicitly, but they are used to define the time scale, $ST$, and also the elementary task executions set, $\{ T_{ij} \}$. If *the periods are fuzzy*, the semantics has to be precise both concerning the *time scale* and the *task executions set*. In fact, we have to specify what a fuzzy period means: for real time system users/developers the most intuitive and valid meanings of periods seem to be defined from two points of view:

- a local value corresponding to the time interval between two task execution occurrences,
- a global value defining the average value of the time intervals between two successive task executions.

For the first point, only the time interval between two time points is needed to be considered and a computation model can be found in [CHE 93]. For the second one, the concrete problem is to define the time scale used to compute the average values. One way would be to use the same time scale definition as for the unfuzzy case. If it gives a practical solution, it has to be validated on real application with the real application users/developers. This point is being studied, and tests are currently under definition in the IRTC project supporting this work at the French Atomic Energy Agency (CEA).

Before, to consider the general case where execution times and constraints (periods) are both fuzzy, it is interesting to discuss a little the simplest case where only execution times are fuzzy. The scale time, $ST$, depends only upon the preset task periods $p_i$. The periods are given precisely thus, $ST$ *remains unfuzzy in this case*. Ready times, $r_{ij}$ and deadlines, $d_{ij}$, are generally fixed and unfuzzy, but when a task is broken, the corresponding new deadlines are computed from the execution time values and thus become fuzzy.

*We will assume, for the moment, that the used task set do not entail such task splitting up.* So we can set, for the first evaluation, that *the ready times ($r_{ij}$) and the deadlines ($d_{ij}$) remain unfuzzy*.

And finally, the current time variable $t$ is managed by adding to it successively the task execution times. Thus, the current time, $t$, *has to be considered as a computed fuzzy variable*.

Two fuzzy operations have to be defined : *addition* of $t$ and $e_i$; *comparison* between the precise ready times ($r_{ij}$) and $t$ or the sum of $t$ and $e_i$. The usual and classical definition of fuzzy addition and comparison can be used and these two operations can be performed in the same way as in the previous paragraphs.

The question is mainly to precise the choice criteria and the effect of the chosen criteria on the schedule. The choice criteria is "*Choose among the ready executions, $r_{ij} \leq t$, the execution, $T_{kl}$, which has the earliest deadline, $d_{kl}$*". Because the current time, $t$, is computed from the fuzzy task execution times, so the comparison, $r_{ij} \leq t$, has to be represented by a couple of measures: its possibility and necessity measures. It means that this criteria has to be redefined by taking into account these two measures.

One simple solution is to select between tasks those which have the maximal necessity value. If all tasks have a necessity value equal to zero, then the selection will be based on the possibility values. And if all the possibility values are also null, then it means that there is absolutely no ready tasks at this moment. After this, the second selection criteria refers to the deadlines which are assumed as unfuzzy and entails no modification on the algorithm.

In fact, if the computing techniques are classical one in Fuzzy set theory, the main problem is to interpret the solution obtained: *which guaranties can be given on its correctness and optimality ? And, first of all, what means in fuzzy context concepts as correctness and optimality ? How the precise criteria used in precise context can be redefined for fuzzy scheduling ?*

Two extreme strategies can be envisaged as in [DUB 89]: *pessimistic* and *optimistic scheduling* corresponding, respectively, to the use of the maximal and the minimal execution time values. In this case, the algorithm has not really to be redefined: it uses respectively the minimal and maximal execution time values as precise execution time values. Moreover, the semantic of the resulting schedules are quite clear to explain to the users/developers. The only bad point is that, in practice, this approach does not exploit really the power of Fuzzy set theory and is similar to the use of the error calculus theory. For this reason, these two strategies are no more developed here by waiting to have a more global proposal covering not only these two extreme cases.

# 4  CONCLUSIONS AND PERSPECTIVES

Real time task scheduling is a fast expanding application domain. It involves more and more complex applications where the program execution time evaluation is very difficult. Moreover, in these applications, even if the real time constraints have to be globally respected, they are not so critical and precise as in the classical in board real time applications. For these reasons, the Fuzzy set theory seems to be very pertinent both for capturing human constraints formulation and imprecise task execution time evaluation.

Real time task scheduling covers two complementary aspects: the definition of schedulability criteria and the execution plan providing. Concerning the schedulability criteria, this paper shows by the analysis of the EDF scheduling algorithm that the usual results in fuzzy arithmetic can be easily applicable and can provide more realistic schedulability evaluation.

Concerning the redefinition of the algorithm itself by the Fuzzy set theory, the difficulty seems to be not in the fuzzy computing techniques, but mostly in the definition of the new algorithms objectives, i.e.: *what kind of solutions is provided, what are its characteristics and what means now correct and optimal ?*

The work presented in this paper can be considered as the first attempt to deal with fuzzy real time constraints and fuzzy execution times in real time scheduling. Tests and development are currently managed in order to obtain next year a fuzzy scheduler integrated into an industrial case tool for real time applications.

# REFERENCES

[BUR 91] A. Burns, "Scheduling hard real-time systems: a review", IEE/BCS Soft. Eng. Jour., USA, May 91.

[CHE 89] H. Chetto, M. Chetto, "Some results of the Earliest deadline scheduling algorithm", IEEE Transaction on Software Engineering, USA, October 1989.

[CHE 90] H. Chetto, M. Silly, T. Bouchentouf, "Dynamic scheduling of real-time tasks under precedence constraints", Revue The Journal of Real-Time Systems, USA, Sept. 1990.

[CHE 91] Z. Q. Chen, F. Terrier, "About Temporal Uncertainty", IEEE/ACM Developing and Managing Expert System Programs (DMESP'91), Sept. 1991, Washington D.C. USA.

[CHE 93] Z. Q. Chen, "Représentation et Gestion de Connaissances Temporelles et Incertaines", Ph.D. dissertation, University of Paris XI, Orsay, France, April 1993.

[DUB 89] D. Dubois, H. Prade, "Processing Fuzzy Temporal Knowledge", IEEE-"Transactions on Systems, Man, and Cybernetics", vol. 19, n°4, July/August 1989.

[DUB 92] D. Dubois, J. Lang, H. Prade, "Timed possibilistic logic", Fundamenta Informaticae Special Issue on Artificial Intelligence (Z. Ras, ed.), 1992.

[DUT 88] S. Dutta, "An event-based fuzzy temporal logic", 18th IEEE International Symposium on Multiple-Valued Logic, Palma de Mallorca, Spain, 1988.

[ISO 91] P. Isomursu, "On the Real-Time aspects of embedded expert systems", Espoo 1991, Technical Research Centre of Finland, Publications 78, Finland, 1991.

[JEN 93] E. D. Jensen, "A scheduling model for scaleable Real-Time computer systems", Proceedings *Real-Time Systems Conference*, Paris, January 1993.

[LIU 73] C. L. Liu, J. W. Layand, "Scheduling Algorithms for multiprogramming in a Hard Real-Time Environment", Journal of the ACM, 20:46-61, 1973.

[LOC 86] C. D. Locke, "Best-Effort Decision Making for Real-Time Scheduling", Doctoral Dissertation, Computer Science Department, Carnegie-Mellon University, 1986.

[SHI 91] W.-K. Shih, J.W.S. Liu, J.-Y. Chung, "Algorithms for scheduling imprecise computations with timing constraints", SIAM Journal on computing, Volume 20, n°3, pp. 537-552, USA, June 1991.

[VAL 89] R. Valette, J. Cardoso, D. Dubois, "Monitoring manufacturing systems by means of Pétri nets with imprecise markings", IEEE International Symposium on Intelligent Control, USA, September 89.

[ZAD 78] L. A. Zadeh, "Fuzzy sets as a basis for a theory of possibility", Fuzzy Sets & Systems, n° 1, 1978.

# Finding the Efficient Solutions to a Multiobjective Linear Program with Fuzzy Constraints

Richard W. Kelnhofer and Xin Feng
Department of Electrical and Computer Engineering
Marquette University
Milwaukee, WI 53233

*Abstract*— **This paper presents a method for obtaining the efficient solutions of a multiobjective linear programming problem over fuzzy linear constraints. The method is based on finding the set of maximally efficient solutions to a multiobjective linear program. This is accomplished by converting the fuzzy constraints into crisp constraints by adding a variable representing the degree of membership in the fuzzy feasible set. In addition, an objective function representing the maximization of this variable is augmented to the objectives of the original problem. An example is provided to illustrate the method.**

## I. INTRODUCTION

In multiobjective mathematical programming, a decision maker is required to maximize two or more objectives simultaneously over a given set of decision variables. Rarely does a single decision optimize all the objectives simultaneously. Instead, a compromise solution is selected from a set of possible solutions by the decision maker. A number of methods exist for finding a compromise solution [5]. These methods include the weighting method, assigning priorities to the objectives, or setting aspiration levels for the objectives. Each method results in a single solution. A different approach, called vector maximization, is the process by which a set of possible solutions that are efficient are generated [4]. Solutions are efficient when any change from a solution in the set of possible solutions that increases one objective will decrease another.

When modelling the problem, it is assumed that the constraints are known with certainty. However, problems exist in which the constraints are not exactly known or are vague in nature. In these instances, fuzzy set theory can be used in the model. Fuzzy set theory was first introduced in a fuzzy decision making process by Bellman and Zadeh [1]. Tanka et al. [7] applied the concepts of fuzzy sets to the decision making processes by considering the objectives as fuzzy goals over the $\alpha$-cuts of a fuzzy constraint set and Zimmermann [9] showed that classical algorithms can be used to solve mutliobjective fuzzy linear programming problems.

There are two limitations to the approaches taken by Zimmermann and Tanaka et al. First, only a single crisp solution is generated. Methods that formulate fuzzy solutions give the decision maker the added flexibility of selecting specific solutions within the fuzzy solution set. Second, the solution obtained is dependent on the membership functions assigned to the goals of the objectives. Setting low aspiration levels can result in decisions that may be suboptimal when compared to the original objectives of the problem.

In this paper we address these problems by proposing a multiobjective linear programming problem with fuzzy liner constraints. In our approach, the objectives remain crisp and only the constraints are fuzzy. The set of maximally efficient fuzzy solutions are found by extending an algorithm originally developed by Ecker et al. [3]. The algorithm generates the solutions by finding all efficient extreme points and efficient faces on a polyhedron representing the fuzzy feasible set. In addition to finding the efficient fuzzy set, the algorithm also generates a set of weight vectors. The weights provide a set of linear programming problems in which the objective function of each problem is a weighted combination of the original objectives. The set of linear programs are optimal for all solutions on the efficient face characterized by the weight vector.

In the next section, the general concepts of multiobjective programming, and linear programming with fuzzy constraints are reviewed. In Section III, the algorithm used to calculate the set of efficient fuzzy solutions is outlined. In Section IV, we illustrate the method by means of example. Finally, some concluding remarks are offered in Section V.

## II. PROBLEM FORMULATION

### A. Multiobjective Linear Programming

Consider the following multiobjective problem consisting of $k$ linear objective functions and $m$ linear constraints,

$$
\begin{aligned}
\text{vmax:} \quad & F(\mathbf{x}) = \mathbf{C}\mathbf{x} \\
\text{subject to:} \quad & \mathbf{A}\mathbf{x} \leq \mathbf{b} \\
& x_i \geq 0, \quad i = 1, \ldots, n,
\end{aligned}
\tag{1}
$$

0-7803-1896-X/94 $4.00 ©1994 IEEE

where the notation, "vmax", represents the maximization of the linear vector function $F : \Re^n \to \Re^k$, $\mathbf{x} \in \Re^n$, $\mathbf{C}$ is a $k \times n$ matrix with rows corresponding to the transpose of the coefficients of the objective functions, $\mathbf{A}$ is an $m \times n$ matrix, and $\mathbf{b} \in \Re^m$. We define some concepts commonly used in multiobjective linear programming.

**Definition 1 (Feasible Solutions)** *A feasible solution of (1) is any* $\mathbf{x} \in X$, *where* $X = \{\mathbf{x} \in \Re^n | \mathbf{A}\mathbf{x} \leq \mathbf{b}, x_i \geq 0, i = 1, \ldots, n\}$, *and* $X$ *is called the feasible set.*

**Definition 2 (Basic Variable)** *Consider the set of linear equations* $\mathbf{A}\mathbf{x} = \mathbf{b}$. *Let* $\mathbf{A}$ *be an* $m \times n$ *matrix with rank* $m \leq n$, *and let* $\mathbf{B}$ *be a non-singular* $m \times m$ *matrix formed by taking any* $m$ *linearly independent columns of* $\mathbf{A}$. *Index these columns by* $j_1, \ldots, j_m$. *Hence,* $\mathbf{B}$ *forms a column basis of* $\mathbf{A}$. $x_{j_k}$ *is equal to the* $k$th *component of the vector* $\mathbf{B}^{-1}\mathbf{b}$ *and is called a basic variable, provided that* $x_{j_k} \geq 0, k = 1, \ldots, m$. *The remaining* $n - m$ *variables are equal to zero and are called nonbasic.*

**Definition 3 (Polyhedron)** *Consider a halfspace given by the set* $\{x | a_1 x + \cdots + a_n x \leq b\}$. *A polyhedron is the intersection of a finite set of halfspaces.*

It is obvious from the above definitions that the feasible set of a multiobjective linear programing problem is a polyhedron formed by the intersection of the halfspaces corresponding to the $m$ constraints.

Because there is no natural ordering to $F(\mathbf{x})$, a single point that is optimal for all objectives rarely exists. There are a number of approaches that are commonly used in finding a compromise solution to (1). One approach is to generate a set of solutions, $X_E$ that is efficient in the sense that any change from a point $\mathbf{x}_E \in X_E$ that increases one objective will decrease another. Such solutions are also called Pareto-optimal or vector maximum [6]. The formal definition for an efficient point is given below.

**Definition 4 (Efficiency)** *A solution* $\mathbf{x}_E$ *is said to be efficient if* $f_j(\mathbf{x}) > f_j(\mathbf{x}_E)$ *implies that* $f_i(\mathbf{x}) < f_i(\mathbf{x}_E)$ *for at least one other index* $i \in \{1, \ldots, k\}$.

In general, $X_E$ will consist of edges and faces of the polyhedron, $X$.

*B. Multiobjective Linear Programming with Fuzzy Constraints*

In multiobjective linear programming, the coefficients are assumed crisp. While this assumption may be valid in many applications, there are problems in which the constraints are based on forecasted projections, or small violations of the constraints are permissible to a certain degree. For these instances,

fuzzy set theory can be used to model the uncertainty in the constraints.

A fuzzy version of (1) is formulated using the concepts of fuzzy numbers as defined by Dubois and Prade [2]. A general fuzzy multiobjective linear programming problem is given by

$$
\begin{aligned}
\text{vmax:} \quad & F(\mathbf{x}) = \tilde{\mathbf{C}}\mathbf{x} \\
\text{subject to:} \quad & \tilde{\mathbf{A}}\mathbf{x} \leq \tilde{\mathbf{b}} \\
& x_i \geq 0, \quad i = 1, \ldots, n,
\end{aligned} \tag{2}
$$

where $\tilde{\mathbf{C}}$ and $\tilde{\mathbf{A}}$ are $k \times n$ and $m \times n$ matrices respectively with coefficients consisting of fuzzy numbers, and $\tilde{\mathbf{b}}$ is an $n$-dimensional vector also with coefficients consisting of fuzzy numbers.

For the remainder of the paper we will use an approach similar to that found in [8], and model the problem with fuzzy numbers in the vector $\tilde{\mathbf{b}}$ and crisp coefficients in the matrices $\mathbf{C}$ and $\mathbf{A}$. However, we will not model the objectives as goals with fuzzy aspiration levels. Instead we will keep the objectives crisp. Proceeding in this fashion we have the problem

$$
\begin{aligned}
\text{vmax:} \quad & F(\mathbf{x}) = \mathbf{C}\mathbf{x} \\
\text{subject to:} \quad & \mathbf{A}\mathbf{x} \leq \tilde{\mathbf{b}} \\
& x_i \geq 0, \quad i = 1, \ldots, n.
\end{aligned} \tag{3}
$$

The fuzzy feasible set, $X_f$, is the intersection of the $m$ fuzzy inequality constraints. The membership functions for the constraints are equal to

$$
\mu_j(\mathbf{x}) = \begin{cases} 1, & \text{if } (\mathbf{A}\mathbf{x})_j \leq b_j \\ \gamma_j(\mathbf{x}), & \text{if } b_j < (\mathbf{A}\mathbf{x})_j \leq b_j + p_j \\ 0, & \text{otherwise} \end{cases} \tag{4}
$$
$$
j = 1, \ldots, m,
$$

where $(\mathbf{A}\mathbf{x})_j$ denotes the $j$th component of the vector $\mathbf{A}\mathbf{x}$, $p_j$ is called the tolerance interval of the $j$th constraint, and $\gamma_j : \Re^n \to [0, 1]$ is monotonically decreasing in $\mathbf{x}$. The membership function of the feasible set is found by taking the minimum of all membership functions of the constraints. Hence,

$$
\mu_{X_f}(\mathbf{x}) = \min\left(\mu_1(\mathbf{x}), \ldots, \mu_m(\mathbf{x})\right). \tag{5}
$$

In [8], Zimmermann has shown that if the membership functions of the constraints are linear, then standard linear programming algorithms can be used to solve (3). Using linear membership functions, (4) can be written as

$$
\mu_j(\mathbf{x}) = \begin{cases} 1, & \text{if } (\mathbf{A}\mathbf{x})_j \leq b_j \\ 1 - \dfrac{(\mathbf{A}\mathbf{x})_j - b_j}{p_j}, & \text{if } b_j \leq (\mathbf{A}\mathbf{x})_j \leq b_j + p_j \\ 0, & \text{otherwise} \end{cases}
$$
$$
j = 1, \ldots, m.
$$

$$\tag{6}$$

The fuzzy feasible set is a polyhedron in $\Re^{n+1}$ characterized by the equations

$$
\begin{aligned}
\mathbf{Ax} + \mathbf{p}\alpha &\leq \mathbf{b} + \mathbf{p} \\
\alpha &\leq 1,
\end{aligned}
\tag{7}
$$

where $\mathbf{p} = (p_1, \ldots, p_m)^T$ and $\alpha$ is in the $(n+1)$th coordinate of $\Re^{n+1}$. The value of the membership function for $X_f$ is on the surface of the polyhedron for which $\alpha$ is maximized. Therefore, the set of fuzzy efficient solutions to (3) can be found by solving for the efficient solutions of the crisp problem

$$
\begin{aligned}
\text{vmax:} \quad & \begin{pmatrix} \mathbf{Cx} \\ \alpha \end{pmatrix} \\
\text{subject to:} \quad \mathbf{Ax} + \mathbf{p}\alpha &\leq \mathbf{b} + \mathbf{p} \\
\alpha &\leq 1 \\
\alpha, x_i &\geq 0, \quad i = 1, \ldots, n,
\end{aligned}
\tag{8}
$$

where $\mathbf{p} = (p_1, \ldots, p_m)^T$.

## III. Algorithm for Finding the Fuzzy Efficient Solution Set

In this section we propose a new approach that uses an algorithm developed by Ecker et al. [3] to solve the complete set of efficient solutions to (8). The algorithm consists of six steps. The first step is an initialization stage. The second step generates an efficient vertex on the feasible set. The third step finds all efficient vertices adjacent to the current vertex. The fourth step finds the set of efficient faces adjacent to the current vertex. The fifth step calculates a weight vector when a new efficient face is found. Finally, in the sixth step the algorithm is either terminated or another iteration is performed.

Before we outline the steps of the algorithm we must modify the problem to facilitate computational requirements. First, arrange the problem such that it has all equality constraints. This is accomplished by adding slack variables corresponding to the inequality constraints in (8) and define the matrix

$$
\bar{\mathbf{A}} = \left( \begin{array}{c|c|c} \mathbf{A} & \mathbf{p} & \\ \hline \mathbf{O}_n & 1 & \mathbf{I}_{m+1} \end{array} \right),
\tag{9}
$$

where $\mathbf{O}_n$ is an $n$-dimensional row vector consisting of all zeroes, and $\mathbf{I}_{m+1}$ is an identity matrix of size $(m+1) \times (m+1)$. Next, define the vectors

$$
\bar{\mathbf{b}} = \left( (\mathbf{b} + \mathbf{p})^T, 1 \right)^T,
\tag{10}
$$

and

$$
\mathbf{y} = \left( \mathbf{x}^T, \alpha, s_1, \ldots, s_{m+1} \right)^T,
\tag{11}
$$

where $s_i$, are the slack variables. Finally, let

$$
\bar{\mathbf{C}} = \left( \begin{array}{c|c} -\mathbf{C} & \mathbf{O}_{m+2} \\ \hline \mathbf{O}_n & (-1, \mathbf{O}_{m+1}) \end{array} \right).
\tag{12}
$$

In (12), a sign change has been introduced to accommodate the minimization problem. The first $k$ rows of $\bar{\mathbf{C}}$ represent the original objectives and the $(k+1)$th row represents the maximization of $\alpha$ over the feasible set formed by the constraints in (8). We now have the crisp multiobjective problem with equality constraints,

$$
\begin{aligned}
\text{vmin:} \quad & \bar{\mathbf{C}}\mathbf{y} \\
\text{subject to:} \quad & \bar{\mathbf{A}}\mathbf{y} = \bar{\mathbf{b}} \\
& y_i \geq 0, \quad i = 1, \ldots, n+m+2.
\end{aligned}
\tag{13}
$$

The steps of the algorithm are as listed.

*Step 1.* Form the initial tableau

$$
\mathcal{T}^{(0)}: \quad \begin{array}{|c|c|} \hline & \mathbf{y} \\ \hline 0 & \bar{\mathbf{C}} \\ \hline \bar{\mathbf{b}} & \bar{\mathbf{A}} \\ \hline \end{array}
\tag{14}
$$

Let $L$ represent the total number of efficient faces the algorithm has found and let $I_l$ equal the set of extreme points adjacent to the $l$th efficient face. Initially, $L = 0$ and $I_l$ is empty.

*Step 2.* Find an extreme point, $\mathbf{y}^{(1)}$, with lowest possible membership. This is accomplished by setting $y_{n+1} = 0$. Note that $\alpha = y_{n+1}$. Let $B_{\mathcal{T}}$ equal the set of indices of all basic variables of the initial efficient extreme point and let $N_{\mathcal{T}}$ equal the set of indices for all nonbasic variables. Compute $\mathcal{T}^{(1)}$ by performing the appropriate pivots in the columns of $\mathcal{T}^{(0)}$ indexed by $B_{\mathcal{T}}$. Let $\bar{\mathbf{C}}_N$ equal the transpose of the matrix formed by taking the nonbasic columns of $\mathbf{C}$ in the current tableau. Note that $\bar{\mathbf{C}}_N$ is an $(n+1) \times (k+1)$ matrix.

*Step 3.* Consider the vector $\mathbf{v} = (v_1, \ldots, v_{k+1})^T$, and a vector $\mathbf{w}$ having indices corresponding to the nonbasic variables. Also consider the linear programming problems, for all $i \in N_{\mathcal{T}}$

$$
\begin{aligned}
\text{min:} \quad & w_i \\
\text{subject to:} \quad & -\bar{\mathbf{C}}_N \mathbf{v} + \mathbf{I}_{(k+1)}\mathbf{w} = \bar{\mathbf{C}}_N \mathbf{e} \\
& w_j, v1, \ldots, v_{k+1} \geq 0, \quad j \in N_{\mathcal{T}},
\end{aligned}
\tag{15}
$$

where $\mathbf{e}$ is a $(k+1)$-dimensional unit vector . Let $J_{\mathcal{T}}$, equal the set of $i$ such that the optimal value for $w_i$ in (15) is zero. $J_{\mathcal{T}}$ contains the set of indices of the nonbasic variables that can be pivotted into the basis and the resulting extreme point is efficient.

*Step 4.* Next let $F_l$ be a maximal subset of $J_{\mathcal{T}}$ such that the set

$$
G(F_l) = \{(\mathbf{v}, \mathbf{w}) \mid -\bar{\mathbf{C}}_N \mathbf{v} + \mathbf{I}_{(k+1)}\mathbf{w} = \bar{\mathbf{C}}_N \mathbf{e},
$$
$$
w_i = 0 \text{ if } i \in F_l\}
\tag{16}
$$

1913

is nonemtpy. The set $F_l$, $l \in \{1, \ldots, L\}$, represents the $l$th efficient face adjacent to the extreme point at the current tableau and contains the indices of nonbasic variables that can be pivotted into the basis in which the resulting point is an efficient vertex adjacent to the $l$th face. Note that $J_T = \bigcup_{l=1}^{L} F_l$ and $G(F_l)$ can be found by means of a tableau. If a new efficient face is found, then Step 5, else Step 6.

*Step 5.* For each new $l$, calculate $\lambda^{(l)} = \mathbf{v}^* + \mathbf{e}$, where $\mathbf{v}^*$ is found from (16). The vector $\lambda^l$ is a weight vector such that the linear program

$$
\begin{aligned}
\text{min:} \quad & \lambda^{(l)} \bar{\mathbf{C}} \mathbf{y} \\
\text{subject to:} \quad & \bar{\mathbf{A}} \mathbf{y} = \bar{\mathbf{b}} \\
& y_i \geq 0, \quad i = 1, \ldots, n+m+2,
\end{aligned}
\tag{17}
$$

is optimal for all efficient solutions on the $l$th face.

*Step 6.* Repeat steps 2, 3, and 4, by pivoting a nonbasic variable $y_j$, $j \in J_T$ into the basis and calculate a new tableau. Let $V$ equal the set of indices of the nonbasic variables corresponding to efficient extreme points that have not been examined. Terminate the algorithm when all efficient extreme points, edges, and faces have been calculated. This occurs when $V$ is empty. The set of efficient solutions is equal to the union of the efficient faces found.

## IV. AN EXAMPLE

In this section we consider an example to demonstrate the algorithm. For illustrative purposes we will list the variables using the original notation. Consider

$$
\text{vmax:} \quad \begin{pmatrix} x_1 + x_2 \\ x_1 \end{pmatrix}
\tag{18.a}
$$

$$
\begin{aligned}
\text{subject to:} \quad -x_1 + 2x_2 &\leq \tilde{2} \\
2x_1 + x_2 &\leq \tilde{8} \\
3x_1 + x_2 &\leq \tilde{12} \\
x_1, x_2 &\geq 0.
\end{aligned}
\tag{18.b}
$$

Using tolerance intervals of 1, 2, and 1 for the three constraints in (18.b) respectively, the initial tableau is given by

|         |       | $x_1$ | $x_2$ | $\alpha$ | $s_1$ | $s_2$ | $s_3$ | $s_4$ |
|---------|-------|-------|-------|----------|-------|-------|-------|-------|
|         |       | $y_1$ | $y_2$ | $y_3$    | $y_4$ | $y_5$ | $y_6$ | $y_7$ |
| $T^{(0)}$: | 0  | -1    | -1    | 0        | 0     | 0     | 0     | 0     |
|         | 0     | -1    | 0     | 0        | 0     | 0     | 0     | 0     |
|         | 0     | 0     | 0     | -1       | 0     | 0     | 0     | 0     |
|         | 3     | -1    | 2     | 1        | 1     | 0     | 0     | 0     |
|         | 10    | 2     | 1     | 2        | 0     | 1     | 0     | 0     |
|         | 13    | 3     | 1     | 1        | 0     | 0     | 1     | 0     |
|         | 1     | 0     | 0     | 1        | 0     | 0     | 0     | 1     |

Beginning with the initial efficient extreme point, $\mathbf{x}^{(1)} = (23/7, 22/7)^T$ and $\mu_{X_J}(\mathbf{x}^{(1)}) = 0$. The corresponding tableau is

|         |       | $x_1$ | $x_2$ | $\alpha$ | $s_1$ | $s_2$ | $s_3$ | $s_4$ |
|---------|-------|-------|-------|----------|-------|-------|-------|-------|
|         |       | $y_1$ | $y_2$ | $y_3$    | $y_4$ | $y_5$ | $y_6$ | $y_7$ |
| $T^{(1)}$: | 45/7 | 0   | 0     | 5/7      | 2/7   | 0     | 3/7   | 0     |
|         | 23/7  | 0     | 0     | 1/7      | -1/7  | 0     | 2/7   | 0     |
|         | 0     | 0     | 0     | -1       | 0     | 0     | 0     | 0     |
|         | 22/7  | 0     | 1     | 4/7      | 3/7   | 0     | 1/7   | 0     |
|         | 2/7   | 0     | 0     | 8/7      | -1/7  | 1     | -5/7  | 0     |
|         | 23/7  | 1     | 0     | 1/7      | -1/7  | 0     | 2/7   | 0     |
|         | 1     | 0     | 0     | 1        | 0     | 0     | 0     | 1     |

The basic variables are indexed by $B_{T^{(1)}} = \{1, 2, 5, 7\}$ and therefore, $N_{T^{(1)}} = \{3, 4, 6\}$. Next we examine

|            |       | $v_1$ | $v_2$ | $v_3$ | $w_3$ | $w_4$ | $w_6$ |
|------------|-------|-------|-------|-------|-------|-------|-------|
| $G^{(1)}(\emptyset)$: | -1/7 | -5/7 | -1/7 | 1     | 1     | 0     | 0     |
|            | 1/7   | -2/7  | 1/7   | 0     | 0     | 1     | 0     |
|            | 5/7   | -3/7  | -2/7  | 0     | 0     | 0     | 1     |

and find that either $w_3$ and $w_4$ can be made nonbasic by pivoting in columns $v_1$ and $v_2$ respectively. It is impossible to make $w_6$ a nonbasic variable such that $w_6 \geq 0$. Hence, $J_{T^{(1)}} = \{3, 4\}$. From the tableau

|            |   | $v_1$ | $v_2$ | $v_3$ | $w_3$ | $w_4$ | $w_6$ |
|------------|---|-------|-------|-------|-------|-------|-------|
| $G^{(1)}(\{3,4\})$: | 0 | 1 | 0 | -1 | -1 | -1 | 0 |
|            | 1 | 0     | 1     | -2    | -2    | 5     | 0     |
|            | 1 | 0     | 0     | -1    | -1    | 1     | 1     |

$G^{(1)}(\{3,4\}) \neq \emptyset$. Therefore, $F_1 = \{3, 4\}$ and $\mathbf{v}^* = (0, 1, 0)$. After the first iteration of the algorithm, $L = 1$, $I_1 = \{(\mathbf{x}^{(1)}, \mu(\mathbf{x}^1))\}$, and $\lambda^{(1)} = (1, 2, 1)^T$. If $y_3$ enters the basis then $y_5$ becomes nonbasic. Likewise if $y_4$ enter the basis, $y_2$ becomes nonbasic and we obtain $V = \{\{4, 5, 6\}, \{2, 3, 6\}\}$.

In the next iteration, $y_3$ enters the basis. The efficient vertex equals $\mathbf{x}^{(2)} = (13/4, 3)^T$ and $\mu_{X_J}(\mathbf{x}^{(2)}) = 1/4$, with tableau

|         |       | $x_1$ | $x_2$ | $\alpha$ | $s_1$ | $s_2$ | $s_3$ | $s_4$ |
|---------|-------|-------|-------|----------|-------|-------|-------|-------|
|         |       | $y_1$ | $y_2$ | $y_3$    | $y_4$ | $y_5$ | $y_6$ | $y_7$ |
| $T^{(2)}$: | 25/4 | 0   | 0     | 0        | 3/8   | -5/8  | 7/8   | 0     |
|         | 13/4  | 0     | 0     | 0        | -1/8  | -1/8  | 3/8   | 0     |
|         | 1/4   | 0     | 0     | 0        | -1/8  | 7/8   | -5/8  | 0     |
|         | 3     | 0     | 1     | 0        | 1/2   | -1/2  | 1/2   | 0     |
|         | 1/4   | 0     | 0     | 1        | -1/8  | 7/8   | -5/8  | 0     |
|         | 13/4  | 1     | 0     | 0        | -1/8  | -1/8  | 3/8   | 0     |
|         | 3/4   | 0     | 0     | 0        | 1/8   | -7/8  | 5/8   | 1     |

The basic variables are indexed by $B_{T^{(2)}} = \{1, 2, 3, 7\}$ and the nonbasic variables by $N_{T^{(2)}} = \{4, 5, 6\}$.

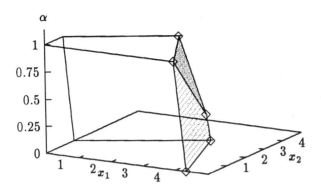

Fig. 1. Efficient Fuzzy solutions for example.

## V. Conclusions

In this paper a method for finding the efficient fuzzy solution set of a multiobjective linear programming problem with fuzzy linear constraints was presented. We have shown that a crisp multiobjective problem can be generated from the fuzzy problem by adding an additional objective and by increasing the dimensionality of the space spanned by the feasible set. In general, methods that generate the efficient solutions still require a decision maker to select an appropriate solution. Therefore, the material presented here is not recommended as a procedure for generating a specific solution to the problem, but more importantly, a tool for verifying the efficiency of solutions obtained using interactive fuzzy goal setting techniques.

## References

[1] R. E. Bellman and L. A. Zadeh. Decision-making in a fuzzy environment. *Management Science*, 17(4):141–164, December 1970.

[2] D. Dubois and H. Prade. Ranking fuzzy numbers in the setting of possibility theory. *Information Sciences*, 30:183–224, 1983.

[3] J. G. Ecker, N. S. Hegner, and I. A. Kouada. Generating all maximal efficient faces for multiple objective linear programs. *Journal of Optimization Theory and Applications*, 30(3):353–381, March 1980.

[4] A. M. Geoffrion. Proper efficiency and the theory of vector maximization. *Journal of Mathematical Analysis and Applications*, 22:618–630, 1968.

[5] R. E. Rosenthal. Concepts, theory, and techniques-principles of multiobjective optimization. *Decision Sciences*, 16:133–152, 1985.

[6] R. M. Soland. Multicriteria optimization: A general characterization of efficient solutions. *Decision Sciences*, 10:26–38, 1979.

[7] H. Tanaka, T. Okuda, and K. Asai. On fuzzy-mathematical programming. *Journal of Cybernetics*, 3,4:37–46, 1974.

[8] H. J. Zimmermann. Description and optimization of fuzzy systems. *International Journal of General Systems*, 2(4):209–215, 1976.

[9] H. J. Zimmermann. Fuzzy programming and linear programming with several objective functions. *Fuzzy Sets and Systems*, (3):45–55, 1978.

We examine

$$G^{(2)}(\emptyset):$$

|  | $v_1$ | $v_2$ | $v_3$ | $w_4$ | $w_5$ | $w_6$ |
|---|---|---|---|---|---|---|
| 1/8 | -3/8 | 1/8 | 1/8 | 1 | 0 | 0 |
| 1/8 | 5/8 | 1/8 | -7/8 | 0 | 1 | 0 |
| 5/8 | -7/8 | -3/8 | 5/8 | 0 | 0 | 1 |

From the previous iteration we know that $\{4,5\} \subseteq J_{T^{(2)}}$ and we see that $w_6$ can be made nonbasic by pivoting in column $v_3$. Hence all pivots will result in efficient extreme points and $J_{T^{(2)}} = \{4,5,6\}$. We next search for efficient faces. It is not necessary to check $G^{(2)}(\{4,5\})$ because $G^{(1)}(\{3,4\}) \neq \emptyset$. However, it remains to check the sets $G^{(2)}(\{4,6\})$ and $G^{(2)}(\{4,5,6\})$.

$$G^{(2)}(\{4,6\}):$$

|  | $v_1$ | $v_2$ | $v_3$ | $w_4$ | $w_5$ | $w_6$ |
|---|---|---|---|---|---|---|
| 0 | -1 | 1 | 0 | 5 | 0 | -1 |
| 1 | -1 | 0 | 0 | 2 | 1 | 1 |
| 1 | -2 | 0 | 1 | 3 | 0 | 1 |

Note that it is impossible to make $w_5$ a nonbasic variable in the above tableau. Therefore we have found another efficient extreme face with $F_2 = \{4,6\}$. After the second iteration, $I_1 = \left\{ \left(\mathbf{x}^{(1)}, \mu(\mathbf{x}^{(1)})\right), \left(\mathbf{x}^{(2)}, \mu(\mathbf{x}^{(2)})\right) \right\}$, $I_2 = \left\{ \left(\mathbf{x}^{(2)}, \mu(\mathbf{x}^{(2)})\right) \right\}$, $\lambda^{(2)} = (1,1,2)$, and $V = \{\{2,3,6\}, \{2,5,6\}, \{4,5,7\}\}$,

Continuing the algorithm in this fashion we find a total of five efficient extreme points and two efficient faces. The weight vectors corresponding to the efficient faces are $\lambda^{(1)} = (1,2,1)^T$ and $\lambda^{(2)} = (1,1,2)^T$. The efficient fuzzy solution set is plotted in Figure 1.

# APPROXIMATION THEORY OF FUZZY SYSTEMS

Xiao-Jun Zeng         Madan G. Singh

Decision Technologies Group
Department of Computation
UMIST, P.O. Box 88
Manchester M60 1QD, U.K.

## 1. INTRODUCTION

In recent years, there have been a number of applications of fuzzy systems theory in various fields, for example, in control systems. In most of these applications, the main design objective is to construct a fuzzy system to approximate a desired control or decision (often experientially based or heuristic control or decision of skilled operators or experts ). From a mathematical point of view, such a design objective is in fact to find a mapping from the input space to the output space which can approximate the desired (control or decision) function within a given accuracy; hence the problems of designing fuzzy systems can be viewed as approximation problems. However up till now, despite the fast development and wide applications of fuzzy system theory, few properties or features of such approximations have been reported. The awareness of such theoretic omission has motivated the research in this paper which discuss the approximation properties of fuzzy systems. Since fuzzy systems can be represented as linear combinations of fuzzy basic functions (FBFs)[1],[2] and the approximation properties of fuzzy systems are closely related to the properties of FBFs, we will firstly give an analysis of FBFs and then present several approximation properties of fuzzy systems: (1)basic approximation property which reveals the basic approximation mechanism of fuzzy systems; (2)uniform convergent property which shows that fuzzy systems with defined approximation accuracy can always be obtained by dividing the input space into finer and finer fuzzy regions; (3) universal approximation property which shows that fuzzy systems are universal approximators and extend some previous results.

## 2. FUZZY SYSTEMS AND FUZZY BASIC FUNCTIONS

In the section, we will give the mathematical formulae of fuzzy systems and fuzzy basic functions. Since a multi-input multi-output(MIMO) fuzzy system can alwaye be seperated into a group of MISO fuzzy systems [3], without loss of generality we assume in this paper that fuzzy systems are multi-input single-output(MISO) systems $f: U \subset R^n \mapsto V \subset R$ where $U = U_1 \times U_2 \times ... \times U_n \subset R^n$ is the input space and $V \subset R$ is output space.

General speaking, fuzzy systems are composed of four principal components: fuzzifier, fuzzy rule base, fuzzy inference engine, and defuzzifier. In this paper we assume that the fuzzifier is the most commonly used *singleton fuzzifer* and that the *fuzzy rule base* consists of $N = \Pi_{j=1}^n N_j$ rules in the following form:

$$R_{i_1 i_2 ... i_n} : IF \ x_1 \ is \ A_{i_1}^1 \ and \ ... \ and \ x_n \ is \ A_{i_n}^n,$$

$$THEN \ y \ is \ C_{i_1 i_2 ... i_n} \qquad (1)$$

$$i_1 = 1, 2, ..., N_1, \ ..., \ i_n = 1, 2, ..., N_n,$$

where $x_j$ (j = 1,2,...,n) are the input variables of the fuzzy system, $y$ is the output variable of the fuzzy system, and fuzzy sets $A_{i_j}^j \subset U_j$ and $C_{i_1 i_2 ... i_n} \subset V$ are linguistic terms characterized by fuzzy membership functions $A_{i_j}^j(x_j)$ and $C_{i_1 i_2 ... i_n}(y)$ respectively.

The fuzzy inference engine is a decision making logic module which employs fuzzy rules from the fuzzy rule base to determine a mapping from the fuzzy sets in the input space $U$ to the fuzzy sets in the output space $V$. Let $A$ be an arbitrary fuzzy set in $U$ and $A(X)$ (where $X = (x_1, x_2, ..., x_n)$ be its membership function, then each rule $R_{i_1 i_2 \ldots i_n}$ of (1) determines a fuzzy set $V_{A \circ R_{i_1 i_2 \ldots i_n}}$ in $V$ based on the sup-star composition [1],[2]:

$$
\begin{aligned}
& V_{A \circ R_{i_1 i_2 \ldots i_n}}(y) \\
= \ & Sup_{X \in U}[A(X) * R_{i_1 i_2 \ldots i_n}(X, y)] \\
= \ & Sup_{X \in U}[A(X) * A_{i_1}^1(x_1) * \ldots * A_{i_n}^n(x_n) \\
& \qquad\qquad * C_{i_1 i_2 \ldots i_n}(y)] \qquad (2)
\end{aligned}
$$

In this paper, we assume that $*$ is the *(algebraic) product*, then the sup-star composition in the fuzzy inference engine becomes a sup-product composition and (2) is simplified to:

$$
\begin{aligned}
& V_{A \circ R_{i_1 i_2 \ldots i_n}}(y) \\
= \ & Sup_{X \in U}[A(X) A_{i_1 i_2 \ldots i_n}(X) C_{i_1 i_2 \ldots i_n}(y)] \quad (3)
\end{aligned}
$$

where

$$
A_{i_1 i_2 \ldots i_n}(X) = A_{i_1}^1(x_1) A_{i_2}^2(x_2) \ldots A_{i_n}^n(x_n). \quad (4)
$$

Further we choose the defuzzifier to be the *centroid defuzzifier* [1],[2],[3] (called center-average defuzzifer in [2]) which is defined as

$$
y = \frac{\sum_{i_1 i_2 \ldots i_n \in I} V_{A \circ R_{i_1 i_2 \ldots i_n}}(y_{i_1 i_2 \ldots i_n}) y_{i_1 i_2 \ldots i_n}}{\sum_{i_1 i_2 \ldots i_n \in I} V_{A \circ R_{i_1 i_2 \ldots i_n}}(y_{i_1 i_2 \ldots i_n})} \tag{5}
$$

where the index set $I$ defined as

$$
I = \{ i_1 i_2 \ldots i_n \mid i_j = 1, 2, \ldots, N_j;\ j = 1, 2, \ldots, n \}, \tag{6}
$$

and $y_{i_1 i_2 \ldots i_n}$ is the point in $V$ at which $C_{i_1 i_2 \ldots i_n}(y)$ achieves its maximum value.

According to [2], for the above fuzzifier, fuzzy rule base, fuzzy inference engine, and defuzzifier, the corresponding fuzzy system can be formulated as follows.

*Lemma 1*[2]. The fuzy systems with singleton fuzzifier, product inference and center-average defuzzifier as well as normal fuzzy sets $C_{i_1 i_2 \ldots i_n}$ $(i_1 i_2 \ldots i_n \in I)$ in the rule base (1) are of the following form:

$$
\begin{aligned}
y &= f(X) = \frac{\sum_{i_1 i_2 \ldots i_n \in I} A_{i_1 i_2 \ldots i_n}(X) y_{i_1 i_2 \ldots i_n}}{\sum_{i_1 i_2 \ldots i_n \in I} A_{i_1 i_2 \ldots i_n}(X)} \\
&= \frac{\sum_{i_1 i_2 \ldots i_n \in I} \left[ \Pi_{j=1}^n A_{i_j}^j(x_j) \right] y_{i_1 i_2 \ldots i_n}}{\sum_{i_1 i_2 \ldots i_n \in I} \left[ \Pi_{j=1}^n A_{i_j}^j(x_j) \right]} \\
&= \sum_{i_1 i_2 \ldots i_n \in I} B_{i_1 i_2 \ldots i_n}(X) y_{i_1 i_2 \ldots i_n} \qquad (7)
\end{aligned}
$$

where $y_{i_1 i_2 \ldots i_n}$ is the point at which $C_{i_1 i_2 \ldots i_n}$ achieves its maximum value and

$$
\begin{aligned}
B_{i_1 i_2 \ldots i_n}(X) &= \frac{A_{i_1 i_2 \ldots i_n}(X)}{\sum_{i_1 i_2 \ldots i_n \in I} A_{i_1 i_2 \ldots i_n}(X)} \\
&= \frac{\Pi_{j=1}^n A_{i_j}^j(x_j)}{\sum_{i_1 i_2 \ldots i_n \in I} \Pi_{j=1}^n A_{i_j}^j(x_j)} \quad i_1 i_2 \ldots i_n \in I, \quad (8)
\end{aligned}
$$

are known as *Fuzzy Basic Functions*(FBFs)[1].

By (7), it is obvious that fuzzy systems can be represented by linear combination of the FBFs. In the rest of this paper, we always assume that the conditions of Lemma 1 are satisfied.

For a SISO fuzzy system with the fuzzy rule base

$$
R_i: \quad IF \ x \ is \ A_i, \ THEN \ y \ is \ C_i, \quad (9)
$$

$$
i = 1, 2, \ldots, N,
$$

the FBFs are given by

$$
B_i(x) = \frac{A_i(x)}{\sum_{i=1}^N A_i(x)}, \quad x \in U \subset R, \quad (10)
$$

$$
i = 1, 2, \ldots, N,
$$

and the fuzzy system can be written as

$$
f(x) = \sum_{i=1}^N B_i(x) y_i, \quad x \in U \subset R. \quad (11)
$$

## 3. PROPERTIES OF FUZZY BASIC FUNCTIONS

In this section, we will firstly present the properties of FBFs of SISO fuzzy systems, and then extend these properties to MISO fuzzy systems.

### A. Properties of FBFs of SISO fuzzy systems

In this subsection, we assume that fuzzy sets $A_1, A_2, \ldots, A_N$ in $U \subset R$ are the linguistic terms in IF-THEN rules base (9) and $A_i(x)$ $(i = 1, 2, \ldots, N)$ are the corresponding fuzzy membership functions.

*Definition 1. Pseudo Trapezoid-Shaped(PTS) Function and PTS Membership Function:* Let $[a, b] \subset U \subset R$. The PTS function is a continuous function in $U$ given by

$$
A(x; a, b, c, d, h) = \begin{cases} I(x) & x \in [a, b] \\ h & x \in [b, c] \\ D(x) & x \in (c, d) \\ 0 & x \in U - [a, b] \end{cases}
$$

Where $a \leq b \leq c \leq d$, $a < d$, $I(x) \geq 0$ is a strictly monotone increasing function in [a, b] and $D(x) \geq 0$ is a strictly monotone decreasing function in (c,d] . When the membership function of a fuzzy set $A$ is a PTS function, it is called the PTS membership function and denoted by $A(x) = A(x; a, b, c, d, h)$. When the fuzzy set $A$ is normal (i.e., h = 1), then its membership function is simply denoted by $A(x) = A(x; a, b, c, d)$.

*Remark 1.* According to Definition 1, obviously trapezoid-shaped (membership) functions and triangle-shaped (membership) functions are special cases of the PTS (membership) functions.

*Definition 2. Complete Partition :* Fuzzy sets $A_i$ $(i = 1, 2, ..., N)$ are said to be a complete partition on $U \subset R$ (simply said complete) if for any $x \in U$, there exists $A_i$ such that $A_i(x) > 0$.

*Definition 3. Consistency:* Fuzzy sets $A_1$, $A_2, ..., A_N$ are said to be consistent if $A_i(x_0) = 1$ for some $x_0 \in U$, then for all $j \neq i$, $A_j(x_0) = 0$.

*Definition 4. Order between Normal Fuzzy Sets:* For two normal fuzzy sets $A$ and $B$ in $U \subset R$, define $A > B$ if $M(A) > M(B)$(i.e., if $x \in M(A)$ and $\hat{x} \in M(B)$, then $x > \hat{x}$), where $M(.)$ is defined by $M(C) = \{x | x \in U$ and $C(x) = 1\}$ for a fuzzy set $C$.

The background to and intuitive explanations of the above concepts and notations are given in [4].

*Lemma 2 [4].* If $A_i$ are consistent and normal fuzzy sets in $U = [a, b] \subset R$ with PTS membership functions $A_i(x) = A_i(x; a_i, b_i, c_i, d_i)$ (i=1,2,...,N), then there exists a rearrangement $\{i_1, i_2, ..., i_N\}$ of $\{1, 2, ..., N\}$ such that

$$A_{i_1} < A_{i_2} < ... < A_{i_N}.$$

According to Lemma 2, without loss of generality we can assume that $A_1 < A_2 < ... < A_N$ when $A_1, A_2, ... , A_N$ satisfy the conditions of Lemma 2, because we can always rearrange $A_1, A_2, ..., A_N$ such that the assumption is satisfied.

*Theorem 1.* Let fuzzy sets $A_i$ in $U = [a, b]$ (i=1,2,...,N) be normal, consistent and complete with PTS membership functions $A_i(x) = A_i(x, a_i, b_i, c_i, d_i)$ (i=1,2,...,N) and $A_1 < A_2 < ... < A_N$. If $B_i(x)$ $(i = 1, 2, ..., N)$ are the corresponding FBFs defined by (10), then the following properties hold.

(a)*Structured Similarity:* the FBFs $B_i(x)$ (i=1,2,...,N) are PTS functions which are given

by

$$B_i(x) = B_i(x; a_i, d_{i-1}, a_{i+1}, d_i)$$

$$= \begin{cases} \frac{A_i(x)}{A_{i-1}(x) + A_i(x)} & x \in [a_i, d_{i-1}) \\ 1 & x \in [d_{i-1}, a_{i+1}] \\ \frac{A_i(x)}{A_i(x) + A_{i+1}(x)} & x \in (a_{i+1}, d_i] \\ 0 & x \in U - [a_i, d_i] \end{cases} \quad (12)$$

$$i = 1, 2, ..., N$$

where $d_0 = a_1$ and $a_{N+1} = d_N$.

(b)*Compatibility:* $B_i = \{(x, B_i(x)) | x \in U\}$ (i=1,2,...,N) are fuzzy sets in $U$ and are normal, consistent and complete with $B_1 < B_2 < ... < B_N$.

(c)*Complementarity:* $\{x \in U \mid 0 < B_i(x) < 1\} = (a_i, d_{i-1}) \cup (a_{i+1}, d_i)$ and

$$B_i(x) + B_{i-1}(x) = 1, \quad x \in (a_i, d_{i-1}),$$
$$B_i(x) + B_{i+1}(x) = 1, \quad x \in (a_{i+1}, d_i),$$

$$i = 1, 2, ..., N.$$

## B. Properties of FBFs of MISO fuzzy systems

In this subsection, we will discuss the properties of FBFs of MISO systems. For this purpose, we will firstly present a seperation property which gives an explicit relationship between MISO FBFs and SISO FBFs and then use this property to extend the above properties of SISO FBFs to MISO FBFs.

*Theorem 2.* Let $B_{i_1 i_2 ... i_n}(X)$ $(i_1 i_2 ... i_n \in I)$ be FBFs defined in (8), then

$$B_{i_1 i_2 ... i_n}(X) = \Pi_{j=1}^n B_{i_j}^j(x_j), \quad i_1 i_2 ... i_n \in I, \quad (13)$$

where $B_1^j(x_j)$, $B_2^j(x_j)$, ... , $B_{N_j}^j(x_j)$ are FBFs generated by fuzzy sets $A_1^j$, $A_2^j$, ... $A_{N_j}^j$ $(j = 1, 2, ..., n)$, i.e.,

$$B_{i_j}^j(x_j) = \frac{A_{i_j}^j(x_j)}{\sum_{i_j=1}^{N_j} A_{i_j}^j(x_j)} \quad (14)$$

$$i_j = 1, 2, ..., N_j; \, j = 1, 2, ..., n.$$

Based on Theorem 1 and Theorem 2 , we give the properties of FBFs of MISO fuzzy systems as follows.

*Theorem 3.* Suppose that (1) fuzzy sets $A_{i_1 i_2 ... i_n} = A_{i_1}^1 \times A_{i_2}^2 \times ... \times A_{i_n}^n$ ( $i_1 i_2 ... i_n \in I$ ) in $U = [a_1, b_1] \times [a_2, b_2] \times ... \times$

$[a_n, b_n] \subset R^n$ have the membership functions $A_{i_1 i_2 \ldots i_n}(X) = \Pi_{j=1}^n A_{i_j}^j(x_j)$ and $A_{i_j}^j(x_j) = A_{i_j}^j(x_j; a_{i_j}^j, b_{i_j}^j, c_{i_j}^j, d_{i_j}^j)$ ( $i_j = 1, 2, \ldots, N_j$; $j = 1, 2, \ldots, n$ ) are PTS membership functions of fuzzy sets $A_{i_j}^j$ in $U_j = [a_j, b_j] \subset R$; (2) fuzzy sets $A_1^j, A_2^j, \ldots A_{N_j}^j$ in $U_j$ are normal, consistent and complete , and $A_1^j < A_2^j < \ldots < A_{N_j}^j$ ( $j = 1, 2, \ldots, n$); (3) $B_{i_1 i_2 \ldots i_n}(X)$ ( $i_1 i_2 \ldots i_n \in I$) are FBFs defined in (8). Then the following properties hold.

(a)*Structured Similarity:* the FBFs $B_{i_1 i_2 \ldots i_n}(X) = \Pi_{j=1}^n B_{i_j}^j(x_j)$ ( $i_1 i_2 \ldots i_n \in I$) and $B_{i_j}^j(x_j)$ are PTS functions given by

$$B_{i_j}^j(x_j) = B_{i_j}^j(x_j; a_{i_j}^j, d_{i_j-1}^j, a_{i_j+1}^j, d_{i_j}^j), \quad (15)$$

$$i_j = 1, 2, \ldots, N_j; \ j = 1, 2, \ldots, n.$$

where $d_0^j = a_1^j, \ a_{N_j+1}^j = d_{N_j}^j$.

(b)*Compatibility:* $B_{i_1 i_2 \ldots i_n} = \{(X, B_{i_1 i_2 \ldots i_n}(X) ) \mid X \in U\} = B_{i_1}^1 \times B_{i_2}^2 \times \ldots \times B_{i_n}^n$ ( $i_1 i_2 \ldots i_n \in I$) are fuzzy sets in $U$ and fuzzy sets $B_1^j, B_2^j, \ldots B_{N_j}^j$ in $U_j$ are normal, consistent and complete ( $j = 1, 2, \ldots, n$) with $B_1^j < B_2^j < \ldots < B_{N_j}^j$ ( $j = 1, 2, \ldots n$), where fuzzy sets $B_{i_j}^j = \{(x_j, B_{i_j}^j(x_j) ) \mid x_j \in U_j\}$ ( $i_j = 1, 2, \ldots, N_j; \ j = 1, 2, \ldots, n$).

(c)*Complementarity:* For any given $X_0 \in U$, there exist at most $2^n$ FBFs $B_{i_1 i_2 \ldots i_n}(X)$ which are not equal to zero at $X_0$ and the sum of these FBFs is equal to 1 at $X_0$ .

*Remark 2.* Another property of FBFs is less fuzzility [4], [5], we omit it here because of the space limitation.

# 4.APPROXIMATION PROPERTIES OF FUZZY SYSTEMS

In this section, we always assume that $g(X)$ is the desired (control or decision) function in $U = [a_1, b_1] \times [a_2, b_2] \times \ldots \times [a_n, b_n] \subset R^n$ and $f(X)$ is the fuzzy system given in (7). Now we will discuss the approximation properties of fuzzy systems.

## A. Basic approximation property and fuzzy approximation mechanism

In order to present the basic approximation property of fuzzy systems in a unified and compact form, we will introduce some notation and provide preliminary results.

Firstly we denote the sets $U_{i_j}^j \subset U_j$ to be

$$U_{i_j}^j = \begin{cases} [a_1^j, d_1^j) & i_j = 1 \\ (a_{i_j}^j, d_{i_j}^j) & 2 \leq i_j \leq N_j - 1 \\ (a_{N_j}^j, d_{N_j}^j] & i_j = N_j \end{cases} \quad (16)$$

$$i_j = 1, 2, \ldots, N_j; \ j = 1, 2, \ldots, n.$$

Secondly define set value mappings $I_{i_j}^j$ from $U_{i_j}^j$ to $I_{N_j}$ (where $I_{N_j}$ is the set which is composed of all subsets of set $\{0, 1, 2, \ldots, N_j + 1\}$ ) as

$$I_{i_j}^j(x_j) = \begin{cases} \{i_j-1, i_j\} & x_j \in (a_{i_j}^j, d_{i_j-1}^j) \\ \{i_j\} & x_j \in [d_{i_j-1}^j, a_{i_j+1}^j] \\ \{i_j, i_j+1\} & x_j \in (a_{i_j+1}^j, d_{i_j}^j) \end{cases} \quad (17)$$

$$i_j = 1, 2, \ldots, N_j; \ j = 1, 2, \ldots, n.$$

Now we introduce a lemma which will be used for the definitions of further notations.

*Lemma 3.* If the conditions of Theorem 3 are satisfied, then

$$U_j = \bigcup_{i_j=1}^{N_j} U_{i_j}^j, \quad U = \bigcup_{i_1 i_2 \ldots i_n \in I} U_{i_1 i_2 \ldots i_n} \quad (18)$$

where $U_{i_1 i_2 \ldots i_n} = U_{i_1}^1 \times U_{i_2}^2 \times \ldots \times U_{i_n}^n \subset U$.

*Remark 3.* It has been proved in [5] that

$$U_{i_1 i_2 \ldots i_n} = S_{A_{i_1 i_2 \ldots i_n}}, \ i_1 i_2 \ldots i_n \in I,$$

where $S_{A_{i_1 i_2 \ldots i_n}}$ is the support of fuzzy set $A_{i_1 i_2 \ldots i_n}$.

Based on Lemma 3, further we define set value mappings $I_{i_1 i_2 \ldots i_n}$ from $U_{i_1 i_2 \ldots i_n}$ to $I_{N_1} \times I_{N_2} \times \ldots \times I_{N_n}$ as

$$I_{i_1 i_2 \ldots i_n}(X) = I_{i_1}^1(x_1) \times I_{i_2}^2(x_2) \times \ldots \times I_{i_n}^n(x_n)$$
$$= \{(j_1, j_2, \ldots j_n) \mid j_k \in I_{i_k}^k(x_k); \ k = 1, \ldots, n\} \quad (19)$$

$$X \in U_{i_1 i_2 \ldots i_n}, \ i_1 i_2 \ldots i_n \in I,$$

and finally define

$$I_U(X) = I_{i_1 i_2 \ldots i_n}(X), \ if \ X \in U_{i_1 i_2 \ldots i_n}. \quad (20)$$

Under the assumptions of Theorem 3, it can be proved[5] that the set value mapping $I_U(X)$ are well defined (set value) function on $U$.

*Theorem 4.* Suppose that the conditions of Theorem 3 are satisfied, then for any $X \in U$,

$$|g(X) - f(X)|$$
$$= \left|g(X) - \sum_{i_1 i_2 \ldots i_n \in I_U(X)} B_{i_1 i_2 \ldots i_n}(X) y_{i_1 i_2 \ldots i_n}\right|$$
$$\leq max\{|g(X) - y_{i_1 i_2 \ldots i_n}| \mid i_1 i_2 \ldots i_n \in I_U(X)\} \quad (21)$$

Consider SISO fuzzy system (11). For any $x \in U = [a, b]$, by Lemma 3 there exists $i$ ($1 \leq i \leq N$) such that

$$x \in U_i = \begin{cases} [a_1, d_1) \\ (a_i, d_i) \\ (a_N, d_N] \end{cases}$$

(it was proved in [4] that $a_1 = a$, $d_N = b$ ). Then from (17), (19), and (20) it is implied for $x \in U_i$ that

$$I_U(x) = \begin{cases} \{i-1, i\} & x \in (a_i, d_{i-1}) \\ \{i\} & x \in [d_{i-1}, a_{i+1}] \\ \{i, i+1\} & x \in (a_{i+1}, d_i) \end{cases} \quad (22)$$
$$i = 1, 2, \ldots, N.$$

Hence from Theorem 4, we have

$$|g(x) - f(x)| = |g(x) - B_{i-1}(x)y_{i-1} - B_i(x)y_i|$$
$$\leq max\{|g(x) - y_{i-1}|, |g(x) - y_i|\}, x \in (a_i, d_{i-1}) \quad (23)$$
$$|g(x) - f(x)| = |g(x) - y_i|, x \in [d_{i-1}, a_{i+1}] \quad (24)$$
$$|g(x) - f(x)| = |g(x) - B_i(x)y_i - B_{i+1}(x)y_{i+1}|$$
$$\leq max\{|g(x) - y_i|, |g(x) - y_{i+1}|\}, x \in (a_{i+1}, d_i). \quad (25)$$

Now we will use (23), (24), and (25) to give a simplified discussion about the approximation mechanism of fuzzy systems. Based on Theorem 1, dividing $U_i = S_{B_i} = \{x \mid B_i(x) > 0\}$ into its certainty subset $C(i) = \{x \mid B_i(x) = 1\} = [d_{i-1}, a_{i+1}]$ and uncertainty subset $U(i) = \{x \mid 0 < B_i(x) < 1\} = (a_i, d_{i-1}) \cup (a_{i+1}, d_i)$, then $y_i$ is the value of fuzzy system $f(x)$ in $C(i)$ (since $f(x) = y_i$ when $x \in C_i$ from (11) and the consistency of Theorem 1(b) ). For any $x \in U = [a, b]$, if $x$ is located in one of the certainty subset $C(i)$, then from (24), the fuzzy system approximates $g(x)$ by $y_i$; If $x$ is located in one of the uncertsinty subset $U(i)$, then $x \in (a_i, d_{i-1})$ or $x \in (a_{i+1}, d_i)$. For $x \in (a_i, d_{i-1})$ (similar for $x \in (a_{i+1}, d_i)$ ), from (23) and Theorem 1(c) the fuzzy system approximates $g(x)$ by a convex weighted sum of $y_{i-1}$ and $y_i$ (the values of the fuzzy system in $C(i-1)$ and $C(i)$). Noting $(a_i, d_{i-1})$ is

the area located between the certainty subsets $C(i-1) = [d_{i-2}, a_i]$ and $C(i) = [d_{i-1}, a_{i+1}]$, we can conclude that the fuzzy system $f(x)$ approximates $g(x)$ in a uncertainty subset by a convex weighted sum of the values of $f(x)$ in the two neighboring certainty subsets. Further from Theorem 1(a) and the definition of PTS functions, the magnitude of the weigh $B_{i-1}(x)$ to $y_{i-1}$ ($B_i(x)$ to $y_i$) depends on the distance of $x$ to $C(i-1)$ ( $C(i)$ ); the shorter the distance, the larger the weight. If we define $y_i = g(x_i)$ for $x_i \in C(i)$, such an approximation mechanism is quite similar to the approximation machanism in mathematical approximation theory. More detailed discussion on this aspect is given in [5].

## B. Uniform convergent property

In practical applications, when the designed fuzzy system does not have the desired approximation accuracy, an intuitive idea to modify the system and improve its performance is to divide the input space into finer partitions and redesign the fuzzy system based on the finer partition of input space.. Some recent examples[6] showed that using such an idea can achieve very satisfactory approximation accuracy. The following convergent property gives a theoretical foundation to this intuitive idea and can be used as a guide for practical design.

In order to give the uniform convergent property, denote

$$\vec{a}_{i_1 i_2 \ldots i_n} = (a_{i_1}^1, a_{i_2}^2, \ldots, a_{i_n}^n) \quad (26)$$
$$\vec{d}_{i_1 i_2 \ldots i_n} = (d_{i_1}^1, d_{i_2}^2, \ldots, d_{i_n}^n) \quad (27)$$
$$\delta_{i_1 i_2 \ldots i_n} = ||\vec{d}_{i_1 i_2 \ldots i_n} - \vec{a}_{i_1 i_2 \ldots i_n}|| \quad (28)$$
$$\delta = max\{\delta_{i_1 i_2 \ldots i_n} | i_1 i_2 \ldots i_n \in I\} \quad (29)$$

where $|| \cdot ||$ is any norm defined in $R^n$. With this notation, we have the following result.

*Theorem 5.* Suppose that the conditions of Theorem 3 are satisfied and

$$Inf_{X \in S_{A_{i_1 i_2 \ldots i_n}}} g(X) \leq y_{i_1 i_2 \ldots i_n}$$
$$\leq Sup_{X \in S_{A_{i_1 i_2 \ldots i_n}}} g(X), \quad i_1 i_2 \ldots i_n \in I. \quad (30)$$

If fuzzy system

$$f_\delta(X) = \sum_{i_1 i_2 \ldots i_n \in I} B_{i_1 i_2 \ldots i_n}(X) y_{i_1 i_2 \ldots i_n}$$

and $B_{i_1 i_2 \ldots i_n}(X)$ ($i_1 i_2 \ldots i_n \in I$) are FBFs given in (13) and (15), then

$$lim_{\delta \to 0} Sup_{X \in U} |g(X) - f_\delta(X)| = 0 \quad (31)$$

Remark 4. If we denote $||.||_p$ $(p = 1, 2, ..., \infty)$ as the $p$-norm in $R^n$, then it is easy to verify that $\delta_{i_1 i_2 ... i_n} = ||\vec{d}_{i_1 i_2 ... i_n} - \vec{a}_{i_1 i_2 ... i_n}||_p = Sup_{X, X' \in S_{A_{i_1 i_2 ... i_n}}} ||X - X'||_p$. Based on this fact, $\delta_{i_1 i_2 ... i_n}$ can be defined as the magnitude of the support $S_{A_{i_1 i_2 ... i_n}}$ and $\delta$ defined in (29) can be viewed as the magnitude of the partition $\{A_{i_1 i_2 ... i_n}; i_1 i_2 ... i_n \in I\}$. Hence Theorem 5 means that the fuzzy system $f_\delta(X)$ converges to the desired (control or decision) function $g(X)$ when the magnitude of the partition $\{A_{i_1 i_2 ... i_n}; i_1 i_2 ... i_n \in I\}$ tends to zero under some appropriate conditions.

### C.Universal approximation property

In this part, we will prove the universal approximation property of fuzzy systems. Similar results on the universal approximation property have been obtained in [1],[7] for some types of fuzzy systems by using the sophisticated Stone-Weierstrass theorem. In the following this property is extended to a general class of fuzzy systems. Further a constructed proof have been given in [5] which presented an idea of how to get the desired fuzzy system $f(X)$ and shows some basic reasons why such a property holds (we omit the proof here because of the space limitation).

*Theorem 6.* Let $g(X)$ be a given continuous function on $U = [a, b]$ and $\epsilon > 0$ be an arbitrary real number, then there exists a fuzzy system $f(X)$ in the form of (7) such that

$$Sup_{X \in U} |g(X) - f(X)| < \epsilon.$$

*Remark 5.* Theorem 6 is different from the results of [1] in the assumptions of membership functions and different from the result of [7] in the defuzzifier of fuzzy systems. From fuzzy control theoretic point of view, the result in [7] in fact reveals that expert fuzzy controllers[8] are universal fuzzy controller whereas the result in [1] means that fuzzy logic controllers with Gaussian membership functions are universal fuzzy controllers and Theorem 6 here means that fuzzy logic controllers with the PTS membership functions are universal fuzzy controllers.

## 5.CONCLUSIONS

In this paper, we discussed the approximation properties of fuzzy systems and presented several basic approximation properties of fuzzy systems.

Because these results are obtained under the assumption that the T-norm in fuzzy implication and inference is a (algebraic) product, further research work is to extend the results to fuzzy systems with other T-norms such as min (intersection), bounded and drastic [3] and compare the results between different T-norms.

## References

[1] L.X. Wang and J.M. Mendel, *IEEE Trans. Neural Networks*, Vol. 3, no. 5, pp. 807-814, 1992.

[2] L.X. Wang, *IEEE Trans. Fuzzy Systems*, Vol.1, no.2, pp.146-155, 1993.

[3] C.C. Lee, *IEEE Trans. Syst. Man, Cybern.*, Vol. 20, no. 2, pp. 404-435, 1990.

[4] X.J. Zeng and M.G. Singh, "Approximation theory of fuzzy systems—SISO case," submitted for publication.

[5] X.J. Zeng and M.G. Singh, "Approximation theory of fuzzy systems—MIMO case," submitted for publication.

[6] L.X. Wang and J.M. Mendel, *IEEE Trans. Syst., Man, Cybern.*, Vol. 22, no. 6, pp. 1414-1427, 1992.

[7] J.J. Buckley, *Automatica*, Vol. 28, no. 6, pp. 1245-1248, 1992.

[8] J.J. Buckley, "Theory of the fuzzy controller: a brief survey," in *Cybernetics and Applied Systems*, C.V. Negoita,Ed. New York: Marcel Dekker, 1992, pp. 293-307.

# Fuzzy reasoning with feature cumulation as cluster analysis

*Wladyslaw Homenda*
*Institute of Mathematics, Warsaw University of Technology*
*pl. Politechniki 1, 00-661 Warsaw, Poland.*

Abstract: *The paper deals with the problem of fuzzy reasoning in cluster analysis. The problem of repetitive fuzzy information as features' estimation is considered. The "cumulative" operators on fuzzy sets is introduced and relevant composition operator of fuzzy relational equations is defined to cumulate repetitive information.*

Keywords: *repetitive information, cluster analysis, feature cumulation, fuzzy reasoning, fuzzy relational equations.*

## 1. Introduction.

A lot of research has already been devoted to fuzzy reasoning and it applications in different fields including fuzzy controllers, diagnosis systems, pattern recognition, object identification, etc. This work was inspired by problems of pattern recognition and cluster analysis [11].

Let's consider a fuzzy reasoning rule of the form:

$$\text{If } F_1 \text{ is } x_{11} \quad \text{and} \quad \dots \quad \text{and} \quad F_k \text{ is } x_{1k} \quad \text{then} \quad Y_1 \text{ is } y_1$$

$$\dots\dots\dots\dots$$

$$(1) \qquad \text{If } F_1 \text{ is } x_{n1} \quad \text{and} \quad \dots \quad \text{and} \quad F_k \text{ is } x_{nk} \quad \text{then} \quad Y_n \text{ is } y_n$$

$$\text{If } F_1 \text{ is } x_1 \quad \text{and} \quad \dots \quad \text{and} \quad F_k \text{ is } x_k$$

---

$$\text{then} \quad Y \text{ is } B$$

where:

- $F_1 , \dots , F_k ,$          are names of features,
- $x_{11} , \dots x_{1k} , \dots x_{n1} , \dots x_{nk}$   are relationships between features and classes, namely - $x_{ij}$ is a relationship between ith feature and jth class,
- $x_1 , \dots x_k, y_1 , \dots y_n$      are certainty factors from [0,1],
- $C_1 , \dots C_n$             are names of classes;
- $[Y_1 , \dots , Y_n], B$     are fuzzy sets from $F(\{C_1 , \dots C_n \})$ space;
- $X, Y$               are variables.

---

*Now the author is with Centro de Investigacion Cientifica y de Education Superior de Ensenada, Km. 107 Carret. Tijuana/Ensenada, Ensenada, B.C., Mexico 22860.*

This form of fuzzy reasoning rule is similar to that given by expert in knowledge acquisition process in, for example, cluster analysis systems. Descriptions obtained from the expert are of the form: if features $F_1$, ... $F_k$ are observed with certainty factors $x_{i1}$, ... $x_{ik}$ respectively then belongingness of object Y to classes $C_1$, ... $C_n$ should be considered with certainty factor $y_1$, ... , $y_n$. Having the given space of features $F_1$, ... ,$F_k$ with certainty factors $x_1$, ... $x_k$, we are faced the problem of concluding relevant space of classes $C_1$, ... $C_n$ and their certainty factors $y_1$, ... $y_n$.

This form of fuzzy reasoning may also be expressed as a fuzzy relational equation

$$(2) \qquad X \circ R = Y$$

where

- R           is a fuzzy relation obtained from above fuzzy reasoning rule,
                               $R = [x_{ij} : i = 1, ... , n, j = 1, ... , k]$
- X           is an observed feature vector,
- Y           is an object classification,
- $\circ$           is composition operator.

Fundamental concept of fuzzy reasoning have been discussed including papers by Baldwin [1, 2], Pedrycz [8,9,10], Sanches [11], Yager [12], Zadeh [13, 14].

Interpreting the composition operator in (2) as a max-min composition, i.e.

$$(2a) \qquad Y(y_j) = \max_k \{ X(x_k) \wedge R(x_k, y_i) \}$$

one could easily check some basic results:
- existence of a solution of an equation (2) with X or R unknown [7, 8, 9, 10],
- structure of the set of solution [6],
- solving algorithm [8].

This natural way of rule formulation has its disadvantages in cluster analysis, especially - but not necessary, when repetitive information is investigated, i.e. when pieces of information concerning this same feature but obtained from different sources (given by a few experts, observed in different experiments, etc.) are processed. To clarify this problem let's consider an example. Let grades of certainty factors of explored object are given by n independent experts and they are equal 0.25, 0.75, 0.75, ...., 0.75. Applying minimum operator (and-like operation) to combine collected values the obtained result will be equal 0.25. The result is not acceptable. The grade of resulting certainty factor should be higher because all but one grades are equal to 0.75. Notice that the result will be even less for any other t-norm modeling and-like operations since inequality

$$\min(a,b) \leq a \, t \, b$$

holds for any t-norm. In case of or-like operators (max or any other s-norm) the problem is quite similar. For more detailed discussion on this subject see [5].

In the following chapters, a different approach to composition operator of fuzzy sets is presented. It allows to avoid the drawback connected with repetitive information processing.

## 2. Preliminaries.

Let's define new operators on fuzzy sets based on those introduced in [5]. Given a mapping

$$f : (-1,1) \to (-\infty, +\infty)$$

called transforming function, which is continuous, increasing and

$$(\forall x, -1 < x < 1)) \; f(x) = -f(-x)$$
$$\lim f(x) = -\infty \quad \text{for } x \to -1$$
$$\lim f(x) = +\infty \quad \text{for } x \to +1$$

we define:

- an additive operator s :

$$(AsB)(x) = f^{-1}(f(A(x)) + f(B(x)))$$

- a multiplicative operator w :

$$(AwB)(x) = f^{-1}(f(A(x)) * f(B(x)))$$

where A, B are fuzzy sets, + and * are arithmetic operators on real numbers. The fuzzy sets forms a ring with both additive and multiplicative operators.

It is necessary to recall that certainty factor is interpreted here as a number belonging to [-1, 1] interval of real numbers rather than [0,1]. The simple linear mapping: x -> 2*x-1 moves certainty factors from [0, 1] to [-1,1] range. As usual, the higher the positive value of c.f. the more certain information of including an element to the fuzzy set. And vice versa, the higher the negative value of c.f. the more certain information of excluding an element from the fuzzy set. The zero value means lack of information whether an element belongs to the fuzzy set or not (see also [5]).

Examples of transforming function are: hyperbolic arc tangent, x/(1-abs(x)), tan($\pi$/2*x) etc. But the problem of choosing appropriate transforming function depends on considered application and features of the application.

In [5] the multiplicative operator was defined as a mapping w : (-$\infty$, +$\infty$) x (-1, 1) -> (-1, 1) and it formed linear space of fuzzy sets with additive operator, whereas in this paper the multiplicative operator is defined as a mapping w : (-1, 1) x (-1, 1) -> (-1, 1) and forms a ring of fuzzy sets with additive operator (studying this structure is of marginal significance in this paper). This kind of operators - ring operators - seems to be more relevant than that defined in [5]. The difference lies in an interpretation of the fuzzy relation: if linear operators were used, elements of fuzzy relation R would be importance factors (real numbers) rather than certainty factors (numbers from [-1,1] interval).

The problems of choosing transforming function in given application, of special treatment of crisp values, of differences between above definition of multiplication operator and that in [5] are of less significance of this paper. They were discussed in [4,5].

# 3. Feature cumulation.

Let us consider the rule of form (1) as an equation with composition operator defined by using of addition-multiplication of fuzzy sets defined in the previous chapter, i.e.

$$(3) \qquad Y(y_j) = \underset{i}{s} ( X(x_i) \; w \; R(x_i, y_j) ) \qquad j = 1, 2, ..., k$$

This approach to fuzzy reasoning and fuzzy relational equations is similar to that of max-min composition since we face again a space of features X and a space of classes Y. Elements of spaces of features and classes are related to each other by means of a fuzzy relation. In other words, $R(x_i, y_j)$ stands for the degree to which the feature $x_i$ is associated with the class $y_j$.

The main difference between these two approaches is that in max-min composition the result fuzzy set of classes reflects only certainty factors of the selected features of objects, while in addition-multiplication composition the resulting class fuzzy set depends on all premises, i.e. all the certainty factors of feature (features) are cumulated in resulting certainty factors of classes. For more detailed discussion on this subject see [4].

Let us introduce modified form of (3) formula to utilize repetitive estimation of features of explored object:

$$(4) \qquad Y(y_l) = \underset{i}{s}(( \underset{j}{s} X_j(x_i)) \; w \; R(x_i, y_l)), \quad l = 1, 2, ..., k$$

The result of features cumulation: $( s \; X_j(x_i))$ reflects expectation for increasing combined certainty if all or at least most of expert opinions are of high or moderately high certainty. A simple example should clarify this problem. Note that left side values are from traditional range [0,1] and right side values are transformed from that range to [-1,1] interval.

Let relation R is as follow:

$$R = \begin{bmatrix} .8 & .8 & .2 \\ .8 & .2 & .8 \\ .2 & .8 & .8 \\ .2 & .2 & .8 \end{bmatrix} \qquad R = \begin{bmatrix} .6 & .6 & -.6 \\ .6 & -.6 & .6 \\ -.6 & .6 & .6 \\ -.6 & -.6 & .6 \end{bmatrix}$$

Let there are five identical estimations of features equal to $X_1$ and one estimation equal to $X_2$:

$$X_1 = [ 0.1 \; 0.1 \; 0.6 \; 0.8 ] \qquad X_1 = [ -0.8 \; -0.8 \; 0.2 \; 0.6 ]$$
$$X_2 = [ 0.5 \; 0.5 \; 0.6 \; 0.6 ] \qquad X_2 = [ 0.0 \; 0.0 \; 0.2 \; 0.2 ]$$

Applying min operator to combine given six feature estimations and then utilizing (2a) formula to obtain classification of explored object and then combining partial result by applying max operator we are given the result [ 0.5 0.6 0.6] (values are from [0,1]). On the other hand formula (4) gives [ -0.95 0.26 0.74] (values are from [-1,1]). The first result does not distinguish between second and third class of object belongingness, while it may be expected that third class should be favored as most of estimations points into last feature - the significant one for third class. And the second result satisfies this expectation.

## 4. Concluding remarks.

The problems concerned in context of information repetitiveness of fuzzy reasoning process are signalized in the paper. High or moderately high repetitive estimation of certainty factor of a premise of fuzzy reasoning rule causes an expectation of increasing of the result certainty factor of this premise. Unfortunately, this expectation is not satisfied in the approach of fuzzy relational equations with max-min composition. This lack of satisfaction is a result of a lattice structure of this approach to fuzzy sets. To avoid such a disadvantage a new composition operator with a cumulative feature is introduced.

It should be noticed that problems under consideration need further investigation. Namely, further studying is needed for formal - not intuitive - definition of cumulation of features, solvability of fuzzy relational equations with cumulative composition, inverse problem, problem of choosing of transforming function. But, first of all, I hope that presented here approach leads to a new aspect of fuzziness and uncertainty and is worth further investigation.

## References

[ 1] J.F. Baldwin,   Fuzzy logic and its application to fuzzy reasoning, in: M.M. Gupta et al., Eds, Advances in Fuzzy Set Theory and Applications, North-Holland, (1979)83-115.

[ 2] J.F. Baldwin,   A new approach to approximate reasoning using fuzzy logic, Fuzzy Sets & Systems 2(1979)309-325.

[ 3] J.C.Bezdek,   Computing with Fuzzy Uncertainty, IEEE Communication Magazine, September 1992.

[ 4] W.Homenda,   Fuzzy relational equations with cumulative composition operator as a form of fuzzy reasoning, Proc. of the Intern. Fuzzy Engin. Symposium, 1(1991)277-285, Yokohama.

[ 5] W.Homenda, W.Pedrycz,    Processing uncertain information in the linear space of fuzzy sets, Fuzzy Sets& Systems, 44 (1991) 187-198.

[ 6] A. Di Nola, S Sessa,    On the set of solutions of composite fuzzy relation equation, Fuzzy Sets & Systems 9(1983)275-286.

[ 7] C.P Pappis, M. Sugeno, Fuzzy relational equations and the inverse problem, Fuzzy Sets & Systems, 15(1985)79-90.

[ 8] W. Pedrycz,   Inverse problem in fuzzy relational equations, Fuzzy Sets & Systems, 36(1990)277-291.

[ 9] W. Pedrycz,   Fuzzy relational equations with triangular norms and their resolutions, BUSEFAL, 11(1982)24-32.

[10] W. Pedrycz,   Numerical and applicational aspects of fuzzy relational equations, Fuzzy Sets and Systems, 11(1983)1-18.

[11] W. Pedrycz,   Fuzzy sets in pattern recognition: methodology and methods, Pattern Recognition 23 (1990) 121-146.

[12] R.R Yager   An approach to inference in approximate reasoning, Internat. J. Man-Machine Stud. 13(1980)323-338.

[13] L.A. Zadeh,   Fuzzy algorithms, Inform. and Control 12(1968)94-102.

[14] L.A. Zadeh,   Similarity relations and fuzzy orderings, Inform. Sci 3(1971)177-200.

# Characterization of fuzzy integrals viewed as aggregation operators

Michel GRABISCH
Thomson-CSF
Central Research Laboratory
Domaine de Corbeville
91404 Orsay Cedex, France
email grabisch@thomson-lcr.fr

## 1  Introduction

Fuzzy logic has brought a fairly large amount of aggregation operators to the field of decision making and reasoning, namely triangular norms and conorms and their extended definition, averaging operators of various kind, etc. Faced with a real problem, a practitioner could be troubled by such a diversity if no guideline is provided to him in order to choose an operator suitable to his problem. A good way to help the practitioner is to try to exhibit all the properties of a given class of operators. A better way, but more difficult, is to try to *characterize* the class of operators, i.e. to isolate the minimal set of properties which are verified by these operators *and only them*. For example, t-norms are operators on $[0, 1]^2$ characterized by commutativity, associativity, monotonicity and certain limit conditions. Inside the class of t-norms, the Frank family is characterized by the property $\top(a, b) + \bot(a, b) = a + b$, where $\top, \bot$ are a pair of dual t-norm/t-conorm.

A particular class of aggregation operators are fuzzy integrals, when they are defined on a finite universe. Fuzzy integrals, introduced by Sugeno [12] in 1974, have been extensively studied on a rather mathematical point of view by several authors, such as Weber [13], Murofushi [7, 8, 9], Wang, to cite a few. However, there has been comparatively few theoretical studies on fuzzy integrals viewed as aggregation operators (see for example the works of Murofushi and Sugeno [10], and Grabisch [3, 4, 6]), although there were used in applications essentially in this way (see a survey of applications in [5]).

As it is known that fuzzy integrals cover a wide range of aggregation operators of the averaging type, such as weighted average, medians, a large part of generalized means [6], weighted minimum and maximum, OWA (Ordered Weighted Averaging) operators [2, 6], it would be of valuable interest to try to characterize them. Recently, Fodor *et al.* [2] have succeeded at characterizing OWA: as OWA is a particular case of fuzzy integral, their results can serve as a basis for the problem addressed here. We will show several results on characterization of different kinds of fuzzy integrals. The proofs of all these results are merely a generalization to the ones presented in [2], and for space limitations, are not all presented here.

## 2  Mathematical background and notations

### 2.1  Notations

We introduce some notations which will be used throughout the paper. Permutations on a given index set $I$ are denoted by $\sigma$. Given a set of real numbers $a_i, i \in I$, the permutation of them will be denoted either by $a_{\sigma(1)}, a_{\sigma(2)}, \ldots$ or by $[a_1, a_2, \ldots]_\sigma$. The particular permutation which ranks the $a_i$ in increasing order is denoted by $(\cdot)$, i.e. $a_{(1)} \leq a_{(2)} \leq \cdots$. We denote by $\mathfrak{S}$ the set of all permutations.

For a given set $X = \{x_1, \ldots, x_m\}$, we define successively $A_i = \{x_i, \ldots, x_m\}$, $A_{(i)} = \{x_{(i)}, \ldots, x_{(m)}\}$, and $A_{\sigma(i)} = \{x_{\sigma(i)}, \ldots, x_{\sigma(m)}\}$ .

### 2.2  Some definitions about operators

We denote $m$-ary aggregation operators by $M^{(m)}(a_1, \ldots, a_m)$. We will drop the superscript $(m)$ when there is no fear of ambiguity. Otherwise indicated, $a_i \in \mathbb{R}, \forall i$.

**Definition 1** *Let* $w_1, \ldots, w_m$ *be a set of weights such that* $\sum_{i=1}^{m} w_i = 1$. *The Ordered Weighted Averaging (OWA) operator is defined by* $\mathrm{OWA}_{w_1, \ldots, w_m}(a_1, \ldots, a_m) = \sum_{i=1}^{m} w_i a_{(i)}$

**Definition 2** *Let* $M^{(m)}$ *be an operator. Then:*

- $M$ *is* commutative or neutral (N) *iff* $M(a_1, \ldots, a_m) = M(a_{\sigma(1)}, \ldots, a_{\sigma(m)})$, $\forall \sigma \in \mathfrak{S}$.

- $M$ *is* idempotent (I) *iff* $M(a, \ldots, a) = a$, $\forall a \in \mathbf{R}$.

- $M$ *has the* ordered linkage property (OL) *[2] iff* $M^{(m+1)}(M^{(m)}(a_{(1)}, \ldots, a_{(m)}), M^{(m)}(a_{(2)}, \ldots, a_{(m+1)}), \ldots$ $\ldots, M^{(m)}(a_{(m+1)}, \ldots, a_{(2m)})) = M^{(m)}(M^{(m+1)}(a_{(1)}, \ldots, a_{(m+1)}), M^{(m+1)}(a_{(2)}, \ldots, a_{(m+2)}), \ldots$ $\ldots, M^{(m+1)}(a_{(m)}, \ldots, a_{(2m)}))$.

- $M$ *has the* ordered linkage property with permutation (OLP) *iff* $M^{(m+1)}([M^{(m)}(a_{(1)}, \ldots, a_{(m)}),$ $M^{(m)}(a_{(2)}, \ldots, a_{(m+1)}), \ldots, M^{(m)}(a_{(m+1)}, \ldots, a_{(2m)})]_\sigma) = M^{(m)}(M^{(m+1)}([a_{(1)}, \ldots, a_{(m+1)}]_\sigma),$ $M^{(m+1)}([a_{(2)}, \ldots, a_{(m+2)}]_\sigma), \ldots, M^{(m+1)}([a_{(m)}, \ldots, a_{(2m)}]_\sigma)$, $\forall \sigma \in \mathfrak{S}$.

- $M$ *is* stable under the same positive linear transformation (SPL) *iff* $M(ra_1 + t, \ldots, ra_m + t) = rM(a_1, \ldots, a_m) + t$, $\forall r > 0, \forall t \in \mathbf{R}$.

- $M$ *is* stable under positive linear transformation with same unit, comonotonic zeroes (SPLUC) *iff* $M(ra_{\sigma(1)} + t_{\sigma(1)}, \ldots, ra_{\sigma(m)} + t_{\sigma(m)}) = rM(a_{\sigma(1)}, \ldots, a_{\sigma(m)}) + T(t_{\sigma(1)}, \ldots, t_{\sigma(m)})$, $\forall a_1 \leq \cdots \leq a_m, \forall r > 0, \forall t_1 \leq \cdots \leq t_m \in \mathbf{R}, \forall \sigma \in \mathfrak{S}$.

Note that SPLUC implies SPL, OLP implies OL.

**Theorem 1** *(Fodor, Marichal, Roubens [2]) The class of OWA operators corresponds to the operators which satisfy neutrality, monotonicity, ordered linkage and stability for the same positive linear transformation.*

**Theorem 2** *(Fodor, Marichal, Roubens [2]) The class of OWA operators corresponds to the operators which satisfy neutrality, monotonicity, idempotency and stability for positive linear transformation, same unit, independent zeroes and ordered values (i.e. SPLUC with $\sigma = Id$ only).*

## 2.3 Fuzzy measures and integrals

Here is a brief introduction to fuzzy measures and integrals, restricted to a finite space $X = \{x_1, \ldots, x_m\}$. We assume also that every subset of $X$ is measurable. More general definitions can be found in e.g. [7, 8, 9, 11].

**Definition 3** *A* fuzzy measure *$\mu$ defined on $X$ is a set function $\mu : 2^X \longrightarrow [0, 1]$ verifying the following axioms :*
(i) $\mu(\emptyset) = 0, \mu(X) = 1$
(ii) $A \subseteq B \Rightarrow \mu(A) \leq \mu(B)$

We turn to fuzzy integrals: as we consider them as aggregation operators, we adopt an operator-like notation.

**Definition 4** *Let $\mu$ be a fuzzy measure on $X$. The* Sugeno integral *of a function $f : X \longrightarrow [0, 1]$ with respect to $\mu$ is defined by :*

$$\mathcal{S}_\mu(f(x_1), \ldots, f(x_n)) \stackrel{\Delta}{=} \bigvee_{i=1}^{n} (f(x_{(i)}) \wedge \mu(A_{(i)})) \tag{1}$$

The definition can be extended, using a t-norm $\top$ instead of $\wedge$: this was done by Weber [13], and considered also by Grabisch *et al.* under the name "quasi-Sugeno" integral [3]. We will use this name subsequently, and denote this integral by $\mathcal{S}_\mu^\top$.

**Definition 5** *Let $\mu$ be a fuzzy measure on $X$. The* Choquet integral *of a function $f : X \longrightarrow \mathbf{R}^+$ with respect to $\mu$ is defined by*

$$\mathcal{C}_\mu(f(x_1), \ldots, f(x_n)) \stackrel{\Delta}{=} \sum_{i=1}^{n} (f(x_{(i)}) - f(x_{(i-1)})) \mu(A_{(i)}) \tag{2}$$

The introduction of t-conorms can generalize the definition of the Choquet integral. We present below such a generalization (see Murofushi and Sugeno [8] for a more complete definition).

**Definition 6** *Let $\mathcal{F} = (\triangle, \perp)$ be a pair of Archimedean t-conorms whose generation functions are respectively $h, g$, with $g(1) = 1$, i.e. a nilpotent t-conorm, and $\mu$ a fuzzy measure on $X$. The* fuzzy t-conorm integral *of a function $f : X \longrightarrow [0, 1]$ based on $\mathcal{F}$ with respect to $\mu$ is defined by $\mathcal{F}_\mu(f) \stackrel{\triangle}{=} h^{-1}[\, C_{g \circ \mu}(h \circ f)\,]$.*

Remark that the Choquet integral is recovered with $\triangle = \perp = \widehat{+}$.

Remark also that the integrand is defined on $[0, 1]$ for the Sugeno and the t-conorm integral, and on $\mathbb{R}^+$ for the Choquet integral. The extension to real numbers is not considered in this paper.

## 2.4 Properties of fuzzy integrals

(for other properties, see [4, 6, 9, 11]). Fuzzy integrals of any type trivially verify idempotency, monotonicity, and are not commutative in general. However, commutative fuzzy integrals are such that the fuzzy measure verify $\mu(A) = \mu(B)$ whenever $|A| = |B|$ [4, 6]. In fact, commutative Choquet integrals *coincide* with OWA operators [2, 6]. A very useful property of the Choquet integral is the following:

**Property 1** *([9]) Let $f$ and $g$ be two comonotonic functions, i.e. $f(x) < f(y) \Rightarrow g(x) \leq g(y)$. Then for any fuzzy measure $\mu$, the Choquet integral is additive: $C_\mu(f(x_1)+g(x_1), \ldots, f(x_m)+g(x_m)) = C_\mu(f(x_1), \ldots, f(x_m)) + C_\mu(g(x_1), \ldots, g(x_m))$.*

Remarking that $C_\mu(\alpha a_1, \ldots, \alpha a_m) = \alpha C_\mu(a_1, \ldots, a_m)$, $\forall \alpha > 0$, and using property 1, it is easy to show that the Choquet integral verifies SPLUC (and consequently SPL), provided $t_i + r a_i \geq 0$, $\forall i$. Similar results exist for other fuzzy integrals; the following can be shown:

**Theorem 3** *Let $\mathcal{F} = (\triangle, \perp)$ define a fuzzy t-conorm integral. Then for every fuzzy measure $\mu$, the following holds:*
**(i)** $\mathcal{F}_\mu(f(x_1) \triangle g(x_1), \ldots, f(x_m) \triangle g(x_m)) = \mathcal{F}_\mu(f(x_1), \ldots, f(x_m)) \triangle \mathcal{F}_\mu(g(x_1), \ldots, g(x_m))$ *whenever $f, g$ are comonotonic.*
**(ii)** $\mathcal{F}_\mu(\alpha \triangledown a_1, \ldots, \alpha \triangledown a_m) = \alpha \triangledown \mathcal{F}_\mu(a_1, \ldots, a_m)$, $\forall \alpha \in [0, 1]$. $\triangledown$ *is a t-norm defined by $a \triangledown b = h^{-1}(h(a) \cdot h(b))$, where $h$ is the additive generator of $\triangle$.*

For showing **(ii)**, we used the fact that $\triangledown$ is the only t-norm which is distributive with $\triangle$, after a result from Bertoluzza [1].

**Theorem 4** *For every fuzzy measure $\mu$, every quasi-Sugeno integral $\mathcal{S}_\mu^\top$ verifies:*
**(i)** $\mathcal{S}_\mu(f(x_1) \vee g(x_1), \ldots, f(x_m) \vee g(x_m)) = \mathcal{S}_\mu(f(x_1), \ldots, f(x_m)) \vee \mathcal{S}_\mu(g(x_1), \ldots, g(x_m))$ *whenever $f, g$ are comonotonic.*
**(ii)** $\mathcal{S}_\mu(\alpha \top a_1, \ldots, \alpha \top a_m) = \alpha \top \mathcal{S}_\mu(a_1, \ldots, a_m)$, $\forall \alpha > 0$.

We will now show that fuzzy integrals verify the ordered linkage property with permutation.

**Theorem 5** *For every fuzzy measure $\mu$, the Choquet integral, every quasi-Sugeno integral, and every fuzzy t-conorm integral with $\mathcal{F} = (\triangle, \triangle)$ verify the ordered linkage property with permutation (OLP).*

**Proof :** (for Choquet integral only: other cases are much the same) using the definition of Choquet integral and rearranging terms, we have:
$M^{(m+1)}([M^{(m)}(a_{(1)}, \ldots, a_{(m)}), \ldots, M^{(m)}(a_{(m+1)}, \ldots, a_{(2m)})]_\sigma) = M^{(m+1)}([$
$a_{(1)}(1 - \mu^{(m)}(A_{(2)})) + a_{(2)}(\mu^{(m)}(A_{(2)}) - \mu^{(m)}(A_{(3)})) + \cdots + a_{(m)}\mu^{(m)}(A_{(m)})$,
$\cdots$
$a_{(m+1)}(1 - \mu^{(m)}(A_{(2)})) + a_{(m+2)}(\mu^{(m)}(A_{(2)}) - \mu^{(m)}(A_{(3)})) + \cdots + a_{(2m)}\mu^{(m)}(A_{(m)})]_\sigma)$.
Remark that we can apply property 1 to the above, and we get:
$(1 - \mu^{(m)}(A_{(2)}))M^{(m+1)}([a_{(1)}, \ldots, a_{(m+1)}]_\sigma) + (\mu^{(m)}(A_{(2)}) - \mu^{(m)}(A_{(3)}))M^{(m+1)}([a_{(2)}, \ldots, a_{(m+2)}]_\sigma) + \cdots + \mu^{(m)}(A_{(m)})M^{(m+1)}([a_{(m)}, \ldots, a_{(2m)}]_\sigma)$
but this nothing more than $M^{(m)}(M^{(m+1)}([a_{(1)}, \ldots, a_{(m+1)}]_\sigma), M^{(m+1)}([a_{(2)}, \ldots, a_{(m+2)}]_\sigma), \cdots$
$\cdots M^{(m+1)}([a_{(m)}, \ldots, a_{(2m)}]_\sigma)$. $\square$

# 3 Characterization of the Choquet integral

**Theorem 6** *(Characterization 1) The class of Choquet integrals corresponds to the operators which satisfy the properties of monotonicity, idempotence, ordered linkage with permutation (OLP), and stability for the same positive linear transformation (SPL) with positive zero.*

**Proof:** We have already established that Choquet integrals verify all these properties. Let us show that operators verifying these properties are necessarily Choquet integrals, in a recursive way.

∗ ($m = 2$). From SPL, we can write:

$$M^{(2)}(a_1, a_2) = \begin{cases} a_1 + (a_2 - a_1)M^{(2)}(0,1), & \text{if } a_1 \leq a_2 \\ a_2 + (a_1 - a_2)M^{(2)}(1,0), & \text{if } a_2 \leq a_1 \end{cases}$$

The two cases are to be distinguished because we need a *positive* linear transformation. Letting $M^{(2)}(0,1) = \mu^{(2)}(\{x_2\})$ and $M^{(2)}(1,0) = \mu^{(2)}(\{x_1\})$, we have $M^{(2)} \equiv \mathcal{C}_{\mu^{(2)}}$.

∗ ($m = 3$). From SPL, we scale $a_1, a_2, a_3$ to the range $[0,1]$. Assuming that $a_1 \leq a_2 \leq a_3$, we have:

$$M^{(3)}(a_1, a_2, a_3) = a_1 + (a_3 - a_1)M^{(3)}\left(0, \frac{a_2 - a_1}{a_3 - a_1}, 1\right)$$

We define $M^{(2)}(0,1) = (a_2 - a_1)/(a_3 - a_1)$, and we apply OLP with $(0,0,1,1)$, $\sigma = Id$. Because of idempotency, we can write $M^{(3)}(0, \frac{a_2-a_1}{a_3-a_1}, 1) = M^{(3)}(M^{(2)}(0,0), M^{(2)}(0,1), M^{(2)}(1,1)) = M^{(2)}(M^{(3)}(0,0,1), M^{(3)}(0,1,1))$ (from OLP).

We define $\mu^{(3)}(\{x_3\}) \triangleq M^{(3)}(0,0,1)$, and $\mu^{(3)}(\{x_2, x_3\}) \triangleq M^{(3)}(0,1,1)$. Then, after some manipulations we get $M^{(3)}(a_1, a_2, a_3) = a_1[1 - \mu^{(3)}(\{x_2, x_3\})] + a_2[\mu^{(3)}(\{x_2, x_3\}) - \mu^{(3)}(\{x_3\})] + a_3\mu^{(3)}(\{x_3\})$.

Recall that this result holds only if $a_1 \leq a_2 \leq a_3$. Consider now the general case, and let $\sigma$ the permutation such that $a_{\sigma(1)} \leq a_{\sigma(2)} \leq a_{\sigma(3)}$. Then:

$$M^{(3)}(a_1, a_2, a_3) = a_{(1)} + (a_{(3)} - a_{(1)})M^{(3)}\left([0, \frac{a_{(2)} - a_{(1)}}{a_{(3)} - a_{(1)}}, 1]_\sigma\right)$$

As before, we define $M^{(2)}(0,1) = (a_{(2)} - a_{(1)})/(a_{(3)} - a_{(1)})$, and we apply OLP with $(0,0,1,1)$ and permutation $\sigma$: $M^{(3)}([0, \frac{a_{(2)}-a_{(1)}}{a_{(3)}-a_{(1)}}, 1]_\sigma) = M^{(3)}([M^{(2)}(0,0), M^{(2)}(0,1), M^{(2)}(1,1)]_\sigma) = M^{(2)}(M^{(3)}([0,0,1]_\sigma), M^{(3)}([0,1,1]_\sigma))$.

We define $\mu^{(3)}(A_{\sigma(3)}) = M^{(3)}([0,0,1]_\sigma)$, and $\mu^{(3)}(A_{\sigma(2)}) = M^{(3)}([0,1,1]_\sigma)$. Substituting as above, we get $M^{(3)}(a_1, a_2, a_3) = a_{(1)}[1 - \mu^{(3)}(A_{(2)})] + a_{(2)}[\mu^{(3)}(A_{(2)}) - \mu^{(3)}(A_{(3)})] + a_3\mu^{(3)}(A_{(3)})$. Taking all possible permutations $\sigma$, we have $M^{(3)} \equiv \mathcal{C}_{\mu^{(3)}}$, and from the hypothesis of monotonicity, all the coefficients $\mu^{(3)}$ form effectively a fuzzy measure.

∗ (general case) similar proof: start from:

$$M^{(m+1)}(a_1, \ldots, a_m) = a_{(1)} + (a_{(m+1)} - a_{(1)})M^{(m+1)}\left([0, \frac{a_{(2)} - a_{(1)}}{a_{(m+1)} - a_{(1)}}, \ldots, \frac{a_{(m)} - a_{(1)}}{a_{(m+1)} - a_{(1)}}, 1]_\sigma\right)$$

and apply OLP with $(\underbrace{0, \ldots, 0}_{m}, \underbrace{1, \ldots, 1}_{m})$. □

Having a look at the proof, we see that we can propose a tighter characterization, since:
1) the OLP property is not used in its full strength: in fact we can replace OLP by the following weaker requirement: $M^{(m+1)}([M^{(m)}(0, \ldots, 0), M^{(m)}(0, \ldots, 0, 1), \ldots, M^{(m)}(0, 1, \ldots, 1), M^{(m)}(1, \ldots, 1)]_\sigma) = M^{(m)}(M^{(m+1)}([0, \ldots, 0, 1]_\sigma), M^{(m+1)}([0, \ldots, 0, 1, 1]_\sigma), \ldots, M^{(m+1)}([0, 0, 1 \ldots, 1]_\sigma), M^{(m+1)}([0, 1, \ldots, 1]_\sigma))$.
2) the idempotence is not necessary, but only $M^{(m)}(0, \ldots, 0) = 0$, $M^{(m)}(1, \ldots, 1) = 1$.

We provide now a second characterization, similarly to the case of OWA.

**Theorem 7** *(Characterization 2) The class of Choquet integrals corresponds to the operators which satisfy the properties of monotonicity, idempotence, and stability for positive linear transformation with same unit and comonotonic positive[1] zeroes (SPLUC).*

---

[1] In fact, it suffices that $t_i + ra_i \geq 0$, $\forall i$

**Proof:** We have already established that Choquet integrals verify all these properties. Let us show that any $M_{(m)}$ verifying these properties is a Choquet integral. Using SPLUC and the fact that $M(0, 0, \ldots, 0) = 0$, we deduce that $T \equiv M$. Let us suppose $r = 1$, and $a_1 = a_2 = \cdots = a_{m-1} = 0$, $a_m = a$, $t_1 = t_2 = \cdots t_{m-1} = 0$, $t_m = t$. Then for every $\sigma \in \mathfrak{S}$, $M([0, \ldots, 0, a+t]_\sigma) = M([0, \ldots, 0, a]_\sigma) + M([0, \ldots, 0, t]_\sigma)$. Let us denote $M([0, \ldots, 0, x]_\sigma)$ by $f_{\sigma(m)}$. The solution to the above functional equation is $f_{\sigma(m)}(x) = Kx$, with $K$ being a positive constant. Let us denote this constant by $\mu(A_{\sigma(m)})$.

In a similar way, we introduce $f_{\sigma(m-1)}, f_{\sigma(m-2)}$, etc, defined by

$$f_{\sigma(i)}(x) = M^{(m)}([\underbrace{0, \ldots, 0}_{i-1}, x, x, \ldots, x]_\sigma)$$

which are solution of the same functional equation, so we define new constants $\mu(A_{\sigma(m-1)}), \mu(A_{\sigma(m-2)})$, etc. Note that $f_{\sigma(1)}(x) = M(x, x, \ldots, x) = x$, so that $\mu(A_{\sigma(1)}) = 1$.

This being done for all $\sigma \in \mathfrak{S}$, we have defined a proper fuzzy measure $\mu$, which is monotonic by monotonicity of $M$. We need only to verify that $M(a_1, \ldots, a_m)$ takes the form of a Choquet integral for any $m$-uple. Consider the particular permutation $\sigma$ such that $M(a_1, \ldots, a_m) = M([a_{(1)}, \ldots, a_{(m)}]_\sigma)$. Applying SPLUC and the definition of $f_{\sigma(i)}$, we have $M([a_{(1)}, \ldots, a_{(m)}]_\sigma) = M([a_{(1)}, \ldots, a_{(1)}]_\sigma) + M([0, a_{(2)} - a_{(1)}, \ldots, a_{(2)} - a_{(1)}]_\sigma) + \cdots + M([0, \ldots, 0, a_{(m)} - a_{(m-1)}]_\sigma) = \sum_{i=1}^{m} (a_{(i)} - a_{(i-1)}) \mu(A_{\sigma(i)})$. $\square$

# 4   Characterization of other fuzzy integrals

An examination of the preceding results shows that there are two key properties for characterizing the Choquet integral, namely the fact that Choquet integral is linear with comonotonic functions (characterizations 1 and 2), and the ordered linkage property with permutations (characterization 1). We have shown in §2.4 that other fuzzy integrals verify OLP too, and that they are still *pseudo*-linear in some sense with comonotonic functions (see Th. 3 and 4). Thus, we should be able to give similar characterizations for the quasi-Sugeno and fuzzy t-conorm integrals. In fact, the following results can be shown (proofs omitted, but similar to previous ones):

**Theorem 8** *The class of quasi-Sugeno integrals $\mathcal{S}_\mu^\top$ corresponds to the operators which satisfy the properties of monotonicity, idempotence, ordered linkage with permutation (OLP), and:*

$$M((r \top a_1) \vee t, \ldots, (r \top a_m) \vee t) = (r \top M(a_1, \ldots, a_m)) \vee t, \quad \forall r, t \in [0, 1]$$

**Theorem 9** *The class of fuzzy t-conorm integrals with $\mathcal{F} = (\triangle, \triangle)$, $\triangle$ being a nilpotent t-conorm, corresponds to the operators which satisfy the properties of monotonicity, idempotence, and stability for positive pseudo-linear transformation with same unit and comonotonic positive zeroes (SPPUC), i.e:*

$$M\left((r \bigtriangledown a_{\sigma(1)}) \triangle t_{\sigma(1)}, \ldots, (r \bigtriangledown a_{\sigma(m)}) \triangle t_{\sigma(m)}\right) = \left(r \bigtriangledown M(a_{\sigma(1)}, \ldots, a_{\sigma(m)})\right) \triangle T(t_{\sigma(1)}, \ldots, t_{\sigma(m)})$$

$\forall 0 \leq a_1 \leq \cdots \leq a_m \leq 1, \forall r \in [0, 1], \forall 0 \leq t_1 \leq \cdots \leq t_m \leq 1, \forall \sigma \in \mathfrak{S}$, *where $\bigtriangledown$ is the associated t-norm defined by $x \bigtriangledown y = h^{-1}(h(x)h(y))$, with $h$ being the generator of $\triangle$.*

The key point in the proof is to show that the only solution to the functional equation $f(x \triangle y) = f(x) \triangle f(y)$ is $f(x) = K \bigtriangledown x$, $K \in [0, 1]$. This comes essentially from the SPPUC property, which implies $f(x \bigtriangledown y) = f(x) \bigtriangledown y$, which gives the desired result taking $x = 1$ (and thus $K = f(1)$).

In a similar way, we can provide a second characterization of quasi-Sugeno integrals:

**Theorem 10** *The class of quasi-Sugeno integrals $\mathcal{S}_\mu^\top$ corresponds to the operators which satisfy the properties of monotonicity, idempotence, and $M\left((r \top a_{\sigma(1)}) \vee t_{\sigma(1)}, \ldots, (r \top a_{\sigma(m)}) \vee t_{\sigma(m)}\right) = \left(r \top M(a_{\sigma(1)}, \ldots, a_{\sigma(m)})\right) \vee T(t_{\sigma(1)}, \ldots, t_{\sigma(m)}), \forall 0 \leq a_1 \leq \cdots \leq a_m \leq 1, \forall r \in [0, 1], \forall 0 \leq t_1 \leq \cdots \leq t_m \leq 1, \forall \sigma \in \mathfrak{S}$.*

# 5  Conclusion

We have found two types of characterization of fuzzy integrals. It happens that the property which is common and assential for fuzzy integrals is ordered linkage with permutations (in fact, its restricted version to 0 and 1 is sufficient). The second important property is stability for certain kinds of transformation: linear for the Choquet integral, based on *min* and t-norms for quasi-Sugeno integrals, and on pair of distributive t-norm/t-conorms for fuzzy t-conorms integrals. This brings an answer to the problem of how to choose between different kind of fuzzy integrals: if we need stability for positive linear transformations, as it is the case with multicriteris evaluation problem, the Choquet integral is the only solution.

Some work remains to do in order to generalize the results presented here for any $m$-uple of real numbers: it seems that it can be easily done in case of Choquet integral, since an extended definition for negative numbers exists. The work could be harder for other fuzzy integrals, which are defined only for integrand whose domain is $[0, 1]$.

# 6  Acknowledgment

The author is very much indebted to Prof. Roubens for invaluable discussions on fuzzy integrals and characterization of OWA operators.

# References

[1] C. Bertoluzza, On the distributivity between t-norms and t-conorms, *2nd IEEE Int. Conf. on Fuzzy Systems*, San Francisco, march 93, 140-147.

[2] J. Fodor, J.L. Marichal, M. Roubens, Characterization of the ordered weighted averaging operators, submitted to *IEEE Tr. on Fuzzy Systems*.

[3] M. Grabisch, T. Murofushi, M. Sugeno, Fuzzy Measure of Fuzzy Events Defined by Fuzzy Integrals, *Fuzzy Sets & Systems* 50, 293-313.

[4] M. Grabisch, On the use of fuzzy integral as a fuzzy connective, *2nd IEEE Int. Conf. on Fuzzy Systems*, San Francisco, march 93, 213-218.

[5] M. Grabisch, A survey of applications of fuzzy measures and integrals, *5th IFSA Congress*, Seoul, july 1993, 294-297.

[6] M. Grabisch, On equivalence classes of fuzzy connectives — The case of fuzzy integrals, submitted to *IEEE Tr. on Fuzzy Systems*.

[7] T. Murofushi, M. Sugeno, An interpretation of fuzzy measure and the Choquet integral as an integral with respect to a fuzzy measure, *Fuzzy Sets & Systems* 29 (1989), 201-227.

[8] T. Murofushi, M. Sugeno, Fuzzy t-conorm integrals with respect to fuzzy measures: generalization of Sugeno integral and Choquet integral, *Fuzzy Sets & Systems* 42 (1991), 57-71.

[9] T. Murofushi, M. Sugeno, A theory of fuzzy measures. Representation, the Choquet integral and null sets, *J. Math. Anal. Appl.* 159 No 2 (1991) 532-549.

[10] T. Murofushi, M. Sugeno, Hierarchical decomposition of Choquet integral systems, *Japanese J. of Fuzzy Theory and Systems* 4 No 4 (1992), 437-442.

[11] H.T. Nguyen, M. Grabisch, E.A. Walker, *Fundamentals of Uncertainty Calculi, with Applications to Fuzzy Inference*, Kluwer Acad., to appear in 1994.

[12] M. Sugeno, *Theory of fuzzy integrals and its applications*, Doct. Thesis, Tokyo Institute of Technology (1974).

[13] S. Weber, $\perp$-decomposable measures and integrals for Archimedean t-conorms $\perp$, *J. Math. Anal. Appl.* 101 (1984), 114-138.

# Understanding Relations between Fuzzy Logic and Evidential Reasoning Methods

Prasad Palacharla and Peter Nelson

*Abstract–* **The world around us is very uncertain and unpredictable. No bivalent logic (also called Boolean or binary logic or law of excluded middle) can possibly solve real world problems without oversimplification. Many approximate reasoning (also called multi-valued, or continuous or fuzzy logic) methods have been proposed and applied to the real world problems. These methods differ in the way they represent uncertainty and handle uncertainty in the real world problems. In this paper we will study, compare and understand relations between fuzzy logic and evidential reasoning methods for application to data fusion problem in transportation engineering.**

## I. INTRODUCTION

SET theory has been found to be at the center of universe for the modern computing world. But this has many unrealistic assumptions. Every element in the world either belongs or does not belong to a set; either member or not a member of a set; either true or false. This is an extreme oversimplification of the real world. It does not consider the fact that an element can partially belong to a set and partially not belong to a set; partially member and partially not member of a set; partially true and partially false. This kind of uncertainty can not be handled by regular set theory. Hence Max Black first introduced vague sets and Zadeh introduced fuzzy sets [1].

Fuzzy set is an extension of regular set in which for each element there is also a degree of membership associated with it. The degree of membership can be any value between 0 and 1. An element whose degree of membership in a set is 0, does not belong to that set at all. An element whose degree of membership in a set is 1, belongs one hundred percent to that set. An element whose degree of membership in a set is 0.8, belongs eighty percent to that set and so on. Also an element can belong to more than one set to various degrees of membership. This provides a powerful scheme for representation of uncertainty. Thus this continuous or fuzzy logic includes conventional or binary logic as a special case and extends beyond that.

The Dempster-Shafer theory of evidence provides another scheme for representation of uncertainty [4]. This theory developed by Dempster and later extended by his student Shafer, deals with weights of evidence and with numerical degrees of support based on evidence. The theory uses a number between 0 and 1 to indicate the degree of support (or a degree of belief) a body of evidence provides for a proposition. Here the proposition could represent a set of multiple hypotheses rather than a single hypothesis. The theory focuses on combination of degrees of belief based on one body of evidence with those based on an entirely different body of evidence.

In this paper we will study, compare and understand relations between fuzzy logic and evidential reasoning methods for application to the data fusion problem in transportation engineering. *Data fusion* is the process of collecting, organizing, and integrating raw and processed data from multiple sources and producing some sort of meaningful inference from these data [9]. Data fusion in advanced traveler information systems is to collect raw

– The authors are with the Department of Electrical Engineering and Computer Science, University of Illinois at Chicago, Chicago, Illinois 60607-7053.

data from various sources such as loop detectors, probe vehicles, historical database, and operator and estimate current travel times and/or traffic congestion levels. There is lot of interest in this problem recently [2, 3, 6, 7, 8].

## II. FUZZY LOGIC

The continuous or fuzzy logic differs from Boolean or binary logic in the way the degree of membership varies from 0 to 1. The membership function in binary logic suddenly jumps from 0 to 1 at a crisp point. Whereas the membership function in fuzzy logic smoothly varies from 0 to 1 and from 1 to 0. Also the membership functions for various fuzzy sets can be overlapping. The main objective of using fuzzy logic for solving real world problems is to capture non-linear relationship between inputs and outputs without much oversimplification.

Loop detectors are sensors beneath the road which continuously measure percent occupancy and volume count of vehicles. As occupancy, rather than flow, is a better explanatory variable of travel time [5], it can be converted to travel time using regression model or a neural network model [2]. The probe vehicles directly measure travel time. The historical database contains travel time estimates from previous observations. The operator data could be a non-numerical data such as high congestion or low congestion. To fuse all these data we need to convert all the data into same units. For this purpose we can use fuzzy logic to convert travel times into traffic congestion levels (Figure 1).

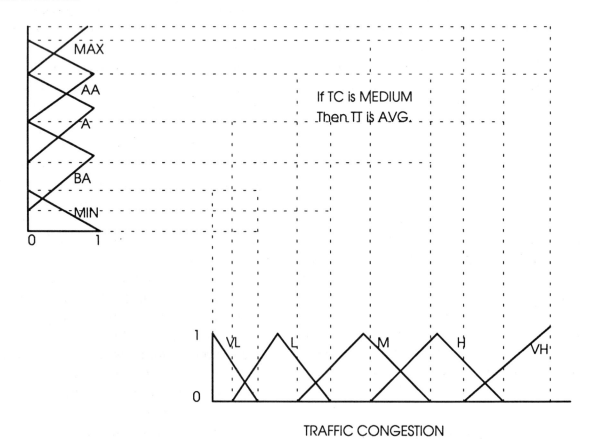

Figure 1 Fuzzy logic implementation for travel time/traffic congestion conversion.

R1) If Travel Time is *Minimum (MIN)*,
     Then Traffic Congestion is *Very Low (VL)*

R2) If Travel Time is Below *Average (BA)*,
     Then Traffic Congestion is *Low (L)*

R3) If Travel Time is *Average (A)*,
     Then Traffic Congestion is *Medium (M)*

R4) If Travel Time is *Above Average (AA)*,
     Then Traffic Congestion is *High (H)*

R5) If Travel Time is *Maximum (MAX)*,
     Then Traffic Congestion is *Very High (VH)*

Figure 2  Fuzzy expert system.

The non-linear relationship between travel times and traffic congestion levels is approximated by fuzzy patches or regions. Each region represents a different fuzzy rule that covers a range of travel times associated with a spectrum of congestion levels. All the fuzzy rules together form a fuzzy expert system (shown in Figure 2). The fuzzy expert system works like a fuzzy associative memory in the sense that rules can be fired in parallel as well as in partial [1]. First the input value is fuzzified by computing its membership values in all fuzzy sets (fuzzification). Fire all rules in parallel to the degree of membership values in the input fuzzy sets. Then aggregate all the corresponding values of the output fuzzy sets and find the centroid of the output regions to get the output value (defuzzification).

## III. EVIDENTIAL REASONING

The problem domain, or *frame of discernment*, $\theta$ (which includes all hypotheses), is always one of the hypotheses sets. Several alternative sets of hypotheses can be derived from a single collection of evidence. Each of these sets has an associated confidence range called a *belief interval* and a basic probability assignment, or *BPA*. As with conventional probabilities the *BPA* ranges from zero to one. But unlike conventional probabilities, the *BPA* represents the chance that the evidence supports the truth of the multiple hypotheses, rather than the chance that a single hypothesis is true [4].

The belief interval describes a particular hypothesis set not only in terms of its believability (the weight of evidence in support of it), but also in terms of its plausibility (the weight of evidence that does not negate it). Belief intervals are calculated for each of the final hypothesis sets after combining all available pieces of evidence. A belief interval takes the form

$$[Bel(H), Pl(H)]$$

where $H$ is some hypotheses set and both $Bel(H)$ and $Pl(H)$ are within the interval $[0, 1]$. $Bel(H)$ is computed by combining the *BPA* associated with $H$ with the *BPA*s of all hypotheses that are subsets of $H$ and as a result implicitly support $H$. $Pl(H)$ is calculated by summing the *BPA*s of all hypothesis sets (including $\theta$) that are not disjoint from $H$ (i.e., those sets that have common members with $H$). $Bel(H)$ is the present believability of $H$ supported by evidence. $Pl(H)$ is the maximum believability of $H$ possible given additional evidence. Hence, $Bel(H)$ is always lower than $Pl(H)$. The larger the distance between $Bel(H)$ and $Pl(H)$, the more room there is for improving $Bel(H)$ with the addition of missing pieces of evidence. A large belief interval implies more uncertainty in a given body of evidence. Hence the belief interval is also a measure of uncertainty.

Since there are no clear boundaries between traffic congestion levels, the travel time ranges are overlapping (Figure 3). The observed travel time is mapped to corresponding traffic congestion level(s) by looking at the link specific conversion graph. If the observed travel time is in just one range then the corresponding traffic congestion is the resultant hypothesis. But if the travel time is in more than one range then the corresponding traffic congestion levels will be a hypothesis set.

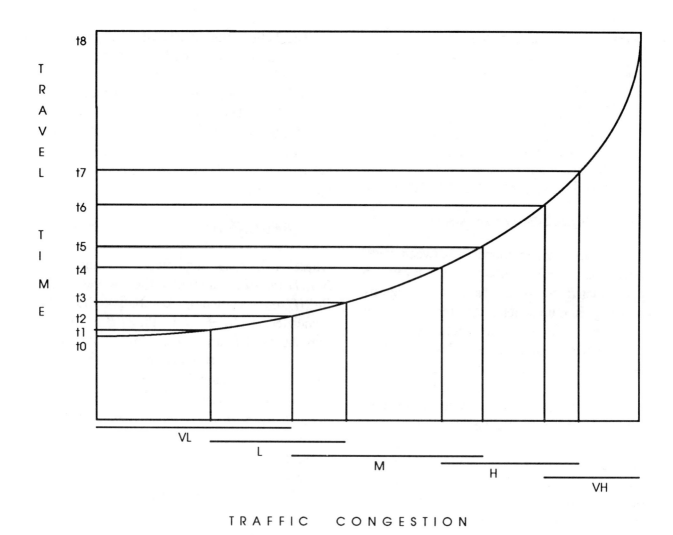

Figure 3. Look-up table for travel time/traffic congestion conversion

Imagine that we have several pieces of evidence or observed data from various sources of traffic congestion data (Table 1). Let $\theta = \{VL, L, M, H, VH\}$ be the frame of discernment. The *BPA* values are system parameters chosen according to the quality of the sources. Our goal is to find the belief intervals of different traffic congestion levels: very low (*VL*), low (*L*), medium (*M*), high (*H*), and very high (*VH*).

Since there is 0.8 belief on subset $\{H, VH\}$ from operator evidence, the remaining belief or disbelief (i.e., $1 - 0.8 = 0.2$) is on the entire set $\theta$. Similarly the beliefs of $\{H\}$ and $\theta$ are 0.6 and 0.4 respectively from probe vehicle evidence. The results of combining these two

pieces of evidence using evidential reasoning are shown in Table 2. The entries in the table are calculated by finding the intersections of rows and columns and multiplying their *BPA* values [3, 4].

Note that there are two occurrences of the hypothesis set $\{H\}$ in the resulting hypotheses. The associated *BPA*s of these two sets will be combined to form a single *BPA* and a single hypothesis set: $BPA_{\{H\}} = 0.48 + 0.12 = 0.60$. Sequentially the results of combining the previous results with next evidences are shown in Tables 3 and 4. Final results are normalized by dividing *BPA* of each of the hypotheses by $(1-k)$ where $k$ is the *BPA* of $\emptyset$ as a result of contradictions (Table 5).

| EVIDENCE | IMPLIES | BPA | RATIONALE |
|---|---|---|---|
| Operator | {H, VH} | 80% | Road repair causes delay |
| Probe Vehicle | {H} | 60% | High travel time recorded |
| Loop Detectors | {M, H} | 60% | Occupancy & volume count |
| Historical Data | {L, M} | 30% | Previous observations |

Table 1. Available pieces of evidence and their basic probability assignments

| | | {H} 0.60 | θ 0.40 |
|---|---|---|---|
| {H, VH} | 0.80 | {H} 0.48 | {H, VH} 0.32 |
| θ | 0.20 | {H} 0.12 | θ 0.08 |

Table 2. Combining operator and probe vehicle evidence

| | | {M, H} 0.60 | θ 0.40 |
|---|---|---|---|
| {H} | 0.60 | {H} 0.36 | {H} 0.24 |
| {H, VH} | 0.32 | {H} 0.192 | {H, VH} 0.128 |
| θ | 0.08 | {M, H} 0.048 | θ 0.032 |

Table 3. Combining previous results with loop detector evidence

| | | {L, M} 0.30 | θ 0.70 |
|---|---|---|---|
| {H} | 0.792 | Ø 0.2376 | {H} 0.5544 |
| {M, H} | 0.048 | {M} 0.0144 | {M, H} 0.0336 |
| {H, VH} | 0.128 | Ø 0.0384 | {H, VH} 0.0896 |
| θ | 0.032 | {L, M} 0.0096 | θ 0.0224 |

Table 4. Combining previous results with historical data evidence

| Hypothesis | Actual BPA | Normalized BPA |
|---|---|---|
| {H} | 0.5544 | 0.770 |
| {H,VH} | 0.0896 | 0.120 |
| {M,H} | 0.0336 | 0.046 |
| {M} | 0.0144 | 0.020 |
| {L,M} | 0.0096 | 0.014 |
| θ | 0.0224 | 0.030 |
| Ø | 0.2760 | – |

Table 5. Results of normalization

We next calculate the belief intervals associated with each of the hypotheses. Since the set {H} is a singleton (a set containing a single hypothesis), it has no subsets. Therefore, the believability of {H} is equal to its *BPA*. Since {H}, {H, VH}, {M, H} and θ are the only sets that are not disjoint from {H}, the plausibility of {H} is calculated as follows:

$Pl(\{H\})$ = Maximum possible believability of {H} with additional pieces of evidence = Sum of BPAs of {H}, {H, VH}, {M, H} and θ = 0.77 + 0. 12 + 0.05 + 0.03 = 0.97. Hence the belief interval of {H} is [0.77, 0.97].

The belief intervals for the remaining hypotheses are calculated in a similar fashion and all intervals are listed below:

| | |
|---|---|
| $\{H\}$ | $[0.77, 0.97]$ |
| $\{H, VH\}$ | $[0.89, 0.97]$ |
| $\{M, H\}$ | $[0.83, 1.00]$ |
| $\{M\}$ | $[0.02, 0.11]$ |
| $\{L, M\}$ | $[0.03, 0.11]$ |

A hypothesis set that is not naturally derived from a given body of evidence, but created to know its belief interval, is called a *synthetic* hypothesis set. Thus belief intervals for the synthetic hypothesis sets $\{VL\}$, $\{L\}$ and $\{VH\}$ are calculated in the same fashion. The final results of applying the D-S theory in our example can best be understood by plotting a graph between belief intervals and the various traffic congestion levels (Figure 4).

| | |
|---|---|
| $\{VL\}$ | $[0.00, 0.03]$ |
| $\{L\}$ | $[0.00, 0.04]$ |
| $\{M\}$ | $[0.02, 0.11]$ |
| $\{H\}$ | $[0.77, 0.97]$ |
| $\{VH\}$ | $[0.00, 0.15]$ |

## IV. CONCLUSION

Both fuzzy logic and evidential reasoning methods provide powerful schemes for representing and handling uncertainty. Fuzzy logic uses a number between 0 and 1 to represent the degree of membership of an element in a fuzzy set. Evidential reasoning uses a number between 0 and 1 to represent the degree of support of truth of multiple hypotheses. Fuzzy logic allows an element to be a member of multiple fuzzy sets to varying degrees. Evidential reasoning allows multiple hypotheses to be derived from multiple evidences. Fuzzy expert system works like a fuzzy associative memory in which rules can be fired in parallel as well as in partial. Evidential reasoning mainly focuses on combination of evidences.

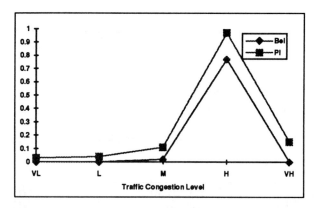

Figure 4. Belief intervals of traffic congestion levels

## V. REFERENCES

[1] Kosko, B. (1993), *Fuzzy Thinking*, Hyperion, New York, NY.

[2] Nelson, P. and Palacharla, P. (1993), "A Neural Network Model for Data Fusion in ADVANCE," *Pacific Rim Transportation Technology Conference Proceedings*, Volume I, pp. 237-243, Seattle, WA.

[3] Palacharla, P. and Nelson, P. (1994), "Evidential Reasoning in Uncertainty for Data Fusion," Submitted to *Fifth International Conference on Information Processing and Uncertainty Methods for Knowledge Based Systems*, Paris.

[4] Shafer, G. (1976), *A Mathematical Theory of Evidence*, Princeton University Press, Princeton.

[5] Sisiopiku, V. (1993), *Travel Time Estimation from Loop Detector Data for Advanced Traveler Information Systems Applications*, Ph.D. Thesis, University of Illinois at Chicago, Chicago, IL.

[6] Sumner, R. (1991), "Data Fusion in Pathfinder and TravTek," *Vehicle Navigation and Information Systems Conference Proceedings*, P-253, Part I, Society of Automotive Engineers, Dearborn, MI.

[7] Tarko, A. and Rouphail, N. (1993), "Travel Time Data Fusion in ADVANCE," *Pacific Rim Transportation Technology Conference*, Volume I, pp. 36-42, Seattle, WA.

[8] Tarko, A., Rouphail, N. and Tsai, J. (1993), "Detection of Traffic Congestion Using Fuzzy Operator Logic in ADVANCE-like Systems," *Intelligent Vehicles Symposium Proceedings*, pp. 189-194, Tokyo.

[9] Thomopoulos, S.C.A. (1990), "Sensor Integration and Data Fusion," *Journal of Robotic Systems*, 7(3), pp. 337-372.

# Accounting for Perception in Possibility Distributions: An Algorithmic Approach to Solving the Problem "How tall is Dan?"

Carol S. Russell

Franklin and Marshall College
Lancaster, PA 17604

## 1 Introduction

Presented with a proposition such as "Dan is tall," the usual approach to determining just how tall Dan might be is to determine the test-score for the semantic proposition:

$$p \overset{\Delta}{=} \text{Dan is tall}$$

As defined in Zadeh [1986], the test-score for p is determined by:

$$\tau \overset{\Delta}{=} \mu\text{Tall}[\text{Height} = \text{Height(Dan)}]$$

where $\mu$Tall is the membership function ranging over the subset of real numbers representing height and Height(Dan) is Dan's actual height. If Dan's actual height is unknown, then proposition p can be converted to a possibility distribution over the real-valued subset of Height as in:

$$\text{Poss}\{\text{Height(Dan)} = x\} = \mu\text{Tall}[\text{Height} = x].$$

This possibility distribution is the same as given in Zadeh [1978] and Kandel [1986]. In effect, this gives a range of solutions, each associated with a possibility measure, for the question: "How tall is Dan?" Unfortunately, this approach does not allow for the use of additional information pertaining to Dan's height, namely the perception of the person who originally made the claim: "Dan is tall."

This paper will present an algorithmic approach to incorporating such perception into the process of establishing the possibility distribution over the set of feasible heights.

## 2 Influence of perception on the choice of height descriptors

In general, choice of words by an individual is grounded in personal experience and perception [Luria 1981, Rommetveit 1968]. Therefore, when a proposition such as "Dan is tall" is uttered, one tends to select the descriptor which most closely represents one's perception of the similarity or difference between one's own height and that of the subject. The perception of the observer establishes a context for the proposition. Yager [1982] addresses the issue of context and the use of linguistic hedges but only explores the concentration and dilution effects on established distributions. It is the author's contention that the perception-context of the observer also imposes a new distribution of fuzzy values on the set of heights which might be quite different from the generic distribution of fuzzy values usually applied in a given case.

For example, the generic distribution of values over the range of heights for the descriptor *Average* may be unimodal with a fairly large bandwidth as given by a $\pi$-distribution (see Appendix I) where the peak is at 5'10" (70") and the effective bandwidth is 8 inches. However, in a sample of 52 men whose heights ranged from 5'8" to 6'1" who labeled themselves as *Average*, while the distribution was still unimodal, the effective bandwidth for the application of the descriptor *Average* to other men tended to be quite a bit smaller (approximately 4") and centered at or near the height of the individual. Thus using the original $\pi$-distribution to establish possibility values for Height(Dan) would yield values that tend to be less accurate than if we accounted for the shift in the peak and the concentration of the bandwidth.

However, when dealing with the descriptor *Tall*, which is usually represented by a monotonically increasing function such as an S-distribution (see Appendix I), the empirical data suggest that not only does the scale shift according to the perception and height of the observer, but also the distribution of the values over the effective range of heights tends to become unimodal rather than increasing. In addition, for values of height less than that of the observer, the fuzzy measures should decrease very rapidly when the observe considers himself *Tall* as well, indicating that the likelihood of those height values being labeled *Tall* is much smaller than would be the case in the generic form. That the distribution changes from a monotonically increasing distribution to a unimodal distribution can be attributed to the fact that as the possible values of height increases, one would tend to use the hedged term *VeryTall* more frequently than the unhedged term *Tall*.

In the empirical study, this change in distribution can be seen by observing the descriptors applied to a male subject who is 6'4" tall. The results are summarized in Table 1. These results can be explained if we consider the degree of difference that is perceived by the observer between his own height and self-perception of height and the height of the subject. It is interesting to note that across all heights of respondents and self-descriptors, a subject with height 6'9" is ranked as *VeryTall* by 97% of the respondents and only 3% ranked the same man as *Tall*. Thus as the height values of the subject increase, the value of the possibility of that height being in the descriptor set *Tall* actually decreases. This is contrary to the notion that the membership function for the descriptor Tall is monotonically increasing. A similar, symmetric case can be seen when dealing with the descriptors *Short* and *VeryShort* with respect to the monotonically decreasing Z-distribution normally applied to those descriptors where $Z(x; \alpha, \beta, \gamma) = 1 - S(x; \alpha, \beta, \gamma)$.

|  | Observer | | | |
|---|---|---|---|---|
| Subject H=6'4" | Short H ≤ 5'9" | Average 5'6" ≤ H ≤ 6'1" | Tall 5'10" ≤ H ≤ 6' | Tall 6' ≤ H ≤ 6'5" |
| Tall | 58% | 76% | 7% | 74% |
| VeryTall | 42% | 24% | 93% | 26% |

Table 1  Height descriptors applied to male subject of height 6'4"

## 3  The algorithm using perception

The degree of perceived difference between the observer and the subject, as mentioned in the previous section, can be represented as an ordered pair (r, d) where r is one of the relations "=", "<" or ">", and d is an arbitrary scalar value.

To illustrate this concept, suppose John considers himself to be short. John makes the statement: "Dan is tall." Then we can conclude:

1.  John is shorter than Dan: r = "<";
2.  The degree to which John is shorter than Dan exceeds the degree to which John is shorter than someone who is *Average*; and
3.  The degree to which John is shorter than Dan is not so great than John perceives Dan to be *VeryTall*.

Based on these observations, Table 2 gives possible values for the degrees of perceived difference.

|  | Subject | | | | |
|---|---|---|---|---|---|
| Observer | VeryShort | Short | Average | Tall | VeryTall |
| VeryShort | (=, 1) | (<, 2) | (<, 3) | (<, 4) | (<, 5) |
| Short | (>, 2) | (=, 1) | (<, 2) | (<, 3) | (<, 4) |
| Average | (>, 3) | (>, 2) | (=, 1) | (<, 2) | (<, 3) |
| Tall | (>, 4) | (>, 3) | (>, 2) | (=, 1) | (<, 2) |
| VeryTall | (>, 5) | (>, 4) | (>, 3) | (>, 2) | (=, 1) |

Table 2  Degrees of perceived difference

Given the values in Table 2, the previous observations and the following assumptions:

1.  A is the OBSERVER;
2.  A's height, h(A), in inches is known;
3.  A's self-descriptor, s(A), is known, where s(A) ∈ {*VeryShort, Short, Average, Tall, VeryTall*};
4.  A makes an observation that "B is X" where X ∈ {*VeryShort, Short, Average, Tall, VeryTall*}; and
5.  the degree of perceived difference is the ordered pair (r, d);

then the following functions determine the possibility values in response to the question: "What is B's height, h(B)?"

For $60 \leq h \leq 84$:

1.  if r = "<", set P = $\pi(h; h(A) + 2kd, 6kd)$.

$$Poss\{h(B) = h\} = \begin{cases} 0 & ; h < h(A) \\ P & ; h(A) \leq h \leq h(A) + 3kd \\ P^2 & ; h > h(A) + 2kd \text{ and } X \neq VeryTall \\ 1 & ; h > h(A) + 2kd \text{ and } X = VeryTall \end{cases}$$

2.  if r = ">", set P = $\pi(h; h(A) - 2kd, 6kd)$.

$$Poss\{h(B) = h\} = \begin{cases} 1 & ; h < h(A) - 2kd \text{ and } X = VeryShort \\ P & ; h < h(A) - 2kd \text{ and } X \neq VeryShort \\ P^2 & ; h(A) - 2kd \leq h \leq h(A) \\ 0 & ; h > h(A) \end{cases}$$

3.  if r = "=", set P = $\pi(h; h(A), 8kd)$

$$Poss\{h(B) = h\} = \begin{cases} 1 & ; h \leq h(A) \text{ and } X = VeryShort \\ P & ; h > h(A) \text{ and } X = VeryShort \\ P & ; h < h(A) + 2kd \text{ and } X = Short \\ Z(h; h(A) - 8kd, h(A), h(a) + 8kd) & ; h > h(A) + 2kd \text{ and } X = Short \\ P^2 & ; X = Average \\ S(h; h(A) - 8kd, h(A), h(A) + 8kd) & ; h \leq h(A) - 2kd \text{ and } X = Tall \\ P & ; h > h(A) - 2kd \text{ and } X = Tall \\ P & ; h < h(A) \text{ and } X = VeryTall \\ 1 & ; h \geq h(A) \text{ and } X = VeryTall \end{cases}$$

The effect of these functions can be seen by comparing the standard possibility distribution for the generic observation "Dan is tall" to the possibility distribution generated when the statement is issued by a person known to be 6'3" tall and considering himself to be tall, or to the same statement made by a man 5'8" tall who considers himself to be average. The three distributions over the range of heights from 60" to 84" can be found in Appendix II.

## 4  Empirical support

A sampling of 139 adult males as described in Section 2 has shown that the algorithm generates possibility values that correspond more closely to the distribution of descriptors when applied to adult males of known heights when k = 1 in the above function than the use of the generic distribution for the corresponding descriptors. Table 3 shows some selected values of height with the corresponding possibility values generated by the generic distribution and by the proposed algorithm from the proposition "Dan is tall." Table 4 shows the empirical results for the same heights. While it is generally accepted [Zadeh 1978] that possibility values do not necessarily correspond to probability values, it is preferable to generate possibility values that do correspond to common sense and practical experience.

This correlation should also hold for the case of female observer to female subject with some appropriate value of k. It is also believed that cross-gender observations can also be validated. These analyses are the subjects of on-going verification.

## 5  Conclusions

The purpose of this work has been to provide a more realistic response to questions such as "How tall is Dan" based on known characteristics of an observer and the descriptors used for both the observer and the subject. Although this paper addressed only possibility distributions for height based on height-perception relations between a male observe and a male subject, it is readily apparent that other meaningful correlations may be possible in other areas of linguistic attributes such as age (old, young, etc.) and girth (fat, thin, etc.) The investigation of corresponding possibility distributions to establishing actual age or weight, for example, should be the subject of future investigation..

| h | Generated by S-distribution with $\alpha = 60$, $\beta = 72$ and $\gamma = 84$ | Generated by algorithm with h(A)=75" and s(A)=*Tall* | Generated by algorithm with h(A)=68" and s(A)=*Average* |
|---|---|---|---|
| 63" | .03 | .00 | .00 |
| 71" | .42 | .13 | .97 |
| 76" | .78 | .97 | .78 |
| 81" | .97 | .13 | .06 |

Table 3  Poss{Height(Dan) = h} for selected values of h corresponding to proposition "Dan is tall"

| h | h(A)=75" s(A)=*Tall* | h(A)=68" s(A)=*Average* |
|---|---|---|
| 63" | 0% | 0% |
| 71" | 0% | 25% |
| 76" | 100% | 75% |
| 81" | 0% | 25% |

Table 4  Percentage of respondents, A, declaring "Dan is tall" when height(Dan)=h

## References

Kandel, Abraham. Fuzzy Mathematical Techniques and Applications. Addison-Wesley Publishing Company, Reading, MA. 1986.

Luria, Alexander R. Language and Cognition (ed. James V. Wertsch). John Wiley & Sons, Inc. New York. 1981.

Rommetveit, Ragnar. Words, Meanings, and Messages: Theory and Experiments in Pyscholinguistics. Academic Press Inc., New York. 1986.

Yager, Ronald R. "Linguistic Hedges: Their Relation to Context and Their Experimental Realization" in *Cybernetics and Systems: And International Journal*, vol. 13, pp.357-374. 1982.

Zadeh, Lotfi A. "Test-Score Semantics as a Basis for a Computational Approach to the Representation of Meaning," originally published in *Literary and Linguistic Computing*, vol. 1, pp.24-35 (1986); reprinted in Fuzzy Sets and Applications: Selected Papers by L.A. Zadeh (edited by R.R. Yager, S. Ovchinnikov, R.M. Tong and H.T. Nguyen). John Wiley & Sons. 1987.

Zadeh, Lotfi A. "Fuzzy Sets as a Basis for a Theory of Possibility," originally published in *Fuzzy Sets and Systems*, vol. 1, pp.3-28 (1978); reprinted in Fuzzy Sets and Applications: Selected Papers by L.A. Zadeh (edited by R.R. Yager, S.Ovchinnikov, R.M. Tong and H.T. Nguyen). John Wiley & Sons. 1987.

## Appendix I

The S-Distribution

$$S(x;\ \alpha,\beta,\gamma) = \begin{cases} 0 & ;\quad x \le \alpha \\[2mm] 2\left(\dfrac{x-\alpha}{\gamma-\alpha}\right)^2 & ;\quad \alpha \le x \le \beta \\[2mm] 1-2\left(\dfrac{x-\gamma}{\gamma-\alpha}\right)^2 & ;\quad \beta \le x \le \gamma \\[2mm] 1 & ;\quad x \ge \gamma \end{cases}$$

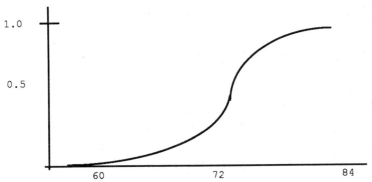

Figure 1  The S-distribution for $_\mu$Tall(x) with $\alpha = 60$, $\beta = 72$ and $\gamma = 84$

The $\pi$-Distribution

$$\pi(x;\beta,\gamma) = \begin{cases} S(x;\ \gamma-\beta,\ \gamma-\dfrac{\beta}{2},\ \gamma); & x \le \gamma \\[3mm] 1-S(x;\ \gamma,\ \gamma+\dfrac{\beta}{2},\ \gamma+\beta); & x \ge \gamma \end{cases}$$

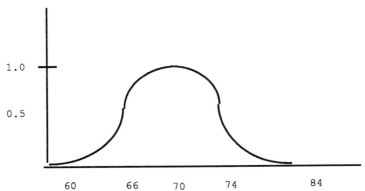

Figure 2  $\pi$-distribution for $_\mu$Average(x) with $\gamma = 70$ and $\beta = 8$

## Appendix II

Poss{Height(Dan)=h} for 60" ≤ h ≤ 84" in response to "How tall is Dan" based on statement "Dan is tall."

| h in inches | Generated by S-Distribution with $\alpha=60$ $\beta=72$ $\gamma=84$ | Generated by algorithm with h(A)=75" and s(A)=*Tall* | Generated by algorithm with h(A)=68" and s(A)=*Average* |
|---|---|---|---|
| 60 | .00 | .00 | .00 |
| 61 | .00 | .00 | .00 |
| 62 | .01 | .00 | .00 |
| 63 | .03 | .00 | .00 |
| 64 | .06 | .00 | .00 |
| 65 | .09 | .00 | .00 |
| 66 | .13 | .00 | .00 |
| 67 | .17 | .00 | .00 |
| 68 | .22 | .01 | .00 |
| 69 | .28 | .03 | .77 |
| 70 | .35 | .07 | .89 |
| 71 | .42 | .13 | .97 |
| 72 | .50 | .20 | 1.00 |
| 73 | .58 | .28 | .99 |
| 74 | .65 | .97 | .94 |
| 75 | .72 | 1.00 | .88 |
| 76 | .78 | .97 | .78 |
| 77 | .83 | .88 | .65 |
| 78 | .88 | .72 | .50 |
| 79 | .91 | .50 | .35 |
| 80 | .94 | .28 | .22 |
| 81 | .97 | .13 | .13 |
| 82 | .99 | .03 | .06 |
| 83 | 1.00 | .00 | .01 |
| 84 | 1.00 | .00 | .00 |

# On "Bold" Resolution Theory

Helmut Thiele
thiele@ls1.informatik.uni-dortmund.de

Stephan Lehmke
lehmke@ls1.informatik.uni-dortmund.de

University of Dortmund, Department of Computer Science 1,
D-44221 Dortmund, Germany

*Abstract*— The research in applying resolution theory to fuzzy logic so far has concentrated on fuzzy logic generated by the 'standard connectives' *min* and *max* for logical *and* and *or*, respectively, and the Łukasiewicz negation $1 - x$ for *not*. The expressive power of this logic, however, is very low.

In this paper, we try to direct interest to resolution in fuzzy logics generated by 'nonstandard connectives' by showing that the resolution rule is correct and complete if *and* and *or* are interpreted by the 'bold' connectives, which is especially valuable because the logic thus generated has very high expressive power.

Keywords: fuzzy logic, resolution principle, nonstandard connectives

## I. Introduction

The extension of resolution theory to fuzzy logic has been studied since the mid-70s, for example by Lee [8], and has led to several "Fuzzy Prolog" implementations (see [5], [9], and [10]).

In most publications, the underlying logic is generated by Zadeh's [14] standard set of logical connectives, where the logical *or* and the logical *and* are interpreted by the functions *max* and *min*, respectively, and where the logical *not* is interpreted by Łukasiewicz negation $1 - x$.

This interpretation offers a variety of advantages. As these logical connectives are idempotent and distributive with respect to each other, it is possible to transform each propositional formula into a conjunctive normal form and to represent these formulas by sets of clauses, thereby reducing the 'administrative overhead'. On the other hand, the logic created by this interpretation of the connectives suffers from a severe lack of expressiveness, an indication of which is the fact that the expressive Łukasiewicz implication cannot be defined in terms of these connectives.

Lee [8] proves a variety of reduction theorems which show that the expressiveness of the logic thus generated is not very far from that of classical two-valued logic.

So it seems promising to focus research on resolution in fuzzy logics generated by different sets of connectives which offer more expressiveness, for example a pair of dual s- and t-norms as an interpretation of logical *or* and logical *and*, respectively. It can be shown, however, that any idempotent t-norm is identical to *min* and any idempotent s-norm is identical to *max* (see, for example, [7]). So the ease of representation is lost once we leave the realm of *min* and *max*, while a raise in expressive power is not guaranteed for all dual t-/s-norm pairs.

Here, we will study a logic generated by the well-known "bold" fuzzy-logical connectives where *or* is interpreted by $min(1, x + y)$ and *and* is interpreted by $max(0, x + y - 1)$, which has much more expressive power than the $min/max$ logic. Especially, Łukasiewicz implication can be defined herein. To avoid the difficulties invoked by the problem of skolemization we restrict ourselves to propositional calculus.

Furthermore, on the level of propositional logic we can recognize, demonstrate and overcome the difficulties caused by the loss of idempotency and distributivity.

For this purpose, in this paper we will consider only clauses of "first level", i. e. disjunctions of literals, whereas in future reports we will show how the resolution rule can be applied to clauses of "higher level", meaning disjunctions of formulas that can be either literals or conjunctions of clauses of higher level.

To simplify our considerations, first of all we use only so-called 1-models. For more general concepts of models, see section VI.

In the next two sections, the syntax and semantics of the logic thus generated are defined. In section IV the resolution rule is introduced and proved correct. In section V it is proved complete. Section VI provides generalizations of the presented concepts and open research problems.

## II. Syntax

We start with a set $PV$ of propositional variables $p, q, r, \ldots$

## Definition 1

1. **Literals** are either propositional variables or negated propositional variables, i. e. the set $LIT$ of all literals is defined by

$$LIT =_{def} PV \cup \{\neg p \,|\, p \in PV\}.$$

2. If $n \geq 1$ is a nonnegative integer and $l_1, \ldots, l_n$ are literals, then $l_1 \oplus \cdots \oplus l_n$ is said to be a **clause**. Furthermore, we introduce the empty clause denoted by $\square$. An arbitrary clause is denoted by $C$.

The nonnegative integer $n$ is called the length of the clause

$$C = l_1 \oplus \cdots \oplus l_n.$$

The empty clause has the length zero. A literal $l$ is considered to be a clause of length one.

## III. Semantics

We use the closed interval $\langle 0, 1 \rangle$ of all real numbers $r$ with $0 \leq r \leq 1$ as the set of all logical values.

An assignment of the propositional variables is a mapping of the form

$$\alpha : PV \to \langle 0, 1 \rangle.$$

By the following definition we introduce the mapping $VAL(C, \alpha)$ which determines the logical value of a clause $C$ with respect to the assignment $\alpha$.

## Definition 2

1. $VAL(\square, \alpha) =_{def} 0$.
2. $VAL(p, \alpha) =_{def} \alpha(p)$ for $p \in PV$.
3. $VAL(\neg p, \alpha) =_{def} 1 - \alpha(p)$ for $p \in PV$.
4. $VAL(l_1 \oplus \cdots \oplus l_n, \alpha)$
$$=_{def} min\left(1, \sum_{\lambda=1}^{n} VAL(l_\lambda, \alpha)\right).$$

The following definition introduces the concepts of *model* and of *semantic consequence* used in the paper presented.

Let $C$ be an arbitrary clause, $X$ a set of clauses, and $\alpha$ an arbitrary assignment.

## Definition 3

1. $\alpha$ is said to be a **model** of $X$ (shortly $\alpha \models X$) $=_{def}$ For every clause $C \in X$, $VAL(C, \alpha) = 1$.

2. $C$ is a **semantic consequence** of $X$ (shortly $X \Vdash C$) $=_{def}$ For every assignment $\alpha : PV \to \langle 0, 1 \rangle$, if $\alpha$ is a model of $X$, then $VAL(C, \alpha) = 1$.

## Remarks

1. The concept of model introduced can be denoted *1-model* because we require $VAL(C, \alpha) = 1$. Several generalizations of this concept are possible; we will discuss this problem at the end of this paper.

2. In the definition of $X \Vdash C$ we have required that $VAL(C, \alpha) = 1$ in the conclusion. This is essential in order to establish the correctness of the resolution rule.

## IV. The resolution rule and its correctness

Since $\oplus$ lacks idempotency, it does not make sense to represent clauses as sets of literals in analogy to the two-valued logic. As using multisets seemed too much effort for the purpose, we will directly manipulate the syntactical structure of the clauses as defined in section II.

So in the following, when we speak of an *occurrence* of a literal $l$ in a clause $C$, we always mean an occurrence of the *character sequence denoting* $l$ in the *character sequence denoting* $C$.[1] Analogous assumptions hold for the removal of literals from clauses.

## Definition 4

By $C \setminus l$ we denote the result of removing the first occurrence of (the character sequence denoting) a literal $l$ from (the character sequence denoting) a clause $C$.

**Case 1.** $l$ does not occur in $C$.
  Then $C \setminus l =_{def} C$.
**Case 2.** $C = l$.
  Then $C \setminus l =_{def} \square$.
**Case 3.** $C = l \oplus D$ for some clause $D$.
  Then $C \setminus l =_{def} D$.
**Case 4.** $C = D \oplus l \oplus E$ for two clauses $D, E$ and $l$ does not occur in $D$.
  Then $C \setminus l =_{def} D \oplus E$.
**Case 5.** $C = D \oplus l$ for some clause $D$ and $l$ does not occur in $D$.
  Then $C \setminus l =_{def} D$.

By $C \,\tilde{\setminus}\, l$ we denote the result of applying $\setminus$ often enough as to remove **all** occurrences of $l$ from $C$.

---

[1] If $p$ denotes a propositional variable, an occurrence of $\neg p$ in $C$ is not considered to be an occurrence of $p$ in $C$.

We use ROBINSON's [11] "classical" resolution rule in the following modified form. If $X$ is an arbitrary set of clauses, $C$, $D$ are arbitrary clauses and $p$ is a propositional variable, then we can express the resolution rule, anticipating the following exact definition, by

$$\frac{X \vdash C \text{ and } p \text{ occurs in } C \qquad X \vdash D \text{ and } \neg p \text{ occurs in } D}{X \vdash ((C \oplus D) \smallsetminus p) \smallsetminus \neg p}^{2}$$

We define the relation $X \vdash^{k} C$ by induction on $k$ as follows.

### Definition 5

*1. $X \vdash^{0} C =_{def} C \in X$ (embedding)*

*2. $X \vdash^{k+1} C$*

$=_{def} X \vdash^{k} C$ *or there are clauses $D$ and $E$ and a propositional variable $p$ such that $p$ occurs in $D$, $\neg p$ occurs in $E$ and*

$$X \vdash^{k} D \qquad \text{and}$$
$$X \vdash^{k} E \qquad \text{and}$$
$$C = ((D \oplus E) \smallsetminus p) \smallsetminus \neg p$$

### Definition 6

$X \vdash C =_{def}$ *There is a nonnegative integer $k \geqq 0$ such that $X \vdash^{k} C$.*

### Theorem 1 (Correctness)

*For every set $X$ of clauses and every clause $C$,*

$$\text{If } X \vdash C, \text{ then } X \Vdash C.$$

**Proof** We show by induction on $k$:

$$\text{If } X \vdash^{k} C, \text{ then } X \Vdash C.$$

The induction basis $k = 0$ is trivial.

In order to prove the induction step it is sufficient to prove the following lemma.

### Lemma 1

*If $p$ occurs in $C$, $\neg p$ occurs in $D$ and if $VAL(C, \alpha) = 1$ and $VAL(D, \alpha) = 1$, then $VAL(((C \oplus D) \smallsetminus p) \smallsetminus \neg p, \alpha) = 1$.*

**Proof** The lemma above follows if we prove, for arbitrary $x, y, z \in \langle 0, 1 \rangle$:

---

²The parentheses are meta-symbols meant to clarify the precedence of the operations.

If $min(1, x + z) = 1$ and $min(1, y + (1 - z)) = 1$, then $min(1, x + y) = 1$.

From the assumptions above we get

$$x + z \geqq 1$$

and

$$y + (1 - z) \geqq 1.$$

We get by addition

$$x + y + 1 \geqq 2,$$

hence

$$x + y \geqq 1,$$

hence

$$min(1, x + y) = 1.$$

Thus, **Lemma 1** and also **Theorem 1** is proved.

## V. THE COMPLETENESS OF THE CORRESPONDING REFUTATION SYSTEM

Following the "philosophy" of ROBINSON's resolution theory, we show a special reversed form of **Theorem 1**:

$$\text{If } X \Vdash \square, \text{ then } X \vdash \square,$$

i. e. we show the following:

### Theorem 2

*If $X$ has no 1-model, then $X \vdash \square$.*

Before we can carry out the proof of theorem 2, we need some preliminary definitions and lemmata.

Let $p$ be a propositional variable and $X$ a set of clauses.

### Definition 7

$$0. \ X_p^0 =_{def} \left\{ C \smallsetminus p \ \middle| \ \begin{array}{l} C \in X \text{ and} \\ \neg p \text{ does not occur in } C \end{array} \right\}$$
$$1. \ X_p^1 =_{def} \left\{ C \smallsetminus \neg p \ \middle| \ \begin{array}{l} C \in X \text{ and} \\ p \text{ does not occur in } C \end{array} \right\}$$

### Lemma 2

*0. If $X_p^0$ has a 1-model, then $X$ has a 1-model.*
*1. If $X_p^1$ has a 1-model, then $X$ has a 1-model.*

### Proof

**ad 0.** Assume that $\alpha^0$ is a 1-model of $X_p^0$ and $C$ is any clause of $X$. We construct an assignment $\alpha$ by defining, for each $q \in PV$,

$$\alpha(q) =_{def} \begin{cases} \alpha^0(q) & \text{if } q \neq p \\ 0 & \text{if } q = p \end{cases}$$

**Case 1.** $C \in X_p^0$.

Then neither $p$ nor $\neg p$ occur in $C$, so

$$VAL(C, \alpha) = VAL(C, \alpha^0) = 1.$$

**Case 2.** $C \notin X_p^0$, but $C \curlyveedownarrow p \in X_p^0$.

Then $\neg p$ does not occur in $C$. As $VAL(p, \alpha) = 0$, obviously

$$VAL(C, \alpha) = VAL(C \curlyveedownarrow p, \alpha)$$
$$= VAL(C \curlyveedownarrow p, \alpha^0) = 1.$$

**Case 3.** $C \notin X_p^0$ and $C \curlyveedownarrow p \notin X_p^0$.

Then $\neg p$ occurs in $C$, so

$$VAL(C, \alpha) \geqq VAL(\neg p, \alpha) = 1.$$

**ad 1.** Can be proved analogously.

Let $p$ be a propositional variable, $C$ a clause and $X$ a set of clauses.

## Lemma 3

0. If $\neg p \in X$ and $X_p^0 \vdash C$, then $X \vdash C$.
1. If $p \in X$ and $X_p^1 \vdash C$, then $X \vdash C$.

## Proof

**ad 0.** In order to prove this assertion we show that $\neg p \in X$ and $X_p^0 \vdash^k C$ implies $X \vdash C$ by induction on $k$.

**Induction basis.** $k = 0$.

This means $C \in X_p^0$.

**Case 1.** $C \in X$.

Then $X \vdash C$ by embedding.

**Case 2.** $C \notin X$.

Then $C$ was constructed from some clause $D \in X$ by removing all occurrences of $p$. But this can be done by applying the resolution rule several times, always choosing $\neg p$ as one of the clauses to be resolved. So $X \vdash C$.

**Induction step.** We assume the assertion holds for $k$. Now, we prove it for $k + 1$.

**Case 1.** $X_p^0 \vdash^k C$.

Then $X \vdash C$ follows directly by the induction hypothesis.

**Case 2.** There are clauses $D$ and $E$ and a propositional variable $q$ such that $q$ occurs in $D$, $\neg q$ occurs in $E$ and

$$X_p^0 \vdash^k D \qquad \text{and}$$
$$X_p^0 \vdash^k E \qquad \text{and}$$
$$C = ((D \oplus E) \smallsetminus q) \smallsetminus \neg q.$$

By the induction hypothesis, from

$$X_p^0 \vdash^k D \text{ and } X_p^0 \vdash^k E$$

we get

$$X \vdash D \text{ and } X \vdash E,$$

hence by the resolution rule

$$X \vdash ((D \oplus E) \smallsetminus q) \smallsetminus \neg q,$$

i. e. $X \vdash C$.

**ad 1.** Can be proved analogously.

## Lemma 4

*If $X$ has no 1-model, then $\square \in X$ or there exists a literal $l$ such that $l \in X$.*

**Proof** We assume $X$ has no 1-model.

We prove this lemma indirectly by assuming that

$$\square \notin X \text{ and there is no literal } l \text{ such that } l \in X.$$

Then every clause in $X$ contains at least two literals connected by $\oplus$. We define an assignment $\alpha$ for every $q \in PV$ by

$$\alpha(q) =_{def} \frac{1}{2}.$$

Then $VAL(l, \alpha) = \frac{1}{2}$ for every literal $l$ and obviously $VAL(C, \alpha) = 1$ for each clause $C$ containing at least two literals.

So $\alpha$ is a 1-model for $X$, contradicting the assumption that $X$ has no such model.

## Lemma 5

*If $X$ has no 1-model, then there is a finite subset $X_{fin}$ of $X$ without 1-model.*

This result has been proved by GOTTWALD [3] for the (full) LUKASIEWICZ propositional calculus, using the theory of compact topological spaces.

**Proof (of Theorem 2)** Let $X$ be a set of clauses without 1-model. By **Lemma 5** it is sufficient to consider a finite subset $X_{fin}$ of $X$. We prove theorem 2 by induction on the number $n$ of different propositional variables occurring in $X_{fin}$.[3]

**Induction basis.** $n = 0$.

If no propositional variables occur in $X_{fin}$, then, by **Lemma 4**, we can conclude that $\square \in X_{fin}$, so

$$X_{fin} \vdash \square$$

by embedding.

---

[3] Obviously, here and in the following a propositional variable $p$ is considered to *occur* in $X$ if either $p$ or $\neg p$ occur in any clause of $X$.

**Induction step.** We assume that $n + 1$ different propositional variables occur in $X_{\text{fin}}$, for some non-negative integer $n \geq 0$.

By **Lemma 4**, we can distinguish two cases:

**Case 1.** $\square \in X_{\text{fin}}$.

In this case $X_{\text{fin}} \vdash \square$ holds by embedding.

**Case 2.** $\square \notin X_{\text{fin}}$ and there is some literal $l$ such that $l \in X_{\text{fin}}$.

**Case 2.0** $l = \neg p$ for some propositional variable $p$.

Now, by **Lemma 2**, we can conclude that $(X_{\text{fin}})_p^0$ has no 1-model.

But, as the propositional variable $p$ does not occur in $(X_{\text{fin}})_p^0$, only a total of $n$ different propositional variables occurs in $(X_{\text{fin}})_p^0$, so it follows from the induction hypothesis that

$$(X_{\text{fin}})_p^0 \vdash \square.$$

But the result of **Lemma 3** means that every clause $C$ resolvable from $(X_{\text{fin}})_p^0$ is also resolvable from $X_{\text{fin}}$, so we can directly conclude

$$X_{\text{fin}} \vdash \square.$$

**Case 2.1** $l = p$ for some propositional variable $p$.

$X_{\text{fin}} \vdash \square$ can be concluded analogously to case 2.0.

## VI. Conclusions and further research

We can generalize our results to the following modifications of the concept of *model*.

Let $\beta$ be an assignment, i. e. $\beta : PV \rightarrow \langle 0, 1 \rangle$.

### Definition 8

1. *Models of type 1.* Assume $s \in \langle 0, 1 \rangle$.

$\beta$ is said to be a model of type 1 of $X$ (or an $s$-model, shortly $\beta \overset{1}{\models} X$)

$=_{def}$ For every $C \in X$, $VAL(C, \beta) \geq s$.

2. *Models of type 2.* Assume $s \in \langle 0, 1 \rangle$ and $s \neq 1$.

$\beta$ is said to be a model of type 2 of $X$ (or a strong $s$-model, shortly $\beta \overset{2}{\models} X$)

$=_{def}$ For every $C \in X$, $VAL(C, \beta) > s$.

3. *Models of type 3.* Assume $\eta : X \rightarrow \langle 0, 1 \rangle$.

$\beta$ is said to be a model of type 3 of $X$ (or an $\eta$-model, shortly $\beta \overset{3}{\models} X$)

$=_{def}$ For every $C \in X$, $VAL(C, \beta) \geq \eta(C)$.

4. *Models of type 4.* Assume $\eta : X \rightarrow \langle 0, 1 \rangle$ with $\eta(C) < 1$ for every $C \in X$.

$\beta$ is said to be a model of type 4 of $X$ (or a strong $\eta$-model, shortly $\beta \overset{4}{\models} X$)

$=_{def}$ For every $C \in X$, $VAL(C, \beta) > \eta(C)$.

### Definition 9

$X \overset{i}{\Vdash} C$

$=_{def}$ For every $\beta : PV \rightarrow \langle 0, 1 \rangle$, if $\beta$ is a model of type $i$ for $X$, then $VAL(C, \beta) = 1$.

**Remark** The relation $X \overset{i}{\Vdash} C$ is not embedding, in general, i. e. we don't have

$$\text{If } C \in X, \text{ then } X \overset{i}{\Vdash} C.$$

Therefore there is no correctness theorem (see **Theorem 1**) like

$$\text{If } X \vdash C, \text{ then } X \overset{i}{\Vdash} C,$$

in general.

### Theorem 3

For $i = 1, 2, 3, 4$,

$$\text{If } X \overset{i}{\Vdash} C, \text{ then } X \Vdash C.$$

**Proof** Trivial, because

$$\beta \text{ is a 1-model of } X$$

implies

$$\beta \text{ is a model of type } i \text{ of } X.$$

### Theorem 4 (Completeness)

If $X$ is finite and $X$ has no model of type $i$, then $X \vdash \square$.

**Proof** By **Theorem 3** and **Theorem 2**.

**Remark** It remains to prove the compactness of the model concepts of type $i$ ($i = 1, 2, 3, 4$), in general. Lemma 5 gives the compactness of models of type 1 only for the special case $s = 1$.

In future reports, the authors will treat the following problems:

1. Extension of the resolution rule presented in this paper to hierarchical expressions of conjunctions of disjunctions which cannot be 'collapsed' to CNF because of the lack of distributivity.

2. Further extension of the resolution rule to arbitrary expressions of a 'bold' fuzzy predicate calculus.

3. Restrictions of resolution, like linear resolution, resolution with horn formulas, and SLD resolution.

*Acknowledgement*

The authors wish to thank Ulrich Fieseler for helpful scientific discussion.

## REFERENCES

[1] Liya Ding, Zuliang Shen, and Masao Mukaidono. The properties of fuzzy logic for fuzzy prolog. In *First Asian Fuzzy Systems Symposium* [4].

[2] Didier Dubois, Henri Prade, and Ronald R. Yager, editors. *Readings in Fuzzy Sets for Intelligent Systems*. Morgan Kaufmann, 1993.

[3] Siegfried Gottwald. Fuzzy propositional logics. *Fuzzy Sets and Systems*, 3:181–192, 1980.

[4] Institute of Systems Science, National University of Singapore. *First Asian Fuzzy Systems Symposium*, Singapore, November 23–26, 1993.

[5] Mitsuru Ishizuka and Naoki Kanai. Prolog-elf incorporating fuzzy logic. In Aravind Joshi, editor, *IJCAI-85 — Proceedings of the Ninth International Joint Conference on Artificial Intelligence*, pages 701–703. Morgan Kaufmann, 18–23 August, 1985.

[6] Hiroaki Kikuchi and Masao Mukaidono. The soundness and completeness of linear resolution for fuzzy logic program. In *First Asian Fuzzy Systems Symposium* [4].

[7] George J. Klir and Tina A. Folger. *Fuzzy Sets, Uncertainty, and Information*. Prentice Hall, 1988.

[8] Richard C. T. Lee. Fuzzy logic and the resolution principle. *Journal of the ACM*, 19(1):109–119, 1972. Reprinted in [2].

[9] T. P. Martin, J. F. Baldwin, and B. W. Pilsworth. The implementation of FPROLOG — a fuzzy prolog interpreter. *Fuzzy Sets and Systems*, 23:119–129, 1987.

[10] Masao Mukaidono, Zuliang Shen, and Liya Ding. Fundamentals of Fuzzy Prolog. *International Journal of Approximate Reasoning*, 3:179–193, 1989. Reprinted in [2].

[11] John Alan Robinson. A machine-oriented logic based on the resolution principle. *Journal of the ACM*, 12:23–41, 1965.

[12] Zuliang Shen, Liya Ding, and Masao Mukaidono. Fuzzy resolution principle. In *ISMVL — The 18th International Symposium on Multiple-Valued Logic*, pages 210–215, 1988.

[13] Guglielmo Tamburrini and Settimo Termini. Towards resolution in a fuzzy logic with Lukasiewicz implication. In *IPMU '92 — International Conference on Information Processing and Management of Uncertainty in Knowledge-Based Systems*, pages 271–277, Mallorca, July 6–10, 1992. Universitat de les Illes Balears.

[14] Lotfi A. Zadeh. Fuzzy sets. *Information and Control*, 8:338–353, 1965.

# A NETWORK MODEL BASED ON FUZZY QUEUEING SYSTEM

Jung Bok JO[*], Yasuhiro TSUJIMURA[**], Mitsuo GEN[**], and Genji YAMAZAKI[*]

[*]Dept. of Engineering Management, Tokyo Metropolitan Institute of Technology,
Hino City, Tokyo 191, JAPAN

[**]Dept. of Industrial and Systems Engineering, Ashikaga Institute of Technology,
Ashikaga, 326, JAPAN

## ABSTRACT

The purpose of this paper is to combine the ability of fuzzy set to represent more realistic situations with the well-established traditional queueing network model problem. We propose an fuzzification of M/M/1 queueing system. We also apply fuzzy set theory to the open central server network model with the fuzzy queues. Thus, we represent the characteristic and performance of open central server network model based on fuzzy queueing system.

## I. INTRODUCTION

Many mathematical models have been developed to solve queueing network problems. Most of these assume probability distribution for arrivals and services[8,9]. When we develop a new system or some information about an existing system is unavailable, data may be represented using the possibility theory rather than probability theory. In this case, we apply to fuzzy number for the arrival and service rates of queueing network system. In actual practice, the service rate is frequently described as 'fast', 'slow', or 'moderately slow', and so on, which are linguistic fuzzy terms and can be best described by using fuzzy sets.

The problem of fuzzy queues has been analyzed by Prade[1] and Li and Lee[5] through the use of the extension principle. Buckley[2] considers elementary queueing systems, with multiple parallel servers with finite or infinite system capacity and calling source, whose arrivals and departures are restricted by arbitrary possibility distributions. Recently Negi and Lee's contribution[4] was shown that their approach can utilize the advantages of both the fuzzy and probability approaches to make the model more realistic and less restrictive. Their approach for F/F/1 system is a single value simulation of a fuzzy variable by the use of two random variables.

In this paper, the fuzzy set theory is introduced into the Little's law, and the performance measures for the fuzzy M/M/1 queue when arrival and service rates are fuzzy variables by using this fuzzy Little's law[13,14]. We also apply fuzzy set theory to the open central server model[6] with a fuzzy queue. Furthermore, the characteristic of open central server network model based on fuzzy queueing system is shown.

## II. PRELIMINARIES

The following notations are used throughout this paper:

$\tilde{N}(t)$ : number of customers in the system at time t

$\tilde{A}(t)$ : number of customers who arrived in the interval[0,t]

$\tilde{N}$ : average number of customers in the system

$\tilde{W}$ : average customer waiting time in the system

$\tilde{N}_Q$ : average number of customers waiting in queue

$\tilde{W}_Q$ : average customer waiting time in queue

$\tilde{P}_n$ : probability of n customers in the system

where $\sim$ denotes a triangular fuzzy number(TFN).

Let $\tilde{A}$ be a TFN with membership function $\mu_{\tilde{A}}(x)$. We represent an interval of confidence of TFN $\tilde{A}$ for all $\alpha$-cut level, that is, $A(\alpha) = [a_L^{(\alpha)}, a_R^{(\alpha)}]$ for $\alpha \in [0,1]$. If $\alpha = 1$, $a_L^{(\alpha)}$ is equal to $a_R^{(\alpha)}$. And the TFN $\tilde{A}$ is specified by triplet $(a_1, a_2, a_3)$ where:

(1) $a_1 < a_2 < a_3$;

(2) $\mu_{\tilde{A}}(x) = 0$ outside $(a_1, a_3)$ and equals one at $a_2$;

(3) $\mu_{\tilde{A}}(x)$ is continuous and monotonically increasing from zero to one on $[a_1, a_2]$;

(4) $\mu_{\tilde{A}}(x)$ is continuous and monotonically decreasing from one to zero on $[a_2, a_3]$.

For any TFN $\tilde{A}$, we can write:

(1) $\tilde{A} \geq 0$ if $a_1 \geq 0$;

(2) $\tilde{A} > 0$ if $a_1 > 0$;

(3) $\tilde{A} \leq 0$ if $a_3 \leq 0$;

(4) $\tilde{A} < 0$ if $a_3 < 0$.

The support of $\tilde{A}$ is $(a_1, a_3)$. Comparisons of two TFNs are assumed to be made by Kaufmann and Gupta[10] methodology using linear ordering of fuzzy numbers.

Dubois and Prade[3] and Zimmermann[7] have defined the fuzzifying function and integration of a fuzzifying function over a crisp interval, as follows:

__Definition 1__  Let X and Y be universes and $\widetilde{P}(Y)$ be the set of all fuzzy sets in Y(power set).

$$\widetilde{f} : X \rightarrow \widetilde{P}(Y) \text{ is a mapping}$$

$\widetilde{f}$ is a fuzzifying function, iff

$$\mu_{\widetilde{f(x)}}(y) = \mu_{\widetilde{R}}(x,y), \ \forall \ (x,y) \in X \times Y$$

where $\mu_{\widetilde{R}}(x,y)$ is the membership function of a fuzzy relation.

__Definition 2__  Let $\widetilde{f}(x)$ be a fuzzifying function from $[a,b] \subseteq \mathbb{R}$ to $\mathbb{R}$ such that, $\widetilde{f}(x)$, $\forall \ x \in [a,b]$ is a fuzzy number. And $\alpha$-level curves $f_\alpha^-(x)$ and $f_\alpha^+(x)$ are the curves which values of membership function of $\widetilde{f}(x)$ are $\mu_{\widetilde{f(x)}}(f_\alpha^-(x)) = \mu_{\widetilde{f(x)}}(f_\alpha^-(x)) = \alpha$ respectively. The integral of $\widetilde{f}(x)$ over [a,b] is then defined to be the fuzzy set as follows:

$$\int_a^b \widetilde{f}(x)dx = \left\{ \left( \left[ \int_a^b f_\alpha^-(x)dx , \int_a^b f_\alpha^+(x)dx \right] , \alpha \right) \right\}.$$

__Definition 3__  For TFNs $\widetilde{A}$ and $\widetilde{C}$, we assume that zero does not belong to the support of $\widetilde{A}$, $\widetilde{A} > 0$ and $\widetilde{C} \geq 0$. By Buckley and Qu[12], the $\alpha$-cut of solution of an fuzzy equation $\widetilde{A} \cdot \widetilde{X} = \widetilde{C}$ is retrieved from

$$\forall \ \alpha \in [0,1]$$

$$X(\alpha) = \left[ \frac{c_L^{(\alpha)}}{a_L^{(\alpha)}}, \frac{c_R^{(\alpha)}}{a_R^{(\alpha)}} \right],$$

iff $a_1 c_2 > c_1 a_2$ and $a_3 c_2 < c_3 a_2$. In order to consider some simplification and obtain an approximate fuzzy number in terms of a TFN, we approximate this $\alpha$-cut of solution by an approximate TFN, that is,

$$\widetilde{X} = \left( \frac{c_1}{a_1}, \frac{c_2}{a_2}, \frac{c_3}{a_3} \right).$$

Thus, we define the operator $\ominus$ as follows:

$$\widetilde{C} \oplus \widetilde{A} = \left( \frac{c_1}{a_1}, \frac{c_2}{a_2}, \frac{c_3}{a_3} \right),$$

*iff $\widetilde{A} > 0$, $\widetilde{C} \geq 0$, $a_1 c_2 > c_1 a_2$ and $a_3 c_2 < c_3 a_2$.*

Fuzzy Little's law

We introduce fuzzy set theory into Little's law. The fuzzy number of customers in the system observed up to time t is as follows:

$$\widetilde{N}_t = \frac{1}{t} \int_0^t \widetilde{N}(\tau)d\tau \qquad (1)$$

which we call the time average of $\widetilde{N}(\tau)$ up to time t, where a fuzzifying function $\widetilde{N}(\tau)$ and a integration of fuzzifying function are defined by definition 1, 2, respectively. Of course, $\widetilde{N}_t$ depend on t, but $\widetilde{N}_t$ tends to a steady-state $\widetilde{N}$ according to increment of t, that is,

$$\widetilde{N} = \lim_{t \to \infty} \widetilde{N}_t. \qquad (2)$$

The fuzzy arrival rate over the interval [0,t] is as follows:

$$\widetilde{\lambda}_t = \frac{\widetilde{A}(t)}{t}. \qquad (3)$$

The steady-state fuzzy arrival rate is defined as

$$\widetilde{\lambda} = \lim_{t \to \infty} \widetilde{\lambda}_t \qquad (4)$$

assuming that the limit exist. The fuzzy average time of the customer delay up to time t is similarly defined as

$$\widetilde{W}_t \backsimeq \frac{1}{\widetilde{A}(t)} \int_0^t \widetilde{N}(\tau)d\tau \qquad (5)$$

that is, the fuzzy average time spent in the system per customer up to time t. The steady-state fuzzy average time of the customer delay is defined as

$$\widetilde{W} = \lim_{t \to \infty} \widetilde{W}_t \qquad (6)$$

assuming that the limit exists. Therefore, we have

$$\widetilde{W} \backsimeq \frac{\widetilde{N}}{\widetilde{\lambda}}. \qquad (7)$$

Eq.(7) is represented using $\alpha$-cut level, as follows:

$$\forall \ \alpha \in [0,1],$$

$$W(\alpha) = \left[ \frac{N_L^{(\alpha)}}{\lambda_R^{(\alpha)}}, \frac{N_R^{(\alpha)}}{\lambda_L^{(\alpha)}} \right]. \qquad (8)$$

By deconvolution of eq.(8) we can compute the $\widetilde{N}$ with interval of confidence as follows:

$\forall \; \alpha \in [0,1],$

$$N(\alpha) = [\lambda_R^{(\alpha)} W_L^{(\alpha)}, \lambda_L^{(\alpha)} W_R^{(\alpha)}].$$  (9)

Similar formulas exist for $\tilde{N}_Q$ and $\tilde{W}_Q$, that is

$$\tilde{W}_Q = \frac{\tilde{N}_Q}{\tilde{\lambda}}.$$  (10)

And we have

$\forall \; \alpha \in [0,1],$

$$N_Q(\alpha) = [\lambda_R^{(\alpha)} W_{QL}^{(\alpha)}, \lambda_L^{(\alpha)} W_{QR}^{(\alpha)}].$$  (11)

## III. THE FUZZY M/M/1 QUEUEING SYSTEM

The fuzzy M/M/1 queueing system consist of a single queueing station with a single server. Customers arrive according to a Poisson process with fuzzy rate $\tilde{\lambda}$, and the fuzzy probability distribution of the service time is exponential with fuzzy mean $1/\tilde{\mu}$ second. From one among of the properties of the Poisson process with fuzzy parameter $\tilde{\lambda}$ [5], we can get:

For every $t \geq 0$ and $\delta \geq 0$,

$$\tilde{P}\{ N(t+\delta) - N(t) = 0 \} = 1 - \tilde{\lambda}\delta + o(\delta)$$  (12)

$$\tilde{P}\{ N(t+\delta) - N(t) = 1 \} = \tilde{\lambda}\delta + o(\delta)$$  (13)

$$\tilde{P}\{ N(t+\delta) - N(t) \geq 2 \} = o(\delta)$$  (14)

where we generically denoted by $o(\delta)$ a function of $\delta$ such that

$$\lim_{\delta \to 0} \frac{o(\delta)}{\delta} = 0.$$

### 1. Fuzzy Markov chain formulation

We could analyze the process $N(t)$ in terms of the discrete-time fuzzy Markov chains. Let us focus attention at the times

$$0, \delta, 2\delta, ..., k\delta, ...$$

where $\delta$ is a small positive number. We denote

$N_k$ : number of customers in the system at time $k\delta$.

Let $\tilde{P}_{ij}$ denote the corresponding transitional fuzzy probabilities

$$\tilde{P}_{ij} = \tilde{P}\{ N_{k+1} = j \mid N_k = i \}.$$

By using eqs.(12) through (14), one can show that

$$\tilde{P}_{00} = 1 - \tilde{\lambda}\delta + o(\delta)$$  (15)

$$\tilde{P}_{ii} = 1 - \tilde{\lambda}\delta - \tilde{\mu}\delta + o(\delta) \qquad i \geq 1$$  (16)

$$\tilde{P}_{i,i+1} = \tilde{\lambda}\delta + o(\delta) \qquad i \geq 0$$  (17)

$$\tilde{P}_{i,i-1} = \tilde{\mu}\delta + o(\delta) \qquad i \geq 1$$  (18)

$$\tilde{P}_{ij} = o(\delta) \qquad i \; and \; j \neq i, \; i+1, \; i-1.$$  (19)

The state transition diagram for the fuzzy Markov chain $\{N_k\}$ is shown in Fig. 1, where we have omitted the terms $o(\delta)$.

### 2. Derivation of the stationary distribution

Consider now the steady-state fuzzy probabilities

$$\tilde{P}_n = \lim_{t \to \infty} \tilde{P}\{ N(t) = n \}.$$

Note that during any time interval, the total number of transitions from state n to n+1 must differ from the total number of transitions from n+1 to n by at most 1. Thus asymptotically, the frequency of transitions from n to n+1 is equal to the frequency of transitions from n+1 to n. Equivalently, the fuzzy probability that the system is in state n and makes a transition to n+1 in the next transition interval is the same as the fuzzy probability that the system is in state n+1 and makes a transition to n, that is,

$$\tilde{P}_n \tilde{\lambda}\delta + o(\delta) = \tilde{P}_{n+1}\tilde{\mu}\delta + o(\delta).$$  (20)

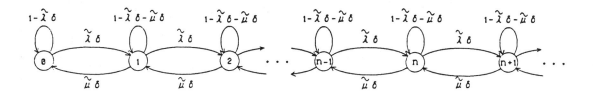

Fig. 1 Discrete-time fuzzy Markov chain for the M/M/1 system.

By taking the limit in this equation as $\delta \to 0$, we obtain

$$\tilde{P}_n \, \tilde{\lambda} = \tilde{P}_{n+1} \tilde{\mu} \; . \tag{21}$$

Eq.(21) can be also written as

$$\tilde{P}_{n+1} = \tilde{\rho} \tilde{P}_n \; , \tag{22}$$

where

$$\tilde{\rho} = \tilde{\lambda} \oslash \tilde{\mu} \; . \tag{23}$$

It follows that

$$\tilde{P}_n \doteq \tilde{\rho}^n \tilde{P}_0 \; . \tag{24}$$

If $\tilde{\rho} < 1$, the probabilities $\tilde{P}_n$ are all positive, so

$$\sum_{n=0}^{\infty} \tilde{P}_n = \sum_{n=0}^{\infty} \tilde{\rho}^n \tilde{P}_0 \doteq \frac{\tilde{P}_0}{1 - \tilde{\rho}} . \tag{25}$$

Combining eq.(24) and eq.(25), we finally obtain

$$\tilde{P}_n \doteq \tilde{\rho}^n (1 \ominus \tilde{\rho}) \sum_{n=0}^{\infty} \tilde{P}_n \; , \tag{26}$$

where $\ominus$ Minkowski subtraction[10].

We can now calculate the fuzzy average number of customers in the system in steady-state:

$$\tilde{N} \doteq \frac{\displaystyle\sum_{n=0}^{\infty} n \tilde{P}_n}{\displaystyle\sum_{n=0}^{\infty} \tilde{P}_n} = \frac{\displaystyle\sum_{n=0}^{\infty} \tilde{P}_n \cdot \sum_{n=0}^{\infty} n \tilde{\rho}^n (1 \ominus \tilde{\rho})}{\displaystyle\sum_{n=0}^{\infty} \tilde{P}_n}$$

$$= \frac{\tilde{\rho} \displaystyle\sum_{n=0}^{\infty} \tilde{P}_n}{(1 - \tilde{\rho}) \displaystyle\sum_{n=0}^{\infty} \tilde{P}_n} \; . \tag{27}$$

The fuzzy average delay per customer is given by fuzzy Little's law,

$$\tilde{W} \doteq \frac{\tilde{N}}{\tilde{\lambda}} \doteq \frac{\tilde{\rho} \displaystyle\sum_{n=0}^{\infty} \tilde{P}_n}{\tilde{\lambda} (1 - \tilde{\rho}) \displaystyle\sum_{n=0}^{\infty} \tilde{P}_n} . \tag{28}$$

If eq.(28) is represented $\alpha$-cut level, we have

$$\forall \; \alpha \in [0,1],$$

$$W(\alpha) = \left[ \frac{\rho_L^{(\alpha)} c_L^{(\alpha)}}{\lambda_R^{(\alpha)} c_R^{(\alpha)} (1 - \rho_L^{(\alpha)})} , \frac{\rho_R^{(\alpha)} c_R^{(\alpha)}}{\lambda_L^{(\alpha)} c_L^{(\alpha)} (1 - \rho_R^{(\alpha)})} \right] .$$

The fuzzy average waiting time in queue $\tilde{W}_Q$, is the fuzzy average delay $\tilde{W}$ less the fuzzy average service time $1/\tilde{\mu}$, so

$$\tilde{W}_Q \doteq \frac{\tilde{\rho} \displaystyle\sum_{n=0}^{\infty} \tilde{P}_n}{\tilde{\lambda} (1 - \tilde{\rho}) \displaystyle\sum_{n=0}^{\infty} \tilde{P}_n} - \frac{1}{\tilde{\mu}} . \tag{29}$$

## IV. FUZZY QUEUEING NETWORK MODEL

One of the key performance concepts used in studying a computer system is the bottleneck device or server. The bottleneck of a system is the first server to saturate as the load on the system is increased. We can identify the bottleneck of the system if we know the service demands $\tilde{D}_k$, for $k = 1, 2, ..., K$. The bottleneck device is device j where j is the integer for which $\tilde{D}_j = \tilde{D}_{max}$, where

$$\tilde{D}_{max} = max \{ \tilde{D}_1, \tilde{D}_2, ..., \tilde{D}_K \} \; . \tag{30}$$

Now, for an application of the proposed fuzzy queueing model, we consider that the open central server system, shown in Fig. 2. The central server is, of course, the CPU. Lavenberg and Sauer in Lavenberg[11] use the central server queueing model to study memory management of a multiprogramming system. They were trying to optimize the performance of operating system. The open central server model

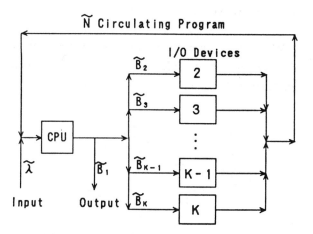

Fig. 2 Open central server system.

can be used to model a system, such as a transaction processing system in which the system is not limited by the multiprogramming level. As shown in Fig. 2, there are K-1 I/O devices, each with its own queue, and each exponentially distributed with average service rate $\tilde{\mu}_k$ (k = 2,3,...,K). The CPU was assumed to have an exponential distribution. The queue discipline is assumed to be first come first serve (FCFS) for the I/O devices. Upon completion of a CPU service, a customer exits the system with probability $\widetilde{B}_1$ or enters service at I/O device k with probability $\widetilde{B}_k$ (k = 2,3,...,K). Upon completion of an I/O service, the customer returns to the CPU queue for another cycle. An incoming arrow leading to the CPU queue is added to indicate the arriving traffic with average rate $\tilde{\lambda}$. The arrival pattern is assumed to have an exponential interarrival time. We assume the system is not overload, so the throughput is also $\tilde{\lambda}$. Therefore, Jackson's theorem can be applied to the system as stated in the following algorithm.

Algorithm 1 (Lavenberg's algorithm)

Consider the open central server system described above. Suppose we are given the average arrival rate $\tilde{\lambda}$, the average service time at each device, $\widetilde{S}_k$ (k = 2,3,...,K), and branching probabilities $\widetilde{B}_k$ (k =2,3,...,K). The service demand, $\widetilde{D}_k$, is the total amount of service that a customer needs at the k-th device. $\widetilde{V}_k$ and $\tilde{\rho}_k$ are visit ratio and utilization at the k-th device, respectively. And $\widetilde{N}_k$ is average number of customer at the k-th device. Then we use Jackson's theorem to calculate the performance measures of the system as follows:

Step 1   [Calculate the demands, $\widetilde{D}_k$]
If the demands are not known, use algorithm 2 to calculate

$$\tilde{D}_k, \quad k = 1,2,...,K.$$

Step 2   [Calculate device performance]
For k = 1,2,...,K, calculate

$$\tilde{\rho}_k \doteq \tilde{\lambda} \cdot \tilde{D}_k ,$$

$$\tilde{N}_k \doteq \frac{\tilde{\rho}_k \sum_{n=0}^{\infty} \tilde{P}_n}{(1-\tilde{\rho}_k) \sum_{n=0}^{\infty} \tilde{P}_n}, \; and \;\; \tilde{W}_k \doteq \frac{\tilde{\lambda} \; \tilde{D}_k \sum_{n=0}^{\infty} \tilde{P}_n}{\tilde{\lambda} \; (1-\tilde{\rho}_k) \sum_{n=0}^{\infty} \tilde{P}_n}.$$

Step 3   [Compute system performance measures]

Set average response time(average time in system) to

$$\tilde{W} = \sum_{k=1}^{K} \tilde{W}_k .$$

Set average number of customer in the system to

$$\forall \; \alpha \in [0,1]$$

$$N(\alpha) = [\lambda_R^{(\alpha)} W_L^{(\alpha)}, \lambda_L^{(\alpha)} W_R^{(\alpha)}].$$

The bottleneck device is device j, where j is the integer for which $\widetilde{D}_j = \widetilde{D}_{max}$, where

$$\tilde{D}_{max} = max \{ \tilde{D}_1, \tilde{D}_2,..., \tilde{D}_K \} .$$

The maximum possible throughput is given by

$$\tilde{\lambda}_{max} \doteq \frac{1}{\tilde{D}_{max}} .$$

Algorithm 2

Consider the open central server system of Fig. 2. This algorithm is constructed the parameters needed to use Labenverg's algorithm.

Step 1   Set the visit ratio $\widetilde{V}_1$ for the CPU to

$$\tilde{V}_1 \doteq \frac{1}{\tilde{B}_1} .$$

Step 2   For k = 2,3,...,K calculate

$$\tilde{V}_k \doteq \tilde{B}_k \cdot \tilde{V}_1.$$

Step 3   [Calculate the demands, $\widetilde{D}_k$]
Set

$$\tilde{D}_k \doteq \tilde{V}_k \cdot \tilde{S}_k , \quad k=1,2,...,K .$$

Now, we show the following numerical example[6] to illustrate fuzzy Lavenberg's algorithm.

## V. NUMERICAL EXAMPLE

We investigate a transaction processing computer system that can be modeled as an open central server model with one CPU and three disks. We find that the TFN of average transaction arrival rate, $\tilde{\lambda}$ is (0.18, 0.2, 0.21) transactions per second with an TFN of average CPU service requirement of (2.8, 3, 3.2) seconds per transaction. The TFNs of average I/O service requirements are (0.9, 1, 1.1), (1.8, 2, 2.1),and

(3.9, 4, 4.1) seconds, respectively.

Let us step through Lavenberg's algorithm. In Step 1 we set $\tilde{D}_1$ = (2.8, 3, 3.2), $\tilde{D}_2$ = (0.9, 1, 1.1), $\tilde{D}_3$ = (1.8, 2, 2.1), and $\tilde{D}_4$ = (3.9, 4, 4.1) seconds. In Step 2 we compute $\tilde{N}_1$ = (0.762, 1.5, 2.732), $\tilde{N}_2$ = (0.145, 0.25, 0.402), $\tilde{N}_3$ = (0.359, 0.667, 1.052), and $\tilde{N}_4$ = (1.767, 4, 8.259) and $\tilde{\rho}_k$, $\tilde{W}_k$ (k = 1, 2, 3, 4) as the following Table 1.

Table 1 The server utilizations and the response times.

| k | $\tilde{\rho}_k$ | $\tilde{W}_k$ |
|---|---|---|
| 1 | (0.504, 0.6, 0.672) | (3.629, 7.5, 15.176) |
| 2 | (0.162, 0.2, 0.231) | (0.690, 1.25, 2.225) |
| 3 | (0.324, 0.4, 0.441) | (1.712, 3.333, 5.844) |
| 4 | (0.702, 0.8, 0.861) | (8.413, 20.0, 45.883) |

In Step 3, from the Table 1 we calculate the average transaction response time

$$\tilde{W} = \tilde{W}_1 + \tilde{W}_2 + \tilde{W}_3 + \tilde{W}_4$$
$$= (\ 14.444,\ 32.083,\ 69.128\ )\ seconds,$$

and the average number of transactions being processed is

$$\tilde{N} \fallingdotseq (\ 3.033,\ 6.417,\ 12.443\ ).$$

The bottleneck is the third disk driver, so the TFN of maximum possible throughput, $\tilde{\lambda}_{max}$ is (0.244, 0.25, 0.256) transactions per second.

## VI. CONCLUSION

In this paper, a fuzzy M/M/1 system that is applied the fuzzy Little's law was proposed, and the performance measures for open central server network model using Lavenberg's algorithm was analyzed. Therefore we can get more realistic solution when some data of network are ambiguous. This analysis was much simple and required less computation.

## REFERENCE

[1] H.M. Prade: An Outline of Fuzzy or Possibilistic Models for Queueing Systems, In : Fuzzy sets, pp.147–153(P.P. Wang and S.K. Chang, editors), Plenum Press(1980).

[2] J.J. Buckley: Elementary Queueing Theory Based on Possibility Theory, Fuzzy Sets and Systems 37, pp.43–52(1990)

[3] D. Dubois and H. Prade: Fuzzy Sets and Systems : Theory and Applications, Plenum Press(1980).

[4] D.S. Negi and E.S. Lee: Analysis and simulation of fuzzy queues, Fuzzy Sets and Systems 46, pp.321–330(1992).

[5] R.J. Li and E.S. Lee: Analysis of Fuzzy Queues, Computers and Mathematics with Applications 17, pp.1143–1147(1989).

[6] A.O. Allen: Probability, Statistics, and Queueing Theory with Computer Science Applications, 2nd ed., Academic Press(1990)

[7] H.J. Zimmermann: Fuzzy Set Theory and Its Applications, 2nd ed., Kluwer Academic Publishers (1991).

[8] D.P. Bertsekas: Data Networks, 2nd ed., Prentice-Hall Inc(1992).

[9] R.W. Wolff: Stochastic Modeling and the Theory of Queues, Prentice-Hall Inc(1989).

[10] A. Kaufmann and M.M. Gupta: Fuzzy Mathematical Models in Engineering and Management Science, Van Nostrand Reinhold(1988).

[11] S.S. Lavenberg, ed.: Computer Performance Modelling Handbook, Academic Press(1983).

[12] J.J. Buckley and Y. Qu: Solving linear and quadratic fuzzy equations. Fuzzy Sets and Systems, 38, pp.43–59(1990).

[13] J.B. Jo, Y. Tsujimura, M. Gen and G. Yamazaki: Performance Evaluation of a Fuzzy Queueing Network, Proc. of Fifth IFSA World Cong., pp.580–583(1993).

[14] J.B. Jo, Y. Tsujimura, M. Gen and G. Yamazaki: A Delay Model of Queueing Network System Based on Fuzzy Sets Theory, Inter. J. of Comp. and Ind. Engg., Vol.25, Nos.1–4, pp.143–146(1993)

# CONSISTENCY CHECKING BASED ON HIGH LEVEL FUZZY PETRI NETS

**HELOISA SCARPELLI**

*UFSCar/CCT/DC - Cx. Postal 676*

*CEP 13565-905 - São Carlos - SP - Brazil*

**FERNANDO GOMIDE**

*Unicamp/FEE/DCA - Cx. Postal 6101*

*CEP 13081-970 - Campinas - SP - Brazil*

### Abstract

The problem of verifying the integrity of fuzzy knowledge bases is discussed. An approach to find potential inconsistencies in fuzzy rule based systems is described. The approach models the knowledge base as a High Level Fuzzy Petri Net and uses the structural properties of the net for verification. Basic notions on approximate reasoning and High Level Fuzzy Petri Nets are also given. The method used for consistency checking is briefly reviewed. Procedures for discovering potential inconsistencies at both local and global levels are described.

## 1 Introduction

The problem of validation of knowledge bases is an important and difficult problem. Ordinary and high level Petri nets have been proposed as knowledge representation formalisms where structural and behavioral properties of the net can be used to prove properties of the system being modeled or to verify the knowledge base integrity [1], [5]. In a fuzzy environment, the issues of concern when talking about integrity checking are the definition of concepts such as inconsistency, redundancy and completeness , and the investigation of suitable similarity measures for comparison of fuzzy propositions [2], [7], [15]. In this paper we suggest a new mechanism to perform inconsistency checking in fuzzy systems, based on the High Level Fuzzy Petri Net (HLFPN) model. The HLFPN model has been proposed by the authors [10], [13] as a powerful formal tool for modeling, designing, and executing fuzzy systems. In the present work the usefulness of this approach in the field of integrity checking is demonstrated. The procedures proposed in this paper are based on the results of applying the *reflection on the input* methodology [15] to several cases of fuzzy inconsistencies analysed in [12]. The organization of the subsequent discussion is as follows. Section 2 introduces some basic notions such as knowledge representation in approximate reasoning and the HLFPN model. The results of applying the method of *reflection on the input* proposed in [15] are summarized in section 3. Section 4 includes descriptions of the algorithms for local and global conflict checking. Conclusions and future work are presented in section 5.

## 2 Preliminary Notions

The most used inference pattern in approximate reasoning [16] contains the premisses

$$P_1 \quad : \quad V \text{ is } A' \tag{1}$$

$$P_2 \quad : \quad \text{IF } V \text{ is } A \text{ THEN } U \text{ is } B \tag{2}$$

where $V$ and $U$ are two variables having base sets $X$ and $Y$ respectively, $A$ and $A'$ are normal fuzzy subsets of $X$ and $B$ and $B'$ are normal fuzzy subsets of $Y$.

$P_2$ translates into a relationship on $X \times Y$ such that $(V, U)$ is $H$ where [15] $H(x, y) = (1 - A(x)) \vee B(y)$ (the symbol $\vee$ denotes *max* operator). We can obtain $U$ *is* $B'$ where $B' = A' \circ H$, that is

$$B'(y) = max_x[A'(x) \wedge ((1 - A(x)) \vee B(y))] \tag{3}$$

Many other methods have been suggested for the representation of $H$ [8] . A generalized fuzzy reasoning method is obtained by using T-operators in addition to the *min* and *max* operators. A general representation of implication function [6]can be denoted by

$$H_{A \to B}(x, y) = f_{\to}(A(x), B(y)).$$

Given the information $A'(x)$, the inferred value for $U$ is obtained as

$$B'(y) = sup_x[A'(x) * H_{A \to B}(x,y)] \quad (* = T - norm) \tag{4}$$

Two net-based structures to represent and to evaluate rules and facts in approximate reasoning, namely the Fuzzy Petri Net (FPN) model [11] [14] and the High Level Fuzzy Petri Net (HLFPN) model [10] [13] have been developed by the authors. The concept of Fuzzy Petri Nets defined in [11] is derived from Petri Nets [9] and from the fuzzy Petri net defined in [3]. The High Level Fuzzy Petri Nets proposed in [10] derive from Predicate/Transition nets (PrT-Net)[4]. A HLFPN consists of:

1. A directed graph, defined by a triple $(P, T, F)$ such that $P$ and $T$ are two disjoint sets of vertices $(P \cap T = \emptyset)$ called places and transitions respectively, and $F$ is a set of directed arcs, each one connecting a place $p \in P$ to a transition $t \in T$ or vice-versa $(F \subset P \times T \cup T \times P)$;

2. A labeling of the arcs with $t$-uples of variables;

3. A set of formulae, used as inscriptions inside some transitions.

The inscriptions inside transitions are typically of three types: expressions used to apply compositional modus ponens, expressions to perform conjunction of multiple antecedents, expressions to aggregate partial results of rules.

Let us define, for each transition $t$, the set of input places $I(t) = \{p \in P \mid \langle p, t \rangle \in F\}$ and output places $O(t) = \{p \in P \mid \langle t, p \rangle \in F\}$. Places of a HLFPN may contain objects called *tokens*. A transition $t \in T$ is enabled whenever each place $p \in I(t)$ contains a token as specified by the label on $\langle p, t \rangle$. An enabled transition $t$ can be fired by removing from each place $p \in I(t)$ a token as specified by the label on the arc $\langle p, t \rangle$ and by adding to each place $p' \in O(t)$ a token specified by the label on the arc $\langle t, p' \rangle$. The token value is evaluated by the inscribed formulae in $t$. The lack of any formula inscribed in a transition means that the token value in $p'$ remains the same as in $p$.

The basic premises in (1) and (2) can be modeled as shown in figure 1(a), where $H_{A \to B}(x_i, y_j) = d_{ij}$. The fact $V$ is $A'$ is represented by putting a token $\langle a'_1, a'_2, \cdots, a'_n \rangle$ in the place $V$ such that $A'(x_i) = a'_i, x_i \in X$.

The conclusion $U$ is $B'$ is reached by firing $t$. The symbol $\gamma$ stands for a T-norm operator. The marking yielded by firing transition $t$ is shown in figure 1(b).

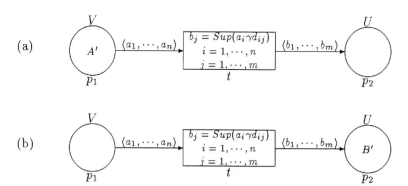

Figure 1: High Level Fuzzy Petri Net for a conditional rule.(a) Before transition firing, where $A' = \langle a'_1, a'_2, \cdots, a'_n \rangle$. (b) After transition firing, where $B' = \langle b'_1, b'_2, \cdots, b'_m \rangle$.

# 3   Inconsistencies in Fuzzy Logic

A methodology called *reflection on the input*, to find potential conflicts in a knowledge base, has been proposed in [15]. The authors analysed [12] in some detail the process of *reflection on the input* when applied to several cases of fuzzy environment not discussed in [15]. In this context, a fuzzy subset $A$ is considered inconsistent if it is subnormal [15]. The *degree of inconsistency* of a fuzzy subset $F$ of a base set $Y$ is defined as[1] $c = Cert(U$ is $\emptyset/U$ is $F)$, where $\emptyset$ denotes the empty set.

To get a degree of inconsistency of $\alpha$ for the variable taking the value given by $F$, it must be the case that $F(y) \leq 1 - \alpha$, $\forall y$.

In each of the cases analysed in [12], some necessary conditions for the existence of potential conflicts have been derived, which provide the basis for the procedures presented in the sequel. The conditions are summarized in Table 1, where $X, Y, Z, X^i$ denote base sets, $V$ is a variable on $X$, $V_i$ is a variable on $X^i$, $U$ is a variable on $Y$, $U_i$ is a variable on $Y^i$ and $W$ is a variable on $Z$. Fuzzy subsets are denoted by $A, B, C, D, G, S$, with or without subscripts.

| KB portion | derived necessary conditions |
|---|---|
| $P_1$ : IF $V$ is $A_1$ THEN $U$ is $B_1$ | $Poss(B_1/B_2) \leq 1 - \alpha < 1$ |
| $P_2$ : IF $V$ is $A_2$ THEN $U$ is $B_2$ | $Poss(A_1/A_2) \leq \alpha > 0$ |
| $P_3$ : $V$ is $D$ | |
| $P_1$ : IF $V_1$ is $A_1$ and $\cdots$ and $V_k$ is $A_k$ THEN $U$ is $B_1$ | $Poss(B_1/B_2) \leq 1 - \alpha$ |
| $P_2$ : IF $V_1$ is $D_1$ and $\cdots$ and $V_l$ is $D_l$ THEN $U$ is $B_2$ | $Poss(A_i/D_i) \geq \alpha, i = 1, \cdots, l.$ |
| IF $V$ is $A$ THEN $U$ s $B_1$ | $Poss(B_1/B_2) \leq 1 - \alpha$ |
| IF $W$ is $D$ THEN $U$ is $B_2$ | |
| $P_1$ : IF $V$ is $A_1$ THEN $U$ is $B_1$ | $Poss(S_2/S_3) \leq 1 - \alpha$ |
| $P_2$ : IF $U$ is $G_1$ THEN $W$ is $S_2$ | $Poss(A_1/A_3) \geq \alpha$ |
| $P_3$ : IF $V$ is $A_3$ THEN $W$ is $S_3$ | $Cert(G_2/B_1) \geq \alpha$ |
| $P_1$ : IF $V_1$ is $A_1$ and $V_2$ is $A_2$ THEN $U_1$ is $B_1$ | $Poss(C_2/C_3) \leq 1 - \alpha$ |
| $P_2$ : IF $U_1$ is $B_2$ and $U_2$ is $B_3$ and $U_3$ is $B_4$ THEN $W$ is $S_2$ | $Poss(A_2/A_3) \geq \alpha$ |
| $P_3$ : IF $V_2$ is $A_3$ and $U_3$ is $B_5$ THEN $W$ is $C_2$ | $Poss(B_4/B_5) \geq \alpha$ |
| | $Cert(B_2/B_1) \geq \alpha$ |

Table 1: Derived necessary conditions for the existence of potential inconsistencies. Lines 1, 2 and 3 refer to local inconsistencies. Lines 4 and 5 deal with global inconsistencies.

# 4   Using HLFPN for Consistency Checking

In this section we must assume that the implication method is the one given by expression $H$ in section 2, the inference uses $max - min$ composition and the aggregation is performed under $min$ operator.

Consider the KB shown in Table 2. The HLFPN model for this KB can be seen in figure 2. The parallel rules 3 and 4 of Table 2 have been denoted in figure 2 by a a compact representation of parallel rules introduced in [14].

| | |
|---|---|
| 1) | IF $V_1$ is $A_1$ and $V_2$ is $C_1$ THEN $V_4$ is $B_1$ |
| 2) | IF $V_4$ is $B_2$ and $V_5$ is $D_1$ and $V_6$ is $E_1$ THEN $V_8$ is $H_1$ |
| 3) | IF $V_2$ is $C_2$ THEN $V_7$ is $G_1$ |
| 4) | IF $V_2$ is $C_3$ THEN $V_7$ is $G_2$ |
| 5) | IF $V_3$ is $F_1$ THEN $V_7$ is $G_3$ |

Table 2: Knowledge Base Example.

We call a *level_1 HLFPN* a net representation obtained from a HLFPN through the definition of special transition and places in a higher level of abstraction such that each place in the net is associated with one and only one variable and each transition corresponds to one and only one rule in the KB (see definition in [12]). The level_1 HLFPN concerning the KB in Table 2 is illustrated in figure 3.

The concept of condition test matrix is essential to the procedure developed here.

---

[1]$Poss(E/F) = max_z[E(z) \wedge F(z)]$ and $Cert(E/F) = 1 - Poss(\overline{E}/F)$.

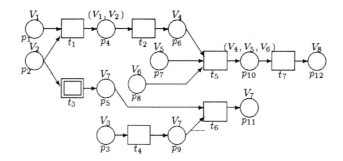

Figure 2: HLFPN model for the Knowledge Base Example.

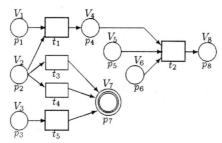

Figure 3: Level_1 HLFPN for the Knowledge Base Example.

**Definition 4.1** *Consider a level_1 HLFPN $\mathcal{N}1$ of a KB. Suppose each transition in the model is associated with a rule $R$ such as $R$ : IF $V_1$ is $A_1$ and $\cdots$ and $V_k$ is $A_k$ THEN $U$ is $B$.*

*The* condition test matrix (CT-matrix) $C$ for $\mathcal{N}1$ *is defined as a* $n \times m$ *matrix such that each row corresponds to a place $p$ in $\mathcal{N}1$ and each column corresponds to a transition $t$ in $\mathcal{N}1$. The entry $c\langle p_i, t \rangle$ is defined as*

$$c\langle p_i, t \rangle = \begin{cases} -A_i & \text{if } p_i \in I(t) \text{ and is associated with } V_i \text{ in } R \\ B & \text{if } p_i \in O(t) \text{ and is associated with } U \text{ in } R \\ 0 & \text{otherwise} \end{cases}$$

For example, the CT-matrix for the level_1 HLFPN of figure 3 is shown in Table 3.

|         |       | $(1)t_1$ | $(2)t_2$ | $(3)t_3$ | $(4)t_4$ | $(5)t_5$ |
|---------|-------|----------|----------|----------|----------|----------|
| $(V_1)$ | $p_1$ | $-A_1$   | 0        | 0        | 0        | 0        |
| $(V_2)$ | $p_2$ | $-C_1$   | 0        | $-C_2$   | $-C_3$   | 0        |
| $(V_3)$ | $p_3$ | 0        | 0        | 0        | 0        | $-F_1$   |
| $(V_4)$ | $p_4$ | $B_1$    | $-B_2$   | 0        | 0        | 0        |
| $(V_5)$ | $p_5$ | 0        | $-D_1$   | 0        | 0        | 0        |
| $(V_6)$ | $p_6$ | 0        | $-E_1$   | 0        | 0        | 0        |
| $(V_7)$ | $p_7$ | 0        | 0        | $G_1$    | $G_2$    | $G_3$    |
| $(V_8)$ | $p_8$ | 0        | $H_1$    | 0        | 0        | 0        |

Table 3: CT-matrix Example.

**Definition 4.2** *Let $A = (a_{ij}), i = 1, \cdots, n, j = 1, \cdots, m$ and $B = (b_{jk}), j = 1, \cdots, m, k = 1, \cdots, l$ be matrices with elements, which are fuzzy subsets over base sets. Then we define the* matrix-composition *such that $A * B = (c_{ik}), i = 1, \cdots, n, k = 1, \cdots, l$, where $c_{ik} = \sum_{j=1}^{m} S(a_{ij} b_{jk})$ and $S(a, b) = Poss(A/B)$ if $A$ and $B$ are no null sets with the same sign and 0 otherwise. The symbol $\sum$ indicates a formal sum of values.*

Assume now we have a consistent knowledge base $K$ modeled in a HLFPN $\mathcal{N}$ and a new rule $R$ is to be added to $K$. We desire to check whether this knowledge update may cause some inconsistency or not.

Suppose $p$ is the place in $\mathcal{N}1$ associated with $U$, the variable in $R$'s consequent. The desired checking is performed through the algorithms 4.1 and 4.2.

**Algorithm 4.1**          Consistency Checking at the Local Level

| | |
|---|---|
| Step 1. | Construct a CT-matrix $C'$ extracting from $C$ all columns $t$ such that $c\langle p,t \rangle \neq 0$ (corresponding to rules with $U$ in the consequent). |
| Step 2. | Construct a test vector $V$ for $R$. |
| Step 3. | Apply matrix composition between $C'$ and $V$, giving $F$ ($F = V * C'$). |
| Step 4. | Analyse the values in $F$ comparingagainst the stipulated consistency grade $\alpha$. |

**Algoritm 4.2**          Consistency Checking at the Global Level

| | |
|---|---|
| Step 1. | Find the CT-matrix $C'$ from $C$, corresponding to the minimal cover for the new rule $R$. |
| Step 2. | Construct a test vector $V$ to rule $R$. |
| Step 3. | Apply matrix composition between $V$ and $C'$, yielding $F$ ($F = V * C'$). |
| Step 4. | Analyse the values in $F$, comparing against the stipulated consistency grade $\alpha$. |
| Step 5. | For those rows in $C'$ with two values with opposite sign (the variable is a link variable), check the compatibility between sets in the consequent and antecedents of chained rules. |

To illustrate the Algorithm 4.1, suppose the rule 6) IF $V_2$ is $C_4$ THEN $V_7$ is $G_4$ is to be added to the KB in Table 2. Variable $V_7$ is associated with place $p_7$ in $\mathcal{N}1$ (figure 3). Matrix $C'$ includes the last three columns of $C$ (Table 3), since rules 3, 4 and 5 in Table 2 have $V_7$ in the consequent.

The test vector for a rule has as many elements as there are variables in the KB. Its entries are defined as in the CT-matrix. For rule 6, the matrix composition yields

$$F = \begin{bmatrix} 0 + Poss(C_4/C_2) + 0 + 0 + 0 + 0 + Poss(G_4/G_1) + 0 \\ 0 + Poss(C_4/C_3) + 0 + 0 + 0 + 0 + Poss(G_4/G_2) + 0 \\ 0 + 0 + 0 + 0 + 0 + 0 + Poss(G_4/G_3) + 0 \end{bmatrix}$$

Each row in the column vector is a formal sum of possibility values where, the last non zero element corresponds to sets in the consequent of rules and all other ones correspond to sets in the antecedent of rules.

To test for global integrity, it becomes necessary to first identify the relevant portion of the global network affected by the rule, that is, the minimal cover for the new rule.(The basic idea is due to [1]).

As an example of applying the Algorithm 4.2, assume that the KB in Table 2 will be updated by inserting the rule IF $V_2$ is $C_5$ and $V_7$ is $G_4$ THEN $V_8$ is $H_2$. The minimal cover for this rule is the subnet that includes the two rules linked by variable $V_4$ : rules 1 and 2. The corresponding CT-matrix $C'$ for the subnet contains only columns $t_2$ and $t_3$ from Table 3.

The matrix composition yields $F$ such that

$$F = \begin{bmatrix} 0 + Poss(C_5/C_1) + 0 + 0 + 0 + 0 + 0 + 0 \\ 0 + 0 + 0 + 0 + 0 + 0 + 0 + Poss(H_2/H_1) \end{bmatrix}$$

In the next step, we find, by inspecting the CT-matrix, that $V_4$ is the link node and verify the condition $Cert(B_1/B_2) \geq \alpha$.

## 5    Conclusions

We have demonstrated the usefulness of HLFPN model in the field of integrity checking, and have proposed algorithms for inconsistency checking both at the local and global levels. Results of the method of reflection on the input in the scope of fuzzy logic were used. The condition test matrix defined here allows one to take advantage of the structural properties of the net to discover the existence of potential conflicts, based on the above method. The verification process proposed here is based on a particular inference method and on the use of *min* operator in aggregation and composition of fuzzy propositions. Our future research will examine methods for consistency checking where different inference methods and different T-operators are used. Other types of anomalies not discussed in this paper such as redundancy and incompleteness of fuzzy rules will also be addressed.

**Acknowledgments:** The first author acknowledges the support of CNPq,for RHAE grant # 460322/91.5 and also the support of FAPESP for grant # 92/2151-8. The second author is grateful to CNPq for grant # 300729/86-3.

# References

[1] AGARWAL, R. and TANNIRU, M. - *A Petri-Net Based Approach for verifying the integrity of production systems*, Int. J. Man-Machine Studies, 36, 1992, pp. 447-468.

[2] CHANG, A. and HALL, L. - *The validation of fuzzy knowledge-based systems*, in Fuzzy Logic and the Management of Uncertainty, L. Zadeh and J. Kacprzyk, eds., John Wiley & Sons, 1992, pp.589-604.

[3] CHEN, S.; KE, J. S. and CHANG, J. - *Knowledge Representation using Fuzzy Petri Nets*, IEEE Trans. Knowledge nad Data Engineering, 2, 3, 1990, pp. 311-319.

[4] GENRICH, H. J. - *Predicate/Transition Nets*, in W. Brauer, W. Reisig and G. Rozenberg(eds.), Petri Nets: Central Models and Their Properties, Lecture Notes in Computer Science 254, Springer-Verlag, 1986, pp.207-247.

[5] GIORDANA, A. and SAITTA, L. - *Modeling Production Rules by Means of Predicate Transition Networks*, Information Sciences, 35,1,1985, pp.1-41.

[6] GUPTA, M. M. and QI, J. - *Theory of T-norms and fuzzy inference methods*, Fuzzy Sets and Systems, 40, 1991, pp. 431-450.

[7] LEUNG, K. S. and SO, Y. T. - *Inconsistency in Fuzzy Rule-based Expert Systems*, Proceedings of the International Conference on Fuzzy Logic & Neural Networks, Japan, 1990, pp.849-852.

[8] MIZUMOTO, M. and ZIMMERMANN, H. - *Comparison of Fuzzy Reasoning Methods*, Fuzzy Stes and Systems, 8, 1982, pp.253-283.

[9] MURATA, T. - *Petri Nets: Properties, Analysis and Applications*, Proceedings IEEE,77, 4, 1989, pp.541-580.

[10] SCARPELLI, H. ; GOMIDE, F. and PEDRYCZ, W. - *Modeling Fuzzy Reasoning using High Level Fuzzy Petri Nets*, Technical Report RT-DCA 023/92, DCA/FEE/UNICAMP, 1992. (submitted for publication)

[11] SCARPELLI, H. and GOMIDE, F. - *Fuzzy Reasoning and Fuzzy Petri Nets*, Fifth IFSA World Congress, Seoul, Korea, July, 5-9, 1993.

[12] SCARPELLI, H.; GOMIDE, F. - *A High Level Net Approach for Discovering Potential Inconsistencies in Fuzzy Knowledge Bases*, Technical Report RT-DCA 006/92, DCA/FEE/UNICAMP, 1993. (submitted for publication)

[13] SCARPELLI, H. and GOMIDE, F. - *Fuzzy Reasoning and High Level Fuzzy Petri Nets*, First European Congress on Fuzzy and Intelligent Technologies, Aachen, Germany, September, 7-10, 1993.

[14] SCARPELLI, H. and GOMIDE, F. - *Fuzzy Reasoning and Fuzzy Petri Nets in Manufacturing Systems Modeling*, Journal of Intelligent and Fuzzy Systems, vol. 1, n. 3, 1993.

[15] YAGER, R. and LARSEN, H. - *On Discovering Potential Inconsistencies in Validating Uncertain Knowledge Bases by Reflecting on the Input*, IEEE Trans. on Systems, Man and Cybernetics, 21, 4, july/august 1991, pp. 790-801.

[16] ZADEH, L. A. - *A Theory of Approximate Reasoning*, Machine Inteligence 9, Hayes, Michie & Kulich, (eds.), 9, 1979, pp.149-194.

# Acquisition of Fuzzy Classification Knowledge Using Genetic Algorithms

Hisao Ishibuchi, Ken Nozaki, Naohisa Yamamoto and Hideo Tanaka
Department of Industrial Engineering, University of Osaka Prefecture
Gakuen-cho 1-1, Sakai, Osaka 593, JAPAN

*Abstract* — This paper proposes a genetic-algorithm-based approach to the construction of fuzzy classification systems with rectangular fuzzy rules. In the proposed approach, compact fuzzy classification systems are automatically constructed from numerical data by selecting a small number of significant fuzzy rules using genetic algorithms. Since significant fuzzy rules are selected and unnecessary fuzzy rules are removed, the proposed approach can be viewed as a knowledge acquisition tool for classification problems. In this paper, first we describe a generation method of rectangular fuzzy rules from numerical data for classification problems. Next we formulate a rule selection problem for constructing a compact fuzzy classification system as a combinatorial optimization problem. Then we show how genetic algorithms are applied to the rule selection problem. Last we illustrate the proposed approach by computer simulations on a numerical example and the iris data of Fisher.

## I. INTRODUCTION

Fuzzy-rule-based control systems have been applied to various problems (for example, see Sugeno[1] and Lee[2]). Fuzzy rules in those control systems were usually derived from human experts. Recently several approaches have been proposed for automatically generating fuzzy rules from numerical data without domain experts[3,4]. Tuning techniques for the membership functions of antecedent and consequent fuzzy sets have been also proposed in many studies. Genetic algorithms[5,6] have been employed for the learning of fuzzy rules in some studies. For example, the membership functions of antecedent and consequent fuzzy sets of fuzzy rules were adjusted in Karr & Gentry[7], fuzzy partitions of input spaces were determined in Nomura *et al.*[8], and an appropriate fuzzy set in the consequent part of each fuzzy rule was selected in Thrift[9]. Appropriate fuzzy sets in the antecedent and consequent parts of each fuzzy rule were also selected by the fuzzy classifier system in Valenzuela-Rendon[10].

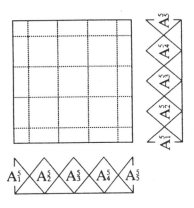

Fig.1 A fuzzy partition by a simple fuzzy grid

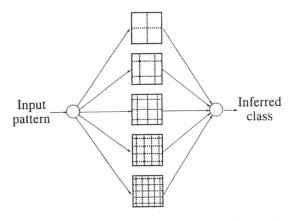

Fig.2 A classification system based on 90 fuzzy rules

In the above-mentioned studies, fuzzy-rule-based systems were mainly applied to control problems. A few methods for generating fuzzy rules have been proposed for classification problems[11-14]. Ishibuchi *et al.*[11] proposed a rule generation method from numerical data based on fuzzy partitions by simple fuzzy grids. Fig.1 shows an example of the fuzzy partition of a two-dimensional pattern space by a simple fuzzy grid. The fuzzy classification method in [11] simultaneously employed all the fuzzy rules generated for several fuzzy partitions of different sizes. In Fig.2, we show a fuzzy classification system based on 90 ($=2^2+3^2+4^2+5^2+6^2$) fuzzy rules generated for five fuzzy partitions. The main drawback of this approach is that the number of fuzzy rules is enormous especially for classification problems in high-dimensional pattern spaces. In order to remove unnecessary fuzzy rules from fuzzy classification systems, Ishibuchi *et al.*[13,14] proposed a genetic-algorithm-based approach that can reduce the number of fuzzy rules in fuzzy classification systems.

Since the above-mentioned fuzzy classification methods were based on fuzzy partitions by simple fuzzy grids as shown in Fig.1 and Fig.2, all the fuzzy rules in those methods had square fuzzy subspaces. Fuzzy rules with rectangular fuzzy subspaces, however, may be more appropriate than square fuzzy rules in many classification problems. As an example, let us consider a two-class classification problem in Fig.3 where closed circles and open circles denote the given patterns in Class 1 and Class 2, respectively. If we try to classify all the given patterns by fuzzy rules with square fuzzy subspaces, fine fuzzy partitions and a large number of fuzzy rules are required. On the contrary, if we use fuzzy rules with rectangular fuzzy subspaces, much less fuzzy rules may be required. This discussion motivates us to propose a genetic-algorithm-based approach to the construction of compact fuzzy classification systems with rectangular fuzzy rules. In the proposed approach, a more appropriate fitness function than in our former work[13,14] is introduced from a point of view of knowledge acquisition.

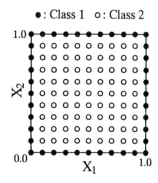

● : Class 1   o : Class 2

**Fig.3** A two-class classification problem

## II. FUZZY CLASSIFICATION SYSTEMS

In this section, we describe a generation method of rectangular fuzzy rules from numerical data. Basically, this method is the same as the generation method of square fuzzy rules in Ishibuchi *et al.*[11] except for fuzzy partitions.

### A. Classification Problems

Let us consider a classification problem in the $n$-di-mensional pattern space $[0,1]^n$. It is assumed that $m$ patterns $x_p = (x_{p1}, x_{p2}, ..., x_{pn})$, $p = 1,2,...,m$, are given as training data from $M$ classes ($C1$: Class 1, $C2$: Class 2, ..., $CM$: Class $M$). Fig.3 shows an example of the classification problem with $n = 2$, $M = 2$ and $m = 121$.

### B. Fuzzy Partition

Let us divide the $i$th axis of the pattern space (*i.e.*, the domain interval of the $i$th attribute value $x_{pi}$) into $J_i$ fuzzy subsets $\{A_1^{J_i}, A_2^{J_i}, ..., A_{J_i}^{J_i}\}$. We can use any type of membership functions (*e.g.*, triangular, trapezoid and exponential) for these fuzzy subsets. In this paper, symmetric triangular membership functions are employed for $A_{j_i}^{J_i}$, $j_i = 1,2,...,J_i$ (see Fig.1 where $J_1 = J_2 = 5$). For the case of $J_i = 1$, let us define the membership function of $A_1^1$ as

$$\mu_1^1(x) = \begin{cases} 1, & \text{if } 0 \le x \le 1, \\ 0, & \text{otherwise.} \end{cases} \tag{1}$$

That is, the fuzzy subset $A_1^1$ is the unit interval $[0,1]$.

The pattern space is partitioned by Cartesian product of the fuzzy partitions for the $n$ axes. In our former work[11-14], we employed the same fuzzy partition for each axis. This led to the fuzzy partition of the pattern space into square fuzzy subspaces as shown in Fig.1 and Fig.2. In this paper, we relax this restriction on the fuzzy partition to generate rectangular fuzzy subspaces.

Since the axis of the $i$th attribute value $x_{pi}$ is divided into $J_i$ fuzzy subsets $\{A_1^{J_i}, A_2^{J_i}, ..., A_{J_i}^{J_i}\}$ for each $i$, the pattern space is partitioned into the following $J_1 \cdot J_2 \cdot ... \cdot J_n$ fuzzy subspaces:

$$\{A_{j_1}^{J_1} \times ... \times A_{j_n}^{J_n} \mid j_1 = 1,2,...,J_1, ..., j_n = 1,2,...,J_n\}. \tag{2}$$

Various fuzzy partitions of the pattern space can be constructed by combining different fuzzy partitions for the $n$ axes of the pattern space. Let us assume that the possible values of $J_i$ are $J_i = 1,2,...,J_{i_{max}}$ for each $i$. Then the total number of possible fuzzy partitions of the pattern space constructed by the combination of the fuzzy partitions for the $n$ axes is $J_{1_{max}} \cdot ... \cdot J_{n_{max}}$. If we use all these fuzzy partitions, the following fuzzy subspaces are generated:

$$\{A_{j_1}^{J_1} \times ... \times A_{j_n}^{J_n} \mid j_1 = 1,2,...,J_1; J_1 = 1,2,...,J_{1_{max}}, ...,$$
$$j_n = 1,2,...,J_n; J_n = 1,2,...,J_{n_{max}}\}. \tag{3}$$

For example, we show all the 36 fuzzy subspaces generated for the case of $n = 2$, $J_{1\max} = 3$ and $J_{2\max} = 3$ in Fig.4.

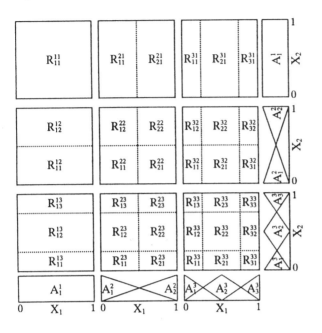

**Fig.4** Generated fuzzy subspaces

### C. Rule Generation

Fuzzy if-then rules corresponding to the generated fuzzy subspaces in (3) are written as follows:

Rule $R_{j_1 \cdots j_n}^{J_1 \cdots J_n}$: If $x_{p1}$ is $A_{j_1}^{J_1}$ and ... and $x_{pn}$ is $A_{j_n}^{J_n}$

$\quad$ then $x_p$ is Class $C_{j_1 \cdots j_n}^{J_1 \cdots J_n}$ with $CF = CF_{j_1 \cdots j_n}^{J_1 \cdots J_n}$,

$$j_1 = 1,2,\dots,J_1; \ J_1 = 1,2,\dots,J_{1\max}, \ \dots,$$
$$j_n = 1,2,\dots,J_n; \ J_n = 1,2,\dots,J_{n\max}, \quad (4)$$

where $R_{j_1 \cdots j_n}^{J_1 \cdots J_n}$ is the label of the fuzzy rule, the consequent $C_{j_1 \cdots j_n}^{J_1 \cdots J_n}$ is one of the $M$ classes and $CF_{j_1 \cdots j_n}^{J_1 \cdots J_n}$ is the certainty of the rule. The consequent and the certainty of each fuzzy rule can be determined from the given numerical data in the same manner as in our former work[11-14]. It should be noted that the fuzzy rule can be generated if at least one pattern is given in the corresponding fuzzy subspace. If there is no given pattern in the fuzzy subspace, the fuzzy rule can not be generated. In this case, we generate a dummy rule by specifying the consequent and the certainty as $C_{j_1 \cdots j_n}^{J_1 \cdots J_n} = \phi$ and $CF_{j_1 \cdots j_n}^{J_1 \cdots J_n} = 0$, respectively. Dummy rules have no effect on the fuzzy reasoning for classifying new patterns[13-14].

Let us denote the set of all the generated fuzzy rules including dummy rules by $S_{ALL}$:

$$S_{ALL} = \{R_{j_1 \cdots j_n}^{J_1 \cdots J_n} \mid j_1 = 1,2,\dots,J_1; \ J_1 = 1,2,\dots,J_{1\max}, \ \dots,$$
$$j_n = 1,2,\dots,J_n; \ J_n = 1,2,\dots,J_{n\max}\}. \quad (5)$$

### D. Classification of New Patterns

Let us denote a subset of the rule set $S_{ALL}$ by $S$. A new pattern $x_p = (x_{p1},\dots,x_{pn})$ is classified by the following procedure using the fuzzy rules in $S$.

**Step 1:** Calculate $\alpha_{CT}$ for $T = 1,2,\dots,M$ as

$$\alpha_{CT} = \max\{CF_{j_1 \cdots j_n}^{J_1 \cdots J_n} \cdot \prod_{i=1}^{n} \mu_{J_i}^{J_i}(x_{pi}) \mid C_{j_1 \cdots j_n}^{J_1 \cdots J_n} = \text{Class } T$$
$$\text{and } R_{j_1 \cdots j_n}^{J_1 \cdots J_n} \in S\}. \quad (6)$$

**Step 2:** Find Class $X$ ($CX$) such that

$$\alpha_{CX} = \max\{\alpha_{C1}, \alpha_{C2}, \dots, \alpha_{CM}\}. \quad (7)$$

If two or more classes take the maximum value in (7) then the classification of $x_p$ is rejected (i.e., $x_p$ is left as an unclassifiable pattern), else assign $x_p$ to Class $X$ ($CX$) determined by (7).

## III. RULE SELECTION BY GENETIC ALGORITHMS

### 3.1 Formulation of a Rule Selection Problem

Our rule selection problem is to find a compact rule set $S$ that has high classification power. Therefore our problem has the following two objectives:

(i) The first objective is to maximize the number of correctly classified patterns by the fuzzy rules in $S$.

(ii) The second objective is to minimize the number of the fuzzy rules in $S$.

By combining these two objectives, we formulate the following problem.

$$\text{Maximize } W_{NCP} \cdot NCP(S) - W_S \cdot |S|, \quad (8)$$
$$\text{subject to } S \subseteq S_{ALL}, \quad (9)$$

where $W_{NCP}$ and $W_S$ are positive weights such that $W_S \ll W_{NCP}$, $NCP(S)$ is the number of correctly classified patterns by $S$, and $|S|$ is the number of the fuzzy rules in $S$. While the same problem was formulated in our former work[13,14], the definition of $S_{ALL}$ is different: $S_{ALL}$ in our former work consisted of only square fuzzy rules, but $S_{ALL}$ in this paper includes rectangular fuzzy rules.

In general, a fuzzy rule in a coarse fuzzy partition is applicable to more patterns than that in a fine fuzzy partition. Therefore the former is more desirable for constructing a compact fuzzy classification system than the latter. A fuzzy rule in a coarse fuzzy partition is also desirable from a point of view of knowledge acquisition because it is a general rule that can be valid in a large subspace of the pattern space. Therefore we modify the rule selection problem (8)-(9) by assigning a different weight to each fuzzy rule.

Let us define an index of the fineness of each fuzzy rule as

$$Fineness(R_{j1\cdots jn}^{J_1\cdots J_n}) = J_1 + \ldots + J_n. \qquad (10)$$

This index can be viewed as the fineness of the fuzzy partition where the fuzzy rule is generated. That is, the finer a fuzzy partition is, the larger the fineness of fuzzy rules in that fuzzy partition is. In order to select fuzzy rules in coarse fuzzy partitions (*i.e.*, those with small fineness values), we modify the objective function of the rule selection problem as follows.

$$\text{Maximize } W_{NCP} \cdot NCP(S) - \sum_{R_{j1\cdots jn}^{J_1\cdots J_n} \in S} Fineness(R_{j1\cdots jn}^{J_1\cdots J_n}).$$
$$(11)$$

### B. Genetic Operations for the Rule Selection Problem

In genetic algorithms, a rule set $S$ is treated as an individual. The value of the objective function in (11) is the fitness value of each individual. That is, the fitness function $f(S)$ is defined as

$$f(S) = W_{NCP} \cdot NCP(S) - \sum_{R_{j1\cdots jn}^{J_1\cdots J_n} \in S} Fineness(R_{j1\cdots jn}^{J_1\cdots J_n}).$$
$$(12)$$

In genetic algorithms, each individual should be represented as a string. In this paper, let us represent a rule set $S$ as $S = s_1 s_2 \ldots s_N$ where $N$ is the number of the fuzzy rules in $S_{ALL}$, and $s_r = 1, -1$ or $0$ denotes the following:

(1) $s_r = 1$ means that the $r$th rule is included in the rule set $S$,

(2) $s_r = -1$ means that the $r$th rule is not included in the rule set $S$,

(3) $s_r = 0$ means that the $r$th rule is a dummy rule.

The index $r$ of each rule can be arbitrary specified.

Since dummy rules have no effect on the fuzzy reasoning for classifying new patterns, they should be excluded from a rule set $S$. Therefore they are represented

as $s_r = 0$ in the coding process in order to prevent $S$ from including them. A string $S = s_1 s_2 \ldots s_N$ is decoded as

$$S = \{R_{j1\cdots jn}^{J_n\cdots J_n} \mid s_r = 1; \; r = 1, 2, \ldots, N\}. \qquad (13)$$

The following genetic operations are employed to generate and handle a set of strings (*i.e.*, a population) in genetic algorithms of this paper (Step 1 through Step 5 are iterated until a prespecified stopping condition is satisfied).

**Step 0 (Initialization):** Generate an initial population containing $N_{pop}$ strings where $N_{pop}$ is the number of strings in each population. In this operation, each string is generated by assigning 0 to dummy rules and randomly assigning 1 or -1 to each of the other rules with the probability of 0.5.

**Step 1 (Selection):** Select $N_{pop}/2$ pairs of strings from the current population. The selection probability $P(S)$ of a string $S$ in a population $\Psi$ is specified as

$$P(S) = \frac{f(S) - f_{\min}(\Psi)}{\sum\limits_{S' \in \Psi} \{f(S') - f_{\min}(\Psi)\}}, \qquad (14)$$

where

$$f_{\min}(\Psi) = \min\{f(S) \mid S \in \Psi\}. \qquad (15)$$

**Step 2 (Crossover):** For each selected pair, randomly choose bit positions. Each bit position is chosen with the probability of 0.5. Interchange the bit values at the chosen positions in the selected pair.

**Step 3 (Mutation):** For each bit value of the generated strings by the crossover operation, apply the following mutation operation:

$$s_r = 1 \; \rightarrow \; s_r = -1 \text{ with the probability } P_m(1 \rightarrow -1),$$
$$s_r = -1 \; \rightarrow \; s_r = 1 \text{ with the probability } P_m(-1 \rightarrow 1).$$

**Step 4 (Elitist strategy):** Randomly remove one string from the $N_{pop}$ strings generated by the above operations, and add the string with the maximum fitness value in the previous population to the current one.

**Step 5 (Termination test):** If a prespecified stopping condition is not satisfied, return to Step 1.

The crossover operation in Step 2 was called the uniform crossover in Syswerda[16]. In Step 3, different mutation probabilities $P_m(1 \rightarrow -1)$ and $P_m(-1 \rightarrow 1)$ are assigned to the mutations from 1 to -1 and from -1 to 1, respectively. A larger probability is usually assigned to

$P_m(1 \rightarrow -1)$ than to $P_m(-1 \rightarrow 1)$ in order to reduce the number of fuzzy rules in each individual. The effect of these biased mutation probabilities was investigated in [14]. The total number of generations is used as a stopping condition in this paper.

## IV. COMPUTER SIMULATIONS

The genetic algorithm described in the last section was applied to the classification problem in Fig.3. The rule set $S_{ALL}$ was generated by specifying $J_{1_{max}} = 6$ and $J_{2_{max}} = 6$. The total number of the fuzzy rules in $S_{ALL}$ is $(1 + 2 + \dots + 6) \cdot (1 + 2 + \dots + 6) = 441$ (see (5)). Therefore the length of each string in the genetic algorithm was also 441. The positive constant $W_{NCP}$ in the fitness function (12) and the population size $N_{pop}$ were specified as $W_{NCP} = 1000$ and $N_{pop} = 50$, respectively. The mutation probabilities were specified as $P_m(1 \rightarrow -1) = 0.01$ and $P_m(-1 \rightarrow 1) = 0.001$. The algorithm was terminated after 2000 populations were generated.

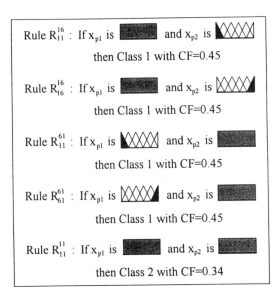

**Fig.5** Selected five fuzzy rules for the classification problem in Fig.3

The selected fuzzy rules by the genetic algorithm with these parameter specifications are shown in Fig.5 where the meshed rectangles represent the antecedent fuzzy sets $A_1^1$ (*i.e.*, the unit interval). Each fuzzy rule in Fig.5 can be interpreted as the following linguistic rules:

(1) Rule $R_{11}^{16}$: If $x_{p2}$ is *very small* then Class 1,

(2) Rule $R_{16}^{16}$: If $x_{p2}$ is *very large* then Class 1,

(3) Rule $R_{11}^{61}$: If $x_{p1}$ is *very small* then Class 1,

(4) Rule $R_{61}^{61}$: If $x_{p1}$ is *very large* then Class 1,

(5) Rule $R_{11}^{11}$: All patterns are Class 2.

If we try to intuitively derive classification rules from the given patterns in Fig.3, we will have similar linguistic rules. Therefore we can conclude that the selected fuzzy rules coincide with our intuitive pattern recognition.

In order to examine the ability of the proposed approach for high-dimensional classification problems, we applied it to the iris data of Fisher (three-class classification problem in a four-dimensional pattern space[15]). We used the same parameter specifications as in the above computer simulation except for the mutation probabilities and the rule set $S_{ALL}$. The mutation probabilities were specified as $P_m(1 \rightarrow -1) = 0.1$ and $P_m(-1 \rightarrow 1) = 0.0001$. The rule set $S_{ALL}$ was specified by $J_{i_{max}} = 4$ for $i = 1, 2, 3, 4$.

The genetic algorithm with these parameter specifications selected only five fuzzy rules in Fig.6 that can correctly classify 149 patterns (99.33% of the given 150 patterns). From the point of view of the number of fuzzy rules, this result outperforms our previous work[14] where ten square fuzzy rules were selected and the same classification rate was obtained.

| $x_1$ $x_2$ $x_3$ $x_4$ | Consequent | CF | Patterns* |
|---|---|---|---|
| | Class 1 | 1.00 | 50 |
| | Class 2 | 0.41 | 8 |
| | Class 2 | 0.41 | 18 |
| | Class 2 | 0.92 | 23 |
| | Class 3 | 0.71 | 50 |

Patterns* : the number of patterns correctly classified by each fuzzy if-then rule

**Fig.6** Selected five fuzzy rules for the iris data

In order to investigate the relation among the three classes of the iris data, the genetic algorithm with the same parameter specifications was also applied to the following three classification problems with two classes (Class I and Class II):

**Problem 1**: Class I: 50 patterns in Class 1
Class II: 100 patterns in Class 2 & 3

**Problem 2**: Class I: 50 patterns in Class 2
Class II: 100 patterns in Class 1 & 3

**Problem 3**: Class I: 50 patterns in Class 3
Class II: 100 patterns in Class 1 & 2

Simulation results for these problems are summarized in Table 1. We can see from the first row of this table that the 50 patterns in Class 1 (*i.e.*, Iris setosa) are entirely separated from the other 100 patterns by the selected three fuzzy rules. This shows that the patterns in Class 1 are easily distinguished from the patterns in the other classes. The selected three fuzzy rules for Problem 1 are shown in Fig.7

**Table 1** Simulation results for the two-class problems

|           | The number of the selected rules | Classification rate |
|-----------|----------------------------------|---------------------|
| Problem 1 | 3                                | 100%                |
| Problem 2 | 5                                | 99.3%               |
| Problem 3 | 5                                | 99.3%               |

| $x_1$ $x_2$ $x_3$ $x_4$ | Consequent | CF   | Patterns* |
|-------------------------|------------|------|-----------|
|                         | Class 1    | 1.00 | 50        |
|                         | Class 2&3  | 1.00 | 99        |
|                         | Class 2&3  | 1.00 | 1         |

Patterns* : the number of patterns correctly classified by each fuzzy if–then rule

**Fig.7** Selected three fuzzy rules for separating Class 1 from the other classes

## V. CONCLUDING REMARKS

In this paper, we proposed a genetic-algorithm-based approach to the construction of compact fuzzy classification systems with rectangular fuzzy rules. In the proposed approach, first a large number of rectangular fuzzy rules were generated from numerical data. Then significant rules were selected from the generated fuzzy rules by genetic algorithms to form a compact fuzzy classification system. By computer simulations on a numerical example, we demonstrated that the proposed approach can select a small number of fuzzy rules that coincide with our intuitive pattern recognition. The ability of the proposed approach was also demonstrated by the application to the iris data. That is, we showed that 149 patterns in the iris data can be correctly classified by only five rectangular fuzzy rules.

## REFERENCES

[1] M.Sugeno, An introductory survey of fuzzy control, *Information Sciences* **36** (1985) 59-83.

[2] C.C.Lee, Fuzzy logic in control systems: fuzzy logic controller, *IEEE Trans. on SMC* **20** (1990) 404-435.

[3] L.X.Wang and J.M.Mendel, Generating fuzzy rules by learning from examples, *IEEE Trans. on SMC* **22** (1992) 1414-1427.

[4] M.Sugeno and T.Yasukawa, A fuzzy-logic-based approach to qualitative modeling, *IEEE Trans. on Fuzzy Systems* **1** (1993) 7-31.

[5] J.H.Holland, *Adaptation in Natural and Artificial Systems*, University of Michigan Press, Ann Arbor, Michigan (1975).

[6] D.E.Goldberg, *Genetic Algorithms in Search, Optimization, and Machine Learning*, Addison-Wesley, Reading, Massachusetts (1989).

[7] C.L.Karr and E.J.Gentry, Fuzzy control of pH using genetic algorithms, *IEEE Trans. on Fuzzy Systems* **1** (1993) 46-53.

[8] H.Nomura, I.Hayashi and N.Wakami, A self-tuning method of fuzzy reasoning by genetic algorithms, *Proc. of 1992 Intern. Fuzzy Systems and Intelligent Control Conference* (Louisville, Kentucky, March 16-18, 1992) 236-245.

[9] P.Thrift, Fuzzy logic synthesis with genetic algorithms, *Proc. of 4th Intern. Conference on GA* (San Diego, California, July 13-16, 1991) 509-513.

[10] M.Valenzuela-Rendon, The fuzzy classifier system: A classifier system for continuously varying variables, *Proc. of 4th Intern. Conference on GA* (San Diego, California, July 13-16, 1991) 346-353.

[11] H.Ishibuchi, K.Nozaki and H.Tanaka, Distributed representation of fuzzy rules and its application to pattern classification, *Fuzzy Sets and Systems* **52** (1992) 21-32.

[12] H.Ishibuchi, K.Nozaki and H.Tanaka, Efficient fuzzy partition of pattern space for classification problems, *Fuzzy Sets and Systems* **60** (1993) in press.

[13] H.Ishibuchi, K.Nozaki and N.Yamamoto, Selecting fuzzy rules by genetic algorithm for classification problems, *Proc. of 2nd IEEE Intern. Conference on Fuzzy Systems* (San Francisco, California, March 28-April 1, 1993) 1119-1124.

[14] H.Ishibuchi, K.Nozaki, N.Yamamoto and H.Tanaka, Genetic operations for rule selection in fuzzy classification systems, *Proc. of 5th IFSA World Congress* (Seoul, Korea, July 4-9, 1993) 15-18.

[15] R.A.Fisher, The use of multiple measurements in taxonomic problems, *Annals of Eugenics* **7** (1936) 179-188.

[16] G.Syswerda, Uniform crossover in genetic algorithms, *Proc. of 3rd Intern. Conference on GA* (George Mason University, June 4-7, 1989), Morgan Kaufmann, San Mateo, California, 2-9.

# Genetic Learning Algorithms for Fuzzy Neural Nets

P. V. Krishnamraju*        J. J. Buckley[†]        K. D. Reilly*        Y. Hayashi[‡]

* Dept. of Computer and Information Sciences, University of Alabama at Birmingham, Birmingham, AL 35294.

† Mathematics Department, University of Alabama at Birmingham, Birmingham, AL 35294.

‡ Department of Computer and Information Sciences, Ibaraki University, Hitachi-shi, Ibaraki 316, Japan.

### Abstract

In this paper we present a genetic learning algorithm for fuzzy neural nets. Illustrations are provided.

## 1   Introduction

This paper is concerned with learning algorithms for fuzzy neural nets. In this section we first introduce the basic notation to be employed and then discuss what we mean by a fuzzy neural net. Then we briefly survey the literature on learning algorithms for fuzzy neural nets. The second section contains a description of our genetic learning algorithm. The third section contains experimental results. The last section has a brief summary and conclusions.

All our fuzzy sets are fuzzy subsets of the real numbers. We place a bar over a symbol if it represents a fuzzy set. So $\bar{A}$, $\bar{B}$, ..., $\bar{W}$, $\bar{V}$, are all fuzzy subsets of the real numbers. The membership function of a fuzzy set $\bar{A}$ evaluated at $x$ is written as $\bar{A}(x)$. The $\alpha$-cut of a fuzzy set $\bar{B}$ is

$$\bar{B}[\alpha] = \left\{ x \mid \bar{B}(x)) \geq \alpha \right\} ,\qquad(1)$$

for $0 < \alpha \leq 1$. We separately specify $\bar{B}[0]$, the support of $\bar{B}$, as the closure of the union of the $\bar{B}[\alpha]$, $0 < \alpha \leq 1$. A triangular fuzzy number $\bar{N}$ is defined by three numbers $a < b < c$ where: (1) $\bar{N}(x) = 0$ for $x \leq a$ and $x \geq c$ and $\bar{N}(x) = 1$ if and only if $x = b$; and (2) the graph of $y = \bar{N}(x)$ is a straight line segment from $(a, 0)$ to $(b, 1)$, $((b, 1)$ to $(c, 0))$ on [a,b] (on [b, c]). If the graph of $y = \bar{N}(x)$ is a continuous, monotonically increasing (decreasing), curve on $[a, b]$ (on $[b, c]$) from $(a, 0)$ to $(b, 1)$ (from $(b, 1)$ to to $(c, 0)$) then we say $\bar{N}$ is a triangular shaped fuzzy number. All our fuzzy sets are either triangular fuzzy numbers, or triangular shaped fuzzy numbers. We abbreviate a triangular fuzzy number, or a triangular shaped fuzzy number, as $\bar{N} = (a \,/\, b \,/\, c)$. We say $\bar{N} \geq 0$ if $a \geq 0$. A measure of fuzziness of $\bar{N}$ is $b$ - $a$. We write $fuzz\,(\bar{N}) = b$ - $a$. We say $\bar{M}$ is more fuzzy than $\bar{N}$ if $fuzz\,(\bar{M}) \geq fuzz\,(\bar{N})$. We employ standard fuzzy arithmetic, based on the extension principle, throughout the paper. If $\bar{M}(x) \leq \bar{N}(x)$ all $x$, then we write $\bar{M} \leq \bar{N}$.

All our fuzzy neural nets are layered (three layers), feedforward, networks. In this paper a fuzzy neural net has to have fuzzy signals and/or fuzzy weights. The basic fuzzy neural net used in this paper is shown in Figure 1. The input $\bar{X}$, the weights $\bar{W}_i$ and $\bar{V}_i$, are all triangular fuzzy numbers. Due to fuzzy arithmetic the output $\bar{Y}$ (and target $\bar{T}$) can be a triangular shaped fuzzy number.

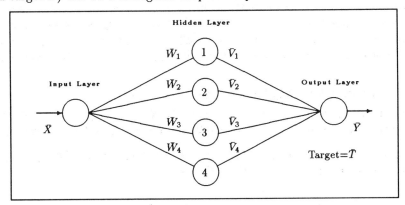

**Figure 1: Fuzzy Neural Net**

All neurons, except the input neuron, have a transfer function $y = f(x)$. This $f$ is assumed to be a continuous, non-decreasing, mapping from $\Re$ into $[-\tau, \tau]$ for $\tau$ some positive integer. The input node acts like the identity mapping so the input to node $i$ in the hidden layer is

$$\bar{I}_i = \bar{X} \cdot \bar{W}_i, \; 1 \leq i \leq 4 \; . \tag{2}$$

Node $i$'s output is

$$\bar{Z}_i = f(\bar{I}_i), \; 1 \leq i \leq 4 \; , \tag{3}$$

evaluated using the extension principle. Therefore, the input to the output node is

$$\bar{I}_0 = \bar{Z}_1 \cdot \bar{V}_1 + ... + \bar{Z}_4 \cdot \bar{V}_4 \; , \tag{4}$$

with final output

$$\bar{Y} = f(\bar{I}_0) \; . \tag{5}$$

Standard fuzzy arithmetic is used to evaluate equations (2)-(5).

The training data for the fuzzy neural net is $(\bar{X}_l, \bar{T}_l)$, $1 \leq l \leq L$, where $\bar{T}_l$ is the desired output when $\bar{X}_l$ is the input. If $\bar{X}_l$ is the input, then let $\bar{Y}_l$ be the actual output, $1 \leq l \leq L$. The learning problem for fuzzy neural nets is to find the weights $(\bar{W}_i, \bar{V}_i)$, so that $\bar{Y}_l$ is close to $\bar{T}_l$, $1 \leq l \leq L$.

In [2, 4, 5, 6, 11, 12, 13] the authors developed a fuzzy backpropagation algorithm for a fuzzy neural net. Actually, what they did was to directly fuzzify the standard delta rule in backpropagation to update the values of the weights. It is interesting to note that this procedure can fail to converge to the correct weights. What they found [4] was that there are values for the weights that make their error measure sufficiently small, but do not make $\bar{Y}_l$ close to $\bar{T}_l$, $1 \leq l \leq L$. The algorithm has been corrected but no new results have been reported.

In a series of papers [14, 15, 16, 17, 18, 19, 20] these authors have also developed a backpropagation based learning algorithm for fuzzy neural nets. They assumed that the inputs $(\bar{X}_l)$, the weights and the bias (they employed a sigmoidal $f$) terms are all symmetric triangular fuzzy numbers with $\bar{X}_l \geq 0$, all $l$. Let $E$ denote their error measure, which is based on $\alpha$-cuts of $\bar{Y}_l$ and $\bar{T}_l$, $1 \leq l \leq L$. Also, let $\bar{W}_i[0] = [w_{i1}, w_{i2}]$, $\bar{V}_i[0] = [v_{i1}, v_{i2}]$. Their algorithm is based on $\partial E/\partial w_{11}, ..., \partial E/\partial v_{42}$. These derivatives are complicated and become more complicated if we allow more general fuzzy sets for $\bar{X}_l$, $\bar{W}_i$, and $\bar{V}_i$. This method does not seem to generalize to more complicated fuzzy inputs and/or weights.

Genetic algorithms [9, 10, 25] are finding more and more applications in fuzzy systems [3, 8]. One of the authors of this paper (with Y. Hayashi) has proposed a genetic algorithm, and a fuzzy genetic algorithm, to train a fuzzy neural network. Until now, no computer experiments have been presented for genetic learning algorithms for a fuzzy neural net. We discuss our method in the next section.

In [23, 24] the authors have: (1) real number signals; (2) monotone increasing membership functions for the fuzzy weights; and (3) a special fuzzy error measure. They employ a learning algorithm, inspired by standard backpropagation, so that the fuzzy neural net can learn the weights. Yamakawa's new fuzzy neuron [28] has a learning algorithm for the weights. The learning algorithm for the real weights is similar to standard backpropagation. The fuzzy neural net in [22, 27] is similar to Yamakawa's fuzzy neuron. These papers have a learning algorithm for both the real weights and the trapezoidal fuzzy numbers.

## 2   Genetic Learning Algorithms

Genetic algorithms are a method of directed random search. We do not present the fundamentals of genetic algorithms in this paper but instead refer the reader to the popular text [10] on genetic algorithms.

The first thing to discuss is the error measure to be minimized. Let $\bar{Y}_l[\alpha] = [y_{l1}(\alpha), y_{l2}(\alpha)]$ and $\bar{T}_l[\alpha] = [t_{l1}(\alpha), t_{l2}(\alpha)]$, for $\alpha$ in the set $\{0.0, 0.1, ..., 0.9, 1.0\}$. Define

$$E_1 = \frac{1}{2} \sum_{\alpha} \sum_{l} (y_{l1}(\alpha) - t_{l1}(\alpha))^2 / L \; , \tag{6}$$

$$E_2 = \frac{1}{2} \sum_{\alpha} \sum_{l} (y_{l2}(\alpha) - t_{l2}(\alpha))^2 / L \; , \tag{7}$$

and

$$E = (E_1 + E_2) .$$ (8)

The genetic algorithm is to search for the fuzzy weights to drive $E$ to zero.

The transfer function $f$, in each hidden neuron and in the output neuron, is

$$f(x) = \begin{cases} -\tau, & x \leq -\tau, \\ x, & -\tau \leq x \leq \tau, \\ \tau, & x \geq \tau. \end{cases}$$ (9)

where $\tau$ is a positive integer. We choose the value of $\tau$ for the application. The value of $\tau$ is always one in the output neuron because all our target fuzzy sets $\bar{T}$ are in the interval [-1, 1].

We employed tournament selection [21, 26] instead of the more familiar roulette wheel selection [10] to choose population members for mating. The values of the parameters (probability of crossover, etc.) in the genetic algorithm can vary slightly from experiment to experiment but their approximate values are: (1) population size = 2000; (2) probability of crossover = 0.80; and (3) probability of mutation = 0.0003.

In this paper the fuzzy weights are assumed to be symmetric triangular fuzzy numbers. Let $\bar{W}_i = (w_{i1}/w_{i2}/w_{i3})$, $\bar{V}_i = (v_{i1}/v_{i2}/v_{i3})$, $1 \leq i \leq 4$. Then $w_{i2} = (w_{i1} + w_{i3})/2$, $v_{i2} = (v_{i1} + v_{i3})/2$, all $i$, and $\bar{W}_i$ ($\bar{V}_i$) is completely known if you know $w_{i1}$, $w_{i3}$ ($v_{i1}$, $v_{i3}$), $1 \leq i \leq 4$. So, the genetic algorithm just needs to keep track of the supports of the fuzzy weights. A member of the population is

$$Z = (w_{11}, w_{13}, ..., v_{41}, v_{43}) ,$$ (10)

coded in binary notation (zeros and ones).

Now we may discuss the results of our computer experiments on genetic learning algorithms for fuzzy neural nets.

## 3 Experiments

The complete experimental design is shown in Table 1. In the Input column real means $\bar{X}_l$ = real number, $1 \leq l \leq L$, and fuzzy means $\bar{X}_l$ = symmetric triangular fuzzy number, all $l$. Recall that the training data is $(\bar{X}_l, \bar{T}_l)$ with $\bar{X}_l$ input and desired output = $\bar{T}_l$. In the output column real means the target output $\bar{T}_l$ = real number, $1 \leq l \leq L$, and fuzzy designates $\bar{T}_l$ = triangular shaped fuzzy number in [-1, 1], all $l$. The mixed case, case 3, has $\bar{X}_l$ = real for some $l$ and $\bar{X}_l$ fuzzy otherwise, and the same for $\bar{T}_l$. In cases 5-7 the more (less) fuzzy in the Output column stands for: (1) $fuzz(\bar{X}_l) < fuzz(\bar{T}_l)$, $1 \leq l \leq L$, is Output = more fuzzy; (2) $fuzz(\bar{X}_l) > fuzz(\bar{T}_l)$, all $l$, is Output = less fuzzy; and (3) $fuzz(\bar{X}_l) < fuzz(\bar{T}_l)$ some $l$ and $fuzz(\bar{X}_l) > fuzz(\bar{T}_l)$ otherwise is Output = more and less fuzzy. In case 5 we have $fuzz(\bar{X}_l) = fuzz(\bar{T}_l)$, $1 \leq l \leq L$.

In [16] the authors conjectured that: (1) if $fuzz(output) \leq fuzz(input)$, then all the weights can be real numbers; and (2) if $fuzz(output) > fuzz(input)$, then the weights are fuzzy. Our experiments were designed to test this conjecture. We discuss the outcome in the last section.

In this paper we report results on cases 4-6. Additional results are available, as subject fare for the conference and as part of a more detailed study for a journal length paper.

**Table 1**: Experimental Design

| Case | Input ($\bar{X}$) | Output ($\bar{T}$) |
|------|------------------|--------------------|
| 1 | real | fuzzy |
| 2 | fuzzy | real |
| 3 | real and fuzzy | real and fuzzy |
| 4 | fuzzy | fuzzy |
| 5 | fuzzy | more fuzzy |
| 6 | fuzzy | less fuzzy |
| 7 | fuzzy | more and less fuzzy |

## 3.1  Case 4

The training data is generated by a fuzzy function $F$. So, we have $\bar{T} = F(\bar{X})$. In this case we must choose $F$ so that $fuzz(\bar{T}) = fuzz(\bar{X})$. In [16] they used the identity mapping and we chose $\bar{T} = -\bar{X} + 1$, for $\bar{X}$ in [-1, 1], so that $\bar{T}$ is in [-2, 2]. The training set is in Table 2. Recall that ($a$ / $b$ / $c$) designates a triangular (shaped) fuzzy number. The squashing function in the hidden neurons, and in the output neuron, used $\tau=2$ and we added a bias term (to be learned) to $f$ in the output neuron so that the net can learn the addition of a one to $-\bar{X}$ to get $\bar{T}$.

**Table 2**: Training Data For Case 4.

| No. | Input ($\bar{X}$) | Desired Output ($\bar{T}$) |
|---|---|---|
| 1 | (-1.00/-0.75/-0.50) | (1.50/1.75/2.00) |
| 2 | (-0.25/0.00/0.25) | (0.75/1.00/1.25) |
| 3 | (0.50/0.75/1.00) | (0.00/0.25/0.50) |

The trained net has real number weights and the results from test data given at the conference.

## 3.2  Case 5

The training data is produced by $\bar{T} = \bar{A} \cdot \bar{X}$ where $\bar{A} = (1.00/1.50/2.00)$, $\bar{X}$ in [-0.5, 0.5], so that $\bar{T}$ is in [-1, 1]. This is an expansive mapping in that $fuzz(\bar{T}) > fuzz(\bar{X})$. The training set is shown in Table 3. $\bar{T}$ is a triangular shaped fuzzy number.

**Table 3**: Training Data For Case 5.

| No. | Input ($\bar{X}$) | $fuzz(\bar{X})$ | Output ($\bar{T}$) | $fuzz(\bar{T})$ |
|---|---|---|---|---|
| 1 | (-0.50/-0.25/0.00) | 0.50 | (-1.00/ -3/8 /0.00) | 1.00 |
| 2 | (-0.25/0.00/0.25) | 0.50 | (-0.50/ 0.00 / 0.50) | 1.00 |
| 3 | (0.00/0.25/0.50) | 0.50 | (0.00/ 3/8 /1.00) | 1.00 |

The value of $\tau$, in the hidden layer and the output layer, was set equal to one. The fuzzy neural net learned the training data perfectly (zero error), with test results given at the conference. All weights were fuzzy in the neural net.

## 3.3  Case 6

The training set comes from $\bar{T} = F(\bar{X}) = 1/\bar{X}$ for $\bar{X}$ in (-∞, -1] or [1, ∞) so that $\bar{T}$ is in [-1, 1]. This is a contraction mapping because $fuzz(\bar{T}) < fuzz(\bar{X})$. We restricted $\bar{X}$ to be in [1, 3] for training and, as in the previous case, $\bar{T}$ is a triangular shaped fuzzy number. The training data is presented in Table 4. The squashing function in the hidden neurons used $\tau=3$.

**Table 4**: Training Data For Case 6.

| No. | Input ($\bar{X}$) | $fuzz(\bar{X})$ | Output ($\bar{T}$) | $fuzz(\bar{T})$ |
|---|---|---|---|---|
| 1 | (1.00/1.25/1.50) | 0.50 | (2/3 / 4/5 /1.00) | 1/3 |
| 2 | (1.50/1.75/2.00) | 0.50 | (1/2 / 4/7 / 2/3) | 1/6 |
| 3 | (2.00/2.25/2.50) | 0.50 | (2/5 / 4/9 / 1/2) | 1/10 |
| 4 | (2.50/2.75/3.00) | 0.50 | (1/3 / 4/11 / 2/5) | 1/15 |

The fuzzy neural net was unable to learn the training data. We felt this was mainly due to the fact that the piece-wise (3-segment) linear squashing function may not be a sufficiently good approximation for learning this non-linear function. We changed the squashing function to a nonlinear mapping, with results given at the conference and available in subsequent publications.

# 4  Summary and Conclusions

In this paper we presented a genetic algorithm for training a fuzzy neural net. We showed that it worked well for modeling the mapping $\bar{T}=-\bar{X}+1$ and $\bar{T}=\bar{A}\cdot\bar{X}$ where $\bar{A}$ and $\bar{X}$ are triangular fuzzy numbers. It did not work well in modeling $\bar{T}=1/\bar{X}$ putatively because the piece-wise linear squashing function we used may not well approximate a nonlinear function. It should work well using a nonlinear squashing function. It has been shown [1, 7] that a (regular) fuzzy neural net (like that in this paper) is not a universal approximator because it is a monotone increasing mapping. What this means is that if $\bar{X} \le \bar{X}'$ are inputs, then $\bar{Y} \le \bar{Y}'$ are the corresponding outputs. All functions that we tried to approximate were monotone increasing so we see no theoretical reasons why the fuzzy neural net should not be able to model these mappings.

Our results show that you do not necessarily get real weights if $fuzz(output) \le fuzz(input)$ [16]. However, we used a different squashing function than the one employed in [16]. Further research is needed on this conjecture.

# References

[1] J.J. Buckley and Y. Hayashi. Can fuzzy neural nets approximate continuous fuzzy functions? *Fuzzy Sets and Systems*. To Appear.

[2] J.J. Buckley and Y. Hayashi. Fuzzy backpropagation for fuzzy neural networks. Unpublished Manuscript.

[3] J.J. Buckley and Y. Hayashi. Fuzzy genetic algorithm and applications. *Fuzzy Sets and Systems*. To Appear.

[4] J.J. Buckley and Y. Hayashi. Fuzzy neural nets: A survey. *Fuzzy Sets and Systems*. To Appear.

[5] J.J. Buckley and Y. Hayashi. Fuzzy neural networks. In R.R. Yager and L.A. Zadeh, editors, *Fuzzy Sets, Neural Networks and Soft Computing*. To Appear.

[6] J.J. Buckley and Y. Hayashi. Fuzzy neural nets and applications. *Fuzzy Systems and AI*, 1:11–41, 1992.

[7] J.J. Buckley and Y. Hayashi. Are regular fuzzy neural nets universal approximators? In *Proc. of International Joint Conference on Neural Networks*, volume 1, pages 721–724, Nagoya, Japan, October 25-29 1993.

[8] J.J. Buckley and Y. Hayashi. Fuzzy genetic algorithms for optimization. In *Proc. of International Joint Conference on Neural Networks*, volume 1, pages 725–728, Nagoya, Japan, October 25-29 1993.

[9] L. Davis. *Handbook of Genetic Algorithms*. Van Nostrand Reinhold, New York, 1991.

[10] D. E. Goldberg. *Genetic algorithms in search, optimization, and machine learning*. Addison-Wesley, Reading, MA, 1989.

[11] Y. Hayashi and J.J. Buckley. Direct fuzzification of neural networks. In *Proc. of First Asian Fuzzy Systems Symposium*. Singapore, November 23-26 1993. In Press.

[12] Y. Hayashi, J.J. Buckley, and C. Czogala. Fuzzy neural network with fuzzy signals and weights. *Inter. J. Intelligent Systems*, 8:527–537, 1993.

[13] Y. Hayashi, J.J. Buckley, and E. Czogala. Direct fuzzification of neural network and fuzzified delta rule. In *Proc. of the Second International Conference on Fuzzy Logic and Neural Networks (IIZUKA'92)*, pages 73–76, Iizuka, Japan, July 17-22 1992.

[14] H. Ishibuchi, R. Fujioka, and H. Tanaka. An architecture of neural networks for input vectors of fuzzy numbers. In *Proc. of IEEE International Conference on Fuzzy Systems (FUZZ-IEEE'92)*, pages 1293–1300, San Diego, September 7-10 1992.

[15] H. Ishibuchi, R. Fujioka, and H. Tanaka. Neural networks that learn from fuzzy if-then rules. *IEEE Transactions Fuzzy Systems*, 1:85–97, 1993.

[16] H. Ishibuchi, K. Kwon, and H. Tanaka. Implementation of fuzzy if-then rules by fuzzy neural networks with fuzzy weights. In *Proc. First European Congress on Fuzzy and Intelligent Technologies*, volume I, pages 209–215, Aachen, Germany, September 7-10 1993.

[17] H. Ishibuchi, K. Kwon, and H. Tanaka. Learning of fuzzy neural networks from fuzzy inputs and fuzzy targets. In *Proc. Fifth IFSA World Congress*, volume I, pages 147–150, Seoul, Korea, July 4-9 1993.

[18] H. Ishibuchi, H. Okada, and H. Tanaka. Interpolation of fuzzy if-then rules by neural networks. In *Proc. of the Second International Conference on Fuzzy Logic and Neural Networks (IIZUKA'92)*, pages 337–340, Iizuka, Japan, July 17-22 1992.

[19] H. Ishibuchi, H. Okada, and H. Tanaka. Learning of neural networks from fuzzy inputs and fuzzy targets. In *Proc. of International Joint Conference on Neural Networks*, volume III, pages 447–452, Beijing, China, November 3-6 1992.

[20] H. Ishibuchi, H. Okada, and H. Tanaka. Fuzzy neural networks with fuzzy weights and fuzzy biases. In *Proc. of IEEE International Conference Neural Networks*, volume III, pages 1650–1655, San Francisco, March 28-April 1 1993.

[21] J.R. Koza. *Genetic Programming: On the Programming of Computers by Means of Natural Selection*. MIT Press, Cambridge, MA, 1993.

[22] K. Nakamura, T. Fujimaki, R. Horikawa, and Y. Ageishi. Fuzzy network production system. In *Proc. of the Second International Conference on Fuzzy Logic and Neural Networks (IIZUKA'92)*, pages 127–130, Iizuka, Japan, July 17-22 1992.

[23] D. Nauck and R. Kruse. A neural fuzzy controller learning by fuzzy error backpropagation. In *Proc. of NAFIPS*, volume II, pages 388–397, Puerto Vallarta, Mexico, December 15-17 1992.

[24] D. Nauck and R. Kruse. A fuzzy neural network learning fuzzy control rules and membership functions by fuzzy error backpropagation. In *Proc. of IEEE International Conference on Neural Networks*, volume II, pages 1022–1027, San Francisco, March 28-April 1 1993.

[25] R. Serra and G. Zanarini. *Complex Systems and Cognitive Processes*. Springer-Verlag, 1990.

[26] R.E. Smith, D.E. Goldberg, and J.A. Earickson. Sga-c: A c-language implementation of a simple genetic algorithm. Technical Report TCGA Report No. 91002, The Univerity of Alabama, The Clearinghouse for Genetic Algorithms, Department of Engineering Mechanics, Tuscaloosa, AL 35487, 1991.

[27] M. Tokunaga, K. Kohno, K. Hashizume, Y. Hamatani, M. Watanabe, K. Nakamura, and Y. Ageishi. Learning mechanism and an application of ffs-network reasoning system. In *Proc. of the Second International Conference on Fuzzy Logic and Neural Networks (IIZUKA'92)*, pages 123–126, Iizuka, Japan, July 17-22 1992.

[28] T. Yamakawa, E. Uchino, T. Miki, and H. Kusanagi. A neo fuzzy neuron and its application to fuzzy system identification and prediction of the system behavior. In *Proc. of the Second International Conference on Fuzzy Logic and Neural Networks (IIZUKA'92)*, pages 477–483, Iizuka, Japan, July 17-22 1992.

# Color Classification Using Fuzzy Inference and Genetic Algorithm

## Masami Sakurai, Yukio Kurihara and Shiro Karasawa

Industrial Research Institute of Kanagawa Prefecture
3173, Showa-machi, Kanazawa-ku, Yokohama 236, Japan

Abstract: A new color classification method using an integrated color sensor has been proposed. This method quickly classifies an unknown color into one of the reference colors by simplified fuzzy inference; the membership functions used for the fuzzy inference are optimized by a genetic algorithm. Simulation of classification between analogous standard colors shows the effectiveness of the proposed method. Furthermore the proposed method classifies low color-difference glass bottles on the basis of their components for recycling.

## 1 Introduction

The color classification technique is applied to discrimination of various objects. In the industrial field, there are many requests to classify colors simply and quickly. A classification method with an integrated color sensor is suited for these applications.

A common integrated RGB color sensor consists of red, green and blue color detectors. The conventional method usually classifies colors in permissible ranges of parameters that are calculated from the output of three detectors. However, highly analogous colors sometimes cannot be classified satisfactorily by this method, because the outputs from the same color samples include fluctuations mainly due to color nonuniformity and light illuminating nonuniformity. The need for a color classification method for highly analogous colors has recently become especially great.

Simple color classification equipment using fuzzy inference is proposed[1]. This equipment classifies colors simply and quickly using an integrated color sensor. However, the equipment has classification limits because shapes and sizes of the membership functions are fixed.

In this paper we present a new color classification method for analogous colors using fuzzy inference and a genetic algorithm[2]. In this method, the membership functions used for classification are optimized by a genetic algorithm. The demonstration is carried out by simulation of classification of highly analogous standard colors. Furthermore, we apply this method to color classification of glass bottles for recycling; the classification ability of this method is compared with that of the conventional method using the permissible ranges.

## 2 Color classification using fuzzy inference

A color classification method with a color sensor is in general use with an artificial light source such as a lamp. The light source illuminates a sample, and the light reflected from the sample or transmitted through it is detected by a RGB color sensor that has red, green and blue detectors. Parameters used for classification, $r$, $g$, $b$ and $s$, are defined from the red, green and blue detector outputs, $V_r$, $V_g$ and $V_b$, respectively, according to

$$V_s = V_r/V_{r0} + V_g/V_{g0} + V_b/V_{b0} \qquad (1)$$

$$r = (V_r/V_{r0}) /V_s \qquad (2)$$

$$g = (V_g/V_{g0}) /V_s \qquad (3)$$

$$b = (V_b/V_{b0}) /V_s \qquad (4)$$

$$s = V_s/V_{s0}, \qquad (5)$$

where $V_{r0}$, $V_{g0}$ and $V_{b0}$ are the output of detectors obtained with a standard white sample or a colorless one and $V_{s0}$ is a constant normalizing $V_s$.

A classification rule to classify colors by simplified fuzzy inference can be described as

if $r=R_i$ and $g=G_i$ and $b=B_i$ and $s=S_i$
    then $COLOR_i$ ($i=1$, $n$), $\qquad (6)$
where formula (6) is the $i$-th fuzzy rule; $R_i$, $G_i$, $B_i$ and $S_i$ are fuzzy variables; $COLOR_i$ is the name of the

i-th reference color; and the classification rule consists of n fuzzy rules. $\mu_i$, the truth value of the i-th premise, is the membership degree of $COLOR_i$ and is calculated by

$$\mu_i = R_i(r)G_i(g)B_i(b)S_i(s). \qquad (7)$$

Membership functions of $R_i$, $G_i$, $B_i$ and $S_i$ determine the classification ability. Thus we propose an optimization method of these membership functions by a genetic algorithm, as shown in Fig.1.

## 3 Optimization of membership functions by genetic algorithm

In this section, we explain the method of optimizing the membership functions by a genetic algorithm. Genetic algorithms are search algorithms based on the mechanics of natural selection and natural genetics[2]. Genetic algorithms have three operations: a) crossover operation, b) mutation operation, c) selection operation.

Each membership function of $R_i$, $G_i$, $B_i$ and $S_i$ forms a trapezoid in the proposed method. One of the membership functions is encoded to the four real numbers defined by the trapezoid. We regard these real numbers as a gene. Figure 2 shows the encoding of the membership functions. Because four membership functions are required by each reference color, the number of genes needed to express the classification rule is four times the number of reference colors. The classification rule is encoded as a string using these genes. First, the strings having random genotypes are taken as the initial population.

In the crossover operation, some pairs of strings are selected as parents. One of the parents is selected according to its classification abilities, and another is selected at random. A crossover position is randomly selected on each pair of strings. Two new offspring strings are made by swapping real numbers after that position.

In the mutation operation, each gene of the offspring is changed for a random real number with an occurrence probability.

Before the selection operation, the strings are evaluated with the training samples. Evaluation is done using an objective function called

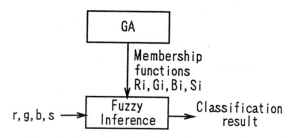

Fig.1 Concept of color classification using fuzzy inference and genetic algorithm.

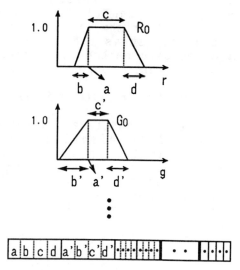

Fig.2 Encoding of the membership functions.

the fitness function:

$$F(k) = w_0E_{0k} + w_1E_{1k} + w_2E_{2k} + w_3E_{3k}, \qquad (8)$$

where $F(k)$ is fitness value of the k-th string; $E_{0k}$ is the classification rate; $E_{1k}$ is the average of the difference from $\mu$ for the correct color to the other $\mu$; $E_{2k}$ is the average of the difference from the membership value for the correct color of each parameter to the same value for the other colors; $E_{3k}$ is the average width of the membership functions; $w_0$, $w_1$, $w_2$ and $w_3$ are the weights for each parameter.

In the selection operation, the strings are selected according to their fitness values. Some strings whose fitness value is low are removed from the population. This operation keeps the size of the population constant.

The proposed method requires a search of a wide solution space because a real number is used as a gene. Therefore, we use the random search operation with the three operations. In the random search operation, each string can search the narrow space around itself before the selection operation. The number of searching trials and the size of the space searched are kept constant. When a string finds a string more suitable than itself, the genes of the string are changed to these of the more suitable string. This operation makes it possible to search a wide space in detail without a large size of the population.

Simulations of classification were performed using data measured by a prototype color sensor whose measuring geometries were 0° illuminated and 45° received. Optimization was performed for classification of five highly analogous standard colors as the references. The following performance of the proposed method was obtained with 200 color datums for each standard color.

classification rate: 100 %,
where
output fluctuation: ±0.2 %,
the minimum color difference: 1.6 ($\Delta L^*$ab).
The same simulations were performed with the conventional method using permissible ranges. As a result,
classification rate: 92.4 %.
From the comparison of these results, we confirmed the effectiveness of the proposed method.

## 4 Application to classification of glass bottles

In order to show the effectiveness of the proposed method, we applied the method to classification of glass bottles, and we compared the classification ability of the proposed method with that of the conventional one. The color of glass bottles to be classified were relatively analogous. Simulations were performed, including fluctuations in sensor output. The output fluctuations are mainly caused by the color nonuniformity and the light illuminating nonuniformity.

Because they are colored by metal ions, classification, and thus recycling, of glass bottles becomes possible. When bottles contain different substances, they must be separated. In the case of highly analogously colored bottles, the classification is difficult because of the color and light nonuniformities. The causes of the nonuniformi-

ties may be as follows:
  a) dirt on the surface,
  b) secular change of the glass,
  c) differing thickness and shape of the glass.
Therefore, the classification of glass bottles is not easy.

Simulations were performed using the sensor outputs $V_r$, $V_g$ and $V_b$ calculated according to

$$V_r = K \sum_\lambda f_r(1_\lambda, \lambda) \tag{9}$$

$$V_g = K \sum_\lambda f_g(1_\lambda, \lambda) \tag{10}$$

$$V_b = K \sum_\lambda f_b(1_\lambda, \lambda) \tag{11}$$

$$1_\lambda = (\phi_\lambda \pm \varepsilon_{\phi\lambda})(\tau_\lambda \pm \varepsilon_{\tau\lambda}), \tag{12}$$

where $\lambda$ is wavelength; $1_\lambda$ spectral power of the light transmitted through a sample glass bottle; $\phi_\lambda$ spectral power of the light source illuminating the sample; $\tau_\lambda$ spectral transmittance of the sample; $\varepsilon_{\phi\lambda}$ a parameter to simulate the light nonuniformity; $\varepsilon_{\tau\lambda}$ a parameter to simulate the color nonuniformity in the sample; K a constant; and $f_r$, $f_g$ and $f_b$ functions expressing characteristics for the RGB color detectors. We used the measured values for $\tau_\lambda$ and $f_r$, $f_g$ and $f_b$ (c.f. Fig. 3). $\phi_\lambda$ was taken according to the CIE standard illuminants[3]. A TiO$_2$/Se integrated color sensor was chosen for the color sensor. TiO$_2$/Se photodiodes have much better sensitivity in the blue region than silicon diodes[4]. A measured spectral response for the chosen color sensor is shown in Fig. 3. $\varepsilon_{\phi\lambda}$ and $\varepsilon_{\tau\lambda}$ were obtained from random numbers determined within a limit that expresses the amount of nonuniformity among the sensor outputs as a ratio to $\phi_\lambda$ or $\tau_\lambda$.

Optimization was performed for classification of four kinds of glass bottles. The minimum color difference between glass bottles was 5.7 ($\Delta L^*$ab). We measured $\tau_\lambda$, the spectral transmittance, for five bottles. Two bottles of them needed to be classified in the same color because they contained the same substances. Figure 4 presents the distribution of measured values for the five bottles on an xy chromaticity diagram[5]. We used 50 datums for each bottle as the training samples. The grades of $\varepsilon_{\phi\lambda}$ and $\varepsilon_{\tau\lambda}$ were set within a limit of 15 %. The parameters for the genetic algorithm were as follows: population size was 100; number of offspring

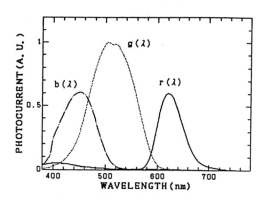

Fig. 3 Spectral responses of the short-cuircuit current for the color sensor.

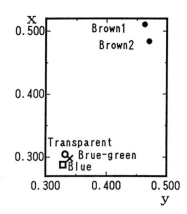

Fig. 4 The chromaticity diagram of glass bottles.

created by crossover operation was 50 for each generation; and occurrence probability of mutation was 5 %. The classification rule that had a maximum fitness value at the 100th generation was evaluated for its classification ability. We examined its classification rates for various grades of $\varepsilon_{\phi\lambda}$ and $\varepsilon_{\tau\lambda}$ by simulations. In the simulations, 100 datums for each bottle were classified.

The same simulations were performed by the conventional method. In this method, the permissible ranges are obtained from a minimum and a maximum value of some classification trials. Therefore, the permissible ranges were yielded by the training samples used for the proposed method. The simulation results were compared with those of the proposed method. The averages of ten simulation results are shown in Table 1. The results show that the proposed method is robust against color nonuniformity and light illuminating nonuniformity.

## 5 Conclusions

In this paper, we proposed a new color classification method using fuzzy inference and a genetic algorithm. We presented the effectiveness of the proposed method in comparison with the conventional method using permissible ranges. Furthermore we applied the proposed method to color classification of glass bottles. When

Table 1 Simulation results.

| $\varepsilon_{\phi\lambda}$ $\varepsilon_{\tau\lambda}$ | Classification rate | |
|---|---|---|
| | Proposed method | Conventional method |
| 0.0% | 1.000 | 0.980 |
| 5.0% | 1.000 | 0.964 |
| 10.0% | 0.990 | 0.933 |
| 15.0% | 0.963 | 0.833 |
| 20.0% | 0.935 | 0.685 |

color and light fluctuations included in sensor output were 5 % and 15 %, respectively, the classification rates in the proposed method were 100 % and 96.3 %, respectively; the rates in the conventional method were 96.4 % and 83.3 %, respectively.

## References

1) Y. Kurihara, K. Kobayashi, S. Karasawa, M. Yamanaka and T. Kuchi, Technical Digest 11th Sensor Sympo., 229(1992).
2) D. E. Goldberg, "Genetic Algorithms in Search, Optimization and Machine Learning," Adison Wesley(1989).
3) JIS, Z8720-1983, Japanese Standards Association(1983).
4) T. Nakata and A. Kunioka, Proc. 5th Sensor Sympo., 257(1985).
5) JIS, Z8701-1982, Japanese Standards Association(1982).

# GA-optimized Fuzzy Controller for Spacecraft Attitude Control

Antony Satyadas and K. Krishnakumar

— Spacecraft attitude control is approached as a non-linear control problem. The robustness quality of Genetic Algorithm (GA)-optimized Fuzzy Logic (FL) Control is demonstrated using a non-linear model of the Space Station Freedom. The main components of the GA-optimized Fuzzy Controller (GAoFC) are the GA (Genetic Algorithm) adaptive learning module, the Objective Function module, and the FC (Fuzzy Control) module. Performance of the non-linear GAoFC is verified using a non-linear simulation of the Space Station. We study cases of manually generated parameters rule set, GA-optimized parameters, and GA-optimized rules and parameters. Results presented substantiate the feasibility of using GA-optimized Fuzzy Controllers in robust non-linear control of spacecraft.

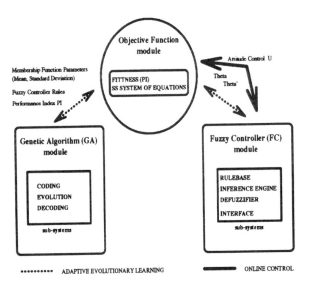

**Figure 1. System Layout**

## I. INTRODUCTION

Spacecraft attitude control has been examined in the past using several approaches such as classical control techniques [1,2] and modern state space techniques [3-5]. With the current interest in an evolutionary approach for constructing a Space Station (SS) in space, there is a need to examine alternate control techniques that can accommodate such evolutionary changes. Techniques reported in references[1-5] use linear approaches for controller designs and make inertia assumptions that are not valid for most proposed configurations. Also, small perturbation assumptions made during linearization will be violated during nominal operations of the SS.

Another substantial deviation from nominal parameters will occur during space shuttle docking and general relative mass motion. These contribute significantly to the changes in moments of inertia and also will introduce transient torque moments. To address some of these problems, recent studies [6-8] have focused on other alternatives. Reference [6] addresses the treatment of changing inertia as an adaptive control problem and uses a linear approach for the same. In reference [7] SS control is approached as a non-linear control problem using Lyapunov's second method for stability analysis and in reference [9] a feedback linearization is used for non-linear control.

This paper presents an approach using Fuzzy Logic and Genetic Algorithms that will provide robust non-linear control (Figure 1.). In the recent past, interests in the working of Fuzzy controllers has brought out many new approaches to solving engineering problems using soft com-

Antony Satyadas is a student IEEE member and Doctoral Student in Computer Science at The University of Alabama, Tuscaloosa, AL 35486-4983 (antony@cs.ua.edu). K. Krishnakumar is IEEE member and Assistant Professor in Aerospace Engineering at The University of Alabama, Tuscaloosa, AL 35487-0280 (kkumar@coe.eng.ua.edu).

puting. There are several benefits in using GA optimized Fuzzy controllers in non-linear control for space applications.

- A GA optimized Fuzzy Controller learns to control a system based on the membership function parameters (and rules) the GA generates through adaptive evolutionary learning. Once GA has learned the appropriate membership function parameters (and rules), the FC module is sufficient to control the spacecraft. This implies that a decentralized controller can be implemented using fuzzy controllers with direct output feedback. Decentralized control, with direct output feedback, is computationally less demanding and more suited when using space-qualified computers that are limited in their processing power.

- The necessity for a learning-type controller for space applications arises due to the conditions of uncertainty in the environment of operation and due to unmodeled dynamics. Fuzzy logic and approximate reasoning have been shown to perform well in problems requiring uncertainty management [8].

The major objective of this study is to highlight the advantages in using fuzzy controllers for spacecraft attitude control. The Space Station (SS) is chosen as the ideal example for demonstrating these advantages. In what follows, we first present the fundamentals of genetic algorithms (GA) and fuzzy control (FC) relevant to this study and outline a

0-7803-1896-X/94 $4.00 ©1994 IEEE

procedure for design and implementation of robust, fuzzy controllers. Next, we present the non-linear SS model used and show the capabilities of fuzzy-control in controlling the pitch attitude of the SS. Three cases are studied. In Case 1, we use a manually generated membership function parameter set and fuzzy logic rules. Case 2 uses GA to learn the membership function parameter values. In Case 3, GA learns the fuzzy rules as well as the membership function parameters. System robustness is tested by analyzing the performance of the system with varying moment of inertia. These study cases demonstrate the art of learning through GA and the synthesis of robust non-linear control laws using the FC. The following section gives a brief overview on the GA concepts used in this paper.

## II. Genetic Algorithm Concepts

Genetic Algorithm (GA) is a biologically inspired highly parallel mathematical search algorithm pioneered by Holland. GA generates entire population of points (typically fixed-length chromosome like character strings) each with associated fitness value, test each point independently, and combines qualities from existing points to form a new population (next generation) containing improved points. The population can be, for example, the fuzzy membership function parameters, fuzzy controller rules, and so on. The fitness value is computed using the information concerning the quality of the solution produced by the members of the population (objective function values). This adaptive evolutionary learning process relates to the evolutionary selection procedure of genetic chromosomes. They rely on Darwinian principle of reproduction and survival of the fittest and naturally genetic operations like mutation. Genetic Algorithm simulates this process, over generations, and identifies the most suitable candidate (for example, membership function parameter, fuzzy controller rule.) Lets look at the mechanics of GA.

The GA module consists of an Encoding, Evolution, and Decoding sub-systems. The population is represented in GA as a character string of 0's and 1's. We will now look at the sub-systems.

The **Encoding** sub-system uses a concatenated, mapped, unsigned binary coding. Let us look at an example. We use Gaussian membership function for the Fuzzy Control module. Two parameters, namely mean and standard deviation, control the Gaussian membership function. The value of each parameter can be encoded as a seven bit binary number (for example, 0000111). There is a membership function associated with every linguistic variable of a state or action in a Fuzzy Control rule. Thus, if there are 24 linguistic variables, a m=(7x2x24) bits string forms one member in the population N (case 2). The population consists of 10 such members. In case 3, GA identifies the fuzzy rules as well as the membership function parameters. Considering 12 rules, 2 states, and 1 action per rule, each member is represented as a m=(7x12x(2+1))+(7x2x24) bits string. Each of these N m-bit strings represent one possible set of parameters (and rules.)

This population of N m-bit strings, is first randomly generated by GA. The **Decoding** sub-system generates the actual values. These values form the output of the GA module. The Objective Function module contains the system of equations, the mathematical model that govern the attitude of the Space Station. The Fuzzy Control module uses the parameter values output by GA module. For each string in the set N, the Objective Function module generates a fitness value (Performance Index) that gives a proportional measure of the extend of control offered by the fuzzy controller with that parameter set. In this study, the performance index passes down penalties for $\dot{\theta}_2$, change of $U_2$, and $\theta_2$ to the GA module. This fitness value is the output of the Objective Function module. This **fitness sequence** is repeated for all the members of the population N. The next step is to use this information to breed the next generation. This is done by the Evolution sub-system.

The **Evolution** sub-system use reproduction, crossover, and mutation for breeding. This relates to copying a string, exchanging portions of strings, and generating random numbers, respectively. **Reproduction** ensures that strings with large fitness values have a larger representation in the next generation (new population). We use tournament selection to select the fittest. Adjacent strings (randomly determined) compete and win based on best fitness value. These reproduced strings are then brought into the mating pool and are ready for crossover. **Crossover** proceeds as follows: Two strings are picked at random from the mating pool. A crossover site on the string is determined at random. Finally, all the characters following the crossing site are exchanged between these two strings. For example, if the strings after reproduction were [11111*111] and [00001*001], crossover at the site marked as * produce [11111*001] and [00001*111]. **Mutation** is the occasional alteration of a value at a particular string position. This alteration ensures that critical pieces of information are not lost over generations. Thus mutation enhances GA's ability to find near-optimal solutions. The **fitness sequence** in the previous paragraph is repeated for the new generation. The best member of the population N is decoded and used as parameters (and rules) for the fuzzy controller. The steps in this paragraph form the **Evolutionary Learning Cycle**. This cycle is repeated for K generations. The final parameters (and rules) are identified as the best available set for the fuzzy controller.

## III. Robust Fuzzy Control Concepts

Fuzzy Control, which combines Fuzzy Logic concepts, approximate reasoning concepts, and optimal control concepts is relatively recent. The following sections present the fuzzy logic, approximate reasoning, and fuzzy-control concepts.

### A. Fuzzy Concepts

A fuzzy set is a class of objects (a set of elements) in which there is no sharp boundary between those objects that belong to the class and those that do not. Each object has associated with it a number which represents its grade of membership in the fuzzy set. According to Zadeh [12], the theory of fuzzy sets represents an attempt at constructing a conceptual framework for a systematic treatment of vagueness and uncertainty in both qualitative and quantitative ways.

### B. Fuzzy Logic

Parallel to fuzzy set theory, Zadeh proposed a fuzzy propositional logic, based on Lukasiewicz's L Aleph 1 multivalued logic [8]. Fuzzy logic allows the degree of membership in a fuzzy set to be interpreted as the truth value of a predicate. Thus the truth value of a proposition can be represented by the fuzzy set *very true, true, more or less true*.

### C. Approximate Reasoning

Approximate reasoning is the process by which a possible imprecise conclusion is deduced from a collection of imprecise premises. We use Generalized Modus Ponens (GMP) for reasoning. The control rules for spacescraft attitude may be expressed as a fuzzy implicational proposition. For example:

**IF** $\theta_2$ is *Very Positive* **AND** $\dot{\theta}_2$ is *Positive*
**THEN** the *attitude control* $U_2$ is *Negative*.

This rule represents the two states *Pitch Angle* $\theta_2$ and $\dot{\theta}_2$, and the action *attitude control* $U_2$. The linguistic terms *very positive Pitch angle* $\theta_2$, *positive* $\dot{\theta}_2$, and *negative attitude control* $U_2$ can be represented by fuzzy sets based on appropriate membership functions. We have used Gaussian membership functions. For a given value of $\theta_2$ and $\dot{\theta}_2$, we can determine their memberships in the various linguistic categories by applying the corresponding membership function. Mean and Standard Deviation are the parameters for the Gaussian membership function. These values are learned using the adaptive evolutionary learning skills of GA. These measures of $\theta_2$ and $\dot{\theta}_2$ are combined using fuzzy conjunction, which amounts to taking the minimum of the membership values. In real life, no single rule match the given input. Therefore any decision about the action (in this case control of spacecraft attitude $U_2$) can be taken only through approximate reasoning. In other words, given a set of rules that relates A and B to C, then given $\acute{A}$ and $\dot{B}$, you have to determine $\acute{C}$. According to Zadehs' theory of approximate reasoning [11], this fuzzy implication proposition can be translated into a fuzzy subset using Lukaziewicz's implication. Lukasiewicz's implication is used as follows:

Given, X is a set of elements; W denote X; $M_w$ is a fuzzy subset of X in which the grade of membership of each of the elements in X is drawn from the unit interval I, [-1,1]; B is a fuzzy subset of the unit interval I; H a fuzzy subset of WxI; H(w,i) is the membership function of $(w,i) \in$ W x I in the fuzzy subset H;

H(w,i) = min [1, (1 - $M_w$ + B(i))] for

w= 1,2 and i = -1.0, -0.9, ..0.0, 0.1, 0,2, .. 1.0.

You get one fuzzy subset per evaluation rule. These series of fuzzy conditions are now combined to draw a conclusion

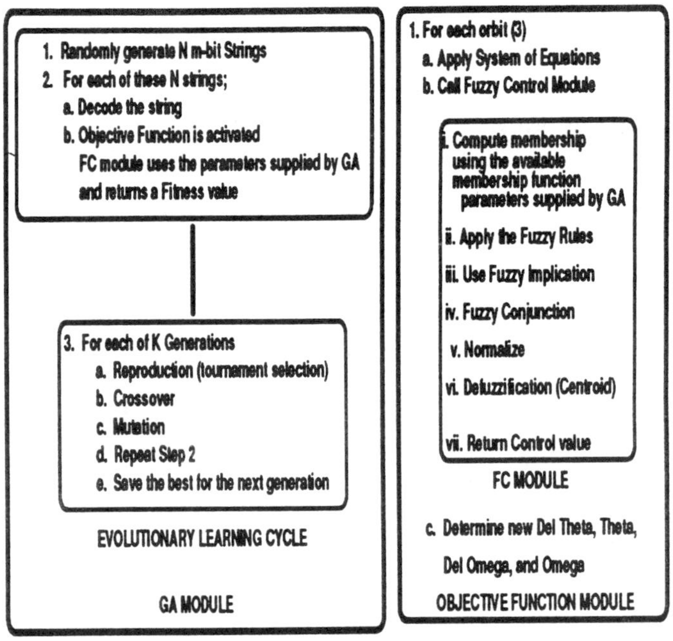

**Figure 2. GA, Objective Function, and FC modules**

by using the compositional rule of inference. The inference is defuzzified and used as control for the spacecraft. In other words, the values of $\theta_2$ and $\dot{\theta}_2$ are adjusted based on their current performance to maintain the stability of the spacecraft. Next we will look at the Fuzzy Control techniques.

### D. Fuzzy Control techniques

The following steps describe the fuzzy control techniques employed for spacecraft attitude control.

1. Identify a mathematical model of the system to be controlled. This includes the system of equations.
2. Identify state and control criteria to be used in the fuzzy controller. In this study, $\theta_2$ and $\dot{\theta}_2$ are the states and attitude control $U_2$ is the action.
3. Define the Fuzzy Controller membership functions, linguistic variables, and number of rules.
4. Identify the appropriate fuzzy inference mechanism, and defuzzification scheme.
5. Design a FC using the off-line supervised controller synthesis technique. This involves Evolutionary learning of Fuzzy Controller Membership Function Parameters (and rules) using Genetic Algorithm.
6. FC on-line simulation using the GA-optimized membership function parameters (and rules).

The mathematical model consists of the system of equations, and the iterative techniques employed to simulate the dynamic environment. The model is implemented as a software package (objective function module) in the computer. The objective of the Fuzzy Controller (FC) module is to use a fuzzy logic controller to achieve optimal spacecraft attitude control ($U_2$) using values of $\theta_2$ and $\dot{\theta}_2$. The fuzzy membership function parameters (and rules) are optimized using the GA module. GA optimization is based on a performance index supplied by the objective function module. Combinations of +/-0.05, and +/-0.00005 are given as the initial values of $\theta_2$ and $\dot{\theta}_2$ respectively, so that the performance index is based on the complete span of state space. The FC module consists of the Rulebase, Interface, Inference Engine, and Defuzzifier sub-systems.

The **Rulebase** sub-system consists of a set of states, actions, and rules. The states and actions have the associated linguistic variables. In this study, the states are *Pitch angle Theta* ($\theta_2$), and $\dot{\theta}_2$. *Attitude control* $U_2$ is the action. The associated linguistic variables are *Negative, Zero, Positive, and so on.* . We have used three linguistic variables each for cases 1 and 2, and eight each for case 3. Each of these linguistic variables, for every state and action, have the corresponding fuzzy membership function. The membership function determines, to what extend $\theta_2$ and $\dot{\theta}_2$ are *Negative, Zero, Positive, and so on.* We have used the Gaussian membership function given by the equation

$$Y = \frac{1}{\sigma\sqrt{2\pi}} e^{\frac{-1}{2}(X-\mu)^2/\sigma^2}$$

where Mean ($\mu$) and Standard Deviation ($\sigma$) are the parameters. The parameters are manually selected for Case 1, and supplied by the GA module for the other two. Combination of the states and action with the associated linguistic variables form the **IF THEN** rules. Thus a typical rule could be:
**IF** $\theta_2$ is *Negative* **AND** $\dot{\theta}_2$ is *Negative*
**THEN** $U_2$ is *Positive*
These rules are user selected for case 1 and 2 (9 each). In case 3, the GA module selects the appropriate set of 12 rules through adaptive evolutionary learning. We use a base set of [-1,+1]. The hedges for the linguistic variables over the base set are also stored in the Rulebase.

The **Inference engine** sub-system apply the rules and membership functions to determine the attitude control $U_2$ for a given value of $\theta_2$ and $\dot{\theta}_2$. First the membership of a given value of $\theta_2$ and $\dot{\theta}_2$ in the states of the fuzzy rules are computed. The Gaussian membership function with GA-optimized parameters are used. The relationship between the states in a rule could be conjunction (AND) or disjunction (OR). In this study, we have limited the relationship to AND. The next step is to apply fuzzy conjunction (this amounts to choosing the minimum) within the states in a rule. We get one value for each rule. This indicates the combined degree of membership for each rule. Now, Lukasiewicz's fuzzy implication is applied to these values to obtain the satisfaction over the base set [-1,1]. The next step is to combine these fuzzy implication sets. Fuzzy conjunction is used. This single curve (fuzzy conjunction curve) is then normalized.

The **Defuzzifier** sub-system computes the centroid of the fuzzy normalized conjunction curve. This crisp value (attitude control $U_2$) is the output of the FC module. $U_2$ is fed back to the Objective Function module to compute new values of $\theta$ and $\omega$. This process is continued for 3 orbits of the space station flight. The GA, Objective Function, and FC modules works together as a closed loop control system

during this adaptive evolutionary learning phase. The 30th generation GA-optimized membership function parameters (and rules in case 3) are identified as the optimal set. The system is now ready for on-line control.

The GA module is delinked from the FC module for on-line control. For case 1, the manually selected fuzzy rules and membership functions parameters are used to control the Space Station. In case 2, GA-optimized membership function parameters and manually selected fuzzy rules are used. In case 3, GA-optimized fuzzy rules and membership function parameters are employed. The FC module ensures attitude control using $\theta_2$ and $\dot{\theta}_2$ as input. Control is provided by determining $U_2$ using the learned fuzzy membership function parameters (and rules). The on-line simulation is thus the interaction between Objective function module and FC module. Details of the Space Station fuzzy control follow.

## III. ROBUST FUZZY CONTROL OF SS

Attitude control of Space Station is challenging due to the varying mass properties, which change the system characterics continuously. One of the proposed control architectures for the attitude control of Space Station consists of an outer loop for the momentum management and an inner loop for the attitude control. The momentum management loop commands the Space Station attitude and the inner loop controls the Space Station to achieve the desired trajectory. A combination of model identification and adaptive control can be utilized for the inner attitude control loop and the reference trajectory is generated by the momentum management system. The study presented in this paper, without any loss of generality, examines only the inner loop pitch attitude control.

### A. Space Station Model

Reference 1 presents the complete non-linear model of the SS. Assumptions of small roll/yaw attitude errors and small products of inertia lead to a simplification of the complete non- linear model. These equations are useful when there is a need for large pitch ($\theta_2$) maneuvers with small roll ($\theta_1$) and yaw ($\theta_3$) maneuvers. With the above simplifications, the equations reduce to:
Attitude Kinematics:

$$\begin{bmatrix} \dot{\theta}_1 \\ \dot{\theta}_2 \\ \dot{\theta}_3 \end{bmatrix} = \begin{bmatrix} 1 & -\theta_3 & 0 \\ 0 & 1 & 0 \\ 0 & \theta_1 & 1 \end{bmatrix} \begin{bmatrix} \omega_1 \\ \omega_2 \\ \omega_3 \end{bmatrix} + \begin{bmatrix} 0 \\ n \\ 0 \end{bmatrix}$$

Space Station Dynamics:

$$\begin{bmatrix} \dot{\omega}_1 \\ \dot{\omega}_2 \\ \dot{\omega}_3 \end{bmatrix} = -[I]^{-1} \begin{bmatrix} 0 & -\omega_3 & \omega_2 \\ 0 & 0 & 0 \\ -\omega_2 & \omega_1 & 0 \end{bmatrix} [I] \begin{bmatrix} \omega_1 \\ \omega_2 \\ \omega_3 \end{bmatrix} + 3n^2$$

$$[I]^{-1} \begin{bmatrix} 0 & -C3 & C2 \\ C3 & 0 & -C_1 \\ -C2 & C1 & 0 \end{bmatrix} [I] \begin{bmatrix} C_1 \\ C_2 \\ C_3 \end{bmatrix} + [I]^{-1} \begin{bmatrix} -U_1 \\ -U_2 \\ -U_3 \end{bmatrix}$$

where
$c_1 \underset{=}{\triangle} -\sin\theta_2$
$c_2 \underset{=}{\triangle} (\sin\theta_2)\theta_3 + \theta_1(\cos\theta_2)$
$c_3 \underset{=}{\triangle} \cos\theta_2$
I = moment of inertia matrix

$$= \begin{bmatrix} 50.28 & 0 & 0 \\ 0 & 10.80 & 0 \\ 0 & 0 & 58.57 \end{bmatrix} \star 10^6 slug.ft^2$$

n = orbital angular velocity = 0.0011 rad/sec
The mathematical model of the system include these system of equations.

### B. Supervised Fuzzy controller Synthesis

A fuzzy controller using the non-linear model is demonstrated in this section. Three cases demonstrate the system capabilities in this study. In case 1, manually generated membership function parameters and fuzzy rules are used (Figure 3.). In case 2, GA-optimized membership function parameters and manually generated rules are used (Figure 4.). In case 3, GA-optimized fuzzy rules and membership function parameters are used (Figure 5.). It may be noted that we have used mirror image of rules in case 3 to ensure that the control is zero when so required. Tables 1 and 2 gives the rules used for these cases and Tables 3-5 gives the membership function parameters. Figure 3-5 gives a graphical summary of the FC performance. Graph A for cases 1-3 shows the performance of the fuzzy controller as a measure of the pitch angle $\theta_2$ with orbits (time). The limitation of manually selected fuzzy rules and membership functions (case 1) is evident. Graph B indicates the membership function curves for the linguistic variables based on the parameters for cases 1-3. For case 3, space constraints limit the curves to those for $U_2$. Graph D indicates the adaptive evolutionary learning performance of the GA-module for cases 2 and 3.

### C. Robustness of the Fuzzy controller

In this phase of the study, the claim that the robustness of the fuzzy controller improves if GA-optimization is used is verified. Two systems, one with GA-optimized membership function parameters (case 2) and the other one with GA-optimized fuzzy rules and membership function parameters (case 3), were tested for their robustness to uncertainties in the moment of inertia of the pitch axis of the SS by varying the moment of inertia from 0 to 25 percent. Comparison of Graph C for cases 2 and 3 highlights the robustness of case 3 where optimization of fuzzy rules and membership functions gives GA a better chance to evolve appropriate solutions.

## IV. Conclusions

This study demonstrated the application of GA-optimized fuzzy-controller to adaptive non-linear control of Space Station. Three capabilities of fuzzy controllers were demonstrated using a non-linear model of the Phase I Space Station Freedom. These capabilities are: (a) Evolutionary learning of the fuzzy rules and membership function parameters using Genetic Algorithms; (b) synthesis of non-linear control laws using fuzzy logic; and (c) Robustness of the GA-optimized Fuzzy Controller.

The fuzzy controller techniques presented seem to be robust based on the successful results obtained using straight forward applications of these techniques. To build upon the ideas presented, fuzzy controllers for the three-axis adaptive fuzzy controller synthesis is currently in progress. This study examined decentralized fuzzy-control to demonstrate the learning and adaptation capabilities of GA-optimized FC on direct output feedback. For space applications, the fixed software/hardware environment provided by fuzzy controller synthesis makes it easy to modify the controller characteristics from a remote site, if there is a need.

Rule interpretation techniques and algorithms for Optimal rule selection from the GA generated combination of fuzzy rules and membership function parameters are currently being explored. Further research in this area is needed to examine the stability and performance robustness of these techniques. Specifically, it will be interesting to examine the theoretical aspects of the fuzzy controller in providing robust control capabilities and the applicability of fuzzy neural controllers. There are many other FC concepts and variations that might improve upon the ideas and results presented in this paper. This study should be seen as constituting a part of the groundwork for future studies in the area of fuzzy-control for Space Station applications.

## References

[1] Wie, B., Byun, K. W., Warren, W., Geller, D., Long, D., and Sunkel, J. W., "New Approach to Attitude/Momentum Control of the Space Station," Journal of Guidance, Control, and Dynamics, vol 12, no. 5, pp. 714-722, 1989.

[2] Yeichner, J., Lee, J., and Barrows, O., "Overview of Space Station Attitude Control System with Active Momentum Management," AAS paper 88-044, February 1988.

[3] Byun, K. W.,Wie, B., Geller, D., and Sunkel, J. W., "Robust H-Infinity Control Design for the Space Station with Structured Parameter Uncertainty," presented at AIAA Guidance, Navigation and Control Conference, Portland, OR, August 20-22, 1990.

[4] Balas, G., Packard, A., and Harduvel, J., "Applications of Mu-Synthesis Techniques to Momentum Management and Attitude Control of the Space Station," presented at AIAA Guidance, Navigation and Control Conference, January 1991.

[5] Parlos, Alexander G, and Sunkel, J. W., "Adaptive Attitude Control and Momentum Management for Large-Angle Spacecraft Maneuvers," Journal of Guidance, Control, and Dynamics, Vol. 15r, no. 4, July - August 1992.

[6] Vadali, S. R. and H. S. Oh, "Space Station Attitude Control and Momentum Management: A nonlinear Look," Journal of Guidance, Control, and Dynamics, May-June 1992.

[7] Bossart, Theodore C. and Singh, Sahjendra N., "Invertibility of Map, Zero Dynamics and Nonlinear Control of Space Station," AIAA-91-2663-CP, 1991.

[8] Satyadas Antony, "Fuzzy Decision Support Expert System," Computer science masters thesis, The University of Alabama, 1992.

[9] Krishnakumar, K., Richard, S., and Bartholomew, S., "Adaptive Neuro-control for the Space Station Freedom," AIAA 93-0407, 1993.

[10] Karr, C. L., and Gentry, E. J.,"Fuzzy Control of pH Using Genetic Algorithms," IEEE Transactions on Fuzzy Systems, Vol. 1, no. 1, pp. 46-53, 1993.

[11] Zadeh, L. A., "A theory of approximate reasoning, Memo "UCB/ERLM 77/58, University of California, Berkeley , 1977.

[12] Zadeh, L. A., "Fuzzy Sets," *Information Control,* Vol. 8, 1965, pp. 338-353.

[13] Lee, M. and Takagi, H.,"Integrating Design Stage of Fuzzy System using Genetic Algorithms", IEEE 2nd Int'l Conf. on Fuzzy Systems, Vol. 1, pp.612-617, 1993.

[14] Holland, J. H.,"Adaptation in Natural and Artificial Systems," Ann Arbor, MI, University of Michigan Press, 1975.

[15] Koza, John, K.,"Genetic Programming," MA, MIT Press, 1992.

[16] Goldberg,"Genetic Algorithms in Search, Optimization, and Machine Learning." MA, Addison-Wesley, 1989.

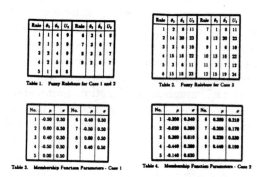

Table 1. Fuzzy Rulebase for Case 1 and 2

Table 2. Fuzzy Rulebase for Case 3

Table 3. Membership Function Parameters - Case 1

Table 4. Membership Function Parameters - Case 2

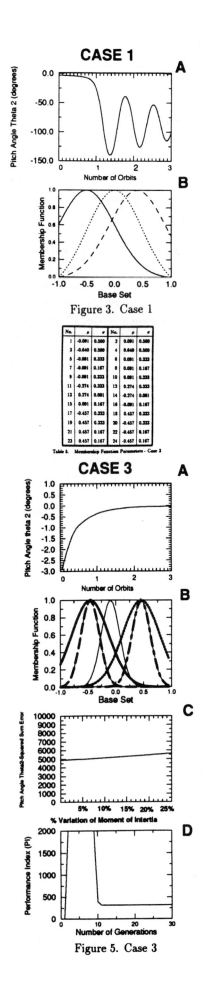

Figure 3. Case 1

Table 5. Membership Function Parameters - Case 3

Figure 5. Case 3

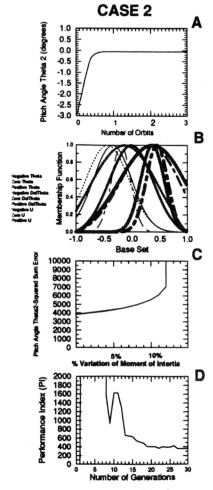

Figure 4. Case 2

# A Genetic Programming Application in Virtual Reality

Sumit Das, Terry Franguiadakis, Michael Papka,
Thomas A. DeFanti, Daniel J. Sandin
Electronic Visualization Laboratory
University of Illinois at Chicago
851 S. Morgan St., Rm. 1120
Chicago, IL, 60607-7053
papka@eecs.uic.edu

## Abstract

Genetic programming techniques have been applied to a variety of different problems. In this paper, the authors discuss the use of these techniques in a virtual environment. The use of genetic programming allows the authors a quick method of searching shape and sound spaces. The basic design of the system, problems encountered, and future plans are all discussed.

## 1 Introduction

### 1.1 Background

Genetic algorithms have been used to solve a wide variety of problems, such as ones discussed in [8, 9, 10]. Used as an optimization technique, genetic algorithms have proven to be an effective way to search extremely large or complex solution spaces. Two such spaces are the vast domains of shape and sound. There exists an infinite variety of shapes, and sounds that can be associated with them. The search for pleasing combinations of the two cannot be carried out exhaustively, and aesthetic values cannot easily be parameterized. Since genetic algorithms do not rely on problem-specific knowledge, they can be used to discover solutions that would be difficult to find by other methods.

A *genotype* is a "blueprint" for a potential solution to a problem. The solution that results from following the directions in the genotype is called the *phenotype*. Genotypes can be altered or combined to create a set of variations. The resulting set of genotypes is used to create a set of phenotypes. A *fitness function* is used to evaluate the relative quality of each phenotype, and the genotypes corresponding to the phenotypes judged "best" are used as the basis for the next generation.

### 1.2 Graphics and Genetic Algorithms

William Latham and Stephen Todd explore the possibilities of shape in their use of *Mutator* [15]. Latham and Todd use basic geometric shapes and a simple set of rules for combining them to construct complex three-dimensional images and animations. The rules are empirically based on organic shapes, such as horns and shells. The use of genetic algorithms simplifies the search for pleasing forms [16]. Providing a selection of shapes, Mutator allows the artist to choose the most pleasing ones, which then survive to produce the next generation. This allows a large variety of shapes to be viewed very quickly. The artist has control over the mutation rate; by raising the rate at the outset and lowering it as the results become more desirable, the search can be optimized, steered, and fine-tuned.

Karl Sims has also used genetic programming to aid in the discovery of new visual images. Sims makes use of symbolic LISP expressions to generate images, shapes, and solid textures [14]. The symbolic expressions are then mutated to generate new images based on aesthetic choice. Both Sims' and Latham/Todd's work make use of the user's aesthetic judgement as the fitness function for the image.

### 1.3 Sound and Genetic Algorithms

The generation of musical sound is usually separated into timbre generation [8, 17] and composition [9, 17]. *Timbre* generation is analogous to selecting or creating instruments to play, while *composition* is the process of deciding what notes the selected instruments will play. There are well known algorithms for timbre generation, such as FM [3] and granular synthesis [17]. Finding a set of parameter values that will satisfy a listener or composer's requirements is a formidable task, and is usually accomplished by trial and error. Horner describes a technique for deriving FM parameters to approximate a desired sound using genetic algorithms [8]. The higher-level requirements of composition are also difficult to quantify and specify. Horner and Goldberg describe the application of genetic algorithms to thematic bridging, a compositional technique in which a repeating initial musical pattern is transformed into a target pattern over a specified duration [9].

### 1.4 Virtual Reality

Virtual reality (VR) is a method of interacting with a computer-simulated environment. One of the major attributes of VR is immersion; that is, to give a user the experience of actually "being there." Sandin defines a VR system as containing a substantial portion of the following: surround vision, stereo cues, viewer-centered perspective, real-time interaction, tactile feedback, and directional sound [13].

The CAVE (CAVE Automatic Virtual Environment) is a virtual reality project being developed at the Electronic Visualization Laboratory (EVL) at the University of Illinois at Chicago (UIC). It makes use of four projectors displaying computer images on three walls and the floor of a ten-foot cube. Images are projected in stereo, so that a user wearing stereo glasses can see the images in true three-dimensional space. The user is tracked using an electromagnetic tracking system, so his/her instantaneous position and orientation are known. This allows the environment to be rendered in correct viewer-centered perspective. The user is able to manipulate objects within the CAVE using a wand, a three-dimensional analog of the mouse of current computer workstations. The user also gets audio feedback through the CAVE's sound system [1]. This includes facilities for spatial localization of sound sources, as well as the ability to play back sampled sounds through a MIDI interface. Facilities for software sound synthesis are under development. A more detailed discussion of the CAVE and its applications can be found in [7, 4, 5, 6, 2].

The ability to view and manipulate three-dimensional objects on a two-dimensional computer screen presents some inherent problems. It can be difficult to comprehend a complex three-dimensional shape on a computer screen. Also, viewing the object from different angles and distances can be cumbersome with mouse and keyboard based interfaces. The three-dimensional nature of the CAVE immediately solves these problems. This paper describes an application of genetic algorithms in virtual reality in which we populate a virtual world with both images and sound. This application, which was demonstrated as part of the CAVE installation at Supercomputing '93, presents one solution for developing a genetically evolving environment of shape and sound.

## 2 Genetic Art in Virtual Reality

### 2.1 System Overview

This application attempts to remedy some of the above problems, inherent in workstation interface technology. Objects are displayed in stereo, so shapes are easier to perceive. Viewers can examine objects in the same ways as in the real world: by walking around and moving relative to the objects, and also by manipulating objects to get a better look at them. Objects can be selected and then acted upon. Depending on the current mode, the selected object can be:

- Manipulated (e.g. rotated and moved)
- Mutated
- Mated to another object (also known as crossover)
- Saved to a file for future reference
- Refined (re-rendered much larger and with higher grid resolution, the significance of the grid resolution is discussed in section 2.3)

In the CAVE, a user is presented with four shapes and a status window which displays information on the current mode of the application and the action available to the user. The wand has three buttons, which are assigned functionality based on common conventions in systems using a three-button mouse. At any time, holding the middle button down allows the selected object to be manipulated. The right button steps through the possible modes. These changes are reflected in the status window, which shows an icon and brief text describing the available action. When the status window displays the desired action, the left button performs the action on the selected shape. Any shape can be selected simply by pointing to it with the wand. While a shape is selected, its associated music plays. Sound is also used in a more traditional way; for example, to provide feedback on choices the user makes and on the internal state of the application program.

### 2.2 System Implementation

The genetic structure is largely based on Sims' work [14], where the genotype is a symbolic mathematical expression. This has two advantages: it allows procedural information as well as parameter data to be encoded in the genotype, and the genotype is not limited to any particular size or structure. The expressions are composed of a small function set containing mathematical and geometric functions:

*plus, minus, times, divide, cylinder, sphere, torus, union, intersection, complement, difference.*

Each function takes a specified number of arguments and returns a real value. The entire expression

$$E(x, y, z)$$

where x, y, and z are interpreted as coordinates in a Cartesian space, also returns a real value. Both mutation and crossover are used to derive the next generation of objects. Mutation is a reproduction technique where a chosen parent's expression is randomly altered, deriving four children which replace the current population (Figure 1). Crossover is the combination of two expressions, with no mutation. Each expression in the new generation contains a part of both parents' expressions (Figure 2). Once again the new population replaces the old population. Both means of reproduction are extensions of techniques discussed in [10, 14], here implemented in C++.

### 2.3 Shape Generation

The shape of the object represents its graphical phenotype. Each shape is created as a polygonal isosurface within a cubical volume embedded in Cartesian space. A threshold value $T$ is chosen (arbitrarily fixed at $T = 1.0$). The expression $E$ is evaluated at discrete points within the volume. These points are arranged in a regular three-dimensional grid. For a point $P = (X, Y, Z)$, if $E(X,Y,Z)$ is within a tolerance $\varepsilon$ of $T$, then $P$ lies on the surface of the shape. The actual polygonization of the surface from the values is accomplished using the marching cubes algorithm [11], which given a three-dimensional grid of values and a threshold, generates a set of triangles representing the isosurface at that threshold.

Figure 1: Mutation. The parent is displayed on the top and one of the four children is displayed on the bottom.

By adjusting the resolution of the grid, the smoothness and accuracy of the surface can be improved at the cost of increased computation and rendering time.

Example: The expression

$$E(x, y, z): X^2 + Y^2 + Z^2$$

with

$$T = 4.0$$

creates a sphere centered at (X, Y, Z) with radius 2. The expression is evaluated at each (X, Y, Z) location corresponding to a grid point in the volume. The resulting surface exists everywhere that

$$-\varepsilon \; < \; E(x, y, z) \; - 4.0 \; < \; \varepsilon$$

so that a polygonal approximation to a sphere is produced.

The accuracy of the approximation depends on the resolution of the grid used. Normally, all four shapes are generated using a 15x15x15 grid. When refined, only the

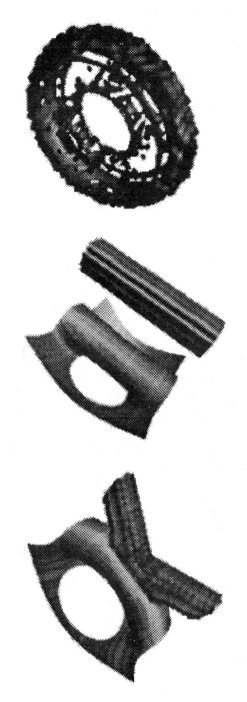

Figure 2: Mating. The parents are displayed on the top and one of the four children is displayed on the bottom

selected shape is displayed, generated using a 60x60x60 grid. When a mutation or mating is performed, all four current objects are replaced by four children resulting from the operation. The initial expressions are generated randomly or by an equation defined by the user, perhaps saved in the course of a previous exploration.

## 2.4 Shape Representation

To make the surfaces more visually interesting, and also to make the shapes more comprehensible, the two sides of each surface are colored differently. One side is left a solid medium gray. Solid noise [12] is used to create the surface texture for the other side. The expression that generates the shape can also be used to generate coloring schemes for one side of the surface.

The function set described above is a subset of the original set. In the authors' opinion, this subset yielded the most interesting shapes. The mutation rate is also important; if it is too high, the children may have little resemblance to the parent(s). If the rate is too low, there may be no perceptible change from one generation to the next.

## 2.5 Sound

The goal of the audio portion of this application is to vary stylistic elements across shape. In effect, each shape has a musical style associated with it. This style is the sonic phenotype. The music is played by a band composed of four MIDI instruments: percussion, bass, marimba, and guitar. The music is generated by manipulating precomposed fragments of music, each four beats long. Each fragment is a musical *phrase* to be played by one of the four instruments. The *phrases* are derived in the following manner: A number of 8-bar (32-beat) *compositions* are created for the band in common musical styles, e.g. reggae, waltz, minimalist. Each composition is separated into four parts, one for each instrument. Each of these parts is then divided into eight four-beat phrases.

For any given shape, each instrument is given eight phrases to choose from. To provide some overall stylistic cohesiveness, these eight phrases can come from at most two compositions. Each instrument chooses the next phrase to play based on a Markov chain, a probabilistic finite state automaton [17]. The probability tables for the Markov chains are generated by repeated evaluations of the expression E, using empirically derived values for the arguments. The state variables that affect the behavior of each instrument and their mappings to the genotype and visual phenotype are as follows:

- Choice of fragments: number of triangles in the isosurface, modulus number of fragments available.

- Markov chain probability tables: derived from expression **E** (see above).

- Musical key to play in: surface area of shape.

- Tempo: number of triangles in isosurface, scaled to available tempo range.

- Probability of rhythmic perturbation of a note: square root of (number of triangles / surface area).

- Probability of pitch perturbation of a note: surface area of shape.

## 3 Conclusion and Future Work

This application is a work in progress; it is constantly being revised and enhanced, and there are still many improvements that can be made. The fact that there are only four shapes visible at any one time is a compromise to allow real-time rendering. Although interesting results are obtained, this limitation makes it difficult to follow a desired evolutionary path, as the genetic pool is so small. This problem can be solved by lowering the grid resolution of the objects or by using more powerful computers. Unfortunately, lower grid resolution results in shapes with less detail.

For purposes of clean implementation and predictability, the function set is severely restricted. A larger population coupled with a richer function set might improve on the visual results.

Many of the restrictions imposed on the sonic phenotype were decided on empirically. The goal throughout is to make the sounds enjoyable and comprehensible to the casual listener, while at the same time making the styles distinct from each other. The music is based on well-known styles of music. Nevertheless, as the phrases combine in various ways, styles emerge that seem distinct from any of the underlying compositions.

Current hardware limitations preclude generating the instrument sounds functionally; instead, sampled sounds are used. Since timbre is one of the major factors of stylistic musical classifications, this is a major limitation. The next version of this application will have facilities to generate at least some of its sounds through direct software synthesis, allowing a much greater range of timbre and interactivity.

The instruments' modifications to the musical fragments are limited to statistically driven perturbations of individual notes. A better solution might be to use a more context-dependent (neighboring notes and states of other instruments), musically-based modification scheme. The musical fragments themselves could also be generated functionally, although encompassing a wider range of styles while remaining "musical" to most listeners would be difficult.

The authors are currently exploring the possibility of extending this work to include behavioral characteristics for the objects, as well as non-user-driven fitness functions. This would necessitate a larger population so that statistical variation would not dominate the actual fitness of genes.

The use of supercomputers for this application is being explored, both for isosurface generation and expression evaluation. The CAVE hardware configuration currently supports a high-speed FDDI/HIPPI connection between the SGI Onyx and the Thinking Machines CM-5, and connections to other supercomputers are being developed. An FDDI/HIPPI prototype connection between the CAVE/SGI Crimson and a 64-processor CM-5, implemented by NCSA, was demonstrated at Supercomputing '93 in Portland, OR.

An interesting area we plan to investigate is *meta-evolution*, such as the evolution of fitness functions, or the evolution of the genotype-to-phenotype mapping. Also, instead of using the entire genotype expression to determine both shape and sound characteristics, we want to explore the possibility of using different parts of the expression to determine these and future characteristics. These may or may not be mutually exclusive parts of the expression.

## 4 Acknowledgments

This material is based upon the work supported by National Science Foundation under Grant No. CDA-9303433, which includes support from the Advanced Research Projects Agency. The authors thank Maxine Brown, Gary Lindahl, Dana Plepys, and Maggie Rawlings for all their help, and all the other members of the Electronic Visualization Laboratory.

## 5 References

[1] R. Bargar and S. Das, "Sound for Virtual Immersive Environments," notes from course #23, *Applied Virtual Reality*, SIGGRAPH 1993, pp 4.1 - 4.18.

[2] R. Bargar and S. Das, "Virtual Sound Composition for the CAVE," (Tokyo, Japan, September 1993), *Proceedings of the 1993 International Computer Music Conference*, International Computer Music Association, September, 1993 (addendum).

[3] J. Chowning, "The Synthesis of Complex Audio Spectra by Means of Frequency Modulation", *Journal of the Audio Engineering Society*, 1973, Vol. 21, pp. 526-534.

[4] C. Cruz-Neira, J. Leigh, M. Papka, C. Barnes, S. Cohen, S. Das, R. Engelman, R. Hudson, T. Roy, L. Siegel, C. Vasilakis, T.A. DeFanti, and D.J. Sandin, "Scientists in Wonderland: A Report on Visualization Application in the CAVE Virtual Reality Environment," *IEEE 1993 Symposium on Research Frontiers in Virtual Reality*, October 1993, pp. 59-66.

[5] C. Cruz-Neira, D.J. Sandin, and T.A. DeFanti, "Surround-Screen Projection-Based Virtual Reality: The Design and Implementation of the CAVE," *Computer Graphics (Proceedings of SIGGRAPH 93)*, ACM SIGGRAPH, August 1993, pp. 135-142.

[6] C. Cruz-Neira, D.J. Sandin, T.A. DeFanti, R.V. Kenyon, and J.C. Hart, "The Cave: Audio Visual Experience Automatic Virtual Environment," *Communications of the ACM*, Vol. 35 No. 6, June 1992, pp. 65-72.

[7] T.A. DeFanti, D.J. Sandin, and C. Cruz-Neira, "A 'Room' with a 'View'," *IEEE Spectrum*, October 1993, pp. 30-33.

[8] A. Horner, J. Beauchamp, and L. Haken, "FM Matching Synthesis with Genetic Algorithms," *Proceedings of the 1992 International Computer Music Conference*, CMA, San Fransisco, 1992.

[9] A. Horner, and D. E. Goldberg, "Genetic Algorithms and Computer-Assisted Music Composition," *Technical report CCSR-91-20*, Center for Complex Systems Research, The Beckman Institute, University of Illinois at Urbana-Champaign, December 1991.

[10] J. R. Koza, *Genetic Programming*, MIT Press, Cambridge, MA, 1992.

[11] W. E. Lorensen and H. E. Cline, "Marching Cubes: A High Resolution 3D Surface Construction Algorithm," *Computer Graphics (Proceedings of SIGGRAPH 87)*, ACM SIGGRAPH, August 1987, pp. 83-91.

[12] K. Perlin, "A Hypertexture Tutorial", notes from course #23, *Procedural Modeling and Rendering Techniques*, SIGGRAPH 1992, pp. 3.1 - 3.28.

[13] D. J. Sandin, "Virtual Reality Technologies: Head Mounted Displays, Booms, Monitor and Projection Based Systems," notes from course #23, *Applied Virtual Reality*, SIGGRAPH 1993, pp 2-1 - 2-5.

[14] K. Sims, "Artificial Evolution for Computer Graphics," *Computer Graphics (Proceedings of SIGGRAPH 91)*, ACM SIGGRAPH, August 1991, pp. 319-328.

[15] S. Todd, and W. Latham, "Artificial Life or Surreal Art?, Towards a Practice of Autonomous Systems," *Proceedings of the First European Conference on Artificial Life*, 1991, MIT Press, Cambridge, Massachusetts, 1991, pp. 504-513.

[16] S. Todd, and W. Latham, *Evolutionary Art and Computers*, Academic Press, New York, 1992.

[17] I. Xenakis, *Formalized Music*, Bloomington: Indiana University Press, 1971.

# A Fuzzy Stop Criterion for Genetic Algorithms Using Performance Estimation

Lee Meyer and Xin Feng

Department of Electrical and Computer Engineering

Marquette University, Milwaukee, Wisconsin, 53233

**Abstract - This article presents a new approach for analyzing the solution performance of Genetic Algorithms (GAs). An adaptive filtering algorithm is combined with a predicting algorithm and memory data from previous GA iterations to estimate the value of the GA's "optimal" solution. If the current GA iteration value is above a certain user-defined acceptance level, the iteration process is stopped and the GA calculates a belief and uncertainty estimations of the found solution. Results indicate this new approach is preferable to the traditional GA iteration approach, in terms of cost/performance and in decreasing the amount of time the GA searches for acceptable solutions.**

## I. INTRODUCTION

Genetic Algorithms have been used extensively for the past several years as effective search methods that rapidly discover and manipulate regularities throughout solution spaces [4]. GAs operate according to the principles of population genetics to evolve solutions for problems with large solution spaces [2] and for problems where no efficient polynomial-time algorithm is known, such as NP-complete problems [1].

However, there are several fundamental concerns regarding conventional GAs: first, GAs do not guarantee that the optimal solution will be found. Second, if the optimal solution is not known, GA performance is difficult to measure accurately. In essence, the ability to measure reliability regarding GA performance is not well-defined. Traditional GAs are programmed to evolve solutions over a fixed number of iterations. In synthetic GAs, where population size decreases, the GA may be structured such that there is one remaining individual, which is taken as the final solution [4], [5]. In either case, reliability of the final solution is still a concern.

In this paper, the concept of a *fuzzy stop criterion* is developed to provide a useful evaluation of the GA's real-time performance. This new approach has several advantages over conventional GAs. First, the fuzzy stop criterion is not based on a fixed value but on achieving a user-defined level of performance for the given problem. Second, data from past performance of the GA is utilized as a *frame of reference* for current GA performance. In addition, the new algorithm calculates *belief* and *uncertainty measures* [7], to provide a secondary reliability measure of the GA's chosen solution.

Coupled with increased control over the iteration process of the GA, the user receives feedback about the GA's real-time performance. At regular intervals the user is updated on the GA's progress, and has the ability to modify solution acceptance parameters *during* the GA's search process. The GA

1990

is able to analyze its real-time performance and compare current performance to past performance values, which aids its estimation process.

## II. FUZZY GOALS AND CRISP GOALS

Problems to be solved may be considered to have two goals: fuzzy goals, and crisp goals. An example is the Traveling Salesperson Problem (TSP) problem., for which there exists a minimum cost path value (although it is not known beforehand). The goal of finding the "minimum cost path" is a crisp goal.

Suppose, however, one were to try and find an approximate solution that is close to the optimal solution. The goal of *this* problem would be a fuzzy goal, due to the vagueness of the term "approximate." This distinction is crucial: in order to solve the TSP problem and guarantee the optimal solution is found, an exhaustive search algorithm must be used. Since GAs do not guarantee finding the optimal solution, when a GA is used to "solve" the TSP problem, it is really solving for the fuzzy goal of "approximate minimum cost path."

Using traditional set theory, a crisp goal may be represented by a crisp set $A$, where:

$A = \{ goal\ state \}, and$

$A' = \{ all\ other\ non\text{-}goal\ states \}$

The mapping function for a crisp goal is thus given as:

$$\mu_A(u) = 1, \Leftrightarrow u \in A;$$
$$0, \Leftrightarrow u \notin A. \qquad (1)$$

A fuzzy goal may be represented as a fuzzy set, where the mapping function is generalized so membership values between (and including) 0 and 1 may be assigned:

$$\mu_A(u):U \to [0,1] \qquad (2)$$

In the context of the TSP problem, a value of "0" represents no membership in the goal state, and a "1" represents total membership in the goal state [7]. By using a GA, the crisp goal of "minimum cost path" or some percentage of the "minimum cost path" is transformed into a fuzzy goal, which is now stated as the "approximate minimum cost path." The user-defined performance value (an $\alpha$-cut) and the corresponding estimated optimal solution de-fuzzifies the fuzzy goal by finding a solution that is acceptable for the crisp goal TSP problem.

The new GA searched for the approximate minimum cost path for a 25-city TSP problem; the chromosome length was set at 25 for the problem. Two chromosomes were designed: one for city order, and the other for holding city distances. The initial populations' city order chromosomes were created at random.

To set the intelligent fuzzy stop criterion for the new GA, an $\alpha$-cut representing the acceptable % optimal solution is defined by the user. When the GA calculates the estimated optimal solution, the fuzzy goal of "approximate minimum cost path" into a crisp goal by using the $\alpha$-cut as the goal cutoff parameter. This $\alpha$-cut represents what "percent" of the optimal solution is an acceptable solution to the given problem.

Belief and Plausibility functions are used to provide the user with some type of confidence in the solution the GA has found. In this article, belief functions will be regarded as representing the evidence that the GA's estimate of the found solution value is reliable [6].

*MOSLOP* (Modified Support Logic Programming) techniques are employed when incorporating belief and uncertainty measures into the new GA [3]. The support pair in MOSLOP consists of a belief and uncertainty measure (*UM*), which can be determined by *Pl(C) - Bel(C)*.

## III. SOLUTION PREDICTION IN GENETIC ALGORITHMS

There are several reasons for incorporating solution prediction into GAs: first, it is difficult for a solution to be determined as "good" after a certain number of arbitrary generations if there is no method established for verifying this "goodness." Second, few engineers are content to blindly guess on setting a variable when little or nothing is known about the range of that variable. If the number of iterations is too low, the GA's final solution will virtually always be sub-optimal. If the number of iterations is too large, the GA wastes valuable computing time for no guaranteed solution improvement.

The new algorithm developed contains several features not found in conventional GAs: an adaptive filtering algorithm, a predicting algorithm, and a reliability checking algorithm.

The GA's learning function consisted of executing the GA a user-specified number of times, to create a "memory record" of how the GA solution process works. Values that are stored in this file are: the minimum solution value for each iteration, the average solution value for each iteration, belief and plausibility values for each iteration, and the order of the predicting equation for each 40-iteration block analyzed. A function called learn_predict() is responsible for determining how many coefficients are used in the GA's predicting equation by determining the number of iterations needed to calculate the smallest solution value in the memory data.

The predicting parts of the GA are found in several functions. The lmsap() function is a least-mean squared adaptive predictor that runs the GA's minimum fitness values for each iteration through an adaptive filter of up to order 40. As processing of the input signals take place, the lmsap() function "learns" the statistics of these inputs, allowing weights to converge to optimum settings [8]. The outputs of the lmsap() function are coefficients for the predicting function. Figure 2 illustrates zproto's predict() function.

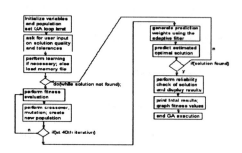

Figure 1. New GA Structure

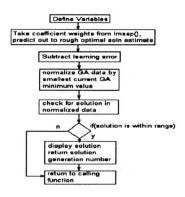

Figure 2. Zproto Predict() Function

The GA makes its initial optimal solution estimate based on information gathered from the current run; then memory data is compared to the current estimate, and the initial solution estimate is updated. All current GA solution points are normalized by this point. The normalized solution values are then analyzed to determine if any values are equal or greater than the $g_{f\alpha}$ cutoff value. The first value greater than or equal to $g_{f\alpha}$ is picked as the solution.

Once a solution has been chosen, a reliability estimate is performed. Seven functions are used to generate an overall belief and uncertainty value, six of whose outputs are combined by MOSLOP rules. The last function calculates relative error of the predicted optimal solution point. The belief value from the MOSLOP calculations and the relative error value yield a final belief and uncertainty measure, in an effort to blend the more subjective belief value with the more objective relative error value.

The three general "rules" these support pairs are based on are:

1. As the number of iterations increase, the minimum solution value improves; the approach of the optimal point indirectly indicates increasing reliability (belief) in performance estimation.

2. As iterations increase, the average solution value improves; again, this is an indication of improving belief in the performance estimate.

3. As iterations increase, the accuracy of calculating the

optimal solution point increases. This is based on the general GA rule which states: while a GA may not guarantee the optimal solution will be found, it does guarantee finding an equal or better solution than its initial solution.

IV. RESULTS

Using belief and uncertainty to test found solutions, the GA must be able to determine which solutions are acceptable, and which are not. There is always the possibility the GA will accept a solution that in reality does not meet user standards (false positives), as well as rejecting valid acceptable solutions (false negatives). Figure 3 shows the results of zproto, using a memory data file that contained a minimum solution value of 1711.

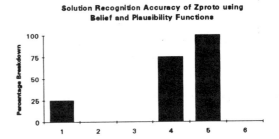

1 - % False Negative Solutions.  2 - % Actual Real Negative Solutions

3 - % False Positive Solutions.  4 - % Actual Real Positive Solutions.

5 - Total % acceptable solutions. 6 - Total % of unacceptable solutions.

Figure 3.  Solution Recognition of Zproto

The above chart was made from 20 separate runs of zproto. There were no false positive solutions recorded, which is very important in a critical application that **must** have the level of performance zproto says exists. In all false negative cases, it was the uncertainty value which was too high; in only one case was belief calculated lower than it should have been.

Figure 4 shows the performance of zproto using memory data vs zproto not using memory data. The $g_{f\alpha}$ value was set at 0.90, belief was set at 0.95 and uncertainty was set to 0.01. Population size was fixed at 80. The final average estimated optimal solution using memory data was 1697, an average of 0.81% off the actual optimal solution (the actual global optimal solution for the 25-city TSP problem programmed into zproto is 1711 [3]).

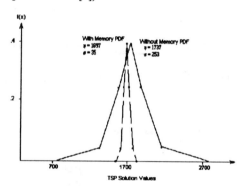

Figure 4. Estimated Optimal Solution with and without memory data

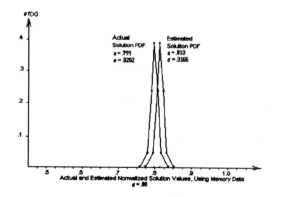

Figure 5. Actual and Estimated Solutions, using memory data, varying

population size

In Figure 5, while population size varied, $g_{f\alpha}$ was fixed at 0.8. In only six out of 41 trials, zproto's estimated optimal solution was higher than the actual solution, meaning zproto generally underestimated the solution it found. The average error between zproto's estimate of the found solution and the actual (real) solution value was 1.358%.

Figure 6 is comprised from trials using different memory data files and different $g_{f\alpha}$ values. Figure 6 illustrates zproto trials that had population size fixed at 80, belief set at 0.95 and uncertainty set at 0.005, and $g_{f\alpha}$ set at 0.9.

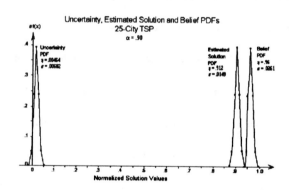

Figure 6. Uncertainty, Estimated Solutions, Belief PDFs

From Figure 6 it can be determined that the uncertainty of zproto in its solution estimations is minimal. The trial data shows zproto has a fairly high degree of belief in its estimated optimal solutions.

Computationally the new algorithm is faster than a conventional fixed-iteration GA. On an Intel 80486-based computer with a clock speed of 33 megahertz (MHz), 256K cache and 4 megabytes (MB) of main memory, and using a population size of 80, 400 iterations required 27.52 seconds of run time. The new algorithm (not including time spent creating a memory data file) averaged 176 generations (12.10 seconds), five iterations of the adaptive filter algorithm (1.10 seconds), five runs of the predicting algorithm (2.40 seconds), and one iteration of the reliability algorithm (0.01 seconds), a total time of 15.61 seconds.

## V. CONCLUSIONS

The concept of an intelligent fuzzy stop criterion has empirical data to support it as a viable, alternative method to the traditional GA fixed-iteration method. The new algorithm provides a higher level of user-GA interaction, allowing the user to request and expect certain levels of performance and reliability; at the same time this method saves the GA time it might needlessly search for the optimal solution.

At first it was believed that an estimation algorithm could be developed without relying on any type of memory data and memory schemes. It was shortly realized thereafter that such a method would never work as well as a method that incorporates empirical memory data into its estimation process. Several sources support this belief: first, natural individuals rely on past information in their prediction processes; two, neural networks, the other artificial intelligent type of program discussed, rely on past values for predicting future results [5]; three, the solution performance of zproto when memory was used in prediction and when memory was not used in prediction.

Several major ideas that will be investigating in future work are: developing adaptive learning and analysis methods the GA builds and updates itself; applying the concept of an intelligent fuzzy stop criterion and estimation algorithms to other problems than TSP problems; and devising a scaling mechanism that would allow for the use of a single memory data file on multiple problems.

## BIBLIOGRAPHY

[1]  G. Brassard and P. Bratley, *Algorithmics: Theory and Practice*, Prentice-Hall, New Jersey, 1988.

[2]  L. Davis, *Handbook of Genetic Algorithms*, Van Nostrand Reinhold, New York, 1991.

[3]  X. Feng and R. Tang, "Uncertainty Measure in Support Logic Programming for Uncertainty Handling in Expert Systems," *IEEE Transactions on Fuzzy Systems*, submitted for publication.

[4]  D. Goldberg, *Genetic Algorithms in Search, Optimization and Machine Learning*, Addison Wesley, 1989.

[5]  S. Grossberg, *Neural Networks and Natural Intelligence*, MIT Press, Cambridge, MA, 1988.

[6]  J. Halpern and R. Fagin, "Two Views of Belief: Belief as Generalized Probability and Belief as Evidence," *Artificial Intelligence*, vol. 54, pages 275-317, 1992.

[7]  G. Klir and T. Folger, *Fuzzy Sets, Uncertainty, and Information*, Prentice Hall, 1988.

[8]  S. Orfanidis, *Optimum Signal Processing*, MacMillan, New York, 1985.

# Fuzzy Vector Quantization Algorithms

Nicolaos B. Karayiannis and Pin-I Pai

Department of Electrical Engineering
University of Houston
Houston, Texas 77204-4793, USA

**Abstract:** This paper presents the development of efficient algorithms employing fuzzy logic for codebook design. These algorithms achieve the quality of vector quantizers provided by computationally demanding approaches, while capturing the advantages of the $k$-means algorithm, such as speed, simplicity, and conceptual appeal. The development of these algorithms is based on effective strategies for the transition from soft to crisp decisions during the clustering process. The uncertainty associated with training vector assignment is quantitatively measured by various families of membership functions, including those used in fuzzy $k$-means algorithms. The application of the proposed algorithms in image compression based on vector quantization provides the basis for evaluating their computational efficiency and comparing the quality of the resulting codebook design with that provided by competing techniques.

## 1  Introduction

Vector quantization methods attempt to subdivide a random set of vectors into subsets, or clusters, which are pairwise disjoint, all nonempty, and reproduce the original set via union. The available vectors, also called training vectors, are divided into clusters on the basis of some optimality criterion, while the cluster centers provide the codebook vectors. In the context of vector quantization, the clustering process is also referred to as the codebook design. Codebook design is usually performed by an iterative algorithm which is known as *k-means* (or *c-means*) algorithm [1]. A variation of the $k$-means algorithm is known in the engineering literature as the LBG (Linde, Buzo, Gray) algorithm [1], [2]. Although the $k$-means algorithm is simple and intuitively appealing, it strongly depends on the selection of the initial codebook, and it can easily be trapped in local minima [1]. The selection of a good initial codebook was attempted by random code, splitting, and pairwise nearest neighbor techniques [1], [2]. However, there seems to be no simple solution to the local minima problem. For a given initial codebook, the algorithm finds the nearest local minimum in the space of all possible codebooks.

Crisp vector quantization techniques assign each training vector to a single cluster and ignore the possibility that this training vector may also belong to other clusters. Based on the concept of fuzzy sets introduced by Zadeh [3], Ruspini developed the first fuzzy clustering algorithm [4]. Dunn considered an alternative formulation of the clustering process and proposed the *fuzzy k-means* algorithm [5]. Bezdek extended Dunn's formulation and produced a family of fuzzy $k$-means algorithms, which includes Dunn's original algorithm as a special case [6]. Fuzzy clustering algorithms consider each cluster as a fuzzy set, while a membership function measures the possibility that each training vector belongs to a cluster. This paper presents efficient algorithms for vector quantizer design, which exploit the advantages offered by fuzzy clustering algorithms while satisfying the requirements imposed by vector quantization.

## 2  Vector Quantization

A vector quantizer $Q$ of dimension $n$ and size $k$ is defined as a mapping from a vector in $\mathcal{R}^n$, into a finite set $\mathcal{Y}$ containing $k$ vectors, called *codebook vectors*. The mapping can be stated as $Q : \mathcal{R}^n \to \mathcal{Y}$, where $\mathcal{Y} = \{\mathbf{y}_1, \mathbf{y}_2, \ldots, \mathbf{y}_k\}$ and $\mathbf{y}_j \in \mathcal{R}^n \; \forall j = 1, 2, \ldots, k$. The set $\mathcal{Y}$ is called the *codebook* of size $k$. The design of a vector quantizer requires a set of *training vectors*, called the *training set*. A training set $\mathcal{X}$ is formed by $M$ training vectors, that is, $\mathcal{X} = \{\mathbf{x}_1, \mathbf{x}_2, \ldots, \mathbf{x}_M\}$, $\mathbf{x}_i \in \mathcal{R}^n \; \forall i = 1, 2, \ldots, M$. A vector quantizer is designed by assigning the $M$ training vectors to $k$ clusters, each represented by a codebook vector. The quality of the codebook design is frequently measured by the following average distortion

$$D = \tfrac{1}{M} \sum_{i=1}^{M} d_{min}(\mathbf{x}_i) = \tfrac{1}{M} \sum_{i=1}^{M} \min_{\mathbf{y}_j \in \mathcal{Y}} d(\mathbf{x}_i, \mathbf{y}_j) \qquad (1)$$

The $k$-means algorithm assigns each training vector to a certain cluster on the basis of *hard* or *crisp* decisions. According to the nearest neighbor condition employed, the training vector $\mathbf{x}_i$ is assigned to the $j$th cluster if $d(\mathbf{x}_i, \mathbf{y}_j) = d_{min}(\mathbf{x}_i) = \min_{\mathbf{y}_j \in \mathcal{Y}} d(\mathbf{x}_i, \mathbf{y}_j)$, where $d(\mathbf{x}_i, \mathbf{y}_j) = \|\mathbf{x}_i - \mathbf{y}_j\|^2$. The nearest neighbor condition can be conveniently described by a *membership function*, which is defined as $u_j(\mathbf{x}_i) = 1$ if $d(\mathbf{x}_i, \mathbf{y}_j) = d_{min}(\mathbf{x}_i)$, and $u_j(\mathbf{x}_i) = 0$ if $d(\mathbf{x}_i, \mathbf{y}_j) \neq d_{min}(\mathbf{x}_i)$. The codebook vectors are evaluated by minimizing a distortion measure defined as

$$J_1 = \sum_{j=1}^{k} \sum_{i=1}^{M} u_j(\mathbf{x}_i)\|\mathbf{x}_i - \mathbf{y}_j\|^2 \tag{2}$$

The minimization of $J_1 = J_1(\mathbf{y}_j, j = 1, 2, \ldots, k)$ with respect to $\mathbf{y}_j$ results in

$$\mathbf{y}_j = \sum_{i=1}^{M} \bar{u}_j(\mathbf{x}_i)\, \mathbf{x}_i \quad \forall\, j = 1, 2, \ldots, k. \tag{3}$$

where $\bar{u}_j(\mathbf{x}_i) = u_j(\mathbf{x}_i)/\sum_{i=1}^{M} u_j(\mathbf{x}_i)$. The codebook vector $\mathbf{y}_j$ defined in (3) is the Euclidean center of gravity or centroid of all the training vectors assigned to the $j$th cluster.

Fuzzy $k$-means algorithms assign to each training vector a membership value between zero and one which indicates to what extent the particular vector belongs to a certain cluster. The derivation of the fuzzy $k$-means algorithms was based on the constrained minimization of [6]

$$J_m = \sum_{j=1}^{k} \sum_{i=1}^{M} u_j(\mathbf{x}_i)^m \|\mathbf{x}_i - \mathbf{y}_j\|^2 \tag{4}$$

where $1 < m < \infty$. For a given set of codebook vectors, the minimization of $J_m = J_m(u_j, j = 1, 2, \ldots, k)$ under the constraint $\sum_{j=1}^{k} u_j(\mathbf{x}_i) = 1 \;\; \forall\, i = 1, 2, \ldots, M$ results in [6]

$$u_j(\mathbf{x}_i)^{-1} = \sum_{\ell=1}^{k} \left(d(\mathbf{x}_i, \mathbf{y}_j)/d(\mathbf{x}_i, \mathbf{y}_\ell)\right)^{\frac{1}{m-1}} \tag{5}$$

where $d(\mathbf{x}_i, \mathbf{y}_j) = \|\mathbf{x}_i - \mathbf{y}_j\|^2$. For a given set of membership functions, the codebook vectors can be evaluated by minimizing $J_m = J_m(\mathbf{y}_j, j = 1, 2, \ldots, k)$ as follows [6]

$$\mathbf{y}_j = \sum_{i=1}^{M} \bar{u}_j(\mathbf{x}_i)^m\, \mathbf{x}_i \quad \forall\, j = 1, 2, \ldots, k. \tag{6}$$

where $\bar{u}_j(\mathbf{x}_i)^m = u_j(\mathbf{x}_i)^m/\sum_{i=1}^{M} u_j(\mathbf{x}_i)^m$. The "fuzziness" of the clustering produced by these algorithms is controlled by the parameter $m$ [6]. As this parameter approaches unity, the partition of the training vector space is a nearly crisp decision-making process. Increasing this parameter tends to degrade membership toward the fuzziest state [6]. To date, there are no reliable criteria for the selection of the optimal parameter for a given set of training vectors.

## 3 A Strategy for Vector Assignment

The strategy proposed in this paper for the assignment of training vectors to various clusters is based on the combination of fuzzy logic with the concept of topological neighborhood used in the ordering of self-organizing feature maps [7], [8].

### 3.1 Transition from Soft to Hard Decisions

Let $\mathcal{I}_i^{(\nu)}$ be the set of the codebook vectors that belong to the hypersphere centered at the training vector $\mathbf{x}_i \in \mathcal{X}$ during the $\nu$th iteration. The training vector $\mathbf{x}_i$ can only be assigned to those clusters whose centers belong to the set $\mathcal{I}_i^{(\nu)}$. When the clustering process begins, each training vector $\mathbf{x}_i$ can be assigned to every cluster. Hence, the corresponding hypersphere includes all the cluster centers, i.e., $\mathcal{I}_i^{(0)} = \mathcal{Y}$. The existence of overlapping hyperspheres centered at the training vectors guarantees the participation of all the training vectors in the formation of the new set of codebooks, therefore reducing the effect of the initial set of codebooks on the codebook design process. The hyperspheres centered at the training vectors are gradually shrinking during the codebook design process, in order to improve the resolution of clustering. When the set of codebooks included in a hypersphere centered at a certain training vector is empty, the training vector is assigned to the cluster with the closest center. The gradual transition from the fuzzy to the crisp mode during the clustering process is based on the following strategy: After the $\nu$th iteration, the hypersphere located at the training vector $\mathbf{x}_i$ includes the vectors $\mathbf{y}_j \in \mathcal{I}_i^{(\nu)}$ whose distance from $\mathbf{x}_i$ is less than or equal to the average of the distances between $\mathbf{x}_i$ and $\mathbf{y}_j \in \mathcal{I}_i^{(\nu)}$, defined by

$$d_{ave}(\mathbf{x}_i) = \mathcal{N}(\mathcal{I}_i^{(\nu)})^{-1} \sum_{\mathbf{y}_j \in \mathcal{I}_i^{(\nu)}} d(\mathbf{x}_i, \mathbf{y}_j) \tag{7}$$

where $\mathcal{N}(\mathcal{I}_i^{(\nu)})$ is the total number of elements in the set $\mathcal{I}_i^{(\nu)}$. The set $\mathcal{I}_i^{(\nu+1)}$ is formed in terms of $\mathcal{I}_i^{(\nu)}$ as $\mathcal{I}_i^{(\nu+1)} = \{\mathbf{y}_j \in \mathcal{I}_i^{(\nu)} : d(\mathbf{x}_i, \mathbf{y}_j) \leq d_{ave}(\mathbf{x}_i)\}$. If $\mathcal{I}_i^{(\nu)}$ is an empty set or contains a single codebook vector, i.e., $\mathcal{N}(\mathcal{I}_i^{(\nu)}) < 2$, the training vector $\mathbf{x}_i$ is transferred from the fuzzy to the crisp mode. In this case, the radius of the hypersphere centered at $\mathbf{x}_i$ reduces to zero.

## 3.2 The Membership Function

The certainty of the assignment of the training vector $\mathbf{x}_i$ to the $j$th class is measured by the membership function $u_j(\mathbf{x}_i)$, which can take values between 0 and 1. The evaluation of $u_j(\mathbf{x}_i), j = 1, 2, \ldots, k$, depends on the number of cluster centers included in the set $\mathcal{I}_i^{(\nu)}$ during the $\nu$th iteration.

If the set $\mathcal{I}_i^{(\nu)}$ is empty, the assignment of $\mathbf{x}_i$ is based on the nearest neighbor condition. In this case, $u_j(\mathbf{x}_i) = 1$ if $d(\mathbf{x}_i, \mathbf{y}_j) = d_{min}(\mathbf{x}_i)$ and $u_j(\mathbf{x}_i) = 0$ if $d(\mathbf{x}_i, \mathbf{y}_j) \neq d_{min}(\mathbf{x}_i)$, where $d_{min}(\mathbf{x}_i) = \min_{\mathbf{y}_j \in \mathcal{I}_i^{(\nu)}} d(\mathbf{x}_i, \mathbf{y}_j)$. After the $\nu$th iteration, $\mathbf{x}_i$ is transferred from the fuzzy mode to the crisp mode, i.e., $\mathbf{x}_i$ is assigned to a single cluster on the basis of the nearest neighbor condition.

If the hypersphere centered at the training vector $\mathbf{x}_i$ includes a single codebook vector, say $\mathbf{y}_{j^*}$, $\mathbf{x}_i$ is assigned to the $j^*$th cluster with value of the membership function equal to one, i.e., $u_j(\mathbf{x}_i) = 1$ if $d(\mathbf{x}_i, \mathbf{y}_j) = d(\mathbf{x}_i, \mathbf{y}_{j^*})$ and $u_j(\mathbf{x}_i) = 0$ if $d(\mathbf{x}_i, \mathbf{y}_j) \neq d(\mathbf{x}_i, \mathbf{y}_{j^*})$. After this iteration, $\mathbf{x}_i$ is transferred from the fuzzy mode to the crisp mode.

If $\mathcal{I}_i^{(\nu)}$ is a nonempty set including more than one codebook vectors, the membership function $u_j(\mathbf{x}_i)$ takes values between zero and one. If $\mathbf{y}_j \notin \mathcal{I}_i^{(\nu)}$, the membership function $u_j(\mathbf{x}_i)$ is zero, i.e., $u_j(\mathbf{x}_i) = 0$. If $\mathbf{y}_j \in \mathcal{I}_i^{(\nu)}$, the membership function $u_j(\mathbf{x}_i)$ measures the certainty of the inclusion of the training vector $\mathbf{x}_i$ into the $j$th cluster. In this case, the membership function depends on the distances between $\mathbf{x}_i$ and $\mathbf{y}_j \in \mathcal{I}_i^{(\nu)}$, that is, $u_j(\mathbf{x}_i) = f(d(\mathbf{x}_i, \mathbf{y}_j), \mathbf{y}_j \in \mathcal{I}_i^{(\nu)})$. The closer the training vector $\mathbf{x}_i$ is to the codebook vector $\mathbf{y}_j$, the higher is the certainty that $\mathbf{x}_i$ belongs to the $j$th cluster. These intuitively reasonable requirements suggest the selection of any membership function $u_j(\mathbf{x}_i)$ which satisfies the following requirements: (i) $u_j(\mathbf{x}_i)$ is a decreasing function of the distance $d(\mathbf{x}_i, \mathbf{y}_j)$, (ii) $u_j(\mathbf{x}_i)$ approaches unity as $d(\mathbf{x}_i, \mathbf{y}_j)$ approaches zero, and (iii) $u_j(\mathbf{x}_i)$ approaches zero if $d(\mathbf{x}_i, \mathbf{y}_j) = d_{max}(\mathbf{x}_i)$, where $d_{max}(\mathbf{x}_i) = \max_{\mathbf{y}_j \in \mathcal{I}_i^{(\nu)}} d(\mathbf{x}_i, \mathbf{y}_j)$.

At the last stages of the transition from the fuzzy to the crisp mode the decrease of the distortion measure (1) becomes very slow. In this phase, the rate at which the total distortion decreases is tested after every iteration. If this rate decreases below a certain threshold after a certain iteration all the training vectors are transferred to the crisp mode. The partition of the training vector space is based in subsequent iterations on the nearest neighbor condition.

# 4 Fuzzy Vector Quantization Algorithms

Given a strategy for the transition from soft to crisp decisions during the codebook design process, the development of specific algorithms depends on the selection of the membership function and the evaluation of the codebook vectors at each iteration [9].

## 4.1 Fuzzy Vector Quantization 1

This algorithm was developed by constructing a family of membership functions satisfying the conditions proposed in section 3. According to these conditions, $u_j(\mathbf{x}_i)$ can be a function of both $d(\mathbf{x}_i, \mathbf{y}_j)$ and $d_{max}(\mathbf{x}_i)$, that is, $u_j(\mathbf{x}_i) = f(d(\mathbf{x}_i, \mathbf{y}_j), d_{max}(\mathbf{x}_i))$. A family of membership functions can be formed by observing that $d(\mathbf{x}_i, \mathbf{y}_j)/d_{max}(\mathbf{x}_i) = 0$ if $d(\mathbf{x}_i, \mathbf{y}_j) = 0$ and $d(\mathbf{x}_i, \mathbf{y}_j)/d_{max}(\mathbf{x}_i) = 1$ if $d(\mathbf{x}_i, \mathbf{y}_j) = d_{max}(\mathbf{x}_i)$. Since $d(\mathbf{x}_i, \mathbf{y}_j)/d_{max}(\mathbf{x}_i)$ is an increasing function of the distance $d(\mathbf{x}_i, \mathbf{y}_j)$, the membership function $u_j(\mathbf{x}_i)$ can be of the form

$$u_j(\mathbf{x}_i) = f(d(\mathbf{x}_i, \mathbf{y}_j), d_{max}(\mathbf{x}_i)) = (1 - d(\mathbf{x}_i, \mathbf{y}_j)/d_{max}(\mathbf{x}_i))^\mu \tag{8}$$

where $\mu$ is a positive integer. The evaluation of the codebook vectors must be consistent with the fact that vector assignment is based on crisp decisions toward the end of the vector quantizer design. This can be guaranteed by the minimization with respect to $\mathbf{y}_j$ of the discrepancy measure $J_1 = J_1(\mathbf{y}_j, j = 1, 2, \ldots, k)$, defined in (2), which results in the formula (3). The direct implication of this selection is that the proposed algorithm reduces to the crisp $k$-means algorithm after all the training vectors have been transferred from the fuzzy to the crisp mode. The combination of the above formulas with the vector assignment strategy proposed in section 3 results in the Fuzzy Vector Quantization 1 (FVQ 1) algorithm [9].

### 4.2 Fuzzy Vector Quantization 2

This algorithm is based on the certainty measures used for training vector assignment by the family of fuzzy $k$-means algorithms. The feasibility of this choice was tested by verifying that the membership function (5) satisfies the conditions proposed in section 3 [9]. The codebook vectors can be evaluated in this case by (6), resulting from the minimization of $J_m = J_m(\mathbf{y}_j, j = 1, 2, \ldots, k)$ and also used in fuzzy $k$-means algorithms. Since the assignment of the training vector is based on hard decisions toward the end of the vector quantizer design, the corresponding membership function $u_j(\mathbf{x}_i)$ takes the values 0 and 1. In this case, $u_j(\mathbf{x}_i)^m = u_j(\mathbf{x}_i)$ regardless of the value of $m$. Therefore, the codebook vectors are evaluated by the same formula used in the crisp $k$-means algorithm toward the end of the vector quantizer design. The combination of the formulas used in fuzzy $k$-means algorithms for evaluating the membership functions and the cluster centers with the vector assignment strategy proposed in section 3 results in the Fuzzy Vector Quantization 2 (FVQ 2) algorithm [9].

### 4.3 Fuzzy Vector Quantization 3

Following the approach that provided the fuzzy $k$-means algorithm, the membership functions result from the constrained minimization of $J_m = J_m(u_j, j = 1, 2, \ldots, k)$. This minimization results in the membership function given in (5). According to the formulation of the crisp $k$-means algorithm, the codebook vectors can be determined by minimizing the objective function $J_1 = J_1(\mathbf{y}_j, j = 1, 2, \ldots, k)$. This minimization results in the codebook vectors given in (3). The Fuzzy Vector Quantization 3 (FVQ 3) algorithm is obtained by evaluating the membership function by (5) and the cluster centers by (3), and assigning training vectors on the basis of the strategy proposed in section 3. Since the codebook vectors are evaluated by (3), the proposed algorithm reduces to the crisp $k$-means algorithm toward the end of the codebook design [9]. The FVQ 3 is expected to perform well when $m$ approaches unity, since in this case $J_m$ approaches asymptotically $J_1$. The computational burden associated with the evaluation of the membership functions can be moderated by requiring that $\frac{1}{m-1} = \lambda$, where $\lambda$ is a positive integer.

## 5 Experimental Results

The application of vector quantization to images is based on the decomposition of one or more sampled images into rectangular blocks which form vectors of fixed size. The available training vectors are divided into clusters, while the cluster centers provide the codebook vectors. Given a codebook, each block of the image can be represented by the binary address of its closest codebook vector. The image is reconstructed by replacing each image block by its closest codebook.

The $256 \times 256$ Lenna image was divided into $4 \times 4$ blocks and the resulting 4096 vectors were used as training vectors for the design of a codebook of size $2^7 = 128$. In this experiment, the quality of the designed codebook was measured by the normalized distortion $\bar{D}^{(\nu)} = D^{(\nu)}/D^{(0)}$, where $D^{(0)}$ is the initial distortion and $D^{(\nu)}$ is the distortion after the completion of the $\nu$th iteration. The codebook design was performed by the proposed FVQ algorithms, the $k$-means and fuzzy $k$-means algorithms. In these experiments, $m$ was evaluated as $m = 1 + 1/\lambda$, where $\lambda$ was taking all integer values between 1 and 5. The quality of the codebook provided by fuzzy $k$-means algorithms depends strongly on the proximity of this parameter to unity, with the lowest distortion corresponding to $m = 1.2$. When the FVQ 1 algorithm was used to design this codebook, the membership function was evaluated according to the formula given in (8) with $\mu = 1, 2, 3, 4, 5$. On the average, the value of $\mu = 2$ led to better codebooks. Regardless of the value of $\mu$, the FVQ 1 algorithm results in codebooks superior to those provided by the family of fuzzy $k$-means algorithms. When the same codebook design was performed by the FVQ 2 and FVQ 3 algorithms, the quality of the resulting codebook was not significantly affected by the parameter $m$. These two algorithms

1999

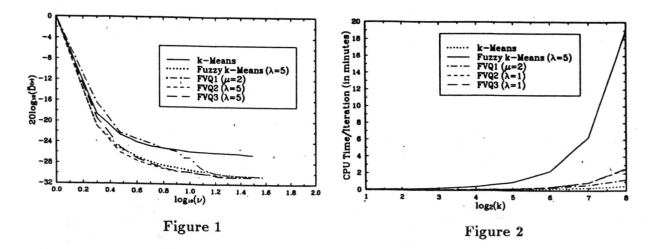

Figure 1                                                                 Figure 2

achieve the best codebook design among all the algorithms tested in this experiment, with the FVQ 3 algorithm resulting in a codebook slightly better than that provided by the FVQ 2 algorithm. The results of this set of experiments are summarized in Figure 1, which shows $20 \log_{10}(\bar{D}^{(\nu)})$ as a function of $\log_{10}(\nu)$ when the same codebook design is performed by the proposed algorithms and the $k$-means algorithms, both crisp and fuzzy.

Figure 2 shows the average CPU time per iteration in minutes required by various algorithms for the design of codebooks of size $k = 2^n, n = 1, 2, \ldots, 8$ as a function of $n = \log_2(k)$. According to Figure 2, the application of fuzzy $k$-means algorithms in the codebook design for vector quantization of images is computationally very demanding. The time required for each iteration of the fuzzy $k$-means algorithm grows exponentially with the number of bits required for representing a $4 \times 4$ image block. Because of their structure, the proposed algorithms are computationally less demanding than the fuzzy $k$-means algorithms. In fact, the FVQ algorithms are closer to the crisp $k$-means algorithm from a computational point of view.

Figure 3 shows the Lenna image of size $256 \times 256$, which was used as the original image in this experiment. The pixels of this image take values between 0 and 255, i.e., the representation of each pixel requires 8 bits. The proposed algorithms and the $k$-means algorithms were used to design codebooks consisting of 256 vectors, which implies the representation of each image block containing $4 \times 4 = 16$ pixels by 8 bits. In this experiment the compression rate was $8/16 = 0.5 \, bits \, per \, pixel$ ($bpp$). The resulting images were evaluated subjectively by the Peak Signal to Noise Ratio (PSNR). Figures 4 and 5 show the images reconstructed from the codebook designed by the $k$-means algorithm and the fuzzy $k$-means algorithm with $\lambda = 5$, respectively. The PSNR achieved by the $k$-means and the fuzzy $k$-means algorithms was 27.16 dB and 29.91 dB, respectively. The PSNR achieved by the FVQ 1, FVQ 2, and FVQ 3 algorithms was 29.67 dB, 29.90 dB, and 30.01 dB, respectively. Figure 6 shows the image reconstructed from the codebook designed by the FVQ 3 algorithm with $\lambda = 1$.

## 6    Conclusions

This paper presented a methodology for incorporating fuzzy logic in vector quantization design. The development of the algorithms was based on a flexible strategy allowing the transition from soft to crisp decisions as the clustering process evolves, thus eliminating the effect of the initial codebook selection on the quality of clustering and also avoiding a priori assumptions regarding the level of fuzziness necessary for a given clustering task. The proposed algorithms were mainly used to perform codebook design in image compression based on vector quantization. However, the range of potential real-world applications of such algorithms is very wide, essentially including any problem involving clustering of random data.

## References

[1]  A. Gersho and R. M. Gray, *Vector Quantization and Signal Compression*, Kluwer Academic Publishers, Boston, MA., 1992.

**Figure 3**

**Figure 4**

**Figure 5**

**Figure 6**

[2] Y. Linde, A. Buzo, and R. M. Gray, "An Algorithm for Vector Quantizer Design", *IEEE Transactions on Communications*, Vol. 28, pp. 84-95, 1980.

[3] L. A. Zadeh, "Fuzzy Sets", *Information and Control*, Vol. 8, pp. 338-353, 1965.

[4] E. H. Ruspini, "A New Approach to Clustering", *Information and Control*, Vol. 15, pp. 22-32, 1969.

[5] J. C. Dunn, "A Fuzzy Relative of the ISODATA Process and its use in Detecting Compact Well-Separated Clusters", *Journal of Cybernetics*, Vol. 3, No. 3, pp. 32-57, 1973.

[6] J. C. Bezdek, R. Ehrlich, and W. Full, "FCM: The Fuzzy c-Means Clustering Algorithm", *Computers and Geosciences*, Vol. 10, No. 2-3, pp. 191-203, 1984.

[7] T. Kohonen, "The Self-Organizing Map", *Proceedings of the IEEE*, Vol. 78, pp. 1464-1480, 1990.

[8] N. B. Karayiannis and A. N. Venetsanopoulos, *Artificial Neural Networks: Learning Algorithms, Performance Evaluation, and Applications*, Kluwer Academic Publishers, Boston, MA, 1993.

[9] N. B. Karayiannis and P.-I Pai, "Fuzzy Vector Quantization Algorithms and Their Application in Image Compression," *Systems, Neural Nets, and Computing Technical Report No. 93-01*, University of Houston, March 1993.

# Extraction of Straight Line Segments using Perceptual Organization and Fuzzy Thresholding Method

In-Cheon Lee and Kyung-Whan Oh

Artificial Intelligence Lab,    Department of Computer Science,    Sogang University,
Seoul 121-742,    KOREA
e-mail : incheon@ailab1.sogang.ac.kr

**Abstract** - Straight-line segments are often used as key features in analyzing scenes. In this paper, we extract straight-line segments using advanced Paton's template masks for straight-line segments detection in real time. To overcome a problem of a complexity of scenes and simplify straight-line segments, straight-line segments are grouped into longer straight-line segments by collinearity of perceptual organization. After the existance of undetected-line segments by parallelism and symmetry of perceptual organization is supposed, undetected-line segments are extracted by applying a new threshold again. Here, feedback control can be seen. Data are integrated by bottom-up and top-down method. To control thresholds in a structural manner, a fuzzy logic controller is used. By extracting straight-line segments of natural scenes and man-made objects, the effectiveness and the advantage of this method are shown. This proposed approach can be applied to a pattern recognition.

## I. Introduction

Straight-line segments are often used as key features in analyzing scenes. In computer vision, extracting straight-line segments in real time and exactly is important. Therefore, many methods for extracting straight-line segments have been studied[1-6,11].

In this paper, Paton's template masks are used in order to reduce a processing time. To detect two end points, edge-like scanning method is used. A start point and an end point during the scanning process are recorded as two end points. In contrast of the Paton's method, to extract line segments of every orientation, collinearity of perceptual organization is used. After extracting line segments over some lengths, these line segments are grouped recursively. Threshold have a great influence on analyzing scenes in computer vision. In perceptual grouping, many thresholds are needed. In order to solve a threshold problem, feedback control is needed. Scenes contain many man-made objects. Generally, this objects are composed of parallelism and symmetry. When straight-line segments of scenes are extracted, they may not be extracted or may be extracted partially because of lighting condition and thresholds. To solve this problem, parallelism and symmetry of perceptual organization are used[7-9]. After supposing the expansion of a less detected straight line segment in a pair of parallel or symmetric line segment, template masks are reused with a lower threshold. Here, feedback control can be seen. To control thresholds in a structural and systematic manner, a fuzzy logic controller is used[10,12]. The outputs of this method are straight line segments which are more structured and grouped.

The results of extracting straight line segments about man-made objects and natural scenes are shown. Also, this approach can be applied to a pattern recognition. In applying to pattern recognition, extended perceptual organization relations can be used.

## II. Perceptual organization and its conditions

In humans, perceptual organization is the ability to readily group elements in an image based on various relationships between them. Here, perceptual organization is implemented as a mechanism to exploit geometrical regularities in the shapes of objects as projected onto images. Straight line segments are grouped by next conditions.

1) Collinearity - Similarity, Proximity, Continuation

2) Parallelism - Similarity, Symmetry, Familiarity

3) Symmetry - Similarity, Symmetry

4) Cotermination - Proximity, Closure

5) Polygon - Proximity, Closure

## III. Straight line extraction using template masks and collinearity

This chapter explains merits and demerits of the Paton's straight line extraction method and the modified straight line extraction method. And the collinearity of perceptual organization is combined with the modified straight line extraction method in order to make scene informations further intergrated.

### 3.1 Paton's straight line extraction method

After multiplying left and right gray levels which are two pixels distant to a middle line by these template masks, if the total value is over threshold, the middle line is recorded as a straight line segment.

### 3.2 Defects of Paton's method and their solutions

Defects of Paton's method and their solutions are as follows.

1) The length of a line is subordinate to template size. To solve this, if a calculated total template vaule is over threshold through the scanning of a scene ( like Sobel's edge detection method) , the point is recorded as a point of a line. After scanning continuously to the same direction, if the length of a line is over threshold, the line is recorded as a straight line segment.

2) Paton's method extracts lines of only eight orientations. To solve this, after finding somewhat short lines, these lines are grouped into a longer line segment through collinearity of perceptual organization. This modified method extracts lines of all directions.

3) By a lighting condition and a template size, endpoints of a juntion may not meet each other. To solve this, we use the cotemination of perceptual organization.

4) There is a posibility that false lines and undetected lines exist. This happens by a lighting condition, a template size, and a template threshold. To solve this, we use a fuzzy thresholding method. This method is explained in the next chapter.

### 3.3 The extraction of straight line segments through collinearity

Collinearity in perceptual organization is a condition which short lines are grouped into a longer line by similarity, proximity, and continuation. To be grouped by collinearity, the next constraints should be satisfied. <Figure 1> represents examples of grouping.

1) The orientation difference between two lines should be below some threshold.

2) The vertical distance between two lines should be below some threshold.

3) The distance between the two endpoints which are the nearest end points of two lines should be below some threshold.

When collinearity is applied, the order of applied angle is important. It happens that the following grouping process depends on the result of the previous grouping processes. Therefore, the adaptation of a threshold plays a important part in the exactness of grouping. Because general scenes mainly contain vertical and horizontal line segments, vertical and horizontal grouping processes preceed other angle grouping processes. Merits of the method using modified template masks and collinearity are time effectiveness, noise tolerance, and abundant line segment extraction. Noise tolerance results from template size. Because the size of template mask is big, noise have little influence on a line segment extraction. And abundant line segments can be extracted, for template masks can be applied to a broader range of a scene. Noise and local change have little influence on extracting a line, too. But because template masks move in a scene through scanning, duplicate line segments may be extracted. Duplicate line segments have a similar orientation. Therefore, duplicate line segments can be grouped by collinearity.

### 3.4 Maximum value orientation expansion method(MVOEM)

Generally, scenes contain roof edges as well as step edges. We have difficulty in a roof edge detection. In a roof edge, though gray level changes slowly, template mask magnitude of that orientation is maximum when compared to other orientation template mask magnitudes. Though a short line is not over threshold, it has a possibility that can be expanded to that orientation. Therefore, a line can be expanded continuously. It makes line segmnts extracted exactly and abundantly. But in natural scenes, foliage and woods may be extracted complicatedly.

## IV. Straight-line extraction using perceptual organization and fuzzy thresholding method(FTM)

To extract more line segments, after the expansion of a undetected line segment is supposed through perceptual organization, a new threshold which is lower is reapplied. In order to control structurely, we use a simple fuzzy logic controller.

### 4.1 Application of perceptual organization

Subjected to a lighting condition and other conditions, undetected line segments may exist. In this case, after the expansion of a undetected line segment is supposed through perceptual organization, a new threshold which is lower is reapplied. Therefore, we can get expanded line segments. Here, we use parallelism and symmetry as conditions of perceptual organization.

1. Parallelism is the characteristic which is mainly not in natural scene but in man-made objects. Two lines which are parallel have similarity and symmetry. If following conditions are satisfied, two lines are parallel. <Figure 2> shows constraints of parallelism.

1) Two line segments have similar orientation.

$$O \text{ (Orientation)} = | \theta_2 - \theta_1 |$$

2) The vertical length between two lines should be within some threshold.

H (Height)

3) The ratio between the length of a short line and the length of a longer line should be above some threshold.

LR (Length Ratio)

4) The ratio between the length of a duplicate part of a short line to a longer line by X and Y axis projection and the length of a longer line should be above some threshold.

VLR (Vertical Length Ratio), XAPR (X Axis Projection Ratio)

YAPR (Y Axis Projection Ratio)

5) The difference between a first line template total mean and a second line template total mean should be below some threshold. It is applied to objects having a familiarity.

D (Difference) = FTTM (First Template Total Mean)

\- STTM (Second Template Total Mean)

Examples which can become parallel lines are given in <Figure 3>.

2. Symmetry has similarity and symmetry. If two lines are parallel, they are symmetric. To be symmetric, except for a orientation condition, other conditions have the same ones as parallelism.

$$180 - \alpha \ <= \ a1 + a2 \ <= \ 180 + \alpha \qquad \alpha : \text{threshold}$$

<Figure 3> shows lines which are symmetric, where a short line was detected with less information owing to lighting condition and other conditions. Then, we can suppose that short line is not completely detected. Therefore, to expand a short line, template masks are reapplied to short line segment.

### 4.2 Fuzzy thresholding method(FTM)

After the expansion of undetected line is supposed, a fuzzy logic controller(FLC) is used in order to adjust a threshold. Merits of a FLC are flexibility and structural representation of rules.

1) Parallelism - Fuzzy membership functions for parallel constraints are shown in <Figure 4>. This functions are made by heuristic experiments. To reduce a processing time, simple fuzzy membership functions are used.

Also, to reduce a processing time, simple max and min operators are used for inference.

$$P = \min \{ O, H, LR, \max \{ VLR, XAPR, YAPR \}, D \}$$

Inference result is transformed into defuzzified threshold value through the next function. This function is calculated through a fuzzy membership function, too.

$$U = \mu^{-1} ( P )$$

For identification of a parallelism, next rules are used.

> if ( P > α ) then " Two Lines are Parallel" Exit
> else if ( P < β ) then " Two Lines are not Parallel" Exit
> else if ( α < P < β ) then {
> > To expand a line, a new threshold is reapplied to a short line.
>
> > threshold = threshold − U ∗ α
> > > α : scaling constant
>
> }

To expand a short line, above process is repeated. If a variation value between each step is small, exit. Because lines extracted by collinearity represent a line information about all orientations, a short line is expanded by Brensenham line drawing algorithm pixel by pixel. Template mask which is the nearest mask to that orientation of a short line is selected as a applied template mask. Then, a new threshold is reapplied to that template mask. Brensenham line drawing algorithm is chosen for time effectiveness.

2. Symmetry − To be symmetric, except for a orientation condition, other conditions have the same fuzzy membership functions as parallelism. Fuzzy membership functions for symmetric constraints are shown in <Figure 4>.

4.3 Comparison to maximum value orientation expansion method(MVOEM)

Compared to maximum value orientation expansion method, fuzzy thresholding method extracts simple and structured lines. In maximum value orientation expansion method, lines are expanded uncondionally to maximum value orientation, Therefore, too complex lines can be extracted.

4.4 Other perceptual organization methods

4.4.1 Cotermination

By a lighting condition and template size, endpoints of a juntion may not meet each other. To solve this, we use the cotemination of perceptual organization. Cotermination should have proximity and closure. Constraints for cotermination are two ones as follows.

1) The difference between angles of two lines should be within some threshold. If too large or too small, two lines are collinear.
2) The distance between two endpoints which are the nearest end points of two lines should be below some threshold.

4.4.2 Polygon

Polygon should have proximity and closure. Of cotermination conditions, lines can be enclosed by enlarging threshold in 2) of cotermination constraints. As groupings go up to high level, thresholds can be given in order to integrate more information, for result of previous processes supports the next process.

V. Experiment Results

We use 256 ∗ 256 images of 256 gray levels. We experiment in 486-PC. <Figure 5, 6> show results of experiment. Jaggies in lower part of a car are a single line representing grouping result.

## VI. Conclusions

In this paper, we extract straight-line segments using advanced Paton's template masks, perceptual organization and fuzzy thresholding method. This proposed method can be applied to pattern recognition. Further study is a curve grouping using perceptual organizaion.

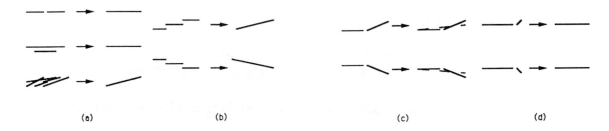

<Figure 1> Examples grouped by collinearity

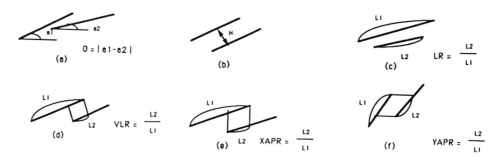

<Figure 2> Constraints of parallelism

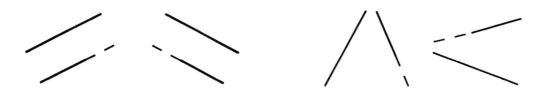

<Figure 3> Examples of undetected parallel and symmetric lines

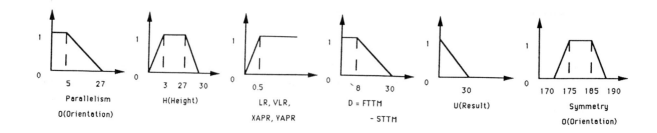

<Figure 4> Fuzzy membership functions for parrallel and symmetric constraints

(a) Raw image

(b) Image using collinearity

<Figure 5> Line segments extraction result using collinearity

(a) Raw image  (b) Line segments extraction result using a FTM

<Figure 6> Example of line segments extraction using a fuzzy thresholding method(FTM)

## References

[1] R. Nevatia and K. R. Babu, "Linear Feature Extraction and Description," *Comput. Graphics Image Proc. 13*, pp. 257-269, 1980.

[2] R. O'Gorman and M. B. Clowes, "Finding Picture Edges Collinearity of Feature Points," *IEEE Trans. Comput. 25*, pp. 449-456, April. 1976.

[3] C. W. Kang, R. H. Park, and K. H. Lee, "Extraction of Straight Line Segments using Rotation Transformation : Generalized Hough Transformation," *Pattern Recognition 24*, pp. 633-641, 1991.

[4] S. A. Dudani and A. L. Luk, "Locating Straight-Line Edge Segments on Outdoor Scenes," *Pattern Recognition 10*, pp. 145-157, 1978.

[5] J. B. Burns, A. R. Hanson, and E. M. Riseman, "Extracting Straight Lines," *IEEE Trans. Patt. Anal. Machine Intell.*,vol. PAMI-8, no. 4, pp. 425-455, July. 1986.

[6] K. Paton., "Line Detection by Local Methods," *Comput. Graphics Image Proc. 9*, pp. 316-332, 1979.

[7] R. Mohan and R. Nevatia, "Perceptual Organization for Scene Segmentation and Description," *IEEE Trans. Patt. Anal. Machine Intell.*, vol. 14, no. 6, pp. 616-635, June. 1992.

[8] H. Q. Lu and J. K. Aggarwal, "Applying Perceptual Organization to the Detection of Man-made Objects in Non-urban Scenes," *Pattern Recognition 25*, pp. 835-853, 1992.

[9] S. Sarkar and Kim L. Boyer, "Integration, Inference, and Mamagement of Spatial Information Using Bayesian Networks : Perceptual Organization," *IEEE Trans. Patt. Anal. Machine Intell.*, vol. 15, no. 3, pp. 256-274, March. 1993.

[10] H. B. Kang and E. L. Walker, "Perceptual Grouping Based on Fuzzy Sets," in *Proc. IEEE Conf. Fuzzy Systems*, vol. 1, pp. 651-659, 1992.

[11] M. Boldt, R. Weiss, and E. Riseman, "Token Based Extraction of Straight Lines," *IEEE Syst. Man Cybern*, vol. 19, no. 6, pp. 1581-1594, Dec. 1989.

[12] K. Miyajima and T. Norita, "Region Extraction for Real Image Based on Fuzzy Reasoning," in *Proc. IEEE Conf. Fuzzy Systems*, vol. 1, pp. 229-236, 1992.

# Fuzzy sensor for the perception of colour

E. Benoit, L. Foulloy, S. Galichet, G. Mauris

LAMII/CESALP
Laboratoire d'Automatique et de MicroInformatique Industrielle
Université de Savoie, BP 806, 74016 Annecy Cedex, France

**Abstract:** This paper describes a fuzzy colour sensor implemented on a single 80C196 micro-controller. This sensor is able to perform linguistic description of precise or imprecise measurements. First the theoretical foundations of fuzzy sensors are presented. Then the application to the description of colours is proposed.

## 1 Introduction

Introduced since several years, *fuzzy symbolic sensors*, also called *fuzzy sensors*, are intelligent sensors which are able to receive, to produce and to handle fuzzy symbolic information [1], [2]. So they can be used as perceptive organs for symbolic controllers like fuzzy controllers and fuzzy expert systems. Furthermore, this kind of sensor can own the most advanced functionalities of intelligent sensors, like auto-adaptation and learning, by using artificial intelligence techniques as proposed in [3].

Fuzzy sensors are specially useful to describe subjective information such as colour, smell, taste, danger, comfort and so on. They are able to reproduce human perception without any numerical model of the phenomenon. This paper is devoted to the perception of colour based on an acquisition with three transducers either considered as precise or imprecise.

Section 1 introduces the basic foundations of fuzzy sensors. Section 2 extends the basic definitions to imprecise measurements. It starts from a crisp subsets approach. Then the links with the possibility and the necessity of fuzzy events are exhibited. Finally, section 4 presents the measurement principle for multi-dimensional spaces and its application to colour sensing.

## 2 Foundations of fuzzy sensors

Fuzzy symbolic sensors are based on the translation of information from a numerical representation to a symbolic one. To perform a symbolic measurement, it is necessary to clearly specify the relation between symbols and numbers. Let $X$ be the universe of discourse associated with the measurement of a particular physical quantity. Denote $x$ any element of $X$. In order to symbolically characterize any measurement over $X$, let $L$ be a set of words, representative of the physical phenomenon. For example, the set $L=\{Close, Medium, Far\}$ could be used to represent a distance. Denote $\mathcal{F}(E)$ the set of the fuzzy subsets of a set $E$.

Introduce an injective mapping $\tau : L \to \mathcal{F}(X)$, called the *fuzzy meaning* of a symbol (see Zadeh [4]). It associates any symbol $L$ of $L$ with a fuzzy subset of $X$. Injectivity guarantees that two different symbols may not have the same meaning. In other words, two synonymous symbols have the same meaning. The fuzzy meaning of a symbol $L$ is characterized, for all $x \in X$, by its membership function denoted $\mu_{\tau(L)}(x)$.

Another mapping $\iota : X \to \mathcal{F}(L)$, called the *fuzzy description* of a measurement over $L$ associates any measurement of $X$ with a fuzzy subset of symbols of $L$. The fuzzy description of a measurement is characterized, for all $L \in L(X)$, by its membership function $\mu_{\iota(x)}(L)$. Any measurement that belongs to the meaning of a symbol, can obviously be symbolically described at least by this symbol. Therefore, the description of a measurement is linked to the meaning of a symbol by the following relation:

$$\mu_{\iota(x)}(L) = \mu_{\tau(L)}(x) \tag{1}$$

It means that if a symbol belongs to the description of a measurement at a grade of membership $\mu_{\iota(x)}(L)$, then the measurement belongs to the meaning of the symbol at the same grade of membership. Let $X = [0, 40]$ be the universe of discourse for the measurement of distances. The fuzzy meanings of *Close*, *Medium*, and *Far* are represented by the membership functions (see Fig. 1). The fuzzy description of the measurement $x = 12$ cm is obtained according to Eq. (1).

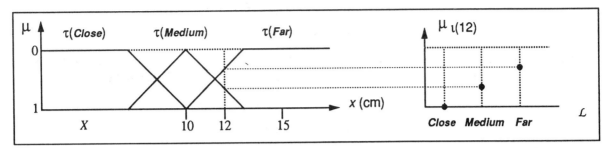

Fig. 1 : Fuzzy meanings and fuzzy description

### 2.1)  Link with the distribution of possibility

Let us start with an example. What could be the exact numerical value of a distance knowing that this distance is medium? From the crisp set point of view, the answer is the following: $\theta$ belongs to a subset of possible values such that each value is at least described by the symbol *Medium*. Formally we have:

$$\theta \in \{ x \mid Medium \in \iota(x) \} \tag{2}$$

Now, when the description is fuzzy, the answer becomes: $\theta$ belongs to a fuzzy subset of possible values. Each possible value $x$ is characterized by its grade of membership which can be interpreted as the degree of possibility that $x$ is the measurement knowing that the distance is *Medium*. A mapping $\pi_\theta : X \to [0, 1]$, called the *distribution of possibility*, can be defined by:

$$\pi_\theta(x) = \mu_{\iota(x)}(Medium) \tag{3}$$

By using Eq. (1) we obtain:

$$\pi_\theta(x) = \mu_{\iota(x)}(Medium) = \mu_{\tau(Medium)}(x) \tag{4}$$

We clearly recover Zadeh's proposal to link fuzzy subsets to possibility distributions [5].

## 3   Fuzzy descriptions of fuzzy subsets

The aim of this section is to present formally the fuzzy descriptions when the measurement is itself a fuzzy subset. First, the problem is presented within the framework of usual set theory. Therefore, the crisp descriptions of crisp subsets are defined. Then these definitions are extended to the fuzzy description of fuzzy subsets. Finally, they are linked to the possibility and the necessity of fuzzy events.

### 3.1)  Crisp descriptions of crisp subsets

The measurement is now a crisp subset $A$ of $X$. A family $(\iota(x))_{x \in A}$ of crisp subsets of $L$ may be obtained from the description of each element of $A$. On the first hand, the least upper bound of the family (i.e. the supremum) is the subset of all the symbols that may describe any element in $A$. On the other hand, the greatest lower bound of the family (i.e. the infinimum) is the subset of common symbols that describes any element in $A$.

The supremum of the family, denoted as $\iota^+(A)$, is called the upper description of the subset $A$. It is defined by:

$$\iota^+(A) = \cup_{x \in A} \iota(x) \tag{5}$$

The infinimum of the family, denoted $\iota^-(A)$, is called the lower description of the subset $A$. It is defined by:

$$\iota^-(A) = \cap_{x \in A} \iota(x) \tag{6}$$

Dubois and Prade [6] proposed a similar approach by considering a relation on $X \times L$ defined by its graph rather than as a mapping from $X$ to $\mathcal{P}(L)$. By using their formalism, the description of a subset is reduced to the upper or lower image of the subset of $X \times L$ induced by the relation.

Remember the characteristic function of a subset $A$ of a set $X$ is a mapping $\chi_A : X \rightarrow \{0,1\}$ such that $\chi_A(x) = 1$ if $x \in A$. By using the characteristic function, equations (5) and (6) can be respectively rewritten as:

$$\chi_{\iota+(A)}(L) = \vee_{x \in A} \chi_{\iota(x)}(L) \tag{7}$$

$$\chi_{\iota-(A)}(L) = \wedge_{x \in A} \chi_{\iota(x)}(L) \tag{8}$$

To prepare the extension to the fuzzy case, Eq. (7) and (8) can be expressed according to the characteristic function of the subset $A$. Let $f$ and $g$ be two mapping from $\{0, 1\} \times \{0, 1\}$ to $\{0, 1\}$. Eq. (7) and (8) become:

$$\chi_{\iota+(A)}(L) = \vee_{x \in X} f(\chi_{\iota(x)}(L), \chi_A(x)) \tag{9}$$

$$\chi_{\iota-(A)}(L) = \wedge_{x \in X} g(\chi_{\iota(x)}(L), \chi_A(x)) \tag{10}$$

Properties of $f$ and $g$ must be:

$$\forall a \in \{0,1\} \ f(a, 0) = 0 \ \text{and} \ g(a, 0) = 1 \tag{11}$$

$$f(a, 1) = a \ \text{and} \ g(a, 1) = a \tag{12}$$

According to their respective properties $f$ is a conjunction and $g$ is an implication.

$$\chi_{\iota+(A)}(L) = \vee_{x \in X} \wedge (\chi_{\iota(x)}(L), \chi_A(x)) \tag{13}$$

$$\chi_{\iota-(A)}(L) = \wedge_{x \in X} \vee (\chi_{\iota(x)}(L), 1 - \chi_A(x)) \tag{14}$$

## 3.2) Fuzzy description of fuzzy subsets

The extension to the fuzzy description of crisp subsets comes straightforwardly. First, we have to consider fuzzy operators instead of crisp ones. Therefore conjunctions will be characterized by triangular norms, disjunction by triangular conorms and implications will be extended to fuzzy implications. They will be respectively denoted T, $\perp$ and $\Rightarrow$. Then, the fuzzy description and the membership function of the fuzzy subset will be used instead of the crisp description and the characteristic function. Assuming $A$ is a fuzzy subset of $X$, the upper description and the lower description of $A$ is given by the following equations.

$$\mu_{\iota+(A)}(L) = \perp_{x \in X} T(\mu_{\iota(x)}(L), \mu_A(x)) \tag{15}$$

$$\mu_{\iota-(A)}(L) = T_{x \in X} \Rightarrow (\mu_{\iota(x)}(L), \mu_A(x)) \tag{16}$$

## 3.3) Links with the possibility and the necessity of fuzzy events

The possibility and the necessity of a fuzzy event were introduced by Zadeh [5]. It allows to find solutions to problems such as: knowing the distance is medium, what is the possibility for the distance to be about 12cm? The fuzzy event is here "about 12 cm". The knowledge "the distance is medium" is represented by a distribution of possibility. According to Zadeh, the possibility of a fuzzy event $A$, knowing the distribution of possibility $\pi$ is given by Eq (17).

$$\Pi(A) = \sup_{x \in X} \min(\mu_A(x), \pi(x)) \tag{17}$$

This equation lets clearly appear the aggregation, by means of the triangular norm *min*, between the grade of membership of the event A and the degree of possibility. When some ambiguity may occur, we will explicitly specify the distribution of possibility. Therefore, $\Pi(A,\pi)$ will denote the possibility of event $A$ knowing the distribution of possibility $\pi$ (see [7] for an equivalent notation).

The necessity, or certainty, is a dual notion of the possibility. An event will be considered as certain if and only if its contrary is impossible. The necessity of the event $A$ is then easily deduced from the possibility of the opposed event, $\neg A$, that is:

$$N(A) = 1 - \Pi(\neg A) \tag{18}$$

Using the definition of the possibility of a fuzzy event given in (17), we finally obtain:

$$N(A) = \inf_{x \in X} \max(\mu_A(x), 1 - \pi(x)) \tag{19}$$

Let us now come back to the definitions of the inferior and superior fuzzy interpretations of fuzzy events. By using the operator *min* as T-norm and the operator *max* as T-conorm, equation (15) may be rewritten as:

$$\mu_{\uparrow+(A)}(L) = \sup_{x \in X} \min (\mu_{\iota(x)}(L), \mu_A(x)) \tag{20}$$

From the equivalence between the meaning and the description (equation (1)), we deduce:

$$\mu_{\uparrow+(A)}(L) = \sup_{x \in X} \min (\mu_{\tau(L)}(x), \mu_A(x)) \tag{21}$$

otherwise formulated by using definition (17) as:

$$\mu_{\uparrow+(A)}(L) = \Pi(\tau(L), A) \tag{22}$$

The grade of membership of any symbol $L$ to the upper description of the fuzzy measurement $A$ is the possibility of the meaning of $L$, knowing the fuzzy measurement $A$. Provided it is normalized, the fuzzy measurement is the distribution of possibility.

Using the same approach, the lower description of a fuzzy measurement can be described in terms of the necessity of the meaning of $L$, knowing the fuzzy measurement $A$.

$$\mu_{\uparrow-(A)}(L) = \inf_{x \in X} \max (\mu_{\iota(x)}(L), 1 - \mu_A(x)) \tag{23}$$

$$\mu_{\uparrow-(A)}(L) = \inf_{x \in X} \max (\mu_{\tau(L)}(x), 1 - \mu_A(x)) \tag{24}$$

$$\mu_{\uparrow-(A)}(L) = N(\tau(L), A) \tag{25}$$

Finally, upper and lower descriptions provide an efficient way for obtaining an interval of the degree of compatibility between the measurement with any symbol, that is, for determining the best symbols characterizing a fuzzy measurement as proposed in [8]. Upper and lower descriptions are implemented in the fuzzy colour sensor described in the next section.

# 4 Description of colours

## 4.1) Introduction

Human beings perceive electromagnetic light beams by four types of photochemical transducers: three cones for day vision and one rod for night vision. The photometric sensors give back information in relation to the received energy. The sensation of colour is generated by the different spectral sensitivities of the cones. The "blue" cones detect short wavelength, whereas the "green" and "red" cones detect respectively medium and long wavelengths.

The artificial sensing (for example the video camera) is based on three photometric transducers that recreate the effects of red, green and blue cones. The responses of the detectors are normalized between 0 and 1, and then the colour space is simply a unit cube (R,G,B). During a colour description in the common language, the luminosity is usually expressed separately. For example we can say "pale blue" or "dark red". In order to introduce this knowledge in the sensor, a non linear mapping is applied on the RGB cube:

$$\tag{26}$$

$$\begin{bmatrix} C'_1 \\ C'_2 \end{bmatrix} = \begin{bmatrix} 1 & -\dfrac{1}{2} & -\dfrac{1}{2} \\ 0 & -\dfrac{\sqrt{3}}{2} & \dfrac{\sqrt{3}}{2} \end{bmatrix} \begin{bmatrix} \dfrac{r}{Y'} \\ \dfrac{v}{Y'} \\ \dfrac{b}{Y'} \end{bmatrix} \qquad Y' = max(r, v, b)$$

Then, luminance and chrominance can be described separately. The measurement set of the luminance is monodimensional, and its description is not developed in this paper. The measurement set of the chrominance is two-dimensional, it is called the chrominance plane of colour and it is used as the measurement set.

## 4.2) Partition of multidimensional spaces

A fuzzy sensor can be used to symbolically describe a measurement characterized by its three coordinates $r, g, b$. This type of fuzzy sensor was first implemented as a software product that can describe points in a colour image [9]. It was also implemented as a single component on a 80C196B micro-controller (fig. 2). In both cases, the fuzzy description of the measurement is based on a partitioning of the chrominance plane using Delaunay's triangulation method.

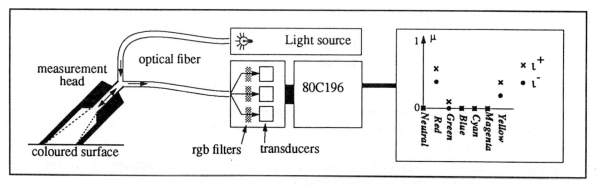

Fig. 2 : Fuzzy colour sensor

Let us present the method used to build fuzzy meanings under a multidimensional measurement set. Let $\mathcal{E}$ be a n-dimensional measurement set. Let $V$ be a small subset $V$ of $\mathcal{E}$ such that the description of any element of $V$ is known and precise. Then the measurement set is partitioned in n-simplexes with the Delaunay's triangulation method. A n-simplex in a n-dimensional space is a polyhedra with n+1 vertices. The points used to perform the triangulation are the elements of the set $V$.

The membership function of the meaning of each symbol is defined on each n-simplex by a multi-linear interpolation. The restriction on a n-simplex of the membership function of the meaning of a symbol $s$ is:

$$\mu_{\tau(s)}(v) = \mu_{\tau(s)}(x_1, x_2, ..., x_n) = a_1x_1 + a_2x_2 + ... + a_nx_n + a_{n+1} \tag{27}$$

Since the value of this function is known for the n+1 vertices of the n-simplex, the n+1 factors $a_i$ can be calculated as follows, where $v_i$ is the $i^{th}$ vertex of the n-simplex, and $x_{i_j}$ is its $j^{th}$ component.

$$A = M^{-1}B \qquad M = \begin{bmatrix} x_{1_1} & ... & x_{1_n} & 1 \\ ... & ... & ... & ... \\ x_{n+1_1} & ... & x_{n+1_n} & 1 \end{bmatrix} \qquad A = \begin{bmatrix} a_1 \\ ... \\ a_{n+1} \end{bmatrix} \qquad B = \begin{bmatrix} \mu_{\tau(s)}(v_1) \\ \mu_{\tau(s)}(v_2) \\ ... \\ \mu_{\tau(s)}(v_{n+1}) \end{bmatrix}$$

This process is performed on each n-simplex and for each symbol. Then the fuzzy meaning of each symbol is defined on $\mathcal{E}$.

The knowledge needed for the configuration of the sensor is very compact. As mentioned before, our method does not take into account all information on the human perception of colour. Missing information are introduced by learning with an operator. He indicates either new symbols and their associated characteristic measurements or extend the meaning of existing symbols to a larger domain in the chrominance colour space.

An example is provided in fig. 3. The initial knowledge is represented by seven crisp points in the chrominance plane which are respectively associated with a colour. Fig. 3.a shows Delaunay's triangulation associated with the initial knowledge. Figure 3.b shows the result of the learning with the operator. Three crisp points have been introduced to extend the definition of the following existing colours (*Red, Green, Blue*). These new points correspond to the human perception of these colours with the respect to the analysed surface. One point has been introduced to define a new symbol *Orange* which corresponds to the operator knowledge. This new symbol was not included in the initial knowledge of the fuzzy sensor. Figure 3.c is the fuzzy meaning of the symbol *Red* according to Delaunay's triangulation given in fig. 3.b.

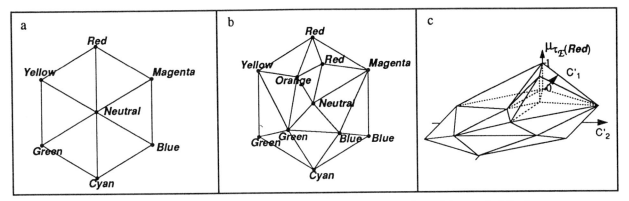

Fig. 3 : Evolution of the sensor knowledge and fuzzy meaning of the symbol *Red*

# 5   Conclusion

This paper has presented a fuzzy colour sensor based on linguistic description of the chrominance of a surface. The interpolative method used to create the meaning of symbols allows the configuration of this sensor without any numerical model of the perceptive process. If a human or a system is able to describe a colour, then this sensor is able to reproduce this perception based on a small set of examples.

The upper and lower descriptions allows the management of imprecise measurements. They can be used to compare several measurement, to extract a more precise description or to compute information about the quality of the sensor. Fuzzy expert systems or a fuzzy controllers can be directly interfaced with fuzzy sensors. Therefore subjective perceptions such as colour, smell or comfort can be used in control systems.

# 6   References

[1]   Foulloy L., Mauris G., *"An ultrasonic fuzzy sensor"*, Proc. of Int. Conf. on Robot Vision and Sensory Control, Zürich, Switzerland, February 1988, pp. 161-170.

[2]   Benoit E., Foulloy L., *"Symbolic sensors"*, Proc. the 8th International Symposium on Artificial Intelligence Based Measurement and Control (AIMaC'91), Kyoto, Japan, September 1991, pp131-136.

[3]   Zingales G., Narduzzi C., *"The role of artificial intelligence in measurement"*, Proc. of 8th International Symposium on Artificial Intelligence based Measurement and Control (AIMaC'91), september 1991, Ritsumeikan University, Kyoto, Japan. pp. 3-12.

[4]   Zadeh L.A., *"Quantitative fuzzy semantics"*, Information Sciences, Vol. 3, 1971, pp. 159-176

[5]   Zadeh L., *"Fuzzy sets as a basis for a theory of possibility"*, Fuzzy sets and systems, 1, pp. 3-28.

[6]   Dubois D., Prade H., *"Upper and lower images of a fuzzy set induced by a fuzzy relation : application to fuzzy inference and diagnosis"*, Information Sciences, 64, pp. 203-232, 1992.

[7]   Prade H., *"A quantitative approach to approximate reasoning in rule-based expert systems"*, Rapport No 225 du LSI, Univ. Paul Sabatier, Toulouse, Mai 1985.

[8]   Dubois D., Prade H., *"Rough fuzzy sets and fuzzy rough sets"*, Int. J. of General Systems, Vol. 17, 1990, pp. 191-209.

[9]   Mauris G., Benoit E., Foulloy L., *"Fuzzy symbolic sensors : From concept to applications"*, to appear in the Int. Journal of Measurement, 1994.

# Generation of Fuzzy Rules Involving Spatial Relations for Computer Vision

Frank Chung-Hoon Rhee and Raghu Krishnapuram
Department of Electrical and Computer Engineering
University of Missouri, Columbia, MO 65211

## ABSTRACT

Rule-based systems are commonly used in computer vision for scene analysis. In this paper, we propose a method for generating fuzzy IF-THEN type rules involving spatial relationships between labeled regions automatically from training data. The proposed method consists of five stages: I) determination of membership functions for representation of spatial relations, II) extraction of training data that represent spatial relations, III) estimation of class relation membership functions, IV) elimination of redundant relations, and finally V) generation of rules. Results are shown for two examples.

## 1. Introduction

Properties (features) of object regions are commonly used in rule-based computer vision systems for tasks such as scene interpretation. In addition to properties, relationships (such as spatial relations) among regions play an important role in an effective rule-based system. For example, a vehicle may be recognized by the fact that it is above a region that has been labeled as "road." An example of a rule that involves spatial relations might be:

IF the relation between an **object** and a **Road region** is above AND its relation to a **Sky region** is below
THEN the object is a **Vehicle**.

Spatial relations such as "above" and "below" defy precise definitions, and they are best modeled by fuzzy sets.

In the literature, there have been several studies [1]-[6] involving spatial relationships between objects. Some of the most important spatial relations between objects for computer vision purposes can be categorized as follows [7]: left-of, right-of, above, below, in-front-of, behind, near, far, inside, outside, and surround. Some of the relations involve angular information [6][8], while others involve area and distance [5]. From the list of spatial relations mentioned above, we will deal only with the relations "left-of," "right-of," "above," and "below" in this paper. However, the methods we employ are quite general, and can be easily extended to cover other types of relations.

In this paper, we present the various steps involved in our proposed method for generating fuzzy IF-THEN type rules involving spatial relationships between labeled regions. The rules are automatically generated from training data that are extracted from angular information. The proposed method consists of the five stages mentioned in the abstract. We use fuzzy connective based aggregation networks [9]-[12] in Stage IV to filter out irrelevant relations (relationships among objects) from the antecedent clauses, and again in Stage V to generate rules. Now, a detailed description for each stage is discussed as follows.

## 2. Stage I: Determination of the Membership Functions for Representation of Spatial Relations

Human perception of spatial positions among objects can be viewed in terms of angular information [6][8]. We now discuss a method for measuring angular information.

Consider an area $A$ that consists of $M$ regions namely $R^1,...,R^M$. For purposes of this discussion, we assume that all the regions are labeled, and there is exactly one region in the area with a particular label. (This assumption is not required for the methods proposed here, however it is used to keep the notation simple. In other words, there could be several regions in the area with the same label, and our methods still apply.) Let the labels of the $M$ regions be $\Lambda^1,...,\Lambda^M$ respectively. Let $r_i^p$ and $r_j^q$ represent points in regions $R^p$ and $R^q$ respectively, and $l_{r_i^p r_j^q}$ denote a line connecting $r_i^p$ and $r_j^q$. Also, let $\theta_{r_i^p r_j^q}$ denote the angle of $l_{r_i^p r_j^q}$ with respect to the horizontal line in a Cartesian coordinate space. Angle $\theta_{r_i^p r_j^q}$ can be regarded as the angular relationship between points $r_i^p$ and $r_j^q$. This is illustrated in Fig. 1.

Now consider the relation between a point $r_i^p$ in region $R^p$ and the entire region $R^q$. Let $r_i^p$ denote the $i^{th}$ point in $R^p$ and $r_j^q$ the $j^{th}$ point in $R^q$ such that $R^p=\{r_1^p,...,r_{N^p}^p\}$ and $R^q=\{r_1^q,...,r_{N^q}^q\}$. Here, $N^p$ and $N^q$ represent the number of points in $R^p$ and $R^q$ respectively. The angle $\theta_{r_i^p \text{REL} R^q}$ representing the relation between point $r_i^p$ and region $R^q$ can be considered as the average angle of all lines $l_{r_i^p r_j^q}, j = 1,...,N^q$ [8]. One approach for calculating this angle is to find unit vectors in the directions of all line segments $l_{r_i^p r_j^q}$, perform vector addition of all unit vectors, and then calculate the angle of the resultant vector.

We need linguistic labels to describe the relation between two regions. In this paper, we consider only one relation called "angular position," i.e., the relation between objects that can be represented in terms of angles. Thus, our domain of discourse can be chosen as $[-\pi,\pi]$. We choose to define four labels over this interval, with four corresponding membership functions. The four labels are referred to as "left-of," "right-of," "above," and "below." A simple way to choose membership functions for the four linguistic labels is given below. More complex definitions may also be used [6].

If the angular information suggests that an object is "left-of" another, the membership function representing "left-of" defined over the domain of angle $\theta$ may be defined as

$$\mu_{\text{left-of}}(\theta, f_l) = \begin{cases} 0 & \pi/2 < |\theta| \le \pi \\ \dfrac{\pi/2 - |\theta|}{\pi/2(1 - f_l)} & f_l \, \pi/2 \le |\theta| \le \pi/2 \\ 1 & |\theta| < f_l \, \pi/2, \end{cases} \qquad (1)$$

where $-\pi \le \theta \le \pi$ and $f_l$ is a fuzzification factor for "left-of" such that $0 \le f_l < 1$. As the value of $f_l$ increases from 0 to 1, the membership values range from most pessimistic to most optimistic. The membership degree to which a point $r_i^p$ in region $R^p$ is "left-of" region $R^q$ ($R^p\text{REL}R^q$) can be obtained by mapping $\theta_{r_i^p \text{REL}R^q}$ through (1). The membership functions for "right-of," "above," and "below" can be defined similarly, with fuzzification factors $f_r$, $f_a$, and $f_b$, respectively. Fig. 2 shows the membership functions for the four spatial relations. The domain of discourse for each of these membership functions is $[-\pi, \pi]$, and they are periodic with a period $T = 2\pi$.

### 3. Stage II: Extraction of Training Data for Representing Relations

For this stage, we consider the relations between one particular class and all other classes at any given time. Let $\Lambda^p$ be the class for which the relations are being considered. For spatially identifying a class (region) label, we need to find all the relations between the class label under consideration ($\Lambda^p$) and the remaining class labels. Hence, there are $M-1$ relations describing class label $\Lambda^p$ which is associated with region $R^p$. We can define two types of angles associated with a point $r_i^p$ in region $R^p$. The first type is represented by the angles $\theta_{r_i^p \text{REL}R^q}$, $q = 1, ..., M$ and $q \ne p$ (see Fig. 3). This set of angles is denoted by $\Theta_k^p$ and correspond to positive (or "1") targets. Each such set $\Theta_k^p$ defines a point in a $M-1$ dimensional cube of size $[-\pi, \pi] \times \cdots \times [-\pi, \pi]$. However, positive targets alone are not sufficient for training. Therefore, we also need angles corresponding to negative (or "0") targets. These angles correspond to all possible angle sets that can occur in the space in the $M-1$ dimensional cube that is not "occupied" by the points $\Theta_k^p$, $k = 1, ..., N^p$. We can randomly sample this empty space and generate angle sets corresponding to "0" targets. We denote such angle sets by $\overline{\Theta}_l^p$. Hence, the training data set for label $\Lambda^p$ consists of $2N^p$ entries, $N^p$ of them of which are $\Theta_k^p$, $k = 1, ..., N^p$, and $N^p$ of which are from $\overline{\Theta}_l^p$.

### 4. Stage III: Estimation of Class Relation Membership Functions

Membership functions representing classes can provide a way of determining the degree of satisfaction of the various criteria (features) used in the antecedent clauses of the rules [9]-[12]. They can also be used to compute the degrees of satisfaction of criteria in the bottom layer of the aggregation networks [11][12]. These ideas can also be applied to relations. In this section, we discuss a histogram-based method for estimating class membership functions for representing relations.

Let $\{\theta_{r_u^p \text{REL}R^j} \mid u = 1, ..., N^p\}$ denote a set of angles describing the relation between region $R^p$ and region $R^j$. There is one such set for each $j$, $j = 1, ..., M$ and $j \ne p$, where $M$ denotes the number of regions. The angles in each of these sets fall in the domain $[-\pi, \pi]$. There is one such domain for each of the relations $R^p\text{REL}R^j$ ($j = 1, ..., M$ and $j \ne p$). The angles are crudely quantized into $Q$ levels and a histogram is constructed. The histogram is then smoothed to obtain $m_j^p(\theta_{r_i^p \text{REL}R^j})$, which denotes the region $R^p$ membership function for relation $R^p\text{REL}R^j$. The membership function is normalized so that the highest value is 1. The type of smoothing and the number of quantization levels used controls the type and extent of the interpolation.

### 5. Stage IV: Elimination of Redundant Spatial Relations

In [11][12], a method to remove redundant features from the antecedent clauses of rules was presented. A similar network for detecting redundant relations can also be designed. For example, if a "tree" appears on the left side of a "house" in some images and on the right side or above or below in others, then the relation between class "tree" and class "house" should be considered redundant. A three layer aggregation network for eliminating redundant relations is shown in Fig. 4. At the input layer, a "relation generator" computes the necessary training data for the $M-1$ relations involving a particular class label $\Lambda^p$ as explained in Section 3 (Stage II). The data are extracted from typical scenes and aggregated at the middle layer. The membership functions computed using the method described in Section 4 (Stage IV) are used as activation functions in the nodes of the middle layer. There are $M-1$ membership functions in this layer where $M$ denotes the number of class labels. The generalized mean fuzzy aggregation operator [11][12] is used at the output nodes. The network is then trained to learn the $M-1$ weights and the attitude parameter associated with the generalized mean. If a particular weight $w_i$ is very small, then the corresponding relation is considered redundant. See [9]-[12] for details.

### 6. Stage V: Generation of Rules

After eliminating redundant relations, the final stage is to obtain a compact set of rules with conjunctive and/or disjunctive antecedent clauses. We achieve this by training an approximate three-layer fuzzy aggregation network. The target values in the training data are chosen to be 1 for the data representing the relation, and 0 for the data representing the non-relation (angles corresponding to "0" target values). The learning is implemented using a modified gradient descent method as mentioned in [9]-[12]. When the training is complete and all the redundant connections are eliminated, the resulting network is interpreted as a set of decision rules. The nodes in the middle and top layers can represent either conjunctive or disjunctive nodes depending on the final values of the attitude parameter of the aggregation function (generalized mean). We now present a method to obtain the initial approximate structure for the three-layer network.

Fig. 5 shows the approximate network structure that is used to train for rules. From the figure, the input layer consists of $K_p^*$ nodes. Each node in the input layer represents a non-redundant relation between the class label $\Lambda^p$ and the remaining labels. It is to be noted that the indexing for the relations has been changed. In other words, after removing the redundant relations between pairs of labels via Stage IV, the non-redundant ones are sequentially reordered. The bottom layer consists of $K_p^*$ groups of nodes, where each group corresponds to a non-redundant relation. The $k^{th}$ group consists of $L_k$ nodes, where $L_k$ is the number of labels for spatial relation $k$. In our case we have four labels for each relation, and they are called "left," "right," "above," and "below." Thus, $L_k=4$. The $j^{th}$ node in the $k^{th}$ group uses $\mu_{kj}$ as the activation function where $\mu_{kj}$ represents the membership function for the $j^{th}$ label for the $k^{th}$ non-redundant relation. The middle layer consists of $K_p^*$ nodes. The $j^{th}$ node in group $k$ in the bottom layer is connected to the corresponding node in the middle layer if the support of $\mu_{kj}$ has a non-empty intersection with the support of $m_k^p()$ (the class $\Lambda^p$ membership function for non-redundant relation $k$). The rationale behind this connection is that if the support of the membership function of a spatial relation has no intersection with the support of the class $\Lambda^p$ membership function, then it cannot appear in the antecedent clause of a rule that describes the class. This connection process is repeated for all the groups in the bottom layer. The nodes in the middle layer are connected to the node in the top layer. All middle and top-layer nodes use a suitable fuzzy aggregation function (such as the generalized mean) as the activation function and all angles of each non-redundant relation are fed to the corresponding group of bottom-layers nodes as input. This process is repeated for all other classes. This completes the construction of the initial approximate aggregation network for generating rules that involves spatial relations. This network is then trained with the training data generated as explained in Section 3 (Stage II) to further eliminate redundant relations and generate rules involving the relation "angular position." We now present two examples to illustrate our proposed method for rule generation.

## 7. Experimental Results
**Example 1:** Natural scene problem

Fig. 6 shows two 200×200 images of natural scenes used in training for rules. The images consist of three regions (classes): road, sky, and vegetation. In total, 100 samples from the class region under consideration (sampled uniformly) were used in computing the "positive" (or "1") entries. Each of the 100 "positive" entries is obtained by calculating the angle representing the relation between a given sample and all the pixels from the remaining class regions. We used 100 "negative" (or "0") entries that were obtained by randomly sampling the space unoccupied by the "positive" entries (see Section 3). This was achieved by moving a square window of size 36°×36° across the entire relation space and collecting angle sets in unoccupied windows. Out of the collected angle sets, 100 were picked randomly. Fig. 7 shows scatter plots of the training data representing the relation of each class label with respect to all other class labels.

Next, the redundancy detection method described in Section 5 was used to eliminate redundant relations. Fig. 8 shows the estimated class membership functions of the two relations describing each class label using the histogram method after mapping each feature into the interval $[-\pi,\pi]$. The domain of the relations was quantized into 32 levels and the resulting histograms were smoothed using a triangular window function with a support of 7 units. The redundancy detection method eliminated the relation between sky and vegetation for class sky, and the relation between vegetation and sky for class vegetation.

Next, we used trapezoidal membership functions such as the ones in Fig. 2 to represent four spatial relations: "left-of (L)," "right-of (R)," "above (A)," and "below (B)." The fuzzification factor for each spatial relation was chosen to be 0.2 (i.e., $f_l=f_r=f_a=f_b=0.2$). These four trapezoidal membership functions were then used as activation functions for the 4 nodes within each group in the bottom layer of the approximate network structures for generating the rules, as shown in Fig. 9. Fig. 10 shows the reduced networks after training. Table 1 shows the final weights (which denote the confidence factors for the relations) and the attitude ($p$) parameter values for the nodes specified in Fig. 10. Values of $p$ greater than 1 denote disjunctive attitudes and values less than 1 denote conjunctive attitudes. The rules obtained as described above for the three class labels from the final network are summarized below.

$R_{road}$: IF the relation between a **region** and a **Sky region** is **below** AND
its relation to a **Vegetation region** is **left-of** AND **below**
THEN the region is **Road**.

$R_{sky}$: IF the relation between a **region** and a **Road region** is **above**
THEN the region is **Sky**.

$R_{veg}$: IF the relation between a **region** and a **Road region** is **right-of** OR **above**
THEN the region is **Vegetation**.

It can been seen that the latter part of the antecedent clause for rule "road" is conjunctive (i.e, ···its relation to a **Vegetation region** is **left-of** AND **below**···). The reason for this is that since the partition of the domain of discourse for the spatial relations was done independently of the training data, the training data may badly straddle the membership functions of the spatial relations. As can be seen in Fig. 7(a), the training data for "road" straddles the supports of the spatial relations "left-of" and "below." This indicates the need for more training data and more than four labels for the spatial relations. It must be noted however, that these rules are correct according to the training data chosen and the spatial relation membership functions picked.

**Example 2:** Airport scene problem

For this next example, we present results involving spatial relations that describe relationships among labeled objects in airport scenes. Fig. 11 shows three synthetically generated 200×200 images of airport scenes used in training. Each training image consists of three objects: airplane, concourse (i.e., airplane docking area), and terminal. We used 50 samples from each image representing the object label under consideration (sampled uniformly). All the pixels from the remaining regions was used while calculating the angles in "positive" entries. A total of 150 samples were selected to represent the "negative" entries using the procedure described in the previous example. The redundancy detection method did not find any redundant relations for any of the object labels. The rules generated by the three layer approximate networks are summarized below.

$R_{air}$:   IF the relation between an **object** and a **Concourse** is **above** OR **left-of** OR **below** AND
        its relation to a **Terminal** is **left-of**
        THEN the object label is a **Airplane**.

$R_{con}$:   IF the relation between an **object** and an **Airplane** is **right-of** OR **above** OR **below** AND
        its relation to a **Terminal** is **left-of**
        THEN the object label is a **Concourse**.

$R_{term}$:   IF the relation between an **object** and an **Airplane** is **right-of** AND
        its relation to a **Concourse** is **right-of**
        THEN the object label is a **Terminal**.

## 8. Summary and Conclusions

In this paper, we introduced a method for automatically generating rules involving spatial relations for rule-based computer vision. Although we have shown results only for four types of relations ("left-of," "right-of," "above," and "below"), the proposed method is general and can be used for other types of spatial relations as well. Our experimental results show that the proposed method is effective if proper case is taken to select the training data.

## 9. References

[1] D. Dubois and M.-C. Jaulent, "A general approach to parameter evaluation in fuzzy digital images," *Pattern Recognition Letters*, vol. 6, pp. 251-259, 1987.

[2] S. K. Pal, "Fuzziness, image information, and scene analysis," in *An Introduction to Fuzzy Logic Applications in Intelligent Systems*, R. R. Yager and L. A. Zadeh, Eds. Boston, MA: Kluwer Academic Publishers, pp. 121-146, 1992.

[3] A. Rosenfeld, "Fuzzy digital topology," *Information and Control*, vol. 40, pp. 76-87, 1979.

[4] A. Rosenfeld, "The fuzzy geometry of image subsets," *Pattern Recognition Letters*, vol. 2, pp. 311-317, 1984.

[5] A. Rosenfeld, "Fuzzy geometry: an overview," *Proc. First IEEE Int. Conf. Fuzzy Systems*, San Diego, CA, March 1992, pp. 113-118.

[6] A. Ralescu and K. Miyajima, "Representation of spatial relations between regions in a 2D segmented image," *Proc. IFSA'93*, Seoul, Korea, July 1993, pp. 1317-1320.

[7] J. Freeman, "The modeling of spatial relations," *Computer Graphics and Image Processing*, vol. 4, pp. 156-171, 1975.

[8] R. Krishnapuram, J. M. Keller, and Y. Ma, "Quantitative analysis of properties and spatial relations of fuzzy image regions," *IEEE Trans. Fuzzy Syst.*, vol. 1, no. 3, pp. 222-233, 1993.

[9] R. Krishnapuram and J. Lee, "Fuzzy-connective-based hierarchical aggregation networks for decision making," *Fuzzy Sets Syst.*, vol. 46, no. 1, pp. 11-27, Feb. 1992.

[10] R. Krishnapuram and J. Lee, "Fuzzy-set-based hierarchical networks for information fusion in computer vision," *The Journal of Neural Networks*, vol. 5, no. 2, pp. 335-350, March 1992.

[11] F. C.-H. Rhee and R. Krishnapuram, "Fuzzy rule generation methods for high-level computer vision," accepted for publication in *Fuzzy Sets Syst.*.

[12] R. Krishnapuram and F. C.-H. Rhee, "Compact fuzzy rule base generation methods for computer vision," *Proc. Second IEEE Int. Conf. Fuzzy Systems*, San Francisco, CA, March 1993, pp. 809-814.

| node | road 1 | road 2 | road 3 | sky 1 | vegetation 1 |
|------|--------|--------|--------|-------|-------------|
| weights | 0.990 | 1.000 | 0.153 | 1.000 | 0.208 |
| | 0.010 | | 0.847 | | 0.792 |
| $p$ | -6.046 | 1.076 | 0.226 | 1.076 | 4.380 |

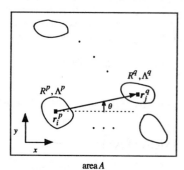

area $A$

Fig. 1. The angle $\theta$ used to represent the angular relationship between point $r_i^p$ and $r_j^q$.

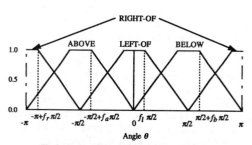

Fig. 2. Membership functions for various spatial relations.

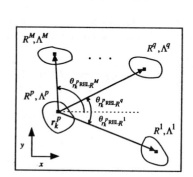

Fig. 3. Angles representing the angular relationship between the class label $\Lambda^p$ and the remaining class labels.

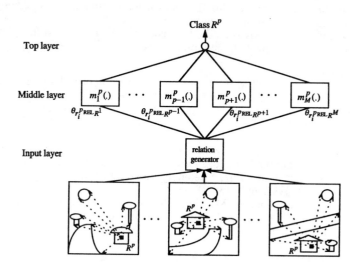

Fig. 4. Aggregation network used for detecting redundant relations for class label $R^p$.

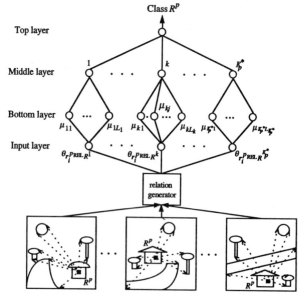

Fig. 5. An approximate network structure for generating rules involving spatial relations.

Fig. 6. Images of natural scenes used in training for rules.

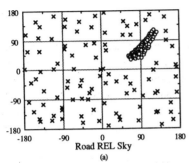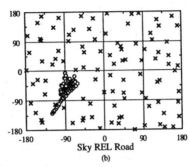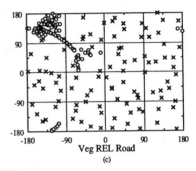

Fig. 7. Scatter plot of relations describing (a) road, (b) sky, and (c) vegetation for the scenes in Fig. 6.
o's represent positive targets and ×'s represent negative targets.

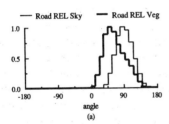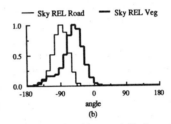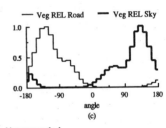

Fig. 8. Estimated class membership functions for the training scenes in Fig. 6 generated by the histogram method
for the two relations describing the class label (a) road, (b) sky, and (c) vegetation.

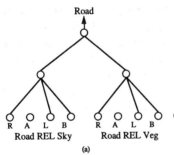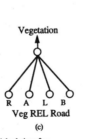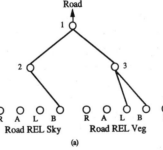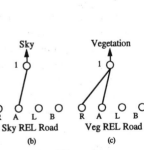

Fig. 9. Approximate network structures for generating rules involving spatial relations for
(a) road, (b) sky, and (c) vegetation.

Fig. 10. Reduced networks after training for (a) road, (b) sky, and (c) vegetation.

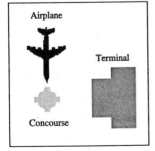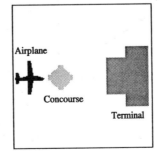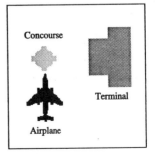

Fig. 11. Synthetic images of airport scenes used for training.

# Fuzzy Measures of Uncertainty in Perceptual Grouping

Ellen L. Walker and Hang-Bong Kang

*Abstract* [*]— **Perceptual grouping is an important phase in image processing for object recognition. When primitive objects are grouped, a measure of certainty of the group is useful both in deciding which of competing groupings to make and in recognizing objects by matching models to the eventual groups. Fuzzy measures are a useful form for describing certainty in grouping. In this paper, we derive three fuzzy measures for primitive grouping, and discuss the extension to fuzzy measures for hierarchical grouping.**

## I. INTRODUCTION

Perceptual grouping is the process of grouping together primitive image elements that belong to the same higher-level object, based on appropriate geometric relationships. In computer vision, perceptual grouping has been used to extract mid-level tokens such as long lines, parallel pairs, and symmetric ribbons from fragmented image data. These structures can then be used as seeds for model-based object recognition.

For a given set of primitive tokens, many alternative groupings can be generated. To choose among these groupings, additional information about their significance, or certainty, is needed. Certainty information is also appropriate for sorting groupings to match to object models. In general, the most certain groupings should be matched first, and the results of this matching fed back to aid in interpretation of the remaining data.

Intuitively, the certainty of a grouping is separate from its geometric description. Figure 1 shows an example from collinear grouping. The original data segments are represented as heavy lines, and the grouped line segments are represented as light lines overlaid on the original data. Although line segments (a) and (b) are geometrically identical, (b) is intuitively more certain because the data segments are closer to collinear with the grouped segment, and there are almost no gaps between data segments.

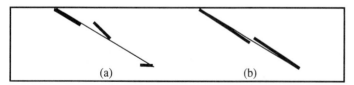

**Figure 1: Grouped line (b) is more certain than grouped line (a)**

This paper will explore the measurement of uncertainty in grouped structures, in particular looking at the applicability of fuzzy measures of uncertainty.

## II. CERTAINTY MEASURES FOR SIMPLE GROUPING

We will discuss two classes of certainty measures for grouping: necessity measures and possibility measures; and consider the application of fuzzy techniques to each. To simplify this discussion, we will emphasize the issue of determining whether or not to group a set of tokens, and will assume that algorithms for geometric grouping are available as needed.

Following the notation of Kruse et. al. [7], we can divide the data space $\Omega$ of all possible interpretations of the given tokens into subsets $A$ and $\Omega - A$, where $A$ is the set of interpretations in which the tokens are grouped, and $\Omega - A$ is the set of interpretations in which the tokens are not grouped. The certainty with which the tokens should be grouped is the same as the certainty that the correct interpretation is in $A$.

This certainty can be bounded by a range (*necessity*, *possibility*). The *possibility* of the correct interpretation being in $A$ is the highest degree of belief of all interpretations that group the tokens, while the *necessity* of the correct interpretation being in $A$ is the negation of the highest degree of belief of all interpretations that do not group the tokens.

Thus, the certainty of grouping a particular token set is related both to the compatibility of the tokens with the grouping criteria and to the chance that the tokens could have arisen by accident (i.e. the compatibility of the tokens with a non-grouped interpretation). Prior work on certainty of grouping has looked at the necessity (or non-accidentalness) approach or the possibility approach, but these have not been considered together before.

### A. Possibility measure

The possibility of grouping is the highest degree of belief that can be assigned to a particular group. This measure depends on the compatibility of the token with the grouping criteria, i.e. how well do the given tokens support the best group that can be made from them? In general, support is measured by considering the degree of approximation that is needed to create the grouped geometric object from the original objects.

A first step in measuring the degree of approximation is to aggregate the geometric distance from all input tokens to the grouped token. This measures the total approximation required in the generation of the grouped token, but does not take into consideration the coverage of the grouped token by the original tokens.

The research described here was supported in part by NSF grant IRI-9011631. This work was performed at Rensselaer Polytechnic Institute, Troy NY, 12180,. Dr. Kang was in the Electrical and Systems Engineering Department and Dr. Walker is in the Computer Science Department.

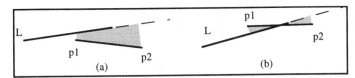

**Figure 2:** Aggregate approximation for collinear grouping is the area of the shaded region

As an example, consider the degree of approximation necessary for a simple case of collinear grouping, where the grouped segment is extended to include the projection of the input segment. In this case, the aggregated distance from the input line segment to the grouped segment is the area of a polygon formed by the input segment and its projection (see Figure 2). When the input segment does not cross the grouped segment, this polygon is a trapezoid, otherwise the polygon is a pair of triangles whose areas are added.

In general, the area of the difference region depends on the average perpendicular distance from the segment to the grouped line and the length of the projection of the segment onto the line. A weakness of this measure is that it is independent of the angle between segments: any pair of segments with the same projection and the same center point will yield the same area. This behavior is counterintuitive. A second weakness is that the measure discriminates against including longer segments in a group, since the longer the segment, the longer the projection, the greater the area of the difference region. This behavior makes it difficult to recursively group increasingly longer segments from a highly fragmented image, however.

Both of these weakness can be avoided by dividing the area measure by the length of the input segment to get a measure of difference area per unit length. This distance ratio now depends on the angle between segments, as well as the average perpendicular distance between the segment endpoints. For a given angle and average perpendicular distance, the measure is independent of segment length. The formula for the measure $A/L$ (difference Area / Length) is:

$$A/L = \begin{cases} \dfrac{|(d_1 + d_2)\cos\theta|}{2}, & \text{if } d_1 d_2 \geq 0 \\[2ex] \dfrac{(d_1^2 + d_2^2)|\cos\theta|}{2(|d_1| + |d_2|)}, & \text{if } d_1 d_2 < 0 \end{cases},$$

where $d_1$ and $d_2$ are the (signed) perpendicular distances from the endpoints of the input segment to the grouped line, and $\theta$ is the angle between the input segment and grouped line.

As the approximation necessary for grouping increases, the compatibility between the new segment and the group decreases. Thus, the compatibility measure, or *possibility* for adding a line segment to a grouped line varies inversely with $A/L$. As formulated, $A/L$ is unbounded. If the measure will be used directly for comparing specific groupings, then its inverse can be used directly as a measure. However, a fuzzy measure

will be easier to combine with the necessity measure, so we can define a possibility measure as:

$$poss_{collinear} = \mu_{A/L}(d_1, d_2, \theta),$$

with $\mu_{A/L}$ defined so that $\mu_{A/L}(0,0,0) = 1$ and $\lim_{d_1, d_2 \to \infty}(\mu_{A/L}(d_1, d_2, \theta)) = 0$. The appropriate mapping between $A/L$ and $\mu_{A/L}$ is application-dependent, so it must be determined empirically.

In general, a possibility measure for any grouping relationship can be determined by finding the aggregate difference between the new token and the grouped token, and mapping this difference through a membership function so that the minimum distance has membership value 1, and the maximum distance (depending on the application) has the membership value 0.

The possibility measure serves as an upper bound on the certainty of grouping. It takes into account only geometric compatibility, ignoring important information such as gaps between tokens that are being grouped. This information is considered, however, in a necessity measure for grouping.

### B. Necessity measure

The necessity of grouping is defined as the possibility of not grouping, i.e. the highest degree of belief of all interpretations that do not group the tokens. This is equivalent to the possibility that the geometric relationships between the tokens arose by accident, i.e. the *accidentalness* of the grouping.

Thus, the necessity of a group is simply the non-accidentalness of the group. Grouping line segments according to non-accidentalness has been analyzed extensively by Lowe [4, 5]. In order to measure the probability of accidental occurrence of a relationship, it is necessary to make assumptions about the background distribution of objects. In Lowe's work, the assumption was that the distribution of segments was uniform with respect to orientation, and that the scale-independent density of endpoints was a unitless constant $D$.

Given these assumptions, Lowe derived the following formula for the expected number of accidental occurrences of segments based on average endpoint distance, segment length, and minimum gap between the input segment and grouped segment (before extending the grouped segment). The resulting formula is:

$$E = \frac{4D}{\pi} \frac{\theta d_{perp}(d_1 + d_2) + (g + l)}{l^2}$$

Although $D$ cannot be directly extracted from the data, the possibility of accidental occurrence is proportional to $E$. A necessity measure for grouping can be derived from $E$ by first mapping $E$ to the interval [0,1] and then taking its inverse, i.e. the necessity measure is:

$$necc_{collinear} = 1 - \mu_E(d_1, d_2, \theta, g)$$

### C. An empirical measure

A single measure of certainty for grouping should lie between the possibility and necessity measures, and should closely approximate human intuition as to what should be grouped.

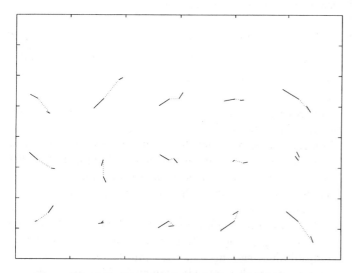

**Figure 3: Segments judged *almost* collinear**

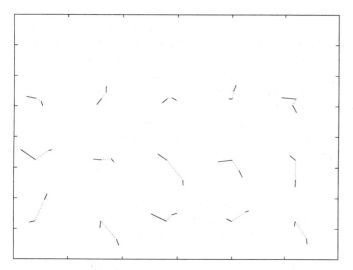

**Figure 4: Segments judged *very roughly* collinear**

Since humans tend to classify compatibility only coarsely (e.g. the five-term set: *exactly, approximately, roughly, very roughly, not*), it makes sense to develop fuzzy membership function based on such classes.

In particular, we define a membership function for each grouping relation that relates measurements on the input tokens to a degree of applicability of the relation. We call this degree of applicability the *goodness value* of the grouping with respect to the grouping criteria.

As an example of membership function extraction, this section describes an experiment in which a goodness measure for collinearity, $\mu_{collinear}$, was determined from classification data.

In an informal experiment, ten graduate students were shown 500 randomly generated line segment pairs, and asked to classify each according to a five-term set: *exactly collinear, almost collinear, roughly collinear, very roughly collinear,* and

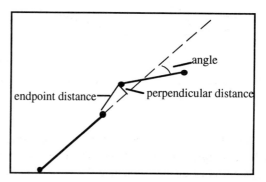

**Figure 5: Geometric constraints for collinearity**

*not collinear.* The segments judged *almost collinear* and *very roughly collinear* are shown in Figures 3 and 4, respectively.

For each segment pair, three geometric constraints were measured (see Figure 5): the angle between the segments (*angle constraint*), the shortest gap between the endpoints of the segments *(endpoint distance constraint),* and the perpendicular distance from an endpoint of the shorter segment to the line containing the longer segment (*perpendicular distance constraint*). Because the significance of a gap of a given size is greater when the segments are shorter, both distance constraints were normalized by the length of the longer segment in the pair.

These constraints are consistent with the measures derived above, as well as with other work in collinear grouping. Both Lowe [5] and Andress and Kak[1] grouped collinear line segments according to angle, endpoint distance, and perpendicular distance, and Mohan[6] grouped co-curvilinear line segments according to endpoint distance and angle.

For each constraint, a linguistic partition into five triangular membership functions (one for each term in the set) was created. The centers of the membership functions were the mean values of the constraint for each classification, and the widths were determined by adapting the K-nearest-neighbors algorithm. Generally, this algorithm minimizes the following objective function with respect to the width $\sigma_i$'s:

$$E = \frac{1}{2}\sum_{i=1}^{K}\left[ \sum_{j \in Knearest}\left( \frac{m_i - m_j}{\sigma_i} \right)^2 - r \right]^2,$$

where $r$ describes the amount of overlap [3]. This objective function assumes that the regions of the partition are symmetric about the mean.

Since the membership functions need not be symmetric, the left width and right width of each membership function were computed separately, considering only the single nearest neighbor in the left (or right) direction. Assuming the $\sigma_i$'s are sorted with $\sigma_1$ as the smallest, then the radii are:

$$\sigma_i(left) = \frac{m_i - m_{i-1}}{r}, \;\; \sigma_i(right) = \frac{m_{i+1} - m_i}{r}$$

As boundary conditions, we assume that $m_0$ and $m_6$ are negative and positive infinity, respectively. The final membership functions representing the linguistic partitions of

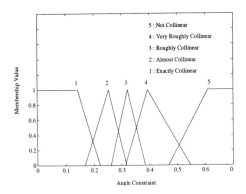

**Figure 6: Angle constraint function**

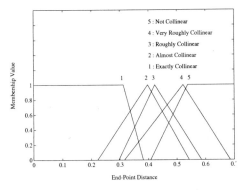

**(b) Figure 7: Endpoint distance constraint function**

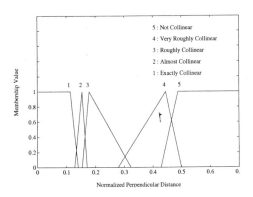

**(c) Figure 8: Perpendicular distance constraint function**

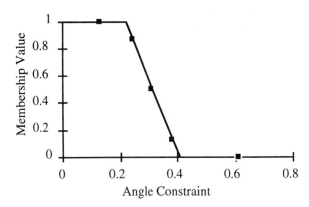

**Figure 9: Membership function for collinearity based on angle**

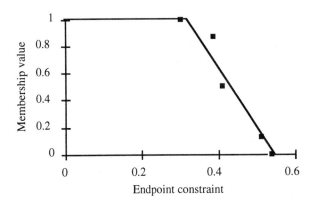

**Figure 10: Membership function for collinearity based on endpoint distance**

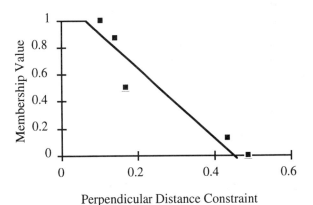

**Figure 11: Membership function for collinearity based on perpendicular distance**

each constraint with respect to the five specified degrees of collinearity are shown in Figures 6 through 8.

To construct a unified membership function that maps each constraint to a degree of collinearity, we used the five-term set from [2] to represent the collinearity membership levels for each linguistic class. The resulting membership function mapped the center point of each term to the membership level for that term. For simplicity, a trapezoidal membership function was fit to the five points, as shown in Figures 9 - 11.

The final membership function was derived by combining the functions from the three constraints. The method of combination was determined by examining sample segment pairs from all collinearity classes. For each sample, the three constraint membership values were compared to the original human classification. This analysis showed that the classification of a segment was consistent with the least collinear classification of its three constraints. Therefore, collinearity requires the conjunction of the three constraint membership functions, so it can be computed by taking their minimum (or $T_3$):

$$\mu_{collinearity} = T_3(\mu_{angle}, \mu_{endpt\_dist}, \mu_{perp\_dist})$$

This membership function is consistent with the intuition that all three of the constraints must be satisfied for a pair of line segments to be collinear. It is also consistent with the forms of the possibility and necessity measures described earlier, i.e. membership functions can be extracted so that

$$necc_{collinearity} \leq \mu_{collinearity} \leq poss_{collinearity}.$$

## III. CERTAINTY MEASURES FOR HIERARCHICAL GROUPING

Most images are too complex to be described by a single level of grouping. Typically, hierarchical grouping is performed at several levels, and often some elements are recursively grouped. As an example, in CANC [6] line segments are recursively grouped to form longer segments, which themselves are grouped to form higher-level structures such as symmetries and ribbons.

Certainty measures for hierarchical grouping must take into account the fact that the input tokens can be the results of prior grouping. Thus, the certainty of input tokens can vary, and this should be taken into account in the determination of the new group's certainty. Once input token certainty is considered, the system can also take into account certainty values for the primitive tokens if they are available. For example, the certainty of a primitive line segment is a function of its length and the contrast across the segment in the image.

A second consideration in recursive grouping is the relative contribution to the overall certainty of each element of the group. Elements that contribute more to the geometry of the group should also contribute more to the certainty of the group. As an example, the certainty contribution of a line segment in collinear grouping should vary with its length relative to the length of the grouped segment.

Finally, the independent contribution of each element should be no more than the possibility of grouping that element, as computed in the previous section. Although other certainty measures can be used (such as the empirical one), the advantage of using necessity as a limit is that it is independent of the other factors (relative size and current certainty) that are included in the hierarchical grouping measure.

In its simplest form, the hierarchical grouping measure is a sum of products: for each token in the group, multiply the contribution of that token to the group by the conjunction of the compatibility of that token with the group and the maximum certainty of that token. This conjunction is computed by a T-norm. For lines, this method produces the following equation:

$$CV = \frac{\sum_i L_i * T(P_i, CV_i)}{L},$$

where $CV$ is the certainty value of the resulting segment, $CV_i$ is the certainty value of the $i$'th input segment, $L_i$ is the length of the $i$'th segment, $P_i$ is the goodness value of the $i$'th segment with respect to the merged segment, and $L$ is the length of the merged segment.

The choice of T-norm depends on the desired relationship between the fuzzy measures of compatibility and certainty value. In the most likely case, these will trade off, so the T-norm will be $T_2$, or multiplication. However, a more conservative interpretation is that certainty should depend on the lesser of prior compatibility and certainty. In that case, $T_3$, or minimum, is the correct conjunction. This determination should be made empirically.

## IV. CONCLUSION

Perceptual grouping is a significant technique for preprocessing in image understanding. Decisions made both in the grouping process itself and in the subsequent object-recognition phase can benefit from certainty information in addition to the geometric description of each group.

This paper has described three measures of certainty for simple perceptual grouping: a possibility measure, a necessity measure, and an empirically-derived measure that lies between the extremes of possibility and necessity. Representing these measures as fuzzy measures allows the application of fuzzy set theory in their combination. Additionally, fuzzy measures are easily related to linguistic descriptions of the geometric relationships that control grouping, and are thus amenable to extraction from human-derived data, either by hand or by a supervised learning procedure.

## REFERENCES

[1] K. Andress and A. Kak, "Evidence Accumulation and Flow of Control in a Hierarchical Spatial Reasoning System," *AI Magazine*, vol. 9, no. 2, 1988.

[2] P.P. Bonissone and K.S. Decker, "Selecting Uncertainty Calculi and Granularity: An Experiment in Trading-off Precision and Complexity," in Uncertainty in Artificial Intelligence, L.N. Kanal and J.F. Lemmer, Eds. Amsterdam: North-Holland, 1986.

[3] Lin and Lee, "Neural-Network-Based Fuzzy Logic and Decision System," *IEEE Trans. Computer*, 1320-1336, December 1991.

[4] D.G. Lowe, *Perceptual Organization and Visual Recognition*. Hingham, MA: Kluwer Academic Publishers, 1985.

[5] D.G. Lowe, "Three-Dimensional Object Recognition from Single Two-Dimensional Images," *Artificial Intelligence*, vol. 31, no. 3, 355-395, March 1987.

[6] R. Mohan and R. Nevatia, "Perceptual Organization for Scene Segmentation and Description," *IEEE Trans. Pattern Anal. Machine Intell.*, vol. 14, no. 6, 616-635, June 1992.

[7] R. Kruse, E. Schwecke, and J. Heinsohn, *Uncertainty and Vagueness in Knowledge Based Systems*. Springer-Verlag, 1991.

# Fixed-Time Life Test Using the Fuzzy Theory

## C. J. Hwang [1] and C. H. Yeh [2]
Yuan - Ze Institute of Technology
Chun-li ,Taiwan, R. O. C.

## Abstract

The single sample plans with type I ( fixed-time ) censoring has been studied for a long time. It is hard to get the required producer's risk $\alpha$ and consumer's risk $\beta$ exactly, because the sample size **n** and the acceptance number $r_0$ must be integers. The design procedures to determine the acceptance numbers and sample sizes for fixed time censoring based on the fuzzy set theory is presented. A simple decision rule for finding the minimization of sample size was considered in the design procedures.


**Key words :** fixed-time life test, fuzzy number, membership function.

## 1. Introduction

In using the traditional single sample plans, the actual producer's or consumer's risk is often different from its specified value. In order to satisfy both the producer's and consumer's risk , the actual risk should be chosen close enough to the required value. Kanagawa et al. treated the sample size as fuzzy numbers, and using membership function to get the required fixed number of failures and acceptance value [1]. It is difficult to compare their test plan with the traditional plan, because the fuzzy plans inevitably include more subjectivity. Also they treated the sample size as fuzzy numbers. The membership function $\mu_n(n)$ represents the grade of satisfaction for the sample size [2]. The minimization of sample size was explicitly considered in design procedure.

The purpose of this paper is to find the minimum sample size and get the required producer's and consumer's risk. A computerized approach is presented which provides the design procedure for fixed time life test.

---

[1]    Department of Computer Engineering and Science
[2]    Department of Electrical Engineering

0-7803-1896-X/94 $4.00 ©1994 IEEE

## 2. Notation

| | |
|---|---|
| **n** | sample size |
| $\alpha$ | specified producer's risk |
| $\beta$ | specified consumer's risk |
| $\alpha'$ | actual producer's risk |
| $\beta'$ | actual consumer's risk |
| **A** | fuzzy number of real **A** |
| $\mu_A$ | membership function of **A** |
| $t_0$ | test time |
| **r** | number of failures during $t_0$ |
| $r_0$ | acceptable number of failures during $t_0$ |
| $\theta_0$ | acceptable good mean life |
| $\theta_1$ | acceptable poor mean life |
| $p_0$ | acceptable quality level $(\mathbf{AQL}) = 1 - e^{\frac{-t_0}{\theta_0}}$ |
| $p_1$ | rejectable quality level $(\mathbf{RQL}) = 1 - e^{\frac{-t_0}{\theta_1}}$ |

## 3. Fixed time sampling plan

Suppose that **n** identical components are placed on life test. Let $\theta_0$ be acceptable good mean life and $\theta_1$ be unacceptable mean life. $p_0$ is acceptable quality level, $p_1$ is rejectable quality level. Assume the sampling distribution is given by the Poisson distribution. The specified producer's risk is $\alpha$, and the consumer's risk is $\beta$. The sample is taken without replacement, then the probability to accept the lot is :

$$p(\, r < r_0 \mid m_0 = np_0) \; = \; 1 - \alpha$$

$$= \sum_{r=0}^{r_0-1} \frac{(np_0)^r e^{-np_0}}{r!} \tag{1}$$

$$p(\ r < r_0 \ | \ m_1 = np_1) \ = \ \beta$$

$$= \sum_{r=0}^{r_0 - 1} \frac{(np_1)^r e^{-np_1}}{r!} \qquad (2)$$

The discrimination ratio is described as following :

$$\frac{m_1}{m_0} = \frac{p_1}{p_0} \qquad (3)$$

We can find the $(r_0, n)$ value from $p_1/p_0$ and Poisson distribution. In the fixed time test plan, if $0 < r < r_0$ then the lot is accepted, otherwise the lot is rejected. The equality in equations (1) and (2) are usually impossible since $r_0$ and $n$ must be integers. In traditional design, if $\beta$ is fixed, the following criterion is adopted:

$$p(\ r < r_0 \ | \ m_0 = np_0\ ) \ \geq \ 1 - \alpha \qquad (4)$$

We choose the $(r_0, n)$ in order that the $\alpha'$ is close to $\alpha$ . Also we can fix $\alpha$, then

$$p(\ r < r_0 \ | \ m_1 = np_1\ ) \geq \ \beta \qquad (5)$$

So, we choose $(\ r_0,\ n\ )$ to get minimum $(\beta' - \beta\ )$. The actual producer's and consumer's risk are usually less than the specified value.

## 4. Fuzzy design

Fuzzy set A can be described as

$$A = \frac{\mu_A(x)}{x}, \quad \mu_A(x) \text{ is membership function of A}$$

The membership function can be found on a theoretical basis. It should be easy to calculate and fit to the problem and can be described by only a few parameters, $\mu_A(x)$ can be defined as following:

$$\mu_A(x) = \left(\frac{x-a}{a-a_i}\right)^k, \quad ; i = 1,2 \ \text{ and } \ a_1 < x < a_2 \qquad (6)$$

The k should be chosen which is satisfied by the producer and consumer. Consider the product set of fuzzy number **A** and **B**, i.e. **A** ∩ **B**, the membership function of **A** ∩ **B** is

$$\mu_{A \cap B}(x) \equiv \min\{\mu_A(x), \mu_B(x)\} \qquad (7)$$

$x'$ which maximizes $\mu_{A \cap B}(x)$ lies between **A** and **B**,
i.e **A** < $x'$ < **B**. Our problem is to find ($r_0$, **n** ) such that
$\mu_{\alpha \cap \beta}(\alpha', \beta') = \min\{\mu_\alpha(\alpha'), \mu_\beta(\beta')\}$ is maximized.
i.e. $\mu_\alpha(\alpha') = \mu_\beta(\beta')$ \qquad (8)

The design procedure for finding the optimum ($r_0$, **n** ) is shown as following:

Assume: $\alpha_1 < \alpha < \alpha_2$, $\beta_1 < \beta < \beta_2$, and $t_0$, $\theta_0$, $\theta_1$ are known.

1). Calculate $p_0 = 1 - e^{\frac{-t_0}{\theta_0}}$, $p_1 = 1 - e^{\frac{-t_0}{\theta_1}}$

2). Define membership function $\mu_\alpha(\alpha)$, $\mu_\beta(\beta)$

3). Find $r_0$ from $p_1/p_0$ and Poisson distribution, two $r_0$ can be obtained.

4). Find ($r_0$,**n**) for each $r_0$ from $\mu_\alpha(\alpha') = \mu_\beta(\beta')$

5). Calculate $\alpha'$, $\beta'$ from equations (1), (2)

6). Find $\mu_\alpha(\alpha')$, $\mu_\beta(\beta')$ for each $r_0$, choose the ($r_0$, **n** ) which has the larger membership value.

7). In order to compare the different test plan, we define the difference from the required risks as following:

$$diff = \sqrt{(\alpha' - \alpha)^2 + (\beta' - \beta)^2} \qquad (9)$$

If the sampling distribution is the Chi-square distribution, the similar procedures shown above will be applied [3,4].

## 5. Example

Consider the fixed time life test. Suppose that $\alpha = \beta = 0.1$, $p_0 = 0.009$, $p_1 = 0.05$. The membership function of $\alpha$, $\beta$ is

$$\mu_\alpha(\alpha') = \sqrt{\frac{\alpha'-0.01}{0.1-0.01}} \qquad\qquad ; \quad 0.01 < \alpha' < 0.1$$

$$= \frac{\alpha'-0.2}{0.1-0.2} \qquad\qquad ; \quad 0.1 < \alpha' < 0.2$$

$$\mu_\beta(\beta') = \sqrt{\frac{\beta'-0.1}{0.1-0.01}} \qquad\qquad ; \quad 0.01 < \beta' < 0.1$$

$$= \frac{\beta'-0.25}{0.1-0.25} \qquad\qquad ; \quad 0.1 < \beta' < 0.25$$

Find the best test plan. Find ( $r_0$, $n$ ) such that

$$\mu_\alpha(\alpha') = \mu_\beta(\beta')$$

The discrimination ratio $p_1/p_0$ = 0.05/0.009 = 5.56 from Poisson distribution we can get $r_0$ between 2 and 3. If $r_0$ =2, we can get $n$ = 69 from equation 1 and 2. The actual risks are $\alpha'$=0.128, $\beta'$=0.142, and $\mu_\alpha(\alpha')$ = $\mu_\beta(\beta')$ = 0.72. If $r_0$ = 3, we can get $n$ =112, and $\alpha'$=0.082, $\beta'$=0.082, $\mu_\alpha(\alpha')$ = $\mu_\beta(\beta')$ = 0.89. we choose ( $r_0$, $n$ )= ( 3,112 ) as the test plan.

## 6. Discussion

If the consumer's risk is fixed, then we choose the test plan ($r_0$, $n$ ) such that the actual producer's risk $\alpha'$ is specified value. Let $\beta$ =0.1, we can get ($r_0$, $n$ ) = (2,78), then $\alpha'$ =0.157. If ($r_0$, $n$ ) = (3,106), then $\alpha'$ = 0.072.

Similarly, if the producer's risk $\alpha'$ is fixed, then we choose the test plan ($r_0$, $n$ ) such that the actual the consumer's risk $\beta'$ is close to the specified value. Let $\alpha'$=0.1, ($r_0$, $n$ ) = (2,59), then $\beta'$ = 0.207. If ($r_0$, $n$ ) = (3,122), then $\beta'$ = 0.058. From table 1 we can find that if one risk is fixed, the difference between actual and specified value will be large. The difference of ($r_0$, $n$ ) = (3,112) has the minimum value. If 5% error is allowed, ($r_0$, $n$ ) = (2,69) will be chosen.

## Table 1.  Fixed-time test plan

| α | β | $r_0$ | n | α' | β' | diff |
|---|---|---|---|---|---|---|
|  | 0.1 | 2 | 78 | 0.157 | 0.1 | 5.7% |
|  | 0.1 | 3 | 106 | 0.072 | 0.1 | 2.8% |
| 0.1 |  | 2 | 56 | 0.1 | 0.207 | 10.7% |
| 0.1 |  | 3 | 122 | 0.1 | 0.058 | 4.2% |
|  |  | 2 | 69 | 0.128 | 0.142 | 5.0% |
|  |  | 3 | 112 | 0.082 | 0.082 | 2.5% |

## 7. Conclusion

A design procedure for the fixed time life test based on Fuzzy sets theory was discussed. The simple decision rule for finding the minimization of the sample size was considered in design procedures.

## References :

[1] A. Kanagawa and H. Ohta, "Fuzzy Design for Fixed-Number Life Tests", IEEE trans. on Reliability, Vol. 39, No. 3, pp.394-397, 1990

[2] A. Kanagawa and H. Ohta, "A Design for Single Sampling Attribute Plan Base on Fuzzy Sets Theory, " Fuzzy sets and Systems, Vol.37, pp. 173-181, 1990.

[3] P. Griffiths and I.D. Hill, Applied Statistics Algorithms, Ellis Horwood Limited, West sussex, England, pp. 157 - 161, 1985.

[4] S.L. Meyer, Data Analysis for Scientists and Engineers, John Wiley & sons, Inc., New York, pp.263 - 273, 1975

# A Fuzzy Approach to Input Variable Identification

Yinghua Lin
Department of Computer Science
New Mexico Institute of Mining and Technology
Socorro, NM 87801

George A. Cunningham, III
Department of Electrical Engineering
New Mexico Institute of Mining and Technology
Socorro, NM 87801

## Abstract

We introduce "fuzzy curves" and use them for input variable identification. We give several examples to show that, in contrast to previously proposed methods, our method is computationally simple and identifies significant input variables easily.

# 1  Introduction

Sugeno and Yasukawa [3] propose a method to create a fuzzy model from input-output data. To identify the significant input variables, they build a one input fuzzy system for every input candidate. The input of the fuzzy system corresponding to the best performance index is chosen as the first significant input. Taking this input and every remaining input candidate, they build a series of two input fuzzy systems. The two inputs corresponding to the best performance measure are considered as two important inputs. They keep adding input variables one at a time until the performance index cannot be improved. For $N$ input candidates the maximum number of fuzzy systems to be tested for input variable identification is $\frac{N(N+1)}{2}$.

Takagi and Hayashi [4] propose a neural network driven fuzzy reasoning system to analyze input-output data. They identify the input variables by eliminating each input variable and checking the performance index. If the performance index does not get worse when an input variable is removed from the system, then that input variable is not significant. They need to train a possible maximum of $\frac{N(N+1)}{2}$ neural nets with this system.

For a system with a large number of input candidates, building $\frac{N(N+1)}{2}$ fuzzy or neural network systems is not practical. To solve this problem, we introduce fuzzy curves and use them to identify the input variables before modeling the system. Our method is computationally simple, and the time complexity of our identification process is linear with respect to the number of input candidates.

The paper's second section introduces fuzzy curves and describes how we identify the input variables. The third and fourth sections give several examples to demonstrate our method.

| No. | $x_1$ | $x_2$ | $x_3$ | $y$ | No. | $x_1$ | $x_2$ | $x_3$ | $y$ |
|-----|------|------|------|-----|-----|------|------|------|-----|
| 1 | 14.9 | 2.0 | 9.3 | 3.8 | 11 | 6.9 | 2.3 | 10.3 | 3.3 |
| 2 | 16.6 | 1.7 | 5.8 | 3.7 | 12 | 7.4 | 1.9 | 9.4 | 2.8 |
| 3 | 21.3 | 2.1 | 9.1 | 3.0 | 13 | 11.3 | 2.0 | 8.4 | 3.0 |
| 4 | 24.3 | 2.9 | 7.0 | 4.7 | 14 | 17.6 | 2.0 | 8.2 | 2.7 |
| 5 | 26.6 | 1.7 | 4.8 | 4.1 | 15 | 19.5 | 2.4 | 6.5 | 3.3 |
| 6 | 23.2 | 1.3 | 4.7 | 3.1 | 16 | 21.6 | 2.5 | 5.1 | 4.9 |
| 7 | 22.2 | 2.5 | 4.5 | 3.1 | 17 | 26.5 | 1.9 | 6.3 | 4.8 |
| 8 | 18.1 | 2.5 | 5.9 | 2.5 | 18 | 26.4 | 2.2 | 8.1 | 4.4 |
| 9 | 13.7 | 2.8 | 7.9 | 3.3 | 19 | 23.1 | 3.8 | 4.9 | 2.2 |
| 10 | 7.7 | 2.6 | 9.3 | 2.3 | 20 | 21.1 | 2.5 | 5.1 | 1.8 |

Table 1: Data points used in An Introduction to Fuzzy Curves.

# 2   An Introduction to Fuzzy Curves

We consider a multiple-input, single-output system. We call the input candidates $x_i$ ($i = 1, 2, \ldots, n$), and the output variable $y$. Assume that we have $m$ training data points available and that $x_{ik}$ ($k = 1, 2, \ldots, m$) are the $i^{th}$ coordinates of each of the $m$ training points. Table 1 shows an example with $n = 3$ and $m = 20$.

For each input variable $x_i$, we plot the $m$ data points in $x_i$–$y$ space. Figure 1 illustrates the data points from Table 1 in the $x_1$–$y$, $x_2$–$y$, and $x_3$–$y$ spaces. For every point in $x_i$–$y$ space, we draw a fuzzy membership function defined by

$$\mu_{ik}(x_i) = \exp\left(-\left(\frac{x_{ik} - x_i}{b}\right)^2\right) \tag{1}$$

We typically take $b$ as about 10% of the length of the input interval of $x_i$. Figure 2 shows the fuzzy membership functions for the points in Figure 1. In Figure 2, the point where $\mu_{ik} = 1$ coincides with $x_{ik}$.

We use centroid defuzzification to produce a fuzzy curve $c_i$ for each input variable $x_i$ by

$$c_i = \frac{\sum_k^m \mu_{ik}(x_i) \cdot y_k}{\sum_k^m \mu_{ik}(x_i)} \tag{2}$$

Figure 3 shows the fuzzy curves $c_1$, $c_2$, and $c_3$ for the data in Table 1. We rank the importance of the input variables $x_i$ according to the range covered by their fuzzy curves $c_i$. The ranges of fuzzy curves in Figure 3 are 1.32 for $c_1$, 1.24 for $c_2$, and 0.89 for $c_3$. Hence, we deduce that $x_1$ is most significant input, $x_2$ is second, and $x_3$ is third.

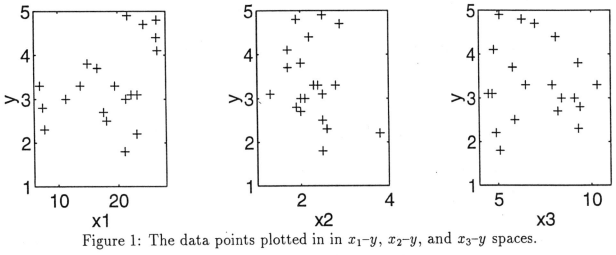

Figure 1: The data points plotted in in $x_1$–$y$, $x_2$–$y$, and $x_3$–$y$ spaces.

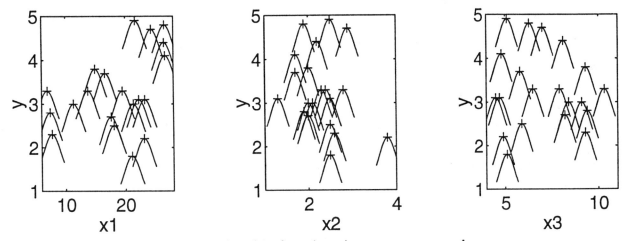

Figure 2: Fuzzy membership functions in $x_1$–$y$, $x_2$–$y$, and $x_3$–$y$ spaces.

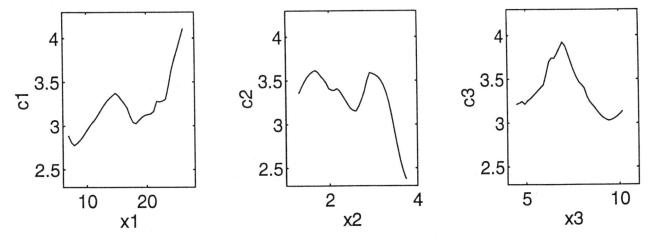

Figure 3: Fuzzy curves $c_1$, $c_2$, and $c_3$.

# 3  Simple Examples

We start with a two input, single output nonlinear system defined by

$$y = \left(1 + x_1^{1.5} - 5\cos(2x_2)\right)^2, \qquad 0 \leq x_1, x_2 \leq 3. \tag{3}$$

We randomly take 100 points from $0 \leq x_1, x_2, x_3, x_4, x_5 \leq 3$. $x_1$ and $x_2$ are used by Equation 3 to obtain 100 input-output data, and $x_3, x_4$, and $x_5$ are dummy inputs used to illustrate input variable identification. We find that the ranges of the fuzzy curves, $c_1$ to $c_5$, are 43.5, 48.3, 11.7, 14.3, and 11.4, respectively. From this, we identify $x_1$ and $x_2$ as the significant input variables for the system.

To determine if we can identify significant inputs with noise, we add 10% random noise to $x_1$, $x_2$, and $y$ after using Equation 3 to obtain the output data. The ranges of fuzzy curves, $c_1$ to $c_5$, are 45.9, 55.7, 21.2, 23.8, and 14.0, respectively. So our method can identify input variables within noise. To determine if it can identify incorporated variables, we use

$$y = \left(1 + x_1^{1.5} - 5\cos(2\sqrt{x_2 x_3})\right)^2, \qquad 0 \leq x_1, x_2, x_3 \leq 3 \tag{4}$$

and

$$y = \left(1 + (x_1 x_4)^{0.75} - 5\cos(2\sqrt{x_2 x_3})\right)^2, \qquad 0 \leq x_1, x_2, x_3, x_4 \leq 3 \tag{5}$$

The ranges of the fuzzy curves(from $c_1$ to $c_5$) for Equation 4 are 40.8, 34.6, 30.4, 15.7, and 18.4, respectively, and the ranges of fuzzy curves for Equation 5 are 27.9, 33.3, 32.5, 27.1, and 17.6, respectively. Hence, our method successfully identifies the significant inputs in all cases shown.

# 4  Comparison to Prior Work

We tested our method using several previously published examples. We used our significant inputs and trained a fuzzy-neural net to produce the given output data. In each example, our performance measure is equal to or better than that given in previously published work.

The Box and Jenkins gas furnace data is taken from [1] and [3]. We used 5 of 10 possible inputs, trained on the first 250 data points, and predicted the last 40 points. The results are shown in Figure 4.

The Chemical Oxygen Demand in Osaka Bay example was originally in [2] and then used in [4]. In contrast to both [2] and [4], we found that only 4 of the 5 inputs are significant, we use the first 33 points as training data, and predict the last 12 points. Our results are shown in Figure 5

The daily stock prices data is from [3] for stock A. It has ten input variables. In contrast to [3], we eliminated 4 input variables, trained on the first 80 data points, and predicted the last 20. The results are shown in Figure 6.

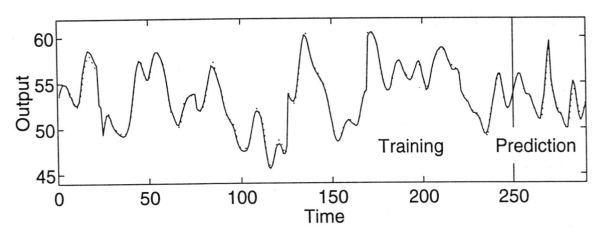

Figure 4: Performance on the Box and Jenkins gas furnace data. We build the model on the first 250 points and predict the next 40. Actual is shown by the solid line —, model by the dotted line ···.

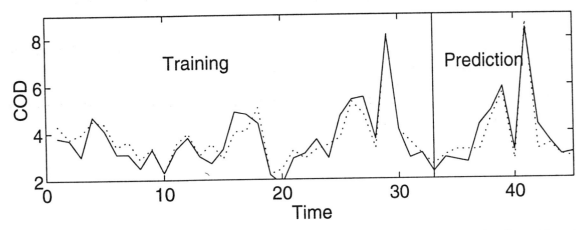

Figure 5: Performance on the Osaka Bay data. We build the model on the first 33 points and predict the next 12. Actual is shown by the solid line —, model by the dotted line ···.

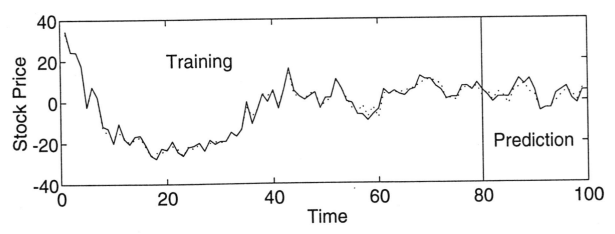

Figure 6: Performance on the stock price data. We build the model on the first 80 points and predict the next 20. Actual is shown by the solid line —, model by the dotted line ···.

# 5 Conclusions

We introduce the concept of a fuzzy curve and use it to identify input variables. Our method is simple to implement, and the time complexity is linear with respect to the number of input candidates. We successfully test our method with several examples.

# References

[1] G. E. P. Box and G. M. Jenkins, *Time Series Analysis, Forecasting and Control*. San Francisco: Holden Day, 1970.

[2] S. Fujita and H. Koi, Application of GMDH to Environmental System Modeling and Management, in *Self-Organizing Methods in Modeling: GMDH Type Algorithms*, (S.J. Farlow, Ed.), Statistics Textbooks Monographs Ser., Vol. 54, Marcel Dekker, New York, 1984, p. 257-275.

[3] M. Sugeno and T. Yasukawa, "A Fuzzy-Logic-Based Approach to Qualitative Modeling," *IEEE Trans. Fuzzy Systems*, vol. 1, no. 1, 1993, 7-31.

[4] H. Takagi, I. Hayashi, "NN-Driven Fuzzy Reasoning," *Int'l. J. Approximate Reasoning*, vol. 5, no. 3, 1991, 191-212.

# THE CONTINUOUS-TIME MARKOVIAN PROCESS
# WITH FUZZY STATES

Salvador Gutiérrez
Inst. Tecnológico de Morelia
Depto. de Computación y Sistemas

Av. Tecnológico No. 1500
C.P. 58120 Morelia, Mich.
México FAX (43) 12-1643

## Abstract

Continuous-time markovian processes have a very wide range of application, but the resulting state sets can be uncomfortable or unwieldy to the decision maker. A way to simplify such state sets introducing fuzzy sets is explored. A process with fuzzy states can be defined over another (underlying) process with crisp or fuzzy states, establishing a relation that can be the basis for a systematic methodology for simplifying this kind of stochastic processes.

## Continuous-Time Markovian Process

Let $\{X(t), x \geq t\}$ be an irreducible Markovian Process with $X(t) \in X \subseteq \mathbb{N}$

$$P\{X(t+s) = j \mid X(s) = i, X(u) = x(u), 0 \leq u < s\}$$
$$= P\{X(t+s) = j \mid X(s) = i\} \quad i,j \in X; \; s,t \in \mathbb{R} \quad \ldots(1)$$

f i $\qquad\qquad P\{X(t+s) = j \mid X(s) = i\}$

is independent of time, then the process is called stationary.

## Process with Fuzzy States

Let $\{F(t): t \geq 0\}$, $F(t) \in F \subseteq \mathbb{N}$, *where* $\#F \leq \#X$ be a process where $F(t)$ are fuzzy states

defined by membership functions $\mu_k(x) \in [0,1]$, $k \in F$, $x \in X$ , we have

**<u>Prop.1</u>** $\quad \begin{aligned} &P\{F(t+s) = k \mid F(s) = l; \; F(v) = f(v); \; 0 \leq v < s\} \\ &= P\{F(t+s) \; k \mid F(s) = l\} \quad \ldots(2) \end{aligned}$

provided there is a function $\phi: X \to F$ such as

$$\{\phi(x) = k^* \mid \mu_{k^*}(x) = \overset{\max}{\underset{k}{}}(\mu_k(x))\}, \text{ where } k, k^* \in F, x \in X \qquad \ldots(3)$$

and $\overset{\max}{\underset{k}{}}(\mu_k(x))$ is the first occurrence of the maximum value of $\mu_k(x)$, scanning for increasing values of k.

**Proof** We get eq. (2) applying $\phi$ to eq. (1). In fact, the markovian property can be induced over the process with fuzzy states if $\phi$ is any ordinary function uniquely assigning one value of $F$ as image to each value of $X$, as long as it does not introduce dependence upon more than one state.

Then, $X$ is known as an underlying process of $F$.

**Lemma 1.** The pre-images (fibers) of $r \in F$ under $\phi$, induce a partition of $X$.
**Proof** Since $\phi$ is a function mapping elements of $X$ on elements of $F$. Then, any $x \in X$ can be in the pre-image of only one f such that $\phi(x) = f$, $f \in F$. Since all x's are assigned, the union of all pre-images of all x's covers $X$. in the past.

Now, if for $k, l \in F$

$$P_{kl}(t) = \sum_{\substack{i \mid \phi(i) = k, \\ j \mid \phi(j) = l}} P_{ij}(t) \qquad \ldots(4) \quad \text{and}$$

$$P_{ll}(t) = \sum_{\substack{i \mid \phi(i) = l, \\ j \mid \phi(j) = l}} P_{ij}(t) \qquad \ldots(5) \quad,$$

we have $P_{kl}(t+s) = \sum_{r=1}^{\infty} P_{kr}(t) P_{rl}(s) \qquad \ldots(6)$ (The Chapman-Kolmogorov eqs. -ref. 1, 2, 3-).

**Proof** Since $X$ is a continuous-time markovian process, the Chapman-Kolmogorov equations hold

for it; i.e., $P_{ij}(t+s) = \sum_{u=1}^{\infty} P_{iu}(t) P_{uj}(s); \ t, s \in \mathbb{R}^+; \ i, j, u \in X \subseteq \mathbb{N} \qquad \ldots(7)$. Then, applying (7)

for all i,j such that $\phi(i) = k$, $\phi(j) = l$; $i, j \in X$, $k, l \in F$, since r running over all the range of $F$ means that u runs over all of $X$, if $\phi(u) = r$.

All the results known for continuous-time markovian processes, from Kolmogorov's backward and forward equations to limit probabilities can be derived in the same way, since the process with fuzzy states becomes just another process on a different variable with a reduced state set (where states are fuzzy sets or representatives of fuzzy sets obtained via a defuzzifying operator).

The underlying process can also be a process with fuzzy states. Then, a process with fuzzy states defined over another process (with crisp or fuzzy states), is called a "coarsening" of the underlying process. We can define a relation $X \blacktriangleright Y$, meaning $Y$ is a coarsening of $X$. This relation is easily proven transitive and defines a hierarchy among these processes. If $X \blacktriangleright Z$ and $Y \blacktriangleright Z$, then $X$ and $Y$ are called equivalently reducible ($X \equiv Y$).

In a practical situation it is possible to find an adequate coarsening of a process, having a manageable number of fuzzy states, with names and meanings more adequate for human decision makers.

## Conclusion

Continuous-time markovian processes have many useful applications, but sometimes the number and interpretation of the states can become uncomfortable for human decision makers. But it is always possible to build an adequate process with fuzzy states which corresponds to the underlying process and is a simplified version of it, with fuzzy names and concepts which turn decision-making easier. Much work remains to be done, specially with other kind of stochastic processes, but it can be rewarding since interesting applications tend to have a complex state set structure.

## References

1.- S. Ross, "Stochastic Processes", John Wiley, NY 1983.

2.- E. Parzen, "Stochastic Processes", Holden-Day, San Francisco, Cal. 1962.

3.- D.R.Cox and H.D. Miller, "The Theory of Stochastic Processes", Methuen, London, 1965.

# A METHOD FOR COMPUTING THE MOST TYPICAL FUZZY EXPECTED VALUE

**Stamatis Vassiliadis** - IBM AWDS, 11400 Burnet Road, Austin, TX, 78758
**George Triantafyllos** - IBM LSCD, 522 South Road, Poughkeepsie NY, 12601-5400
**Gerald G. Pechanek** - IBM Microelectronics, 203 Silicon Drive, Research Triangle Park, NC

**ABSTRACT:** A new method for computing the most typical fuzzy expected value of a membership function in a fuzzy set is described. The fuzzy expected value computed by this method denoted as the Clustering Fuzzy Expected Value is based on grouping of individual responses, that meet certain criteria, into clusters. Each cluster is considered a "super response" and contributes to the result proportional to its relative size and the difference in opinion from the mean of the entire sample. In so doing, the Clustering Fuzzy Expected Value represents the opinion of the majority of the population, but it also respects the opinion of the minority. A comparison is made with existing schemes, such as the Fuzzy Expected Value and the Weighted Fuzzy Expected Value, also intended to compute the most typical value in fuzzy sets. The advantages of Clustering Fuzzy Expected Value are demonstrated by examples for cases where other methods fail to perform.

## I. INTRODUCTION

The Fuzzy Expected Value (FEV) [1] has been introduced by A. Kandel to indicate the most "typical" grade of membership of a given fuzzy set [1-5] and it was meant to replace, in that capacity, the mean value and the median [6]. The FEV has been defined by the following [1] and [6]:

Consider a fuzzy relation defined over the universe X which yields a fuzzy set A with a membership function $\chi_A$ that satisfies $0 \leq \chi_A(x) \leq 1$, $x \in X$. Let $\mu$ be a fuzzy measure defined over the subsets of X and let $\xi_T = \{x \mid \chi_A(x) \geq T\}$, $0 \leq T \leq 1$. Then, the fuzzy expected value, FEV, of $\chi_A$ over A, is define by

$$FEV \equiv \sup_{T \in [0,1]} \{ \min[T, \mu(\xi_T)] \} \qquad (1)$$

where, $\mu(\xi_T) = f_A(T)$ is a function of the threshold T. The $FEV(\chi_A)$ may be calculated as the median of the set $\{\chi_1, \ldots, \chi_m, \mu_1, \ldots, \mu_{m-1}\}$ Where, $\chi_i = \chi_A(x)$, $1 \leq i \leq m$, $0 \leq \chi_1 \leq \chi_2 \leq \ldots \leq \chi_m \leq 1$, and, $\mu_i = 1/N \sum_{j=i+1}^{m} n_j$, $1 \leq i \leq m-1$, $N = \sum_{i=1}^{m} n_i$, and $n_i$ is a finite population.

M. Friedman, M. Schneider and A. Kandel [6] suggested that the FEV "may occasionally generate improper results". They attributed such a failure to "its insensitivity to population's density", and proposed the weighted fuzzy expected value (WFEV), to "replace the FEV whenever it fails to function" [6], defined by the following: Let $\omega(x)$ be a non-negative monotonically decreasing function defined over the interval $[0,1]$ and $\lambda$ a real number greater than 1. The solution s of

$$s = \frac{\chi_1 \omega(|\chi_1 - s|)n_1^\lambda + \ldots + \chi_m \omega(|\chi_m - s|)n_m^\lambda}{\omega(|\chi_1 - s|)n_1^\lambda + \ldots + \omega(|\chi_m - s|)n_m^\lambda} \qquad (2)$$

is the WFEV and is denoted $WFEV(\omega, \lambda)$. Where, $\omega(x) = e^{-\beta x}$, $\beta > 0$, $\lambda$ is a real number $> 1$, n is the number of values in a set, and s is the weighted fuzzy expected value of order $\lambda$.

In this paper we reexamine the FEV and WFEV and investigate their capabilities. The main contributions of our work can be summarized by the following:

- We demonstrate that the FEV always fails to perform satisfactorily under certain circumstances.
- We show that under certain conditions WFEV fails to represent the most typical value.
- We propose a new quantity denoted as the Clustering Fuzzy Expected Value (CFEV) that appears promising in attributing the most "typical" grade of membership of a given fuzzy set, and demonstrate its performance, with examples, including the cases where both FEV and WFEV clearly fail to produce satisfactory results.

The paper is organized as follows: In the next section, we present the benchmarks needed for evaluation purposes. Following that, we present an evaluation of both, FEV and WFEV. We proceed in the next sections with the presentation of the Clustering Fuzzy Expected Value (CFEV), and the evaluation of the performance of the CFEV, (a comparison is made with the mean, median, CFEV, FEV and WFEV). Finally, we conclude the presentation with some remarks.

## II. THE FRAMEWORK AND THE EVALUATION BENCHMARKS

In this presentation, we assume that for a set $A$ of membership values $\chi_i$ corresponding to the populations $n_i$, the most representative membership value is a value $x$ "close" to the majority of the population, but that it also considers the membership grades of the minority of the populations in the fuzzy set. For example, consider the three populations, $n_1 = n_2 = 10$ and $n_3 = 45$, having the membership values $\chi_1 = 0.2$, $\chi_2 = 0.6$ and $\chi_3 = 0.7$, respectively. In this example, the "most representative" value should be close to the "majority" of the opinions (0.6 and 0.7), but also representative of the "minority" (0.2). In essence the value produced by a proposed measure that presumably correspond to the fuzzy expected value for the set should produce a value in the interval 0.6 - 0.7 with some "fluctuation" due to the existence of the value 0.2. To establish the goodness of the methods proposed by [1], [6] and our proposal we use a set of examples. This is the only approach used thus far to establish the validity of proposals for fuzzy expected values, see for example [1, 6]. Our benchmarks are described in Figure 1 In Figure 1, Examples #1 - #7 are reported in [6], and the remainder of the examples were developed-for our evaluations. Each example in Figure 1 contains the population group with the same opinion ($i$), the opinion of population group i ($\chi_i$), the size of population group i having opinion $\chi_i$ ($n_i$), and the percentage of population group i having opinion $\chi_i$ based on the total population ($p_i$).

| exmpl | i | Xi | Ni | Pi |
|---|---|---|---|---|
| #1 | 1 | 0.400 | 1 | 0.08 |
|  | 2 | 0.500 | 3 | 0.25 |
|  | 3 | 0.550 | 4 | 0.33 |
|  | 4 | 0.600 | 2 | 0.17 |
|  | 5 | 1.000 | 2 | 0.17 |
| #2 | 1 | 0.100 | 5 | 0.83 |
|  | 2 | 1.000 | 1 | 0.17 |
| #3 | 1 | 0.050 | 70 | 0.70 |
|  | 2 | 0.300 | 30 | 0.30 |
| #4 | 1 | 0 | 50 | 0.50 |
|  | 2 | 0.50 | 50 | 0.50 |

| exmpl | i | Xi | Ni | Pi |
|---|---|---|---|---|
| #5 | 1 | 0.200 | 35 | 0.35 |
|  | 2 | 0.300 | 25 | 0.25 |
|  | 3 | 0.600 | 40 | 0.40 |
| #6 | 1 | 0.050 | 90 | 0.90 |
|  | 2 | 1.000 | 10 | 0.10 |
| #7 | 1 | 0.125 | 7 | 0.07 |
|  | 2 | 0.375 | 19 | 0.19 |
|  | 3 | 0.625 | 31 | 0.31 |
|  | 4 | 0.875 | 43 | 0.43 |
| #8 | 1 | 0.100 | 5 | 0.05 |
|  | 2 | 0.150 | 25 | 0.25 |
|  | 3 | 0.200 | 15 | 0.15 |
|  | 4 | 0.550 | 55 | 0.55 |

| exmpl | i | Xi | Ni | Pi |
|---|---|---|---|---|
| #9 | 1 | 0.100 | 69 | 0.69 |
|  | 2 | 0.300 | 31 | 0.31 |
| #10 | 1 | 0.500 | 51 | 0.51 |
|  | 2 | 0.800 | 40 | 0.40 |
|  | 3 | 0.900 | 9 | 0.09 |
| #11 | 1 | 0.100 | 37 | 0.37 |
|  | 2 | 0.150 | 25 | 0.25 |
|  | 3 | 0.400 | 38 | 0.38 |
| #12 | 1 | 0.500 | 49 | 0.49 |
|  | 2 | 0.950 | 11 | 0.11 |
|  | 3 | 0.975 | 40 | 0.40 |

Figure 1. Benchmark Examples

## III. AN EVALUATION OF FEV AND WFEV

It has been suggested that FEV "may occasionally generate improper results" and the failure has been associated with "its insensitivity to population's density" [6]. While it is true that the FEV generates improper results, it can be suggested that its failure may not be considered as neither occasional nor attributed to its insensitivity to the population density. The FEV formulation, regarding the attribution of the most typical value, suggest a "direct" association of the population with the grades of membership, and such an association, if any association exist, is at best unclear. To clarify, consider Example #3 in Figure 1. The definition of the FEV suggests the following computations: $N = \sum_{i=1}^{2} n_i = 30 + 70 = 100$ and $\mu_1 = (1/100) \sum_{i=1}^{2} n_2 = 0.30$ Then, $FEV(\chi_A)$ is the median of the set $\{0.05, 0.30, 0.30\}$ i.e. indicates that the FEV value for the example is 0.3. Given that $n_2$ and $n_1$ are "populations" and $\mu_1$ is in effect a membership grade the FEV proposes a direct relationship among them.

In assuming that the "most typical expected value" should somehow include the minority, it can be claimed that the FEV has produced an improper value [6]. This can be explained because the computation of FEV suggests that the "opinions" of people ($\chi_1, \chi_2$) and the "percentages" of people (implicitly incorporated in the $\mu_i$ computation) having those opinions are related by the definition of the fuzzy measure of Equation (1) to produce a "typical opinion". In fact, the definition of FEV implies that this measure will always produce a non "typical" value under certain condition described by the following two properties proved in [7].

*Property 1:* For a given set of populations, there exists at least one set of membership grades such that the FEV will always fail to represent the most typical value for the set $\chi_i$.

*Property 2:* For a given set of membership grades, there exists a set of populations such that FEV will always fail to represent the most typical value for the set.

We begin our evaluation of WFEV by considering Example #7 in Figure 1. In this example, $\chi = \{0.125, 0.375, 0.625, 0.875\}$ and $P = \{0.07, 0.19, 0.31, 0.43\}$. To compute the WFEV values, the parameters $\lambda$, $\beta$ must be chosen. Choosing $\lambda = 1$, $\beta = 2$ as suggested in [6], Equation (2) converges after 4-5 iterations to $s = 0.745$. Since 74% of the population answered either 0.625 or 0.875, the most typical value is expected to fall somewhere between 0.625 and 0.875. Indeed, all methods give an answer in this interval:

(FEV = 0.625, mean = 0.65, median = 0.625, WFEV = 0.745). What should be, however, the best value? M. Friedman, M. Schneider and A. Kandel suggested that *"WFEV is clearly the best choice"* because *"The majority of the elements have membership grades either 0.625 or 0.875. The most typical value is expected to fall somewhere in the middle, but to also indicate the existence of the few elements with memberships 0.375 and 0.125."* While it may be claimed that the WFEV value falls in between the interval (0.625, 0.875) it is not clear why such value is "clearly the best choice". Indeed, it can be argued that the median (0.65) is also a good choice as it falls in between the interval (0.625, 0.875), and it considers "few" elements that fall outside that interval. Furthermore, it can be argued that the mean is a better choice then WFEV because of the following: The "few elements" are 26% of the total population, thus if we assume that 1 over 4 in a population is hardly "few", the value should be closer 0.625 than to 0.875. Additionally, most of the elements, i.e. 57% of them, are between 0.125 and 0.625 indicating that while the most representative value should be in between 0.625 - 0.875 (74% of the population), the 57% of the population should have a bigger influence than just 0.13 from the majority having opinion 0.875. Consequently, the median, which also falls in the proper interval and has a 0.225 "deviation" from 0.875, should be considered to be the better choice not the WFEV. Clearly, which value can be considered as the most typical may be application dependent. What can not be suggested however is that the WFEV value "constitutes clearly the best choice" without taking into account the application as the previous argumentation suggests.

So far, the WFEV was considered using the $\lambda = 2$ and $\beta = 1$ as given in [6]. However, as M. Friedman, et. al. pointed out, the choice of the best $\lambda, \beta$ may require experimentation [6], or even the development of an expert system that will chose the best $\lambda, \beta$ for a particular application based on a user's supplied knowledge base but, as the choice is application dependent, they provide no guidelines of how to choose those parameters. Since the behavior of WFEV depends on good values for $\lambda$ and $\beta$, the problem becomes how to chose those values. Clearly, this additional requirement may create a difficulty and potential problems may arise as it may lead to unpredictable situations. To quantify, consider, for example, the following case:

If we assume that Example #7 in Figure 1 is our knowledge base and that WFEV = 0.745 is a good answer then we need to choose $\lambda, \beta$ such that the desired result for Example #7 is obtained. If the WFEV algorithm is applied to Example #7, the following is obtained:

```
WFEV(λ = 2, β = 1)  =  0.74508
WFEV(λ = 1, β = 6)  =  0.74418
```

Both sets of values for $\lambda, \beta$ yield almost the same results for the example under considerations, and there may be other sets of values that yield the same answer, as well. Let us now consider Example #2 in Figure 1 and apply the two sets of values found above to calculate WFEV. The choice of $\lambda = 2, \beta = 1$ yields $WFEV = 0.11483$ which may be considered as typical fuzzy expected value for Example #2. However, the second choice for $\lambda = 1, \beta = 6$ yields $WFEV = 0.97263$ which is anything but a typical value for the given example.

Several other problems may arise with WFEV depending on the data being used. For example, as it is demonstrated in [7] WFEV may produce different values depending on the number of significant digits of the membership grades. Consider the following example in Figure 2 which is a modification of Example #7, with the modifications being trivial (added more precision). Given that the two examples are "close" and the fact that for Example #13, the majority of elements is between 0.624600 and 0.875513, we expect, as also indicated in [6], to obtain an expected value between those two numbers. A computation of WFEV however produces the WFEV value 0.599 clearly outside of what has been considered in [6] as the proper range. In fact, if we assume that the values of Example #13 were the outcome of intermediate floating point computations, then several cases may be cited, depending on the method used to compute the final value (truncation, rounded up and truncation, etc.) [7] where the WFEV will produce incorrect or unpredictable results [7].

## IV. DESCRIPTION OF THE CLUSTERING FUZZY EXPECTED VALUE (CFEV)

The CFEV, introduced here as an alternative to WFEV, can be described by the following: First, the elements of a fuzzy set are clustered into separate groups, such that the membership grades in the same cluster are closely associated. The clustering of membership grades is based on two parameters, "s" and "d", "s" being the maximum allowable distance between the first and last elements in a cluster, and it serves as a measure of the size of a cluster, and "d" is the maximum allowable distance between two consecutive elements within a cluster, and it represents a measure of the closeness of the elements within a cluster. Both, "s" and "d", serve as the cut-point for establishing the clusters. The elements of the formed clusters are confined to the following two inequalities.

$$| C_{i,j-1} - C_{i,j} | < d \ (gap) \quad (3)$$

$$| C_{i,1} - C_{i,n} | < s \ (size) \quad (4)$$

In the Inequalities (3) and (4) $C_{i,j-1}$ and $C_{i,j}$ are two consecutive elements in cluster $i$, and $C_{i,1}$ and $C_{i,n}$ are, respectively, the first and last elements in cluster $i$.

| i | Xi | Pi | i | Xi | Pi | i | Xi | Pi | i | Xi | Pi | i | Xi | Pi |
|---|----|----|---|----|----|---|----|----|---|----|----|---|----|----|
| 1 | 0.125091 | 2 | 7 | 0.624709 | 3 | 13 | 0.625300 | 3 | 19 | 0.874503 | 3 | 25 | 0.875109 | 3 |
| 2 | 0.125001 | 5 | 8 | 0.624803 | 3 | 14 | 0.625400 | 3 | 20 | 0.874604 | 3 | 26 | 0.875210 | 3 |
| 3 | 0.375000 | 6 | 9 | 0.624904 | 3 | 15 | 0.625500 | 3 | 21 | 0.874705 | 3 | 27 | 0.875311 | 3 |
| 4 | 0.375001 | 7 | 10 | 0.625000 | 3 | 16 | 0.874200 | 4 | 22 | 0.874806 | 3 | 28 | 0.875412 | 3 |
| 5 | 0.375090 | 6 | 11 | 0.625100 | 3 | 17 | 0.874301 | 3 | 23 | 0.874907 | 3 | 29 | 0.875513 | 3 |
| 6 | 0.624600 | 4 | 12 | 0.625300 | 3 | 18 | 0.874402 | 3 | 24 | 0.875008 | 3 | | | |

Figure 2. Example 13.

The two "distance" parameters, "s" and "d", have been introduced to indicate two properties of membership values. The first, associated with "d", relies on the closeness of the "responses", i.e. if two membership values are "d" apart, than they may be considered to have a "close relationship". Clearly, "being close" may not be respected alone with the distance between two consecutive elements in the set, and the "s" parameter is required to establish the maximum distance between two arbitrary elements that can be considered close. Clearly, the parameters and algorithms are application related, as for some applications for example, 0.1 may be considered a close distance between consecutive elements, but not for other applications. By way of example only, we consider that clustering follows the following assumption:

1. Maximize the size of each cluster. Since all elements in the original cluster are closely associated, according to Inequality (3), the resulting clusters should have as many elements as possible.
2. Minimize the number of clusters which follows from the previous requirement.

Clearly, depending on the application, other assumptions are possible and not excluded a priori. The clusters are formed by applying the following algorithm (other algorithms are also possible):

**Clustering Algorithm**

- Step 1: Select the "s" and "d" values.
- Step 2: Cluster the elements within the fuzzy set based on the "d" value, i.e. select consecutive elements into a cluster in a manner that satisfies Inequality (3).
- Step 3: If the first and last elements of the formed cluster satisfy Inequality (4), the cluster qualifies, and the process is completed if there are no more clusters to be formed. Otherwise, if there are more clusters to be formed, go to step 2. If the formed cluster does not qualify, proceed to step 4.
- Step 4: In the non-qualifying cluster,
  1. Select a new value $d'$ for the parameter $d$ by taking the greatest distance between two consecutive elements of the set.
  2. Divide the original cluster into sub-clusters using $d'$ as the criteria indicate.
  3. For each sub cluster that does not satisfy Inequality (4) repeat sub-steps 1 and 2.
  4. Merge the resulting clusters to find all possible solutions that satisfy Inequality (4).
  5. From all possible solutions produced in sub-step 4 select the one that maximizes the following function:

$$\prod_{i=1}^{n} s_i/n = \max \qquad (5)$$

where, $s_i$ is the size of cluster i (distance between the smallest and biggest number), and n is the number of clusters for each solution.

## V. COMPUTATION OF THE CLUSTERING FUZZY EXPECTED VALUE (CFEV)

When the clusters are formed by the mean of an appropriate algorithm, the choices of s and d, representing an application, the mean of each cluster, $W_{A_i}$, and the mean of the entire set, $W_A$, are computed. Based on the means and the number of observations of the entire fuzzy set and that of the individual clusters, the CFEV is represented as follows:

$$CFEV = W_A + \sum_{i=1}^{m} \left( \frac{N_i}{N} \right)^x (W_{A_i} - W_A) \qquad (6)$$

$$x \geq 1$$

In Equation (6), the mean of the entire fuzzy set is adjusted based on the population sizes and the means of the formed individual clusters. In this process, each cluster shifts the mean of the entire fuzzy set by a factor proportional to the distance between the mean of the cluster and that of the entire fuzzy set, the ratio of the population size of the cluster and that of the entire fuzzy set. Thus, a cluster with a mean smaller than that of the entire fuzzy set reduces the mean of the fuzzy set by a factor proportional to the distance between the two means and the ratio of the population size of the cluster to that of the entire fuzzy set. On the other hand, a cluster with a mean larger than that of the entire fuzzy set increases the mean of the fuzzy set by a factor which is also based on the distance and ratio of the cluster and the fuzzy set.

Assuming the data to be analyzed with CFEV represents a sample of membership grades, whose values are in the interval [0, 1], it can be proven than the CFEV will always produce a result that is in the same interval, .i.e. $0 \leq CFEV \leq 1$ [7].

Given that the CFEV value depends on the value x to determine the "deviation" from the mean, it is of interest to present some of the "behavior" of CFEV when different x values are considered. Assuming two fuzzy sets (A) with $n = \{5,10,5\}$ and corresponding $\chi_i = \{0.5, 0.8, 0.3\}$ and (B) with $n = \{10,5,5\}$ and corresponding $\chi_i = \{0.95, 0.6, 0.7\}$, Figure 3 shows the value of the CFEV as a function of the parameter x.

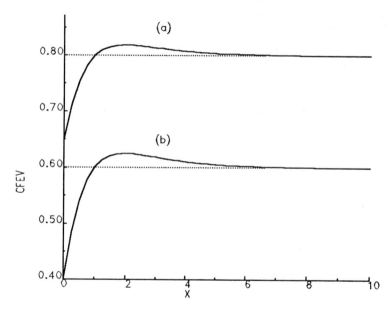

Figure 3. CFEV of examples (A) and (B).

## VI. EVALUATION

In order to evaluate the proposed measure we choose $x = 2$ in the following CFEV equation and evaluated the benchmarks of Figure 1 with the addition of example 13.

$$CFEV = W_A + \sum_{i=1}^{m} \left( \frac{N_i}{N} \right)^2 (W_{A_i} - W_A) \qquad (7)$$

We report the complete evaluation in Figure 4. The evaluation shown in Figure 4 includes the CFEV value (for d = 0.10 and s = 0.20), the mean value, the median, the FEV and $WFEV(\omega = 2, \lambda = 1)$ for the reported examples. We note that the CFEV produces satisfactory results in all cases for the assumed definition of what constitutes the "most typical" value while the other quantities do not. In addition to the circumstances indicating that the FEV and WFEV will fail, as discussed in the previous sections, we note that where the mean and median have been reported to fail [6], the CFEV has produced a satisfactory answer. For example, as suggested in Example #1 [6], while the mean will provide an incorrect "typical value" (0.608), the "correction" in CFEV provides a deviation from the mean that is enough to "pull" the value in the proper range. Furthermore, where the median produces an unacceptable answer, e.g. Example #6 [6], the CFEV again produces the correct value. The evaluation of the CFEV performance when compared to the other approaches suggest that the proposed approach can be considered to be a better choice for establishing the most typical value in a fuzzy set. A more extensive evaluation may be found in [7] where we compare CFEV with WFEV using several examples in which WFEV fails and show that CFEV produces satisfactory results.

| Example | CFEV | Mean | Median | FEV | WFEV |
|---------|------|------|--------|-----|------|
| #1 | 0.582 | 0.608 | 0.550 | 0.550 | 0.576 |
| #2 | 0.167 | 0.250 | 0.100 | 0.167 | 0.115 |
| #3 | 0.104 | 0.125 | 0.050 | 0.300 | 0.083 |
| #4 | 0.250 | 0.250 | 0.250 | 0.500 | 0.250 |
| #5 | 0.391 | 0.385 | 0.300 | 0.400 | 0.402 |
| #6 | 0.077 | 0.145 | 0.050 | 0.100 | 0.054 |
| #7 | 0.677 | 0.650 | 0.625 | 0.625 | 0.745 |

| Example | CFEV | Mean | Median | FEV | WFEV |
|---------|------|------|--------|-----|------|
| #8 | 0.385 | 0.375 | 0.550 | 0.550 | 0.478 |
| #9 | 0.146 | 0.162 | 0.100 | 0.300 | 0.130 |
| #10 | 0.654 | 0.656 | 0.500 | 0.500 | 0.614 |
| #11 | 0.211 | 0.226 | 0.150 | 0.380 | 0.232 |
| #12 | 0.742 | 0.739 | 0.950 | 0.510 | 0.686 |
| #13 | 0.677 | 0.650 | 0.625 | 0.625 | 0.599 |

Figure 4. The Evaluation of the Reported Examples

## VII. CONCLUDING REMARKS

In this paper we described a method for computing the most typical fuzzy expected value of a membership function in a fuzzy set. The fuzzy expected value computed by this method, which is referred to as the Clustering Fuzzy Expected Value (CFEV), is based on grouping of the individual responses into clusters, and it represents the opinion of the majority of the population, but it also respects the opinion of the minority.

It was shown that:

- the FEV and WFEV have failed to perform satisfactorily in some applications,
- the advantages of the CFEV over the FEV and CFEV were demonstrated by examples for cases where other methods fail to perform satisfactorily, thus, eliminating the need for experimentation to chose the best method, or the development of an expert system to chose the best $\lambda, \beta$ to be applied in generating the CFEV value, as suggested in [6].

The final suggestion of the presentation is that the CFEV represents a promising methodology in establishing the "most typical" value. CFEV was used as part of a question answering system [8] to compute the most typical value of the responses to interview questions.

## VIII. REFERENCES

[1]   A. Kandel,"Fuzzy Mathematical Techniques with Applications," *Addison-Wesley*, Reading, Massachusetts, pp. 72-101, 1986.
[2]   G. Klir and T. Folger,"Fuzzy Sets, Uncertainty, and Information," *Prentice Hall*, Englewood Cliffs, New Jersey, 1988.
[3]   D. Dubois and H. Prade,"Fuzzy Sets and Systems: Theory and Applications," *Academic Press*, New York, New York, 1980.
[4]   R. Yager, S. Ovchinnikov, R. Tong and H. Nguyen (eds.),"Fuzzy Sets and Applications: Selected Papers by L.A.Zadeh," *John Wiley and Sons*, New York, New York, 1987.
[5]   A. Kaufmann and M. Gupta,"Introduction to Fuzzy Arithmetic: Theory and Applications," *Van Nostrand Reinhold Company Inc.*, New York, New York, 1985.
[6]   M. Friedman, M. Schneider and A. Kandel,"The Use of Weighted Fuzzy Expected Value (WFEV) in Fuzzy Expert Systems," *Fuzzy Sets and Systems Vol. 31*, pp. 37-45, May 1989.
[7]   S. Vassiliadis, G. Triantafyllos, and W. Kobrosly,"CFEV: The Clustering Fuzzy Expected Value," *IBM Technical Report TR 01.C729, Endicott, NY, p:34*, April 1993.
[8]   S. Vassiliadis, G. Triantafyllos, and W. Kobrosly,"A Fuzzy Reasoning Database Question Answering System," *IBM Poughkeepsie, TR 00.3749*, pp. 38, September 1993.

# Reliability structure functions based upon fuzzy numbers

Vincenzo CUTELLO
*Department of Mathematics*
*University of Catania*
*Catania, Italy*

Javier MONTERO
*Department of Statistics and O.R.*
*Complutense University*
*Madrid, Spain*

*Abstract*— Abstract: The main aim of this paper is to establish a basis for reliability structure functions where the space of states for the system and its components is assumed to be a family of fuzzy numbers. A particular non-probabilistic uncertainty can be in this way modeled, generalizing classical crisp binary and multistate models in Reliability Theory. The importance of an underlying order relation defined on the given family of fuzzy numbers is also stressed.

**Key words:** Reliability Theory, Structure Functions, Fuzzy Numbers.

## I. INTRODUCTION.

Theoretical research works on Reliability always refer to a basic mathematical model, which establish how the real system is understood. The system is then represented in terms of a *structure function*, that is, a mapping assigning a system state to each possible profile of component states. Classical Reliability Theory develops probabilistic properties of binary systems, in such a way that all uncertainty is assumed to be only of a random nature, and every component -and the system itself- is just allowed to be either in complete failure state or perfect functioning state. Since these two assumptions relative to the states and their uncertainty appear to be too restrictive for many applications, different alternative approaches can be proposed, both to add possibilistic uncertainty and to include intermediate states between those two extreme states (see [22]). A general model relative to the space of states has been proposed in [18], by means of structure functions $\phi : L^n \to L_0$, where $L$ and $L_0$ are two arbitrary complete lattices and $n$ is the number of components (see also [6,20]). In this way, classical binary systems $\phi : \{0,1\}^n \to \{0,1\}$ ([1,3]), multistate systems $\phi : \{0,1,\ldots,K\}^n \to \{0,1,\ldots,K\}$ ([9,8,11,21]) and continuum systems $\phi : [0,1]^n \to [0,1]$ ([2,4,19]) are all trivially generalized.

In this paper we shall explore a model where the space of states are fuzzy numbers, leading to a particular system

$$\phi : L^n \to L$$

L being a lattice of fuzzy numbers.

## II. THE FUZZY-STATE ASSUMPTION.

It is a usual assumption in reliability theory to focus theoretical studies onto systems which are *isotonic*, in the sense that such the associated structure function is non-decreasing (that is, system performance does not improve when more components fail). Obviously, such an isotonicity (non-decreasingness) depends on the particular partial order associated to the space of states.

Fuzzy-state assumptions are justified when the system or its components are simultaneously to some extent in a fuzzy success and/or in a fuzzy failure state.

Such a situation is common when dealing with complex systems or components, since some states that can be identified are not easily ranked. Defining an appropriate crisp valuation set may not be clear at all, and such a difficulty may be due to some inherent fuzziness.

A fuzzy number is represented by a membership function on the real line, that is, a mapping

$$\mu : \Re \rightarrow [0,1]$$

which associates a degree of membership $\mu(x)$ to each real number $x \in \Re$. Assuming that states are modeled according to fuzzy numbers, structure functions in this context should be defined as mappings

$$\phi : N^n \rightarrow N$$

where $N$ represents an appropriate subset of fuzzy numbers. Notice that performance under Kaleva's approach [12] is defined as a fuzzy number on the unit real interval where probabilities take values.

A standard assumption is to restrict the model to the set of normal convex fuzzy numbers, that is, those fuzzy numbers $\mu : \Re \rightarrow [0,1]$ such that

- there exists a "mean value" $x \in \Re$ such that $\mu(x) = 1$, and

- $\mu(y) \geq min\{mu(x), \mu(z)\}$ whenever $x < y < z$.

Second condition -convexity- is equivalent to impose that every level set $\{x/\mu(x) \geq \alpha\}$ is convex, for any $\alpha \in [0,1]$ (see [13]). Convex fuzzy numbers are sometimes named single-caved fuzzy numbers in the context of group decision making (see [17]). It is easy to check that a fuzzy number $\mu$ verifies normality -that is, the above first condition- together with convexity if and only if there exists a "mean value" point $x \in \Re$ with $\mu(x) = 1$ such that $\mu(z) \leq \mu(y)$ whenever $z < y < x$ or $x > y > z$. Frequent additional assumptions to these normal convex fuzzy numbers are continuity (i.e., its membership function $\mu$ is continuous) and uniqueness of the mean value point. Sometimes it is also assumed that $lim_{x \rightarrow \infty}\mu(x) = lim_{x \rightarrow -\infty}\mu(x) = 0$. In some contexts

it is also efficient the restriction to triangular fuzzy numbers (see, e.g., [13]).

Least upper bound and greatest lower bound can be both defined for any pair of normal convex fuzzy numbers, by applying Zadeh's extension principle (see e.g., [7]) to the canonical order of the real line. In fact, Mizumoto and Tanaka [15,16] already proved that normal convex fuzzy numbers form a distributive lattice under the "∨" (sup, join) and the "∧" (inf, meet) operations. Given two fuzzy numbers $\mu, \nu$, the least upper bound $\mu \vee \nu$ and the greatest lower bound $\mu \wedge \nu$ are respectively defined by the fuzzy numbers

$$
\begin{aligned}
(\mu \vee \nu)(t) &= sup_{max(x,y)=t} min\{\mu(x), \nu(y)\} \text{ and} \\
(\mu \wedge \nu)(t) &= sup_{min(x,y)=t} min\{\mu(x), \nu(y)\}.
\end{aligned}
$$

A partial order relation $\leq$ such that $\mu \leq \nu$ if and only if $\mu \vee \nu = \nu$ (or equivalently, $\mu \wedge \nu = \mu$) was therefore defined on the set of fuzzy numbers. Given two normal convex fuzzy numbers $\mu, \nu$ it can be proven that $\mu \leq \nu$ if and only if there exists $y \in \Re$ such that $x \leq y \leq z$ for some $x, z$ with $\mu(x) = 1$ and $\mu(z) = 1$ in such a way that $\mu(a) \leq \nu(a)$ for all $a > y$ and $\mu(a) \geq \nu(a)$ for all $a < y$.

Practical considerations suggest that the concept of perfect functioning and complete failure must always be clearly defined. Hence, it seems appropriate to assume that our fuzzy numbers must be all defined on the same closed interval $I$ of the real line (which can as well be the compact real line). Let us then assume hereafter that $N_I$ is the set of normal convex fuzzy numbers $\mu$ such that $\mu(x) = 0 \ \forall x \notin I$, where $I$ is a previously fixed closed real interval. Given an arbitrary family $\{\mu_i, i \in M\}$ of normal convex fuzzy numbers in $N_I$, the least upper bound $\vee\{\mu_i, i \in M\}$ and the greatest lower bound $\wedge\{\mu_i, i \in M\}$ will respectively be given by the following fuzzy numbers:

$$
\begin{aligned}
(\vee\{\mu_i, i \in M\})(t) &= sup_{sup\{x_i\}=t} inf\{\mu_i(x_i), i \in M\} \\
(\wedge\{\mu_i, i \in M\})(t) &= sup_{inf\{x_i\}=t} inf\{\mu_i(x_i), i \in M\}
\end{aligned}
$$

The above least upper bound $\vee\{\mu_i, i \in M\}(t)$ can be alternatively obtained by taking

$$sup\{\mu_i(t), i \in M\}$$

if for all i there exists $x \leq t$ such that $\mu_i(x) = 1$; or

$$inf\{\mu_i(t), i \in M, \mu_i(x) < 1 \ \forall x \geq t\}$$

otherwise.

Analogously, we have that the above greatest lower bound $\wedge\{\mu_i, i \in M\}(t)$ can be alternatively obtained by taking

$$sup\{\mu_i(t), i \in M\}$$

if for all $i$ there exists $x \geq t$ such that $\mu_i(x) = 1$; or

$$inf\{\mu_i(t), i \in M, \mu_i(x) < 1 \ \forall x \leq t\},$$

otherwise.

Hence, we obtain that the set $N_I$ of normal convex fuzzy numbers vanishing outside a fixed closed interval $I$ is a complete lattice.

Finally, let us denote by $U = N_I$ the subset of normal convex fuzzy numbers vanishing outside the unit interval $I = [0, 1]$ (i.e., $\mu(x) = 0$ for all $x \notin [0, 1]$). An interesting structure function based upon fuzzy numbers as states can be naturally derived from continuum systems. Given a continuum system

$$\phi : [0, 1]^n \rightarrow [0, 1]$$

being a nondecreasing and continuous mapping such that $\phi(1, 1, ..., 1) = 1$ and $\phi(0, 0, ..., 0) = 0$, then it can be proven that the mapping

$$\psi : U^n \rightarrow U$$

obtained by applying Zadeh's extension principle, that is,

$$\psi(\mu_1, ..., \mu_n)(t) = \nu(t),$$

where

$$\nu(t) = sup_{\phi(x_1, ..., x_n) = t} min\{\mu_1(x_1), ..., \mu_n(x_n)\}$$

is well defined. Moreover, it is nondecreasing with respect to the above partial ordering defined on $U$. These systems derived by extension from a given continuum system will require special attention in future research.

Obviously, other partial order relations -different from the one considered above- can be more appropriate in particular problems. Alternative partial orders

between fuzzy numbers will lead to different lattice state spaces. The complete lattice of fuzzy numbers considered here in order to get the basic comparision between different states gives in fact not much information. Many practical problems will require a more informative order relation. For example, sometimes it is defined a ranking function mapping each fuzzy set into the real line, where a natural order exists (see [5]). A specific order structure may be more appropriate when we restrict the model to a smaller class of normal convex fuzzy numbers, or just because we found it more accurate or more operative in practice. A more general approach can be developed if a fuzzy binary relation is defined over the set of fuzzy numbers. Comparison of fuzzy numbers by means of two dimensional fuzzy preference relations (see, e.g., [7,5,23]) have not been considered in this paper.

## III. HYBRID SYSTEMS

Hybrid failure diagnosis Systems (as described in [24]) represent a particular class of decision systems which work on different modes according to the circumstances. Whenever the problem is well defined they operate on the basis of a probabilistic risk analysis of the different failure situations accounting for interdependicies and external events. Otherwise the system is designed so to shift to a mode in which heuristics (in the form of production rules) are used. Our proposed model lends itself quite naturally to a fuzzy extension of such (crisp) Hybrid Systems, because it provides both a natural way to handle partial failures of multi-component systems and to formalize experts heuristics in terms of fuzzy IF-THEN rules.

Fuzzy numbers can be described in terms of *set of confidence*, i.e. the support of the fuzzy number and *levels of presumption*, i.e. the $\alpha$-levels of the fuzzy number. As a consequence, we can say that the bigger is the set of confidence the lesser is the certain information we have. In this way, we can mathematically model the *well-definition* of the problem and provides an algorithmic way to decide when to shift from one mode to the other in the Hybrid System at hand.

## IV. FINAL COMMENTS

Crisp non-binary systems and some fuzzy-state based models indeed meet within the context of lattice theory. Complexity in real state spaces appears as a basis for fuzzy-state based reliability theories. Mizumoto and Tanaka [16] already showed that fuzzy sets of type 2 whose grades are normal convex fuzzy numbers are L-fuzzy sets in the sense of Goguen [10]. Following [20], it was therefore expected a deep link between some Fuzzy-State based Reliability and Lattice Reliability. Whenever a (complete) lattice structure is assumed over the set of fuzzy states defining a given space of states, all considerations on general (complete) lattices do apply to such fuzzy models. In particular, it has been shown in the previous section that normal convex fuzzy numbers vanishing ouside a fixed complete support provide a complete lattice state space on the basis of a natural partial order relation defined on those fuzzy numbers. Moreover, a particular type of structure function derived from crisp continuum systems and Zadeh's extension principle has been defined, showing also its isotonicity.

**Acknowledgment:** This research has been partially supported by Dirección General de Investigación Científica y Técnica (Spanish national grant PB91-0389).

## REFERENCES

[1] R.E. Barlow and B. Proschan. *Statistical Theory of Reliability and Life Testing.* Silver Spring, MD (1981).

[2] L.A. Baxter. Continuum structures I. *Journal of Applied Probability* 21 (1986), 802–815.

[3] Z.W. Birnbaum, J.D. Esary and S.C. Saunders. Multi component systems and structures. *Technometrics* 3 (1961), 55–77.

[4] H.W. Block and T.H. Savits. Continuous multistate structure functions. *Operations Research* 32 (1984), 703–714.

[5] G. Bortolan and R. Degani. A review of some methods for ranking fuzzy subsets. *Fuzzy Sets and Systems* 15 (1985), 1–19.

[6] B. Cappelle Multistate structure functions and possibility: an alternative approach to reliability. In: E.E. Kerre (Ed.), *Introduction to the Basic Principles of Fuzzy Set Theory and some of its Applications.* Communication and Cognition, Gent (1991), 252–293.

[7] D. Dubois and H. Prade. Decision-making under fuzziness. In: M.M. Gupta, R.K. Ragade and R.R. Yager (Eds.), newblock *Advances in Fuzzy Set Theory and Applications.* North Holland, Amsterdam (1979), 279–302.

[8] E. El-Neweihi, F. Proschan and F. Sethuraman. Multistate coherent systems. *Journal of Applied Probability* 15 (1978), 675–293.

[9] E. El-Neweihi and F. Proschan. Degradable systems: a survey of multistate system theory. *Communication Statistical Theory and Methods* 13 (1984), 405–432.

[10] J.A. Goguen. L-fuzzy sets. *Journal of Mathematical Analysis and Applications* 18 (1967), 145–174.

[11] W.S. Griffith. Multistate reliability models. *Journal of Applied Probability* 17 (1980), 735–744.

[12] O. Kaleva. Fuzzy performance of a coherent system. *Journal of Mathematical Analysis and Applications* 117 (1986), 234–246.

[13] A. Kaufmann and M.M. Gupta. *Introduction to Fuzzy Arithmetic.* Van Nostrand Reinhold, New York (1985).

[14] G.J. Klir and T.A. Folger. *Fuzzy sets, Uncertainty and Information.* Prentice Hall, Englewood Cliffs, NJ, 1988.

[15] M. Mizumoto and K. Tanaka. Some properties of fuzzy sets of type 2. *Information and Control* 31 (1976), 312–340.

[16] M. Mizumoto and K. Tanaka. Some properties of fuzzy numbers. In: M.M. Gupta, R.K. Ragade and R.R. Yager (Eds.), *Advances in Fuzzy*

*Set Theory and Applications.* North Holland, Amsterdam (1979), 153–164.

[17] J. Montero. Single-peakedness in weighted aggregation of fuzzy opinions in a fuzzy group. In: J. Kacprzyk and M. Fedrizzi (Eds.), *Multiperson Decision Making using Fuzzy Sets and Possibility Theory.* Kluwer, Dordrecht (1990), 163–171.

[18] J. Montero, J. Tejada and J. Yáñez. General structure functions. *Kybernetes.* To appear.

[19] J. Montero, J. Tejada and J. Yáñez. Structural properties of continuum systems. *European Journal of Operational Research* 45 (1990) 231-240.

[20] J. Montero, B. Cappelle and E. Kerre. The usefulness of complete lattices in reliability analysis. In: T. Onisawa and J. Kacprzyk (Eds.), *Fuzzy sets and possibility theory in reliability and safety analysis.* Springer-Verlag, Heidelberg (1994).

[21] B. Natvig. Two suggestions of how to define a multistate coherent system. *Advances in Applied Probability* 14 (1982), 434–455.

[22] T. Onisawa and J. Kacprzyk (Editors). *Fuzzy sets and possibility theory in reliability and safety analysis.* Springer-Verlag, Heidelberg (1994).

[23] S.V. Ovchinnikov. Transitive fuzzy orderings of fuzzy numbers. *Fuzzy Sets and Systems* 30 (1989), 283–295.

[24] E. Paté-Cornell and H. Lee. Hybrid systems for failure diagnosis. In: D.E. Brown, and C.C. White (Eds.), *Operations Research and Artificial Intelligence: the integration of problem solving strategies.* Kluwer Academic Publishers, Boston (1990), 141–165.

# DIAGNOSIS BY APPROXIMATE REASONING ON DYNAMIC FUZZY FAULT TREES

**MIHAELA ULIERU**

Institute of Automatic Control
Technical University of Darmstadt
Landgraf-Georg Str. 4, 64283 - Darmstadt, Germany

**Abstract.** An *approximate-reasoning model* for diagnosis of continuous dynamic systems is introduced based on a previously developed *fuzzy extension* of the fault tree analysis and synthesis approach. The concept of *dynamic fuzzy fault tree* naturally emerges from the act of *matching* the *fuzzified fault tree* with the *dynamic symptoms*. Management of the *incipient fault dynamics* via fuzzy information processing is illustrated on a simple example.

**Key Words.** Approximate-Reasoning Model, Dynamic Symptoms, Dynamic Fuzzy Fault Tree, Similarity Evaluation by Possibility Measure, Incipient Fault Dynamics, Dynamic Fuzzy Information.

## 1. INTRODUCTION

The concept of *fault tree* has been widely used in systems reliability evaluation [1]. By modeling fault's propagation to its manifestations (symptoms) through deterministic chains of undetectable effects (events) Fig 1 [2] it attempts to determine the *probability of fault's occurrence* from the a-priori determined conditional probabilities of the events and symptoms. While adequate to *discrete* systems for which *either* a normal *or* a faulty behaviour can be clearly stated, the static fault tree analysis and synthesis approach is inappropriate for the diagnosis of continuous dynamic systems [3;4]

The performances of *continuous dynamic systems* are in many cases characterized by a *gradual transition* from the range of *acceptable* behaviours to that of *unacceptable* ones [5]. This defines an *unsharp boundary* between *faulty/non-faulty* behaviour which has to be adequately managed especially when the fault develops *incipiently*. In this case the fault shows up *gradually*, via *dynamic symptoms* developing continuously in time as increasing or decreasing *process variables* (measurable signals, parameters, states), Fig. 2. Sometimes these symptoms occur intermittently, Fig.3. An important role in incipient fault detection is played by those symptoms detected by *subjective human perceptions* (e.g. a *strange sound* or the *smell of burning*) which cannot be monitored but can be integrated with the process variables as fuzzy subsets [6].

The application of the classical static frame of the fault tree approach for managing the dynamics of *incipient faults* with a fast development may lead to wrong diagnostic decisions. This is due to the inability of the induced probabilistic reasoning strategy to access and process on-line the ambiguous (fuzzy) information brought by the dynamic and subjective symptoms [7] which may have different *simultaneous* developments (e.g. some symptoms may occur

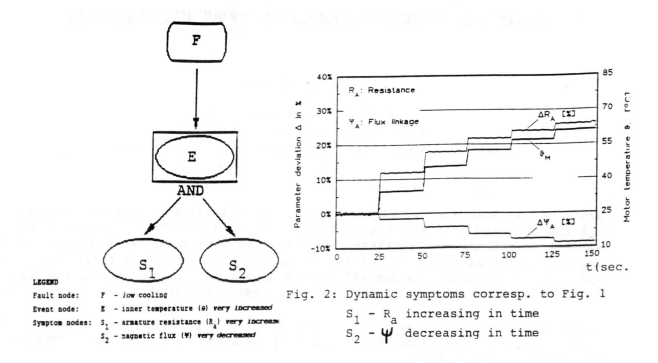

Fig. 2: Dynamic symptoms corresp. to Fig. 1
$S_1$ - $R_a$ increasing in time
$S_2$ - $\Psi$ decreasing in time

LEGEND

Fault node:     F - *low cooling*

Event node:     E - *inner temperature ($\theta$) very increased*

Symptom nodes:  $S_1$ - *armature resistance ($R_a$) very increase*
                $S_2$ - *magnetic flux ($\Psi$) very decreased*

Fig.1: Fault tree example

intermittentely, others may have a slow development while others may occur abbruptely.) This paper proposes a *possibilistic reasoning approach* for processing the fault trees, which makes them suitable for diagnosing complex faults in continuous dynamic systems.

## 2. DYNAMIC FUZZY FAULT TREES

### 2.1. Capturing the Fuzzy Information

The first step towards accessibility to the fuzzy information is the *fuzzification* of the fault trees, explained in detail in [2] on the same example given in Fig.1. In short, by fuzzifying the variables represented by each node (as the fuzzification in fuzzy control) the *meaning* of each link will be captured by a fuzzy relation of the product space of the universes symbolized by the connected nodes, Fig. 4.

If a *symmetrical* fuzzy relation is choosen, (this can be achieved by defining it via any *commutative* operator as for example minimum or product) the link becomes an *association* of the elements represented by its nodes [4]. This fairly models the representation of the diagnostic knowledge existing in the expert's mind. Indeed, with gathering experience, the diagnostician builds in his/her mind *associations* mapping symptoms with faults. When confronted with a concrete diagnostic task he tries to find *similarities* with the known associations in order to make the diagnostic *decision*.

In the following it will be shown how the process of *approximate reasoning by similarity evaluation* [8] (which emulates the diagnostician's decision-making) can be modeled via processing the fuzzy fault trees by possibilistic reasoning.

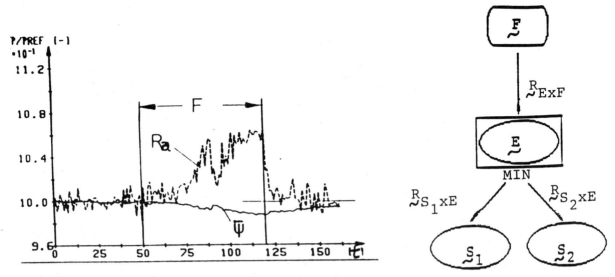

Fig. 3: Intermittent symptom          Fig.4: Fuzzy fault tree

## 2.2. Possibilistic Reasoning on the Fuzzy Fault Tree

The process of *similarity evaluation by possibility measures* is detailed in [2], [9] and [10]. First the tree's information is *compressed* by fuzzy composition to *direct symptoms-fault associations* $R_{\sim S_i x F}$ (for the example in

Fig. 4: $R_{\sim S_i x F} = R_{\sim S_i x E} \; o \; R_{\sim E x F}$, i=1,2.). By *matching* them with the *detected*

values $S_i'$ of the symptoms (in our case: i=1,2) the possibility distributions $F_{\sim S_i} = S_i' o R_{\sim S_i x F}$ (i=1,2), acting as *local fault possibilities* brought by the

detection every symptom, are determined. These are distinct *pieces fuzzy information* by which each detected symptom reflects fault's occurrence. By aggregating them on the tree's structure, a *global fault possibility distribution* $GP_{\sim F}$ is obtained (for the example in Fig. 4, $\mu_{GP}(f) = \underset{f \in F}{MIN} \{\mu_{F_{S1}}(f),$

$\mu_{F_{S2}}(f)\}$, f∈F.)

In fact, by the act of matching the *similarity* between the *tree's information* (reflecting the situation when the corresponding fault has certainly occurred) and the information brought by the *detected symptoms* (reflecting the concrete diagnostic situation) is evaluated. As measures of similarity *possibility measures* were choosen [10]. Finally as similarity measure $PM = \underset{f \in F}{sup} \; GP_{\sim F}$ is

computed as basic information concerning fault's occurence.

## 2.3. Emergence of the Dynamic Fuzzy Fault Trees

Consider now the dynamic evolution in time of symptoms $S_1$ and $S_2$ ($s_1(t)$ $s_2(t)$)

detected as in Fig. 2 as a reflection of the fast development of the incipient fault F. By *matching* $s_1(t)$ and $s_2(t)$ with the approximate reasoning model induced by the fuzzified fault tree, the possibility distributions $F_{\sim R_a}(t)$ and

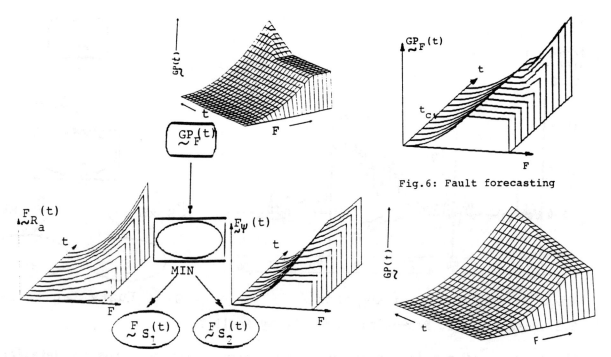

Fig.6: Fault forecasting

Fig.5: Dynamic fuzzy fault tree

Fig.7: Dynamic fault possibility by product

$F_{\sim\psi}(t)$ by which the evolution in time of the symptoms brings information concerning fault's gradual development, are determined, (in Fig. 5 attached to the bottom nodes). The *fault dynamics* will be reflected through the evolution in time of its global possibility distribution $GP_{\sim}(t)$ (illustrated in Fig. 5 at the top node) obtained by aggregation of $F_{\sim R_a}(t)$ and $F_{\sim\psi}(t)$ via the MIN connective of the tree. In this way a *dynamic fuzzy fault tree*, which models the fault dynamics via processing dynamic fuzzy information on the structure of the crisp fault tree, naturally emerges.

## 3. DIAGNOSIS OF DYNAMIC FAULTS VIA FUZZY INFORMATION PROCESSING ON THE DYNAMIC FUZZY FAULT TREE

### 3.1. Forecasting Dynamic Faults

By further estimating the future evolution of $F_{\sim R_a}(t)$ and $F_{\sim\psi}(t)$ via *statistical prediction techniques* [11] , the time moment $t_c$ when the fault becomes critical can be forecasted, Fig. 6. For this e.g. a time series analysis of the symptoms as ARMA models can be used [12].

### 3.2. Incipient Faults Dynamics

Figure 7 illustrates the dynamics of fault's possibility for F in our example obtained by making the aggregation via the norm *product* for the evolutions of the detected dynamic symptoms $R_a(t)$ and $\Psi(t)$ as in Fig. 2. By comparison with the dynamics of fault's possibility distribution obtained via the MIN connective, which is illustrated at the top node in Fig. 5 for the same detected symptoms, it appears clear that the MIN connective (and in general

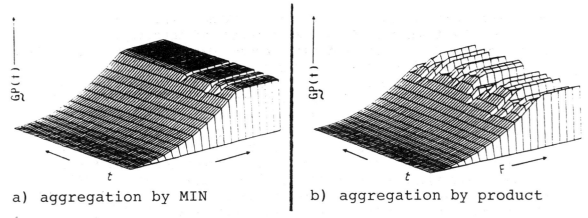

a) aggregation by MIN                    b) aggregation by product

Fig.8: Dynamic possibility distribution of an intermittent fault

the *possibilistic* MIN/MAX pair), due to its great nonlinearity, makes a sharper discrimination, by being therefore more adequate for *incipient faults detection*.

### 3.3. Monitoring Intermittent Faults

The dynamic of the fault possibility when the intermittent symptom illustrated in Fig. 3 was detected is presented in Fig. 8 - a) under the MIN connective and b) under the *product*. The MIN connective is less sensitive than the product, by allowing only the symptom generating the *smallest* local fault possibility to contribute in fault's monitoring. Therefore the *product* (and in general the *probabilistic* ·/⊕ pair), due to its effect of *interactivity*, allowing *all* the symptoms to contribute with information in monitoring the dynamics, is more adequate for *diagnosing intermittent faults*.

### 4. CONCLUSIONS

A dynamic approach for fuzzy fault trees processing has been developed. It determines, based on similarity evaluation, the *possibility distribution* of fault's occurence rather than its probability. The possibilistic approach allows the on-line processing of the *dynamic fuzzy information* gathered from the on-line detected dynamic and subjective symptoms. By this the dynamic fuzzy fault tree is very effective as an *approximate reasoning model* for diagnosis of continuous dynamic systems especially when complex situations as incipient linear and/or intermittent faults are dealt with. Hints concerning the use of the right connectives depending on the contextual diagnostic task have been extracted by experiments.

### REFERENCES

[1] Lee, W.S., Grosh, D.L., Tillman, F.A. and Lie, C.H., "Fault Tree Analysis, Methods and Applications - A Review", *IEEE Transactions on Reliability*, Vol. R-34, No. 3, 1985.

[2] Ulieru, M., "Processing fault trees by approximate reasoning," *Proceedings of the 12th IFAC World Congress*, Vol. 9 pp. 221-224, Sydney, Australia Oxford: Pergamon Press, 1993.

[3] Ulieru, M., "From Boolean Probabilistic to Fuzzy Possibilistic Fault

Tree Processing in Diagnostic Decision Making", *Proceedings of EUFIT'93*, First European Congress on Fuzzy and Intelligent Technologies, Vol. 1, pp. 408-414, Aachen, Germany, Sept. 7-10, 1993.

[4] M. Ulieru, "From fault trees to fuzzy relations in managing heuristics for technical diagnosis," *Proceedings of IEEE/SMC'93 Conference*, Vol. 1, pp. 743-748, Le Touquet, France, October 1993.

[5] Leitch, R.R. and Quek, H.C., "A Behaviour Classification for Integrated Process Supervision", Proceedings of *IEE International Conference on Control*, Vol. 2, , pp. 127-133, Publ. No 332, 1991.

[6] R. Isermann, "On the integration of fault detection methods via unified fuzzy symptom representation," *Proceedings of EUFIT'93*, Vol. 1, pp. 400-407, Aachen, Germany, September 7-10, 1993.

[7] Ulieru, M., "Coping with Complex Diagnostic Tasks via Fuzzy Information Processing", *Int. Conf. on Information Processing and Management of Uncertainty - IPMU'94*, Paris, July 4-7, 1994.

[8] Ruspini, E.H., "Approximate Reasoning: Past, Present, Future", *Information Sciences*, 57-58/1991, pp. 297-317.

[9] Ulieru, M., "A Unified Fuzzy Approach for Diagnostic Decision by Causal Networks", *Proceedings QUARDET'93 - III IMACS International Workshop on Qualitative Reasoning and Decision Technologies*, pp. 672-682, Barcelona, Spain, June 16-18, 1993.

[10] Ulieru, M. and Isermann, R., "Design of a Fuzzy -Logic-Based Diagnostic Model for Technical Processes", *Int. Journal of Fuzzy Sets and Systems*, Vol. 58, Nr. 3/1993, pp. 249-271.

[11] Box, G. and Jenkins, G., *Time Series Analysis - forecasting and control -*, pp. 126-170, Holden-Day, San Francisco, 1970.

[12] Isermann, R., "Fault Detection Methods for the Supervision of Technical Procasses", *Process Automation*, pa/1981, pp. 36-44.

# An Improved Fuzzy Connectivity Diagnosis Method

William J. Yakowenko

Dept. of Computer Science, UNC-CH, Chapel-Hill NC, 27599-3175

yakowenk@cs.unc.edu

Hiroyuki Watanabe

Dept. of Electrical and Computer Engineering, Univ. of Missouri, Columbia MO, 65211

watanabe@risc1.ecn.missouri.edu

### Abstract

The connectivity diagnosis method proposed by Norris et al [NPB87] produced disappointing results in our experiments. In exploring the reasons for this, we designed a new algorithm in the same spirit as the original, and which is both more intuitive and more effective. By making use of the frequency of occurrence of symptoms in the training set, the new algorithm is better at selecting representative syndromes, and can recognize more than one syndrome for each disease.

## 1  The Original Algorithm

The NPB connectivity algorithm generates a pair of tables from a set of training data of patients having a particular disease, and then diagnoses each new patient by pattern-matching his symptoms with these tables. One of the two tables represents *positive* evidence of a disease; the other, *negative*. So, a strong diagnosis is a result of matching a *large* amount of positive evidence and a *small* amount of negative evidence. Both are handled by fuzzy set membership, and the results averaged.

The tables themselves are generated by measuring the similarity of symptoms' occurrence vectors. That is, each symptom has a vector describing which patients exhibited it, and these vectors are compared to each other. The result is a fraction, in which the numerator is the number of patients exhibiting both symptoms, and the denominator is the number exhibiting one or both symptoms. Since frequency of occurrence is not taken into account, even rare symptoms can match each other perfectly, as long as they occur in exactly the same sets of patients. At levels of similarity ranging from 0.0 to 1.0, symptoms are grouped together. At level zero, all symptom vectors match each other, so a single group is formed from the entire symptom set. At increasing levels of similarity, groups from lower levels are broken into smaller and smaller pieces; symptoms which matched at some low level of similarity might no longer match at a higher level. Each symptom occurrence vector matches itself perfectly, so even if there are no perfect matches between distinct symptoms, each symptom forms a group by itself at the level of 1.0. But not every symptom is characteristic of the disease, so some of these must be removed. To accomplish this, only one group is kept at each level: the largest one that has at least two symptoms and is a subset of the group at the next lower level. The result forms a single "mountain peak", with one group at each level; the largest at the bottom including all symptoms, the smallest at the top, and each being a subset of the one at the next lowest level. See figures 1 and 2. Here, symptom set 1, 2, 3 forms the mountain peak.

Diagnosis is done by comparing a patient's symptoms with those in each group in the table, noting the fraction of the group's symptoms which the patient has. These fractions are then weighted by their levels, and summed. After normalization so that a perfect match at all levels would come out to be 1.0, this is the evidence for the disease.

There are two fundamental problems with this method. First, the similarity measure used does not take frequency of occurrences into account at all. If two symptoms are exhibited by identical sets of patients,

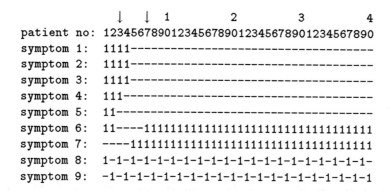

```
                    ↓   ↓  1         2         3         4
patient no: 1234567890123456789012345678901234567890
symptom 1:  1111------------------------------------
symptom 2:  1111------------------------------------
symptom 3:  1111------------------------------------
symptom 4:  111-------------------------------------
symptom 5:  11--------------------------------------
symptom 6:  11----1111111111111111111111111111111111
symptom 7:  ----111111111111111111111111111111111111
symptom 8:  1-1-1-1-1-1-1-1-1-1-1-1-1-1-1-1-1-1-1-1-
symptom 9:  -1-1-1-1-1-1-1-1-1-1-1-1-1-1-1-1-1-1-1-1
```

**Figure 1:** A training set showing two syndromes. The two patients marked by arrows will be used in later examples. Note the two syndromes: symptoms 1, 2, 3, and 4 (and 5 and 6 to a lesser extent); and symptoms 6 and 7 (and 8 and 9).

their similarity is 100%, even if the training set is large, and these symptoms rare. Therefore, similarity of occurrence vectors is not by itself a very good indicator of which symptoms to group together as a syndrome of a disease, particularly in the presence of other diseases. Second, a disease may have more than one syndrome, but only one can be encoded in a single peak.

The first problem with the existing algorithm is that it is sensitive to "noise"; symptoms of other diseases. Mountain peaks occur not for those symptom patterns which most frequently indicate the disease, but for those which occur in the *same subsets* of patients. This is particularly problematic for diseases which do not occur in isolation. A patient having more than one disease will typically exhibit symptoms for each, and it is not a simple matter to automatically decide which diseases caused which symptoms. So, it happens that some subset of patients with the disease we are diagnosing also has some other disease. In this subset, the symptoms caused by this other diseases have some degree of self-similarity; they form groups and peaks just as those caused by the disease we are considering. Since frequency of occurrence is not taken into account, these groups and peaks have as much chance as any others of being chosen for use in the final table. This is somewhat compensated for by the later averaging-in of negative evidence; such an ill-chosen positive symptom set will very likely score high in the negative table. But although this weakens the strength of a diagnosis based on such poorly chosen symptoms, it does nothing to strengthen the diagnosis of patients exhibiting discarded but truly representative symptom patterns.

A related problem is that similarity is reflexive; every symptom vector is self-similar at the level of 1.0. This means that every symptom forms its own peak, whether or not other symptoms join it in doing so. Therefore, some criteria must be used to "weed-out" some of these, as only some subset of symptoms is truly indicative of any disease. The criteria used was the size of the group; only the largest were kept. But just as similarity of occurrence vectors is not by itself a good indicator of which symptoms to group, these group sizes are not good indicators of which groups to keep. The actual syndromes of a disease may be quite small, and so may be easily out-numbered by "noise" as described above. Furthermore, a disease may in fact have more than one syndrome, but only one can be represented by a single peak. So discarding groups this way may produce a table that is either incomplete or simply wrong. Figure 2 shows the connectivity table resulting from the given training set. Although most patients in the training set have symptoms number six and seven, this peak is entirely discarded in favor of one having a larger number of symptoms, though fewer patients.

## 2  Our Modified Algorithm

Intuitively, our approach is first to connect symptoms not only by similarity of occurrence vectors, but also frequency of co-occurrence. Specifically, this means we divide the number of patients having both symptoms by the total number of patients in the training set, rather than dividing by the number of patients having one

| similarity level | symptoms | | | | | | | | | | patient # 3 | 7 |
|---|---|---|---|---|---|---|---|---|---|---|---|---|
| 100% | { 1 | 2 | 3 } | { *4* } | { *5* } | { *6* } | { *7* } | { *8* } | { *9* } | | 3/3 | 0/3 |
| 90% | { 1 | 2 | 3 } | { *4* } | { *5* } | { *6* } | { *7* } | { *8* } | { *9* } | | 3/3 | 0/3 |
| 80% | { 1 | 2 | 3 } | { *4* } | { *5* } | { *6* | *7* } | { *8* } | { *9* } | | 3/3 | 0/3 |
| 70% | { 1 | 2 | 3 | 4 } | { *5* } | { *6* | *7* } | { *8* } | { *9* } | | 4/4 | 0/4 |
| 60% | { 1 | 2 | 3 | 4 | 5 } | { *6* | *7* } | { *8* } | { *9* } | | 4/5 | 0/5 |
| 50% | { 1 | 2 | 3 | 4 | 5 } | { *6* | *7* } | { *8* } | { *9* } | | 4/5 | 0/5 |
| 40% | { 1 | 2 | 3 | 4 | 5 } | { *6* | *7* | *8* | *9* } | | 4/5 | 0/5 |
| 30% | { 1 | 2 | 3 | 4 | 5 } | { *6* | *7* | *8* | *9* } | | 4/5 | 0/5 |
| 20% | { 1 | 2 | 3 | 4 | 5 } | { *6* | *7* | *8* | *9* } | | 4/5 | 0/5 |
| 10% | { 1 | 2 | 3 | 4 | 5 } | { *6* | *7* | *8* | *9* } | | 4/5 | 0/5 |
| 0% | { 1 | 2 | 3 | 4 | 5 | 6 | 7 | 8 | 9 } | | 5/9 | 3/9 |
| | | | | | | | | | weighted sum: | 4.49 | 0.0 |

Figure 2: The connectivity table produced by the original algorithm, from the training set shown in Figure 1. All groups shown in italics are discarded, including those representing the most common syndrome. The pattern-matching fraction is shown at the right for the single group retained at each level for two selected patients, as is the weighted sum for each patient.

or both symptoms. This eliminates the problem of "noise" coming from the presence of other diseases; any other single disease is likely to be less frequent, and so its symptoms will no longer group together at such high levels. Then, rather than keeping only the largest group at each level, forming a single peak, we keep *all* groups, and so form many such peaks. This is possible because our connectivity measure is not reflexive; a symptom vector might be similar to itself only up to some intermediate level, so the initial connectivity table does not contain all symptoms at all levels, and no weeding-out rule is necessary. During diagnosis, each peak accumulates weighted evidence from all groups of which it is a subset. The strength of our diagnosis is then the highest of that for any single peak. This eliminates the problem of ignoring all syndromes but one.

Using this modified algorithm, let us re-consider the previous example, in which some small set of symptoms was more common, while some larger set was rarer. The symptoms in the smaller group will match each other only at some low level, corresponding to their frequency; if one in ten patients in the training set exhibited these symptoms, they will form a peak only at the 0.1 level. The more common symptoms, on the other hand, will peak much higher; perhaps they are exhibited by the other 90% of the patients in the training set, and so might peak at the level of 0.9. Thus all of the disease's possible syndromes are represented, and the rarer ones peak much lower, since they might actually be just noise. Figure 3 shows the new connectivity table, in which all syndromes are represented, and the more common one has a higher peak.

If a disease has more than one syndrome represented in the training set, each one will have to peak below 100%, because none will occur in all patients. For instance, if two syndromes are each represented by half of the patients, each one will peak at or below 50%. This puts diseases having multiple syndromes at a disadvantage, compared to those having only one. To correct for this, and make the diagnosis strengths of different diseases comparable to each other, we normalize the weighted sums so that a patient perfectly matching the highest peak will always get the same result, no matter what level that peak is. This leaves rarer syndromes at a disadvantage to more common ones, an arguable position since we cannot automatically differentiate them from noise.

# 3 Experimental Results

Our method was developed because of the disappointing success rate of the NPB algorithm in this situation. In our tests, with various training sets, and symptom data, the original algorithm achieved from 30% to

| similarity level | symptoms | | | | | | | | | | patient # 3 | 7 |
|---|---|---|---|---|---|---|---|---|---|---|---|---|
| 100% | | | | | | | | | | | | |
| 90% | | | | | | | { 6 } { 7 } | | | | $[\frac{0}{1}\ \frac{0}{1}]$ | $[\frac{1}{1}\ \frac{1}{1}]$ |
| 80% | | | | | | | { 6   7 } | | | | 0/2 | 2/2 |
| 70% | | | | | | | { 6   7 } | | | | 0/2 | 2/2 |
| 60% | | | | | | | { 6   7 } | | | | 0/2 | 2/2 |
| 50% | | | | | | | { 6   7 } { 8 } { 9 } | | | | $[\frac{0}{2}\ \frac{1}{1}\ \frac{0}{1}]$ | $[\frac{2}{2}\ \frac{1}{1}\ \frac{0}{1}]$ |
| 40% | | | | | | | { 6   7   8   9 } | | | | 1/4 | 3/4 |
| 30% | | | | | | | { 6   7   8   9 } | | | | 1/4 | 3/4 |
| 20% | | | | | | | { 6   7   8   9 } | | | | 1/4 | 3/4 |
| 10% | { 1   2   3 } | | | | | | { 6   7   8   9 } | | | | $[\frac{3}{3}\ \frac{1}{4}]$ | $[\frac{0}{3}\ \frac{3}{4}]$ |
| 0% | { 1   2   3   4   5 | | | | | | 6   7   8   9 } | | | | 4/9 | 3/9 |
| | | | | | | max weighted sum: | | | | | 0.75 | 4.25 |

**Figure 3:**  The connectivity table produced by our algorithm. No groups are discarded, so both syndromes in the training set are represented. There are five distinct peaks, and the more common patterns of symptoms form higher peaks. Pattern-matching fractions are again shown at the right. For levels having more than one group, fractions are listed for each in braces. Each peak has a weighted sum; only the highest one is shown for each patient.

55% true positive (*TP*) diagnosis rates, with 19% to 28% false positive (*FP*). Our modified algorithm, by contrast, produced 47% to 70% TP, with only 17% to 25% FP, on the same data. Furthermore, using the best (prototypical - see [WYK93]) training set, our rates jumped to 60%-70% TP, and 17%-20% FP; while no consistently better rates were found for any training set with the NPB algorithm (improvements for some symptom sets were offset by worsening of others, or TP increases offset by FP increases).

# 4   Conclusion

Although the original NPB connectivity algorithm was based on a good idea, that of recognizing co-occurring patterns of symptoms, its implementation was prone to misdiagnosis in the presence of multiple diseases. This was corrected primarily by recognizing that the frequency of occurrence of a symptom is valuable information, and incorporating it into our similarity measure. This in turn made it straightforward to allow for diseases having multiple syndromes, rather than only recognizing one per disease. Though there is perhaps still room for improvement, these changes alone have produced remarkable performance gains.

# References

[Atk74]  Atkin R.H., *Mathematical Structure in Human Affairs,* Crane, Russak & Co, New York, 1974

[NPB87]  Norris, Pilsworth, Baldwin, "Medical Diagnosis from Patient Records - A Method using Fuzzy Discrimination and Connectivity Analyses," *Fuzzy Sets & Systems,* 23 (1987) 73-87

[WYK93]  Watanabe, Yakowenko, Kim, Anbe, Tobi, "Application of a Fuzzy Discrimination Analysis for Diagnosis of Valvular Heart Disease," *IEEE Transactions on Fuzzy Systems,* to appear

# GOAL-DRIVEN REASONING FOR FUZZY KNOWLEDGE-BASED SYSTEMS USING A PETRI NET FORMALISM

A. Bugarín and S. Barro

Departamento de Electrónica e Computación. Facultade de Física.

Universidade de Santiago de Compostela. 15706 Santiago de Compostela (Spain).

E-mail: senen@gaes.usc.es

## ABSTRACT

In this paper we describe a Goal-Driven reasoning algorithm over Fuzzy Knowledge Bases. The approach to the problem permits characterizing the chaining between rules without requiring a complete identification between the propositions. The certainty degree of the rules is implemented by means of linguistic truth values and we include different considerations on the goals. A simple and adequate description of the process is achieved through the use of a Petri Net formalism for the representation and manipulation of Fuzzy Knowledge Bases.

## 1. INTRODUCTION

The introduction of Fuzzy Production Systems (FPSs) with chaining between the rules in its Fuzzy Knowledge Base (FKB) has allowed their application to problems which presented a higher level of complexity than those usually found in control. In this type of Systems it is interesting to structure the set of rules prior to the process of executing the FKB, specially in those cases in which a large number of variables and/or rules participate. In order to carry out this task several formal representation methods have been described. These methods permit a convenient representation of the rules of a FKB and the relationships they establish between the propositions (implication, conjunction and disjunction), facilitating the design of strategies for the execution of the FKB. We can classify most of the solutions of this type found in the literature into two groups: those that make use of a matrix representation [1-2] and those based on the Petri Net (PN) formalism. [3-8]

The Petri Net formalism permits the implementation in a simple and highly descriptive manner of the chaining relationships which hierarchize the FKB and establish a temporal order for its execution. It is also a valuable assistance for the systematic formulation of the different processes that are carried out over the FKB (evaluation according to different strategies, inconsistency detection, FKB depuration, ...). Most of the proposals, [3-6] however, do not make an intensive use of these great representative potentials. They present a high level of rigidness in the mechanism for the qualification of the rules (numerical and interval degrees are suggested) and with respect to the type of chaining that is allowed (a coincidence between variable and value is required in chained propositions), which leads to a merely symbolic processing, thus significantly restricting the possibility of performing fuzzy comparisons. In other cases, [7-8] the chaining mechanisms are more flexible (only a coincidence at the variable level is required), but they do not carry out an optimum approximation to the calculation of the inferences, computationally penalizing the process of executing the FKB. The dynamic processes described in the PN generally corresponds to the execution of the FKB following a particular strategy (Data-Driven), although in some cases alternative strategies are presented (Goal-Driven). [5,7-8] In this context we frame the proposal we make in the present paper: using a flexible and generic representation in the PN formalism we propose the description of a Goal-Driven inference algorithm over the FKB. With this objective, we briefly describe the characterization of the FKB in the formalism and we carry out an analysis of the Goal-Driven inference process in FKBs with chaining between rules.

## 2. EXECUTION OF A FKB IN THE PN FORMALISM

We will consider that the r-th rule $R^r$ in the FKB of a FPS is the following:

$$IF \; X_1^r \; IS \; A_1^r \; AND...AND X_{M_r}^r \; IS \; A_{M_r}^r$$
$$THEN \qquad\qquad (1)$$
$$X_{M_r+1}^r \; IS \; B_1^r \; AND...AND X_{M_r+N_r}^r \; IS \; B_{N_r}^r \quad (\tau^r)$$

where $X_{m_r}^r, X_{M_r+n_r}^r, A_{m_r}^r, B_{n_r}^r$, $m_r=1,...,M_r$, $n_r=1,...,N_r$, are, respectively, different linguistic variables and their corresponding values. $M_r$, $N_r$, $r=1,...,R$ represent the number of antecedents and consequents of rule $R^r$ and $\tau^r$, $r=1,...,R$, are linguistic values of the truth variable, [9] expressing the linguistic degree of certainty of the rules. Finally, $R$ represents the number of rules in the FKB.

In our approach [10] we represent this FKB by means of a Petri Net $\{P,T,A\}$ where $P=\{p_{m_r}^r\}$, $m_r=1,...,M_r+N_r$ is a finite set of **places**, $T$ is a finite set of **transitions** and $A\subseteq\{T\times P\}\cup\{P\times T\}$ is a set of **directed arcs**. We identify the places in the PN with single propositions in the FKB by means of the assignment function $\alpha:P \rightarrow PR$ defined as

$$\alpha(p_{m_r}^r) = \begin{cases} "X_{m_r}^r \; IS \; A_{m_r}^r" & if \; m_r \leq M_r \\ "X_{m_r}^r \; IS \; B_{m_r-M_r}^r" & if \; m_r > M_r \end{cases} \quad (2)$$

where $m_r=1,...,M_r+N_r$ and $r=1,...,R$. We denote by $PR$ the set of **propositions** in the FKB.

We also define the transition set $T$ as $T=T^R\cup T^C=\{t^1,...,t^R,t^{R+1},...,t^{R+C}\}$. Subset $T^R=\{t^1,...,t^R\}$ includes the transitions associated with each of the rules in the FKB, whereas subset $T^C=\{t^{R+1},...,t^{R+C}\}$ includes the transitions that are associated with the chaining variables in the FKB. The **input** and **output** functions over set $T$ have a different interpretation depending on the subset of $T$ in which they are

considered: if $t^j\in T^R$, $p_i\in I(t^j) \Leftrightarrow \alpha(p_i)$ in the antecedent part of $R^j$ and $p_i\in O(t^j) \Leftrightarrow \alpha(p_i)$ in the consequent part of $R^j$. If $t^j\in T^C$, $p_i\in I(t^j)$, $p_k\in O(t^j) \Leftrightarrow \alpha(p_i)$ is linked with $\alpha(p_k)$, i.e., $\alpha(p_i)$ and $\alpha(p_k)$ assert on the same variable.

The graphic representation of the PN is a directed arc graph, with circles representing places and vertical lines representing transitions. Only directed arcs in $A = \bigcup_{t^j\in T} \{t^j\times O(t^j)\}\cup\{I(t^j)\times t^j\}$ are permitted. Figure 1 shows an example of the PN formalism we have just described.

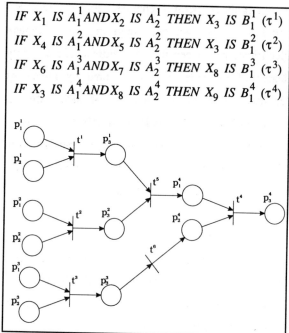

Fig. 1: Representation of a KB in the PN formalism.

The representation of the dynamic process of evaluating the FKB is carried out by the **Marking function**, which associates tokens to the places in the PN. Marking functions in this formalism take binary values ($\{1,0\}$) that indicate if the Degree of Fulfillment (DOF) of each proposition is or not known at a given moment. A **fulfillment function** $g:P \rightarrow [0,1]$ is defined for quantifying the DOF of the proposition associated to a given place, i.e., $g(p)=DOF(\alpha(p))$, $p\in P$.

Applying the sup-min compositional rule of inference, the DOF of a proposition in the consequent part of rule $R^r$ can be obtained from the DOF of the propositions in the antecedent part. Taking the previous definitions into account, this implication process can be represented by

$$g(p_{M_r+n_r}^r) = \bigwedge_{p_{m_r}^r \in I(t^r)} g(p_{m_r}^r), \quad n_r = 1,...,N_r, \quad r = 1,...,R \quad (4)$$

Also, if we consider the chaining situation represented by the transition $t^j \in T^C$, we will have [11] $\forall p_{m_t}^t \in O(t^j)$,

$$g(p_{m_t}^t) = \bigvee_{p_{M_s+n_s}^s \in I(t^j)} \tau^s \Big( g(p_{M_s+n_s}^s) \Big) \circ^s \mu_{p_{M_s+n_s}^s, p_{m_t}^t} \quad (5)$$

with $s, t \in \{1,...,R\}$, $m_t \in \{1,...,M_t\}$, $n_s \in \{1,...,N_s\}$ and operator $\circ^s$ defined as

$$\circ^s = \begin{cases} \wedge & \text{if } \tau^s \text{ is monotonically increasing} \quad (6) \\ \vee & \text{if } \tau^s \text{ is monotonically decreasing} \end{cases}$$

The numerical values $\mu_{p_{M_s+n_s}^s, p_{m_t}^t} \in [0,1]$, that can be calculated in the moment the FKB is defined, quantify the DOF between the possibility distributions $\tau^s(B_{n_s}^s)$ and $A_{m_t}^t$.

Applying expression (5) to the example of figure 1, we obtain that, for example, the chaining of variable $X_3$ is represented by $t^5$, and that, consequently, $g(p_1^4) = (\tau^1(g(p_3^1)) \circ^1 \mu_{p_3^1, p_1^4}) \vee (\tau^2(g(p_3^2)) \circ^2 \mu_{p_3^2, p_1^4})$. In a similar way, $g(p_2^4) = \tau^3(g(p_3^3)) \circ^3 \mu_{p_3^3, p_2^4}$.

We will now see how from the analysis of expressions (4) and (5) an inference process following a Goal-Driven (GD) strategy can be described. In it we will present the problem of confirming an initial hypothesis. The formulation of a set of sufficient conditions for the hypothesis to be verified will constitute the GD reasoning.

## 3. GD REASONING OVER THE PN REPRESENTATION OF A FKB

We present the reasoning process according to the GD strategy as the formulation of a set of necessary conditions for the confirmation of a given initial hypothesis. Using the formalism that was described

before, we can formulate the starting hypothesis as

$$"g(p_k^r) > \eta" \quad (7)$$

with $k \in \{1,...,M_r+N_r\}$, $r \in \{1,...,R\}$, $\eta \in [0,1]$. From (7) we will obtain the necessary conditions established over the DOFs of other propositions in the FKB. In order to do this, the representation of the FKB in the PN formalism will be very useful, as it will permit expressing these conditions in a very simple way. Given the fact that the operation associated with the implication process (expression (4)) is different to that carried out in the chaining process (expression (5)) we will differentiate these two situations in the analysis of DO reasoning. Finally, we will present the results corresponding to situations in which the starting hypotheses are $"g(p_k^r) < \eta"$ and $"g(p_k^r) = \eta"$.

### 3.1 GD reasoning in implication

In this case, $p_k^r \in O(t^r)$, $t^r \in T^R$, $k \in \{M_r+1,...,M_r+N_r\}$, $r \in \{1,...,R\}$. Expression (4) is satisfied if

$$g(p_i^r) > \eta \quad \forall p_i^r \in I(t^r) \quad (8)$$

### 3.2 GD reasoning in chaining

In this case, $p_k^j \in O(t^j)$, $t^j \in T^C$, $k \in \{1,...,M_r\}$, $j \in \{R+1,...,R+C\}$. Expression (5) is satisfied if

$$\exists p_i^s \in I(t^j) / \tau^s(g(p_i^s)) \circ^s \mu_{p_i^s, p_k^r} > \eta \quad (9)$$

• If $\tau^s$ is increasing, $\circ^s \equiv \wedge$. Consequently (9) is transformed in

$$\exists p_i^s \in I(t^j) / \tau^s(g(p_i^s)) \wedge \mu_{p_i^s, p_k^r} > \eta \quad (10)$$

As the values $\mu_{p_{M_s+n_s}^s, p_{m_t}^t} \in [0,1]$ are calculated in the moment of defining the FKB, we will have

$$\begin{cases} \tau^s(g(p_i^s)) > \eta & \text{if } \mu_{p_i^s, p_k^r} > \eta \\ \text{IMPOSSIBLE} & \text{otherwise} \end{cases} \quad (11)$$

and, if we denote the inverse of the linguistic truth value $\tau^s$ as $\tau_{INV}^s(x) \triangleq {\tau^s}^{-1}(x)$ $x \in [0,1]$, we will have

$$\begin{cases} g(p_i^s) > \tau_{INV}^s(\eta) & \text{if } \mu_{p_i^s,p_k^r} > \eta \\ \text{IMPOSSIBLE} & \text{otherwise} \end{cases} \qquad (12)$$

- If $\tau^s$ is decreasing, $\circ^s \equiv \vee$. Consequently, (9) becomes

$$\exists p_i^s \in I(t^j) \;/\; \tau^s(g(p_i^s)) \vee \mu_{p_i^s,p_k^r} > \eta \qquad (13)$$

Following an analogous procedure we obtain

$$\begin{cases} g(p_i^s) \in [0,1] & \text{if } \mu_{p_i^s,p_k^r} > \eta \\ g(p_i^s) < \tau_{INV}^s(\eta) & \text{otherwise} \end{cases} \qquad (14)$$

As only the values of the linguistic truth variable that are bijective functions present any intuitive meaning, [9] the existence of a single inverse function $\tau_{INV}^s$ is guaranteed. However, the implementation of certain rule qualification mechanisms that are widely used in control (for instance, numerical degrees of certainty) may require the use of values that do not verify this property. In these cases, it is enough to provide an adequate definition of the corresponding inverse function in order to maintain what we have presented here.

Tables I and II summarize the results corresponding to expressions (12) and (14), together with those obtained in the same way when we consider initial hypothesis (7) with operators '=' and '<' ("$g(p_k^r) = \eta$" and "$g(p_k^r) < \eta$"). These tables display the necessary conditions for the hypothesis made in each case to be verified. From its analysis we can deduce that in some cases (those labelled as IMPOSSIBLE), the starting hypothesis is proven as false, without any need for additional analysis. In other cases (condition $g(p_i^s) \in [0,1]$) the initial hypothesis is undoubtedly true, whereas in the remaining cases, it is translated into a new set of conditions established over other variables. The conditions over the compound operator '$\geq$' will be solved as two conditions established over the corresponding simple operators ('=' and '>') linked by means of the classical 'or' logic operator.

The confirmation of the new conditions (and, as a consequence, the verification of the starting hypothesis that generates them) can be carried out through external methods (consultation of a fact base, direct interpellation to the user, ...) or by iterating the process. This last possibility cannot be applied indefinitely, as it is limited by the number of chaining variables in the FKB.

We will now formulate the algorithm that permits carrying out the process we have just described.

## 4. GD REASONING ALGORITHM

For the sake of clarity we have divided the description of the algorithm into three blocks, corresponding to the three operator types we have considered for the initial hypothesis ('>', '=' and '<'). Furthermore, we have used a compact notation ($C_{HO,RC}^{\circ^s}$) in order to refer to the conditions described in Tables I and II. In it, superindex $\circ^s$ refers to table I ($\circ^s = \wedge$) or table II ($\circ^s = \vee$). Subindex $HO \in \{<,>,=\}$ represents the operator described in the initial hypothesis and locates the column in the corresponding table. Finally, subindex $RC \in \{<,>,=\}$, which characterizes the appropriate row, describes the result from the relative comparison between the value of $\eta$ described in the hypothesis and the corresponding DOF $\mu_{p_i^s,p_k^r}$. This way, the GD reasoning algorithm can be described as:

**Block 1.** Initial hypothesis:
$C_>$: "$g(p_k^r) > \eta$", $\quad p_k^r \in P$, $\eta \in [0,1]$
- If $p_k^r \in O(t^r)$
  - If $\forall p_i^r \in I(t^r)$, $g(p_i^r) > \eta$, then $C_>$ is verified.
  - Otherwise $C_>$ is not verified.
- If $p_k^r \notin O(t^r)$
  - If $\exists t^j \in T^C \;/\; p_k^r \in O(t^j)$
    - If $\exists p_i^s \in I(t^j) \;/\; C_{>,RC}^{\circ^s}$ is verified, then $C_>$ is verified.
    - Otherwise, $C_>$ is not verified.
  - Otherwise, $C_>$ must be verified externally.

**Block 2.** Initial hypothesis:

$C_<$: "$g(p_k^r)<\eta$", $\quad p_k^r \in P$, $\eta \in [0,1]$

- If $p_k^r \in O(t^r)$
  - If $\exists p_i^r \in I(t^r)$ / $g(p_i^r)<\eta$, then $C_<$ is verified.
  - Otherwise, $C_<$ is not verified.
- If $p_k^r \notin O(t^r)$
  - If $\exists t^j \in T^C$ / $p_k^r \in O(t^j)$
    - If $\forall p_i^s \in I(t^j)$, $C_{<,RC}^{o^s}$ is verified, then $C_<$ is verified.
    - Otherwise, $C_<$ is not verified.
  - Otherwise, $C_<$ must be verified externally.

**Block 3.** Initial hypothesis:

$C_=$: "$g(p_k^r)=\eta$", $\quad p_k^r \in P$, $\eta \in [0,1]$

- If $p_k^r \in O(t^r)$
  - If $\forall p_i^r \in I(t^r)$, $g(p_i^r)\geq\eta$, and, besides, $\exists p_i^r \in I(t^r)$ / $g(p_i^r)=\eta$, then $C_=$ is verified.
  - Otherwise, $C_=$ is not verified.
- If $p_k^r \notin O(t^r)$
  - If $\exists t^j \in T^C$ / $p_k^r \in O(t^j)$
    - If $\forall p_i^s \in I(t^j)$, $C_{\leq,RC}^{o^s}$ is verified and, besides, $\exists p_i^r \in I(t^j)$ / $C_{=,RC}^{o^s}$, then $C_=$ is verified.
    - Otherwise, $C_=$ is not verified.
  - Otherwise, $C_=$ must be verified externally.

## 5. EXAMPLE

We will consider the FKB and its representation in the PN formalism described in Figure 1. We will also assume that the linguistic truth values that qualify the rules are $\tau^1(x)=x^2$, $\tau^2(x)=\sqrt{x}$, $\tau^3(x)=1-x$, $\tau^4(x)=0.75x$, $x\in[0,1]$, and that $\mu_{p_3^1,p_1^4}=0.4$, $\mu_{p_3^2,p_1^4}=1$ and $\mu_{p_3^3,p_2^4}=0.85$. Over this FKB we will try to confirm the initial hypothesis $C_1$: "$g(p_3^4)<0.7$". Consequently, we apply block 2 of the algorithm:

1• As $p_3^4 \in O(t^4)$, the necessary condition for $C_1$ to be verified is {"$g(p_1^4)<0.7$" or "$g(p_2^4)<0.7$"}, which will be the new hypotheses $C_2$ and $C_3$ that have to be verified.

2• $C_2$: "$g(p_1^4)<0.7$". As $p_1^4 \in O(t^5)$, $t^5 \in T^C$ and $p_3^1$, $p_3^2 \in I(t^5)$, the necessary condition will be { $C_{<,<}^{\wedge}$ and $C_{<,>}^{\wedge}$ }, because $\tau^1$ and $\tau^2$ are increasing,

$\mu_{p_3^1,p_1^4}<0.7$ and $\mu_{p_3^2,p_1^4}>0.7$. As $C_{<,<}^{\wedge}$ is always verified (S. Table I), this condition will be equivalent to $C_{<,>}^{\wedge}$, which becomes the new hypothesis: $C_{2.1}$: "$g(p_3^2)<0.7^2=0.49$".

As $p_3^2 \in O(t^2)$, the necessary condition is $C_{2.2}$: {"$g(p_1^2)<0.49$" or "$g(p_2^2)<0.49$"}, whose confirmation can only be obtained through external methods.

3• $C_3$: "$g(p_2^4)<0.7$". As $p_2^4 \in O(t^6)$, $t^6 \in T^C$ and $p_3^3 \in I(t^6)$, the necessary condition will be $C_{3.1}$: $C_{<,>}^{\vee}$, as $\tau^3$ is decreasing and $\mu_{p_3^3,p_2^4}>0.7$. As $C_{3.1}$ is not verified in any case (S. Table II), the hypothesis is always false.

Consequently, the necessary conditions for the initial hypothesis "$g(p_3^4)<0.7$" to be verified are {$g(p_1^2)<0.49$ or $g(p_2^2)<0.49$}. Thus, if, for example, in the fact base we find the evidences $g(p_1^2)=0.5$ and $g(p_2^2)=0.8$, we would have that the initial condition $C_1$: "$g(p_3^4)<0.7$" is not verified.

## 6. CONCLUSIONS

In this work we have described a GD reasoning process over a FKB with chaining between rules. The solution we propose is highly flexible in several aspects: the chaining mechanism only requires the repetition of the variable, the qualification of the rules is carried out by means of linguistic values of the truth variable and the formulation of the initial hypothesis permits confirming expressions with operators '>', '=' and '<'. This way, the approach is more general than that of other solutions. The use of the PN formalism simplifies the development of the expressions and permits a simple algorithmic description of the process. In addition, it systemizes the description of any FKB with the characteristics we mention and facilitates, in any case, the description of the processes that are carried out over it.

## 7. REFERENCES

[1] C. G. Looney, "Expert Control Design with Fuzzy Rule Matrices," *Int. J. Expert Syst.: Res. Appl.*, vol. 1, no. 2, pp. 159-168, 1988.

[2] S.-M. Chen, J.-S. Ke and J.-F. Chang, "An Inexact Reasoning Technique based on Extended Fuzzy Production Rules," *Cybern. Syst.*, vol. 22, pp. 151-171, 1991.

[3] S.-M. Chen, J.-S. Ke and J.-F. Chang, "Knowledge Representation Using Fuzzy Petri Nets," *IEEE Trans. Syst. Man Cybern.*, vol. SMC-2, no. 3, pp. 311-319, 1990.

[4] C. G. Looney, "Fuzzy Petri Nets for Rule-Based Decisionmaking," *IEEE Trans. Syst. Man Cybern.*, vol. SMC-18, no. 1, pp. 178-183, 1988.

[5] M.G. Chun and Z. Bien, "Fuzzy Petri Net Representation and Reasoning Methods for Rule-Based Decision Making Systems," *IEICE Trans. Fundamentals*, vol. E76-A, no. 6, 974-983, 1993.

[6] M. L. Garg, S.I. Ashon and P.V. Gupta, "A Fuzzy Petri Net for knowledge representation and reasoning," *Inf. Process. Letters*, 39, pp. 165-171, 1991.

[7] H. Scarpelli and F. Gomide, "Fuzzy Reasoning and Fuzzy Petri Nets," *Proc. 5th IFSA World Congress*, 1326-1329, 1993.

[8] H. Scarpelli and F. Gomide, "Fuzzy Reasoning and High Level Fuzzy Petri Nets," *Proc. EUFIT '93 Conference*, vol. 2, pp. 600-605, 1993.

[9] J. F. Baldwin, "A New Approach to Approximate Reasoning using a Fuzzy Logic," *Fuzzy Sets and Systems*, vol. 2, pp. 309-328, 1979.

[10] A. Bugarín and S. Barro, "Fuzzy Reasoning supported by Petri Nets," *IEEE Trans. Fuzzy Systems*. Accepted to be published.

[11] A. Bugarín, S. Barro and C. V. Regueiro, "Saving Computation Time through Compaction of Chained Fuzzy Rules," in *Proc. EUFIT '93 Conference*, vol. 3, pp. 1543-1549, 1993.

## ACKNOWLEDGEMENTS

This work was supported by the Xunta de Galicia and the Spanish CICyT under grants XUGA20601B92 and TIC91-0997.

*Table I: Necessary conditions for each initial hypothesis ($\tau^s$ increasing).*

| $C_{HO,RC}^{\wedge}$ | INITIAL HYPHOTESIS | | |
|---|---|---|---|
| | $g(p_k^r) > \eta$ | $g(p_k^r) = \eta$ | $g(p_k^r) < \eta$ |
| $\mu_{p_i^s, p_k^r} > \eta$ | $g(p_i^s) > \tau_{INV}^s(\eta)$ | $g(p_i^s) = \tau_{INV}^s(\eta)$ | $g(p_i^s) < \tau_{INV}^s(\eta)$ |
| $\mu_{p_i^s, p_k^r} = \eta$ | IMPOSSIBLE | $g(p_i^s) \geq \tau_{INV}^s(\eta)$ | $g(p_i^s) < \tau_{INV}^s(\eta)$ |
| $\mu_{p_i^s, p_k^r} < \eta$ | IMPOSSIBLE | IMPOSSIBLE | $g(p_i^s) \in [0,1]$ |

*Table II: Necessary conditions for each initial hypothesis ($\tau^s$ decreasing).*

| $C_{HO,RC}^{\vee}$ | INITIAL HYPHOTESIS | | |
|---|---|---|---|
| | $g(p_k^r) > \eta$ | $g(p_k^r) = \eta$ | $g(p_k^r) < \eta$ |
| $\mu_{p_i^s, p_k^r} > \eta$ | $g(p_i^s) \in [0,1]$ | IMPOSSIBLE | IMPOSSIBLE |
| $\mu_{p_i^s, p_k^r} = \eta$ | $g(p_i^s) < \tau_{INV}^s(\eta)$ | $g(p_i^s) \geq \tau_{INV}^s(\eta)$ | IMPOSSIBLE |
| $\mu_{p_i^s, p_k^r} < \eta$ | $g(p_i^s) < \tau_{INV}^s(\eta)$ | $g(p_i^s) = \tau_{INV}^s(\eta)$ | $g(p_i^s) > \tau_{INV}^s(\eta)$ |

# Medical Expert System
# with Elastic Fuzzy Logic

H. Chris Tseng and Dennis W. Teo
Intelligent Control Laboratory
Electrical Engineering Department
Santa Clara University, Santa Clara, CA 95053, USA

## Abstract

We investigate the use of elastic fuzzy logic in a typical medical diagnosis. Fuzzy rules are formulated using common symptoms as antecedents and diagnoses as consequents. Different weighting factors on antecedents are assigned in each rule. This is done to better reflect the fact that in most diagnoses, there are major symptoms among all related symptoms. We also formulated a geometry-mean fuzzification scheme in issuing final decisions. A mutiple-pass interactive scheme to retrieve symptom descriptions from patients is used to capture more realistic diagnosis information. An internal medicine expert system with the proposed framework is used to illustrate our design

## 1.    Introduction

Ever since the introduction of MYCIN [6], the first medical expert system in 1975, expert systems have become a very popular approach to medical problem solving. Numerous expert systems including EMERGE [4], utilize fuzzy logic to better represent the uncertainty. Initiated by Zadeh in his 1965 paper [11], fuzzy logic and fuzzy set theory have in general progressed rapidly, and are finding their way into many medical applications [1,2,8] , expert systems [3,12] and fuzzy expert systems [5].

The development of our medical expert system, a rule-based medical expert system for the analysis of common illness addressed the uncertainty of partial information and also the importance of the presence of key antecedents in a fuzzy rule base.

Our system uses a multiple-pass scheme to retrieve symptom information, i.e., after the initial run of the algorithm, the system will determine any suspected missing antecedents, followed by a confirmation by the end-user and then a final evaluation.

## 2    Fuzzy Logic in Uncertainty Management

Fuzzy logic works through the use of *fuzzy sets* [9] -- collection of objects which are less rigid then traditional sets. In traditional sets, an object is either in the set or not in the set. As such, there is no in between.

The fuzzy logic rules supplied by a human expert are conditional statements normally expressed as *if-then* rules. A typical fuzzy if...then rule is as follows:

If (Fever is Low) and (Headache is Mild) and (Muscle Ache is Somewhat Severe) and (Fatigue is Severe) then (Cold/Influzenza is Somewhat Severe).

A rule is also regarded as an ordered pair of symbol strings, with a left hand side (antecedents) and a right hand side (consequent). The antecedents and consequents are either conventional or fuzzy set. The antecedents part expresses some condition on the state of the data base and whether it is satisfied at any given point. The consequent part specifies changes to be made to the data base whenever the rule is satisfied.

## 2.1. Gaussian Membership Function

Like any other expert system, the membership functions is use to determine the degree of uncertainty. The degree to which an element belongs to a set is given by its degree of membership. The membership function can be represented in a number of way, e.g., triangle, trapezoid or a Gaussian curve. The membership functions for our system are represented by the Gaussian curve which is governed by the following equation:

$$\mu = \exp\left[-\frac{1}{2}\left(\frac{x-\tilde{x}}{\sigma}\right)\right] \qquad (2.1)$$

where $\tilde{x}$ = mean of the Gaussian curve
$x$ = degree of symptom (0 to 5)
$\mu$ = degree of membership
$\sigma$ = width of the Gaussian Curve.

A typical Gaussian membership function curve is as shown below in Figure 1.

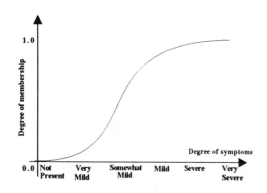

Figure 1  Gaussian Membership Curve

Unlike conventional system such as MYCIN [6] where the values of the degree of symptoms are left as values with no meaning. That is, the data:

The temperature is high.

Togai and Watanabe [7] suggested using the idea of fuzzy subsets that was introduced by Zadeh, where the *degree of symptoms*, i.e. high, is characterized by the grade of membership or membership function. The Gaussian membership curve of figure 1 illustrate this..

## 2.2. Medical Diagnosis with Fuzzy Logic

A typical fuzzy rule is shown below:

Rule $i$:  If $S_1^i$ is $A_1^i$, $S_2^i$ is $A_2^i$, ... , $S_n^i$ is $A_n^i$, then $C_i$

and the equation for calculating the score is:

$$SCORE_i = \left(\prod_{j=1}^{n_i} \mu_{ij}\right)^{\frac{1}{n_i}} \qquad (2.2)$$

where $\mu_{ij}$ = Degree of the fuzzy membership of the $j$th antecedent of rule $i$.
$n_i$  = Number of antecedents in rule $i$.

In the above rule, $S_1^i ... S_n^i$ are the antecedents, $A_1^i ... A_n^i$ are the degrees and $C_i$ is the consequent for rule $i$. For a rule to be executed, all the symptoms associated with that particular rule must be satisfied. From there on, we can calculate the score of uncertainty. However, in our studies of determining the overall degree membership of symptoms, we have concluded that taking just the product of the various degrees of membership will lead to a smaller probability score. Furthermore, as the number of symptoms increases, the score decreases. To counteract this trend, the final degree of membership is be raised to its appropriate reciprocal of the number of symptoms, thus providing a more meaningful result. Therefore, this is the geometry-mean of the score. The score is simply the product of all the degree of membership associated with a particular rule, raised to the reciprocal of the number of symptoms of that rule. Figure 2 shows an example of the degree associated with muscle ache, which is a symptoms for cold and influenza.

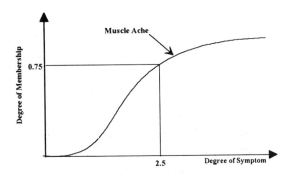

Fig 2 Membership function for Muscle Ache

From the figure above, the degree associated with muscle ache is 0.75. Like wise, the degree of membership for fever, headache and fatigue, which are the other remaining symptoms, can be obtain from their respective membership curves. We propose the geometry-mean defuzzification scheme to compute the score count as follows. Assume that the degrees of membership for fever, headache and fatigue are 0.3, 0.35 and 0.85 respectively, then the score for cold and influenza can be calculated from equation (2.2) as:

$$SCORE_i = (0.3 \times 0.35 \times 0.85 \times 0.75)^{\frac{1}{4}} \quad (2.3)$$

However, this tabulation is only feasible if all the antecedents are present for that rule. In the case where one of the antecedent is missing from a rule, the algorithm can be modified as shown in section 4, to accommodate this lack of information.

## 3. Elastic Fuzzy Logic

In a conventional fuzzy logic theory, the fuzzy rules, derived from the fuzzy set principle by Zadeh [11], are usually in the form:
"If (fuzzy or crisp condition) and (fuzzy or crisp condition) then (fuzzy action)"

where the "fuzzy or crisp condition" are words or phrase such as "fever is high" The score of rule $i$ with $j$ antecedents is computed form [10] as shown below:

$$SCORE_j = \prod_{j=1}^{n_i} \mu_{ij} \quad (3.1)$$

where $\mu_{ij}$ is the degree of membership function of the $j$th antecedent in rule $i$ and $n_i$ is the number of antecedents for the ith rule. However, there are cases where certain antecedents will stress more importance than others. As such, the calculation of the fuzzy *if...then* rules would have to be modified to reflect this. Suggested by Werbos [10], a typical elastic fuzzy rule formulation for this purpose would be in the form of:

$$SCORE_i = \prod_{j=1}^{n_i} \mu_{ij}^{\gamma_{ij}} \quad (3.2)$$

where $\gamma_{ij}$ is the weight of the parameter, or the degree of importance of the $j$th antecedent in the $i$th fuzzy rule. For example, a $\gamma_{ij}$ of 2 is simply equivalent to having the antecedent appear twice in the rule. For each rule that has to be tabulated, $\gamma$ is initially set to 1. But if a antecedent stresses a greater emphasis, then $\gamma$ is raised accordingly. In order to incorporate this important aspect of elastic fuzzy logic, our equation (2.2) has been modified to incorporate this concept of elastic fuzzy logic. The equation is shown below:

$$SCORE_i = \left( \prod_{j=1}^{n_i} \mu_{ij}^{\frac{1}{\gamma_{ij}}} \right)^{y_i} \qquad (3.3)$$

$$\text{where } y_i = \sum_{j=1}^{n_i} \frac{1}{\gamma_{ij}} \qquad (3.4)$$

where $\gamma_{ij}$ are the weight or importance of the $j$th antecedent in the $i$th rule. Our equation is modified to counteract the trend in which the score decreases in value as the number of antecedents increases.

Our equation (3.3) is also different from EMERGE [4] expert system. EMERGE uses the equation

$$score_i = \sum_{j=1}^{m_i} a_i(j)d_i(j) \qquad (3.5)$$

$$\text{where } \sum_{j=1}^{m_i} a_i(j) = 1 \qquad (3.6)$$

and $m$ is the total number of antecedents, $a_i(j)$ is the weighting factor and $d_i(j)$ is the degree of membership of a symptom. Unlike equation (3.5) that EMERGE uses, our equation is more flexible and more realistic. And if all the elastic weights are 1, then equation (3.3) becomes a typical equation use in a traditional expert system.

## 4. Multiple Pass for Missing Antecedent Detection Scheme.

In order for a rule to be issued as a conclusive statement, all the required antecedents of a particular rule must be fulfilled. As such, if one or more antecedents of a rule is missing, the rule will not be used as the final decision. To solve the problem where the user accidentally omits an antecedent/symptom or more, we propose an algorithm that would take this partial absence of antecedents into account. For example, the following rule

If (fever) and (fatigue) and (muscle-ache) and (headache), then (cold/influenza)

where fever, fatigue and headache is keyed into the system by the user and muscle-ache is left out.

The multiple-pass performs as follows. After the first pass calculation, the system will analyze and determine scores of any rules that are not fully fulfilled. This is only applied to rules with one missing antecedent. The algorithm is as follows:

**Step 1:** Calculate $\mu_{ij}$ and $score_i$ with equation (3.2) for all the rules. Then the system determines if there is any missing antecedent. If yes, then go to step 2. If no, then go to step 5.
**Step 2:** The system determines what are the missing antecedents for the rules involved and prompt the user for a possible input.
**Step 3:** The user confirms the missing antecedent(s) and also the degree(s), if available.
**Step 4:** Go to step 1.
**Step 5:** The system list the consequents of the top 5 score count and presents them to the user.

The score with one missing antecedent is calculated as shown:

$$SCORE_i = \left( \prod_{j=1}^{n_i-1} \mu_{ij} \right)^{y_i} \qquad (4.1)$$

Where $y_i$ is defined in equation (3.4). Among all the calculated scores with the missing antecedent, the highest score is picked and the missing antecedent associated with that score is brought to the attention of the user. The user will have to confirm the presence of that symptom. If that symptom is indeed left out the user will have to provide the degree associated with it.

With this additional information supplied to the system, a final calculation is made again with the equation (3.2). And the final analysis is done.

With our multiple-pass scheme, we are able to obtain a more realistic information for the medical diagnosis purpose.

## 5. Conclusion

The use of fuzzy logic in dealing with decision problems that are vague in nature is demonstrated in our propose medical expert system. Elastic fuzzy logic not only models better information but also allows possible adaptation learning with neural networks. This framework initiates the way to integrate experience and practitioners' knowledge with fuzzy expert rules. Our medical expert system for example, though not universal, is typical in many uncertainty management problems. The schemes proposed here can certainly be useful for other related problems.

## References

1. **Adlassing, K. P.**, A fuzzy logical model of computer-assisted medical diagnosis, *Math. Inform. Med.*, 19(3), 141, 1980.

2. **Esogbue, A. O. and Elder, R. C.**, Management and valuation of a fuzzy mathematical model for medical diagnosis, in *Fuzzy Sets System*, 10, 223,

3. **Gupta, M. M. and Sanchez, E., Eda.**, *Approximate Reasoning in Decision Analysis*, North-Holland, Amsterdam, 1992.

4. **Hudson, D. L. and Cohen M.E.**, The role of approximate reasoning in a medical expert system, *Fuzzy Expert System*, 11, 165, 1991. CRC, Florida.

5. **Kandel, A.**, The evolution from expert systems sto fuzzy expert system, *Fuzzy Expert Systems*, 1, 4, 1991. CRC, Florida.1

6. **Shortliffe, E.H.**, *Computer-Based Medical Consusltation, MYCIN*, Elsevier/North-Holland, New York, 1976.

7. **Togai, M. and Watanabe.**, Expert system on a chip: an engine for approximate reasoning, *Fuzzy Expert Systems*, 18, 259, 1991. CRC, Florida.

8. **Vila, M. A. and Delgado, M.**, On medical diagnosis using possibility measures, in *Fuzzy Sets Systems*, 10, 211, Elsevier/North-Holland, Amsterdam, 1983.

9. **Welch, C. A.**, *Fuzzy logic: from software to silicon a primer*, Togai Infralogic, 1992.

10. **Werbos, P. J.**, Elastic fuzzy logic: A better way to combine neural and fuzzy capabilities, *World Congress on Neural Networks*, Vol II, 1993.

11. **Zadeh, L. A.**, Fuzzy Sets, *Inf. Control*, 8,338, 1965.
Elsevier/North-Holland, Amsterdam, 1983.

12. **Zadeh, L. A.**, The role of fuzzy logic in the management of uncertainty in expert systems, *Fuzzy Sets Systems*, 11, 199, 1983.

# A Fuzzy Logic Approach to Intelligent Alarms in Cardioanesthesia

K. Becker, H. Käsmacher[*]), G. Rau, G. Kalff[*]), H.-J. Zimmermann[**])

Helmholtz-Institute for Biomedical Engineering, [*]) Clinic of Anesthesiology, Medical Faculty,
[**]) Institute of Operations Research, Aachen University of Technology
Pauwelsstraße 20, 52074 Aachen, Germany, Fax: ++241 8089416,
e-mail: becker@ergo.hia.rwth-aachen.de

## Abstract

One of the most important tasks of the anesthetist performing anesthesia of a patient undergoing cardiac surgery is the evaluation of the patient's hemodynamic state. During cardiac operation when the heart comes to work again after hours of cardioplegia the vital parameters (blood pressure, heart rate etc.) can reach extreme values and fluctuations. Current alarm facilities are triggered yielding an alarm cascade and often turned off because the permanent noise level in the operating theatre disturbs the surgical team.

An intelligent alarm system to support the anesthetist in this situation has been developed. The system works online, gathering all required data from a general anesthesia information system. Based on the decision making process of the anesthetist, the most important vital parameter constellations are evaluated using a fuzzy inference approach, and are presented graphically on the user interface.

## I. Introduction

During open-heart surgery, management of narcosis and stabilisation of the patients circulation are the most important tasks of the anesthetist. Currently, he is supported by monitoring devices which present vital parameters like blood pressures, temperatures and blood gases. Conventional monitoring devices are equipped with threshold alarm facilities which means that if any parameter exceeds a predetermined, fixed threshold, an acoustical and optical alarm signal is generated (Fig. 1). The limitation of this approach is the difficulty of determining appropriate thresholds for each signal and the fact that deterioration of one hemodynamic parameter like loss of blood volume can lead to a change of all blood pressures i.e. arterial pressure, venous pressure, atrial pressure etc., each triggering a different alarm signal [4].

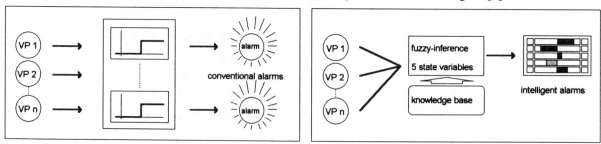

Fig. 1: Alarm function of monitoring devices: conventional alarms versus intelligent alarms [7]

Our approach is an intelligent alarm system to support the anesthetist in monitoring and evaluating the patients hemodynamic state during open-heart surgery. Based on the decision making process of the anesthetist, the most important vital parameter constellations are evaluated using a fuzzy inference approach. The evaluation yields an estimation of five hemodynamic state variables for which a continuos alarm visualisation is presented on the user interface. A graphical explanation of the system inference is available on request.

## II. Knowledge representation: The hemodynamic state variable model

The anesthetist estimates about the patient state recognising various invasively measured blood pressures, blood gases and visual impressions of the _____ blood gases and visual impressions of the patient. So catheterisation makes it easier to determine the patient's hemodynamic state but their application is not without risk [9].

Experienced cardioanesthetists base their therapy decisions on the state of certain "hemodynamic state variables" i.e. "blood volume", "myocardial contractility", "afterload", "heart frequency" and "depth of anesthesia" [8]. The hemodynamic state variable model provides a higher level concept of the patients state than the measured vital parameters because the state variables can be influenced directly by the application of drugs or blood volume.

The minimum information an anesthetist needs for an estimation of the hemodynamic state variables are the current systolic arterial pressure, left atrial pressure, heart frequency and the knowledge about administered anesthetic drugs and the course of operation (Fig. 2).

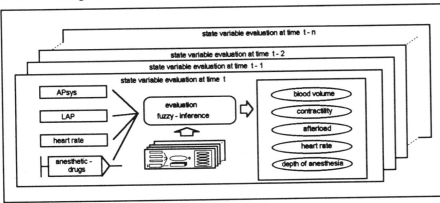

Fig. 2:
State variable
evaluation by the
cardioanesthetist

## III. Structure of the intelligent alarm system

During complicated surgery the anesthetist has little time to spare for computer interaction, therefore the intelligent alarm system gathers all information needed from the Anesthesia Information System AIS [3] (Fig. 3). The anesthetist is not forced to interact with the system and can concentrate on patient care.

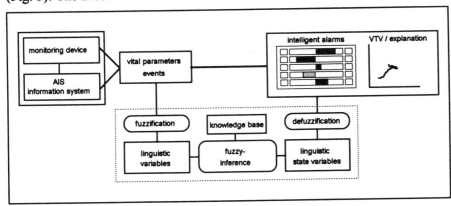

Fig. 3:
Structure of the intelligent
alarm system

For processing of expert knowledge in the form of diagnostic rules a fuzzy-inference [10] approach has shown to be well suited. By choosing compensatory operators [11] for fuzzy-inference, ill-defined problems are solvable by compensatory reasoning and faulty parameters are eliminated by increasing consideration of other more plausible parameters.

For this approach, the patient's vital parameters are fuzzified and their constellations evaluated in a fuzzy inference engine using fuzzy-rules. The inference results are presented graphically on the user interface. An explanation of the fuzzy inference process is available upon request in combination with a permanent trend visualisation.

## IV. Knowledge acquisition: Membership functions

Ten anesthetists who have been working in cardio-anesthesia for between 3 and 15 years individually rated their meaning on the bandwidth of terms like: very low, a little too low, good, a little too high and very high of the measured vital parameters. For every term each anesthetist checked a part of the measurement scale. Each of the n (=10) anesthetists received a weight w = 1/n (1/10) and the weights were summed up for each term.

These findings were used to define a set of linear and symmetric membership functions for the linguistic input variables which describe the vital parameters heart rate, systolic arterial pressure ($AP_{sys}$ Fig. 4), left atrial pressure and central venous pressure.

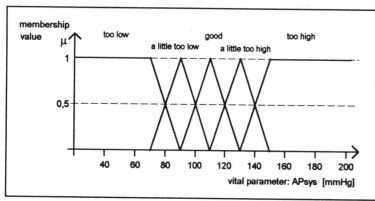

Fig. 4:
Membership functions for the linguistic variable: "arterial systolic pressure" $AP_{sys}$

### Fuzzy knowledge base

In order to determine a fuzzy knowledge base for the evaluation of the hemodynamic state variables sets of questionnaires were created. For each hemodynamic state variable 25 stimuli were presented graphically using virtual analogue scales for the parameter representation in order to avoid bias caused by linguistic uncertainty. These stimuli covered the whole hemodynamic state space [2]. Thirteen experienced cardioanesthetists marked their evaluation of the state variable for each stimuli on the given analogue scale. This estimation was transformed into fuzzy rules about the state variable (Fig. 5).

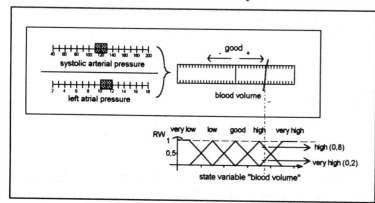

Fig. 5:
Extraction of the fuzzy rule weights. The extracted rule for this example would be:
**If systolic arterial pressure is good and left atrial pressure is high**
then **blood volume** is with a **plausibility of 0.8 high**
and with a **plausibility of 0.2 very high**.

In this way we acquired 50 rules for the evaluation of each state variable. These rules were implemented in the knowledge base of the intelligent alarm system.

## V. User interface: Profilogram

The intelligent alarms are presented on the left section of the user interface in a so-called profilogram [6]. It displays information about the state variables related to the evaluation of the vital parameter constellations. The icons in Fig. 6 (right side) represent the state variables "blood volume", "myocardial contractility", "afterload", "heart rate" and "narcotic level". In a rectangular field beside each icon the result of the evaluation is redundantly coded by colour and shape. Bars to the left side of the centre line indicate a state variable becoming "too low"; in the same way a bar to the right indicates a state variable becoming "too high". When the length of the bar increases, the colour of the bar changes from green to red.

### Vital Trend Visualisation - VTV

In the right section of the user interface the trend of the most important hemodynamic vital parameters, LAP and $AP_{sys}$ is presented in one co-ordinate system (Fig. 6 left side). Plotting the vital parameter combinations in a time series yields the Vital-Trend-Visualisation (VTV) [1]. The VTV trajectory enables the anesthetist to recognise pathologic trends in the patient's circulatory functions and enables him to prevent deterioration of the patient's state.

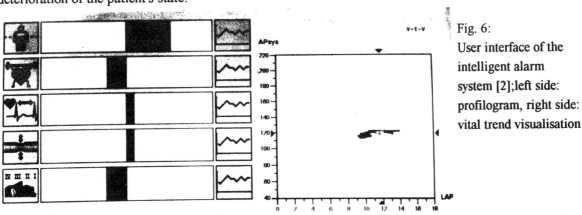

Fig. 6:
User interface of the intelligent alarm system [2];left side: profilogram, right side: vital trend visualisation

### Explanation module

The explanation module of the system can be accessed by touching and activating one of the state variable icons. Then the VTV trend trajectory of the most important vital parameters LAP and APsys is combined with a complete state plane plot of the inference for the activated state variable (Fig. 7). The black line represents the vital parameter trend of the last ten minutes. The underlying contour plot of the inference state plane is coloured similar to the profilogram. This visualisation allows the anesthetist to comprehend the system evaluation and to predict the behaviour of the state variable after blood volume or drug application by visually extrapolating the trend curve.

## VI. Conclusion

An intelligent alarm system based on fuzzy inference has been implemented to support the anesthetist during cardiac surgery. The knowledge base of the system has been acquired by thirteen experienced cardioanesthetists. The system evaluates the current vital parameter constellations to get an estimate of five so-called hemodynamic state variables.

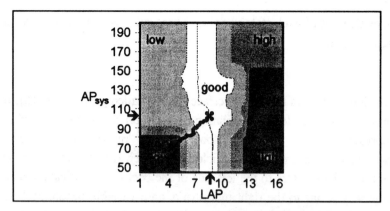

Fig. 7:
Inference explanation module:
State variable "blood volume"

This estimation is presented on a graphic user interface which has been especially adapted to the situation in the operating theatre so that it is possible to 'catch the situation at a glance'. The knowledge base has been validated in laboratory using simulation techniques [5] and is currently undergoing validation with real data from cardiac operations. Field testing in the operating theatre is planned for the future.

## Acknowledgement

This work has been supported by grant Ka 251/3-1 of the DFG (German Research Foundation).

## List of References

[1] Becker, K., Käsmacher, H., Juffernbruch, K., Rau, G., Kalff, G., Zimmermann, H.-J.
An Intelligent Alarm System using Fuzzy-Inference: Knowledge Acquisition and Implemention
Proc 2nd Europ Conf Eng Med, p. 57-58, 1993, Issn. 0928-7329, Elsevier Publishers Amsterdam

[2] Becker, K., Kaesmacher, H., Juffernbruch, K., Rau, G., Kalff, G., Zimmermann, H.-J.
An Intelligent Alarm System for Decision-Support in Cardioanaesthesia: Knowledge Base and User Interface, Proc 1st Europ Cong Fuzzy Intelligent Techn, EUFIT'93, 1023-1026 , Augustinus Verlag, Aachen (1993)

[3] Klocke, H., Trispel, S., Rau, G., Hatzky, U., Daub, D.: An Anesthesia Information System for Monitoring and Record Keeping during Surgical Anesthesia. J Clin Mon, 2, 4, 246-261, October 1986

[4] Loeb, R.G., Jones, B.R., Leonard, R.A., Behrman, K.: Recognition Accuracy of Current Operating Room Alarms, Anaesth Analg, 75, 499-505, 1992

[5] Popp, H.-J., Schecke, Th., Rau, G., Käsmacher, H., Kalff, G.:An interactive computer simulator of the circulation for knowledge acquisition in cardio-anesthesia, Int J Clin Mon Comp, 8, 151-158, 1991

[6] Rau, G., Schecke, Th., Langen M.: Visualization and Man-Machine Interaction in Clinical Monitoring Tasks, First Conference on Visualization in Biomedical Computing, May 22-25 1990, Atlanta Georgia

[7] Schecke Th., Rau, G., Popp, H.-J., Käsmacher, H., Kalff, G., Zimmermann, H.-J.: A Knowledge-Based Approach to Intelligent Alarms in Anesthesia, IEEE Eng Med Biol, Vol 10, 4, 38-43, 1991

[8] Schecke, Th., Langen, M., Popp, H.-J., Rau, G., Käsmacher, H., Kalff, G.: Knowledge-based decision support for patient monitoring in cardioanesthesia, Int J Clin Mon Comp, 9: 1-11, 1992

[9] Shah, K.B., Rao, T.L.K., Laughlin, S., El-Etr, A.A..
A Review of Pulmonary Artery Catheterisation in 6245 Patients, Anesthesiology, 61:271-275, 1984

[10] Zadeh, L.A.: Outline of a New Approach to the Analysis of Complex Systems and Decision Processes IEEE Trans Syst Man Cybern, Vol.SMC-3, 28-44, 1973

[11] Zimmermann, H.-J., Zysno, P.: Latent connectives in human decision making:
Fuzzy Sets and Systems, 4, 37-51, 1980

# The Fuzzy Knowledge-Based System CLINAID: A Correspondence Between its Relational and Neural Net Structures

Ladislav J. KOHOUT, Wyllis BANDLER, Song GAO, and John ANDERSON

## I. INTRODUCTION: CLINAID GENERIC ARCHITECTURE

The paper describes in some detail the structure of inference in a generic family of multi-context Knowledge-Based Systems of CLINAID architectural type. The system is designed for knowledge-based decision making with risk and under uncertainty, operating in a multi-environmental situation and making decisions within a multiplicity of contexts [8]. Because risk situations depending on a rare combination of risk factors as well as on the context in which the data are used appear not only in medicine, our approach has wider applicability than just the domain of medicine. We have shown [8] that similar circumstances may also occur in insurance underwriting and computer assisted manufacturing. Thus the empty shell of CLINAID obtained by removing the deep knowledge structures pertinent to medicine may find significant applications in commercial non-medical applications. The paper also pays attention to the translation of fuzzy relational structures into fuzzy neural network representation.

Since 1980 we have been engaged in a continuous development of fuzzy relational architectures for knowledge based systems. A notable development based on the theory of fuzzy triangle products [1],[2],[3] has been described in a series of papers and a monograph [8].

The basic architecture consists of the following co-operating units (basic shell substratum):

1. Diagnostic Unit (comprised of several parallel co-operating centres)
2. Treatment Recommendation Unit.
3. Patient Clinical Record Unit.
4. Co-ordination and Planning Unit.

The *Clinaid* approach significantly differs from more conventional AI approaches in its potential capabilities. Unlike fuzzy CLINAID, the majority of extant medical expert systems deal with a limited medical context, the largest domain of knowledge being just a single medical field, e.g. internal medicine in CADUCEUS. The inherent limitation of such medical expert systems is conceptual and logical: their knowledge bases and inference engines cannot easily mix the knowledge from several fields without adverse effects. CLINAID successfully deals with this problem by introducing a *multi-centre architecture* in the Diagnostic Unit by which the multiple contexts are handled concurrently.

An important extension of the reasoning capabilities of CLINAID is *interval-based* inferential structure [12], [7]. The semantic justification of logics used for this purpose is provided by a theoretical device called the *checklist paradigm* [10]. This paradigm not only gives a sound epistemological justification for interval-valued inference, but also supports knowledge and data retrieval techniques utilizing information structures with interval credibility weights.

The multi-centre diagnostic unit has been implemented in several versions: in Lisp (cf. [8], chapter 10), in Pascal (by H. Kalantar); a statistical simulator for testing various types of fuzzy logic was implemented San Andres Gutierrez in Modula 2, based on the previous Lisp architecture of CLINAID. All these version used triangle product based fuzzy inference, in which a number of implication operators have been compared. The LISP architecture contains a learning unit utilizing neural network that was used to adapt the relational knowledge structure to different medical environments. The learning unit can monitor the changes in the nature of diagnostic and other clinical data, the rate of misdiagnosis of the system, due to changes in the nature of the environments, and other factors, and can provide recommendations to the clinician concerning changes of the contents of the knowledge base [5]. Another extension of CLINAID architecture is the design blueprint for DEXTERON, a real-time fuzzy expert system with deep knowledge for assessment of movement performance of neurological patients by classification of the dynamic features of their handwriting [11].

Affiliations: LJK and WB – Department of Computer Science, Florida State University, Tallahassee, Florida 32306-4019, USA; SG – Centre for Remote Sensing and GIS, University of NSW, Australia; JA – King's College School of Medicine and Dentistry, University of London, U.K.

## II. Knowledge Representation and Inference in Clinaid

### A. Triangle and Square Types of Relational Products

Where $R$ is a relation from $X$ to $Y$, and $S$ a relation from $Y$ to $Z$, a *product relation* $R * S$ is a relation from $X$ to $Z$, determined by $R$ and $S$. There are several types of product used to produce product-relations [4], [9]. Each product type performs a **different logical action** on the intermediate sets, as *each logical type* of the product enforces a *distinct specific meaning* on the resulting product-relation $R * S$.

$$(R \circ S)_{ik} = \bigvee_j (R_{ij} \wedge S_{jk})$$
$$(R \lhd S)_{ik} = \bigwedge_j (R_{ij} \rightarrow S_{jk})$$
$$(R \rhd S)_{ik} = \bigwedge_j (R_{ij} \leftarrow S_{jk})$$
$$(R \square S)_{ik} = \bigwedge_j (R_{ij} \equiv S_{jk})$$

The logical symbols *AND, OR*, both *implications* and the *equivalence* in the above formulas represent the connectives of some many-valued logic, **chosen** according to the properties of the products required. The details of choice of the appropriate many-valued connectives are discussed in [1], [2],[3]. It is important to distinguish what we call *harsh* fuzzy products (defined above) from a different family, the family of *mean* products. Given the general formula $(R@S)_{ik} ::= \#(R_{ij} * S_{jk})$, a mean product is obtained by replacing the outer connective $\#$ by $\sum$ and normalizing the resulting product appropriately. The mean products provide an effective inference mechanism in Clinaid applications.

Interval inference is based on the semantics provided by the checklist par-adigm. The checklist paradigm generates *pairs of* distinct connectives of the same logical type that determine the end points of intervals, thus providing formally and epistemologically justified systems of *interval-valued* approximate inference. The inequality restricting the possible values of measure m(F) expressing the logical values that fall within the interval, is written in its general form as: $contop \geq m(F) \geq conbot$. In this paper we employ $m(F) = m_1(F) = 1 - (\alpha_{10}/n)$. This choice yields the bounds [1]:

$$min(1, 1 - a + b) \geq m_1(PLY) \geq max(1 - a, b)$$

where the Lukasiewicz implication represents one bound while the other bound (plybot) is the Kleene-Dienes implication operator. Choosing for *con* the connective type **AND** we get for the same measure the bounds $min(a,b) \geq m_1(AND) \geq max(0, a+b-1)$ [6].

### B. Dealing With the Inferential Context in CLINAID

The diagnostic unit of Clinaid deals with a *multiplicity* of contexts provided by the several body systems and other factors. Deep knowledge concerning *individual contexts* takes into account cross-contextual similarities and differences. We distinguish: a) the *context of different body systems*; b) the *context of different diagnostic levels* within the super-context of a specific body system.

The inferential process involves several levels. The medical meaning (semiotic descriptors) of all sets entering the relations at these levels is as follows:

| | | |
|---|---|---|
| B | .... | a set of body systems. |
| G | .... | a set of general diseases. |
| D | .... | a set of specific diseases. |
| I | .... | a set of investigations. |
| P | .... | a set of patients. |
| S | .... | a set of symptoms and physical signs. |
| Y | .... | a set of syndromes. |

The strategy of clinical inference used over this diagnostic hierarchy is realized by means of fuzzy triangle products [4]. Within each specific body system of Clinaid, (e.g. cardiac or respiratory, etc.) the inferential levels have to be related appropriately. At each level of the diagnostic process, different fuzzy relations are used to perform the inference, because the *purpose* of each fuzzy relation is different.

The relations are created, represented and computed dynamically. Only the items of information which are relevant for reaching the conclusion of a specific level are utilized when each level is activated. The relations are made redundant after the process has passed through that diagnostic level. In this way the complexity of the relevant knowledge structure and the processing time are sharply reduced.

The activation of levels within the specific body system represented in a centre of the diagnostic unit is ordered by the following sequence:

(body system) $\Rightarrow$ syndromes $\Rightarrow$ general diseases $\Rightarrow$ working diagnoses.

Signs and symptoms of a patient entered into the system are matched with the knowledge structures distributed within the individual levels of a specific centre. The stream of computations passes through the levels of the above sequence, in an iterative manner. The main contribution of interval reasoning is to provide the limits by which the degrees of plausibility of logical conclusion must be bound. Below, we present a concrete example of actual computations that clearly demonstrates the merit of our method.

## C. Relationships Over the Basic Relational Structures of Clinaid

Global inferential activity of CLINAID consists of matching the relevant empirical findings with medical theories and normative facts ([8] Chapter 10 section 4, pp. 181-184). Empirical findings obtained through communication with patients (syndromes), direct observations of patient by a clinician (signs) and by laboratory tests and investigations. Medical theories and normative facts are held in the knowledge bases of CLINAID and are represented mostly as many-valued logic based (fuzzy) relations. The match of the two kinds of information is performed by relational compositions which are called circle and square products, and triangle sub- and super-products as described above. Some relational formulas employed in CLINAID are listed and their conceptual meaning explained below.

Name:     A set of:

B     ....     body systems.
D     ....     specific diseases.
V     ....     variability index of observation events.
P     ....     patients.
S     ....     symptoms and physical signs.

Relation $PVS \in \mathcal{R}(P \times V \times S)$ is a relaation between *patients, variability index (time) of observation events* and *signs and symptoms*. $VBS \in \mathcal{R}(V \times B \times S)$ is between *variability index of observation events* and *signs and symptoms*. $PS \in \mathcal{R}(P \leadsto S)$ relates *patients* to *symptoms and physical signs*. $PD \in \mathcal{R}(P \leadsto D)$ is a relaation from *patients* to *specific diseases*. $SBR(S \leadsto B)$ relates *symptoms and physical signs* to *body systems*. Finally, $SD \in \mathcal{R}(S \leadsto D)$ is a relation from *symptoms and physical signs* to *specific diseases*.

Given are two ternary relations, $PVS$ and $VBS$, from which the composed relations $PBS$ and $PB$ are computed. $PVS$ is a ternary (3-place) relation relating the set of patients $P$ with the set of all time instances $V$ at which the observations of relevant signs and symptoms $S$ are made. This relation belongs to the family of relations capturing the relevant empirical findings. $VBS$ is a ternary relation relating the variability index of observational events $V$ with the set of body systems $B$ and the set of signs and symptoms $S$. The $VBS$ relation captures the relationship between signs and symptoms and body systems. The actual observations of patients captured by the $PVS$ relation have to be matched with the knowledge contained in the relation $VBS$ over the set $V$. This yields the computation

$$PBS_{imk} = (PVS \triangleleft VBS)_{imk} = (1/cardV) \sum_{j=l} (PVS_{ijk} \rightarrow VBS_{lmk})$$

$$PB_{im} = (PBS \square^* PBS)_{im} = (1/cardS) \sum_k (PBS_{imk} \wedge (PBS_{imk} \equiv PBS_{imk}))$$

where *card V, card S* are the cardinalities (sizes) of the sets of relevant elements, over which the relational composition (matching) is performed. The second product yielding the relation $PB$ computes the degree of involvement of each body system relative to the composition over available symptoms and signs (and generally also over other observables) characterising individual patients. The connectives used by this method are divided into two classes: TOPCON and BOTCON, computing the top and bottom point of the interval of the possible values of the products. We have used the following connectives in the examples of the next section:

$TOPCON: \ a \rightarrow_5 = min(1, 1 - a + b)$;
$a \equiv b = min(1 - a + b, 1 - b + a)$;
$BOTCON: \ a \rightarrow_6 b = max(1 - a, b)$;
$a \equiv b = max(a + b - 1, 1 - (a + b))$.

We present a sample result of working diagnoses computed from real clinical data [7]. The first part of the example shows how the interval-based inference helps in selecting the relevant body system in CLINAID, while the second part presents the actual working diagnoses within the Cardiac body system.

Let us assume that our clinical case is a patient that exhibits the following set S of signs and symptoms. The set S is a fuzzy set consisting of $s_1 = SOB/.8$ (Shortness of breath), $s_2 = AS/.6$ (Ankle swelling), $s_3 = HM/.7$ (Hepatomegaly), $s_4 = ANX/.6$ (Anxiety), $s_5 = JVPH/.8$ (Jugular venous pressure high), $s_6 = CMRM/.7$ (Cardiac murmurs), $s_7 = PLP/.7$ (Palpitations) and $s_8 = RVEP/.75$ (Right ventricular enlargement). From this set, and using the clinical knowledge structure of the Diagnostic unit one computes the relation $PB$ that suggests the involvement of individual body systems. This relation is computed by the products $PVS \triangleleft VBS$ and $PBS \square^* PBS$ listed above. The computation shows that the top candidate is the *cardio-vascular* body system.

The following syndromes within the cardio-vascular system are identified as the most plausible for patient $p_1$: *Acute cardiac failure (ACF), Left sided heart failure (LHF), Right sided heart failure (RHF), Murmurs (MRM)* and *Dysrythmia (DSR)*. The intervals of the possibility of their occurrence are listed in decreasing order as the fuzzy interval set of syndromes $Y$:

$$Y = \{ACF/[.72, .57], DSR/[.63, .43],$$
$$MRM/[.56, .48], RHF/[.54, .47],$$
$$LHF/[.38, .36]\}.$$

The fuzzy triangle product $PYD = PS \lhd SYD$ from patients to diseases (indexed by syndromes) is given by the formula

$$PYD_{ikm} = (PS \lhd SYD)_{jkm} =$$
$$(1/cardS) \sum_j (PS_{ij} \rightarrow SYD_{jkm})$$

It provides the final working diagnosis. This is listed in Table 1 below. *Congestive Heart Failure* is the correct diagnosis.

**Table 1:**
Working Diagnoses for syndrome
Acute Cardiac Failure ACF/[0.72, 0.57].

| Disease | Possibility of occurrence |
| --- | --- |
| Congestive Heart Failure | [1.0, 0.76] |
| Heart Block | [0.89, 0.64] |
| Myocardial Infarction | [0.73, 0.59] |
| Malignant Hypertension | [0.72, 0.59] |
| Cardiomyopathy | [0.73, 0.58] |

### III. Choice of Special Implication Operators for Applicative Implementation of CLINAID with Qualitative Information

In the previous sections, interval-based fuzzy reasoning using checklist paradigm based pairs of many-valued logic systems were employed. These systems assume that *fuzzy relations* describing medical kowledge as well as *fuzzy membership* functions for clinical data are available. This is the case for some body systems of CLINAID as documented in the references. But such full fuzzy knowledge bases are scarce at present. Indeed, difficulty of eliciting any numerical degrees apart form relative ranking of concepts is a recurrent problem with the development of decision support systems in medicine. There is, however, evidence indicating that the establishing linguistic relationships and identification of factors which are relevant to a problem is more important than the precise numerical values used. This problem has been successfully dealt with in our applicative (i.e. functional) realization of CLINAID (see [8] chapter 10 and 11).

When one uses a mixture of fuzzy and crisp, or entirely crisp items of information with mean and square products of relations, this still yields fuzzy relations with well defined membership functions. Thus, one is generating well-defined graded (i.e. non-crisp, fuzzy) statements from qualitative crisp infor-

mation. In this case the interval logic does not operate in its full generality, but a pair of fuzzy interval logic connectives collapses into (coincides with) a single point-based logic fuzzy connective.

Note the following equivalences of the values of some many-valued logic connectives. For fuzzy $a \in [0, 1]$ and crisp $b \in \{0, 1\}$: (i) The interval pair of values of Lukasiewicz ($\rightarrow_5$) and Kleene-Dienes ply ($\rightarrow_6$) operators collapse into a point defined by the intuitionistic logic *Standard Star S*$*$ ply operator ($\rightarrow_3$). (ii) The interval pair of values of fuzzy *OR* and of bold *OR* will collapse into the Skolem ply $a \Rightarrow b$, which is interestingly related to $\rightarrow_3$. More formally, the following theorem holds.

**Theorem 1**: *collapse of interval-based many-valued logics (MVL) into point-based MVL.*

For fuzzy $a \in [0, 1]$ and crisp $b \in \{0, 1\}$ the following equalities hold:
(1) $a \rightarrow_5 b = a \rightarrow_6 b = a \rightarrow_3 b = 1$ if $a \leq b$; **otherwise** it will attain the **value** $b$ for $a > b$.
(2) $min(1, a + b) = max(a, b) = \neg(\neg a \leftarrow_7 b) = \neg[(\neg a \wedge b) \vee \neg b] = a \Rightarrow b = 0$ if $a \leq \neg b$; **otherwise** it will attain the **value** $a$ for $a > \neg b$.
(where $a \rightarrow_5 b = 1 - a + b$ (Lukasiewicz), $a \rightarrow_6 b = max(1 - a, b)$ ( Kleene-Dienes), $a \rightarrow_7 b = (a \wedge b) \vee \neg b$ (Early Zadeh EZ), and $a \rightarrow_3 b$ is the intuitionistic Standard Star $S*$ ply operator.) ¶

### IV. Relational Neural Network Representation of a Version of CLINAID

The sets, relations and formulas listed below describe one layer of the cascade of the relational network. For example, we had three layers in the example of the previous section: (i) identification of the body system, (ii) identification of relevant syndromes and (iii) computation of working diagnoses.

Name: A set of:

A .... input objects (e.g. patients, syndromes).
B .... characteristic attributes of objects.
C .... output objects (e.g. diseases, syndromes)
D .... domain of questions

| Name: | Definition: | Relation |
| --- | --- | --- |
| V .... | $\mathcal{R}(A \rightsquigarrow B)$ | input data rel. |
| W .... | $\mathcal{R}(B \rightsquigarrow C)$ | knowledge base rel. |
| R .... | $\mathcal{R}(A \rightsquigarrow C)$ | result of inference rel. |
| Q .... | $\mathcal{R}(C \rightsquigarrow D)$ | questioning strategy rel. |

An iterative formula for an arbitrary level is as follows:

$$R_{ik}^{\sigma} = (1/card_{over\, j} W_{jk}) \sum_k (V_{ij}^{\sigma} \Rightarrow W_{jk})$$

where $TOP_R^\sigma = \bigwedge_k R_{ik}^\sigma$, and $T_R^\sigma = TOP_R^\sigma - \Delta$. $TOP_R^\sigma$ is a variable (floating) treshold which defines the limiting operation, which is an analogy to the S-function of the conventional neural networks. This operation filters out the irrelevant information. It computes the output values $R\Theta_{ik}^\sigma$ of the layer according to the formula

$$R\Theta_{ik}^\sigma = R_{ik}^\sigma \Rightarrow (1 - T_R^\sigma)$$

It can be seen that the treshold may vary during the actual computation as the result of feedback (more input data) from the clinician. Thus our relational structure represents a variable relational structure, that changes due to the changed context of the computation. In conventional neural networks the S-function is usually fixed. Our system is therefore more universal than the conventional neural nets.

Each layer, when active, may provide feedback to the clinician, suggesting missing items (e.g. signs or symptoms) that the clinician either accepts or rejects. When there is sufficient information, the iterative computation in the active layer is completed and a subsequent layer is activated. The relation providing the feedback is given by the formula: $Q_{ij}^\sigma = R_{ik}^\sigma \circ W_{kj}^T$, where $\circ$ is the usual max-min circle fuzzy product of relations.

The neural net translation presented in this section is a direct transcription of the algorithms previously implemented in LISP. The LISP architecture used a subset of LISP without the cut operator. Hence, it is an applicative architecture based on functional programming methods. This architecture and its fuzzy neural network translation has demonstrated its practical utility. Considering the known correspondences between applicative computer languages, lambda calculus, combinatory logic and Cartesian closed categories, the neural network translation may also offer new theoretical insights when properly analysed.

## REFERENCES

[1] W. Bandler and L.J. Kohout. Semantics of implication operators and fuzzy relational products. *Internat. Journal of Man-Machine Studies*, 12:89–116, 1980. Reprinted in Mamdani, E.H. and Gaines, B.R. eds. *Fuzzy Reasoning and its Applications*. Academic Press, London, 1981, pages 219-246.

[2] W. Bandler and L.J. Kohout. A survey of fuzzy relational products in their applicability to medicine and clinical psychology. In L.J. Kohout and W. Bandler, editors, *Knowledge Representation in Medicine and Clinical Behavioural Science*, pages 107–118. An Abacus Book, Gordon and Breach Publ., London and New York, 1986.

[3] W. Bandler and L.J. Kohout. Fuzzy implication operators. In M.G. Singh, editor, *Systems and Control Encyclopedia*, pages 1806–1810. Pergamon Press, Oxford, 1987.

[4] W. Bandler and L.J. Kohout. Relations, mathematical. In M.G. Singh, editor, *Systems and Control Encyclopedia*, pages 4000 – 4008. Pergamon Press, Oxford, 1987.

[5] Song Gao. *Design and Inplementation of Diagnostic Unit of Medical Knowledge-Based System* CLINAID. Thesis, Dept. of Computer Science, Brunel University, U.K., 1986.

[6] L. J. Kohout and W. Bandler. Interval-valued systems for approximate reasoning based on checklist paradigm semantics. In P. Wang, editor, *Advances in Fuzzy Set Theory and Technology, vol. 1*, pages 167–193. Bookwrights Press, Durham, NC, 1993.

[7] L. J. Kohout, I. Stabile, W. Bandler, and J. Anderson. CLINAID: Medical Knowledge-Based System based on fuzzy relational structures. In Cohen and D. Hudson, editors, *Comparative Approaches in Medical Reasoning*, in press. Springer, New York, 1994.

[8] L.J. Kohout, J. Anderson, and W. Bandler. *Knowledge-Based Systems for Multiple Environments*. Ashgate Publ. (Gower), Aldershot, U.K., 1992.

[9] L.J. Kohout and W. Bandler. Fuzzy relational products in knowledge engineering. In V. Novák et al., editor, *Fuzzy Approach to Reasoning and Decision Making*, pages 51–66. Academia and Kluwer, Prague and Dordrecht, 1992.

[10] L.J. Kohout and W. Bandler. Use of fuzzy relations in knowledge representation, acquisition and processing. In L.A. Zadeh and J. Kacprzyk, editors, *Fuzzy Logic for the Management of Uncertainty*, pages 415–435. John Wiley, New York, 1992.

[11] L.J. Kohout and M. Kallala. Evaluator of neurological patients' dexterity based on relational fuzzy products. In *Proc. of Second Expert Systems International Conference (London, October 1986)*, pages 1–12. Learned Information Inc., New Jersey, USA and Oxford, UK, 1986.

[12] L.J. Kohout and I. Stabile. Interval-valued inference in medical knowledge-based system Clinaid. *Interval Computations*, in press, 1993.

# Inference Method for Natural Language Propositions involving Fuzzy Quantifiers in FLINS

## Wataru OKAMOTO, Shun'ichi TANO,
## Toshiharu IWATANI and Atsushi INOUE

**Laboratory for International Fuzzy Engineering Research**
**Siber Hegner Building 3FL., 89-1 Yamashita-cho**
**Naka-ku Yokohama-shi 231 JAPAN**
**E-mail: wataru@fuzzy.or.jp**

## Abstract

We propose an inference method necessary for constructing a natural language communication system called FLINS (Fuzzy LINgual System). We show that for natural language propositions involving fuzzy quantifiers, for example "Most tall men are heavy", we can infer a modified proposition "Many tall men are *very* heavy", where the fuzzy quantifier "Many" in the inferred proposition can be resolved analytically. Generally, for natural language propositions involving three types of quantifiers, a monotone nonincreasing type (FEW, ...), a monotone nondecreasing type (MOST, ...) and a triangular type (SEVERAL, ...), we can resolve fuzzy quantifiers analytically for inferred propositions.

## 1. Introduction

At the Laboratory for International Fuzzy Engineering Research (LIFE), we are developing FLINS which we can teach and question, and answer using natural language[1].

To develop such a Teach, Question and Answer (TQA) system able to handle natural language, including ambiguity in semantics, we divide the inference engine into a three-layered structure: non-fuzzy reasoning in the first layer, fuzzy reasoning in the second layer, and fuzzy case-based reasoning in the third. This paper describes the fuzzy reasoning in the second layer, especially as it relates to the inference method for natural language proposition (NLP) involving fuzzy quantifiers (FQs).

Zadeh has proposed an inference method for NLPs that involves fuzzy quantifiers[2,3]. This method proved helpful but it is insufficient for constructing FLINS.

Zadeh showed how to infer using NLPs involving FQs when the quantifier is MOST and the modifier is VERY, for example,

Most foreigners are tall

⇔ *Many* foreigners are *very* tall

But for more general NLPs, for example, which involve other types of quantifiers (FEW, SEVERAL,...) and modifiers (rather,...) we cannot use Zadeh's method. Therefore, we need to extend

Zadeh's method to handle inferences such as

Few heavy foreigners are tall

⇔ *A* few heavy foreigners are *rather* tall

## 2. State of the Art
### 2.1 Goal of inferences based on Fuzzy NLPs

In FLINS we set the syntax of NLPs as follows:

QA is mF

We will use the same notations as that of Zadeh[3]: Q, Q' for a fuzzy quantifier, A for a fuzzy quantity, m for a modifier, F for a fuzzy predicate, $k1$ for knowledge, $q1$ for a question and $a1$ for an answer. In FLINS, knowledge is given by $k1$, and a question is given by $q1$ as follows:

$k1$: QA is F
$q1$: How A is mF?

To answer the above question, we need to infer

$k1$: QA is F ⇔ Q'A is mF    (1)

Finally we can make an answer as follows:

$a1$: Q'A is mF

For example, we consider the knowledge expressed by the following NLP:

Most heavy foreigners are tall

The problem is how to answer the question

How many heavy foreigners are *very* tall?

If we can infer

Most heavy foreigners are tall

⇔ Many heavy foreigners are *very* tall

we can make an answer

Many heavy foreigners are very tall

As a result, we need method to infer (1).

### 2.2 Zadeh's Inference Method and its Limitations

Zadeh showed that for special cases, inference (1) can be resolved, so for the case in Eq. (1), Q=MOST, m=VERY, A is a non-fuzzy quantity. For example,

Most foreigners are tall

⇒ Q' foreigners are *very* tall

In the above inference, Q' can be resolved.

Next, we show how to determine Q'. Zadeh showed that when a database containing height data for N foreigners exists (for i=1,...,N we denote the degree of membership to the fuzzy set by $\mu_{tall}(i) = \mu_i$ ), the fuzzy quantifier Q' is given by the following relation[2,3].

$$\mu_{Q'}(\gamma) = MAX_\mu \left[ \mu_{MOST}\left(\frac{1}{N}\sum_{i=1}^{N} \mu_i\right)\right], \text{ for } \gamma = \frac{1}{N}\sum_{i=1}^{N} \mu_i^2$$

As we assume that the membership function $\mu_{MOST}$ is monotone nondecreasing, the MAX operator can be replaced inside the parenthesis as follows:

$$\mu_{Q'}(\gamma) = \mu_{MOST}\left(MAX_\mu\left(\frac{1}{N}\sum_{i=1}^{N} \mu_i\right)\right), \text{ for } \gamma = \frac{1}{N}\sum_{i=1}^{N} \mu_i^2$$

As we will show later $MAX(\ ) = \sqrt{\gamma}$, $\mu_{Q'}$ can be expressed as follows (Note that $\mu_{Q'}$ depends neither on N nor on F):

$$\mu_{Q'}(\gamma) = \mu_{MOST}\left(\sqrt{\gamma}\right) \qquad (2)$$

This method has not yet been used for other types of quantifiers and modifiers such as Q=FEW (monotone nonincreasing function form), SEVERAL (triangular function form) and m = rather. Therefore we need to extend the method to handle the various types of NLPs.

### 2.3 Our Extended Inference Method

Here we briefly overview our results. For further details, refer to our paper[4].

### 2.3.1 Inference in the case of a monotone quantifier Q, such as MOST or FEW

In this section, we extend Zadeh's method to more general cases, such as the cases of the following quantifiers Q and modifiers m in inference (1).

$$m \begin{cases} 0 < m < 1 \text{ (dilation; rather,.....)} \\ \\ m > 1 \text{ (concentration; very,...)} \end{cases}$$

$$Q \begin{cases} \text{monotone non-decreasing (most, many,...)} \\ \\ \text{monotone non-increasing (few, little,...)} \end{cases}$$

The possible combinations are shown in cases 1 to 4 in Table 1. Case1 includes Zadeh's case.

Table 1: Characteristics of CASES

| Q \ m | m > 1 (very, ....) | 0 < m < 1 (rather,..) |
|---|---|---|
| monotone nondecreasing ( MOST,...) | (CASE 1) | (CASE 2) |
| monotone nonincreasing (FEW,...) | (CASE 3) | (CASE 4) |

We assume that A is non-fuzzy in Eq. (1).

Again we assume we have a database for N (for i=1,...,N we denote by $\mu_F(i) = \mu_i$ the membership degree to the fuzzy set F for the i-th term).

The possibility value for QA is F is $\mu_Q\left(\frac{1}{N}\sum_{i=1}^{N} \mu_i\right)$

The possibility value for Q'A is mF is $\mu_{Q'}\left(\frac{1}{N}\sum_{i=1}^{N} \mu_i^m\right)$

As the possibilities for both propositions in Eq. (1) are equal, we get

$$\mu_{Q'}(\gamma) = \mu_Q\left(\frac{1}{N}\sum_{i=1}^{N} \mu_i\right), \text{ for } \gamma = \frac{1}{N}\sum_{i=1}^{N} \mu_i^m$$

Since there are many sets of values $\mu_1,...,\mu_N$ that lead to the same value of $\gamma$ ($0 \leq \gamma \leq 1$), many $\mu_{Q'}$s exist. Therefore, it is impossible to determine a unique distribution for $\mu_i$. The main results relative to the optimal distribution for $\mu_i$ are given in the following theorem (The proof is omitted because of limited space).

[THEOREM]

The value of $\frac{1}{N}\sum_{i=1}^{N} \mu_i$ is maximum (or minimum)

for $\mu_1 = \mu_2 = ... = \mu_N = \sqrt[m]{\gamma}$

when $\gamma = \frac{1}{N}\sum_{i=1}^{N} \mu_i^m$, $0 \leq \mu_1,...,\mu_N \leq 1$

Furthermore, the following inequalities hold:
(1) m>1

$$\frac{1}{N}\sum_{i=1}^{N} \mu_i \leq MAX\left(\frac{1}{N}\sum_{i=1}^{N} \mu_i\right) = \sqrt[m]{\gamma}$$

(2) 0<m<1

$$\frac{1}{N}\sum_{i=1}^{N} \mu_i \geq MIN\left(\frac{1}{N}\sum_{i=1}^{N} \mu_i\right) = \sqrt[m]{\gamma}$$

This theorem shows that, following the functional form of $\mu_Q$ in Table 1, $\mu_{Q'}$ has MAX or MIN value. We apply the theorem to CASE1.

As the membership function for $\mu_{MOST}$ is monotone nondecreasing and m=2 (very), using the definition of $\mu_{Q'}(\gamma)$, the following inequality holds.
(CASE1)

$$\mu_{MOST}\left(\frac{1}{N}\sum_{i=1}^{N} \mu_i\right) \leq MAX\left[\mu_{MOST}\left(\frac{1}{N}\sum_{i=1}^{N} \mu_i^2\right)\right] = \mu_{MOST}\left(\sqrt{\gamma}\right)$$

$$\Rightarrow \mu_{Q'}(\gamma) \leq \mu_{MOST}\left(\sqrt{\gamma}\right)$$

As mentioned in Sec. 2.2, for CASE1, Zadeh uses the MAX operator [2,3]. He states that, according to the extension principle, the MAX value must be taken. But this is not sufficient to get a unique distribution of $\mu_{Q'}(\gamma)$ for the four cases.

**As the inferred values are determined definitely and the membership function for Q' has a value different from that of Q, we have proposed that the extreme values could be taken as a result of the fuzzy inference for the above four cases.**

This interpretation includes Zadeh's method, because we get Eq. (2) by setting Q=MOST and m=2(VERY) in CASE1 as follows:

$$\mu_{Q'}(\gamma) = MAX_\mu\left[\mu_{MOST}\left(\frac{1}{N}\sum_{i=1}^{N} \mu_i\right)\right] = \mu_{MOST}\left(\sqrt{\gamma}\right)$$

In CASE4 the extreme values are obtained as follows:
(CASE 4) m=1/2 (rather), Q=FEW

$$\mu_{Q'}(\gamma) = MAX_\mu\left[\mu_{FEW}\left(\frac{1}{N}\sum_{i=1}^{N} \mu_i\right)\right] = \mu_{FEW}\left(\gamma^2\right) \qquad (4)$$

We summed the characteristic features (to which extent $\mu_{Q'}$ is different from $\mu_Q$) in Table 2. In the matrices we show from top to bottom "case number", "effects on $\mu_Q$ to get $\mu_{Q'}$", "functional expression for $\mu_{Q'}$", "MAX or MIN operation to get $\mu_{Q'}$".

Table 2: Characteristic features of $\mu_{Q'}$

| m ＼ Q | m > 1 (very, ....) | 0 < m < 1 (rather,..) |
|---|---|---|
| monotone nondecreasing ( MOST,....) | (CASE 1) dilation $\mu_Q\left(\sqrt[m]{\gamma}\right)$ MAX | (CASE 2) concentration $\mu_Q\left(\sqrt[m]{\gamma}\right)$ MIN |
| monotone nonincreasing (FEW,...) | (CASE 3) concentration $\mu_Q\left(\sqrt[m]{\gamma}\right)$ MIN | (CASE 4) dilation $\mu_Q\left(\sqrt[m]{\gamma}\right)$ MAX |

In CASE4, as we concentrate the fuzzy predicate F, the quantifier Q' is dilated to hold Poss{QA is mF} = Poss{Q'A is mF } because $\mu_{FEW}$ has monotone nondecreasing functional form.

In addition, $\mu_{Q'}$ has the maximum value, so it is our choice as the inferred result in Eq. (1).

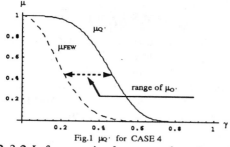

Fig.1 $\mu_{Q'}$ for CASE 4

### 2.3.2 Inference in the case of a triangular quantifier Q, such as SEVERAL

When Q is a triangular-type fuzzy quantifier, such as SEVERAL, first we divide SEVERAL into two quantifiers Q1 and Q2 (it is important that Q1 is monotone nondecreasing and Q2 is monotone nonincreasing, and Q = Q1 AND Q2, using an AND operator). We then infer that Q1 ⇒ Q1' and Q2 ⇒ Q2', following the results of Sec. 2.3.1, and we get the inferred Q' from Q1' AND Q2'. The membership functions for Q1, Q2 and SEVERAL are shown in Fig. 2.

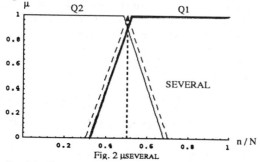

Fig. 2 $\mu_{SEVERAL}$

In this figure, the perpendicular axis is the grade value of membership functions and the parallel axis is the proportion (natural number n to total number N of elements of the universe of discourse).

For example, the knowledge "Several foreigners are tall" is transformed as follows:

Several foreigners are tall

$$\Leftrightarrow \begin{cases} Q1 \text{ foreigners are tall} \\ \qquad AND \\ Q2 \text{ foreigners are tall} \end{cases}$$

$$\Rightarrow \begin{cases} Q1' \text{ foreigners are very tall} \\ \qquad AND \\ Q2' \text{ foreigners are very tall} \end{cases}$$

$$\Leftrightarrow Q' \text{ foreigners are very tall}$$

When m>1, the membership functions for Q1', Q2', Q', SEVERAL are shown in Fig. 3.

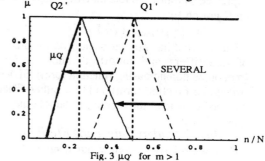

Fig. 3 $\mu_{Q'}$ for m > 1

As a result, the membership function for Q' is similar to that of SEVERAL, but is shifted to the left (as shown by the thick arrows in the figure). For m (0<m<1), the same discussion holds except that the shift direction is reversed.

### 2.4 Problem

In Sec. 2.3, we showed the inference method for NLPs limited to the case of non-fuzzy A. Therefore, we cannot handle the following case by using this method. For example, we consider the knowledge expressed by the following NLP:

Most <u>heavy</u> foreigners are tall

The problem is how to answer the question below

How many <u>heavy</u> foreigners are *rather* tall?

More generally, how do we answer question q2 when knowledge k2 is given and A is a fuzzy quantity?

k2: QA is F

q2: How A are mF?

To answer this question, we need to infer a fuzzy quantifier Q' which satisfies

k2: QA is F ⇔ Q'A is mF

## 3. Inference Method for NLPs with fuzzy A

### 3.2 Inference Method: monotone type

We now extend our method of determining Q' when A is fuzzy and Q is a monotone function. As before, we assume that we have a database for N (for i=1,...,N we denote as $\mu_F(i) = F_i$, $\mu_A(i) = A_i$ the extent of membership in the fuzzy sets F and A for the i-th term).

Accordingly, if we denote MIN as $\wedge$ and $A = \sum_{i=1}^{N} A_i$, the possibility for "QA is F" is $\mu_Q\left(\dfrac{1}{A}\sum_{i=1}^{N} F_i\wedge A_i\right)$,

the possibility for "Q'A is mF" is $\mu_{Q'}\left(\dfrac{1}{A}\sum_{i=1}^{N} F_i\wedge A_i^m\right)$.

Because the possibilities for both propositions must be equal, we have Eq. (5) as follows:

For $\gamma = \dfrac{1}{A}\sum_{i=1}^{N} A_i\wedge F_i^m$, $A=\sum_{i=1}^{N} A_i$

$$\mu_{Q'}(\gamma)=\mu_Q\left(\dfrac{1}{A}\sum_{i=1}^{N} A_i\wedge F_i\right) \qquad (5)$$

When $A_1,...,A_N$ are given, there are many sets of values $F_1,...,F_N$ that lead to the same value of $\gamma$ ( $0\leq\gamma\leq 1$ ). Therefore, we know that many $\mu_Q$'s exist. As before, we propose taking the extreme values resulting from the fuzzy inference. The kind of extreme values (MAX or MIN) correspond to the result in the Table 2. First we consider the case of N=2 and then we consider the general case of N.

### 3.2.1 Case of N=2

First we consider the simple case (N=2). Then we get the expression for $\mu_{Q'}$ from Eq. (5) as follows:

For $\gamma = \dfrac{1}{A}\sum_{i=1}^{2} A_i\wedge F_i^m$, $A=\sum_{i=1}^{2} A_i$

$$\mu_{Q'}(\gamma)=\mu_Q\left(\dfrac{1}{A}\sum_{i=1}^{2} A_i\wedge F_i\right) \qquad (6)$$

Here we assume Q=MOST and m=1/2 (rather) then we take the minimum value of Eq. (6) following CASE2 in Table 2. So we get

$$\mu_{Q'}(\gamma)=\text{MIN}_\mu\left[\mu_{\text{MOST}}\left(\dfrac{1}{A}\right)\sum_{i=1}^{2} A_i\wedge F_i\right]; \text{ for } \gamma = \dfrac{1}{A}\sum_{i=1}^{2} A_i\wedge F_i^{1/2}$$

This result agrees with human intuition because in the inference

Most heavy foreigners are tall

$\Rightarrow$ Q' heavy foreigners are rather tall

we predict that Q' has concentrated functional form compared with membership function of MOST.

Next we try to get a mathematical expression for $\mu_{Q'}$. Following the functional form of $\mu_{\text{MOST}}$,

$$\mu_{Q'}(\gamma)=\mu_{\text{MOST}}\left[\text{MIN}\left(\dfrac{1}{A}\sum_{i=1}^{2} A_i\wedge F_i\right)\right]; \text{ for } \gamma = \dfrac{1}{A}\sum_{i=1}^{2} A_i\wedge F_i^{1/2}$$

So we must get MIN() for a value of $\gamma$. We set

$$\gamma = \dfrac{A_1\wedge\sqrt{F_1}+A_2\wedge\sqrt{F_2}}{A_1+A_2}, \; y = \dfrac{A_1\wedge F_1+A_2\wedge F_2}{A_1+A_2} \qquad (7)$$

When $A_1$ and $A_2$ are given, we may change the values of $F_1$, $F_2$ randomly under the above conditions. And because Eq. (7) includes the MIN operator, we assign 3 intervals for each $F_i$ (so for $F_1$ and $F_2$, there are 3*3=9 intervals). According to these intervals, Eq.(7) is as follows (we set $A = A_1 + A_2$, $A_1 \leq A_2$ ):

(1) $F_1\geq A_1$, $F_2\geq A_2$         (2) $F_1\geq A_1$, $A_2\geq F_2\geq A_2^2$

$\gamma = \dfrac{A_1+A_2}{A}=1$, $y=\dfrac{A_1+A_2}{A}=1$    $\gamma = \dfrac{A_1+A_2}{A}=1$, $y=\dfrac{A_1+F_2}{A}$

(3) $F_1\geq A_1$, $F_2\leq A_2^2$         (4) $A_1\geq F_1\geq A_1^2$, $F_2\geq A_2$

$\gamma = \dfrac{A_1+\sqrt{F_2}}{A}$, $y=\dfrac{A_1+F_2}{A}$    $\gamma = \dfrac{A_1+A_2}{A}=1$, $y=\dfrac{F_1+A_2}{A}$

(5) $A_1\geq F_1\geq A_1^2$, $A_2\geq F_2\geq A_2^2$    (6) $A_1^2\leq F_1\leq A_1$, $F_2\leq A_2^2$

$\gamma = \dfrac{A_1+A_2}{A}=1$, $y=\dfrac{F_1+F_2}{A}$    $\gamma = \dfrac{A_1+\sqrt{F_2}}{A}$, $y=\dfrac{F_1+F_2}{A}$

(7) $F_1\leq A_1^2$, $F_2\geq A_2$         (8) $F_1\leq A_1^2$, $A_2\leq F_2\leq A_2$

$\gamma = \dfrac{\sqrt{F_1}+A_2}{A}$, $y=\dfrac{F_1+A_2}{A}$    $\gamma = \dfrac{\sqrt{F_1}+A_2}{A}$, $y=\dfrac{F_1+F_2}{A}$

(9) $F_1\leq A_1^2$, $F_2\leq A_2^2$

$$\gamma = \dfrac{\sqrt{F_1}+\sqrt{F_2}}{A}, \; y=\dfrac{F_1+F_2}{A}$$

Of all 9 cases, case(9) is the most important. The minimum value of y in case(9) is less than or equal to all values of y in other cases for the same value of $\gamma$. For example, we compare case(9) with case(3) as follows:

We set $F_1 = A_1^2$ for case(9) and get

$$\gamma = \dfrac{A_1+\sqrt{F_2}}{A}, \; y=\dfrac{A_1^2+F_2}{A}$$

Now the above $\gamma$ of case(9) becomes equal to the $\gamma$ of case(3). Then we compare y of the modified case(9) and y of case(3).

$$\dfrac{A_1^2+F_2}{A} \leq \dfrac{A_1+F_2}{A} \quad ( A_1\leq 1 )$$

So, for the same value of $\gamma$, the value of y in case(9) is less than or equal to that of y in case(3). For the other 7 cases, we can show the same result through the same procedure. From now on we consider only case(9). Next, we will get the minimum value of y in case (9).

For a given value of $\gamma$, y has minimum value in case that $F_1 = F_2 = F$ (you can easily get this by drawing a graph of $\gamma$ and y).

Here we divide the range of the case(9) into the following two intervals.

(1) $0\leq F_1\leq A_1^2$, $0\leq F_2\leq A_1^2$

$$\gamma = \dfrac{2\sqrt{F}}{A}, \; \text{MIN}(y)=\dfrac{2}{A}F$$

$$\therefore \; \text{MIN}(y)=\dfrac{A}{2}\gamma^2 \text{ for } 0\leq\gamma\leq\dfrac{2A_1}{A} \qquad (8)$$

(2) $0\leq F_1\leq A_1^2$, $A_1^2\leq F_2\leq A_2^2$

$$\gamma = \dfrac{A_1+\sqrt{F_2}}{A}, \; \text{MIN}(y)=\dfrac{A_1^2+F_2}{A}$$

$$\therefore \; \text{MIN}(y)=\dfrac{A_1^2}{A}+A\left(\gamma-\dfrac{A_1}{A}\right)^2 \text{ for } \dfrac{2A_1}{A}\leq\gamma\leq 1 \quad (8\text{-}1)$$

This result can be achieved by Fig. 4. Because of the condition that $0\leq F_1\leq A_1^2$, $A_1^2\leq F_2\leq A_2^2$, $\sqrt{F_1}$, $\sqrt{F_2}$ take values on the broadest line in Fig. 4. So y has minimum value when $\sqrt{F_1}=A_1$. We can easily extend the above discussion to the case of 0<m<1.

(1) 0<m<1

$$y_1=\dfrac{A}{2}\gamma^{1/m} \qquad\qquad \text{for } 0\leq\gamma\leq\dfrac{2A_1}{A} \quad (9)$$

$$y_2=\dfrac{A_1^{1/m}}{A}+A\left(\gamma-\dfrac{A_1}{A}\right)^{1/m} \text{ for } \dfrac{2A_1}{A}\leq\gamma\leq 1 \quad (9\text{-}1)$$

And the same procedure can be applied to the case of m>1. In this case, we get the maximum value for the case $F_1 \leq A_1$, $F_2 \leq A_2$ as follows:

(2) m>1

$$y_3 = \sqrt[m]{\frac{2\gamma}{A}} \qquad \text{for} \quad 0 \leq \gamma \leq \frac{2A_1^m}{A} \qquad (10)$$

$$y_4 = \frac{A_1}{A} + \sqrt[m]{\frac{1}{A}\left(\gamma - \frac{A_1^m}{A}\right)} \qquad \text{for} \quad \frac{2A_1^m}{A} \leq \gamma \leq \frac{A_1^m + A_2^m}{A} \qquad (10\text{-}1)$$

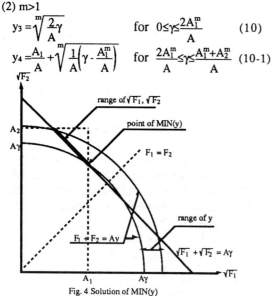

Fig. 4 Solution of MIN(y)

So, using the above extreme values of y, we get inferred $\mu_{Q'}$ from Eq. (11).

$$\mu_{Q'}(\gamma) = \mu_Q(\text{extreme value of } y) \qquad (11)$$

We summarize the result in Table 3.

Table 3: Characteristic features of $\mu_{Q'}$

| Q \ m | m > 1 (very, ....) | 0 < m < 1 (rather,..) |
|---|---|---|
| monotone nondecreasing ( MOST,...) | (CASE 1) dilation $\mu_Q(y_3)$, $\mu_Q(y_4)$ MAX | (CASE 2) concentration $\mu_Q(y_1)$, $\mu_Q(y_2)$ MIN |
| monotone nonincreasing (FEW,...) | (CASE 3) concentration $\mu_Q(y_3)$, $\mu_Q(y_4)$ MIN | (CASE 4) dilation $\mu_Q(y_1)$, $\mu_Q(y_2)$ MAX |

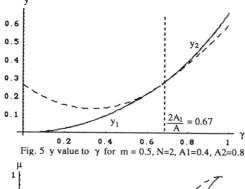

Fig. 5 y value to $\gamma$ for m = 0.5, N=2, A1=0.4, A2=0.8

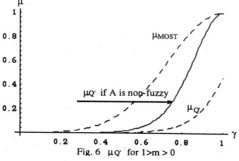

Fig. 6 $\mu_{Q'}$ for 1>m > 0

We show MIN(y) in Fig. 5. For 2 intervals of $\gamma$, y has 2 curves. $\mu_{Q'}$ is shown in Fig. 6 when Q=MOST, m=rather. It can be seen that $\mu_{Q'}$ has concentrated functional form compared with $\mu_{MOST}$. And compared with the case that A is non-fuzzy, $\mu_{Q'}$ has more concentrated functional form. The maximum grade value of $\mu_{Q'}$ is less than 1 because A is a fuzzy quantity in Eq. (1). But when A=2 (so $A_1 = A_2 = 1$), $\mu_{Q'}$ is $\mu_Q(\gamma^2)$ and equal to the result for the case that A is non-fuzzy.

### 3.2.2 General Case of N

We can apply the same method as N=2 to the general case. Following the procedure of N=2, we proceed as follows. As before, we assume we have a database for N (for i=1,...,N we denote as $\mu_F(i) = F_i$, $\mu_A(i) = A_i$ the extent of membership in the fuzzy set F and A for the i-th term). Further we assume that Q is monotone nondecreasing and 0<m<1. Following the functional form of $\mu_Q$ and value of m, we choose the minimum value in Table 2 and get the following.

$$\mu_{Q'}(\gamma) = \mu_Q\left[\text{MIN}\left(\frac{1}{A}\sum_{i=1}^N A_i \wedge F_i\right)\right]; \text{ for } \gamma = \frac{1}{A}\sum_{i=1}^N A_i \wedge F_i^m$$

So we must get MIN(y) for a given value of $\gamma$, by setting

$$\gamma = \frac{1}{A}\sum_{i=1}^N A_i \wedge F_i^m , \quad y = \frac{1}{A}\sum_{i=1}^N A_i \wedge F_i$$

When $A_1,...,A_N$ are given, we may change values of $F_1,...,F_N$ randomly under these conditions, and because the expression includes MIN operator($\wedge$), we divide 3 intervals for each Fi (so for $A_1,...,A_N$, there are $3^N$ intervals). The notation is as follows: (we set $A = A_1 +,..., +A_N$, $A_1 \leq,...,\leq A_N$ )

(1) $F_i \geq A_i$; i=1,...,N

$$\gamma = \frac{1}{A}\sum_{i=1}^N A_i = 1, \quad y = \frac{1}{A}\sum_{i=1}^N A_i = 1$$

⋮       ⋮

(I) $F_i \leq A_i^{1/m}$; i=1,...,N

$$\gamma = \frac{1}{A}\sum_{i=1}^N F_i^m, \quad y = \frac{1}{A}\sum_{i=1}^N F_i$$

Of all $3^N$ cases, case(I) is the most important. We can show that the minimum value of y for case(I) is less than or equal to all the values for other cases (proof is omitted here). From now on we consider only case(I). Next, we get the minimum value of case(I) as follows. For a given value of $\gamma$, y has minimum value for $F_1 = ,..., = F_N$ if 0<m<1[4]. So the following Eq. (12-1) holds true. For Eq. (12-2) to Eq.(12-N), we get the following expressions through the same procedure as for N=2 (proof is omitted here).

(1) $0 \leq \gamma \leq \frac{NA_1}{A}$ $\qquad\qquad$ $y_1 = \frac{A}{N}\gamma^{1/m}$ $\qquad$ (12-1)

(2) $\frac{NA_1}{A} \leq \gamma \leq \frac{A_1 + (N-1)A_2}{A}$

$$y_2 = \frac{A_1^{1/m}}{A} + \frac{A}{N-1}\left(\gamma - \frac{A_1}{A}\right)^{1/m} \qquad (12\text{-}2)$$

$$\vdots \qquad \vdots \qquad \vdots$$

$$(n)\ \frac{1}{A}\left(\sum_{i=1}^{n-2} A_i + (N-n+2)A_{n-1}\right) \lesssim \gamma \lesssim \frac{1}{A}\left(\sum_{i=1}^{n-1} A_i + (N-n+1)A_n\right)$$

$$y_n = \frac{1}{A}\sum_{i=1}^{n-1} A_i^{1/m} + \frac{A}{N-n+1}\left(\gamma - \frac{1}{A}\sum_{i=1}^{n-1} A_i\right)^{1/m} \qquad (12\text{-}n)$$

$$\vdots \qquad \vdots \qquad \vdots$$

$$(N)\ \frac{1}{A}\left(\sum_{i=1}^{N-2} A_i + 2A_{N-1}\right) \lesssim \gamma \lesssim \frac{1}{A}\sum_{i=1}^{N} A_i = 1$$

$$y_N = \frac{1}{A}\sum_{i=1}^{N-1} A_i^{1/m} + A\left(\gamma - \frac{1}{A}\sum_{i=1}^{N-1} A_i\right)^{1/m} \qquad (12\text{-}N)$$

So, y has N curves for N intervals.

We show the results in Fig. 7 for the case of N=3 and m=1/2 (rather). According to $y_n$, we can easily get $\mu_{Q'}$ from Eq. (11). For m>1, we get mathematical expressions like Eq. (12-1) to Eq. (12-N) through the same procedure.

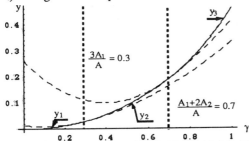

Fig. 7 y value to $\gamma$ for m = 0.5, N=3, A1=0.1, A2=0.3, A3=0.6

### 3.3 Inference Method: triangular type

When Q is a triangular-type fuzzy quantifier, we get $\mu_{Q'}$ in the same way as in Sec. 2.3.2.

For example,

Several heavy foreigners are tall

$$\Leftrightarrow \begin{cases} Q1 \text{ heavy foreigners are tall} \\ \qquad \text{AND} \\ Q2 \text{ heavy foreigners are tall} \end{cases}$$

$$\Rightarrow \begin{cases} Q1' \text{ heavy foreigners are very tall} \\ \qquad \text{AND} \\ Q2' \text{ heavy foreigners are very tall} \end{cases}$$

$$\Rightarrow Q' \text{ heavy foreigners are very tall}$$

We show the inferred result in Fig. 8 when N=2.

It can be seen that $\mu_{Q'}$ is shifted to the left compared with $\mu_{SEVERAL}$, and more shifted compared with the case that A is a non-fuzzy quantity.

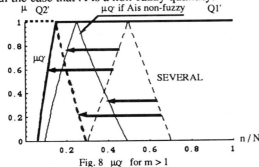

Fig. 8 $\mu_{Q'}$ for m > 1

### 3.4 Inference Method: when * is used instead of MIN as AND operator

Next we consider the inference when we use the * operator instead of $\wedge$. In this case we can prove the following theorem (proof is omitted here)

[THEOREM]

Conditions:

$$\gamma = \frac{1}{A}\sum_{i=1}^{2} A_i * F_i^m,\quad A = \sum_{i=1}^{2} A_i,\ 0 \leq A_i, F_i \leq 1$$

Under these conditions, $\frac{1}{A}\sum_{i=1}^{N} A_i * F_i$ has extreme values when $F_1 = ,..., = F_N$ and these extreme values are as follows:

(1) m>1

$$MAX\left(\frac{1}{A}\sum_{i=1}^{N} A_i * F_i\right) = \sqrt[m]{\gamma}$$

(2) 0<m<1

$$MIN\left(\frac{1}{A}\sum_{i=1}^{N} A_i * F_i\right) = \sqrt[m]{\gamma}$$

Following the above theorem and our proposition to choose extreme values as inferred results, we can show that we get the same $\mu_{Q'}$ as in the case of non-fuzzy A in Sec. 2.3. So the results are the same as in Table 2. For triangular Q, we can apply the same procedure as in Sec. 3.3 to this case and easily get inferred $\mu_{Q'}$.

## 4. Conclusion

We have shown that in the inference "QA is F $\Rightarrow$ Q'A is mF", Q' can be determined analytically for three types of quantifiers (monotone nonincreasing types, monotone nondecreasing types, and triangular types) when A is a fuzzy quantity. We have also shown that the inferred result Q' is the same as the case that A is a non-fuzzy quantity if we use the mathematical product as an AND operator.

The inferred result, when we use * as an AND operator, may match human intuition more closely because the inferred result is the same regardless of whether A is fuzzy or non-fuzzy.

We are also currently trying to achieve an analytical resolution for Q' when "QA is F $\Rightarrow$ Q' (mA) is F" is inferred.

## References

[1] S. Tano, W. Okamoto and T. Iwatani, *New Design Concepts for the FLINS- Fuzzy Lingual System:Text-based and Fuzzy-centered Architectures*, Proc. of Int'l Symposium on Methodologies for Intelligent Systems, pp. 286-294 (1993).

[2] L. A. Zadeh, *A Theory of Approximate Reasoning*, Machine Intelligence, **Vol. 9**, New York: Halstead Press, pp. 149-194 (1979).

[3] L. A. Zadeh, *A Computational Approach to Fuzzy Quantifier in Natural Languages*, Comp. and Maths. with Appls., **Vol. 9**, pp. 149-184 (1983).

[4] W. Okamoto, S. Tano, T. Iwatani, *Estimation of Fuzzy Quantifiers in FLINS*, Proc. of 1'st Asian Fuzzy Systems Symposium (1993).

# ACQUISITION OF FUZZY RULES BY DATA ANALYSIS OF BIOKINETICS

B. Ruggeri and G. Sassi*

Dip. di Chimica Inorganica, Chimica Fisica,Chimica dei Materiali
Università di Torino via P. Giuria, 7; 10100 - TORINO (Italy)
*Dip. di Scienza dei Materiali e Ingegneria Chimica
Politecnico di Torino c.so Duca Abruzzi, 24; 10129 - TORINO (Italy)

## Summary
This paper deals with the use of fuzzy mathematical approach in the description of a kinetics in a bioreactor. An original procedure to handle a set of experimental data to build up a fuzzy algorithm is presented. The information is stored in IF...THEN statements able to describe the experienced dominium.

## 1. Sources of data uncertainty in biokinetics

Recently a great number of paper dealing with the micro-organism modelling appeared in many journals. Nevertheless many situations involving biomass activity in biotechnology are mismodelled or no-modelled at all (Ruggeri and Sassi, 1993A).

A possible (reductive!) explanation of the encountered difficulties are founding in the not adequate (actually) experimental set-up and devices which are able to capture only such "holistic" (Ruggeri and Sassi, 1993B) aspects of the cellular phenomena because of the different relaxation times (Röels, 1982) of microbial mechanisms and those of investigating devices.

More strictly specking, the uncertainty of experimental data is coming from different sources hence also the modelling and the "conclusions" obtained by these ones inherit this uncertainty.

The first and the most important source of uncertainty streams from the variability of data, the micro-organisms behaviour in equal or comparable situations does not show identical trajectories. This, typically, is founded on data taken from living systems and reflects the rich variability of their nature (an example is the difficulty of obtaining the same inocule activity in a bioreactor).

The second source of uncertainty is connected with the arbitrary level of precision of experimental investigation, which is due not only to the power of sensors, but to the "measure context" too, which determines the observations themselves. As an example the determination of the degree on ATP of a microbial population is a measure of the "activity" of energized reactions as a whole, but not of the energized bioreaction pathways.

The third source of uncertainty is originated by the use of natural language (usually done in life sciences) in order to describe observations on biology and microbiology. As an example, for the reader not introduced in the field, it is possible to mention the mechanism of facilitate transport across a biological membrane: <<a transport protein is "usually" specific for a "certain" molecule or group of structurally "similar" one; however "sometimes" a single protein is able to transport two "different" species across the membrane>> (Horton et al., 1993).

All the above recalled sources of uncertainty, with the combination of corpuscular nature of micro-organism, the degree of ignorance about biological mechanisms, the degree of accuracy and precision of experimental tools, determine the vagueness of experimental data. But the bioreactor science needs to have a mathematical tool in order to handle precise details (bioreactor dimensions, mixing power, oxidation flow rate etc.) together with imprecise ones linked mainly to the biocatalyst.

Recently Ruggeri and Sassi (1993C) used a fuzzy approach to take into account the hydrodynamic shear stress on mold activity. The relatively encouraging results obtained, were forcing the Authors to stress the use of fuzzy approach in the biotechnology and bioreactor dominia.

This paper candidates a fuzzy data analysis procedure to capture the knowledge on the kinetics of a bioreaction. Fuzzy approach is used in order to infer several experimental data to obtain an inferential motor of IF...THEN nature able to perform a fuzzy modelling of kinetics.

## 2. Knowledge dominia

In the present case the goal is to have a suitable representation of a bioreaction kinetics by a fuzzy algorithm in order to take into account such parameters which otherwise are not computable by traditional Constitutive Equations of classical approach (Himmemblau and Bishoff, 1968). To this purpose the micro-organism behaviour is described by a verbal parameter, the so called "Physiological state" (PS) (Malek, 1975; Konstantinov and Yoshida, 1991). To test the proposed fuzzy methodology the citric acid production by a strain of *Aspergillus niger*

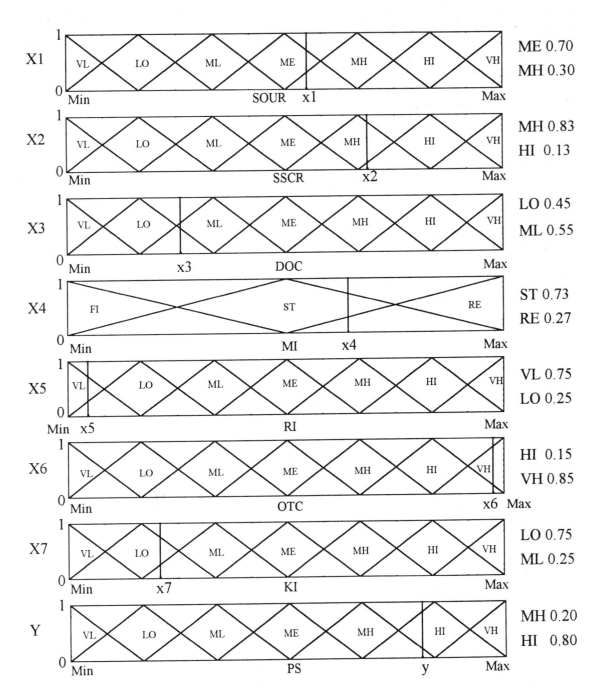

Figure 1 - Procedure to set up the inferential rules

was used. This fermentation is one of the most complex one because of its strong aerobic conditions dependence and the nature of the mold which is susceptible on the shear stress generated by turbulence magnitude induced in order to increase mass transfer rate (Röhr et al., 1983).

The system under study can be structured following an holistic approach in three sub-systems: the biotic phase, the abiotic phase and the subsystem constituted by their mutual interactions. Owing the ill-defined nature of the "real world" which must be modelled, the experimental data must be generated by monitoring such parameters in order to have information about all the sub- systems, in accordance with the aim, the finalities and the scope of the representation (modelling purposes).

The scope of modelling was the evaluation of the ability of the micro-organism to produce citric acid hence: "PS: specific citric acid production", is the consequent variable.

The following parameters were selected as affecting the PS, in fuzzy word they are the antecedent variables:

SOUR    Specific Oxygen Uptake Rate (Biotic phase)
SSCR    Specific Substrate Consumption Rate (Biotic phase)
DOC     Dissolved Oxygen Concentration (Abiotic phase)
RI      Rheological Index (Interactions)
MI      Morphological Index (Biotic phase)
OTC     Oxygen Transfer Coefficient (Interactions)
KI      Kolmogoroff's Index (Interactions)

Inside the brackets the sub-system of interest is reported. For all the antecedents, as well as for the consequent, the dominium was experimentally evaluated; details on experimental set-up together with data may be found in the previous reports (Sassi, 1990; Sassi et al., 1991).

### 3. Fuzzy rules and fuzzy algorithm set-up

One considers after experimentation to answer the following questions: What was it happen? What does data means?

Holistically speaking the answer will be pursued by undertaking a logical relationships (rules) set describing the system. Traditionally specking the question is how to set-up a model in order to conduct a descriptive analysis. The Authors renounce to try to correlate the above parameters invoking several mechanisms but they are viewed as "holistic" markers giving information on the whole system, i.e., the bioreactor.

The present procedure can be considered as equivalent to a classical kinetic study in the Chemical Reaction Engineering sense. The time variation of the consequent is obtained by looking the variation of the antecedents vs. time before entering in the fuzzification procedure.

Recalling in mind the sources of uncertainty of section 1, the experimental data can be regarded as a fuzzy vector coming from pseudo-exact data which give a grey tone picture of reality.

The fuzzification of data was reached by formulating labels and their membership functions. For each antecedent and consequent, apart the Morphological Index (MI) for which the experimental investigation (a microscopic observation of mold morphology) suggests to use not more than three labels, the labels used were the following: VL, very low; LO, low; ML, medium low; ME, medium; MH, medium high; HI, high; VH, very high. The labels for MI were: FI, filamentous; ST, standard; RE, retarded; which indicate the experienced situations where the mold was: "not able to produce citric acid", "able to produce citric acid" and "able to produce citric acid with difficulties" for FI, ST, RE respectively. In the following the antecedents are indicated with X and the consequent with Y; their numerical values with x and y respectively. The selected membership functions were of triangular shape for all the labels as shown in Figure (1); the support of each of them was obtained dividing the experimented ranges homogeneously. In the fuzzification of MI parameter an arbitrary scale [0,1] was selected to convert the microscopic image in a score scale.

In order to set-up the fuzzy rules the following procedure was used: the vector of experimental data was constituted by elements representing a crisp situation, each one is:

$$x^\circ = (x_1, ..., x_7, y) \qquad /1/$$

entering with $x^\circ$ on the membership function as shown in Figure (1) each experimental data forming the crisp situation is converted in a fuzzy set, i.e., every point is coupled with a label and its grade of membership as shown in the right field of Figure (1).

The crisp situation /1/ is converted in a set of fuzzy conditional statements of the following type:

IF $X_1$ is LO AND $X_2$ is ML AND $X_3$ is MH AND $X_4$ is ST AND $X_5$ is LO
AND $X_6$ is VH AND $X_7$ is ME AND Y is ME THEN "$r_j$ is a rule" /2/

Owing the membership function used each of 8 parameters in /1/ belongs to 2 labels hence $2^8=256$ fuzzy conditional statements are possible. If one considers that each of 6 experimental runs have 24 points, the possible fuzzy conditional statements are 36864 which are a little part of the 17294403 possible combinations, but in any case a big number to handle. At this point a decision structure in order to accept the conditional statement as a representative rules of the system under study needs.

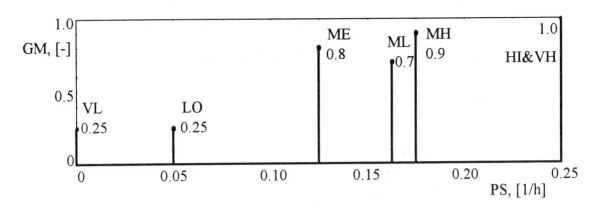

Figure 2 - Singletons of the consequent PS

Considering the rule set {R} as a subset of the possible combination as a graph, in particular {R} is a fuzzy graph, each element $r_j$ is associated to a grade of membership; this grade of possibility that "$r_j$ is a rule" is evaluated by applying the following (Zadeh, 1973):

$$m_R(r_j) = \sup_{k=1,u} \min(m(x_1), ..., m(x_7), m(y)) \qquad /3/$$

where u is the number of equal conditional statements, obtained processing different crisp situations. The /3/ reduce the number of possible rules if a threshold of acceptance of $m_R(r_j)$ is introduced; changing the threshold value the rule number can be modulated.

The last criterion of decision structure is of heuristic nature: the rules number is selected in accordance with software able to perform fuzzy inference using the stated rules. In the present application a PC-Board FP3000 (Omron, 1991) was employed, it is able to handle until 128 rules. The above procedure (implemented on a PC) was applied to all the 144 experimental situations with a threshold value of 0.5, it gave 85 rules (See Table I).

The structure of the fuzzy algorithm was constituted by the 85 rules in which PS is the consequent; $A_{i,j}$ ($B_j$) are the label of the i-th antecedent (consequent) in the j-th rule (they belongs to {VL,LO,ML,ME,MH,HI,VH}, only $A_{4,j}$ belong to {FI,ST,RE}):

$$r_j: \text{ IF } X_1 \text{is } A_{1,j} \text{ AND...AND } X_7 \text{is } A_{7,j} \text{ THEN PS is } B_j \qquad j=1,...,85 \qquad /4/$$

To compute the fuzzy algorithm /4/ the same membership functions of antecedents used in the rules creation procedure was used, while that of the consequent PS were defined as singletons which abscissa were defined by a trial and error method (see Figure (2)).

## 4. Results and comments

In order to evaluate the efficiency of fuzzy algorithm /4/, to the modelling of mold citric acid production, the logical relations are used to evaluate the possibility that PS has a certain numerical value for a given crisp situation of antecedents.

To this goal the inferential motor (Bandemer and Näther, 1992) was constituted by min-max Zadeh's norm:

$$GM_l = \sup_{k=1,v} \min_{i=1,7} (m_{A_{i,k}}(x_i)) \qquad /5/$$

where $GM_l$ is the grade of membership of the l-th label of the PS and v is the number of rule with the same consequent label. In order to squeeze out the final conclusion and get a deterministic value of the specific citric acid production rate the center of gravity method (CG-method) was selected as defuzzification procedure.

In Figures (3), a comparison of deterministic values evaluated by fuzzy simulation and that experimentally determined of specific citric acid production rate are shown, the agreement is good.

As a comment the Authors highlight the peculiarity of the fuzzy approach as compared with the traditional one. If one looks at rules /4/ as a traditional "kinetics" in (bio)chemical engineering it is remarkable that they are able to

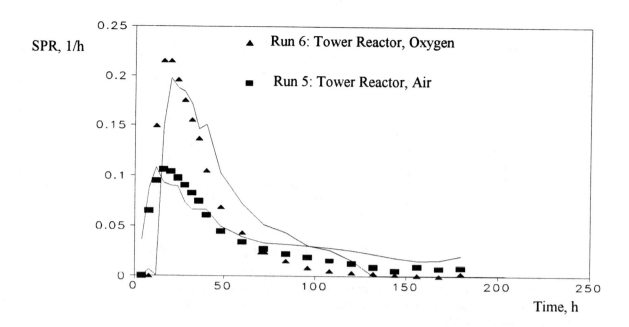

Figure 3 - Experimental (symbols) and predicted (lines) data. Specific Citric Acid Production Rate (SPR) by *A. niger B60* on banana extract medium.

take into account several parameters different than traditional ones encountered in usual kinetic expressions (such as OTC, $k_L a$ in traditional terms, KI, MI) and of different nature: numerical and linguistic ones (MI) which are impossible to take into account by a Constitutive Equation approach to kinetics.

Other important aspect concerns the data used to set up the rules, i.e., the "knowledge domain". They can be obtained not only by experimental investigations, but also and mixing together information coming from test runs, literature and/or theoretical findings. As an example, in the specific case it is possible to take into account data coming from different researchers on citric acid production by *A. niger* taking care to harmonize them by adding some other antecedent, but using the same framework, i.e. the model can be more and more entangled but the mathematical complexity of fuzzy algorithm remain the same.

## References

Bandemer, H. and Näther, 1992. *Fuzzy data analysis*, Kluwer Academic Publ., London.

Himmemblau, D.M. and Bishoff, R.B., 1977. *Process analysis and simulation, deterministic systems*, J. Wilej & Son, New York.

Horton, H.R., Morau, L.A., Ochs, R.S., Rown, J.D. and Scringeour, K.G., 1993. *Principles of biochemistry*, Printice Hall Inc., New York.

Kostantinov, K. and Yoshida, T., 1989. Physiological state control of fermentation processes, *Biotechnol. Bioengng.*, **33**, 1145-1156.

Malek, I., 1975. *Physiological state of continuous growth microbial cultures*, Proc. 6th Int. Symp. Contin. Cult. Microorg., Oxford.

Omron, 1991. *Fuzzy guide book, Cat. P30-1-1, Fuzzy Inference board FP3000 Processor*, Shimokaiinji, Nagaokakio-city, Kyoto.

Röels, J.A., 1982. *Energetics and kinetics in biotechnology*, Elsevier Biomed. Publ., Amsterdam.

Röhr, M., Kubicek, C.P. and Kominek, J., 1983. Citric Acid. In: Rehem, H.J. and Reeds, G., *Biotechnology*, v.3, pp. 419-454, Verlag Chimie, Basel.

Ruggeri, B. and Sassi, G., 1993A. On the modelling approaches of biomass behaviour in bioreactor, *Chem. Engng. Comm.*, **122** 1-56.

Ruggeri, B. and Sassi, G., 1993B. Macro-approach and fuzzy modelling of bioreactors. In: *Trends in biotechnology*, J. Menon, Ed., New Deli (In press).

Ruggeri, B. and Sassi, G., 1993C. Hydrodynamic shear stress on biocatalysts: a fuzzy approach opportunity. In: *Bioreactor and bioprocess fluid dynamics*, A.W., Nienow, Ed., BHR Group, Cambridge, UK.

Sassi, G., 1990. *Produzione di acido citrico da scarti agroalimentari*, PhD Thesis, Politecnico di Torino, Torino.

Sassi, G., Ruggeri, B., Specchia, V. and Gianetto, A., 1991. Citric acid production by *A. niger* with banana extract, *Bioresource Technol.*, **37**, 259-269.

Zadeh, L.A., 1973. Outline of a new approach to the analysis of complex systems and decision processes, *IEEE Trans. Syst. Man Cybern.*, **3**, 28-44.

### Table I - The 85 rules resulting from a threshold value of 0.5

| X1 | X2 | X3 | X4 | X5 | X6 | X7 | Y | | X1 | X2 | X3 | X4 | X5 | X6 | X7 | Y | | X1 | X2 | X3 | X4 | X5 | X6 | X7 | Y |
|----|----|----|----|----|----|----|----|---|----|----|----|----|----|----|----|----|---|----|----|----|----|----|----|----|----|
| LM | VH | LO | ST | ME | LO | VL | VL | | VL | VL | VL | FI | HI | VL | HI | VL | | LM | HM | LO | ST | LM | LO | VL | LM |
| LM | HM | LO | ST | ME | LO | VL | VL | | VL | VL | VL | FI | HI | VL | VH | VL | | LM | ME | LO | ST | LM | LO | VL | LM |
| VL | VL | LO | ST | VL | VL | VL | VL | | LM | ME | HI | ST | ME | VH | VL | VL | | LM | LM | LO | ST | LO | LO | VL | LM |
| VL | VL | LO | ST | VL | LO | VL | VL | | HM | HM | HI | ST | ME | VH | VL | VL | | ME | ME | LO | ST | LM | LM | VL | LM |
| VL | VL | LO | FI | VL | LO | VL | VL | | HI | HI | HI | ST | LM | HI | VL | VL | | LO | LM | LO | ST | LO | LO | VL | LM |
| LM | HI | LO | ST | ME | LM | VL | VL | | VL | VL | HI | ST | VL | LM | VL | VL | | ME | ME | LO | ST | ME | LM | VL | LM |
| VL | VL | VL | FI | LO | VL | VL | VL | | VL | VL | VH | ST | VL | LM | VL | VL | | LO | ME | LO | ST | LM | LO | VL | LM |
| VL | VL | LO | FI | LO | VL | VL | VL | | VL | VL | VH | ST | LO | LM | VL | VL | | LO | LM | LO | ST | LM | LO | LO | LM |
| VL | VL | LO | FI | LM | VL | LO | VL | | VL | VL | VH | ST | LO | LM | LO | VL | | ME | ME | LO | ST | LM | LM | LO | LM |
| VL | VL | VL | FI | HM | VL | LM | VL | | VL | VL | VH | ST | LO | LO | LO | VL | | ME | HM | LO | FI | HI | LM | LO | LM |
| VL | VL | VL | FI | HI | VL | LM | VL | | VL | VL | VH | FI | LO | LO | LO | VL | | VL | LO | HI | RE | LO | LM | VL | LM |
| HM | VL | LO | ST | ME | LM | VL | VL | | LM | ME | LO | ST | LM | LM | VL | ME | | ME | HM | LO | ST | LM | LM | VL | HM |
| ME | LM | LO | ST | ME | LM | VL | VL | | LM | HM | LO | ST | ME | LM | VL | ME | | ME | HI | LO | ST | HM | LM | LO | HM |
| VL | LO | LO | FI | ME | VL | LO | VL | | HM | HM | LO | ST | ME | LM | VL | ME | | HM | HM | HI | RE | LM | HI | VL | HM |
| VL | LO | VL | FI | HM | VL | LM | VL | | HI | HM | LO | ST | ME | LM | LO | ME | | LO | LM | HI | RE | LO | ME | VL | HM |
| VL | VL | VL | FI | HM | VL | ME | VL | | ME | HM | LO | ST | HI | LM | LO | ME | | ME | HM | LO | ST | ME | ME | VL | HI |
| VL | VL | VL | FI | HI | VL | ME | VL | | LO | LO | HI | RE | LO | ME | VL | ME | | ME | HI | LO | ST | HM | ME | LO | HI |
| VL | VL | VL | FI | HI | VL | HM | VL | | LO | LO | LO | ST | LO | LO | VL | LO | | HM | ME | HI | RE | LM | HM | VL | HI |
| LO | LM | LO | FI | HM | LO | LM | VL | | VL | LO | VL | ST | LO | VL | VL | LO | | LM | LM | HI | RE | LO | HM | VL | HI |
| LO | LM | VL | FI | HI | VL | LM | VL | | VL | LO | VL | ST | VL | VL | VL | LO | | VL | VL | VL | ST | VL | VL | VL | VL |
| LO | LO | VL | FI | HI | VL | LM | VL | | LO | LO | VL | ST | LO | VL | VL | LO | | LM | ME | LO | ST | LM | LO | VL | ME |
| VL | LO | VL | FI | VH | VL | HM | VL | | LO | LO | LO | FI | ME | LO | LO | LO | | ME | HM | LO | ST | LM | LM | VL | ME |
| VL | VL | VL | FI | VH | VL | HI | VL | | LM | LM | LO | ST | LM | LO | LO | LO | | VL | VL | VL | ST | VL | VL | VL | LO |
| LO | LM | LO | FI | HI | LO | LM | VL | | LO | LM | LO | FI | ME | LO | LO | LO | | LM | HI | LO | ST | ME | LM | VL | LO |
| LO | LO | VL | FI | HI | LO | LM | VL | | LO | LM | LO | FI | HI | LO | LO | LO | | LO | LM | LO | ST | LM | LO | LO | LO |
| LO | LO | VL | FI | HI | LO | ME | VL | | VL | LO | HI | RE | VL | LM | VL | LO | | LO | ME | LO | ST | LO | LO | VL | LM |
| VL | LO | VL | FI | HI | VL | ME | VL | | VL | VL | HI | RE | VL | LM | VL | LO | | ME | HM | LO | ST | LM | LM | VL | LM |
| VL | LO | VL | FI | HI | VL | HM | VL | | LO | LO | HI | RE | LO | ME | VL | HM | | LO | LM | HI | RE | LO | ME | VL | HI |

# Systematic Design Approach for Multivariable Fuzzy Expert System[*]

Bin-Da Liu      Chun-Yueh Huang

Department of Electrical Engineering
National Cheng Kung University, Tainan, Taiwan, R. O. C.

## Abstract

An algorithm to improve the Gupta's structure is proposed to design the multivariable fuzzy expert system. By this algorithm, the mapping problem is solved but the properties of systematic analysis and the regular design procedure are preserved. Besides, a new scheme, which can save great amount of storage, in implementing the relation matrices is presented. A famous example of inverted pendulum in fuzzy system is used to explain this algorithm. Taking the advantage of the composition rule of fuzzy inference, the inference speed of this structure is fast and it is suitable for real time processing.

## I. Introduction

Fuzzy logic has been successfully used in several expert systems [1-3]. In the real world, the applications of fuzzy logic in the expert systems are multivariable ( multi-input multi-output ) systems. Recently, several researches are devoted to develop the methodology to analyze and synthesize the multivariable fuzzy expert system [4-7]. Gupta [4] proposed a systematic and regular design methodology for multivariable fuzzy systems. Based on his methodology, the design procedure is easier and flexible. However, the accuracy of inferential result is limited because of the simplification in multivariable structure and the lack of good mapping property. In this paper, an algorithm based on the Gupta's structure is proposed, which not only solves the mapping problem but also preserves the properties of systematic analysis and regular design procedure. Besides, a new scheme, which can save great amount of storage, is proposed to implement the relation matrices.

## II. The multivariable fuzzy expert system

Figure 1 shows the general structure of the fuzzy expert system, which includes the fuzzifier module, the defuzzifier module, the inference engine, and the knowledge base. The fuzzifier module is used to translate the input signal to the fuzzy representation and the defuzzifier module is used to convert the inferential fuzzy logic results into the crisp solution variable to the output

---

[*] This work is supported by the National Science Council, Republic of China, under Grant NSC-83-0404-E-006-029.

of the fuzzy expert system. Generally, an expert system is used to simulate the expert of human, so the semantic variable descriptions for the production rules in the expert system are more appreciate. In a fuzzy expert system, the fuzzy production rules, whose conditions of the premise and the conclusion are represented by the linguistic variables, are used to represent the decision-making procedure of the knowledge base of late. Taking the advantage of the linguistic variables or natural languages to represent the decision-making procedure in the fuzzy production rules, the imprecise information and the system uncertainty can be dealt with. Therefore, the inference engine has to be equipped with the computational capabilities to analyze and conclude the reasonable inference result from the premises with uncertain information.

Based on the composition rule of inference, the relation of the production rules in the knowledge base can be computed before the inference process. With this methodology, the inference speed is fast and the inferential structure is suitable for real time applications. Figure 2 shows the block diagram of the fuzzy expert system that is implemented by the methodology of the composition rule of inference. During the inference process, the input signal is translated into the degree of the membership of the distinct fuzzy sets. The results are then inferred according to the degree of membership that belongs to the premise of the fuzzy production rules. The higher the degree of the premise conforms, the more the conclusion of inference contributes.

Consider the 3-input 3-output fuzzy expert system shown in Fig. 3. Therefore, the fuzzy production rules to describe the decision-making procedure of the fuzzy expert system have the form of

$$\text{IF } X_{1(i)} \text{ and } X_{2(i)} \text{ and } X_{3(i)} \text{ THEN } Y_{1(i)} \text{ and } Y_{2(i)} \text{ and } Y_{3(i)}, \quad i = 1, ..., N \quad (1)$$

where $X_{k(i)}$ is the fuzzy linguistic label of the $k$th input variable defined in the universe of discourse $X^k$, $k=1, 2, 3$; $Y_{j(i)}$ is the fuzzy linguistic label of the $j$th output variable defined in the universe of discourse $Y^j$, $j=1, 2, 3$. Let the domains of the universe of discourse $X^k$ be $p_k$ and $Y^j$ be $q_j$, respectively. The overall fuzzy relation $R$ of the system is expressed as follows

$$R = \bigvee_{i=1}^{n} \{ X_{1(i)} \wedge X_{2(i)} \wedge X_{3(i)} \wedge Y_{1(i)} \wedge Y_{2(i)} \wedge Y_{3(i)} \}, \quad (2)$$

where $\vee$ and $\wedge$ are MAX operation and MIN operation of fuzzy logic, respectively, and $\dim[R] = p_1 \times p_2 \times p_3 \times q_1 \times q_2 \times q_3$.

Suppose that the current inputs are $X_1$, $X_2$, and $X_3$. The following compositional rule of inference is used to obtain the present output $Y$.

$$Y = X_1 \circ X_2 \circ X_3 \circ R. \quad (3)$$

The result of this composition is a compound fuzzy set $Y$ in the universe $Y_1 \times Y_2 \times Y_3$, where $\dim[Y] = q_1 \times q_2 \times q_3$.

Since the relation (2) and the fuzzy output (3) are multidimensional, a direct analysis and synthesis of the multivariable fuzzy system is difficult to obtain. Gupta [4] proposed an algorithm to solve this problem and to obtain a more tractable structure. Based on his algorithm, the individual outputs can be calculated by the projection of the compound output fuzzy set (3) on the respective universes

$$Y_1 = Y_{11} \wedge Y_{21} \wedge Y_{31}$$
$$Y_2 = Y_{12} \wedge Y_{22} \wedge Y_{32} \tag{4}$$
$$Y_3 = Y_{13} \wedge Y_{23} \wedge Y_{33}.$$

where

$$Y_{kj} = X_k \circ R_{kj}, \qquad k=1, 2, 3; j=1, 2, 3 \tag{5}$$

and

$$R_{kj} = \overset{n}{\underset{i=1}{\vee}} \{ X_{k(i)} \wedge Y_{j(i)} \} \text{ for } k = 1, 2, 3; j = 1, 2, 3. \tag{6}$$

It should be noted that the nine-dimensional fuzzy relation matrix $R$ and the three-dimensional fuzzy output matrix $Y$ are decomposed into nine two-dimensional fuzzy sub-relation matrices $R_{11}, ..., R_{33}$ and the final results are three one-dimensional fuzzy output matrices $Y_1, Y_2,$ and $Y_3$. Now the decomposed system can be represented by the block diagram as shown in Fig. 4. Since the relationship representation is too simple, the mapping problems exist in the inference process. These problems will induce an incorrect inferential result, although the condition of the premise is clearly defined in the rules of the knowledge repository. Furthermore, the mapping problem will also seriously affect the system performance when the number of input variables is large.

## III. The proposed algorithm

In order to solve the mapping problem, we propose an algorithm to modify the relation matrices. For considering the system implementation and easily understanding the mapping problem, a famous example of inverted pendulum [8] in fuzzy system is given. Originally, the inverted pendulum requires seven rules. The state-action diagram representation of the control rules for the inverted pendulum system and the membership function of the distinct linguistic label for the rod angle ($\theta$), the angular velocity of the rod ($\dot{\theta}$), and the amount of drive of the dolly ($\omega$) are shown in Fig.5. Because the production rules in the knowledge base of the expert system are clearly defined by the linguistic variables, the domain of the input variable can be defined by the number of linguistic labels, and so does the output variable. In order to realize the system by using the composition rule of fuzzy inference, the dimensions of the relation matrices can be defined by the product of the domains of the input variable and the output variable. In the relation matrices, only those pairs whose members are defined in the rules to each other will be assigned to be a value 1, which indicates that they are clearly defined in the knowledge base. Based on the relation matrix structure, the fuzzy inference procedure for system uncertainty is mainly concerned in the fuzzifier module and defuzzifier module. This methodology can save a lot of memory spaces in system implementation without violating the concepts of fuzzy theory. In this system, we can obtain the internal relationship between $\theta$ and $\omega$ and the relationship between $\dot{\theta}$ and $\omega$ in the form of matrices according to Eq. (6), as shown in Figs. 6(a) and 6(b) respectively.

Frequently, there are many-to-one mapping problems between the premise and the conclusion in the relation matrices. For example, when the inputs are $\theta=$**NS** and $\dot{\theta}=$**NS**, the inferential result of the output is $\omega=$**ZE** according to Eq. (5). This inferential result is consistent with the definition of rules in the knowledge base. But when the inputs are $\theta=$**NS** and $\dot{\theta}=$**PS**, the inferential result of the output $\omega$ is between **NS** and **ZE**, which is inconsistent with the definition of rules. However, the inferential error caused by the mapping problem of the relationship can be solved by the *MatrixExtend* algorithm as follows:

***Algorithm***: *MatrixExtend*

Given:   $N$ rules of the knowledge base.
            Assume there are $k$ input variables and $j$ output variables.
            Let $p_k$ be the number of the linguistic label of $k$th input variable.
            Let $q_j$ be the number of the linguistic label of $j$th output variable.
            Let **S** represent linguistic label.

Step1:   Establish the internal relation matrices $R_{kj}$

$$R_{kj} = \bigvee_{i=1}^{N} \{ X_{k(i)} \wedge Y_{j(i)} \}$$

Step2:   For $u = 1$ to $j$    /* Scan every internal relation matrix */
           {            /* $j$ sets relation matrices */
            i = 0           /* Initialize the index number */
          For $v = 1$ to $k$     /* Regard $k$ relation matrices as one set */
            {
              For w $= 1$ to $q_j$ /* Scan every linguistic label of the output variable */
                {
                IF
                there are many-to-one mapping conditions in the linguistic label $\mathbf{S}_w$
                THEN
                   rename the linguistic label by adding footnote $\mathbf{S}_{wi}$
                   and shift the whole row relationship of the same set relation
                   for eliminating the many-to-one mapping situations.
                   i=i+1   /* Increase the index number */
              }
            $\mathbf{S}_w = \wedge_i \mathbf{S}_{wi}$
            }
          }

The relation matrices, Figs. 6(a) and 6(b), of the fuzzy system after modification by the *MatrixExtend* algorithm are shown in Figs. 6(c) and 6(d), respectively. By using these matrices and Eq. (5), when the inputs are $\theta$=NS and $\dot{\theta}$=PS, the inferential result $\omega$=NS is obtained which is consistent with the definition of rules.

Thus, we have constructed a fuzzy expert system with its inferential scheme the same as that shown in Fig. 4 while the internal relation matrices of the system are established by the *MatrixExtend* algorithm. The inferential result of the system is obtained from Eqs. (4) and (5). Since the system structure is regular and the analysis procedure is systematic, the fuzzy expert system is suitable for VLSI implementation.

## IV. Conclusions

In this paper, an algorithm to improve the Gupta's structure has been proposed. By this algorithm, the mapping problem is solved but the systematic analysis and the regular design procedure for the fuzzy expert system are preserved. Besides, great amount of storage is saved by using the new scheme to implement the relation matrices. A famous example of inverted pendulum in fuzzy system has been used to explain this algorithm. With this design procedure, designing the multivariable fuzzy expert system is easier and more accurate. Taking the advantage of the composition rule of fuzzy inference, the inference speed of this structure is fast and it is suitable for real time applications. Moreover, this proposed structure is suitable for VLSI implementation.

## References

[1]  P. J. King and E. H. Mamdani, "The application of fuzzy control systems to industrial processes," *Automatica*, vol. 13, pp.235-242, 1977.

[2]  W. J. M. Kickert and H. R. Van Nautha Lemke, "Application of a fuzzy controller in a warm water plant," *Automatica*, vol. 12, pp.301-308, 1976.

[3]  L. J. Huang and M. Tomizuka, "A self-placed fuzzy tracking controller for two-dimensional motion control ," *IEEE Trans. Syst., Man., Cybern.*, vol. 20, No. 5, pp.1115-1124, 1990.

[4]  M. M. Gupta, J. B. Kiszka and G. M. Trojan, "Multivariable structure of fuzzy control systems," *IEEE Trans. Syst., Man, Cybern.*, vol. 16, No. 5, pp.638-656, 1986.

[5]  A. D. Nola, W. Pedrycz, and S. Sessa, "When is a fuzzy relation decomposable in two fuzzy sets?" *Fuzzy Sets and Systems,* vol. 16, pp.87-90, 1985.

[6]  C. W. Xu, "Decoupling fuzzy relational systems - an output feedback approach," *IEEE Trans. Syst., Man., Cybern.*, vol. 19, No. 2, pp.414-418, 1990.

[7]  C. W. Xu and Y. Z. Lu, "Decouping in fuzzy systems: a cascade compensation approach," *Fuzzy Sets and Systems,* vol. 29, pp.177-185, 1989.

[8]  T. Yamakawa, "Stabilization of an inverted pendulum by a high-speed fuzzy logic controller hardware system," *Fuzzy Sets and Systems,* vol. 32, pp.161-180, 1989.

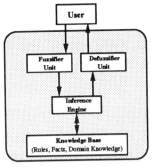

Fig. 1 The general structure of the fuzzy expert system.

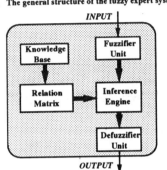

Fig. 2 The block diagram of the fuzzy expert system by using the composition rule of inference.

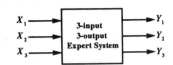

Fig. 3 The 3-input 3-output fuzzy expert system.

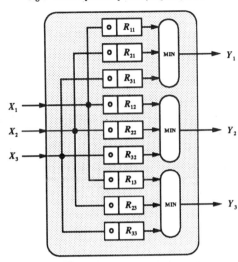

Fig. 4 The block diagram of the decomposed 3-input 3-output fuzzy expert system.

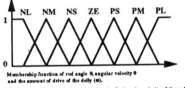

Membership function of rod angle $\theta$, angular velocity $\dot\theta$ and the amount of drive of the dolly ($\omega$).

NL ( Negative Large )   Input : Rod angle $\theta$, Angular velocity of the rod $\dot\theta$
NM ( Negative Medium )  Output : The amount of drive of the dolly ($\omega$).
NS ( Negative Small )
ZE ( Approximately Zero )
PS ( Positive Small )
PM ( Positive Medium )
PL ( Positive Large )

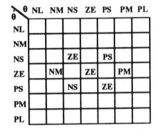

Fig. 5 State-action diagram representation of the control rules for inverted pendulum system.

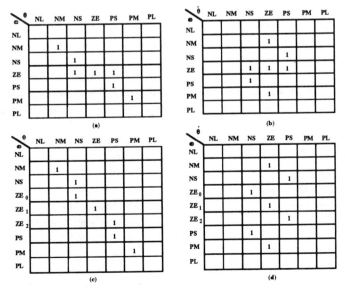

Fig. 6 Internal relations of the inverted pendulum system in the form of matrix.

# DCS-1: A Fuzzy Logic Expert System for Automatic Defect Classification of Semiconductor Wafer Defects

Dr. Marc Luria, Maty Moran, Dr. Dan Yaffe
Galai Laboratories, Migdal Haemek, Israel
Jim Kawski, ADE Corporation, Boston, Mass.

## Abstract

DCS-1, a Defect Classification System for classifying defects on a semiconductor wafers using a builtin fuzzy logic expert system is described. DCS-1 is advantageous over manual classification because it is both objective and fast. Defect classes are defined using DDL, Defect Description Language, a language of fuzzy predicates which was developed for this application. The paper describes a unique approach to fuzzy logic and image processing where fuzzy features are computed using traditional image processing techniques and classifications are decided using fuzzy logic rules.

## 1. Introduction

In manufacturing, the classification of images is still considered a bottleneck. While much of the manufacturing process has been automated using very robust tools, a human operator often needs to visually inspect the manufactured item. Galai, a company with long experience in image processing tools, realized that using traditional software techniques would not suffice to classify images. As a result, we chose a hybrid approach: traditional algorithms for feature extraction are combined with a fuzzy logic rule interpreter for classification.

Our first major attempt at classification was in the semiconductor industry. In semiconductor wafer manufacturing, classification of defects is an important step toward understanding malfunctions in the manufacturing process. Every FAB (semiconductor FABrication) engineer designs a classification strategy which will help clarify what went wrong in the manufacturing process.

## 2. Defect Classification -- Current Technology

Currently, defects are classified in the following manner. The defects are:

A. Detected using automatic detection tools
B. Classified by a human operator

## 2.1. Defect Detection

The detection of patterned wafer defects is a well developed technology. There are a number of commercially available automatic inspection systems which analyze the inspected wafer and determine defect location in an automatic manner, e.g., Tencor 7500, KLA, and OSI. Within a few minutes, these detection tools can scan an entire 8 inch wafer and report the defect locations with relatively good accuracy. They simply compare one die on the wafer to its neighboring die. The detection tools assume that all dies are similar and it is unlikely to have a defect in the neighboring die. These systems produce a defect map - a list of defect coordinates on a wafer.

## 2.2. Manual Defect Classification

The wafer is then inspected using a review station which displays each defect to a human operator. The task of the operator is to classify the defects according to the defects defined by the engineer for a particular level in the wafer.

There are a number of competing systems (Leica, Kinetic, etc.). These systems consist of a microscope, a stage, and the apparatus for handling the wafers. The review station also includes computer (e.g. 486) that handles the interface with the detection tool and also with the human operator. After loading the defect map from the defect detection tool, the review station moves the microscope field of view to the first defect. The human operator examines the defect, and enters the classification into the review station. After the human operator has classified all the defects on the wafer, the defect classification information is passed to a FAB database system. The database is used by the FAB engineer to make decisions regarding changes in the process line.

The process of defect classification is often time-consuming, monotonous, and costly. In practice, there may be up to 4000 defects in a single 8" wafer. The monotonous work causes fatigue and eye-strain which in turn cause errors in classification. In addition, human decision making is directly affected by the individual experience base of the operator. Thus, a different shift, with a different operator may produce different results.

No adequate statistical information is available with respect to the speed and accuracy of the human operators. One FAB engineer told us that the operators examine only 100 defects on each wafer. Each wafer takes the human operator 45 minutes (including frequent breaks). Another engineer told us that he found that when using more than 4 defect classes on a wafer, inter-operator reliability deteriorated considerably. Using four or fewer classes on a wafer, inter-operator reliability is about 80%.

Due to the time required and the cost of a human operator, FAB engineers often decide to concentrate on defect classification for a particular level further along the fabrication process. Automatic defect classification using DCS-1 avoids these problems. The results are reproducible, accurate, and fast.

3.      Fuzzy Logic in DCS-1.

In order to test DCS-1 in the real world, the DCS-1 design team worked with a number of wafer fabrication plants around the world. FAB engineers told us that they wanted total control over the classification process. There are two reasons for this demand:

A.      The defect classes produced during fabrication are some of the FAB's best kept secrets;

B.      Defect class structure may change suddenly due to temporary changes in production. A defect may appear for a week and then disappear.

To meet the demand of total control over the classification process, an expert system, with rules defined locally by the FAB engineer, determines the class of each defect. The descriptions, however, of defect classes are never exact. For example, a

particular defect class might be described as being circular. Traditional logic and programming methods would stipulate that a defect is circular or not circular (true or false). However, when the FAB engineer states that a particular defect is circular, he does not mean that a slight aberration in the perimeter would prevent classification of the defect as a particular defect class. Instead, he means that the defect is more-or-less circular. Often, he has noted the circularity in order to differentiate the defect from other similar but non-circular defects.

In order to translate these inexact characterizations of a class into a language that the computer could understand, Fuzzy Logic techniques are employed.

For example, *Circular* is a fuzzy concept. Instead of stating that a defect object is either circular or not circular, when using Fuzzy Logic one can refer to the full range of truth values between true and false. Traditionally, fuzzy truth values range between 0 and 1, where 0 is false and 1 is true. Thus, the following values might be assigned to the following images:

### IMAGE VALUE

| Image | Value |
| --- | --- |
| Quarter | .99 |
| Protozoan | .8 |
| Car Tire at Rest | .7 |
| American Football | .5 |
| Pencil | .01 |

Similarly, one can use fuzzy linguistic quantifiers [3] to say that a quarter is very circular, a protozoan moderately circular, and a pencil is not circular. Fuzzy concepts can be connected using logical connectives. The fuzzy truth value of a conjunctive is the minimum of the conjuncts and the truth value of a disjunctive is the maximum of the disjuncts.

4.      DDL - Defect Description Language

In order to facilitate fuzzy descriptions of defect classes that can be defined locally by the FAB engineer, we have developed DDL - Defect Description Language. DDL was designed in such a way that the FAB engineer who has only limited knowledge of fuzzy logic could easily describe the defect classes in a way similar to natural language. FAB engineers are very busy people who want a tool that they can learn and master quickly. Therefore, giving FAB engineers a generic fuzzy logic expert system shell would not be appropriate.

Furthermore, we also assume that FAB engineers have only a limited knowledge of image processing. As discussed in the previous section, Visual Predicates are based on Visual Features. For example, the Visual Predicate "Elongated" is based on the Visual Feature "Aspect Ratio". Since we cannot assume that the FAB engineer will understand the significance of "Aspect Ratio", we cannot assume that he could design fuzzification parameters for the relationship between Elongated and Aspect Ration in a generic Fuzzy Logic expert system shell. DDL is composed of Visual Predicates, which are fuzzy terms which describe the visual characteristics of the object. For example, Scattered, Circular, and Elongated are all visual predicates which can be used in defining a specific defect class.

Defect classes are defined by using the Visual Predicates. Each class is defined as an if-then rule, where the description of the class is the antecedent and the class name is the consequent. For example, the defect class, *Scratch* might be defined as:

**Rule #8:**
>    *Scratch* <----
>    Elongated &
>    Size > 60 &
>    Scattered

In other words, the Scratch defect class is identified when the defect object is Elongated, its size is greater than 60 units, and the defect image is scattered across the field of view. Therefore, if a particular defect was determined to be:

>    Elongated: .89
>    Size > 60: 1.00 (e.g., it size is 100)
>    Scattered .83

The truth value of the defect being *Scratch* would be .83.

Furthermore, the DDL rules can include fuzzy modifiers, such as Somewhat, Very, and Not. For example, suppose that in Rule #8 the engineer would have specified instead that Scratch is *Very Elongated*. Although the defect is .89 when classified as *Elongated*, it might be only .70 under the Visual Predicate of *Very Elongated*. Thus, the truth value of the defect being Scratch would only be .70.

For each level of the wafer, one rule such as Rule #8 is defined for each class. Although the same class is defined at different levels, a new rule is needed for each level, since the visual manifestations of that defect may be different at each level. These rules are then tested on a set of images.

Although the classes themselves are not pre-defined, the Visual Predicates in DDL are. This is due to the fact that each of the Visual Predicates is defined using image processing algorithms for that particular Visual Predicate. Actually, the truth value of each Visual Predicate is determined through the use of visual features. These features are converted to the fuzzy Visual Predicates by using a feature fuzzification process. We currently computer the Fuzzy Value using both trapezoidal numbers and S functions. [1]

5.    DCS-1 Algorithm

Once the rule set has been defined, the DCS-1 automatically classifies all the defects on the wafer. DCS-1 is interfaced to a commercially available review station. DCS-1 uses a defect map transferred from a previous 100% inspection system (such as Tencor 7500, KLA, OSI). Each of the defects in the defect map is then analyzed and classified according to the algorithm described in the following sections:

5.1.    Image Extraction:

The Defect image is extracted using image comparison techniques since the defect map provided from the review station is not sufficient. The image is compared to a reference image (the same image on another part of the wafer) and the two images are compared. Currently, the comparison is done using a T-805 transputers.

5.2.    Feature Extraction:

A wide range of visual features, both spatial and textural are computed from the defect image, using advanced image processing algorithms. Only those features which are needed for a particular level are computed. Currently, these features are computed on a 486-based PC compatible running under OS/2. However, since many of these features are independent, we are currently implementing these features in parallel on the transputer discussed in Section 5.4.

## 5.3. Fuzzification:

The fuzzy predicates are computed using the features extracted from the image. As described in the previous section, the feature fuzzification is done using a trapezoidal numbers or S-functions.

## 5.4. Rule Application

Each of the rules, which corresponds to one of the classes in a particular level, is evaluated using the Visual Predicates computed in the previous step. This evaluation is done using special-purpose code developed for the DCS-1. Rules are loaded once when a particular wafer is loaded into the review station. Since the rules will be used many times, they are pre-compiled for better performance.

## 5.5. Defect Classification Decision

After all the rules have been evaluated, the classification with the highest truth value is displayed on the engineer's console and appended to a classification log file.

## 6. Results

The DCS-1 was tested on two different sets of real data collected from two wafer manufacturing sites. The data sets were composed of video images taken using a review station connected to the DCS-1. In addition, each of the defect images was classified by the FAB engineer. We wrote the classification rules, but they were approved by the FAB engineers.

Success was measured by the number of times that the engineer's classification was equivalent to that of DCS-1 divided by the number of images.

|         | # of Images | Success | # of Defect Classes |
|---------|-------------|---------|---------------------|
| Site #1 | 32          | 89%     | 6                   |
| Site #2 | 69          | 85%     | 7                   |

The system is currently being tested in three beta sites.

## 7. DCS Knowledge Acquisition

In this section, we describe the development of advanced knowledge acquisition tools for DCS-1. Knowledge acquisition is the stage of acquiring knowledge from the expert for the expert system. In DCS-1, the task is to provide an environment where the expert (FAB engineer) can easily define new defect classes or modify old ones. These rules can then immediately be tested on real data (defect images). We are currently developing semi-automatic and automatic tools that will determine rules in an automatic fashion. Such systems are called adaptive fuzzy systems, since they adapt the fuzzy rules.

## 7.1. DCS Knowledge Acquisition Tools

We are currently developing a tool for the DCS called the DCSKA for DCS Knowledge Acquisition System. This system provides an easy-to-use interface for the user to define Fuzzy Logic rules. It is assumed that the user has only a limited background in Fuzzy Logic.

All the fuzzy rules are defined using the DCSKA. The DCSKA interface allows the user to enter new classes into the system and define rules for those classes. The development of new rules is done in a structured manner. New rules are defined by pointing with the mouse at parts of the language in a dialogue box. All visual predicates are documented on-line. In the future, the documentation for each element will comprise a short description and a long description. The long description may include visual information. For example, the description of *Circular* may include example pictures of *Circular*, *Very Circular*, *Somewhat Circular*, and *Not Circular*.

## 7.2. Testing in the DSCKA

The other major part of the DCSKA is the testing component. Testing is done on real images. The testing procedure currently works as follows. The user saves a number of representative images while using the DCS. He also manually enters the classification for each of the defect images. He does this using a special Learn Mode which calculates Visual Predicates for all of the potential predicates in the DCS. (In normal mode, only those predicates necessary for classification are calculated.)

The user can see the outcome of all these predicates as well as the classification data for each of these images. This information can be used to make changes in the defect classification rules.

In addition, an overview of the results is provided. The user can examine the classification success rate for each of the defect classes and the total success rate for the classification.

## 8. Analysis

DCS-1 classifies defects using the rules defined in the Defect Description Language. This classification is reproducible, fast, and accurate. The DDL allows FAB engineers who have little experience in programming or expert systems to accurately and quickly describe a defect class which can then be tested immediately on real data.

### 8.1. Standardization of Classification

In order to use DDL, the FAB engineers must specify exactly the definitions of the defects in the wafer level. Thus, a side-effect of the use of the DCS-1 is the standardization of defect classes. Similar benefits have been realized in other applications where an expert system was used. The introduction of the expert system has forced companies to standardize the running of a business. In wafer production, standardization is particularly important since large manufacturers may produce the same wafer at multiple sites. After the FAB engineer has developed the classification for a particular level, the DDL description of that classification can be distributed around the world. The FAB engineer at the local production site can thus use the expertise of the FAB engineer where the classification is developed.

### 8.2. Galai's Approach to Image Processing and Fuzzy Logic

In this section, we wish to reiterate our basic approach and discuss how it differs from other approaches to image processing and Fuzzy Logic. Galai has developed a large number of sophisticated image-processing algorithms. Therefore, even though it was recognized within the company that a Fuzzy Logic approach was necessary for intelligent tasks such as classification, we did not want to discard that developed technology.

The approach we developed was to adapt our traditional image processing algorithms. The output of these algorithms is fuzzified and then used in Fuzzy Logic rules. We believe that this approach in some ways mimics human behavior. Image processing algorithms that compute visual features correspond to unconscious behavior while fuzzy logic rules correspond to conscious decision-making behavior. It is clear that people compute visual features in some very automatic unconscious fashion which they have trouble expressing. On the other hand, classification is a conscious task that people can express. In our system, feature extraction is done in an automatic fashion by using C code. Decision-making, a more conscious process is accomplished using a fuzzy logic expert system.

Thus, we have not had to replace our algorithms and use a completely different approach to low-level image processing, such as neural networks, or low-level Fuzzy Logic. Instead, we have modified the traditional image-processing methods which have been developed over the last decade to provide information to a high-level Fuzzy Logic expert system. We believe that by using this approach we have two major advantages.

A. The ability to build on pre-existing in-house developed image-processing technology for feature extraction

B. The ability to make complex user-defined classifications using state-of-the-art Fuzzy Logic expert system capabilities.

As such, this approach is different from earlier approaches to image processing using fuzzy logic.

There have been a number of other systems using Fuzzy Logic techniques that are relevant to classification of objects using visual inspection techniques. Most of the previous works done on the application of Fuzzy Logic to Image Processing have dealt with the fuzzification of known "crisp" concepts. Rosenfeld [2] developed an extension of the topological concepts of connectedness and surroundness. In [3], he continued this direction and fuzzified various visual terms. The fuzzy concepts of adjacency, convexity, starshapedness, area, perimeter, and compactness of fuzzy objects were described.

Pal and Rosenfeld [4] dealt with the problem of segmentation of ill-defined images. Incorporating fuzziness to the description of regions provided more meaningful results than by considering fuzziness in the grey level alone.

Dellepiane, et. al. [5] described the IBIS system which deals with the processing and interpretation of images in the biomedical and remote sensing areas. The system deals with model generation and model matching in image data. These tasks, performed using fuzzy restrictions [6], give satisfactory results in the analysis of topographic images.

Pedrycz et.al [7] describe a system for recognizing electro-cardiographic signals from an electro-cardiogram. They do not use a rule-based approach, but instead compute 16 features for every image. Each of the features is evaluated with respect to its fuzzy value for 3 features: *Normal, Borderline,* or *Abnormal.* Each diagnosis is defined using a 3 x 16 matrix that identifies the fuzzy values for each of the 16 features. The diagnosis that best matches the image is identified as the image classification. This approach is in some ways more robust then that of the DCS-1 since more information about the class is included in the description. However, this solution is unacceptable in the wafer classification domain for two reasons. Firstly, most defect classes are characterized by a relatively small number of feature values. For example, dust that drops on the wafer may be of any size, any color, and any shape. Thus, most features for a particular defect class are "don't cares." Secondly, this approach demands much more input from the expert (FAB engineer) and a greater understanding of Fuzzy Logic. A rule-based system is easier for the FAB engineer to understand.

9. Conclusion

The DCS-1 provides FAB engineer with an important tool for defect classification. Using this tool, FAB engineers have already been able to classify defects and work towards increasing yields in their semiconductor manufacturing facilities. They have found the DDL, Defect Description Language, easy to

use, and have written rules using the DCSKA Knowledge Acquisition system.

*References:*

[1] Zadeh, L.A. (1965), "Fuzzy Sets," *Information and Control*, 8, 338-353.

[2] A. Rosenfeld, "Fuzzy Digital Topology," Information Control, vol. 40, No. 1, 1979, pp. 76-87.

[3] A. Rosenfeld, "The Fuzzy Geometry of Image Subsets," *Pattern Recognition Letters*, vol. 2, 1984, pp. 311-317

[4] S.K. Pal, A. Rosenfeld, "Image Enhancement and Thresholding by Optimization of Fuzzy Compactness," *Pattern Recognition Letters*, vol. 7, 1988, pp. 77-86.

[5] S. Dellepiane, G. Venturi, G. Vernazza, "Model Generation and Model Matching of Real Images by a Fuzzy Approach," *Pattern Recognition*, vol. 25, No. 2, 1992, pp. 115-137.

[6] L.A. Zadeh, "Calculus of Fuzzy Restrictions," *Fuzzy Sets and Their Applications to Cognitive and Decision Processes*, Academic Press, 1975.

[7] W. Pedrycz, G. Bortolan, and R. Degani, "Classification of Electrocardiographic Signals: A Fuzzy Pattern Matching Approach," *Artificial Intelligence in Medicine 3*, 1991, pp. 211-226.

and fuzzified various visual terms. The fuzzy concepts of adjacency, convexity, starshapedness, area, perimeter, and compactness of fuzzy objects were described.

Pal and Rosenfeld [4] dealt with the problem of segmentation of ill-defined images. Incorporating fuzziness to the description of regions provided more meaningful results than by considering fuzziness in the grey level alone.

Dellepiane, et. al. [5] described the IBIS system which deals with the processing and interpretation of images in the biomedical and remote sensing areas. The system deals with model generation and model matching in image data. These tasks, performed using fuzzy restrictions [6], give satisfactory results in the analysis of topographic images.

Pedrycz et.al [7] describe a system for recognizing electro-cardiographic signals from an electro-cardiogram. They do not use a rule-based approach, but instead compute 16 features for every image. Each of the features is evaluated with respect to its fuzzy value for 3 features: *Normal, Borderline,* or *Abnormal*. Each diagnosis is defined using a 3 x 16 matrix that identifies the fuzzy values for each of the 16 features. The diagnosis that best matches the image is identified as the image classification. This approach is in some ways more robust then that of the DCS-1 since more information about the class is included in the description. However, this solution is unacceptable in the wafer classification domain for two reasons. Firstly, most defect classes are characterized by a relatively small number of feature values. For example, dust that drops on the wafer may be of any size, any color, and any shape. Thus, most features for a particular defect class are "don't cares." Secondly, this approach demands much more input from the expert (FAB engineer) and a greater understanding of Fuzzy Logic. A rule-based system is easier for the FAB engineer to understand.

## 9. Conclusion

The DCS-1 provides FAB engineer with an important tool for defect classification. Using this tool, FAB engineers have already been able to classify defects and work towards increasing yields in their semiconductor manufacturing facilities. They have found the DDL, Defect Description Language, easy to use, and have written rules using the DCSKA Knowledge Acquisition system.

*References:*

[1] Zadeh, L.A. (1965), "Fuzzy Sets," *Information and Control*, 8, 338-353.

[2] A. Rosenfeld, "Fuzzy Digital Topology," Information Control, vol. 40, No. 1, 1979, pp. 76-87.

[3] A. Rosenfeld, "The Fuzzy Geometry of Image Subsets," *Pattern Recognition Letters*, vol. 2, 1984, pp. 311-317

[4] S.K. Pal, A. Rosenfeld, "Image Enhancement and Thresholding by Optimization of Fuzzy Compactness," *Pattern Recognition Letters*, vol. 7, 1988, pp. 77-86.

[5] S. Dellepiane, G. Venturi, G. Vernazza, "Model Generation and Model Matching of Real Images by a Fuzzy Approach," *Pattern Recognition*, vol. 25, No. 2, 1992, pp. 115-137.

[6] L.A. Zadeh, "Calculus of Fuzzy Restrictions," *Fuzzy Sets and Their Applications to Cognitive and Decision Processes*, Academic Press, 1975.

[7] W. Pedrycz, G. Bortolan, and R. Degani, "Classification of Electrocardiographic Signals: A Fuzzy Pattern Matching Approach," *Artificial Intelligence in Medicine 3*, 1991, pp. 211-226.

# IMPACT OF FUZZY NORMAL FORMS ON KNOWLEDGE REPRESENTATION

## I.B. Türkşen

Department of Industrial Engineering
University of Toronto
Toronto, Ontario, M5S 1A4

## ABSTRACT

Fuzzy normal forms can be generated with an application of "Normal Form Generation Algorithm" on fuzzy truth tables. This takes place at the third level of knowledge representation, i.e., propositional level. It is shown that at least three distinct sets of normal forms can be generated depending on the axioms one is willing to impose on the propositional fuzzy set and logic theories. All are conjunctive-disjunctive and complement based De Morgan Logics with the following three classes of axioms that identify each general class of fuzzy normal forms in order of least to most restrictive set of axioms in the following sense: (i) boundary and monotonicity; (ii) boundary, monotonicity, associativity and commutativity; (iii) boundary, monotonicity, associativity, commutativity and idempotency.

## 1. INTRODUCTION

Recently, we have proposed that knowledge representation should be expressed at least in terms of four levels (Türkşen, 1993). These are identified as linguistic, meta-linguistic, propositional, and computational levels. The meta-linguistic expressions are symbolic representations of linguistic expressions in a natural language. Generally, there is more than one interpretation associated with each meta-linguistic expression that corresponds to a linguistic expression. For example, the conjunction of two linguistic concepts is expressed as "A AND B"; but there are at least two interpretations of this as (a) NON-COMMUTATIVE and (b) COMMUTATIVE, "AND." Another example is Implication, which is expressed as "A→B"; but there are at least two interpretations of this as (a) NOT A OR B, known as S-implication, and (b) OR{X|A AND X ⊆ B}, known as R-implication, etc.

On the other hand, it is known that there are two distinct canonical forms known as Disjunctive and Conjunctive Normal Forms, DNF, CNF, respectively, in two-valued logic. In a previous investigation, it was shown that the fuzzification of these canonical forms generated a containment, i.e., $DNF_i(.) \subseteq CNF_i(.)$, i=1, . . . , 16, for all of the sixteen possible propositional expressions (Türkşen, 1986). Furthermore, recently, it was shown that the "latent connectives"

$$\mu^{(1)}{}_{A\Theta B} = \mu^{(1-\gamma)}_{A \cap B} \mu^{\gamma}_{A \cup B} \quad \text{and}$$

$$\mu^{(2)}{}_{A\Theta B} = (1-\gamma)\mu_{A \cap B} + \gamma\mu_{A \cup B},$$

defined by Zimmermann and Zysno (1980) as a result of their experimental studies, are contained within DNF(A AND B) and CNF(A OR B) (Türkşen, 1992).

However, these results did not answer three important questions. They are stated as follows: 1) Are there fuzzy normal forms? 2) Can they be obtained from fuzzy truth tables? 3) If there are fuzzy normal forms, what is their relationship to the fuzzified extensions of the classical normal forms (Türkşen, 1986)? In order to respond to these questions, fuzzy normal forms are discussed in the next section.

## 2. Fuzzy Normal Forms

In a recent study, it was shown that fuzzy normal forms can be generated from the fuzzy truth table directly (Türkşen, 1993). Here, the results of that study are reviewed briefly.

The fuzzy truth table is shown in Table 1 (Kaufmann, 1975) with a new interpretation. It should be noted that:

(1) there are eight entries here as opposed to four entries that are in the classical truth table. (2) There is an order among the degrees of truth, $\tau(.)$ and falsehood $\phi(.)$ assigned to meta-linguistic variables A and B, as shown in Table 1. This is required due to monotonicity axiom. (3) Fuzzy normal forms can be derived using the Normal form Generation Algorithm (Türkşen, 1993). However, the degrees of truth

must be assigned to a particular meta-linguistic expression under its column heading. (4) A truth or falsehood assignment is made to a particular meta-linguistic expression such as "A AND B" by an interpretation of all four entries shown in a row of Table 1.

For example, the assignment of the degrees of truth or falsehood that are made to the meta-linguistic expression "A AND B" is shown in Table 1 in accordance with what Rescher (1969, pp.122-123) calls an "orthodox" interpretation. *That is, a conjunction (∩) takes the "falsest" and a disjunction (∪) takes the "truest" true value.*

Furthermore, the corresponding primary conjunctions in the fuzzy truth table, Table 1, are written on the basis of the last two columns such that $\tau(.)$ is interpreted as affirmation and $\phi(.)$ is interpreted as complementation in order to be able to apply the normal form generation algorithm. The reason for this is that the "falsest" and the "truest" values are in the last two columns of Table 1. (Recall the order). Finally, $\tau(.)$'s identify "Fuzzy Disjunctive Normal Forms," FDNF's, conjunctions and $\phi(.)$'s identify "Fuzzy Conjunctive Normal Forms," FCNF's conjunctions in the column of the Truth Assignments made to a meta-linguistic expression. Furthermore, $\tau(.)$'s specify the

affirmation and $\phi(.)$'s specify the complementation of the sets in the last two columns. When we apply the normal form generation algorithm to the particular assignments and interpretations made in Table 1, we get the fuzzy disjunctive and conjunctive normal forms, FDNF and FCNF as follows.

$$FDNF(A \text{ AND } B) = (B \cap A) \cup (A \cap B)$$

$$FCNF(A \text{ AND } B) = (B \cup A) \cap (c(B) \cup A) \cap (B \cup c(A)) \cap (A \cup B) \cap (c(A) \cup B) \cap (A \cup c(B))$$

It should be noted that no axioms were specified in the derivation except that $(\cap, \cup, c)$ is a De Morgan Logic aside from the axioms of boundary and monotonicity. Depending on what set of axioms we impose, we get at least three different classes of fuzzy logics. These are:

i)     If we assume $(\cap, \cup, \mathbf{c})$ is a De Morgan Logic such that only the boundary and the monotonicity conditions are to be applicable, then we have FDNF and FCNF expressions associated with the pseudo conjunction-disjunction, and complementation based fuzzy logics. Let us identify these as $FDNF^{(1)}(.)$ and $FCNF^{(1)}(.)$. It should be noted that these are the ones derived from the fuzzy truth table directly as shown above.

ii) If we assume $(\cap, \cup, c)$ is a De Morgan Logic such that boundary, monotonicity, associativity, and commutativity conditions are to be applicable, then we have $FDNF^{(2)}(.)$ and $FCNF^{(2)}(.)$ expressions associated with t-norm, s-norm and complement based fuzzy logics. In this case, the expressions derived above can be re-arranged by applying commutativity and associativity as follows:

$$FDNF^{(2)}(A\ AND\ B) = (A \cap B) \cup (A \cup B),$$

$$FCNF^{(2)}(A\ AND\ B) = (A \cup B) \cap (A \cup c(B)) \cap (c(A) \cup B) \cap (A \cup B) \cap (A \cup c(B)) \cap (c(A) \cup B).$$

iii) Finally, if we assume $(\cap, \cup, c)$ is a De Morgan Logic such that boundary, monotonicity, associativity, and commutativity and idempotency conditions are to be applicable, then we have $FDNF^{(3)}(.)$ and $FCNF^{(3)}(.)$ expressions associated with idempotent t-norm, s-norm and complement based fuzzy logics. These are equivalent to the fuzzified extensions of the classical normal forms as investigated in our previous study (Türkşen, 1986); i.e.,

$$FDNF^{(3)}(A\ AND\ B) = A \cap B$$
$$= DNF(A\ AND\ B)$$

$$FCNF^{(3)}(A\ AND\ B) = (A \cup B) \cap (A \cup c(B)) \cap (c(A) \cup B) = CNF(A\ AND\ B).$$

Therefore, $FDNF^{(3)}(A\ AND\ B) \subseteq FCNF^{(3)}(A\ AND\ B)$.

In a similar manner, we can obtain $FDNF(A\ OR\ B)$ and $FCNF(A\ OR\ B)$ for those three classes of fuzzy logics as follows.

i) For $(\cap, \cup, c)$ that have boundary and monotonicity conditions, we get:

$$FDNF^{(1)}(A\ OR\ B) = (A \cap B) \cup (A \cap c(B)) \cup (c(A) \cap B) \cup (B \cap A) \cup (B \cap c(A)) \cup (c(B) \cap A)$$

$$FCNF^{(1)}(A\ OR\ B) = (A \cup B) \cap (B \cup A)$$

ii) For $(\cap, \cup, c)$ that have boundary, commutativity, associativity and monotonicity, by applying commutativity and associativity, we get:

$$FDNF^{(2)}(A\ OR\ B) = (A \cap B) \cup (A \cap B) \cup (A \cap c(B)) \cup (A \cap c(B)) \cup (c(A) \cap B) \cup (c(A) \cap B)$$

$$FCNF^{(2)}(A\ OR\ B) = (A \cup B) \cap (A \cup B)$$

iii) For $(\cap, \cup, c)$ that have boundary, commutativity, associativity, monotonicity and idempotency, by applying idempotency, we get:

$FDNF^{(3)}(A \text{ OR } B) = (A \cap B) \cup (A \cap c(B)) \cup (c(A) \cap B)$

$FCNF^{(3)}(A \text{ OR } B) = (A \cup B)$

Again in this case, we obtain the same result that was obtained by the fuzzification of the classical normal forms (Türkşen, 1986). With these results, we are faced with an open question. That is, "are there idempotent t-norms, s-norms with appropriate complements that form De Morgan Logics that are idempotent but not distributive or absorptive?" Recall that Weber (1983) shows that distributivity implies absorption which in turn implies idempotency and which in turn implies min-max operators.

## 3. CONCLUSIONS

It is shown that fuzzy normal forms are derived from the fuzzy truth table with a particularly orthodox interpretation. Now, we have fuzzy normal forms for three basic classes of conjunction-disjunction complement based De Morgan Logics. These we have identified as: (i) the class-one fuzzy normal forms, $FDNF^{(1)}$ and $FCNF^{(1)}$ build on only the axioms of boundary and monotonicity. (ii) the class-two normal forms

$FDNF^{(2)}$ and $FCNF^{(2)}$ build on only the axioms of boundary, monotonicity, associativity and commutativity. (iii) the class-three normal forms $FDNF^{(3)}$, $FCNF^{(3)}$ build on only the axioms of boundary, monotonicity, associativity, commutativity and idempotency.

This third class of fuzzy normal forms is equivalent in form to the classical normal forms. This class of fuzzy normal form did not require the assumptions of distributivity or absorption. This poses the possibility of De Morgan Logics that are in existence without the requirement of distributivity and absorption axioms. The third class gives the interval-valued fuzzy sets proposed by Türkşen (1986) with the containment, i.e., $FDNF^{(3)} \subseteq FCNF^{(3)}$.

The result of the current study (Türkşen, 1993) shows that we should apply $FDNF^{(3)} \subseteq FCNF^{(3)}$ only to Zadeh's De Morgan Logic, i.e., min-max-standard complement triple.

Table 1.  Interpretation of the Fuzzy Truth Table

| Degrees of Truth and Falsehood Assignments to the Meta-Linguistic Variables $\subseteq$ $\subseteq$ $\subseteq$ | | | | Degrees of Truth and Falsehood Assignments to the Meta-Linguistic Expression "A AND B" | Primary Fuzzy Conjunctions |
|---|---|---|---|---|---|
| $\tau(A)$ | $\tau(B)$ | $\phi(B)$ | $\phi(A)$ | $\phi(A)$ | $c(B) \cap c(A)$ |
| $\tau(A)$ | $\phi(B)$ | $\tau(B)$ | $\phi(A)$ | $\phi(A)$ | $B \cap c(A)$ |
| $\phi(A)$ | $\tau(B)$ | $\phi(B)$ | $\tau(A)$ | $\phi(B)$ | $c(B) \cap A$ |
| $\phi(A)$ | $\phi(B)$ | $\tau(B)$ | $\tau(A)$ | $\tau(A)$ | $B \cap A$ |
| $\tau(B)$ | $\tau(A)$ | $\phi(A)$ | $\phi(B)$ | $\phi(B)$ | $c(A) \cap c(B)$ |
| $\tau(B)$ | $\phi(A)$ | $\tau(A)$ | $\phi(B)$ | $\phi(B)$ | $A \cap c(B)$ |
| $\phi(B)$ | $\tau(A)$ | $\phi(A)$ | $\tau(B)$ | $\phi(A)$ | $c(A) \cap B$ |
| $\phi(B)$ | $\phi(A)$ | $\tau(A)$ | $\tau(B)$ | $\tau(B)$ | $A \cap B$ |

## REFERENCES

[1] N. Rescher, Many-Valued Logics, McGraw-Hill, New York, 1969, 122-123.

[2] I.B. Türkşen, "Fuzzy Normal Forms," (to appear) (Invited Special Issue, 1993).

[3] I.B. Türkşen, "Interval-Valued Fuzzy Sets and 'Compensatory 'AND'," Fuzzy Sets and Systems, 52, 2(1992) 295-307.

[4] I.B. Türkşen, "Interval-Valued Fuzzy Sets Based on Normal Forms," Fuzzy Sets and Systems, 20(1986) 191-210.

[5] H.J. Zimmerman and P. Zysno, "Latent Connectives in Human Decision Making," Fuzzy Sets and Systems, 4(1980) 37-51.

# Fuzzy Decision Trees by Fuzzy ID3 Algorithm and Its Application to Diagnosis Systems

Motohide Umano[1]   Hirotaka Okamoto[2]   Itsuo Hatono[1]   Hiroyuki Tamura[1]
Fumio Kawachi[3]   Sukehisa Umedzu[3]   Junichi Kinoshita[3]

1: Department of Systems Engineering, Osaka University
1-1, Machikaneyama-cho, Toyonaka, Osaka 560, Japan
Tel: +81-6-844-1151 ext.4629, Fax: +81-6-857-7664
Internet: umano@sys.es.osaka-u.ac.jp
2: Department of Precision Engineering, Osaka University, Japan
Currently he is at Kawasaki Steel Corporation, Japan
3: Kansai Technical Engineering Co., Ltd., Japan

**Abstract:** A popular and particularly efficient method for making a decision tree for classification from symbolic data is ID3 algorithm. Revised algorithms for numerical data have been proposed, some of which divide a numerical range into several intervals or fuzzy intervals. Their decision trees, however, are not easy to understand. We propose a new version of ID3 algorithm to generate a understandable fuzzy decision tree using fuzzy sets defined by a user. We apply it to diagnosis for potential transformers by analyzing gas in oil.

## 1. Introduction

Knowledge acquisition from data is very important in knowledge engineering. A popular and efficient method is ID3 algorithm proposed by J.R. Quinlan [1],[2] in 1979, which makes a decision tree for classification from symbolic data.

The decision tree consists of nodes for testing attributes, edges for branching by values of symbols and leaves for deciding class names to be classified. ID3 algorithm applies to a set of data and generates a decision tree which minimizes the expected value of the number of tests for classifying the data.

For numerical data, its revised algorithms have been proposed, which divide a numerical range of attribute into several intervals. To make a decision tree flexible, some algorithms are proposed to fuzzify the interval [3],[4]. Their decision trees, however, are not easy to understand because

(1) we cannot know how a range of attribute is divided into intervals,

(2) a range of attribute may be divided into different intervals on different test nodes,

(3) one attribute may appear more than one time in one sequences of tests.

Moreover, we need a long sequence of tests since the decision tree is binary.

As for numerical data, many tuning methods for fuzzy rules are also proposed in the research of fuzzy control, which is called neuro-fuzzy technique [5]. Since it generates rules that contain all combinations of all fuzzy sets in attributes, it has several difficulties as follows:

(1) so many fuzzy rules are generated,

(2) fuzzy sets in the rules are not understandable since they are tuned for fitting the training data,

(3) the more the number of attributes are, the less convergent the error is.

Thus we propose a new algorithm to generate a fuzzy decision tree from data using fuzzy sets defined by a user. And we apply it to diagnosis of potential transformers by analyzing gas in oil in them [6].

## 2. Fuzzy ID3 Algorithm

ID3 algorithm [1],[2] applies to a set of data and generates a decision tree for classifying the data. Our algorithm, called fuzzy ID3 algorithm, is extended to apply to a fuzzy set of data (several data with membership grades) and generates a fuzzy decision tree using fuzzy sets defined by a user for all attributes. A fuzzy decision tree consists of nodes for testing attributes, edges for branching by test values of fuzzy sets defined by a user and leaves for deciding class names with certainties. An example of fuzzy decision trees is shown in Fig. 1.

Our algorithm is very similar to ID3, except ID3 selects the test attribute based on the information gain which is computed by the probability of ordinary data but ours by the probability of membership values for data.

Assume that we have a set of data $D$, where each data has $\ell$ numerical values for attributes $A_1$, $A_2$, ..., $A_\ell$ and one classified class $C =$ $\{C_1, C_2, \ldots, C_n\}$ and fuzzy sets $F_{i1}, F_{i2}, \ldots, F_{im}$ for the attribute $A_i$ (the value of $m$ varies on every attribute). Let $D^{C_k}$ to be a fuzzy subset in $D$ whose class is $C_k$ and $|D|$ the sum of the membership values in a fuzzy set of data $D$. Then an algorithm to generate a fuzzy decision tree is in the followings:

1. Generate the root node that has a set of all data, i.e., a fuzzy set of all data with the membership value 1.

2. If a node $t$ with a fuzzy set of data $D$ satisfies the following conditions:
   (1) the proportion of a data set of a class $C_k$ is greater than or equal to a threshold $\theta_r$, that is,
   $$\frac{|D^{C_k}|}{|D|} \geq \theta_r,$$
   (2) the number of a data set is less than a threshold $\theta_n$, that is,
   $$|D| < \theta_n,$$
   (3) there are no attributes for more classifications,
   then it is a leaf node and assigned by the class name (more detailed method is described below).

3. If it does not satisfy the above conditions, it is not a leaf and the test node is generated as follows:

   3.1 For $A_i$'s $(i = 1, 2, \ldots, \ell)$, calculate the information gains $G(A_i, D)$, to be described below, and select the test attribute $A_{max}$ that maximizes them.

   3.2 Divide $D$ into fuzzy subsets $D_1, D_2, \ldots, D_m$ according to $A_{max}$, where the membership value of the data in $D_j$ is the product of the membership value in $D$ and the value of $F_{max,j}$ of the value of $A_{max}$ in $D$.

   3.3 Generate new nodes $t_1, t_2, \ldots, t_m$ for fuzzy subsets $D_1, D_2, \ldots, D_m$ and label the fuzzy sets $F_{max,j}$ to edges that connect between the nodes $t_j$ and $t$.

   3.4 Replace $D$ by $D_j$ $(j = 1, 2, \ldots, m)$ and

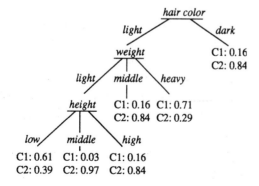

Fig. 1. Fuzzy decision tree.

repeat from 2 recursively.

The information gain $G(A_i, D)$ for the attribute $A_i$ by a fuzzy set of data $D$ is defined by

$$G(A_i, D) = I(D) - E(A_i, D), \qquad (1)$$

where

$$I(D) = -\sum_{k=1}^{n}(p_k \cdot \log_2 p_k), \qquad (2)$$

$$E(A_i, D) = \sum_{j=1}^{m}(p_{ij} \cdot I(D_{F_{ij}})), \qquad (3)$$

$$p_k = \frac{|D^{C_k}|}{|D|}, \qquad (4)$$

$$p_{ij} = \frac{|D_{F_{ij}}|}{\sum_{j=1}^{m}|D_{F_{ij}}|}. \qquad (5)$$

As for assigning the class name to the leaf node, we propose three methods as follows:

(a) The node is assigned by the class name that has the greatest membership value, that is, other than the selected data are ignored.

(b) If the condition (a) in step 2 in the algorithm holds, do the same as the method (a). If not, the node is considered to be empty, that is, the data are ignored.

(c) The node is assigned by all class names with their membership values, that is, all data are taken into account.

Now, we will illustrate one cycle of the algorithm. In the beginning of some cycle, we have a fuzzy set of data $D$:

| $\mu$ | height | weight | hair color | class |
|---|---|---|---|---|
| 1 | 160 | 60 | blond | $C_1$ |
| 0.8 | 180 | 80 | black | $C_2$ |
| 0.2 | 170 | 75 | black | $C_2$ |
| 0.7 | 175 | 60 | red | $C_1$ |
| 1 | 160 | 75 | black | $C_2$ |
| 0.3 | 175 | 60 | red | $C_2$ |
| 1 | 165 | 60 | blond | $C_2$ |
| 0.5 | 180 | 70 | blond | $C_1$ |

Note that this includes inconsistent data, say, the fourth and sixth data. Fuzzy sets *low*, *middle* and *high* in the attribute *height*, fuzzy sets *light*, *middle* and *heavy* in *weight*, and fuzzy sets *light* and *dark* in *hair color* are defined as

$$
\begin{aligned}
low &= \{1/160, 0.8/165, 0.5/170, 0.2/175\}, \\
middle &= \{0.5/165, 1/170, 0.5/175\}, \\
high &= \{0.2/165, 0.5/170, 0.8/175, 1/180\}, \\[4pt]
light &= \{1/60, 0.8/65, 0.5/70, 0.2/75\}, \\
middle &= \{0.5/65, 1/70, 0.5/75\}, \\
heavy &= \{0.2/65, 0.5/70, 0.8/75, 1/80\}, \\[4pt]
light &= \{1/blond, 0.3/red\}, \\
dark &= \{0.6/red, 1/black\}.
\end{aligned}
$$

Note that we can define fuzzy sets of the continuous membership functions.

First, we calculate the information $I(D)$. Since we have $|D| = 5.5$, $|D^{C_1}| = 2.2$ and $|D^{C_2}| = 3.3$, we have

$$
\begin{aligned}
I(D) &= -\frac{2.2}{5.5}\log_2\frac{2.2}{5.5} - \frac{3.3}{5.5}\log_2\frac{3.3}{5.5} \\
&= 0.971(\text{bits}).
\end{aligned}
$$

Next, we calculate the expected information for all $A_i$'s. For *height*, using the step 3.2 in the algorithm, we have the fuzzy sets of data $D_{height,low}$, $D_{height,middle}$ and $D_{height,high}$:

| low | mid | high | H | W | HC | C |
|---|---|---|---|---|---|---|
| 1 | 0 | 0 | 160 | 60 | blond | $C_1$ |
| 0 | 0 | 0.8 | 180 | 80 | black | $C_2$ |
| 0.1 | 0.2 | 0.1 | 170 | 75 | black | $C_2$ |
| 0.14 | 0.35 | 0.56 | 175 | 60 | red | $C_1$ |
| 1 | 0 | 0 | 160 | 75 | black | $C_2$ |
| 0.06 | 0.15 | 0.24 | 175 | 60 | red | $C_2$ |
| 0.8 | 0.5 | 0.2 | 165 | 60 | blond | $C_2$ |
| 0 | 0 | 0.5 | 180 | 70 | blond | $C_1$ |

The membership value is calculated by the product of $\mu$ in $D$ and the membership value of the fuzzy sets *low*, *middle* and *high* of the value of the *height* in $D$.

Then for *low*, we have

$$|D_{height,low}| = 3.1,$$
$$|D_{height,low}^{C_1}| = 1.14,$$
$$|D_{height,low}^{C_2}| = 1.96$$

and

$$I(D_{height,low})$$
$$= -\frac{1.14}{3.1}\log_2\frac{1.14}{3.1} - \frac{1.96}{3.1}\log_2\frac{1.96}{3.1}$$
$$= 0.949(\text{bits}).$$

For *middle*, we have $|D_{height,middle}| = 1.2$, $|D_{height,middle}^{C_1}| = 0.35$, $|D_{height,middle}^{C_2}| = 0.85$, and $I(D_{height,middle}) = 0.871(\text{bits})$.

For *high*, we have $|D_{height,high}| = 2.4$, $|D_{height,high}^{C_1}| = 1.06$, $|D_{height,high}^{C_2}| = 1.34$, and $I(D_{height,high}) = 0.990(\text{bits})$.

Now we can calculate the expected information after testing by the *height* as

$$E(height, D) = \frac{3.1}{6.7} \times 0.949$$
$$+ \frac{1.2}{6.7} \times 0.871 + \frac{2.4}{6.7} \times 0.990$$
$$= 0.950(\text{bits}).$$

Thus we have the information gain for the attribute *height* as

$$G(height, D) = I(D) - E(height, D)$$
$$= 0.971 - 0.950$$
$$= 0.021(\text{bits}).$$

By similar analysis for *weight* and *hair color*, we have

$$G(weight, D) = 0.118,$$
$$G(hair\ color, D) = 0.164.$$

Since we select the attribute that maximize the gain, we have the *hair color* as the test attribute. Now we have a subtree shown in Fig. 2. For the fuzzy sets of data $D_1$ and $D_2$, we apply the same process until it hold the leaf condition

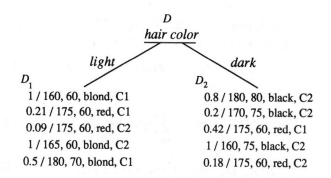

Fig. 2. Generated subtree.

(1), (2) or (3) in the step 2 in the algorithm. For this data, we have the fuzzy decision tree shown in Fig. 1.

The fuzzy decision trees generated by this algorithm is very easy to understand, because an attribute appears at most once in any paths from the root to leaf nodes and the test for an attribute is the same even when it appears at different nodes, i.e., branching by the fuzzy sets defined by a user.

## 3. Inference by Fuzzy Decision Tree

Inference in an ordinary decision tree is executed by starting from the root node and repeating to test the attribute at the node and branch to an edge by its value until reaching at a leaf node, a class attached to the leaf being as the result.

On the other hand, in a fuzzy decision tree we must branch more than one edge with certainty. An example is shown in Fig. 3, where $A_i$ stands for the attribute to be tested, $F_{ij}$ on the edge a fuzzy sets and the numerical value on the edge is a membership value of $F_{ij}$ for the value of the attribute.

Now We must decide three operations. First, for the operation to aggregate membership values for the path of edges, we adopt the multiplication from many alternatives. Second, for the operation of the total membership value of

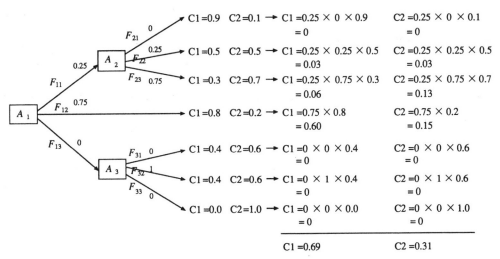

Fig. 3. Fuzzy Reasoning in Fuzzy Decision Tree

the path of edge and the certainty of the class attached to the leaf node, we also adopt the multiplication. Finally, for the operation to aggregate certainties of the same class from the different paths of edges, we adopt the addition from several alternatives. When the total certainty value exceeds the unity, we can normalize them. In Fig. 3, we have the results "$C_1$ with 0.69 and $C_2$ with 0.31".

We newly formulated this inference method. However, we have found that this is essentially the same as a reasoning method of fuzzy rules for fuzzy control, so called "×-×-+ method".

## 4. Application to Diagnosis

We apply our algorithms to diagnosis of potential transformers which contain oil [6]. When a part of it is damaged, several kinds of gases are generated and dissolved in oil. Then a little amount of oil is sampled out and analyzed. Thus we have amounts of 10 gases, $CH_4$, $C_2H_6$, $C_2H_4$, $C_3H_8$, $C_3H_6$, $C_2H_2$, $H_2$, $CO$, $CO_2$ and other gases (we think other gases as a kind of gas).

Several potential transformers are destroyed and checked their causes a year. Now we have 220 data that have been checked already. We have two classes of causes, one is rough (4 classes) and the other detailed (17 classes). Note that the checked transformer were varied in size, makers, usage condition and degree of checking. Really several data is inconsistent with each other.

We divide them in half, one half is for generating a decision tree and the other half is for checking it. We apply our algorithm to these data with fuzzy sets on each gas which are defined by experts and shown in Fig. 4, to generate two fuzzy decision tree for rough causes and detailed causes.

Then we check the results of above three methods (a), (b) and (c). The results of rough causes is shown in Table 1 and that of detailed one Table 2, where the threshold $\theta_r$ ranges from 0.6 to 1 by 0.1 and $\theta_n$ is fixed to 1. In Table 2, (a–n) means to check the inferred classes with the greatest $n$ certainties.

We have the best result by the method (c). We have the best correction rates 87.0% for rough causes by checking the inferred class with the greatest degree and 63.5% for detailed causes by checking the inferred classes with the greatest 3 certainties.

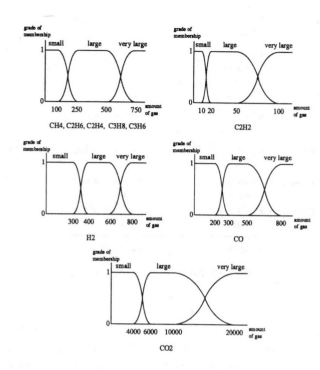

Fig. 4. Fuzzy sets for each gas.

Table 1. Correction rates for rough causes

| $\theta_r$ | 0.6 | 0.7 | 0.8 | 0.9 | 1.0 |
|---|---|---|---|---|---|
| (a) | 69.6% | 74.8% | 73.0% | 64.4% | 49.6% |
| (b) | 69.6% | 74.9% | 77.4% | 80.0% | 77.4% |
| (c) | 68.7% | 78.3% | 82.6% | 85.2% | 87.0% |

Table 2. Correction rates for detailed causes

| $\theta_r$ | 0.6 | 0.7 | 0.8 | 0.9 | 1.0 |
|---|---|---|---|---|---|
| (a–1) | 34.7% | 32.2% | 28.7% | 24.4% | 18.3% |
| (a–2) | 40.9% | 38.3% | 33.9% | 27.8% | 20.0% |
| (a–3) | 40.9% | 38.3% | 34.8% | 28.7% | 20.9% |
| (b–1) | 41.7% | 41.7% | 40.9% | 41.7% | 41.7% |
| (b–2) | 52.2% | 51.3% | 49.6% | 50.4% | 49.6% |
| (b–3) | 54.8% | 53.9% | 52.2% | 53.0% | 52.2% |
| (c–1) | 39.1% | 40.9% | 40.9% | 42.6% | 44.4% |
| (c–2) | 55.7% | 57.4% | 54.8% | 54.8% | 56.5% |
| (c–3) | 62.6% | 63.5% | 62.6% | 63.5% | 63.5% |

rithm, which will be described in the forthcoming paper.

## 5. Conclusion

We proposed a new algorithm to generate a fuzzy decision tree from numerical data using fuzzy sets defined by a user. Next, we formulated an inference method for such a fuzzy decision tree. Finally, we applied it to diagnosis of potential transformers by analyzing gas in oil.

Since we can easily transform a fuzzy decision tree to a set of fuzzy rules, this can be considered as a method to generate fuzzy rules from a set of numerical data. But this is not so good in the sense of the number of fuzzy rules. We have already formulated another better rule generation method based on the fuzzy ID3 algo-

## References

[1] J.R. Quinlan (1979) : "Discovering Rules by Induction from large Collections of Examples," in D. Michie (ed.): *Expert Systems in the Micro Electronics Age*, Edinburgh University Press.

[2] J.R. Quinlan (1986) : "Induction of Decision Trees," *Machine Learning*, Vol.1, pp.81–106.

[3] T. Tani and M. Sakoda (1991) : "Fuzzy Oriented Expert System to Determine Heater Outlet Temperature Applying Machine Learning," *7th Fuzzy System Symposium* (Japan Society for Fuzzy Theory and Systems), pp.659–662 (in Japanese).

[4] S. Sakurai and D. Araki (1992) : "Application of Fuzzy Theory to Knowledge Acquisition," *15th Intelligent System Symposium* (Society of Instrument and Control Engineers), pp.169–174 (in Japanese).

[5] H. Ichihashi (1993) : "Tuning Fuzzy Rules by Neuro-Like Approach," *Journal of Japan Society for Fuzzy Theory and Systems*, Vol.5, No.2, pp.191–203 (in Japanese).

[6] F. Kawachi and T. Matsuura (1990) : "Development of Expert System for Diagnosis by Gas in Oil and Its Evaluation in Practice Usage," *Technical Meeting on Electrical Insulation Material* (The Institute of Electrical Engineers of Japan), EIM-90-40 (in Japanese).

# FUZZ-IEEE Authors Index

# FUZZ-IEEE Authors Index

# FUZZ-IEEE Authors Index